Proceedings

21st International Conference on VLSI Design

Held jointly with
7th International Conference on Embedded Systems

Hyderabad, India
January 4-8, 2008

Technical Co-Sponsorship
IEEE Circuits and Systems Society
IEEE Solid State Circuits Society
IEEE Electron Devices Society

Sponsored by
VLSI Society of India

Sister Conference
IEEE/ACM Design Automation Conference

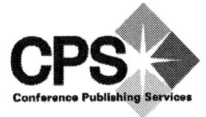

Los Alamitos, California

Washington • Tokyo

Copyright © 2008 by the Institute of Electrical and Electronic Engineers, Inc
All Rights Reserved

Copyright and Reprint Permissions: Abstracting is permitted with credit to the source. Libraries are permitted to photocopy beyond the limit of U.S. copyright law for private use of patrons those articles in this volume that carry a code at the bottom of the first page, provided the per-copy fee indicated in the code is paid through Copyright Clearance Center, 222 Rosewood Drive, Danvers, MA 01923.

For other copying, reprint or republication permission, write to IEEE Copyrights Manager, IEEE Service Center, 445 Hoes Lane, Piscataway, NJ 08854. All rights reserved.

***This publication is a representation of what appears in the IEEE Digital Libraries.Some format issues inherent in the e-media version may also appear in this print version.**

IEEE Catalog Number:	CFP08041-PRT
ISBN 13:	978-1-4244-3039-0
ISSN	1063-9667

Additional Copies of This Publication Are Available From:

Curran Associates, Inc
57 Morehouse Lane
Red Hook, NY 12571 USA
Phone: (845) 758-0400
Fax: (845) 758-2633
E-mail: curran@proceedings.com

Table of Contents

VLSI Design 2008

Message from the General Chairs
Message from the Program Chairs
Conference Steering Committee
Conference Committee
Program Committee
Reviewers
VLSI Design 2007 Fellowship Recipients
VLSI Design 2007 Best Paper Awards
VLSI Design Conference History
Embedded Systems Design Conference History
Plenary Invited Keynote Speakers

Tutorials

Gateway to Chips: High Speed I/O Signalling and Interface 3
Nidhir Kumar, Senthil N. Velu, and Rajan Verma

DFM / DFT / SiliconDebug / Diagnosis 5
Srikanth Venkataraman and Nagesh Tamarapalli

Oversampling Analog-to-Digital Converter Design 7
Shanthi Pavan and Nagendra Krishnapura

Programming and Performance Modelling of Automotive ECU Networks 8
Samarjit Chakraborty and S. Ramesh

Architecture Exploration for Low Power Design 10
Vinod Kathail and Tom Miller

Memory Design and Advanced Semiconductor Technology 12
D. Harame, Subramanian S. Iyer, Josef S. Watts, Rajiv Joshi, and John E. Barth Jr.

Scan Delay Testing of Nanometer SoCs 13
Adit D. Singh

Cross-Layer Approaches to Designing Reliable Systems Using Unreliable Chips 14
Fadi Kurdahi, Nikil Dutt, Ahmed Eltawil, and Sani Nassif

OpenSPARC — A Scalable Chip Multi-Threading Design 16
Dwayne Lee

Implementing the Best Processor Cores 17
Vamsi Boppana, Rahoul Varma, and S. Balajee

SESSION A1: Fault Tolerance

A Power Efficient Approach to Fault-Tolerant Register File Design...21
Mojtaba Amiri-Kamalabad, Seyed Ghassem Miremadi, and Mahdi Fazeli

Reconfiguring CMOS as Pseudo N/PMOS for Defect Tolerance in Nano-Scale CMOS.............................27
Maryam Ashouei, Adit D. Singh, and Abhijit Chatterjee

Single Error Correcting Finite Field Multipliers Over GF(2^m)..33
Jimson Mathew, A. Costas, A.M. Jabir, H. Rahaman, and D.K. Pradhan

A Robust Architecture for Flip-Flops Tolerant to Soft-Errors and Transients
from Combinational Circuits 39
Aditya Jagirdar, Roystein Oliveira, and Tapan Jyoti Chakraborty

Energy-Efficient Soft-Error Protection Using Operand Encoding and Operation Bypass 45
Kaushal R. Gandhi and Nihar R. Mahapatra

SESSION B1: Wireless/Communication

Retimed Decomposed Serial Berlekamp-Massey (BM) Architecture for High-Speed
Reed-Solomon Decoding 53
Shahid Rizwan

Exploring the Processor and ISA Design for Wireless Sensor Network Applications...........................59
Shashidhar Mysore, Banit Agrawal, Frederic T. Chong, and Timothy Sherwood

Concurrent Multi-Dimensional Adaptation for Low-Power Operation in Wireless Devices.....................65
Rajarajan Senguttuvan, Shreyas Sen, and Abhijit Chatterjee

Adaptive Signal Scaling Driven Critical Path Modulation for Low Power Baseband
OFDM Processors 71
Muhammad Mudassar Nisar, Rajarajan Senguttuvan, and Abhijit Chatterjee

Fault Tolerant Dynamic Antenna Array in Smart Antenna System
Using Evolved Virtual Reconfigurable Circuit...77
D. Dhanasekaran and K. Boopathy Bagan

SESSION C1: Embedded Systems

Multimedia Tools and Architectures for Hardware/Software Co-Simulation
of Reconfigurable Systems 85
Valery Sklyarov, Iouliia Skliarova, Bruno Pimentel, and Manuel Almeida

A Modeling of a Dynamically Reconfigurable Processor Using SystemC 91
Junji Kitamichi, Koji Ueda, and Kenichi Kuroda

A Scalable and Reconfigurable Coprocessor for Image Composition...97
Jalaj Jain

Predictable Implementation of Real-Time Applications on Multiprocessor
Systems-on-Chip...103
Alexandru Andrei, Petru Eles, Zebo Peng, and Jakob Rosen

An Approach to Software Performance Evaluation on Customized Embedded Processors......................111
Soumyajit Dey, Monu Kedia, and Anupam Basu

SESSION D1: Technology

Compact Modeling of Suspended Gate FET 119
  Yogesh Singh Chauhan, D. Tsamados, N. Abelé, C. Eggimann, M. Declercq,
  and A.M. Ionescu

Optimal Dual-V_T Design in Sub-100 Nanometer PDSOI and Double-Gate Technologies 125
  Aditya Bansal, Jae-Joon Kim, Keunwoo Kim, Saibal Mukhopadhyay, Ching-Te Chuang,
  and Kaushik Roy

Recursive Statistical Blockade: An Enhanced Technique for Rare Event Simulation
with Application to SRAM Circuit Design 131
  Amith Singhee, Jiajing Wang, Benton H. Calhoun, and Rob A. Rutenbar

NBTI Degradation: A Problem or a Scare? 137
  Kewal K. Saluja, Shriram Vijayakumar, Warin Sootkaneung, and Xaingning Yang

On-Chip Process Variation Detection Using Slew-Rate Monitoring Circuit 143
  Amlan Ghosh, Rahul M. Rao, Jae-joon Kim, Ching-Te Chuang, and Richard B. Brown

SESSION A2: Testing/DFT

On Common-Mode Skewed-Load and Broadside Tests 151
  Irith Pomeranz, Sudhakar M. Reddy, and Sandip Kundu

Testing Flash Memories for Tunnel Oxide Defects 157
  Mohammad Gh. Mohammad and Kewal K. Saluja

On the Detection of Missing-Gate Faults in Reversible Circuits by a Universal Test Set 163
  Hafizur Rahaman, Dipak K. Kole, Debesh K. Das, and Bhargab B. Bhattacharya

Memory Yield Improvement through Multiple Test Sequences and Application-Aware
Fault Models 169
  Aman Kokrady, C.P. Ravikumar, and Nitin Chandrachoodan

Design-for-Testability for Improved Path Delay Fault Coverage of Critical Paths 175
  Irith Pomeranz and Sudhakar M. Reddy

Design-for-Testability for Synchronous Sequential Circuits that Maintains
Functional Switching Activity 181
  Irith Pomeranz and Sudhakar M. Reddy

A Partitioning Based Physical Scan Chain Allocation Algorithm that Minimizes
Voltage Domain Crossings 187
  Nilabha Dev, Sandeep Bhatia, Subhasish Mukherjee, Sue Genova, and Vinayak Kadam

SESSION B2: Interconnects

Wiring-Area Efficient Simultaneous Bidirectional Point-to-Point Link for Inter-Block
On-Chip Signaling 195
  Charbel J. Akl and Magdy A. Bayoumi

Energy-Aware Interconnect Optimization for a Coarse Grained Reconfigurable Processor 201
  Andy Lambrechts, Praveen Raghavan, Murali Jayapala, Francky Catthoor,
  and Diederik Verkest

Integrated TIA-Equalizer for High Speed Optical Link 208
  Saurav Bandyopadhyay, Pradip Mandal, Stephen E. Ralph, and Kenneth Pedrotti

Single Edge Clock (SEC) Distribution for Improved Latency, Skew, and Jitter Performance 214
  Jeff Mueller and Resve Saleh

Threshold Voltage Control through Multiple Supply Voltages for Power-Efficient
FinFET Interconnects..220
 Anish Muttreja, Prateek Mishra, and Niraj K. Jha

Analysis of Delay Variation in Encoded On-Chip Bus Signaling under Process Variation........................228
 Sampo Tuuna, Ethiopia Nigussie, Jouni Isoaho, and Hannu Tenhunen

Exploiting Variable Cycle Transmission for Energy-Efficient On-Chip Interconnect Design....................235
 T. Venkata Kalyan, Madhu Mutyam, and P. Vijaya Sankara Rao

SESSION C2: Architecture

Dynamic Aggregation of Virtual Addresses in TLB Using TCAM Cells..243
 Rupak Samanta, Jason Surprise, and Rabi Mahapatra

Continuous Frequency Adjustment Technique Based on Dynamic Workload Prediction........................249
 Hwisung Jung and Massoud Pedram

Recursive versus Iterative Algorithms for Solving Combinatorial Search Problems
in Hardware 255
 Iouliia Skliarova and Valery Sklyarov

Exhaustive Enumeration of Legal Custom Instructions for Extensible Processors........................261
 Nagaraju Pothineni, Anshul Kumar, and Kolin Paul

An Optimal Multi-Functional Unit Dynamic Instruction Selection Logic
at Submicron Technologies 267
 Terrell Bennett and Rama Sangireddy

A 2.1GHz 6.5mW 64-bit Unified PopCount/BitScan Datapath Unit for 65nm
High-Performance Microprocessor Execution Cores 273
 Rajaraman Ramanarayanan, Sanu Mathew, Vasantha Erraguntla, Ram Krishnamurthy,
 and Shay Gueron

Dynamic Error Detection for Dependable Cache Coherency in Multicore Architectures....................279
 Hui Wang, Sandeep Baldawa, and Rama Sangireddy

SESSION D2: Analog

Mismatch Aware Analog Performance Macromodeling Using Spline Center
and Range Regression on Adaptive Samples 287
 Shubhankar Basu, Balaji Kommineni, and Ranga Vemuri

An Input Stage for the Implementation of Low-Voltage Rail to Rail Offset
Compensated CMOS Comparators..294
 Jaime Ramirez-Angulo, Lalitha Mohana Kalyani-Garimella, Annajirao Garimella,
 Sri Raga Sudha Garimella, Antonio Lopez-Martin, and Ramon Gonzalez Carvajal

Highly Linear Wide Dynamic Swing CMOS Transconductance Multiplier
Using Source-Degeneration V-I Converters..300
 Sri Raga Sudha Garimella

Chaos-Modulated Ramp IC for EMI Reduction in PWM Buck Converters—
Design and Analysis of Critical Issues..305
 Rupam Mukherjee, Amit Patra, and Soumitro Banerjee

A Fast Settling 100dB OPAMP in 180nm CMOS Process with Compensation
Based Optimisation 311
 Amal Kumar Kundu, Subho Chatterjee, and Tarun Kanti Bhattacharyya

VLSI Implementation of a Digitally Tunable G_m-C Filter with Double CMOS Pair 317
  S. Ramasamy, B. Venkataramani, and K. Anbugeetha

A 9 bit 400 MHz CMOS Double-Sampled Sample-and-Hold Amplifier 323
  Sounak Roy and Swapna Banerjee

SESSION A3: Physical Design/CAD

A New Approach for Estimation of On-Resistance and Current Distribution
in Power Array Layouts..331
  *Jyotirmoy Ghosh, Siddhartha Mukhopadhyay, Amit Patra, Barry Culpepper,
  and Tawen Mei*

An Elitist Non-Dominated Sorting Based Genetic Algorithm for Simultaneous
Area and Wirelength Minimization in VLSI Floorplanning..337
  Pradeep Fernando and Srinivas Katkoori

Fast Congestion Aware Routing for Pin Assignment...343
  Shashank Prasad

A Novel Approach to Compute Spatial Reuse in the Design of Custom Instructions 348
  Nagaraju Pothineni, Anshul Kumar, and Kolin Paul

Addressing the Challenges of Synchronization/Communication and Debugging Support
in Hardware/Software Cosimulation..354
  Banit Agrawal, Timothy Sherwood, Chulho Shin, and Simon Yoon

SESSION B3: Low Power - I

Incorporating PVT Variations in System-Level Power Exploration
of On-Chip Communication Architectures..363
  Sudeep Pasricha, Young-Hwan Park, Fadi J. Kurdahi, and Nikil Dutt

Energy-Efficient, High Performance Circuits for Arithmetic Units..371
  Sundeepkumar Agarwal, Pavankumar V K, and Yokesh R

Delay and Energy Efficient Design of On-Chip Encoded Bus with Repeaters 377
  Qingli Zhang, Jinxiang Wang, and Yizheng Ye

A Robust Level-Shifter Design for Adaptive Voltage Scaling...383
  Ankur Gupta, Rajat Chauhan, Vinod Menezes, Vikas Narang, and Roopashree H.M.

Low Power Hardware Architecture for VBSME Using Pixel Truncation....................................389
  Asral Bahari, Tughrul Arslan, and Ahmet T. Erdogan

SESSION C3: NoC/SoC

MoCSYS: A Multi-Clock Hybrid Two-Layer Router Architecture and Integrated
Topology Synthesis Framework for System-Level Design of FPGA Based
On-Chip Networks 397
  Arun Janarthanan and Karen A. Tomko

MPSoC Communication Architecture Exploration Using an Abstraction Refinement Method 403
  Hao Shen and Frédéric Pétrot

An NoC Test Strategy Based on Flooding with Power, Test Time
and Coverage Considerations..409
  Mahshid Sedghi, Elnaz Koopahi, Armin Alaghi, Mahmood Fathy, and Zainalabedin Navabi

High-Level Modeling Approach for Analyzing the Effects of Traffic Models
on Power and Throughput in Mesh-Based NoCs 415
  Somayyeh Koohi, Mohammad Mirza-Aghatabar, Shaahin Hessabi, and Masoud Pedram

PTSMT: A Tool for Cross-Level Power, Performance, and Thermal Exploration
of SMT Processors 421
  Deepa Kannan, Aseem Gupta, Aviral Shrivastava, Nikil D. Dutt, and Fadi J. Kurdahi

SESSION D3: Nano

Single Event Upset: An Embedded Tutorial 429
  Fan Wang and Vishwani D. Agrawal

Fault-Tolerant Computing Using a Hybrid Nano-CMOS Architecture 435
  Muzaffer O. Simsir, Srihari Cadambi, Franjo Ivančic, Martin Roetteler, and Niraj K. Jha

Analysis and Robust Design of Diode-Resistor Based Nanoscale Crossbar PLA Circuits............441
  Rajat Subhra Chakraborty, Somnath Paul, and Swarup Bhunia

A New Threshold Voltage Model for Omega Gate Cylindrical Nanowire Transistor 447
  Biswajit Ray and Santanu Mahapatra

Design of Reversible Finite Field Arithmetic Circuits with Error Detection 453
  Jimson Mathew, Hafizur Rahaman, Babita R. Jose, and Dhiraj K. Pradhan

SESSION A4: Verification

Exploiting Circuit Reconvergence through Static Learning in CNF SAT Solvers 461
  Yinlei Yu, Cameron Brien, and Sharad Malik

Efficient Linear Macromodeling via Discrete-Time Time-Domain Vector Fitting 469
  Chi-Un Lei and Ngai Wong

Formal Verification of a Public-Domain DDR2 Controller Design..475
  Abhishek Datta and Vigyan Singhal

Enhanced TED: A New Data Structure for RTL Verification 481
  Pejman Lotfi-Kamran, Mehran Massoumi, Mohammad Mirzaei, and Zainalabedin Navabi

Simulation Acceleration with HW Re-Compilation Avoidance 487
  Kyuho Shim, Kesava Talupuru, Maciej Ciesielski, and Seiyang Yang

A Module Checking Based Converter Synthesis Approach for SoCs 492
  Roopak Sinha, Partha S. Roop, and Samik Basu

SESSION B4: Low Power - II

Energy Reduction in SRAM using Dynamic Voltage and Frequency Management.........................503
  Mohammed Shareef I, Pradeep Nair, and Bharadwaj Amrutur

Unified $V_{dd} - V_{th}$ Optimization Based DVFM Controller for a Logic Block.................................509
  Kannan S.A., Sreeram N.S., and Bharadwaj S. Amrutur

Temperature and Process Variations Aware Power Gating of Functional Units 515
  Deepa Kannan, Aviral Shrivastava, Vipin Mohan, Sarvesh Bhardwaj, and Sarma Vrudhula

A Robust Top-Down Dynamic Power Estimation Methodology for Delay Constrained
Register Transfer Level Sequential Circuits 521
  Sriram Sambamurthy, Jacob A. Abraham, and Raghuram S. Tupuri

Total Power Minimization in Glitch-Free CMOS Circuits Considering Process Variation..........527
  Yuanlin Lu and Vishwani D. Agrawal

Power Reduction of Functional Units Considering Temperature and Process Variations 533
  Deepa Kannan, Aviral Shrivastava, Sarvesh Bhardwaj, and Sarma Vrudhula

SESSION C4: Architecture/Arithmetic

Stall Power Reduction in Pipelined Architecture Processors 541
  Pejman Lotfi-Kamran, Amir-Mohammad Rahmani, Ali-Asghar Salehpour, Ali Afzali-Kusha, and Zainalabedin Navabi

A Novel Carry-Look Ahead Approach to a Unified BCD and Binary Adder/Subtractor..........547
  Sreehari Veeramachaneni, Kirthi Krishna M, Prateek G V, Subroto S, Bharat S, and M.B. Srinivas

Memory Architecture Exploration Framework for Cache Based Embedded SOC 553
  T.S. Rajesh Kumar, C.P. Ravikumar, and R. Govindarajan

A 100MHz to 1GHz, 0.35V to 1.5V Supply 256 x 64 SRAM Block Using Symmetrized
9T SRAM Cell with Controlled Read 560
  Satish Anand Verkila, Siva Kumar Bondada, and Bharadwaj S. Amrutur

A Novel Approach to Design BCD Adder and Carry Skip BCD Adder..........566
  Ashis Kumer Biswas, Md. Mahmudul Hasan, Moshaddek Hasan, Ahsan Raja Chowdhury, and Hafiz Md. Hasan Babu

A Merged Synthesis Technique for Fast Arithmetic Blocks Involving Sum-of-Products
and Shifters..........572
  Sabyasachi Das and Sunil P. Khatri

SESSION D4: Design/MEMS/Optical

A Jitter Reduction Circuit Using Autocorrelation for Phase-Locked Loops
and Serializer-Deserializer (SERDES) Circuits 581
  Hari Vijay Venkatanarayanan and Michael Lee Bushnell

GyroCompiler: A Soft IP Model Synthesis and Analysis Framework
for Design of MEMS Based Gyroscopes 589
  Jairam S and Navakanta Bhat

Behavioral Modeling of a CMOS Compatible High Precision MEMS Based
Electron Tunneling Accelerometer..........595
  T.K. Bhattacharyya and Anandaroop Ghosh

An Optical Reconfiguration System with Four Contexts..........601
  Naoki Yamaguchi and Minoru Watanabe

An Acceleration and Optimization Method for Optical Reconfiguration 607
  Minoru Watanabe and Naoki Yamaguchi

0.35μ, 1 GHz, CMOS Timing Generator Using Array of Digital Delay Lock Loops 613
  S. Balaji, Vinay B. Chandratre, and Menka Tewani

SESSION A5: Synthesis

Variability-Tolerant Register-Transfer Level Synthesis 621
   Anish Muttreja, Srivaths Ravi, and Niraj K. Jha

A Galois Field Based Logic Synthesis Approach with Testability 629
   J. Mathew, H. Rahaman, A.K Singh, A.M. Jabir, and D.K Pradhan

A Timing-Driven Synthesis Technique for Arithmetic Product-of-Sum Expressions 635
   Sabyasachi Das and Sunil P. Khatri

Clock Period Minimization with Iterative Binding Based on Stochastic
Wirelength Estimation during High-Level Synthesis..641
   Vyas Krishnan and Srinivas Katkoori

On the Use of Hash Tables for Efficient Analog Circuit Synthesis.........................647
   Almitra Pradhan and Ranga Vemuri

An Inversion-Based Synthesis Approach for Area and Power Efficient Arithmetic
Sum-of-Products 653
   Sabyasachi Das and Sunil P. Khatri

SESSION B5: Low Power - III

A Low Voltage, Low Ripple, on Chip, Dual Switch-Capacitor Based Hybrid
DC-DC Converter 661
   Kaushik Bhattacharyya and Pradip Mandal

Voltage and Temperature Scalable Standard Cell Leakage Models Based on Stacks
for Statistical Leakage Characterization 667
   Janakiraman Viraraghavan, Bishnu Prasad Das, and Bharadwaj Amrutur

Self-Sleep Buffer for Distributed MTCMOS Design 673
   Charbel J. Akl and Magdy A. Bayoumi

Power Management of Interactive 3D Games Using Frame Structures....................679
   Yan Gu and Samarjit Chakraborty

Voltage and Temperature Scalable Gate Delay and Slew Models
Including Intra-Gate Variations..685
   *Bishnu Prasad Das, Janakiraman V. Bharadwaj Amrutur, H.S. Jamadagni,
   and N.V. Arvind*

SESSION C5: Security

Single Chip Encryptor/Decryptor Core Implementation of AES Algorithm 693
   Monjur Alam, Santosh Ghosh, Dipanwita RoyChowdhury, and Indranil Sengupta

Reduced Complementary Dynamic and Differential Logic: A CMOS Logic Style
for DPA-Resistant Secure IC Design...699
   Srividhya Rammohan, Vijay Sundaresan, and Ranga Vemuri

Power Attack Resistant Efficient FPGA Architecture for Karatsuba Multiplier..........706
   Chester Rebeiro and Debdeep Mukhpodhyay

Watermarking Video Clips with Workload Information for DVS..............................712
   Yicheng Huang, Samarjit Chakraborty, and Ye Wang

Throughput Efficient Parallel Implementation of SPIHT Algorithm 718
   Anilkumar V. Nandi and R.M. Banakar

SESSION D5: Invited Special Session: Standards in EDA

Organizer: Dr. Nagi Naganathan, LSI Corp

Standards in EDA: An Introduction..727
  Nagi Naganathan, LSI Corp

Industry Standards from Accellera 728
  Shrenik Mehta, SUN Microsystems

IEEE Market-Oriented Standards Process and the EDA Industry..729
  Dennis Brophy, Mentor Graphics

Design Automation Standards: The IP Providers Perspective...730
  Dr. John Goodenough, ARM

Driving Analog Mixed Signal Verification through Verilog-AMS...731
  Sri Chandra, Freescale

VSI Standards, Current Status and Future Work...732
  Kathy Werner, Freescale

Author Index 733

Message from the General Chairs

Srimat Chakradhar   J A Chowdary   Dasaradha Gude

Welcome to Hyderabad! This historic city is once again playing host to the 21st International Conference on VLSI Design and the 7th International Conference on Embedded Systems that will be held jointly during January, 4-8, 2008. The conference is being held at India's largest convention center, the Hyderabad International Convention Center (HICC). With a seating capacity of over 4000 (and inbuilt flexibility to expand to 6500), HICC is also billed as South Asia's first world-class convention center.

VLSI Design 2008 begins with a thought provoking two-day tutorial program that showcases the latest trends in technology. This year, our tutorial chairs Atul Jain and Preeti Ranjan Panda have assembled a world-class tutorial program that covers many hot and relevant topics like chip-level multiprocessing, design of reliable systems using unreliable chips, new memory design and I/O trends. Inaugural ceremony for the conference is scheduled for the evening of January 5th, 2008. This session is sponsored by AMD, the Platinum Sponsor for 2008. The inaugural session includes a keynote talk by Dirk Meyer, President and COO of AMD who will make a case for the VLSI industry to re-orient innovation around the customer rather than the traditional drivers of transistor density and clock frequencies.

The three-day technical program starts on January 6th with a keynote from Wally Rhines, CEO of our premium Gold sponsor Mentor Graphics. He takes a unique and unconventional stance. He postulates that experience can be a bottleneck, and new players like India will emerge as strong forces to reckon in the world VLSI industry. FPGAs are emerging as viable alternative to ASICs in many market segments. Ivo Bolsens, CTO of our premium Gold sponsor Xilinx, will articulate how FPGAs will revolutionize future IT platforms. Program Chairs N. Ranganathan and V. Visvanathan have assembled an outstanding technical program. VLSI Design 2008 offers a pre-competitive forum that promotes sharing of research and technology development costs, complete with an unprecedented no-strings-attached free rapid review and feedback cycle. This year, there are four parallel tracks with over 100 thought provoking technical presentations. The technical program, keynotes, panels, and tutorials have been carefully selected to give you an insight into the latest research and development trends in VLSI. To promote excellence in the design of electronic systems in universities and educational establishments, Design Contest Chairs H. Parameswaran and N. Krishnapura have solicited design and implementation submissions from full-time students. Top design entries will be showcased in the technical program.

Attendees of VLSI Design 2008 will experience a delicate balance of technology principles and industrial practices, all conveniently located in a single location! VLSI Design conference has run a stellar exhibits program for over two decades. This year, the Exhibits Chairs Naresh Malipeddi and Sanjeev Aggarwal have assembled an enviable panorama of a record number of over 70 exhibits. The list of exhibitors includes all the major industry players, as well as numerous innovative startups. The three-day Exhibits Program will run concurrently with the technical program. Industry Forum chairs Pradip Dutta, Venkat Rajaraman and P. L. Narayana have added an exciting fifth, parallel track to the technical program. This track provides the attendees a unique opportunity to synchronize with the latest trends in industrial practices. All speakers in this track are from the industry. Exhibits Program, along with the Industry Forum, will bring you up to speed with leading edge industrial practices in all aspects of VLSI Design.

A corporation creates a link with an outside issue or event, hoping to influence the audience by the connection. Sponsorships are an important part of the marketing mix of corporations. Thanks to valiant efforts by our dynamic Sponsorship Chair Uma Mahesh, well known first tier VLSI Design and Embedded systems companies vied enthusiastically to sponsor various events in the conference.

VLSI Design Conference has a long tradition of a large fellowship programme. These fellowships support interested faculty and students who are not in a position to arrange for their own funds to attend the conference. The fellowships range from registration fee waiver to full travel support including registration fee. Fellowship Chairs K. Subbarangaih and K. Lal Kishore were instrumental in selecting over 200 fellows.

Expressions of gratitude do not cancel an indebtedness anymore than a promissory note cancels an account. We would be remiss if we did not thank the numerous volunteers who generously gave of their time and wisdom. The Organizing Chairs, V. Gopi Krishna and Sandeep Ramineni, accepted the difficult challenge of coordinating all activities of the conference, including local arrangements necessary to engage an audience of over 1000 professionals for five days. We sincerely thank them for undertaking this daunting task. Finances for the 2008 conference were managed by Prem Nivasa, and we are grateful to him for navigating us through the complex world of trusts, accounting, taxation and cash flow. Advertising still ranks as a prime necessity. Importance of publicity in gaining recognition and establishing credibility cannot be overemphasized. We are indebted to our Publicity Gurudutt Bansal, C. S. Rao and Shashidhar Reddy for providing us with superb global publicity. Nagi Naganathan and C. P. Ravikumar once again stepped up to the plate and ensured timely publication of the conference proceedings in hardcopy and CD-ROM formats.

Understanding the trends, feeling the pulse of the industry, forging new connections and relationships can open doors to new opportunities. It is the random opportunity to meet people you would not have otherwise met, the chaotic, the unexpected, the unplanned discovery of new thoughts wherein lies the hidden value of the conference. The 2008 conference, with its generous breaks, banquets, lunches and dinners, is a tastefully orchestrated artistic endeavor of hundreds of volunteers, a form of performance, a sublime experience where we all are challenged, enlarged, and made wiser. You will look back on the conference as a turning point in your thinking, your career, and your life.

Welcome again to the historic city of Hyderabad!

With warm regards,
Srimat Chakradhar, J A Chowdary, Dasaradha Gude

Message from the Program Chairs

Nagarajan Ranganathan   Vish Visvanathan

Welcome to the 21st International Conference on VLSI Design and the 7th International Conference on Embedded Systems taking place during January 4-8, 2008 at HICC, Hyderabad, India. This joint-conference is a forum for researchers and designers to present and discuss various aspects of VLSI design, electronic design automation (EDA), enabling technologies, and embedded systems. It covers the entire spectrum of activities in the two vital areas of VLSI and embedded systems, which underpin the semiconductor industry. This joint annual conference continues to attract a large number of researchers and engineers representing academia, industry and government from around the world.

The five-day program will consist of regular paper sessions, special sessions, banquets, plenary keynote talks, panel discussions, industrial exhibits and two days of tutorials. The 2008 conference has a particular emphasis on the challenges and opportunities in nanoelectronics, as the CMOS technology continues to scale further in the nanometer dimension. The new paradigms in design, EDA and system implementation will be presented and discussed in the 2008 edition of this conference.

This year we received 336 paper submissions from around the world and we had a challenging task ahead of us of getting a rigorous review process accomplished. The international technical program committee (TPC) consisted of 59 members from both academia and industry. There were two program committee meetings held simultaneously in New Jersey and Bangalore. Our target was to obtain about three to four reviews for each paper and make sure that each paper was reviewed by at least one tpc member. On papers with only 2 reviews, if we found the reviews to be concurring, we went with the recommendation and if we found reviews to be conflicting, then we sought additional reviews. We are extremely thankful to the TPC members for their efficient handling of the reviewer assignments and the relentless pursuit of the reviews. As usual, most of the reviews trickled in during the last week prior to the TPC meeting, however, thanks to huge team of reviewers, acknowledged by their names in the following proceedings pages, we received sufficient number of reviews to make good decisions. We accepted a total of 107 refereed papers and included one invited paper. The distribution of the number of reviews for the papers is summarized below:

No. of papers with 5 or more reviews -   102
No. of papers with exactly 4 reviews -   109
No. of papers with exactly 3 reviews -   102
No of papers with 2 reviews -   23
Total number of paper submissions -   336
Total number of reviews received -   1298
Total number of papers accepted -   107

We want to congratulate and thank everyone who served as a reviewer or as a member of the TPC for accomplishing the huge job of getting enough quality reviews. Thus, we are proud to say we have an outstanding technical program once again this year. It is unfortunate that we could not include several other good papers in the program, due to the limited number of sessions that could be accommodated.

This year, we have an invited special session titled, "Standards in EDA", thanks to the efforts by Nagi Naganathan of LSI Corporation. The invited speakers include Shrenik Mehta of SUN Microsystems, Dennis Brophy of MentorGraphics, John Goodenough of ARM, Sri Chandra and Kathy Warner of Freescale providing their views on various efforts in EDA standards.

A key aspect of this year's technical program is a set of eminent speakers who will deliver keynote speeches in plenary sessions and the banquets:

Dirk Meyer, President and COO, AMD
Ivo Bolsens, Vice President and CTO, Xilinx
Marc Tremblay, CTO, SUN Microsystems
Wally Rhines, CEO, MentorGraphics
Anantha Chandrakasan, Professor and Director of Microsystems Technology Laboratories, MIT
Alain Artieri, Senior System Architect, STMicroelectronics
Michael Campbell, Senior Vice President of Engineering, Qualcomm
Ramesh Senthinathan, Senior Director of Engineering, Broadcom
Santanu Das, President and CEO, TranSwitch

Further, we have three panels scheduled one on each day of the technical program. The first panel on Jan 6th is proposed and coordinated by A.P.K. Sreekumar of Rambus and the panel is titled, "Memory Requirements of Next Generation Chips and Systems" with a diverse set of panelists from Rambus, Intel, TI, Marvell etc are being invited. The second panel is being coordinated by Vivek De of Intel and the panel is titled, "New Frontiers in Energy Efficient Computing with Integrated Multi-Core Platforms in Nanoscale CMOS" with panelists from both industry and academia. The third panel is being organized by C. P. Ravikumar of Texas Instruments India and the panel is titled, "Growing the Right Talent for a Growing Semiconductor Industry" with several invited panelists. We sincerely thank A.P.K. Sreekumar, Vivek De and C.P. Ravikumar for volunteering their time and services in organizing the panels.

We would like to express our appreciation to the tutorial chairs, Atul Jain and Preeti Ranjan Panda for conducting the review process and selecting a wonderful set of 10 tutorials for attendees to choose from for attending during the first two days of the conference.

Our gratitude and thanks are due to many members of the organizing team – first to the General Co-chairs, Srimat Chakradhar (Chak), J.A. Chowdary, Dasaradha Gude for allowing us the freedom to structure the technical program and helping us with numerous logistics. Of course, Chak has been spearheading several aspects of this conference in a big way and it is a real pleasure to work with him. We cannot thank enough S. Uma Mahesh for his help and support with every aspect of the conference and the program. His promptness in correspondence as well as getting things done has been amazing. We express our appreciation to Nagi Naganathan and C.P. Ravikumar for assembling the proceedings material and organizing the CD production. Another quiet contributor behind the scenes is Sandeep Ramineni who has done a wonderful job managing the conference secretariat. We are greatly indebted to Lisa O'Conner and Tom Baldwin of IEEE CS Press for being accommodative and getting the proceedings published in time for the conference. We thank Mike Bushnell for hosting the TPC meeting in Rutgers University, New Jersey and C.P. Ravikumar for the meeting in Bangalore. Another person who has always been there to provide us wisdom and mentorship is Vishwani Agrawal and so, thank you Vishwani, as always. Last, but not the least, we would like to thank IEEE CAS, SSC and EDS societies as well as VSI and ISA for providing the technical co-sponsorship.

We sincerely hope that you will enjoy a memorable technical program and a unique overall experience at the 21st International Conference on VLSI Design and the 7th International Conference on Embedded Systems.

Nagarajan Ranganathan and Vish Visvanathan

Conference Steering Committee

Vishwani D. Agrawal, Chair
Jaswinder Ahuja
M. Balakrishnan
Srimat T. Chakradhar
Partha Pratim Das
Apurva Kalia
Bobby Mitra
A. Prabhakar
N. Ranganathan
C. P. Ravikumar

Conference Committee

Steering Committee Chair

Vishwani Agrawal

General Chairs

Srimat Chakradhar
NEC Labs

General Chairs

J A Chowdary
NVIDIA Graphics Pvt. Ltd.

General Chairs

Dasaradha Gude
AMD, India

Program Chairs

N. Ranganathan
University of South Florida

Program Chairs

V. Visvanathan
Texas Instruments

Organizing Chair

Venigalla Gopi Krishna
AMD, India

Publication Chairs

Nagi Naganathan
LSI Corp, USA

Publication Chairs

C. P. Ravikumar
TI, India

Publicity Chairs

Gurudutt Bansal
Cadence, India

Publicity Chairs

C.S. Rao
AP Invest

Publicity Chairs

Shashidhar Reddy
Qualcomm India Pvt. Ltd.

Tutotrial Chairs

Atul Jain
TI, India

Tutorial Chairs

Preeti Ranjan Panda
IIT, Delhi

Sponsorship Chair

S. Uma Mahesh
Indrion Technologies

Exhibit Chairs

Naresh Malipeddi
Conexant Systems Inc.

Exhibit Chairs

Sanjeev Aggarwal
Cadence, India

Industry Forum Chairs

Pidugu Lakshmi Narayana
Cypress Semiconductor
Technology India Pvt. Ltd.

Industry Forum Chairs

Pradip K. Dutta
Synopsys India Pvt. Ltd.

Industry Forum Chairs

Venkat Rajaraman
NVIDIA Graphics Pvt. Ltd.

Design Contest Chairs

Harindranath Parameswaran
Cadence, India

Design Contest Chairs

Nagendra Krishnapura
IIT, Chennai

Finance Chair

Prem Nivasa
Mentor Graphics

Organizing Committee

Sandeep Ramineni
AMD, India

IEEE Liaison

N. Ranganathan

SSCS Liaison

Sreedhar Natarjan

ACM/SIGDA Liaison

Sharad Seth

VSI Liaison

Biswadeep Mitra

ISA Liaison

Poornima Shenoy

Fellowship Chairs

K. Subbarangaih
VEDA IIT

Fellowship Chairs

K. Lal Kishore
JNTU

Program Committee

Nagarajan Ranganathan, **Program Chair**
Vish Visvanathan, **Program Chair**

Vishwani Agrawal
Bharadwaj Amrutur
Swapna Banerjee
Shabbir Batterywala
Sanjukta Bhanja
Navakanta Bhat
Mike Bushnell
Chaitali Chakrabarti
Srimat Chakradhar
Vikas Chandra
Nitin Chandrachoodan
Bernard Courtois
Vijay Degalahal
Nikil Dutt
Vidyasagar Ganesan
Rajesh Gupta
Narender Hanchate
Joerg Henkel
Vikram Iyengar
Atul Jain
Ahmed Jerraya
Niraj K. Jha
V. Kamakoti
Mahmut Kandemir
Bhaskar Karmakar
Srinivas Katkoori
Anshul Kumar
Radu Marculescu

Vinod Menezes
Prabhat Mishra
Durgamadhab Misra
Saraju Mohanty
Nagi Naganathan
Vijaykrishnan Narayanan
Zain Navabi
David Pan
Harindranath Parameswaran
Rubin Parekhji
Amit Patra
Shanthi Pavan
Irith Pomeranz
Preeti Ranjan Panda
Srivaths Ravi
CP Ravikumar
Subir Roy
Kewal Saluja
Sachin Sapatnekar
Sharad Seth
Li Shang
Adit Singh
S. Srinivasan
Vinoo Srinivasan
Pradip Thaker
Vinita Vasudevan
G.S Visweswaran
Yuan Xie

Reviewers

Afshin Abdollahi
Abhijit Mukund Abhyankar
Jacob A. Abraham
Jais Abraham
Andrea Acquaviva
Chaitanya Adapa
Ali Afzali-Kusha
Vishwani D. Agrawal
Vinay Agrawal
Waleed K Al-Assadi
Jins Davis Alexander
Mythri Alle
Nicholas Allec
Stelian Alupoaei
Kokrady Aman
Ahmed Amine
Bharadwaj S Amrutur
Federico Angiolini
Arvind NV
Iuliana Bacivarov
Amit Yashawant Badole
Maryam Shojaei Baghini
R Bahl
Brian Bailey
Shankar Balachandran
M Balakrishnan
Lakshmanan Balasubramaniam
Ashok Balivada
Neal Kumar Bambha
Chaitanya Bandi
Nirmalya Bandyopadhyay
Ansuman Banerjee
Swapna Banerjee
Subhadeep Banik
Kausar Banoo
Mohit K Bansal
Steven Bartling
Diganta Baruah
Kanad Basu
Prasenjit Basu
Shabbir Batterywala
Kia Bazargan

Leonid Belostotski
Luca Benini
Shaleen Jain Bhabu
Sanjukta Bhanja
Sarvesh Bhardwaj
Lava Bhargava
Ashish Bhargave
Anand Bhat
Navakanta Bhat
Bhargab B. Bhattacharya
Koustav Bhattacharya
Sambuddha Bhattacharya
Tarun Kanti Bhattacharyya
Basabi Bhaumik
Swarup Kumar Bhunia
Ravindra Bidnur
David M Binkley
Navin Bishnoi
Santosh Biswas
Partha Biswas
Clive Bittlestone
Subash Chandra Bose
P. Oscar Boykin
Rainer M Buchty
Sanjay Burman
Michael L. Bushnell
Yu Cao
Amlan Chakrabarti
Chaitali Chakrabarti
Samarjit Chakraborty
Supratik Chakraborty
Arun N. Chandorkar
Vikas Chandra
Nitin Chandrachoodan
Sreeram Chandrasekar
Frank MC Chang
Hongliang Chang
Chip Hong Chang
Abhijit Chatterjee
Shouribrata Chatterjee
Subhomoy Chattopadhyay
Charlie Chung-Ping Chen

Guangyu Chen
Guilin Chen
Mingsong Chen
Kameshwar Rao Chesetti
Madhav Y Chikodikar
Chen-Ling Chou
Prohor Choudhury
Sonali Chouhan
D Roy Chowdhury
Joel Dylan Joel Coburn
Bernard Courtois
Manoj Kumar Dadhich
Balasaheb S Darade
Bishnu Prasad Das
Debesh Das
Shamik Das
Subrangshu Kumar Das
Pallab Dasgupta
Ramyanshu Datta
Rakesh Gnana David J
Azadeh Davoodi
Vijay Sai Degalahal
V.R. Devanathan
Ish Dham
Harsh Dhand
Nagu Dhanwada
Anindya Sundar Dhar
Li Ding
Abdulkadir Utku Diril
Alok S. Doshi
David E Duarte
Nikil Dutt
Arijit Dutta
Basant Kumar Dwivedi
Soumya Eachempati
Prakash Easwaran
Peter Ehlig
Mohamed Mostafa Elsayed
Piet Engelke
Eric Ericson Fabris
Yan Feng
Pradeep R Fernando
Goerschwin Fey
Kannan Gaddam
Vidyasagar Ganesan

Anup Gangwar
Rajat Garg
Vivek Garg
Aman Gayasen
George E. Georgiou
Dhruva Ghai
Praveen Ghanta
Georges G.E. Gielen
Abhijit Giri
Girishankar G
Surace Giuseppe
Ananth Somayaji Goda
Neeraj Goel
Chandramouli Gopalakrishnan
Raja Gopalan
Maziar Goudarzi
Hillary Grimes
Peter G Grun
Krishnaiah Gummidipudi
Rajesh K Gupta
Aseem Gupta
Pallav Gupta
Rajat Gupta
Rajesh Gupta
Shyam Sunder Gupta
Suvodeep Gupta
Upavan Gupta
Vishal Gupta
Guoling Han
Narender Hanchate
Andreas Hansson
Harish B. P.
David Money Harris
Mohd. Hasan
John P Hayes
Joerg Henkel
Sebastian Herbert
Lakshmikantha V Holla
Shengyan Hong
Jie Hu
Mohammed Imdad Hussain
Kevin Maurice Irick
Mary Jane Irwin
Vikram Iyengar
Indira Iyer

Ashok Jagannathan
Palkesh Jain
Praveen Jain
Maxey Jay
Murali Jayapala
Niraj K Jha
Rajiv Vasant Joshi
Santiram Kal
Sumant D Kale
Kamakoti V.
Mahmut T Kandemir
Kalyana R Kantipudi
Jung-Chun Kao
Rohit Kapur
Sougata Kumar Kar
Vinod Kariat
Bhaskar J. Karmakar
Ravishankar Karthikeyan
Srinivas Katkoori
Srinivas Katkoori
Arun Kejariwal
Igor Keller
Sunil P Khatri
Ajay Khoche
Parveen Khurana
Vida Kianzad
Chris H. Kim
Jung Sub Kim
Ganesh Kiran
Dong-Ik Ko
Hakduran Koc
Amit Kohli
Aman Kokrady
Heon-Mo Koo
Makesh Kothandaraman
Elias Kougianos
Sumitha Krishnamurthi
Gopalakrishnan Perur Krishnan
Ramakrishnan Krishnan
Vyas Krishnan
Nagendra Krishnapura
Bruce Krogh
Bram X Kruseman
Sudhir S Kudva
Anshul Kumar

Shashi Kumar
Abhaya Kumar
Anil KV Kumar
Ashok Kumar A
Phani Sudheendra Kumar
Ranjith Kumar
Sandip Kundu
Prasad kuppa
Tomas Lang
Erik Karl Axel Larsson
Kusum Lata
Kyoungwoo Lee
Feihui Li
Hong Li
Lin Li
Peng Li
Zheng Li
Ingchao Lin
Loganathan Lingappan
Karthikeyan Lingasubramanian
Greg M Link
Jing-Jia Liou
Jinfeng Liu
Yongpan Liu
Yung-Hsiang Lu
Anuj Madan
Patrick H Madden
Gabor Madl
Paolo Maffezzoni
Rabi N Mahapatra
Santanu Mahapatra
Kingsuk Maitra
Ashis Maity
Ananta K Majhi
Deyasini Majumdar
Franco Maloberti
Mahesh Mamidipaka
Pradip Mandal
Prasanth Mangalagiri
Stefan Mangard
Rajit Manohar
Sujan Manohar
Radu Marculescu
Radu Marculescu
Yehia Massoud

Sunil H Matange
Jimson Mathew
Mahesh M Mehendale
Dinesh P Mehta
Vinod Menezes
Salvador Mir
Prabhat Mishra
Tania Mishra
Durga Misra
Debasis Mitra
Raj S Mitra
Raja Mitra
Bijitendra Mittra
Siamak Mohammadi
Kartik Mohanram
Saraju P Mohanty
Paras Lal Mohla
Rajat Moona
A. El mourabit
Matthieu Moy
Jayanta Mukherjee
Debdeep Mukhopadhyay
Satyakiran Munaga
Amol Jitendra Mupid
Rajeev Murgai
Venkatesan Muthukumar
Madhu Mutyam
Rajiv M Nadig
Nagi Naganathan
Krishnaswamy Nagaraj
Kiran Nagaraja
Veerapaneni Nagbhushan
Srinath Robin Naidu
Anindya Sundar Nandi
Soumitra Kumar Nandy
Harihar Narayanan
Vijay Narayanan
Sudha Natarajan
Venkatesh Natarajan
Zainalabedin Navabi
Michael Nicolaidis
Nicola Nicolici
Chrysostomos A. Nicopoulos
Michael Thaddeus Niemier
Arthur Nieuwoudt

Rishiyur S Nikhil
Mehrdad - Nourani
Nagaraj NS
Adrian Nunez
Nahmsuk Oh
Satoshi Ohtake
Hidetoshi Onodera
Marco Ottavi
Ozcan Ozturk
Ravi R Pai
David Z Pan
Preeti Ranjan Panda
Rajendran Panda
Pankaj Pandekar
Jagdish Narayan Pandey
Harindranath Parameswaran
Sri Parameswaran
Rubin A. Parekhji
Chetan D Parikh
Sudeep Pasricha
Nirav Patel
Sachin Patkar
Amit Patra
Kolin Paul
Shanthi Pavan
Chung-Ching Peng
Matthew Pirretti
Ilia Polian
Irith Pomeranz
Salvatore Pontarelli
P. G. Poonacha
Nagaraju Pothineni
Nachiketh Potlapally
Anil Prabhakar
Ramya Prabhakar
Kanti Prasad
Rajendra Pratap
Rudra Pratap
Razvan Racu
Damu Radhakrishnan
Arvind Raghavan
Raghavendra R G
Vijay Raghunathan
Hafizur Rahaman
Karthik Rajagopal

Subramanian Rajagopalan
S. Ramachandran
Rajaraman Ramanarayanan
Sornavalli Ramanathan
S. Ramesh
Parvinder Rana
Nagarajan Ranganathan
Abhishek Ranjan
Jagdish C Rao
Madhu Rao
K.R.K. Rao
Shyam Rapaka
Srivaths Ravi
C. P. Ravikumar
Harish M Rawlani
Arijit Raychowdhury
Sudhakar M. Reddy
Syam Sundar Reddy
Harinath Renukamurthy
Matteo Sonza Reorda
Andrew Jonathan Ricketts
Rituparna
Peng Rong
Garrett S. Rose
Aurobinda Routray
Matt Rowley
Soumyaroop Roy
Subir Kumar Roy
Shanq-Jang Ruan
Goutam Saha
Kaushik Saha
Aryabartta Sahu
Debapriya Sahu
Vineet Sahula
Kewal K. Saluja
Sanjeev Saluja
Padmini Sampath
Raja Kiran Kumar Reddy Sandireddy
Juan Carlos Martinez Santos
Sachin S Sapatnekar
Hendra Saputra
Smruti Ranjan Sarangi
B. K. Sarkar
Adil Sarwar
Prashant Saxena

Jayashree Saxena
Peter C.S. Scholtens
Michael J. Schulte
Henry Selvaraj
Siddhartha Sen
Mainak Sen
Dimitrios Serpanos
Sharad C. Seth
Kavish Seth
Dharin N Shah
Kalpesh Amrutlal Shah
Vishal Shah
Li Shang
Akbar Shareef
Anmol Sharma
Dinesh Sharma
Mohit Sharma
Nithin Shastri
Farhana Sheikh
Rupesh S Shelar
Shuo Sheng
Khushboo Umeshkumar Sheth
Zhijie Shi
Sean X. Shi
Aviral Shrivastava
Sachin Shrivastava
Shagufta Siddique
Biplab K Sikdar
B.V.N. Silpa
Montek Singh
Adit D. Singh
Amarjeet Singh
Jawar Singh
Jaya Singh
Raj Singh
Shashank Singh
Virendra Singh
Vigyan X Singhal
Vipul Kumar Singhal
Amith Singhee
Rajaram Sivasubramanian
Alastair Smith
Milind A. Sohoni
Vassos Soteriou
Ramalingam Sridhar

Sridharan K
Shekhar S Srikantaiah
Srinivasan Srinath
Jithendra Srinivas
Srinivasa R. STG
H. C. Srinivasaiah
MR Hariharan Srinivasan
S. Srinivasan
Srinivasan Raj
Suresh Srinivasan
Venkataraman Srinivasan
Vinoo N Srinivasan
Ankur Srivastava
Saket Srivastava
Michiel Steyaert
Subbarangaiah K
Mukherjee, Subhashish
Sivaramakrishnan Subramanian
Jairam Sukumar
Fei Sun
Sudhakar Surendran
Susmita Sur-Kolay
Emil Talpes
Siddhartha V. Tambat
Baris Taskin
Mohammad Tehranipoor
Pradip A Thaker
Samit Thange
Theocharis Theocharides
Brandon Thompson
Narayanan V Thondugulam
Praveen Tiwari
Saurabh Kumar Tiwary
Andrea M Tonello
Kim Yaw Tong
Nur A Touba
Patrick Traynor
Udayakumar H.
Nachiket Urdhwareshe
Balaji Vaidyanathan
Venkat Rao Vallapaneni
Ankush Varma
Vinita Vasudevan
Kamakoti Veezhinathan
Venkataraghavan S K

Mahalingam Venkataraman
Ravi Krishnan Venkatesan
R Venkatraman
Ajay K. Verma
Vijaykrishnan N.
Vikram K N
Janakiraman Viraraghavan
Anant Vishnoi
V. Visvanathan
Natarajan Viswanathan
G. S. Visweswaran
Srinivas kumar Vooka
Duncan M. Walker
Xiaofang Wang
Fan Wang
Feng Wang
Liang-Kai Wang
Yu Wang
Yu Wang
Lan Wei
Xiaoqing Wen
Gunter Winkler
Kai-Chiang Wu
Kaijie Wu
Xiaoxia Wu
Hans-Joachim Wunderlich
Yuan Xie
Jiang Xu
Gefu Xu
Rajesh Yadav
Haihua Yan
Aditya Yanamandra
Yonghong Yang
Chunhua Yao
Nitin V Yogi
Wenjian Yu
Xiaoming Yu
Morteza Saheb Zamani
Nicholas H Zamora
Jindrich Zejda
Wei Zhang
Danella Zhao
Changyun Zhu
Jianwen Zhu
Wang, Zuoding

VLSI Design 2007
Fellowship Recipients

Menaka Devi T	Adhiyamaan College of Engineering
Durga Devi	Anna University
P. Sakthivel	Anna University
N. Ramadass	Anna University
S. Natarajan	Anna University
J. Raja Paul Perinbam	Anna University
M Srinivasa Rao	B.M.S College Of Engineering
S. Kaja Mohideen	B.S.A Crescent Eng. College
E.N. Ganesh	B.S.A Crescent Eng. College
Prasun Ghosal	BECT, Sibhpur, WB
Hafizur Rahaman	Bengal Eng. College Shibpur
Dr. Hemangee K. Kapoor	Dhirubhai Ambani Inst of ICT
Dhirendra Mathur	Govt. Eng College, Ajmer
K. Kishore Kumar	ICFAI Hyderabad
Madhu Mutyam	IIIT Hyderabad
Ujjwal Maulik	Jadavpur university
Sheetal Umesh Bhandari	IIIT Pune
S. Dasgupta	IIT Roorkee
S. Ramanarayan Reddy	IPU Delhi
Sanghamitra B.	ISI Kolkata
Chandan Sarkar	Jadavpur University Kolkata
D. Mukhopadhyay	Jadavpur University.kolkata
Reema Agarwal	Jaipur Engineering College
M.Shanthi	Kumaraguru College of technology
D.Boolchandani	Malaviya National Institute of Technology
S.M. Rezaul Hasan	Massey University, New Zealand
Udipi Cholayya Niranjan	MIT Karnataka
Ghanshyam Choudhary	MNIT, Jaipur
Srikanta	National Institute of Eng, Mysore
Vimal P. Singh Thoudam	NER Inst of Science and Tech
A.K. Panda	NIST Orissa
G.V. Kiran Kumar	NIST Orissa
M. Suresh	NIST Orissa
Suchismita Roy	NIT Durgapur
Thavasi Raja G	NIT Trichy
Sreehari Rao Patri	NIT Warangal
Partha Sarkar	Orissa Engineering College
Bhuvaneswari M C	PSG College, Coimbature
A. Natarajan	PSG College, Coimbature

Rajkumar Nalliah	Ramakrishna Eng. College, Coimbature
S. Jayanthy	Ramakrishna Eng. College, Coimbature
UMA B.V.	RVCE, Bangalore
K.V.Padmaja	RVCE, Bangalore
Venugopal C.R	S J College of Eng., Mysore
M Laxminarasimha	S.R Eng. College, Warangal
A. Muruganandam	SCT, Salem
Perala Prasada Rao	SERC Warangal
S.R.Biradar	Sikkim Manipal Institute of Tech
Sudhakar Reddy	SKIT, Srikalahasti
Jagadeeswari	Sri Ramakrishna Engineering College
V. Vaithinathan	SSN College of Engineering
Poornima G.R	SVCE Bangalore
Vekata Siva Reddy	SVCE Bangalore
Harpet Vohra	Thapar Institute of Engg and Tech
Vijayan s	Travancore Engineering College, Kerala
Rajat Kumar Pal	University of Calcutta
Samrat Lagnajeet Sabat	University of Hyderabad
Rakesh K Vaid	University of Jammu
K.S. Gurumurthy	UVCE Bangalore
P.Sakthivel	Vellalar College of Engg. Technology
S. Karthik	VIT, Vellore
S.Sivanantham	VIT, Vellore
D. S. Harish Ram	VLB Janakiammal College of Eng
Pachkawade Vinayak Ashok	VNIT Nagpur
B.Gopala Krishnayya	VNR VJIT Hyderabad
Srinivas Rao M	VNR VJIT Hyderabad
Balaji N	VNR VJIT Hyderabad
Roji Morjorie	VNR VJIT Hyderabad
C.D. Naidu	VNR VJIT Hyderabad
Shaila Subbaraman	Walchand College of Engineering, Sangli
Amlan Chakrabarti	University of Calcutta
Khalil I. Mahmoud	Anna University
Nitin Yogi	Auburn University
Sudip Ghosh	BESU Shibpur
P.Rangababu	Hyderabad Central University
Mamatha S	IIIT Hyderabad
J.V.R Ravindra	IIIT Hyderabad
Deepak G.C	IISc Bangalore
Siva Rama Krishna	IISc Bangalore
Balaji Jayaraman	IISc Bangalore
M.R. Arulalan	IISc Bangalore
Basavaraj Talwar	IISc Bangalore
Subhasis Banerjee	IISc Bangalore
Bishnu Prasad Das	IISc Bangalore

C. Venkatesh	IISc Bangalore
Satyam Dwivedi	IISc Bangalore
Kusum Lata	IISc Bangalore
Vrajesh D. Maheta	IIT Bombay
Angad B Sachid	IIT Bombay
Sudhakar Shankarro Mande	IIT Bombay
Ramesh R Navan	IIT Bombay
Pankaj Jha	IIT Bombay
Brajesh Pandey	IIT Bombay
Marshnil Dave	IIT Bombay
Sonali Chouhan	IIT Delhi
Lava Bhargava	IIT Delhi
Nagaraju Pothineni	IIT Delhi
Rama Shankar Sharma	IIT Delhi
Sukhendu Deb Roy	IIT Delhi
Neeraj Goel	IIT Delhi
Aryabartta Sahu	IIT Delhi
Anant Vishnoi	IIT Delhi
Thimmappa S.D	IIT Delhi
Saurabh Chaudhury	IIT Kharagpur
Chandan Giri	IIT Kharagpur
Ansuman Banerjee	IIT Kharagpur
Tapas Kumar Maiti	IIT Kharagpur
Sushanta K. Mandal	IIT Kharagpur
Soumya Pandit	IIT Kharagpur
Gopal Paul	IIT Kharagpur
Santosh Biswas	IIT Kharagpur
P. Rakesh Babu	IIT Kharagpur
Santanu Kundu	IIT Kharagpur
Subrat Kumar Panda	IIT Kharagpur
Pravanjan Choudhury	IIT Kharagpur
Shyamala Ravi	IIT Madras
Noor Mahamad S.K	IIT Madras
Santosh Kumar Vishvakarma	IIT Roorkee
Balwinder Raj	IIT Roorkee
Najeeb K	IIT, Madras
Pritha Banerjee	ISI Kolkata
Swarup Kumar Das	ISI Kolkata
Nagaraj Reddy	Jadavpur University
Anup D.	Jadavpur University
Nuka Siva Sankara Reddy	JNTU Hyderabad
Ulhas Deshmukh	MNIT Jaipur
Ashis Kumar Mal	NERIST Nirjuli
Shanthala S	NMAM IT
Omid Kaveh	NUI Iran

Rajamani Sethuram	Rutgers
V.S.Kanchan Baskaran	SSN col of Eng
Shu-Ming Chang	Tatung University
Rajesh Garg	Texas A&M University
Aseem Gupta	UCI
Karthik Baddam	University of Southampton
Chitranjan K Singh	University of Texas, Dallas
Sushma Honnavara prasad	University of Texas, Dallas
Vishnu vimjam	Virginia Tech Univ
Srihari Veeramachaneni	IIIT Hyderabad
M Sudhakar	IIIT Hyderabad
R.V Kamala	IIIT Hyderabad
Pratap Kumar Das	IISc Bangalore
Viveka K.R	IISc Bangalore
Jagdish Narayan Pandey	IISc Bangalore
Abhilasha Kawle	IISc Bangalore
Sathe Chaitanya	IISc Bangalore
Rakesh Nalluri	IIT Delhi
Sujan Kundu	IIT Kharagpur
Pradipta Patra	IIT Kharagpur
Debashis Mandal	IIT Kharagpur
Monu Kedia	IIT Kharagpur
Sayak Ray	IIT Kharagpur
Samiran Ganguly	Indian School of Mines, Dhanbad
Deblina Sarkar	Indian School of Mines, Dhanbad
Sanjiv Mangal	VNIT Nagpur
Abhishek Chaudhary	VNIT Nagpur
Shantanu A. Bhalerao	VNIT Nagpur
Rahul M. Badghare	VNIT Nagpur
J.V.R Akshaya	B.S.A Crescent Eng. College
Nasrullah Habeeb M	GCE Kannur
Mohd Abubakr	GRIET Hyderabad
Robin Singh	IIIT Hyderabad
Pendyala Krishna Seshu	IIIT Pune
Arpit Garg	MNIT Jaipur
Garima Khandpur	MNIT Jaipur
I. Radhika	NIT Warangal
Amandeep Singh	PEC, chandigarh
Aaswin Sreenivasan	PSG College, Coimbature
A. Andrew Arul Rose	SSN College of Engineering
Kumara K. Arjuna Chirayu	UVCE Bangalore
Varaprasad Gandi	VIT Vellore
Mohammed Fahmitha	B.S.A Crescent Eng. College
Suleesh R.	GCE Kannur
Swapnil Mishra	IIIT Hyderabad

Sravanthi Mantha	IIIT Pune
Parul Jindal	MNIT Jaipur
Ravinder Kumar Bansal	MNIT Jaipur
Rajan Arora	PEC, chandigarh
Shankar Raj. V	SSN College of Engineering
Suresh Kumar	UVCE Bangalore
Mohamed Yousuff C	AHCET Vellore
Balaji. V	Anna University
Ramalatha Marimuthu	Anna University
Junaid Yousuf	BIT Pilani
Darshan Shah	D.Y. PIET Pimpri, Pune
Vivek Jha	ICFAI Hyderabad
Lingamneni Avinash	IIIT Hyderabad
Kiran T. Nathan	IIIT Hyderabad
Mayank Agarwal	IIIT Hyderabad
Raghunandan C.	IIIT Hyderabad
Rishov Biswas	IIIT Kolkata
Nayan B. Patel	IISc Bangalore
Thejas	IISc Bangalore
Arvind M	IISc Bangalore
Madhusudan Srinivasan	IISc Bangalore
Rakesh Gana David J	IISc Bangalore
Shuaeb Fazeel H M D	IISc Bangalore
Raja Reddy Patukuri	IISc Bangalore
Nageswararao Pedapati.	IIT Bombay
Srikanth Kurra	IIT Delhi
Santosh Kumar	IIT Delhi
Pushkal Tripati	IIT Delhi
Tushar Khadtare	IIT Delhi
Rakesh G	IIT Delhi
Rahul Nagarajan	IIT Delhi
Prashanth Reddy Gade	IIT Guwahati
Kshitij Yadav	IIT Kharagpur
Rahul Bhattacharya	IIT Kharagpur
Baidurya Chatterjee	IIT Kharagpur
Vishnu Konda	IIT Madras
Shyam Shroff	IIT Madras
Chandra Sekhar Yapara	IIT Madras
N. Karthikeyan	Madras Institute of Technology
Srinivas R	MCIS Manipal
Yogesh Agarwal	MGM JNU, Maharashtra
Kaushik Vaidyanathan	MIT Chennai
Shiva Shankar	NIT Surathkal
Lakshmi Lalitha Pragada	RIT, Yanam
Saurabh Kotiyal	SIT Jaipur

Satyam Trivedi	TIET Patiala
Ritesh Garg	University of Texas, Austin
Krishna Kunchaparthi	VEDA IIT
K. Kranthi Kiran	VEDA IIT
Rosni Basu	West Bengal University
N H Satyanarayana Manyam	NIT Trichy
Mukesh P.R	SCE Salem
Raghavendra Pavan M	NITK Surathkal
Krishna Chaitanya C	NITK Surathkal
Ashish Choudhary	RGPV Bhopal
Devendra Kumar	ZHCET, AMU
M.VijayaRaju	Hyderabad Central University
Abhijit Hazra	JIS College of Engineeering
Susanta Chakrabarti	University of Shibpur
Basavaraj Gorguddi	BVB College of Eng., Hubli
Harish Padaki	BVB College of Eng., Hubli
Prafulla Patil	Walchand College of Engineering, Sangli
Anup Kulkarni	Walchand College of Engineering, Sangli
Nixon Vincent	GECP, Kerala
Anu Asokan	GECP, Kerala
Santanu Kapat	IIT Kharagpur
Gaurav	IIT Bombay
C. Y. Gopinath	BIT Bangalore
Chandra Sekharappa K	BIT Bangalore
Gururaj V. Naik	IISc Bangalore
Koganti Venkata Harikrishna	IISc Bangalore
B. Satish Anand	IISc Bangalore
S.A. Kannan	IISc Bangalore
Mohammed Shareef	IISc Bangalore
Lakshmipathi	IISc Bangalore
Manodipan Sahoo	IISc Bangalore
Janakiraman V	IISc Bangalore
SriLakshmi.N.B	UVCE Bangalore
Bhanumathi S M	UVCE Bangalore
Prathibha B S	UVCE Bangalore
Shilpa Soman	UVCE Bangalore
Pradeep P A	UVCE Bangalore
Sangnamitra Ghosh	JIS College of Engineeering
Rupan Mukharjee	IIT Kharagpur

VLSI Design 2007
Best Paper Awards

Prof. Arun Kumar Chaudhury Best Paper Award (Tie between two papers):
"Defect-Aware Synthesis of Droplet-Based Microfluidic Biochips"
     by Tao Xu, Krishnendu Chakrabarty (Duke University)

"Dynamically Optimizing FPGA Applications by Monitoring Temperature and Workloads"
     by Phillip H. Jones, Young H. Cho ,John W. Lockwood (Washington University at St. Louis)

Best Student Paper Award:
"Low Power Pipelined TCAM Employing Mismatch Dependent Power Allocation Technique"
    by K.R.Viveka, Abhilasha Kawle and Bharadwaj S Amrutur (IISc Bangalore)

Honorable Mention Award:
"Spectral RTL Test Generation for Microprocessors"
     by Nitin Yogi and Vishwani D. Agrawal (Auburn University)

Design Contest Winners:
Youngil Ahn, Jinsub Park, Young-Dae Kim, Jonghwa Choi, Seungyoul Kim & Younggap You
from Dept. of Information Communication Eng, Chungbuk National University, Korea for their
design titled
"A Security PDA Systems based on Crypto-Processor"

and the second place has been awarded to
Md Safiullah, S. Chakrabarti & K. S. Dasgupta from G. S. S. School of Telecommunications,
Indian Institute of Technology, Kharagpur, India for their design titled,
"FPGA Implementation of a Burst QPSK Modem for Satellite Applications"

VLSI Design Conference History

Meeting Sequence	Place	Dates	Number of Papers	Number of Posters	Number of Tutorials	Proceedings Pages
First	Madras, India	Dec. 26-28, 1985	29	0	1	193
Second	Bangalore, India	Dec. 15-18, 1988	26	21	4	496
Third	Bangalore, India	Jan. 6-9, 1990	30	22	4	390
Fourth	New Delhi, India	Jan. 4-8, 1991	45	16	9	315
Fifth	Bangalore, India	Jan. 4-7, 1992	57	24	4	378
Sixth	Bombay, India	Jan. 3-6, 1993	70	9	6	371
Seventh	Calcutta, India	Jan. 5-8, 1994	87	0	6	448
Eighth	New Delhi, India	Jan. 4-7, 1995	77	6	6	456
Ninth	Bangalore, India	Jan. 3-6, 1996	75	16	6	480
Tenth	Hyderabad, India	Jan. 4-7, 1997	84	18	6	608
Eleventh	Chennai, India	Jan. 4-7, 1998	98	0	6	624
Twelfth	Goa, India	Jan. 7-10, 1999	103	0	6	682
Thirteenth	Calcutta, India	Jan. 3-7, 2000	93	0	6	590
Fourteenth	Bangalore, India	Jan. 3-7, 2001	77	0	9	592
Fifteenth	Bangalore, India	Jan. 7-11, 2002	109	0	8	834
Sixteenth	New Delhi, India	Jan. 4-8, 2003	84	0	6	622
Seventeenth	Mumbai, India	Jan. 5-9, 2004	120	44	8	1132
Eighteenth	Kolkata, India	Jan. 3-7, 2005	113	23	9	922
Nineteenth	Hyderabad, India	Jan. 3-7, 2006	136	0	11	880
Twentieth	Bangalore, India	Jan. 6-10, 2007	147	0	15	990
Twenty First	Hyderabad, India	Jan. 4-8, 2008	108	0	10	780

Embedded Systems Conference: History

Meeting Sequence	Place	Dates	Number of Papers	Proceedings Pages
First	New Delhi, India	Jan. 2-4, 2002	8	70
Second	New Delhi, India	Jan. 4-8, 2003	84	622
Third	Mumbai, India	Jan. 5-9, 2004	120	1132
Fourth	Kolkata, India	Jan. 3-7, 2005	113	922
Fifth	Hyderabad, India	Jan. 3-7, 2006	136	880
Sixth	Bangalore, India	Jan. 6-10, 2007	147	990
Seventh	Hyderabad, India	Jan. 4-8, 2008	108	780

VLSI Design 2008 Plenary Invited Keynote Speakers

Saturday, January 5, 2008

Inaugural Keynote	**Dirk Meyer** **President and COO, AMD**

Sunday, January 6, 2008

Plenary Keynote I	**About new paradigm with/in India - why experience is sometimes a bottleneck** **Wally Rhines** **CEO, Mentor Graphics**
Plenary Keynote II	**System design challenges and solutions for future nanoscale process technologies** **Alain Artieri** **Senior System Architect, STMicroelectronics**
Banquet Keynote	**TBA**

Monday, January 7, 2008

Plenary Keynote I	**Design considerations for next generation micro-power systems** **Anantha Chandrakasan** **Professor and Director, Microsystems Tech. Labs, MIT**
Plenary Keynote II	**Partnerships for success: DFM, DFT and the value of accelerated yield learning the IFM world** **Michael Campbell** **Senior Vice President of Engineering, Qualcomm**
Banquet Keynote	**FPGA: The future platform for transforming, transporting and computing** **Ivo Bolsens** **Vice President and CTO, Xilinx**

Tuesday, January 8, 2008

Plenary Keynote I	**Embedded processor design challenges & opportunities for converging multi-media and communications applications** **Ramesh Senthinathan** **Senior Director of Engineering, Broadcom**
Plenary Keynote II	**Trends in telecommunication and their impact on VLSI** **Santanu Das** **President and CEO, TranSwitch**
Plenary Keynote III	**Marc Tremblay** **CTO, SUN Microsystems**
Plenary Keynote IV	**Emerging technologies for information and signal processing** **Pinaki Mazumder** **Program Director, Emerging Models and Technologies, National Science Foundation, USA**
Plenary Session	**The EKA Supercomputer Plenary Session:** **Supercomputers: The Eka experience** **Sunil Sherlekar, TCS** **Anatomy of the Eka supercomputer** **N. Seetha Rama Krishna, TCS** **Applications of supercomputing to nanoelectronics** **Rajendra Patrikar, TCS**

Tutorials

21st International Conference on VLSI Design

Gateway to Chips: High Speed I/O Signalling and Interface (Full Day)

Nidhir Kumar - (Project Leader) *ARM Embedded Technologies.* nidhir.kumar@arm.com
Senthil N. Velu - (Technical Lead Engineer) *ARM Embedded Technologies.*
senthil.velu@arm.com
Rajan Verma - (Senior Design Engineer) *ARM Embedded Technologies.* rajan.verma@arm.com
Contact Address: *ARM Embedded Technologie*s, Salarpuria Hall Mark, Level 3. Marthahalli,
Bangalore – 560087

Abstract

The design of inputs and outputs to integrated circuits has traditionally been a straightforward task involving procurement of a specification and its implementation. In the past few technology generations design and implementation of integrated circuit I/O's have become much more complex. Just as Moore's Law predicts that functions per chip will double every 1.5 – 2 years to keep up with consumer demand, there is a corresponding demand for processing electrical signals at progressively higher rates. The International Technology Roadmap for Semiconductors (ITRS) predicts the I/O bandwidth (Gb/s) for high performance ASICs to be 30 Gb/s by the year 2015. Adding to the complexity is the need to conform to a plethora of emerging I/O specifications and continued focus on reliability regarding Electro Static Discharge (ESD) and Simultaneous Switching Noise (SSN), and the circuit designer has about as much challenges as one can stand. This tutorial presents the techniques and methods employed to build a low power, high bandwidth, highly reliable I/O. It covers the popular signaling standards like LVDS, DDR, XAUI and PCI-Express. Also to be covered are concepts of ESD and Signal Integrity. This section of the tutorial will cover the origins of ESD failures in chips, circuit and layout guidelines to avoid ESD failures and ESD testing procedures. Finally, the tutorial will give a detailed architectural overview of various emerging I/O's such as the DDR, LVDS, and the USB-PHY.

Speaker Biographies

Senthil N. Velu received a Bachelors degree in Electrical and Electronics Engineering from Madras University in 1998 and a M.S. degree in Electrical and Computer Engineering from North Carolina State University in 2002. He joined the Centre for Circuit and Systems Solutions at Carnegie Mellon University as a member of Research Staff in 2002. At Carnegie Mellon he was involved in building a analog automation tool to optimize multiple design objectives in PLL and ADC design. In 2004 he joined ARM Physical IP as Technical Lead of the Analog Mixed Signal – I/O Circuit Design Group. His thesis and subsequent research has involved design and optimization of analog / mixed signal circuits. He has authored and presented several papers in IEEE and RFIC conferences and journals on this topic. His work has received an Honorable Mention Award at the RFIC conference (2000). His currents interests include Analog circuit design automation and high speed interface circuit design.

Nidhir Kumar received a Bachelors degree in Electrical Engineering from Delhi University in 1997. He was involved in High Speed I/O design at S.T. Microelectronics between 1998 and 2000. Since 2000, he has been working extensively in the area of IO and mixed signal domain at ARM Physical IP. He was instrumental in leading the design teams of high speed IOs including DDR3 and GDDR3 at 1.6Gbps. He also has prior experience in PLL design. At ARM, he leads the design and development activities in high speed IO design.

Rajan Verma received a Bachelors degree in Electronics Engineering from Punjabi University. He Joined Semiconductor Complex Limited, a govt. of India organization, and played a key role in design and development of several analog and mixed signal products. He also made a significant contribution in development of High speed data PHYs like USB2.0 and 10/100 Ethernet PHY. Currently he is working as senior design engineer at ARM and his current interests include analog and mixed signal design and high speed interface design.

21st International Conference on VLSI Design

DFM / DFT / SiliconDebug / Diagnosis (Full Day)

Srikanth Venkataraman, *Intel Corporation*
Nagesh Tamarapalli, *AMD India Design Center*

Abstract

Semiconductor yield has traditionally been limited by random particle-defect based issues. However, as the feature sizes reduced to 0.13 micron and below, systematic mechanism-limited yield loss began to appear as a substantial component in yield loss. In addition, it is becoming clear that ramping yield would take longer and final yields would not reach historical norms. A key factor for not reaching previously attained yield levels is the interaction between design and manufacturing. Yield losses in the newer processes include functional defects, parametric defects and issues with testing. Each of these sources of yield loss needs to analyzed and understood by designers and tool developers. In addition, new techniques and methods must be devised to minimize the impact of these yield loss mechanisms. After an introduction of the issues involved in the first section, the second section covers Design-for-Manufacturing (DFM) techniques to analyze the design content, flag areas of design that could limit yield, and make changes to improve yield. However, once the changes are made it is necessary to quantify their impact so that knowledge about yield contribution of different features can be fed back to design and DFM tools. Test presents an opportunity to close the loop by crafting test patterns to expose the defect prone features during automatic test pattern generation (ATPG) and by analyzing silicon failures through diagnosis to determine the features that are actually causing yield loss and their relative impact. The third section covers design techniques (DFX) to improve testability, debuggability and diagnosability, and DFM and defect aware test generation to both meet product quality and expose yield issues at test. Section four covers the basic concepts and theoretical aspects of debug and diagnosis including algorithmic IC diagnosis, scan chain diagnosis, critical path based techniques and diagnosis of delay defects. The applications of the basic concepts and techniques for silicon debug are covered in section five. Section six covers the application of statistical diagnosis techniques to determine the features that are actually causing yield loss and their relative impact. Finally, in section seven, future trends, challenges and directions are covered.

Speaker Biographies

Srikanth Venkataraman is a Principal Engineer at Intel Corporation in Hillsboro, OR. He manages an R&D group responsible for developing CAD tools for diagnosis, debug and test quality applications in the Design and Technology Solutions group. He has successfully developed and deployed several tools in test and diagnosis used all across Intel. His research interests include the areas of VLSI Test (product design for testability and test CAD), Fault diagnosis, Design Verification and Debug, CAD for VLSI, S/W Engineering and Development. He received his Ph.D. in Electrical and Computer Engineering from the University of Illinois at Urbana-Champaign. He has worked at Texas Instruments and View Logic Systems (Sunrise Test System). He has over 60 publications, 3 patents issued and 2 patents pending. He received the best paper award at IEEE VLSI Test Symposium 2000, top 10 papers at IEEE International Test Conference 2000 and the best panel at the IEEE VLSI Test Symposium 99. Intel awards include an Intel Achievement Award (2006), five Divisional Recognition Award (2000, 2002, and 2004), Technical Recognition Award (2002), Excellence Award (2001), Discover Award (2000), best papers at Intel Design and Test Technology Conference (2002, 2003). He has presented a tutorial on diagnosis, DFM, test and debug at the IEEE Design Automation Conference 2006, ISQED 2007, VLSI Test Symposium 2006, 2004 and 2003, IEEE International Test Conference 2004, Design Automation and Test in Europe 2004, European Test Symposium 2006, VLSI Design Conference 2006 and International Symposium on Testing and Failure Analysis 2006, 2004 and 2003. He is a member of IEEE, IEEE Computer Society and ACM.

978-1-4244-3039-0/08 $25.00 © 2008 IEEE

Nagesh Tamarapalli is with AMD India Design Center in Bangalore, India, where he manages a group engaged in DFT and manufacturing test development for the next generation microprocessors. Prior to AMD, he was with Mentor Graphics DFT group where he worked on logic BIST, test compression and diagnosis tools. He has published in leading test conferences such as International Test Conference, Asian Test Symposium and journals such *IEEE Transactions on CAD*. A paper he co-authored at International Test Conference 1999 on logic BIST has been selected for Honorable Mention Award. This and another paper he co-authored at International Test Conference have been selected for "significant papers from the past 35 years". He is the co-inventor of 11 approved/pending US patents in the area of testing. He has delivered DFT seminars in India and USA at several venues including VLSI Design conference 2006 and ISQED 2007. He holds MS in Electrical Engineering from Indian Institute of Technology, Kharagpur, India and PhD in Electrical Engineering from McGill University, Montreal, Canada.

21st International Conference on VLSI Design

Oversampling Analog-to-Digital Converter Design (Full Day)

Shanthi Pavan, *IIT, Madras*
Nagendra Krishnapura, *IIT, Madras*

Abstract

Analog-to-Digital converters (or Sigma-Delta) converters have now become routine aspects of high-performance signal processing, ranging from precision audio to RF transceivers. In this tutorial, we will present, in a systematic fashion, the basics and design aspects of delta-sigma data converters, along with a case study of a high performance ADC designed for digital audio. The intended audience is analog/mixed signal designers with limited prior exposure to over sampling converters and graduate students. Anyone interested in designing, simulating and testing such converters should benefit greatly by attending this tutorial.

Speaker Biographies

Shanthi Pavan is an Assistant Professor of Electrical Engineering at the Indian Institute of Technology, Madras in Chennai. He obtained the B.Tech degree in Electronics and Communication Engg from the Indian Institute of Technology, Madras in 1995 and the masters and doctoral degrees from Columbia University, New York in 1997 and 1999 respectively. From 1997 to 2000, he was with Texas Instruments in Warren, New Jersey, where he worked on high speed analog filters and data converters. From 2000 to June 2002, he worked on microwave ICs for data communication at Bigbear Networks in Sunnyvale, California. Since July 2002, he has been with the Electrical Engineering Department of the Indian Institute of Technology, Madras, where he teaches, conducts research and consults for several companies in the areas of high speed analog circuit design and signal processing. Dr. Pavan serves on the editorial board of the *IEEE Transactions on Circuits and Systems: Part II Express Briefs.* Apart from having taught several short courses in industries (like Texas Instruments, ST Microelectronics, National Semiconductor, Genesys Microsystems and many others), Dr. Pavan has offered a tutorial on SystemLevel Aspects of A/D Converter Design in the VLSI Design Conference in Hyderabad in 2006. He is also involved in increasing analog mixed signal competence in India through the KRK Foundation and its activities.

Nagendra Krishnapura is an Assistant Professor of Electrical Engineering at the Indian Institute of Technology, Madras in Chennai. He obtained the B.Tech degree in Electronics and Communication Engg from the Indian Institute of Technology, Madras in 1996 and the masters and doctoral degrees from Columbia University, New York in 1998 and 2000 respectively. Between 2000 and 2005, he worked as a senior design engineer at Celight, Inc. and Multilink (later Vitesse Semiconductor) where he designed integrated circuits for high speed communications. At the Electrical Engineering Department of the Indian Institute of Technology, Madras he teaches, conducts research and consults for several companies in the areas of high speed analog circuit design and signal processing.

Dr. Krishnapura has been an adjunct faculty at Columbia University, New York, where he taught several advanced courses in the analog/mixed signal area. He is also involved in increasing analog mixed signal competence in India through the KRK Foundation and its activities.

978-1-4244-3039-0/08 $25.00 © 2008 IEEE

Programming and Performance Modelling of Automotive ECU Networks (Half Day)

Samarjit Chakraborty, *National University of Singapore*
S. Ramesh, *General Motors R&D, India Science Laboratory, Bangalore*

Abstract

The last decade has seen a phenomenal increase in the use of electronic components in automotive systems, resulting in the replacement of purely mechanical or hydraulic-implementations of different functionalities. Today, in high-end cars, it is common to have around 70 electronic control units (ECUs), each consisting of programmable processors, one or more microcontrollers and a set of sensors and actuators. Different functionalities (e.g. adaptive cruise control or anti-lock braking) are then implemented in a distributed fashion with parts of a task being mapped onto one or more ECUs and these ECUs exchanging messages and signals via high-speed communication buses. The heterogeneity and the distributed nature of these implementations, coupled with the emergence of new standards and protocols for the automotive domain have given rise to new challenges – both in terms of programming large-scale ECU networks, as well as in evaluating their performance and timing properties. This tutorial will provide a comprehensive overview of the recent developments in this domain and also highlight some of the challenges facing embedded systems designers and programmers. The topics covered will include time-triggered architectures for implementing safety-critical applications, emerging protocols for the automotive domain such as FlexRay, techniques for performance and timing analysis of FlexRay-based ECU networks, and languages and tools for developing distributed implementations of automotive functionality around FlexRay and other related protocols. Apart from discussing the relevant protocols, languages and modelling/analysis techniques, the tutorial will also cover practical case studies and some commercially available tools and their functionality.

Speaker Biographies

Samarjit Chakraborty is an Assistant Professor of Computer Science at the National University of Singapore. He obtained his Ph.D. in Electrical and Computer Engineering from ETH Zurich in 2003. For his Ph.D. thesis, he received the ETH Medal and the European Design and Automation Association's "Outstanding Doctoral Dissertation Award" in 2004. His work has also received Best Paper Award nominations at DAC 2005, CODES+ISSS 2006 and ECRTS 2007. Samarjit's research interests are primarily in the area of system-level design and analysis of real-time and embedded systems. He has extensively published in major research forums on this topic including DAC, DATE, CODES+ISSS, ASP-DAC, RTSS and RTAS, and has also served on the technical program committees of many of these conferences. He is currently serving as the TPC Co-Chair of the Hardware/Software Co-design track of the 2007 IEEE Real-Time Systems Symposium (RTSS), TPC Chair of the Hardware/Software Co-design track of the 2007 International Conference of Embedded and Ubiquitous Computing (EUC) and the TPC Co-Chair of the 2007 IEEE Workshop on Embedded Systems for Real-Time Multimedia (ESTIMedia). Over the last few years he has been working on various problems related to performance modelling and analysis of real-time and embedded systems and teaches a graduate-level course on this topic at NUS. He is also interested in studying various models, protocols and architectures in the context of automotive systems and collaborates with General Motors R&D in this area. He has given invited talks on various topics related to design, modelling and analysis of embedded systems at various universities and industrial labs, including UC Berkeley, MIT, CMU, Philips, General Motors and Creative Technology Labs. His experience with conducting tutorials include (i) a tutorial at the IEEE International Conference on Multimedia & Expo (ICME) at Amsterdam in July 2005, entitled *"Multimedia Processing on Multiprocessor SoC Platforms: What should Multimedia System Developers know about Architectural Design, Performance Analysis and Platform Management?"* (jointly with Radu Marculescu from CMU and Paul Stravers from Philips

Research), (ii) a half-day solo tutorial at the ACM Multimedia Conference (MM) at Santa Barbara in October 2006 on *"Flexible Modelling and Performance Debugging of Real-Time Embedded Multimedia Systems"*, (iii) a tutorial at the VLSI Design Conference at Bangalore in January 2007 on *"Performance Debugging of Complex Embedded Systems"* (jointly with Abhik Roychoudhury from NUS), and (iv) a tutorial at the ARTIST2 Winter School on Modelling, Testing, and Verification for Embedded Systems (MOTIVES) at Trento, Italy in February 2007 on *"Interactive Performance Debugging of Real-Time Systems"*.

S. Ramesh is a Technical Fellow at the General Motors India Science Laboratory in Bangalore, India. As a technical fellow, he plays a key role in setting up a Centre of Excellence in rigorous control software engineering for automotive embedded systems. As part of this Centre, he is involved in directing a group of young researchers in devising and developing rigorous methods and tools for model-based development and verification of distributed embedded software. S. Ramesh has more than 20 years of research experience in the areas of high level language design, validation and verification of distributed systems, embedded software and hardware designs. Earlier, he was a Professor in the Dept of Computer Science and Engineering, Indian Institute of Technology, Bombay. At IIT Bombay, he was also the head of the Centre for Formal Design and Verification of Software which he co-founded. As part of the Center activities, he carried out many industry-sponsored projects on verification of hardware and embedded software. S. Ramesh has also been actively involved in many national and international collaborative research projects. He has published more than sixty papers in international conferences and journals, co-edited a few conference and workshop proceedings, special issues in international journals, been panelists and on the program committee for many international conferences, and refereed several papers for many international journals and conferences.

Architecture Exploration for Low Power Design (Half Day)

Vinod Kathail, *Synfora, Inc.*
Tom Miller, *Sequence Design, Inc.*

Abstract

Increasingly SoC design is driven by integrated mobile devices such as cell phones, music players and hand-held game consoles. These devices rely on standard algorithms such as H.264, 802.11n, or JPEG2000, which allow room for innovative implementations that can result in differentiated products. Designers of these devices are constantly on a tread-mill to integrate more and more features within tight time-to-market constraints. They also play a very delicate balancing act to meet the aggressive goals for power, area and performance.

Power consumption, rather than area, is an increasingly key design parameter for these designs, and architectural and micro-architectural choices at the high level have a significant impact (sometimes an order of magnitude) on the power consumption. For example, reducing the operating frequency and increasing the parallelism to maintain constant throughput reduces the power consumption, but at the expense of additional area. Currently, designers have no practical method to perform such architectural exploration for low power design. Instead, they rely on their intuitions and past experiences — implementation of multiple design alternatives is not a viable option because of the high cost of manual design.

Application engine synthesis, for example as provided by Synfora's PICO (Program-In Chip-Out) product, enables the design of a complete subsystem such as H.264 encoder from an un-timed C algorithm. A designer can explore multiple design alternatives by changing the design constraints or the C code.

An ESL design-flow that integrates application engine synthesis with an industry-leading RTL power estimation technology, such as Sequence Power Theater, enables a designer to explore multiple algorithms and architectures with different power profiles to determine the optimal algorithm-architecture combination in a very short period of time. As an example, it took less than a day to perform four different designs for an imaging pipeline for a digital camera. The power consumption for these designs ranged from 24mW to 8mW, a 3X change in power consumption.

This tutorial will describe in detail and demonstrate an ESL design flow for architectural exploration to determine low power designs. Participants in this tutorial will learn:

1. The impact of architectural and micro-architectural choices on power consumption

2. How architectural exploration reduces power and how to examine power/performance/area tradeoff using PICO Express and Power Theater

3. How Exploring algorithmic variations reduces power consumption

4. Integration into an existing SoC design methodology

We will use real-life examples such as an imaging pipeline for a camera to illustrate the process

Targeted Audience and Prerequisites: SoC architects, SoC project managers and RTL design engineers interested in low power design, architectural exploration and moving to a higher level of abstraction for design to reduce time to market.

Speaker Biographies

Dr. Vinod Kathail is CTO and one of the founders of Synfora, Inc. Previously, he was the R&D Program Manager and a principal scientist at Hewlett-Packard Laboratories and had overall responsibility for the PICO project. At HP Labs, he was one of the architects of a new style of parallel computer architecture, now known as EPIC, and he invented many architectural concepts and compiler techniques for EPIC architectures. He has more than 10 patents and numerous publications in prestigious journals in the area of processors architecture and high-performance compilers. Vinod received his Sc.D. degree in Electrical Engineering and Computer Science from MIT.

Tom Miller is General Manager, Logical Business Unit and VP of R&D for Sequence Design, with over 22 years experience in the EDA industry. He was one of the original founders of Sente Inc. While at Sente Inc., Mr. Miller was the Vice President of Engineering focusing on IC power estimation products. Prior to his position at Sente, Mr. Miller was the Vice President of Engineering for the System Physical Group at Cadence Design Systems, dealing with the physical design and electrical analysis of printed circuit boards. He also held the position of Vice President of Engineering for the PCB Group at Valid Logic Systems. Mr. Miller holds a BS in both Mathematics and Computer Science, as well as a MS in Computer Science from the University of Illinois. He has published papers on software development techniques and hardware description languages, and holds a patent for RTL power estimation methodologies and techniques.

Memory Design and Advanced Semiconductor Technology

D. Harame, *Systems and Technology Group, IBM* (dharame@us.ibm.com)
Subramanian S. Iyer, *Systems and Technology Group, IBM* (ssiyer@us.ibm.com)
Josef S. Watts, *Systems and Technology Group, IBM,* (jswatts@us.ibm.com)
Rajiv Joshi, *Systems and Technology Group, IBM,* (rvjoshi@us.ibm.com)
John E. Barth Jr., *Systems and Technology Group, IBM*

Abstract

This tutorial will provide a bottom-up view of the changes in semiconductor memory design as we move into the nanometer regime. We begin by discussing the breakdown of scaling and the power problem. As innovation replaces classical scaling we investigate the use of stress engineering to improve device level performance. Technology challenges in lithography and interconnects are addressed. The consequences of innovation and scaling on RF/Analog characteristics must also be considered. The scaling of memory presents yet another challenge.

We proceed to discuss the modeling of these effects for the circuit designer including discussion of the many new and traditional sources of variation. We describe how these are characterized how they can be controlled by layout rules and how the remaining variation can be describe in the model to enable Statistical Timing and other advanced circuit techniques.

At the circuit level we consider in detail embedded DRAM and SRAM design for both bulk and SOI. We discuss the benefits and challenges of advanced technologies including methods for creating robust designs in the presence of manufacturing variation. We also discuss the design innovations required to utilize advanced technologies for overcoming the "memory wall", "power wall" and "ILP wall".

Speaker Biographies

David Harame received the PhD in Electrical Engineering from Stanford University in 1984. He has authored or co-authored over 154 articles and holds 16 patents. He is an IBM Fellow of the IBM Corporation and IEEE Fellow.

Subramanian S. Iyer is Distinguished Engineer and Chief Technologist for the Semiconductor Research and Development Center, IBM Systems & technology Group. He obtained his B.Tech in Electrical Engineering at the Indian Institute of Technology, Bombay, and his M.S. and Ph.D. in Electrical Engineering at the University of California at Los Angeles.

Josef Watts received the BS degree in nuclear engineering from the University of Wisconsin, Madison in 1975, the AM degree in physics from Dartmouth College, Hanover, NH in 1987 and the PhD degree in electrical engineering from Purdue University, West Lafayette, IN in 1995. He is currently on assignment with IBM in Bangalore, India and is a senior technical staff member.

Dr. Rajiv V. Joshi is a research staff member at T. J. Watson research center, IBM. He received his B.Tech degree from Indian Institute of Technology (Bombay, India), M.S degree from Massachusetts Institute of Technology and Doctorate in Eng. Science from Columbia University, USA.

John Barth is developing SOI embedded DRAM macros for high performance microprocessor cache applications. He received his BSEE degree from Northeastern University Boston, MA in 1987, and MSEE degree from National Technological University (NTU) Fort Collins, CO in 1992.

978-1-4244-3039-0/08 $25.00 © 2008 IEEE

Scan Delay Testing of Nanometer SoCs (Full Day)

Dr Adit D. Singh, *Auburn University* (adsingh@eng.auburn.edu)

Abstract

Delay defects that degrade performance and cause timing related reliability failures are emerging to be a major concern in nanometer technologies. Extensive at-speed functional testing to screen out such defects can be prohibitively expensive. Scan based structural delay tests are being pursued as a possible cost effective solution to this problem. However, recent research indicates that several formidable challenges must be overcome before such an approach can be fully effective. These include poor delay test coverage, and inaccuracies in the observed circuit timing due to false paths, power supply noise, clock stretching etc. This tutorial aims at a comprehensive discussion of these challenges and proposed solutions, aided by data from recently published industrial studies from Intel, IBM. TI, Freescale, LSI Logic, and universities.

Speaker Biography

Adit D. Singh is James B. Davis Professor of Electrical and Computer Engineering at Auburn University, where he directs the VLSI Design and Test Laboratory. His technical interests span all aspects of VLSI test and reliability. He has published one hundred fifty research papers, served as a consultant for several major semiconductor companies, and holds international patents that have been licensed to industry. He has held leadership roles at international test conferences, including serving as General Chair of the 2000 IEEE VLSI Test Symposium, the 2003 IEEE Defect Based Test Workshop, and the 2004 IEEE Memory Test Workshop. He also currently serves on the editorial boards of *IEEE Design and Test Magazine*, and JETTA. Dr. Singh is a Fellow of IEEE, and is presently Vice Chair of the IEEE Test Technology Technical Council. He can be reached at email: adsingh@auburn.edu.

21st International Conference on VLSI Design

Cross-Layer Approaches to Designing Reliable Systems using Unreliable Chips (Full Day)

Fadi Kurdahi, *UCI*
Nikil Dutt, *UCI*
Ahmed Eltawil, *UCI*
Sani Nassif, *IBM*

Abstract

The design for manufacturing and yield (DFM&Y) is fast becoming an indispensable consideration in today's SoCs. Most current flows only consider manufacturability and yield at the lowest levels: process, layout and circuit. As such, these metrics are treated as an afterthought. With advanced process nodes, it has become increasingly expensive—and soon prohibitive—to guarantee bit level error free chips. The challenge now is to design reliable systems using chips that may have some faults. This has lead to approaches that consider DFM&Y at the system level where more benefit can be reaped, and to consider the problem across the design layers. This tutorial covers cross layer approach to design for DFM&Y spanning from the application all the way to manufacturing, overviews various techniques being explored today, and demonstrates its effectiveness on key applications including wireless, multimedia and imaging. We believe that this tutorial will benefit a large percentage of the attendees at VLSI Design 2008, and should elicit an excellent response at the VLSI Design 2008 conference. he tutorial is intended for application designers, chip architects, managers, CAD tool developers, researchers and students interested in System-on-Chip design, platform-based design methodologies, and trends in design for manufacturing and yield at the system level. Attendees should have basic (undergraduate-level) knowledge of VLSI Design and SoC design flows. Familiarity with architectural concepts such as IP based design, and applications such as wireless and multimedia is desirable, but not required. No specific knowledge of CAD tools or modeling languages is required for this tutorial.

Speaker Biographies

Fadi Kurdahi received his PhD from the University of Southern California in 1987. Since then, he has been a faculty at the Department of Electrical & Computer Engineering at UCI, where he conducts research in the areas of Computer Aided Design of VLSI circuits, high-level synthesis, and design methodology of large scale systems, and serves as the Associate Director for the Center for Embedded Computer Systems (CECS). He was Associate Editor for *IEEE Transactions on Circuits and Systems II* 1993-1995, Area Editor in *IEEE Design and Test* for reconfigurable computing, and served as program chair, general chair or on program committees of several workshops, symposia and conferences in the area of CAD, VLSI, and system design. He received the best paper award for the *IEEE Transactions on VLSI* in 2002, the best paper award in 2006 at ISQED, and other distinguished paper awards at DAC, EuroDAC and ASP-DAC. He has delivered several successful tutorials covering cross-layer design and reconfigurable computing in the past at DAC (1990), IEEE/ICM'93, IEEE/ICM'94, DATE (2003), and IEEE/ICPS 2004. He is a Fellow of the IEEE.

Nikil Dutt received a Ph.D. in Computer Science from the University of Illinois at Urbana-Champaign in 1989, and is currently a Chancellor's Professor in the Center for Embedded Computer Systems at the University of California, Irvine, with academic appointments in the CS and EECS departments. His research interests are in embedded systems, electronic design automation, computer architecture, optimizing compilers, system specification techniques, and distributed systems. He is a coauthor of five books, numerous conference and journal papers, and has received 6 Best Paper Awards (CHDL'89, CHDL'91, VLSI Design 2003, CODES+ISSS 2003, CNCC 2006, ASPDAC 2006). He serves as Editor-in-Chief for *ACM TODAES* and as Associate Editor for *ACM TECS* and *IEEE TVLSI*, and has served in

978-1-4244-3039-0/08 $25.00 © 2008 IEEE

several organizational and technical capacities for the leading IEEE and ACM conferences in design automation and embedded systems. He has organized and delivered successful tutorials on Embedded Memories (ICCAD-98, DAC-99, DATE-2000 and VLSI Design 2001) and on communication architectures (ASPDAC-2006, DATE-2006, VLSI Design-2007).

Sani Nassif received his PhD from Carnegie-Mellon University in the eighties. He worked for ten years at Bell Laboratories on various aspects of design and technology coupling including device modeling, parameter extraction, worst case analysis, design optimization and circuit simulation. He joined the IBM Austin Research Laboratory in January 1996 where he is presently managing the tools and technology department, which is focused on design/technology coupling and includes activities in: model to hardware matching, simulation and modeling, physical design, statistical modeling, statistical technology characterization and similar areas.

Ahmed M. Eltawil received his B. Sc and M. Sc. degrees in Communications and Electrical Engineering from Cairo University, Egypt in 1997 and 1999 respectively. He received his doctorate degree from the University of California, Los Angeles in 2003 with a focus on VLSI architectures for wideband wireless communications. From January 2001 till August 2003 he was director of ASIC Engineering at Innovics Wireless, where he led the development of the first reported diversity enabled third generation W-CDMA mobile station. In January 2005, he joined the University of California, Irvine as an assistant professor in the Electrical Engineering and Computer Science Dept. His current research interests are in advanced digital circuit and signal processing techniques for communication systems. Dr Eltawil holds several awards in his field including being the first Henry Samueli faculty fellow of Engineering at the University of California, Irvine as well as being a co-recipient of the 2006 ISQED best paper award.

OpenSPARC - A scalable Chip Multi-Threading Design (Half Day)

Dwayne Lee, *SUN Microsystems*

Abstract

This tutorial is about OpenSPARC and provides details on the first Chip Multi-Threading 64-bit, 32-thread microprocessor made available as open source under the GNU General Public License (GPL) Audience: Any person who would like to know more about the architecture of this state of the art microprocessor and possibility using it as a teaching vehicle or as a basis for a project

Speaker Biography

Not available

Implementing the Best Processor Cores (Half Day)

Dr. Vamsi Boppana, Senior Director, *Technology Open-Silicon, Inc.* 490 N McCarthy Blvd Suite 220 Milpitas, CA 95035 vamsi.boppana@open-silicon.com

Rahoul Varma, Manager, *Processor Division Core Implementation Bangalore ARM Embedded Technologies* Pvt. Ltd. Level 3, Block B, Salarpuria Hall Mark, Marthahalli - Sarajapur Outer Ring Road, Varthur Hobli, Bangalore 560087 Rahoul.Varma@arm.com

S. Balajee, *Processors Implementation Group Wireless Design Team Texas Instruments* India Bangalore balajee@ti.com

Abstract

It is well-known that varying architectural, technological and implementation aspects of embedded microprocessors, such as ARM, can produce widely differing performance and power specifications. Frequency specifications of high-end realizations are often nearly 2x-3x over vanilla flows. Power optimization techniques used in high-end processor designs have also been reported to have the potential to produce 3x-10x improvements in power over standard flows. This tutorial reviews high-end processor design challenges, techniques and presents state-of-the-art flows for implementing embedded processors. These techniques include processor and architecture selection, verification, selection of technology node/process, selection of macros, selection and optimization of standard cell libraries, design/architecture and power planning, advanced timing and power optimization, design closure, design integration, variability-tolerance, and design-for-manufacturability. The tutorial arms the audience with the best techniques, tools and methodologies to select and achieve the best Silicon for state-of-the-art embedded processors.

Speaker Biographies

Vamsi Boppana received the B.Tech (Hons) in computer science and engineering from the Indian Institute of Technology, Kharagpur, India, in 1993 and the M.S. and Ph.D. degrees from the department of electrical and computer engineering at the University of Illinois at Urbana-Champaign in 1995 and 1997, respectively. He is the Senior Director of Technology for Open-Silicon, Inc, a leading fabless ASIC vendor. Prior to Open-Silicon, Dr. Boppana was the Co-founder and Vice-President of Engineering of Zenasis Technologies, a VLSI design company creating leading-edge automated transistor-level optimization technologies. He has authored or co-authored over 40 technical papers and has six granted patents. He has served on the program committee and as a session chair for several computer-aided design and test conferences, including the International Conference on Computer-Aided Design. His current research interests include all aspects of VLSI design, test and verification. Dr. Boppana received the Indian Institute of Technology TCS Best Project Award in 1993, the University of Illinois Van Valkenburg Fellowship for demonstrated excellence in research in 1995, the Best Paper Award at the VLSI Test Symposium in 1997, and a Fujitsu Laboratories of America Intellectual Property Contribution Award in 1999.

Rahoul Varma BEng(Hons) MCIM graduated in Microelectronics System Design from Brunel University, UK in 1997 and has been with ARM Ltd (Cambridge, UK) for 10 years; Since mid-05 Rahoul has led a team of processor implementation engineers in ARM Bangalore. Before his arrival in India he was the technical lead of the architecture and development of a L2 Cache Controller for ARM11 based products. Rahoul has three patents in Cache design, Trace and Security. He earlier worked in the ARM SoC consultancy group and delivered 5 device level tape outs delivering in the areas from spec to layout. He was also involved in the foundry program where he was technically responsible for the addition of many productized ARM hard macros and silicon qualification projects.

S. Balajee graduated with a Masters degree in Computer Science from BITS, Pilani in 1991. He has been with Texas Instruments for the past fourteen years, contributing in various technical and management roles. He is currently responsible for the processors development group in Wireless India Design Team. Balajee was also elected as Senior Member Technical Staff in 2003.

SESSION A1:
Fault Tolerance

21st International Conference on VLSI Design

A Power Efficient Approach to Fault-Tolerant Register File Design[*]

Mojtaba Amiri-Kamalabad, Seyed Ghassem Miremadi, Mahdi Fazeli
Dependable Systems Laboratory
Department of Computer Engineering
Sharif University of Technology
Tehran, Iran
{mojtaba_amiri, m_fazeli}@ce.sharif.edu, miremadi@sharif.edu

Abstract

Recently, the trade-off between power consumption and fault tolerance in embedded processors has been highlighted. This paper proposes an approach to reduce dynamic power of conventional high-level fault-tolerant techniques used in the register file of processors, without affecting the effectiveness of the fault-tolerant techniques. The power reduction is based on the reduction of dynamic power of the unaccessed parts of the register file. This approach is applied to three transient fault-tolerant techniques: Single Error Correction (SEC) hamming code, duplication with parity, and Triple Modular Redundancy (TMR). As a case study, this approach is implemented on the register file of an OpenRISC 1200 processor. The experimental calculation of the power consumption shows that the proposed approach saves about 67%, 62%, and 58% power for TMR, duplication with parity, and SEC hamming code, respectively.

1. Introduction

Reliability and power consumption are two important factors in many embedded systems applications. Examples of such applications are pacemakers [1], smartcards [2], and defibrillators. Since these portable systems are battery operated with limited battery life, power is of decisive importance in design of such systems.

Trends in CMOS technology are resulting in circuits with higher susceptibility to transient faults or single event upsets (SEUs) (Bit-flips due to the impact of particles, such as neutrons and alpha particles, on flip-flops) [3, 4]. Most of the fault-tolerant techniques that are used in embedded processors use spatial redundancy such as triple modular redundancy (TMR) and duplication [5], or informational redundancy such as parity check and single error correction (SEC) hamming code [5, 6]. All these techniques suffer from high power consumption due to the use of redundancy; therefore cannot be used in embedded applications where power consumption is the major concern. One way to reduce power consumption is by using low power design techniques such as dynamic voltage scaling (DVS) [7]. However, such technique increases the susceptibility of circuits to SEU [8, 9]. Based on the above discussion, low power consumption and high reliability are conflicting objectives in the design of a power efficient reliable embedded processor. Hence, new design techniques are required to address both objectives, simultaneously [2, 10].

One of the most vulnerable components to SEU in a microprocessor is the register file. Studies show that SEUs in the register file mainly lead to data error [11, 12]. In addition, it has been concluded that the register file represents a substantial portion of the power consumption in modern microprocessors [13].

In this paper, we propose an approach that decreases the power consumption of fault-tolerant techniques used in the register file without disrupting the effectiveness of the fault-tolerant techniques. The power reduction is based on clock gating of the register file. The power reduction includes two steps: 1) clock gating of the whole register file when the register file is not accessed at all, and 2) partial clock gating of the register file when a register is accessed. This approach is implemented in the register file of an OpenRISC 1200 microprocessor. The implementation is based on the RTL description and hence is portable to different technology and CAD tools.

The rest of this paper is organized as follows: section 2 provides background on power consumption components and SEU-tolerant techniques. Section 3

---

[*]This work was partially supported by a grant from Iran Telecommunication Research Center (ITRC).

978-1-4244-3039-0/08 $25.00 © 2008 IEEE

describes the proposed approach and implementation scenarios for TMR, Hamming code, and duplication with parity in the register file. Section 4 presents experimental system and results. Finally, Section 5 concludes the paper.

2. Background

2.1. Power consumption components

The power dissipation of COMS digital circuit can be expressed as equation 1. :

$$P_{Total} = P_{dynamic} + P_{Static} \qquad (1)$$

The dominant part of power is currently the dynamic power, although with technology scaling the contribution of static power is no longer negligible [14]. The dynamic power can be express as fallow:

$$P_{dynamic} = \alpha C_L V_{DD}^2 f_{clk} + I_{SC} V_{DD} \qquad (2)$$

where f_{clk} is the clock frequency, α is the average switching activity, V_{DD} is the supply voltage. The first term in equation 2 is switching power that is from charging and this charging of the capacitive load C_L of the circuit. The second term is short circuit power is from current I_{SC} that arises when NMOS and PMOS transistors are simultaneously on. The first term is dominant part of dynamic power [14].

2.1. SEU-tolerant techniques

Continuous technology scaling, smaller devices and lower operating voltages are resulting in circuits with higher susceptibility to transient faults or single event upsets (SEUs) [3, 4]. Traditionally, SEUs where mentioned as a major concern only for space applications. Due to technology shrinkage SEUs have become the major concern, even, at ground level [4]. Mostly a single SEU causes a Single-Bit Upset (SBU). In addition to SBU a possibility exist for Multi-Bit Upsets (MBUs) caused by a single particle. It has been observed that 1-10% SEU can affect multiple adjacent memory cells and cause MBUs in set of satellite experiments [15, 16]. Also, this phenomenon has been seen in FPGAs [17] and integrated circuits with memory structure [18, 19].

The SEU-tolerant techniques utilize some form of redundancy; variations include informational redundancy (redundant data structure), spatial redundancy (redundant hardware), and temporal redundancy (redundant sequential operations) [3]. These techniques have different performance, area, and power overheads. Due to critical applications of embedded processors, techniques with low performance overhead are desirable; hence

informational and spatial redundancies are effectively employed in these processors.

Spatial or modular redundancy, such as TMR and duplication are widely used in embedded processors to protect against SEUs. Although, spatial redundancy involves high area and power overheads, but can effectively tolerate MBUs.

The increase in hardware content in informational redundancy is typically less than spatial modular redundancy hence informational redundancy, inherently, introduce less power and area overheads. Examples of this redundancy are Hamming code and parity check that effectively employed to tolerant SBUs in data structure of embedded processors [5, 6].

Recently, new fault-tolerant techniques for the register file have proposed that considered power consumption. In [12], a duplication-based approach proposed that employed unused registers as duplication of actively-used registers. Although this approach significantly decreases power overhead in comparison to the methods that protect the entire register file, but only 78% register file accesses are to the registers with duplicates. The method proposed in [20], consumes about 80% of the power of a register file with full ECC and could not protect the registers against all the double-bit errors.

3. Proposed Approach

As mentioned in previous section, spatial and informational redundancies are widely used in the register file of microprocessors. The power overhead of these redundancies is a problem. In order to address this problem, DVS as a low power technique is not suitable because it increases the rate of SEU. Another effective approach to reduce dynamic power in low

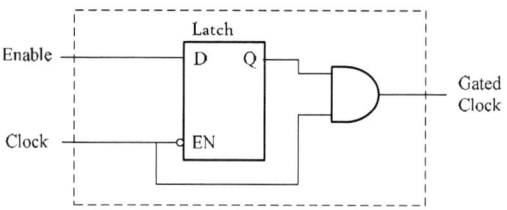

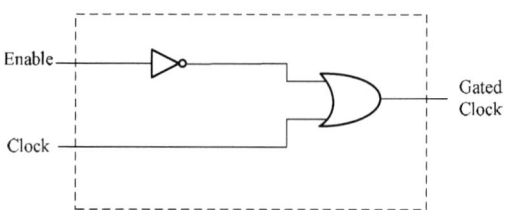

Figure 1. Different clock gating styles

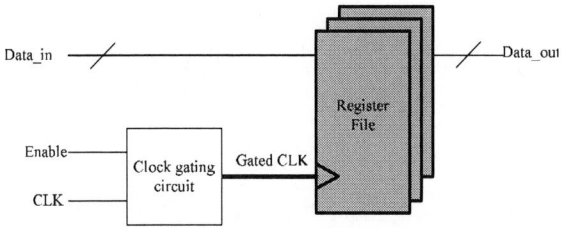

Figure 2. Coarse-grained clock gating in the register file

power design is RTL clock gating [21, 22, 23] that don't affect the effectiveness of the fault-tolerant techniques.

The basic concept in clock gating is that, registers are not clocked when they don't have to write new data. Two styles are generally proposed for clock gating (figure 1). One uses transparent latch, while second uses an OR gate. Although the second style is more area and power efficient, to prevent timing error, that could not be acceptable in reliable design, first style has been used in proposed approach [21]. In traditional clock gating of the register file when data should be write in a register, all registers in the register file will be clocked. This happens because the enable of all registers are equal.

In order to reduce the power consumption overhead of fault-tolerant techniques used in the register file, such as TMR and duplication, partial clock gating

approach is proposed for the register file. In traditional RTL clock gating for the register file, only the register enable signal is used so it gate the clock in gorse grain (figure 2). By this, power could save significantly [21], but it could be better if it be fine grain. The proposed approach demonstrates this. Traditional approach can be preformed automatically by tools such as Synopsys Power Compiler, but our custom approach guarantees better results.

If register write address is used to determine the register enable signal, much power could be saved. We use the register write address lines to gate clock (figure 3). This partial clock gating approach for the register file can partition the register file to different grain parts, from using only most significant bit of write address to the whole address. There is an efficient selection here for power reduction that will be discussed.

The power saving due to clock gating insertion can be expressed by equation 3[21]:

$$P_{Saved} = V_{DD}^2 f_{clk} \left[C_{gated}(1 - \alpha) - C_{add} \right] \qquad (3)$$

where C_{gated} is the total input capacitance of gated circuit, C_{add} is capacitance of additional circuit due to clock gating, α is the switching activity of gated clock. In conventional clock gating for register file C_{gated} is capacitance of all registers of register file. In the case of using register write address used to determine the register enable signal the total power saving for a 32

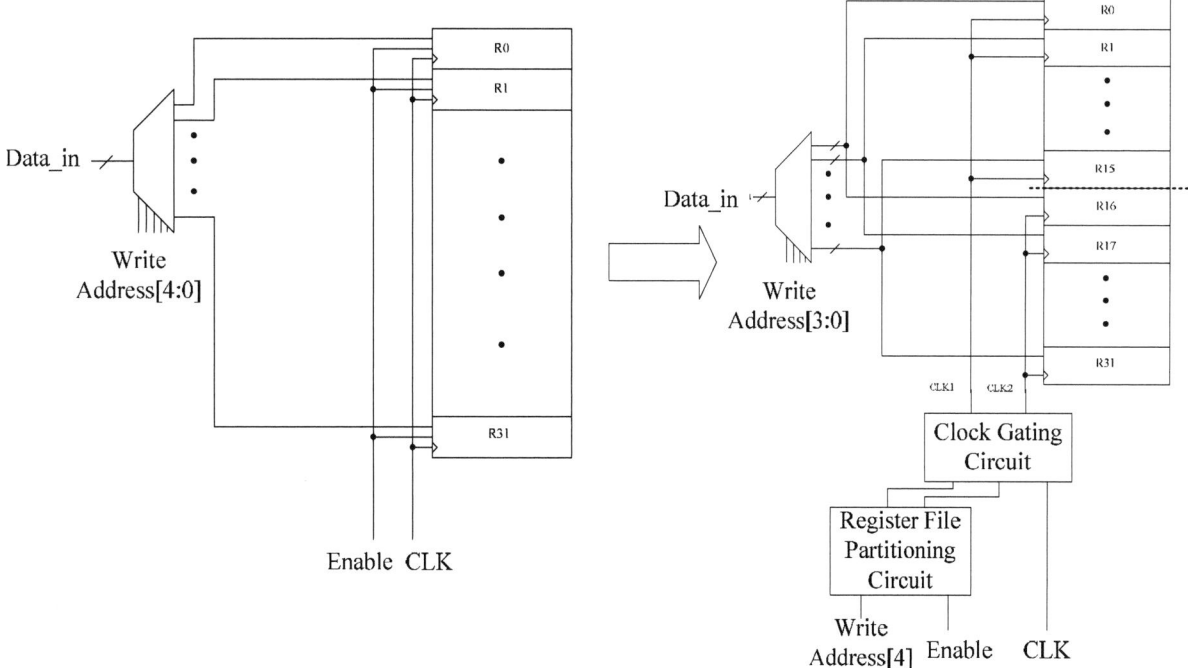

Figure 3. Partial clock gating of the register file

register file can be expressed by equation 4:

$$P_{Saved} = \overbrace{V_{DD}^2 f_{clk} \left[(2^n - 1) \frac{2^n}{32} C_{gated}\alpha - C_{add_n} \right]}^{saved\ by\ custom\ clock\ gating} +$$

$$\overbrace{V_{DD}^2 f_{clk} [C_{gated}(1-\alpha) - C_{add}])}^{saved\ by\ conventional\ clock\ gating} =$$

$$V_{DD}^2 f_{clk}([C_{gated}((1-\alpha) + (2^n - 1)2^{5-n})\alpha) - C_{add_{total}}]) \quad (4)$$

where n is the number of register write address bits that used to determine the register enable signal, C_{add_n} is capacitance of additional circuit due to our custom clock gating because in this method although we add capacitance by adding clock gating we reduce the complexity and capacitance of register write address decoder, $C_{add_{total}}$ is summation of C_{add_n} and C_{add}.

The proposed approach is applied to the register file that employs three effective fault-tolerant techniques i.e. TMR, duplication with parity, and SEC hamming code.

For a TMR implementation of a register file there are two possible manners. One directly implements TMR from three identical register files, another only tripling the storage cells. The latter exploits the fact that the address decoding is equal among all three register file. To save power and area, latter manner is used.

Two parity bits have been used for implementation of the duplication with parity, so that one of them uses the data odd bits for generation of the parity bit, and other one uses the data even bits for generation of the parity bit. As discussed in [24], this method detects at least three bit faults. Here, as discussed about the TMR, duplication of storage cells has been used. Thus,

Table 1. Power and area overheads of conventional techniques in the register file

Overhead	Power			
Benchmark Technique	Bubble sort	Quick sort	Matrix Multiplication	Area
SED Hamming code	35%	37%	36%	28%
Duplication with parity	101%	109%	105%	113%
Triple Modular Redundancy	203%	203%	202%	201%

the duplicates will be used after fault detection.

The SEC hamming code implementation of the register file is as discussed in [25]. In addition, in this method, we have a power efficient substitution because a fine grain enable generation has an overhead itself. So, we could not use all register write address lines to generate enable signals.

4. Experimental system and results

For the experiments, we used OpenRISC 1200 microprocessor [26] as a case study and Synopsys Power Compiler for power results on TSMC 0.18um process. This embedded processor model has been successfully implemented on FPGA and ASIC technologies. In evaluations three benchmarks, quick sort, bubble sort, and matrix multiplication, have been used. The register file of the OpenRISC 1200 microprocessor has two read ports and one write port.

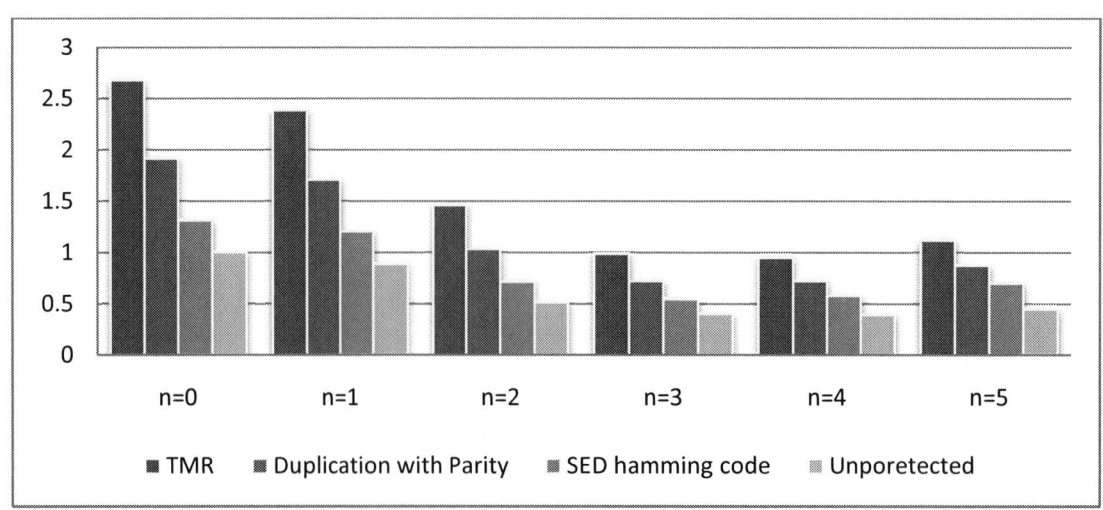

Figure 4. Power reduction achievements with respect to different number of write address bit usage

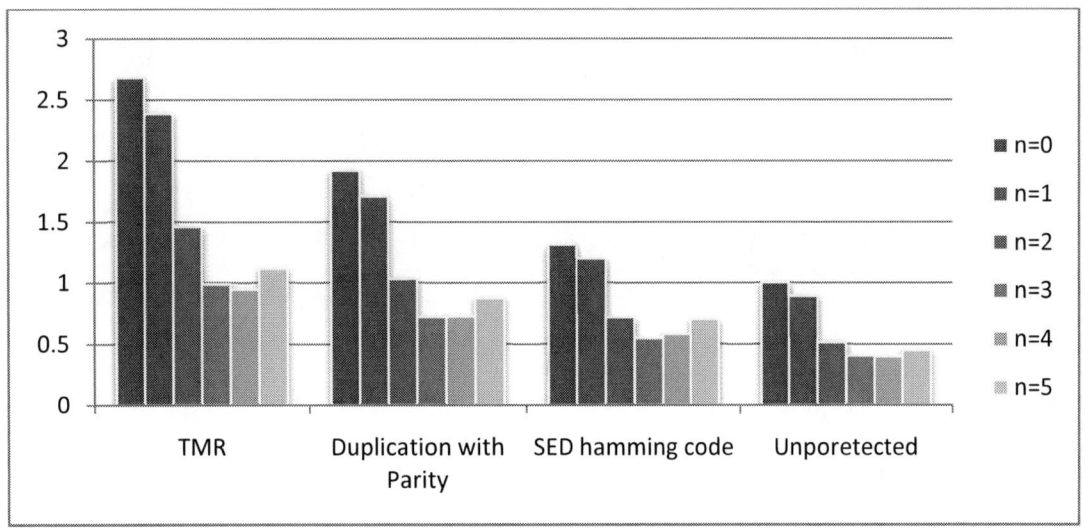

Figure 5. Power reduction achievements for each fault-tolerant technique with respect to different number of write address bit usage

The comparison of power and area overheads of the mentioned fault-tolerant techniques is showed in table 1. This table indicates that SEC hamming code lead to lower overhead. But, as motioned in section 2, it cannot tolerate MBUs that are more susceptible by CMOS technology scaling.

As seen in table 1, the used benchmarks are relatively the same in power consumption. This happens because that they relatively access the register file as the same as each other. So the averages of results of the benchmarks are shown in figure 4 and 5.

The proposed approach is implemented on those fault-tolerant techniques different number of write address usage (n). The results of the figure 4 and 5 are normalized to power consumption of an unprotected register file that does not use our approach (n=0). Figure 4 illustrates that using 3 or 4 write address bits of the register file lead to efficient results. According to equation 4, this is explainable.

As illustrated in figure 5, the proposed approach save more power in TMR and duplication with parity than two others. As mentioned in equation 4, this is achieved due to lager capacitance of these techniques. This is desirable because these fault-tolerant techniques can tolerate MBUs.

5. Conclusions

High fault-tolerance against SEUs and low power consumption are key objectives in design of embedded processors. Since, it has been shown that these two objectives are at odds, i.e. the utilization of fault-tolerance techniques increases power consumption and the utilization of low power techniques increase SEU

rate, new design approaches are required to achieve both objectives simultaneously. Toward this goal, this paper has proposed an approach to reduce dynamic power of the conventional fault-tolerance techniques that used in register file. The power reduction is based on partial clock gating of unaccessed parts of the register file. Experimental results have shown that proposed approach save more power in triple modular redundancy which can tolerate MBUs. This approach can be employed in other data structure of embedded processor with negligible modifications.

6. References

[1] J.C. Knight, "Safety Critical Systems: Challenges and Directions", *Proceedings of the 24th International Conference on Software Engineering*, Orlando, Florida, USA, May 2002, pp. 547-550.

[2] A. Maheshwari, W. Burleson, and R. Tessier, "Trading off Transient Fault Tolerance and Power Consumption in Deep Submicron (DSM) VLSI Circuits", *IEEE Transactions on Very Large Scale Integration Systems*, May 2004, vol. 12, no. 3, pp. 299-311.

[3] W. Heidergott, "SEU Tolerant Device, Circuit and Processor Design", *Proceedings of the 42nd Design Automation Conference (DAC'05),* Anaheim, California, USA, Jun. 2005, pp. 5-10.

[4] C. Constantinescu, "Impact of Deep Submicron Technology on Dependability of VLSI Circuits", *Proceedings of the International Conference on Dependable Systems and Networks (DSN'02)*, Bethesda, Maryland, USA, Jun. 2002, pp. 205-209.

[5] J. Gaisler, "Evaluation of a 32-bit Microprocessor with Built-in Concurrent Error-Detection", *Proceedings of the 26th Annual International Symposium on Fault-Tolerant Computing (FTCS-27)*, Seattle, Washington, USA, Jun. 1997, pp. 42-46.

[6] J. Gaisler, "A Portable and Fault-Tolerant Microprocessor Based on the SPARC V8 Architecture", *Proceedings of the IEEE/IFIP International Conference on Dependable Systems and Networks (DSN'02)*, Bethesda, Maryland, USA, Jun. 2002, pp. 409-415.

[7] A. Ejlali, B. M. Al-Hashimi, M. T. Schmitz, P. Rosinger, and S. G. Miremadi, "Combined Time and Information Redundancy for SEU-Tolerance in Energy-Efficient Real-Time Systems", *IEEE Transactions on Very Large Scale Integration Systems,* Apr. 2006, vol. 14, no. 4, pp. 323-335.

[8] D. Zhu, R. Melhem, and D. Mosse, "The Effects of Energy Management on Reliability in Real-Time Embedded Systems", *Proceedings of the International Conference on Computer Aided Design (ICCAD'04)*, San Jose, California, USA, Nov. 2004, pp. 35-40.

[9] P.Hazucha, and C. Svensson, "Impact of CMOS Technology Scaling on the Atmospheric Neutron Soft Error Rate", *IEEE Transactions on Nuclear Science*, 2000, vol. 47, no. 6, pp. 2586-2594.

[10] R. Melhem, D. Mosse, and E. Elnozahy, "The Interplay of Power Management and Fault Recovery in Real-time Systems", *IEEE Transactions on Computers*, Feb. 2004, vol. 53, no. 2, pp. 217-231.

[11] G.P. Saggese, N.J. Wang, Z.T. Kalbarczyk, S.J. Patel, and R.K. Iyer, "An Experimental Study of Soft Errors in Microprocessors", *IEEE Micro,* Nov. 2005, vol. 25, no. 6, pp. 30-39.

[12] G. Memik, M.T. Kandemir, and O. Ozturk "Increasing Register File Immunity to Transient Errors", *Proceedings of Design, Automation and Test in Europe (DATE'05),* Munich, Germany, Mar. 2005, vol. 1, pp. 586-591.

[13] V. Zyuban, and P. Kogge, "The Energy Complexity of Register Files", *Proceedings of the International Symposium on Low Power Electronics and Design (ISLPED'98)*, Monterey, California, USA, Aug. 1998, pp. 305-310.

[14] L. Benini, G. De Micheli, E. Macii, "Designing Low-Power Circuits: Practical Recipes", *IEEE Circuits and Systems Magazine,* 2001, vol. 1, no. 1, pp. 6-25.

[15] M.K. Oldfield, and C.I. Underwood, "Comparison between Observed and Theoretically Determined SEU Rates in the TEXAS TMS4416 DRAMs and On-Board the UoSAT-2 Microsatellite", *IEEE Transactions on Nuclear Science*, Jun. 1998, vol. 45, no. 3, pp. 1590-1594.

[16] C.I. Underwood, "The Single-Event-Effect Behavior of Commercial-Off-The-Shelf Memory Devices – A Decade in Low-Earth Orbit", *Proceedings of the 4th European*

Conference on Radiation and Its Effects on Components and Systems, Palm Beach, Cannes, France, Sep. 1997, pp. 251-258.

[17] R. Koga, J. George, G. Swift, C. Yui, L. Edmonds, C. Carmichael, T. Langley, P. Murray, K. Lanes, and M. Napier, "Comparison of Xilinx Virtex-II FPGA SEE Sensitivities to Protons and Heavy Ions", *IEEE Transactions on Nuclear Science*, Oct. 2004, vol. 51, no. 5, pp. 2825-2833.

[18] G. M. Swift, and S. M. Guertin, "In-Flight Observations of Multiple-Bit Upset in DRAMs," *IEEE Transactions on Nuclear Science*, Dec. 2000, vol. 47, no. 6, pp. 2386-2391.

[19] R. Koga, K. B. Crawford, P. B. Grant, W. A. Kolasinski, D. L. Leung, T. J. Lie, D. C. Mayer, S. D. Pinkerton, and T. K. Tsubota, "Single Ion Induced Multiple-Bit Upset in IDT 256K SRAMs", *Proceedings of the Second European Conference on Radiation and its Effects on Components and Systems (RADECS)*, St. Malo, France, Sep. 1993, pp. 485-489.

[20] P. Montesinos, W. Liu, and J. Torrellas, "Using Register Lifetime Predictions to Protect Register Files against Soft Errors*", Proceedings of the 37th Annual IEEE/IFIP international Conference on Dependable Systems and Networks (DSN'07)*, Edinburgh, UK, Jun. 2007, pp. 286-296.

[21] F. Emnett, and M. Biegel, "Power Reduction through RTL Clock Gating," *Synopsys User Group*, San Jose, Mar. 2000.

[22] N. Raghavan, V. Akella, and S. Bakshi, "Automatic Insertion of Gated Clocks at Register Transfer Level", *Proceedings of the Twelfth International Conference on VLSI Design*, Goa, India, Jan. 1999, pp. 48-54.

[23] F. Iozzi, S. Saponara, A.J. Morello, and L. Fanucci, "8051 CPU Core Optimization for Low Power at Register Transfer Level", *PhD Research in Microelectronics and Electronics*, Lausanne, Switzerland, Jul. 2005, vol. 2, pp. 178-181.

[24] A. J. Ricketts, M. Mutyam, N. Vijaykrishnan, and M. J. Irwin, "Investigating Simple Low Latency Reliable Multiported Register Files", *IEEE Computer Society Annual Symposium on VLSI (ISVLSI '07)*, Porto Alegre, Brazil, May 2007, pp. 375-382.

[25] R. Naseer, R. Z. Bhatti, and J. Draper, "Analysis of Soft Error Mitigation Techniques for Register Files in IBM Cu-08 90nm Technology", *Proceedings of the 49th IEEE International Midwest Symposium on Circuits and Systems*, San Juan, Puerto Rico, Aug. 2006, vol. 1, pp. 515-519.

[26] OpenRISC 1200 Specification, OPENCORES.ORG, http://www.opencores.org/cvsget.cgi/or1k/or1200

978-1-4244-3039-0/08 $25.00 © 2008 IEEE

21st International Conference on VLSI Design

Reconfiguring CMOS as Pseudo N/PMOS for Defect Tolerance in Nano-Scale CMOS

Maryam Ashouei	Adit D. Singh	Abhijit Chatterjee
Georgia Institute of Technology	Auburn University	Georgia Institute of Technology
ashouei@ece.gatech.edu	adsingh@eng.auburn.edu	chat@ece.gatech.edu

ABSTRACT

End-of-the-roadmap nanoscale CMOS is expected to suffer from significant defectivity due to manufacturing defects, random process variations, and wear-out during normal operational. To ensure acceptable yield and reliable operation of the circuit during its life-time, future circuits must be equipped with significant defect-tolerance capabilities. Traditional defect-tolerance approaches are too expensive to be applied to general purpose circuits. In this paper, we propose a defect-tolerant CMOS logic gate architecture that exploits the inherent functional redundancy in static CMOS. This is accomplished by reconfiguring the CMOS logic gate to a pseudo-NMOS-like gate in the presence of a defect. The resulting defect-tolerant logic architecture incurs only a modest area overhead. The proposed gate design can tolerate defects in either the pull-up or pull-down network of the gate. The architecture can tolerate multiple defects across the logic gates of a CMOS logic circuit. The effectiveness of the proposed defect tolerance technique and its impact on circuit delay and power is studied. It is shown that the technique imposes little delay overhead (less than 6%) but incurs power dissipation overhead (less than 20%) in the presence of defects.

Keywords

Nanoscale CMOS, defect-tolerant, pseudo-NMOS

1. INTRODUCTION

The continued scaling of transistor channel lengths to the limits of CMOS technology over the next decade is projected to require complex new transistor structures such as FinFETS and the trigate transistors, along with specialized high-k gate dielectric materials, and possibly metal gates. Reliably fabricating billions of such complex devices on a die is expected to be a huge challenge, requiring large designs to employ a significant defect tolerance capability [1],[2]. This capability must be designed not only to achieve acceptable yield in the face of numerous manufacturing defects, but also to recover from field failures due to latent defects and wear-out, which are also expect to be a major problem. Current defect-tolerance approaches for random logic have high overhead and can typically handle only a few defects; they are, therefore, generally not practical for this application.

In this paper we present a novel low overhead defect-tolerance approach for CMOS designs capable of efficiently recovering from dozens of defects. Our approach takes advantage of the inherent functional redundancy in static CMOS. In a traditional CMOS gate, the function is in fact implemented twice: once by the N-transistor network and again by the P-transistor network. Our standard-cell-based defect-tolerance approach affects recovery in the presence of faulty transistors in the pull-up P-network by reconfiguring the logic gate implemented by the cell into a pseudo-NMOS gate, replacing the faulty P-network with a single (properly sized) P transistor, which acts as a resistive pull up. Similarly, a fault in the N-network is recovered from by replacing it by a properly sized single N transistor acting as a resistive pull down. Thus, each gate (primitive standard cell) can tolerate any number of defects in either the P-network or the N-network, but not both at the same time. The cost of this extensive defect tolerance capability is a modest increase in cell area, and also increased cell leakage current and delay in both the normal mode, and the reconfigured defect-tolerant mode. We study these trade-offs in this paper through simulation experiments on benchmark circuits.

Note that in this paper our primarily focus is on making the active components of the chip (the standard cells) defect tolerant; we do not explicitly address defectivity in the inter-cell interconnection network. The assumption here is that transistor reliability will be limiting as we approach end-of-the-roadmap CMOS. While interconnect reliability can be enhanced to some measure through conservative design and/or the use of redundancy, it remains an important challenge to be addressed in future work.

The rest of the paper is organized as follows. Section 2 describes prior work in logic circuit defect tolerance design. Section 3 describes the proposed standard cell based defect-tolerance approach. Section 4 presents the overall chip architecture that supports our new approach, including a grid based routing methodology for control signals to support reconfiguration in the presence of defects. In Section 5, simulation results are presented. In Section 6, diagnosis of faulty gates for the purpose of reconfiguration is discussed. Conclusions and future work are discussed in Section 7.

This work was supported by NSF NSF Grant CCR-0635016 and GSRC MARCO 2003-DT-660.

978-1-4244-3039-0/08 $25.00 © 2008 IEEE

2. PREVIOUS WORK

While considerable research has been done on memory built-in self-repair designs [3]-[5], there is not much work on built-in redundancy and repair of logic circuitry. Recently, there were two work proposed on the subject [6]-[7]. Authors in [6] propose to add transistor-level redundancy by having parallel transistors as permanent back-up (active redundancy) similar to what is shown in Figure 1(a) for an inverter. The extra parallel transistors increase the diffusion capacitance of the gate. It also acts as a larger load for the driving gate. These two effects increase the propagation delay of paths with the proposed redundant architecture. At the same time, the redundancy provides higher current drive in absence of defects, which in turn reduces the propagation delay. The simulation results in [6] shows 17% increase in the delay of an inverter chain of 2 when both parallel sets of transistors are working. If one of the sets is defective the delay will increase by 42%. The proposed technique also increases the area and power by 33% and 45% respectively. To minimize the large delay, power, and area overhead associated with the technique, the authors propose to use such redundancy for gates on non-critical paths and with lower switching activity.

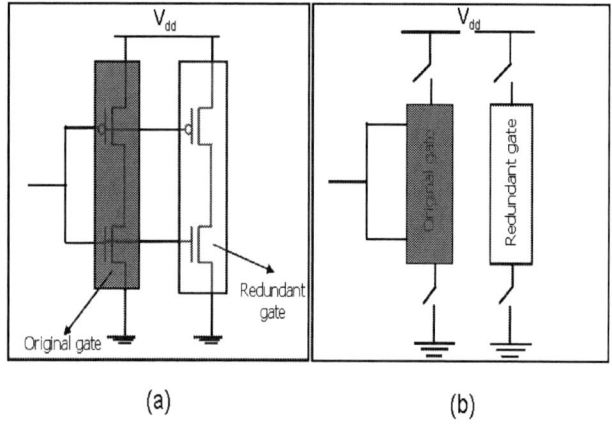

Figure 1. Previously proposed defect tolerant techniuqes

Another drawback of the technique in [6] is that it can only handle very limited defect types. Mainly, it is suitable for open defects. On the positive side, the technique is very simple and does not require any defect diagnosis and built-in self-repair reconfiguration circuitry.

To address some of the issues associated with the above technique, the authors in [7] propose a passive redundancy technique shown in Figure 1 (b). The main advantage of this technique is that it covers more defect types including shorts in each of the parallel branches so long as the gates of the transistors are not shortened and the cell output is not stuck-at. It should be noted that this technique, similar to [6], also suffer from large delay, power, and area overheads. Furthermore, the technique requires extra switches, which will add to the overall overhead.

Additionally, the technique also requires some sort of diagnosis mechanism in order to bypass the defective gates.

To address the need of defect tolerance, we propose a novel low-overhead technique. It imposes relatively low delay and power overhead as compared to previous techniques. Similar to the technique in [7], our proposed technique also requires a defect diagnosis method to identify the defective gates for reconfiguration. The details of the proposed technique are presented in the next section.

3. FAULT-TOLERANT DESIGN OF CMOS GATES

In this section, the concept of defect-tolerant CMOS gates is introduced. Such gates have the capability to endure any number of defects as long as the defects are either in the P-network or in the N-network.

3.1 OVERALL ARCHITECTURE

The architecture of the proposed defect-tolerant CMOS gate is shown in Figure 2. The proposed gate operates in the following way: During the normal operation, i.e. when there is no fault, $switch_1$ and $switch_2$ are connected to V_{DD} and GND respectively. In this case, $switch_3$ ($switch_4$) connects the gate of the pull-up (pull-down) transistor to the V_{DD} (GND) to turn it off.

During the defect-tolerant operation, if the defect(s) is in the P-network, $switch_1$ is turn off to disconnect the faulty P-network from the power supply and $switch_3$ connects the gate of pull-up transistor to the GND. This converts the cell to a pseudo-NMOS cell. The setting of $switch_2$ and $switch_4$ remains the same as the normal operation. Conversely, for a defect in the N-network $switch_2$ is turn off and $switch_4$ turns on the pull-down transistor. Analogous to the pseudo-NMOS case, we call this mode of operation pseudo-PMOS.

The extra resistance contributed by $switch_1$ and $switch_2$ along with the extra junction capacitance contributed by the two pull-up and pull-down transistors, imposes an extra delay during the normal operation of the gate. Such overhead is shown in Table 1 for NAND2 and NOR2. The table also shows the delay and leakage overhead in the case of faulty P-network or faulty N-network. To obtain this data from SPICE, the pull-up and pull-down transistors are assumed to be weak transistors (one-fifth the size of transistors in the N/P-network) and $switch_1$ and $switch_2$ are reasonably sized up. The overhead values were computed by comparing the proposed fault-tolerant gates to their traditional CMOS counter-parts. The delay overhead is moderately low for the defect-free gates, but increases for defective gates.

28

978-1-4244-3039-0/08 $25.00 © 2008 IEEE

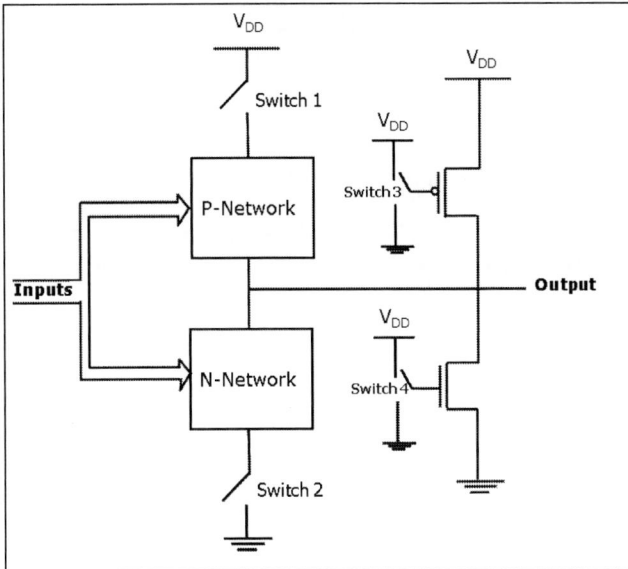

Figure 2. Architecture of the defect-tolerant CMOS gate

Table 1. Delay and leakage overhead of the fault-tolerant cells

Gate Type	Delay Overhead			Leakage Overhead		
	No Fault	P-net Faulty	N-net Faulty	No Fault	P-net Faulty	N-net Faulty
NAND2	7.4 %	37 %	27 %	12%	120 X	135 X
NOR 2	5.9 %	3.8 %	43 %	14.5%	145 X	155 X

The table also shows that there is a large leakage overhead when the cell is reconfigured as a pseudo-NMOS or pseudo-PMOS in the presence of a defect. This is because pseudo-N(P)MOS gates, unlike standard CMOS dissipate power in steady state when the output is low (high). The leakage overhead could be reduced by using higher-threshold-voltage pull-up/pull-down transistors. The delay and leakage overhead of using higher-V_t transistors compared to a traditional CMOS gates are shown in Table 2. The advantage of using a slower transistor in the single pull-up/down branch is that it does not increase the delay overhead in the absence of defects. This can be easily seen by comparing (the second columns of) Table 1 and Table 2. For both cases the delay overhead is 7.4% and 5.9% for the NAND2 and NOR2 respectively. On the other hand, the leakage overhead in the absence of a defect reduces from 12% to 7% for NAND2 and from 14.5% to 9% for NOR2 gates.

When there is a defect in the system, using the higher V_t increases the delay penalty. At the same time, such a higher V_t transistor severely reduced the leakage in the fault-tolerant mode, a reduction of 22% and 18% when the P-network and N-network are faulty respectively. Nevertheless, the leakage increase is still huge (of the order of 100X). We argue that such a huge increase in the leakage

is tolerable as long as there are not many defective cells on a single chip. For example, even if 0.1% of cells are defective, then with the assumption of 100X leakage increase in the defect-tolerant mode, the overall chip leakage will increase by less than 10%. This is because the leakage increase caused by the few defective cells amortizes across the whole chip. The yield increase obtained by salvaging the defective dies through reconfiguring the defective cells into the fault-tolerant mode makes this approach attractive and the increase in the leakage bearable. It is clear that as the defect density (percentage of defective cells) increases, the leakage overhead also increases. For a 1% cell defect ratio, the leakage almost doubles. However, in practice, we expect to target a few hundred defects in designs with several million gates, a cell defect ratio of well under 0.01%.

Before presenting the simulation results, in the next section we discuss the overall framework and architecture for employing the proposed defect-tolerant CMOS standard cell and, in particular, issues related to the layout and routing of the reconfiguration control signals.

Table 2. Delay and leakage overhead of the fault-tolerant cell with high V_t on the pull-up and pull-down transistors

Gate Type	Delay Overhead			Leakage Overhead		
	No Fault	P-net Faulty	N-net Faulty	No Fault	P-net Faulty	N-net Faulty
NAND2	7.4 %	52 %	48 %	7 %	94 X	110 X
NOR 2	5.9 %	16 %	61 %	9 %	113 X	128 X

4. RECONFIGURATION CONTROL ARCHITECUTRE

A major concern with any defect-tolerance scheme is the overhead associated with the reconfiguration control signals, since they must be routed to each reconfigurable cell. For our standard cell based defect tolerant design, we propose grid-based two-dimensional reconfiguration architecture comprised of X and Y control signals. Control signals X_i and Y_j address the standard cell located at (i,j) in the two-dimensional array of standard cells as shown in Figure 3. While for simplicity we assume that all cells are the same size, and that each row has the same number of cells, it is not difficult to extend the scheme to cells with different sizes by designing the grid based on the smallest quantized cell size. Larger cells are assigned multiple column addresses. Thus a cell twice as wide as the basic cell in any row would have two column control signals routed through it, either one of which can be used to address it. While the X and Y grid control signals can address a desired cell for reconfiguration, at least one bit of additional information is needed to distinguish between the two cases where the P network is defective and where the N network is defective. This can be achieved by making at least one of the two control signals two bit wide.

29

978-1-4244-3039-0/08 $25.00 © 2008 IEEE

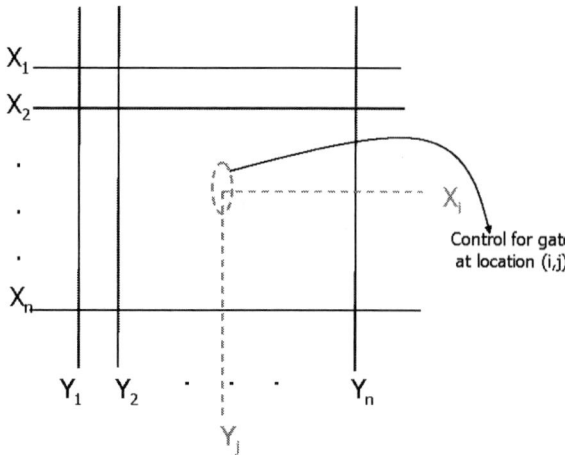

Figure 3. Grid-based reconfiguration control architecture

The main advantage of the 2-dimention technique is that it reduces the routing overhead from n to $n^{0.5}$. One disadvantage of the proposed routing architecture is that it may results in unnecessary reconfiguring some defect-free gates. For example if the gates in location (X_1, Y_1) and (X_2, Y_2) of Figure 3 are defective, with the grid-based routing technique, the gates in location (X_1, Y_2) and (X_2, Y_1) are also considered as defective and will be reconfigured. While unnecessarily reconfiguration gates as pseudo N/PMOS gates will not affect the functionality, it may hurt the power consumption.

An attractive option for implementing the control grid signals is the polysilicon layer in the layout. Note that there are no switching performance constraints on the control signals. These are not signals that switch during normal operation. Thus the reconfiguration control grid can be laid out in the polysilicon layer, which is too slow to be used for the functional interconnect. Just as a fixed pitch metal1 rail is used in standard cell layouts to route the power and ground signals, we propose using horizontal and vertical polysilicon traces to route the reconfiguration control grid, with short metal 1 (or even diffusion) "jumper" connections to allow crossover. *Such a strategy allows the reconfiguration control signals to be implemented with virtually no adverse impact on the availability of the metal layers for use for the functional interconnect.* A regular grid structure in polysilicon is, in fact, consistent with the requirement of ensuring an even trace distribution in each mask layer to guarantee the planarity. In advanced processes, each process layer is checked for trace density and unused areas filled with dummy traces to ensure proper planarization by chemical mechanical polishing (CMP). Our control signal grid can be designed to ensure the same for the polysilicon layer.

Ongoing related research is investigating layout designs for placing this control grid in standard cells. The goal is a design that automatically realizes the reconfiguration control structure in polysilicon for any arrangement of the standard cells during the placement step in physical design. This will allow routing to be performed by the metal layers without any constraints whatsoever. Current placement and routing tools (indeed all design tools) could be used without the need for them to be aware of the defect tolerance capability incorporated in the designs.

5. EXPERIMENTAL RESULTS

To evaluate the yield improvement achievable by the proposed reconfiguration architecture, circuits from the ISCAS'85 benchmark circuits was synthesized using Synopsys Design Compiler with a library of 2-input to 4-input NAND and NOR gates and inverters. The synthesized circuits were used with SPICE 70nm models [7] to compute the propagation delay and power using a look-up table method similar to [9]. Similar SPICE-based look-up table was also generated for the proposed defect-tolerant gates. All gates had a transistor channel length of 70nm, V_{DD} of 1V, V_t of 0.2V. It is assumed that a high threshold of 0.3 V is used for the transistors in the pull-up and pull-down transistors.

For analyzing the delay and the leakage overhead of the proposed defect-tolerant cell architecture, all gates in the ISCAS circuits were replaced by their defect-tolerant counter-parts. The delay and leakage were evaluated for the case with no defect and for the case with 1% defect density. A defect density of X means that on average X% of cells in a circuit are defective. The defect density distribution is assumed to have a Gaussian distribution with the mean X. The defects are assumed to be totally random, i.e. no cluster defects are considered. Table 3 shows the delay, leakage power, and dynamic power overhead for different benchmark circuits. The overhead in the column "No defect" is calculated by comparing the delay/power of the defect-free defect-tolerant circuit and the regular circuit (with no defect-tolerant cells). It can be seen that by making circuit defect tolerant, delay increases by an average of 7%. The leakage and dynamic power increases by 4.7% and 1.7% respectively. Table 3 also shows the delay/power overhead for the case of 1% defect density. The data is computed comparing the average of delay/leakage/dynamic power with the ones of defect-free but defect-tolerant circuit. It can be seen that the delay overhead is very low, always below 6% and the leakage overhead is always below 20%. The dynamic power overhead is negligible. The technique proposed here have much less overhead compared to the ones in [6] and [7]. It should be mentioned that although the delay is not affected significantly in the presence of defects, this result might change if the circuit has too many critical or close to critical paths.

The distribution of delay and leakage increase for different benchmark circuit is shown in Figure 4 and Figure 5. From the delay distribution, one can see that most of the dies have delay increase of less than 5%. In some cases where the defects are mostly on the critical paths, the delay increase is large, but the cases are rare as they are only in the tail of the delay increase distribution. Such dies can perhaps be discarded as bad dies because of their excessive delay.

Table 3. Average percentage of delay and power increase

Circuit	No defect (compared to non-defect-tolerant circuit)			1% defect density (compared to defect-free defect-tolerant circuit)		
	Delay	Leakage	Dynamic	Delay	Leakage	Dynamic
C432	7.8	4.7	3.4	2.6	19.7	2.7
C1908	3.9	5.4	0.1	3.4	17.9	2.7
C2670	7.0	4.9	2.0	1.4	17.3	2.2
C3540	7.1	4.3	1.9	5.9	16.8	2.6
C5315	7.5	4.4	1.2	3.4	16.7	2.8
C7552	8.2	4.7	1.8	2.8	17.1	2.7

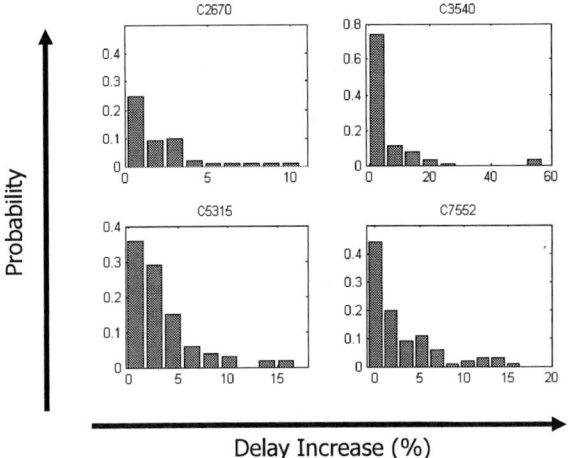

Figure 4. Distribution of the percentage of delay increase

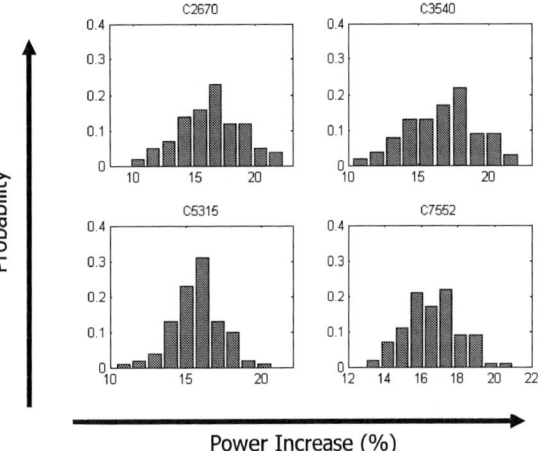

Figure 5. Distribution of the percentage of leakage increase

6. DIAGNOSIS

The key to the success of the proposed defect tolerance approach is effective diagnosis of the defects in the chip. Defect diagnosis is an extensively researched problem [10][11], particularly in the context of physical failure analysis (FA) where the objective is to predict the physical location of a defect through electrical analysis, so that the site can be de-layered and examined for the source of the defect. Experience indicates that while predicting the exact site of the defect on the layout is extremely challenging, sources of faults can generally be located to within small sets of electrical nodes. While this level of resolution may often not be good enough for FA, it may be sufficient for our purposes. Observe that the proposed defect tolerance methodology is clearly best suited to design using complex CMOS gates as standard cells so as to minimize the reconfiguration overhead. Often many of the candidate fault sites indicated by fault diagnosis may fall within the same standard cell, requiring the same reconfiguration action for recovery. Furthermore, if there are multiple defect candidate sites in different gates, the reconfiguration capability also allows repeated trial and testing of potential recovery reconfigurations until the fault effects are eliminated.

It is important to keep in mind that yield recovery is statistical. Not all defective parts may be recoverable. However, we project that just as current memory chips today are capable of recovery from up to a dozen defects, the proposed approach will allow general purpose circuits to tolerate dozens, and perhaps even hundreds of defects, making it viable to implement large complex designs in future technologies that cannot be fabricated free of defects. In the simulation results in this paper, we assume that the diagnosis technique identifies the defective gates with perfect accuracy.

7. CONCLUSION AND FUTURE WORK

In this paper a novel defect-tolerant architecture for standard CMOS cells was presented. Such gates could be used to perform post-manufacture circuit reconfiguration to recover circuits with dozens, even hundreds, of manufacturing defects in nanoscale CMOS technology, where defect free fabrication of large designs is expected to be virtually impossible. The proposed defect-tolerant cell design could also be used to reconfigure the circuit during its life time as defects appear because of the aging of active components. In this first paper, we have shown the potential of this novel approach; how with low delay overhead and modest increase in leakage, dies with a significant number of random defects could be reconfigured for full functionality.

Several areas remain to be addressed in future research. A detailed defect tolerant cell design, with control signal routing to the reconfiguration switches, needs to be fully developed. Using an estimate of the area overhead of such a cell, yield models can be developed to more accurately evaluate the applicability of the proposed approach. The assumption in this research so far has been that the defective cells could be identified using some diagnosis techniques already developed by other researchers. How the accuracy of such diagnosis technique may affect the quality of the reconfigured defect-tolerant circuit is also the subject of future work. Clearly, the area overhead of the proposed cell architecture becomes even more acceptable if the technique is applied to more complex gates rather than the primitive standard cells used in this research. This will also be further studied.

8. REFERENCE

[1]. M. A. Breuer, et. al., "Defect and Error Tolerance in the Presence of Massive Numbers of Defects", IEEE Design and Test of Computers", Vo. 21, No. 3, May-June 2004, pp. 216 – 225.

[2]. Y. Zorian, D. Gizopoulos, "Design for Yield and Reliability" (Guest Editor's Introduction), IEEE Design and Test of Computers, Vol. 21, No. 3, May/June 2004, pp. 177 – 182.

[3]. M. Nicolaidis, et. al., "A Memory Built-In Self-Repair for High defect Densities Based on Error Polarities", IEEE International Symposium on Defect and Fault Tolerance in VLSI Systems 2003, pp.459 – 466.

[4]. Y. Zorian, "Embedded Memory Test & Repair: Infrastructure IP for SOC Yield", International Test Conference 2002, pp. 340 – 349.

[5]. A. Tanabe, et. al, "A 30-ns 64-Mb DRAM with Built-In Self-Test and Self-Repair Function", IEEE Journal on Solid State Circuits, Vol.27. No 11, Nov. 1992, pp. 1525 – 1533.

[6]. N. Sirisantana, et. al, "Enhancing Yield at the End of the Technology Roadmap", IEEE Design & Test of Computers, Vol. 21. Issue 6, Dec. 2004, pp. 563 – 571.

[7]. R. Kothe, et. al., "Embedded Self Repair by Transistor and Gate Level Reconfiguration", IEEE Design and Diagnostics of Electronic Circuits and systems, April 2006, pp. 208 – 213.

[8]. Y. Cao, et. al., "New paradigm of predictive MOSFET and interconnect modeling for early circuit simulation," CICC 2000, pp. 201 –204.

[9].Y. H. Dhillon, et. al. "Soft-Error Tolerance Analysis and Optimization of Nanometer Circuits," DATE 2005, pp. 288 – 293.

[10]. S. Y. Kuo and W. K. Fuchs, "Fault diagnosis and spare allocation for yield enhancement in large reconfigurable PLAs," IEEE Tran. on Computers, Feb. 1992, pp. 221-226.

[11]. P.G. Ryan, W. K. Fuchs, "Dynamic fault dictionaries and two-stage fault isolation," IEEE Tran. On VLSI, March. 1998, pp. 176 – 180.

Single Error Correcting Finite Field Multipliers Over $GF(2^m)$

J. Mathew, A. Costas, A. M. Jabir*, H. Rahaman, D. K. Pradhan
Department of Computer Science, University of Bristol, UK
Oxford Brookes University, UK

Abstract: This paper presents a new method for designing single error correcting Galois field multipliers over polynomial basis. The proposed method uses multiple parity prediction circuits to detect and correct logic errors and gives 100% fault coverage both in the functional unit and the parity prediction circuitry. Area, power and delay overhead for the proposed design technique is analyzed. It is found that compared to the traditional Triple Modular Redundancy (TMR) techniques for single error correction the proposed technique is very cost efficient.

Index Terms: **Error Correcting Codes, Galois Field Multiplier, Cryptography, VLSI.**

1. Introduction

As the integrated circuit density increases, high performance integrated circuits, characterised by high operating frequencies, low voltage levels and small noise margins will be increasingly susceptible to temporary faults. With reducing feature sizes, failures due to causes such as radiation can severely impact field-level product reliability, not only for memory, but for logic as well [3]. On chip error masking techniques such as error correction could be one of the options to mitigate logic error.

Furthermore, it has recently been shown that for many digital signature and identification schemes an attacker can inject faults into the hardware and these incorrect outputs can completely expose the secret signatures [13]. Subsequently, [14] presents a technique showing how the presence of faults in the public parameters of an elliptic curve cryptosystem may expose the secret key. Motivated by the above problems, a Single Error Correcting (SEC) Galois filed multiplier is considered in this paper.

The Galois field multiplier design is one of the most well-researched and widely investigated topics, having great impact on the solvability of large class of design problems in cryptography, coding theory, Galois switching theory and digital signal processing. Over the years, many solutions have been proposed for efficient multiplier design, such as bit-serial, digit-serial and bit parallel. For error detection in finite field multipliers, a number of schemes have been proposed in the literature [1, 7]. One way to detect errors in a finite field multiplier is to use the parity prediction technique [7]. A second approach is to scale the inputs of the multiplier by a factor and at the end of the multiplication, the correctness of the result is checked by one or two divisions [15]. The main techniques that can be used for single error correction are (1) error detection and retry, and (2) error masking. Error detection and retry involves using concurrent error detection (CED) circuitry [10] that monitors the outputs of a circuit for the occurrence of an error. If an error is detected, the system recovers through rollback and retry, thereby preventing a failure. Error masking involves using circuitry that masks (i.e. corrects) errors using schemes such as the TMR.

The technique of [7] has considered detection of single stuck at faults in polynomial basis Galois Field multipliers. They used simple parity prediction technique for error detection. The main problem with their approach is that for a low complexity bit parallel multiplier the delay overhead is 69.2% (Table 1, [11]). For performance critical applications this delay overhead may be critical. Our technique fundamentally differs from this technique in two critical issues. Firstly, our technique addresses the problem of single error correction. Secondly, and more importantly, since our parity prediction circuit runs parallel with the multiplier, delay penalty comes only in the decoding and correction logic. While most of the previous work provides fault detection techniques, this work proposes error correction techniques and investigates the hardware cost and performance.

978-1-4244-3039-0/08 $25.00 © 2008 IEEE

To date, error correction techniques in Galois field multiplier have not been properly explored. The contribution of this paper is a novel single error correcting architecture for Galois field multipliers. Area, power and delay overhead for the proposed design technique are analyzed. It is found that compared to traditional TMR techniques for single error correction the proposed technique is very cost efficient.

The rest of this paper is organized as follows. In the next section, we review the basics of Galois field multiplications. Then, in Section 3, we show how single error correction is achieved. Section 4 presents the experimental results.

2. Background

Let $GF(N)$ denote a set of N elements, where N is a power of a prime number, with two special elements 0 and 1 representing the additive and multiplicative identities respectively, and two operators addition '+' and multiplication '·'. $GF(N)$ defines a Galois Field (also called *finite field*), if it forms a *commutative ring* with identity over these two operators in which every element, apart from 0, has a multiplicative inverse. Finite fields can be generated with primitive polynomials of the form $p(x) = x^{m-1} + \sum_{i=0}^{m} c_i x^i$, where $c_i \in GF(2)$ [5]. It is conventional to represent the elements of $GF(2^m)$ as a power of the *primitive element* α, where α is the root of $P(x)$, i.e. $p(\alpha)=0$. The set $\{1, \alpha, ..., \alpha^{m-1}\}$ is referred to as the polynomial basis (PB). Each element $A \in GF(2^m)$ can be expressed with respect to the PB as a polynomial of degree m over $GF(2)$, i.e. $A(x) = \sum_{i=0}^{m-1} a_i x^i$, where $a_i \in GF(2)$. Given A, B $\in GF(2^m)$, the PB multiplication over $GF(2^m)$ can be defined as $C(x)=A(x).B(x) \bmod P(x)$. Details can be found in [2,6].

Mastrovito has proposed an algorithm, along with its hardware architecture, for PB multiplication [5], which is popularly known as the Mastrovito algorithms/multipliers. The formulation for PB multiplication based on the Mastrovito algorithms is summarized below for completeness [4].

Consider a multiplier with A and B inputs where $A = [a_0, a_1, a_2, ..., a_{m-1}]$ and $B = [b_0, b_1, b_2, ..., b_{m-1}]$. The a_i, and b_i are the coordinates of A and B respectively where $0 \le i \le m-1$. The multiplication outputs are given as follows:
$$c = d + Q^T e \qquad (1)$$

$$d = L \times b \qquad (2)$$
$$e = U \times b \qquad (3)$$
where $b = B^T = [b_0, b_1, b_2, ..., b_{m-1}]^T$.

$$L = \begin{bmatrix} a_0 & 0 & 0 & 0 ... 0 \\ a_1 & a_0 & 0 & 0 ... 0 \\ a_2 & a_1 & a_0 & 0 ... 0 \\ \vdots & & & \\ a_{m-2} & a_{m-3} ... a_1 & a_0 & 0 \\ a_{m-1} & a_{m-2} ... a_2 & a_1 & a_0 \end{bmatrix} \quad U = \begin{bmatrix} 0 & a_{m-1} & a_{m-2} ... a_2 & a_1 \\ 0 & 0 & a_{m-1} & a_3 & a_2 \\ \vdots & & & \\ 0 & 0 ... 0 & a_{m-1} & a_{m-2} \\ 0 & 0 ... 0 & 0 & a_{m-1} \end{bmatrix}$$

L is an $m \times m$ lower triangular matrix and U is an $(m-1) \times m$ upper triangular matrix. Matrix Q is derived from the technique described in [14].

We have derived the Q matrix for the multiplier over $GF(2^4)$ from the first principles in example 1. The general architecture for this implementation is shown in Fig. 1. This structure is divided into two parts: the *Inner Product (IP)-network* and the *Q-network*. The IP-network, which has m blocks, generates the d and e in Eq. (1). For $\{0 \le i \le m-2\}$, each block constitutes two inner product cells, namely, $IP(i+1)$ and $IP(m-i-1)$. However the last block constitutes only one such cell, $IP(m)$. The Q-network takes d and e as inputs and generates c. It constitutes m binary trees of EXOR gates (BTX_0, BTX_1, ..., BTX_{m-1}). For the multiplier structure shown in Fig. 1, the IP-network has a total of m^2 AND gates and $(m-1)^2$ EXOR gates. The maximum number of EXOR gates required for the Q-network depends on the Q-matrix. The multiplier structure is similar to the multiple output Positive Polarity Reed-Muller (PPRM)-like form.

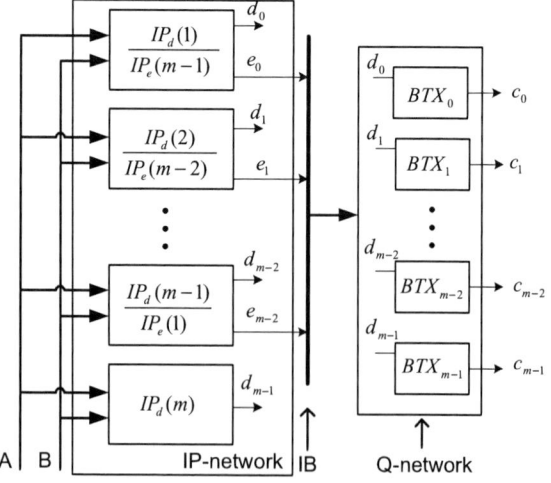

Fig.1 Architecture of the BP Multiplier over GF(2^m) [4]

The number of lines on the interconnection bus (IB) is fixed and is equal to the number of e_js, i.e. m-1. There are three buses, A, B, and IB. For the Q-network, the maximum number of EXOR gates needed can be determined by the technique presented in [14].

Example 1: A multiplier structure over *GF(16)* defined by the primitive polynomial $P(x) = x^4 + x^3 + 1$ is shown in Fig. 2.

The two inputs of the multiplier are $A = (a_0, a_1, a_2, a_3)$ and $B = (b_0, b_1, b_2, b_3)$. The polynomial representation of $GF(2^4)$ elements is as follows.

$A(x) = a_0 + a_1 x + a_2 x^2 + a_3 x^3$, $B(x) = b_0 + b_1 x + b_2 x^2 + b_3 x^3$, where A, B $\in$ GF(2^4). The product $C(x) = A(x) \times B(x)$.

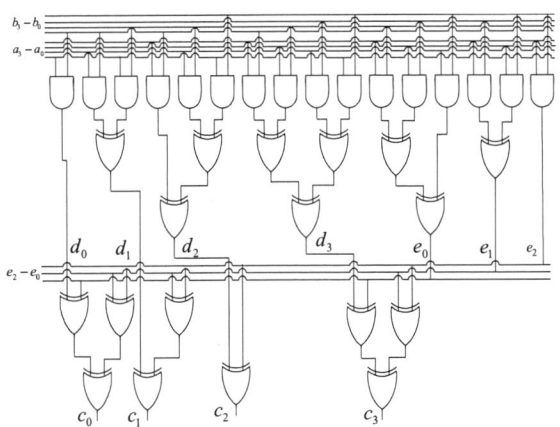

Fig.2 Architecture of the BP Multiplier over GF(2^4)

Now, $C(x) = (a_0 + a_1 x + a_2 x^2 + a_3 x^3) \times (b_0 + b_1 x + b_2 x^2 + b_3 x^3) = a_0 b_0 +$
$(a_0 b_1 + a_0 b_1)x + (a_0 b_2 + a_1 b_1 + a_2 b_0)x^2 + (a_0 b_3 + a_1 b_2 + a_2 b_1 + a_3 b_0)x^3 + (a_1 b_3 + a_2 b_2 + a_3 b_1)x^4 + (a_2 b_3 + a_3 b_2)x^5 + a_3 b_3 x^6$.

Let us denote the lower order m coefficients as $d_0, d_1, \dots, d_{m-1}$ and the higher order ($m - 1$) coefficients as $e_0, e_1, \dots, e_{m-2}$. Then $C(x)$ can be expressed as shown in the following.

$$C(x) = d_0 + d_1 x + d_2 x^2 + d_3 x^3 + e_0 x^4 + e_1 x^5 + e_2 x^6 \quad (4)$$

Here, we define the product over the primitive polynomial $P(x) = x^4 + x^3 + 1$ as $A(x)\ B(x)\ mod\ P(x)$. Hence, we have,

$x^4 = x^3 + 1$, $x^5 = x(x^3 + 1) = x^4 + x = x^3 + x + 1$, $x^6 = x(x^5) = x(x^3 + x + 1) = x^4 + x^2 + x = x^3 + 1 + x^2 + x = x^3 + x^2 + x + 1$.
Substituting the power of x^4, x^5, and x^6 and simplifying we get,

$A(x)\ B(x)\ mod\ P(x) = C = (e_0 + e_1 + e_2 + d_0) + (e_1 + e_2 + d_1) x + (e_2 + d_2)x^2 + (e_0 + e_1 + e_2 + d_3) x^3$

The above modulo reductions can be represented in the matrix form as given below.

$$\begin{bmatrix} c_0 \\ c_1 \\ c_2 \\ c_3 \end{bmatrix} = \begin{bmatrix} 1 & 1 & 1 \\ 0 & 1 & 1 \\ 0 & 0 & 1 \\ 1 & 1 & 1 \end{bmatrix} \begin{bmatrix} e_0 \\ e_1 \\ e_2 \end{bmatrix} + \begin{bmatrix} d_0 \\ d_1 \\ d_2 \\ d_3 \end{bmatrix}$$

$c = Q^T e + d$, Where

$$Q^T = \begin{bmatrix} 1 & 1 & 1 \\ 0 & 1 & 1 \\ 0 & 0 & 1 \\ 1 & 1 & 1 \end{bmatrix}, \quad e = \begin{bmatrix} e_0 \\ e_1 \\ e_2 \end{bmatrix}, \quad d = \begin{bmatrix} d_0 \\ d_1 \\ d_2 \\ d_3 \end{bmatrix}$$

We can also derive d, and e from the equations (1), (2).

3. Proposed Design Technique

The basic implementation of the Galois field multiplier is shown in Figure 1. The classical bit parallel Galois Field multiplier is designed by the method described in Section II. The modified single error correcting architecture is shown in Figure 3. Apart from the main functional multiplier, it consists of the parity prediction unit, and the output parity generation, comparison and decoding logic. For error detection and correction we use the parity prediction, which is based on Hamming's principles. First, the predicted parity bits are tabulated and these parity bits generated using parity prediction logic as shown in Figure 3. The basic design steps can be explained by the following example.

Consider for example a GF(2^8) multiplier that corrects single errors. Let c_i be the jth bit of ith bits of the multiplier outputs. Therefore, the predicted parity bits are as follows.

$$P_0 = c_0 \oplus c_1 \oplus c_3 \oplus c_4 \oplus c_6 \quad (5)$$
$$P_1 = c_0 \oplus c_2 \oplus c_3 \oplus c_5 \oplus c_6 \quad (6)$$
$$P_2 = c_1 \oplus c_2 \oplus c_3 \oplus c_7 \quad (7)$$

35

978-1-4244-3039-0/08 $25.00 © 2008 IEEE

$$P_3 = c_4 \oplus c_5 \oplus c_6 \oplus c_7 \qquad (8)$$

From the above equations we generate the truth tables. The output parity bits are generated based on the inputs A and B. At the other end, at the outputs C, the parity bits are generated and compared against the predicted parity bits. The comparator outputs, called syndrome bits, are used to locate the bit error position. This is similar to the basic hamming code based error correction technique. The main difference between traditional Hamming code applied to memory error correction and the technique proposed here is that, in the conventional (memory) method an encoder is used to generate the parity bits, whereas here we use parity prediction logic based on the input operands. Intuitively, the basic principle is based on the theory of Hamming codes, however, since we are using the same principle for the logic circuits, we had to use a parity predictor to generate the parity bits from the input operands.

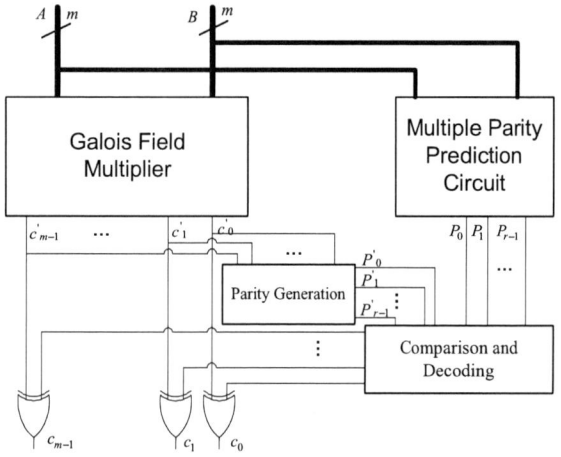

Fig.3: Proposed Architecture of the BP GF Multiplier with SEC

Hamming codes are the simplest of a group of codes known as the linear block codes [2]. The advantage of the Hamming codes is that the number of parity bits grows logarithmically as the number of output bits increases; however, the complexity of the encoding and decoding logic grows linearly as the number of data bits increases in memory based error correction. In the proposed approach, it may not be linear. In Hamming code applied to memory encoding, encoding the bits requires the use of a tree of EXOR operations (gates). The exact number of EXORs operations is difficult to calculate in a closed form, but an upper bound for this number is easily defined.

But in this proposed approach instead of encoders we have parity prediction circuits.

In addition, in this approach we need to encode the output data bits and compare against predicted parity bits. Encoding requires that each of the r parity bits be obtained by the EXOR of k of the m output bits, where m is the number of output bits and k is determined by the parity check equation. Thus, for encoding (for the worst case) there are $r(2^{r-2} - 1)$ two-input EXOR gates. The computation of the ith error location requires r 2-input XOR gates. A binary decoder is needed to decode the error location and one XOR per output bit is needed for data correction. Therefore the total number of two-input XOR gates is $r(2^{r-2} - 1) + 2 m = r(2^{r-2} - 1) + 2m$, apart from decoder circuitry. Since r is a logarithmic function of m, the number of XOR gates is bounded by a linear function of the number of output bits.

For large circuits the size of the parity bits is small in relation to the size of the entire circuit, but the size of the parity prediction and decoding logic can be prohibitive, worst case it could same as that of the multiplier. Consequently, the overall hardware requirement is the sum of parity prediction circuit plus the above encoding, decoding and correcting parts. Let T_e, T_c, T_d and T_{cor} be delays in the encoder, comparison, decoding and correction circuitry respectively. Then the increase in critical path for the new SEC design would be $T = T_e + T_c + T_d + T_{cor}$. However, this delay is comparatively much less than the critical path of the multiplier. As shown in the experimental results, the critical path of the parity prediction circuit is always less than the critical path of the Galois field multiplier and, therefore, it will not be in the critical path of the overall structure.

4. Experimental Results

We provide experimental results for the multipliers over the fields GF(2^k), for $k = 5$ to 8 generated with all the primitive polynomials over these fields. Both the multipliers and the parity prediction hardware were generated with programs written in C++ (Gnu C++) under the Linux operating system in the PLA format. The PLA files were synthesized with the Synopsys[TM] design compliers with the help of the UMC technology libraries in the 0.18 micron CMOS technology for estimation of the actual area, delay, and power (Synopsys[TM] power compiler), which we present in Figs. 4-6, and in Table 1. However, the technique can be easily extended to larger circuits. In Figs 4-6, the x-axis shows the decimal representation of the primitive polynomials. For example, the primitive polynomial $P(x) = x^7 + x^5 + x^4 + x^3 + x^2 + x^1 + 1$ can be represented as $(10111111)_b$ and its decimal equivalent is 191. As expected, the overhead varies depending on the size of the circuits. For instance, as the number of output bit increases, that is for larger multipliers, the overhead comes down.

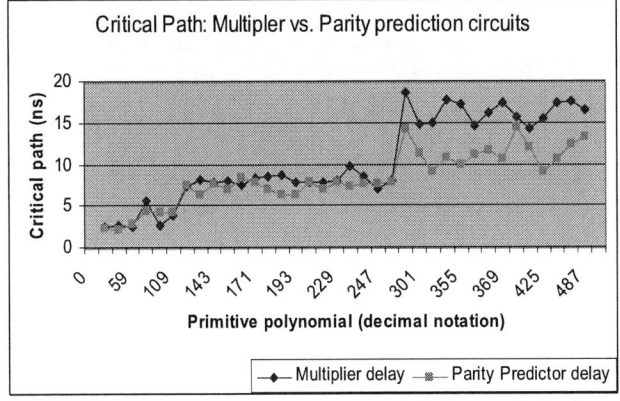

Fig. 4: Critical path analysis GF Multiplier and Parity prediction logic.

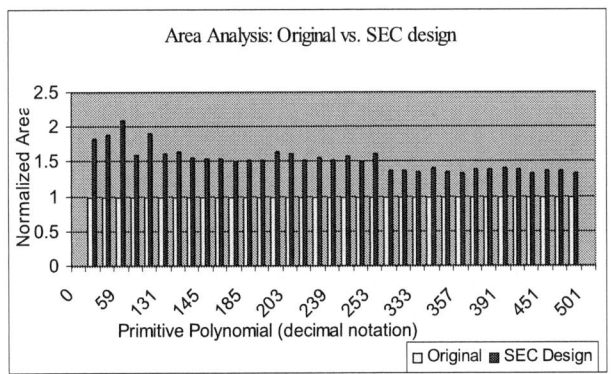

Fig. 5: Area analysis classical GF Multiplier and SEC

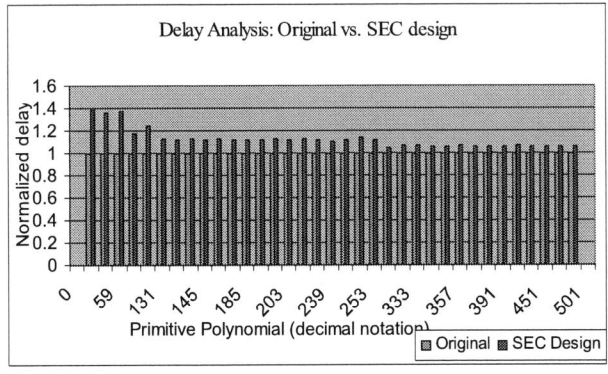

Fig. 6: Delay analysis classical GF Multiplier and SEC

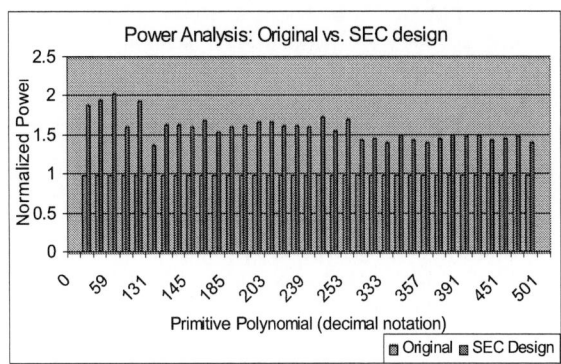

Fig. 7: Power analysis classical GF Multiplier and SEC

For the various primitive polynomials considered in this paper, Fig. 4 shows the critical path of the parity prediction hardware and the functional multipliers. As can be seen, the critical path of the parity prediction logic is always significantly less than that of the multipliers. The area overhead may vary from 112% to 50% depending on the parity prediction logic and a similar variation in power overhead as well (see Figure 5 and Figure 7). It is also interesting to note that there is only a 12.4% increase in critical path, on average, as shown in Figure 6. This overhead could be justified considering the increased reliability of the new design. Furthermore, it gives 100 % fault coverage both in the functional parts and the parity prediction circuits.

5. Conclusions

Error correction is an effective way to mitigate fault attacks in cryptographic hardware intensive applications. Commonly, higher level mechanisms are adapted to protect hardware. This paper proposes an alternate hardware architecture compared to the existing approaches. Experiments carried out on bit-parallel $GF(2^m)$ multipliers show significant performance advantages, as it does not affect the overall delay too much, over conventional designs. More importantly, this is the first time the single error correction architectures are proposed with hardware overhead about 100%, whereas TMR based single error correction techniques take more than 200 % area/power. Consequently, the proposed architecture can be effectively applied in attack tolerant cryptographic architectures.

REFERENCES

1. S. Fenn, M Gossel, M Benaissa, and D. Taylor, "On-line Error Dection for Bit-seial Multipliers in GF(2^m)", J. Electronic Testing: Theory and Applications, vol. 13,pp. 29-40,1998.

2. W. Hamming. "Error Detecting and Error Correcting Codes." *Bell Systems Tech. Journal*, pp. 147-160, 1950

3. Mitra S., Seifert N., Zhang M., Shi Q and Kim K. : "Robust System Design with Built-In Soft Error Resilience," *IEEE Computer*, Vol. 38, Number 2, pp. 43-52, Feb. 2005..

4. A. Reyhani-Masoleh, and M. A. Hasan, "Low Complexity Bit Parallel Architectures for Polynomial Basis Multiplication over GF(2^m)", *IEEE Trans. Computers*, vol.53, no.8, pp.945-959, 2004.

5. E.D. Mastrovito, "VLSI Architectures for Computation in Galois Fields," *PhD thesis, Linkoping Univ., Sweden,* 1991.

6. D. K. Pradhan, "A Theory of Galois Switching Functions", *IEEE Trans. Computers,* vol. 27, no. 3, pp.239-248, Mar. 1978.

7. A. Reyhani-Masoleh, and M. Anwar Hasan, "Fault Detection Architectures for Field Multiplication Using Polynomial Bases",IEEE Transactions on Computers, vol. 55, No. 9, Sept. 2006.

8. H. Rahaman, J. Mathew, A. M. Jabir and D. K. Pradhan, "Easily Testable Implementation for Bit Parallel Multipliers in GF (2^m)", HLDVT 2006.

9. A. M. Jabir D. K Pradhan and J. Mathew , "An Efficient Technique for Synthesis and Optimization of Polynomials in GF(2^m)". IEEE International Conference on Computer Aided Design, November 2006, California, pp. 151–157.

10. M. Nicolaidis and Y. Zorian, "Online Testing for VLSI – A Compendium of Approaches," Jnl. Electronic Testing: Theory and Applications, Vol. 12, Nos. 1/2, pp. 7-20, Feb.-Apr. 1998.

11. G. L. Smith, "Model for Delay Faults Based Upon Paths", *ITC85*, pp. 342-349.

12. C. J. Lin and S. M. Reddy, "On Delay Fault Testing in Logic Circuits", *IEEE TCAD*, pp. 694-703, Sept. 1985.

13. D. Boneh, R. A. DeMillo, and R. J. Lipton. "On the Importance of Eliminating Errors in Cryptographic Computations". Journal of Cryptology, 14:101–119, 2001.

14. M. Ciet and M. Joye. "Elliptic Curve Cryptosystems in the Presence of Permanent and Transient Faults". Designs, Codes and Cryptography, 36(1):33–43, 2005.

15. G. Gaubatz and B. Sunar. Robust finite field arithmetic for fault-tolerant public-key cryptography. presented in the 2ndvWorkshop on Fault Tolerance and Diagnosis in Cryptography (FTDC), 2005.

Table 1: Galois Multiplier vs. Multiple Parity Prediction Circuit

primitive polynomial	GF Multiplier			Multiple Parity Prediction circuit		
	area	delay	power	area	delay	power
59	4909.35	2.66	6.455	4989.98	2.91	6.17
91	18047.18	5.7	22.88	10541.2	4.46	13.38
109	11118.61	3.94	14.45	9621.94	4.29	12.96
131	30865.70	7.51	72.23	18818.13	7.73	26.4
137	30356.09	8.31	39.91	19192.2	6.44	25.07
143	33375.21	7.99	40.26	18176.2	7.87	25.12
145	33468.76	8.13	42.27	17708.52	7.21	24.9
167	34978.34	7.72	39.73	18947.14	8.59	26.85
171	39023.25	8.5	49.07	19037.46	7.96	26.27
185	34797.71	8.71	42.1	18118.17	7.16	25.08
191	34842.87	8.8	40.95	17885.96	6.37	25.26
193	30320.50	7.95	38.82	19063.26	6.41	25.9
203	33875.21	8.02	41.96	20459.95	7.99	27.99
213	33488.12	7.96	38.93	17389.18	7.13	23.8
229	34578.37	8.19	41.32	19311.64	7.96	25.1
239	34123.54	9.86	41.46	17631.115	7.51	24.78
241	32297.87	8.69	36.7	18450.40	7.78	26.31
247	36704.02	7.19	45	18447.18	7.87	24.69
253	33781.64	8.26	41.27	20527.68	8.08	28.39
285	141693.95	18.75	130.93	52925.58	14.41	58.326
301	132597.70	15.02	125.3	49006.48	11.54	56.62
333	140900.46	15.16	129.26	49680.62	9.14	53.52
351	135110.45	17.9	121.51	55173.80	10.93	61.2
355	143200.25	17.44	126.03	50380.56	10.1	55.11
357	144477.65	14.77	133.48	48406.52	11.25	54.37
361	135229.85	16.27	125.55	52715.91	11.86	58.4
369	132407.46	17.63	130.15	52622.36	10.76	64.88
391	130578.51	15.91	126.52	53164.296	14.66	62.55
397	139997.2	14.37	122.33	53661	12.11	62.41
425	141884.2	15.59	118.94	47367.86	9.29	52.96
451	139616.65	17.62	123.93	50512.83	10.86	55.84
463	141232.67	17.79	122.87	53261.04	12.45	59.48
487	150709.5	16.74	138.84	51354.72	13.45	57.15

978-1-4244-3039-0/08 $25.00 © 2008 IEEE

21st International Conference on VLSI Design

A Robust Architecture for Flip-flops Tolerant to Soft-Errors and Transients from Combinational Circuits

Aditya Jagirdar
E.C.E. Dept.
Rutgers University
New Brunswick, NJ, U.S.A.
jagirdar@eden.rutgers.edu

Roystein Oliveira
A.M.D., Inc.
Austin, TX, U.S.A.
roystein.oliveira@amd.com

Tapan J. Chakraborty
Alcatel-Lucent, Inc.
Whippany, NJ, U.S.A
tchakraborty@alcatel-lucent.com

Abstract

Soft-errors are a leading cause of reliability issues during field operations. High-energy particles, either from cosmic rays or from impurities in the packaging material can disrupt charge stored on the internal node capacitances leading to a malfunction of the device. Although this is usually a temporary effect, it may lead to Silent Data Corruption(SDC) when not detected in time. SDC may be detrimental to many real-time commercial applications of the device and demands an effective solution that is cheap in terms of various design overheads. In this paper, we propose two novel flip-flop designs aimed at detecting and correcting soft-errors and transients from combinational circuits.Each design is optimized for a different set of constraints and they have area overheads of 40% and 21% as compared to the standard industrial design of a scan flip-flop.

1. Introduction

Soft errors may occur in a chip when exposed to radiation (Cosmic rays, α-particle, Neutrons etc.). This may cause an erroneous change in state of a system, thus affecting its operation. The effect of the error is temporary and if detected, the system can be restored to its normal operation by resetting the system or by re-execution, if possible. These errors are of concern in mission critical operations (space, medical) where a downtime or delay for system recovery is not affordable.

Exposure to radiations brings the devices on the chip under constant strike by charged particles. A high energy particle striking a node in these devices generates electron hole pairs. In the presence of strong electric field, these charged carriers move toward their respective device contacts. If the collected charge is greater than the threshold value required

for determining a logic level for a transistor, an erroneous value is registered [23]. This minimum threshold value is also known as critical charge.

Earlier, soft error was a concern only in space applications. The altitude at which these systems operate exposes them to strong cosmic rays [26] which corrupts logic, thus affecting systems. The cosmic ray flux incident at sea level (terrestrial cosmic rays), consisting mainly of neutron particles loses most of its energy as it passes through the atmosphere [26] and would cause a negligible change in the threshold charge value to result in a *Single-Event Upset*(SEU). The problem of SEU arising due to α-particle radiation from minute traces of radioactive elements in packaging material is also significant and can be addressed by ensuring a higher degree of purity of packaging materials [6]. Recently, the problem of SEU has become severe and recurrent with shrinking device geometry. Shorter channel lengths means lesser number of charge carriers and a smaller value of threshold charge [1]. The problem of SEU which was initially associated with small and densely packed memories is now commonly observed in combinational circuits and latches [25][2]. SEU's in memories has been addressed by *Error Detecting and Correcting codes* (EDC). In sequential circuits, a flip-flop (FF) input may observe transients caused by SEU's at the FF input node or due to SEU's in the *Combinational Logic Blocks* (CLB) being propagated to the FF input. A FF is also vulnerable to a SEU that causes bit flips during the opaque cycle. Figure 1 shows the *Window of Vulnerability* (WoV) [16] for a FF. The latches in a FF are vulnerable to SEU's during the latch setup time (T_{setup}) and a period T_{tf}. T_{tf} is the time interval during which a transient fault can occur. T_{nvd} is the period when the latch is transparent; T_{prop} is the time required for resulting fault to propagate to the output. In this paper we present two schemes for a FF, both of which provide SEU and *Single-Event Transient* (SET) tolerance over the *WoV*. Both schemes have minimal transistor and clock to Q delay

39

978-1-4244-3039-0/08 $25.00 © 2008 IEEE

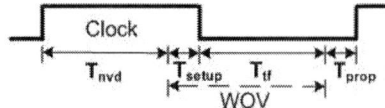

Figure 1. Window of vulnerability (WOV) [16]

overhead.

2. Prior work and motivation

Traditional techniques use *Triple Modular Redundancy* (TMR), where the circuit is triplicated and a *Majority Voter* (MV) is employed[15, 24]. Various combinations of space and time redundancies have been explored to immunize latches from soft-errors[14, 10, 1, 16, 3, 8]. Designs that reuse scan hardware to recover from soft-errors have relatively lower area overheads[12, 7, 5, 4]. They, however, provide only limited tolerance with respect to the *WoV*. Other schemes such as *Concurrent Error Detection* use an output characteristic predictor to verify circuit operation[18, 11]. Online error correction or error-masking techniques use dynamic storage of charge (keeper, node capacitance etc.) to reduce soft-error rate[20, 9]. Various other approaches in terms of gate-sizing, pulse-spreading have been proposed as well[13, 19, 21]. Time-redundancy based architectural approaches have been investigated as well[22, 17]. Although, significant work has been done to deal with single-event upsets, these techniques do not protect the registers during the entire *WoV*. They also have single-points of failure, where a particle-strike can invalidate the corrective operation of the design. In this paper we propose two designs for soft error tolerance with an aim to provide design flexibility. The designs make use of the concept of time redundancy to achieve error masking. The first design provides a simplified clocking scheme with almost no performance degradation. The second design has a low transistor overhead and has a tolerable performance degradation. Our designs make use of the scan hardware in the functional mode to store samples. The decision circuitry is a majority voter that avoids soft errors over the entire *WoV*. Both designs are also tolerant to transients from the *Combinational Logic Blocks*.

3. Our proposed designs

Our designs make use of the basic scan FF (BSFF) as described in [12]. It facilitates scan based structural testing using automated test pattern generation tools, for functional testing using signature analysis, and for efficient post silicon debug [12]. It has two modes of operation, the Functional mode and the Test mode.

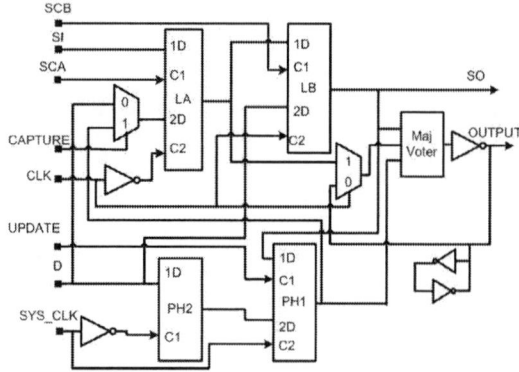

Figure 2. The XSEUFF 1

3.1. XSEUFF 1

The block diagram for the XSEUFF 1 is shown in Figure 2. This design can provide tolerance over the *WoV* from

- SEU in FFs.

- SET due to SEU in combinational blocks and due to crosstalk.

Latches PH2 and PH1 are the master and slave system latches. LA and LB are the master and slave scan latches. The FF captures the data on the input D by loading latches PH2, LA and LB. Latches LA and LB are clocked using $\overline{CLK}$ and *CLK*. *CLK* is delayed by Δ_1 to obtain *SYS_CLK*. Latches PH2 and PH1 are clocked using $\overline{SYS\_CLK}$ and *SYS_CLK*. The clocking scheme is shown in Figure 3. The delay Δ_1 is given by the equation.

$$\Delta_1 = t_{Hold\_LA} + W_{MTT} + t_{Setup\_PH2} \qquad (1)$$

Where W_{MTT} is the *Width of Maximum Tolerable Transient*. It is the design variable that gives the tolerance of the designed circuit to an SEU pulse. The *CLK* can be distributed using the existing scan clock routing structure in a circuit, thus simplifying the task of the designer. A keeper is used at the output node of the FF. The output of the keeper is multiplexed with the output of LA with the *CLK* signal as the select to the multiplexer. The output of the multiplexer, LB and PH1 are inputs to the majority voter. The FF is tolerant to errors over the entire window of vulnerability.

When *TESTBAR* is high, the system is in functional mode. Latch LA loads the data from D when *CLK* is low and latch LB loads the data from D when *CLK* is high. Latch PH2 loads the data from D when *SYS_CLK* is low and latch PH1 loads from PH2 when *SYS_CLK* is high. LA and LB act as the shadow of the system latches, capturing data at different time intervals for comparison. Δ_1 is selected as in Equation (1), to ensure that a SEU is never captured by LA and PH2 at the same time. The voter votes on LA,

978-1-4244-3039-0/08 $25.00 © 2008 IEEE

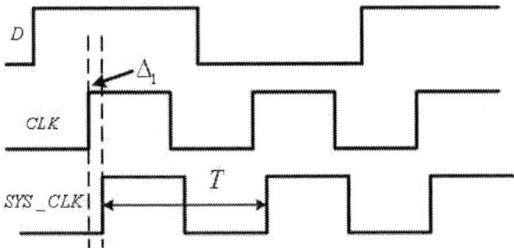

Figure 3. Clocking scheme for the proposed design

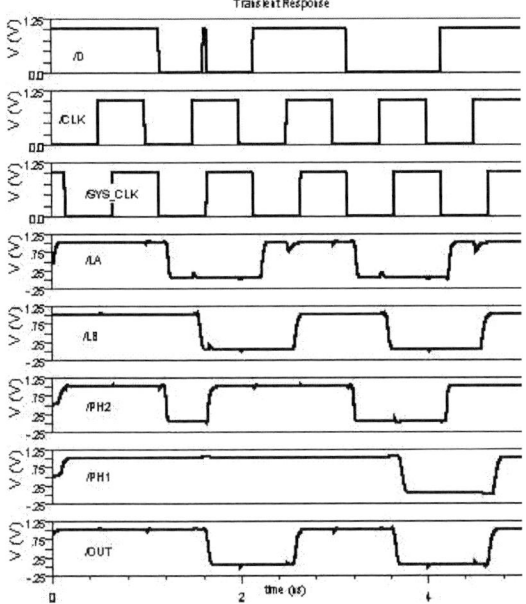

Figure 4. Functional mode simulation

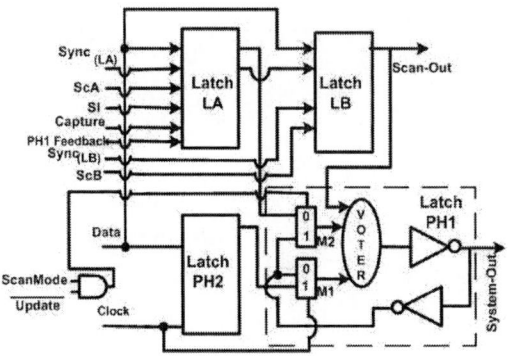

Figure 5. The XSEUFF 2

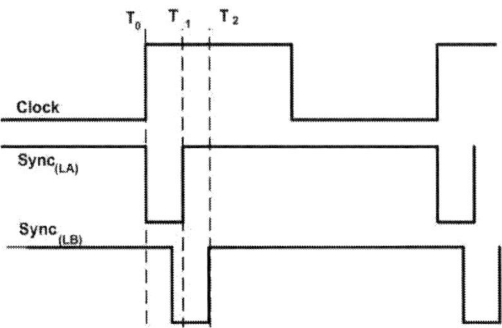

Figure 6. Waveforms for $Sync_{LA}$ and $Sync_{LB}$

LB and PH1 when *SYS_CLK* goes high. This value is captured by the keeper at the output node. When *CLK* goes low, the voter votes on keeper, LB and PH1. This arrangement prevents data corruption at the output due to a SEU in LA or PH1, during latching or during their opaque cycle. As shown in Figure 4, LA latches the value from D. LB goes transparent before PH2 latches in the input data. PH2 latches in an incorrect data due to a SEU from the CLB and is propagated to PH1. The majority voter computes the correct output on the rising edge of *SYS_CLK*. When *CLK* goes low, the voter computes the output based on the value stored on the keeper instead of LA. The design is tolerant to only one latch being corrupted by an SEU over the *WoV*.

During the testing phase, when *TESTBAR* signal is low, the system is in test mode. The scan latches LA and LB are clocked using the scan clocks *SCA* and *SCB* respectively. LA receives the scan data input *SI*. Alternate pulsing of *SCA* and *SCB* loads LA and LB with the test vector and this en-

ables scanning through the scan chain. The vector is then loaded into PH1 by the pulsing of the *UPDATE* signal. The system is switched from test mode to functional mode by asserting *TESTBAR*. The response of the *Circuit Under Test* (CUT) to the applied vector is captured after one functional clock cycle. PH2 is loaded with the response of the previous CUT and is propagated to PH1 at the end of the cycle. *CAPTURE* is then asserted and the final output of the applied vector is captured in LA. The system is then switched to test mode and the captured response is scanned out by alternate assertion of *SCB* and *SCA* for fault analysis.

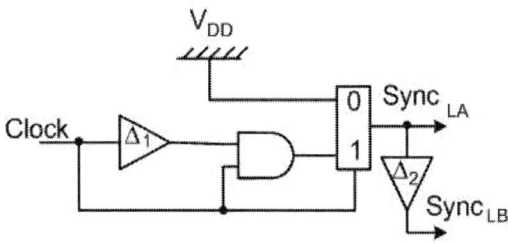

Figure 7. Synchronous signal generator for the XSEUFF 2

978-1-4244-3039-0/08 $25.00 © 2008 IEEE

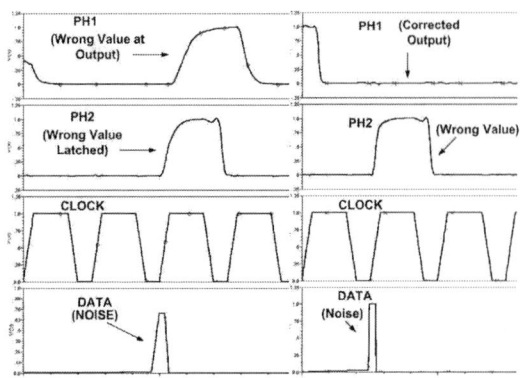

Figure 8. Response of a BSFF vs. XSEUFF 2 to noise pulses

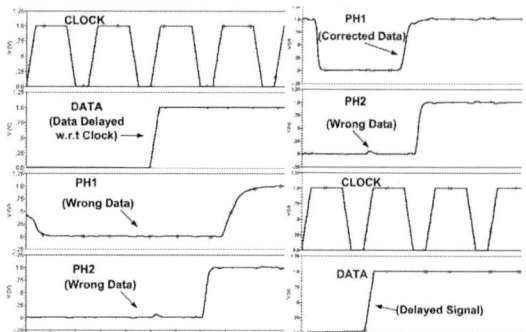

Figure 9. Response of a BSFF vs. XSEUFF 2 to signal delays

3.2. XSEUFF 2

The XSEUFF 2 shown in Figure 5. This design provides tolerance over the *WoV* from

- SEU in FFs.

- SET due to SEU in combinational blocks and due to crosstalk.

- Signal delays arising due to crosstalk.

Temporal sampling of the data line is carried out at the active clock edge, using clocks $Sync_{LA}$ and $Sync_{LB}$ as shown in Figure 6. This filters out any incoming noise that may cause a setup and/or hold time failure. Latches PH2, LA and LB are transparent when their gating signals are low and opaque when they are high. Latch PH1 has been modified to contain the majority voter circuit. In functional mode, PH2, LA and LB get data from the combinational circuit. The pulses $Sync_{LA}$ and $Sync_{LB}$ can be generated by using the circuit shown in Figure 7. Delay parameters Δ_1 and Δ_2 control the relative timing of these synchronous signals. The temporal sampling is carried out by reuse of the scan latches LA and LB. $Sync_{LA}$ and $Sync_{LB}$ are used as gating signals for LA and LB respectively. Pulses $Sync_{LA}$ and $Sync_{LB}$ are driven low after the active clock. This allows LA and LB to sample copies of the expected data in PH2. When Clock goes high, the majority voter gets data from PH2, LA and LB thus ensuring that the corrected data is always latched into it. When PH1 goes opaque (Clock = 0), the redundant values stored in LA and LB ensure that an SEU hit on any of them does not affect the value at the FF output. The FF output changes only after the arrival of the next active clock edge. Temporal sampling at the active clock edge combined with the unique design of PH1 ensures complete immunity of the XSEUFF 2 over the *WoV*.

During simulations, the XSEUFF 2 was subjected to noise pulses and signal delays of varying magnitudes. Its response to a worst case transient pulse is shown in Figure 8. In Figure 8, we compare the behavior of the BSFF [12] and the XSEUFF 2. Latch PH1, in the BSFF receives wrong data from PH2 after the arrival of active clock edge. In the case of XSEUFF 2, time-redundant samples are evaluated by the majority voter and PH1 receives corrected data at the end of the clock cycle. Figure 9 shows the response of the BSFF and the XSEUFF 2 to signal delays. In Figure 9, although PH2 latched in a wrong value, PH1 received the corrected value from the majority voter circuit. Although the XSEUFF 2 is able to transmit corrected data to PH1 in both cases, its tolerance is limited and depends on several circuit parameters. We define the limits for the width of a noise pulse and magnitude of signal delay that this FF can tolerate. If T_{XPulse} is the maximum tolerable width of a single event transient, then Equation (2) gives the relationship between the various sampling instants, the setup time of the FF t_{setup}, the hold-time t_{hold} and the delays of the multiplexers at the input of latches LB and PH1. For instance $t_{cd(LB),r}$ represents the combinational delay of a rising transition at the input of LB. Similarly $t_{cd(LB),f}$ represents the combinational delay of a falling transition at the input of LB. Equation (3) gives the maximum tolerable delay (Δ_{max}) of the XSEUFF 2.

$$T_{XPulse} = \{T_2 - t_{setup} + max(t_{cd(LB),r}, t_{cd(LB),f})\} - \{T_0 + t_{hold} - min(t_{cd(PH2),r}, t_{cd(PH2),f})\} \quad (2)$$

$$\Delta_{max} = \{T_1 - t_{setup} + max(t_{cd(LA),r}, t_{cd(LA),f})\} - \{T_0 - t_{setup}\} \quad (3)$$

The *ScanMode* signal is analogous to the inverted *TESTBAR* in XSEUFF 1. During Scan operation, LA and LB become part of a large scan-chain. LA gets data from *SI*, which is connected to the output of a previous FF or the

Table 1. Normalized transistor count, clock to q and power consumption ratios

Circuit type	Normalized transistor count	Normalized Clock to Q delay	Normalized power consumption
BSFF [12]	1.00	1.00	1.00
EBSFF [12]	1.13	1.23	2.77
ETSFF [12]	1.15	1.01	2.13
EBSHFF [7]	0.91	1.25	1.72
XSEUFF 1	1.54	1.00	2.70
XSEUFF 2	1.37	1.25	1.95

Table 2. Area overhead comparison for the proposed designs

Flip-Flop	BSFF[14]
XSEUFF1	1.40
XSEUFF2	1.21

Table 3. ISCAS '89 benchmark overheads

Circuit	XSEUFF 1 Transistor Overhead (%)	XSEUFF 2 Transistor Overhead (%)
s5378	29.3	20.2
s9234	21.8	15.0
s13207	32.1	22.1
s15850	27.3	18.9
s35932	31.3	21.6
s38417	29.9	20.6
s38584	27.0	18.6
Average	28.4	19.6

tester pin. *SCA* and *SCB* control data transfer between LA and LB during the test mode. *UPDATE* and *CAPTURE* are used to apply the test vector and capture the circuit response respectively. The XSEUFF 2 can be tested for stuck-at-0 and stuck-at-1 faults on all its internal nodes by enabling the scan mode of operation and applying the appropriate vector. A stuck-at fault present on any of the input nodes of the majority voter can be tested by scanning in opposite values in latches LA and LB and disabling the signal generator, so that value latched in PH2 can be used to sensitize this fault.

4. Limitations

XSEUFF 1 and XSEUFF 2 are proposed with the assumption that there can be only one SEU affecting the FF system in a clock cycle. The schemes of XSEUFF 1 and 2 would fail if the width of the maximum tolerable transient exceeds the designed value which causes incorrect values to be latched in more than one FF in a given cycle. XSEUFF 1 increases the setup time of the FF due to the pre-sampling by *CLK* and requires the data to be steady for at least half a clock period.

5. Results

Table 1 gives the transistor counts of different standard cell FF designs, normalized to the transistor count of the BSFF. For each design in this comparison we have only considered transistors local to a cell. All globally routed signals are considered to be inverted at the cell level. The EBSFF [12] and ETSFF [12] designs have 13% and 15% overhead respectively. The EBSHFF [7]

has a lower transistor count as compared to the BSFF. The transistor overhead for XSEUFF 1 is 54%. The transistor overhead for the XSEUFF 2 is 37%. Area overheads for XSEUFF1 and XSEUFF2 were evaluated by comparing the layouts. Although XSEUFF1 has a higher transistor/area overhead, the ease of generation of the additional synchronous signals requires lower design effort as compared to the XSEUFF2. This may be an advantage for designs operating at a higher frequency. However, if it is feasible to generate synchronous signals to a higher degree of precision, then XSEUFF2 provides a greater advantage in terms of area optimization. Table 2 indicates the result of this comparison. The normalized values of Clock to Q delay for different designs are shown in Table 1. For each design, the transistor sizes were maintained for equal rise and fall times. The ETSFF [12] and the XSEUFF 1 have clock to Q delay overhead comparable to the BSFF. The EBSFF [12] and the EBSHFF [7] have a clock to Q delay overhead of about 25%. The XSEUFF 2 has a clock to Q delay overhead of 25 %. The simulations for power were carried out on *Nanosim* tool using the 70 nm BPTM with supply of 1 V at 1 GHZ clock frequency for five clock cycles. The average power consumption values are compared for the functional mode of operation. The normalized values of power consumption for different designs are shown in Table 1.

The transistor overhead for the seven largest ISCAS '89

benchmark circuits is shown in Table 3. The comparison is with reference to the BSFF. All FFs were replaced with the XSEUFF 1 and XSEUFF 2 designs for each benchmark circuit comparison. The average transistor overhead is approximately 28% for XSEUFF 1. The XSEUFF 2 has a overhead of approximately 20%

6. Conclusion

Two designs that provide tolerance to SEUs and transients from the combinational circuit have been proposed in this paper. Transistor, power and timing overheads for the same have been analyzed. With decreasing feature sizes, transistor designs have become more susceptible to soft errors. This increase mandates use of designs that are not just immune to them but also capable of correcting data at run-time, thus ensuring fault-free operation of the entire system. The XSEUFF 1 and 2 offer a convenient trade off between reliability and various overheads.

References

[1] Y. Arima, T. Yamashita, Y. Komatsu, T. Fujimoto, and K. Ishibashi. Cosmic-ray immune latch circuit for 90nm technology and beyond. In *Proc. of Intl. Solid-State Circuits Conference*, pages 492 – 493, February 2004.

[2] R. Baumann. The Impact of Technology Scaling on Soft Error Rate Performance and Limits to the Efficacy of Error Correction. In *Proc. of the Int'l. Electron Devices Meeting (IEDM)*, pages 329–332, 2002.

[3] H. Cha and J. H. Patel. Latch Design for Transient Pulse Tolerance. In *Proc. of the ACM Int'l. Conf. on Computer Design*, pages 385–388, October 1994.

[4] S. DasGupta, R. G. Walther, T. W. Williams, and E. B. Eichelberger. An Enhancement to LSSD and Some Applications of LSSD in Reliability, Availability and Serviceability. In *Proc. of the Int'l. Symp. on Fault Tolerant Computing*, page 289, June 1981.

[5] A. J. Drake, A. J. KleinOsowski, and A. K. Martin. A Self-Correcting Soft Error Tolerant Flop-Flop. In *Proc. of the 12th NASA Symp. on VLSI Design*, October 2005.

[6] J. F. Z. et al. IBM Experiments in Soft Fails in Computer Electronics. *IBM J. of Research and Development*, 40(1):3–18, January 1996.

[7] A. Goel, S. Bhunia, H. Mahmoodi, and K. Roy. Low-overhead Design of Soft-error-tolerant Scan Flip-flops with Enhanced-scan Capability. In *Proc. of the Asia and South Pacific Design Automation Conf.*, January 2006. 6 pp.

[8] H. K.J., G. J.W., W. B., and Z. M. Mitigating single event upsets from combinational logic. In *Proc. of 7th NASA Symposium on VLSI Design*, 1998.

[9] S. Krishnamohan and N. R. Mahapatra. Combining Error Masking and Error Detection Plus Recovery to Combat Soft Errors in Static CMOS Circuits. In *Proc. of the Int'l. Conf. on Dependable Systems and Networks*, pages 40–49, July 2005.

[10] D. Mavis and P. Eaton. Soft Error Rate Mitigation Techniques for Modern Microcircuits. In *Proc. of the Int'l. Reliability Physics Symp.*, pages 216–225, 2002.

[11] S. Mitra and E. J. McCluskey. Which Error Detection Scheme to Choose? *Proc. of the Int'l. Test Conf.*, pages 985–994, October 2000.

[12] S. Mitra, M. Zhang, N. Seifert, Q. Shi, and K. S. Kim. Robust System Design with Built-In Soft-Error Resilience. *Computer*, 38(2):43–52, February 2005.

[13] B. M.P. and B. S.P. Attenuation of single event induced pulses in CMOS combinational logic. *IEEE Transactions on Nuclear Science*, 44:22172223, December 1997.

[14] M. Nicolaidis. Time Redundancy Based Soft-error Tolerance to Rescue Nanometer Technologies. In *Proc. of the VLSI Test Symp.*, pages 86–94, April 1999.

[15] R. Oliveira, A. Jagirdar, and T. J. Chakraborty. A Soft Error Tolerant, Low Overhead TMR Design for Flip-Flops. In *Proc. of the North Atlantic Test Workshop*, pages 97–104, 2006.

[16] M. Omana, D. Rossi, and C. Metra. Novel Transient Fault Hardened Latch. In *Proc. of the Int'l. Test Conf.*, pages 886–892, 2003.

[17] E. P., P. K., and S. R. Time redundancy based scan flip-flop reuse to reduce SER of combinational logic. In *Proc. of Intl. Symposium on Quality Electronic Design*, 2006.

[18] D. Pradhan. *Fault-Tolerant Computer System Design*. Prentice Hall, Englewood Cliffs, NJ, 1986.

[19] Z. Q. and M. K. Cost-Effective Radiation Hardening Technique for Combinational Logic. In *Proc. ACM/IEEE International Conf. Computer-Aided Design*, pages 100 – 106, November 2004.

[20] K. S. and M. N.R. A highly-efficient technique for reducing soft errors in static CMOS circuits. In *Proc. of ACM International Conf. Computer Design (ICCD)*, pages 126 – 131, October 2004.

[21] K. S. and M. N.R. Analysis and Design of Soft-Error Hardened Latches. In *Proc. 15th ACM Great Lakes symposium on VLSI*, pages 328–331, April 2005.

[22] M. S., W. C., E. J., R. S.K., and A. T. A systematic methodology to compute the architectural vulnerability factors for a high-performance microprocessor. In *Proc. of Intl. Symposium on Microarchitecture*, pages 29 – 40, December 2003.

[23] H. H. K. Tang. Nuclear Physics of Cosmic Ray Interaction with Semiconductor Materials: Particle-induced Soft Errors from a Physicists Perspective. *IBM J. of Research and Development*, 40(1):91–108, January 1996.

[24] J. von Neumann. Probabilistic Logics and the Synthesis of Reliable Organisms from Unreliable Components. In C. Shannon and J. McCarthy, editors, *Automata Studies – Annals of Mathematical Studies*, number 34, pages 43–98. Princeton University Press, 1956.

[25] R. Wilson and D. Lammers. Soft Errors Become Hard Truth for Logic. *EE Times*, May 2004.

[26] J. F. Ziegler. Terrestrial Cosmic Rays. *IBM J. of Research and Development*, 40(1):19–39, Jan. 1996.

978-1-4244-3039-0/08 $25.00 © 2008 IEEE

21st International Conference on VLSI Design

Energy-Efficient Soft-Error Protection
Using Operand Encoding and Operation Bypass

Kaushal R. Gandhi and Nihar R. Mahapatra
Department of Electrical & Computer Engineering
Michigan State University
East Lansing, MI 48824-1226, USA
E-mail: {gandhika, nrm}@egr.msu.edu

Abstract

As designs scale further into the nanometer regime, the vulnerability of logic circuits in commodity products to soft errors is increasing and their contribution towards total chip soft-error rate (SER) is predicted to be as high as 60%, much more than that of memory. We employ a value-aware framework that enables operation bypass in combinational circuits to simultaneously reduce both energy consumption and SER in them. Unlike techniques that reduce SER in combinational logic with very high performance and/or energy overheads (since they usually employ significant explicit spatial or temporal redundancy), our technique dynamically exploits operand values by shutting off portions of combinational circuits, thus reducing vulnerable area and energy consumption with minimal performance overheads. On the average across the SPEC CPU2k benchmark suite, we obtain 60% SER reduction and 24% energy savings with minimal impact on performance.

Keywords: low-power design, operand encoding, operation bypass, soft error.

1 Introduction

Soft errors are changes in logic state resulting from the latching of single-event transients (SETs) caused by radiation or electrical noise. The non-ideal scaling of device parasitics and supply/threshold voltages (leading to increased leakage), coupling in high-frequency devices and interconnects, and process variability have all resulted in soft-error tolerance becoming a critical design metric. Among the overall design challenges, identified by ITRS, power reduction and soft-error tolerance feature as key objectives in technologies beyond 90 nm [1].

Transient faults resulting from particle strikes or electrical noise at the outputs of logic gates are expected to become prevalent in future technologies. In particular, the soft-error rate (SER) of combinational logic is projected to increase by nine orders of magnitude from 1992 to 2011, i.e., from technology nodes 600 nm through 50 nm, and at 50 nm it is expected to become comparable to that of unprotected memory [20]. The SER of CMOS products fabricated in advanced technologies is projected to exceed 50,000 failures in time (FIT) per chip, which is higher than those due to all other (defect and manufacturing related) failure mechanisms combined (1-500 FIT per chip) [3].

Soft Errors in Logic Circuits: Due to technology scaling, reduced capacitance of gate output nodes, and lower supply voltages, the amount of charge representing a logic state is diminishing. Consequently, the smallest charge (required at a gate output node) to create an SET that may result in a soft error, also referred to as the *critical charge* (Q_{crit}), is also reducing. This makes logic circuits more susceptible to strikes by cosmic particles and electrical noise. The masking effects in logic circuits that prevent the propagation of SETs and their manifestation as errors are also reducing. With higher clock frequencies, the setup and hold times of pipeline registers/latches become a larger fraction of the clock cycle, and hence the probability that an SET will be latched increases, i.e., the *latching-window masking* effect reduces. Fewer logic levels to accommodate higher clock frequencies results in reduced attenuation of SETs and less *electrical masking*. Finally, *logical masking* (the absence of a functionally-sensitized path to propagate an SET to a primary output) effects decrease as pipeline depth and clock frequencies increase, as the number of logic levels per pipeline stage reduce.

SER Reduction Using Redundancy: A majority of previous logic SER mitigation techniques introduce explicit hardware to add time or computation redundancy. Techniques that rely on redundant threads incur performance overheads of 10-30% and energy overheads of almost 100% (in the sphere of replication, i.e., in components that are protected), since each duplicate instruction incurs the same energy cost as the original instruction [9, 17, 15, 16]. Other techniques that use hardware redundancy, such as triple modular redundancy, incur area and energy overheads of 200%; more refined schemes reduce these overheads but they are still close to 80-90% [13, 14]. This makes them unsuitable for the overhead-sensitive and highly-competitive commercial off-the-shelf (COTS) processor market. So, there is an acute need for design techniques that reduce logic SER but do not compromise power or performance.

Energy-Efficient SER Reduction: Instead of introducing explicit time or computation redundancy, we rely on inherent redundancy present in data values and exploit it to achieve SER reduction. Our approach encodes data values and opportunistically performs operation bypass in por-

45

978-1-4244-3039-0/08 $25.00 © 2008 IEEE

tions of functional units. This bypass protects computational hardware from soft errors.

The key idea is to split an operation (e.g., addition) into two phases. The first phase deals with the input operand values on which arithmetic operations are performed. Encoding the data value helps reduce the number of vulnerable bits. The encoding is such that inherent redundancy in data values is exploited to reduce the number of vulnerable bits in a data word. A data word is statically partitioned into *subwords* (SWs) and each subword is classified as *special* (if all its constituent bits are identical, i.e., an all-0 or all-1 bit pattern) or *regular* (all other bit patterns of the subword), otherwise. Each subword is encoded using an additional *special value bit* (SV bit), which is set ($SV = 1$) if it is special-valued and reset ($SV = 0$), otherwise. Now, a special-valued SW, (i.e., an all-0 or all-1 bit pattern) can be represented using a single bit of the SW and its SV bit, whereas all the SW bits are required to represent a regular value. As explained later in Sec. 2, the overhead per data word is reduced to by embedding the SV bits in a special-valued subword.

In the first phase of our technique, operand subwords fetched from memory are encoded as explained above (i.e., as either special or regular) depending upon their values. Since a special-valued SW is represented using only two bits (its LSB and its SV bit), errors in the remaining bits do not have any impact, i.e., the average number of vulnerable bits across a data word is reduced. This provides protection in structures where the data value may reside (storage components such as the register file, pipeline registers, the issue queue or re-order buffer, load value queue, store queue) or during its communication (in ALU buses). It also carries over to any operation that consumes the data word as explained next.

The second phase deals with the actual operation. In this, for certain predetermined input SW value combinations, the operation in a part of a functional unit (FU) (to which our technique is applied) is bypassed, i.e., to determine the output SW for that combination of input SWs, a very simple alternative method is used instead of the original, more complex hardware. Such *exploitable* input SW combinations can be detected by examining the SV bits of the input SWs. In our work, in addition to energy reduction, we use the operation bypass framework to reduce the vulnerability of FUs.

Most of the overhead of our technique is incurred in encoding in the first phase. But the overhead is low because: (a) encoding data subwords is a very simple operation and employs a very simple zero- or one-detect circuitry and is performed only once when they are read from the data cache or from the immediate field of an instruction; (b) since operands are fetched/read more often than written/updated (typically by a ratio of 2:1) [10], operand encoding occurs much less frequently than operation bypass. This is because, in our design, an operation on encoded operands produces an encoded result, which is preserved as such while it resides in the register file. Subsequent operations on the same encoded operands or their results do not incur encoding overheads. The second phase yields the benefits in terms of SER reduction and energy savings from operation bypass in storage, communications, and computational components in the datapath.

The remainder of the paper is organized as follows. We present an overview of operation bypass in Sec. 2. Next, we present an SER model to estimate soft errors in combinational logic circuits. Following this, we discuss our design and simulation methodology and also mention how we optimally partition the hardware to obtain maximum SER reduction in Sec. 4. The impact of our encoding technique for data words and error hardening of overhead circuitry and simulation results are presented in Sec. 5. Finally, we discuss related work in Sec. 6 and conclude in Sec. 7.

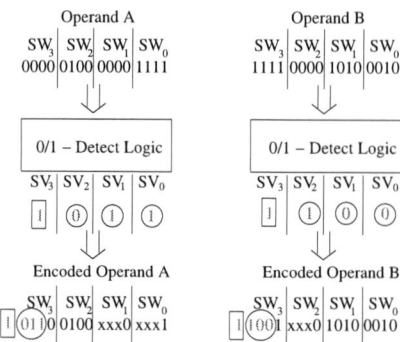

Figure 1. Encoding operands: A special-valued subword can be denoted using only two bits, its least significant bit and a special-value bit. The SV bits can be clubbed together within the most significant SW to reduce encoding overhead. All the other bits of a special-valued SW retain values from a previous cycle to prevent unnecessary switching activity.

2 Operation Bypass Framework

Due to space constraints, we provide a concise description of operation bypass here. More details can be found in [7]. Our two-phase approach of identifying special values and exploiting them in the combinational circuit is best explained using an example. Consider a 16-bit datapath with two operands, Operand A and Operand B, as illustrated in Fig. 1. For simplicity, each operand is partitioned into four-bit SWs, but can be partitioned in any other manner. The 0-1 detect unit determines whether a SW is special-valued or not and outputs the SV bit for each SW, which is 1 if all the bits in the SW are identical and 0, otherwise. Therefore, we choose to encode an operand word only if its most significant SW is special valued. While encoding the operand, all special-valued subwords retain their least significant bit values and the other SW bits retain values from the previous cycle (denoted by x in Fig. 1). This reduces unnecessary switching activity, and thus power, in the flip-flops as well as the combinational logic associated with these other SW bits. The SV bits of the low-order SWs are clubbed together in the most-significant SW (indicated by the circled bits). An additional bit called the *status bit* (indicated by boxed bits) indicates whether a data word is encoded or not. If the most significant SW is not special valued, then the status bit = 0, and all SWs retain their original bit values regardless of special values. This way only a single additional bit per data word is used, thereby minimizing area and energy overheads. This is an effective strategy because we find that a low-order SW is unlikely to have a special value if the most significant SW is not special-valued.

The second phase, which involves an operation on the

978-1-4244-3039-0/08 $25.00 © 2008 IEEE

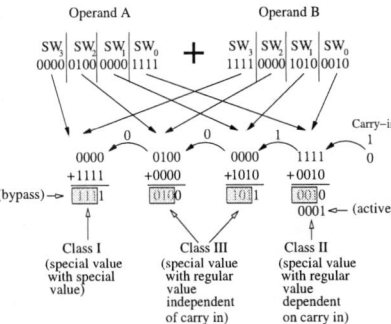

Figure 2. **Exploitable subword combinations:** Three different classes of exploitable combinations for addition are shown. Class I represents combinations of special values with special values. Class II represents combinations of special values with regular or special values but is exploitable depending upon the carry in from a previous SW. Class III is a similar combination but independent of the carry in from a previous group.

SWs, can be modified to exploit combinations of special values. In this, an FU is partitioned so that each functional subunit (FSU) operates on operand SWs. For addition, we identify three different classes of exploitable combinations (see Fig. 2): (1) Class I: combinations of special SW values with special SW values; (2) Class II: class I combinations plus combinations of special SW values with special or regular SW values but depending on input from adjacent FSU; and (3) Class III: class II combinations plus combinations of special SW values with special or regular SW values and independent of input from adjacent FSU. For each of these classes of combinations, the output can be determined by copying one of the SWs or letting the least significant bit perform the operation and extend its output as that of the complete SW. Finally, the output SW is encoded and combined with other encoded output SWs to form the encoded output. This is done in order to facilitate operation bypass in any subsequent operation on the encoded output word.

3 Soft-Error Rate Analysis Model

We develop an SER model to evaluate the SER of the original combinational logic block (CLB) and the two-phase SER reduction approach. Our generation and propagation model of an SET is based on the model proposed in [20], but with a few enhancements. First, we determine the critical charge (Q_{crit}) at each gate output node i in the CLB, where $i \in I$, and I is the set of all gate output nodes susceptible to particle strikes in the CLB. We then enumerate all unique shortest path lengths $L_{i,j}$ (in terms of the number of logic gates) from every node i to every primary output j. If more than one path exists between the pair (i, j), the shorter path length is used. This approximation of critical charge per pair (i, j) (instead of per unique path) is made to reduce the complexity for reconvergent paths, which can be exponential depending upon the topology of the CLB. Next, an FO4 NAND gate chain of length l, for each unique $l \in L_{i,j}$, is used to determine $Q_{crit_{i,j}}$. The error pulse widths at the end of the NAND gate chain, corresponding to all charges ranging from $Q_{crit_{i,j}}$ to Q_{max}, in discrete increments, are also stored in the form of a lookup table (LUT). Q_{max} is the maximum amount of charge that can be induced. The LUT

models electrical derating and aids in determining whether an SET will be generated by a particle strike and the corresponding error pulse width when it reaches a primary output.

Assuming that the probability of an error pulse of width d arriving at a primary output is uniformly distributed across a clock period (T) and the output latch has a latching window of width w, the probability that the error pulse will be latched is given by:

$$P_{latch}(Q) = \begin{cases} 0, & \text{if } d(Q) < w; \\ \frac{d(Q) - w}{T}, & \text{if } w \leq d(Q) \leq T + w; \\ 1, & \text{if } d(Q) > T + w, \end{cases} \quad (1)$$

where Q is the charge induced at a gate output node and $d(Q)$ is the width of the error pulse corresponding to charge Q. The above model describes the latching probability of: (1) a *single* error pulse propagating along a *single* path to a primary output. However, there may be other situations where: (2) a *single* error pulse propagates along *multiple* paths to a primary output (reconvergent paths) and (3) a *single* error pulse propagates to *many different primary outputs* through *multiple* independent paths. The above model under-estimates latching-window masking and over-estimates SER, since it does not distinguish between these three scenarios [20]. The distinction is necessary because the error pulses are correlated in time (i.e., in scenarios (2) and (3)) and the latching-window masking probability must be determined for all the error pulses taken together (*error pulse waveform*) instead of considering each pulse individually (as done in previous work). We enhance the model to consider these scenarios.

In reconvergent paths (scenario (2)), depending upon the logically sensitized paths, error pulses may overlap, and the effective error pulse width is not the sum of widths of the individual pulses but the width of the combination of the pulses, which may overlap in time. Second, an error pulse may propagate along different paths to multiple primary outputs (scenario (3)). In this case, a single SET at a gate output affects multiple primary outputs. Again, the error pulses may overlap in time depending upon their relative arrivals at the different primary outputs. For scenarios (2) and (3), we modify the model in the following way. To determine the relative arrival of error pulses at all primary outputs on which sensitized paths from gate node i (where the error pulse was generated are incident) we first construct a *combined* error pulse waveform by combining error pulses corresponding to different primary outputs.

The sensitization of a path depends upon the primary inputs to the circuit. The combination of error pulses (that constitute an error pulse waveform) varies depending upon the input to the CLB. It is not computationally feasible to determine the error pulse waveform on an input by input basis for 100s of millions of input vectors over which simulation is required to characterize the model. So, we simulate a structural description of the circuit with representative input vectors from a standard benchmark suite and determine the probability of an error propagating to at least one primary output, assuming that an error occurs at a gate output node i. This probability is distributed equally amongst all the candidate paths along which the error pulse could have propagated. Although, this distribution of probabilities along candidate paths is not a completely accurate representation of the error pulse waveform, it is much more accurate compared to considering error pulses along each path without accounting for the relative delay between error pulses, as done in previous soft-error models. Based on this error-pulse waveform, we determine the latching-window masking for scenarios (2) and (3).

Finally, logical derating is modeled as follows. The same structural description of the circuit is used to perform a logical simulation over input vectors from a representative simulation window of SPEC CPU2000 benchmarks [19]. During simulation of the structural description of the CLB, for each input vector, we introduce an error at a gate output node i and observe whether a primary output is affected or not. If at least one primary output changes due to this, we record this as a soft error for node i. This is repeated for every node i in the circuit. Therefore, for every gate output node i we determine the logical masking probability as:

$$1 - P_{logical}(\text{error at gate } g_i) =$$
$$1 - \frac{\text{No. of soft errors observed}}{\text{No. of input vectors}}. \quad (2)$$

To compute the total SER of a circuit, we first determine the soft-error contribution of a single gate. The soft error rate is expressed as a function of the charge induced (Q or higher) at a gate node with vulnerable area A:

$$SER(Q) = K \times F \times A \times e^{\frac{-Q}{Q_s}}, \quad (3)$$

where K is a technology-independent constant, F is the neutron flux, A is the sensitive device area, Q is the induced charge, and Q_s the charge collection slope for the technology, which is strongly dependent on doping and supply voltage. The soft-error contribution of a gate is then defined as:

$$SEC(g_i) =$$
$$P_{logical}(g_i) \times \sum_{k=Q_{crit}}^{Q_{max}} \{(SER_{Q_k} - SER_{(Q_k+q)}) \times P_{latch}(Q_k)\}, \quad (4)$$

where q is size of the discrete charge intervals into which the charge range is divided, $P_{latch}(Q_k)$ is the probability that an error-pulse waveform resulting from induced charge Q_k will be latched, and $P_{logical}(g_i)$ is the probability that a functionally-sensitized path exists between gate output node i and at least one primary output. The summation over k is performed to determine SER due to a particle strike inducing a charge of Q_{crit} or higher. Since the error pulse width varies with the charge induced, we divide the range of induced charges into finite intervals and weight each interval by the probability of the error pulse being latched. The total SER of the circuit is given by:

$$SER_{total} = \sum_{\forall i} SEC(g_i) \quad (5)$$

4 Simulation Methodology

Our baseline design for evaluating SER and power reduction is one without any operation bypass technique implemented on it. We apply our design techniques to a 64-bit carry lookahead tree adder (CLA) using benchmark inputs. We consider integer addition in the execute stage of an out-of-order superscalar processor. To compare and evaluate our design schemes, we use a basic configuration where the original CLA does not have any operation bypass technique implemented on it. We determine its performance, power, area, and SER, and these serve as the comparison points after we implement operation bypass on it. Although we implement operation bypass on the CLA, it has been shown to be applicable in the context of functional units and so the operation bypass framework is not limited to a single type of operation [8]. Encoding of special values and operands is identical to the strategy proposed in [7] (as discussed earlier in Sec. 2).

We used a modified version of SimpleScalar [5] to simulate a 64-bit single-core superscalar processor using 64-bit binaries to trace operand values. We simulated a structural description of a 64-bit carry-lookahead tree adder (CLA) with these traced input vectors to generate the logical masking data for the SER model. We also used circuit-level simulations to generate the latching-window and electrical masking models. For energy modeling, we used circuit-level simulations for a 64-bit CLA using the traced inputs from architectural simulation. The adder was designed as a static CMOS circuit. We used 64-bit binaries for SPEC CPU2K benchmarks to generate the adder inputs. The circuits were designed using Cadence tools and 0.18 μm TSMC technology libraries due to the availability of power characterized standard cells at the typical corner for that technology. Since logic SER would increase as technology scales down further, the SER per bit of logic would increase and hence operation bypass to reduce SER would be more effective because it would reduce more soft errors per bit for the same rate of operation bypass. Therefore, our technique would improve with current and future technologies.

Next, we discuss results. The reduced vulnerability of data due to our encoding scheme is presented for byte-wise partitioning scenario below. This is only to illustrate the effectiveness of our encoding scheme. We actually use a dynamic programming approach to partition the datapath in order to maximize SER reduction using our SER analysis model and traced input operands.

5 Results and Discussion

In this section, we first discuss the effect our encoding scheme has on the reduction in vulnerable bits in communication (ALU buses) and storage (register file, pipeline registers) components. Following that, we present *net* SER and energy reduction in functional units (i.e., after accounting for all overheads). In this paper, we have analyzed a 64-bit CLA, but as shown previously [4, 6], operation bypass can be extended to the whole processor pipeline. Therefore, similar benefits can be achieved across the processor pipeline.

Reduced Vulnerable Bits Using Operand Encoding: An operand in the encoded format is shown in Fig. 3. While

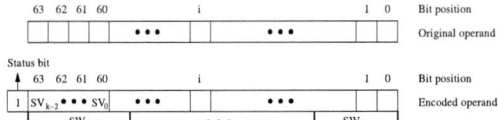

Figure 3. Encoding scheme: Operand in the original and encoded formats. In the encoded format, special value bits (SV bits) are stored in the most significant subword and an additional status bit is used to indicate whether an operand is encoded or not. Note that special values are represented using the SV bit and each subword's least significant bit. The other bits of that subword are not required. An operand is encoded only if its most significant subword is special valued, i.e., all its constituent bits are identical.

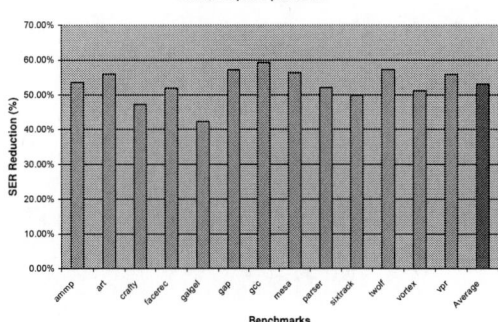

Figure 4. General-purpose optimization: One half of the benchmarks are randomly chosen as the training set for the DP algorithm to determine partitions for minimizing SER. These partitions are applied to the remaining test benchmarks on which operation bypass is implemented, which yields 53% reduction in SER and 22% reduction in energy on average.

loading a data value from the data cache or reading an immediate value, the operand is encoded only if its most significant subword is special valued, i.e., all its constituent bits are identical; in the example illustrated, each SW is a byte. The encoding strategy was discussed earlier in Sec. 2.

Now, in the encoded format, a bit is vulnerable only if: (1) the subword it belongs to is a regular value; (2) it is the status bit; (3) it is a special value bit; or (4) it is the least significant bit of a special value. Here, for simplicity, we assume a byte-wise partitioned datapath, but as we show later, it can be partitioned in any other manner to optimize SER reduction. Consider an operation on two operands. From our analysis of the traced input operands, we find that the least significant byte of Operand 1 is special valued 30.38% of the time. So, its 7 most significant bits are vulnerable 69.62% of the time, while its least significant bit is vulnerable 100% of the time. So, on average, the number of vulnerable bits in the least significant byte is $(0.6962 \times 7 + 0.3038 \times 1.0) = 5.87$, when operation bypass is applied. In the original format, all 8 bits are used and are thus vulnerable. For the operand as a whole, we find that, on average, in Operand 1, only 33.56 bits are vulnerable during an operation, whereas in the original format all 64 bits are vulnerable. Note that the most significant subword and the status bit are always vulnerable. This represents a 47.6% reduction in the number of vulnerable bits for Operand 1. For Operand 2, a similar analysis reveals that, on average, only 32.93 bits are vulnerable as opposed to 64 in the unencoded format, representing a 48.5% reduction in the number of vulnerable bits.

SER Reduction and Energy Savings: Our experiments were conducted for a 64-bit CLA. The results presented here represent SER reduction in the CLA alone. However, this can be easily extended to other functional units where similar SER reduction and energy savings can be expected.

Using operation bypass, we are able to prevent the generation of soft errors and their propagation to primary outputs in different ways. Consider an FU with two adjacent FSUs: FSU_0 and FSU_1. In the original design of the CLA (without operation bypass), an SET could be generated in FSU_0 and propagate to: (1) primary output of FSU_0; (2) primary output of FSU_1; (3) or both. After implementing our technique, cases (1), (2), and (3) can be avoided if the operation is bypassed in FSU_0 or case (2) can be avoided if the operation in FSU_1 is bypassed. Also, a particle striking the alternative computation circuitry in a bypassed FSU will (effectively) not result in an error because that portion of the hardware is error hardened.

As mentioned previously, we use a dynamic programming (DP) algorithm to optimally partition a word into SWs (and an FU into corresponding FSUs) based on a training set of operand inputs so that SER is minimized. To demonstrate the effectiveness of our approach, we consider three increasing degrees of customization. In general-purpose optimization, the training set consists of all add operations in the standard SimPoint window from 13 randomly-chosen benchmarks. The partitions obtained from the DP algorithm are applied to the remaining 13 test benchmarks; we obtain an SER reduction of 53% on average for a 64-bit CLA (see Fig. 4).

In workload-specific optimization, the training set is comprised of the first half of all addition operations from all benchmarks and the test set is comprised of the latter half of all addition operations from all benchmarks. The optimal partitioning is obtained from the training set and it is applied to the test set to achieve SER reduction of 58% and energy savings of 21% (see Fig. 5).

In program-specific optimization, we use the first half of the addition operations as the training set for each benchmark. Using the training set, we determine the optimal partitioning for each benchmark and apply it to the remaining half of the addition operations for the same benchmarks to achieve SER reduction of 60% and energy savings of 24% (see Fig. 6).

The performance overheads are minimal: an increase of 4% in the latency of the FU. But, often the bottleneck in the microprocessor system is the issue queue, and hence it may not directly affect performance. If it does severely affect performance, then our technique can be restricted to functional units in the execute stage and other radiation hardening or SER reduction techniques can be applied to other simpler functional units such as, address-generation hardware. Also, since operation bypass introduces simpler, alternative, and faster means of computation when an FSU is bypassed, the slack generated as a result can be exploited to reduce any performance overheads. We defer exploration of this to future study.

6 Related Work

At the circuit level, an error can be prevented by either mitigating the generation of an SET, preventing its propa-

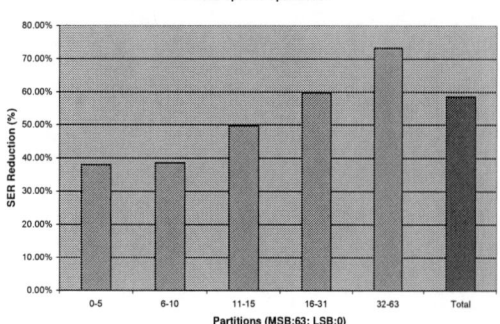

Figure 5. Workload-specific optimization: One half of the operations from each benchmark is used as the training set and the remaining operations are treated as the test set. The partitions are determined using the DP algorithm to determine the optimal partitioning for minimum SER. SER reduction of 58% achieved on average.

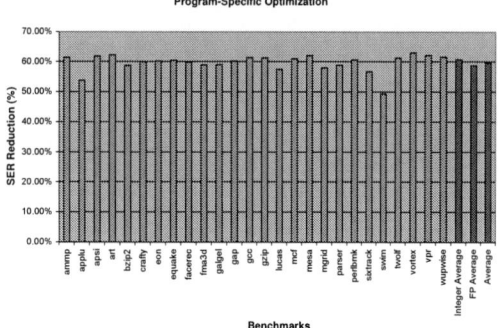

Figure 6. Program-specific optimization: A partitioning is determined for each benchmark individually. One half of the operations from a benchmark is chosen as the training set and the remaining operations are chosen as the test set. The partitioning across each benchmark is program-specific and this yields the best results for that given benchmark. Average SER reduction of 60% and energy reduction of 24% are obtained across all benchmarks.

gation, preventing it from latching at the output, or masking/correcting it before/after it is latched. Typically, design techniques employ either time redundancy (error detection and correction through rollback and retry [16]) or hardware redundancy (mask the propagation of errors through schemes such as triple modular redundancy (TMR)) to achieve one or more of the above [13]. Other techniques such as radiation hardening [25, 22], body biasing [23, 12], or error-tolerant latches [11] have also been proposed. At the architecture level, hardware structures are replicated (the Tandem Computer [2], Compaq NonStop Himalaya [24], and IBM S/390 [21] explicitly use hardware replication) or redundant threads are used to ensure correct execution of a program [18, 15]. The performance overhead using redundant threads can be as much as 30% and incurs energy overhead close to 80-100%. But the area overhead is minimal. On the other hand, hardware redundancy incurs overheads of 100% and similar energy overheads and with performance overheads.

Our technique, in contrast to the above, does not incur heavy overheads as is characteristic for any technique that introduces redundancies. Whether it is temporal or spatial

redundancy, the power overhead is always significant, at least 30% for microarchitectural techniques. In contrast to value-based low-power techniques [4, 6, 8, 7], we propose options for reducing SER using a similar framework.

7 Conclusion

In this paper, we presented operation bypass techniques to reduce both SER and energy consumption. These differ from previous operation bypass works that focus only on energy savings. Further, compared to approaches that use explicit temporal and spatial redundancy, we exploit the inherent redundancy present in data values to not only obtain SER reduction but also to *simultaneously* reduce energy consumption. This is the most important contribution of our work, since these two metrics are often in conflict with each other—optimizing one generally adversely affects the other. Previous approaches typically provide SER reduction with substantial energy, performance, and/or area overheads, whereas our approach simultaneously reduces both SER and energy, with minimal performance and area overheads. We also presented an SER model that captures latching-window masking more accurately by taking the relative arrival times of error pulses into consideration. To maximize the effectiveness of our operation bypass approach, we employed a DP algorithm to optimally partition data words and FUs to minimize SER. As a result, we obtained average SER reduction of 60% and energy savings of 24% using program-specific optimization across the SPEC CPU2k benchmark suite for a 64-bit CLA. Similar results can be expected for other functional units. In future work, we will exploit the slack arising from operation bypass to further enhance its effectiveness.

References

[1] Semiconductor Industry Association. The International Technology Roadmap for Semiconductors. URL:http://public.itrs.net/Common/2005ITRS/Design.pdf, 2005.

[2] J. Bartlett, J. Gray, and B. Horst. Fault tolerance in Tandem computer systems. Technical Report 86.2, Tandem Computers, March 1986.

[3] Robert Baumann. Soft Errors in Advanced Computer Systems. *IEEE Design and Test of Computers*, 22(3):258–266, May 2005.

[4] D. Brooks and M. Martonosi. Value-based clock gating and operation packing: dynamic strategies for improving processor power and performance. In *ACM Transactions on computer systems*, May 2000.

[5] D. Burger and T.M. Austin. The SimpleScalar Tool Set, version 2.0. *Computer Architecture News*, pages 13–25, June 1997.

[6] R. Canal, A. Gonzalez, and J. Smith. Very low power pipelines using significance compression. In *Proceedings of the Annual ACM/IEEE International Symposium on Microarchitecture*, 2000.

[7] K. Gandhi and N. Mahapatra. Exploiting data-dependent slack using dynamic multi-VDD to minimize energy consumption in datapath circuits. In *Proc. 9th Design, Automation and Test in Europe (DATE 2006), Munich, Germany*, March 2006.

[8] K. R. Gandhi and N. R. Mahapatra. Dynamically exploiting frequent operand values for energy efficiency in integer functional units. In *Proceedings of the 18th. International Conference on VLSI Design, Kolkata*, 2005.

[9] M. Gomaa and T. N. Vijaykumar. Opportunistic transient-fault detection. In *Proc. of the 32nd Annual International Symposium on Computer Architecture*, pages 172–183, 2005.

[10] J.L. Hennessy and D.A. Patterson. *Computer Architecture: A Quantitative Approach, Third edition*. Morgan Kaufmann Publishers, 2003.

[11] Srivathsan Krishnamohan and Nihar R. Mahapatra. Combining error masking and error detection plus recovery to combat soft errors in static CMOS circuits. In *Proc. ACM International Conference on Dependable Systems and Networks*, pages 40–49, 2005.

[12] S. Mitra, T. Karnik, N. Seifert, and M. Zhang. Logic soft errors in sub-65nm technologies design and CAD challenges. In *Proceedings of Design Automation Conference*, June 2005.

[13] K. Mohanram and N. A. Touba. Cost-effective approach for reducing soft error failure rate in logic circuits. In *Proc. International Test Conference*, 2003.

[14] K. Mohanram and N. A. Touba. Partial error masking to reduce soft error failure rate in logic circuits. In *Proc. International Symposium on Defect and Fault Tolerance in VLSI Systems*, pages 433–440, 2003.

[15] Shubhendu S. Mukherjee, Michael Kontz, and Steven K. Reinhardt. Detailed design and evaluation of redundant multithreading alternatives. In *Proceedings of the 29th annual international symposium on Computer architecture*, pages 99–110, 2002.

[16] M. Nicolaidis. Time redundancy based soft-error tolerance to rescue nanometer technologies. In *17th IEEE VLSI Test Symposium*, 1998.

[17] N. Oh, P. Shirvani, and E. McCluskey. ED4I: Error detection by diverse data and duplicated instructions. *IEEE Transactions on Computers*, 51:180–199, 2002.

[18] Eric Rotenberg. AR-SMT: A microarchitectural approach to fault tolerance in microprocessors. In *Symposium on Fault-Tolerant Computing*, pages 84–91, 1999.

[19] T. Sherwood, E. Perelman, and B. Calder. Automatically characterizing large scale program behavior. In *Proc. of International Conference on Parallel Architectures and Compiler Techniques*, September 2001.

[20] Premkishore Shivakumar, Michael Kistlerand, Stephen W. Keckler, Doug Burger, and Lorenzo Alvisi. Modeling the effect of technology trends on the soft error rate of combinational logic. In *Proc. ACM International Conference on Dependable Systems and Networks*, pages 389–398, June 2002.

[21] T. J. Slegal et al. IBM's/390 G5 Microprocessor Design. *IEEE Micro*, pages 12–23, March 1999.

[22] T. Karnik et. al. Selective node engineering for chip-level soft error improvement. In *Proceedings of Symposium on VLSI Circuits, Digest of Technical Papers*, pages 204–205, 2002.

[23] T. Karnik et. al. Impact of body bias on alpha- and neutron-induced soft error rates of flip-flops. In *Proceedings of Symposium on VLSI Circuits, Digest of Technical Papers*, pages 324–325, 2004.

[24] A. Wood. Data integrity concepts, features, and technology. White paper, Tandem division, Compaq Computer Corporation.

[25] Q. Zhou and K. Mohanram. Cost-Effective Radiation Hardening Technique for Combinational Logic. In *Proc. ACM/IEEE International Conf. Computer-Aided Design (ICCAD)*, November 2004.

SESSION B1:
Wireless/Communication

21st International Conference on VLSI Design

Retimed Decomposed Serial Berlekamp-Massey (BM) Architecture for High-Speed Reed-Solomon Decoding

Shahid Rizwan

Korea Advanced Institute of Science and Technology, Republic of Korea

shahidrizwan06@hotmail.com

Abstract

This paper presents a retimed decomposed inversion-less serial Berlekamp-Massey (BM) architecture for Reed Solomon (RS) decoding. The key idea is to apply the retiming technique into the critical path in order to achieve high decoding performance. The standard basis irregular fully parallel multiplier is separated into partial product generation (PPG) and partial product reduction (PPR) stages to implement the proposed modified decomposed inversion-less serial BM algorithm. The proposed RS (255,239) decoder is implemented in verilog HDL and synthesized with 0.18 μm CMOS std130 standard cell library. The proposed architecture achieves almost 76 % increase in speed and throughput, and can be used in high-speed and high-throughput applications such as DVD, optical fiber communications, etc.

1. Introduction

Among a large number of error correction coding developed so far, linear block codes such as Reed-Solomon and BCH have a lot of applications in many digital areas because of their powerful error correction capability and efficient encoding and decoding procedures [1]. These applications include DVD, DTV, satellite communications, wireless systems, optical communications, etc.

Due to increasing demand for high capacity of optical communications, the high-speed and high-throughput implementations of RS decoders are desirable to meet higher data rate requirements. However all existing RS decoder architectures have limitations in terms of speed and throughput. Among the decoding steps in RS decoding, the second block i.e. the key equation solver (KES) block, which computes the coefficients of the error location

polynomial has been known to be the most responsible for the performance against high-speed, small-area, low-power because of its complexity and its delay and thus has been the bottleneck to achieve high-speed decoding. Either the Berlekamp-Massey (BM) algorithm or the Euclidian algorithm can be used to solve the key equation for an error locator polynomial and an error evaluator polynomial. The Berlekamp-Massey (BM) algorithm results in more efficient software and hardware implementations. If we investigate the literature, we would come up with three distinct BM architectures i.e. inversion-less parallel BM architecture [5], inversion-less serial BM architecture [7] and the inversion-less decomposed serial BM architecture [6]. The parallel BM architecture has a shorter latency but a lower decoding speed. The serial BM architecture has a higher latency but possesses a higher decoding speed. Decomposed serial BM structure has a medium latency in comparison to the above two architectures and also the highest decoding speed. So this is the fastest architecture amongst the present BM architectures. In the decomposed serial BM algorithm, the critical path includes a multiplier, an adder and a multiplexer. The critical path has a feedback nature so it could not be pipelined easily. A retimed decomposed inversion-less serial BM architecture is proposed in this paper to break the critical path to achieve high-speed and high-throughput Reed Solomon (RS) decoding. The rest of this paper is organized as follows. Section 2 explains the Reed-Solomon decoder, section 3 gives an insight into the retiming technique and section 4 explains the proposed retimed BM architecture. Pipelined error evaluator block has been given in section 5 and the concluding remarks are made in section 6.

2. Reed-Solomon Decoder

A general RS code denoted by (n, k) can correct up to $t = \lfloor n - k / 2 \rfloor$ symbol errors [7], where n and k

978-1-4244-3039-0/08 $25.00 © 2008 IEEE

represent the length of a block and the length of the information symbols, respectively.

A syndrome-based RS decoder consists of three components [1]. First part is a syndrome calculator. It generates a syndrome polynomial that is used in the second component for solving a key equation (KES). The Berlekamp-Massey (BM) algorithm is used to solve the key equation for an error locator polynomial and an error evaluator polynomial because it is considered to be the one with the least hardware complexity. The BM algorithm computes the error locator polynomial in 2t iterations [2]. Computation of error locator and error evaluator polynomial in parallel results in less latency (merged architecture). Then in the third component, these two polynomials are used to compute the error locations and the corresponding error values according to the Chien search and Forney's algorithm. In addition, delay elements are used in parallel to these three components in order to buffer the received symbols according to the latency of these components. The distinct blocks in RS decoder block usually operate in parallel (pipelined mode) and are shown in **Figure 1**. The main objective of this paper is the improvement of the KES block to improve the speed of the Reed-Solomon decoding. It is hard to pipeline the KES stage because of the presence of feed back signals. The use of a specialized technique i.e. retiming results in a high-throughput and high-speed Reed Solomon decoding architecture.

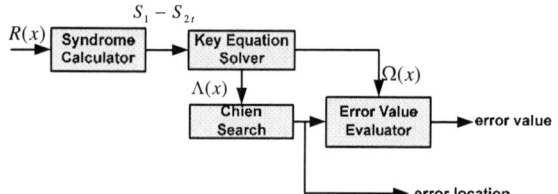

Figure 1   RS Decoder

3. Retiming Technique

Retiming is a transformation technique used to change the locations of the delay elements in a circuit without affecting the input/output characteristics of the circuit. Retiming maps a circuit G to a retimed circuit G_r. A retiming solution is characterized by a value r (V) for each node V in the graph. Let w (e) denotes the weight of the edge e in the original graph G, and let w_r (e) denote the weight of the edge e in the retimed graph Gr. The weight of the edge $U \xrightarrow{e} V$ in the retimed graph is computed from the weight of the edge in the original graph using

$$w_r\ (e) \quad = \quad w\ (e) + r\ (V) - r\ (U)$$
(A)

For example consider a filter circuit in graphical from in **Figure 2**(a), and the retimed filter drawn in **Figure 2**(b). The retiming values r (1) = 0, r (2) = 1, r (3) = 0, and r (4) = 0 can be used to obtain the retimed data flow graph (DFG) in **Figure 2**(b) from the data flow graph (DFG) in **Figure 2**(a). For example, the edge $3 \xrightarrow{e} 2$ in the retimed DFG contains

$$
\begin{aligned}
w_r(3 \xrightarrow{e} 2) \quad &= \quad w(3 \xrightarrow{e} 2) + r(2) - r(3) \\
&= \quad 0 + 1 - 0 = 1
\end{aligned}
$$

delay, and the edge $2 \xrightarrow{e} 1$ contains

$$
\begin{aligned}
w_r(2 \xrightarrow{e} 1) \quad &= \quad w(2 \xrightarrow{e} 1) + r(1) - r(2) \\
&= \quad 1 + 0 - 1 = 0
\end{aligned}
$$

delays.

A retiming solution is feasible if $w_r(e) \geq 0$ holds for all edges. The solution that maps **Figure 2**(a) to **Figure 2**(b) is feasible because all of the edges in Fig. 2(b) have nonnegative weights.

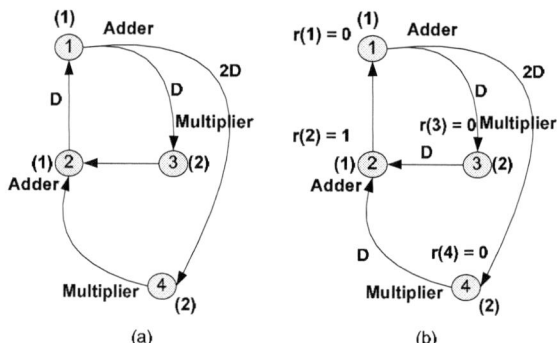

(a)   (b)

Figure 2   Data Flow Graph (DFG)

Although the filters in **Figure 2**(a) and **Figure 2**(b) have delays at different locations, these filters have the same input/output characteristics. These 2 filters can be derived from one another using retiming. The critical path of the filter in **Figure 2**(a) passes through 1 multiplier and 1 adder and has a computation time of 3 time units. The retimed filter in **Figure 2**(b) has a critical path that passes though 2 adders and has a computation time of 2 time units. By retiming the filter in **Figure 2**(a) to obtain the filter in **Figure 2**(b), the clock period has been reduced from 3 to 2 or by 33% [3].Retiming has many applications in synchronous circuit design. The applications include reducing the clock period of the circuit, reducing the number of registers in the circuit, reducing the power consumption of the circuit, and logic synthesis.

Cutset retiming is a useful technique that is a special

54

978-1-4244-3039-0/08 $25.00 © 2008 IEEE

case of retiming. A cutset is a set of edges that can be removed from the data flow graph (DFG) to create 2 disconnected sub-graphs. Cutset retiming only affects the weights of the edges in the cutest. It is often used in combination with slow-down. The procedure is to first replace each delay in the DFG with N delays to create an N-slow version of the DFG and then to perform cutest retiming on the N-slow DFG. In an N-slow system, $N-1$ null operations (or 0 samples) must be interleaved after each useful signal sample to preserve the functionality of the algorithm [3].

4. Retimed Decomposed Inversion-less Berlekamp-Massey (BM) Architecture

Modified decomposed inversion less BM algorithm is given as

$$\Lambda_j^{(i)} = \left\{ \begin{array}{ll} \delta.\Lambda_0^{(i-1)} & for\, 0 \leq j \leq 1 \\ \delta.\Lambda_j^{(i-1)} + \Delta^{(i)}\tau_{j-1}^{(i-1)} & for\, 2 \leq j \leq 2t+1 \end{array} \right\} \quad (1)$$

$$\Delta_j^{(i+1)} = \left\{ \begin{array}{ll} 0 & for\, 0 \leq j \leq 1 \\ \Delta_{j-1}^{(i+1)} + S_{i-j+3}.\Lambda_{j-1}^{(i)} & for\, 2 \leq j \leq 2t+1 \end{array} \right\} \quad (2)$$

where $\Lambda^{(i)}(x) = \Lambda_0^{(i)} + \Lambda_1^{(i)}x + ... + \Lambda_t^{(i)}x^t$ is the error locator polynomial, $\tau_j^{(i)}$'s are the coefficients of $\tau^{(i)}(x)$, and Δ_j^{i+1}'s are the partial results in computing discrepancy $\Delta^{(i+1)}$.

We can decompose the ith iteration into 2t + 2 cycles. $\Lambda_j^{(i)}$ requires at most two Finite Field Multipliers (FFMs) and $\Delta_j^{(i+1)}$ requires only one FFM. Standard basis irregular fully parallel multiplier with separate partial product generation (PPG) and partial product reduction (PPR) stages has been used in our design [4]. At first cycle partial product generation (PPG) operation is done and in the next cycle partial product reduction (PPR) is done.

For implementing the above idea we propose a retimed decomposed inversion-less BM architecture. The steps used to modify the original decomposed BM structure to the retimed structure are as follows.

i.   Identifying the nodes and the delay elements in the original circuit and drawing a data flow graph of the original circuit. (**Figure 3**)

ii.   Making a 2-slow version of the original circuit. (**Figure 4**)

iii.   Applying retiming technique by giving retiming value of -1 to all the multiplier nodes and then adjusting the delays of different branches to retain the original operation of the circuit. Then shifting the registers to inside of the multipliers to break the critical path. (**Figure 5**)

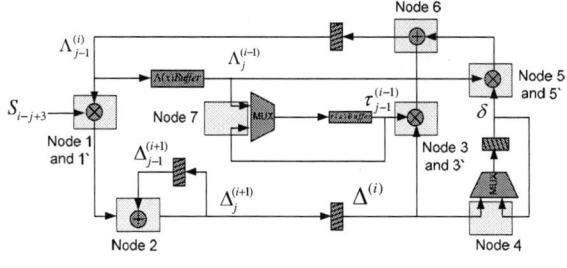

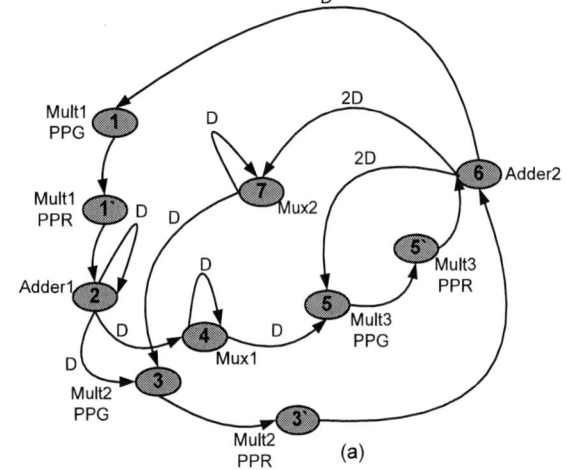

Figure 3   DFG for decomposed BM

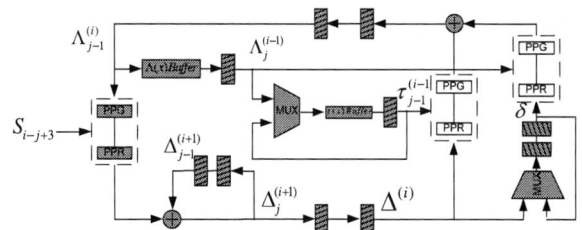

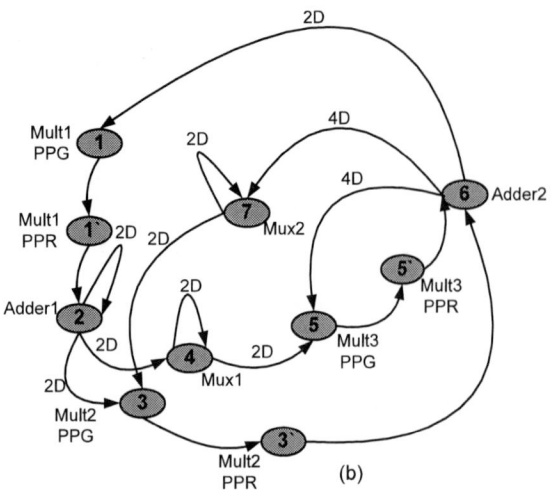

Figure 4   DFG for 2-slow version

iv.   Using register minimization technique to minimize the number of registers in the circuit. (**Figure 6**)

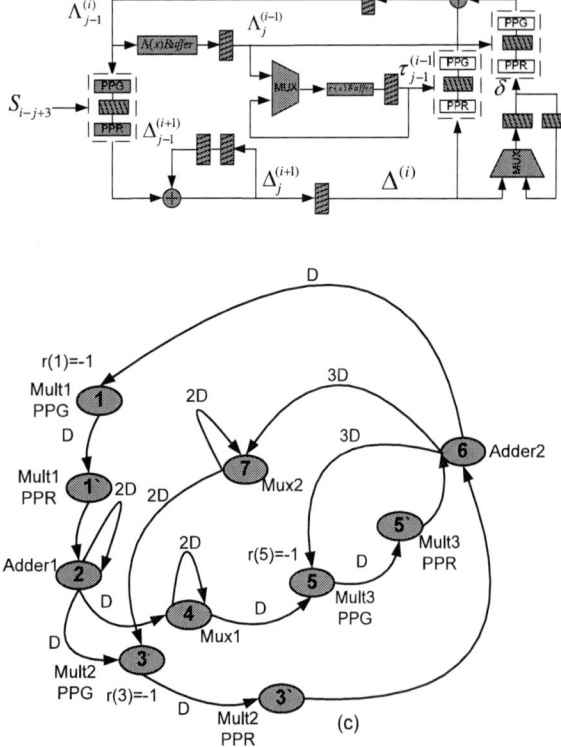

Figure 5   DFG after retiming

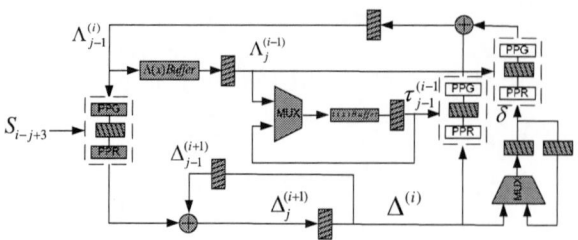

Figure 6   Register Minimization

The retimed decomposed serial BM architecture for RS (255, 239) decoder for error locator polynomial computation is shown in **Figure 6**. All operations are done is standard basis so there is no need to convert from dual basis to standard basis and vice versa as was done in [6]. A similar architecture in parallel to this architecture could be used to compute the error evaluator polynomial.

Table 1   Performance comparison

t = 8 case	Parallel BM	Serial BM	Decomp Serial BM	Retimed Decomp Serial BM
Cycle Time(ns)	4.71	2.99	2.70	1.53
Speed (MHz)	212	334	370.4	653.6
Through put (Gbps)	1.69	2.67	2.96	5.23
Latency (cycles)	16	288	144	288

Table 1 shows the performance comparison for different architectures for RS (255, 239) decoder. Compared to the previously proposed decomposed serial architecture [6] our architecture improves the speed from 370 MHz to 653 MHz. In addition the throughput increases from 2.96 Gbps to 5.23 Gbps. All these improvements are achieved at the expense of an additional latency plus some additional registers. In the proposed architecture, one iteration of the BM algorithm takes 2t + 2 clock cycles which is equivalent to the serial architecture's latency. Our proposed architecture achieves almost 76 % increase in speed and throughput.

5. Pipelined Error Evaluator Block

The basic error evaluator block given in literature has a delay of about 2.70 ns[6]. The critical path consists of an inverter, a multiplier, an adder and a multiplexer.

978-1-4244-3039-0/08 $25.00 © 2008 IEEE

Addition of two pipeline stages results in considerable speed improvement (**Figure 7**). Synthesis results are given in **Table 2**.

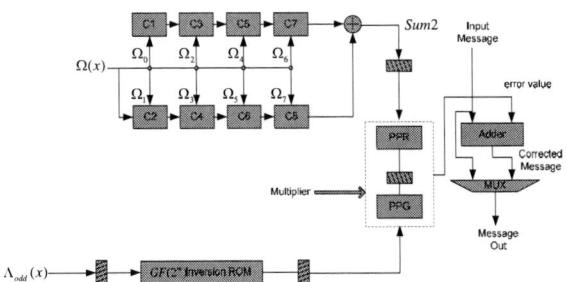

Figure 7   Pipelined Error Evaluator

Table 2   Synthesis Results

t = 8 case	Basic Error corrector block	Two stage pipeline	Three stage pipeline
Cycle Time(ns)	2.70	2.13	1.53
Speed (MHz)	370.4	469.5	653.6
Critical path	Inverter + multiplier+ adder multiplexer	Inverter + multipli er (PPG)	Inverter

6. Conclusions

In this paper, we have proposed a retimed decomposed inversion less serial BM architecture that uses the irregular fully parallel multiplier. In the original decomposed inversion less serial BM algorithm, the critical path includes a multiplier, an adder and a multiplexer. The retimed architecture breaks this critical path using retiming technique and results in considerable speed as well as throughput improvements. This improvement is achieved at the expense of a few extra registers. In addition, pipelining the error evaluator block results in considerable speed improvement. The proposed architecture was implemented in Verilog HDL and synthesized with UMC $0.18\,\mu m\ std130$ standard cell library. Our proposed architecture achieves almost 76 % increase in speed and throughput, which could be used in high speed and high throughput applications such as optical fiber communication. Layout was done using Milkyway and Apollo tool and is shown in **Figure 8**. The chip size is $2.0\,\mu m$ x $1.9\,\mu m$.

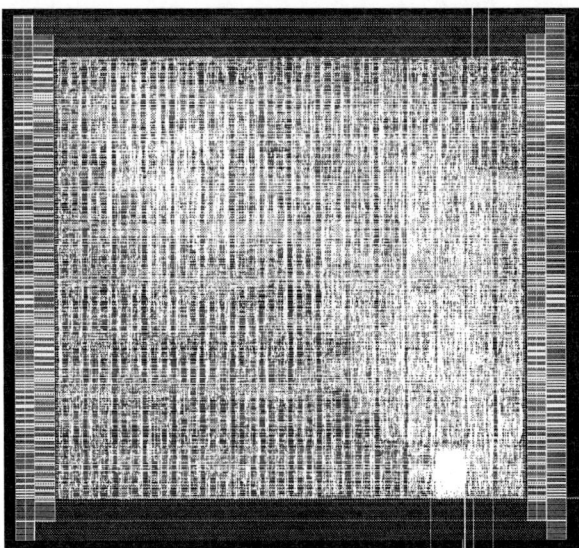

Figure 8   RS Decoder Layout

10. References

[1]   S. B. Wicker and V. K. Bhargava, "Reed Solomon Codes and Their Applications," IEEE PRESS, 1994.

[2]   Richard E. Blahut, *"Theory and Practice of Error Control Codes",* Addison-Wesley Publishing Company, 1983.

[3]   Keshab K. Parhi, "VLSI Digital Signal Processing Systems, Design and Implementation", A Wiley-Interscience Publication, 1999.

[4]   *Lijun Gao; Parhi, K.K.,* "Custom VLSI design of efficient low latency and low power finite field multiplier for Reed-Solomon codec", Circuits and Systems, 2001. ISCAS 2001.

[5]   *Reed, I.S.; Shih, M.T.,* "VLSI design of inverse-free Berlekamp-Massey algorithm", Computers and Digital Techniques, IEE Proceedings-E, Volume: 138, Issue: 5, Sept. 1991.

[6]   *Hsie-Chia Chang; Shung, C.B.; Chen-Yi Lee,"* A Reed-Solomon product-code (RS-PC) decoder chip for DVD applications", Solid-State Circuits, IEEE Journal of, Volume: 36, Issue: 2, Feb. 2001.

[7]   *Hyeong-Ju Kang; In-Cheol Park,* "A high-speed and low-latency Reed-Solomon decoder based on a dual-line structure", Acoustics, Speech, and Signal Processing, 2002. Proceedings. (ICASSP '02).

[8]   *Hanho Lee; Meng-Lin Yu; Leilei Song, "*VLSI design of Reed-Solomon decoder architectures ", Circuits and Systems, 2000. Proceedings. ISCAS 2000 Geneva.

21st International Conference on VLSI Design

Exploring the Processor and ISA Design for Wireless Sensor Network Applications

Shashidhar Mysore, Banit Agrawal, Frederic T. Chong, and Timothy Sherwood
Department of Computer Science
University of California, Santa Barbara
{shashimc, banit, chong, sherwood}@cs.ucsb.edu

Abstract

Power consumption, physical size, and architecture design of sensor node processors have been the focus of sensor network research in the architecture community. What lies at the foundation for these research is the hardware-level design which determines the boundaries for achievable utility and performance. Architecture design and evaluation, however, cannot be accomplished independent of the applications and software that run on these sensor nodes. On one hand, some researchers have proposed architectures that can cater to a variety of application classes while trading off on some performance improvements. On the other hand, a set of application-specific architectures have been proposed which perform certain operations extremely well but are not versatile enough to run a variety of applications.

This paper provides a design space exploration and optimizations platform to characterize the processor and ISA design tailored for a particular application or a class of applications. We collect a wide variety of sensor network applications to create a comprehensive benchmark suite called the WiSeNBench. We then present a careful profiling of these benchmark applications using an ARM simulator to identify some of the key characteristic behaviors. This also opens up avenue for a possible re-look at the classes of applications that could be supported on next-generation sensor networks and efficient architectural designs to enable these applications.

1 Introduction

Sensor network applications include environmental monitoring, structural sensing, battlefield communication, traffic, health, security monitoring, and other automation techniques. Consequently, sensor network research involves architecture, application optimization, communication protocol design, and developing efficient communication hardware. Small form factor, low-power budget, low-resource availability, and real-time requirements are some of the characterizing factors of sensor nodes. Each of the above is an additional constraint imposed on the designers of such networks. Given these various sites of improvement, in this paper we focus on sensor network applications, their characteristics, and explore ways in which they can influence the design of the underlying architecture.

Knowledge of the underlying hardware aids in efficient software development. A top-down approach towards software development may be well suited in scenarios where the processor architecture is already well defined and scope for optimization, if any, is just on the software front. With sensor applications and sensor network research, however, the scenario is quite different. Sensor applications require a special purpose hardware suitable to cater to a different set of requirement. Medical applications for example, need highly non-intrusive tiny sensors which are usually harmless to the human body as foreign-bodies. Whereas sensors spread on a military terrain have to be more tolerant to physical impacts and wide operating temperature range. These physical characteristics in which the sensors are placed and the difference in their utility makes it important to do research on sensor networks on a case by case basis.

Unique set of findings for a sensor application is good, but the sheer number of applications sensor networks are finding today rules out the feasibility of developing processors unique to each and every application. This brings us to a point where we will have to trade between the amount of customization available on a processor and the performance-cost benefits one would like to achieve. We believe that this aspect needs to reflect on the research focus in this area. The major contributions of this paper are a) To study some of the most important applications for sensor networks, including those in TinyBench [8] and SenseBench [27]. This includes a careful profiling of application behavior and its microarchitectural implications. b) To characterize the workload that is prominently visible among sensor network applications. c) To characterize the workload that is unique to a class of sensor network applications. d) To propose optimizations to existing sensor network architectures based on the observations made in (a,b,c).

In order for the characterization to be useful, the set of applications used for the purpose should be representative of the domain being analyzed. To this end, we present a thorough survey on sensor network applications, classify them based on the core functionality an application serves, and characterize each class independently.

Microarchitectural characteristics of programs serves as an important aspect in making architectural design decisions. The code size of an application influences critical decisions

978-1-4244-3039-0/08 $25.00 © 2008 IEEE

made based on the footprint of a program. The execution pattern and the bottleneck region optimizations improve the response time. The memory access patterns (both spatial and temporal) provide avenues for various memory placement and memory design related optimizations. Studying the composition of the dynamic instruction stream aids in instruction set architecture optimization. Dynamic instruction execution sequences help architects not only understand the behavior of a program but also help in improving functional unit design to reduce the overall area or even increase transistor utility.

In this paper, we architecturally characterize all sensor network applications and compare the results for different class of sensor network applications. First, we collect all representative set of applications from TinyOS benchmarks [10], TinyBench [8], and from [28] and we build some of our own simple applications. We then classify them into various classes of applications to compare the architectural findings for different classes. Specifically, we study the following architectural characterizations and optimizations.

- Find the most frequently executed instructions
- Find the most frequently executed pair and triple of instructions
- Instruction-set and footprint optimization by combining frequently executed pair of instructions
- Memory behavior of the applications

2 Related Works

Wireless sensor network (WSN) has identified its applications in many disciplines including environmental engineering, military/security applications, and civil engineering. The main challenge is to make the sensor node a low-energy device so that it can scavenge energy from various sources. From an architectural perspective, the processor used inside should be designed to consume less energy while coordinating with other components in a manner which minimizes the total energy consumption. While initial microprocessors used in a Mote are of ATMEL (AVR) family, they were synchronous processors and not specifically designed for the sensor node applications. Past research [5, 9] have shown that an asynchronous processor design is the ideal choice for microprocessor to save energy, whereas in synchronous design energy could be wasted in clocking the synchronous processor and other components. ARM cores and variations of ARM cores such as StrongARM and XScale have also been used to see the energy-performance tradeoffs for sensor node applications.

Ekanayake et al. [5] have designed a low-energy asynchronous processor that only takes 24 pJ/instruction, whereas ATMel or ARM family processors takes energy in the order of nJ/instruction. They design a new ISA, new coprocessors which includes timers, radio units, and processor core for low energy design. But they do not provide any motivation or reasoning behind selecting this instruction set that could help the architecture community to understand the ISA design better in tandem with sensor network benchmarks. Similarly, Hemstead et al. [9] have also designed an event processor along with some hardware accelerators to improve the performance and energy consumption. Nazhandali et al. [28] have designed a sub-threshold sensor network processor that

can run at very low voltage and hence at very low frequency. This sensor network processor is a CISC architecture and it consumes 1.6 pJ/instruction.

While these processors are very well optimized for all sensor network applications, some key insights on designing a particular ISA for these architectures would be very helpful. Some past benchmarks such as MediaBench [25] designed for multimedia applications, NetBench [26] , and MiBench [7] for embedded applications, do exhibit an architectural characterization and provide some insight/platform to build an optimized architecture. Ideally, ISA design or other components design should result from the characterization of all sensor network applications on a base processor. There are two benchmarks for sensor network applications TinyBench [8] and SenseBench [27] for this purpose. In TinyBench, all the applications are targeted for TinyOS and does not scale well for a general study of architecture exploration. While SenseBench provides a set of generalized benchmarks, it does not cover all the applications and the architectural characterization is limited to code size, energy per benchmark and real-time performance requirement. Instead we make our benchmark more comprehensive by extensively scanning research literature and also performing a large set of architectural characterization. Architectural characterization will also vary from different class of sensor node applications as we move from security/military applications to environmental/structural monitoring applications where the computational requirements are different. Therefore, we present a complete architectural characterization of the all sensor network applications (which we call *WiSeNBench*) and we then group them into different classes of applications and compare our findings for these different classes of applications.

3 WiSeNBench: Wireless Sensor Network Benchmark

In this section, we describe our benchmark suite *WiSeNBench* in detail. *WiSeNBench* consists of a large spectrum of sensor network applications and core algorithms that are mainly used inside sensor network applications. Identifying and collecting this set of applications required non-trivial efforts due to a plethora of wireless sensor network applications and many more applications which are not yet explored. To also make sure that these applications cover many different classes of applications, we had to rigorously scan the research literature in different domains of research in wireless sensor network. Specifically, we look for various cryptographic applications [24], security protocols [29], digital signal processing (DSP) applications [3], hashing techniques [36], message digest [31], random number generator [33], compression techniques [30], routing [4], applications related to computational geometry [23], some basic algorithms [27], and many pertinent survey papers [1, 22].

Based on this study, we identify the potential applications that will run on the sensor network processor and collect the optimized code for these applications. The main problem in the collection phase was to find the optimized code for one generic language instead of code written in a specialized language (such as nesC [6]) or targeted for a very specific architecture [8]. While we could find the optimized code

978-1-4244-3039-0/08 $25.00 © 2008 IEEE

Table 1: WiSeNBench: Different classes of sensor network applications with its brief description and reference.

Class of applications	Application [reference]	Brief Description
Compression/ Hash/Digest	CRC [14]	Cyclic redundancy check (CRC) generates a checksum to correct errors for a block of data.
	RLE [15]	Run length encoding (RLE) is a simple data compression technique which scans the data and then stores the data with associated count.
	Hash algorithms [19]	A set of hash algorithms to produce a fix-length data for indexing and better search.
	Bloom filter [18]	Bloom filter consists of a set of hash algorithms and a hash table to resolve containment queries.
	MD5 [31]	Message digest algorithm 5 (MD5) is a powerful hash function to create a 128-bit key for integrity checking.
Routing/Radio	SMAC [37]	S-MAC is an energy-efficient medium access control (MAC) protocol.
	Ad-hoc routing [34]	A routing technique in a distributed multihop wireless network with a shared wireless channel.
	EnergyEff routing [32]	Its an unidirectional level-hop routing algorithm to assign each intermediate nodes a level to reach the sink node.
Cryptography/ Security	RC5 [16]	RC5 is a fast block cipher for RSA data security and has variable key size and rounds. We consider both encryption and decryption in this case.
	TEA [20]	Tiny encryption algorithm (TEA) is a block cipher which is very simple to design and code size is also very small.
	Crypto [13]	Crypto3 is a cryptographic technique to encrypt password in an Unix based system.
	RC6 [17]	RC6 is an advanced version of RC5 for data security.
	SPINS [29]	SPINS is a security protocol for sensor network which has two components (1) SNEP (2) TESLA
Computational geometry	Voronoi diagrams [12]	Voronoi diagrams is a special decompostion of metric space using a set of distinct points.
	Delaunay triangulation [11]	Delaunay triangulation for a set of points is the triangulation of points with some specific property.
	Localization [21]	Localization algorithms for sensor node to approximate its position.
Digital signal processing	FFT	Fast fourier transform (FFT) algorithm is a fast and efficient algorithm to calculate DFT and IDFT.
	FIR	Finite impulse response (FIR) is a type of digital filter and its response finally settles down to zero.
	IIR	Infinite impulse response (IIR) filter has non-zero response over a long period of time.
	Speech filtering	Some specialized filters such as Kalman filters for speech processing.
Basic core algorithms	sorting algos	This application contains 7 types of sorting algorithms with different runtime complexity and implementation.
	Fibonacci	Fibonacci numbers are special numbers from a well defined recurrence relation.
	MatrixMul	This is a simple matrix multiplication algorithm for small sizes of matrix.
	Binary search	Binary search algorithm is a typical search algorithm for a sorted list.
	Min-max finder	Finds the minimum and maximum values in a list of values.
	majority consensus	Finds the majority value in a list of values.
	Top10	Finds the top 10 values in a list of values.
	sum-array	Provides the sum of all values present in a list.

for many applications (about 70% of those in the suite), we also develop optimized code from scratch for various applications (about 30%) for which we could not find optimized code written in a generic language (such as C).

After the identification and collection phase, we accomplish a good representative set of a wide variety of sensor network applications. To characterize it better, we categorize these benchmarks into various classes: 1) Compression 2) Routing 3) Security 4) Computational geometry 5) DSP 6) Basic algorithms. Table 1 shows different classes of benchmarks with a brief description and the reference if applicable.

Although all the benchmarks can be run as a separate standalone binary, we preferred to combine all the applications into one single binary and create a unified framework. The motivation behind creating a this framework is to enable centralized control of inputs to these benchmarks, a simpler simulation platform, and easier statistics collection.

4   Experimental Setup

In this section, we explain our experimental setup which includes the compilation of the unified benchmark, simulation, and the statistics collection. We use the ARM SimpleScalar simulator [35], which was extended from the original SimpleScalar simulator [2]. We setup cross-gcc suites for ARM processor to compile the benchmark and make a single static binary as SimpleScalar ARM would not handle the dynamic binary file with shared object files. Since we are only interested in the architecture of a very simple RISC processor with no out-of-order execution, and no caches, we conservatively use sim-safe simulator for our experimentation. We modify the code of sim-safe simulator to extract the related statistics which is explained in detail below along with other implementation issues.

Using the gcc cross-compilation suites we first create the binary and then make sure that we collect statistics based on

978-1-4244-3039-0/08 $25.00 © 2008 IEEE

various functions in the binary . Although Some benchmarks use multiple functions, we combined the results of all related functions. We use ARM disassembler in *binutils* toolset to disassemble the generated binary and feed this disassembled file to a PERL script which parses the disassembled file and generates a C header file containing a large structure with all the function initialization. Each function initialization mainly consists of following three entries: <*function-name,start-pc,end-pc*>. This header file is used with the sim-safe simulator. Since we intend to do architectural characterization based on each function, we identify every function range in the sim-safe execution loop and do relevant processing required to collect the statistics. At the end of simulation, the simulator prints all the related statistics to a file. Specifically, we consider the following statistics:

- Codesize - This is the footprint of the application and a crucial factor in the design of a resource constrained sensor node processor.
- Memory accesses - We characterize the memory access behavior by finding out number of load/stores instructions executed by a particular application.
- Loads - Percentage of loads in the memory accesses is another important factor in terms of energy consumption for sensor network device.
- Frequent instructions - To make some instructions power-conscious during their execution, it is important to understand the distribution of frequently executed instruction. Architects can optimize these instructions for energy savings and performance.
- Frequent pair of instructions - We also find the frequent pair of instructions while executing a specific function. This will give us an idea about the quantitative improvement we can achieve in energy savings and performance. This can also be done statically to improve the code size by combining frequent instruction pairs into a single instruction.

5 Results

Having described the complete benchmark suite in Section 3 and the experimental setup in Section 4, we now present some characterization results based on the parameters discussed in the previous section.

5.1 CodeSize

Codesize or the footprint of a program is a very important parameter as it directly signifies the amount of memory required for a particular application. We statically compute the codesize for ARM ISA and present the results in Figure 1. We can see that most of the DSP applications have much larger code size, whereas most of other applications have code size less than 500 bytes except MD5 and RC5.

5.2 Memory Accesses

A whole host of program optimization techniques are aided by the knowledge of memory access behavior of a program. We compute the total memory accesses to signify the memory intensive behavior of each benchmark. We present the results as percentage of memory accesses (load/stores) as a percentage of total executed instructions. Figure 2 shows the percentage of memory accesses for all of the benchmarks.

We can see that most of applications have 40-60% memory accesses, while some of the applications such as DSP applications, Fibonacci numbers accesses memory through more than 60% of the executed instructions.

5.3 Load accesses

To characterize the memory accesses further, we calculate the percentage of loads in the memory accesses. As we can clearly see from the Figure 3, percentage of loads is larger than 50% for almost all applications which signifies that there are more loads than stores (which is also intuitive). We also find that basic algorithms have higher percentage of loads (specially sorting algorithms), whereas for DSP applications loads and stores are almost evenly distributed.

5.4 Frequent instructions

We find the frequent instructions for each application in *WiSeNBench* to get an idea if any particular instruction is suitable for optimization in terms of energy or performance. Although we gather results for all the benchmarks, due to space constraints we only present the results for two applications. These two applications represent a class of applications: 1) TEA - from cryptographic class 2) FIR - DSP class. We present the frequent instructions as the percentage of total instructions. For TEA, the results are shown in the graph on the left in Figure 4 and we find that almost 7 frequent instructions account for 95% of instructions, with load and add instructions at the top of the list. Similar results are presented for FIR application in graph on the right in Figure 4. We find that for FIR applications instructions are widely distributed with load and branch instructions at the top of the list with about 28% and 10% respectively of total instructions .

5.5 Frequent pairs of instructions

We also find the frequent pairs of instructions to further characterize the application behavior to seek any possible optimization of combining frequent pair of instructions. Although there is a tradeoff in combining the instructions, it can certainly result in lower code size, possibly lesser energy consumption and possibly improved performance. Once again, for this analysis we present the results for only TEA and FIR applications and they are shown in Figure 5. We see a very similar behavior found in frequent instructions analysis. For TEA applications, frequent pairs are attributed to a small number of pairs, whereas for FIR applications it is distributed to many pairs of instructions. Interestingly, the 2nd and 3rd frequent pairs < *mov, load* > and < *load, mov* > are same in both FIR and TEA applications. In TEA, we find that < *add, load* > instruction pair is found to be more frequent (may be due to array accesses), whereas in FIR < *load, br* > is found to be more frequent at about 9%.

6 Conclusion

Sensor networks, though classified under one umbrella, have varied requirements and utilities. To efficiently design protocols, architectures, and applications, it is important to characterize the applications and categorize them based on their effects at the microarchitectural-level. We present a new set of comprehensive benchmarks called the WiSeNBench in a unified framework. We show that WiSeNBench effectively

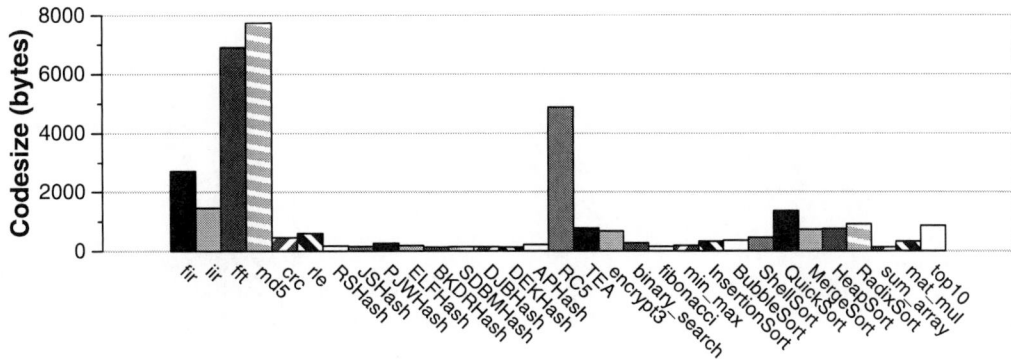

Figure 1: Footprint of all application in our benchmark. The y-axis shows the code size in bytes for each benchmark on x-axis.

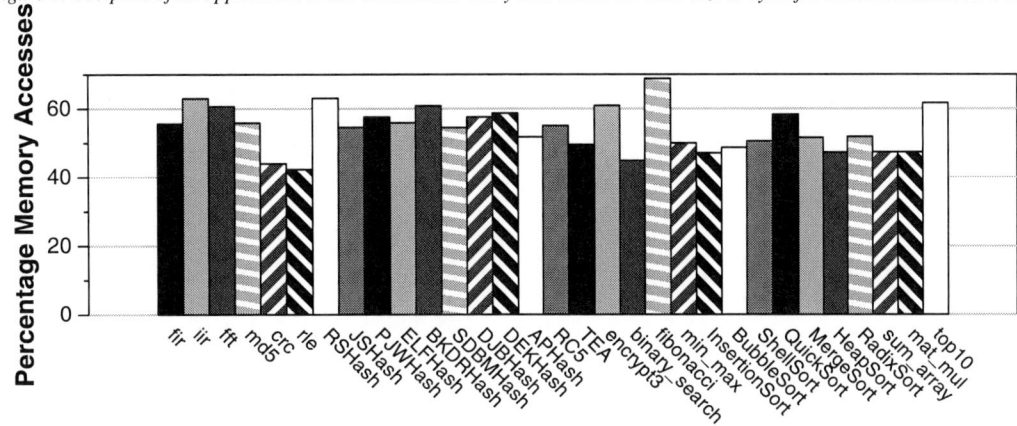

Figure 2: Memory access behavior of all applications in WiSeNBench. The y-axis shows the memory accesses as a percentage of total instructions for each benchmark on x-axis.

characterizes the myriad application set sensor network often deal with and provides insights into the behavior of each of these. Architectural characterization was performed using ARM SimpleScalar Simulator. We find that the code size of MD5, RC5 and DSP applications is larger compared to other categories. On the instruction stream composition front, we find that the set of basic algorithms execute larger percentage of loads and that the DSP applications. Also, we believe that there is a potential for further research on ISA design based on the results presented on frequent instructions and frequent pairs of instructions.

References

[1] I. Akyildiz, W. Su, Y. Sankarasubramaniam, and E. Cayirci. A survey on sensor networks. 2002.

[2] Douglas C. Burger and Todd M. Austin. The simplescalar tool set, version 2.0. Technical Report CS-TR-1997-1342, 1997.

[3] C. Chiasserini and R. Rao. The concept of distributed digital signal processing in wireless sensor networks, 2002.

[4] Douglas S. J. De Couto, Daniel Aguayo, John Bicket, and Robert Morris. A high-throughput path metric for multi-hop wireless routing. In *MobiCom '03: Proceedings of the 9th annual international conference on Mobile computing and networking*, pages 134–146, New York, NY, USA, 2003. ACM Press.

[5] Virantha Ekanayake, IV Clinton Kelly, and Rajit Manohar. An ultra low-power processor for sensor networks. In *ASPLOS-XI: Proceedings of the 11th international conference on Architectural support for programming languages and operating systems*, pages 27–36, New York, NY, USA, 2004. ACM Press.

[6] David Gay, Philip Levis, J. Robert von Behren, Matt Welsh, Eric A. Brewer, and David E. Culler. The nesc language: A holistic approach to networked embedded systems. In *PLDI*, pages 1–11. ACM, 2003.

[7] M. R. Guthaus, J. S. Ringenberg, D. Ernst, T. M. Austin, T. Mudge, and R. B.Brown. Mibench: A free, commercially representative embedded benchmark suite. In *In IEEE Workshop in workload characterization*, 2001.

[8] Mark Hempstead, David Brooks, and Matt Welsh. TinyBench: The Case For A Standardized Benchmark Suite for TinyOS Based Wireless Sensor Network Devices. In *Proceedings of the First IEEE Workshop on Embedded Networked Sensors(EmNets'04)*, November 2004.

[9] Mark Hempstead, Nikhil Tripathi, Patrick Mauro, Gu-Yeon Wei, and David Brooks. An ultra low power system architecture for sensor network applications. In *ISCA '05: Proceedings of the 32nd Annual International Symposium on Computer Architecture*, pages 208–219, Washington, DC, USA, 2005. IEEE Computer Society.

[10] Jason Hill, Robert Szewczyk, Alec Woo, Seth Hollar, David E. Culler, and Kristofer S. J. Pister. System architecture directions for networked sensors. In *Architectural Support for Programming Languages and Operating Systems*, pages 93–104, 2000.

[11] http://mathworld.wolfram.com/DelaunayTriangulation.html. Delaunay triangulation, 1999.

[12] http://mathworld.wolfram.com/VoronoiDiagram.html. Voronoi diagrams, 1999.

[13] http://michael.dipperstein.com/. Crypto3: Unix password cryptography algorithm, 2004.

[14] http://michael.dipperstein.com/. Cyclic redunancy check (crc) implementation, 2004.

[15] http://michael.dipperstein.com/rle/index.html. Run length encoding (rle) discussion and implementation, 2004.

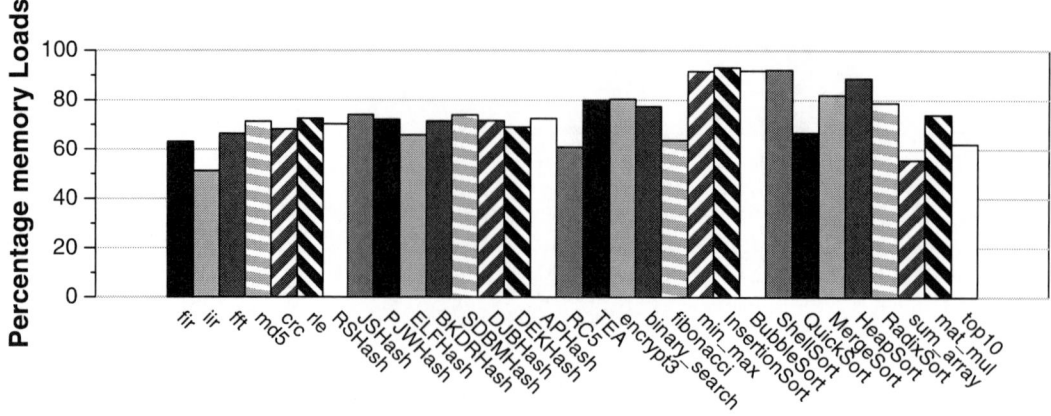

Figure 3: Percentage of loads in the memory accesses is shown on y-axis for each application in WiSeNBench. This gives an idea about the load/store contribution and scope of read optimization if percentage of loads is higher.

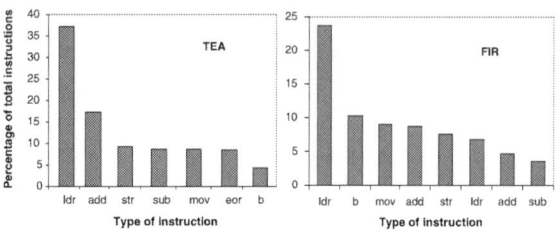
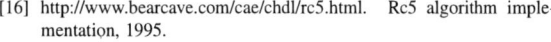

Figure 4: Frequent instructions for the benchmarks is shown on the x-axis. The frequency of these instructions is shown as the percentage of the total instruction executed on the y-axis.

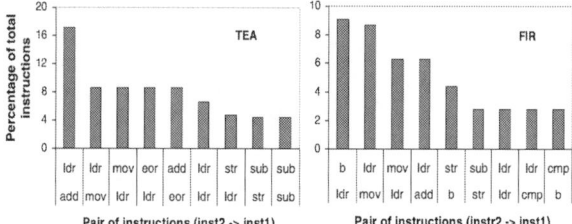

Figure 5: Frequent pairs of instructions is shown on the x-axis. The frequency of these pairs of instructions is shown as the percentage of the total instruction executed on the y-axis.

[16] http://www.bearcave.com/cae/chdl/rc5.html. Rc5 algorithm implementation, 1995.

[17] http://www.codeproject.com/cpp/hexenc.asp. Rc6 encryption and decryption, 2002.

[18] http://www.partow.net/programming/hashfunctions/index.html. Bloom filters implementation, 2006.

[19] http://www.partow.net/programming/hashfunctions/index.html. General purpose hash function algorithms, 2006.

[20] http://www.simonshepherd.supanet.com/tea.htm. Tiny encryption algorithm (TEA), 2006.

[21] Lingxuan Hu and David Evans. Localization for mobile sensor networks. In *MobiCom '04: Proceedings of the 10th annual international conference on Mobile computing and networking*, pages 45–57, New York, NY, USA, 2004. ACM Press.

[22] Holger Karl and Andreas Willig. A short survey of wireless sensor networks, 2003.

[23] Brad Karp and H. T. Kung. Gpsr: greedy perimeter stateless routing for wireless networks. In *MobiCom '00: Proceedings of the 6th annual international conference on Mobile computing and networking*, pages 243–254, New York, NY, USA, 2000. ACM Press.

[24] Y. W. Law, J. M. Doumen, and P. H. Hartel. Benchmarking block ciphers for wireless sensor networks. Technical report, Centre for Telematics and Information Technology, Univ. of Twente, The Netherlands, February 2005. Imported from DIES.

[25] C. Lee, M. Potkonjak, and W. Mangione-Smith. Mediabench: A tool for evaluating and synthesizing multimedia and communicatons systems. In *International Symposium on Microarchitecture (Micro-30)*, pages 330–335, 1997.

[26] Gokhan Memik, William H. Mangione-Smith, and Wendong Hu. Netbench: a benchmarking suite for network processors. In *ICCAD '01: Proceedings of the 2001 IEEE/ACM international conference on Computer-aided design*, pages 39–42, Piscataway, NJ, USA, 2001. IEEE Press.

[27] L. Nazhandali, M. Minuth, and T. Austin. Sensebench: toward an accurate evaluation of sensor network processors. In *In Proceedings of the IEEE International Workload Characterization Symposium*, October 2005.

[28] Leyla Nazhandali, Bo Zhai, Javin Olson, Anna Reeves, Michael Minuth, Ryan Helfand, Sanjay Pant, Todd M. Austin, and David Blaauw. Energy optimization of subthreshold-voltage sensor network processors. In *ISCA*, pages 197–207. IEEE Computer Society, 2005.

[29] Adrian Perrig, Robert Szewczyk, Victor Wen, David E. Culler, and J. D. Tygar. SPINS: security protocols for sensor netowrks. In *Mobile Computing and Networking*, pages 189–199, 2001.

[30] D. Petrovic, R.C. Shah, K. Ramchandran, and J. Rabaey. Data funneling: Routing with aggregation and compression for wireless sensor networks. In *Sensor Network Protocols and Applications (SNPA'03)*, May 2003.

[31] R. Rivest. The MD5 Message-Digest algorithm, April 1992. RFC 1321.

[32] C. Schurgers and M. Srivastava. Energy efficient routing in wireless sensor networks. 2001.

[33] Deva Seetharam and Sokwoo Rhee. An efficient pseudo random number generator for low-power sensor networks. In *LCN*, pages 560–562. IEEE Computer Society, 2004.

[34] R. Shah and J. Rabaey. Energy aware routing for low energy ad hoc sensor networks. In *In Proc. IEEE Wireless Communications and Networking Conference (WCNC)*, Orlando, FL, 2002.

[35] The SimpleScalar-Arm Power Modeling Project. www.eecs.umich.edu/ tnm/power/, 2000.

[36] Volker Turau and Christoph Weyer. Location-aware in-network monitoring in wireless sensor networks. In *Proceedings of the Sensor Networks Workshop at Informatik 2004*, Ulm, Germany, September 2004.

[37] W. Ye, J. Heidemann, and D. Estrin. An energy-efficient mac protocol for wireless sensor networks. In *INFOCOMM*, 2002.

978-1-4244-3039-0/08 $25.00 © 2008 IEEE

21st International Conference on VLSI Design

Concurrent Multi-Dimensional Adaptation for Low-Power Operation in Wireless Devices

Rajarajan Senguttuvan, Shreyas Sen, Abhijit Chatterjee

Georgia Institute of Technology, Atlanta, GA, USA, 30332

{rajs,shreyas.sen,chat}@ece.gatech.edu

Abstract[1]

In this paper, we develop a multi-dimensional adaptive power management approach for wireless systems that optimally trades-off power vs. performance across temporally changing operating conditions by concurrently tuning control parameters in the RF and digital baseband components of the wireless device. A key contribution of this paper is the development of a multi-dimensional optimal control law that determines how the various control parameters should be concurrently tuned to guarantee minimum power consumption across changing channel conditions without compromising overall system bit error rate. Simulation results indicate significant power savings in the receiver RF front end using the proposed approach in addition to power savings in the baseband processor itself.

1. Introduction

The issue of handset power consumption and heat dissipation is critical in future designs of multi-mode multi-standard wireless devices. Several techniques for low power operation of wireless circuits have been proposed in the literature [1][2]. At the physical layer, low-power design methodologies for digital systems have been studied extensively [3][4] but relatively fewer techniques have been proposed for radio frequency (RF) systems. There is a need for co-optimizing RF front-end and baseband power consumption metrics as systems become more complex. Some of the design techniques proposed for power minimization in RF circuits include bias current reuse [5], controlled positive feedback [6], high impedance interfaces and sub-threshold biasing [7]. The issue with these approaches is that design margins are incorporated into the circuits to account for worst-case estimates of process variability, and thermal effects in the RF front-end, and channel conditions corresponding to different modes of transmission.

However, for a significant portion of the time a wireless device is powered up for operation, it is not in a worst case environment. Hence, the circuit operation is not optimal from a power consumption standpoint under the majority of operating conditions. There also exist MAC and network-level dynamic power management schemes [1] that strive to conserve power by adapting the data rate (modulation and coding rates) based on certain channel quality metrics derived from the analysis of training symbols. Present-day wireless devices also feature high-power, low-power and shut-down modes, that are correspondingly exercised depending on the prevailing operating conditions. Though these approaches are effective in reducing the power consumption levels, they do so in a few discrete steps, and hence do not fully exploit the built-in design margins for lowest power operation while satisfying performance requirements.

2. Motivation

To satisfy the demands for future generation of wireless systems, aggressive power management strategies should be employed at the *system-level* that adapts not only the RF and digital circuits, but also takes into account end-to-end cross layer interactions. In a recently published work [8], an energy scalable RF transmitter was proposed, where the front-end is dynamically tuned (supply, bias, resistances) for each data rate modulation set by the higher-level link layer protocol. The effectiveness of the approach described in [9] depends on the accuracy of channel estimation procedure, and the pre-set transmitter calibration coefficients obtained during the design phase. We present an alternate approach for a dynamically tunable RF front-end that tunes the circuitry based on feedback from an *adaptation metric* that is computed during run time. The key innovations of the presented work are outlined as follows

- Development of a methodology for *concurrently adapting multiple RF parameters along with digital baseband signal processing algorithms* of a wireless device under dynamically changing channel conditions.

---

[1] This work was funded in part by NSF ITR award CCR-0325555 and GSRC/FCRP 2003-DT-660.

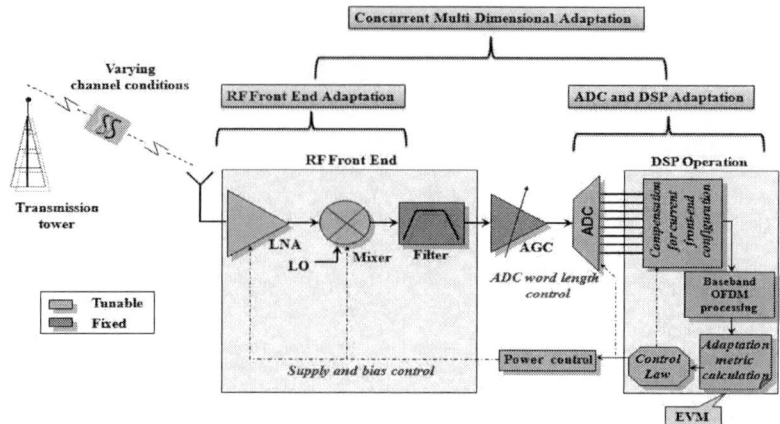

Figure 1 Illustration of the proposed approach on a wireless receiver

- Development of a *multi-dimensional control law* for adaptation of the above parameters such that *minimum power is consumed* without compromising system-level bit-error rate (BER).

We presented an adaptive RF front-end for low power operation in [9]. In this paper a multidimensional control law that adaptively fine-tunes the bias/supply voltages of the RF front-end, and ADC resolution to reduce the overall power consumption is presented. The DSP control also intelligently adapts the analog-to-digital converter (ADC) resolution (wordsize) whenever possible by monitoring channel conditions (feedback from adaptation metric). Thus, design margins in the wireless device are fully utilized for saving *more power* under favorable channel conditions in a standard-compliant manner. While power/ performance trade-off studies of RF transmitters have been investigated, no comparable work exists for RF receivers. In this paper we use a WLAN orthogonal frequency division multiplexing (OFDM) receiver chain (RF-ADC-DSP) as the test vehicle to study and evaluate the effectiveness of the proposed system-level adaptation approach.

3. Proposed Approach for Minimum Power Consumption

3.1 An Overview

An illustration of the proposed approach for minimum power operation of the receiver is shown in Figure 1. The tunable supply/bias voltages for LNA and mixer, along with the tunable ADC word length serve as control 'knobs' for system. Thus, the receiver can be operated in several different configurations based on the settings of these control knobs. During the run time operation of the receiver, the performance is periodically evaluated through the computation of a suitable adaptation metric (described later). To do this,

the received data at the baseband is first compensated to calibrate-out the effects induced by the RF front-end in its configuration. The control block then uses this information to set the new configuration of the front-end for optimal power consumption while ensuring that the system-level performance metrics are not violated. This circuit-level energy flexibility comes at limited increase in the area cost and complexity. The development of receiver system used for this work is described next.

3.2 Wireless Receiver System

This section first presents the circuit-level implementation of the LNA and mixer. Circuit-level ADS simulations are performed for various supply and bias settings to evaluate their performance. The data from these simulations is then used to develop behavioral models which are used for system-level study of the proposed approach.

A. LNA and Mixer circuit design

A wideband LNA (Figure 2) is designed in CMOS 0.18μ technology. A common gate (CG) input stage provides good input matching over a wide range, the common source (CS) intermediate stage provides the required gain, and the class A output stage provides good signal swing. The operating range of the designed LNA is from 2-7 GHz, and the gain is observed to be 16 dB at 2.4GHz for a nominal supply voltage of 1.8V and bias of 0.8V. The LNA has NF of about 3dB, and the total DC power consumption of the LNA is about 25.6mW. For the mixer, a double-balanced Gilbert cell architecture was chosen. A balun is used for single-to-double ended conversion. The mixer provides a gain of 11.8 dB, NF of 8 dB. The operating range of the mixer is from 2-7GHz, and the power consumption of 6.6mW at a nominal supply voltage of 1.8V and bias of 0.8V.

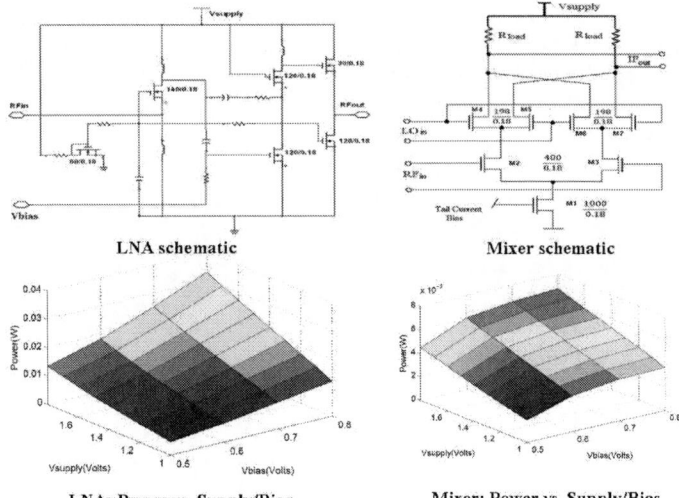

Figure 2 LNA and Mixer: Schematics and power dissipation profiles (power vs. supply/bias)

B. Performance tradeoff with supply and bias voltages:

The power consumption of the LNA and mixer decreases with the lowering of supply and bias voltages, but trades-offs with gain, non-linearity and NF specifications. This, in turn affects the system-level performance. The modeling of the LNA and mixer is realized through a non-linear transfer function of the

$$y(t) = \alpha_0 + \alpha_1 x(t) + \alpha_2 x^2(t) + \alpha_3 x^3(t) + \alpha_4 x^4(t) + \alpha_5 x^5(t) \qquad (1)$$

where α_0 is the DC offset, α_1 is the small signal gain, and α_2, α_3, α_4, α_5 are non-linearity coefficients. The coefficients define the linear (gain) and non-linear (harmonics and inter-modulation terms) effects of the amplifier. Transistor-level simulation data for the LNA and mixer obtained from the Agilent ADS tool is used to extract the coefficients in Equation (1). The NF is factored-in through addition of white noise to the output signal. For the mixer, in addition to the non-linear transfer function implementation, the frequency mixing operation is realized by a simple multiplication operation as given in Equation (2).

$$Y(t) = C \times x_1(t) \times x_2(t) \qquad (2)$$

where, C represents conversion gain of the mixer.

The choice of supply and bias voltages of the LNA and mixer constitute a total of 400 different settings. Behavioral parameters of the RF front-end are extracted for each of these 400 configurations for use in the simulation study. It should be noted here that many of these 400 configurations are not optimal from a power consumption standpoint. Therefore,

- The set of configurations are pruned down to a limited set of optimal values using a multi-dimensional optimization approach in the design phase (described later).

- The reduced set of supply and bias voltages are used to dynamically configure the RF front-end during run time for minimizing power across changing environmental conditions using an optimal control law.

C. Channel modeling

The effect of the channel includes the sum contribution of the air-antenna interface effect at the transmitter and receiver, respectively, and the effect of the physical medium (air) between the transmit and receive antennas.

(1) Propagation loss and noise addition are modeled as a simple attenuation of the received signal followed by addition of white noise to the received signal.

(2) The multipath and fading effect are modeled using an FIR filter. The length of the FIR filter determines the maximum delay spread in the channel.

(3) Interference is modeled as a combination of adjacent channel interference and microwave interference. Adjacent channel interferer is modeled as an OFDM signal in a neighboring band. Emissions from this band affect the signal in the band of interest (2.4GHz). The adjacent channel interferer is given by

$$i\_adj(t) = f(A(t), f_c(t))_{OFDM} \qquad (3)$$

where A(t) is the time varying amplitude and $f_c(t)$ is the time varying frequency. The microwave interferer is modeled as an AM-FM source that allows the characterization of frequency wander based on the work presented in [10].

3.3. Choosing a Suitable Adaptation Metric

The key issues to be considered in developing a dynamic feedback-driven for wireless systems include defining a *suitable adaptation metric* and the *bounds* on this metric for satisfactory operation. For feedback, an adaptation metric must be chosen such that it provides the best indication of the system performance under all possible environmental conditions. In this paper, error vector magnitude (EVM) specification is used as the adaptation metric. EVM quantifies the difference between the transmitted and received modulated data, thus capturing the cumulative effect (non-idealities, noise etc.) of all the blocks in the transmit-receive chain. The choice of EVM as an adaptation metric in a tunable word length OFDM receiver was explored in [11].

The quality of service (QoS) metrics for wireless systems includes specifications such as bit error rate (BER) or packet error rate (PER) [12]. We develop a relationship between EVM and BER specifications of the wireless device under consideration [9]. Figure 3 plots the EVM vs. BER for a wireless system under different channel conditions and system non-idealities for two modulation schemes (QPSK and 16-QAM). An upper bound on the BER specification can then be translated into an upper bound on the EVM. For example, if BER bound is set at 1e-3, the corresponding mean EVM bound for the QPSK and 16-QAM cases can be approximated to about 35% and 12.5%, respectively.

3.4. Design Phase Optimization and Development of Control Law

Figure 4(a) illustrates the overall approach. During the design phase, a set of tunable control 'knobs' that trade-off the device performance with power consumption are first identified. Here, the supply and bias voltages of the LNA and the mixer, and the ADC wordsize are fixed as the control 'knobs'. The environmental conditions (channel) encountered by the device can vary from good to bad constituting an infinite set of possibilities. Since it is not possible to characterize a device for all these possibilities, it is imperative to carefully model *a finite set of channel conditions* that adequately span the range ('good' to 'bad'). The channel conditions are obtained by perturbing the different channel parameters (noise, multipath components, and interference sources). For each of these channel conditions, optimal settings for control 'knobs' are computed through a multi-dimensional co-optimization for the power and EVM.

In [9], we presented a heuristic optimization procedure for determining the optimum RF front-end supply and bias voltages. In this work, we extend the optimization by tune the ADC wordsize along with the RF front-end voltages. The output of the optimization procedure is a *'maximum EVM-minimum power locus'* that gives the relationship between the tunable parameters of the system for minimum power operation while satisfying

the maximum EVM criterion. Furthermore, digital compensation is performed in the baseband as described below.

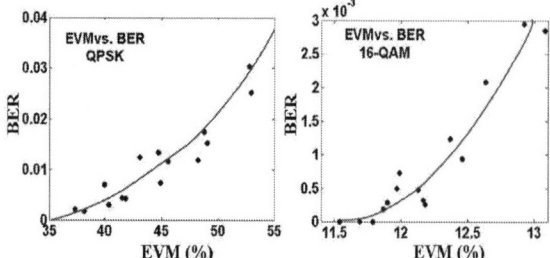

Figure 3 (a) EVM vs. BER relations (b) EVM threshold and guard band estimation for QPSK

The performance of the LNA and mixer (gain, non-linearity and distortion parameters) trade-off with power consumption as the supply and bias voltages are scaled. If the relation between these circuit metrics and voltages are characterized during design optimization phase, this information can be used to perform compensation during run time for dynamically changing front-end effects. In this work, inverse transfer function characteristics for LNA and mixer are extracted from ADS circuit simulations for the optimized (pruned) set of supply and bias voltages along the locus. They are stored in a look-up table in the DSP for digital compensation during run time operation. The procedure is illustrated below in Figure 4(b).

3.5. Run Time Operation of the Device

During run time, the control law continuously strives to reduce power consumption of the device for every data rate under all environmental conditions while simultaneously ensuring that the performance metrics (EVM, BER) are met. This is done by moving up or down the optimal locus of voltage values obtained from the optimization procedure. The threshold and guard band for EVM (feedback) is set based on the estimated channel quality and current data rate. For example, a poor channel would have a larger guard band and vice versa. Wireless communication protocols have built-in strategies for data rate control at the transmitter end through feedback from the receiver. In our approach, the voltage scaling is performed independently for each data rate in a standard-compliant manner. By allowing the system to dynamically adapt this way, the device power consumption hovers around the lowest possible value for which received signal quality meets the required specification. Thus the design margins are fully exploited by operating the system close to error boundary, in the process significantly saving power.

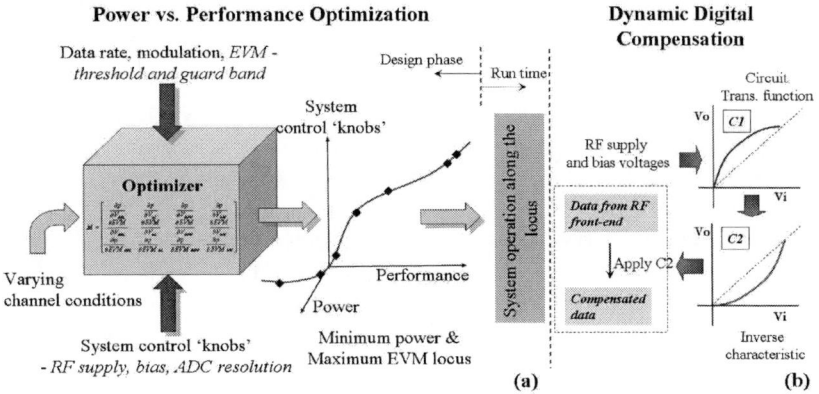

Figure 4 Figure 4 (a) Design phase optimization and run time operation (b) Digital compensation

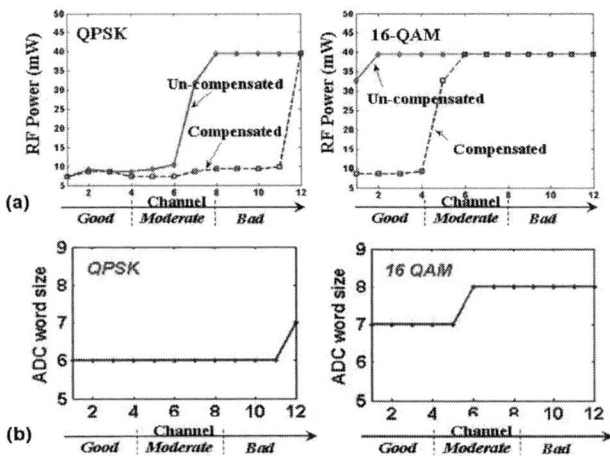

Figure 5 (a) RF power consumption under different channel conditions (compensated and uncompensated) for QPSK and 16-QAM modulations (b) Optimum ADC wordsize

4. System Simulations and Inferences

The power-EVM optimizer was simulated to obtain a set of optimal control 'knobs' under a set of channels *(good-moderate-bad)*. The optimizer was run for two different modulations: QPSK and 16-QAM. From the computed voltage set, it is observed that for a majority of the channel conditions, the optimal control voltages were lower than nominal voltage values yielding significant savings in device power consumption. It was also observed that the optimal voltage values and the associated power consumption is lower for QPSK-modulated signals compared to 16-QAM. This is due to the tighter requirements on the signal quality and SNR for a 16-QAM signal (higher data rates).

Next, run time operation (QPSK and 16-QAM) is simulated by running the optimal control law, first without digital compensation, and then with digital compensation. It is observed that for all four cases, the control law operates the device at lower than nominal power (40mW) whenever the channel conditions are good. Furthermore, it is observed that when digital compensation is performed, more power is saved under moderate and bad channel conditions compared to the uncompensated case. The plots shown in Figure 5 highlight the key advantage of the proposed approach – *save more power while the channel conditions are good.*

In wireless standards such as WLAN, the higher-level radio link control (RLC) protocol dynamically changes the data rate (modulation and coding) based on the channel conditions. The control law operates within the framework of RLC protocol by operating the device near the threshold for each data rate. Figure 6 shows the computed upper and lower bounds of EVM specification for QPSK modulation are 35% and 12.5%, respectively, for a BER compliance of 10^{-3}.

69

978-1-4244-3039-0/08 $25.00 © 2008 IEEE

Within QPSK modulation, there exits an EVM margin of 23%-Δ_1- Δ_2, taking into account the guard band limits. This existing margin is exploited to save power in our approach. It should be noted here that the EVM margins are different for each modulation. Hence the amount of realizable power saving is dependent on the modulation format the device is operating under.

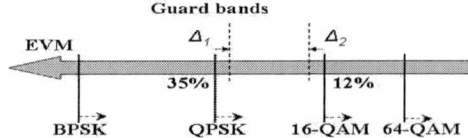

Figure 6 Exploiting the EVM margins for each data rate

As observed from Figure 5(b), preliminary simulations indicate that sufficient margin exists for ADC word size (maximum of 8 bits) pruning under favorable channel conditions. Up to 2 bits of resolution can be sacrificed for QPSK modulation under majority of the channel conditions. Though the margin is lower in the case of 16-QAM modulation, the system budget allows for a bit drop under good channel conditions.

5. Key Issues

The proposed multi-dimensional adaptation can be extended to include multiple control knobs (ADC word size, and sampling rate etc.) leading to a comprehensive system-level optimization problem. The software/hardware and power overhead for implementing the proposed feedback control must be considered carefully. The proposed approach will be effective in increasing the battery life of the future multi-mode, multi-standard radios with multiple RF front-ends and power hungry DAC/ADCs, while consuming relatively little power and software/hardware overheads. The design of the power control circuitry is crucial in the proposed approach, as it determines the power savings that can be obtained under varying channel conditions. Under fast varying channel conditions, low-power operability would be limited by the response and settling time of the feedback control circuitry.

6. Conclusions

A multi-dimensional approach for dynamic adaptation of a wireless device for reduced power consumption without violating system-level performance metrics was presented. The proposed approach aggressively exploits the built-in margins in wireless systems to enable minimum power operation. We also propose the use of DSP-based digital compensation for additional power savings by calibrating-out the front-end effects. The framework

was implemented on an OFDM RF receiver system and analyzed under different channel conditions.

10. References

[1] H.Woesner, J.P. Ebert, M. Schlager, and A. Wolisz, "Power saving mechanisms in emerging standards for wireless LANs: The MAC layer perspective", IEEE Personal Communication Systems, 5(3): 40-48, 1998.

[2] Tasic, A., Serdjin, W.A., Long, J.R., "Adaptive multi-standard ciruits and systems for wireless communications", IEEE Circuits and Systems Magazine, Vol 6., Issue 1., pp. 29-37.

[3] D. Ernst, S. Das, S. Lee, D. Blaauw, T. Austin, T. Mudge, N. S. Kim, K. Flautner, "RAZOR: Circuit-Level Correction Of Timing Errors For Low-Power Operation," IEEE Micro, Vol. 24, Issue 6, Nov-Dec 2004 pp:10 – 20.

[4] Burd, T.D., Pering, T.A., Stratakos, A.J., Brodersen, R.W., "A dynamic voltage scaled microprocessor system," IEEE Journal of Solid-State Circuits, Volume 35, Issue 11, Nov. 2000 Page(s):1571 – 1580

[5] A. N. Karanicolas, "A 2.7-V 900-MHz CMOS LNA and Mixer," IEEE Journal of Solid-State Circuits, Vol. 31, No. 12, December 1996, pp. 1939-1944.

[6] T. Kawamura, M. Suzuki and H. Ichino, "An Extremely Low-power Bipolar Current-mode 1/0 Circuit for Multi-Gbit/s Interfaces," Proceedings of 1994 Symposium on VLSl, June 1994, pp: 31-32.

[7] B. G. Perumana, S. Chakraborty, C. H. Lee,and J. Laskar, "A Fully Monolithic 260-_W, 1-GHz Subthreshold Low Noise Amplifier," IEEE Microwave And Wireless Components Letters, Vol. 15, No. 6, June 2005, pp. 428-430.

[8] Debaillie, B. Bougard, B., Lenoir, G., Vandersteen G., Catthoor, F., "Energy-scalable OFDM transmitter design and control", 43rd IEEE Design Automation Conference, july 24-28, pp. 536-541.

[9] R. Senguttuvan, S. Sen, A. Chatterjee, "VIZOR: Virtually zero-margin adaptive RF for ultra low-power wireless communication", IEEE International Conference on Computer Design, Lake Tahoe, USA, 2007.

[10] Zhao, Y., Agee, B.G., Reed, J.H., "Simulation and measurement of microwave oven leakage for 802.11 WLAN interference management", Microwave, Antenna, Propagation and EMC Technologies for Wireless Communications, vol 2, Aug 8-12, 2005, pp.1580-1583.

[11] Yoshizawa, S., Miyanaga, Y., "Tunable wordlength architecture for a low power wireless OFDM demodulator", IEICE Transaction Fundamentals, Vol E89-A, No. 10, October 2006.

21st International Conference on VLSI Design

Adaptive Signal Scaling Driven Critical Path Modulation for Low Power Baseband OFDM Processors

Muhammad M. Nisar, Rajarajan Senguttuvan, Abhijit Chatterjee

School of Electrical and computer Engineering

Georgia Institute of Technology, Atlanta, GA, USA

{mnisar, rajs, chat@ece.gatech.edu}

Abstract

Modern wireless communication systems are designed for worst-case channel noise and interference conditions. This results in circuits that over-perform and consume more power than necessary most of the time when channel conditions are not worst-case. In this paper, we propose a power savings methodology that allows "graceful degradation" of baseband system performance when channel conditions are good without compromising overall bit-error rate. This adaptation is achieved by "dynamically modulating the active circuit critical paths via signal scaling" and by simultaneously modulating the supply voltage of the baseband signal processing circuitry. The proposed approach uses continuous monitoring of the error vector magnitude (EVM) of the demodulated signal to drive the adaptation procedure. The proposed architecture can reduce power consumption in the baseband processor of an OFDM receiver by as much as 30%. As opposed to other schemes, the proposed technique requires no modification of existing signal processing hardware.

1. Introduction

Power consumption is a major design driver for remote handheld DSP applications. In baseband signal processing of wireless communication systems, the noise performance of specific signal processing algorithms for signal demodulation and symbol decoding can be traded off for power. When the received signal is of high quality (i.e. there is little signal interference, channel noise or fading), the noise performance (noise figure degradation) of the underlying signal processing algorithms can be reduced without compromising the overall bit-error rate of the decoded symbols. Conversely, when the received signal is of poor quality, the baseband signal processing algorithms must operate with high fidelity to maintain the same prescribed bit-error rate. In operating conditions where the worst case channel is seen infrequently, significant power can be saved by reducing the performance of the baseband signal

processing algorithms (trade off performance for power) when channel conditions are not worst-case and vice-versa.

In prior research, simulation based techniques [1]-[4] have been proposed to find the optimal wordlength for digital signal processing algorithms in wireless communications and filtering applications. Such algorithms optimize the wordlength according to predetermined system-level performance metrics. Often, the resulting digital circuits are implemented with large wordlength values. A method for tuning the wordlength of digital baseband OFDM signal processing algorithms [5] has been developed that allows dynamic adjustment of the wordlengths of digital filtering and FFT operations driven by continuous monitoring of the error vector magnitude EVM of the demodulated signal (similar to the monitoring technique proposed in this paper). However, the method relies on the use of *gated clocks* for wordlength adaptation. This in turn, requires significant modifications to be made to the computer arithmetic hardware to enable use of the proposed technique for reducing power consumption.

In this paper, we present an adaptive power savings technique that scales the input data and correspondingly lowers the supply voltage while keeping the signal quality (injected noise power due to reduced wordlength) above a minimum value needed to maintain a prescribed maximum bit error rate. The approach exploits the fact that the critical circuit path lengths of the underlying arithmetic units (multipliers and adders) are reduced by input data scaling. This reduction in circuit critical path length allows the correct operation of the arithmetic units at lower supply voltage without incurring additional bit errors. Since scaling of the input data results in dropping of the LSB bits of the data only, the performance degradation of the OFDM modulation and demodulation algorithms is graceful. This allows gradual system-level performance degradation and allows robust and fine-grained performance vs. power control driven by the monitored EVM metric (the proposed EVM metric is discussed in Section 4). To adapt to changes in the input signal quality, a PID feedback controller is implemented that

---

This work was supported in part by NSF Grant CCR-0635016 and by GSRC MARCO under award 2003-DT-660.

978-1-4244-3039-0/08 $25.00 © 2008 IEEE

dynamically scales the input data and controls the DSP supply voltage. A key feature of the proposed approach is that it can be applied to existing baseband DSP processors *without any architectural changes.*

The rest of the paper is organized as follows. The next section gives an overview of the proposed power savings methodology for wireless systems. Next, the proposed input scaling concept is presented. Then scaling and supply voltage feedback control is discussed along with EVM metric. This is followed by experimental results and a discussion of future research in this area.

2. Overview

In wireless communication systems, existing power management schemes [6]-[8] rely on channel quality metrics to modulate the data communication rate (radio-link control). These metrics are derived from the analysis of received "training symbols" embedded in each packet of transmitted data. Training symbols are short bit sequences that are transmitted prior to the data in each packet to enable the receiver to characterize the channel, and to calibrate the receiver for current channel conditions (noise, attenuation, etc.). A range of channel conditions can be accommodated for each specified data rate of communication that the wireless system can support. If the channel conditions become inferior to the worst channel quality for a specific data rate, then the wireless system switches to the next *lower* possible data communication rate. On the other hand, if the channel condition improves significantly, then the data rate is increased to the next *higher* data rate possible.

For any given data rate, the wordlength of the baseband signal processing system is always (statically) set in such a way that the system bit error rate (BER) is less than the maximum allowed BER value for the wireless communication protocol being used for communication. This wordlength therefore corresponds to the worst channel quality that the data rate can accommodate. Given the fact that most of the time in which the mobile device operates at that data rate, it will not be working in a worst case environment; the effective wordlength can be reduced when the channel is not worst-case, saving power while keeping the system BER within quality of service (QoS) requirements. A simple example of the above is when a wireless device is used close to a transmission tower vs. far from any transponder. In the following section, we discuss a communication system capable of maximum power savings by using a feedback system that adapts to wireless channel quality.

2.1 Communication System with Signal Quality Feedback

In order to adjust the signal processing circuit operating conditions (voltage, data word size etc), a signal quality metric is needed, which quantifies the cumulative sum of the quality of transmission, the channel quality degradation and the quality of signal reception. Such a metric, called the *adaptation metric* is computed by the baseband signal processor in real-time (online). When this adaptation metric has a "high" value, the quality of signal processing in the baseband DSP can be traded off (degraded) for power consumption within specific limits and vice versa. In our presented technique, the signal is degraded by scaling down the input (dropping LSB bits) and correspondingly adjusting the supply voltage. This dynamic adjustment is done to save power while keeping the adaptation metric of the demodulated signal below a specified upper limit.

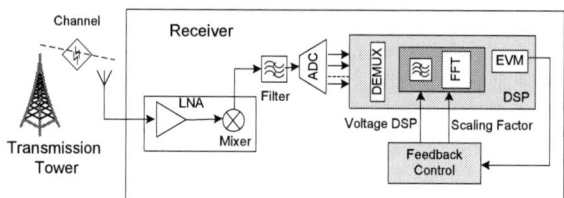

Figure 1: Block Diagram of the Overall Proposed Architecture

In our approach, the EVM of the demodulated signal is used as the adaptation metric. Such a metric is chosen because the EVM of the demodulated signal is seen to have strong statistical correlation with BER for fading as well as AWGN channels [5]. Our own studies indicate that the EVM vs. BER correlation holds true when channel conditions as well baseband signal processing wordlength are changed concurrently (see Figure 5). The feedback system of Figure 1 generates the control signals for adaptive scaling of signals in the baseband DSP. Such a system always strives to operate at the lowest power consumption levels (lower but acceptable performance) for any specified data rate through adaptive control of DSP wordlength. The use of such a feedback control technique to adapt to dynamically changing air channel conditions saves significant amount of power whenever the channel quality is not worst-case.

3. Input Scaling Approach and Voltage Adjustment

Dynamic voltage scaling benefits from the margins in the circuit design, workload and latency constraints and allows power to be traded for the quality in the circuit. Due to the quadratic relationship between power and supply voltage, significant power savings can be achieved if the circuit operates at lower supply voltages. If we define *"critical voltage"* as the voltage required for the correct operation of the circuit then in LSB first architectures, reducing supply voltage lower than the critical supply voltage degrades the circuit performance catastrophically because of errors in the MSB bits. This drastic drop in circuit performance impedes efforts to try to operate the circuit below the critical supply voltage levels. If gradual performance degradation can be achieved with voltage scaling below the critical supply voltage, then the DSP circuit can be made to operate at an acceptable performance level but with much lower power consumption. In the following, we explain how signal scaling reduces the chances of critical path excitation in DSP arithmetic circuits.

Any downward signal scaling by a factor that is a power of 2 causes the data to be shifted right by an appropriate number of bits. As long as sign extension is performed correctly in 2's complement arithmetic, the effect is to reduce the circuit critical path by the number of least significant bits that is "shifted out" of the arithmetic computation. This has two implications: (a) the supply voltage can now be reduced by an amount that is proportional to the reduction in the circuit critical path length (the critical path corresponding to a given scaling factor is called the *active critical path* for that scaling factor) and (b) the performance of the DSP circuit is degraded gracefully since only the least significant bits of DSP computation are eliminated to trade off performance for lower power consumption.

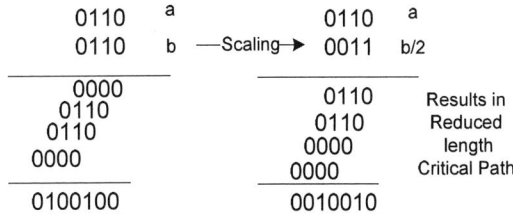

Figure 2: Critical Path reduction with Scaling

For a simple illustration of the proposed concept, consider multiplication of two 4-bit binary numbers as shown in the Figure 2. The critical path of such a multiplication is 2^{n-1}, where n is the number of multiplier/multiplicand bits. Now if multiplicand b is scaled by a factor 2 i.e. a right shift of one bit, the active critical path of such scaled multiplication will reduce to $2^{n-1}-1$. Clearly, there is reduction in critical path length of the multiplier with input scaling. This concept holds in booth multipliers as well (most commonly implemented multiplier architecture in DSP) as the number of partial products that need to be added for the correct output will reduce by signal scaling.

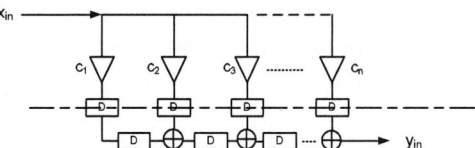

Figure 3: Transposed Form pipelined FIR filter

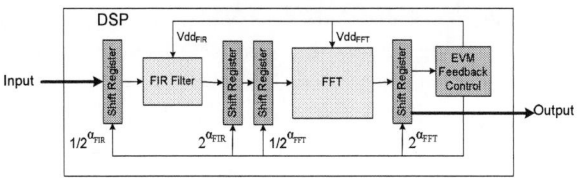

Figure 4: Input Scaling, Voltage adjustment via EVM feedback controller

Figure 3 shows an n-tap FIR filter implemented in transpose-form. The critical path of the pipelined FIR filter is T_M, where T_M is the multiplier time delay. The filter coefficients C_n are fixed in such a filter. The input to the filter is scaled down by a factor 2^α (shifting right and dropping α LSB bits), therefore reducing the active critical path lengths in the filter. This allows lowering the operating voltage for the circuit. Since, input scaling by 2^α reduces the active critical path by α bits, the supply voltage can only be lowered till we do not exceed the timing requirements of the new active critical path, as this will ensure that no MSB bits get corrupted because of the reduced supply voltage. The data out of the filter is scaled up by the same scaling factor 2^α. The system performance degradation due to the dropping of LSB bits is determined using the error vector magnitude (EVM) of the demodulated signal. If signal quality is good (EVM is low), the noise performance of the filter is intentionally degraded by increasing input signal scaling. If the received signal quality is poor (large EVM), the EVM block starts up-scaling the input and adjusts system supply voltage accordingly. In Section 4.1, the EVM metric is discussed in greater detail and its choice as a metric for circuit adaptation is justified.

The complete scaling approach for the DSP block along with feedback is shown in Figure 4. Separate input scaling and voltage control for FIR filter and FFT ensures finer control of EVM degradation and allows circuit power savings to be maximized. Abrupt degradation in signal quality might result in corrupted DSP data because of the response time of the feedback system. However, in wireless communication systems at every protocol layer, robust techniques ensure that only correct data is retained for further processing. In the worst case scenario, this will result in a packet error, thereby, mandating a retransmission request of the transmitted data.

978-1-4244-3039-0/08 $25.00 © 2008 IEEE

4. Scaling and Supply Voltage Control

The implemented feedback control continuously monitors the quality of the demodulated signal by measuring EVM. In order to operate the receiver within the prescribed limits of bit error rate, the receiver operates within a given range of EVM values determined by extensive simulation of the wireless transmission protocol. If the EVM value is below the prescribed threshold, the feedback control degrades system performance by input scaling. For every scaling level, the filter is set to a lower voltage level determined by lookup table that is obtained via prior simulation and calibration experiments. Since, we know the effect of input scaling on the length of active critical paths and the effect of voltage scaling on the delay characteristics of the critical paths, a lookup table is constructed that determines different supply voltage levels as per the timing requirements of the requisite active critical paths. The advantage of such a lookup table is that it ensures that no MSB bits get corrupted while signal scaling and supply voltage levels are modulated by the control mechanism. In the next section, the adaptation metric EVM is explained along with its relationship to BER.

4.1. Adaptation Metric – EVM

The performance of a communication system is normally defined in terms of the BER and as the name suggests, it gives the rate at which errors occur during communication. Typical BER values for wireless systems are in the order of 10^{-3}-10^{-4}. Measuring the BER of a system typically incurs a long time as several thousand bits are transmitted and received. Clearly, using BER as the adaptation metric is not feasible due to the dynamic nature of the problem at hand. We use EVM for characterizing system performance as it can be computed fairly quickly for each data frame of the received signal and for its strong correlation with BER [5]. The EVM is given by the sum of the vector differences between the received and ideal signals and is computed as

$$EVM = \sqrt{\frac{1}{N}\frac{\sum_{1}^{N}\left\|y_i - x_i\right\|^2}{\left\|y_{max}\right\|^2}} \qquad (1)$$

where, y_i and x_i are the received and ideal complex modulated data $(I+jQ)$, y_{max} is the outermost data point in the constellation, and N is the number of data points used for computation. In case of quadrature phase shift keying (QPSK), x_i gives one of "1+j", "1-j", "-1-j" and "-1+j" as known data. System-level simulations were performed to determine the correlation between the EVM and BER. Figure 5 plots the EVM and BER for a wireless system under different channel conditions and system non-idealities. Two different modulation schemes used in practice – QPSK and 16-QAM were employed for evaluation purposes and about 10^5 bits were transmitted and received. The different values of EVM and BER are obtained by perturbing channel conditions. From the plots it is observed that in general an increase in EVM is associated with an increase in BER, and vice versa. A relationship between the two specifications can therefore be established. An upper bound on the BER specification can then be translated into an upper bound on the EVM. For example, if BER bound is set at 5e-4, the corresponding mean EVM bound for the QPSK and 16-QAM cases can be approximated to about 35% and 12%, respectively.

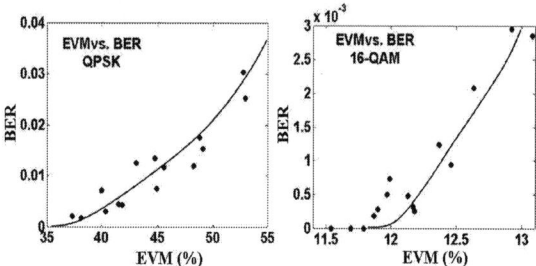

Figure 5: EVM vs. BER relation: System-level simulations

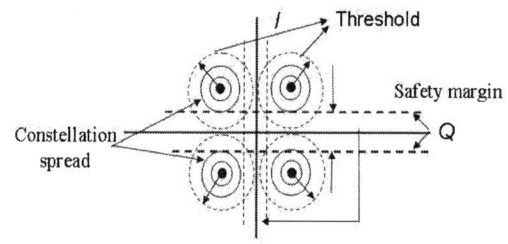

Figure 6: QPSK constellation: Setting the EVM threshold

A graphical illustration of the QPSK encoded symbols is shown in Figure 6. As the channel conditions and receiver performance degrade, the constellation points for each symbol lie inside circles of increasing size, corresponding to increasing EVM. When the circles cross the horizontal and vertical constellation boundaries, the received symbols are decoded incorrectly and bit errors occur. The objective of the EVM based feedback control is that the receiver will operate within the boundary of this threshold value (largest circle).

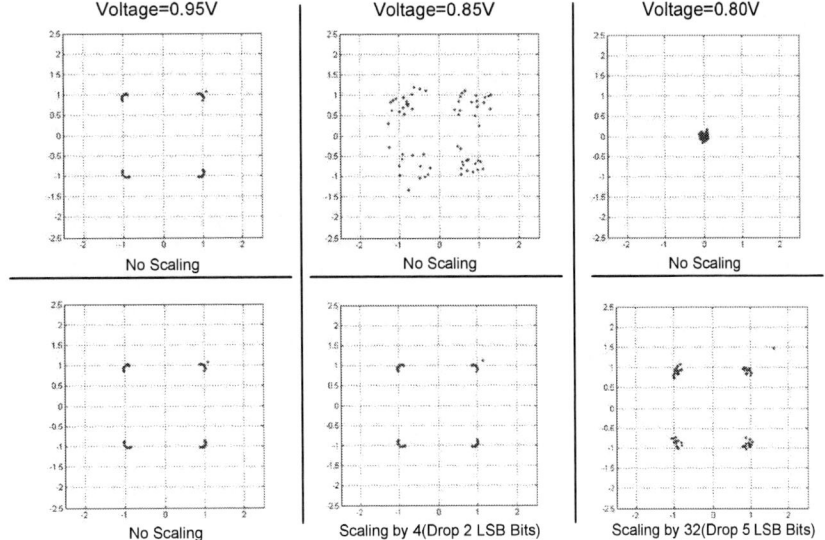

Figure 7: Constellation Degradation with Voltage Scaling and Recovery with Input Scaling. (Columns represent different voltage levels; top row has constellation points without Input Scaling; bottom row shows constellation points with Adaptive Input Scaling)

5. Experimental Results

An OFDM receiver model using QPSK modulation was implemented in MATLAB. A 9-tap low pass FIR filter and 128-point FFT in Radix-2 Single Path Delay Feedback [11] format were implemented at binary level to observe the effects of input scaling and adaptive voltage scaling on the system EVM. Bit level functions were implemented for ripple carry adder and array multiplier [12]. For the time delay and power estimations of the circuit, a full-adder was implemented in HSPICE with 70nm BPTM device model [13]. The time delay of the full-adder at 1 volt was 40.07ps and average power consumption was 8.161µw. Taking full-adder as the basic building block, the delay of the n-bit adders or n-bit multipliers is computed; based on number of full-adders in the critical path of the adder or multiplier. The clock speed of the FIR filter is equal to the critical path length, which is the time delay of a 12-bit array multiplier. The input to the 12-bit system is in Q6.6 format, where the MSB bit represents the sign bit, 5 bits represent the integer portion of the data and the remaining 6 bits represent the fractional values. Negative numbers are in two's complement form.

As evident from Figure 7, system performance degrades catastrophically as the system voltage is reduced on FIR-filter without any input scaling. However, the system EVM improves significantly for lower voltage levels in presence of adaptive input

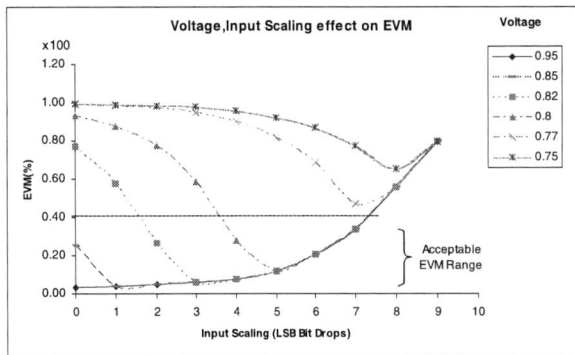

Figure 8: EVM improvement at low voltages with Input Scaling.

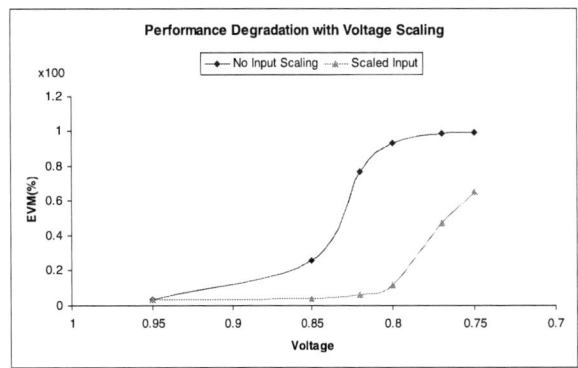

Figure 9: Drastic EVM Degradation with Voltage Overscaling vs. Graceful EVM Degradation with Adaptive Signal Scaling.

scaling. This phenomenon is also evident from the voltage curves of Figure 8, showed at different scaling levels. The plots show that using the proposed approach, for the given input it is possible to lower the supply voltage to 0.8 volt from 1 volt while keeping the system within acceptable EVM range. It is also shown in Figure 9 that system performance degradation curve is much better for adaptive input scaling as compared to regular voltage overscaling technique.

Table 1: Power Dissipated in FIR filter and FFT at different voltage levels

Voltage (volts)	Power (mW)	Power (mW)
	FIR	FFT
1	18.672	28.2
0.95	15.686	23.7
0.9	13.055	19.7
0.85	10.771	16.3
0.80	8.936	13.5

The 9-tap FIR filter used in the simulations has 9 multipliers and 8 adders. Average power consumption in the multipliers and adders was based on the number of full adders in those circuits. The power consumption of the filter and FFT at different voltage levels is shown in Table 1. In the experiments, voltage is reduced in either the FIR-filter or the FFT block at a given time. For the given input and circuit, it was noted that supply voltage can be lowered to 0.9v on FIR filter or FFT without any adverse effect on system performance. Although, this aggressive voltage overscaling is not possible without some feedback control while being within the acceptable performance limits, however to make a conservative estimate of the power savings we compared the power consumption at 0.8 volts to that of 0.9 volts, instead of 1 volt. The resulting power savings is approximately 31%. Since EVM block will be used for adaptive control of the whole OFDM receiver (RF-front end, DSP) its cost on the power is not considered.

6. Conclusion and Future Work

In this paper we presented an adaptive signal scaling scheme to reduce the power consumption of the OFDM receiver. The scheme ensures that system operates within acceptable performance limits while minimizing system power consumption. The proposed scheme can be applied to existing DSP systems without any significant architectural changes. The power cost of the EVM feedback network and a co-optimization procedure for FIR and FFT relative voltage settings to achieve optimal system power savings is left for future work.

7. References

[1]. H. Choi and W.P. Burleson, "Search-based wordlength optimization in VLSI/DSP synthesis", VLSI Signal Processing Workshop VII, pp.198-207, Oct.1994.

[2]. A. Oppenheim and R. Schaffer, Discrete-time Signal Processing, Prentice Hall, 1998.

[3]. K. Han and B.L. Evans, "Wordlength optimization with complexity and distortion measure and its applications to broadband wireless demodulator design", Proc. IEEE ICASSP2004, vol.5,pp.37-40, May 2004.

[4]. A.G. Dempster and M. D. Macleod, "Variable statistical wordlength in digital filters", IEE Proc.-Vis. Image Signal Process., Vol. 143, No. 1, Feburary 1996.

[5]. S. Yoshizawa and Y. Miyanaga "Tuneable wordlength Architecture for a Low Power Wireless OFDM demodulator", IEICE Trans. Fundamentals, Vol.E89-A, No.10, October 2006.

[6]. Debaillie, B. Bougard, B., Lenoir, G., Vandersteen G., Catthoor, F., "Energy-scalable OFDM transmitter design and control", 43rd IEEE Design Automation Conference, july 24-28, pp. 536-541.

[7]. Tasic, A., Serdjin, W.A., Long, J.R., "Adaptive multi-standard ciruits and systems for wireless communications", IEEE Circuits and Systems Magazine, Vol 6., Issue 1., pp. 29-37.

[8]. Abidi, A., Pottie, G.J., Kaiser, W.J., "Power-conscious design of wireless circuits and systems", Proceedings of the IEEE, vol 88,Issue 10, Oct 2000, pp. 1528-1545.

[9]. T. Pering, T. Burd and R. Broderson, " The Simulation and Evaluation of Dynamic Voltage Scaling Algorithms," Proc. Of Int'l Symp. on Low Power Electronics and Design(ISLPED 98), ACM Press, 1998, pp. 76-81.

[10]. J.G. Proakis, Digital Communications, 2nd Edition, McGraw-Hill, 1989.

[11]. H. Wold and M. Despain, "Pipeline and Parallel-Pipeline FFT Processors for VLSI Implementations" IEEE Transcations on Computers, Vol. C-33, No. 5, May 1984.

[12]. P. Denyer and D. Renshaw, "VLSI Signal Processing: A Bit-Serial Approach," Addison-Wesley, 1985.

[13]. BPTM 70nm: Berkley predictive technology model.

21st International Conference on VLSI Design

FAULT TOLERANT DYNAMIC ANTENNA ARRAY IN SMART ANTENNA SYSTEM USING EVOLVED VIRTUAL RECONFIGURABLE CIRCUIT

*D.Dhanasekaran, **K.Boopathy Bagan

*Assistant Professor, SVCE, Pennalur,Sriperumbudur-602108.

**Assistant Professor, Madras Institute of Technology, Chrompet, Chennai-44.

nddsekar@svce.ac.in,nddsekar@gmail.com

ABSTRACT

A majority of applications require cooperation of two or more independently designed, separately located, but mutually affecting subsystems. In addition to good behavior of each of the subsystems, an effective coordination is very important to achieve the desired overall performance. However, such a co-ordination is very difficult to attain mainly due to the lack of precise system models and/or dynamic parameters. In such situations, the evolvable hardware (EHW) techniques, which can achieve the sophisticated level of information processing the brain is capable of, can excel. In this paper, a new virtual reconfigurable circuit based drive circuit for array elements in smart antenna using the techniques of evolved operators is presented. The idea of this work is to develop a system that is tolerant to array element failure (fault tolerance) by utilizing phased array input programmer connected to a programmable VLSI chip. The approach chosen here is based on functional level evolution whose architecture contains many nonlinear functions and uses an evolutionary algorithm to evolve the best configuration. The system is tested for its effectiveness by choosing a real-time phase control in three element array of smart antenna with three input phases and introducing different element failures such as: element fails as open circuit, sensor fails as short circuit, noise added to individual element, multiple element failure etc.. In each case the mean square error is computed and used as the performance index.

Keywords: Virtual Reconfigurable circuit, Evolvable hardware, element validation.

I. INTRODUCTION

On systems that perform real-time processing of data, performance is often limited by the processing capability of the system. Providing a high processing speed is therefore often a crucial factor to be considered when implementing real-time systems. Also there is a need to have flexible systems that can be changed according to new specifications. Systems based on software are flexible, but often suffer from insufficient processing capability. Alternately, dedicated hardware can provide the highest processing performance, but is less flexible for changes.

Reconfigurable hardware [1] devices offer both the flexibility of computer software, and the ability to construct custom high performance computing circuits. Thus, in any cases they make a good compromise between software and hardware solutions. The structure of a reconfigurable hardware device can be changed any number of times by downloading into the device a software bit string called configuration bits. Field Programmable Gate Arrays (FPGA) and Programmable Logic Devices (PLD) are typical examples of reconfigurable hardware devices. FPGAs are hardware devices whose architecture can be determined by downloading a binary string, called architecture bits.

In this paper, a fault tolerant system using virtual reconfigurable circuit and evolved operators designed by coordinating the data from multiple element of smart antenna array, (that are commonly present in a smart antenna for maintaining the steering beam angle for various element failures and environmental conditions) is used, to handle exceptions such as element faults or extreme situations incorrectly handled by the less sophisticated conventional systems. The proposed method finds application mainly in the area of element validation, control engineering and other related fields to estimate the true variations of the signal during the failure period of an element.

II. EVOLVING CIRCUITS

Evolvable Hardware (EHW) is a new concept in the development of online adaptive machines. In contrast to conventional hardware where the structure is irreversibly fixed in the design process, EHW is designed to adapt to changes in task requirements or changes in the environment through its ability to reconfigure its own hardware structure online and autonomously [2]. The capacity for adaptation is achieved through evolutionary algorithms such as Genetic Algorithm (GA).

Virtual reconfigurable hardware is the combination of Genetic Algorithms and the software reconfigurable devices. The structure of the reconfigurable device can be determined by downloading binary bit strings called the architecture bits [1]. The architecture bits are treated as chromosomes in the population by the GA, and can be downloaded to the reconfigurable device resulting in changes to the hardware structure. The modified functionality of the device can then be evaluated and the fitness of the chromosome is calculated. The performance of the device is improved as the population is evolved by GA

978-1-4244-3039-0/08 $25.00 © 2008 IEEE

according to fitness. This process is repeated till the desired performance is achieved or particular number of generations is reached. To design conventional hardware, it is necessary to prepare all the specifications of the hardware functions in advance. In contrast to this, VRC can reconfigure

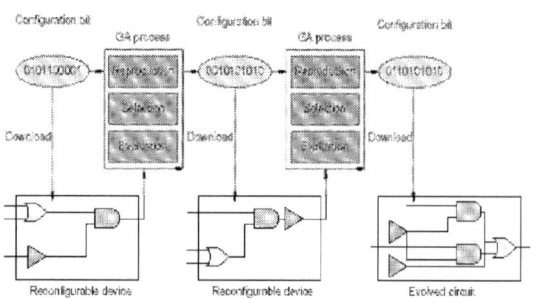

Figure 1 Basic concept of evolving hardware configurations

III. GENETIC ALGORITHM

Genetic Algorithm [6] determines how the hardware structure should be reconfigured whenever a new hardware structure is needed for a better performance. GA was proposed to model adaptation of natural and artificial systems through evolution, and is well known as one of the most powerful search procedures. The canonical GA has a population of chromosomes; each of them is obtained by encoding a point in the search space. Usually, they are represented by the strings of binary characters. The sequence of operations performed by the GA is shown in Figure 2. At an initial state, chromosomes in the population are generated at andom, and processed by many operations, such as itself without such specifications and can be considered as an online adaptive hardware. The basic concept of evolving circuits is shown in figure 1. The GA implementation configures the evolving design by placing individuals in Random Access Memory (RAM). The fitness value is calculated by the GA from the feedback signals originating in the evolving design. Since the GA and the evolving design are implemented on the same chip, the evolution process may continuously observe the evolving design.

The proposed architecture has many advantages over traditional hardware and software systems. First, it can automatically improve its performance by changing its hardware configuration according to the GA. Second, the reconfigurable devices can change evaluation, selection, crossover and mutation. The latter three operations are called the genetic operations, and one cycle of the evaluation and the genetic operation is counted as a generation. The evaluation assigns the fitness values to the chromosomes, which indicates how well the chromosomes perform as solutions of the given problem.

According to the fitness values, the selection determines which chromosomes can survive into the next generation. The crossover chooses some pairs of chromosomes, and exchanges their sub-strings at random. Finally, the mutation randomly picks some positions in the chromosome and flips their values. The major advantages of GA are its robustness and superior search performance in problems without a prior knowledge.

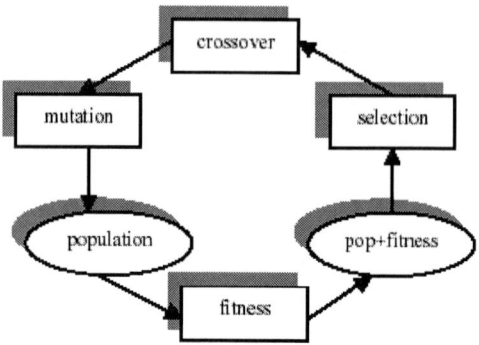

Figure 2 Operations in a Genetic unit

IV. VIRTUAL RECONFIGURABLE CIRCUITS IN FPGA'S

Although various evolvable systems have been implemented as Application Specific Integrated Circuits (ASIC), this solution is relatively expensive [5]. Hence a great effort is invested to designing evolvable systems at the level of FPGAs. These solutions can be divided into two groups: (i) FPGA is used for evaluation of circuits produced by evolutionary algorithm, which is executed in software. (ii) The whole evolvable system is implemented in the FPGAs. This type of implementation integrates a hardware realization of evolutionary algorithm and a reconfigurable device. The typical feature of these approaches is that the chromosomes are transformed to configuration bit stream and the configuration bit stream is uploaded into the FPGA. Most families of FPGAs can be configured externally (i.e. from an external device connected to the configuration port).

The approach utilizing VRC offers many benefits, such as

a). It is relatively inexpensive, because the whole evolvable system is realizable using an ordinary FPGA.

b). The architecture of the reconfigurable device can be designed exactly according the needs of a given problem.

c). Since, the complete evolvable system is available at the level of Hardware Description Language (HDL) code it can easily be modified and synthesized for various target platforms (FPGA families).

The VRC consists of 8 programmable elements realized on top of an ordinary FPGA. Slices have to implement a new array of programmable elements, new routing circuits and new configuration memory. The virtual circuit can be configured internally or from FPGA's I/O pins if new configuration memory is connected to them. The VRC is shown in figure 3.

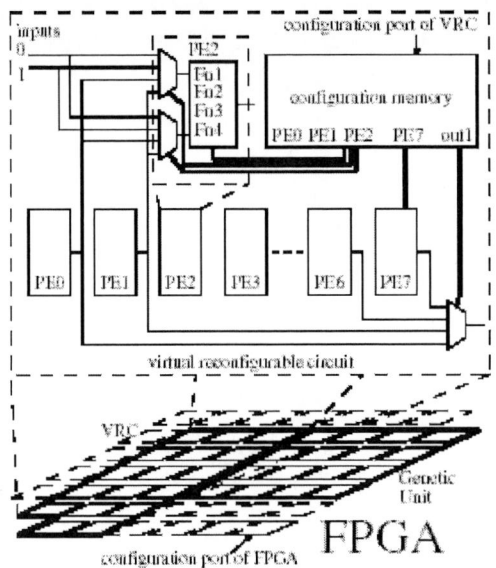

Figure 3 VRC module used in EHW

The main advantage of the VRC is that the array, the routing circuits and the configuration memory can be designed exactly according to the requirements of a given application. Furthermore, style of reconfiguration and granularity of new virtual reconfigurable circuit can exactly reflect the needs of a given application.

V. DETAILS OF EHW CHIP

The proposed EHW chip along with the VRC unit is shown in figure 4. It consists of the inputs routed through the communication component module. The configuration memory in the VRC is controlled by the genetic unit. The failure detection mechanism detects the failure of a sensor and accordingly the VRC is reconfigured. The evolvable system is composed of basic modules; input buffer, virtual reconfigurable circuit, pseudo random number generator, population memory, selection unit, mutation unit, fitness evaluator and output buffer and is shown in Figure 5. Both the Genetic unit and the

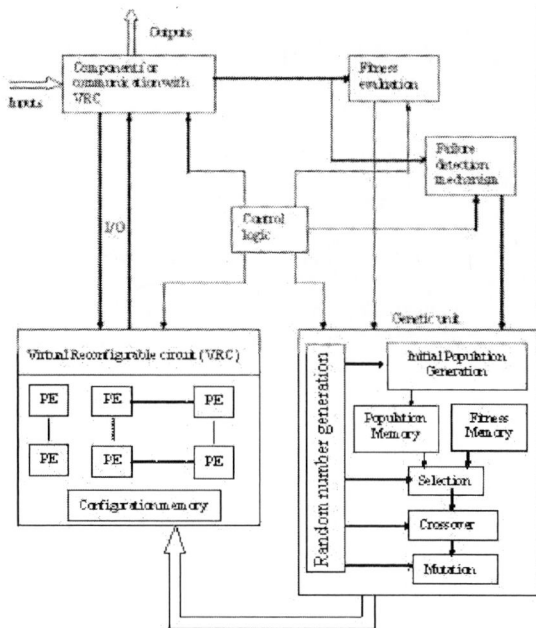

Figure 4 Proposed evolvable hardware chip

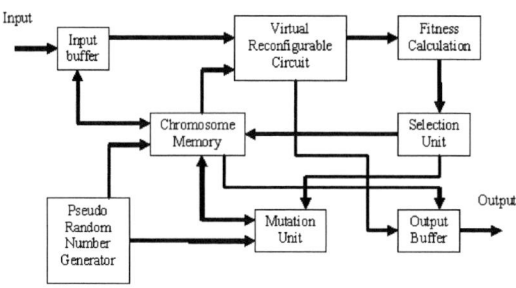

Figure 5 Block diagram of the VRC and the Genetic unit

A. Pseudo Random Number Generator

The Pseudo Random Number Generator (PRNG) [4] is used in two of the major steps in GA. Firstly, during initial population creation, and secondly to select individuals for crossover and mutation. One of the most common PRNG for FPGA implementation is a Linear Feedback Shift Register (LFSR). In this work, a word size of 12 is chosen. It is important to choose a good polynomial to ensure that the PRNG can generate a maximal sequence of 2_n-1 random numbers, while keeping the number of taps to a minimum for efficiency. For the chosen 12 bit word the polynomial

$$x_{12} \text{ (xnor) } x_6 \text{ (xnor) } x_4 \text{ (xnor)} x_1$$

is used. The block diagram of the LFSR is shown in Figure 6.

79

978-1-4244-3039-0/08 $25.00 © 2008 IEEE

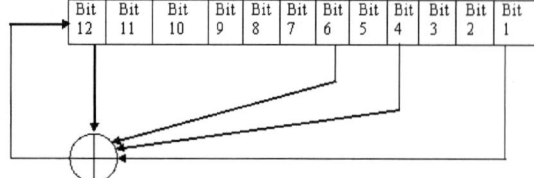

Bit 12	Bit 11	Bit 10	Bit 9	Bit 8	Bit 7	Bit 6	Bit 5	Bit 4	Bit 3	Bit 2	Bit 1

Figure 6 Pseudo random noise generator

The PRNG is designed so that a random number is generated in every clock. The 12th bit is taken as the random bit. Initial seed value is loaded in to the PRNG. To create a 10 bit random number, 10 single bit PRNG are combined in parallel.

B. Input Buffer

Input buffer consists of RAM. Original and distorted images are read from the file and stored in the input buffer. During runtime pixels are given as input to the VRC from the input buffer.

C. Initial Population Generation

During initial population generation, a 250 bit chromosome is created using 10 bit random number generator in 25 clock cycles. Chromosomes are stored in the Block RAM of FPGA. The initial population size is taken as 16. Totally 16x25 clock cycles are needed for initial population generation.

D. Fitness Calculation

MDPP fitness function is used to evaluate the chromosome. The original and filtered images are taken from the memory and the absolute difference between the corresponding pixel values is added and the fitness is evaluated.

E. Selection Unit

Selection unit selects the chromosome which has highest fitness as the best chromosome and it is retained for subsequent generations.

F. Mutation Unit

The chromosome which has highest fitness is selected for mutation. Using PRNG, bit by bit mutation is done for the creation of Childs. Fifteen new Childs are created in every generation and stored in the population memory.

G. Output Buffer

After the specified number of generations the evolution is completed and the best chromosome is stored in the memory. The fitness value and filtered image signal are calculated and stored in the output buffer.

VI. PROGRAMMING THE PE'S IN VRC

The logical configuration of the circuit is defined by a set of 250 bits, 10 bits for each one of the 25 PEs in the reconfigurable architecture as shown

111 000 0000 010 001 1100 110 010 0011 111
011 0010 110 111 0100 ………… 101 000 1000
110 010 1010

The first six bits of each ten bits represent the source of inputs to the PE (sel1& sel2). The other four bits of each ten bit (sel3) indexes the function from Table 1 to be applied by the PE. The configuration word contains details about the interconnection between the processing elements (PE) of the VRC and the functional operations performed within each PE. In this work, the interconnection between the PE's is restricted to its nearest two neighbors. The configuration word is updated the moment a sensor failure is detected. The genetic unit controls the configuration word in the VRC module. The genetic unit is programmed to give the best chromosome and using this, the initial configuration of the VRC is chosen. The reconfiguration of the circuit is required once a sensor failure is detected by the failure detection mechanism. In the present work, each PE except the first stage is assumed to receive inputs from any of the previous two stages. A total of 25 PE's is used in the VRC. The logical configuration of the circuit is defined by a set of 25 integer triplets, one for each of the 25 PEs in the reconfigurable architecture. The first two integers of each triplet represent the source of inputs to the PE (sel1& sel2) and the third integer of the triplet (sel3) indexes the function to be applied by the PE. A total of 16 functions are used in each PE and these are listed in Table-I.

Table 1 List of Functions used in the Processing Element

Function code	Function	Function code	Function
0000	X >> 1	1000	X & 0x0F
0001	X	1001	X & xF0
0010	~ X	1010	X	0x0F
0011	X & Y	1011	X	0x F0
0100	X	Y	1100	Min(X,Y)
0101	X ^ Y	1101	Max(X,Y)
0110	(X+Y)>>2	1110	Y<<1
0111	(X+Y) >>1	1111	X+Y

The two inputs to the PE's are selected using two multiplexers as shown in figure 7. The function performed in each of the PE depends on the status of the three select lines. The routing circuits are implemented using multiplexers. The

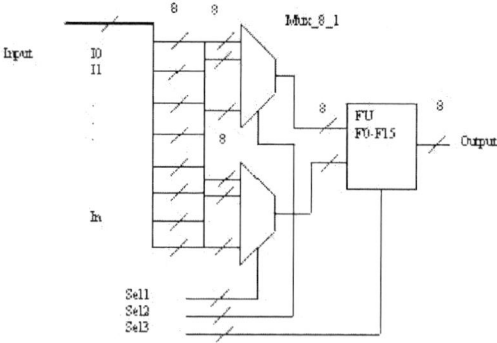

Figure 7 Selection of PE's using MUX logic

VII. DISCUSSION OF RESULTS

Case study –I: All three elements are faultless and noiseless

The results obtained by using the evolvable hardware chip on a real-time plant for this case configuration memory is composed of flip-flops. All bits of the configuration memory are connected to multiplexers that control routing and selection of functions in PEs. are shown in figure 8. The first three plots show the ariations of the three faultless sensors. The fourth plot is the average of the three sensor output and the last plot is the output of the reconfigured VRC.

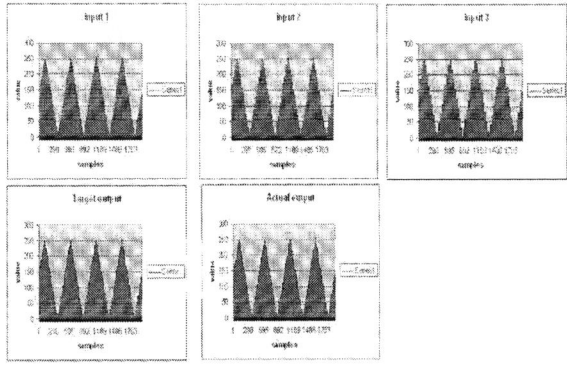

Figure 8 Experimental results obtained by using the EHW chip

Case study –II First element input is noiseless and noise is added to element 2 and 3

The results obtained by using the evolvable hardware chip on a real-time plant for this case are shown in figure 9. The ability of the circuit to

Case study –III element 3 is open circuit and element 1 and 2 are faultless

The VRC output captured using the Model-Sim package corresponding to this condition is shown reconfigure itself and filter the noise is demonstrated in this study. Gaussian noise was added to the element 2 and 3 and the VRC output was found to match the required output.

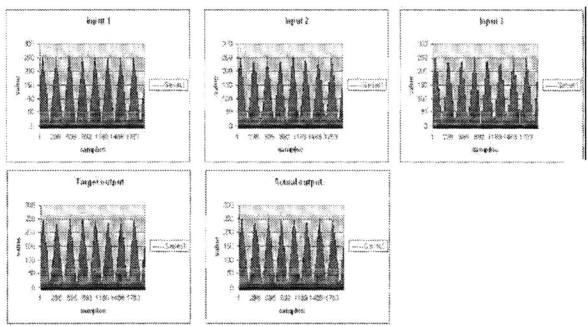

Figure 9 Experimental results obtained by using the EHW chip with noise added to sensors 2 and 3 alone

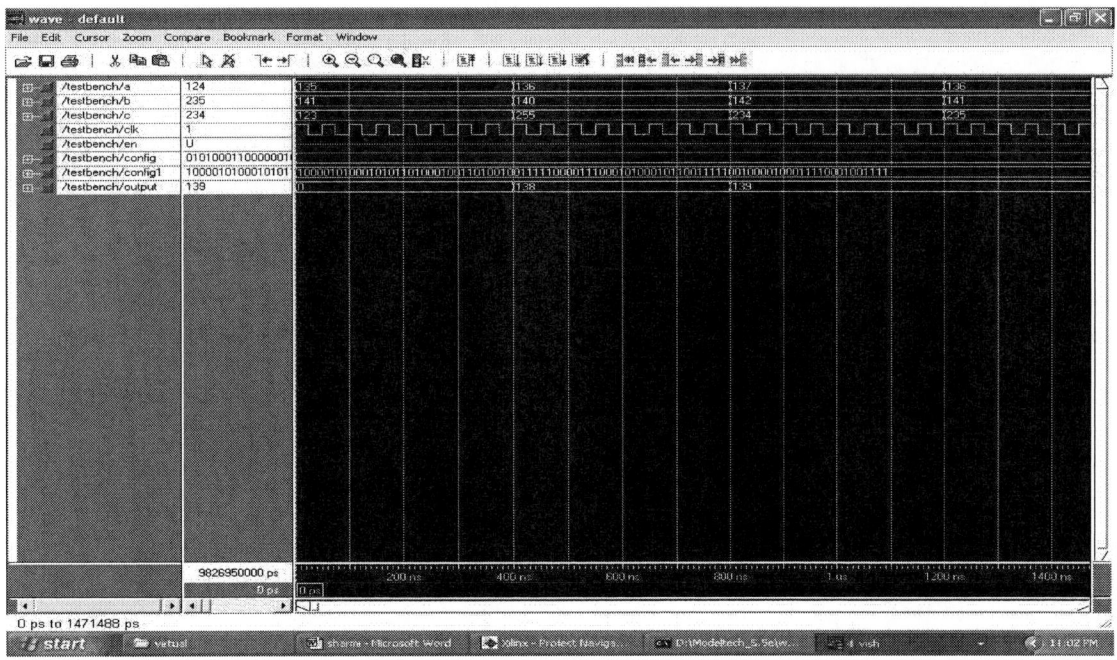

Figure 10 VRC output corresponding to element 3 open and element 1 and 2 faultless

978-1-4244-3039-0/08 $25.00 © 2008 IEEE

in figure 10. The VRC has reconfigured itself to reject the element 3 value and gives an output which closely matches with the average of the sensor 1 and 2 readings. The performance of the evolved chip in each of the case is measured in terms of the MSE value and is given by

$$MSE = \sum_{k=1}^{N} [actual\ output(k) - target\ output(k)]^2$$

where actual output(k) and target output(k) are the VRC circuit output and ideal output respectively and N is the total number of samples.

VIII. CHIP IMPLEMENTATION RESULTS

The evolved filter is the result of the evolution of an array of 4x6 PEs with one PE at the output. Number of generations is 3000. The coding was done in VHDL and simulations were performed in ModelSim 6. The hardware evolution took 2 minutes on Xilinx Virtex FPGA xcv800 running at 49 MHz. This compares favorably with software simulations run previously which it took approximately 6 hours (Pentium III/800 MHz system) to achieve the best chromosome, the speed has been increased by 180 times and the evolution time has been greatly reduced. Number of generations was taken as 3000. Hardware evolution took 12 minutes in Xilinx VirtexE FPGA xcv2000e. This compares favorably with software simulations run previously which took approximately 6 hours to give the best chromosome. The VRC takes 1754 slices of the Xilinx Virtex FPGA xcv800 (9408 slices) and the whole evolvable system including the GA takes 3204 slices. Since small amount of the resources are only used i.e only 34% of the resources, chromosomes can be operated in parallel and the processing time can be further reduced. The synthesis report obtained is given in Table-2 below.

Table 2 Synthesis Report - Device Utilization Summary
(Population Size = 16, Chromosome Length = 250)

Target information: Vendor: Xilinx Family: Virtex Device: v800fg680 Speed: -6 Optimization Goal: Speed		
Number of Slices	3204 out of 9408	34%
Number of Slice Flip Flops	1087 out of 18816	5%
Number of 4 input LUTs	6200 out of 18816	32%
Number of bonded IOBs	79 out of 516	15%

Number of BRAMs	8 out of 28	28%
Number of GCLKs	1 out of 4	25%
Minimum period	20.160ns	
Maximum Frequency	49.603MHz	
Minimum input arrival time before clock	27.706ns	
Maximum output required time after clock	6.887ns	

IX. CONCLUSION

The work has presented a novel approach to overcome the element failure of smart antenna that may be installed in space station based on the technique of evolvable hardware. FPGA model for the function level evolvable hardware is analyzed and associated with the evolutionary algorithms employed. The evolution time has been greatly reduced by implementing the evolutionary algorithm in hardware. The EHW architecture evolves circuits without a priori information and reconfigures itself and is tolerant to different fault conditions. The EHW system outperforms conventional ones in terms of computational effort (greatly reduced), robustness (improved) and implementation cost (reduced significantly).

X. REFERENCES

[1]. Gerald R. Clark (1999), "A Novel Function-Level EHW Architecture within Modern FPGAs", Proceedings of the Congress on Evolutionary Computation (CEC 99), IEEE.
[2]. Hollingworth G, Smith S and Tyrrell A (2000), "Safe Intrinsic Evolution of Virtex Devices", Proceedings of the Second NASA/DoD Workshop on Evolvable Hardware, IEEE, pp. 195-202.
[3]. Hollingworth G, Smith S and Tyrrell A (1999), "Design of Highly Parallel Edge Detection Nodes using Evolutionary Techniques", Proceedings of the 7th Euromicro Workshop on Parallel and Distributed Processing, IEEE, pp. 35 – 42.
[4]. Gallagher, J.C.; Vigraham, S.; Kramer, G (2004) "A family of compact genetic algorithms for intrinsic evolvable hardware.",IEEE transactions on Evolutionary computation Volume 8, Issue 2, April Page(s):111 - 126

XI. BIOGRAPHIES

1) Mr. D.Dhanasekaran is working as an Assistant Professor, ECE dept. In Sri Venkataswara College Engg. College, Pennalur,Sriperumbudur,affiliated to the Anna university. His areas of interest include Evolvable Computing, reconfigurable computing,VLSI signal processing and neural networks.
2) Dr. K.Boopathy Bagan completed his doctoral degree from Anna university . He is presently working as a professor, ECE dept. In Madras Institute of Technology, Chrompet, Chennai. His areas of interest include VLSI signal processing, Genetic Algorithms and evolvable hardware.

SESSION C1:
Embedded Systems

21st International Conference on VLSI Design

Multimedia Tools and Architectures for Hardware/Software Co-Simulation of Reconfigurable Systems

Valery Sklyarov, Iouliia Skliarova, Bruno Pimentel, Manuel Almeida

Department of Electronics, Telecommunications and Informatics, IEETA
University of Aveiro, 3810-193 Aveiro, Portugal
skl@ua.pt, iouliia@ua.pt, brunopimentel@ua.pt, manuel.almeida@ua.pt

Abstract

The paper describes novel multimedia tools and architectures for hardware/software co-simulation of reconfigurable systems. The main contributions are provided in the following three areas: 1) multimedia tools making it possible to manage animated graphical objects for virtual simulation of real world physical objects in the scope of reconfigurable system design; 2) a remotely accessible prototyping system, which is very helpful for both solving the problems of hardware design and supporting multimedia systems which can be used in vast varieties of practical applications, the most important of which are engineering training and education; 3) design methodology based on physical circuits and virtual objects. A number of illustrative examples demonstrating capabilities of the proposed approach are presented and discussed.

1. Introduction

The information revolution has focused considerable attention on multimedia tools related to engineering and education. One important application from this scope is hardware/software co-simulation with general architecture illustrated in Fig. 1, where two primary components (a host computer and an FPGA-based reconfigurable prototyping board) communicate with each other.

The developed system is considered to be partially implemented in software of the host computer and partially in hardware of the prototyping board. The software/hardware parts interact through an interface and they are assembled in such a way that we can explore systems with either more hardware or more software. In other words the boundary between software and hardware is rather flexible.

One of the most important components of the hardware/software co-simulator shown in Fig. 1 is a virtual visual sub-system, which enables the designer to

verify the functionality of the developed system in a visual mode using the host computer monitor (see the left-hand part of Fig. 1). Let us consider an example taken from [1] and shown in Fig. 2.

A visual interface with a plotter is presented on the screen (see the left-hand part of Fig. 2). There are four sensors shown in form of filled circles. The states of sensors are indicated by different fill color for the corresponding circle. Any motion of the pen unit is shown on the screen. We can control the plotter either from software of the host PC computer or from physical FPGA-based circuits in the attached hardware (prototyping board). Drawing can be performed using any of three pens of different colors. The pens and other details can be either shown or hidden. A multimedia library enables the circuit designers to construct different visual objects on a monitor screen and to establish proper communication mechanisms between software of the host computer and the prototyping board.

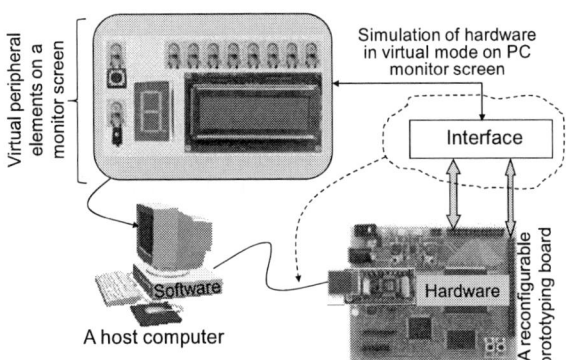

Figure 1. General idea of hardware/software co-simulation

This type of hardware/software co-simulators is very promising in numerous areas such as: virtual design space exploration, rapid dissemination of different models and methods in the scope of hardware design, comparing alternative and competitive circuit implementations, education, engineering training, etc. It is important to note that the considered technique permits to get valuable assistance from different types of multimedia, such as the Internet, mobile video conferencing, interactive tools for intensive education, etc.

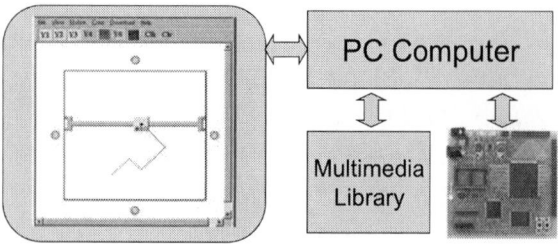

Figure 2. Hardware/software co-simulation of plotter functionality

Further steps in hardware/software co-simulation can be done in the following directions:
1. Using reconfigurable hardware (based on FPGAs), which enables a soft approach for developing both hardware and software to be applied by the circuit designers.
2. Establishing remote wireless interactions between software (host computer) and hardware (reconfigurable board) components.
3. Applying the strategy of remote design space exploration using the Internet facilities.

It is known that good results (in particular, in the scope of education) can be achieved by combining real hand-on experiments with simulation, and tests in remote/virtual laboratories (in accordance with suggestions done in [2] for different domains, including engineering, natural sciences, education, and psychology). Examples of such laboratories are discussed in [3,4] and the majority of them deal with fixed experiments, where the users are only able to control a limited number of parameters and to observe visually the behavior of a system and/or to receive the resulting measurement data for subsequent processing and analysis. The proposed here methods and tools have a number of distinctive features. The most important of them are remote prototyping and virtual design. The latter enables physically implemented circuits and circuits modeled in software to be combined and explored. All the required details will be described in the subsequent sections.

The remainder of the paper is organized in 6 sections. Section 2 presents basic ideas and architecture of the proposed hardware/software co-simulator and evidences the importance of multimedia tools, which permit the solution of many design problems to be simplified. Section 3 discusses some practical applications. Section 4 describes the developed multimedia tools. Section 5 considers a remote collaboration of system components. Section 6 presents an example. The conclusion is given in section 7.

2. Architecture of hardware/software co-simulator

The hardware/software co-simulator is composed of an FPGA-based reconfigurable prototyping core board (DETIUA-S3) [5] communicating with a host computer through either wired (USB) or wireless (Bluetooth) interface. Two types of software/hardware (reconfigurable hardware) co-simulation with the host computer have been examined and they allow to model:
1. Functionality of peripheral devices, such as dipswitches, pushbuttons, LEDs, LCD panels, segment displays, etc. Some examples are shown in Fig. 1.
2. Functionality of application-specific devices, such as the considered above plotter, combinatorial accelerators, etc.

The DETIUA-S3 functions in two mutually compatible modes. The first one is an interaction with physical hardware through extension buses, such as that shown in Fig. 3 for communications with a VGA monitor; a mouse; and a keyboard. The second mode enables the designers to interact with virtual hardware modeled in software of the host computer through either USB or Bluetooth interface. Dependently on the target requirements we can choose either more hardware or more software and this feature makes the considered architecture rather flexible and appropriate for a vast variety of practical applications.

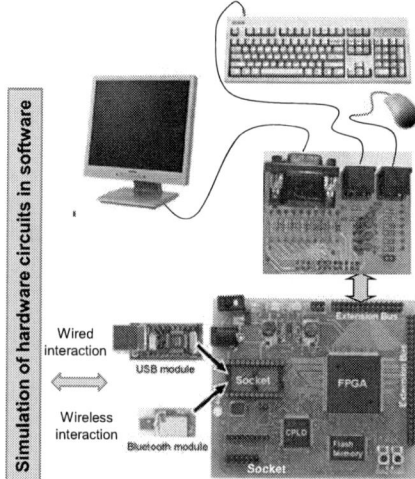

Figure 3. Architecture of software/hardware co-simulator

Configuration is achieved through uploading a bit-stream to the FPGA static memory. It is done in two steps. At the first step the FPGA is configured in such a way that enables the host computer to load up to 7 bitstreams to a flash memory available on the board. The flash memory is divided into three logical sections. The first section contains a bitstream that has to be pre-loaded to FPGA in order to allow the following set of operations to be performed: 1) transferring an application-specific bitstream to the second section; 2) erasing flash memory sectors; 3) transferring data to the third section of the flash memory and vice versa. The second logical section is used to store the default application-specific bitstream for subsequent quick loading into the FPGA during the second step. The third memory section enables the designer to store up to 6 additional bitstreams for FPGA reconfiguration. If required, the second and the third sections can be updated at any time. Since there are totally 7 different user bitstreams available in the flash memory, the logic capacity of FPGA can be multiplied by 7.

The board has two extension buses (see Fig. 3) with 80 pins totally. The latter can be used to connect physical external devices (see Fig. 4). The developed software (a prototyping board manager - PBM in the host computer) and hardware (an agent module) make it possible to replace physical devices with virtual devices as shown in Fig. 4. The pre-designed and implemented in FPGA agent module establishes connections between the user circuits and either physical or virtual devices. In the second case inputs and outputs of the user circuits are converted to special codes that can be transmitted through the selected interface (USB or Bluetooth). These codes are recognized by the PBM in the host computer and they are handled by software in a virtual mode much like it is done in physical devices.

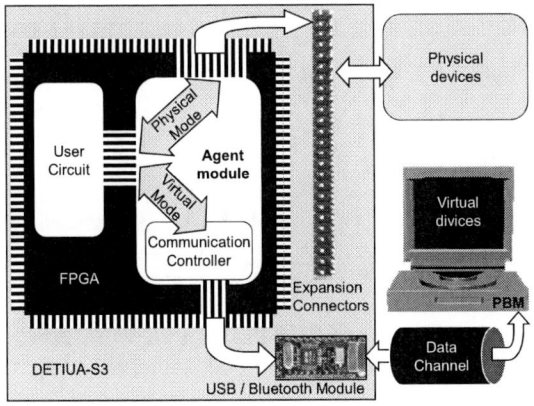

Figure 4. Interaction of user circuits with physical and virtual devices

3. Practical applications

The developed hardware/software tools can be employed in different areas enabling the designers to partition the developed system in such a way that one part of the system will be implemented in reconfigurable

hardware and another part will be modeled in software. Such technique is especially interesting in the scope of engineering training and education.

Let us consider an example. Suppose we would like to design a reprogrammable finite state machine (FSM), which implements different algorithms over ternary vectors whose elements have one of three possible values: 1, 0 and – (don't care). Different algorithms permit to execute such operations as: testing if the given vector contains N successive 1s (0s, don't cares); if the vector does not have values 1 (0s, don't cares); if the number of 1s in the vector is greater than the number of 0s, etc. Such operations are frequently required for numerous combinatorial search problems [6] and we would like to examine execution time for different algorithms and the ability of FSM to be efficiently reprogrammed (using, for example, methods [7]).

Suppose, an initial vector has to be entered from either pushbuttons or DIP switches and the results of the selected operation together with the execution time have to be displayed on an LCD (see Fig. 5). The designed circuit includes the following two primary blocks (see Fig. 5): the reprogrammable FSM and the handler making it possible to customize the FSM in such a way that enables the desired algorithm to be realized. Suppose at the beginning that the FSM is implemented in FPGA and the handler is modeled in software of a host computer. After the FSM has been tested, both blocks (i.e. the FSM and the handler) can be implemented in FPGA. Thus, the considered technique enables the circuit to be designed incrementally. Dependently on availability of peripheral devices (such as the LCD shown in Fig. 5) either physical or virtual interaction with such devices can be employed.

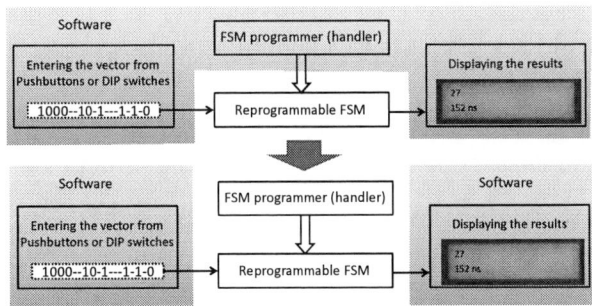

Figure 5. Different modes of hardware/software co-simulation

The considered tasks are very interesting for education because the students can start with examining software modules and then continue with incremental conversion of the tested software modules to the relevant hardware implementations. These tasks are also very appropriate for design space exploration. Indeed, the proposed technique enables the designers to concentrate on the most critical components of the explored system (implementing them in hardware) and to abstract from less significant components (modeling them in software).

4. Multimedia tools

The developed multimedia tools are shown in Fig. 6. They are organized in libraries of C++ and C# classes for *visual objects*, which provide support for visualization of different components on a monitor screen (of the host computer), and for *hardware objects* such as memories, FSMs, etc. The latter can be seen as software models of the relevant hardware circuits.

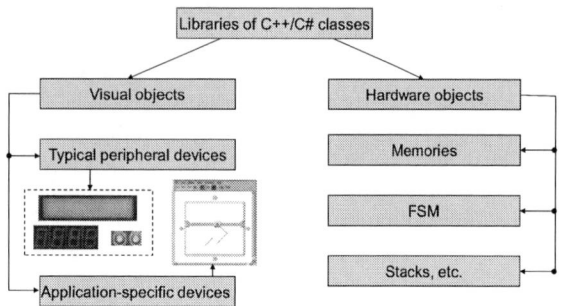

Figure 6. The structure of libraries

Fig. 7 depicts an example of a visual object, which includes three groups of functions that are 1) private functions providing the desired functionality; 2) public functions enabling a visual manipulation of the object on the monitor screen; 3) public functions establishing external interface with the object. Let us assume that it is required to display an LCD on a monitor screen (see the left-top corner of Fig. 7). In this particular case, inputs and outputs of the relevant object model inputs and outputs of a physical microcontroller, such as [8]. Dependently on target requirements either full or reduced functionality of the microcontroller can be provided.

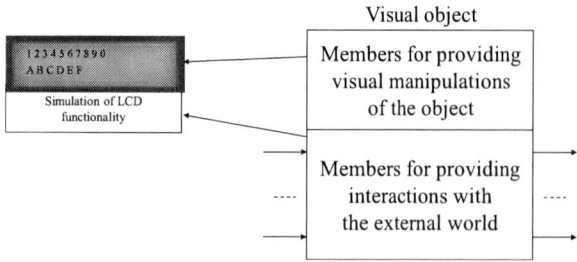

Figure 7. The structure of a visual object

Fig. 8 depicts a subset of the developed classes emulating the functionality of peripheral devices, namely: pushbuttons (the class vPushButton), DIP switches (the

class vDipSwitch), LEDs (the class vLED), segment displays (the class vSevenSegmentDisplay), and LCD (the class vLCD).

All classes for typical peripheral devices are derived from the class UserControl (from the .NET library) providing functions for graphical visualization and implement the interface vPeripheral (see Fig. 8) introducing names for common interface operations (such as GetName(), GetOutputPins(), etc.).

There are two special classes named *vPin* and *vConnector*, which permit to establish interface between visual objects (see Fig. 7) and an input/output module. The latter serves the data channel from the side of the host computer and it is considered to be a part of the PBM, which also provides support for the following functions:

- Uploading user bitstreams to the second section of the board's flash memory (see section 2);
- Transferring data from a host computer to the third section of the board's flash memory and vice versa;
- Run-time data exchange between the user and the board.

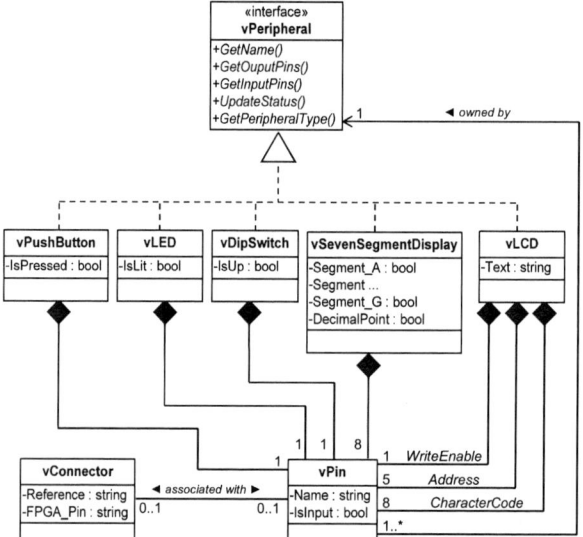

Figure 8. A set of C# classes developed for virtual peripheral devices

Objects of the classes *vPin* and *vConnector* (see Fig. 8) correspond to physical pins of the relevant peripheral devices and the DETIUA-S3 extension buses respectively. The association shown in Fig. 8 stands for establishing connections between the extension buses and virtual peripheral devices. Class attributes indicate pin numbers (*FPGA_pin*), connector names (*Reference*), etc. The input/output module periodically updates the values of *vPin* (for FPGA outputs) and *vConnector* (for FPGA inputs). This is needed to set up data exchange between the user circuits, implemented in FPGA, and virtual devices, modeled in the host computer. Finally, data of *vConnector* and the agent module (see Fig. 4) are exchanged enabling virtual devices to behave much like real peripheral devices. The agent module implements interface between FPGA pins and user circuits (see Fig. 4). Thus, this module (considered to be a part of a library) has to be included to any FPGA circuit interacting with virtual hardware modeled in the host computer.

Let us characterize now hardware objects. They are instances of C++ classes modeling functionality of different circuits. Fig. 9 depicts an example of a hardware object, which includes three groups of functions, that are 1) private functions providing the desired functionality; 2) public functions supplying visual facilities in such a way that the selected states of the object can be displayed on a monitor screen and these states can be changed by users if required; 3) public functions establishing external interface with the object. Suppose a hardware object implements the functionality of a dual-port memory (see the left-hand part of Fig. 9). The latter can be read/written either by rows or by columns.

In this particular case, functions from the first group read/write either horizontal or vertical memory vectors. Functions from the second group enable the users to communicate with the object through the visual interface (i.e. to test visually any row/column entering the necessary addresses or to write any row/column once again specifying the necessary addresses). Functions from the third group establish an interaction between the considered hardware object and other hardware objects and/or user circuits implemented in FPGA.

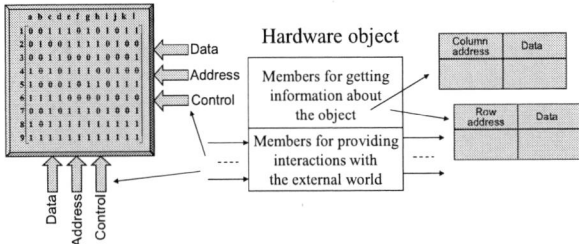

Figure 9. Structure of a hardware object

5. Remote collaboration

The developed software tools provide remote users with the most of the PBM functionality through the Internet (see Fig. 10). The DETIUA-S3 board is connected to a server through either wired (USB) or wireless (Bluetooth) interface. In addition, co-simulation tools have been developed enabling remote users to construct digital systems in such a way that they are partially implemented in FPGA and partially modeled in software of a user computer.

The software management system enables only one user to work with the board within the specified time slot. As soon as physical communication is terminated the board becomes available for the next user. Such system is being developed for educational process. Thus, the users are students; the time slots are determined by teachers. The objectives are the following:

• The remote prototyping board (the DETIUA-S3) has to implement the circuits developed by the students;

• Virtual peripheral devices have to provide interfaces with the circuits;

• The proposed software and hardware tools have to be used for co-simulating the developed and virtual components.

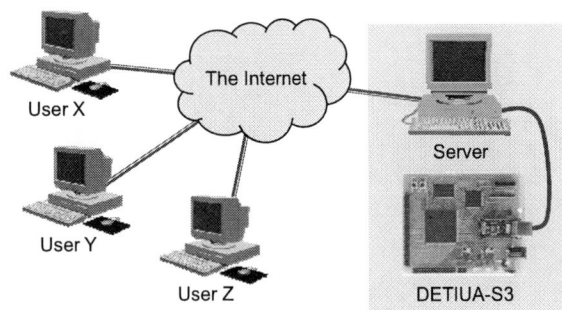

Figure 10. Interaction between the DETIUA-S3 and remote users

6. An example

Suppose we have to design a combinatorial accelerator shown in Fig. 11. The main idea is to verify if this accelerator can be reused for solving different combinatorial problems formulated over Boolean and ternary matrices (such as that have been discussed in [6]). The part shown with grey background is projected to be reusable and the control unit is intended to be reprogrammable in such a way that allows implementing different combinatorial algorithms (such types of combinatorial accelerators are described in detail in [6]).

Let us model the reusable part (i.e. the part shown with grey background) in software involving hardware objects (such as that shown in Fig. 9) and implement in FPGA the control unit. Suppose that a request for reprogramming the control unit has to be done from a host computer, which knows a particular problem that must be solved. Examples of such problems might be the Boolean satisfiability, binary matrix covering, etc. To model the considered reusable circuit in the host computer it is necessary to develop a program using the library of hardware classes, which permit to define the relevant hardware objects, such as matrices, stacks, registers (see Fig. 11), etc. Suppose the control unit is modeled by a reprogrammable FSM whose functionality can be changed through reloading the FSM's RAM blocks. Methods of synthesis of reprogrammable FSMs are proposed in [7]. Interaction between software and hardware parts can be provided with the aid of the developed interface components (the agent module and the PBM), which establish links between the designated inputs and outputs of the circuit implemented in hardware (in FPGA) and hardware objects modeled in software (see Fig. 4). Finally, we can test the circuit for a particular algorithm (for example for the Boolean satisfiability) with the aid of hardware object functions supplying visual facilities.

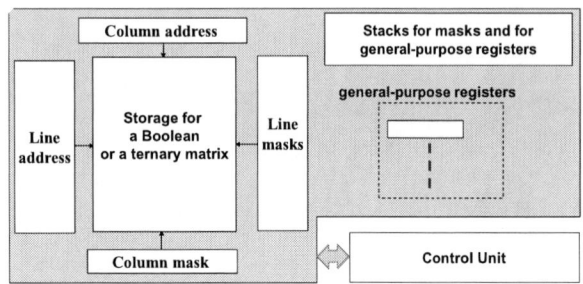

Figure 11. Structure of a combinatorial accelerator

Now let us consider how to change the circuit functionality in order to examine different algorithms. For such purposes it is necessary to implement a handler, which is able to alter the algorithm of the control unit (we assume that the execution unit, shown with grey background, is exactly the same). Thus, we have to apply the same technique that is demonstrated in Fig. 5. At the beginning the handler can be modeled in software. After the handler has been tested it can be implemented in FPGA. Incrementally other blocks of the explored system (see Fig. 11) might be converted from software to hardware. This technique gives vast opportunities for hardware/software co-simulation and consequently for the design space exploration. Obviously this task is very interesting and helpful for students.

The presented above system has not been completely finished yet, although the majority of basic components (such as the DETIUA-S3 board, C++/C# libraries for virtual peripheral devices and for numerous hardware objects, remote interaction with the board through the Internet) have been implemented and tested. The results of testing demonstrate good capabilities of the developed components for remote monitoring and design of reconfigurable systems.

7. Conclusion

The proposed tools for hardware/software co-simulation of reconfigurable systems, including remote monitoring and design, are very promising in a vast varieties of practical applications, such as virtual design space exploration, rapid dissemination of different models and methods in the scope of hardware design, comparing alternative and competitive circuit implementations using the Internet facilities, education, engineering training, etc. These tools possess the following distinctive features:

- Reconfiguration of the prototyping board through either wired (USB) or wireless (Bluetooth) interface;
- Remote monitoring of prototyping board and remote design of reconfigurable systems;
- Software/reconfigurable hardware co-simulation through the developed interfaces supported by the relevant hardware projects and software tools.

8. References

[1] V. Sklyarov, "Hardware/Software Modeling of FPGA-based Systems", *Parallel Algorithms and Applications*, vol. 17, no. 1, 2002, pp. 19-39.

[2] J. Ma and J. V. Nickerson, "Hand-On, simulated, and remote laboratories: A comparative literature review", *ACM Computing Surveys*, vol. 38, no. 3, article 7, Sept. 2006.

[3] Z. Nedic and J. Machotka, "Remote Laboratory NetLab for Effective Teaching of 1st Year Engineering Students", *iJOE International Journal of Online Engineering*, vol. 3, no. 3, article 8, 2007.

[4] D. Z. Deniz, A. Bulancak, and G. Özcan, "A Novel Approach to Remote Laboratories", Proc. 33rd ASEE/IEEE Frontiers in Education Conference, CO, USA, November 2003, vol. 1, pp. T3E-8 - T3E-12.

[5] M. Almeida, B. Pimentel, V. Sklyarov, and I. Skliarova, "Design Tools for Rapid Prototyping of Embedded Controllers", Proc. 3rd International Conference on Autonomous Robots and Agents - ICARA'2006, Palmerston North, New Zealand, December 2006, pp. 683-688.

[6] I. Skliarova and V. Sklyarov, "Design Methods for FPGA-based Implementation of Combinatorial Search Algorithms", Proc. Int. Workshop on SoC and MCSoC Design - IWSOC'2006, 4th Int. Conf. on Advances in Mobile Computing and Multimedia - MoMM'2006, Yogyakarta, Indonesia, December 2006, pp. 359-368.

[7] V. Sklyarov, "Reconfigurable models of finite state machines and their implementation in FPGAs", *Journal of Systems Architecture*, 47, 2002, pp. 1043-1064.

[8] HD44780U Dot Matrix Liquid Crystal Display Controller/Driver, Hitachi, 1998.

21st International Conference on VLSI Design

A Modeling of a Dynamically Reconfigurable Processor using SystemC

Junji Kitamichi[†]   Koji Ueda[‡]   Kenichi Kuroda[†]

[†] School of Computer Science and Engineering,
The University of Aizu,
Tsuruga, Ikki-machi, Aizu-Wakamatsu,
Fukushima 965-8580, JAPAN
{kitamiti,kuroda}@u-aizu.ac.jp

[‡] NEC Electronics Corporation
1753, Shimonumabe,
Nakahara-Ku, Kawasaki,
Kanagawa 211-8668, JAPAN

Abstract

Recently, Dynamically Reconfigurable Processors (DRPs) have been proposed. In this paper, we describe a model of a DRP using a Dynamic Module Library (DML), which we have developed for the modeling of general-purpose dynamically reconfigurable systems. The DML is an extended SystemC library and enables the modeling of the dynamic generation and elimination of modules, ports and channels and the dynamic connection and dispatch between port and channel. Using the DML, we can model the DRP naturally. The architecture of the proposed DRP is based on an MIPS-type architecture and supports the instructions, which are for the dynamically reconfigurable operational units and for their generation and elimination. We describe the proposed DRP model and its evaluation results.

1 Introduction

Recently, many types of Dynamically Reconfigurable Processors (DRPs) based on FPGA technology have been proposed[1] [2] [3]. DRPs are implemented on a unique Dynamically Reconfigurable Architecture (DRA), and a specialized design environment is provided for the DRP. In the case of the system design for a new application-specific Dynamically Reconfigurable System (DRS), the existing design methods and CAD systems can not deal with this system design. In the case of a new DRP architecture design, a design environment that is independent of the specific DRA will be needed at the system level, which is the beginning of system design.

The system level design[4] is more abstract that the conventional Register Transfer (RT) level design, and system specification languages, such as SystemC [5], SpecC[6],

and SystemVerilog[7], are proposed.

We have been developing a Dynamic Module Library (DML) [8] for the modeling of general DRSs from the system level to the RT level, which provides the dynamic generation and elimination of a module and the dynamic connection and dispatch of the port and channel. In SystemC2.1.v1, a dynamic process is introduced, which enables the dynamic process generation and elimination. However, using this process, the modules, ports, and channels cannot be generated and eliminated dynamically and cannot be connected and dispatched dynamically after the simulation has started. Using the DML, the dynamic generation and elimination of a module and the dynamic connection and dispatch of the port and channel are implemented and modeled naturally.

In this paper, we describe a DRP model at the RT level using the DML and its evaluation results. The proposed processor is based on an MIPS-type processor which supports the dynamically reconfigurable operational units that are dynamically generated and eliminated as well as the instructions for dynamic reconfiguration and their operations.

2 Background

2.1 System Modeling using SystemC

SystemC[5] is a system description language. SystemC has the ability for system modeling using a class library of C++. System models from the abstraction level to the concrete RT level can be described in SystemC, and they can be simulated at the different levels.

The behavior of the system is described in the modules. In the module, the definitions of input and output ports and the definitions and initializations of processes, such as the behavior of the module, the generations of instances of sub-modules, the connection with ports in the sub-modules

978-1-4244-3039-0/08 $25.00 © 2008 IEEE

to channels, and the behavior of processes are described. The modules are defined by inheriting the *sc_module* class, a class of SystemC. The channels are the communication paths between modules, and the ports are connected to channels and the port of modules. For example, *sc_in* and *sc_fifo* are types of an input port and a FIFO channel respectively.

In the older version of SystemC than 2. 1.v1, processes are initialized only at the start of a simulation and are never generated or eliminated during a simulation. The idea of a dynamic process is added in SystemC 2. 1.v1. With the dynamic process, it is possible to activate processes after the simulation has started. However, each port and channel is defined as static and is never generated or eliminated dynamically during the simulation in SystemC 2. 1.v1.

2.2 Dynamically Reconfigurable Architectures

Recently, Dynamically Reconfigurable Architectures (DRAs)[9] have been proposed. A DRA is based on FPGA, whose configuration memory is SRAM, and can be reconfigured while the system is operating. There are many types of DRAs[1][2][10][11], such as the multi-context type / partially reconfigurable type and the fine-grain / course-grain types. The reconfigurable units of fine-grain architectures are constructed by Look-Up Tables (LUTs) that use memory technology. LUTs can be constructed as any logic function, so every region can be efficiently used. On the other hand, ALUs construct reconfigurable units of coarse-grain architectures; therefore, they can compute numerical operations faster than fine-grain architectures. However, some reconfigurable units cannot be utilized, and the usage of the area is worse.

Multi-context DRA[1] and partially DRA[10] [11] make up one classification of DRAs. Multi-context DRAs have some configuration data memories, and the circuits are reconfigured by switching the configuration data. The switching of the configuration data takes place at once. The contents of unused configuration memories can be rewritten. In the partially DRA, the reconfiguration of circuits corresponds to the rewriting of the configuration data in some areas.

2.3 Dynamic Module Library

We describe a modeling method of the systems that includes the dynamic generation and elimination of modules and interface at the system level. In the DRS, modules and an interface are generated or eliminated while the system is running. In order to express these behaviors naturally at the system level, modules, ports and channels have to be generated or eliminated dynamically. However, their generation and elimination cannot be performed after the simula-

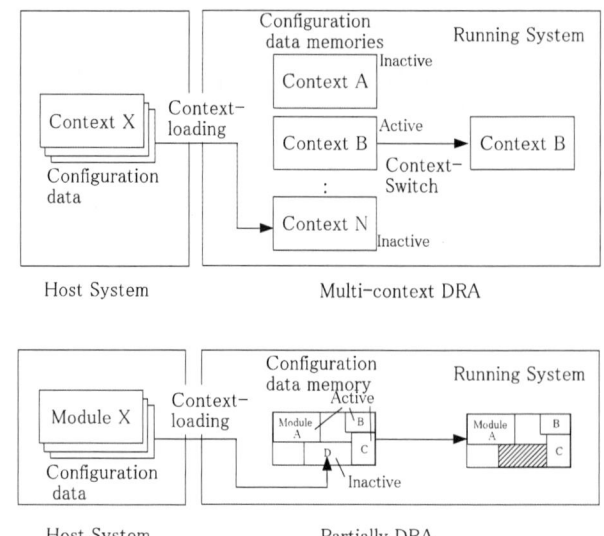

Figure 1. Outline of two types of DRAs

tion has started, even if SystemC 2.1.v1 is used. We have proposed a Dynamic Module Library(DML) [8] including dynamic modules, dynamic ports, and dynamic channels to solve these problems.

If the DML is used in modeling, the dynamic module class *DynamicModule*, which collects functions, such as *newmod()*, *new* and *delmod()*, for the dynamic module, port and channel is used rather than the *sc_module* class. In a dynamic module, the usual ports, such as *sc_port* and channels, such as *sc_fifo* cannot be used for their definition. In addition, the *sc_port* and *sc_fifo* have a binding method to connect with each other, but they do not have any method to dispatch. The dynamic port, such as *dc_port* and dynamic channel, such as *dc_fifo* are used in a dynamic module and have methods to connect and dispatch.

The DML has a channel pool, which is a set of static channels, and manages the connection and dispatch of the channel and port in the channel pool. The dynamic ports are implemented and managed in the same way as the dynamic channels.

The dynamic module, ports and channel are implemented by the specific hardware resources on the adopted DRA. In the case of Multi-context DRA, one or more dynamic modules may be mapped into the context data, and dynamic channel may be mapped into the store operation into the memory outside DRA and the reload operation from the outside memory.

If the dynamic modules were used in the design, the simulation time could be increased because of the overhead of the operations of the DML. Therefore, several acceleration methods[8] are built in the DML.

Some modeling methods for DRSs using SystemC have

been proposed. The method proposed in [12] has some constraints on the modeling, such that all modules must be on same level of hierarchy and instantiated in th same component. The target modeling class in [13] and [14] is more limited type of DRAs than our proposed method.

3 Proposed Processor Architecture

We describe a proposed DRP architecture. The proposed processor has some operational units, which are generated and eliminated dynamically. It consists of a processor core and an Extended Functional Unit(EFU), which contains some Dynamically Reconfigurable Operational Units(DROUs). The overview of the proposed processor is shown in Figure 2. In this paper, we deal with the modeling independently of a specific DRA. In the modeling, we use SystemC and the DML.

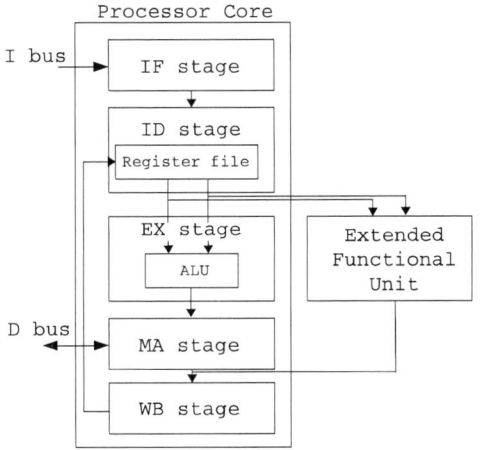

Figure 2. Block Diagram of the Proposed Processor

3.1 Processor Core

The proposed processor is a 32-bit pipelined processor that executes the subset of instructions of MIPS[15] in the 5-stage pipeline. The instruction set consists of about 50 instructions, such as integer/logical operations, Load/Store, and Jump/Branch, and each instruction is executed in order. The processor has an integer- and a floating-point register files, and has the data transfer instructions between each other.

In the execution of instructions, the proposed processor executes the Instruction Fetch(IF), the Instruction Decode(ID), the EXecution(EX), the Memory Access(MA) and the Write Back(WB) stages in order. In some instructions, the processor skips some stages. For example, in the case of integer arithmetic instructions, it skips the MA stage

and forwards the result to the ID stage or the EX stage. The processor has an integer ALU in the EX stage as the static operational unit, and executes the Load/Store instructions, Addition/Subtraction/logical instructions, and Multiplication/Division instructions at 5, 4 and about 35 clocks.

The instructions on the dynamic reconfiguration are the Dynamically Generation (DG) instructions of the DROU and the OPerational (OP) instructions using the DROU. The DROUs are generated in the EFU, execute the operation for the data received from the ID stage, and send the results to the WB stage.

A PRISC[16] is developed with the DRP closely connected with the RF, as in our proposed processor.

3.2 EFU and its Interface to the Processor Core

The EFU has two input and one output ports and contains several DROUs. The EFU contains the values of the Configuration Address (CA) and Block Address (BA) in the table in order to manage the status of the reconfiguration.

The DG and OP instructions are also 32-bit width. The DG instructions contain a 6-bit CA field and a 3-bit BA field. The value of the CA field denotes the type of DROUs, and the value of the BA field denotes the location of the dynamic reconfiguration blocks in the EFU.

We can describe more abstract specification of the construction of instructions and the type of operations using the parameters and abstract data types. In the proposed model, for the simplicity, we adopted the fixed construction of instructions and the fixed types of operations.

We assume that the number of clocks for the generation, elimination of the dynamic reconfiguration blocks and the execution of the operation is defined according to the types of DROUs and can be given as the parameter. The ID stage is stalled during the cycles specified by these parameters.

The OP instructions consist of 2 or 3 operands. A 3-bit address field, which is one of the operands, specifies the location of dynamic reconfiguration blocks. The operational results are sent to the WB stage.

The execution of these instructions is shown in Figure 3. In the case of the DG instructions, the ID stage sends a signal *Config_en*, which indicates the start of dynamic reconfiguration, and a signal *Config_adr*, which indicates the type of operational units to be dynamically reconfigured, to the EFU. The EFU spends the time *Creating_time* when the through DROU must be reconfigured dynamically. If the DROU has been configured and this DROU has to be eliminated for new unit, the EFU spends the necessary time *Deleting_time* and then issues the signal *Genend* to the ID stage.

In the case of the OP instructions, the ID stage sends source operands *FP_data* or *INT_data*, i.e., floating point data or integer data, respectively, a signal *Drou_op*, which is

93

978-1-4244-3039-0/08 $25.00 © 2008 IEEE

the type of operations, and and a signal *Drou_adr*, which is the location of dynamic reconfiguration blocks at the EFU. The execution time at the DROU is specified as the parameter *Running_time*. The EFU sends the execution results to the WB stage and sends a signal *Exend* to the ID stage. By receiving the signal *Exend*, the ID stage restarts the issue of instructions.

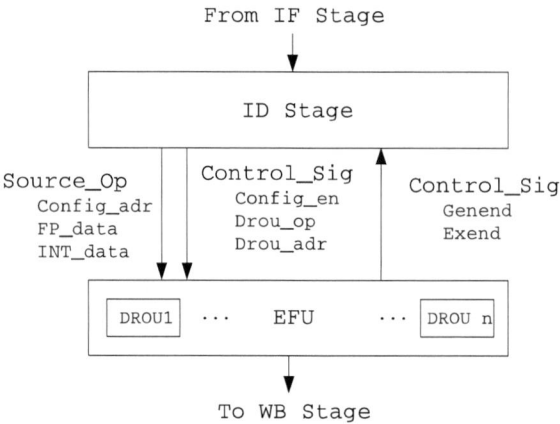

Figure 3. Interface of EFU and Processor Core

3.3 Modeling of the Proposed DRP

We describe the proposed processor model using the DML. We describe the IF, ID, MA, and WB stages as the static modules in the original description style of SystemC at the RT level. Then we omit these modeling. These are described at the RT level, Hereafter, we describe the description about the EX stage, which concerns the dynamically reconfiguration.

(a)Dynamically Reconfigurable Execution Unit

We modeled 4 types of modules; addition, subtraction, multiplication, and division for the floating point as the DROUs. The numbers of clocks for the generation and elimination of modules and operational clocks are given as parameters.

The part of the description of a multiplication DROU is shown in Figure 4. The *Dynamic_Mulf* class is defined by inheriting the *DynamicModule* class. This class has the dynamic ports *dc_in* and *dc_out* and user processes, such as *running*.

The *running* process executes the floating point multiplication using the values from the input ports *fsin* and *ftim* and sends the results through the output port *fout*. The behavior of the *running* process, which is executed through the internal variables *fs_signTmp*,*ft_signTmp* and etc., is described in the original description style of SystemC. *mulf_etime* is used as the parameter, which is the period of the operation, and

the *running* process outputs the result *sign1_tmp*, *exp5_tmp*, and *man5_tmp* through the port *fout* and waits for the next operation . The description of the multiplication process is omitted in Figure 4.

```
#include "DynamicModule.h"

class Dynamic_Mulf: public DynamicModule
{
public:
  dc_in<sc_uint<DWORD> > fsin,ftin;
  dc_out<sc_uint<DWORD> > fout;
  dc_in<bool> dc_en;

  SC_HAS_PROCESS(Dynamic_Mulf);
  Dynamic_Mulf(const char *name, int ctime,
               int dtime, sc_time_unit tunit):
    DynamicModule(name, ctime, dtime, tunit)
  {

  }
  ~Dynamic_Mulf(){};

  void running();
  void creating();
  void deleting();
};

void Dynamic_Mulf::running(){
    :
  while(true){
    while(dc_en.read()==false){
      if( dwait(dc_en.value_changed_event()) )return;
    }

    fs_signTmp = fsin.read()[31];
    ft_signTmp = ftin.read()[31];
    fs_expTmp =  fsin.read().range(30,23);
    ft_expTmp =  ftin.read().range(30,23);
    fs_manTmp =  fsin.read().range(22,0);
    ft_manTmp =  ftin.read().range(22,0);

      :

    dwait(mulf_etime, SC_NS);
    fout.write( (sign1_tmp, exp5_tmp, man5_tmp) );
    dwait(mulf_waits, SC_NS);
  }
}
```

Figure 4. Description of DROU

(b)Control of Dynamic Reconfiguration in EFU

We describe the model of dynamic reconfiguration control in the EFU. Part of the description of the model is shown in Figure 5. This control unit has 4 states;*Idle*, *Creating*, *Running*, and *Deleting*. **(1)**The control unit transits the *Idle* state by the *Reset* signal. **(2)**It transits the *Creating* state by the *Config_en* signal from the ID stage. In this state, it stalls for the *Creating_time*. **(3)**Then it sends the *Genend* signal to the ID stage and transits the *Running* state. **(4)** and**(5)**In the *Running* state, the OP instruction is executed. It stalls for the *Running_time*, sends the executed results to the WB stage, sends the *Exend* signal to the ID stage, and waits for the next instruction issue. **(6)**By the arrival of the *Config_en* signal, it transits the *Deleting* state and stalls for the *Delet-*

ing_time. **(7)**To reconfigure a new DROU, it transits the *Creating* state.

Creating_time and *Deleting_time* are the parameters, which are defined according to the scale of module, adopted DRA, and etc.

The transitions in **(6)** and **(7)** are the behavior in the case that the DROU has been configured and this operational unit should be eliminated for the new unit.

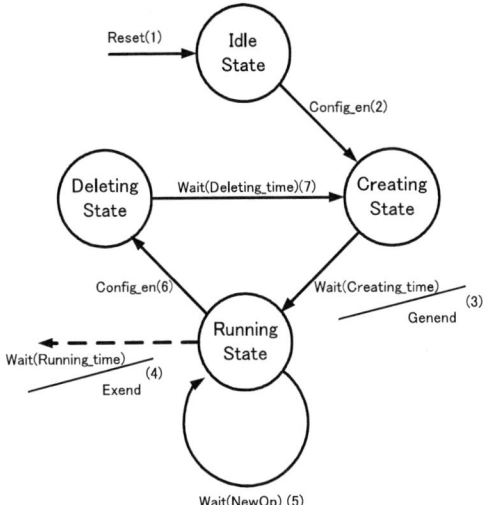

Figure 5. State Diagram in EFU

We describe the behavior for the dynamic reconfiguration. This description, which is the part of step **(6)** reported above, is shown in Figure 6. The behavior of the reconfiguration for a new DROU *Dynamic_Addf* at the *unit[i]*, which denotes the specified block in the EFU by the issued instruction is described. The controller executes following procedures.

(1) With the method *detach*, the channels *dina*, *dinb*, and *dout* and the ports *fsin*, *stim*, and *fout* are dispatched, respectively. **(2)**With the method *delmod*, the module is eliminated, and the time to eliminate the module is passed. **(3)**With the method *new*, a new DROU is generated. **(4)**With the method *bind*, the channels *dina*, *dinb*, and *dout* and new generated ports are connected, respectively. **(5)**The time *addf_ctimeto* to generate the module is passed.

Other channels *dina*, *dinb*, and *dout* are static and connected to the outer ports of the EFU.

4 Experimental Results

We show the experimental results of the proposed processor model, which is described using SystemC 2.1.v1 and the DML. The experimental environment is a Xeon 3.2GHz, Mem 4GB, Linux 2.6.9, GCC 3.4.6, and SystemC 2.1.v1 by OSCI.

```
// If unit[i] is used and reconfigured to Dynamic_Addf
        unit[i]->fsin.detach(dina);
        unit[i]->ftin.detach(dinb);
        unit[i]->fout.detach(dout);
        unit[i]->dc_en.detach(ex_en);
        unit[i]->delmod();
        wait(unit[i] -> dtime, SC_NS);
        unit[i] = new Dynamic_Addf("addf",
                    addf_ctime, addf_dtime, SC_NS);
        unit[i]->fsin.bind(dina);
        unit[i]->ftin.bind(dinb);
        unit[i]->fout.bind(dout);
        unit[i]->dc_en.bind(ex_en);
        unit[i]-> dtime = addf_dtime;
        wait(addf_ctime, SC_NS);
    }
                              :

//Connection Input Port and Channel
void Dynamic_Control::data_in() {
  while(true) {
    wait();
    dina = in0.read();
    dinb = in1.read();
  }
}

//Connection Channel and Output Port
void Dynamic_Control::data_out(){
  while(true) {
    wait();
    if(exend_tmp==true){
      out.write(dout.read());
    }
  }
}
```

Figure 6. Description of the Control for DROUs

The Mux model, which is modeled in the original description style of SystemC, is prepared as a comparison with the proposed processor using the DML. In the Mux model, the generation and elimination of a module are described using some multiplexers and de-multiplexers.

The lengths of the codes after eliminating comments, blank lines, and macros for C++ with the preprocessor of *gcc* are shown in Table 1.

Table 1. Code lengths of the models

	S Modules	D Modules	Control
Mux model	8248	787	507
Proposed method	8248	751	588

In Table 1, "S Modules" denote the static modules of data path and control units and the common description in the proposed model and the Mux model. "D Modules" denote the descriptions of DROUs in the proposed model and the descriptions of static DROUs, multiplexers, and de-multiplexers in the Mux model. "Control" denotes the description of the control of the dynamic reconfiguration. In the proposed model, this includes the descriptions of the generation and elimination of modules, ports, and channels and the connection and dispatch of ports and channels.

In the "D Modules," the description of the Mux model is slightly longer, and, in the "Control," that of the proposed model is longer. The difference in the total descriptions is slight.

In the simulation, the times of 1,000 executions of DG and OP instructions are measured. As a result, the time of the Mul model is 6.13 sec., that of the proposed method without the acceleration method is 6.90 sec., and that of the proposed method with acceleration methods is 6.72 sec. The Mux model is the fastest, although the difference among them is slight. In the case of a low frequency of reconfiguration, faster simulation of the proposed method is possible[8].

The memory usage of the DML depends on the number of dynamic module types, and is fixed during the simulation. In the experiment, it is dominated mainly by the user processes of the static modules, especially in the ID stage.

5 Conclusion

We described a DRP model which can dynamically generate and eliminate the operational units using the DML. We can easily and naturally model the generation and elimination of the DROUs in the proposed processor.

The proposed model can be described as having a similar size as that of the Mux model, which uses multiplexers and de-multiplexers for switching the function, and similar speed simulation is possible. We showed the variability of DML for the modeling of general-purpose DRPs.

In the proposed model, the processor core is modeled using SystemC at the near-RT level, and, on the contrary, the parts of the dynamic reconfiguration are modeled at the abstracted level. We will refine the description level of the total system from the system level to the RT level and apply our method for other DRP types. We research the system partitioning from system model into static devices and dynamically reconfigurable devices and the technology mapping of dynamic modules and channels to a concrete DRA.

Acknowledgment

This research has been supported by the Kayamori Foundation of Information Science Advancement.

References

[1] H. Nakano, T. Shindo, T. Kazami, and M. Motomura, "Development of Dynamically Reconfigurable Processor Lsi," NEC Technical Journal,Vol.56,No.4,pp.99-102, 2003.

[2] T.Toyo, H.Watanabe, and K.Shiba. "Implementation of Dynamically Reconfigurable Processor DAP/DNA-

2," Proc, of IEEE VLSI-TSA Int. Sympo. on VLSI Design, Automation & TEST, pp.321-322, Apr.2005.

[3] J.E.Carrillo and P.Chow, "The Effect of Reconfigurable Units in Superscalar Processors," Proc. of the 2001 ACM/SIGDA 9th Int. Sympo. on Field Programmable Gate Arrays, pp.141-150, Feb.2001.

[4] R. Bergamaschi and W. Lee, "Designing Systems-on-Chip Using Cores," Proc. of the 37th Conf. on Design Automation, pp.425-430, Jun.2000.

[5] Open SystemC Initiative, "SystemC 2.1 Language Reference Manual," http://www.systemc.org.

[6] SpecC Technology Open Consortium, http://www. specc.org.

[7] Accellera Organization, Inc., http://www. systemverilog.org.

[8] K. Asano, J. Kitamichi, and K. Kuroda, "Proposal of Dynamic Module Library for System Level Modeling and Simulation of Dynamically Reconfigurable Systems,"Proc. of 20th Int. Conf. on VLSI DESIGN(VLSI DESIGN 2007),pp. 373-378, 2007.

[9] M. Gokhale and P. S. Graham," Reconfigurable computing : Accelerating computation with field-programmable gate arrays," SPRINGER,2005.

[10] Xilinx,"Vertex-4 Configuration Guide,"http://direct. xilinx.com/bvdocs/userguides/j_ug071.pdf, UG071 (v1.3), 2005.

[11] T. Shiozawa, N. Imlig, K. Nagami, K. Oguri, A. Nagoya, and H. Nakada, "An Implementation of Longest Prefix Matching for IP Router on Plastic Cell Architecture," Proc. of FPL 2000, LNCS 1896, pp.805-809, 2000.

[12] A. Pelkonen, K. Masselos, and M Cupák, "System-Level Modeling of Dynamically Reconfigurable Hardware with SystemC,"Proc. of Int. Paralles and Distributed Processing Symp.(IPDPS'03),pp.174-181,2003.

[13] A. V. de Brito, E. U. K. Melcher, and W. Rosas, "An open-source tool for simulation of partially reconfigurable systems using SystemC,"Proc. of 2006 Emerging VLSI Technologies and Archtectures(ISVLSI'06), pp.434-435,2006.

[14] P. A. Hartmann, A. Schallenberg, F. Oppenhaimer, and W. Nebel, "OSSR+R:Simulation and Synthesis of Self-adaptive Systems,"Proc. of Field Programmable Logic and Applications(FPL2006),pp.177-182, 2006.

[15] G. Kane, "MIPS RISC Architecture-R2000/R3000-," Prentice-Hall, 1988.

[16] R.Razdan and M.Smith, "A High-Performance Microarchitecture with Hardware-Programmable Functional Units," Proc. of 27th the Annual Int. Sympo. on Microarchitecture,pp.172-180, Nov.1994.

21st International Conference on VLSI Design

A Scalable and Reconfigurable Coprocessor for Image Composition

Jalaj Jain
LSI Research and Development Pune Pvt. Ltd., Pune, India
jalajiitm@yahoo.com

Abstract

Image composition is an important post processing step in graphics sub system, video sub system and emerging MPEG-4 audio-visual standard. Image composition is achieved by rendering image elements independently with each element has an associated converge information "Alpha". Moving from one application to another i.e. graphics to video or vice versa, hardware architecture for image composition has to change accordingly. Therefore, in this paper, we propose scalable and reconfigurable coprocessor for image composition. We also calculate the operating clock frequency, required system data bus width and number of planes that can be processed in real time for video, graphics and MPEG-4 applications. Verilog implementation and synthesis for 90nm process shows an estimate of 400MHz achievable clock frequency and 90k gates which results in 0.25 mm² silicon area for composition of 3 high definition planes. Simulation model shows that proposed coprocessor can compose 15 high definition planes of size 1920×1080 in real time for 64 bit data transfer on system bus.

1. Introduction

Cost and time-to-market are important factors in consumer electronics product development cycle. However with increasing number of consumer products and emerging complex multimedia standards, it is becoming difficult to target time-to-market factor with reduced development cost. Video sub system and graphics sub system are the mandatory modules in multimedia system on chip (SOC). Video sub system is mainly composed of memory controller, transport/program stream demultiplexer, image composition unit and MPEG-2/4 decoder. While main components of graphics sub system are 3D/2D graphics accelerator (raster-engine, texture engine and pixel engine) and image composition unit. In MPEG-4 standard [1], which is object based approach for description and coding for multimedia contents,

audiovisual scenes are decomposed in multiple audio and video objects. At the decoder end, output frame is reconstructed by composition of multiple video object planes (VOP). Hardware accelerators for existing standards like H.263 [2] or MPEG-2[3] have built in support for video/audio decoding and display, except for the compositing stage which is very specific to MPEG-4. Therefore, image composition unit has different requirement for each application. So, in this paper, we propose scalable coprocessor that can be used as a hardware accelerator in multimedia SOC and MPEG-4, as shown in figure 1.

The organization of the paper is as follows. Section 2 introduces image composition arithmetic equations. Proposed scalable coprocessor architecture is detailed in the section 3. Performance analysis and implementation results are discussed in section 4. This paper is then concluded in Section 5.

2. Compositing Arithmetic

The basic idea for compositing operations is to compute contributions of each input plane at each pixel. For this purpose, a mixing factor "Alpha" is required at each pixel.

Let's assume 2 image planes A and B. One of these planes can be used as background plane whereas other plane as a foreground plane. Each pixel of plane A and plane B is represented by four elements vector $PA[i]$ and $PB[i]$, representing the alpha, red, green and blue components, where

$$0 \le i \le 3$$
$$0 \le PA[0] \le 1, 0 \le PB[0] \le 1$$
$$0 \le PA[1] \le 255, 0 \le PB[1] \le 255$$
$$0 \le PA[2] \le 255, 0 \le PB[2] \le 255$$
$$0 \le PA[3] \le 255, 0 \le PB[3] \le 255$$

The composite color and alpha can be computed on component basis by adding the color of plane A times it's alpha to the color of plane B times its alpha. It can be formulated as follows.

978-1-4244-3039-0/08 $25.00 © 2008 IEEE

$$C[w,h,i]_{i\in[1,3]} = (PA[w,h,i] \times \overline{ALPHA\_A(w,h)}) + \quad (1)$$
$$(PB[w,h,i] \times \overline{ALPHA\_B(w,h)})$$
$$\overline{ALPHA\_C(w,h)} = \overline{ALPHA\_A(w,h)} + \overline{ALPHA\_B(w,h)}$$

Where w is the width of plane A and B, h is the height of plane A and B, C is the resultant composite plane color value, $\overline{ALPHA\_A}$ alpha associated with plane A, $\overline{ALPHA\_B}$ alpha associated with plane B and $\overline{ALPHA\_C}$ alpha associated with plane C.

Alpha associated with plane A and plane B depends on the composition operators as listed in table I [4], [5]. For composition of N planes, we can generalize equation (1) as follows.

$$C[w,h,i] = (\overline{P\_1[w,h,i]} \times \overline{ALPHA\_1(w,h)}) + \quad (2)$$
$$\sum_{K=2}^{N} (\overline{P\_K[w,h,i]} \times \overline{ALPHA\_K(w,h)})$$
$$ALPHA\_C[w,h] = (\overline{ALPHA\_1(w,h)}) + \sum_{K=2}^{N} \overline{ALPHA\_K(w,h)})$$

Where $\overline{P\_K}$ is the associate color defined as $(PA[0] \times P\_K)$ and N is the number of planes.

Hardware implementation of equation (2) will result in inefficient VLSI architecture. Therefore, we propose to decompose the equation (2) in multiple 3 planes composition. 3 planes composition is again decomposed in 2, 2 planes composition. This decomposition process is formulated in equation (3) and equation (4).

$$C[w,h,i] = \sum_{j=0}^{j=N} \sum_{k=0}^{k=2} \overline{P\_(k+(j\times2))(w,h,i)} \times \overline{ALPHA\_(k+(j\times2))(w,h)} \quad (3)$$

$$\sum_{k=0}^{k=2} (\overline{P\_(k+(j\times2))(w,h,i)} \times \overline{ALPHA\_(k+(j\times2))(w,h)}) = \quad (4)$$
$$\sum_{k=0}^{k=1} (\overline{P\_(k+(j\times2))(w,h,i)} \times \overline{ALPHA\_(k+(j\times2))(w,h)}) +$$
$$(\overline{P\_(2+(j\times2))(w,h,i)} \times \overline{ALPHA\_(2+(j\times2))(w,h)})$$

Further, 2 planes composition is decomposed in following common basic math operations.
- Alpha to pixel multiplication
- Alpha to alpha multiplication
- Alpha inverse operation

3. Proposed Scalable and Reconfigurable Coprocessor

The block diagram of proposed coprocessor is shown in figure 2. It consists of direct memory access (DMA), command processing unit (CPU) and processing unit. Processing unit consists of control unit, plane reorder unit, common math operation unit, blending unit and mixing unit.

Coprocessor performs the data read and write operation from SDRAM by AXI bus system. Application software can initiate commands by writing command field in SDRAM buffers. Separate buffers are allocated for commands, source planes and destination plane. The command processing unit is responsible for managing, fetching, decode and execution of precompiled list of commands, and generates all necessary control signals for DMA and processing unit. Command is terminated by setting command completion flag. DMA fetches the source planes from SDRAM and write the composite plane to SDRAM. Therefore, the data operation is memory to memory. To pre fetch source planes, separate read memory buffers are available in DMA. Processing unit is based on equation (4). It is responsible for blending 3 source planes and mixing their alpha values as defined in the command code.

DMA read channels and processing unit are scalable in nature. The number of read channels and processing unit entirely depends on the end application. As an example, video sub system needs 3 read channels, one each for main video plane, sub video plane and graphics plane.

3.1 Command Description

Coprocessor command consists of following fields.
- First source plane address pointer
- Second source plane address pointer
- Third source plane address pointer
- Source plane z order
- Alpha selection for first, second and third source plane
- Destination plane address pointer
- Source plane height
- Source plane width
- Command code
- Command compilation flag

Command code is used to specify the one of the 11 possible, 2 planes blending operation as listed in table I. It has single bit flag indicating if 3 source planes are involved. If this flag is set to one, additional fields are used to specify the one of the 11 possible, 2 blend operations for 3rd source plane.

3.2 Processing Unit

Processing unit consists of plane reorder unit (figure 3), common math operation unit (figure 3), alpha blending unit (figure 4) and alpha mixing unit (figure

5). In the figure 3, 4 and 5, we assume top plane as a foreground plane and bottom plane as a background plane.

Planes reorder unit changed the z order of source planes as specified in command code and select between the plane A and plane B as a background plane and foreground plane. Common math unit pre calculates the all possible valid combinations for alpha and color. The possible valid combinations for alpha are as follows:

$$\alpha^{top}, 1 - \alpha^{top}, \alpha^{bottom} \text{ and } 1 - \alpha^{bottom}$$

Possible valid combination for pixel and alpha are as follows:

$$(P^{top} * \alpha^{top}), (P^{top} * (1 - \alpha^{top})), (P^{bottom} * \alpha^{bottom}) \text{ and } (P^{bottom} * (1 - \alpha^{bottom}))$$

In alpha blending and alpha mixing units, we calculate all the possible combinations of alpha values and color values for foreground and background plane. Then we select the 2 alpha and 2 color values, as specified in command code, and perform multiply-add operations. Table II and III lists the configuration signals (as defined in figure 4 and 5) for alpha blending unit and alpha mixing unit respectively. For an example, suppose we want to blend the 3 planes in following z order

A _top_ B _out_ C

So we decompose the above equation and configure the alpha blending unit as follows.

2 plane blending output = A _top_ B
Program SELA00 to 100
Program SELA01 to 011
Program SELC0 to 00
Program SELC1 to 10
3 plane blending output = (2 plane blending output) _out_ C
Program SELA00 to 101
Program SELA01 to 000
Program SELC0 to 00
Program SELC1 to 10

Same concept can be used for the blending multiple planes in any specified z order.

There are 3 parallel units, one for each of the color component. All 3 parallel units share the common math operation unit. Figure 6 shows, how 2 plane composition units can be used to composite the 3 planes.

3.3 Coprocessor Configuration for Multiple Plane Composition

To blend multiple planes in specified order, application software can issue multiple commands to coprocessor. Hence, in multiple pass, N planes can be blended. However this approach may not be suitable for high speed applications, for example, 3 D graphics application.

Since proposed coprocessor architecture is scalable in nature, we can reuse the processing unit and DMA data buffers to support composition of multiple planes as shown in figure 7.

4. Performance Analysis and Implementation Results

In this section, we discuss the real time performance and implementation results of proposed coprocessor for video, graphics and MPEG-4 applications. We assume ideal conditions on system bus and double buffer storage mechanism in DMA. Total number of clock cycles required to composite the final plane depends on the SDRAM to DMA read operation, alpha blending & mixing operation and DMA to SDRAM write operation.

4.1 Video Sub System

In video sub system, image composition is used for picture-in-picture (PIP) in which, main video plane, sub video plane and graphics plane are blended to create final plane to be displayed. Figure 8 shows the graph between total clock cycles requirement vs. bus data width for following parameters.

- Main video, sub video and graphics plane size: 1920×1080

- Pixel data size: 32 bit

- Frame rate: 25 frames/second

- Number of planes: 3

4.2 Graphics Sub System

In graphics sub system, there may be a need to composite the multiple planes. Multiple processing units can be used in parallel to reduce the required clock cycles. However beyond certain number of parallel processing units, there is no significant performance improvement as shown in figure 9. Simulation results are derived for following parameters.

- Graphics plane size: 1920×1080

- Pixel data size: 32 bit

- Frame rate: 25 frames/second

- Data width: 64 bit
- Number of planes: 15

4.3 MPEG-4

For MPEG-4, video object planes, that can be processed, increase linearly with bus data width for 300 MHz clock frequency as shown in figure 10. Simulation results are derived for following parameters.

- Video object plane size: 704 × 576
- Pixel data size: 32 bit
- Frame rate: 25 frames/second
- Clock frequency: 300 MHz
- Maximum number of VOP: 512

To summarize, for video sub system, 3 high definition planes can be composed in real time for 64 bit data transfer on system bus and 200 MHz operating clock frequency. For Graphics sub system, proposed coprocessor can process 15 high definitions planes for 64 bit data transfer on system bus and 450MHz operating clock frequency. For MPEG-4, composition of 115 video object planes of size 704 × 576 is achieved in real time for 64 bit system data bus width and 650MHz clock frequency.

4.4 Implementation Results

The proposed coprocessor has been described in verilog hardware description language and synthesized with magma flow for 90nm TSMC library at 400MHz. Each processing unit requires 12, 8 bit multiplier, 6 adder and 4 rounding/saturation units which results in approximately 75k gates with control circuitry. DMA buffers are implemented as a synchronous RAM of size 16 ×32, with 11k gates. Total gate count is 90k which results in 0.25 mm^2 and it is proportional to the number of processing units and DMA read buffers. For example, 5 parallel processing units and 5 sets of DMA read buffers will result in total area of 1.25 (0.25 × 5) mm^2.

5. Conclusion

The scalable and reconfigurable coprocessor architecture is presented which can be easily integrated in high definition multimedia SOC, thus significantly reducing the workload of the main decoding unit. Scalability aspect of coprocessor makes it an attractive choice and provides common platform for composition of high definition planes in different multimedia applications. Simulation model shows the real time

performance of proposed coprocessor for composition of high definition video and graphics planes. There is a need of low power architecture, which is still an open problem for future research. Further investigation is also needed to optimize the area by custom adders and multiplication units.

6. References

[1] ISO/IEC JTC/SC29/WG11, "MPEG-4 Video Verification Model V8.0", MPEG96/N1796 July 1997.

[2] ITU-T Draft Recommendation H.263, "Video Coding for Low Bit rate Communication".

[3] ISO/IEC 13818-2 "Generic Coding of Moving Pictures and Associated Audio", (MPEG-2), Part2, Video, November 1993.

[4] James Blinn, "Compositing, Part 1: Theory", IEEE Computer Graphics and Application Staff, Volume 14, Issue 5, September 1994, pp. 83-87.

[5] T.Porter and T. Duff, "Compositing, Digital Images", Computer Graphics (Proc. Siggraph), Volume 18, Issue 3, July 1984, pp. 253-259.

Table I
2 Plane composition operators

	Operation	$\overline{ALPHA\_A}$	$\overline{ALPHA\_B}$
1	Clear	0	0
2	A	1	0
3	B	0	1
4	A over B	1	$1-PA[0]$
5	B over A	$1 - B[0]$	1
6	A in B	$PB[0]$	0
7	B in A	0	$PA[0]$
8	A out B	$1- PB[0]$	0
9	B out A	0	$1 - PA[0]$
10	A top B	$PB[0]$	$1 - PA[0]$
11	B top A	$1- PB[0]$	$PA[0]$
12	A xor B	$1- PB[0]$	$1 - PA[0]$

Table II
Reconfiguration signals for alpha blending unit

SEL00/SEL01	A0/A1	SELC0/SELC1	P0/P1
000	0	00	P^{top}
001	1	01	$P^{top} \times \alpha^{top}$
010	α^{top}	10	P^{bottom}
011	$1 - \alpha^{top}$	11	$P^{bottom} \times \alpha^{bottom}$
100	α^{bottom}		
101	$1-\alpha^{bottom}$		

Table III
Reconfiguration signals for alpha mixing unit

SELA10 / SELA11	A0/1
0000	$(\alpha^{top} \times \alpha^{bottom})$
0001	$(1 - \alpha^{top}) \times (1 - \alpha^{bottom})$
0010	$(1 - \alpha^{top}) \times \alpha^{bottom}$
0011	$\alpha^{top} \times (1 - \alpha^{bottom})$
0100	α^{top}
0101	α^{bottom}
0110	$(1 - \alpha^{top})$
0111	$(1 - \alpha^{bottom})$
1000	0
1001	1

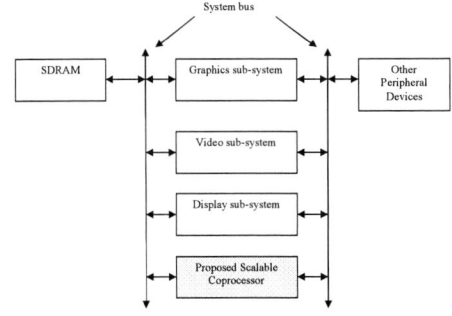

Figure 1: An example of system diagram for multimedia system on chip

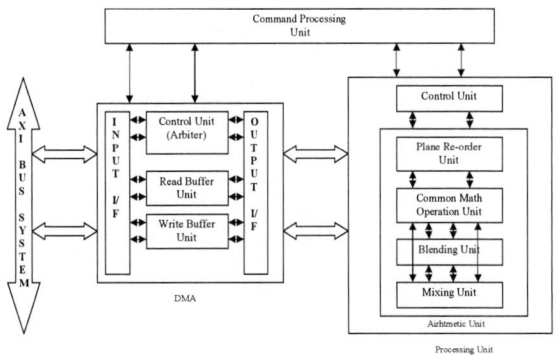

Figure 2: Coprocessor block diagram

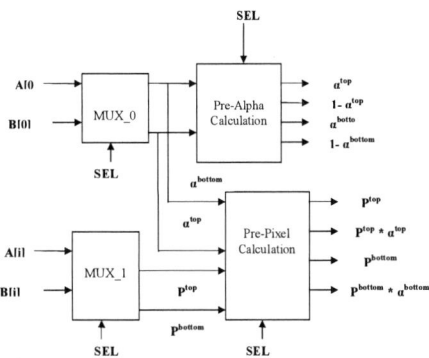

Figure 3: Plane reorder and common math operation unit

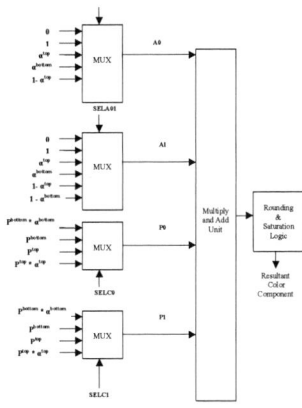

Figure 4: Alpha blending unit

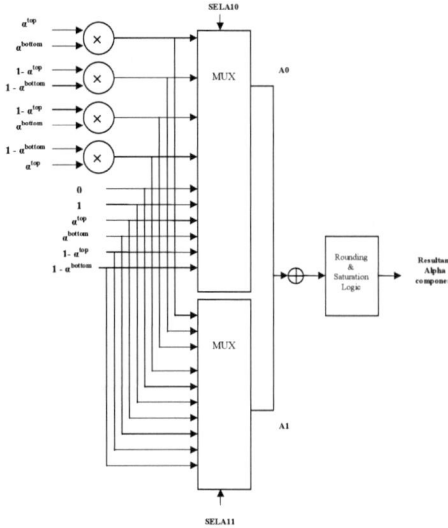

Figure 5: Alpha mixing unit

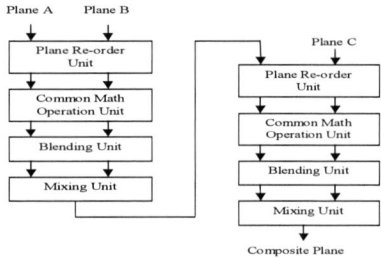

Figure 6: 3 Plane composition using two, 2 plane composition units

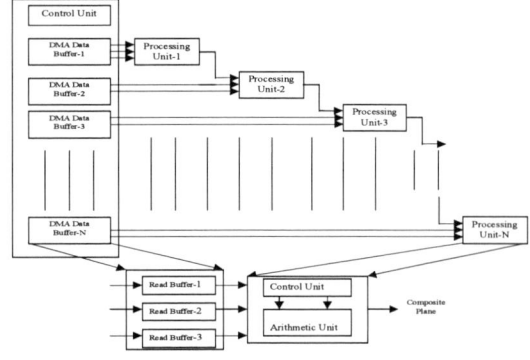

Figure 7: Multiple plane composition

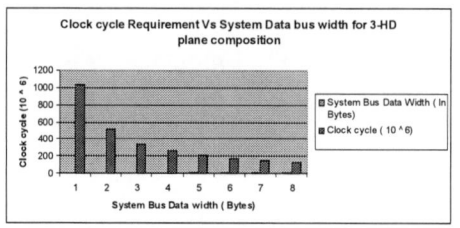

Figure 8: Simulation model results for 3 HD video planes composition

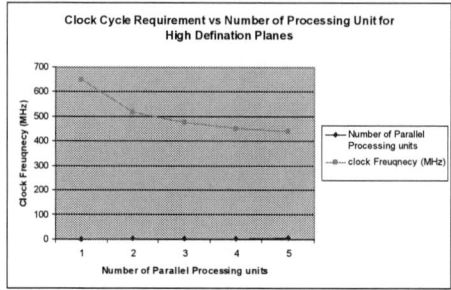

Figure 9: Simulation model results for high definition graphics plane composition

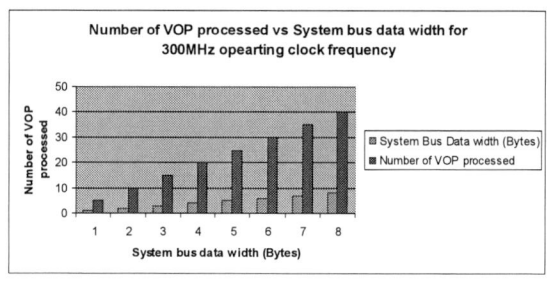

Figure 10: Simulation model results for MPEG-4 video object planes composition

21st International Conference on VLSI Design

Predictable Implementation of Real-Time Applications on Multiprocessor Systems-on-Chip

Alexandru Andrei, Petru Eles, Zebo Peng, Jakob Rosen

Deptartment of Computer and Information Science
Linköping University, Linköping, S-58183, Sweden

Abstract—Worst-case execution time (WCET) analysis and, in general, the predictability of real-time applications implemented on multiprocessor systems has been addressed only in very restrictive and particular contexts. One important aspect that makes the analysis difficult is the estimation of the system's communication behavior. The traffic on the bus does not solely originate from data transfers due to data dependencies between tasks, but is also affected by memory transfers as result of cache misses. As opposed to the analysis performed for a single processor system, where the cache miss penalty is constant, in a multiprocessor system each cache miss has a variable penalty, depending on the bus contention. This affects the tasks' WCET which, however, is needed in order to perform system scheduling. At the same time, the WCET depends on the system schedule due to the bus interference. In this context, we propose, for the first time, an approach to worst-case execution time analysis and system scheduling for real-time applications implemented on multiprocessor SoC architectures.

I. INTRODUCTION AND RELATED WORK

We are facing an unprecedented development in the complexity and sophistication of embedded systems. Emerging technologies provide the potential to produce complex multiprocessor architectures implemented on a single chip [24]. Embedded applications, running on such highly parallel architectures are becoming more and more sophisticated and, at the same time, will be used very often in applications for which predictability is very important. Classically, these are safety critical applications such as automotive, medical or avionics systems. However, recently, more and more applications in the multimedia and telecommunications area have to provide guaranteed quality of service and, thus, require a high degree of worst-case predictability [5]. Such applications impose strict constraints not only in terms of their logical functionality but also with concern to timing. The objective of this paper is to address, at the system-level, the specific issue of predictability for embedded systems implemented on current and future multiprocessor architectures. Providing predictability, along the dimension of time, should be based on scheduling analysis which, itself, assumes as an input the worst case execution times (WCETs) of individual tasks [10], [14]. While WCET analysis has been an investigation topic for already a long time, the basic driving force of this research has been, and still is, to improve the tightness of the analysis and to incorporate more and more features of modern processor architectures. However, one of the basic assumptions of this research is that WCETs are determined for each task in isolation and then,

in a separate step, task scheduling analysis takes the global view of the system [22]. This approach is valid as long as the applications are implemented either on single processor systems or on very particular multiprocessor architectures in which, for example, each processor has a dedicated, private access to an exclusively private memory.

The main problems that researchers have tried to solve are (1) the identification of the possible execution sequences inside a task and (2) the characterization of the time needed to execute each individual action [15]. With advanced processor architectures, effects due to caches, pipelines, and branch prediction have to be considered in order to determine the execution time of individual actions. There have been attempts to model both problems as a single ILP formulation [11]. Other approaches combine abstract interpretation for cache and pipeline analysis with ILP formulations for path analysis [21], or even integrate simulation into the WCET analysis flow [12], [23]. There have been attempts to build modular WCET estimation frameworks where the particular subproblems are handled separately [4], while other approaches advocate a more integrated view [7]. More recently, preemption related cache effects have also been taken into consideration [16], [20].

The basic assumption in all this research is that, for WCET analysis, tasks can be considered in isolation from each other and no effects produced by dependencies or resource sharing have to be taken into consideration (with the very particular exception of some research results regarding cache effects due to task preemption on monoprocessors, [20]). This makes all the available results inapplicable to modern multiprocessor systems in which, for example, due to the shared access to sophisticated memory architectures, the individual WCETs of tasks are depending on the global system schedule. This is pointed out as one major unsolved issue in [22] where the current state of the art and future trends in timing predictability are reviewed. The only solution for the above mentioned shortcomings is to take out WCET analysis from its isolation and place it into the context of system level analysis and optimization. In this paper we present the first approach in this direction.

A framework for system level task mapping and scheduling for a similar type of platforms has been presented in [2]. In order to avoid the problems related to the bus contention, they use a so called additive bus model. This assumes that task execution times will be stretched only marginally as an effect of bus contention for memory accesses. Consequently,

978-1-4244-3039-0/08 $25.00 © 2008 IEEE

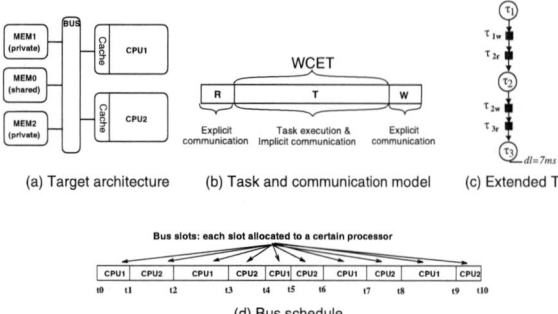

(a) Target architecture (b) Task and communication model (c) Extended TG

(d) Bus schedule

Fig. 1. System and task models

they simply neglect the effect of bus contention on task execution times. The experiments performed by the authors show that such a model can be applied with relatively good approximations if the bus load is kept below 50%. There are two severe problems with such an approach: (1) In order for the additive model to be applicable, the bus utilization has to be kept low. (2) Even in the case of such a low bus utilization, no guarantees of any kind regarding worst-case behavior can be provided.

The remainder of the paper is organized as follows. Preliminaries regarding the system and architecture model are given in Section II. Section III outlines the problem with a motivational example and is followed in Section IV by the description of our proposed solution. In Section V we present the approach used for the worst-case execution time analysis. The bus access policy is presented in Section VI. Experimental results are given in Section VII.

II. SYSTEM AND APPLICATION MODEL

In this paper we consider multiprocessor system-on-chip architectures with a shared communication infrastructure that connects processing elements to the memories. The processors are equipped with instruction and data caches. Every processor is connected via the bus to a private memory. All accesses from a certain processor to its private memory are cached. A shared memory is used for inter-processor communication. The accesses to the shared memory are not cached. This is a typical, generic, setting for new generation multiprocessors on chip, [9]. The shared communication infrastructure is used both for private memory accesses by the individual processors (if the processors are cached, these accesses are performed only in the case of cache misses) and for interprocessor communication (via the shared memory). An example architecture is shown in Fig. 1(a).

The functionality of the software applications is captured by task graphs, $G(\Pi, \Gamma)$. Nodes $\tau \in \Pi$ in these directed acyclic graphs represent computational tasks, while edges $\gamma \in \Gamma$ indicate data dependencies between these tasks (explicit communications). The computational tasks are annotated with deadlines dl_i that have to be met at run-time. Before the execution of a data dependent task can begin, the input data must be available. Tasks mapped to the same processor are communicating through the cached private

memory. These communications are handled similarly to the memory accesses during task execution. The communication between tasks mapped to different processors is done via the shared memory. Consequently, a message exchanged via the shared memory assumes two explicit communications: one for writing into the shared memory (by the sending task) and the other for reading from the memory (by the receiving task). Explicit communication is modeled in the task graph as two communication tasks, executed by the sending and the receiving processor, respectively as, for example, τ_{1w} and τ_{2r} in Fig. 1(c). During the execution of a task, all the instructions and data are stored in the corresponding private memory, so there will not be any shared memory accesses. The reads and writes to and from the private memories are cached. Whenever a cache miss occurs, the data has to be fetched from the memory and a cache line replaced. This results in memory accesses via the bus during the execution of the tasks. We will refer to these as implicit communication. This task model is illustrated in Fig. 1(b). Previous approaches that are proposing system level scheduling and optimization techniques for real-time applications only consider the explicit communication, ignoring the bus traffic due to the implicit communication [18]. We will show that this leads to incorrect results in the context of multiprocessor systems.

In order to obtain a predictable system, which also assumes a predictable bus access, we consider a TDMA-based bus sharing policy. Such a policy can be used efficiently with the contemporary SoC buses, especially if QoS guarantees are required, [17], [13], [5].

We introduce in the following the concept of bus schedule. The bus schedule contains slots of a certain size, each with a start time, that are allocated to a processor, as shown in Fig. 1(d). The bus schedule is stored as a table in a memory that is directly connected to the bus arbiter. It is defined over one application period, after which it is periodically repeated. An access from the arbiter to its local memory does not generate traffic on the system bus. The bus schedule is given as input to the WCET analysis algorithm. When the application is running, the bus arbiter is enforcing the bus schedule, such that when a processor sends a bus request during a slot that belongs to another processor, the arbiter will keep it waiting until the start of the next slot that was assigned to it.

III. MOTIVATIONAL EXAMPLE

Let us assume a multiprocessor system, consisting of two processors CPU_1 and CPU_2, connected via a bus. Task τ_1 runs on CPU_1 and has a deadline of 60 time units. Task τ_2 runs on CPU_2. When τ_2 finishes, it updates the shared memory during the explicit communication E_1. We have illustrated this example system in Fig. 2(a). During the execution of the tasks τ_1 and τ_2, some of the memory accesses result in cache misses and consequently the corresponding caches must be refilled. The time interval spent due to these accesses is indicated in Fig. 2 as M_1, M_3, M_5 for τ_1 and M_2, M_4 for τ_2. The memory accesses are executed by the implicit

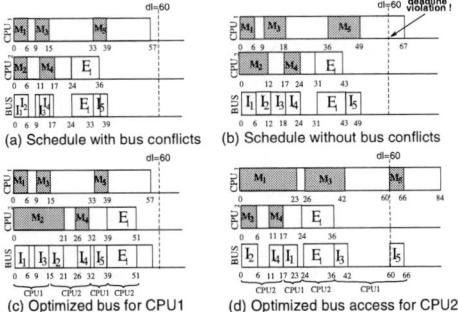

Fig. 2. Schedule with and without bus conflicts

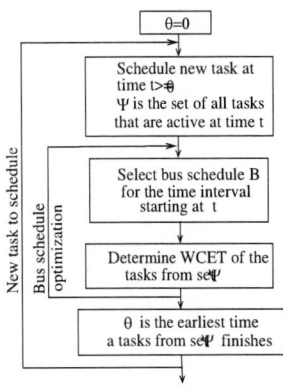

Fig. 3. Overall Approach

bus transfers I_1, I_2, I_3, I_4 and I_5. If we analyze the two tasks using a classical worst-case analysis algorithm, we conclude that task τ_1 will finish at time 57 and τ_2 at 24. For this example, we have assumed that the cache miss penalty is 6 time units. CPU_2 is controlling the shared memory update carried out by the explicit message E_1 via the bus after the end of task τ_2. This scenario is illustrated in Fig. 2(a).

A closer look at the execution pattern of the tasks reveals the fact that the cache misses may overlap in time. For example, the cache miss I_1 and I_2 are both happening at time 0 (when the tasks start, the cache is empty). Similar conflicts can occur between implicit and explicit communications (for example I_5 and E_1). Since the bus cannot be accessed concurrently, a bus arbiter will allow the processors to refill the cache in a certain order. An example of a possible outcome is depicted in Fig. 2(b). The bus arbiter allows first the cache miss I_1, so after 6 time units needed to handle the miss, task τ_1 can continue its execution. After serving I_1, the arbiter grants the bus to CPU_2 in order to serve the miss I_2. Once the bus is granted, it takes 6 time units to refill the cache. However, CPU_2 was waiting 6 time units to get access to the bus. Thus, handling the cache miss I_2 took 12 time units, instead of 6. Another miss I_3 occurs on CPU_1 at time 9. The bus is busy transferring I_2 until time 12. So CPU_1 will be waiting 3 time units until it is granted the bus. Consequently, in order to refill the cache as a result of the miss I_3, task τ_1 is delayed 9 time units instead of 6, until time 18. At time 17, the task τ_2 has a cache miss I_4 and CPU_2 waits 1 time unit until time 18 when it is granted the bus. Compared with the execution time from Fig. 2(a), where an ideal, constant, cache miss penalty is assumed, due to the resource conflicts, task τ_2 finishes at time 31, instead of time 24. Upon its end, τ_2 starts immediately sending the explicit communication message E_1, since the bus is free at that time. In the meantime, τ_1 is executing on CPU_1 and has a cache miss, I_5 at time 36. The bus is granted to CPU_1 only at time 43, after E_1 was sent, so τ_1 can continue to execute at time 49 and finishes its execution at time 67 causing a deadline violation. The example in Fig. 2(b) shows that using worst-case execution time analysis algorithms that consider tasks in isolation and ignore system level conflicts leads to incorrect results.

In Fig. 2(b) we have assumed that the bus is arbitrated using a simple First Come First Served (FCFS) policy. In order to achieve worst-case predictability, however, we use a TDMA bus scheduling approach, as outlined in Section II. Let us assume the bus schedule as in Fig. 2(c). According to this bus schedule, the processor CPU_1 is granted the bus at time 0 for an interval of 15 time units and at time 32 for 7 time units. Thus, the bus is available to task τ_1 for each of its cache misses (M_1, M_3, M_5) at times 0, 9 and 33. Since these are the arrival times of the cache misses the execution of τ_1 is not delayed and finishes at time 57, before its deadline. Task τ_2 is granted the bus for its cache misses at times 15 and 26 and finishes at time 31, resulting in a longer execution time than in the ideal case (time 24). The explicit communication message E_1 is started at time 39 and completes at time 51.

While the bus schedule in Fig. 2(c) is optimized according to the requirements from task τ_1, the one in Fig. 2(d) eliminates all bus access delays for task τ_2. According to this bus schedule, while τ_2 will finish earlier than in Fig. 2(c), task τ_1 will finish at time 84 and, thus, miss its deadline.

The examples presented in this section demonstrate two issues:

1) Ignoring bus conflicts due to implicit communication can lead to gross subestimations of WCETs and, implicitly, to incorrect schedules.

2) The organization of the bus schedule has a great impact on the WCET of tasks. A good bus schedule does not necessarily minimize the WCET of a certain task, but has to be fixed considering also the global system deadlines.

IV. ANALYSIS, SCHEDULING AND OPTIMIZATION FLOW

We consider as input the application task graph capturing the dependencies between the tasks and the target hardware platform. Each task has associated the corresponding code and potentially a deadline that has to be met at runtime. As a first stage, mapping of the tasks to processors is performed. Traditionally, after the mapping is done, the WCET of the tasks can be determined and is considered to be constant and known. However, as mentioned before, the basic problem is that memory access times are, in

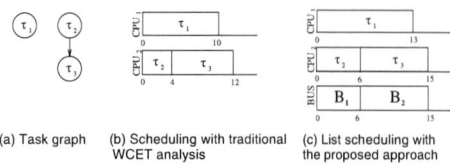

(a) Task graph   (b) Scheduling with traditional WCET analysis   (c) List scheduling with the proposed approach

Fig. 4. System level scheduling with WCET analysis

principle, unpredictable in the context of the potential bus conflicts between the processors that run in parallel. These conflicts (and implicitly the WCETs), however, depend on the global system schedule. System scheduling, on the other side, traditionally assumes that WCETs of the tasks are fixed and given as input. This cyclic dependency is not just a technical detail or inconvenience, but a fundamental issue with large implications and which invalidates one of the basic assumptions that support current state of the art. In order to solve this issue, we propose a strategy that is based on the following basic decisions:

1) We consider a TDMA-based bus access policy as outlined in Section II. The actual bus access schedule is determined at design time and will be enforced during the execution of the application.

2) The bus access schedule is taken into consideration at the WCET estimation. WCET estimation, as well as the determination of the bus access schedule are integrated with the system level scheduling process (Fig. 3).

We will present our overall strategy using a simple example. It consists of three tasks mapped on two processors, as in Fig. 4.

The system level static cyclic scheduling process is based on a list scheduling technique [8]. List scheduling heuristics are based on priority lists from which tasks are extracted in order to be scheduled at certain moments. A task is placed in the ready list if all its predecessors have been already scheduled. All ready tasks from the list are investigated, and that task τ_i is selected for placement in the schedule which has the highest priority. We use the modified partial critical path priority function presented in [14]. The process continues until the ready list is empty.

Let us assume that, using traditional WCET estimation (considering a given constant time for main memory access, ignoring bus conflicts), the task execution times are 10, 4, and 8 for τ_1, τ_2, and τ_3, respectively. Classical list scheduling would generate the schedule in Fig. 4(b), and conclude that a deadline of 12 can be satisfied.

In our approach, the list scheduler will choose tasks τ_1 and τ_2 to be scheduled on the two processors at time 0. However, the WCET of the two tasks is not yet known, so their worst case termination time cannot be determined. In order to calculate the WCET of the tasks, a bus configuration has to be decided on. This configuration should, preferably, be fixed so that it is favorable from the point of view of WCETs of the currently running tasks (τ_1 and τ_2, in our case). Given a certain bus configuration, our WCET-analysis will determine the WCET for τ_1 and τ_2. Inside an optimization loop, several alternative bus configurations are considered. The goal is to

reduce the WCET of τ_1 and τ_2, with an additional weight on reducing the WCET of that task that is assumed to be on the critical path (in our case τ_2).

Let us assume that B1 is the selected bus configuration and the WCETs are 12 for τ_1 and 6 for τ_2. At this moment the following is already decided: τ_1 and τ_2 are scheduled at time 0, τ_2 is finishing, in the worst case, at time 6, and the bus configuration B1 is used in the time interval between 0 and 6. Since τ_2 is finishing at time 6, in the worst case, the list scheduler will schedule task τ_3 at time 6. Now, τ_3 and τ_1 are scheduled in parallel. Given a certain bus configuration B, our WCET analysis tool will determine the WCETs for τ_1 and τ_3. For this, it will be considered that τ_3 is executing under the configuration B, and τ_1 under configuration B1 for the time interval 0 to 6, and B for the rest. Again, an optimization is performed in order to find an efficient bus configuration for the time interval beyond 6. Let us assume that the bus configuration B2 has been selected and the WCETs are 9 for τ_3 and 13 for τ_1. The final schedule is illustrated in figure 4.

The overall approach is illustrated in Fig. 3. At each iteration, the set ψ of tasks that are active at the current time t, is considered. In an inner optimization loop a bus configuration B is fixed. For each candidate configuration the WCET of the tasks in the set ψ is determined. During the WCET estimation process, the bus configurations determined during the previous iterations are considered for the time intervals before t, and the new configuration alternative B for the time interval after t. Once a bus configuration B is decided on, θ is the earliest time a task in the set ψ terminates. The configuration B is fixed for the time interval (t, θ), and the process continues from time θ, with the next iteration.

In the above discussion, we have not addressed the explicit communication of messages on the bus, to and from the shared memory. As shown in Section II, a message exchanged via the shared memory assumes two explicit communications: one for writing into the shared memory (by the sending task) and the other for reading from the memory (by the receiving task). Explicit communication is modeled in the task graph as two communication tasks, executed by the sending and the receiving processor, respectively (Fig. 1 in section II). A straightforward way to handle these communications would be to schedule each as one compact transfer over the bus. This, however, would be extremely harmful for the overall performance, since it would block, for a relatively long time interval, all memory access for cache misses from active processes. Therefore, the communication tasks are considered, during scheduling, similar to the ordinary tasks, but with the particular feature that they are continuously requesting for bus access (they behave like a hypothetical task that continuously generates successive cache misses such that the total amount of memory requests is equal to the worst case message length). Such a task is considered together with the other currently active tasks in the set Ψ. Our algorithm will generate a bus configuration and will schedule the communications such that it efficiently accommodates both the explicit message communication as

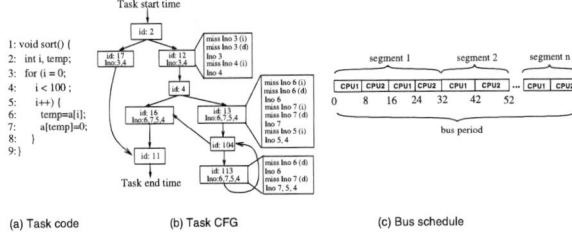

(a) Task code (b) Task CFG (c) Bus schedule

Fig. 5.   Example task WCET calculation

well as the memory accesses issued by the active tasks.

It is important to mention that the approach proposed in this paper guarantees that the worst-case bounds derived by our analysis are correct even when the tasks execute less than their worst-case. In [1] we have formally demonstrated this. The intuition behind the demonstration is the following:

1) Instruction sequences terminated in shorter time than predicted by the worst-case analysis cannot produce violations of the WCET.

2) Cache misses that occur earlier than predicted in the worst-case will, possibly, be served by an earlier bus slot than predicted, but never by a later one than considered during the WCET analysis.

3) A memory access that results in a hit, although predicted as a miss during the worst-case analysis, will not produce a WCET violation.

4) An earlier bus request issued by a processor does not affect any other processor, due to the fact that the bus slots are assigned exclusively to processors.

In the following sections we will address two aspects: the WCET estimation technique and bus access policy.

V. WCET ANALYSIS

We will present the algorithm used for the computation of the worst-case execution time of a task, given a start time and a bus schedule. Our approach builds on techniques developed for "traditional" WCET analysis. Consequently, it can be adapted on top of any WCET analysis approach that handles prediction of cache misses. Our technique is also orthogonal to the issue of cache associativity supported by this cache miss prediction technique. The current implementation is built on top of the approach described in [23], [19] that supports set associative and direct mapping.

In a first step, the control flow graph (CFG) is extracted from the code of the task. The nodes in the CFG represent basic blocks (consecutive lines of code without branches) or control nodes (capturing conditional instructions or loops). The edges capture the program flow. In Fig. 5(a) and (b), we have depicted an example task containing a for loop and the corresponding CFG, extracted from this task. For the nodes associated to basic blocks we have depicted the code line numbers. For example, node 12 (id:12) captures the execution of lines 3 ($i = 0$) and 4 ($i < 100$). A possible execution path, with the for loop iteration executed twice, is given by the node sequence 2, 12, 4 and 13, 104, 113, 104, 16, 11. Please note that the for loop was automatically

unrolled once when the CFG was extracted from the code (nodes 13 and 113 correspond to the same basic block representing an iteration of the for loop). This is useful when performing the instruction cache analysis. Intuitively, when executing a loop, at the first iteration all the instruction accesses result in cache misses. However, during the next iterations there is a chance to find the instructions in the cache.

We have depicted in Fig. 5(b) the resulting misses obtained after performing instruction (marked with an "i") and data (marked with an "d") cache analysis. For example, let us examine the nodes 13 and 113 from the CFG. In node 13, we obtain instruction cache misses for the lines 6, 7 and 5, while in the node 113 there is no instruction cache miss. In order to study at a larger scale the interaction between the basic blocks, data flow analysis is used. This propagates between consecutive nodes from the CFG the addresses that are always in the cache, no matter which execution path is taken. For example, the address of the instruction from line 4 is propagated from the node 12 to the nodes 13, 16 and 113.

Let us consider now the data accesses. While the instruction addresses are always known, this is not the case with the data [19], [16]. This, for example, is the case with an array that is accessed using an index variable whose value is data dependent, as in Fig.5(a), on line 7. Using data dependency analysis performed on the abstract syntax tree extracted from the code of the task, all the data accesses are classified as predictable or unpredictable [19]. For example, in Fig.5(a) the only unpredictable data memory access is in line 7. The rest of the accesses are predictable. All the unpredictable memory accesses are classified as cache misses. Furthermore, they have a hidden impact on the state of the data cache. A miss resulted from an unpredictable access replaces an unknown cache line. One of the following predictable memory accesses that would be considered as hit otherwise, might result in a miss due to the unknown line replacement. Similar to the instruction cache, dataflow analysis is used for propagating the addresses that will be in the cache, no matter which program path is executed.

Until this point, we have performed the same steps as the traditional WCET analysis that ignores resource conflicts. In the classical case, the analysis would continue with the calculation of the execution time of each basic block. This is done using local basic block simulations. The number of clock cycles that are spent by the processor doing effective computations, ignoring the time spent to access the cache (hit time) or the memory (miss penalty) is obtained in this way. Knowing the number of hits and misses for each basic block and the hit and miss penalties, the worst case execution time of each CFG node is easily computed. Taking into account the dependencies between the CFG nodes and their execution times, an ILP formulation can be used for the task WCET computation, [19], [16], [11].

In a realistic multiprocessor setting, however, due to the variation of the miss penalties as a result of potential bus conflicts, such a simple approach does not work. The main difference is the following: in traditional WCET analysis it is sufficient for each CFG node to have the total number of misses. In our case, however, this is not sufficient in order to take into consideration potential conflicts. What is needed is, for each node, the exact sequence of misses and the worst-case duration of computation sequences between the misses. For example, in the case of node 13 in Fig. 5(b), we have three instruction sequences separated by cache misses: (1) line 6, (2) line 7 and (3) lines 5 and 4. Once we have annotated the CFG with all the above information, we are prepared to solve the actual problem: determine the worst-case execution time corresponding to the longest path through the CFG. In order to solve this problem, we have to determine the worst-case execution time of a node in the CFG. In the classical WCET analysis, a node's WCET is the result of a trivial summation. In our case, however, the WCET of a node depends on the bus schedule and also on the node's worst-case start time.

Let us assume that the bus schedule in Fig. 5(c) is constructed. The system is composed of two processors and the task we are investigating is mapped on $CPU1$. There are two bus segments, the first one starting at time 0 and the second starting at time 32. The slot order during both segments is the same: $CPU1$ and then $CPU2$. The processors have slots of equal size during a segment.

The start time of the task that is currently analyzed is decided during the system level scheduling (see section IV) and let us suppose that it is 0. Once the bus is granted to a processor, let us assume that 6 time units are needed to handle a cache miss. For simplicity, we assume that the hit time is 0 and every instruction is executed in 1 time unit.

Using the above values and the bus schedule in Fig. 5(c), the node 12 will start its execution at time 0 and finish at time 39. The instruction miss (marked with "i" in Fig. 5(b)) from line 3 arrives at time 0, and, according to the bus schedule, it gets the bus immediately. At time 6, when the instruction miss is solved, the execution of node 12 cannot continue because of the data miss from line 2 (marked with "d"). This miss has to wait until time 16 when the bus is again allocated to $CPU1$ and, from time 16 to time 22 the cache is updated. Line 3 is executed starting from time 22 until 23, when the miss generated by the line 4 requests the bus. The bus is granted to $CPU1$ at time 32, so line 4 starts to be executed at time 38 and is finished, in the worst case, at time 39.

In the following we will illustrate the algorithm that performs the WCET computation for a certain task. The algorithm must find the longest path in the control flow graph. For example, there are four possible execution paths (sequences of nodes) for the task in Fig. 5(a) that are captured by the CFG in Fig. 5(b):(1) 2, 17, 11, (2) 2, 12, ,4, 16, 11, (3) 2, 12, 4, 13, 104, 16, 11 and (4) 2, 12, 4, 13, 104, 113, 104, ..., 104, 16, 11. The execution time of a particular node in the CFG can be computed only after the execution times of all its predecessors are known. For example, the execution time of node 16 can be computed only after the execution time for the nodes 4 and 104 is fixed. At this point it is interesting to note that the node 104 is the entry in a CFG loop (104, 113, 104). Due to the fact that the cache miss penalties depend on

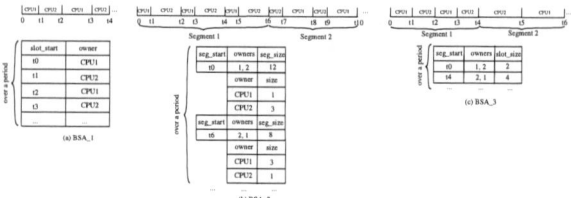

Fig. 6. Bus Schedule Table (system with two CPUs)

the bus schedule, the execution times of the loop nodes will be different at each loop iteration. Thus, loop nodes in the CFG must be visited a number of times, given by the loop bound (extracted automatically or annotated in the code). In the example from Fig. 5(b), the node 113 is visited 99 times (the loop is executed 100 times, but it was unrolled once). Each time a loop node is visited, its start time is updated and a new end time is calculated using the bus schedule, in the manner illustrated above for node 12. Consequently, during the computation of the execution time of the node 16, the start time is the maximum between the end time of the node 4 and node 104, obtained after 99 iterations. The worst-case execution time of the task will be the end time of the node 11.

The worst-case complexity of the WCET analysis is exponential (this is also the case for the classical WCET analysis). However, in practice, the approach is very efficient, as experimental results presented in Section VII show.[1]

VI. BUS SCHEDULE

The approach described in section V relies on the fact that during a time slot, only the processor that owns the bus must be granted the access. The bus arbiter must take care that the bus requests are granted according to the bus schedule table.

The assignment of slots to processors is captured by the bus schedule table and has a strong influence on the worst-case execution time. Ideally, from the point of view of task execution times, we would like to have an irregular bus schedule, in which slot sequences and individual slot sizes are customized according to the needs of currently active tasks.Such a schedule table is illustrated in Fig. 6(a) for a system with two CPUs. This bus scheduling approach, denoted as BSA_1, would offer the best task WCETs at the expense of a very complex bus slot optimization algorithm and of a large schedule table.

Alternatively, in order to reduce the controller complexity, the bus schedule is divided in segments. Such a segment is an interval in which the bus schedule follows a regular pattern. This pattern concerns both slot order and size. In Fig. 6(b) we illustrate a schedule consisting of two bus segments. This bus scheduling approach is denoted BSA_2.

The approach presented in Fig. 6(c) and denoted BSA_3 further reduces the memory needs for the bus controller. As opposed to BSA_2, in this case, all slots inside a segment have the same size.

---

[1]In the classical approach WCET analysis returns the worst-case time interval between the start and the finishing of a task. In our case, what we determine is the worst-case finishing time of the task.

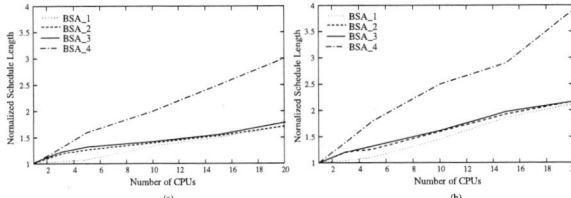

Fig. 7.   Experimental results

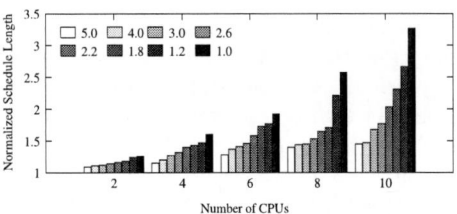

Fig. 8.   Experimental results

In the final approach, BSA_4, all the slots in the bus have the same size and repeated according to a fix sequence.

The bus schedule is a key parameter that influences the worst-case execution time of the tasks. As shown in Section IV the bus schedule is determined during the system scheduling process. Referring to Fig. 3, successive portions of the bus schedule are fixed during the internal optimization loop. The aim is to find a schedule for each portion, such that globally the worst-case execution time is minimized. In order to find an efficient bus schedule for each portion, information produced by the WCET analysis is used in the optimization process. In particular, this information captures the distribution of the cache misses along the detected worst case paths, for each currently active task (for each task in set Ψ). We have deployed several bus access optimization algorithms, specific to the proposed bus schedule alternative (BSA_1, BSA_2, BSA_3, BSA_4).

In the case of BSA_1, each portion of the bus schedule (fixed during the internal optimization loop in Fig.3) is determined without any restriction. For BSA_2 and BSA_3, each portion of the bus schedule corresponds to a new bus segment, as defined above. In case of BSA_2, the optimization has to find, for each segment, the size of the slots allocated for each processor, as well as their order. The search space for BSA_3 is reduced to finding for each bus segment, a unique slot size and an order in which the processors will access the bus.

In the case of BSA_4, the slot sequence and size is unique for the whole schedule. Therefore, the scheme in Fig. 3 is changed: a bus configuration alternative is determined before system scheduling and the list scheduling loop is included inside the bus optimization loop.

The bus schedule optimization algorithms are based on simulated annealing. Due to space limitations, the actual algorithms are not presented here. They will be described in a separate paper. Nevertheless, their efficiency is demonstrated by the experimental results presented in the next section.

VII. EXPERIMENTAL RESULTS

The complete flow illustrated in Fig. 3 has been implemented and used as a platform for the experiments presented in this section. They were run on a dual core Pentium4 processor at 2.8 GHz.

First, we have performed experiments using a set of synthetic benchmarks consisting of random task graphs with the number of tasks varying between 50 and 200. The tasks are mapped on architectures consisting of 2 to 20 processors. The tasks are corresponding to CFGs extracted from various C programs (e.g. sorting, searching, matrix multiplications, DSP algorithms). For the WCET analysis, it was assumed that ARM7 processors are used. We have assumed that 12 clock cycles are required for a memory access due to a cache miss, once the bus access has been granted.

We have explored the efficiency of the proposed approach in the context of the four bus scheduling approaches introduced in Section VI. The results are presented in Fig. 7. We have run experiments for configurations consisting of 2, 4, 6, ... 20 processors. For each configuration, 50 randomly generated task graphs were used. For each task graph, the worst-case schedule length has been determined in 5 cases: the four bus scheduling policies BSA_1 to BSA_4, and a hypothetical ideal situation in which memory accesses are never delayed. This ideal schedule length (which in practice, is unachievable, even by a theoretically optimal bus schedule) is considered as the baseline for the diagrams presented in Fig. 7. The diagram corresponding to each bus scheduling alternative indicates how many times larger the obtained bus schedule is relative to the ideal length. The diagrams correspond to the average obtained for the 50 graphs considered for each configuration.

A first conclusion is that BSA_1 produces the shortest delays. This is not unexpected, since it is based on highly customized bus schedules. It can be noticed, however, that the approaches BSA_2 and BSA_3 are producing results that are close to those produced by BSA_1, but with a much lower cost in controller complexity. It is not surprising that BSA_4, which restricts very much the freedom for bus optimization, produces very low quality results.

The actual bus load is growing with the number of processors and, implicitly, that of simultaneously active tasks. Therefore, the delays at low bus load (smaller number of processors) are close to the ideal ones. The deviation from the ideal schedule length is growing with the increased bus load due to the inherent delays in bus access. This phenomenon is confirmed also by the comparison between the diagrams in Fig. 7(a) and (b). The diagrams in Fig. 7(b) were obtained considering a bus load that is 1.5 times higher (bus speed 1.5 times smaller) than in Fig. 7(a). It can be observed that the deviation of schedule delays from the ideal one is growing faster in Fig. 7(b).

The execution times for the whole flow, in the case of the largest examples (consisting of 200 tasks on 20 processors) are as follows: 125 min. for BSA_1, 117 min. for BSA_2, 47 min. for BSA_3, and 8 min. for BSA_1.

The amount of memory accesses relative to the computations has a strong influence on the worst case execution time.

We have performed another set of experiments in order to asses this issue. The results are presented in Fig. 8. We have run experiments for configurations consisting of 2, 4, 6, ... 10 processors. For each configuration, 50 randomly generated task graphs were used. For each task graph, we have varied the ratio of clock cycles spent during computations and memory accesses. We have used eight different ratios: 5.0, 4.0, 3.0, 2.6, 2.2, 1.8, 1.4, 1.0. A ratio of 3.0 means that the number of clock cycles spent by the processors performing computations (assuming that all the memory accesses are cache hits) is three time higher than the number of cache misses multiplied with the cache miss penalty (assuming that each cache miss is handled in constant time, as if there are no conflicts on the bus). So, for example, if a task spends on the worst case CFG path 300000 clock cycles for computation and 100000 cycles for memory accesses due to cache misses (excluding the waiting time for bus access), the ratio will be 3.0. During this set of experiments we have assumed that the bus is scheduled according to the BSA_3 policy. Similar to the previous experiments from Fig. 7, the ideal schedule length is considered as the baseline for the diagrams presented in Fig. 8. Each bar indicates how many times larger the caculated worst case execution is relative to the ideal length. For example, on an a architecture with six processors and a ratio of 5.0, the worst case execution time is 1.28 times higher than the ideal one. Using the same architecture but with a smaller ratio (this means that the application is more memory intensive), the deviation increases: for a ratio of 3.0, the worst case execution time is 1.41 times the ideal one, while if the ratio is 1.0 the worst case execution time is 1.92 times higher than the ideal one.

In order to validate the real-world applicability of this approach we have analyzed a smart phone. It consists of a GSM encoder, GSM decoder [3] and an MP3 decoder [6], that were mapped on 4 ARM7 processors (the GSM encoder and decoder are mapped each one on a processor, while the MP3 decoder is mapped on two processors). The software applications have been partitioned into 64 tasks. The size of one task is between 1304 and 70 lines of C code in case of the GSM codec and between 2035 and 200 lines in case of the MP3 decoder. We have assumed a 4-way set associative instruction cache with a size of 4KB and a direct mapped data cache of the same size. The results of the analysis are presented in table I, where the deviation of the schedule length from the ideal one is presented for each bus scheduling approach.

BSA_1	BSA_2	BSA_3	BSA_4
1.17	1.33	1.31	1.62

TABLE I

RESULTS FOR THE SMART PHONE

VIII. CONCLUSIONS

In this paper, we have presented the first approach to the implementation of predictable RT applications on multiprocessor SoCs, which takes into consideration potential conflicts between parallel tasks for memory access. The approach comprizes WCET estimation and bus access optimization in the global context of system level scheduling. Experiments have shown the efficiency of the approach.

REFERENCES

[1] A. Andrei. *Energy Efficient and Predictable Design of Real-Time Embedded Systems*. PhD thesis, Linkoping University, Sweden, 2007.
[2] D. Bertozzi, A. Guerri, M. Milano, F. Poletti, and M. Ruggiero. Communication-aware allocation and scheduling framework for stream-oriented multi-processor systems-on-chip. In *DATE*, pages 3–8, 2006.
[3] Jutta Degener and Carsten Bormann. GSM 06.10 lossy speech compression. Source code available at http://kbs.cs.tu-berlin.de/~jutta/toast.html.
[4] J. Engblom, A. Ermedahl, M. Sjodin, J. Gustafsson, and H. Hansson. Worst-case execution-time analysis for embedded real-time systems. *International Journal on Software Tools for Technology Transfer*, 4(4):437–455, 2003.
[5] K. Goossens, J. Dielissen, and A. Radulescu. AEthereal Network on Chip: Concepts, Architectures, and Implementations. *IEEE Design & Test of Computers*, 2/3:115–127, 2005.
[6] Johan Hagman. mpeg3play-0.9.6. Source code available at http://home.swipnet.se/~w-10694/tars/mpeg3play-0.9.6-x86.tar.gz.
[7] R. Heckmann, M. Langenbach, S. Thesing, and R. Wilhelm. The influence of processor architecture on the design and the results of WCET tools. *Proceedings of the IEEE*, 91(7):1038–1054, 2003.
[8] E.G. Coffman Jr and R.L. Graham. Optimal Scheduling for two processor systems. *Acta Inform.*, 1:200–213, 1972.
[9] I. A. Khatib, D. Bertozzi, F. Poletti, L. Benini, and et.all. A multiprocessor systems-on-chip for real-time biomedical monitoring and analysis: Architectural design space exploration. In *DAC*, pages 125–131, 2006.
[10] H. Kopetz. *Real-Time Systems-Design Principles for Distributed Embedded Applications*. Kluwer Academic Publishers, 1997.
[11] Y.T.S. Li, S. Malik, and A. Wolfe. Cache modeling for real-time software: Beyond direct mapped instruction caches. In *IEEE Real-Time Systems Symposium*, pages 254–263, 1996.
[12] T. Lundqvist and P. Stenstrom. An Integrated Path and Timing Analysis Method based on Cycle-Level Symbolic Execution. *Real-Time Systems*, 17(2/3):183–207, 1999.
[13] S. Pasricha, N. Dutt, and M. Ben-Romdhane. Fast exploration of bus-based on-chip communication architectures. In *CODES+ISSS*, pages 242–247, 2004.
[14] P. Pop, P. Eles, Z. Peng, and T. Pop. Analysis and Optimization of Distributed Real-Time Embedded Systems. *ACM Transactions on Design Automation of Electronic Systems*, Vol. 11:593–625, 2006.
[15] P. Puschner and A. Burns. A Review of Worst-Case Execution-Time Analysis. *Real-Time Systems*, 2/3:115–127, 2000.
[16] H. Ramaprasad and F. Mueller. Bounding Preemption Delay within Data Cache Reference Patterns for Real-Time Tasks. In *Real-Time and Embedded Technology and Applications Symposium*, pages 71–80, 2005.
[17] E. Salminen, V. Lahtinen, and T. Hamalainen K. Kuusilinna. Overview of bus-based system-on-chip interconnections. In *ISCAS*, pages 372–375, 2002.
[18] S. Schliecker, M. Ivers, and R. Ernst. Integrated analysis of communicating tasks in mpsocs. In *CODES+ISSS*, pages 288–293, 2006.
[19] J. Staschulat and R. Ernst. Worst case timing analysis of input dependent data cache behavior. In *ECRTS*, 2006.
[20] J. Staschulat, S. Schliecker, and R. Ernst. Scheduling analysis of real-time systems with precise modeling of cache related preemption delay. In *ECRTS*, pages 41–48, 2005.
[21] H. Theiling, C. Ferdinand, and R. Wilhelm. Fast and Precise WCET Prediction by Separated Cache and Path Analysis. *Real-Time Systems*, 18(2/3):157–179, 2000.
[22] L. Thiele and R. Wilhelm. Design for Timing Predictability. *Real-Time Systems*, 28(2/3):157–177, 2004.
[23] F. Wolf, J. Staschulat, and R. Ernst. Associative caches in formal software timing analysis. In *DAC*, pages 622–627, 2002.
[24] W. Wolf. *Computers as Components: Principles of Embedded Computing System Design*. Morgan Kaufman Publishers, 2005.

978-1-4244-3039-0/08 $25.00 © 2008 IEEE

21st International Conference on VLSI Design

An Approach to Software Performance Evaluation on Customized Embedded Processors

Soumyajit Dey, Monu Kedia, Anupam Basu
Indian Institute of Technology Kharagpur
Kharagpur-721 302, INDIA
{soumyajit.dey, anupambas, monu.kedia}@gmail.com

Abstract

Evaluation of software performance on a given customized embedded processor is an important step in the design space exploration of embedded system architectures. Such evaluations help system designers in taking early design decisions regarding the hardware architecture most suitable for the target application. Simulation based performance evaluations, although very accurate, can be prohibitively slower. In this paper, we present a novel hybrid approach consisting of an initial simulation run (one time) followed by analysis of intermediate level (IR) application code by an evaluation engine. Our results show that the evaluation engine can accurately (more than 95%) estimate the excecution cycles of application or application task on a given customized embedded processor while it is at least an order of magnitude faster in terms of time taken.

1. Introduction

Instruction set customized processors are evolving as a viable solution addressing the needs of flexibility and performance in the domain of embedded systems. These customized cores try to deliver the performance close to Application Specific Integrated Circuits (ASICs) while retaining the flexibility of a General Purpose Processor (GPP). Recent results like [9], [12], [15], [19] have shown that by extending a base processor core with some intelligently selected instructions, the performance of the processor for the application or application domain can be improved significantly. Such design tools are marketed by vendors like CoWare [1], Tensilica [2], ARC [3] etc. But, keeping in view the shrinking time-to-market window in today's embedded system design and existence of several customized processor cores in the market from silicon vendors, a system designer is often tempted to adopt an off-the-self processor core, rather than designing and synthesizing one. However, doing a performance evaluation of the available off-the-self processor cores for the target application by cycle accurate simulations of the application for each of the available processors is tedious and time consuming.

In this paper, we present a hybrid approach consisting of an initial simulation run followed by analysis of IR-level application code using an evaluation engine to predict the execution time statistics on any given instruction set customized processor. In the present work, we study the behavior of our evaluation engine both in terms of the accuracy of the predicted execution time and also in terms of how fast it is as compared to the simulation based estimations.

The methodology has been implemented in the Tensilica (Xtensa) [2] design platform. An obvious application of the proposed approach becomes the estimation of task-level execution times which are the inputs to any Design Space Exploration (DSE) algorithms performing the application-to-architecture mapping [10], [6], [11] in heterogeneous MPSoC architectures [17]. In a multi-processor platform, identifying the most suitable processor for an application task is non trivial and necessitates the performance evaluation of each application task on each of the PEs in the platform. It is easy to see that in a M processor and N tasks system, doing such evaluations by cycle accurate simulation will result in a performance evaluation phase consisting of M*N simulations. Using our approach the application to architecture mapping during an MPSoC design can be done using simulation of each tasks on the base processor (M Simulations) and then M*N estimations by our evaluation engine.

Along with the prediction of execution cycles, our evaluation engine is also capable of automatically augmenting the application tasks with the Custom Instructions (CIs) available in the processor hardware and generating scheduled code. This can be seen as important step in the system design-flow using Tensilica platform

111

978-1-4244-3039-0/08 $25.00 © 2008 IEEE

where the CIs have to be manually embedded in the application code for porting into an architecture that consists of different extensions of the base Xtensa core.

Our results show that the evaluation engine is at least an order of magnitude faster than simulation based evaluation techniques. Further, the predicted execution times are more than 95% accurate in all the test cases.

Rest of the paper is organized as follow : In section 2 we do a detailed survey of the related work. In section 3 we present our hybrid approach for performance evaluation and in the subsequent section we present the detailed architecture of the evaluation engine. Section 5 presents the results of the case studies and in section 6 we come to the conclusions with a focus on possible future works in this direction.

2. Related Work

The evaluation of the processor for an application can be classified into two broad approaches.

1.**Simulation/Execution Based Approach**: In this approach, the application or application task is compiled for the target processor and either executed on an evaluation platform or simulated using a cycle accurate simulator. Several vendors [1], [2] provide such simulation infrastructures. In [14], a phase-accurate simulation strategy with delay models and trace has been reported for evaluation of pipelined architectures. The time taken to simulate the application at cycle accurate level is prohibitively high for large design spaces. However, we perform the simulation only on an identified base processor architecture and after that the execution time on different processors is predicted by our evaluation engine. So, we do not need the simulation infrastructures and compilers for all different processors.

2.**Analytical Approach**: In this kind of approach, the application is not simulated or executed. The application is rather analyzed with respect to the instruction set architecture of the processor. [18] present an approach to evaluate the instruction set for an application based on efficient graph covering and task scheduling method. In [16], a covering problem is used for finding out the optimal compilation of the application on special architectures like DSPs. Though this kind of technique can be used for execution time analysis, but they tend to be slower. Methodologies investigating the more commonly addressed problem of CI selection can be found in [5], [7] and [8]. In [13], a methodology of processor evaluation and application performance estimation has been reported. However, the impact of custom instructions have not been emphasized. The focus has been more on architectural parameters like register file-size, memory band-width etc. Our hybrid approach is similar to the graph covering based approach of [18] but we consider only a set of instructions defined in a separately written parameter file to carry out graph covering. The uncovered portion of the application graph will be covered by instructions of the base processor core. The effect of control-flow constraints in identifying instructions across basic blocks have been studied in [19]. However, no integrated approach for predicting architecture specific execution times along with the generation of CI enabled code for the different architectures have been reported in any of these works. This is precisely the novelty of the present work.

3. Our Approach

We assume that we have several customized embedded processors which differ in their ISAs as architectural design alternatives. We identify a base processor configuration which consists of instructions doing basic arithmetic, logical, comparison and jump instructions. Our goal is to evaluate the performance of the given application or application tasks on these different customized processors. Our approach to-wards this consists of two steps. As the first step, execution time statistics of the application or each of the application tasks is obtained via cycle accurate simulation using the simulator for base processor core. This is the only simulation run in our approach. The execution time statistics obtained via this initial simulation run forms the input to our second step. In the second step, an evaluation engine predicts the execution time statistics for each of the customized embedded processors. The inputs to our evaluation engine are as follows:

1. Performance statistics obtained from the simulation on the identified base processor.

2. A parameter file which captures the architectural differences in terms of the instruction set of the customized processor core in comparison with the identified base processor.

This parameter file contains the definition of all extra instructions contained in a customized processor. An instruction definition consists of the number of input/output parameters, latency information and dependency relations. Essentially the evaluation engine tries to estimate how these architectural differences will effect the execution time of the application or application tasks.

4. Architecture of the Evaluation Engine

The architecture of our evaluation engine is shown in Fig 1. Fig 2 shows the transformation steps performed on an example code by the evaluation engine.

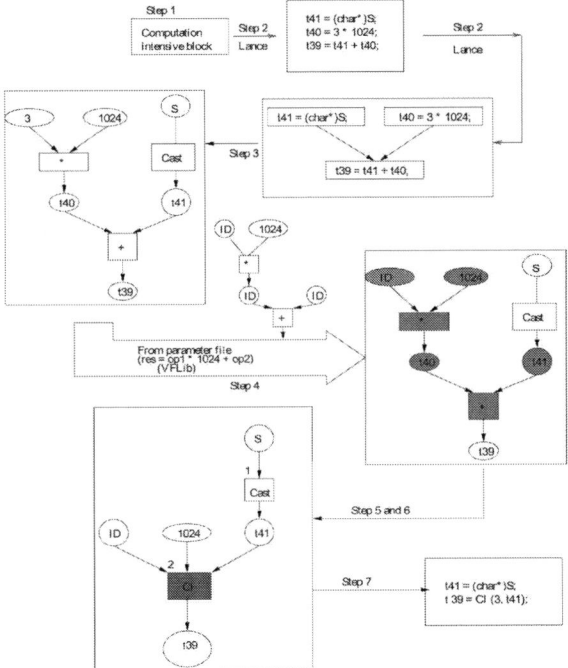

Figure 1. The Evaluation Engine

Figure 2. An Example of transformation steps

The details of each of the steps of the evaluation engine are as follows:

Application analysis and DFG generation (Step 1-3):
The application is simulated and profiled on the base processor using the simulator and profiler available in Tensilica [2] platform and the performance statistics are noted. We also note the number of times each basic block is executed and than identify the computation intensive basic block(s) which are the bottlenecks in the

application. The argument in favor of considering only such blocks is that if the architectural differences of the custom processor (*with the base processor*) does not effect (*improves*) the execution time of the computation intensive basic blocks, it is less likely that the custom processor will perform any better from the base processor for the concerned application task. In step 1, the DFG of these computation intensive basic blocks are generated using a compiler front-end tool called Lance [4]. These DFGs are converted to the operator/operand graph (OOG) in step 2.

Identification of sub-graphs in OOG (Step 4):
In this step we read each of the instructions given in the parameter file. We generate an equivalent DFG for those instructions and then using a graph/sub-graph matching library called VFLib, we identify the places where each of the instruction graphs occur in the OOG. We apply the following semantics for any node of an instruction graph to match with a node of the OOG :
1. An operator node of instruction graph can match only the operator node of OOG which have same type.
2. An operand node of instruction graph which is not constant can match with an operand node of OOG which is either constant or is an identifier.
3. An operand node of instruction graph which is a constant can match with an operand node of OOG which is constant and have same type and value.

Though the algorithm employed in VFLib is still exponential in the worst case, the incorporation of these semantic rules to guide the matching gives practical run-time. Apart from these semantic rules we also check the input/output locations of the instruction graph and the sub-graph of the OOG which matches it. If the input/output locations are not the same we do not consider that as a valid match. At the end of this step we have a list of sub-graphs of OOG which matches with some instruction given in the parameter file. It is important to note here that some of these sub-graphs may have common nodes. It is allowed to have input nodes of two (or more) instructions to match with the same node in the OOG. In all other cases where two sub-graphs of the OOG (which matches some instruction graph corresponding to the instruction given in the parameter file) intersect we can select only one of them for this OOG. This forms the motivation for optimally selecting the matching sub-graphs of OOG to be replaced by the corresponding instruction of the parameter file in order to estimate the performance.

Selection and Collapsing (Step 5):
The input to this step is a set M in which each element is a sub-graph of the OOG, which matches the DFG corresponding to some instruction defined in the parameter file. This set is essentially obtained from the previous

113

978-1-4244-3039-0/08 $25.00 © 2008 IEEE

step. The output is a set S (S being a subset of M) such that any two sub-graphs in M can intersect only in the nodes which form input nodes in the corresponding instruction DFG. The set S is identified by using an integer linear programming (ILP) based formulation with the objective of maximizing the performance gain obtained by replacing the nodes in those sub-graphs by the corresponding instruction. The ILP formulation is discussed later in a separate sub-section.

Scheduling and Code Generation (Step 6,7):
After the ILP based selection of the custom patterns in the OOG, the corresponding code segments in the IR are replaced by the CIs (instructions in the parameter file) in our evaluation infrastructure. The modified OOG after the collapsing of selected sub-graphs is scheduled by topological sorting and a scheduled IR level code is generated. Topological sort schedules the IR level code (augmented with CIs after optimal selection by the ILP solver) thereby ensuring the correctness of the data dependencies in the IR level code. Further code optimizations are left to be done by the Tensilica compiler.

After this step, the application code becomes automatically back-annotated with the identified CIs and it is ready to be compiled and executed/simulated in the customized processor. This automation along with the proposed performance estimation methodology in the Tensilica design platform is an important novelty of the present work. This is one important reason behind using the Lance compiler front end as it generates the IR level code in a subset of C along with the DFG. Hence, this code can be augmented with the CIs and compiled for the final mapping in the customized processors.

4.1. The ILP Formulation

The input/output constraints and objective function of the ILP formulation is described as below :

1.**Input**: Consider a set M, such that each element m_i of M is a sub-graph of an OOG such that m_i matches some instruction graph in the parameter file. This set is identified in step 4. We also consider a function ϕ, such that $\phi(m_i)$ gives us the instruction corresponding to m_i.

2.**Output**: Set S $\subseteq$ M, such that the sub-graph corresponding to each element s_i of S will be replaced by the instruction $\phi(s_i)$ described in the parameter file. The goal of ILP is to find the set S, so as to maximize the performance obtained by using the custom instructions. The set S must also adhere to the constraints discussed in the subsequent paragraph.

3.**ILP Decision Variables**: The problem of identifying the optimal set S is formulated using several boolean variables. We consider a boolean variable b_i corresponding to each element $s_i \in$ M. If b_i is true than $s_i \in$ S oth-

erwise $s_i \notin$ S. The values of these decision variables are found using ILP.

4.**ILP Objective Function**: The objective function for ILP optimization problem is defined to minimize the execution time by selecting an optimal set S $\subseteq$ M. Each element $m_i \in$ M corresponds to a sub-graph in OOG which can be replaced by an instruction described in the parameter file of the custom processor. Such replacement may result in the decrement of cycle count. Let $SU(m_i)$ denotes the decrement in cycle count obtained by replacing m_i with the corresponding instruction in the parameter file. The value of $SU(m_i)$ is obtained as the difference between the sum of latency information of the basic operations (which are involved in the sub-graph corresponding to m_i in the OOG) on the base processor and the latency information of $FI(m_i)$ available in the parameter file. The objective function OF is represented mathematically in terms of ILP decision variables as follow :

$OF = maximize\ (\ \sum SU(m_i) * b_i\ )\ \forall\ m_i \in S$

5.**ILP Constraints**: The constraints ensure that any two subgraphs s_i, $s_j \in$ S do not intersect in the nodes other than the input nodes of the corresponding instruction graphs. Fig 3 explains this point graphically. In this figure, s_i corresponds to the match in OOG with the instruction graph for *Result = (Op1 + Op2)* Op3* and s_j corresponds to the match in OOG with the instruction graph for *Result = (Op1 * Op2) + Op3*. Since both these graphs intersect in their operator nodes we can select any one of them for this OOG. Let IN(m_i) $\subseteq$ M, such that $\forall$ $m_j \in$ IN(m_i), m_j intersect with m_i in nodes other the input nodes in the corresponding instruction graphs. Such non-intersection constraint can be mathematically represented as follow :

$\forall m_i \in M\ (\forall\ m_j \in IN(m_i), 0 \leq \sum b_j \leq 1).$

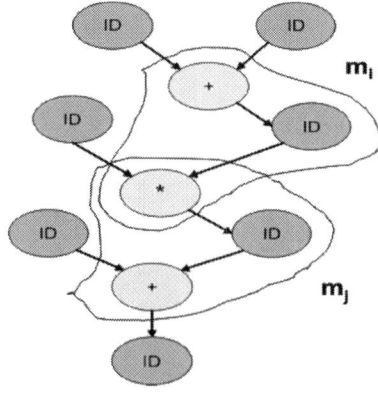

Figure 3. Intersection of m$_i$ and m$_j$ in OOG

The solutions of formulated ILP have been generated by a popular Linear Programming solver called GLPK.

978-1-4244-3039-0/08 $25.00 © 2008 IEEE

Once the sub-graphs are selected, each of them are collapsed as a single node and labeled with the matching instruction of the parameter file.

5. Results

To study the applicability of our approach we start with a base processor core (LX1.0) used in the Tensilica Xtensa [2] configurable processor platform. We consider three custom processor configurations whose CIs are as shown in figure 4. On the application side, we study our approach on three cryptographic benchmarks namely, Blowfish, DES and Gost. Blowfish is a symmetric block cipher which can take variable length key (up to 448 bits). The heart of this algorithm is a function named F (the Feistel function), which consumes more than 75% of the total execution time. DES is also a block cipher in which the block size is 64 bits. The important part of this application is the round function which accounts for around 78% of the total execution time. Gost is a 64-bit block cipher with 256 bits key. Like Blowfish, Gost spend most of its execution time (around 80%) inside the F function. For each application, we feed the following inputs to our evaluation engine :

1.Cycle count taken on the base processor core.
2.Number of times the basic block of the computation intensive function is executed during the initial simulation run.
3.C code for the computation intensive basic block.
4.Parameter file describing the processor core for which the execution time has to be predicted by the evaluation engine.

Processor	Extra Instructions in comparison of LX1.0 Instruction Set Architecture (ISA)
P1	1. Result = 4 * Op1 + Op2
	2. Result = 1024 * Op1 + Op2
P2	1. Result = 4 * Op1 + Op2
	2. Result = (Op1 >> 8) & Op2
	3. Result = (Op1 >> 16) & Op2
	4. Result = (Op1 >> 24) & Op2
P3	1. Result = (Op1 >> Op2) & 255
	2. Result = (Op1 << 8)	Op2
	3. Result = (Op1 << 16)	Op2
	4. Result = (Op1 << 24)	Op2

Figure 4. Processor Con gurations

Table 1 shows the execution time of each application on each of the processors as predicted by the evaluation engine and also as obtained by simulation. As can be seen, the evaluation engine predicts the cycle count for each of the processor configurations quiet accurately (

more than 95% in all the case). The accuracy of 100% obtained for Gost with processor configuration P1 and for DES and Blowfish with processor configuration P3 needs some explanation. This basically reflects that the chosen processor configuration will perform no better than the base processor for this application. So, the execution time is predicted to be the same as obtained from the initial simulation, resulting in 100% accuracy. This implies that the CI patterns for the chosen processor did not occur in the OOGs of the respective the applications. Also, the prediction by evaluation engine is more than an order of magnitude faster than simulating the application on that processor. An important decision that can be taken from the results of Table 1 is that, processor P1 is best suited for Blowfish, P2 for DES and P3 for Gost benchmark. The obvious reason behind such decisions is the the reduction in execution times offered by the respective processors. Such decisions are consistent with both the execution time statistics produced by our evaluation engine and one obtained by simulating the application on the respective processor configurations. This validates the use of our approach for selecting the ideal processor for an application. It is also evident that such accurate execution time data generated by our evaluation engine can be used for mapping the application tasks optimally onto heterogeneous MPSoC platforms.

Table 1. Results

Appli-cation	Base Processor (K cycles)	P1			
		Evaluation Engine (Predicted No. of Kcycles)	Simulation (Actual No. of Kcycles)	Accuracy	Relative time taken (Sim /Eval)
Blowfish	965.56	888.24	909.00	97.71%	32.90
DES	1560.97	1471.38	1445.32	98.22%	18.16
Gost	1912.86	1912.86	1912.86	100%	52.56
		P2			
Blowfish	965.56	926.90	929.37	99.73%	25.15
DES	1560.97	1417.62	1383.96	97.56%	17.53
Gost	1912.86	1818.36	1763.50	96.98%	54.39
		P3			
Blowfish	965.56	965.56	965.56	100%	29.89
DES	1560.97	1560.97	1560.97	100%	30.43
Gost	1912.86	1708.18	1764.41	96.81%	57

6. Conclusion and Future Work

In this paper we have looked into the problem of estimation of the software performance on an instruction set customized embedded processor. We have presented a novel hybrid approach consisting of one time simulation of the application on an identified base processor architecture and then prediction of execution time on a customized processor by our evaluation engine. Along with the prediction of execution time, the application tasks are automatically augmented with CIs optimally for execution in the heterogeneous processors. Our approach also finds use in the application to architecture mapping during MPSoC design. Our work basically turns around the classical instruction selection problem for code generation. However, the presented framework is significant in the context of performance estimation for a given processor architecture and corresponding code-generation.

In the current work our evaluation engine can predict only the execution time for a custom processor. Refinements for improving the prediction accuracy is a natural extension to the present work. Another interesting future work can be augmenting the evaluation engine with cache performance estimation due to CIs and the effect of other micro-architectural parameters.

References

[1] Lisatek. http://www.coware.com/.

[2] Xtensa, http://www.tensilica.com/.

[3] Arc, http://www.arc.com/.

[4] Lance, http://www.lancecompiler.com/.

[5] K. Atasu, L. Pozzi, and P. Ienne. Automatic application-specific instruction-set extensions under microarchitectural constraints. In *DAC*, 2003.

[6] C. Brandolese, L. Pomante, F. Salice, and D. Sciuto. Affinity-driven system design exploration for heterogeneous multiprocessor soc. *IEEE Transactions on Computers*, 55(5), May 2006.

[7] N. Clark, H. Zhong, and S. Mahlke. Processor acceleration through automated instruction set customization. In *MICRO-36*, 2003.

[8] N. Clark, H. Zhong, W. Tang, and S. Mahlke. Automatic design of application specific instruction set extensions through dataflow graph exploration. *IJPP*, 31(6), Dec 2003.

[9] N. T. Clark, H. Zhong, and S. A. Mahlke. Automated custom instruction generation for domain-specific processor acceleration. *IEEE Transactions on Computers*, 54(10), October 2005.

[10] C. Erbas, S. C. Erbas, and A. D. Pimentel. A multiobjective optimization model for exploring multiprocessor mappings of process networks. In *CODES+ISSS*, 2003.

[11] C. Erbas, S. C. Erbas, and A. D. Pimentel. Multiobjective optimization and evolutionary algorithms for the application mapping problem in multiprocessor system-on-chip design. *IEEE Transactions on Evolutionary Computation*, 10(3), June 2006.

[12] S. Fei, S. Ravi, A. Raghunathan, and N. K. Jha. Custom-instruction synthesis for extensible-processor platforms. *IEEE Transactions Computer-Aided Design of Integrated Circuits and Systems*, February 2004.

[13] T. V. K. Gupta, P. Sharma, M. Balakrishnan, and S. Malik. Processor evaluation in an embedded systems design environment. In *VLSI Design*, January 2000.

[14] J. K. Kim and T. G. Kim. Trace-driven rapid pipeline architecture evaluation scheme for asip design. In *ASP-DAC*, 2003.

[15] J. Lee, K. Choi, and N. Dutt. Efficient instruction encoding for automatic instruction set design of configurable asips. In *ICCAD*, 2002.

[16] S. Liao, S. Devadas, K. Keutzer, and S. Tjiang. Instruction selection using binate covering for code size optimization. In *IEEE/ACM international conference on Computer-aided design*, 1995.

[17] G. Martin. Overview of the mpsoc design challenge. In *ACM IEEE Design Automation Conference*, 2006.

[18] M. Masuda and K. Ito. Rapid and precise instruction set evaluation for application specific processor design. In *IEEE International Symposium on Circuits and Systems*, May 2005.

[19] P. Yu and T. Mitra. Characterizing embedded applications for instruction-set extensible processors. In *DAC*, 2004.

SESSION D1:
Technology

21st International Conference on VLSI Design

Compact Modeling of Suspended Gate FET

Y. S. Chauhan*, D. Tsamados*, N. Abelé*,†, C. Eggimann‡, M. Declercq* and A. M. Ionescu*

*Ecole Polytechnique Fédérale de Lausanne (EPFL), Lausanne,1015, Switzerland

†ST Microelectronics, Crolles, France

‡Center for Integrated Systems, CISX 312, Stanford University, Stanford, CA 94305-4075, USA

E-mail: (yogeshsingh.chauhan@epfl.ch, yogeshsingh.chauhan@in.ibm.com), adrian.ionescu@epfl.ch

Abstract—**For the first time, a compact model for Suspended Gate (SG) FET valid for entire bias range is proposed. The model is capable of simulating both pull-in and pull-out effects, which are the two important phenomena of this device. A novel hybrid numerical simulation approach combining ANSYS Multiphysics and ISE-DESSIS in a self-consistent system is developed. The model is then validated on this numerical device simulation of SGFET. The model shows excellent performance over the entire drain and gate voltage range. The model has been implemented in Verilog-A code and tested on ELDO and Spectre simulators, which makes it useful for circuit simulations using SGFET devices.**

I. INTRODUCTION

The Suspended Gate FET (SGFET) [1]–[4] has been under study for the last two decades. Fig. 1 shows the device architecture of SGFET. The gate of the device is hanging over the oxide with a gap, which could be air or vacuum. Till last decade, the main attention in the literature has been on the use of SGFET for sensing applications [4], [5]. Ionescu et al. first time showed the application of SGFET for abrupt switching applications [1]. Fig. 2 shows the typical $I_D - V_G$ characteristics of SGFET. As V_G starts increasing, the beam starts moving down due to electrostatic attraction and I_D increases. But as the small capacitance of the gap is in series with the oxide capacitance, I_D is very small. At a specific gate bias, the electrostatic force cannot be compensated by the mechanical restoring force anymore, and the beam collapses on the oxide. This is called pull-in effect as shown in Fig. 2. After pull-in, increase in I_D with V_G is similar to the standard MOSFET. If V_G is decreased from some high value, then I_D starts decreasing. At certain value of V_G, the system becomes unstable due to combined electro-mechanical force and beam is pulled-out. This causes sudden decrease in I_D due to large decrease in capacitance. This effect is called pull-out effect as shown in Fig. 2. These unique effects have found great importance for memory and switching applications [6],

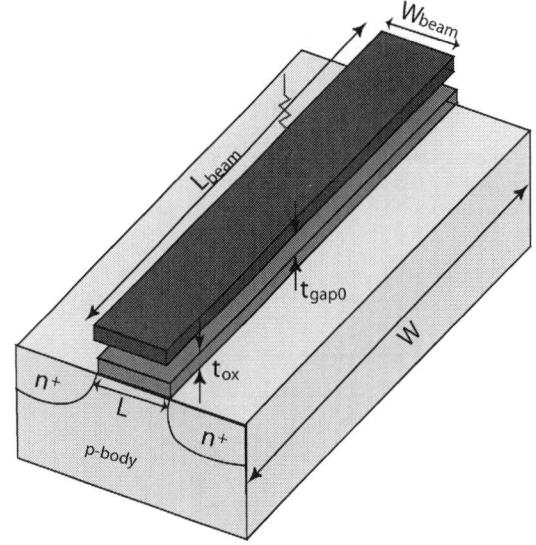

Fig. 1. Schematic representation of N-type Suspended Gate FET (SGFET) [6], [7]. The gate is suspended above oxide with an air or vacuum gap in between gate and oxide.

[7]. Recently SGFET based Single Electron Transistor (SET) has also been demonstrated [8].

As the application of this device in the past has been mainly focused on sensing applications, very few models have been proposed for this device. Most of these models [4] are valid only in the strong inversion region. Also none of these models have discussed about pull-in and pull-out effects, which are very important for switching and memory applications [6]. Ionescu et al. [1] proposed a simple charge based model valid for entire gate bias range but only for linear region of output characteristics. This model successfully reproduced the pull-in effect and elucidated the importance of SGFET for abrupt switch applications. Here we will develop an advanced charge based compact model for SGFET, valid for the entire voltage operation range. The model will be capable of modeling both the pull-in and pull-out effects. The model has been coded in Verilog-A code and is being used for circuit simulation based on SGFET devices [9].

119

978-1-4244-3039-0/08 $25.00 © 2008 IEEE

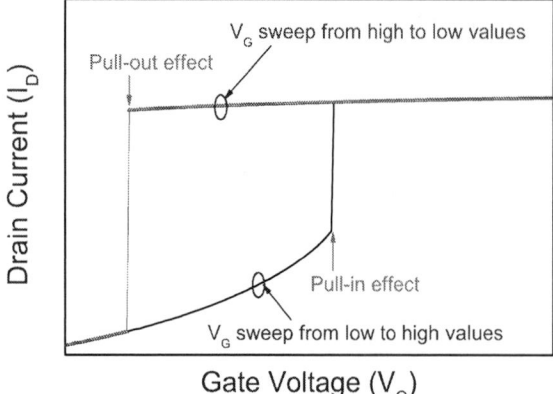

Fig. 2. Typical $I_D - V_G$ characteristics of SGFET. The pull-in effect occurs when V_G is increased from low to high values while pull-out effect occurs when V_G is decreased from high to low values. Note that the SGFET has dynamic threshold voltage (V_{th}), which varies with the beam position. V_{th} decreases when beam moves downwards with increase in V_G and vice-versa.

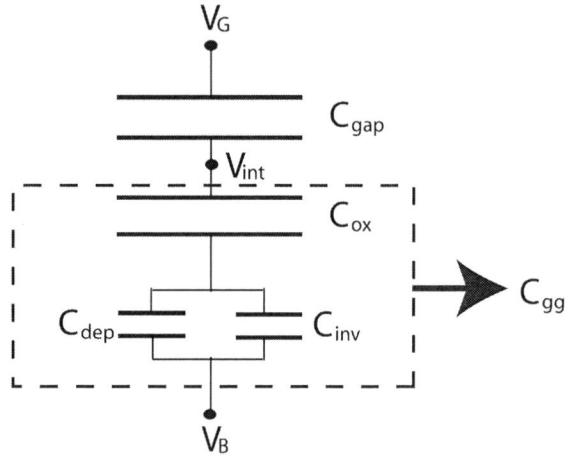

Fig. 3. Equivalent capacitive divider of SGFET. C_{gap}, C_{ox}, C_{dep}, C_{inv} and C_{gg} represents gap, oxide, depletion, inversion and gate-to-gate capacitances respectively.

II. SUSPENDED GATE FET MODEL

The modeling strategy for SGFET in this work is based on charge based EKV formalism [10], [11]. The SGFET can be considered as a capacitive divider between C_{gap} and C_{gg} as shown in Fig. 3, where C_{gap} and C_{gg} are the capacitance of the gap and input capacitance of MOS, respectively. The potential at the internal node (V_{int}) can be expressed as

$$V_{int} = V_G \frac{C_{gap}}{C_{gap} + C_{gg}} \qquad (1)$$

where C_{gg} is derived at the end of this section. The $C_{gap} = \frac{\epsilon_0 W_{beam} L_{beam}}{t_{gap0} - \Delta_y}$ is the capacitance of the gap.

The initial gap is denoted as t_{gap0} and Δ_y is the displacement of the beam from its initial position. From (1), it is clear that V_{int} is a function of known applied biases, geometrical parameters and the unknown variable Δ_y. The displacement of the beam Δ_y is obtained by balancing the electrical and mechanical forces on the beam as,

$$\frac{1}{2} \epsilon_0 W L \left(\frac{V_G - V_{int}}{t_{gap0} - \Delta_y} \right)^2 = k_{beam} \Delta_y \qquad (2)$$

where k_{beam} is the lumped linear spring constant of the beam as depicted in Fig. 1. Please note that the RHS of (2) should be replaced by the $m \frac{d^2 \Delta_y}{dt^2} + b \frac{d\Delta_y}{dt} + k_{beam} \Delta_y$ for dynamic simulation, where m is the dynamic mass of the beam and b is the damping factor. Eq. (1) and (2) can be solved together for both V_{int} and Δ_y. The solution of these coupled equations is valid only for $V_G - V_{int} < V_{PI}$ [3], [5]. To impose this condition, we can formulate the modified internal node potential as

$$V_{mid}^{PI} = \begin{cases} V_{int}, & \text{if } V_G < V_{int} + V_{PI}; \\ V_G, & \text{otherwise.} \end{cases} \qquad (3)$$

Using similar strategy as discussed above, the internal node potential for the pull-out case can be written as,

$$V_{mid}^{PO} = \begin{cases} V_{int}, & \text{if } V_G < V_{PO}; \\ V_G, & \text{otherwise.} \end{cases} \qquad (4)$$

where pull-in voltage V_{PI} and pull-out voltage V_{PO} are expressed as

$$\begin{aligned} V_{PI} &= \sqrt{\frac{8k_{beam}}{27\epsilon_0 W_{beam} L_{beam}} t_{gap0}^3} \\ V_{PO} &= \sqrt{\frac{2W L k_{beam} t_{gap0} \epsilon_0}{C_{gg}^2}} \end{aligned} \qquad (5)$$

Now as we know the internal node potential for both pull-in (V_{mid}^{PI}) and pull-out case (V_{mid}^{PO}), we can use standard EKV formalism to derive the C_{gg} and finally, drain current (I_D) for both cases as described below.

The well-known drift-diffusion current expression using first order mobility model [12], [13] is given by

$$I_D = \frac{\mu_v}{1 + \frac{\mu_v}{v_{sat}} \left| \frac{d\Psi_s}{dx} \right|} W \left(-Q_i \frac{d\Psi_s}{dx} + U_T \frac{dQ_i}{dx} \right) \qquad (6)$$

where $\mu_v (= \rho_v \mu_0)$, Q_i, and Ψ_S are the vertical field dependent mobility, inversion charge, and surface potential, respectively, at any position x in the channel. U_T and ρ_v are the thermal voltage and parameter for vertical field dependence [13], respectively. Using inversion charge linearization relation ($-Q_i = n_q C_{total}(\Psi_P - \Psi_S)$) from EKV formalism [11], Eq. (6) for $\frac{d\Psi_s}{dx} > 0$ can be rewritten as

$$\frac{dq}{d\xi} = -\frac{i_d}{\rho_v(1 + 2q - \delta_{sat}i_d)} \quad (7)$$

where variables are normalized [11] as follows

$$q = \frac{-Q_i}{2n_q C_{total} U_T}, \quad i_d = \frac{I_D}{2n_q \frac{W}{L}\mu_0 C_{total} U_T^2}$$
$$\delta_{sat} = \frac{2\mu_0 U_T}{v_{sat}L}, \quad \xi = \frac{x}{L} \quad (8)$$

Note that q, i_d, δ_{sat} and, ξ are all dimensionless quantities. The n_q and Ψ_P are the slope factor and pinch-off surface potential, respectively [11], [13]. The C_{total} is the total capacitance of series combination of C_{gap} and C_{gg}. Integrating (7) along the channel from $\xi = 0$ to $\xi = 1$, the normalized drain to source current i_d and total normalized inversion charge density q_c in the channel are obtained as

$$i_d = \frac{\rho_v}{1 + \rho_v \delta_{sat}(q_s - q_d)}\left[(q_s^2 + q_s) - (q_d^2 + q_d)\right] \quad (9)$$

$$q_c = \frac{1}{2}\left\{q_s + q_d + \frac{1}{3}\frac{(q_s - q_d)^2}{(1 + q_s + q_d)}[1 + \rho_v \delta_{sat}(q_s - q_d)]\right\} \quad (10)$$

where q_d and q_s are the normalized charge densities at drain and source, respectively [10], [11], [13]. The final expression for i_d including velocity saturation from (9) is given by,

$$i_d = \begin{cases} p_1\left[(q_s^2 + q_s) - (q_d^2 + q_d)\right], & \text{if } q_d > q_{dsat}; \\ p_1\left[(q_s^2 + q_s) - (q_{dsat}^2 + q_{dsat})\right], & \text{otherwise} \end{cases} \quad (11)$$

$$q_{dsat} = (p_2 + q_s) - \sqrt{p_2^2 + \frac{1}{\rho_v \delta_{sat}}(2q_s + 1)} \quad (12)$$

where $p_1 = \frac{\rho_v}{1 + \rho_v \delta_{sat}(q_s - q_d)}$ and $p_2 = \frac{1}{2} + \frac{1}{\rho_v \delta_{sat}}$. Using (10), the expression for normalized C_{gg} can be written as,

$$C_{gg} = \left\{1 - \frac{1}{n_q} + d_{qs} + d_{qd} + \right.$$
$$\frac{1}{3}\frac{(2(q_s - q_d) + 3\rho_v \delta_{sat}(q_s - q_d)^2)[d_{qs} - d_{qd}]}{1 + q_s + q_d}$$
$$\left. - \frac{1}{3}\frac{\{[(q_s - q_d)^2 + \rho_v \delta_{sat}(q_s - q_d)^3](d_{qs} + d_{qd})\}}{(1 + q_s + q_d)^2}\right\} \quad (13)$$

where

$$d_{qs} = \frac{(1/n_q)}{2 + (1/q_s)}, \quad d_{qd} = \frac{(1/n_q)}{2 + (1/q_d)} \quad (14)$$

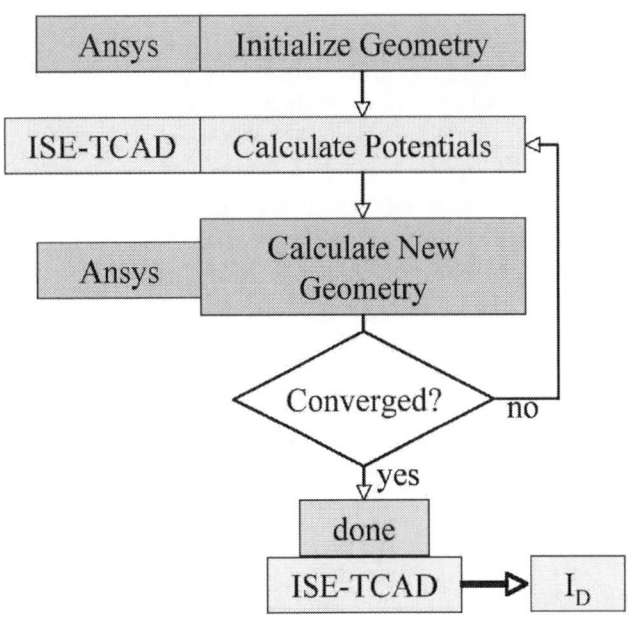

Fig. 4. Flow chart for the finite element analysis of the electromechanical SGFET based on *hybrid* ANSYS-ISE (iteration loop for given bias voltages).

III. NUMERICAL DEVICE SIMULATION OF SGFET

The finite element analysis (FEA) of a SGFET requires a complex multiphysics solver capable of handling the electrostatic domain with semiconductor materials and the structural (micromechanic) domain. In this work a *hybrid* FEA coupling two different tools, ANSYS Multiphysics and ISE-DESSIS in a self-consistent system, is demonstrated [9]. The first one is used for calculations in the coupled electrostatic and structural physics environments while the second one provides the correct boundary conditions for the electrostatic domain after taking into account semiconductor physics.

In Fig. 4, the procedure of the FEA with the two tools is presented in a flow chart. The great advantage of using two different FEA tools, with one completely dedicated to semiconductor devices, is that one can easily add physical models for more complicated phenomena taking place in the semiconductor material e.g. dielectric charging etc.

A. Coupling of FEA software

The *communication* of the electro-mecahnical and semiconductor simulators, by default incompatible software, is implemented and performed through a script written in Perl scripting language. All the parameters needed for building the 2D models in both simulation environments are incorporated in this Perl code. The first step is performed by ANSYS in order to initialize the geometry of the problem, which is then transferred to ISE DESSIS. DESSIS resolves the problem with the proper boundary conditions (bulk, source, drain and gate biases) for the initial position of the movable structure (suspended-gate in our case). Once the solution converged, the resulting boundary conditions (potential at the air-oxide interface) are passed to ANSYS for the electromechanical calculation of this first iteration that produces the new geometry. The new geometry is introduced to DESSIS with the same boundary conditions (biases are assumed constant). The loop of calculations and of boundary conditions transfer between ANSYS and ISE is stopped once the convergence criterion is satisfied. For our case the relative change of the gate's position from the last iteration is chosen as such criterion.

B. ANSYS electromechanical model

The geometry of the 2D model is kept as simple as possible for the first phase of validation. A standard enhancement-mode bulk silicon MOSFET structure with SiO2 dielectric is 2D-described in the simulator. Over the channel region an air (or vacuum) region is introduced, which contains the suspended gate. In the simple case presented here, a metallic gate is assumed and its work function and mechanical properties (Young modulus, Poisson coefficient) can be chosen according to the material used. In ANSYS, the only regions of interest are the air and the metallic suspended gate (the substrate is considered fixed). The ANSYS multiphysics solver is used for the coupling of the two physical environments with morphing activated for the electrostatic domain. Contact elements are used for the detection of the mobile gate pull-in and pull-out. These contact elements are positioned far after the theoretical $\frac{2}{3}$ of the gap and slightly before the air-oxide interface in order to avoid excessive distortion of the electrostatic elements. The contact is assumed *bonded* meaning that once the gate collapses it stays pulled-in until the end of the ANSYS iteration. This solution permits a quick convergence of the solution without influencing the result of that iteration. The pinball region is adapted to the gap dimension in order to avoid accidental closing and pull-in. Spring elements are implemented for simulating the restoring force of the beam.

C. ISE-DESSIS model

Regarding the ISE part, standard long channel MOSFET semiconductor-physics models are applied. At the end of each DESSIS iteration, the currents and potentials

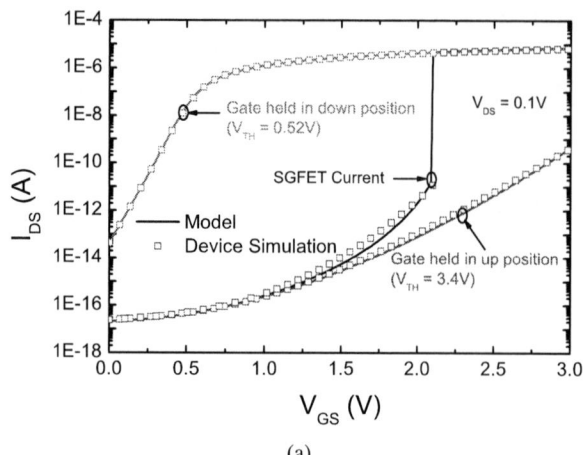

(a)

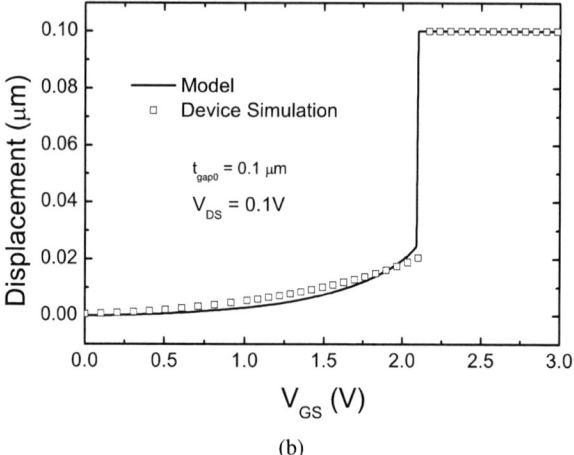

(b)

Fig. 5. (a) Plot of $I_D - V_G$ of SGFET at V_D=0.1V (model- lines and device simulation - symbols). Plots of $I_D - V_G$ for gate held in the up position and gate held in the down position are also shown. (b) Displacement of the beam with the increasing gate voltage. The dimensions of the device are: $L_{beam} = 60\mu m$, $W_{beam} = 2\mu m$, $W = 1\mu m$, $L = 2\mu m$.

are exported, on one hand for sending the later to ANSYS and on the other hand for saving the $I_D - V_G$ (or $I_D - V_D$) characteristics of the SGFET.

Note that the pull-in/pull-out in our device simulation may occur before/after the theoretical limits due to fringing field.

IV. RESULTS AND DISCUSSION

The proposed model is validated on the numerical device simulation of SGFET. The source and body are grounded for all results. Fig. 5 shows the current and displacement variation with gate voltage for a large device ($L_{beam} = 60\mu m$, $W_{beam} = 2\mu m$, $W = 1\mu m$, $L = 2\mu m$). The model shows good accuracy over entire gate bias range. It is interesting to note that SGFET

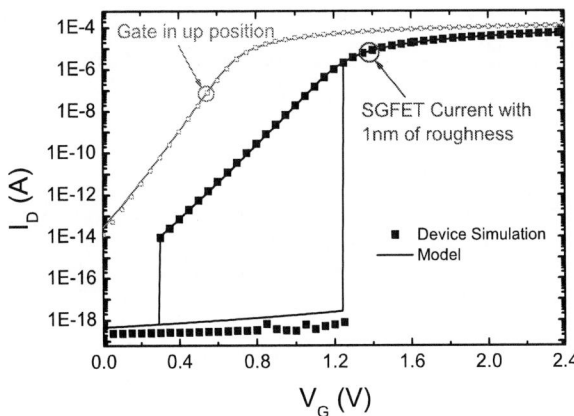

Fig. 6. Plot of $I_D - V_G$ of SGFET at V_D=0.05V (model- lines and device simulation - symbols). It is assumed that the device has 1nm of gap even after pull-in, which is evident in the experimental devices [6]. Plot of $I_D - V_G$ for gate held in the down position (without any roughness) is also shown. The dimensions of the device are: $L_{beam} = 2\mu m$, $W_{beam} = 0.09\mu m$, $t_{gap0} = 17nm$, $W = 1\mu m$, $L = 0.13\mu m$.

shows super-exponential behavior just before pull-in, which was also demonstrated experimentally by Abelé et al. [2].

Fig. 6 shows the $I_D - V_G$ characteristics of a small device ($L_{beam} = 2\mu m$, $W_{beam} = 0.09\mu m$, $W = 1\mu m$, $L = 0.13\mu m$). For this device, we assumed $1nm$ of air gap at the contact between gate and dielectric to simulate the roughness of the two contacting surfaces [6]. It should be noted, this is the first time, pull-out behavior of SGFET has been shown by any compact model and device simulation. An interesting observation of this result is that pull-in occurs around threshold voltage of the MOSFET, which is also the most desired results for switching applications. The device remains in the off condition with low off current and suddenly switches on at pull-in and enters into strong inversion. Even though the leakage current shown in the device simulation and model is very small, it would be higher in the real devices. Still the off current will be limited by the diode leakage and not by the subthreshold current of conventional MOSFET.

Fig. 7 shows the output characteristics ($I_D - V_D$) of the same device as in Fig. 6. As expected, the drain current changes by several orders of magnitude before and after pull-in. It should be noted that this is the first time output characteristics has been simulated by any SGFET model.

Based on the proposed model, the SGFET behavior is further investigated for different geometries. Fig. 8 shows the $I_D - V_G$ characteristics for different spring constants. As K_{beam} increases, the pull-in voltage also increases

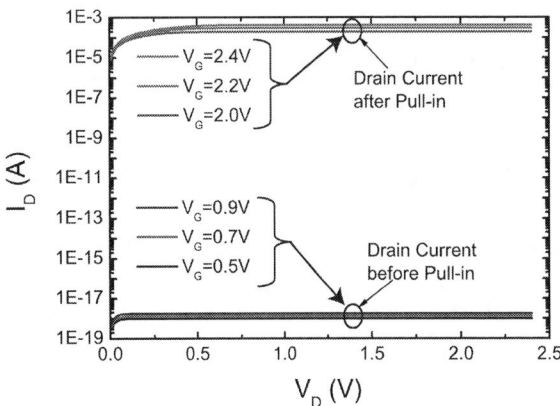

Fig. 7. Plot of $I_D - V_D$ of SGFET for V_G=0.5, 0.7, 0.9 and 2.0, 2.2, 2.4V. It is assumed that the device has 1nm of gap even after pull-in, which is evident in the experimental devices [6]. The output characteristics have several order of difference in magnitude of current before and after pull-in. The dimensions of the device are: $L_{beam} = 2\mu m$, $W_{beam} = 0.09\mu m$, $t_{gap0} = 17nm$, $W = 1\mu m$, $L = 0.13\mu m$.

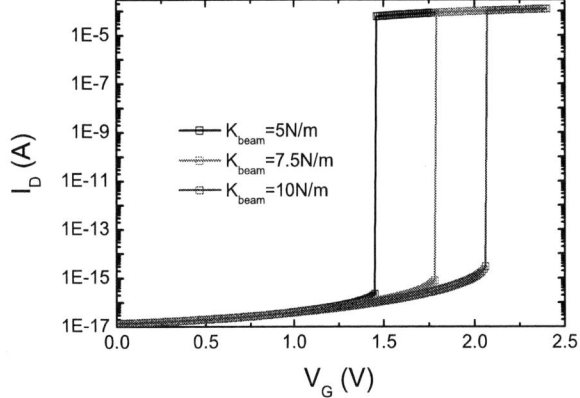

Fig. 8. Plot of $I_D - V_G$ of SGFET at V_D=0.05V for spring constants of K_{beam}=5, 7.5 and 10N/m. The dimensions of the device are: $L_{beam} = 2\mu m$, $W_{beam} = 0.09\mu m$, $t_{gap0} = 17nm$, $W = 1\mu m$, $L = 0.13\mu m$.

(also evident from (5)), which suggests that pull-in voltage can be set to the convenient value by carefully designing the structure of beam. Fig. 9 shows the $I_D - V_G$ characteristics for different beam lengths. The pull-in voltage of the SGFET is inversely proportional to the beam length and thus decreases as L_{beam} increases (see (5)).

Fig. 10 shows the transfer characteristics for different initial gaps. As initial gap decreases, the pull-in voltage decreases. As discussed earlier, the most suitable value for pull-in voltage would be around threshold voltage of the MOSFET, thus giving characteristics close to the ideal switch.

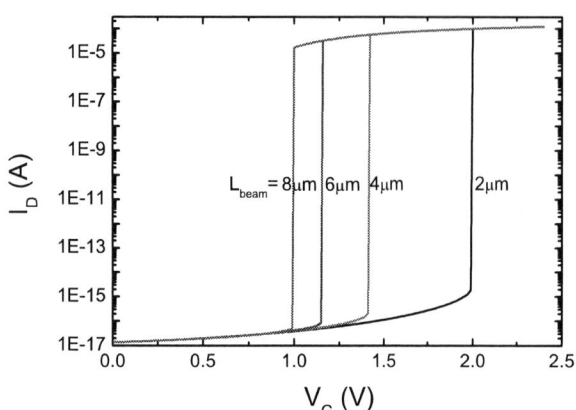

Fig. 9. Plot of $I_D - V_G$ of SGFET at V_D=0.05V for different beam lengths. The dimensions of the device are: $W_{beam} = 0.09\mu m$, $t_{gap0} = 17nm$, $W = 1\mu m$, $L = 0.13\mu m$.

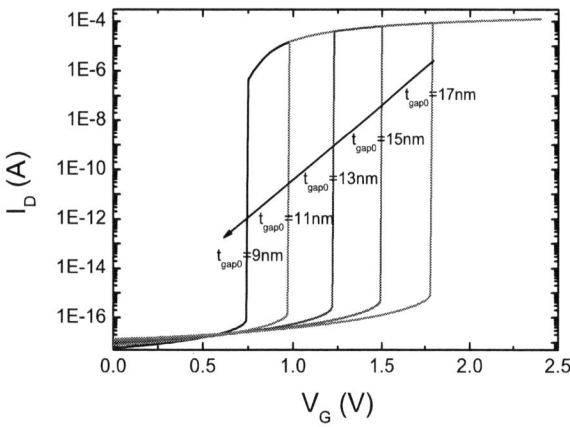

Fig. 10. Plot of $I_D - V_G$ of SGFET at V_D=0.05V for different initial gaps. The dimensions of the device are: $L_{beam} = 2\mu m$, $W_{beam} = 0.09\mu m$, $W = 1\mu m$, $L = 0.13\mu m$.

V. CONCLUSION

A compact model for Suspended Gate FET has been reported here. The model is valid for entire voltage operation range and has been implemented in the Verilog-A code. The model has been validated on the 2D simulation of SGFET and shows good accuracy including pull-in and pull-out effects. The model can be used for the circuit simulation using SGFET e.g. SGFET based resonator detection, Sleep transistor logic, memory and switching applications.

ACKNOWLEDGMENT

The authors would like to thank A. Gangadharaiah, M. Mazza for interesting discussions.

REFERENCES

[1] A. Ionescu, V. Pott, R. Fritschi, K. Banerjee, M. J. Declercq, P. Renaud, C. Hibert, P. Fluckiger, and G. Racine, "Modeling and design of a low-voltage SOI suspended-gate MOSFET (SG-MOSFET) with a metal-over-gate architecture," in *IEEE International Symposium on Quality Electronic Design*, 2002, pp. 496– 501.

[2] N. Abelé, R. Fritschi, K. Boucart, F. Casset, P. Ancey, and A. M. Ionescu, "Suspended-gate MOSFET: bringing new MEMS functionality into solid-state MOS transistor," in *IEEE International Electron Devices Meeting*, Dec. 2005, pp. 479– 481.

[3] H. Kam, D. Lee, R. Howe, and T.-J. King, "A new nano-electro-mechanical field effect transistor (NEMFET) design for low-power electronics," in *IEEE International Electron Devices Meeting*, Dec. 2005, pp. 463– 466.

[4] H. Mahfoz-Kotb, A. Salaun, F. Bendriaa, F. L. Bihan, T. Mohammed-Brahim, and J. Morante, "Sensing sensibility of surface micromachined suspended gate polysilicon thin film transistors," *Sensors and Actuators B: Chemical*, vol. 118, no. 1-2, pp. 243–248, Oct. 2006.

[5] J.-M. Sallese and D. Bouvet, "Principles of space-charge based bi-stable MEMS: The junction-MEMS," *Sensors and Actuators A: Physical*, vol. 133, no. 1, pp. 173–179, January 2007.

[6] N. Abelé, A. Villaret, A. Gangadharaiah, C. Gabioud, P. Ancey, and A. M. Ionescu, "1T MEMS memory based on suspended gate MOSFET," in *IEEE International Electron Devices Meeting*, Dec. 2006, pp. 509–512.

[7] T. Nagami, H. Mizuta, N. Momo, Y. Tsuchiya, S. Saito, T. Arai, T. Shimada, and S. Oda, "Three-dimensional numerical analysis of switching properties of high-speed and nonvolatile nano-electromechanical memory," *IEEE Transactions on Electron Devices*, vol. 54, no. 5, pp. 1132–1139, May 2007.

[8] B. Pruvost, H. Mizuta, and S. Oda, "3-d design and analysis of functional nems-gate mosfets and sets," *IEEE Transactions on Nanotechnology*, vol. 6, no. 2, pp. 218–224, March 2007.

[9] D. Tsamados, Y. S. Chauhan, C. Eggimann, K. Akarvardar, H. P. Wong, and A. M. Ionescu, "Numerical and analytical simulations of Suspended-Gate FET for ultra-low power inverters," in *accepted in IEEE European Solid-State Device Research Conference (ESSDERC)*, Sept. 2007.

[10] Y. S. Chauhan, F. Krummenacher, R. Gillon, B. Bakeroot, M. Declercq, and A. M. Ionescu, "Compact Modeling of Lateral Non-uniform doping in High-Voltage MOSFETs," *IEEE Transactions on Electron Devices*, vol. 54, no. 6, pp. 1527–1539, June 2007.

[11] J.-M. Sallese, M. Bucher, F. Krummenacher, and P. Fazan, "Inversion charge lineariazation in MOSFET modeling and rigorous derivation of the EKV compact model," *Solid-State Electronics*, vol. 46, no. 11, pp. 677–683, April 2003.

[12] M. Bucher, "Analytical MOS Transistor Modelling for Analog Circuit Simulation," Ph.D. dissertation, EPFL, 1999, thesis No. 2114.

[13] Y. S. Chauhan, F. Krummenacher, C. Anghel, R. Gillon, B. Bakeroot, M. Declercq, and A. M. Ionescu, "Analysis and Modeling of Lateral Non-Uniform Doping in High-Voltage MOSFETs," in *IEEE International Electron Devices Meeting*, Dec. 2006, pp. 8.3.1 – 8.3.4.

978-1-4244-3039-0/08 $25.00 © 2008 IEEE

21st International Conference on VLSI Design

Optimal Dual-V_T Design in Sub-100 Nanometer PDSOI and Double-Gate Technologies

Aditya Bansal, Jae-Joon Kim*, Keunwoo Kim*, Saibal Mukhopadhyay*, Ching-Te Chuang* and Kaushik Roy

School of Electrical and Computer Engineering, Purdue University, West Lafayette, IN 47907, USA

*IBM T. J. Watson Research Center, Yorktown Heights, NY 10598, USA

Email: bansal@us.ibm.com

Abstract

Dual-V_T CMOS is an effective way to reduce leakage power in high-performance VLSI circuits. In this paper, we explore the technology design space for dual-threshold voltage transistor design in deep sub-100nm technology nodes. We propose a technique of achieving high-V_T devices – longer gate sidewall offset spacers to increase the channel length without increasing the printed gate length. Effectiveness of all the dual-V_T technology options – increasing channel doping, increasing gate length and proposed technique of increasing spacer thickness – are analyzed at transistor to basic logic gate level. Results indicate that the proposed technique yields lower dynamic power consumption and lower performance penalty compared with longer gate length and high body doping devices. Our proposed technique, however, incurs extra fabrication mask similar to high-V_T by increasing body doping.

1. Introduction

In high-performance circuits, multi-V_T CMOS is an effective method for reducing leakage power [1], [2]. It requires a technology with at least two different transistor threshold voltages (V_T). Lower-V_T devices are used in critical paths to maintain performance, while higher-V_T devices are used in non-critical paths to reduce leakage power by trading off the available timing slack with leakage power. Dependence of performance and leakage power on V_T can be given as [3]

$$Performance,\ I_{on} = k\ (V_{gs} - V_T)^{\alpha} \qquad (1)$$

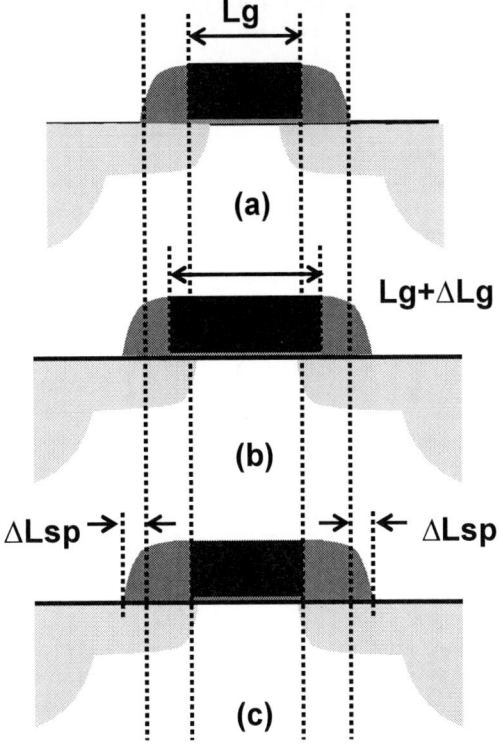

Fig. 2 Schematic of (a) regular-V_T transistor, (b) gate stretched transistor, and (c) spacer stretched transistor. Note that the distance between source/drain junctions is same with gate stretch and spacer stretch. V_T can also be increased by increasing channel doping in regular-V_T FET.

$$Leakage,\ I_{sub} = k'\ exp(q(V_{gs} - V_T)/mkT) \qquad (2)$$

where, k and k' are constants. Above expressions show that increasing V_T reduces the leakage exponentially while performance impact varies as α power (α is approx. 1 in scaled technologies).

Conventionally, high-V_T is achieved by engineering body doping profiles and densities. Increasing the body doping increases total depletion charge in the device channel, thereby increasing V_T. Recently, a microprocessor physical design which uses long channel devices for high-V_T option was demonstrated in [4]. The reason why longer channel device can be used to achieve high-V_T is that V_T rolls off with gate length scaling because of short-channel-effect. Fig. shows the V_T roll-off curve in a sub-90nm PDSOI technology. As can be seen, the V_T of nominal device lies to the left of the knee of the V_T roll-off curve, thus making longer channel length a viable option to

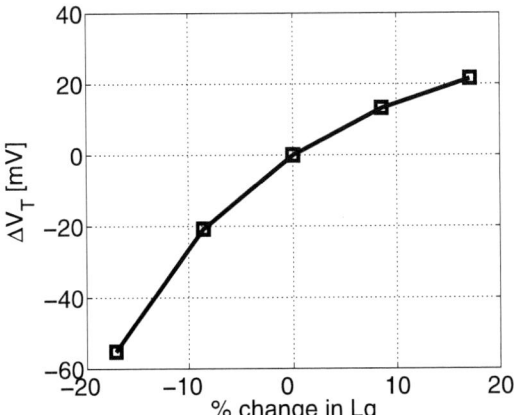

Fig. 1 V_T roll-off curve for NFETs in a sub-90nm PDSOI technology node. Nominal gate length is 35nm.

978-1-4244-3039-0/08 $25.00 © 2008 IEEE

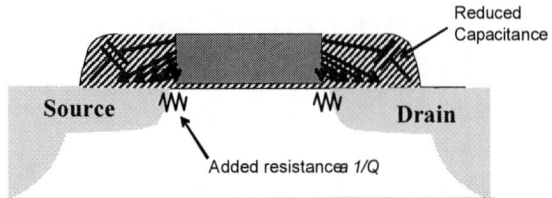

Fig. 3 Schematic of a device with stretched spacers. Geometry is exaggerated for illustration purposes. Increasing spacer, increases parasitic S/D resistance and reduces parasitic gate-to-S/D capacitance.

increase the threshold voltage. Longer channel length can be achieved by increasing the printed gate length, L_g (or gate stretch, Fig. 2(b)).

With increase in gate length, delay penalty is incurred because of reduced on-current as well as increased capacitive loading of the driving stage. Further, gate capacitance (approx. $\varepsilon_{ox}WL_g/t_{ox}$) also increases with increase in channel length, giving simple approximation for gate delay as

$$CV/I = k'' L_g^2 / (Vgs-V_T)^\alpha \qquad (3)$$

where k'' is a constant. Note that increased gate capacitance will also increase the dynamic power consumption and gate leakage current because of increased gate area.

In this work, we propose a novel dual-V_T technology option which uses the increase in V_T similar to gate stretch without increasing the capacitance. In particular

- we propose to increase V_T in dual-V_T design by increasing the gate sidewall offset spacer thickness resulting in increased channel length
- we show that the proposed technique gives similar static power savings as increasing printed gate length along with reduced dynamic power dissipation and lower performance penalty
- we analyze the effectiveness of proposed technique in PD/SOI and double-gate technologies considering the tradeoffs due to different device structures
- we further analyze the impact of process variations in printed gate length and spacer thickness

Drift-diffusion solver [5] is used for mixed-mode device simulations. Quantum confinement in ultrathin silicon body double-gate devices is accounted for by solving 1D Schrodinger equation.

2. Proposed Dual-V_T Design: Spacer Stretch

In deep sub-micron technologies, V_T increases with the increase in channel length because of short-channel-effect. Conventionally, channel length is increased by stretching the printed gate. We propose to increase the channel length, and hence V_T, by increasing the gate sidewall offset spacer thickness. In extremely scaled gate length transistors ($L_g<50nm$), source/drain extension (SDE) doping is controlled at the outer edges of the spacers to achieve the desired channel length (L_{ch}) [6] because of lateral diffusion of dopants. Hence, increasing the spacer

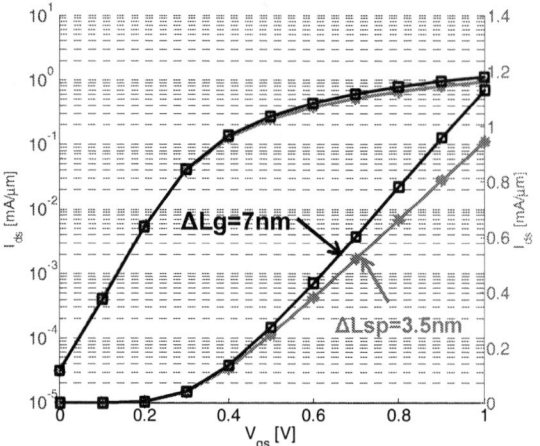

Fig. 4 Id-Vg plot for gate stretch (20%) and equivalent spacer stretch NFETs. Nominal gate length is 35nm.

thickness will effectively increase the channel length. Note that the printed gate length or bottom dimension of the gate remains same as regular-V_T.

Fig. 3 shows the schematic of a transistor with stretched gate sidewall spacers. Stretching the spacer results in

1) weak control of non-overlap region of the channel by the gate sidewall fringing field. This reduces the carrier density under the spacers resulting in increased channel resistance.
2) reduced gate to SDE overlap resulting in reduced overlap and inner/outer fringe capacitances [8].
3) reduced gate to SDE electric field resulting in reduced gate edge direct tunneling leakage.

3. Comparison of Spacer Stretch with Gate Stretch and Increasing Body Doping

Required channel length increase from regular-V_T device depends on either the leakage or performance target. To meet the leakage target, certain minimum increase in body doping or gate stretch or spacer stretch is required. On the other hand, based on available timing slack in non-critical paths, certain delay penalty is affordable. *We target approximately 3X reduction in sub-threshold leakage with the constraint that delay penalty should not increase more than 30%.*

High-V_T transistors, obtained by increasing body doping (Nb), are designed to meet the desired leakage reduction target of 3X. Since gate stretch and spacer stretch techniques result in increased channel length and utilize V_T roll-off to increase V_T, we analyze/compare the impact of these techniques on leakage currents, drive current and gate capacitance.

3.1. Currents (on-current and sub-threshold)

978-1-4244-3039-0/08 $25.00 © 2008 IEEE

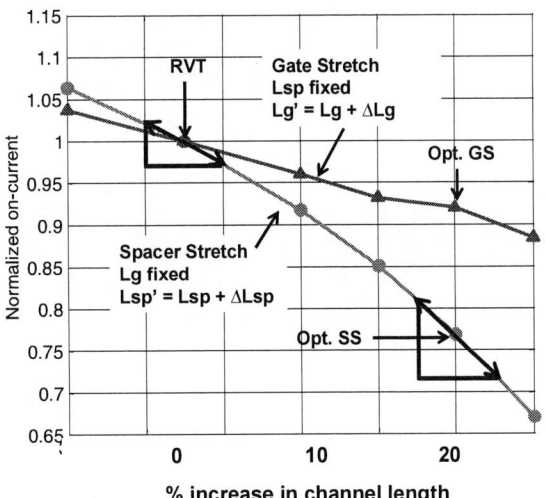

Fig. 5 On-current vs % increase in channel length. Increase in channel length from regular-V_T (0%) is obtained by stretching either gate or spacer. Nominal gate length: 35nm.

Fig. 4 shows the I_d-V_g plots for *20%* of gate stretch over regular-V_T and equivalent spacer stretch in an NFET. It can be seen that the sub-threshold characteristics are almost same in both the cases. In sub-threshold, current is dominated by diffusion and depends upon the channel length. Hence, for iso sub-threshold leakage reduction with gate or spacer stretch, increase in channel length should be same. We increase spacer thickness by an amount so that channel length increases by the same amount as with gate stretch. In inversion, the region under the stretched spacers is weakly controlled by the gate edge electric field. This results in low charge density in these regions compared with strongly controlled channel region directly under the gate. Reduced inversion charge density increases the resistance in the stretched spacer region resulting in reduced on-current. Fig. 5 shows NFET on-current vs % increase in channel lengths obtained by stretching the gate or spacer (L_{sp}). It can be seen that for a given channel length, on-current is lower in stretched spacer transistor compared with the gate stretch because of increased parasitic S/D resistance.

Further note that leakage between gate-drain and gate-source due to edge direct tunneling through the thin gate dielectric reduces with increase in spacer thickness because of reduced gate-to-drain/source overlap [7].

3.2. Gate Capacitance

Stretching the spacer reduces the gate to SDE overlap capacitance and can also result in non-overlap or underlap of gate to SDE. Hence, unlike gate stretch, the gate capacitance does not increase with increase in channel length. Fig. 6 shows the C-V plots obtained for 20% gate stretch over regular-V_T NMOSFET and equivalent spacer stretch. It can be seen that the gate capacitance in inversion (@$V_{gs}=V_{dd}$) is the same in regular-V_T and spacer stretched

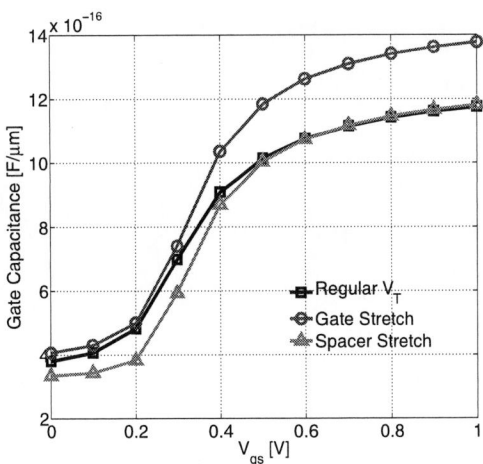

Fig. 6 C-V plots for regular-V_T, gate stretch and equivalent spacer stretch NFETs.

FET. This is because the bottom dimension of gate is same in these cases; however, it increases with gate stretch. Moreover, with spacer stretch also reduces the effective charge accumulation during a voltage swing at the gate which determines the effective gate capacitance (C_{geff}). The effective gate capacitance, computed by integrating the area under the C-V plot reduces by *18%* in equivalent spacer stretched transistor compared to gate stretched one. This results in reduced dynamic power consumption in the preceding stages of stretched spacer gate.

3.3. Performance

Stretching the spacer reduces the on-current because of increased channel and parasitic S/D resistance. Also, parasitic overlap and fringe capacitances reduce with increase in spacer stretch. Thus, net impact on the delay depends on the tradeoff between on-current reduction and capacitance reduction. Fig. 7 shows the trade off between gate delay (*CV/I*) and sub-threshold leakage reduction

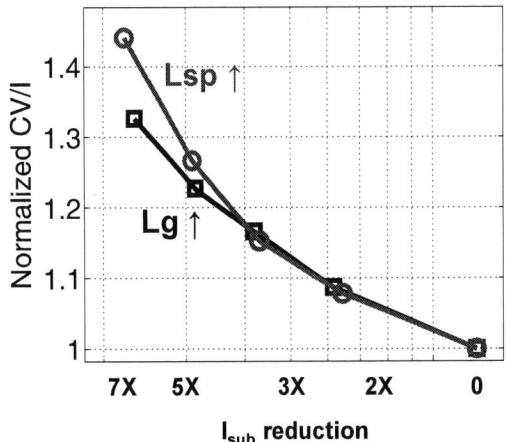

Fig. 7 Gate delay (CV/I) vs. sub-threshold leakage by varying gate length and by varying spacer thickness. Delay and leakage are normalized to regular-V_T NFET.

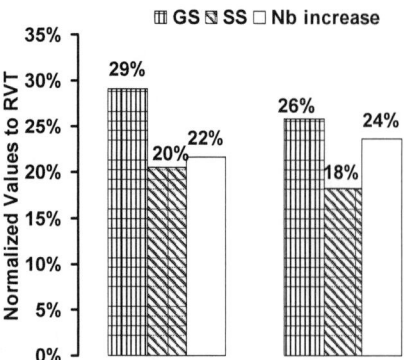

Fig. 8 Mixed-mode simulation results for INV and 2-input NAND gate delay with gate stretch (GS), spacer stretch (SS) and by increasing body doping. All the techniques yield iso leakage reduction (3X). (RVT: regular-V_T).

obtained by increasing channel length using two techniques – stretching the gate (L_g) or stretching the spacer (L_{sp}). It can be seen that till certain spacer stretch, stretching the gate or spacer result in similar performance penalty for given sub-threshold leakage reduction. For leakage reduction target of 4X or higher, large spacer stretch is required resulting in degraded performance.

Fig. 8 shows the propagation delays for INV and 2-input NAND gates. It can be seen that the delay of spacer stretched devices is lower than the gate stretched and high-V_T (by increased Nb) devices at iso off-state leakage reduction (3X). This is not entirely in agreement with the simple gate delay measure CV/I (Fig. 7), where we expected gate stretch and spacer stretch to have similar delay penalties. This is because with increase in spacer thickness, drain to gate capacitance reduces resulting in lower net drain capacitance at the internal nodes. This implies that the net capacitance reduction due to spacer stretch is higher than the on-current reduction.

3.4. Impact of Process Variations

We analyzed the impact of inter-die variations in printed gate length on delay distribution of high-V_T (Nb), gate stretch, and spacer stretch designs (Fig. 9). We first computed the sensitivity of circuit delay to several gate lengths around nominal value. Then assuming Gaussian distribution of gate lengths across the dies, random gate lengths are generated using $\pm 3\sigma = 15\%$ of poly length. Fig. 10 shows the delay variation in a 2-input NAND chain. Gate stretch and spacer stretch are more robust compared to

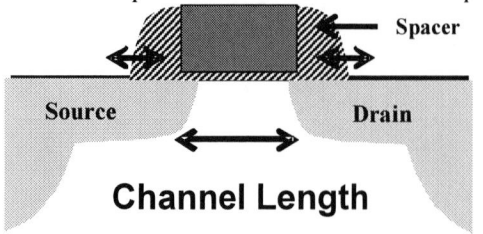

Fig. 9 Thickness of the spacer can vary because of process variations resulting in varying channel lengths. Note that the inside spacer edge variation will vary printed Lg.

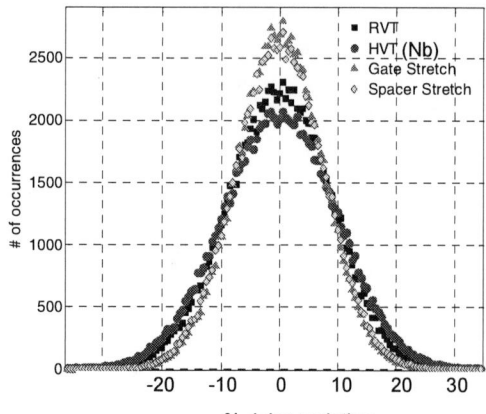

Fig. 10 Inter-die Lg variation in a 2-input NAND chain: $\sigma_{RVT}=8.8\%$, $\sigma_{HVT(Nb)}=9.7\%$, $\sigma_{GS}=7.3\%$, $\sigma_{SS}=7.6\%$. 0% delay variation corresponds to the nominal case.

regular-V_T and high-V_T (Nb) because of increased channel length. Spacer stretch performs almost same as gate stretch because of similar channel lengths. Spacer stretch however, is prone to variation in spacer thickness. Fig. 5 shows the variation in on-current with spacer thickness. For increased channel length of *20%* (with spacer stretch), on-current is more sensitive to channel length variation compared to the nominal channel length (in regular-V_T).

It should be noted that, spacer stretch and high-V_T (Nb) require extra fabrication masks as the two different spacers or body dopings (for regular-Vt and high-Vt) can not be realized in the same mask step. On the other hand, the devices with different gate length can be realized using a single mask. Hence, the spacer stretch has a fabrication overhead compared to gate stretch.

.

4. Double-Gate MOSFETs: Is Spacer Stretch Applicable?

Ultra-thin body (UTB) double-gate devices are promising candidate in sub-50nm nodes because of better scalability and higher on-current. In this section, we analyze the effect of spacer stretch on the resistance under the spacers in double-gate MOSFETs. Fig. 11 shows the schematic of a double-gate MOSFET used for simulation. Original gate sidewall spacer thickness is *5nm*, which is obtained by mutually optimizing the overlap capacitance and on-current to get minimum delay. In the simulations, we have used devices with intrinsic body. Fig. 12 shows the trade off between sub-threshold leakage and gate delay

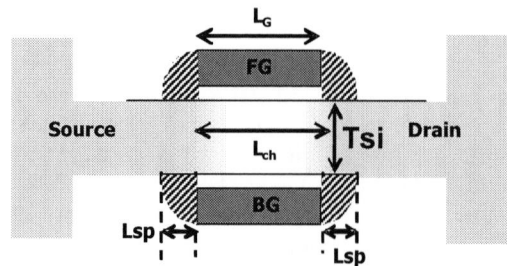

Fig. 11 Schematic of the double-gate MOSFET used for analysis. *Lg=20nm, Lsp=5nm* and *Tsi=9nm, undoped body.*

128

978-1-4244-3039-0/08 $25.00 © 2008 IEEE

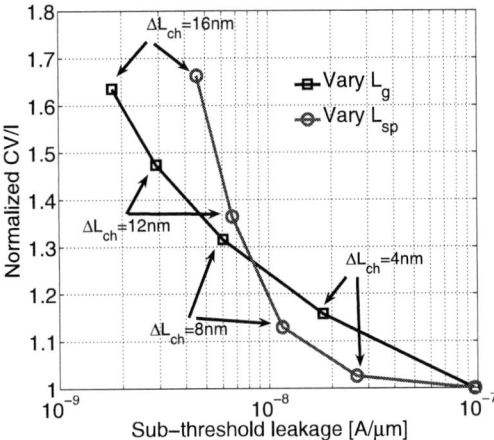

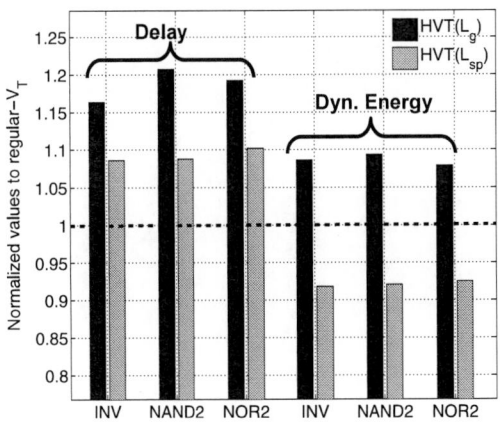

Fig. 12 CV/I vs. Isub for varying *Lg* and *Lsp*. Values are normalized to regular-V_T case with *100nA/μm* of leakage. Body is intrinsically doped, hence, high-V_T by increasing body doping is not considered.

(CV/I_{on}) for two dual-V_T device design techniques in double-gate technology. Sub-threshold leakage reduction target is normally around *3X-5X*. As can be seen from Fig. 12, for sub-threshold leakage reduction target of *10X* or less, higher L_{sp} yields lower gate delay in high-V_T devices compared to L_g. For example, for leakage reduction target of *5X*, L_{sp} should be increased to *8nm* (original *5nm*) or L_g should be increased to *24nm* (original *20nm*). Fig. 13 shows the delay and dynamic energy dissipation in three basic logic blocks – INV, NAND and NOR. HVT(L_{sp}) yields *15-20%* lower dynamic energy compared to HVT(L_g) at lower delay penalty. From Fig. 12 and Fig. 13, we can conclude that, the dual-spacer is an effective method for designing dual-Vt devices even in double gate technology.

5. Discussions: How Device Parameters Impact Effectiveness of Dual-Spacer?

It is interesting to note that, dual-V_T design using spacer thickness is applicable in reducing static leakage only if increasing channel length increases V_T. In other words, the impact of short-channel effect should be high resulting in a sufficient V_T-roll-off with reduction in channel length. Hence, dual-spacer technique is expected to be more effective in devices with higher short-channel effect (SCE). On the other hand, the effectiveness of dual-spacer compared to dual-channel length requires that, the degradation of "on" current with increase in spacer thickness should be as small as possible. This will result in better delay for high-V_T devices in dual-spacer compared to dual-gate length (as explained earlier). The effect of spacer thickness on device current depends on how well the gate edge field can control the inversion region under stretched spacer. This leads us to an interesting discussion on how the different design parameters for the regular-V_T device impact the effectiveness of dual-spacer technique. To analyze this, we considered the double gate devices and

Fig. 13 Delay and dynamic energy in three basic logic gates. Values are normalized to corresponding regular-V_T values. HVT(L_g) has ΔL_g=4nm and HVT(L_{sp}) has ΔL_{sp}=3nm on each side of the gate. Note that both HVT(L_g) and HVT(L_{sp}) have approx. 5X leakage reduction (Fig. 12). Since device channels are undoped, HVT (Nb) case is not considered.

studied the impact of body-doping and silicon thickness on effectiveness of dual spacer technique. This is particularly important for double-gate, as they are known for their better SCE immunity. Hence, the effectiveness of dual-spacer will be determined by the degradation of "on" current with spacer thickness.

5.1. Effect of Doping

To analyze the effect of channel doping (N_b) on the resistance under the stretched spacers, we vary N_b from 1e15cm^{-3} (undoped) to 5e18cm^{-3}. Channel doping increases the depletion charge, resulting in increased resistance under the stretched spacers during inversion. Thus, the use of un-doped channel for UTB double-gate FET will reduce the net channel resistance and resistance increase under space stretch. V_T increases with increase in body doping because of increased depletion charge. V_T also increases with increase in L_{sp} because of increased channel length (V_T roll-off). I_{on} can be given as eq. (1) (on page 1), where $k=\mu C_{ox}W/L$. Mobility (μ) reduces with increase in N_b because of increased gate electric field. To capture the effect of mobility and resistance with increase in N_b and L_{sp}, respectively, we compute the drive current at iso gate overdrive V_{gs}-V_T. Fig. 14 shows the drive-current at iso gate-overdrive and V_{ds}=V_{dd} for varying body doping and spacer thicknesses. As can be seen, higher doping results in large drive-current degradation with increase in spacer thickness. We can infer that increase in resistance due to spacer thickness has more severe impact with increase in body doping.

5.2. Effect of Silicon Thickness

Let us now analyze the impact of spacer stretch in devices with varying silicon thicknesses. Thinning the silicon body results in quantum confinement of charge [9].

129

978-1-4244-3039-0/08 $25.00 © 2008 IEEE

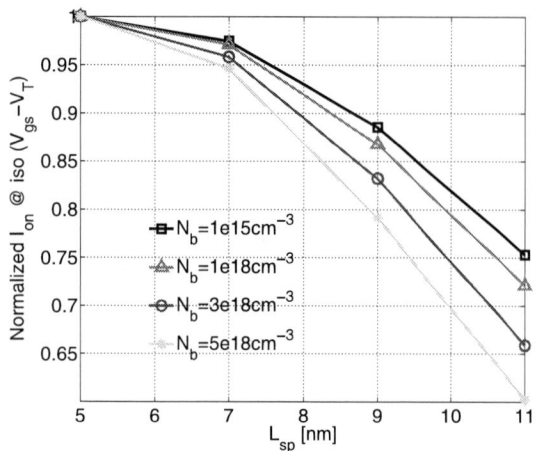

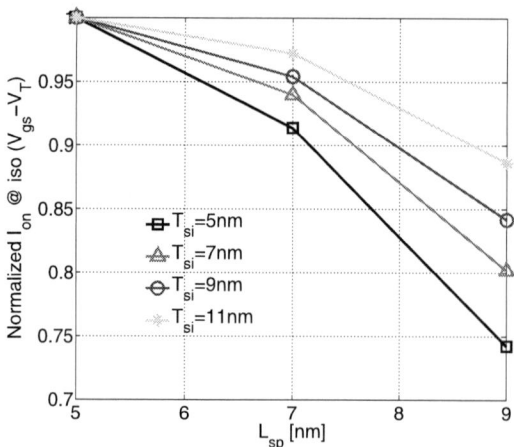

Fig. 14 On-current degradation, at *iso gate overdrive* (V_{gs}-V_T), with increase in N_b and spacer thickness. N_b=$1e15cm^{-3}$ is referred as undoped case. For each body doping, on-current is normalized to its respective value at Lsp=5nm.

This results in peak of inversion charge moving away from the Si-SiO$_2$ interface towards the center of the silicon body. In stretched spacer devices, gate sidewall electric field weakly controls the inversion charge under the spacer regions. As the charge moves away from the surface, the control of gate field on the charge weakens resulting in reduced charge density and increased resistance. Since, V_T varies with T_{si} and L_{sp}, we measure the drive-current at iso gate overdrive (V_{gs}-V_T). Fig. 15 shows on-current at iso gate overdrive for varying silicon thicknesses (T_{si}) and spacer thicknesses (L_{sp}). As expected, for thicker silicon body, on-current degradation is less with increase in spacer thickness.

From the above discussion we can conclude that, the effectiveness of the dual-spacer strongly depends on design of the regular-Vt device. In general, higher SCE in devices improves the leakage reduction with dual-spacer. On the other hand, better control of gate on the inversion charge in the stretched spacer region (lower body doping in that region and/or lower distance of the inversion charge centroid from surface) reduces the on-current degradation with spacer stretch.

6. Conclusions

Dual-V_T technology is an effective way of reducing static power dissipation in high-performance circuits. Increasing channel length by gate stretch is becoming effective as dual-V_T option because of increased short-channel-effect in deep sub-100nm technology nodes. However, gate stretch increases the dynamic power consumption and gate-to-channel tunneling current because of increased gate area. We propose to increase the gate sidewall offset spacer thickness to increase the effective channel length without increasing the gate area. Hence, this does not increase dynamic power and gate-to-channel tunneling current. The reduction in drain/source parasitic capacitances with spacer stretch compensates for the "on" current degradation. This results in lower delay in the high-V_T logic gates designed with spacer stretch FETs compared

Fig. 15 On-current degradation, at *iso gate overdrive* (V_{gs}-V_T), with increase in silicon and spacer thicknesses. For each Tsi, on-current is normalized to its respective value at Lsp=5nm.

to gate-stretch FETs. Our analysis shows that the dual-spacer technique is an effective dual-Vt technology option for sub-65nm technologies both for single-gate PD/SOI and double gate technologies. Fabrication penalty incurred is similar to increasing higher body doping for high-V_T FETs.

7. Acknowledgments

Jae-Joon Kim, Keunwoo Kim, and Ching-Te Chuang are partially supported by the Defence Advanced Research Project Agency (DARPA) contract NBCH30390004 for this work.

8. References

[1] S. Mutoh, T. Douseki, Y. Matsuya, T. Aoki, S. Shigematsu and J. Yamada, "1-V Power Supply High-Speed Digital Circuit Technology with Multithreshold-Voltage CMOS," *IEEE Jrnl. of Solid-State Ckts*, Aug. 1995, pp. 847-854.

[2] L. Wei, Z. Chen and K. Roy, "Design and Optimization of Dual Threshold Circuits for Low Voltage, Low Power Applications," *IEEE Trans. On VLSI Systems*, vol. 17, no. 1, 1999, pp. 16-24.

[3] Y. Taur and T. H. Ning, "*Fundamentals of Modern VLSI Devices*," Cambridge, MA.

[4] S. Rusu, S. Tam, H. Muljono, D. Ayres and J. Chang, "A Dual-Core Multi-Threaded Xeon^{TM} Processor with 16MB L3 Cache," *Digest of Tech. Papers, ISSCC*, 2005, pp. 315-324.

[5] Taurus Device Simulator, Synopsys Inc., 2004.

[6] H.-J. Gossmann, A. Agarwal, T. Parrill, L.M. Rubin and J.M. Poate, "On the FinFET Extension Implant Energy," *IEEE Trans. on Nanotech.*, 2003, pp. 285-290.

[7] A. Bansal, S. Mukhopadhyay and K. Roy, "Device-Optimization Technique for Robust and Low-Power FinFET SRAM Design in NanoScale Era ," *IEEE Trans. on Elec. Dev.*, June 2007, pp. 1409-1419.

[8] A. Bansal, B. C. Paul and K. Roy, "Modeling and Optimization of Fringe Capacitance of Nanoscale DGMOS Devices," *IEEE Trans. on Elec. Dev.*, Feb. 2005, pp. 256-262.

[9] Q. Chen, L. Wang and J. D. Meindl, "Quantum Mechanical Effects on Double-Gate MOSFET Scaling," *Proc. IEEE International SOI Conf.*, 2003, pp. 183-184.

21st International Conference on VLSI Design

Recursive Statistical Blockade: An Enhanced Technique for Rare Event Simulation with Application to SRAM Circuit Design

Amith Singhee[1], Jiajing Wang[2], Benton H. Calhoun[2], Rob A. Rutenbar[1]

[1]Dept. of ECE, Carnegie Mellon University,
Pittsburgh, PA 15213, USA
{asinghee,rutenbar}@ece.cmu.edu

[2]ECE Dept., University of Virginia,
Charlottesville, VA 22903, USA
{jjwang,bcalhoun}@virginia.edu

Abstract

Circuit reliability under statistical process variation is an area of growing concern. For highly replicated circuits such as SRAMs and flip flops, a rare statistical event for one circuit may induce a not-so-rare system failure. The authors of [1] proposed Statistical Blockade as a Monte Carlo technique that allows us to efficiently filter—to *block*—unwanted samples insufficiently rare in the tail distributions we seek. However, there are significant practical problems with the technique. In this work, we show common scenarios in SRAM design where these problems render Statistical Blockade ineffective. We then propose significant extensions to make Statistical Blockade practically usable in these common scenarios. We show speedups of 10^2+ over standard Statistical Blockade and 10^4+ over standard Monte Carlo, for an SRAM cell in an industrial 90nm technology.

1. Introduction

Circuit reliability under statistical process variation is an area of growing concern. Designs that add excess safety margin, or rely on simplistic assumptions about "worst case" corners no longer suffice. Worse, for critical circuits such as SRAMs and flip flops, replicated across 10K - 10M instances on a large design, we have the new problem that statistically rare events are magnified by the sheer number of these elements. In such scenarios, an exceedingly rare event for one circuit may induce a not-so-rare failure for the entire system.

Monte Carlo analysis (MC) [2] remains the gold standard for the required statistical analysis. Standard Monte Carlo techniques are, by construction, most efficient at sampling the statistically likely cases. When used for simulating statistically unlikely or rare events, these techniques are extremely slow. For example, to simulate a 5σ event, 100 million circuit simulations would be required, on average.

One avenue of attack is to abandon Monte Carlo. Several analytical and semi-analytical approaches have been suggested to model the behavior of SRAM cells [3][4][5] and digital circuits [6] in the presence of process variations. All suffer from approximations necessary to make the problem tractable.

[4] and [6] assume a linear relationship between the statistical variables and the performance metrics (*e.g.* static noise margin), and assume that the statistical process parameters and resulting performance metrics are normally distributed. This can result in gross errors, especially while modeling rare events, as we shall show later. When the distribution varies significantly from Gaussian, [4] chooses an F-distribution in an *ad hoc* manner. [3] presents a complex analytical model limited to a specific transistor model (the transregional model) and further limited to only static noise margin analysis for the 6T SRAM cell. [5] again models only the static noise margin (SNM) for SRAM cells under assumptions of independence and identical distribution of the upper and lower SNM, which may not always be valid.

A different avenue of attack is to modify the Monte Carlo strategy. [7] shows how Importance Sampling can be used to predict failure probabilities. Recently, [8] applied an efficient formulation of these ideas for modeling rare failure events for single 6T SRAM cells, based on the concept of *Mixture Importance Sampling* from [9]. The approach uses real SPICE simulations with no approximating equations. However, the method only estimates the exceedence probability of a *single* value of the performance metric. A re-run is needed to obtain probability estimates for another value. No complete model of the tail of the distribution is computed. The method also combines all performance metrics to compute a failure probability, given *fixed* thresholds. Hence, there is no way to obtain separate probability estimates for each metric, other than a separate run per metric. Furthermore, given that [7] advises against importance sampling in high dimensions, it is unclear if this approach will scale efficiently to large circuits with many statistical parameters.

The authors of [1] presented Statistical Blockade (SB), a general and efficient MC method that addresses both problems previously described: very fast generation of samples—rare events—with sound models of the tail statistics for any performance metric. The method imposes almost no *a priori* limitations on the form of the statistics for the process parameters, device models, or performance metrics. The key observation behind Statistical Blockade is that *generating* each sample is not expensive: we are merely creating the parameters for a circuit.

131

978-1-4244-3039-0/08 $25.00 © 2008 IEEE

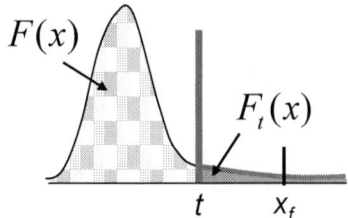

FIGURE 1. Example distribution of a circuit performance metric (e.g., SRAM write time). The solid region is the *tail* region. SB focuses the sampling to this region for fast sampling of rare events.

Evaluating the sample is expensive, because we simulate it. The paper developed a method to quickly *filter* these samples, and *block* those that are unlikely to fall in the low-probability tails we seek. It used techniques from *data mining* [10] to build *classifier* structures, from a small set of Monte Carlo training samples, to create the necessary blocking filter. Given these samples, it showed how to use the rigorous mathematics of *Extreme Value Theory* (EVT [11]) to build sound models of these tail distributions. The paper successfully applied SB to a variety of circuits with dimensionality ranging up to 403, with speedups of up to 2 orders of magnitude over Standard Monte Carlo.

SB can, however, completely fail for certain commonly seen SRAM metrics (e.g., data retention voltage) because of the presence of conditionals in the formulation of the metric. Also, if rare samples with extremely low probability (e.g. 5σ and beyond) are required, SB can still become prohibitively expensive. In this work, we extend the SB technique in two significant ways: 1) we propose a solution to solve the problem of SB failing for certain common SRAM metrics, and 2) we develop a recursive strategy to achieve further speedups of orders of magnitude, while simulating extremely rare events (5σ and beyond).

This paper is organized as follows. Section 2 reviews the Statistical Blockade filtering technique from [1]. Section 3 presents the first problem with the formulation, circuit metrics with conditionals, and proposes a solution. Section 4 presents the second problem with SB, sampling extremely rare events, and proposes a solution. Section 5 presents experimental results and Section 6 offers concluding remarks.

2. Background: Statistical Blockade Filtering

Fig. 1 shows an example distribution $F(x)$ of a circuit metric; e.g., SRAM write time. As an example, consider a 1Mb cache, where the SRAM cell has a failure probability of 1ppm, given a failure threshold, x_f. In such a case we would need to simulate 1 million MC samples to generate one such failure event and made any prediction about the failure probability. In fact, we would need many more to generate sufficient failure events to ensure statistical confidence of the prediction. This approach would become much worse for lower failure

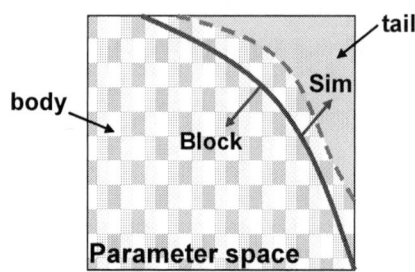

FIGURE 2. The classifier in statistical parameter space is shown as a solid boundary. The dashed line is the exact tail region boundary. The relaxed classification boundary allows us to *block* most non-tail points.

probabilities. This scenario is common in today's SRAM designs.

SB was proposed in [1] to significantly speed up the simulation of rare events and prediction of low failure probabilities. SB defines a *tail threshold* (for example, the 99% point) t, as shown in Fig. 1. Without loss of generality, the part of the distribution greater than t is called the tail. The key idea is to identify that region in the parameter (process variable) space that yields circuit performance values (e.g., SRAM write time) greater than t. Once this is known, those MC samples that do not lie in this tail region are not simulated, or *blocked*. Only those MC samples that lie in the tail region are simulated. Hence, the number of simulations can be significantly reduced. For example, if t is the 99-th percentile, only 1% of the MC samples will be simulated, resulting in an immediate speedup of 100x over standard MC.

To build this model of the boundary of the tail region a small MC sample set (1,000 points) is used to train a classifier. A classifier is an indicator function that allows us to determine set membership for complex, high-dimensional, nonlinear data. Given a data point, the classifier reports true or false on the membership of this point in some arbitrary set. For Statistical Blockade, this is the set of parameter values *not* in the tail region we seek. However, it is difficult, if not impossible, to build an exact model of the tail region boundary. Hence, we relax the requirement to allow for classification error. This is

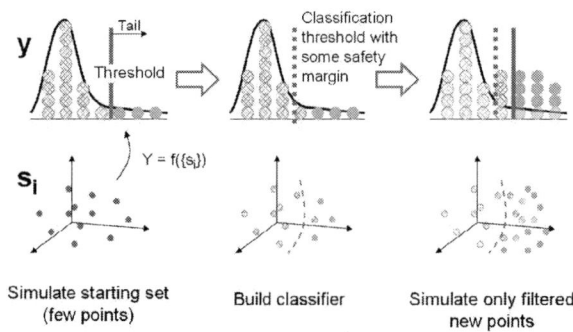

FIGURE 3. Classification based sampling

978-1-4244-3039-0/08 $25.00 © 2008 IEEE

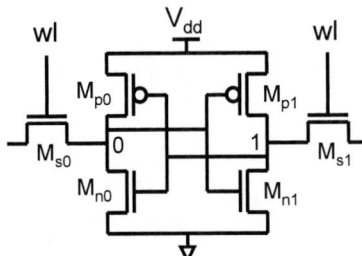

FIGURE 4. A standard 6-T SRAM cell.

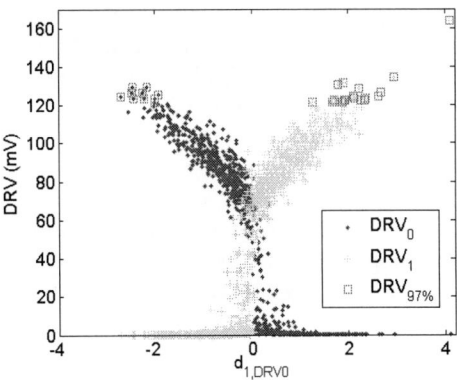

FIGURE 5. Behavior of DRV_0 and DRV_1 along the direction of maximum variation in DRV_0. The worst 3% DRV values are shown as squares, clearly showing the disjoint tail regions (along this direction in the parameter space).

done by building the classification boundary at a *classification threshold* t_c that is less than the tail threshold t. Fig. 2 shows this relaxed classification boundary in the parameter space. The dashed line is the exact boundary of the tail region for the tail threshold t, and the solid line is the relaxed classification boundary for the classification threshold t_c.

SB filtering is then accomplished in three steps (Fig. 3):

1) *Perform initial sampling* to generate data to build a classifier. This initial sampling can be standard Monte Carlo or importance sampling.

2) *Build a classifier* C using a classification threshold t_c. To minimize false negatives (tail points classified as non-tail points), choose $t_c < t$.

3) *Generate more samples* using MC, following the CDF F, but simulate only those that are classified as tail points.

From the simulated samples, some will be in the tail region and some will be in the non-tail region. [1] shows how to use Extreme Value Theory to fit a parametric distribution (the Generalized Pareto Distribution) to these tail points to generate an analytical model for the failure probability, given any failure threshold $x_f > t$.

In the rest of this paper, we will focus on the classifier building and the sample filtering parts of this framework. We will show how they can fail for certain common scenarios and present effective solutions.

3. Classifier failure: Conditionals

3.1 The problem

Consider the 6-T SRAM cell shown in Fig. 4. With scaling reaching nanometer feature sizes, subthreshold and gate leakage become very significant. Particularly for the large memory blocks seen today, the standby power consumption due to leakage can be intolerably high. Supply voltage (V_{dd}) scaling [12] is a powerful technique to reduce this leakage, whereby the supply voltage is reduced when the memory bank is not being accessed. However, lowering V_{dd} also makes the cell unstable, ultimately resulting in data loss at some threshold value of V_{dd}, known as the *Data Retention Voltage* or DRV. Hence, DRV is the lowest supply voltage that still preserves

the data stored in the cell. DRV is computed as follows.

$$DRV = max(DRV_0, DRV_1) \qquad (1)$$

where DRV_0 is the DRV when the cell is storing a 0, and DRV_1 is the DRV when it is storing a 1. If the cell is balanced (symmetric), then $DRV_0 = DRV_1$. However, if there is any mismatch due to process variations, they become unequal. This creates a situation where the standard SB classification technique would fail. We will explain this in more detail now.

Suppose we run a 1,000 sample MC, varying all the mismatch parameters in the SRAM cell according to their statistical distributions. This would give us distributions of values for DRV_0, DRV_1 and DRV. In certain parts of the mismatch parameter space $DRV_0 > DRV_1$, and in other parts $DRV_0 < DRV_1$. This is clearly illustrated in Fig. 5. Using SiLVR, from [13], we extracted the direction in the parameter space that has maximum impact on DRV_0 (maximum variation), called *latent variable* in the paper. The figure plots the simulated DRV_0 and DRV_1 values along this direction ($d_{1, DRV0}$). We can clearly see that they are inversely related: one decreases as the other increases. Now let us take the *max* as in (1), and choose the classification threshold t_c for DRV as the 97-th percentile. Then we pick out the worst 3% points from the classifier training data and plot them against the same

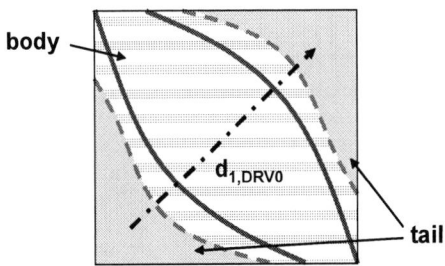

FIGURE 6. Parameter space with two disjoint tail regions for the same circuit metric (e.g. DRV).

133

978-1-4244-3039-0/08 $25.00 © 2008 IEEE

direction $d_{1, DRV0}$, in Fig. 5. We can clearly see that these points (squares) lie in two disjoint parts of the parameter space. Since the tail region defined by a tail threshold $t > t_c$ would be a subset of the classifier tail region (defined by t_c), it is obvious that the tail region consists of two disjoint regions of the parameter space. This is illustrated with a 2-D example in Fig. 6. The figure also shows the direction vector for $d_{1, DRV0}$. The solid tail regions on the top-right and bottom-left corners of the parameter space correspond to the large DRV values shown as squares in Fig. 5.

In such a situation the SB classifier is unable to create a single boundary to separate the tail and non-tail regions. The problem stems from the *max* operation in (1), since it combines subsets of the tail regions of DRV_0 and DRV_1 to generate the tail region of DRV. The same problem occurs for any other such metric (e.g., Static Noise Margin) with a conditional operation. We now propose a solution to this problem.

3.2 Solution

Instead of building a single classifier for the tail of DRV in (1), we will build two separate classifiers, one for the the 97-th percentile ($t_c(DRV_0)$) of $DRV0$, and another for the 97-th percentile ($t_c(DRV_1)$) of DRV_1. The generated MC samples will then be filtered through both these classifiers: points classified as non-tail by *both* the classifiers will be blocked, and the rest will be simulated. In the general case, if the circuit metric y is given as

$$y = max(y_0, y_1, ...)\qquad(2)$$

the resulting algorithm is as follows.

1) Perform initial sampling to generate data to build a classifier and estimate tail and classification thresholds.

2) For each argument y_i of the conditional (2), build a classifier C_i at a classification threshold $t_c(y_i)$ that is less than the tail threshold $t(y_i)$.

3) Generate more samples using MC, but block the samples classified as *non-tail* by *all* the classifiers. Simulate the rest and compute y for the simulated points.

Hence, in the case of Fig. 6, we build a separate classifier for each of the two boundaries. From the simulated points, those with $y > t$ are chosen as tail points for further analysis [1]. Also note that this same algorithm can be used for the case of multiple metrics. Each metric would have its own thresholds and its own classifier, just like each argument in (2).

4. Simulating Extremely Rare Events

4.1 The problem

Consider a 10 Mb memory, with no redundancy or error correction. Even if the failure probability of each cell is as low as 0.1 ppm, every such chip will still fail on average. Hence, the worst case (largest) DRV from a 10 million MC should, on

```
1.for each argument y_i in y = max(y_i)
2.    n = n_0 = 1000
3.    y_{i,tail} = Simulate(MCarlo(n))
4.    while (n < N)
5.        y_{i,tail} = GetWorst(n_0, y_{i,tail})
6.        t = Percentile(y_{i,tail}, 99)
7.        t_c = Percentile(y_{i,tail}, 97)
8.        C = BuildClassifier(y_{i,tail}, t_c)
9.        n = n * 100
10.       y_{i,tail} = Simulate(Filter(C, MCarlo(n))
11.   end
12.y_{tail} = max(y_{0,tail} ∪ y_{1,tail} ∪ ...)
```

FIGURE 7. A recursive formulation for Statistical Blockade for simulating extremely rare events, that can also handle conditionals.

average, be below the standby voltage. To estimate this at least 10 million MC samples have to be run. If we want to reduce the *chip* failure probability to less than 1%, we need to look at the worst case DRV from a 1 billion MC run. This is equivalent, approximately, to the 6σ value of DRV -- the 6σ point from a standard normal distribution has the same failure probability. Using Statistical Blockade, we can reduce the number of samples, using a classification threshold $t_c = $ 97-th percentile. This would reduce the number of simulations from 1 billion to 30 million, which is still very large. Even with a perfect classifier, where we can choose $t_c = t = $ 99-th percentile, the number of simulations would still be 10 million. Moving t_c to higher percentiles will help reduce this further, but many more initial samples will be needed for a believable estimate of t_c and for training the classifier. Now we describe a recursive formulation that reduces the simulation count drastically.

4.2 Solution

Let us first assume that there are no conditionals. For a tail threshold equal to the α-th percentile, let us represent it as t^α, and the corresponding classification threshold as t_c^α. Using the algorithm from Section 3.2, build a classifier C^α and generate sufficient points with $y > t^\alpha$, so that a higher percentile (t^β, t_c^β, $\beta > \alpha$) can be estimated. For this new, higher threshold a new classifier C^β is trained and a new set of tail points ($y > t^\beta$) are generated. This new classifier will block many more points than C^α, significantly reducing the number of simulations. This procedure is repeated to push the threshold out more until the tail region of interest is reached. The complete algorithm is shown in Fig. 7.

Now, we will describe the functions used in the algorithm. In line 1, we repeat lines 2-10 for each argument of the conditional. If there is no conditional, lines 2-10 are not repeated. The conditional *max* is used without loss of generality. N is the total number of MC samples that would be needed to reach the tail regions required; e.g., $N = 1$ billion for reaching 6σ. The function MCarlo(n) generates n samples, and the function Simulate() actually simulates the samples passed to it.

978-1-4244-3039-0/08 $25.00 © 2008 IEEE

The returned vector consists of both the input parameter sets for simulation and the corresponding circuit metrics computed for each sample. The function GetWorst(n_0, x) returns the n_0 worst samples from the set x. BuildClassifier(x, t_c) builds a classifier using training points x. Hence in line 8, C is a classifier. The function Filter(C, x) blocks the samples in x classified as non-tail by C and returns the samples classified as tail points. The function Percentile(x, p) computes the p-th percentile of the output values in the set x. Line 12, takes all the tail points generated for all the arguments in line 1, and computes the overall metric y using the *max* operator.

The basic idea is to use a tail threshold (and its corresponding classification threshold) that is very far out in the tail, so that the simulations are restricted to the very rare events we are interested in. This is being done in a recursive manner by estimating lower thresholds first and using them to estimate the higher threshold without having to simulate a large number of points. For example, if we want to use the 99.9999 percentile as the tail threshold $t^{99.9999}$, we first estimate the 99.99 percentile threshold $t^{99.99}$. To estimate this in turn, we first estimate the 99 percentile threshold t^{99}. At each stage we use a classifier corresponding to that threshold to reduce the number of simulations for estimating the next higher threshold. The next section will present experimental results.

5. Experimental Results

The techniques described in this paper were applied to a standard 6T SRAM cell, for the case of DRV. The cell was implemented in an industrial 90nm process and all the mismatch statistical parameters were varied as per the industrial process design kit (PDK). We used a Support Vector Machine classifier [14], similar to [1].

5.1 An analytical model for DRV

The authors in [15] develop an analytical model for predicting the Cumulative Density Function (CDF) of the DRV, that uses not more than 5,000 MC simulations. The CDF is given as

$$F_{DRV}(x) = 1 - erfc\left(\frac{\mu_0 + k(x - V_0)}{\sqrt{2}\sigma_0}\right) + \frac{1}{4}\left(erfc\left(\frac{\mu_0 + k(x - V_0)}{\sqrt{2}\sigma_0}\right)\right)^2$$
(3)

where x is the DRV value. k is the sensitivity of Static Noise Margin (SNM) to the supply voltage, computed using a DC sweep. μ_0 and σ_0 are the mean and standard deviation of the SNM (SNM_0) distribution for the circuit, for a user-defined supply voltage V_0. SNM_0 is the SNM of the cell while storing a 0. These are computed using a short Monte Carlo run. Complete details regarding this analytical model are provided in [15]. The q-th quantile (e.g., the 6σ point) can be estimated as

$$DRV(q) = \frac{1}{k}(\sqrt{2}\sigma_0 erfc^{-1}(2 - 2\sqrt{q}) - \mu_0) + V_0$$
(4)

Hence, $DRV(q)$ is the supply voltage V_{dd} such that

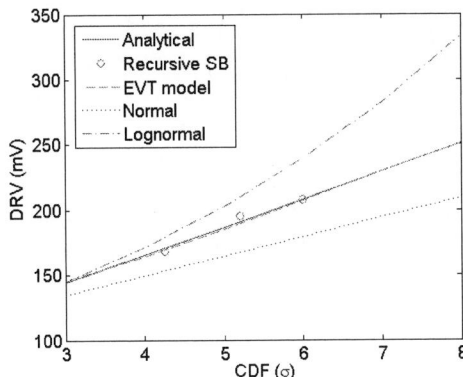

FIGURE 8. The worst case DRV values from RSB closely match the model in (5). Fitting an EVT model as per [1] to the data from RSB also shows close match with (5). Normal and log-normal fits are inaccurate.

$$P(DRV \le V_{dd}) = q$$
(5)

We compare the worst case DRV values from our technique, for a given number of MC samples, with the value predicted by (5) for the corresponding quantile. For example, we can compute the 4.5σ DRV value from (5) and compare it with the worst case DRV from a 1 million sample MC run: 1 ppm is the failure probability of the 4.5σ point.

5.2 Results

Fig. 8 shows a graphical comparison of five different methods:

1) **Analytical**: The 3σ to 8σ DRV values (quantiles) predicted by equation (5).

2) **Recursive SB**: The algorithm in Fig. 7 was run for $N = 1$ billion: lines 5-10 were run three times, corresponding to 100,000, 10 million and 1 billion MC samples, respectively. The worst case DRV from these 3 recursion stages are estimates of the 4.26σ, 5.2σ and 6σ points, respectively.

3) **EVT model**: The tail points from the last recursion stage (1 billion MC) are used to fit a Generalized Pareto Distribution (GPD), as per [1]. This GPD is then used to predict the 3σ to 8σ DRV values.

4) **Normal**: A normal distribution is fit to data from a 1,000 sample MC run, and used to predict the same DRV values.

5) **Lognormal**: A lognormal distribution is fit to the same 1,000 MC samples, and used for prediction.

From the plots, we can immediately see that Recursive SB estimates are very close to the estimates from the analytical model. Table 1 shows the number of circuit simulations performed at each of the three recursion stages, along with the initial 1,000 sample MC run. The total number of simulations used is a very comfortable 41,721, resulting in a speedup of 4 orders of magnitude over standard MC and 700 times over SB.

135

978-1-4244-3039-0/08 $25.00 © 2008 IEEE

Also, we can extend the prediction power to 8σ without any additional simulations, by using the GPD model. Standard MC would need over 1.5 quadrillion points to generate an 8σ point. For this case the speedup over standard MC is extremely large. The normal and lognormal fits show significant error compared to the analytical model. The normal fit is unable to capture the skewness of the actual DRV distribution, while the lognormal distribution has a heavier tail than the true DRV distribution and, hence, over-estimates the skewness.

Stage	Num. simulations
Init	1,000
1	11,032
2	14,184
3	15,505
Total	**41,721**
Speedup over MC	**23,969x**
Speedup over SB[1]	**719x**

TABLE 1. Number of circuit simulations run per recursion stage to generate a 6σ DRV sample

A final point to highlight is that recursive SB is a completely general technique to estimate rare events and their tail distributions. In the case of the SRAM cell DRV experiment, we were lucky enough to have an extremely recent analytical result against which to compare performance. Obviously, if one has such analytical models available, one should use them. Unfortunately, in most cases, one does not, and one must fall back on some sort of Monte Carlo analysis. In such scenarios, recursive Statistical Blockade has three attractive advantages:

1) it is circuit-neutral, by which we mean that any circuit that can be simulated can be attacked with the technique;

2) it is metric-neutral, by which we mean that any circuit performance metric that can be simulated can be analyzed with the technique;

3) as seen in our SRAM DRV experiments, it is extremely efficient, faster usually by several orders of magnitude than simple-minded brute-force Monte Carlo algorithms.

6. Conclusions

Statistical Blockade was proposed in [1] for 1) efficiently generating samples in the tails of distributions of circuit performance metrics, and 2) deriving sound statistical models of these tails. However, the method has some practical shortcomings: it fails for the case of circuit metrics with conditionals, and it requires prohibitively large number of simulations while sampling extremely rare events. This paper presents a recursive formulation of SB that overcomes both these issues efficiently. This new technique was applied to an SRAM cell in an industrial 90nm technology to obtain speedups of up to 4 orders of magnitude over standard Monte Carlo and 2 orders

of magnitude over standard SB.

Acknowledgements: The authors acknowledge the support of the Focus Center for Circuit & System Solutions (C2S2, http://www.c2s2.org), one of five research centers funded under the Focus Center Research Program, a Semiconductor Research Corporation program.

References

[1] A. Singhee, R.A. Rutenbar, "Statistical Blockade: A Novel Method for Very Fast Monte Carlo Simulation of Rare Circuit Events, and its Application", *Proc. DATE*, 2007.

[2] G.S. Fishman, "A First Course in Monte Carlo", Duxbury Press, Oct. 2005.

[3] A.J. Bhavnagarwala, X. Tang, J.D. Meindl, "The Impact of Intrinsic Device Fluctuations on CMOS SRAM Cell Stability", *J.Solid State Circuits*, 26(4), pp 658-665, Apr. 2001.

[4] S. Mukhopadhyay, H. Mahmoodi, K. Roy, "Statistical Design and Optimization of SRAM Cell for Yield Enhancement", *Proc. ICCAD*, 2004.

[5] B.H. Calhoun, A. Chandrakasan, "Analyzing Static Noise Margin for Sub-threshold SRAM in 65nm CMOS", *Proc. ESSCIRC*, 2005.

[6] H. Mahmoodi, S. Mukhopadhyay, K. Roy, "Estimation of Delay Variations due to Random-Dopant Fluctuations in Nanoscale CMOS Circuits", *J. Solid State Circuits*, 40(3), pp 1787-1796, Sep. 2005.

[7] D.E. Hocevar, M.R. Lightner, T.N. Trick, "A Study of Variance Reduction Techniques for Estimating Circuit Yields', *IEEE Trans. CAD*, 2(3), July, 1983.

[8] R. Kanj, R. Joshi, S. Nassif, "Mixture Importance Sampling and its Application to the Analysis of SRAM Designs in the Presence of Rare Failure Events", *Proc. DAC*, 2006.

[9] T.C. Hesterberg, "Advances in Importance Sampling", PhD Dissertation, Dept. of Statistics, Stanford University, 1988, 2003.

[10] T. Hastie, R. Tibshirani, J. Friedman, "The Elements of Statistical Learning", Springer Verlag, 2003.

[11] A.J. McNeil, "Estimating the Tails of Loss Severity Distributions using Extreme Value Theory", *ASTIN Bulletin*, 27(1), pp 117-137, 1997.

[12] R. K. Krishnamurthy et al., "High-performance and low-power challenges for sub-70 nm microprocessor circuits," *Proc. CICC*, 2002.

[13] A. Singhee, R. A. Rutenbar, "Beyond Low-Order Statistical Response Surfaces: Latent Variable Regression for Efficient, Highly Nonlinear Fitting", *Proc. DAC*, 2007.

[14] T. Joachims, Making large-Scale SVM Learning Practical. Advances in Kernel Methods - Support Vector Learning, B. Schölkopf and C. Burges and A. Smola (ed.), MIT-Press, 1999.

[15] J. Wang, A. Singhee, R.A. Rutenbar, B.H. Calhoun, "Statistical Modeling for the Minimum Standby Supply Voltage of a Full SRAM Array", *Proc. ESSCIRC*, 2007.

21st International Conference on VLSI Design

NBTI Degradation: A Problem or a Scare?

Kewal K. Saluja, Shriram Vijayakumar, Warin Sootkaneung, and Xaingning Yang
Department of Electrical and Computer Engineering
University of Wisconsin-Madison
Madison, WI 53706, USA
saluja@ece.wisc.edu, vijayakumar@wisc.edu, sootkaneung@wisc.edu, greg.yang@gmail.com

Abstract

Negative Bias Temperature Instability (NBTI) has been identified as a major and critical reliability issue for PMOS devices in nano-scale designs. It manifests as a negative threshold voltage shift, thereby degrading the performance of the PMOS devices over the lifetime of a circuit. In order to determine the quantitative impact of this phenomenon an accurate and tractable model is needed. In this paper we explore a novel and practical methodology for modeling NBTI degradation at the logic level for digital circuits. Its major contributions include i) A SPICE level simulation to identify stress on PMOS devices under varying input conditions for various gate types and ii) a gate level simulation methodology that is scalable and accurate for determining stress on large circuits. We validate the proposed logic level simulation methodology by showing that it is accurate within 1% of the reference model. Contrary to many other papers in this area, our experimental results show that the overall delay degradation of large digital circuits due to NBTI is relatively small.

1. Introduction

Negative Bias Temperature Instability (NBTI) affects the p-MOSFET transistors. The degradation process caused by the generation of traps and partial recovery associated with reduction in traps is explained in [6], [8], [13]. NBTI degradation worsens at high temperatures, causing a lager shift in the threshold voltage. Further, over long period of time this threshold voltage shift can potentially cause a significant increase in delay of the p-MOSFET devices [4], [6].

NBTI degradation and its impact on circuit reliability and performance have become a key issue due to the continuous decrease of the transistor dimensions. A number of studies have been conducted to investigate the effect of NBTI on both digital and analog circuits [11], [2], [10], [7], [9]. Besides, many studies have also developed several design-time and run-time techniques to

cope with the NBTI degradation [3], [17], [7], [14]. These studies include the use of CAD tools for managing transistor degradation mechanism [3], the use of dynamic voltage scaling (DVS)[17], the use of data flipping to recover the static noise margin of the SRAM [7], and the use of device parameter tuning (Vdd, Vth and gate-size) in digital circuits [14].

Since, in the device community, NBTI is difficult to design around, an efficient and accurate model for predicting the performance degradation due to NBTI is urgently needed to design the circuits based on design-for-degradation techniques. Essentially, in digital circuits, which are more likely to suffer NBTI effects, several models of NBTI degradation in the literature have been used to capture the degradation behavior of all PMOS components as a function of their parameters and bias-voltages [11], [10], [14]. However, for large scale circuits, these device level NBTI stress models require extensive circuit simulation of every PMOS transistor in the circuit [17], [15], thus making them intractable and non-ideal. Design techniques that are meant to offset the effect of NBTI degradation using various methods, such as gate sizing [16], [5], require accurate prediction of degradation for gate or higher level designs of the logic circuits.

In this paper, a gate level simulation methodology which can model NBTI degradation of digital circuits accurately is developed. It employs a previously proposed device-level model [14] which has been experimentally verified to determine the degradation of p-MOSFET transistors in a gate over a period of given time. Our simulation technique can efficiently model V_{th} shift and delay of the devices for industrial scale benchmark circuits (ISCAS'85 and MCNC'91). More importantly, to identify the dynamic behavior of NBTI degradation due to the bias of a PMOS transistor which depends on the position of the PMOS transistor in the pull-up stack and also the logic input value, we use SPICE simulation to determine the input conditions and the extent of the stress for every PMOS transistor in the library of gates mapped onto the design. Our method can

978-1-4244-3039-0/08 $25.00 © 2008 IEEE

also handle arbitrary stress pattern for any workload conditions. Hence it provides a better and more realistic prediction of the NBTI degradation in digital circuits. However, the model proposed in this paper is under the condition that all transistors in mapping gates have paths to ground so that we can construct the gate level model to predict the large scale degradation from practical SPICE simulation. In this phase, since we mainly focus on the effect of NBTI on circuit delay, other impacts such as voltage scaling, divergent paths, and power domain analysis are not considered.

The section 2 below provides an overview of the previously proposed NBTI modeling techniques and how they relate to our method. Section 3 gives the details of the proposed method and in Section 4 the experimental results are presented for numerous benchmark circuits. Finally, we conclude this research in Section 5.

2. Related Work

A comprehensive device-level predictive model covering both static and dynamic NBTI degradation is proposed by Vattikonda et. al. in [14]. This device level model is very accurate having been verified with the experimental data. The model calculates the amount of V_{th} shift due to NBTI degradation over time using the following equations:

Stress Phase:

$$\Delta V_{th} = \sqrt{K_v^2 (t - t_0)^{1/2} + \Delta V_{th0}^2} + \delta_v \qquad (1)$$

Recovery Phase:

$$\Delta V_{th} = (\Delta V_{th} - \delta_v) \left(1 - \sqrt{\eta (t - t_0)/t}\right) \qquad (2)$$

where,

$$K_v = A.t_{ox}. \sqrt{C_{ox} (V_{gs} - V_{th})}. \left(1 - \frac{V_{gs}}{V_{gs} - V_{th}}\right)$$
$$.exp\left(\frac{E_{ox}}{E_o}\right).exp\left(\frac{E_a}{kT}\right) \qquad (3)$$

A detailed explanation of this predictive model and the technology specifications are dealt within the original paper [14]. Using these equations, V_{th} shift of every PMOS transistor in the digital circuit can be determined on a cycle-by-cycle basis. After simulating the circuit operation for certain time period, such as the specified lifetime, the increase in delay of each transistor can be calculated from the obtained V_{th} shift using the alpha-power law [1] model, which expresses the delay dependence on the threshold voltage as:

$$Delay \propto \frac{V_{dd}}{(V_{dd} - V_{th})^\beta} \qquad (4)$$

Finally, the overall circuit delay can be obtained by using any timing analysis scheme based on the increased delay of

each transistor. The drawback of the above methodology is that the modern digital circuits operate at hundreds or even thousands of *MHz* and NBTI is a long term degradation process occurring over the lifetime of the unit which could be several years. Hence, simulating the circuit in a cycle-by-cycle style, till the end of specified lifetime, is difficult, if not impossible, due to the tremendous computation need.

An architecture level model proposed in [12] is based on the circuit area and offers no insight into the logic level performance degradation. As a result this is of little value to the circuit designers who want to mitigate the effects of NBTI during the design phase. At the device and logical level, effects of NBTI degradation were studied by Reddy et al. [11] but the study was limited to SRAM cells. Paul et al. [10] provide a predictive model for circuit level degradation, but it only considers static NBTI effects. Authors in [14] also propose a model for predicting the long-term degradation but in this model the PMOS transistors are assumed to be stressed periodically. Whereas in reality, the input patterns to the circuit are not periodical. Thus the assumption is prone to producing erroneous results.

The simulation method proposed in this paper builds upon the device level model from [14] (Eq(1) and Eq(2)). It is capable of handling any specified or arbitrary workload efficiently. The fact that the original model of [14] has been verified with the experimental data, we argue that the proposed method is rather accurate. Indeed the method proposed by us simulates the NBTI degradation on a cycle-by-cycle basis and the assumption of periodic stress is relaxed. Finally, with a curve-fitting scheme used in our proposed method, it is capable of handling large benchmark circuits.

3. Methodology

3.1. Transistor Level Simulation

Use of the predictive model of degradation and recovery [14] to calculate the amount of threshold voltage shift of each transistor requires the knowledge of technology variables, temperature, and bias voltages (Vdd and Vgs). However, for a particular technology generation and operating temperature, the stress/recovery conditions are solely dependent on the V_{gs} bias of the PMOS transistors. The device is considered to be under *stress* if its V_{gs} is less than the threshold voltage V_{th}, otherwise it is assumed to be in *recovery*. If V_{gs} for a transistor can be inferred from the inputs to a gate, then the conditions for NBTI degradation can be easily deduced for any gate.

In this study, we look at two different models. The first is a simple model that is intuitive for logic level circuit simulations. This allows the PMOS gate to be in one of the two different states, namely *stress* or *recovery*. A PMOS gate is said to be under *stress* when its gate input is 0 and in recovery when its gate input is V_{DD}. This model holds true for PMOS gates analyzed in isolation or in case of inverters and NAND gates which contain PMOS transistors in parallel in the pull-up stack. In these gates, the sources of all the PMOS transistors are tied to the supply voltage V_{DD} and hence V_{gs} is always either $-V_{DD}$ or 0. But for NOR gates with series stacked PMOS devices, the bias conditions depend on the transistor's input and also the states of other transistors in the stack. Extending the *2-state model* to NOR gates would lead to overly pessimistic calculations for threshold voltage degradation as some devices do not experience the maximum possible V_{gs} voltages.

In the extended *3-state model*, we aim to overcome the drawbacks of the simplistic model. SPICE level simulations reveal the status of all PMOS transistors in the pull-up network and their individual exact bias conditions. Details of these evaluations for all possible gate input combinations for various gate types are discussed in conjunction with Tables 1, 2 and 3 in the Results section. However, we explain below the basics for a three input NOR gate shown in Figure 1. Consider the PMOS transistor M_B in a stack. This can be in three different states as follows:

- *Stress*: In Figure 1a) the transistor M_A is ON and as a result M_B experiences a V_{gs} close to $-V_{DD}$ and hence is under complete *stress*.

- *Recovery*: There are two conditions under which a device can be in recovery.

  - Device input is V_{DD}: In Figure 1b) input B is high, hence V_{gs} for M_B is greater than the threshold voltage, therefore it is said to be in *recovery*. Here the stress condition is independent of the states of other transistors in the PMOS stack.

  - Device input is 0: In Figure 1d) transistor M_A is OFF, while M_B and M_C are ON. The sources of both the ON transistors have a direct path to ground and as a result the source node voltages are pulled below V_{th} and as a result the V_{gs} for both these devices (M_B and M_C) are greater than the threshold voltage. Consequently both M_B and M_C are in recovery.

- *Moderate Stress*: In Figure 1c) M_B is ON with M_A and M_C OFF. The source of M_B does not experience the complete supply voltage V_{DD}. Also the source has no discharge path to the ground and hence has a significant voltage but it is still less than V_{DD}. Here M_B is assumed to be under *moderate stress*.

The important deduction here is that the state of a device can be determined by absence or presence of direct paths to the supply lines. These in turn can be determined by the gate inputs, providing us a method to evaluate stress conditions accurately from current gate inputs. To check for con-

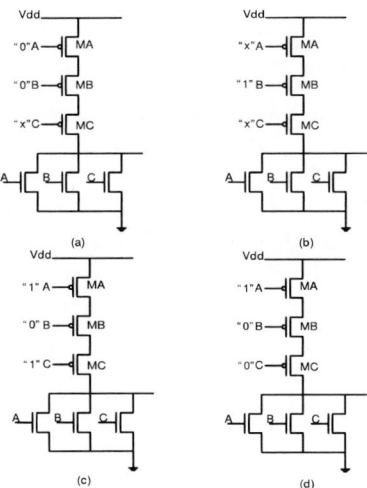

Figure 1. States for a PMOS transistor

sistency and validity of the extended model, SPICE simulations were performed for the 90nm, 65nm, 45nm and 22nm nodes. These nodes were selected because NBTI degradation is dominant only for sub-130nm nodes. The SPICE models for these nodes were based on the Predictive Technology used in [14]. The Vdd values used were 1.2V for 90nm and 65nm nodes, and 1V for the 45nm and 22nm nodes. Simulations were performed for 2, 3 and 4 input NOR gates by feeding the gate inputs through inverters and also loading the gate output with an inverter representing a fanout-factor of 3. In our experiments, each NOR gate was applied all possible combinations and only one signal was changed (swing signal). It was observed that the stress, moderate stress and recovery conditions for the PMOS transistors are consistent across the technology generations.

3.2. Gate Level Simulation

We designed and implemented a simulator that uses the *3-state model* described in the transistor level simulation subsection. The simulator can calculate PMOS transistor NBTI degradation under stress, moderate stress, or recovery phase, in large digital circuits. Further, to determine the long term threshold voltage degradation we use regression method in which we express the degradation by the power law given below:

$$\Delta V_t (t) = \alpha t^n \qquad (5)$$

One could argue to use regression for the complete circuit degradation but, as demonstrated in many of the previous works [3], [17], [14], the circuit degradation is difficult to quantify using closed form approximations. Note that in the above equation, α and *n* are functions of various operating parameters like temperature, supply voltage, probability of stress and device parameters like gate oxide thickness and threshold voltage. The power law gives us a way of estimating the V_t degradation over the lifetime if stable values for α and *n* can be found.

Figure 2 shows the simulation methodology. It consists of two phases

- *Cycle accurate degradation simulation*: This phase carries out a cycle accurate simulation of NBTI degradation of all PMOS devices over a small fraction of the lifetime of the circuit. It is used to determine the values of α and n for all the PMOS devices statistically, by curve fitting, using the *least square fitting* method. In practice the new values need to be calculated on per cycle basis. Once we have the stable values of α and n for all PMOS devices, this phase is terminated. Another salient feature of this method is that the simulation can be carried out for any workload condition.

- *Long Term V_{th} degradation estimation*: Values of α and n obtained in the previous phase are used to estimate the long term V_{th} degradation using Eqn (5). This helps avoid cycle-by-cycle simulation over the entire lifetime of the circuit and makes the use of the simulator for large circuits feasible. The degradation in device delay can be estimated by using Eqn(4).

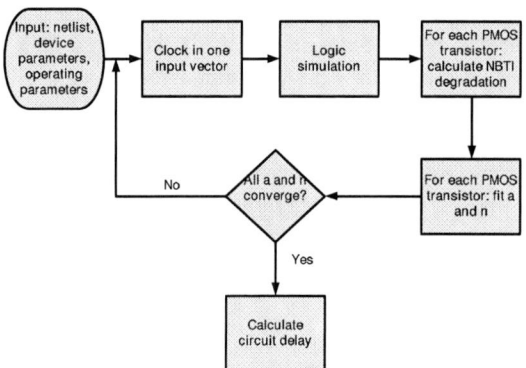

Figure 2. Simulation Methodology Flowchart

Finally, the path delay degradation can be calculated by analyzing the paths and these are taken as an indication of performance degradation of the circuit.

4. Results

The stress models and simulation methodology were evaluated for various ISCAS'85 and MCNC'91 circuits. These circuits were mapped to 2, 3, and 4 input NOR and NAND gates, and inverters. Unless stated otherwise, the simulations were carried out for operating temperature of $100°C$ and $f=1GHz$ and the device parameters with $V_{DD} = 1.2V$, $V_{th} = 200mV$ using the 65nm technology nodes.

4.1. Transistor Level Simulation Results

Tables 1, 2, and 3 show the results of SPICE simulations and bias voltages for the 2, 3, and 4 input NOR gates for the 65nm nodes. In these tables M1, M2, M3, etc denote the

PMOS transistors in the pullup stack. Further, the transistor M1 is assumed to be closest to the power supply. All voltage levels of less than 10mv in magnitude have been represented as zero. From the bias conditions it is apparent that the application of a 2-state model to identify stress/recovery conditions might lead to pessimistic evaluations. It is also seen that for the 3-state model, a device with gate input of 0 experiences significant reverse bias when its paths to V_{DD} and GND are cutoff, and almost negligible bias when cutoff is only from V_{DD}.

Table 1. V_{gs} Values for a 2-Input NOR Gate

INPUT LOGIC LEVEL		V_{gs} (V)	
M1	M2	M1	M2
0	0	-1.2	-1.2
0	1	-1.2	0.12
1	0	0	-0.13
1	1	0	0.36

Table 2. V_{gs} Values for a 3-Input NOR Gate

INPUT LOGIC LEVEL			V_{gs} (V)		
M1	M2	M3	M1	M2	M3
0	0	0	-1.2	-1.2	-1.2
0	0	1	-1.2	-1.2	0
0	1	0	-1.2	0	-0.13
0	1	1	-1.2	0	0.26
1	0	0	0	-0.15	-0.07
1	0	1	0	-0.76	0.44
1	1	0	0	0.25	-0.11
1	1	1	0	0.03	0.29

Table 3. V_{gs} Values for a 4-Input NOR Gate

INPUT LOGIC LEVEL				V_{gs} (V)			
M1	M2	M3	M4	M1	M2	M3	M4
0	0	0	0	-1.2	-1.2	-1.2	-1.2
0	0	0	1	-1.2	-1.2	-1.2	0
0	0	1	0	-1.2	-1.2	0	-0.13
0	0	1	1	-1.2	-1.2	0	0.20
0	1	0	0	-1.2	0	-0.15	-0.09
0	1	0	1	-1.2	0	-0.85	0.34
0	1	1	0	-1.2	0	-0.14	-0.11
0	1	1	1	-1.2	0	0.02	0.23
1	0	0	0	0	-0.17	-0.12	-0.09
1	0	0	1	0	-0.81	-0.81	0.39
1	0	1	0	0	-0.86	0.33	-0.10
1	0	1	1	0	-1.06	0.13	0.23
1	1	0	0	0	0.20	-0.12	-0.03
1	1	0	1	0	0.02	-0.80	0.39
1	1	1	0	0	0.01	0.04	-0.01
1	1	1	1	0	0.01	0.04	0.26

The difference in V_{th} degradation estimations between the 2-state and 3-state model can be seen from Table 4 which compares degradations for 3 and 4 input NOR gates. While computing the values give in this table, we assumed an average value of $V_{gs} = 700mv$ for the moderate stress conditions. It is apparent from this table that the 2-state model overestimated the degradation in all 3 cases. The application of this model to large circuits and the related results, including comparison of two and there states model, are presented in a later subsection.

Table 4. ΔV_{th} for 2- and 3-State Models

Circuit	ΔV_{th} DEGRADATION (mV)					
	2-state model			3-state model		
	Max	Min	Avg.	Max	Min	Avg.
NOR3	11.55	11.35	11.44	11.22	7.59	9.60
NOR4	11.47	11.34	11.40	11.48	6.70	9.57

4.2. Validation of Simulation Methodology

To verify our proposed simulation methodology, we compare our model with the reference model from [14]. We conducted experiments for 45nm, 65nm and 90nm technology nodes. All parameters are obtained from the predictive model of [14]. The NBTI degradation of an inverter is simulated for 10 years of operation. Since the model in [14] can only handle periodical stress case, input vectors generated are patterns of equal 1's and 0's, signifying a duty cycle of 50%. However, We must add that the proposed simulation methodology can handle arbitrary patterns.

Table 5. Param. Dependence and Rel. Error

Parameters	Maximum Error $(x10^{-3})$		
	90nm	65nm	45nm
Baseline			
$T=100^{\circ}C$, duty cycle=50%	7.89	8.23	6.32
Vary Temperature			
$T=50^{\circ}C$	7.52	3.43	2.38
$T=75^{\circ}C$	8.65	3.94	2.89
$T=125^{\circ}C$	5.12	4.69	3.75
Vary duty cycle			
Duty cycle=33%	8.37	7.45	1.41
Duty cycle=66%	2.57	2.15	1.41
Duty cycle=75%	9.26	8.07	6.02
Vary T_{ox}			
T_{ox}=1.5nm	9.04	8.13	6.29
T_{ox}=2.5nm	7.22	6.00	4.19

The results of the comparison for various parameters are shown in Table 5. We use the 3-state model for simulation purposes, but for an inverter there will not exist any divergence from the 2-state model due to the absence of any stacking effect. This table shows the relative error in V_{th} shift of PMOS transistors using proposed model and the reference model. The relative error is defined as follows:

$$RelError = \frac{|\Delta V_{th,reference} - \Delta V_{th,proposed}|}{\Delta V_{th,reference}} \quad (6)$$

It is seen that the relative error between the reference and proposed model is less than 1% in all cases. We conclude that the proposed method gives an accurate estimation of NBTI degradation.

4.3. V_{th} Degradation of Benchmark Circuits

We now present the results of V_{th} degradation for various ISCAS'85 and MCNC'91 circuits using both the 2-state and 3-state models of stress/recovery conditions with $\beta = 1.3$ [1]. Table 6 shows the comparative values. We observe that generally the 2-state model provides a more pessimistic estimation of the threshold shift. We also note that significant variations between the two models are absent. This is so because variations are to be expected only in circuits that have a considerable percentage of 3 or 4 input NOR gates, whereas the mapping was initially done to favor NAND usage. However, we note that for the two circuits, i3 and i4, the deviation is substantial and we found that it is due to the fact that these circuits are NOR gate intensive and hence more prone to stacking effect.

Table 6. ΔV_{th} for 2-State and 3-State Models

Circuit	ΔV_{th} DEGRADATION (mV)					
	2-state model			3-state model		
	Min	Max	Avg.	Min	Max	Avg.
c1355	2.33	21.98	11.54	2.26	21.87	10.09
c1908	1.86	23.88	11.22	3.47	22.43	10.80
c432	6.36	19.85	10.54	6.37	18.52	10.18
c499	2.32	21.79	11.53	2.29	21.92	11.00
i1	2.07	26.46	11.40	2.03	25.94	10.87
i2	4.96	29.07	11.99	4.92	28.59	11.22
i3	4.43	23.42	11.45	3.55	22.60	10.14
i4	4.89	27.86	11.51	4.84	22.20	10.82
i5	4.83	17.09	10.46	5.01	17.13	10.46
i6	5.73	11.59	10.16	5.75	11.54	10.14
i7	5.77	22.35	11.38	5.73	22.34	10.60
i8	2.69	78.57	12.52	2.86	74.90	10.94
i9	4.17	26.90	11.37	4.17	26.83	10.26

4.4. Circuit Delay Changes

The threshold voltage degradation affects the performance by increasing delay of the gates (in the critical path). To quantify this effect, we used Eqn(4) to determine the delay degradation of all gates in a circuit after computing the threshold voltage change using 3-state model.

The table 7 shows the percentage delay degradation in the long path of the circuits. We used three different values of β in Eqn(4). All circuits present a delay degradation close to 1-2%. It should be noted that a V_{th} degradation of 10% over a 10 year period manifests itself as only a 2% degradation in gate delay with β=1.3. For larger values of β the degradation is larger. Another and equally important observation we made is that for some circuits, like c1355 and c499, the critical paths changed between the original circuit and the NBTI degraded circuit. This shows that overdesign for NBTI cannot be targeted at critical paths alone but must consider other long paths in the circuit.

5. Conclusion

In this paper a *3-state* stress/recovery model for NBTI degradation is proposed and its necessity is substantiated

Table 7. Delay Degradation along Critical Path

Circuit	% Delay Degradation		
	$\beta = 1.3$	$\beta = 1.5$	$\beta = 2.0$
c1355	1.83	2.11	2.83
c1908	2.02	2.34	3.14
c432	1.60	1.85	2.48
c499	1.87	2.17	2.90
i1	1.70	1.96	2.63
i2	2.03	2.34	3.14
i3	1.76	2.04	2.73
i4	1.80	2.08	2.79
i5	1.66	1.92	2.56
i6	1.50	1.73	2.32
i7	1.84	2.12	2.85
i8	1.83	2.12	2.83
i9	1.86	2.16	2.89

by SPICE simulations. A simulation methodology, which is accurate, efficient and tractable, is developed that uses the 3-state stress/recovery model. The results of the two models are compared. There is only a 1% divergence between the proposed and the reference simulation methodologies. We also used the 3-state model to quantify the delay degradation of benchmark circuits. The results show the overall delay degradation to be small. Consequently, the effect of NBTI on timing degradation in large circuits is still diminutive compared to the time period a consumer prefers to use the hardware which sharply decreases as new technologies become available.

Our present methodology does not take into consideration the dependence of stress and recovery conditions on the past inputs. The future work will consider refining the stress/recovery model to include the effects of the current and the previous gate inputs.

References

[1] T. S. abd A. R. Newton. Alpha power law MOSFET model and its application to CMOS inverter and other formulas. *IEEE Journal of Solid State Circuits*, 25(2):584–594, April 1990.

[2] G.Chen, K.Y.Chuah, M.F.Li, D. Chan, C.H.Ang, J.Z.Zheng, Y.Jin, and D.L.Kwong. Dynamic NBTI of PMOS transistors and its impact on device lifetime. In *RPSP '03: Proc. of the 41st annual symposium on Reliability Physics*, pages 196–202, Dallas, Texas, 2003.

[3] A. S. Goda and G. Kapila. Design for degradation: Cad tools for managing transistor degradation mechanisms. In *ISQED '05: Proc. of the 6th International Symposium on Quality of Electronic Design*, pages 416–420, Washington, DC, 2005.

[4] B. Kaczer, V. Arkhipov, R. Degraeve, N. Collaert, G. Groeseneken, and M. Goodwin. Temperature dependence of the negative bias temperature instability in the framework of dispersive transport. *Applied Physics Letters*, 86(14):143506, 2005.

[5] K. Kang, H. Kufluoglu, M. A. Alam, and K. Roy. Efficient transistor-level sizing technique under temporal performance degradation due to NBTI. In *Proc. of the IEEE International Conference on Computer Design*, pages 431–436, San Jose, Ca, 2006.

[6] S. V. Kumar, C. H. Kim, and S. S. Sapatnekar. An analytical model for negative bias temperature instability. In *ICCAD '06: Proc. of the 2006 IEEE/ACM International Conference on Computer-Aided Design*, pages 493–496, New York, 2006.

[7] S. V. Kumar, C. H. Kim, and S. S. Sapatnekar. Impact of NBTI on SRAM read stability and design for reliability. In *ISQED '06: Proc. of the 7th International Symposium on Quality Electronic Design*, pages 210–218, Washington, DC, 2006.

[8] S. Mahapatra, M. A. Alam, P. B. Kumar, T. R. Dalei, D. Varghese, and D. Saha. Negative bias temperature instability in CMOS devices. *Microelectron. Eng.*, 80(1):114–121, 2005.

[9] N.K.Jha, P.S.Reddy, D.K.Sharma, and V.R.Rao. NBTI degradation and its impact for analog circuit reliability. *IEEE Trans. on Electron Devices*, 52(12):2609–2615, 2005.

[10] B. C. Paul, K. Kang, H. Kufluoglu, M. A. Alam, and K. Roy. Temporal performance degradation under NBTI: Estimation and design for improved reliability of nanoscale circuits. In *DATE '06: Proc. of the Conference on Design, Automation and Test in Europe*, pages 780–785, Leuven, Belgium, 2006.

[11] V. Reddy, J. Carulli, A. Krishnan, W. Bosch, and B. Burgess. Impact of negative bias temperature instability on product parametric drift. In *ITC '04: Proc. of the International Test Conference*, pages 148–155, Washington, DC, 2004.

[12] J. Srinivasan, S. V. Adve, P. Bose, and J. A. Rivers. Lifetime reliability: Toward an architectural solution. *IEEE Micro*, 25(3):70–80, 2005.

[13] L. Tsetseris, X. J. Zhou, D. M. Fleetwood, R. D. Schrimpf, and S. T. Pantelides. Physical mechanisms of negative-bias temperature instability. *Applied Physics Letters*, 86(14):142103, 2005.

[14] R. Vattikonda, W. Wang, and Y. Cao. Modeling and minimization of PMOS NBTI effect for robust nanometer design. In *DAC '06: Proc. of the 43rd Annual Conference on Design Automation*, pages 1047–1052, New York, 2006.

[15] W. Wang, S. Yang, S. Bhardwaj, R. Vattikonda, S. Vrudhula, F. Liu, and Y. Cao. The impact of NBTI on the performance of combinational and sequential circuits. In *DAC '07: Proc. of the 44rd Annual Conference on Design Automation*, pages 364–369, New York, 2007.

[16] X. Yang and K. K. Saluja. Combating NBTI degradation via gate sizing. In *ISQED '07: Proc. of the 8th International Symposium on Quality of Electronic Design*, pages 47–52, San Jose, Ca, 2007.

[17] X. Yang, E. F. Weglarz, and K. K. Saluja. On NBTI degradation process in digital logic circuits. In *International Conference on VLSI Design*, pages 723–730, Bangalore, India, 2007.

21st International Conference on VLSI Design

On-Chip Process Variation Detection using Slew-Rate Monitoring Circuit

Amlan Ghosh, *Rahul M. Rao, *Jae-joon Kim, *Ching-Te Chuang and Richard B. Brown

University of Utah, Salt Lake City, UT 84112

* IBM TJ Watson Research Center, Yorktown Heights, NY 10598

{aghosh, brown@eng.utah.edu}, *{raorahul, jjkim2, ctchuang@us.ibm.com}

Abstract

The need for efficient and accurate detection schemes to mitigate the impact of process variations on the parametric yield of integrated circuits has increased in the nm design era. In this paper, a new variation detection technique is presented that uses slew as a metric along with delay to determine the mismatch between the drive strengths of NMOS and PMOS devices. The importance of considering both of these metrics is illustrated and a new slew-rate monitoring circuit is presented for measuring slew of a signal from the critical path of a circuit. Design considerations, simulation results and characteristics of the slew-rate monitor circuitry in a 45 nm SOI technology are presented, and a sensitivity of 1MHz/ps is achieved. This scheme can detect the threshold voltage variation in the order of mV, with a sensitivity of 0.95MHz/mV.

1. Introduction

The significance of parametric variations in manufacturing processes has increased in recent technologies due to their enhanced effects on power and performance characteristics of integrated circuits. Newer sources of variation due to nano-scale effects (such as Line-Edge Roughness) and larger variations in critical parameters have become major concerns for parametric yield. This has necessitated various compensation schemes that optimize the operating characteristics of the designs post-fabrication, and enable a greater percentage of manufactured chips to meet the required frequency and power consumption targets.

Several compensation schemes have been reported in literature. A typical compensation scheme consists of a sensor block to determine the extent of variation, followed by a compensation circuit that alters the operating point of the design appropriately. A correction scheme that senses variation in critical path delay and generates a suitable bi-directional body bias was presented in [1]-[4]. In [5], the authors used a combination of power and delay monitoring blocks to optimally adjust the supply and threshold voltage (V_{TH}) of devices in various modes of operation.

Most of these schemes are primarily based only on delay of the critical path of the circuit. However, purely delay-based compensation schemes can result in sub-optimal design in certain scenarios, wherein V_{TH} mismatches between NMOS and PMOS devices result in nearly identical delay but inferior power characteristics compared to a circuit having devices with typical parameters. Power monitors measure the total switching or

leakage power of the system and hence may also fail to identify the presence of such mismatches.

In this paper, we illustrate the problem of using delay as the only metric for detection and compensation of process variations in Section II. We then propose the use of signal slew as a metric to be used in combination with delay (and power) to determine the adjustments to be applied to the design. We also provide an analytical framework to substantiate the use of slew as a metric. In Section III, we briefly describe the design of a slew-rate monitoring circuit. The simulation results and characteristics of the slew-rate monitor circuitry are shown in Section IV. The conclusions of the paper are presented in Section V.

2. Slew as a Device Mismatch Metric

The threshold voltages (V_{TH}) of PMOS and NMOS devices are among the most important parameters affecting circuit characteristics. The difference in V_{TH} as well as their absolute value is important. A balanced V_{TH} between the devices enables the use of lower operating voltage and provides a symmetric static noise margin. Any mismatch in V_{TH} between P and N devices degrades the operating margin, and performance or power of the circuit [6].

A simple 33-stage ring oscillator with FO4 load (representative of a critical path) was used as a test vehicle to measure the effect of variation in device V_{TH} in a state-of-the-art 45nm SOI technology. Normalized delay of the ring oscillator is shown in Fig. 1 for a V_{TH} variation of 50mV to -50mV for both types of devices. In this graph, the x-axis represents the shift in NMOS V_{TH} with the various curves corresponding to different shifts in PMOS V_{TH}. In Fig. 1, point A corresponds to a condition with fast (and leaky) NMOS devices with slow PMOS devices, whereas point B represents slow NMOS devices with fast PMOS devices. As seen in the figure, the delay values at A and B are nearly identical to those of point C, which corresponds to the nominal (i.e. intended) operating point of the circuit in the absence of any V_{TH} variation. Similarly, the region inside the dotted oval in Fig. 1 shows different V_{TH} variation conditions that exhibit a delay very close to nominal. Thus it may not be possible to detect the mismatch between the device types with delay as the only metric, and hence a purely delay-based compensation scheme would not generate any adjustments to well or body bias and/or supply voltage in such scenarios. However, with the V_{TH} of one type of device being lower than the nominal value, its leakage current would be significantly higher, resulting in a substantial increase in power consumption of the circuit. In addition, noise margins of the

143

978-1-4244-3039-0/08 $25.00 © 2008 IEEE

circuit are also degraded, rendering it more susceptible to noise failures.

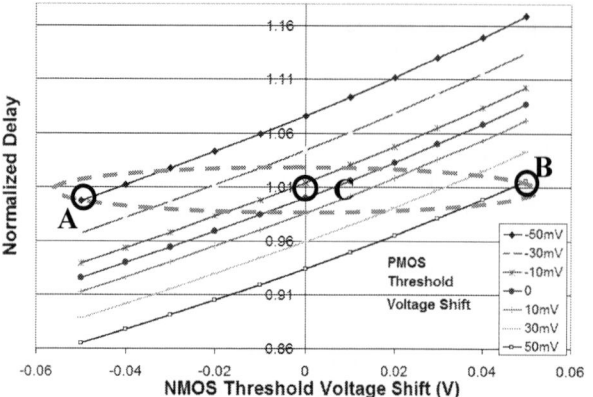

Fig. 1: Normalized delay with NMOS and PMOS threshold voltage.

Let us consider a variation of ΔV_{thn} and ΔV_{thp} in the threshold voltages of NMOS and PMOS devices. The propagation delay of a single stage, t_{pd}, can be computed [7]-[8] as a function of threshold voltages and gate-source voltages (V_{GS}) of both types of devices as

$$t_{pdnom} + \Delta t_{pd} = k\left(V_{GSn}^{-2} + V_{GSp}^{-2} + \frac{2V_{thn}}{V_{GSn}^{3}} + \frac{2V_{thp}}{V_{GSp}^{3}} + \frac{2\Delta V_{thn}}{V_{GSn}^{3}} - \frac{2\Delta V_{thp}}{V_{GSp}^{3}}\right) \quad (1)$$

which can be represented as

$$t_{pdnom} + \Delta t_{pd} = k_0 \Delta V_{thn} - k_1 \Delta V_{thp} + k_2 \quad (2)$$

Here k_2 is representative of t_{pdnom}, the nominal propagation delay of a single stage of the ring oscillator. A similar approximation can be derived using a more accurate alpha-power model [9]-[10]. As can be seen in these equations, the propagation delay is a function of the variation in both device types. Hence, it is difficult to decouple their mismatch from delay variations when NMOS and PMOS V_{TH} variation are of opposite polarities. As an illustration, the variation in oscillator delay from Fig. 1 can be expressed as a function of NMOS and PMOS V_{TH} as

$$\frac{\Delta t_d}{t_{pdnom}} = 1.74 \times \Delta V_{thn} - 1.52 \times \Delta V_{thp} \quad (3)$$

It can be seen that the oscillator delay will be identical to the nominal delay t_{pdnom} for $\frac{\Delta V_{thn}}{\Delta V_{thp}} = 0.877$.

Thus, while delay metric is necessary, it is not sufficient to detect all combinations of variations in device parameters. Delay-based compensation schemes perform adequately only when the device parameters of NMOS and PMOS devices vary in the same way, i.e., when both devices become either equally slower or faster. However, it would fail to find the optimum operating condition when V_{TH} variations for NMOS and PMOS are in opposite directions or of unequal magnitudes, as illustrated in Fig. 1.

The signal slew at any node on the ring oscillator (or critical path) can be used to identify the mismatch between variations in the two types of devices. The rise time at the output of any CMOS gate is determined by the pull-up network, and the fall time by the pull-down network. Hence, the difference between the rise and fall slew is indicative of the relative mismatch between the strength of the NMOS and PMOS devices. The normalized difference of rise and fall slew for a 33-stage ring oscillator, with both NMOS and PMOS V_{TH} shifted from -0.1V to 0.1V from their nominal values, is plotted in Fig. 2. As can be seen, point A represents the condition with fast NMOS devices and slow PMOS devices. This results in a fast fall time and slow rise time, which causes an identifiable negative change in the difference of rise and fall slew. Similarly, in the case of fast PMOS and slow NMOS devices, represented by point B, there exists an identifiable positive change in the slew difference. Thus, the slew difference can be used to suitably characterize the mismatch between the two types of devices. It should be noted that when the parameters of both device types vary in the same way (i.e., both devices are either slow or fast), the impact on slew difference is small. In such cases, a delay-based monitor can be used to measure the extent of variation.

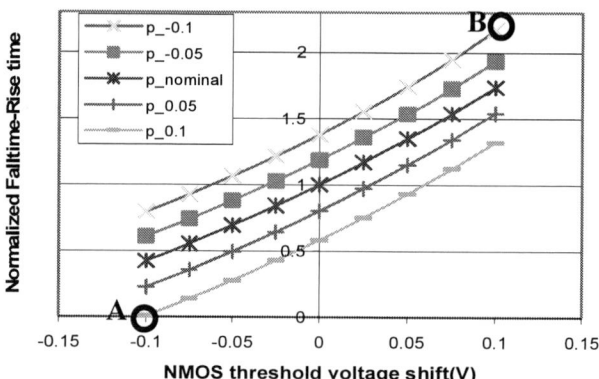

Fig. 2: Normalized difference of fall and rise slew with NMOS and PMOS threshold voltage.

Rise and fall times can also be modeled as a function of V_{TH} of the two device types. For a given set of variations in NMOS and PMOS V_{TH}, the corresponding change in the difference of rise and fall time can be represented as

$$\Delta t_{rise-fall} = k_3\left(\frac{2\Delta V_{thp}}{V_{GSp}^{3}} - \frac{2\Delta V_{thn}}{V_{GSn}^{3}}\right) \quad (4)$$

which can also be written as

$$\Delta t_{rise-fall} = k_4 \Delta V_{thn} - k_5 \Delta V_{thp} \quad (5)$$

As an illustration, the variation in normalized difference of rise and fall time of the oscillator of Fig. 2 can be expressed as a function of NMOS and PMOS V_{TH} as

$$\frac{\Delta t_{rise-fall}}{t_{rise-fall\_nom}} = 6.78 \times \Delta V_{thn} - 3.74 \times \Delta V_{thp} \quad (6)$$

978-1-4244-3039-0/08 $25.00 © 2008 IEEE

3. Design of Slew-Rate Monitor

In this section, the design issues and challenges of a slew-rate monitor are described. Measuring the slew-rate of a signal from the critical path requires very high-speed and precise dynamic apparatus with sensitivity in the pico-second range. A basic block diagram of the slew rate monitor is shown in Fig. 3. The Signal under Test (SUT) is connected to two comparators, A and B, designed using thick-oxide long channel devices. Comparator-A compares the SUT level to a reference voltage equal to 80% of the supply voltage (V_{DD}) while Comparator-B compares the SUT level to a reference voltage equal to 20% of the supply voltage. The reference voltages of 20% and 80% were chosen to provide sufficient noise margin against supply noise on the input signal. This slew rate monitor topology is applicable for any set of two reference voltages. In a low-noise system, 10%-90% reference voltages could be used in order to improve the output sensitivity.

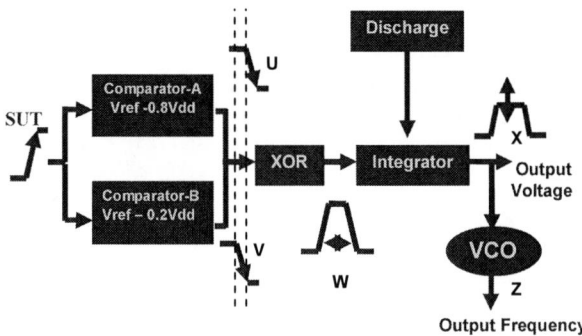

Fig. 3. Block diagram of the slew-rate monitor.

The output of one comparator switches before the output of the other comparator, depending on the slew of the signal and the direction of transition. These two comparator output signals (U and V in Fig. 3) are fed to an exclusive-or (XOR) gate that generates a pulse at its output. The width of this pulse (W in Fig. 3) is a direct representation of the slew of the input. This pulse is integrated over time to generate an output voltage proportional to the pulse width of the XORed output, with the integrator essentially functioning as a time-to-voltage converter. Thus the output of the integrator (X in Fig. 3) is proportional to the slew of the SUT. The discharge circuit resets the integrator before a new input signal is applied.

This output voltage (X) controls the frequency of a voltage controlled oscillator (VCO) to enable easy characterization of the slew monitoring circuit. The VCO has been designed such its frequency varies almost linearly with the control voltage over the desired range. The output voltage (X) can also be used directly to determine the appropriate bias adjustments to cancel the mismatch between the device types. The output frequency of the VCO can be directly monitored, or it can be easily converted to a digital value representative of the device

mismatch, by using it to clock an N-bit counter for a fixed time.

3.1. Comparator

The sensitivity of the slew rate monitor is a function of the performance and voltage offset of the comparators. Two CMOS differential latched comparators with additional diode-connected transistors were designed to achieve the required high speed and accuracy, while operating with chosen reference voltages. Fig. 4 shows the implementation details for Comparator-B, designed primarily using PMOS devices, with a reference voltage of to 20% of.V_{DD}. In the input pre-amplifier stage, P3 and P4 are diode-connected [11]. When the input voltage is less than the reference voltage, an additional current flows through P4 (and in the other case, the extra current flows through P3). This additional current increases the voltage gain of the preamplifier and enhances the speed of the comparator.

A regenerative latch is incorporated as the decision element to achieve a high gain [12]. It uses positive feedback from the cross-gate connection of P6 and P8. To understand its operation, assume i_{N4}, the current flowing through device N4, is much larger than i_{N3}, current through N3. In that case, P7 and P6 are *on*, and P5 and P8 are *off*. If i_{N3} is increased until it is greater than i_{N4}, the drain-source voltage of P6 will be large enough to switch P8 *on*. P8 will draw current from P7 which will reduce the drain-source voltage of P7. This in turn will switch P6 *off*. This regenerative process accelerates the comparison.

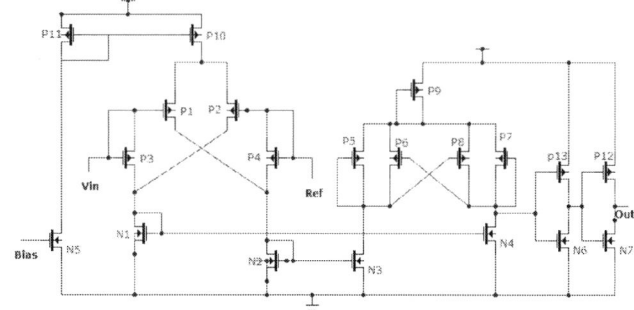

Fig. 4: Schematic diagram of Comparator-B.

P9 is used to shift the level of the output to enable a rail-to-rail output voltage. The voltage drop across it is maintained at nearly V_{THP} by suitably sizing the device. The drive strength of a single latch may not be sufficient in scenarios where a significant amount of load capacitance is to be driven in a relatively short time. Hence, an buffer stage comprising (P13, N6) and (P12, N7), is included as the output stage [9]. A complementary comparator-A using NMOS at the input stage and output latch has been designed with an intended reference voltage of 80% of V_{DD}.

A rising-edge transition of 5 pico-seconds slew was applied as an input to the comparators to determine their dynamic characteristics. The performance characteristics of the two comparators are tabulated in Table 1 and their DC transfer characteristics are shown in Fig. 5. Comparators A and B show DC gains of 231 and 186, respectively.

Table1: Dynamic characteristics of the comparators

	Propagation delay(ps)	Slew(ps)
Comparator-A (Vref = 0.8 Vdd)	73	19
Comparator-B (Vref = 0.2 Vdd)	66	18

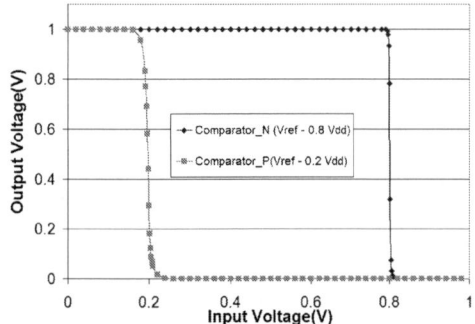

Fig. 5: Transfer characteristics of the Comparators.

3.2. Integrator

A simple charge pump based integrator was designed as shown in Fig. 7. The pulse generated from the XOR gate allows the current to flow through P2, charging the integrating capacitor C1 which has been implemented using on-chip decoupling capacitors. The integrating time constant is dependent on the capacitance (C1) and the integrating current, and hence can be tuned as required. In this design, a reasonably large W/L ratio is chosen for device P2 to obtain a small time-constant while ensuring that the integrator output will not saturate to supply rail.

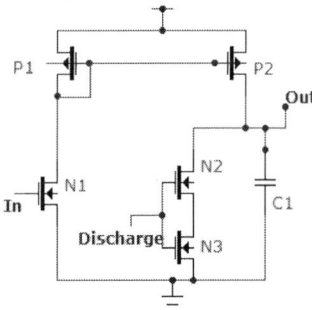

Fig. 6: Schematic block diagram of integrator.

A voltage proportional to the width of the input pulse is generated on the output capacitor. The operating range of the integrator circuit can be tailored to different input slew ranges by keeping multiple current sources with different drive strength and appropriately choosing between them based on the application and input slew range.

A discharge path is provided through the pair of series-connected NMOS transistors (N2 and N3). The discharge signal is asserted to reset the output voltage, and de-asserted just prior to the application of the input to the slew monitoring system. The output voltage of the integrator is shown in Fig. 7 as a function of the input pulse

width. This figure confirms the linear relationship of the output voltage with respect to the pulse width over the range of interest.

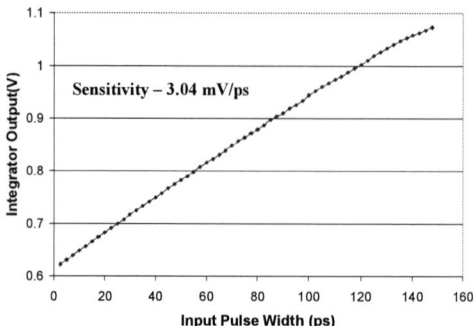

Fig. 7: Output characteristics of the integrator.

3.3. Voltage Controlled Oscillator (VCO)

A single stage of the 33-stage voltage controlled oscillator is shown in Fig. 8. The output voltage of the integrator controls the current through N1, modulating the stage delay, and thereby, the output frequency of the VCO. The output frequency exhibits a linear relationship with the control voltage within the 0.5 to 1V range as shown in Fig. 9. The sensitivity of the VCO is ~1.17MHz/mv in this range.

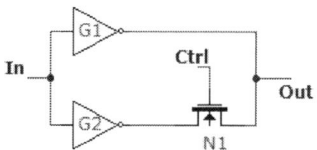

Fig. 8: Schematic block diagram of VCO.

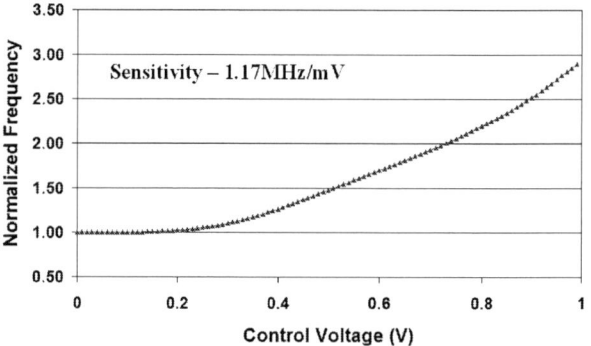

Fig. 9: Normalized frequency characteristics of VCO.

4. Simulation Results and Discussion

The slew-rate monitor circuit described in the previous section has been designed and implemented in a 45 nm SOI technology in an 45um×35um area. Thick oxide long channel devices were used in the design to minimize the impact of variation in the manufacturing process on the sensitivity of the monitor.

The input slew is varied from 25ps to 250ps. The output of the integrator is plotted in Fig. 10. As can be seen,

146

978-1-4244-3039-0/08 $25.00 © 2008 IEEE

the slew-rate monitor exhibits an output sensitivity of 0.5mV/ps. The frequency of the VCO has also been plotted in Fig. 10. The output sensitivity of the oscillator w.r.t the input signal slew is 1MHz/ps as can be seen in Fig. 10.

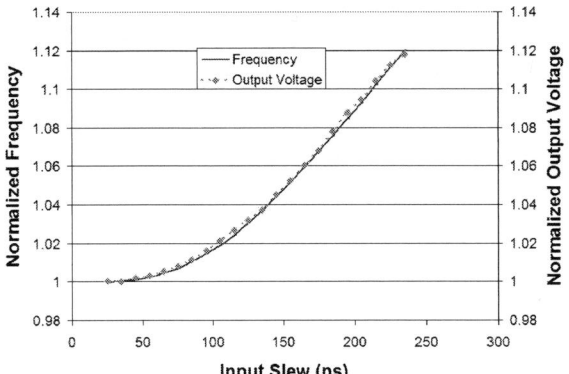

Fig.10: Normalized output frequency of VCO and output voltage of slewrate monitor with input signal slew.

To demonstrate the effect of mismatch between NMOS and PMOS V_{TH}, a 33 stage inverter chain with FO4 load at each node was used as a testbed circuit. The output of an intermediate stage in the chain was fed to the input of the slew-rate monitor. The inverter chain was simulated for various V_{TH} mismatches between N- and P-type devices. When a falling edge transition is applied as the input, the circuit monitors NMOS V_{TH} shift, while applying the rising edge at the input captures the variation in PMOS V_{TH} shift.

Normalized output frequency for differences between rise and fall input transitions of the slew monitor system are shown in Fig. 11 for various mismatches in the threshold voltages of NMOS and PMOS devices. As demonstrated, the slew-rate monitor can detect the NMOS and PMOS variation accurately, with a sensitivity of 0.95MHz/mV.

5. Conclusion

We have illustrated the deficiency of pure delay-based compensation schemes in measuring the mismatch between P- and N-type devices. This has been validated using an analytical framework for impact of V_{TH} variation on circuit performance. We have proposed the use of a slew metric in combination with delay to determine the appropriate supply and body-bias voltages. These voltages can be applied to optimize the power-efficiency and/or performance of designs, resulting in improved parametric yield.

A novel slew-rate monitoring circuit has been designed and implemented in an 45um×35um area. This circuit can be used to measure slew of any signal by setting the reference voltages to the desired points. In a 45 nm SOI technology, the slew-rate monitor shows a sensitivity of 1MHz/ps to the slew of a signal from a representative critical path circuit. This circuit is capable of detecting a V_{TH} mismatch in the order of mV between the two device types with a sensitivity of 0.95MHz/mV.

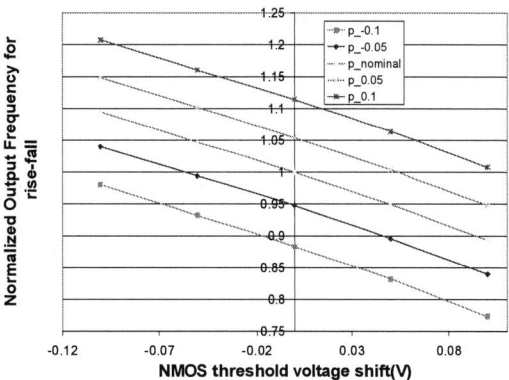

Fig. 11: Normalized output of the integrator with NMOS and PMOS V_{TH} variation,

Acknowledgements:
J. J. Kim and C. T. Chuang are partially supported by the DARPA contract NBCH30390004 for this work. A. Ghosh and R. B. Brown were supported in part by the Engineering Research Centers Program of the National Science Foundation under Award Number EEC-9986866.

References
[1] Tschanz et. al., "Adaptive body bias for reducing impact of Die-to-Die and Within-Die parameter variations on microprocessor frequency and leakage", *IEEE JSSC, Vol. 37, no. 11, pp. 1396-1402, Nov. 2002.*

[2] Fetzer,E.S., "Using Adaptive Circuits to Mitigate Process Variations in a Microprocessor Design", *IEEE Design & Test of Computers,Vol. 23, No. 6, pp. 476-483, June 2006*

[3] Zhou, B. et.al., "Measurement of delay mismatch due to process variations by means of modified ring oscillators", *Proc. ISCAS, pp. 5246 – 5249, Vol. 5, May 2005*

[4] Bassi, A., et. al., "Measuring the effects of process variations on circuit performance by means of digitally-controllable ring oscillators", *Proc .of ICMTS, pp. 214 – 217, March 2003.*

[5] Nomura et. al., "Delay and power monitoring schemes for minimizing power consumption by means of supply and threshold voltage control in active and standby modes", *IEEE JSSC, Vol. 41, no. 4, pp. 805-814, April 2006.*

[6] Rao R, et. al., "Parametric yield analysis and constrained-based supply voltage optimization", *Proc. of ISQED, pp. 284- 290 Mar 2005*

[7] Hamoui A.A. et. al. ,"An analytical model for current, delay, and power analysis of submicron CMOS logic circuits"; *IEEE Trans. on CAS II: Analog and Digital Signal Processing, vol. 47, no. 10, pp:999 – 1007, Oct 2000.*

[8] Tae-Yong C. et. al., "A simple CMOS delay model for wide applications", *Proc. of Asia Pacific Conference Circuits and Systems, pp. 77 – 80, 1996.*

[9] Bisdounis L. et. al., "Analytical transient response and prop-agation delay evaluation of the CMOS inverter for short-channel devices", *IEEE JSSC Vol. 33, no. 2, pp. 302 – 306, Feb 2006.*

[10] Rabaey J. M. et. al., "Digital Integrated Circuits (2nd Edition)", 2002

[11] Park S. et. al., "Low-power Transistor-String and new rail-to-rail comparator in A/D converter", *Proc. of MWSCAS , Vol 1, pp. 194 - 197, 1999*

[12] Baker R.J. et.al. "CMOS Circuit Design, layout, and Simulation: Chap 26", *IEEE Press, 1996.*

SESSION A2:
Testing/DFT

21st International Conference on VLSI Design

On Common-Mode Skewed-Load and Broadside Tests

Irith Pomeranz[1]
School of ECE
Purdue University
W. Lafayette, IN 47907, U.S.A
pomeranz@ecn.purdue.edu

Sudhakar M. Reddy[2]
ECE Dept.
University of Iowa
Iowa City, IA 52242, U.S.A
reddy@engineering.uiowa.edu

Sandip Kundu
ECE Dept.
University of Massachusetts
Amherst, MA 01003, U.S.A.
kundu@ecs.umass.edu

Abstract

Two-pattern tests for delay faults in standard scan circuits can be of one of two types: skewed-load or broadside. Each type of tests creates different conditions during test application due to the different way in which scan mode and functional mode are interleaved. Therefore, tests that are applicable both as skewed-load tests and as broadside tests are useful for comparing the two types of tests with respect to properties such as defect coverage or overtesting. In this work we investigate the possibility of generating tests that are applicable under both test application schemes. We refer to two-pattern tests that are applicable as both skewed-load and broadside tests as common−mode tests. We show that most benchmark circuits have sufficient numbers of common-mode tests to make them an interesting class of tests. Moreover, we show that the use of multiple scan chains increases the number of common-mode tests.

1. Introduction

Two-pattern tests for delay faults in standard scan circuits can be applied under one of two test application schemes. We denote a two-pattern test by $<s_1,v_1;s_2,v_2>$, where s_i is a state and v_i is a primary input vector, for $i = 1,2$. The state s_1 and the primary input vectors v_1 and v_2 are not limited by the various test application schemes. The state s_1 is always scanned in at the beginning of the test, and v_1 and v_2 are applied in two consecutive clock cycles. Depending on the test application scheme, the state s_2 is determined as follows.

Under skewed-load tests [1], after the state s_1 of the first pattern of every test is scanned in, the state s_2 of the second pattern is obtained by a single additional shift of the scan chain(s). Thus, s_2 is determined from s_1 and the scan-in value(s), and the circuit is switched from scan mode to functional mode during the application of the second pattern. For a circuit with n primary inputs, k

---

1. Research supported in part by SRC Grant No. 2004-TJ-1244.
2. Research supported in part by SRC Grant No. 2004-TJ-1243.

state variables and c scan chains, the number of skewed-load tests is 2^{2n+k+c} (two arbitrary input vectors contribute 2^{2n} options, an arbitrary s_1 contributes 2^k options, and c scan-in values add 2^c options).

Under broadside tests [2], the state s_2 of the second pattern is obtained by latching the next-state obtained under the first pattern, and using it as part of the second pattern. Thus, s_2 is determined from s_1 and v_1, and the circuit is switched from scan mode to functional mode during the application of the first pattern. For a circuit with n inputs and k state variables, the number of broadside tests is 2^{2n+k} (two arbitrary input vectors contribute 2^{2n} options, and an arbitrary s_1 contributes 2^k options).

Skewed-load tests are known to result in a higher coverage of delay faults than broadside tests. This is related to the fact that there are more skewed-load tests than broadside tests, and increasing the number of scan chains increases the number of skewed-load tests. Nevertheless, broadside tests are known to detect some faults that cannot be detected by skewed-load tests. Both types of scan-based tests were shown to result in overtesting that may lead to unnecessary yield loss [3]-[5]. Overtesting is a result of the fact that scan makes it possible to bring the circuit to states that it cannot visit during functional operation. In addition, switching the circuit from scan mode to functional mode during the application of a test results in mode switching delays that do not occur during functional operation. In skewed-load tests, the switch from scan mode to functional mode occurs during the application of the second pattern, where the circuit delay is also measured. Thus, the delays related to mode switching have a direct effect on the measured circuit delays. In broadside tests, the switch from scan mode to functional mode occurs earlier, when the first pattern is applied. However, this advantage of broadside tests does not eliminate overtesting completely.

Based on the discussion above, part of the difference between skewed-load and broadside tests is due to the fact that different states may be possible for s_2 under the two types of tests, resulting in different two-pattern tests. This results in the ability to assign different signal-

151

978-1-4244-3039-0/08 $25.00 © 2008 IEEE

transitions to circuit lines under the two types of tests. Another part of the difference has to do with the operation conditions, especially the point where the switch from scan mode to functional mode occurs. When considering overtesting due to operation conditions and the switch from scan mode to functional mode, it is important to be able to compare defect detection under skewed-load tests with that under broadside tests when the signal-transitions under both types of tests are identical. This requires the use of two-pattern tests such that the same test can be applied under both test application schemes. With such a test, the values throughout the (fault free) circuit are the same for both test application schemes. Different defects may then be detected due to the different operation conditions created by the test under different test application schemes. With such tests it is also possible to check whether overtesting occurs to a larger extent with skewed-load tests due to the more extensive use of scan, and the later switching from scan mode to functional mode. Moreover, if the mode switching delay can be quantified by using tests that are applicable under both test application schemes, it may be possible to use skewed-load tests to achieve a higher fault coverage, while accounting for the effects of mode switching delays.

We refer to a two-pattern test that can be applied under both the skewed-load and broadside test application scheme as a *common−mode* test. Our goal in this work is to demonstrate the existence of common-mode tests.

A common-mode test must satisfy the constraints imposed by both test application schemes on the state s_2. Consequently, the number of common-mode tests is lower than the number of skewed-load or broadside tests, and the number of faults detectable by common-mode tests is lower than the number of faults detectable by skewed-load or broadside tests. Nevertheless, as we show later, most benchmark circuits have sufficient numbers of common-mode tests to make it interesting to study this class of tests. In addition, as we show later, the use of multiple scan chains increases the number of common-mode tests and the number of faults they detect.

In a common-mode test, s_2 is completely-specified. It is also possible to generate *partial* common-mode tests, where s_2 is partially specified. The specified values of s_2 are the same whether the test is applied as a skewed-load or as a broadside test. The unspecified values depend on the test application scheme. Only faults that are detectable using the partially specified s_2 should be considered for the comparison of skewed-load and broadside tests. Partial common-mode tests can be used when common-mode tests do not exist for certain faults. We do not consider the generation of partial common-mode tests in this work.

The paper is organized as follows. In Section 2 we demonstrate the existence of common-mode tests by considering exhaustive test sets for benchmark circuits with small numbers of primary inputs and state variables. In Section 3 we describe a procedure for deriving common-mode tests, which is applicable to larger circuits. We also present experimental results. In Section 4 we discuss the effect of the number of scan chains on the ability to generate common-mode tests. Since a higher number of scan chains allows more flexibility in the generation of skewed-load tests, it can result in more common-mode tests. We include in Section 4 results for benchmark circuits with multiple scan chains.

2. Existence of common-mode tests

In this section we demonstrate the existence of common-mode tests for benchmark circuits with small numbers of inputs and state variables. We assume a single scan chain connected such that flip-flop j drives flip-flop $j+1$, for $0 \leq j \leq k-2$, where k is the number of state variables. For $i = 1,2$, let $s_i = s_i(0)s_i(1) \cdots s_i(k-1)$, where s_i is a state and $s_i(j)$ is the value of state variable j under s_i. The state s_2 under a skewed-load test $<s_1,v_1;s_2,v_2>$ satisfies the condition $s_2(j) = s_1(j-1)$ for $1 \leq j \leq k-1$. The value $s_2(0)$ is equal to the scan-in value.

For circuits with small numbers of primary inputs and state variables, we obtain an exhaustive common-mode test set as follows. For every combination of s_1, v_1 and v_2, we compute s_2 such that $s_2(0) = x$, and $s_2(j) = s_1(j-1)$ for $1 \leq j \leq k-1$. s_2 is appropriate for a skewed-load test, with $s_2(0)$ still to be determined. We then compute the next state p_2 obtained under s_1 and v_1. p_2 is appropriate for the second pattern of a broadside test. We set $s_2(0) = p_2(0)$. If $p_2 = s_2$ is obtained, the test $<s_1,v_1;s_2,v_2>$ is a common-mode test since it satisfies the conditions for both skewed-load and broadside tests. In this case we add it to the test set. We denote the resulting test set by E_{common}.

For comparison, we obtain exhaustive skewed-load and broadside test sets as follows. For an exhaustive skewed-load test set, we consider every combination of vectors s_1, v_1 and v_2, and every scan-in value a. We compute s_2 such that $s_2(0) = a$, and $s_2(j) = s_1(j-1)$ for $1 \leq j \leq k-1$. We add the test $<s_1,v_1;s_2,v_2>$ to the test set, denoted by E_{skewed}. For an exhaustive broadside test set, we consider every combination of vectors s_1, v_1 and v_2. We compute s_2 as the next state obtained under s_1 and v_1. We add the test $<s_1,v_1;s_2,v_2>$ to the test set, denoted by E_{broad}.

Table 1: Exhaustive test sets

circuit	in	sv	flts	skewed-load tests	det	broadside tests	det	common tests	det
bbara	4	4	270	8192	262	4096	246	576	184
bbsse	7	4	452	524288	427	262144	415	30720	210
bbtas	2	3	112	256	109	128	109	24	38
beecount	3	3	202	1024	195	512	185	136	136
cse	7	4	742	524288	687	262144	676	29696	221
dk14	3	3	434	1024	419	512	415	160	381
dk15	3	2	292	512	290	256	287	120	279
dk16	2	5	1106	1024	1013	512	1055	36	508
dk17	2	3	248	256	241	128	239	24	200
dk27	1	3	122	64	108	32	116	12	81
dk512	1	4	230	128	211	64	208	10	133
ex2	2	5	620	1024	580	512	603	44	424
ex3	2	4	300	512	284	256	296	28	213
ex4	5	4	334	32768	302	16384	307	1920	92
ex5	2	3	280	256	266	128	258	28	153
ex6	5	3	452	16384	433	8192	430	1920	254
ex7	2	4	296	512	278	256	277	28	178
firstex	2	4	138	512	119	256	74	60	47
keyb	7	5	980	1048576	876	524288	884	43136	357
lion	2	2	68	128	68	64	65	36	56
lion9	2	3	110	256	107	128	96	36	86
mark1	4	4	378	8192	336	4096	334	768	144
mc	3	2	130	512	127	256	119	160	94
opus	5	4	364	32768	345	16384	357	1952	249
shiftreg	1	3	52	64	49	32	49	8	37
sse	7	4	452	524288	427	262144	415	30720	210
tav	4	2	114	2048	112	1024	109	512	96
train11	2	4	182	512	177	256	174	40	143
train4	2	2	68	128	67	64	59	40	59

We simulate each test set under the set of transition faults. The results are shown in Table 1. After the circuit name we show the number of inputs, the number of state variables, and the number of transition faults. For each test set we show the number of tests and the number of detected faults under the corresponding column.

For example, *bbara* has 8192 skewed-load tests that detect 262 transition faults, 4096 broadside tests that detect 246 transition faults, and 576 common-mode tests that detect 184 transition faults.

Table 1 demonstrates that benchmark circuits have non-negligible numbers of common-mode tests that detect non-negligible numbers of transition faults. In the next section we describe a procedure for generating common-mode tests. The procedure is applicable to circuits with larger numbers of primary inputs and state variables.

3. Generation of common-mode tests

In this section we describe a procedure for generating common-mode tests. A deterministic test generation procedure can be used instead. For example, a test generation procedure for broadside tests can be modified to accommodate the additional constraints imposed on s_2 by the skewed-load test application scheme.

We consider circuits with single scan chains. As before, we assume that a scan chain is connected such that flip-flop j drives flip-flop $j+1$, for $0 \leq j \leq k-2$.

In the proposed procedure, for each test, we initially use a weighted-random vector for s_1. The weighted-random vector is selected such that the probability of a 1 in s_1 is varied among the values 1, 1/2, 1/3, 2/3, 1/4, 3/4, $\cdots$, 1/10, 9/10 with each test.

We initially leave v_1 and v_2 unspecified. We compute s_2 and modify s_1 until certain necessary conditions for a common-mode test are satisfied, or until a limit M on the number of modifications is reached. If the necessary conditions are satisfied, we consider the specification of v_1 and v_2 so as to obtain a common-mode test.

In an iteration of the modification process, we have a fully-specified vector for s_1. We compute the next state s_2 obtained under s_1 with a fully-unspecified primary input vector v_1. Thus, s_2 satisfies the conditions for a broadside test $<s_1,v_1;s_2,v_2>$. Since v_1 is unspecified, s_2 may be incompletely-specified.

We modify s_1 such that the specified values of s_2 would be obtainable by a single shift of s_1, i.e., so as to satisfy the condition $s_1(j-1) = s_2(j)$ for every j, $1 \leq j \leq k-1$, where $s_2(j)$ is specified. Thus, s_1 is modified so as to satisfy the conditions on a skewed-load test

$<s_1,v_1;s_2,v_2>$. An iteration ends after s_1 is modified, without recomputing s_2. In the next iteration we consider the modified state s_1, and we compute a new next state s_2 based on s_1. This continues for a constant number M of iterations, or until an iteration where s_1 does not need to be modified in order to match s_2 under a skewed-load test.

We demonstrate the modification process by considering $s_1 = 00001000001001$ for ISCAS-89 benchmark $s298$.

In iteration 1, we have $s_1 = 00001000001001$. We compute the next state $s_2 = \text{x00000000110xx}$ for s_1 and $v_1 = \text{xxx}$. We then modify s_1 into $s_1 = 00000000110001$ in order to match it to s_2 under a skewed-load test. Note that we only modify bits of s_1 that conflict with corresponding bits of s_2. The three states, s_1 before modification, s_2, and s_1 after modification, are shown in Table 2(a).

Table 2: Example of common-mode test generation

	iter	state	
(a)	1	s_1 before	00001000001001
		s_2	x00000000110xx
		s_1 after	00000000110001
(b)	2	s_1 before	00000000110001
		s_2	x000x0011000xx
		s_1 after	00000011000001
(c)	3	s_1 before	00000011000001
		s_2	x000x0011000xx
		s_1 after	00000011000001

In iteration 2, we have $s_1 = 00000000110001$. We compute the next state $s_2 = \text{x000x0011000xx}$ for s_1 and $v_1 = \text{xxx}$. We then modify s_1 into $s_1 = 00000011000001$ in order to match s_1 to s_2 under a skewed-load test. The three states are shown in Table 2(b).

In iteration 3, we have $s_1 = 00000011000001$. We compute the next state $s_2 = \text{x000x0011000xx}$ for s_1 and $v_1 = \text{xxx}$. The states s_1 and s_2 match the conditions of a skewed-load test, and the modification process terminates. The three states are shown in Table 2(c). The test $<s_1,v_1;s_2,v_2>$ with $s_1 = 00000011000001$, $v_1 = \text{xxx}$, $s_2 = \text{x000x0011000xx}$ and $v_2 = \text{xxx}$ is a common-mode test.

In order to obtain a fully-specified test, we fill the unspecified values in v_1 one at a time. Let $v_1(j)$ be the value of input j under v_1. For $0 \leq j \leq n-1$, we specify $v_1(j)$ randomly. We then check whether the test $<s_1,v_1;s_2,v_2>$ satisfies the conditions for a common-mode test. If not, we complement $v_1(j)$ and check the conditions again. If the test is still not a common-mode test, we do not consider any additional primary inputs. If a common-mode test is obtained, we randomly specify v_2.

We repeat this process a constant number of times denoted by F or until a common-mode test is obtained.

In the example of $s298$ we obtain the common-mode test $<00000011000001, 100; 00000001100000, 101>$ after the first attempt to specify v_1.

We repeat the test generation process N times, for a constant N, in order to generate a common-mode test set T_{common} with at most N tests. The test generation process is summarized in Procedure 1.

Procedure 1: Generating a common-mode test set

(1) Set $T_{common} = \phi$. Set $count\,1 = 0$.

(2) Select a random vector s_1. Let v_1 and v_2 be fully-unspecified. Set $count\,2 = 0$.

(3) Compute the next state s_2 obtained under s_1 and v_1.

(4) Set $compl = 0$. For every state variable j such that $1 \leq j \leq k-1$, if $s_2(j)$ is specified and $s_1(j-1) \neq s_2(j)$, complement $s_1(j-1)$ and set $compl = 1$.

(5) If $compl = 1$:
Set $count\,2 = count\,2+1$. If $count\,2 < M$, go to Step 3.

(6) If $compl = 0$:
  (a) Set $count\,3 = 0$.
  (b) Unspecify v_1. Set $j=0$.
  (c) Select a random value for $v_1(j)$.
  (d) If $<s_1,v_1;s_2,v_2>$ is not a common-mode test:
        Complement $v_1(j)$. If $<s_1,v_1;s_2,v_2>$ is not a common-mode test:
          Set $count\,3 = count\,3+1$. If $count\,3 < F$, go to Step 6b. Else, go to Step 7.
  (e) Set $j = j+1$. If $j < n$, go to Step 6c.
  (f) Specify v_2 randomly. Add $<s_1,v_1;s_2,v_2>$ to T_{common}.

(7) Set $count\,1 = count\,1+1$. If $count\,1 < N$, go to Step 2.

We applied Procedure 1 to benchmark circuits using $M = 1000$, $F = 100$ and $N = 100,000$. The results are shown in Table 3.

After the circuit name we show the number of inputs, the number of state variables, and the number of transition faults. Under column *common* we show the number of common-mode tests, and the number of transition faults they detect.

From Table 3 it can be seen that even without performing deterministic test generation it is possible to find significant numbers of common-mode tests for many of the benchmark circuits considered. The tests detect significant numbers of transition faults.

978-1-4244-3039-0/08 $25.00 © 2008 IEEE

Table 3: Generated common-mode test sets

circuit	in	sv	flts	common tests	det
s208	11	8	416	89280	86
s298	3	14	596	128	112
s344	9	15	688	52409	167
s382	3	21	764	64	56
s400	3	21	800	64	56
s420	19	16	840	89922	90
s510	19	6	1020	33107	229
s526	3	21	1052	192	130
s641	35	19	1280	39847	660
s820	18	5	1640	80992	532
s953	16	29	1906	41706	263
s1196	14	18	2392	2004	1854
s1423	17	74	2846	11	254
s5378	35	179	10590	0	0
s9234	19	228	18468	0	0
s13207	31	669	26358	0	0
s15850	14	597	31694	0	0
s35932	35	1728	71864	4875	14258
s38584	12	1452	76864	0	0
b03	5	30	768	1029	68
b04	12	66	2284	56479	364
b06	3	9	332	96	186
b09	2	28	678	37	174
b10	12	17	870	52602	308
b11	8	30	1830	30409	197
b14	33	247	17180	9569	319

4. Circuits with multiple scan chains

In Section 3 we considered circuits with single scan chains. As a result, the condition on a skewed-load test was $s_2(j) = s_1(j-1)$ for $1 \leq j \leq k-1$. The value of $s_2(0)$ was determined by the scan-in value and therefore could be set arbitrarily.

When a circuit has c scan chains, there are c flip-flops that receive their values from the scan inputs, and these values can be set arbitrarily under a skewed-load test. As a result, the circuit has more skewed-load tests, and potentially additional common-mode tests. Broadside tests are not affected by the number of scan chains.

Considering a circuit with c scan chains, we denote the index of the first flip-flop of scan chain i by $f(i)$, for $0 \leq i \leq c-1$. We continue to assume that scan chains are connected such that flip-flop j drives flip-flop $j+1$ (unless flip-flop j is the last flip-flop of a scan chain). Under these conditions, a skewed-load (and a common-mode) test must satisfy the condition $s_2(j) = s_1(j-1)$ for $f(i)+1 \leq j \leq f(i+1)-1$ and $0 \leq i \leq c-1$. Here, $f(c) = k$.

Procedure 1 can be used with this condition to generate common-mode tests for circuits with multiple scan chains. For illustration, we show a common-mode test generated for $s382$ with three scan chains that start at flip-flops 0, 4 and 17. The test generation process starts with $s_1 = 0000$- 1000001001000- 0100. We separate dif-

ferent scan chains by a dash to make it easier to identify the first flip-flop of every scan chain.

In iteration 1, we have $s_1 = 0000$- 1000001001000- 0100. We compute the next state $s_2 =$ xx01- xxxx00x00x000- 0x0x for s_1 and $v_1 =$ xxx. We then modify s_1 into $s_1 = 0010$- 1000000000000- 0000 in order to match it to s_2 under a skewed-load test as shown in Table 4(a). Note that we only modify bits of s_1 that conflict with corresponding bits of s_2, and that the value of the first flip-flop of every scan chain under s_2 does not cause a conflict since it can be determined arbitrarily under a skewed-load test.

Table 4: Example of common-mode test generation with multiple scan chains

	iter		state
(a)	1	s_1 before	0000-1000001001000-0100
		s_2	xx01-xxxx00x00x000-0x0x
		s_1 after	0010-1000000000000-0000
(b)	2	s_1 before	0010-1000000000000-0000
		s_2	xx01-1110000000000-000x
		s_1 after	0010-1100000000000-0000
(c)	3	s_1 before	0010-1100000000000-0000
		s_2	xx01-1110000000000-000x
		s_1 after	0010-1100000000000-0000

In iteration 2, we have $s_1 = 0010$- 1000000000000- 0000. We compute the next state $s_2 =$ xx01- 1110000000000- 000x for s_1 and $v_1 =$ xxx. We then modify s_1 into $s_1 = 0010$- 1100000000000- 0000 in order to match it to s_2 as shown in Table 4(b).

In iteration 3, we have $s_1 = 0010$- 1100000000000- 0000. We compute the next state $s_2 =$ xx01- 1110000000000- 000x for s_1 and $v_1 =$ xxx as shown in Table 4(c). The states s_1 and s_2 match the conditions of a skewed-load test, and the modification process terminates.

After specifying the primary input vectors we obtain the common-mode test <0010- 1100000000000- 0000, 011; 0001- 1110000000000- 0000, 011>.

We partition the flip-flops of benchmark circuits into multiple scan chains by selecting c indices $f(0), f(1), \cdots, f(c-1)$ as follows. We consider values of c that are multiples of two, i.e., $c = 1,2,4,8,\cdots,c_{max}$, where c_{max} is the largest multiple of two that is smaller than $k/2$. For $c = 1$, we use $f(0) = 0$ to obtain a single scan chain as defined earlier. For $c > 1$, we start from $f(0), f(1), \cdots, f(c/2-1)$ selected for $c/2$ scan chains. To this selection we add $c/2$ randomly selected flip-flop indices that are not already included in the selection. For example, for a circuit with 25 flip-flops we start from $f(0) = 0$ for $c = 1$. For $c = 2$ we select one index ran-

155

978-1-4244-3039-0/08 $25.00 © 2008 IEEE

domly, say index 9, to obtain $f(0) = 0$ and $f(1) = 9$. For $c = 4$ we select two additional indices randomly, say indices 5 and 18, to obtain $f(0) = 0$, $f(1) = 5$, $f(2) = 9$ and $f(3) = 18$. We continue up to $c_{max} = 8$ in this case.

With this selection process, the set of first flip-flops of $c/2$ scan chains is contained in the set of first flip-flops of c scan chains. This containment ensures that a common-mode test computed for c_1 scan chains is also applicable when c_2 scan chains are considered, for $c_2 > c_1$. For example, the common-mode test <0010-1100000000000- 0000, 011; 0001- 1110000000000- 0000, 011> computed for $s382$ with $c = 3$ scan chains such that $f(0) = 0$, $f(1) = 4$ and $f(2) = 17$ is also applicable with $c = 4$ scan chains such that $f(0) = 0$, $f(1) = 4$, $f(2) = 10$ and $f(3) = 17$. In this case, the test would be written as <0010- 110000- 0000000- 0000, 011; 0001- 111000- 0000000- 0000, 011>, and the scan-in value of scan chain 2 under the second pattern is equal to the scan-out value of scan chain 1 under the first pattern (this condition is always satisfied when a scan chain is partitioned into two scan chains and the tests obtained before the partitioning are used).

Experimental results for several circuits are shown in Table 5. The case of $c = 1$ is the same as the case of a single scan chain considered in Table 3. We report on values of c that improved the ability to generate common-mode tests relative to the case where $c = 1$, as well as the case of $c = 1$ for comparison.

From Table 5 it can be seen that increasing the number of scan chains increases the number of common-mode tests generated and the number of faults they detect. For the larger circuits the use of multiple scan chains is necessary for the generation of common-mode tests. Only for very few circuits, including $s5378$ and $s15850$, we could not find any common-mode tests even with high numbers of scan chains.

5. Concluding remarks

We defined a common-mode scan-based test as a two-pattern test that can be applied under both the skewed-load and the broadside test application schemes. We demonstrated that benchmark circuits have non-negligible numbers of common-mode tests that detect non-negligible numbers of faults. Common-mode tests can be used for comparing the two test application schemes under identical fault free values, including the study of defect coverage and overtesting due to the test application scheme.

Table 5: Generated common-mode test sets for multiple scan chains

circuit	c	common tests	det
s1423	1	11	254
s1423	2	16	258
s1423	4	32	365
s1423	8	87	432
s1423	16	132	480
s1423	32	5351	1008
s9234	1	0	0
s9234	8	15	1269
s9234	16	35	1736
s9234	32	36	1736
s9234	64	41	1913
s13207	1	0	0
s13207	128	8	3530
s13207	256	76	6677
s38584	1	0	0
s38584	512	5019	11542
b03	1	1029	68
b03	2	2294	181
b03	4	6574	186
b03	8	20021	459
b04	1	56479	364
b04	2	73933	370
b04	4	102145	436
b04	8	126586	496
b04	16	145872	599
b04	32	172293	1193
b14	1	9569	319
b14	2	11029	322
b14	4	12057	330
b14	8	13490	345
b14	16	15112	379
b14	32	17528	632
b14	64	19549	844

References

[1] J. Savir and S. Patil, "Scan-Based Transition Test", IEEE Trans. on Computer-Aided Design, Aug. 1993, pp. 1232-1241.

[2] J. Savir and S. Patil, "Broad-Side Delay Test", IEEE Trans. on Computer-Aided Design, Aug. 1994, pp. 1057-1064.

[3] J. Rearick, "Too Much Delay Fault Coverage is a Bad Thing", in Proc. Intl. Test Conf., Oct. 2001, pp. 624-633.

[4] J. Saxena, K. M. Butler, V. B. Jayaram, S. Kundu, N. V. Arvind, P. Sreeprakash and M. Hachinger, "A Case Study of IR-Drop in Structured At-Speed Testing", in Proc. Intl. Test Conf., 2003, pp. 1098-1104.

[5] J. Rearick and R. Rodgers, "Calibrating Clock Stretch During AC Scan Testing", in Proc. Intl. Test Conf., Nov. 2005, pp. 266-273.

21st International Conference on VLSI Design

Testing Flash Memories for Tunnel Oxide Defects

Mohammad Gh. Mohammad
Computer Engineering Department
Kuwait University
P.O. Box 5969, Safat, 13060, Kuwait
Email: mohammad@eng.kuniv.edu.kw

Kewal K. Saluja
Electrical and Computer Engineering Department
University of Wisconsin
Madison, Wisconsin, USA
Email: saluja@ece.wisc.edu

Abstract— Testing non volatile memories for tunnel oxide defects is one of the most important aspects to guarantee cell reliability. Defective tunnel oxide layer in core memory cells can result in various disturb faults. In this paper, we study various defects in the insulating layers of a 1T flash cell and analyze their impact on cell performance. Further, we present a test methodology and test algorithms that enable the detection of tunnel oxide defects in an efficient manner.

I. INTRODUCTION

Non volatile memories (NVM) in general and flash memories in particular are becoming the number one memory of choice in applications ranging from cell phones to complex systems such as System-on-Chip (SOC). The market share of flash memory is expected to double in the next few years to reach 20 billion dollars. One of the main reasons for this explosive growth is the low cost of flash memories. However, one of the major concerns in manufacturing such memories is the cost of test. With every new generation, the cost of test is increasing due to the introduction of new failure mode(s) which must be considered to properly test such memories. In today's semiconductor manufacturing, more than 30% of final product cost is due to test requirements [1]. Hence, methods to reduce the test cost are likely to be key factors in the continued growth of flash memories and to maintain its competitiveness in the semiconductor memory market.

Since the introduction of flash memories, substantial research has been directed towards the development of efficient tests [2], [3], [4]. In [2] the authors proposed a logical fault model to model various disturb faults and developed efficient tests for their detection. The authors of [3], [4] expanded the modeled faults to include some traditional faults that are present in other type of memories as well as disturb faults that are described by the IEEE Standard Definitions and Characterization of Floating Gate Semiconductor Arrays [5]. In [3], Built-in Self-Test for flash memories was proposed to reduce the cost of testing. In all these papers, special attention was paid to develop efficient test algorithms in detecting all known faults applicable to flash memories.

In this paper, we analyze various defects at different locations in the core memory cell that are responsible for disturb faults and study their impact on cell performance using a 2D device simulator named Atlas [6]. After analyzing the behavior of the defective cells, we determine fault excitation conditions that allow fast and reliable identification of faulty cells. Using these excitation conditions, efficient tests for testing NOR type flash memories are developed.

In section II of this paper we review the previously developed models for flash memory faults. Section III discusses the experimental setup for defect injection and the study of their impact. Findings of simulation studies are provided in section IV. Test algorithms and their detection capabilities are discussed in section V. Section VI concludes the paper.

II. FLASH MEMORY PROGRAM DISTURB FAULT MODELS

Semiconductor memories, including all forms of NVMs such as flash memories, suffer from defects that could occur during the manufacturing process. The characteristics of the defects, whether large or small, can alter the correct behavior of the memory cells. However, the manifestation and detection of these defects depends on their physical characteristics. For example, when a defect is large and it results in a short/open between a line and the power supply (V_{cc}) or ground (GND), then it can manifest itself as a stuck-at (SAF) fault [7]. Alternatively, if the short is between two lines and neither of the lines is V_{cc} or GND, then such a short may result in a coupling fault (CF) or address decoder fault (AF) depending on which two memory lines are shorted together. In other instances, the defects may be too small to result into shorts or opens and may manifest as resistive defects which result in more complex faulty behavior, such as Incorrect-Read-Fault (IRF) or Write-Disturb-Fault (WDF) [8], [9]. Even though tests to detect these faults exist, they can be expensive in terms of test time unless careful attention is paid to the derivation of such tests.

In NVMs, particularly in flash memory literature, the faults are classified into two major categories. The first category consists of those faults that are common between flash and all other type of semiconductor memories such as DRAMs and SRAMs. These faults include SAF, SOF, AF, and CF_{st} (state-coupling) faults [3], [4]. The second category consists of faults that are specific only to NVMs, including flash memories, and do not conform to the traditional faults known to occur in other types of volatile memories such as DRAMs. These faults are known as disturb faults and can be classified as either program or read disturbs [10], [11], [12]. A disturb fault is caused by defects in the insulating layer of a core memory element or it may be induced by electric stress conditions during the different modes of operations.

157

978-1-4244-3039-0/08 $25.00 © 2008 IEEE

The most common memory cell used in today's flash memories is the 1T floating gate transistor (FG). The structure of the 1T transistor is similar to the traditional MOS transistor with an additional floating polysilicon gate that is completely insulated by dielectric from all other conducting terminals. The floating gate is insulated from the substrate by a layer of high quality oxide, called tunnel oxide, whereas it is insulated from the top by a oxide-nitride-oxide (ONO) layer, an interpoly insulator [13]. The logical state of the memory cell is represented by the charge on the floating gate. When there is no charge present on the floating gate, the cell is referred to as erased or containing a logic "1" value. On the other hand, when the floating gate has a negative charge, it is referred to as programmed or having a logic "0" value.

1T flash cells are organized in an array to constitute a memory module. The most common array organizations for flash memories are NOR and NAND array organizations. The NOR array organization is shown in Figure 1. In NOR array organization, Channel Hot Electron Injection (CHEI) mechanism is used to accumulate charge on the floating gate whereas FN-tunneling is used to extract the charge, thus erasing the cell [13]. When accumulating or extracting charge of the floating gate, higher than normal voltages are applied on the various terminals of the memory cell to create high electric field across the tunnel oxide to enable the transfer of charge. These high fields that are used to program and erase the cell pose reliability concerns as well. Due to the way the memory cells are organized in a typical memory array, many of the unselected cells experience the same high electric fields as the selected cell(s). For example, when addressing a cell (i,j), where "i" is the row address and "j" is the column address of a cell, all cells on row i will experience the voltage applied on that row. Similarly, all cells in column j (whether selected or not) will experience the voltage on that column.

Designers are aware of these high fields and they realize that these could result in disturb behavior. As a result, they consider these effects during memory device characterization and they design operating voltages, array organization, and the cell structure to meet the targeted design specifications and minimize the effect of disturbs [13]. But, some of the issues still remain. For example, the quality of the tunnel oxide must be flawless and the manufacturing process must be nearly perfect to ensure the reliability of the memory cell. However, due to the nature of the manufacturing process, contaminants and other anomalies are unavoidable. When these contaminants or defects are located in the insulating layers of the 1T cell, they pose a reliability concern and amplify the unwanted behavior of disturbs. The *IEEE Standard Definitions and Characterization of Floating Gate Semiconductor Arrays* defines nearly all disturbs for all possible memory array organizations and cell structures [5]. In a set of recent papers [2], [3], [4], [14] the most common disturb behaviors were modeled as logical faults. The following is a short description of four most common program disturbs as well as their fault models for NOR type flash memories. Other disturbs which are not very common or which are applicable to other memory array

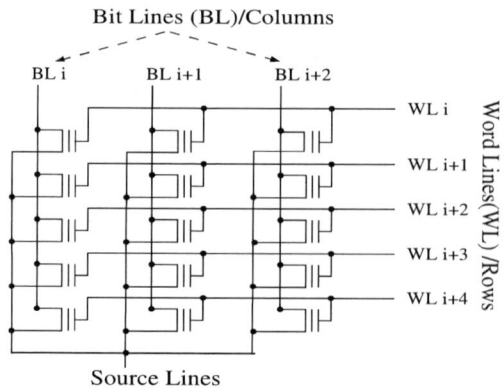

Fig. 1.   NOR Array Organization

organizations can be found in [5], [14]. The description below uses the notation $<S_a;S_v/F/R>$ which is commonly used in representing static coupling faults [15]. In this notation: $S_a \in \{$ w0,w1,r0,r1$\}$, is the *sensitizing operation sequence* (SOS) consisting of read/write operations on the state of the aggressor cell, S_v is a SOS for the state on the victim (before fault excitation) cell, "F" $\in \{0,1\}$ is the state of the faulty cell (after excitation), and R $\in \{0,1,-\}$ is the output of the read operation. The value "-" in field R is used in case of write operation.

Word-line erase disturb (WED)

> exists when a cell being programmed (selected cell) causes another unprogrammed cell (unselected cell), sharing the same word-line, to be erased. This fault is modeled as $<$1w0;0/1/-$>$ fault.

Word-line program disturb (WPD)

> occurs when a cell being programmed causes another unprogrammed cell, sharing the same word-line, to be programmed and is modeled as $<$1w0;1/0/-$>$.

Bit-line erase disturb (BED)

> which is modeled as $<$1w0;0/1/-$>$ fault takes place when a cell being programmed causes another unprogrammed cell, sharing the same bit-line, to be erased.

Bit-line program disturb (BPD)

> arises when a cell being programmed causes another unprogrammed cell, sharing the same bit-line, to be programmed $<$1w0;1/0/-$>$.

Low electric field stresses, such as read disturbs, are also caused by the physical defects in the insulating layer. Thus by identifying defects that cause program disturb faults, we can also detect read disturb faults.

III. 2-D DEVICE SIMULATION SETUP

Two different 2D device simulation tools, namely Athena and Atlas [6], [16], were used to investigate the various defects in 1T cell. Using Athena, we constructed a fault-free $1\mu \times 1\mu$ 1T structure with a tunnel oxide of Å105 and combined ONO stack thickness of Å400. The cell is designed to be programmed by the CHEI using a $1\mu s$ pulse and erased using FN-tunneling with 10ms erase pulse. The bias conditions required to accomplish these operations and the resulting threshold voltages (V_t) are shown in Table I. In this table, labels CG, D, S, and B represent the voltages applied at control gate, drain, source, and base terminals of the memory cell respectively (values given in volts (V)). The erase operation described in Table I is the most common approach used for erasure and is known as *negative gate erase* (NGE). This technique results in lower power consumption as well as better cell reliability [13], [17] when compared to *source side erase* technique which uses high voltage on the source terminal while grounding the control gate of the cell. The

TABLE I

PROGRAM/ERASE BIASES AND THRESHOLD VOLTAGES

Operation	CG	D	S	B	V_t
Program	10	6	GND	GND	7.95
Erase	-8	Floating	7	GND	1.11
Read	3.3	0.5	GND	GND	

number of cells in each row/column in the array organization define the worst case gate/drain stress that a cell can undergo. The duration of the worst case stress can be calculated by multiplying the program time with the number of cells in a row/column (i.e. $T_{stress} = T_P \times (N-1)$), where T_P and N are program time and number of cells in a row/column respectively (for an NxN array organization) [18]. Common stress time found in today's flash memories varies from 0.1ms-2ms depending on the program time and array organization used. In our experiments, we assume a memory array organized as a 128 x 128 grid. This implies that the duration of the worst case gate/drain stress (for a cell) in such an array is $127\mu s$ for a program time of $1\mu s$.

Defect injection was accomplished by injecting a defect in a particular region of the memory cell while maintaining the same processing steps as the fault-free cell. The locations of the defects were limited to the various oxide layers in the structure, namely the tunnel oxide or oxide layers in ONO stack. Five different locations were identified for possible defect injection and corresponding defective devices were created. Thus each defective device had one injected defect in the structure, and these are shown in Figure 2. Defects in the tunnel oxide are located in the oxide area above the diffusion region of the 1T structure (i.e. drain/source overlap) or in the channel region. The ONO layer defects were limited to the bottom (ONO_B) or top (ONO_T) oxide layers.

The defects were characterized by specifying the effective oxide thickness at the defect site. Thus, if the fault free value of the tunnel oxide thickness was Å106, a defect will result in an oxide thickness smaller than this value. For example, a *Å 100 defect* results into effective oxide thickness of Å100 instead of the design value of Å106. All remaining characteristics of the defect size, such as "x" (length) and "z" (along the width of transistor), are kept constant. A total of 11 defective devices were created for simulation as shown in table II. Two studies were carried out on defective cells. The first study was to

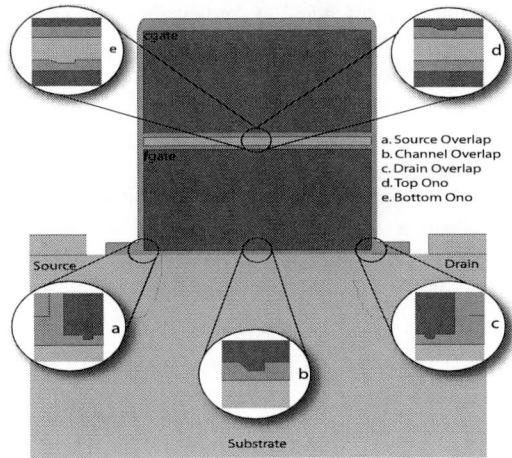

Fig. 2. Defects and Their Locations

determine the impact of the shape of the defects on the cell performance to identify characteristics of killer defects. In the second study, killer defects were injected in different regions of the insulating layers of the 1T cell and their impact on cell operation during normal as well as stress modes were analyzed. The results of these are given below.

TABLE II

TUNNEL OXIDE OF VARIOUS DEFECTS

Defect location	Effective thickness (Å)
ONO_B	67
ONO_T	71
Drain Overlap	100, 80, 60
Source overlap	100, 80, 60
Channel	100, 80, 60

IV. RESULTS AND DISCUSSION

The simulation studies targeted three different objectives. 1) To identify the size of the defects that can adversely impact the reliability of the cell under any mode of operation (i.e. normal operation or stress conditions). 2) To assess the effectiveness of various stress tests on detecting defects. 3) To develop efficient test(s) that detect(s) all defects being investigated in this paper.

For each simulation study, the defect is first injected in the specified region (e.g. drain overlap), and then an operation is performed on the cell (e.g. gate stress). After that, the I-V characteristics are measured and the threshold voltage is extracted. The process is repeated for all combinations of different operations and defect locations.

A. Cell Performance of Cells With Defects

In order to analyze the effect of defect size on cell performance, defects of different shape/size (but of identical total volume) were simulated. Instead of conducting experiments for all defect locations, we inserted defects only in the drain overlap region to carry out these studies. The outcomes are shown in Table III. In this table, the first column represents the operation performed and the remaining columns represent the resulting threshold voltages for three different defects (labeled by their effective tunnel oxide thickness) under various modes of operations. Moreover the rows labeled "Gate/Drain Stress$_E$" in the table represent cells that underwent a $127\mu s$ of stress when they were assumed to be initially erased cells. Rows labeled "Gate/Drain Stress$_P$" correspond to cells that underwent stress assuming they were initially programmed cells. Also, the erase operation in this study was performed using the commonly used NGE method. It is apparent from

TABLE III
DRAIN DEFECT SIZE CHARACTERIZATION

Defect / Operation	Threshold Voltage (V)			
	Fault Free	Å99	Å81	Å49
Program	7.9581	6.3197	7.1271	6.4617
Erase	1.1066	0.36456	-1.9736	-8.0953
Gate Stress$_E$	1.1066	0.36456	-1.9736	-1.5950
Drain Stress$_E$	1.1066	0.36456	-1.9736	-1.7851
Gate Stress$_P$	7.9581	6.3197	7.1271	6.4617
Drain Stress$_P$	7.9577	6.3181	7.1026	3.7307

the table that the oxide thickness of the defect plays a major role in determining the performance (in this case threshold voltage) of the cell. In the case of Å81 and Å49 defects, the cells were depleted when they were erased, depicting the well known over-erase phenomenon [13], [19], [17]. Furthermore, the Å49 defect was the only defect that was identified as a detectable defect using the *Drain Stress$_P$* experiment. As for all other cells, no significant shift in the threshold voltage before and after stress experiment was noticed. In particular, the Å49 defect shifts the threshold voltage of the cell from a programmed V$_t$ of 6.4 to 3.7, i.e. in excess of a 2 volts shift, whereas all other defects cause only a marginal and insignificant shift in the threshold voltage. Interestingly the 2 volts threshold shift does not result in value flip in the memory cell and hence it will not be detected using normal read operation. However, other techniques, such as margin reads [18], can detect such defects. The main conclusion of this set of experiments is that a defect which reduces the effective tunnel oxide thickness to Å80 or less represents a killer defect and such defects must be properly excited and detected.

B. Tunnel Oxide Defects

The second set of experiments was carried out on defective cells with effective tunnel oxide thickness (at defect site) to be Å80, and located in one of the various tunnel oxide regions as specified in section III. For all these cases, we used the NGE erase approach discussed in section III. The experiments were carried out on tunnel oxide defects in source/drain overlaps and in the channel regions. The cells were simulated to see the impact of each defect on the cell program/erase characteristics as well as their behavior under stress conditions. Table IV summarizes the findings. It is evident from this table that only the source overlap defect can potentially be identified in these experiments as explained below. For this defect, the erase

TABLE IV
NEGATIVE GATE ERASE EXPERIMENTS

Defect / Operation	Threshold Voltage (V)			
	Fault Free	Source	Drain	Channel
Program	7.9581	7.9094	7.1271	7.9764
Erase	1.1195	-0.3050	1.1627	1.4534
Gate Stress$_E$	1.1195	-0.3050	1.1627	1.4534
Drain Stress$_E$	1.1195	-0.3050	1.1627	1.4534
Gate Stress$_P$	7.9581	7.9094	7.1271	7.9764
Drain Stress$_P$	7.9577	7.9091	7.1026	7.9761

operation resulted in a depleted cell (cell with negative V$_t$), causing a faulty cell behavior, whereas for all other defects, the threshold shift is only marginal. Thus in the case of source overlap defect, when reading a cell in the same column as the depleted cell, the read data will be corrupted if the addressed cell has a logic "0" value. The stress time for all defective cells was increased to five time ($5 \times 127\mu s$) the worst case stress duration in order to see if an undetected defect would become detectable. However, the defects did not show different behavior under such stress conditions (only a minor shift in V$_t$ for Drain Stress$_P$ experiment). Even though the drain overlap and channel defects do not seem to impact the performance of the memory cell, they will pose reliability concerns. Further, these undetected defects most likely will be excited by cycling and will result in an in-field failure, one of the major concerns in flash memory reliability.

The above study suggests that the stress experiments may not be the most efficient way to detect tunnel oxide defects. In particular, we notice that stress tests do not result in any noticeable shift in the I-V characteristics for any of the defects. Further, it is clear that source overlap defect does impact cell performance (cell becomes depleted after erase). This suggests that the erase operation could be the key to efficient detection of tunnel oxide defects. In the next section we describe a method that can be used to detect these defects.

C. Channel Erase and Tunnel Oxide Defect Detection

In the previous section, it was shown that the drain-overlap and channel defects could not be detected by neither stress condition nor erase/program operation, thus they may remain undetected. The source overlap defect, on the other hand, could be detected because it resulted in a depleted cell when the cell was erased. Investigating further for the reason for this behavior, we found that during erase operation, the overlap area undergoes high electric field stress due to the biases applied to the gate/source terminals of the memory cell. Therefore, we felt that in order to excite and detect defects in any of the tunnel oxide regions, appropriate electric field stress must be present in every region that needs to be tested. One possible approach that offers the opportunity to stress all regions of the cell is the channel erase operation discussed below.

Unlike the commonly used NGE operation, which utilizes negative-gate-positive-source bias condition, channel erase concept results in a uniform electric field stress in all regions of the tunnel oxide. The NGE technique restricts the high electric field region to only the source overlap area. The channel erase approach is accomplished by biases applied either to the control gate only, or by using gate and substrate biases (i.e. in triple well technology). In our study, we chose to use the method where only control gate is biased with -20V while grounding the substrate and floating the source and drain terminal. Further, for the cell structure created in our study, a channel erase approach would require 70ms erase time, which is substantially longer time compared to the NGE approach which requires only 10ms.

In order to analyze the effectiveness of channel erase technique in identifying tunnel oxide defects, we ran the same experiments that were performed previously, but this time using the channel erase approach. The results of this study are shown in Table V. It is evident that every defect in this case results in a depleted cell. These observations suggest that the channel erase technique is far more effective and superior in detecting all tunnel oxide defects.

TABLE V
CHANNEL ERASE EXPERIMENTS

Operation \ Defect	Threshold Voltage (V)			
	Fault Free	Source	Drain	Channel
Program	7.9581	7.9094	7.1271	7.9764
Erase	1.1066	-1.4758	-1.9736	-2.3726
Gate Stress$_E$	1.1066	-1.4758	-1.9736	-2.3726
Drain Stress$_E$	1.1066	-1.4758	-1.9736	-2.3724
Gate Stress$_P$	7.9581	7.9094	7.1271	7.9764
Drain Stress$_P$	7.9577	7.9091	7.1026	7.9761

D. ONO Defects and Impact of Cell Performance

Next we expanded our investigation to study defects in the ONO layer. We used the same approach as before and faulty cells with ONO defects were constructed and simulated. Two type of defects were simulated. First, a defect is created in the bottom oxide layer of the ONO layer (see Figure 2e). The fault free value of this layer was approximately Å96 and in the presence of a defect it has an effective thickness of Å67. Second, a defect in the top oxide layer of the ONO layer (see Figure 2d) was also created and simulated. The fault free thickness in this case was Å100 and that of the defective cell was Å71. The results of the study of these defects are compared to the fault free case in Table VI. It is apparent that defects in the ONO layers do not impact the performance of the flash memory cell and hence can be ignored. This finding supports what was previously argued and suggested in [2], [14] using logical reasoning only.

E. Simulation Summary

We summarize the important findings about various defects as follows. These findings are used to develop efficient tests for various defects in the 1T cell based flash memories.

- **Defect Excitation**: Stress tests are not very effective in defect excitation. It was shown that the erase operation is a more effective way to excite tunnel oxide defects.

TABLE VI
ONO DEFECTS SIMULATION

Operation \ Defect	Threshold Voltage (V)		
	Fault Free	ONO$_B$	ONO$_T$
Program	7.9581	8.5420	8.5437
Erase	1.1066	1.1765	1.1434
Gate Stress$_E$	1.1066	1.1765	1.1434
Drain Stress$_E$	1.1066	1.1765	1.1434
Gate Stress$_P$	7.9581	8.5420	8.5437
Drain Stress$_P$	7.9577	8.5413	8.5431

- **Fault Detection**: Channel erase technique is superior in detecting all defects compared to NGE method.
- **Depleted Cell Behavior**: All tunnel oxide defects result into depleted threshold voltages when channel erase technique is used. Therefore, a test for depleted cell, rather than erased/programmed cell, as previously suggested in [2], [3], [4], is likely to be a more efficient method for detecting such defects.
- **ONO Defects**: No single defect in the ONO layer will result in faulty behavior. Hence, tests for ONO defects can be simplified by removing those patterns.

V. TEST ALGORITHMS

After considering the above findings, we conclude the following for testing flash memories: 1) we must consider all tunnel oxide defects and 2) we must utilize channel erase technique to excite the various defects in tunnel oxide region. We also conclude that ONO defects can be ignored and ONO layer can be assumed to be fault-free. We further conclude that to develop a test to detect tunnel oxide defects, the following conditions must be met:

Programmed Initial State
> All cells to be tested must be programmed (i.e. set to logic "0" state).

Channel Erase Fault Excitation
> Programmed cells to be tested must be erased using channel erase technique.

Figure 3 gives a new test procedure called Flash-CE test, which can be used to detect all defects in the tunnel oxide layer of 1T cells organized in a NOR array. Since most disturb faults are assumed to be caused by defects in the tunnel oxide, we can claim that this algorithm can detect all disturb faults. In Figure 3, n and m represent the number of rows and columns in the memory array, respectively. The working of the algorithm is as follows. Step 1 initializes the array to a programmed state. Step 2 erases all cells in the array using channel erase approach, hence exciting all tunnel oxide defects. The third step programs each cell in the first row ($i = 0$) of the array and reads each element of that row, expecting a value of "0". In case there is any depleted cell in any column (defective cell), the read operation will fail and the value will be read as "1". This is so because of the excessive depletion of the defective cell and as a result the column containing the defective cell will read a logic 1 value.

```
1. For (i=0; i < n; i++)          \* Initialize array        *\
      For (j=0; j < m; i++)
      (w0)_{i,j}
2. For (i=0; i < n; i++)          \* Erase array             *\
      For (j=0; j < m; i++)       \* using channel erase     *\
      (w1)_{i,j}
3. For (j=0; j < m; j++)          \* Program then read row 0  *\
      (w0,r0)_{i=0,j}
4. For (i=1; i < n; i++)
      For (j=0; j < m; j++)       \* Program remaining cells  *\
      (w0)_{i,j}
5. For (i=0; i < n; i++)          \* Erase array             *\
      For (j=0; j < m; i++)       \* using channel erase     *\
      (w1)_{i,j}
6. For (j=m-1; j ≥ 0; j- -)       \* Program then read row n-1 *\
      (w0,r0)_{i=n-1,j}
```

Fig. 3.   Algorithm Flash-CE

After this step, the only cells that remain to be tested are those in the first row. Steps four, five and six initialize, excite, and detect these remaining faults in a similar manner.

In recent years "March tests" have gained popularity and have been used in many test algorithms for testing flash memories [4], [3], [14]. This is due to their simplicity, fault detection capability, and ease of implementation. We have developed an efficient (minimum length) march algorithm, called March$_{CE}$, which can detect all tunnel oxide defects and is as follows:

$$March_{CE} = < \Uparrow w0; E_{Ch}; \Uparrow (w0, r0); E_{Ch}; \Downarrow (w0, r0) >$$

In this algorithm, the term E$_{Ch}$ represents a "w1" on the whole array since a selective "w1" in flash memories is not permissible. Further, the subscript "Ch" in the erase operation signifies the fact that the erase operation uses channel erase instead of the conventional source side or NGE operation. March-CE algorithm is inefficient in detecting other types of faults, such as SAF and SOF (only 50% of SAF, 0% of SOF). However, by adding few additional read operations, the algorithm March$_{CERR}$ given below, can detects 100% of SAF, AF, SOF, TF, and CF$_{st}$ faults.

$$March_{CERR} =$$
$$< \Uparrow w0; E_{Ch}; \Uparrow (r1, w0, r0); E_{Ch}; \Downarrow (r1, w0, r0) >$$

The detection capabilities were computed using RAMSES [20] memory simulator assuming a 1-bit wide memory. In order to implement the channel erase approach in the March algorithms proposed, the design of the memory array may need to be modified. The modification requires the addition of new high voltage switches to the row decoders and additional control logic and the discussion of such design for testability (DFT) concepts is beyond the scope of this paper.

VI. CONCLUSION

In this paper we first studied different defects that are responsible for disturb faults in 1T flash cell using a 2D device simulator. It was found that stress tests are not efficient when it comes to detecting tunnel oxide defects. Oxide-Nitride-Oxide

layer defects were found to be benign and did not result into faulty behavior and hence they can be ignored. Efficient tests based on channel erase techniques were developed to detect tunnel oxide defects (hence disturb faults) as well as other type of faults such as SAF and AF faults.

ACKNOWLEDGMENT

The authors would like to thank Faisal Al-Hasawi of the Public Authority of Applied Education and Training-Kuwait for his help in preparing this manuscript. This work was support by Kuwait University research grant No. [EO01/04].

REFERENCES

[1] M. L. Bushnell and V. D. Agrawal, *Essentials of Electronic Testing for Digital, Memory and Mixed-Signal VLSI Circuits*, 1st ed. Norwell, Massachusetts: Kluwer Academic Publishers, 2000.

[2] M. Mohammad, K. Saluja, and A. Yap, "Fault Models and Test Procedures for Flash Memory Disturbances," *Journal of Electronic Testing: Theory and Applications*, vol. 17, no. 6, pp. 495–508, December 2001.

[3] J.-C. Yeh, C.-F. Wu, K.-L. Cheng, , Y.-F. Chou, C.-T. Huang, and C.-W. Wu, "Flash Memory Built-In-Self-Test Using March-Like Algorithms," 1st *international workshop on Electronic Design, Test, and Applications (DELTA'02)*, pp. 137–141, 2002.

[4] S.-K. Chiu, J.-C. Yeh, C.-T. Huang, and C.-W. Wu, "Diagonal Test and Diagnostic Schemes for Flash Memories," *Proc. International Test Conference*, pp. 37–46, 2002.

[5] *IEEE 1005 Standard Definitions and Characterization of Floating Gate Semiconductor Arrays*, 2nd ed. Piscataway: IEEE Standard Department, 1999.

[6] *ATLAS User's Manual: Device Simulation Software*, Silvaco International, April 1997.

[7] A. van de Goor, *Testing Semiconductor Memories: Theory and Practice*. Gouda, The Netherlands: ComTex Publishing, 1998.

[8] Z. Al-Ars and A. van de Goor, "Test Generation and Optimization for DRAM Cell Defects Using Electrical Simulation," *IEEE Trans. Computer-Aided Design of Integrated Circuits and Systems*, vol. 22, no. 10, pp. 1371–1384, October 2003.

[9] M. Azimane and A. L. Ruiz, "A Special March Test to Detect Delay Coupling Faults for RAMs," 8th *IEEE International Conference on Electronics, Circuits and Systems*, vol. 2, pp. 995–999, 2001.

[10] P. Pavan, R. Bez, P. Olivo, and E. Zanoni, "Flash Memory Cells-An Overview," *Proc. of IEEE*, vol. 85, no. 8, pp. 1248–1271, August 1997.

[11] S. Aritome, R. Shirota, G. Hemink, T. Endoh, and F. Mausouka, "Reliability Issues of Flash Memory Cells," *Proc. of IEEE*, vol. 81, no. 5, pp. 776–787, May 1993.

[12] A. Brand, K. Wu, S. Pan, and D. Chin, "Novel Read Disturb Mechanism Induced by Flash Cycling," *International Reliability Physics Symposium*, pp. 127–132, 1993.

[13] P. Cappelletti, C. Golla, P. Olivo, and E. Zanoni, *Flash Memories*, 1st ed. Norwell, Massachusetts: Kluwer Academic Publisher, 1999.

[14] M. Mohammad and L. Al-Terkawi, "Techniques for Disturb Fault Collapsing," *Journal of Electronic Testing: Theory and Applications*, vol. 23, no. 4, pp. 263–268, August 2007.

[15] A. van de Goor and Z. Al-Ars, "Functional Memory Faults: A formal Notation and a Taxonomy," 18th *IEEE VLSI Test Symposium*, pp. 281–289, 2000.

[16] *ATHENA User's Manual: 2D Process Simulation Software*, Silvaco International, November 1998.

[17] A. Sharma, *Advanced Semiconductor Memories: Architectures, Designs, and Applications*, 1st ed. IEEE Press, 2003.

[18] M. Mohammad and K. Saluja, "Optimizing Program Disturb Faults Test Using Defect-Based Testing," *IEEE Trans. Computer-Aided Design of Integrated Circuits and Systems*, vol. 24, no. 6, pp. 905–915, June 2005.

[19] A. Chimenton, P. Pellati, and P. Olivo, "Overerase Phenomena: An Insight Into Flash Memory Reliability," *Proceedings of The IEEE*, vol. 91, no. 4, pp. 617–626, April 2003.

[20] C.-F. Wu and C.-W. Wu, "RAMSES: A Fast Memory Fault Simulator," *IEEE International Workshop on Defect and Fault Tolerance in VLSI Systems*, pp. 165–173, 1999.

21st International Conference on VLSI Design

On the Detection of Missing-Gate Faults in Reversible Circuits by a Universal Test Set

Hafizur Rahaman[1], Dipak K. Kole[1], Debesh K. Das[2], Bhargab B. Bhattacharya[3]

[1]IT Dept., Bengal Engg. & Sc. University, Howrah – 711 103, India; email: *rahaman_h@it.becs.ac.in*
[2]Dept. of Comp. Sc. & Engg., Jadavpur University, Kolkata – 700 032, India; email: *debeshd@hotmail.com*
[3]ACM Unit, Indian Statistical Institute, Kolkata – 700 108, India; email: *bhargab@isical.ac.in*

ABSTRACT

Logic synthesis with reversible circuits has received considerable interest in the light of advances recently made in quantum computation. Implementation of a reversible circuit is envisaged by deploying several special types of quantum gates, such as k-CNOT. Newer technologies like ion trapping or nuclear magnetic resonance are required to emulate quantum gates. Although the classical stuck-at fault model is widely used for testing conventional CMOS circuits, new fault models, namely, single missing-gate fault (SMGF), repeated-gate fault (RGF), partial missing-gate fault (PMGF), and multiple missing-gate fault (MMGF), have been found to be more suitable for modeling defects in quantum k-CNOT gates. In this paper, it is shown that in an (n × n) reversible circuit implemented with k-CNOT gates, addition of only one extra control line along with duplication each k-CNOT gate yields an easily testable design, which admits a universal test set of size (n +1) that detects all SMGFs, RGFs, and PMGFs in the circuit.

Keywords: Missing-gate faults, quantum computing, reversible logic, testable design, universal test set

1. INTRODUCTION

Reversible logic can be employed to design information lossless circuits, and therefore, has the potential of reducing power consumption drastically [1-4]. It provides a basis for the newly emerging paradigm of quantum computing [5-7]. Since quantum gates or circuits satisfy "no-cloning" behavior, and are information lossless, they do not permit fanout, and ought to have an equal number of inputs and outputs. An *n*-input, *m*-output Boolean function *F* is said to be reversible if and only if $m = n$, and *F* is one-to-one.

A reversible combinational circuit must be fanout free, acyclic, and should consist of only reversible gates, which themselves implement reversible functions; such gates need to be specially designed, e.g., by using Toffoli gates. Reversible circuits have numerous applications to optical computing, digital signal processing, communication, cryptography, nanotechnology, quantum computing, DNA technology, and low-power CMOS design [5-13]. Conventional logic gates such as AND, OR, or EXOR used in digital design are not reversible. To design a reversible circuit, only reversible gates can be used, for

example, the controlled-not (CNOT) gate proposed by Feynman [14], Toffoli [16], or Fredkin [15] gates. Many techniques for the synthesis of reversible logic circuits are known [17-23, 35].

Recently, several researchers [24, 27-30, 32] have studied the problem of fault modeling and testing of reversible logic circuits. The online testability in reversible circuits was studied [25-26, 29]. Universal testability of reversible logic circuits designed with *k*-CNOT gates under the stuck-at fault model (both single and multiple) has also been investigated [30]. A test generation scheme detecting bridging faults in reversible circuits is also reported [32, 33]. However, new fault models, namely, single missing-gate fault (SMGF), repeated-gate fault (RGF), partial missing-gate fault (PMGF), and multiple missing-gate fault (MMGF), have been found to be more suitable for modeling defects in quantum *k*-CNOT gates [27, 34]. They capture better representation of physical failures in reversible logic, particularly for quantum technologies. A *k*-CNOT based circuit can be implemented using trapped-ion technology, where the ions interact with laser pulses [5, 6, 34].

Determination of a universal test set for fault detection in reversible circuits has been studied for a few fault models [27, 30]. It has been shown that by adding one extra control line and a few 1-CNOT gates, any reversible circuit designed with *k*-CNOT gates can be tested for all SMGFs just by applying one test vector. All the irredundant RGFs are also detectable by the same test.

In this paper, we investigate the problem of detecting PMGFs by a universal test set. A PMGF in a *k*-CNOT gate may be of first or higher order depending on the number of partially misaligned or mistuned gate pulses in its quantum implementation [34]. We show that it is always possible to transform an (n × n) reversible circuit implemented with *k*-CNOT gates, by adding only one extra control line and by duplicating each *k*-CNOT gate,

978-1-4244-3039-0/08 $25.00 © 2008 IEEE

so that the modified design becomes easily testable; it then admits a universal test set of size $(n+1)$ that detects all PMGFs of any order, in the circuit. Since the test set is universal, no test generation by ATPG (Automatic Test Pattern Generation) is required. The original functionality of the circuit can be restored by setting the extra control line to logic 0. In addition, all the SMGFs, and detectable RGFs are also tested by the same test set.

2. PRELIMINARIES

Reversible logic: A reversible function has equal number of inputs and outputs, and simply induces a permutation on the set of input vectors to produce an output vector. Therefore, given an input vector, its output vector is unique, and for an output vector, its corresponding input vector can be uniquely restored. Further, in a circuit implementation with reversible gates, no fanout is allowed. In conventional non-reversible logic design, the above restrictions are not imposed. However, a non-reversible Boolean function can always be implemented by a reversible circuit after appropriately transforming it to a reversible one by adding garbage lines and reversible gates [35].

Example: The function $\{x, y \rightarrow x.y\}$ denoting AND operation is not reversible. By adding one extra input and two outputs, a modified but reversible function $\{x, y, z \rightarrow x, y, z \oplus x.y\}$ can be constructed. The AND function can be realized at the output $z \oplus x.y$, by setting the input z to constant zero; the circuit will have two "garbage" outputs. The Toffoli gate [16] realizes this function. By *garbage* is meant the number of extra outputs required to realize the given function.

Reversible gates: The basic CNOT type reversible gates used for synthesis are the following: (i) (1×1) NOT $(x_1 \rightarrow \bar{x}_1)$; (ii) (2×2) controlled NOT (CNOT) gate: $(x_1, x_2) \rightarrow (x_1, x_1 \oplus x_2)$; and (iii) (3×3) Toffoli gate $(x_1, x_2, x_3) \rightarrow (x_1, x_2, x_1 x_2 \oplus x_3)$.

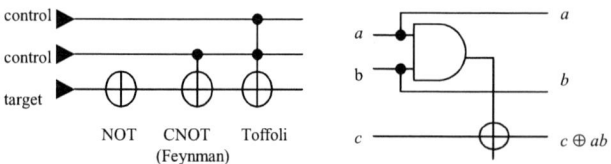

Fig. 1a: NOT, CNOT and Toffoli gates Fig. 1b: Behavior of a Toffoli gate

A generalized Toffoli gate has a set of control inputs C, a target input set T, and has the form $TOF(C;T)$, where $C = (x_{i1}, x_{i2},..., x_{ik})$ $T = \{x_j\}$ and $C \cap T = \varnothing$. It maps an input vector $(x^0_1, x^0_2,..., x^0_n)$ to $(x^0_1, x^0_2,..., x^0_{j-1}, x^0_j \oplus (x^0_{i1}.x^0_{i2}.....x^0_{ik}), x^0_{j+1},..., x^0_n)$. Thus, a NOT gate is $(TOF(x_j))$,

a generalized Toffoli gate which has no controls. The CNOT gate is $(TOF(x_i; x_j))$, a generalized Toffoli gate with one control bit [14, 21]; this is also known as the Feynman gate. The simple (3×3) Toffoli gate is a generalized Toffoli gate with two controls [16]. These three gates are shown in Fig 1. A k-CNOT gate has k control inputs $x_1, x_2, ..., x_k$ and one target input t. It maps the input vector $(x_1, x_2, ..., x_k, t)$ to the output vector $(x_1, x_2, ..., x_k, t \oplus x_1. x_2 ... x_k)$. In other words, a k-CNOT gate has $k+1$ inputs and $k+1$ outputs; the first k outputs follows the respective inputs, and it inverts the target at the $(k+1)$-th output if and only if all the k control inputs are 1. Any reversible function can be realized as a cascade of k-CNOT gates.

3. FAULT DETECTION IN A REVERSIBLE CIRCUIT

Testing of a reversible circuit, in general, turns out to be relatively simpler compared to that of non-reversible logic because of the inherent *ease of controllability* of logic states and *observability* of errors [1]. Another important property that expedites the test generation process is the fact that backtracing is straightforward and always yields a unique vector at the input.

3.1 Fault model

Several new fault models for k-CNOT based reversible circuits were introduced earlier [27, 34]. These are single missing-gate fault (SMGF), the repeated gate fault (RGF), partial missing-gate fault (PMGF), and the multiple missing-gate fault (MMGF). In this section, we briefly explain with examples, the nature of these faults.

Single missing-gate fault (SMGF): This model corresponds to the case when one k-CNOT gate completely disappears from the circuit. In the presence of this fault, the CNOT gate behaves as a simple wire connection, i.e., the pulse implementing the gate operation is short, missing, misaligned or mistuned.

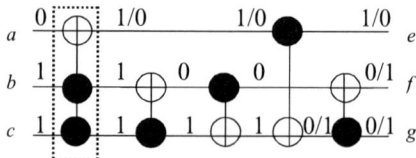

Fig. 2: Single missing-gate fault (SMGF)

Fig. 2 shows an SMGF marked by the dotted box in the reversible circuit ham3\design#1 benchmark, where the first 2-CNOT gate is missing.

An SMGF is detected by setting logic 1 value on all the control inputs of the gate, and any value either 0 or 1 on the target input as well as on the wires not connected to the gate. In the example of Fig. 2, if we apply {a, b, c} = {0, 1, 1} at the input of the circuit, the normal output would be {e, f, g} = {1, 0, 0}, whereas, in the presence of the SMGF fault marked by the dotted box, the output will be {e, f, g} = {0, 1, 1}. The number of possible SMGFs is equal to the number of gates in the circuit.

Repeated-gate fault (RGF): A repeated-gate fault (RGF) is an unwanted replacement of a *k*-CNOT gate by several instances of the same gate [34]. An RGF may be needed to model the occurrence of long or duplicated pulses. Fig. 3 shows an example, where first gate is repeated in the circuit ham3\design#1. The effect of this fault is thus same as that of an SMGF at the first 2-CNOT gate in the original circuit.

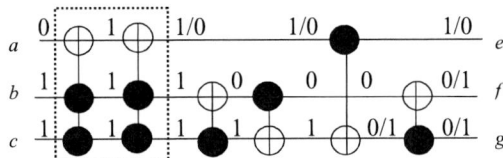

Fig. 3: RGF in ham3\design#1 reversible circuit

If we apply {a, b, c} = {0, 1, 1}, the normal output would be {e, f, g} = {1, 0, 0}, whereas, in the presence of the above RGF marked by the dotted box, the output will be {e, f, g} = {0, 1, 1}. Hence, it is detected by the vector {a, b, c} = {0, 1, 1}.

It is clear that if a RGF replaces a gate by even number of instances of the same gate, its effect is similar to the effect of the SMGF with respect to the same gate. If the RGF replaces a gate by odd number of instances of the same gate, the fault is redundant, i.e., it does not change the function of the circuit. Further, it has been shown that any SMGF test set detects all detectable RGFs [34].

Partial missing-gate fault (PMGF): This is used to model the defects resulting from the partially misaligned or mistuned gate pulses [34]. It changes a *k*-CNOT gate into a *p*-CNOT gate, with *p* < *k*. The corresponding fault is called as $(k - p)^{th}$ *order* PMGF. Fig. 4 shows a first-order PMGF affecting the second control input of the leftmost gate. An SMGF can be seen as a 0-order PMGF.

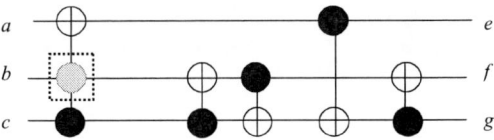

Fig. 4: PMGF in ham3\design#1

In the circuit of Fig. 4, if we apply {a, b, c} = {1, 0, 1}, the normal output would be {e, f, g} = {1, 1, 0}, whereas, in the presence of the first order PMGF fault as shown, the output will be {e, f, g} = {0, 0, 1}. Hence, the vector {a, b, c} = {1, 0, 1} detects this fault. It has been shown [34] that a PMGF (of first or higher order) is detected when a 0 is applied to at least one of the affected control inputs and a 1 to all other control inputs. Thus, a higher order PMGF is detected by a test vector for a first order PMGF, the affected control input of which is one of those affected in the higher order PMGF. Hence, it is sufficient to consider first order PMGFs only.

Multiple missing-gate fault (MMGF): This is defined as complete disappearance of two or more *consecutive k*-CNOT gates from the circuit.

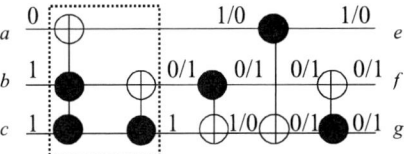

Fig. 5: MMGF in *ham3\design#1*

In the circuit of Fig. 5, it is shown that the circuit has an MMGF marked by the dotted box. This fault is detected by the vector {a, b, c} = {0, 1, 1}.

3.2 Testable design for detecting PMGFs

An exact ATPG scheme has been reported earlier [34] that generates test vectors for various types of missing-gate faults discussed above. To detect all PMGFs by a universal test, the original reversible circuit is augmented by adding one wire and duplicate *k*-CNOT gates.

A first-order PMGF affecting the jth control input can be detected by setting 0 at the jth control input and 1 at all the other control inputs. For such a vector, the fault-free and the faulty gate will produce different values on the target node. Therefore, to detect all first order PMGFs in a *k*-CNOT gate as shown in Fig. 6a, we will have to apply the following *k* test vectors {$x_1 x_2 ... x_j .. x_k$ .. *t*}: (0 1... 1...1 .. X), (1 0... 1...1 .. X), ..., (1 1... 1...0 .. X) at the input level, where X, applied to the target input, may be 0 or 1.

165

978-1-4244-3039-0/08 $25.00 © 2008 IEEE

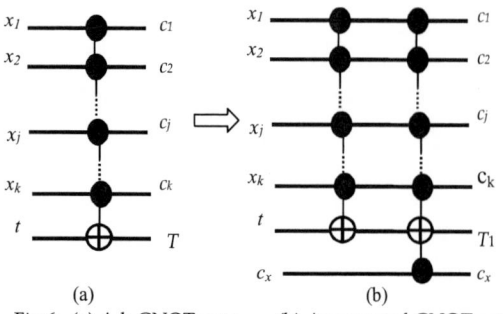

(a)　　　　　　　　(b)

Fig.6: (a) A k-CNOT gate　(b) Augmented CNOT gate.

An $(n \times n)$ reversible circuit R of depth d is built with a cascade of k-CNOT gates. While the above k test vectors applied at the inputs to R are guaranteed to detect all PMGFs at the first CNOT gate, they may not detect a PMGF at a CNOT gate lying at a subsequent level, as the vectors change when they propagate through various levels. However, if we are able to produce the same k patterns at the inputs of each CNOT gate lying at all other levels, then all PMGFs of first order can be detected in the reversible circuit. To restore the same test patterns at each level, we augment a k-CNOT gate as shown in Fig. 6b. The same k-CNOT gate is repeated consecutively, and one additional control input (c_x) is added.

Lemma 1: The target output T_1 of the augmented gate is equal to the target input t when $c_x = 1$.

Proof: The output of the target line $T = t \oplus (x_1.x_2\ldots x_j.\ldots x_k)$. After augmentation, the target output T_1 when $c_x = 1$, is given by:

$$T_1 = T \oplus (1.\, x_1.x_2\ldots x_j.\ldots x_k)$$
$$= T \oplus (x_1.x_2\ldots x_j.\ldots x_k)$$
$$= t \oplus (x_1.x_2\ldots x_j.\ldots x_k) \oplus (x_1.x_2\ldots x_j.\ldots x_k)$$
$$= t$$

Hence the proof follows.

Therefore, it is possible to restore the same test pattern (which is applied at the input level), at the output level. Repeating this augmentation procedure for every k-CNOT gate with a common additional control line (c_x), an $(n \times n)$ reversible circuit (Fig. 7) is modified as in Fig. 8.

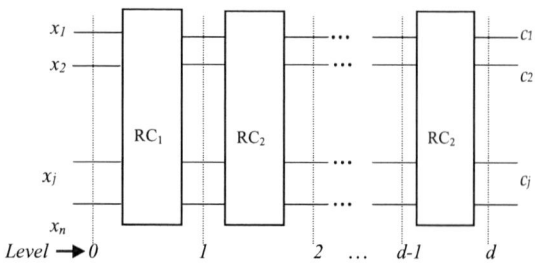

Fig. 7: An $(n \times n)$ reversible circuit of depth d

In other words, a CNOT gate is inserted between every j^{th} and $(j+1)^{th}$ level, where the inserted gate is the same as the one at the jth level with one extra control input.

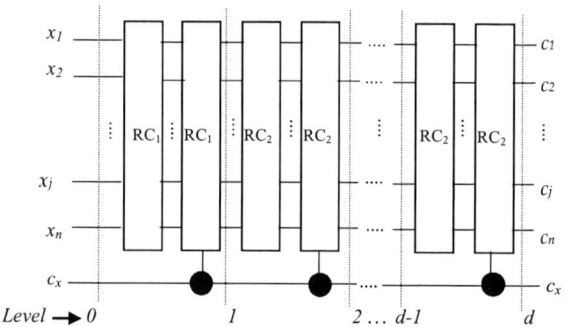

Fig. 8: Augmented reversible circuit

The augmented circuit implements the normal function when the control line c_x is set t 0.

Example: The circuit ham3\design#1 is shown in Fig. 9. The augmented circuit is shown in Fig. 10. For this circuit, the test set: $S\{a, b, c, c_x\} = \{0111, 1011, 1101, 1110\}$, detects all possible first order PMGFs. Since first order PMGFs dominate all other higher order PMGFs [34], this test set detects all PMGFs in general. Further, this is universal in the sense that for all (3×3) reversible circuits, the same test set will work.

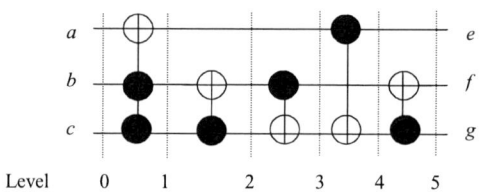

Fig. 9: ham3\design#1 benchmark reversible circuit

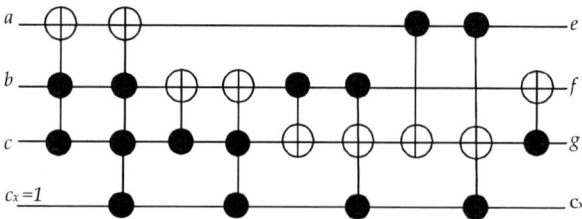

Fig. 10: Testable design for ham3\design#1

The general result stated below, now easily follows.

Theorem 1: In the testable design as shown in Fig. 8, the following universal test S_u of length $(n +1)$ is sufficient to detect all PMGFs of any order ≥ 1.

$$S_U = \begin{pmatrix} x_1 & x_2 & \ldots\ldots x_n & c_x \\ 0 & 1 & \ldots\ldots 1 & 1 \\ 1 & 0 & \ldots\ldots 1 & 1 \\ & & \ldots\ldots\ldots\ldots & \\ 1 & 1 & \ldots\ldots 0 & 1 \\ 1 & 1 & \ldots\ldots 1 & 0 \end{pmatrix}$$

Lemma 2: The test set S_U is also sufficient to detect all the SMGFs in the circuit.

Proof: In a k-CNOT gate, if we apply pattern $\{x_1\ x_2\ ..., x_i... x_k\ t\} = (1\ 1\ ..1\ ...\ 1\ X)$, where X denotes don't care, the target output T becomes the complement of t. Since the reversible circuit is implemented with only k-CNOT gates, only one of the n input lines of the reversible circuit is the target line t for a given CNOT gate. If we apply 0 on the target line and 1 on the remaining lines, then it is able to detect an SMGF on that gate. Clearly at the input level, such a test pattern belongs to the set S_U. For other CNOT gates at all subsequent levels, the required vector reappears in the testable design of Fig. 8, as discussed earlier. Hence, the test set S_U is sufficient to detect all SMGFs in the augmented circuit.

Lemma 3: All detectable RGFs are detected by the test set S_U.

Proof: Follows from the fact that any SMGF test set detects all detectable RGFs [34].

Thus, it follows that the above universal test S_U of length $(n+1)$ is sufficient to detect all SMGFs, all detectable RGFs, and all PMGFs in the augmented reversible circuit. The test set depends only on n and is independent of the functionality of the reversible logic.

4. EXPERIMENTAL RESULTS

We have studied several examples of reversible benchmark circuits [36], the results of which are shown in Table 1. Column 1 shows the circuit name, and column 2 denotes gate count (N), column 3 presents the input size (n). The number of tests for detecting all PMGFs obtained by running the ATPG [34], is shown in column 4. The size of the universal test set as per the proposed method is just ($n+1$) and is shown in column 5. The universal test set can be directly found without the need of running an ATPG. However, the augmentation procedure doubles the gate cost.

Table 1: Comparison of the test set for the PMGF model

Circuit	N	n	# of tests as in [34]	# of tests as in the proposed method
2of5d1	18	6	8	7
4_49tc1	16	4	5	5
hwb4tc	17	4	5	5
hwb5tc	56	5	9	6
hwb6tc	126	6	15	7
hwb7tc	291	7	24	8
rd53d1	12	7	8	8
rd53rcmg	30	7	8	8

5. CONCLUSIONS

This paper presents a design-for-testability technique for testing missing-gate faults in a reversible circuit. The technique derives a universal test set of length ($n+1$) for detecting all partial missing-gate faults (PMGF) along with all single missing-gate faults (SMGF), and all detectable repeated-gate faults (RGF). in an ($n \times n$) reversible combinational circuit designed with k-CNOT gates. The test set also detects a large number of multiple missing-gate faults (MMGF). However, for detection of all MMGFs, additional tests and/or further augmentation may be needed. These would require further investigation.

References

1. R. Landauer, "Irreversibility and heat generation in the computing process," *IBM Research and Development*, vol. 5, pp. 183-191, 1961.
2. C. H. Bennett, "Logical reversibility of computation," *IBM J. Research and Development*, vol. 17, pp. 525-532, Nov. 1973.
3. C. H. Bennett, "Notes on the history of reversible computation," *IBM J. Research and Development*, vol. 32, pp. 16-23, January 1988.
4. R. P. Feynman, *Feynman Lectures on Computation.* (A. J. G. Hey and R. W. Allen, Ed.), Perseus Books, USA, 1996.
5. M. Nielsen and I. Chuang, *Quantum Computation and Quantum Information.* Cambridge University Press, 2000.
6. N. Gershenfeld and I. L. Chuang, "Quantum computing with molecules," *Scientific American*, June 1998.
7. J. Preskill, *Lecture Notes in Quantum Computing.* Technical Report, (http://www.Theory.caltech.edu/~preskill/ph229).
8. W. C. Athas and L. J. Svensson, "Reversible logic issues in adiabatic CMOS," *Manuscript.*
9. P. Picton, "Optoelectronic, multivalued, conservative logic," *International Journal of Optical Computing*, vol. 2, pp. 19-29, 1991.
10. P. Picton, "A universal architecture for multiple-valued reversible logic," *MVL Journal*, vol. 5, pp. 27-37, 2000.

11. R. C. Merkle and K. E. Drexler, "Helical logic," *Nanotechnology*, vol. 7, pp. 325-339, 1996.

12. R. C. Merkle, "Reversible electronic logic using switches," *Nanotechnology*, vol. 4, pp. 21-40, 1993.

13. R. C. Merkle, "Two types of mechanical reversible logic," *Nanotechnology*, vol. 4, pp. 114-131, 1993.

14. R. Feynman, "Quantum mechanical computers," *Optics News*, vol. 11, pp. 11-20, 1985.

15. E. Fredkin and T. Toffoli, "Conservative logic," *Int. J. of Theoretical Physics*, vol. 21, pp. 219-253, 1982.

16. T. Toffoli, "Reversible computing," Tech memo - MIT/LCS/TM-151, MIT Lab for Comp. Sci., 1980.

17. V. V. Shende, A. K. Prasad, I. L. Markov, and J. P. Hayes, "Synthesis of reversible logic circuits," *IEEE Transactions on CAD*, vol. 22, pp. 723-729, June 2003.

18. K. Iwama, Y. Kambayashi, and S. Yamashita, "Transformation rules for designing CNOT-based quantum circuits," In *Proc. Design Automation Conference*, June 2002.

19. A. Mishchenko and M. Perkowski, "Logic synthesis of reversible wave cascades," In *Proc. International Workshop on Logic Synthesis*, pp. 197-202, June 2002.

20. M. Perkowski, P. Kerntopf, A. Buller, M. Chrzanowska-Jeske, A. Mishchenko, X. Song, A. Al-Rabadi, L. Joswiak, A. Coppola, and B. Massey, "Regularity and symmetry as a base for efficient realization of reversible logic circuits," In *Proc. International Workshop on Logic Synthesis*, pp. 245-252, 2001.

21. A. Mischenko and M. Perkowski, "Reversible Maitra cascades for single output functions," In *Proc. International Workshop on Logic Synthesis*, pp. 197-202, 2002.

22. D. M. Miller, "Spectral and two-place decomposition techniques in reversible logic," In *Proc. Midwest Symposium on Circuits and Systems*, Aug. 2002.

23. D. M. Miller and G. W. Dueck, "Spectral techniques for reversible logic synthesis," In *Proc. 6th International Symposium on Representations and Methodology of Future Computing Technologies*, pp. 56-62, March 2003.

24. K. N. Patel, J. P. Hayes, and I. L. Markov, "Fault testing for reversible circuits," In *Proc. VTS*, pp. 410-416, 2003.

25. J. C. Bertrand, N. Giambiasi, and J. J. Mercier, "Sur la recherche de l'inverse d'un automate," *RAIRO*, pp. 64– 87, Apr. 1974.

26. J. C. Bertrand, J. J. Mercier, and N. Giambiasi, 'Sur la recherche de l'inverse d'un circuit combinatoire," *RAIRO*, pp. 21–44, Jul. 1974.

27. J. P. Hayes, I. Polian, B. Becker, "Testing for missing-gate faults in reversible circuits," In *Proc. Asian Test Symposium*, pp. 100-105, 2004.

28. K. Ramasamy, R. Tagare, E. Perkins and M. Perkowski, "Fault localization in reversible circuits is easier than for classical circuits," In *Proc. of ATS 04*.

29. D. P. Vasudevan, P. K. Lala and J. P. Parkerson, "A novel approach for on-line testable reversible logic circuit design," In *Proc. Asian Test Symposium*, 2004.

30. A. Chakraborty, "Synthesis of reversible circuits for testing with universal test set and C-testability of reversible iterative logic arrays," In *Proc. VLSI Design*, pp. 249-254, 2005.

31. N. Scott, Reversible Circuit Viewer v1.14 Simulation.

32. H. Rahaman, D. K. Kole, D. K. Das, and B. B. Bhattacharya, "Detection of bridging faults in a reversible circuit", In *Progress in VLSI Design and Test* (Ed. C. P. Ravikumar), Elite Publishing, New Delhi, 2006, pp. 384-392.

33. H. Rahaman, D. K. Kole, D. K. Das, and B. B. Bhattacharya, "Minimal test set for bridging fault detection in reversible circuits", In *Proc. Asian Test Symp.*, Bejing, China, pp. 125-128, 2007.

34. I. Polian, J. P. Hayes, T. Fienn and B. Becker, "A family of logical fault models for reversible circuits", In *Proc. Asian Test Symp.*, Kolkata, India, 2005, pp. 422-427.

35. D. Maslov, *Reversible Logic Synthesis*. Ph.D Thesis, The University of New Brunswick, Canada, September 2003.

36. D. Maslov, G. Dueck, and N. Scott, *Reversible Logic Synthesis Benchmarks Page*. www.cs.uvic.ca/~dmaslov/, 2004.

21st International Conference on VLSI Design

Memory Yield Improvement through Multiple Test Sequences and Application-aware Fault Models

Aman Kokrady[†], C.P. Ravikumar[†], Nitin Chandrachoodan[Π]

[†]*Texas Instruments India,* [Π]*IIT Madras*

{koko,Ravikumar}@ti.com, nitin@ee.iitm.ac.in

Abstract

In this paper, we propose a way to improve the yield of memory products by selecting the appropriate test strategy for memory Built-in Self-Test (BIST). We argue that by testing the memory through a sequence of test algorithms which differ in their fault coverage, it is possible to bin the memory into multiple yield bins and increase the yield and product revenue. Further, the test strategy must take into consideration the usage model of the memory. Thus, a number of video and audio buffers are used in sequential access mode, but are overtested using conventional memory test algorithms which model a large number of defects which do not impact the operation of the buffers. We propose a binning strategy where memory test algorithms are applied in different order of strictness such that bins have a specific defect / fault grade. Depending on the applications some of these bins need not be discarded but sold at a lower price as the functionality would never catch the fault due to its usage of memory. We introduce the notion of a *test map* for the on-chip memories in a SoC and provide results of yield simulation on two specific test strategies called "Most Strict First" and "Least Strict First". Our simulations indicate that significant improvements in yield are possible through the adoption of the proposed technique. We show that the BIST controller area and run-time overheads also reduce when information about the usage model of the memory, such as sequential access, is exploited.

1 Introduction

System-on-chip designs (SoC) are increasingly becoming memory dominated [1]. Memories are more likely to fail than random logic. Traditionally memories have been screened for memory cell faults, cell coupling faults, address decoder faults, and faults in read/write circuitry. As the transistor dimensions shrink further, subtler defect mechanisms, those that impact the speed of operation, can affect memories [2], [3] and testing becomes more complex. With 200+ memory instances in a design and with the introduction of more expensive memory test algorithms, test application times become longer. BIST can help address the test application time issue, but it comes with overheads of area and power. Determination of BIST architecture for a SoC is therefore a complex optimization problem which must simultaneously address several concerns. Figure 1 shows a symbolic diagram of BIST representation in modern day SOCs. Two commonly used BIST techniques are - 1) multiple BIST controllers, each controller testing a group of local memories [28], 2) Single large programmable controller which interacts with memories through local and global combiners [27]. In sub-100nm technologies, manufacturability and variability concerns have negatively impacted the yield, and design for yield is becoming important [4]. In this paper, we consider yield in the optimization of memory BIST architecture.

Yield of an integrated circuit directly determines its profitability and is defined as the percentage of usable chips in the population of the manufactured chips. With subtle timing-related defects in integrated

circuits, "usability" cannot be simply black or white. With several

Figure 1- Single Controller BIST architecture

forms of process variability, such as lot-to-lot, wafer-to-wafer, die-to-die, and on-the-die, it is necessary to adopt a general notion of "acceptability" and grade the chips into bins. Binning has been used with microprocessors and digital signal processors which are produced in high volume. Wrongly classifying a chip that belongs to a high-speed bin into a low-speed bin due to a limitation of the testing procedure is considered a "yield loss." In platforms for multimedia applications, it is common to find microprocessors, digital signal processors and memory cores ranging from small and sparse register files to large and dense SRAM/DRAM blocks. It is not uncommon to find 200+ memory instances in a typical ASIC and even 1000+ memories in unusual cases, including single-port, two-port, and multi-port SRAMs, ROMs, DRAMs, CAMs, and TCAMs in such SoC.

While catastrophic faults in logic or memory will be a cause for rejecting the chip, a chip with speed faults and other "tolerable" faults may be deemed "acceptable" in some applications. For example, even cell faults in a video RAM may be acceptable since the results of the computation may not be noticeable. In this paper, we will consider the following definition of acceptability for a chip. Let L be the digital logic on the chip, and M be the set of memories. We assume that the logic is tested using a test schedule that consists of stuck-at-fault testing, followed by speed testing. Detection of a fault in stuck-at fault testing is sufficient reason to reject the chip. On the other hand, speed fault testing is repeated at different frequencies and a suitable speed bin is found. Just as logic is tested by a sequence of test algorithms, we propose that a sequence of test algorithms be used to test memories. Let A be the set of memory test algorithms such as galloping pattern test, March 13N, and so on. There is extensive work on the defect coverage of the memory test

169

978-1-4244-3039-0/08 $25.00 © 2008 IEEE

algorithms [5] [6] [7]. A test algorithm such as galloping pattern test will be referred to as a "strict" test algorithm since it detects a larger set of faults. March 13N is a relatively "lenient" algorithm since it does not detect resistive PMOS faults; however, such a fault may be "acceptable" in many applications such as audio or video where the memory access patterns are not random. In fact, even some cell faults may not cause an appreciable change in the audio or video quality. It is therefore feasible to define a *test map* for the chip which is defined as a mapping from M to A. Thus, memory M_j in M will be tested using the algorithm $map(M_j)$. The map will be defined by the product engineer who is aware of the application for which the memory is being targeted and the characteristic of the memory. For example, if the data memory of a video application has in-built error correction, a lenient algorithm may be selected for testing such a memory. This would insure that even if there is a fault, it can be tolerated as during the functional mode, these faults would be corrected. Similarly, a control memory that is non-repairable, where the access patterns are usually random for the same application, should be tested using a strict algorithm.

	A_j	A_k
M1	Pass	Pass
	Pass	Fail
	Fail	Pass
	Fail	Fail
M2	Pass	Pass
	Pass	Fail
	Fail	Pass
	Fail	Fail

Table1 – *Possible outcomes of testing a two-memory SoC using two algorithms – A_j & A_k*

In a formal sense, an algorithm A_j is strict compared to algorithm A_k if A_j models a larger set of faults. Note that it is possible that A_k detects some faults that A_j cannot – however, the set of faults that A_k detects are not "crucial" from the view point of application. Let D_{ij} be defined to be a 0-1 variable which is set to 1 if fault type i is modeled by algorithm j. (See [19] for an example of such a fault coverage matrix D.) Let w_i be the cruciality index for fault type i – where the index is a positive integer such that $w_i > w_j$ implies fault i is more crucial than fault j. Two algorithms j and k can be compared on the strictness metric by comparing the weighted sums $\Sigma w_i D_{ij}$ and $\Sigma w_i D_{ik}$ - the algorithm that has the larger metric will be deemed more strict.

We can generalize the notion of map and define it to be a mapping from M to $A \times A \times \ldots \times A$. For instance, consider a map from M to $A \times A$, where every memory will be tested by a *sequence* of two algorithms A_j, A_k. This will be suitable when the product engineer is not aware of the target application and wishes to segregate memories into categories. The engineer may select a "Most Strict First" (MSF) or a "Least Strict First" (LSF) strategy. Consider that M has only two memories M1 and M2 and let us look at the outcomes (Table1 above) assuming a most strict first strategy.

If both memories pass both A_j and A_k, then the chip would be in a "Premium" bin. If either of the memories fails both A_j and A_k, the chip will be rejected. In the remaining cases, the product engineer can place the chip in the "Non-Premium" bin. More bins can be defined if necessary; for example, if we know that more errors in M2 can be tolerated than in M1, we can define finer bins.

4.2 Previous Work

Breuer first pointed out that error tolerance is a way to cope with the decreasing yield of chips manufactured in deep submicron technologies [8]. He distinguished between fault-tolerance (techniques such as error

correction codes, redundant rows/columns of memory, which can be used to mask the effect of faults) and error tolerance (robustness in the algorithm or design implementation due to which a certain number of errors can be tolerated with some degradation in performance). We apply the concept of error tolerance to improve the yield of memories.

The concept of realistic faults has been used to optimize memory test algorithms and grade the algorithms on realistic coverage earlier [10, 11], the papers only consider the memory layout and the defect density of the memories and not the functional access and usage of the memories. In our work we use the functional usage scenario of the memory to discount certain fault mechanisms which can't occur during functional modes. Work on SOC yield improvement by improving memory yield has so far been concentrated on usage of Repair [24], ECC [25], Repair efficiency [24] and bit-cell architecture fine tuned for DFM [26]. All these are generic methods but require additional cost of area or performance to increase yield. Our work decreases BIST area and test time apart from increase in yield and can give benefit over and above these yield improvement techniques.

Several memory fault/defect simulators have been discussed in the literature [9,10,11]. In [9] the authors use an equation based technique to simulate a fault in different memory architectures. In [10] and [11] the authors use realistic fault models and compare different algorithms on the basis of this defect coverage. The authors of [12] describe an $O(N^2)$ memory fault simulator, where N is the number of bits in the memory. While the earlier fault simulators used simple fault models such as cell stuck-at and cell coupling faults, more complex fault models such as static linked faults (Faults consisting of two or more simple faults) have been described [13], [14]. In this paper, we use results from the literature to associate the list of faults that are targeted by memory test algorithms. There is a tradeoff between the time complexity of the test algorithms, the fault coverage, and the complexity of implementing the test algorithm. We explore this tradeoff by classifying the test algorithms as lenient and strict, and then by using a sequence of memory test algorithms to optimize the area overhead, run-time, and yield. In this paper, our goal is not to build a full memory fault simulator. We are interested in seeing how use of different test algorithms can result in different binning strategy and not in analyzing how effective a given algorithm is in screening faults.

1.2 Contribution of the paper

To the best of our knowledge, ours is the first attempt to look at memory BIST optimization which concurrently addresses the issues of yield, area and test time. Most of the designs today follow identical BIST strategy for memories of the same type. In this paper we argue that the yield, BIST area and even test time can be improved if the access pattern and functionality are considered while testing and creating BIST algorithms. We exploit the ability to test an on-chip memory using more than one algorithm through programmable BIST and propose a way to reduce yield loss through "overtest." We argue that awareness of the target application can be exploited to avoid overtesting.

The paper is organized as follows. In Section 2, we introduce yield as a new metric for memory test and various memory test strategies for yield improvement. In Section 3, we develop the formal framework for yield and fault-level analysis, which are useful in formulation the memory BIST architectural optimization. Section 4 gives the results of adopting our test strategy on an example SoC. Conclusions are presented in Section 5.

2 Yield-aware Memory Test

Self-testing of a memory requires an integrated BIST controller which can generate test patterns according to a specific algorithm and compare the response from the memory with the expected values. A memory "collar" multiplexes the address, data, and control signals from the controller with those from the functional unit. A SoC with a large number of memories will require the use of several BIST controllers to avoid routing congestion. An alternate solution is to have a single programmable BIST controller with distributed data path. A controller is typically intended to test similar memory types, although controllers can also be adopted to test a larger number of memory types. Determining the number of BIST controllers and assigning memories to controllers is a complex optimization problem and the architecture impacts routing congestion, area, performance, and power overheads. BIST controllers have been conventionally hardwired, but with increasing number of fault-types in memories, there are distinct benefits in postponing the decision of the test algorithm to the post-silicon phase. Therefore, modern BIST methodologies promote the use of hardware programmable or software programmable controllers [15]. The overheads of a controller increase as we move from hardwired organization to hardware programmability to software programmability.

With increased focus on mobile entertainment applications, many SoC today are intended for multimedia applications. Interestingly, algorithms for video and audio applications often access data in a strictly increasing or strictly decreasing order of logical addresses. There are many video and audio buffers which use sequential access i.e. they generate address patterns of the form a_0, a_1, ..., a_{N-1}, a_0, a_1, ... Often, the data in some of these buffers is used in a "WORM" fashion i.e. *write once and read multiple times*. Many data caches and data buffers also have known sequences for addresses. We analyzed two sub-modules in a video processor SoC and found that there is over 300KB of memory which uses purely sequential access. For one of the sub-modules, Table 2 shows the instances of single-port and dual-port memories where strictly sequential access was used. It is sufficient if the memory works correctly for a specified access pattern which is far from random.

Memory Type	Subsystem	Size
Single Port	Video interface, streaming IO, Graphic Engine, FIFO	80KB
Dual Port	FIFO	60KB
Two Port	Video interface, Graphic Engine, FIFO	20KB

Table 2 – Memories with strictly Sequential Access

Some memory defects/faults which will not impact sequential access but may impact random access functionality are:

- coupling faults in address decoders
- coupling faults in column muxes
- bit line coupling faults
- IR Drop defects due to address bus switching

Memory test algorithms such as galloping pattern or even MARCH algorithms can "overtest" such memory instances since they test the memory for full random access. What is more, in video and audio applications, an occasional read or write error may not impact the quality of the end application. Overtesting can negatively impact yield. For example, at-speed testing of logic can excite non-functional paths and downgrade good chips [16]. Similarly, reliance on a scan path for test application may cause up to 20% more switching activity than the functional operation and lead to speed test failures.

We argue that memory test algorithms can similarly overlook the functional behavior of the RAM and contribute to yield loss. We propose that a binning strategy be adopted for memory test by associating a *test map* with the on-chip memories and testing the memories using more than one test algorithm. Run-time programmable BIST architectures, which have found wide-spread adoption in the industry today, are useful in implementing such a strategy. Although there are many possible ways to select and order the algorithms in a *test map*, we consider two important ones for purposes of experimentation and illustration, namely, the "Most Strict First" (MSF) and the "Least Strict First" (LSF). Figure 2 illustrates these two strategies. In Figure 2(a), A0 is a stricter algorithm than A1. Chips that A0 classifies as "good" are suitable for all markets and are sold at a premium price; chips that A0 rejects are tested using A1. Any chip that A1 accepts is suitable for a narrower market and is sold at a lower price; chips that both A0 and A1 reject are classified into the "Reject" bin. In Figure 2(b), A1 is a lenient algorithm as compared to A0. Thus, chips that A1 rejects are put into "Reject" bin. Chips that A1 will accept are further tested by A0. If A0 also accepts a chip, the latter is sold at a premium price. Chips accepted by A1 but rejected by A0 are sold for a lower price for a different market. Note that the prices associated with the bins will be different in the two cases. For MSF strategy, these will be monotonically decreasing and for LSF, the prices are monotonically increasing. The test application time will depend on the fault probability in the batch of chips being screened. Intuitively, when the fault probabilities are low, the strictest-first strategy will work best, and vice versa.

3 Analysis of Yield and DPM

When "functionality-aware" testing is used, there is a chance that a "good" memory (one that may have non-crucial faults) gets into the bin of faulty chips, leading to yield loss. We refer to the probability

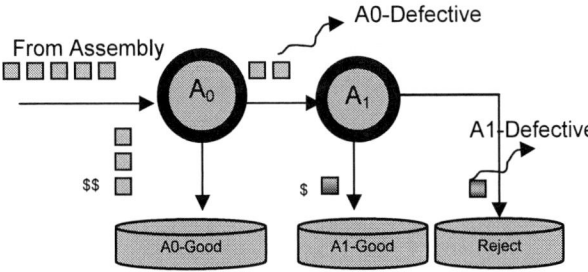

(a) Most Strict First Algorithm, A0 strict

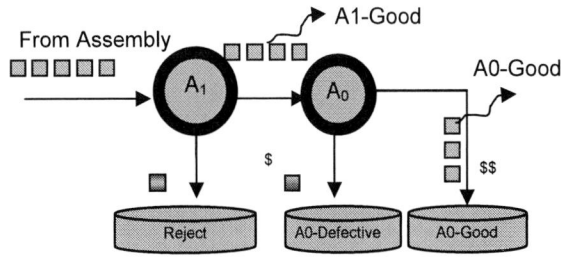

(b) Least Strict First Algorithm, A1 lenient

of rejecting a good memory as P_{rg}. Alternately; a memory with a crucial fault may enter the bin of "good" memories, contributing to the fault level of the shipped parts. The probability of accepting a faulty memory is denoted P_{ad}. Let n be the number of *fault types* which are modeled, of which k are crucial for the target application and the remaining $(n-k)$ are non-crucial. Assume that a sequence of test algorithms applied to the memory can catch $m+l$ fault types, of which m are crucial and l are non-crucial. As the algorithm strictness increases the value of $m+l$ increases. Let p be the probability of a fault in the memory – for simplicity, we shall assume this to be identical for all fault types. In a large sample of chips, the probability that a selected chip is fault-free is $(1-p)^k$. Since only the critical faults will contribute to a fault in functional mode, the probability of the chip being fault free is independent of non-crucial faults $(n-k)$. Using an analysis similar to [17], we can show that

Figure 2: Illustration of Two Test Strategies

$$P_{rg} = \left(1-p\right)^k - \left(1-p\right)^{l+k}$$

The fault level, which is the fraction of defective parts that are shipped to the number of shipped parts can be shown to be

$$FaultLevel = \frac{P_{ad}}{P_{ship}} = 1 - \left(1-p\right)^{k-m}$$

We can note here that fault level is only dependent on k and m and is independent of how many faults are in the non crucial category (l). This means that any of the omitted faults will not lead to higher fault Level. We have omitted the detailed derivations of the two equations above for brevity. The "yield hit" due to classification of a part without crucial faults as a "bad part" is

$$YieldHit = \frac{P_{rg}}{P_{defectfree}} = 1 - \left(1-p\right)^l$$

It is important to note that the fault level does not go up with reducing the test algorithms, as the faults we aren't testing for are not going to occur in the functional mode and hence will not cause any customer returns. The latent defects which can crop up due to non-testing of any particular defect type can arise because of unknown defects we aren't testing for. This makes the defect level a function of crucial faults only and all of them are tested. The crucial set of faults is dependent on the application and usage of the device.

Let Y_0 be the yield when the conventional method is used to screen parts i.e. a strict algorithm is employed to screen the memories. The shipped parts are sold at the premium price, say S_0 leading to a profit of $Y_0 S_0$. When the proposed method is used, we will have bins of the form A_1-Good; A_1-rejected, A_2-good; A_1-rejected, A_2-rejected, A_3-good, and so on. Let the sizes of these bins be Y_0, ΔY_1, ΔY_2 and so on. Assume that the selling prices for these bins are set at S_0, S_1, etc. Then the revenue is

$$S_0 Y_0 + S_1 \Delta Y_1 + S_2 \Delta Y_2 + \dots \ S_{n-1} \Delta Y_{n-1}$$

where n is the number of test algorithms used. Thus the increase in revenue in the proposed system will be

$$S_1 \Delta Y_1 + S_2 \Delta Y_2 + \dots \ S_{n-1} \Delta Y_{n-1}$$

4 Experimental Results

4.1 Lenient Testing Algorithms for Sequential Access Memories
As mentioned in the previous section, when testing a memory which is mainly used as sequential access memory (SAM), we can be lenient in terms of the fault model and get yield improvements. The memory BIST algorithms and architectures can also be simplified when targeting purely sequential access memories since we do not need address scrambling or complex address generators. This is illustrated in Table 3, where we compare the area and BIST run-time for RAM-BIST and SAM-BIST and find that for larger memories, we can get around 20% improvement in area and around 60% improvement in run-time. Given that there can be several such controllers, the overall reduction in real estate and test application time can be significant.

4.2 Yield Improvement

In order to understand the improvement in yield possible through the use of the proposed test strategy, we considered a SoC device with 100 on-chip memories. A yield simulator was developed using the following methodology. One million instances of the SoC were simulated. The simulations run were behavioral where faults were "injected" into an instance of the chip as follows. About 30 memory fault types were considered [18], [19] and [20]. The value of p (fault probability) was treated as a parameter and is shown in number of faults per million. The value of p was determined by the fabrication unit and can vary based on different parameters like technology node, process used and even fabrication unit used. p has been used before in context of realistic faults [10]. Seven memory test algorithms were selected; in the increasing order of strictness, these are PI-PO, DN, DX, MC, PR, MARCH13n, and GALPAT. The criterion for strictness is based on the kind of faults each algorithm catches. Gross defects and hard failures are the most catastrophic and the algorithms which catch these faults are stricter than the algorithms which catch speed defects which in turn are stricter than the algorithms which catch defect due to a particular access pattern. The fault coverage information for these algorithms is shown in Table 4. GALPAT and MARCH13n tend to catch most of gross defects and bit faults. Among the two GALPAT catches a bigger set of faults making it stricter. PR and MC mostly are used to catch speed defects due to precharge circuitry or bit line faults. These catch a subset of gross faults as well but most of them are subset of

Memory Type	BIST area (Random Access Memories)	BIST area (Sequential Access Memories)	Area Improvement	Random BIST test time	Sequential BIST test time	Test Time Improvement
Single Port 8192X133 (Sliced)	20895	17435	18%	7918528	2986188	63%
Single Port 2048X57	11807	10942	8%	598933	227396	63%
Single Port 2048X7	7513	6816	10%	616427	248900	60%
Two Port 16X107	9516	9157	5%	14933	6928	54%
Single Port 128X38	11815	10690	10%	32192	14276	56%
ROM 4096X32	2412	1900	25%	81944	65552	20%
Single Port 8192X128 (Non-sliced)	20604	16992	19%	774291	295949	63%
Table 3 – BIST Area and BIST time improvement from SAM-BIST						

172

978-1-4244-3039-0/08 $25.00 © 2008 IEEE

	GALPAT	MARCH13n	PR[21]	MC[21]	DX[22]	DN	PI-PO[22]
Bit Defects/Faults	1	1	1	1	1	1	1
Column Defects/Faults	1	1	0	1	1	0	0
Sense Amp Defects/Faults	1	1	1	0	0	1	0
Address Decoder Defects/Faults	1	0	0	0	0	0	1
Column Multiplexer Defects/Faults	1	0	0	0	1	0	0
Speed Defects/Faults in Precharge	1	1	1	0	0	0	0
Speed Defects/Faults in Bit-Lines	1	0	1	1	0	0	0
IR-Drop induced Defects/Faults	0	0	0	0	0	1	0

Table 4 – Fault Coverage Matrix for 7 Memory Test Algorithms and 14 fault types (1 = detected, 0 = not detected)

the faults caught by GALPAT or MARCH13n. DX, DN and PI-PO are used to catch specific faults which are usually access pattern dependent. Among the lot DX catches most faults followed by DN and finally PI-PO. Since the order of strictness is only dependent on the tolerance of faults in an application, the order of algorithm execution would give different results. We experimented with both the MSF and LSF strategies which were introduced in the previous section. Eight bins will be created by either of the test strategies. The yield simulator considers one SoC instance at a time, and for each memory instance on the chip, it applies the memory test algorithm. The memory instance is marked "good" or "bad". If all memories on the instance of the SoC are fault free, then the SoC instance is placed in the appropriate bin. The yield simulator was developed in Perl and required several days of run-time on a Sun SPARC station.

The results from the use of MSF and LSF strategies are shown in Figure 3(a) and 3(b) respectively. In Figure 3(a), we show the yield after the application of each of the algorithms. In this experiment, we assumed that all 100 memories would be used in "full random access" mode. GALPAT is the strictest of the algorithms and gives a yield of about 38% when p = 0.0005. When MARCH13n is used to screen the chips that are rejected by GALPAT, the yield improves to about 39%. If all the 7 bins that contain "good" chips are taken into account, the yield improves to about 60%. Similarly, when p =

0.0002, the yield improves from about 70% to 82%. Note that PR algorithm gives the highest incremental improvement in yield for sequential access memories. Such improvements are very significant when we note that a chip is deemed fault-free if *all the 100 memories are fault-free*. In Figure 3(b), we show the simulated yield for the LSF strategy is plotted. While the general trend in the results is similar, note that for larger fault probabilities, the improvement in the yield is not as high as those achievable from the MSF strategy.

Figure 4 shows the number of memory BIST runs that were required during the yield simulation for a single instance of the SoC when the two test strategies are adopted. For p < 0.00045, the MSF strategy will have a shorter run-time and will apply a fewer number of memory test runs. For larger fault densities, we see that LSF has a better run-time performance. In figures 3 and 4, we show that the improvement in yield is not just limited to sequential access memories but any application that still make random access, but is not affected by some types of faults. A classic case being fault in precharge circuitry allows the memory to operate at lower frequencies but fail at high frequencies.

Figure 5 shows the results of applying the memory BIST algorithms when all the 100 memories are assumed to be used in sequential access mode. Notice that a further 5% improvement in yield is possible when this assumption is made. Notice also that the lenient algorithms give better yield improvement than the strict algorithms.

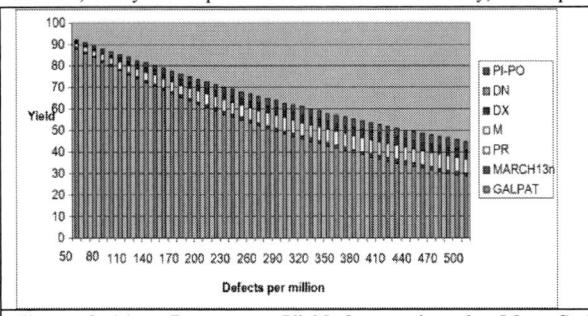

Figure 3 (a) – Percentage Yield from using the Most Strict Algorithm first

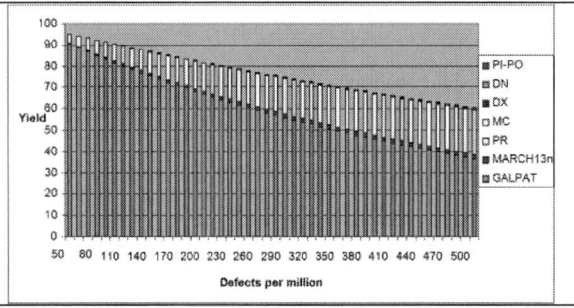

Figure 3 (b) – Percentage Yield from using the Least Strict Algorithm first

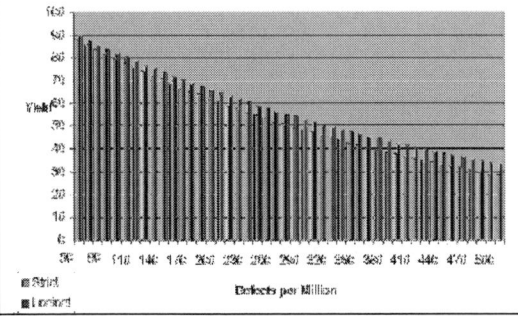

Figure 4 – Number of times memories are tested, as a function of *p*

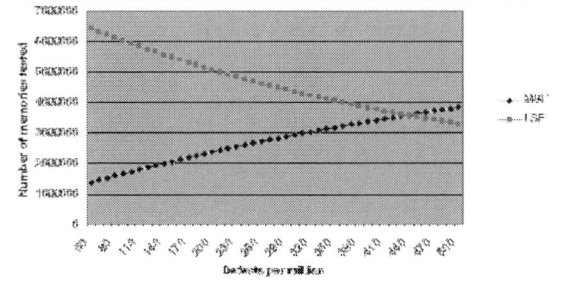

Figure 5 - Results of yield improvement for SAM-BIST

5 Conclusions

In the sub-100 nm regime, yield has become a major concern for semiconductor manufacturers. The major portion of the real-estate in a modern SoC consists of memories. In this paper, we proposed a new memory BIST scheme that can result in significant improvements to product yield without impacting the fault level of the product. We achieve this yield improvement through the use of two techniques:

(a) Repeated testing of memories through a mix of lenient and strict algorithms which have different fault coverage levels. Our memory BIST architecture therefore relies on the use of run-time programmability.

(b) Awareness of the end-usage of the memory – for example, knowledge of the purely sequential access pattern in the functional mode of operation.

We have introduced the notion of a test strategy and in particular, we defined the Most Strict First and Least Strict First strategies. Our results on an example SoC indicate that dramatic improvements in yield, BIST area overhead, and BIST run-time are possible by adopting our proposed methodology. MSF strategy tends to accept a product that later (more lenient) tests may reject, and typically offers a higher yield than LSF. LSF strategy takes a more pessimistic, but safer, approach.

References

[1] International Technology Roadmap for Semiconductors, *International technology roadmap for semiconductors 2004 Update*, http://public.itrs.net/Home.htm, 2004.

[2] Ad. J. van de Goor, Ivo Schanstra, *Address and Data Scrambling: Causes and Impact on Memory Tests*, IEEE International Workshop on Electronic Design, Test and Applications, Page: 128-136, Christchurch, 2002

[3] Dong -Chual Kang, Sung Min Park, and Sang-Bock Cho. *An Efficient Built-In Self-Test Algorithm for Neighborhood Pattern- and Bit-Line-Sensitive Faults in High-Density Memories*, ETRI Journal, vol.26, No.6, Dec. 2004, pp.520-534.

[4] Yervant Zorian, *Optimizing Manufacturability by Design for Yield*, IEEE Electronics Manufacturing Technology Symposium, California, page(s):255 – 258, 2004

[5] Ad. J. van de Goor, *Testing semiconductor memories: theory and practice*, John Wiley & Sons, Chichester, England, 1991

[6] Vonkyoung Kim, Tom Chen, *Assessing Defect Coverage of Memory Testing Algorithms*, Ninth Great Lakes Symposium on VLSI, 1999, Page(s): 340-341

[7] Alvin Jee, Jonathon E, V. Swamy, Mukesh Puri, *Optimizing Memory Tests by Analyzing Defect Coverage*, IEEE International Workshop on Memory Technology, Design and Testing, 2000, Page(s): 20-25

[8] Breuer, M.A., Zhu, H.H., *Error-Tolerance and Multi-Media*, IEEE International conference on Intelligent Information Hiding and Multimedia Signal Processing, 2006. IIH-MSP '06, page(s): 521-524, 2006

[9] Kuo-Liang Cheng, Chih-Wea Wang, Jih-Nung Lee, Yung-Fa Chou, Chih-Tsun Huang, Cheng-Wen Wu, *FAME: A Fault-Pattern Based Memory Failure Analysis Framework*, International Conference on Computer Aided Design, Page(s): 595-598 2003

[10] Jee, A., Ferguson, F.J., *Carafe: an inductive fault analysis tool for CMOS VLSI circuits*, Eleventh VLSI Test Symposium, 1993, Atlantic City, NJ page(s): 92-98

[11] Venkatesh R., Kumar S., Philip J., Shukla S., *A fault modeling technique to test memory BIST algorithms*, IEEE International Workshop on Memory Technology, Design and Testing, 2002, Page(s): 109- 116

[12] Chi-Feng Wu, Chih-Tsun Huang, Cheng-Wen Wu, *RAMSES: a fast memory fault simulator*, International Symposium on Defect and Fault Tolerance in VLSI Systems, 1999, Albuquerque, page(s): 165-173

[13] A. Benso, S. Di Carlo, G. Di Natale, P. Prinetto, *Memory Fault Simulator for Static-Linked Fault*, Proceedings of the 15th Asian Test Symposium 2006, Page(s): 31-36

[14] A. Benso, S. Di Carlo, G. Di Natale, P. Prinetto, *Specification and design of a new memory fault simulator*, 11th IEEE Asian Test Symposium, 2002. Page(s): 92 – 97

[15] Xiaogang Du, Nilanjan Mukherjee, Wu-Tung Cheng, Sudhakar M. Reddy, *Full-Speed Field-Programmable Memory BIST Architecture*, International Test Conference 2005, paper 45.3

[16] Aman Kokrady, C. P. Ravikumar, *Static Verification of Test Vectors for IR Drop Failure*, ICCAD 2003, 760-764

[17] T.W. Williams, N.C. Brown, *Defect Level as a function of Fault Coverage*, IEEE transactions on computers, Vol. C-30, No. 12, December 1981

[18] R. Dekker, F. Beenker, and L. Thijssen, *Fault modeling and test algorithm development for static random access memories*, In Proceedings of International Test Conf. (ITC), pages 343-352, 1988.

[19] Von-Kyoung Kim, Chen, T., *On comparing functional fault coverage and defect coverage for memory testing*, IEEE Transactions on Computer-Aided Design of Integrated Circuits and Systems, page(s): 1676-1683, Vol. 18, Issue 11, 1999

[20] Jee A., *Defect-oriented analysis of memory BIST tests*, IEEE International Workshop on, Memory Technology, Design and Testing, 2002, page(s): 7- 11

[21] Rei-Fu Huang, Yan-Ting Lai, Yung-Fa Chou, Cheng-Wen Wu, *SRAM delay fault modeling and test algorithm development*, Proceedings of the ASP-DAC 2004, Issue , 27-30 Jan. 2004 Page(s): 104 - 109

[22] Theo J. Powell, Wu-Tung Cheng, Joseph Rayhawk, Omer Samman, Paul Policke, Sherry Lai, *BIST for Deep Submicron ASIC Memories with High Performance Application*, International Test conference, ITC 2003, Volume: 1, Page(s): 386- 392

[23] Aadsen D., Fenstermaker L., Higgins F., Ilyoung Kim, Lewandowski J., Nagy J.J., *Test Algorithm for Memory Cell Disturb Failures*, IEEE International Workshop on Memory Technology, Design and Testing, 1998, Page(s): 53-56

[24] Samvel Shoukourian, Valery Vardanian, and Yervant Zorian, *SoC Yield Optimization via an Embedded-Memory Test and Repair Infrastructure*, IEEE Design & Test, Volume 21 , Issue 3 (May 2004), Pages: 200 - 207

[25] Chin-Lung Su, Yi-Ting Yeh, and Cheng-Wen Wu, *An Integrated ECC and Redundancy Repair Scheme for Memory Reliability Enhancement*, Proceedings of the 20th IEEE International Symposium on Defect and Fault Tolerance in VLSI Systems, 2005, Pages: 81 - 92

[26] Jitendra B. Khare, *Memory Yield Improvement - SoC Design Perspective*, International Test conference, ITC 2004, Page(s): 1445

[27] A. Benso, S. Di Carlo, G. Di Natale, P. Prinetto, Bodoni, M.L., *Programmable built-in self-testing of embedded RAM clusters in system-on-chip architectures*, IEEE communications magazine, Sept. 2003, Pages: 90 - 97

[28] Allen C. Cheng, *Comprehensive Study on Designing Memory BIST: Algorithms, Implementations and Trade-offs*, Project Report, Advanced Computer Architecture Lab, Department of Electrical Engineering and Computer Science, The University of Michigan Ann Arbor, 2002

21st International Conference on VLSI Design

Design-for-Testability for Improved
Path Delay Fault Coverage of Critical Paths

Irith Pomeranz[1]
School of Electrical & Computer Eng.
Purdue University
W. Lafayette, IN 47907, U.S.A.
pomeranz@ecn.purdue.edu

and

Sudhakar M. Reddy[2]
Electrical & Computer Eng. Dept.
University of Iowa
Iowa City, IA 52242, U.S.A.
reddy@engineering.uiowa.edu

Abstract

The path delay fault coverage achievable for a circuit may be low even when enhanced scan is available and only faults associated with critical paths are considered. To address this issue we describe a design-for-testability (DFT) approach that targets the critical (or longest) paths of the circuit. In a basic step of the proposed procedure, a fanout branch that is not on a longest path is disconnected from its stem, and driven from a new input in order to reduce the dependencies between off-path inputs of a target path delay fault. We present experimental results to demonstrate the increase in fault coverage of faults associated with longest paths as the number of new inputs is increased. We also discuss the implementation of the DFT approach in the context of scan design.

1. Introduction

The path delay fault model [1] is used for generating tests that detect small delay defects as well as verify that the delay of the circuit is within its specified limit determined by the clock speed. For this purpose, critical paths are typically targeted during path delay fault test generation. However, many paths in a circuit including critical paths may be unsensitizable, implying that path delay faults associated with these paths are untestable. Test generation procedures for path delay faults and procedures for identifying untestable path delay faults [2]-[5] report large percentages of untestable path delay faults in benchmark circuits. Moreover, many of the faults associated with critical paths in these circuits are untestable. Since untestability of a path delay fault does not imply that the path cannot affect the operation speed of the circuit, it is desirable to test the path delay faults associated with all the critical paths of the circuit (critical paths that are known to be false paths may be eliminated from consideration).

Several synthesis-for-testability (*SFT*) [6]-[14] and design-for-testability (*DFT*) [15]-[18] approaches were proposed for path delay faults considering the combinational logic of the circuit (i.e., assuming that the inputs and outputs of the combinational logic are controllable and observable, respectively). *SFT* approaches for multi-level circuits reduce the number of paths and/or increase the testability of the existing paths by resynthesizing the circuit. A reduction in the number of paths improves the circuit testability mainly because the fault coverage denominator (i.e., the total number of faults) is reduced.

DFT approaches insert test-points so as to disconnect circuit lines for the purpose of test application, and thus achieve one of two goals. The first is to eliminate dependencies that prevent path delay faults from being detected [15], [18]. For example, consider two AND gates along a path with off-path

inputs h_1 and h_2. To propagate a transition through the path it is necessary to set $h_1 = h_2 = 1$. However, if h_1 and h_2 are correlated such that they cannot be set to 1 simultaneously, the path delay fault will be untestable. The procedure of [15] disconnects selected fanout branches, whose stems are *inputs* of the circuit, and connects them to new inputs of the circuit. In this way, it removes the dependencies that exist between lines that are driven by different fanout branches of the same inputs, and potentially increases the path delay fault testability. The choice of considering inputs is motivated by two-level circuits where these are the only lines that have fanout branches. The procedure of [15] targets all the path delay faults in the circuit. The procedure of [18] targets primitive path delay faults, and inserts test-points in order to ensure that the maximum size of a primitive set that needs to be considered would not be larger than two.

The second goal of *DFT* approaches is to partition the circuit paths into subpaths [16]-[17]. The procedures of [16]-[17] use test-points to partition the paths such that each subpath can be tested independently of any other subpath. When a test-point is inserted on a line g, the subpaths ending at g are tested through an observation point on g, and a control-point on g is used for controlling the source of the subpaths that start at g. The partitioning is done such that the number of distinct subpaths is significantly smaller than the number of paths. In these approaches, each subpath must be tested using a fast clock cycle that corresponds to the expected delay of the subpath. Thus, testing of subpaths requires the use of several different fast clock cycles [16]-[17], [19]-[20].

DFT techniques developed for transition faults in scan circuits [21]-[26] can be applied to path delay faults as well. These methods are aimed at providing better control over the state variables of the circuit when broadside [27] or skewed-load [28] tests are used, and will at most provide the ability to detect every path delay fault that is detectable under enhanced scan [29]. However, the testability of path delay faults is low even under enhanced scan, as indicated by the results in [2]-[5].

In this work we describe a *DFT* approach for path delay faults considering the combinational logic of the circuit and targeting specifically the path delay faults associated with the critical (or longest) paths of the circuit. The proposed approach is more flexible in selecting lines that will be disconnected than the one in [15]. As a result, it allows 100% fault coverage of faults associated with longest paths to be achieved for circuits where the procedure from [15] did not improve the fault coverage at all. Unlike [18], the faults are tested individually and there is no need to consider subsets of path delay faults. By targeting only critical paths, and ensuring that lines on critical paths are not disconnected, the new approach does not partition paths that need to be tested. Consequently, it does not create shorter subpaths that need to be tested. It thus removes the need for several different fast clock cycles, which are required in [16]-[17].

---

1. Research supported in part by SRC Grant No. 2004-TJ-1244.
2. Research supported in part by SRC Grant No. 2004-TJ-1243.

978-1-4244-3039-0/08 $25.00 © 2008 IEEE

We define the type of tests used throughout this work in Section 2. We also discuss the reason for path delay faults being untestable under enhanced scan, i.e., dependencies that exist between values of off-path inputs.

In Section 3 we describe a procedure for removing dependencies that result in untestable path delay faults. The procedure is based on disconnecting a fanout branch g_i^j from its stem g_i for the purpose of test application, and driving the disconnected fanout branch g_i^j from a new input of the circuit. This modification eliminates the dependencies between lines that are driven by g_i through g_i^j, and lines driven by g_i through its other fanout branches. It thus has the potential of improving the path delay fault coverage achievable for the circuit. By performing this modification only for fanout branches that are not on critical paths (or close-to-critical paths), it is possible to ensure that the number of paths to be tested will not change, each longest path of the original circuit has a corresponding path in the modified circuit, and the longest path lengths in the modified circuit are equal to the corresponding path lengths in the original circuit. By allowing g_i^j to be a fanout branch of an internal stem, we obtain a large number of candidates for modifying the circuit, thus allowing significant increases in fault coverage to be obtained. In addition, by targeting only untestable faults associated with longest paths we keep the number of additional inputs low. We discuss the overheads of the modification, how it fits with scan design, and describe the general *DFT* insertion procedure in Section 3.

Experimental results for benchmark circuits are presented in Section 4. The results demonstrate that 100% fault coverage can be achieved for faults associated with longest paths even when none of these faults is testable in the original circuit.

Throughout this work we measure the length of a path by the number of multi-input gates along the path. Other definitions of path length can be accommodated.

2. Preliminaries

A path is said to be sensitized when its off-path inputs carry non-controlling values. Path sensitization is a necessary condition for the detection of a path delay fault under any type of test (robust, non-robust, etc.). In this work we consider a path delay fault as detected by a two-pattern test that assigns the required transition to the source of the path, and sensitizes the path under the second pattern of the test. The following example demonstrate this detection condition.

We consider ISCAS-89 benchmark circuit $s27$ shown in Figure 1. We consider the path delay fault associated with the path $p = 1$-8-13-14-16-19-20-21-23-26 and the $1 \to 0$ transition at its source. Line numbers are shown on the left in Figure 1 followed by values under the second pattern of a two-pattern test. The path under consideration is shown in bold. To sensitize the path it is necessary to assign the value 1 to off-path input 6, the value 0 to off-path input 4, the value 1 to off-path input 18, and the value 0 to off-path input 5. To detect the path delay fault, it is necessary to assign the $1 \to 0$ transition to line 1, and sensitize the path under the second pattern of the test. The vector shown in Figure 1 sensitizes the path and assigns the required final value to the source of the path. If the vector of Figure 1 is preceded by a vector that assigns the value 1 to line 1, a two-pattern test that detects the path delay fault will be obtained.

In general, we denote a path by $p = g_0\text{-}g_1\text{-}\cdots\text{-}g_{L-1}$ and a transition at its source by $v \to v'$. We denote the corresponding path delay fault by $(p, v \to v')$. We denote the set of off-path inputs of path p by $OFF(p) = \{h_0, h_1, \cdots, h_{m-1}\}$. We denote the non-controlling value of $h_i \in OFF(p)$ by $nc(h_i)$. We denote by $DET(p, v \to v')$ the set of values that the second pattern of a test for the path delay fault $(p, v \to v')$ must assign. The set

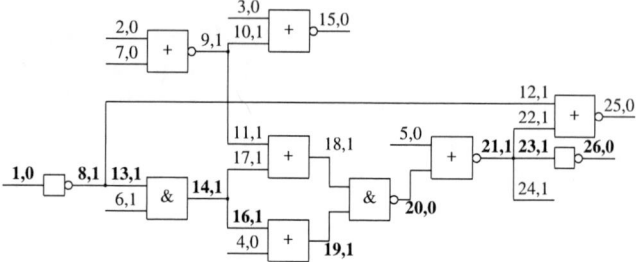

Figure 1: ISCAS-89 benchmark circuit s27

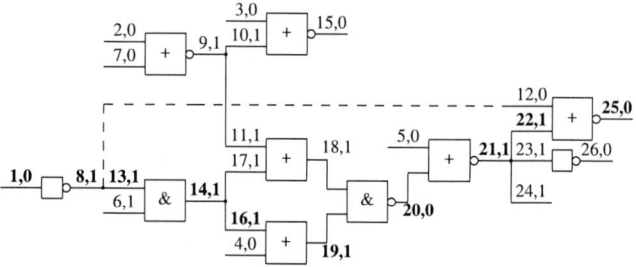

Figure 2: Modified ISCAS-89 benchmark circuit s27

$DET(p, v \to v')$ includes every line $h_i \in OFF(p)$ with its non-controlling value $nc(h_i)$, and in addition it includes the line g_0 with the value v'.

For the example path delay fault in Figure 1 we have $OFF(p) = \{4, 5, 6, 18\}$ and $DET(p, 1 \to 0) = \{(1,0), (4,0), (5,0), (6,1), (18,1)\}$. An entry of $DET(p, 1 \to 0)$ is a pair consisting of a line and its required value.

If t_2 is a vector that assigns all the values in $DET(p, v \to v')$, it is possible to obtain a two-pattern test for $(p, v \to v')$ by using $<t_1, t_2>$ such that t_1 is the complement of t_2. This can be seen as follows. If t_2 assigns all the values in $DET(p, v \to v')$, then in particular it assigns $g_0 = v'$. The only requirement on t_1 is that it would assign $g_0 = v$. Since t_2 assigns $g_0 = v'$, its complement is guaranteed to assign $g_0 = v$ as required for t_1. For simplicity, we assume that t_1 will be the complement of t_2 in every two-pattern test.

A path delay fault $(p, v \to v')$ is untestable when the values in $DET(p, v \to v')$ cannot be assigned by a single vector. An example is the $1 \to 0$ path delay fault on the path $p = 1$-8-13-14-16-19-20-21-22-25 in $s27$. For this fault, $DET(p, 1 \to 0) = \{(1,0), (4,0), (5,0), (6,1), (12,0), (18,1)\}$. Fault detection requires that lines 1 and 12 be set to 0 by the same vector. This is impossible since line 12 always assumes a value that is the complement of the value of line 1. Therefore, the fault is untestable.

3. Removing line dependencies

In this section we describe a method for removing dependencies between lines in the circuit that result in untestable path delay faults associated with longest paths.

We first present an example. We then discuss the overheads of the modification and how it fits with scan design. We then describe the general procedure.

3.1. Example

As shown at the end of Section 2, the dependence that exists between the values of lines 1 and 12 in $s27$ causes the $1 \to 0$ path delay fault on the path $p = 1$-8-13-14-16-19-20-21-22-25 to be untestable. It is possible to

978-1-4244-3039-0/08 $25.00 © 2008 IEEE

remove this dependency without affecting the path p (and without affecting any other longest path) by disconnecting line 12 from line 8 for the purpose of test application, and connecting line 12 to a new input of the circuit. Such a modified circuit is shown in Figure 2. The connection from line 8 to line 12 that was removed is shown by a dashed line. We also show in Figure 2 the second vector of a test that detects the $1 \rightarrow 0$ path delay fault on the path $p = 1$-8-13-14-16-19-20-21-22-25.

The example of Figure 2 demonstrates a case where the dependence that causes a path delay fault to be untestable is between the source of the path (line 1) and one of the off-path inputs of the path (line 12). In the example, the off-path input (line 12) is a fanout branch of a fanout stem on the path (line 8). In general, a dependence that makes a path delay fault untestable can be caused by an off-path input that is driven from a fanout stem on the path through one or more gates.

A dependence between off-path inputs can also cause a path delay fault to be untestable. Such a dependence is caused by a fanout stem g_i whose branches g_i^{j1} and g_i^{j2} drive two off-path inputs as demonstrated by Figure 3. The dashed lines indicate that there may be additional logic along these lines. The target path is the one at the bottom of the figure. Disconnecting one of the fanout branches, either g_i^{j1} or g_i^{j2}, from the stem will remove the dependence between the off-path inputs driven by g_i^{j1} and g_i^{j2}, and improve the testability of the path.

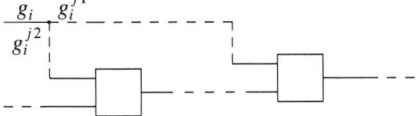

Figure 3: Dependence between off-path inputs

3.2. Modification overheads

The following points should be noted with respect to the modification of Figure 2.

A fanout branch g_i^j that is disconnected from its fanout stem g_i as in Figure 2 requires a multiplexer with a control input s and the following modes of operation. During functional operation ($s = 0$), g_i^j must be driven from g_i. During test application ($s = 1$), g_i^j must be driven from a new input.

When a multiplexer is placed on a fanout branch g_i^j, the delay of any path through it will increase. For this reason we do not allow a fanout branch on a critical path to be disconnected. Additionally, the increased delay due to the placement of multiplexers may increase the delay of a non-critical path so that the path would become critical. This implies that the multiplexers should only be placed on fanout branches such that the increased delay does not create new critical paths, or any new critical paths are also made testable. To simplify the discussion we assume that any fanout branch that is not on a critical path may be disconnected. This assumption can be easily removed by excluding fanout branches on close-to-critical paths.

We note that it is sometimes possible to use existing circuit inputs to drive disconnected fanout branches. Consider a path p that ends at an output z of an n-input circuit. Let the input cone of z consist of m inputs. In this case, only m inputs need to be specified in order to detect a path delay fault associated with p. The remaining $n-m$ inputs can be used for driving fanout branches that are disconnected in order to make the path delay fault testable. We do not consider this option in this work.

In the context of scan design, the new inputs added in order to disconnect fanout branches from their stems can be driven from flip-flops, and these flip-flops can be added to the scan chains of the circuit. The extra flip-flops are similar to the dummy flip-flops used in [22]. The difference is that in [22] they are used to reach the fault coverage of enhanced scan, whereas in the method proposed here they are used to exceed this fault coverage by disconnecting fanout branches from their stems.

After *DFT* insertion, to allow every two-pattern test to be applied to the inputs of the circuit, it is possible to use enhanced scan. Alternatively, any one of the *DFT* approaches described in [21]-[26] for scan circuits can be used to ensure that path delay faults that are testable considering the combinational logic of the circuit can be detected using skewed-load or broadside tests.

The control input s of the multiplexers is common to all the multiplexers, and it needs to be held at 1 during test application.

3.3. Basic modification step

Next, we describe the circuit modification procedure that selects fanout branches to be disconnected so as to improve the path delay fault coverage. We consider a circuit with n inputs, $a_0, a_1, \cdots, a_{n-1}$.

A basic step of the procedure consists of selecting a single fanout branch g_i^j of a stem g_i, disconnecting g_i^j from g_i, and connecting g_i^j to a new input a_n through a multiplexer. The selection of g_i^j is done based on a given set U of untestable (and hard-to-detect) path delay faults. The set U is obtained by performing test generation for the set of target path delay faults P in the circuit. Test generation results in a test set T and a set of undetected path delay faults U. After the circuit is modified, the set of target path delay faults remains the same since fanout branches on critical paths are not disconnected. However, the circuit has new inputs that can be used for detecting path delay faults. Therefore, a new test set is computed for the modified circuit. The modification process can then be repeated for the modified circuit with a new test set T and a new set of undetected path delay faults U until no additional modifications are possible.

The following example demonstrates some of the issues to be considered when selecting a fanout branch g_i^j, such that g_i^j will be disconnected from its stem and connected to a new input. We consider ISCAS-89 benchmark circuit $s27$ shown in Figure 1.

A complete test set T for the path delay faults in $s27$ leaves the path delay faults shown in Table 1 undetected. These path delay faults are untestable. For each untestable path delay fault $(p_i, v \rightarrow v')$ we show in Table 1 the path p_i, the transition $v \rightarrow v'$ and the set $DET(p_i, v \rightarrow v')$.

Table 1: Untestable path delay faults of s27

i	p_i	$v \rightarrow v'$	$DET(p_i, v \rightarrow v')$
0	6-14-16-19-20-21-22-25	$1 \rightarrow 0$	(4,0) (5,0) (6,0) (12,0) (13,1) (18,1)
1	6-14-16-19-20-21-22-25	$0 \rightarrow 1$	(4,0) (5,0) (6,1) (12,0) (13,1) (18,1)
2	1-8-13-14-16-19-20-21-22-25	$1 \rightarrow 0$	(1,0) (4,0) (5,0) (6,1) (12,0) (18,1)
3	6-14-17-18-20-21-22-25	$1 \rightarrow 0$	(5,0) (6,0) (11,0) (12,0) (13,1) (19,1)
4	6-14-17-18-20-21-22-25	$0 \rightarrow 1$	(5,0) (6,1) (11,0) (12,0) (13,1) (19,1)
5	1-8-13-14-17-18-20-21-22-25	$1 \rightarrow 0$	(1,0) (5,0) (6,1) (11,0) (12,0) (19,1)

The circuit has four fanout branches that are not on critical paths, 10, 12, 23 and 24. Based on the information in Table 1 we find that line 12 appears in the set $DET(p_i, v \rightarrow v')$ of all six path delay faults in U. Lines 10, 23 and 24 do not appear in the set $DET(p_i, v \rightarrow v')$ of any path delay fault in U. Moreover, lines 10, 23 and 24 do not drive any off-path input of a critical path. Since line 12 appears in the largest number of sets, we prefer to select line 12 and perform the modification shown in Figure 2 based on line 12. With this modification, all the path delay faults become detectable. Placing a multiplexer on line 12 does not create any new longest paths.

In general, we denote by $n_{01}(g_i^j)$ the number of times a fanout branch g_i^j of a stem g_i appears in a set $DET(p,v \rightarrow v')$ such that $(p,v \rightarrow v') \in U$. When a choice exists, we will prefer to select the fanout branch with the highest value of $n_{01}(g_i^j)$ in order to modify the circuit.

We will allow a fanout branch g_i^j with $n_{01}(g_i^j) = 0$ to be selected. This is needed since such a fanout branch may drive an off-path input of an untestable path delay fault through several levels of gates. It is thus possible that disconnecting such a fanout branch will remove dependencies in the circuit, which cause path delay faults to be untestable or hard-to-detect. However, we will exclude a fanout branch g_i^j with $n_{01}(g_i^j) = 0$ if we can verify that it cannot improve the testability of a target path delay fault, as explained later.

We select fanout branches for modification by considering the path delay faults in U one at a time. When a target path delay fault $(p,v \rightarrow v') \in U$ is considered, we select fanout branches based on the fault until the fault becomes detectable, or no additional options exist for making the fault detectable. The latter case may occur since we do not allow fanout branches on critical paths to be disconnected. We select fanout branches for $(p,v \rightarrow v')$ in three phases. Different criteria are used in each phase as described next.

In Phase 1, we consider fanout branches that must be disconnected from their stems in order to make $(p,v \rightarrow v')$ testable. We consider two cases. In the first case, a fanout stem g_i on the path p has a fanout branch g_i^j such that $g_i^j \in OFF(p)$. More generally, g_i^j may drive an off-path input h of p through single-input gates. Suppose that the value of g_i is w when the source of the path is set to v' and the path is sensitized. If w implies the controlling value on h, the connection between h and g_i through g_i^j will prevent the sensitization of the path. Therefore, g_i^j needs to be disconnected in order to make the path delay fault testable.

The second case we consider in Phase 1 is the case where two fanout branches g_i^{j1} and g_i^{j2} of a stem g_i drive two off-path inputs h_1 and h_2 of the path p through single-input gates, such that neither $g_i = 0$ nor $g_i = 1$ will result in non-controlling values on both h_1 and h_2. In this case, either g_i^{j1} or g_i^{j2} needs to be disconnected in order to make the path delay fault testable.

Using structural analysis, we identify the cases above one at a time and disconnect a fanout branch corresponding to each case if it is not on a critical path.

Phases 2 and 3 are identical in the criteria they use to select a fanout branch to be disconnected. The difference between them is that in Phase 2 we undo a modification that does not result in the target fault being detected, while Phase 3 retains all the modifications. For Phases 2 and 3, we exclude from consideration a fanout branch g_i^j that does not drive any line of the path p. In addition, we exclude a fanout branch g_i^j of a stem g_i if g_i does not have another fanout branch that drives p. In this case, g_i^j cannot contribute to any conflicts related to the detection of $(p,v \rightarrow v') \in U$, and it does not need to be disconnected. Among the remaining fanout branches, we select the fanout branch g_i^j with the highest value of $n_{01}(g_i^j)$. This ensures that g_i^j drives as many undetectable path delay faults as possible, and it is thus likely to help render as many of them as possible testable (including the target fault).

The procedure for selecting a fanout branch for circuit modification is referred to as Procedure 1. There are three variations of Procedure 1 corresponding to Phase 1, 2 and 3.

3.4. Overall procedure

The overall circuit modification procedure is given below as Procedure 2. In every iteration, Procedure 2 selects a path

delay fault $(p,v \rightarrow v') \in U$. The criterion used for selecting the next target path delay fault is based on the number of fanout branches that will be disconnected in Phase 1. Since these fanout branches must be disconnected to make the path delay fault detectable, we prefer to select a fault with a higher number of such fanout branches earlier.

Once a target path delay fault $(p,v \rightarrow v')$ is selected, Procedure 2 calls the version of Procedure 1 that corresponds to Phase 1 (followed by Phases 2 and 3) repeatedly to select a fanout branch g_i^j for modification. This continues until $(p,v \rightarrow v')$ can be detected, or no additional modifications based on $(p,v \rightarrow v')$ are possible.

If a modification based on g_i^j does not make $(p,v \rightarrow v')$ testable, one of two steps is taken depending on the phase of the procedure, as explained next.

In Phase 1, a modification that does not make $(p,v \rightarrow v')$ testable is retained. This is important since each case identified by Procedure 1 is sufficient by itself to make a path delay fault untestable.

In Phase 2, a modification that does not make $(p,v \rightarrow v')$ testable is undone. This prevents the number of inputs from growing unnecessarily. The flag $select(g_i^j)$ prevents a fanout branch from being selected again if an earlier modification based on it failed.

In Phase 3, a modification that does not make $(p,v \rightarrow v')$ testable is retained. This is important for the case where $(p,v \rightarrow v')$ is untestable due to dependencies between multiple subsets of off-path inputs. In this case, it may be necessary to disconnect several fanout branches before the path delay fault becomes testable.

Although Procedure 2 maintains the flag $select(h)$ for every line throughout all the phases, only Phase 2 requires it. When a fanout branch g_i^j is disconnected, g_i^j is connected directly to a new primary input, and will not be considered again.

If Procedure 2 succeeds in detecting a target path delay fault $(p,v \rightarrow v')$, it performs test generation for all the target faults in case additional path delay faults become detectable together with $(p,v \rightarrow v')$. All the detected path delay faults are removed from U.

Procedure 2: Circuit modification
(1) Let the set of inputs be $a_0,a_1,\cdots,a_{n-1}$. Let the set of target path delay faults be P.
(2) Generate a test set T for the faults in P and find the set of undetected path delay faults U.
(3) If $U = \varnothing$, stop.
(4) Select a path delay fault $(p,v \rightarrow v') \in U$. Set $ph = 1$.
(5) Set $select(h) = 0$ for every line h.
(6) Call the version of Procedure 1 that corresponds to Phase ph to select a fanout branch g_i^j. If no fanout branch is selected:
    If $ph < 3$, set $ph = ph+1$ and go to Step 5. Else:
        Mark the fault $(p,v \rightarrow v')$ and remove it from U.
        Go to Step 3.
(7) Set $select(g_i^j) = 1$.
(8) Modify the circuit such that g_i^j would be driven from a new input a_n. Set $n = n+1$.
(9) Attempt to generate a test for $(p,v \rightarrow v')$. If a test cannot be generated:
    (a)   If $ph = 2$, undo the modification of Step 8.
    (b)   Go to Step 6.
(10) Go to Step 2.

Procedure 2 continues to modify the circuit until U becomes empty. If no faults are marked in Step 6, this implies that the fault coverage for the circuit is 100%. It is possible to stop Procedure 2 before this point when the number of extra inputs reaches a preselected limit. Faults that are marked during Step 6 of Procedure 2 may require fanout branches that are on critical paths to be disconnected. Since we do not allow such fanout branches to be disconnected, the faults remain untestable. Nevertheless, other fanout branches may be disconnected in an attempt to make such a fault testable before Procedure 2 identifies that the fault cannot be made testable. This may cause the number of modifications (and the number of extra inputs) to be larger than necessary. To address this issue, if Procedure 2 marks some faults as untestable after modification, we run Procedure 2 again, this time without considering the marked faults of the first run.

We note that the set of path delay faults does not change due to the application of Procedure 2, since fanout branches on critical paths are not disconnected. However, every time the circuit is modified, it has one additional input that does not exist in the circuit before modification. To take advantage of tests generated for the original circuit and for modified circuits obtained earlier during the modification process, we use the following approach.

Let T be the test set for the circuit before modification. Suppose that T is defined over n inputs. After a modification that adds an input a_n to T, we compute a modified test set $\hat{T}$ for the new circuit as follows.

Let $t = <t_1,t_2>$ be a test in T defined over n inputs of the circuit before modification. After modification, it is possible to extend $t = <t_1,t_2>$ into a test for the modified circuit by assigning the value 0 or 1 to a_n under t_1 (and the complement value under t_2). The resulting tests are $t^0 = <t_10,t_21>$ and $t^1 = <t_11,t_20>$. We include both tests in $\hat{T}$. It is possible to show that $\hat{T}$ will detect the same faults detected by T and possibly additional faults. Additional tests can be added to $\hat{T}$ in order to detect additional faults in U.

4. Experimental results

We applied Procedure 2 to the combinational logic of ISCAS-89 and ITC-99 benchmarks.

Before applying Procedure 2 we identify path delay faults that will remain untestable even after modifying the circuit by checking if any of the fanout branches that would be identified in Phase 1 as necessary to be disconnected are on critical paths. During Procedure 2 we do not consider paths that were identified as untestable in a modified circuit.

We use a simulation-based test generation procedure when new tests need to be generated.

The results of Procedure 2 are shown in Tables 2 and 3. We show several rows for every circuit. The first row corresponds to the original circuit. The following rows correspond to modified circuits obtained after making different target path delay faults testable. We place an asterisk following the circuit name if the first run of Procedure 2 marked some faults that were not identified in advance as untestable after modification, and a second run was performed without considering such faults.

In every row, after the circuit name we show the index of the path delay fault targeted by Procedure 2, where -1 corresponds to the original circuit. We then show the number of inputs including extra inputs added by the proposed procedure. In parentheses we show the number of extra inputs. Next, we show the maximum length of a path, the number of path delay faults in the target set P, the number of path delay faults in P that are detected by test generation, and the path delay fault coverage with respect to P.

Table 2: Results for ISCAS-89 benchmarks

circuit	path	inp		len	paths	detect	f.c.
s1423	-1	91	(0)	55	64	24	37.50
s1423	4	94	(3)	55	64	40	62.50
s1423	0	95	(4)	55	64	48	75.00
s1423	16	96	(5)	55	64	64	100.00
s5378	-1	214	(0)	11	1952	1466	75.10
s5378	6	215	(1)	11	1952	1474	75.51
s5378	54	216	(2)	11	1952	1484	76.02
s5378	82	217	(3)	11	1952	1488	76.23
s9234*	-1	247	(0)	29	16384	0	0.00
s9234*	0	272	(25)	29	16384	6144	37.50
s13207	-1	700	(0)	30	480	0	0.00
s13207	38	762	(62)	30	480	24	5.00
s13207	0	763	(63)	30	480	48	10.00
s13207	1	764	(64)	30	480	334	69.58
s13207	12	765	(65)	30	480	384	80.00
s13207	13	766	(66)	30	480	432	90.00
s13207	15	767	(67)	30	480	480	100.00
s35932	-1	1763	(0)	19	34816	0	0.00
s35932	4608	1785	(22)	19	34816	128	0.37
s35932	4610	1807	(44)	19	34816	256	0.74
s35932	4612	1829	(66)	19	34816	384	1.10
s35932	4614	1851	(88)	19	34816	512	1.47
s35932	4616	1873	(110)	19	34816	640	1.84
s35932	4618	1895	(132)	19	34816	768	2.21
...							
s35932	4770	2899	(1136)	19	34816	6976	20.04
s35932	4926	3971	(2208)	19	34816	14080	40.44
s35932	5168	5025	(3262)	19	34816	20944	60.16
s35932	1256	5140	(3377)	19	34816	27856	80.01
...							
s35932	18880	5199	(3436)	19	34816	34816	100.00
s38417*	-1	1664	(0)	28	256	0	0.00
s38417*	1	1690	(26)	28	256	128	50.00
s38584*	-1	1464	(0)	26	13248	0	0.00
s38584*	6998	1478	(14)	26	13248	4608	34.78
s38584*	7	1483	(19)	26	13248	6048	45.65
s38584*	349	1485	(21)	26	13248	6144	46.38
s38584*	352	1487	(23)	26	13248	6240	47.10
s38584*	354	1489	(25)	26	13248	6336	47.83
s38584*	0	1490	(26)	26	13248	6816	51.45
s38584*	40	1491	(27)	26	13248	6960	52.54
s38584*	439	1492	(28)	26	13248	7440	56.16
s38584*	514	1493	(29)	26	13248	7824	59.06
s38584*	674	1494	(30)	26	13248	8064	60.87
s38584*	675	1495	(31)	26	13248	8256	62.32
s38584*	687	1496	(32)	26	13248	8400	63.41
s38584*	690	1497	(33)	26	13248	8544	64.49
s38584*	691	1498	(34)	26	13248	8688	65.58
s38584*	692	1499	(35)	26	13248	8832	66.67
s38584*	693	1500	(36)	26	13248	8976	67.75

We include the length of the longest path and the number of path delay faults associated with longest paths on every row to stress that these numbers are not affected by the proposed *DFT* insertion procedure. The following points can be seen from Tables 2 and 3.

For all the circuits where the fault coverage of faults associated with longest paths is initially low, the proposed modification procedure improves the fault coverage significantly. In many cases, all such faults are detectable in the modified circuit.

For s5378, 464 faults were identified as untestable even in the modified circuit, and these are the only faults that were left untestable after modification. For s35932, a large number of modifications was made to bring the fault coverage from 0% to 100%. We include the results for all the modifications until the fault coverage reached 2%. After that, we include the results only for the modifications that increased the fault coverage to approximately $10r\%$, for $r = 2,4,\cdots,10$.

For most of the circuits, the number of extra inputs grows slowly with every additional path delay fault considered by Procedure 2. When this is not the case, it is possible to remove from consideration faults that require large numbers of extra inputs at

Table 3: Results for ITC-99 benchmarks

circuit	path	inp		len	paths	detect	f.c.
b04	-1	78	(0)	28	32	16	50.00
b04	0	79	(1)	28	32	32	100.00
b05	-1	36	(0)	32	768	156	20.31
b05	0	83	(47)	32	768	572	74.48
b05	57	84	(48)	32	768	612	79.69
b05	4	85	(49)	32	768	636	82.81
b05	59	86	(50)	32	768	645	83.98
b05	76	87	(51)	32	768	664	86.46
b05	78	88	(52)	32	768	691	89.97
b05	122	89	(53)	32	768	722	94.01
b05	127	90	(54)	32	768	745	97.01
b05	167	91	(55)	32	768	747	97.27
b05	169	92	(56)	32	768	765	99.61
b05	216	93	(57)	32	768	766	99.74
b05	456	94	(58)	32	768	767	99.87
b05	740	95	(59)	32	768	768	100.00
b07	-1	53	(0)	26	16	6	37.50
b07	0	54	(1)	26	16	12	75.00
b07	3	73	(20)	26	16	14	87.50
b07	9	74	(21)	26	16	15	93.75
b07	11	75	(22)	26	16	16	100.00
b11	-1	38	(0)	32	112	0	0.00
b11	0	54	(16)	32	112	62	55.36
b11	1	55	(17)	32	112	72	64.29
b11	4	56	(18)	32	112	79	70.54
b11	5	62	(24)	32	112	84	75.00
b11	6	63	(25)	32	112	87	77.68
b11	7	64	(26)	32	112	94	83.93
b11	26	65	(27)	32	112	106	94.64
b11	29	66	(28)	32	112	111	99.11
b11	37	67	(29)	32	112	112	100.00
b12	-1	127	(0)	15	800	160	20.00
b12	0	128	(1)	15	800	320	40.00
b12	40	129	(2)	15	800	440	55.00
b12	120	130	(3)	15	800	640	80.00
b12	160	133	(6)	15	800	800	100.00

the cost of reduced fault coverage. It is also possible to reconsider every extra input and check whether it can be removed (after reconnecting the fanout branch it drives to its original fanout stem) without reducing the fault coverage. We did not implement these options.

5. Concluding remarks

We described a *DFT* approach for path delay faults associated with longest paths. In a basic step of the proposed procedure, a fanout branch that is not on a critical path is disconnected from its stem, and driven from a new input. This reduces dependencies between values of off-path inputs of target paths and increases the number of testable path delay faults. Due to the introduction of new inputs, the method allows path delay faults that are untestable under enhanced scan to be detected. Due to the use of fanout branches that are not on critical paths, the method requires a single fast clock cycle equal to the functional clock cycle for test application. We presented experimental results to demonstrate the increase in fault coverage of faults associated with longest paths as the number of inputs introduced is increased. We also discussed this approach in the context of scan design.

References

[1] G. L. Smith, "Model for Delay Faults Based Upon Paths", in Proc. 1985 Intl. Test Conf., pp. 342-349.

[2] K. Fuchs, F. Fink and M. H. Schulz, "DYNAMITE: An Efficient Automatic Test Pattern Generation for Path Delay Faults", IEEE Trans. on Computer-Aided Design, Oct. 1991, pp. 1323-1335.

[3] K.-T. Cheng and H.-C. Chen, "Delay Testing for Non-robust Untestable Circuits", in Proc. Intl. Test Conf., Oct. 1993, pp. 954-961.

[4] U. Sparmann, D. Luxenburger, K.-T. Cheng and S. M. Reddy, "Fast Identification of Robust Dependent Path Delay Faults", in Proc. Design Autom. Conf., June 1995, pp. 119-125.

[5] S. Kajihara, K. Kinoshita, I. Pomeranz and S. M. Reddy, "A Method for Identifying Robust Dependent and Functionally Unsensitizable Paths", in Proc. 1997 VLSI Design Conf., Jan. 1997, pp. 82-87.

[6] K. Keutzer, S. Malik and A. Saldanha, "Is Redundancy Necessary to Reduce Delay?", IEEE Trans. on Computer-Aided Design, April 1991, pp. 427-435.

[7] S. Devadas and K. Keutzer, "Synthesis of Robust Delay-Fault-Testable Circuits: Practice", in IEEE Trans. on Computer-Aided Design, March 1992, pp. 277-300.

[8] H. Hengster, R. Drechsler and B. Becker, "Testability Properties of Local Circuit Transformations with respect to the Robust Path-Delay-Fault Model", in Proc. 7th Int. Conf. on VLSI Design, Jan. 1994, pp. 123-126.

[9] I. Pomeranz and S. M. Reddy, "On Synthesis-for-Testability of Combinational Logic Circuits", in Proc. 32nd Design Autom. Conf., June 1995, pp. 126-132.

[10] H. Hengster, R. Drechsler and B. Becker, "On Local Transformations and Path Delay Fault Testability", in Journal of Electronice Testing - Theory and Applications, Dec. 1995, pp. 173-191.

[11] I. Pomeranz and S. M. Reddy, "On the Number of Tests to Detect All Path Delay Faults in Combinational Logic Circuits", IEEE Trans. on Computers, Jan. 1996, pp. 50-62.

[12] A. Krstic and K.-T. Cheng, "Resynthesis of Combinational Circuits for Path Count Reduction and for Path Delay Fault Testability", in Proc. 1996 Europ. Design & Test Conf., March 1996, pp. 486-490.

[13] I. Pomeranz and S. M. Reddy, "On Cancelling the Effects of Logic Sharing for Improved Path Delay Fault Testability", in Proc. Intl. Test Conf., Oct. 1996, pp. 357-366.

[14] M. A. Gharaybeh, V. D. Agrawal, M. L. Bushnell and C. G. Parodi, "False-Path Removal using Delay Fault Simulation", Journal of Electronic Testing: Theory and Applications, Oct. 2000, pp. 463-476.

[15] X. Xie and A. Albicki, "Bit-Splitting for Testability Enhancement in Scan-Based Design", in Proc. Intl. Conf. on Computer Design, Oct. 1993, pp. 155-158.

[16] I. Pomeranz and S. M. Reddy, "Design-for-Testability for Path Delay Faults in Large Combinational Circuits Using Test-Points", 31st Design Autom. Conf., June 1994, pp. 358-364.

[17] P. Uppaluri, U. Sparmann and I. Pomeranz, "On Minimizing the Number of Test Points Needed to Achieve Complete Robust Path Delay Fault Testability", in Proc. 14th VLSI Test Symp., April 1996, pp. 288-295.

[18] A. Krstic, K.-T. Cheng and S. T. Chakradhar, "Primitive Delay Faults: Identification, Testing, and Design for Testability", in IEEE Trans. on Computer-Aided Design, June 1999, pp. 669-684.

[19] W.-W. Mao and M. D. Ciletti, "A Variable Observation Time Method for Testing Delay Faults", in Proc. Design Autom. Conf., 1990, pp. 728-731.

[20] V. S. Iyengar and G. Vijayan, "Optimized Test Application Timing for AC Test", IEEE Trans. on Computers, Nov. 1992, pp. 1439-1449.

[21] W. Mao and M. D. Ciletti, "Arrangement of Latches in Scan-Path Design to Improve Delay Fault Coverage", in Proc. 1990 Intl. Test Conf., Sept. 1990, pp. 387-393.

[22] J. Savir and R. Berry, "At-Speed Test is Not Necessarily an AC Test", in Proc. 1991 Intl. Test Conf., Oct. 1991, pp. 722-728.

[23] N. A. Touba and E. J. McCluskey, "Applying Two-Pattern Tests Using Scan-Mapping", in Proc. VLSI Test Symp., April 1996, pp. 393-397.

[24] I. Pomeranz and S. M. Reddy, "On Achieving Complete Coverage of Delay Faults in Full Scan Circuits using Locally Available Lines", in Proc. 1999 Intl. Test Conf., Oct. 1999, pp. 923-931.

[25] S. Wang, X. Liu and S. T. Chakradhar, "Hybrid Delay Scan: A Low Hardware Overhead Scan Based Delay Test Technique for High Fault Coverage and Compact Test Sets", in Proc. Design Autom. and Test in Europe Conf., 2004, pp. 1296-1301.

[26] N. Devtaprasanna, A. Gunda, P. Krishnamurthy, S. M. Reddy and I. Pomeranz, "Methods for Improving Transition Delay Fault Test Coverage Using Broadside Tests", in Proc. Intl. Test Conf., Nov. 2005, pp. 256-265.

[27] J. Savir and S. Patil, "Broad-Side Delay Test", IEEE Trans. on Computer-Aided Design, Aug. 1994, pp. 1057-1064.

[28] J. Savir and S. Patil, "Scan-Based Transition Test", IEEE Trans. on Computer-Aided Design, Aug. 1993, pp. 1232-1241.

[29] S. Dasgupta, R. G. Walther, T. W. Williams and E. B. Eichelberger, "An Enhancement to LSSD and Some Applications of LSSD in Reliability, Availability and Serviceability", in Proc. 11th Fault-Tolerant Computing Symp., 1981, pp. 880-885.

978-1-4244-3039-0/08 $25.00 © 2008 IEEE

21st International Conference on VLSI Design

Design-for-Testability for Synchronous Sequential Circuits that Maintains Functional Switching Activity

Irith Pomeranz[1]
School of Electrical & Computer Eng.
Purdue University
W. Lafayette, IN 47907
pomeranz@ecn.purdue.edu

and

Sudhakar M. Reddy[2]
Electrical & Computer Eng. Dept.
University of Iowa
Iowa City, IA 52242
reddy@engineering.uiowa.edu

Abstract

Design-for-testability (DFT) approaches that allow a synchronous sequential circuit to enter states that it cannot enter during functional operation improve the fault coverage achievable for the circuit. However, non-functional operation during test application may result in switching activity that is significantly higher than under functional operation. This may lead to unnecessary yield loss due to supply voltage droops that slow the circuit but will not occur during functional operation. To address this issue we describe a DFT approach and a test generation procedure that improve the fault coverage by slowing down the state transitions of certain state variables relative to others. Unlike approaches that are based on holding values of state variables stable for unlimited numbers of clock cycles, the proposed approach resumes functional operation every limited number of clock cycles. This is shown to result in maximum switching activity that is in most cases lower than that obtained under the application of a functional test sequence, and never needs to exceed it.

1. Introduction

Test generation for synchronous sequential circuits [1]-[5] is known to have a high computational complexity. Therefore, design-for-testability (*DFT*) is used to facilitate test generation. Scan as well as non-scan *DFT* approaches typically improve the controllability of the state variables of the circuit [6]-[17]. As a result, they allow the circuit to enter states that it cannot enter during its functional operation. Non-functional operation during test application may result in switching activity that is significantly higher than under functional operation. As shown in [18], this may lead to unnecessary yield loss due to supply voltage droops that slow the circuit, causing it to produce incorrect output responses. The yield loss is unnecessary since a similar slow down will not occur during functional operation [18]. Insertion of observation points as well as *DFT* approaches that do not affect the set of states that the circuit can enter alleviate this prob-

---

1. Research supported in part by SRC Grant No. 2004-TJ-1244.

2. Research supported in part by SRC Grant No. 2004-TJ-1243.

lem. However, the improvement in fault coverage that they can provide is limited.

In this work we investigate a *DFT* approach that allows a synchronous sequential circuit to enter states that it cannot enter during functional operation, thus improving the fault coverage achievable for the circuit. However, in doing so, the proposed approach attempts to ensure that the maximum switching activity would not increase. A test generation procedure for this *DFT* approach is also described, which improves the fault coverage without increasing the maximum switching activity relative to a functional test sequence (i.e., a test sequence applied under functional operation conditions). The proposed *DFT* approach is based on slowing down the state transitions of certain state variables relative to others. For example, for a circuit with two state variables y_0 and y_1, during a specific phase of test application, y_0 may change state every clock cycle, while y_1 may change state every two clock cycles. We capture the frequencies by which the state variables change in a vector called the *frequency vector* and denoted by Φ. For the example above, $\Phi = \phi_0\phi_1 = 12$. The use of different frequencies for different state variables achieves two goals.

(1) Consider the circuit state at a time unit u. By holding certain state variables at their values from time unit u while others are allowed to change, the circuit may enter a new state at time unit $u+1$ that it cannot enter during functional operation. Thus, the ability to control the circuit state is improved, and a higher fault coverage may be obtained relative to a functional test sequence.

(2) A value stored in a state variable is not allowed to stay stable indefinitely. In the example above, every second clock cycle both state variables are allowed to change. Although this is not guaranteed to take the circuit back into a state that the circuit can reach during functional operation, as a heuristic, it is likely to prevent the circuit state from deviating significantly from the states it can reach during functional operation.

The first property above exists in other *DFT* approaches based on holding or modifying the values of state variables relative to the values obtained during functional operation [8]-[14], [16]. However, the second property is unique, and it attempts to ensure that the circuit is

978-1-4244-3039-0/08 $25.00 © 2008 IEEE

returned to functional operation soon afterwards in order to prevent the switching activity from deviating significantly from that during functional operation. We will present experimental results to demonstrate that this heuristic fulfills its goal, while unlimited holding of state variable values may result in large increases in maximum switching activity. Moreover, the test generation procedure for the proposed *DFT* approach does not have to consider the switching activity during test generation. In most cases, a test sequence generated while slowing down certain state variables as proposed here will have a maximum switching activity that is lower than that obtained during the application of a functional test sequence.

In Section 2 we introduce the proposed *DFT* approach and discuss the assumptions under which it is developed. In Section 3 we describe a procedure for selecting frequency vectors and performing test generation. Experimental results are presented in Section 4.

2. *DFT* approach

For a circuit with k state variables, we denote the present state variables by $y_0, y_1, \cdots, y_{k-1}$, and the next state variables by $Y_0, Y_1, \cdots, Y_{k-1}$. A frequency vector Φ has k entries, $\Phi = \phi_0 \phi_1 \cdots \phi_{k-1}$. The entry ϕ_i for y_i indicates the frequency by which y_i is allowed to change. With a frequency ϕ_i, y_i is allowed to change at the beginning of clock cycles ϕ_i, $2\phi_i$, $3\phi_i$, $\cdots$, where we start numbering time units from zero. For example, with $\phi_i = 1$, y_i is allowed to change every clock cycle starting from clock cycle one; with $\phi_i = 2$, y_i is allowed to change at the beginning of clock cycles $2, 4, 6, \cdots$; and so on.

To generalize the discussion, we define a *frequency sequence* S, which can be computed based on a frequency vector Φ. For an input sequence T, S contains an entry corresponding to every time unit u of T and every state variable y_i. This entry is denoted by $S(u,i)$. We have $S(u,i) = 1$ if y_i is allowed to change its value at the beginning of time unit $u+1$; otherwise $S(u,i) = 0$. When $S(u,i) = 1$, the value of the next-state variable Y_i at time unit u becomes the value of y_i at time unit $u+1$. When S corresponds to a frequency vector Φ, the relationship between Φ and S is given by the following equation for every time unit u and state variable y_i:

$$S(u,i) = \begin{cases} 1 & \text{if } u \bmod \phi_i \text{ is equal to } \phi_i - 1 \\ 0 & \text{otherwise} \end{cases}$$

We show an example for a circuit with four primary inputs and three state variables in Table 1. For every time unit u, we show the input vector under column T, the vector of S under column S, the present-state under column PS, and the next-state under column NS. The frequency sequence S corresponds to a frequency vector $\Phi = 211$.

At time unit $u = 0$, the present-state is xxx and the next-state is 0x0. Since $S(0,0) = 0$, y_0 retains its x value from time unit 0 to time unit 1. For the other state vari-

Table 1: Example input sequence

u	T	S	PS	NS
0	0010	011	xxx	0x0
1	0100	111	xx0	0x1
2	1001	011	0x1	101
3	0010	111	001	000
4	0001	011	000	010
5	1000	111	010	100
6	0001	011	100	000
7	0000	111	100	000
8	0001	011	000	010
9	0000	111	010	010

ables, the next-state values at time unit 0 become the present-state values at time unit 1. At time unit $u = 1$, the present-state is xx0 and the next-state is 0x1. Since $S(1,i) = 1$ for every i, the next-state at time unit 1 becomes the present-state at time unit 2. At time unit $u = 2$, the present-state is 0x1 and the next-state is 101. Since $S(0,0) = 0$, y_0 retains its 0 value from time unit 2 to time unit 3.

It is possible to define any frequency sequence S so as to maximize the fault coverage. In particular, it is possible to construct S during test generation to maximize the fault coverage of the input sequence T generated. In this case, S does not have to be periodic as in the case where S is based on a frequency vector Φ. Under the proposed *DFT* approach, we restrict S in the following ways in order to reduce the hardware and memory overheads associated with S, and to ensure that the circuit returns to functional operation often enough to prevent increases in the maximum switching activity.

(1) Consider a state variable y_i. If $S(u,i) = 1$ for every u, y_i is allowed to change its value every clock cycle. This is identical to functional operation, and the logic around y_i does not need any modification. However, if $S(u,i) = 0$ for one or more time units u, the structure shown in Figure 1 (or a structure similar to it) must be used to prevent the value of Y_i from being latched in the flip-flop when $S(u,i) = 0$. We attempt to minimize the number of different state variables that require this structure by attempting to assign $S(u,i) = 1$, for every u, for as many state variables y_i as follows. In this way we minimize the hardware overhead as well as the delay overhead of the proposed approach.

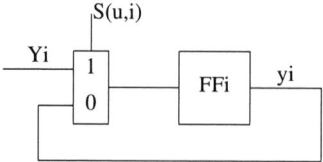

Figure 1: Implementing $\phi_i = 2$

(2) We only allow periodic frequency sequences S that are based on frequency vectors Φ as described earlier. Consequently, only Φ needs to be stored and S can be computed from Φ. For example, the sequence $S(u,i)$ for $\phi_i = 2$ is $010101 \cdots$, the sequence $S(u,i)$ for $\phi_i = 3$ is

$001001 \cdots$, and so on. S can be computed from Φ on a tester, or on-chip using counters. A single counter for every distinct frequency larger than one is sufficient since all the state variables with the same frequency have the same sequence $S(u,i)$.

(3) We limit all the frequencies in Φ to be 1 or 2. This ensures that the circuit returns to functional operation every other clock cycle. For on-chip implementation, this requires a single one-bit counter.

We allow several different frequency vectors to be used for a circuit. Each frequency vector is used for generating an input sequence to detect faults that are not detected under other vectors.

Next, we demonstrate that an arbitrary frequency sequence S may result in a higher switching activity than when an input sequence is applied under functional operation conditions. We consider a test sequence T generated by deterministic test generation. We simulate T under the all-1 frequency sequence, denoted by S_0. This corresponds to application of T under functional operation conditions. We then modify the frequency sequence in steps with the goal of obtaining frequency sequences under which T will detect additional faults. A frequency sequence S_j that improves the fault coverage over S_0 is likely to take the circuit through non-functional operation conditions in order to detect additional faults. Therefore, its switching activity may not be typical of functional operation.

We compute a frequency sequence S_j from the frequency sequence S_{j-1}, for $j = 1,2,\cdots$, as follow. Simulating T under S_{j-1}, we find the number of time units where state variable y_i assumes the value 0 (1), denoted by $n_0(i)$ [$n_1(i)$], for $0 \le i < k$. We compute the maximum difference $\max_{01} = \max\{|n_1(i)-n_0(i)|:0\le i<k\}$. For every state variable y_i, if $n_0(i) < n_1(i)$ and $n_1(i)-n_0(i) > \max_{01}/2$, we consider y_i to have a low number of 0's under S_{j-1} and we set $low(i) = 0$. If $n_1(i) < n_0(i)$ and $n_0(i)-n_1(i) > \max_{01}/2$, we consider y_i to have a low number of 1's under S_{j-1} and we set $low(i) = 1$. Otherwise, we set $low(i) = 2$. We select a time unit u and a state variable y_i such that $low(i) \ne 2$, $S_{j-1}(u,i) = 1$, y_i assumes the value $low(i)$ at time unit u, and y_i assumes the value $\overline{low(i)}$ at time unit $u+1$. We copy S_{j-1} into S_j and set $S_j(u,i) = 0$. In this way, we ensure that y_i will stay at the value $low(i)$ for one additional clock cycle, potentially taking the circuit into new states and detecting new faults.

After obtaining S_j from S_{j-1}, we find the switching activity of T under S_j. The switching activity for a pair of time units $u,u+1$ is the number of $0\to1$ or $1\to0$ transitions that occur between time units u and $u+1$ considering all the lines in the combinational logic of the circuit. We also perform fault simulation of T under S_j. Fault simulation is done to find the cumulative number of detected

faults of T under S_0 and S_j.

In Table 2 we report on S_0 as well as one of the frequency sequences that improved the cumulative number of detected faults and resulted in the highest maximum switching activity.

Table 2: Arbitrary frequency sequences

circuit	seq	hold	det	max
s344	0	0	329	172
s344	62	62	334	203
s382	0	0	364	198
s382	262	262	367	233
s1423	0	0	1414	739
s1423	91	91	1421	769
b03	0	0	334	157
b03	234	234	408	194
b04	0	0	1168	574
b04	616	616	1194	593

In Table 2, we show for every test sequence its index j, the number of 0's in the sequence S_j (i.e., the number of times the value of a state variable is held for one clock cycle), and the number of faults detected by the test sequence when it is simulated under S_0 and S_j. In the last column we show the maximum switching activity for the test sequence T under S_j.

It can be seen that when the frequency sequence S_j is allowed to be an arbitrary sequence, the maximum switching activity may increase significantly while increasing the number of detected faults. This is the motivation for limiting the frequency sequences.

3. Test generation procedure

In this section we describe a test generation procedure that selects frequency vectors to improve the fault coverage achievable for a circuit under the constraints described in Section 2. These constraints require that a frequency sequence S_j would be defined by a frequency vector Φ_j with entries out of the set $\{1,2\}$. In addition, they require to minimize the number of state variables y_i that are assigned a frequency 2 under any frequency vector in order to minimize the number of state variables for which the modification of Figure 1 is needed.

We modified a test generation procedure to accept a given frequency sequence S_j and generate a test sequence T_j for a set of target faults F under S_j. The test generation procedure we use performs only forward-time processing. In this process, S_j affects the way the present-state at time unit $u+1$ is updated given the next-state computed at time unit u. If $S(u,i) = 1$, the next-state value of Y_i at time unit u is copied to y_i at time unit $u+1$; otherwise the present-state value of y_i at time unit u is copied to y_i at time unit $u+1$. Similar rules can be derived for a procedure that uses backward-time processing.

Let the set of target faults for the circuit be F. Let Φ_0 be the all-1 frequency vector. In Phase 0, we perform test generation under Φ_0 for the faults in F. The resulting test sequence is denoted by T_0. Faults detected by T_0

under Φ_0 are dropped from F. We also compute the maximum switching activity under T_0 and Φ_0. We denote it by max_switch_0.

In Phases 1,2, $\cdots$, we generate test sequences T_j based on frequency vectors Φ_j that have at least one frequency equal to 2. The test generation process we use to generate a test sequence T_j for a frequency vector Φ_j does not include any heuristics for limiting the switching activity of the generated test sequence. Consequently, it is possible that a frequency vector we consider will lead to a higher maximum switching activity than the maximum possible during functional operation. We use the maximum switching activity max_switch_0 of T_0 under Φ_0 as an upper bound on the maximum switching activity allowed for any test sequence T_j and frequency vector Φ_j. If a test sequence T_j generated for a vector Φ_j results in a higher maximum switching activity than T_0 under Φ_0, we do not accept T_j and Φ_j. Note that we discard the complete test sequence T_j after it is generated. Since this happens rarely, we can afford to perform the check as a postprocessing step. In most cases, the selection of Φ_j will result in a maximum switching activity that is lower than that of T_0 under Φ_0.

In Phase 1, we consider all the frequency vectors where a single state variable is assigned a frequency equal to 2. For example, for a circuit with four state variables, we consider the frequency vectors $\Phi_1 = 2111$, $\Phi_2 = 1211$, $\Phi_3 = 1121$ and $\Phi_4 = 1112$. For every frequency vector Φ_j, we perform test generation under Φ_j for the faults left in F to obtain a test sequence T_j. We then compute the maximum switching activity max_switch_j of T_j and Φ_j. If $max\_switch_j \le max\_switch_0$, we accept T_j and Φ_j, and we remove their detected faults from F. For a circuit with NSV state variables, we consider NSV frequency vectors in Phase 1.

In Phase p, where $2 \le p \le P_{MAX}$ for a constant P_{MAX}, we consider frequency vectors where exactly p state variables are assigned frequencies equal to 2. The state variables are selected randomly. For a circuit with NSV state variables, we consider NSV frequency vectors in Phase p. For every frequency vector Φ_j we perform test generation under Φ_j for the faults left in F. We then compute the maximum switching activity max_switch_j of the generated test sequence T_j and Φ_j. If $max\_switch_j \le max\_switch_0$, we accept T_j and Φ_j, and we remove their detected faults from F.

During Phases 0, 1, $\cdots$, P_{MAX}, if a frequency vector Φ_j results in a test sequence T_j that improves the fault coverage, the state variables with frequencies equal to 2 under Φ_j are added to a set $FREQ2$. At the end of Phase P_{MAX}, $FREQ2$ contains the state variables that require the modification shown in Figure 1. Since the state variables in $FREQ2$ already require the modification, we attempt to use them for the generation of additional test sequences in

the next phase, Phase P_{MAX}+1. In a frequency vector considered in this phase, n_2 state variables are assigned a frequency equal to 2. The value of n_2 is selected randomly from the range 2 to $|FREQ2|$. A set V of n_2 state variables is then selected randomly out of $FREQ2$. A frequency vector is defined by assigning frequencies equal to 2 for the n_2 state variables in V, and frequencies equal to 1 for the remaining $k-n_2$ state variables. This is repeated NSV times. For every frequency vector Φ_j defined in this way we perform test generation under Φ_j for the faults left in F. We then compute the maximum switching activity max_switch_j of the generated test sequence T_j and Φ_j. If $max\_switch_j \le max\_switch_0$, we accept T_j and Φ_j, and we remove their detected faults from F.

4. Experimental results

We applied the procedure described in Section 3 to ISCAS-89 and ITC-99 benchmark circuits. We used $P_{MAX} = 5$. Every test sequence T_j is generated starting from a completely-unspecified initial state.

We set strict limits on the computational effort expended during test generation. This is done in order to demonstrate that test generation under the proposed DFT approach does not require a complete sequential test generation procedure. To search for a test subsequence for a target fault f, the test generation procedure we use constructs a tree of fault-free/faulty states reachable from the final states obtained under the current test sequence (the final states are completely-unspecified if the current test sequence is empty). The height of the tree (the length of a test subsequence for f) is limited to 64, and the width of the tree (the number of state pairs at any level of the tree) is limited to 32. The number of input vectors considered for every state pair in the tree in order to construct the next level is at most 128. If a test subsequence for f is not found under these constraints, test generation for f is aborted. Otherwise, the test subsequence is concatenated to the current test sequence.

The limits above are not always sufficient for finding a test sequence that detects every detectable fault of the circuit without DFT. To better demonstrate the effects of the proposed DFT approach on the detection of faults that are undetectable under functional operation conditions without increasing the maximum switching activity, we add another phase following Phase 0. In this phase, we simulate a deterministic test sequence T_D under the all-1 frequency vector Φ_0. We remove detected faults from F. We also update max_switch_0 if the maximum switching activity under T_D is higher than under T_0.

The results are shown in Tables 3 and 4. We show several rows for every circuit corresponding to the different phases. The phase is shown under column p, where $p = D$ corresponds to the deterministic test sequence T_D applied under the all-1 frequency vector, and

$p = p_0$ corresponds to phase p_0, for $0 \le p_0 \le 6$. We omit the results for a phase $p_0 \ge 1$ if the fault coverage is not improved in this phase.

Table 3: Results for ISCAS-89 benchmarks

circuit	p	seq tot	eff	len	detected tot	f.c.	max switch	freq2	time
s298	0	0	0	161	263	85.39	160	0/14	1.00
s298	D	0	0	117	265	86.04	183	0/14	0.01
s298	1	14	1	41	282	91.56	133	1/14	3.83
s298	2	28	2	54	283	91.88	175	3/14	7.59
s298	4	56	4	101	287	93.18	175	7/14	15.63
s298	6	84	5	139	296	96.10	175	7/14	22.88
s344	0	0	0	221	329	96.20	202	0/15	1.00
s344	D	0	0	57	329	96.20	172	0/15	0.01
s344	1	15	2	67	336	98.25	167	2/15	9.73
s382	0	0	0	474	357	89.47	195	0/21	1.00
s382	D	0	0	516	364	91.23	198	0/21	0.01
s382	1	21	2	69	368	92.23	110	2/21	6.53
s382	3	63	3	208	374	93.73	141	4/21	17.15
s382	4	84	4	251	375	93.98	141	7/21	21.13
s400	0	0	0	522	372	88.36	213	0/21	1.00
s400	D	0	0	611	380	90.26	271	0/21	0.01
s400	1	21	2	69	384	91.21	111	2/21	6.69
s400	3	63	3	208	390	92.64	147	4/21	17.76
s400	4	84	4	251	391	92.87	147	7/21	22.01
s420	0	0	0	115	179	41.63	119	0/16	1.00
s420	D	0	0	110	179	41.63	116	0/16	0.03
s420	3	48	2	40	181	42.09	106	5/16	19.11
s420	4	64	3	60	184	42.79	106	8/16	25.33
s526	0	0	0	937	452	81.44	290	0/21	1.00
s526	D	0	0	1006	454	81.80	294	0/21	0.00
s526	1	21	2	302	471	84.86	244	2/21	3.22
s526	2	42	4	605	494	89.01	244	5/21	5.64
s526	4	84	5	752	500	90.09	244	8/21	9.10
s526	5	105	6	823	507	91.35	244	11/21	10.53
s526	6	126	7	856	508	91.53	244	11/21	11.37
s641	0	0	0	218	403	86.30	381	0/19	1.00
s641	D	0	0	101	404	86.51	311	0/19	0.00
s1196	0	0	0	1934	1236	99.52	642	0/18	1.00
s1196	D	0	0	238	1239	99.76	634	0/18	0.00
s1196	1	18	2	32	1242	100.00	557	2/18	0.12
s1423	0	0	0	1468	1311	86.53	709	0/74	1.00
s1423	D	0	0	1024	1414	93.33	739	0/74	0.00
s1423	1	74	4	185	1422	93.86	651	4/74	3.37
s1423	4	296	5	241	1423	93.93	651	7/74	11.95
s1423	5	370	7	339	1426	94.13	651	13/74	14.72
s5378	0	0	0	866	3002	65.22	1782	0/179	1.00
s5378	D	0	0	646	3639	79.06	2426	0/179	0.01
s5378	1	179	2	76	3645	79.19	1426	3/179	13.65
s5378	2	358	4	116	3646	79.21	1426	5/179	27.22
s5378	3	537	4	116	3646	79.21	1426	5/179	40.64
s5378	4	716	5	164	3647	79.23	1426	9/179	54.12
s5378	5	895	6	228	3657	79.45	1427	13/179	67.50
s5378	6	1074	7	269	3659	79.49	1427	13/179	80.74

Table 4: Results for ITC-99 benchmarks

circuit	p	seq tot	eff	len	detected tot	f.c.	max switch	freq2	time
b03	0	0	0	154	312	69.03	118	0/30	1.00
b03	D	0	0	130	334	73.89	157	0/30	0.01
b03	1	30	5	217	383	84.73	148	5/30	14.01
b03	2	60	8	279	391	86.50	148	8/30	23.53
b03	3	90	12	365	407	90.04	148	14/30	32.77
b03	5	150	15	464	415	91.81	152	20/30	47.57
b04	0	0	0	362	1168	86.78	624	0/66	1.00
b04	D	0	0	168	1168	86.78	574	0/66	0.00
b04	1	66	37	880	1293	96.06	616	37/66	12.87
b04	2	132	53	1242	1321	98.14	619	47/66	15.44
b04	3	198	62	1439	1332	98.96	619	53/66	16.61
b04	4	264	64	1510	1334	99.11	619	54/66	17.43
b04	5	330	65	1526	1335	99.18	619	55/66	18.16
b05	0	0	0	223	754	41.52	496	0/34	1.00
b05	D	0	0	223	1075	59.20	466	0/34	0.06
b05	1	34	11	299	1110	61.12	432	11/34	7.23
b05	2	68	15	551	1351	74.39	481	15/34	15.32
b05	3	102	16	643	1367	75.28	487	17/34	20.17
b05	4	136	18	770	1451	79.90	487	21/34	25.45
b05	5	170	26	990	1529	84.20	487	29/34	30.48
b05	6	204	29	1074	1556	85.68	487	29/34	34.18
b07	0	0	0	621	834	70.50	294	0/51	1.00
b07	D	0	0	380	837	70.75	282	0/51	0.05
b07	1	51	21	1786	955	80.73	292	21/51	7.10
b07	2	102	31	2344	972	82.16	292	31/51	13.07
b07	3	153	35	2696	990	83.69	292	33/51	18.79
b07	4	204	37	2790	992	83.85	292	34/51	24.78
b07	5	255	45	3266	1005	84.95	292	44/51	31.12
b07	6	306	48	3395	1008	85.21	292	44/51	38.30
b08	0	0	0	473	462	94.48	231	0/21	1.00
b08	D	0	0	415	463	94.68	219	0/21	0.00
b08	1	21	3	230	478	97.75	201	3/21	4.91
b08	2	42	5	278	482	98.57	201	6/21	7.25
b08	3	63	6	368	484	98.98	201	8/21	9.19
b08	4	84	7	390	485	99.18	201	11/21	10.30
b09	0	0	0	511	306	72.86	143	0/28	1.00
b09	D	0	0	269	339	80.71	234	0/28	0.01
b09	1	28	17	660	380	90.48	152	17/28	4.25
b09	2	56	20	754	383	91.19	152	19/28	6.08
b09	3	84	22	862	389	92.62	193	21/28	7.74
b09	4	112	23	877	390	92.86	193	23/28	8.49
b09	5	140	25	972	396	94.29	193	24/28	9.52
b09	6	168	26	1011	397	94.52	193	24/28	9.90
b10	0	0	0	311	467	91.21	231	0/17	1.00
b10	D	0	0	190	467	91.21	233	0/17	0.00
b10	1	17	10	253	493	96.29	216	10/17	5.95
b10	2	34	11	271	495	96.68	216	11/17	8.54
b10	3	51	12	287	496	96.88	216	12/17	11.24
b10	5	85	14	366	500	97.66	216	14/17	16.31
b10	6	102	15	377	501	97.85	216	14/17	17.46
b11	0	0	0	554	933	85.67	408	0/30	1.00
b11	D	0	0	675	997	91.55	416	0/30	0.00
b11	1	30	10	721	1057	97.06	414	10/30	3.73
b11	4	120	12	826	1059	97.25	414	14/30	9.29
b11	6	180	13	840	1060	97.34	414	14/30	12.12

For Phase p_0, where $1 \le p_0 \le 6$, under column *seq* we show the total number of test sequences generated in Phases 1 to p_0, and the number of test sequences that were effective in detecting new faults. For Phases 0 and D these numbers are zero.

Under column *len*, for $p = 0$ and $p = D$ we show the length of the test sequence T_0 and T_D, respectively. For $p = p_0$, where $1 \le p_0 \le 6$, we show the total length of all the test sequences generated in Phases 1 to p_0 that were effective in detecting new faults.

Under column *detected*, for $p = p_0$, we show the total number of detected faults in all the phases up to Phase p_0, followed by the corresponding fault coverage.

Under column *max switch*, for $p = 0$, we show the maximum switching activity of T_0 under Φ_0. For $p = D$

we show the maximum switching activity considering both T_0 and T_D under Φ_0. For $p = p_0$, where $1 \le p_0 \le 6$, we show the maximum switching activity considering all the test sequences generated in Phases 1 to p_0 that were effective in detecting new faults.

Under column $freq2$, for $p = p_0$, we show the total number of state variables with frequency 2 under any frequency vector that was useful in increasing the fault coverage up to Phase p_0 (the size of the set $FREQ2$ at the end of Phase p_0). This is followed by the number of state variables in the circuit to the right of the slash.

Under column *time*, for $p = 0$, we show the run time in Phase 0. For $p = D$, we show the run time for simulating T_D under Φ_0. For $p = p_0$, where $1 \le p_0 \le 6$, we show the cumulative run time for Phases 1 to p_0. All the run times are normalized to the run time of Phase 0.

For example, for $s298$, T_0 has a length of 161, it detects 263 faults under Φ_0, and it has a maximum switching activity of 160. The deterministic test sequence T_D increases the number of detected faults to 265, and it has a maximum switching activity of 183. In Phase 1, 14 test sequences are generated. Only one of them is effective in improving the fault coverage. The length of the sequence is 41, and it increases the number of detected faults to 282. Its maximum switching activity is 133. The set $FREQ\,2$ includes a single state variable. In Phase 2, 14 additional test sequences are generated for a total of 28 test sequence. One additional test sequence is effective in detecting new faults, for a total of two effective test sequences. The length of these sequences is 54. The number of detected faults increases to 283, and the maximum switching activity is 175. A total of three state variables are included in $FREQ\,2$.

From Tables 3 and 4 it can be seen that the limits set on the test generation process are low such that they do not always allow all the detectable faults to be detected in the circuit without DFT. Nevertheless, faults that cannot be detected in the circuit without DFT are detected by the same reduced-complexity test generation procedure after DFT insertion.

For all but one of the circuits considered, the frequency vectors Φ_j, for $j > 0$, result in test sequences that detect additional faults. These faults are detected while keeping the maximum switching activity below its level during the application of T_0 or T_D under Φ_0.

For all the circuits considered, the total number of state variables that need to be modified as shown in Figure 1 is smaller than the number of state variables. A static test compaction process can be used to further reduce this number. To eliminate the need to modify y_i, it is necessary to remove every test sequence T_j such that $\Phi_j(i) = 2$. If this is possible without reducing the fault coverage (i.e., the faults detected by these test sequences are also detected by other test sequences), then the need to modify y_i can be eliminated.

5. Concluding remarks

We described a DFT approach and a test generation procedure that improve the fault coverage achievable for a circuit by allowing the circuit to enter states that it cannot enter during functional operation, while ensuring that the maximum switching activity would not increase. The DFT approach was based on slowing down the state transitions of certain state variables relative to others. Specifically, in every phase of test application, there is a subset of state variables that are allowed to change every clock cycle, and a subset of state variables that are allowed to change every second clock cycle. This was

represented by frequency vectors with entries out of $\{1,2\}$. Test generation was done for frequency vectors with increasing numbers of 2's to identify state variables that are most effective in increasing the fault coverage when they are slowed down. Since only frequencies of 1 and 2 were allowed, functional operation was resumed every two clock cycles. This resulted in maximum switching activity that was in most cases lower than that obtained under the application of a functional test sequence, and never needed to exceed it.

References

[1] R. Marlett, "An Effective Test Generation System for Sequential Circuits", in Proc. Design Autom. Conf., 1986, pp. 250-256.

[2] W.-T. Cheng and T. J. Chakraborty, "Gentest: An Automatic Test Generation System for Sequential Circuits", IEEE Computer, April 1989, pp. 43-49.

[3] T. P. Kelsey and K. K. Saluja, "Fast Test Generation for Sequential Circuits", in Proc. Intl. Conf. on Computer-Aided Design, 1989, pp. 354-357.

[4] T. Niermann and J. H. Patel, "HITEC: A Test Generation Package for Sequential Circuits", in Proc. European Design Autom. Conf., 1991, pp. 214-218.

[5] X. Lin, I. Pomeranz and S. M. Reddy, "Techniques for Improving the Efficiency of Sequential Circuit Test Generation", in Proc. Intl. Conf. on Computer-Aided Design, 1999, pp. 147-151.

[6] M. J. Y. Williams and J. B. Angell, "Enhancing Testability of Large Scale Integrated Circuits via Test Points and Additional Logic", IEEE Trans. on Computers, 1973, pp. 46-60.

[7] E. B. Eichelberger and T. W. Williams, "A Logic Design Structure for LSI Testability", in Proc. 14th Design Autom. Conf., June 1977, pp. 462-468.

[8] M. Abramovici, P. S. Parikh, B. Mathew and D. G. Saab, "On Selecting Flip-Flops for Partial Reset", in Proc. Intl. Test Conf., 1993, pp. 1008-1012.

[9] V. Chickermane, E. M. Rudnick, P. Banerjee and J. H. Patel, "Non-Scan Design-for-Testability Techniques for Sequential Circuits", in Proc. 30th Design Autom. Conf., June 1993, pp. 236-241.

[10] I. Pomeranz and S. M. Reddy, "On the Role of Hardware Reset in Synchronous Sequential Circuit Test Generation", IEEE Trans. on Computers, Sept. 1994, pp. 1100-1105.

[11] P. Parikh and M. Abramovici, "On Combining Design for Testability Techniques", in Proc. 1995 Intl. Test Conf., Oct. 1995, pp. 423-429.

[12] D. K. Das and B. B. Bhattacharya, "Testable Design of Non-Scan Sequential Circuits using Extra Logic", in Proc. Asian Test Symp., Nov. 1995, pp. 176-182.

[13] I. Pomeranz and S. M. Reddy, "On the Use of Fully Specified Initial States for Testing of Synchronous Sequential Circuits", IEEE Trans. on Computers, Feb. 2000, pp. 175-182.

[14] X. Dong, X. Yi and H. Fujiwara, "Non-Scan Design for Testability for Synchronous Sequential Circuits Based on Conflict Analysis", in Proc. Intl. Test Conf., 2000, pp. 520-529.

[15] H. Fujiwara, "A New Class of Sequential Circuits with Combinational Test Generation Complexity", IEEE Trans. on Computers, Sept. 2000, pp. 895-905.

[16] M. Abramovici, X. Yu and E. M. Rudnick, "Low-Cost Sequential ATPG with Clock-Control DFT", in Proc. 39th Design Autom. Conf., June 2002, pp. 243-248.

[17] Y. Xiaoming and M. Abramovici, "Sequential Circuit ATPG using Combinational Algorithms", in IEEE Trans. on Computer-Aided Design, Aug. 2005, pp. 1294-1310.

[18] J. Saxena, K. M. Butler, V. B. Jayaram, S. Kundu, N. V. Arvind, P. Sreeprakash and M. Hachinger, "A Case Study of IR-Drop in Structured At-Speed Testing", in Proc. Intl. Test Conf., 2003, pp. 1098-1104.

21st International Conference on VLSI Design

A Partitioning based Physical Scan Chain Allocation Algorithm that minimizes Voltage Domain Crossings

Nilabha Dev[1], Sandeep Bhatia[2], Subhasish Mukherjee[1], Sue Genova[2], Vinayak Kadam[1]

[1]*Cadence Design Systems, Noida, India*
[2]*Cadence Design Systems, San Jose, USA*

Abstract

In this paper we present an algorithm for allocating scan flops to scan chains based on the placement information of flops. The objective of the algorithm is to reduce the scan wire length, the number of level shifters and the number of lockup latches in the scan path. The algorithm uses a novel partitioning based approach to allocate and order the scan flops for a particular scan chain. The scan flops are allocated to a number of partitions. These partitions are then ordered for the scan chains based on the physical location of the scan-in and scan-out pins of the scan chain and the aspect ratio of the layout. To the best of our knowledge this is the first algorithm that explicitly attempts to reduce the number of level shifters in the scan data path. Experimental results obtained for some industrial circuits show that the number of level shifters in the scan data path is halved for some industrial test cases over approaches that are not multiple supply voltage aware. Scan wire lengths too are reduced by up to 45% over previous approaches for the above designs. Additionally we obtain upto 90% scan wire length reductions over previous approaches [4],[11] for some ISCAS-89 benchmark circuits.

1. Introduction

Most modern VLSI designs incorporate some form of Design-For-Testability (DFT) logic in them. This is done to aid in the process of manufacturing test. The most commonly used DFT technique is scan testing [6]. In this technique flops are replaced by their scan equivalents and then connected into scan chains. During manufacturing tests, test vectors are applied to these scan chains to uncover manufacturing defects.

The scan replacement and scan connection processes happen during the logic synthesis phase of the design life cycle. The order in which flops are configured into the scan chain is independent of the physical location of the flops. Often times it is based on the alphabetical ordering of the flop names. This leads to routing congestion because of long wire lengths and also longer test application times [12].

Power consumption is one of the big concerns of modern chip design today. A number of techniques have been proposed in the literature for reducing power consumption. Among them is the technique of using multiple supply voltages for different parts of a chip. Since signals may pass from one voltage domain to another, special cells called level shifters are required that transform the voltage level of signals as they cross voltage domain boundaries. If a scan chain passes from one voltage domain to another then level shifters will be required in the scan data path also.

In this paper we present an algorithm to solve the problem of finding a configuration of scan flops to scan chains so as to minimize the scan wire length. A simultaneous goal is the reduction in the number of level shifters and lockup latches in the scan data path. Our algorithm is different from first allocating flops to chains and then reordering them based on their physical location as an afterthought.

The structure of the paper is as follows. In Section 2 we present an overview of the existing work. Section 3 explains our algorithm in detail. The results are given in Section 4. Finally Section 5 summarizes the paper and gives directions for future work.

2. Previous work

Most existing work deal with the problem of reordering flops allocated to a scan chain based on placement information. They propose Traveling Salesman Problem (TSP) based algorithms to the problem [2], [8]. In these methods the scan chain wire length reduction problem is modeled as an instance of the TSP where the flops are the cities, the Manhattan distance between the flops is the length of the path between the cities and the objective is to find a minimum cost tour of the cities. This minimum cost tour corresponds to the minimum total scan wire length. This approach is suboptimal because the initial allocation of the flops to the scan chains was not done based on their physical locations.

978-1-4244-3039-0/08 $25.00 © 2008 IEEE

Berthelot et al. in [1] propose a two step process. In the first step a clustering algorithm is applied in which the flops are grouped in a top-down fashion and a TSP algorithm is applied recursively to each of the groups. The ordered flops are then grouped into a single chain. In the second step the chain obtained in the previous step is divided into sub chains of size $p + 1$ where p is a user specified partition size and the TSP algorithm is reapplied to this problem of size $p + 1$. The drawback of their solution is the fact that to allocate the flops to different scan chains they need to split the one single scan chain that they have formed and this may break scan chains arbitrarily resulting in some long wires.

Guiller et al. [7] propose a scan chain allocation strategy that groups flops by their clock domains and then based on their physical location allocates them to different chains. However their algorithm takes one flop at a time and depending on the current configuration of the chain (i.e. the flops that have already been allocated to a chain) allocates the current flop to the chain. This approach is greedy because it uses a local improvement criterion at each step of the allocation procedure. It is likely that the configurations produced by this algorithm are suboptimal.

Rahimi and Soma in [12] present a stable marriage based algorithm to the problem of allocating scan flops to multiple scan chains. An ideal scan chain is drawn between the scanin and scanout of the chain. The flops are allocated to these ideal chains based on the mutual preference of the flops for the chains and vice versa using a graph matching algorithm. The optimal ordering of flops in the scan chain is obtained by simulated annealing. However their approach may lead to some scan chains with long wire lengths. It may so happen that most flops are in a certain region of the chip but there are an insufficient number of ideal scan chains that pass through this region. Also the runtime of their algorithm is $O(n^2)$ where n is the number of flops. This is expensive if n is very large.

The problem of selecting partial scan flops based on physical location so as to reduce the scan wirelength has been studied in [5]. In this the problem is formulated as an instance of a graph matching problem because of which the runtime is expected to be very large.

The authors of [3] propose a scheme that uses scan cell reordering to reduce scan power. This reordering is done based on the test vectors. [4] extends this scheme to reorder the flops based on placement information. The usefulness of the above two techniques are limited because there exists numerous other techniques to reduce scan power without having to reorder the scan cells and hence the structure of the design after the test vectors have been generated.

Genetic algorithm based approaches [11] have also been proposed for the scan reordering problem. However as with all genetic algorithms the runtime of the procedure is expected to be very large.

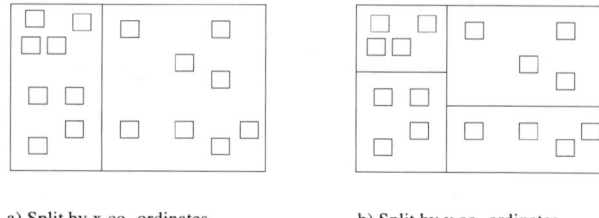

a) Split by x co−ordinates   b) Split by y co−ordinates

Figure 1. An example with 16 flops and 4 partitions

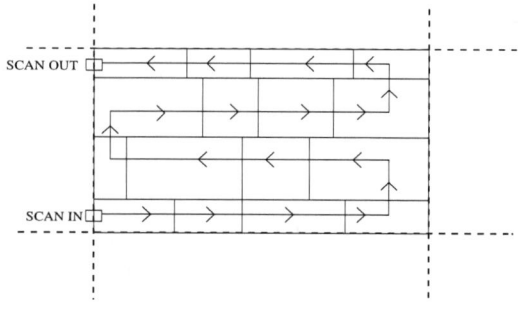

Figure 2. Order of gathering the partitions for given location of scan in and scan out and aspect ratio less than 1

3. Partitioning based scan chain allocation algorithm

In this section we describe the procedure for doing scan chain connection based on placement information. The inputs to the procedure are the physical location of the flops, the number of scan chains and the physical location of the scan in and scan out of the chains.

In the first step of the algorithm the flops are first segregated based on test clock domain information. For our purpose a test clock domain is a set of test clocks that are driven by the same root source pin from the boundary of the chip or an on-chip PLL. Each scan chain is allocated flops only from compatible test clock domains. We then further refine these groupings by their voltage domains. This is done to reduce the number of voltage domain crossings and hence the number of level shifters in the scan data path. Within these subgroups we group the flops by their test clock and test clock edges to minimize the number of lockup elements.

The heart of our algorithm is a partitioning process that at this stage partitions the flops gathered in the above step and groups them based on physical proximity.

Let n be the total number of flops that we take up for partitioning. Assuming a rectangular layout, the flops are partitioned into a grid of size $p \times q$. p and q are determined from the aspect ratio of the layout. Let k be the partition size i.e. the number of flops in a partition. This is a user

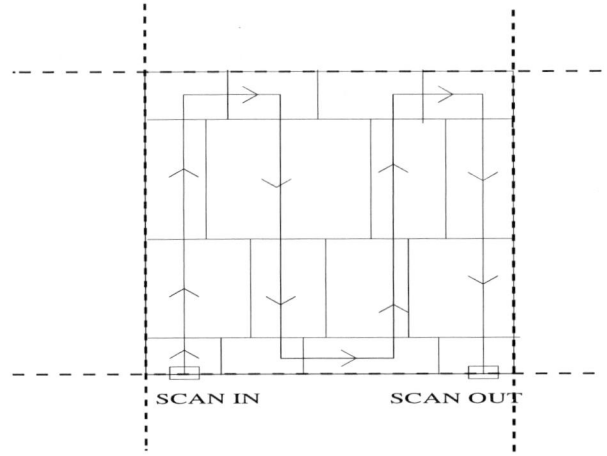

SCAN IN   SCAN OUT

Figure 3. Order of gathering the partitions for given location of scan in and scan out and aspect ratio greater than 1

tunable value. Let the aspect ratio of the design be A. Then p is assigned to be $\sqrt{n/(k \times A)}$ and q is assigned to be $\sqrt{(n \times A)/k}$. The aspect ratio A of the layout acts as a scaling factor for p and q.

The partitioning of the flops into $p \times q$ groups is achieved by first sorting the flops based on their $x - coordinate$ locations and then dividing the flops equally into p groups. Next, the flops in each of the p groups are partitioned into q groups of equal size by sorting on their $y-coordinates$. An example with 16 flops and 4 partitions is shown in Figure 1. In part(a) the layout has been divided into 2 partitions with 8 flops each. In part(b) the flops have divided into 4 partitions with 4 flops each.

The resultant set of $p \times q$ partitions may have different layout area, but each has the same number of flops, i.e; $n/(p \times q)$. In this way we have $p \times q$ partitions of (maybe) unequal layout area but with the same number of flops in each partition. Given $p \times q$ partitions we now gather the partitions necessary for a scan chain. This is determined by the physical location of the scanin and scanout of the chain. We find the partition in which the scanin and scanout are physically located. This can be done by simply comparing the last flop in a partition with x (and y) locations of the scanin(or scanout). These are our starting and ending partitions. Next based on the aspect ratio of the layout and the number of flops that are required for a scan chain, we gather the partitions contiguous to these two starting partitions that give us the required number of flops for the scan chain. An example of a scan chain that spans 16 partitions is shown in figure 2. The diagrams show how the partitions are gathered for a scan chain depending on the aspect ratio of the layout and the location of the scan in and scan out of the chain. The overall algorithm is presented in **CONFIGURE_FLOPS_TO_CHAINS**. The partitioning phase of the algorithm is presented in algorithm **PARTITION_FLOPS**.

CONFIGURE_FLOPS_TO_CHAINS

/* Configure scan flops to chains to minimize wire length and reduce the number of lockup latches and level shifters in the scan data path */

Input : n flops with their clock domains clk_i , supply voltages v_k and physical location $< x, y >$ specified. The location of the scanin and scanout of the chains are also specified. It is assumed that the numbers of scan chains necessary to accommodate all the flops have been allocated. The clock domain to which a scan chain is assigned is also known.

Output : Scan chains with the scan flops arranged so that the number of lockup latches and level shifters in the scan path are minimized and the scan wire length is minimal.

begin

(1) Find the clock domain to which the scan chain is assigned.

(2) Find all the candidate flops for this scan chain. The candidate flops for a scan chain are those that belong to the same clock domain as the scan chain.

(3) Sort the set of candidate flops by voltage domain. This step is performed to reduce the number of level shifters in the scan chain.

(4) Sort the sets of candidate flops found in the earlier step by test clocks. This is done to reduce the number of lockup latches in the scan chain

(5) For each of the subsets found in the earlier step do a partitioning step based on physical location using procedure **PARTITION_FLOPS**. Also depending on the number of flops that are required for the chain only return as many partitions as are necessary in order.

(6) For each of the partitions returned in the previous step the local order of the flops in the scan chain can be improved by running a Traveling Salesman Problem algorithm. For the problem being solved in this section we use a nearest neighbor heuristic to determine the order. The order is improved by using a 2-opt heuristic [9] to remove long edges in the tour. Hold time violations in the scan path are taken care of by specifying a minimum spacing criterion for the TSP routine. The TSP routine then orders the flops so that no two flops which are less than this minimum spacing apart are adjacent to each other in a scan chain.

end

PARTITION_FLOPS

/* Partition flops based on physical location */

Input : A set of flops with their physical co-ordinates given. Also a scan chain with the physical location of its scanin and scanout. Also the number of partitions required given as $p \times q$.

Output : A partition of the flops into $p \times q$ groups. Also an ordering of the partitions based on the location of the scanin and scanout of the chain.

begin

(1) Sort the flops by their $x - coordinates$ and divide them into m groups.

(2) Sort each of the m groups above by their $y - coordinates$ and divide them into n groups.

(3) Find the starting partition for the scanin of the chain, call this $start_x$ and $start_y$. Find the ending partition for the scanout of the chain, call this end_x and end_y

(4) **If** the aspect ratio be less than 1

    (i) **If** $|start_x - end_x| \leq |start_y - end_y|$ then gather flops as in Figure 2. using procedure **GATHER_FLOPS_ROW_WISE**

    (ii) **else if** $|start_x - end_x| > |start_y - end_y|$ then gather flops as in Figure 3. using procedure **GATHER_FLOPS_COLUMN_WISE**

  else if aspect ratio be greater than 1

    (i) **If** $|start_x - end_x| \leq |start_y - end_y|$ then gather flops as in Figure 3. using procedure **GATHER_FLOPS_COLUMN_WISE**

    (ii) **else if** $|start_x - end_x| > |start_y - end_y|$ then gather flops as in Figure 2. using procedure **GATHER_FLOPS_ROW_WISE**

end

---

procedure GATHER_FLOPS_ROW_WISE /* gather the partitions row wise from $start_y$ to end_y */

begin

  (i) in each even row gather the partitions from $start_x$ to end_x

  (ii) in each odd row gather the partitions from end_x downto $start_x$

end

Procedure **GATHER_FLOPS_COLUMN_WISE** is similar to **GATHER_FLOPS_ROW_WISE** where the partitions are gathered column wise from $start_x$ to end_x.

Steps 1 and 2 of algorithm **PARTITION_FLOPS** can also be implemented using a *k-d tree* data structure in case $p = q$. However in our approach we allow the creation of $p \times q$ partitions where $p! = q$. Hence we did not use *k-d tree* based partitioning. Another method of grouping flops based on physical location could be by using a clustering algorithm like *k-means* where the flops would be grouped into k clusters or partitions based on physical location. However we did not consider *k-means* as its best case running time is $\Theta(n^2)$ and the worst case order is $O(2^{\Theta(\sqrt{n})})$. We chose a simple sorting based scheme based on guaranteed worst case runtimes. The results presented in Section 4 prove that we are able to achieve upto 50% wire length decrease using our approach. Also the method we use for gathering the partitions for our scan chain would not work with the irregular clusters that a *k-means* kind of algorithm would produce.

The method of gathering the partitions for a scan chain as explained in Step 4 is dependent on both the aspect ratio of the design and the location of the scanin and scanout of the chain. If the $x - coordinates$ of the scanin and scanout are closer than their $y - coordinates$ then more number of adjacent flops would be gathered if the partitions are gathered row wise rather than column wise if the aspect ratio is less than one. A similar argument holds if the $y - coordinates$ are closer than the $x - coordinates$ of the scanin and scanout and the aspect ratio is greater than one.

The running time of the above algorithm is dominated by the sorting steps in the overall algorithm and during the partitioning phase of the algorithm. The initial sorting steps to divide the flops by clock domain and voltage domain take $O(nlogn)$ time where n is the total number of flops under consideration. The partitioning phase also takes $O(nlogn)$ time mainly because of the sorting step where we sort the flops first by their $x - coordinates$ and then by their $y - coordinates$. Gathering the partitions based on the location of the scanin's and scanout's of the chains is linear in the number of partitions required. In the worst case this could be $p \times q \times s$ where $p \times q$ is the number of partitions and s is the number of scan chains. However p,q and s are all much smaller in comparison to n and do not contribute much to the runtime. The runtime for the TSP based ordering for each of the partitions is of $O(r)$ if each partition were to have r elements. This is because we use a nearest neighbour heuristic for the TSP algorithm. In the worst case all of the $p \times q$ partitions have to be ordered using TSP and the runtime for the TSP routines is then $O(n)$. Therefore the complexity of our scan chain configuration algorithm is $O(nlogn)$.

4. Results

In this section we describe our results and experimental setup. We present the improvements found in scan wire-length and the reduction in the number of level shifters and

Table 1. Characteristics of the designs

Features	A	B	C	D
No of voltage domains	1	1	5	3
No of clock domains	1	36	1	6
No of flops	12000	80000	8000	32000
No of scan chains	8	8	1	6

Table 2. Comparision of our results with the conventional approach on four industrial circuits

Design	Conv, Alloc	Scan Wire length(μm)	Runtime (sec)	#level shifter	#lockup
A	conv	6E+5	4	0	0
	alloc	3.65E+5	28	0	0
B	conv	1.14E+6	38	0	35
	alloc	1.08E+6	74	0	35
C	conv	7.5E+4	1.1	4	0
	alloc	7.1E+4	6.16	2	0
D	conv	4.24E+5	6	10	5
	alloc	3.57E+5	12	7	6

Table 3. Comparison of scan wirelength, in microns between our algorithm (alloc) and ([4]) (with 64 clusters) and ([11]) on selected ISCAS benchmark circuits with 1 and 4 scan chains

Design	#chains	[4]	[11]	(alloc)
s5378	1	1.8E+4	4.3E+4	3.28E+3
	4	1.8E+4	4.3E+4	3.52E+3
s9234	1	3.5E+4	4.5E+4	2.63E+3
	4	3.5E+4	4.5E+4	2.88E+3
s15850	1	1E+5	3.4E+5	8.52E+3
	4	1E+5	3.3E+5	8.87E+3
s13207	1	1.1E+5	3.2E+5	9.65E+3
	4	1.1E+5	3.8E+5	1.02E+4
s38584	1	-	1.1E+6	2.44E+4
	4	-	9.1E+5	2.51E+4
s35932	1	4.2E+5	-	3.06E+4
	4	4.2E+5	-	3.11E+4

lockup latches in the scan data path.

We compare the results of our algorithm against those produced by an algorithm that first arbitrarily allocates flops to scan chains and then re-orders the flops on a per chain basis using a TSP solver. We call this latter approach the conventional (*conv*) approach in the rest of this section. We call our algorithm the allocation based (*alloc*) approach for the rest of this section.

We ran our algorithm on four industrial designs. These designs have multiple clock domains and multiple supply voltage domains. In Table 1 we present the characteristics of these designs. In Table 2 we present the comparative results between the conventional versus our partitioning based algorithm that minimizes the number of voltage domain crossings. We present the total scan wirelength, the number of level shifters and the number of lockup latches used by the two algorithms.

The scan wirelength numbers that we present for both algorithms are those that are reported by Cadence SOC Encounter tool. So the conventional wirelength numbers that are reported are those obtained by arbitrarily allocating scan flops to scan chains and then using SOC Encounter's scan reordering capability to minimize scan wirelength on a per chain basis. The wirelengths reported for our algorithm are those that are obtained by using it to allocate scan flops to scan chain using our partitioning based approach and then reordering on a per chain basis using SOC Encounter. Our results which show a decrease in wirelength and in the number of level shifters and lockup latches prove that allocation

of flops to scan chains is better than arbitrarily allocating flops to chains and then reordering them.

The runtime numbers for the conventional approach in Table 2 is simply the running time of allocating flops to scan chains arbitrarily. The runtime for our algorithm is the execution time of allocating the flops to the scan chains and includes the time for partitioning, gathering the partitions and doing a per chain improvement of scan wirelength by running a Nearest Neighbour based TSP algorithm.

For comparision with other works we have compared the wirelengths obtained in [4] and [11] with our work. Like the above two papers we too use a 0.25μ process. We show order of magnitude improvements in the scan wirelength However we must mention that the scan wirelength obtained is also a function of the placement algorithm used originally to place the design. In our case we used Cadence SOC Encounter to do the original placement . Table 3 shows the results of the above experiments.

In order to show the reduction in the number of level shifters and lockup elements in the scan path on some of the ISCAS89 benchmark circuits we perform an experiment wherein we instantiate eight copies of the same circuit in a design and assign different voltage domains to each copy. Then for different number of scan chains we run our algorithm (alloc) and compare it with (conv), the conventional approach. The results are presented in Table 4. Our algorithm succeeds in reducing the number of level shifters by half and also reduces the scan wirelength.

As a visual comparison we provide the layout of one scan chain for Design A mentioned above without and with using our algorithms in Figures 4 and 5 respectively.

978-1-4244-3039-0/08 $25.00 © 2008 IEEE

Table 4. Comparison of scan wirelength(in microns) and number of level shifters between (conv) and (alloc)

Design*	# chains	(conv) wire-length	(conv) # lvl shifters	(alloc) wire-length	(alloc) # lvl shifters
s5378	1	2.9E+4	7	2.95E+4	2
	4	3.05E+4	5	2.83E+4	3
s9234	1	2.5E+4	7	2.5E+4	2
	4	2.59E+4	5	2.28E+4	3
s15850	1	8.31E+4	7	8.15E+4	2
	4	8.38E+4	5	8.0E+4	3
s13207	1	1.04E+5	7	9.24E+4	2
	4	1.05E+5	5	9.12E+4	3
s38584	1	2.34E+5	7	2.03E+5	2
	4	2.37E+5	5	2.03E+5	3
s35932	1	2.9E+5	7	2.56E+5	2
	4	2.61E+5	5	2.54E+5	3

* Each ISCAS benchmark circuit was instantiated 8 times. Each instantiation was assigned to a different voltage domain.

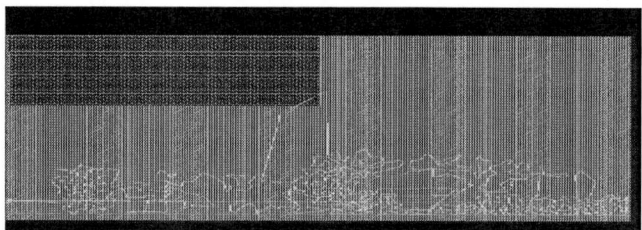

Figure 4. Layout of one scan chain without using our algorithm

5. Conclusions and future work

We presented an algorithm to allocate scan flops to scan chains based on physical location of the flops, the aspect ratio of the layout and the location of the scanin and scanout of the chains. Our algorithm uses a partitiong based scheme for this purpose. In addition to minimizing scan chain wire lengths our algorithm attempts to minimize the number of level shifters and lockup latches in the scan datapath.

Our algorithm has a guaranteed worst case runtime of $O(nlogn)$. We presented results where we compared the performance of our algorithm with that of an approach where the flops are allocated arbitrarily to chains and re-ordered on a per chain basis. For some cases we obtained upto 50% better scan wirelength reduction with our algorithm. We demonstrated the advantages of apriori allocating the correct flops to the correct scan chains over an approach

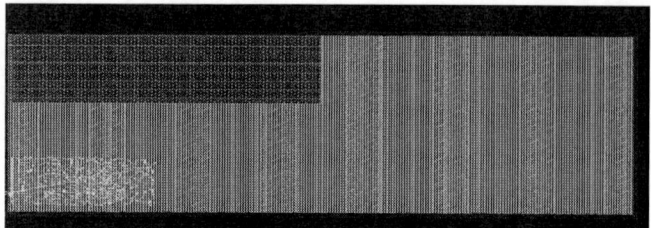

Figure 5. Layout of one scan chain by using our algorithm

where flops are allocated arbitrarily to scan chains.

The future of this work lies in extending it to hierarchical scan segments [10]. In this work the challenge consists of allocating the correct segments to the correct chains without violating any user defined scan chain length restrictions. Also scan segments cannot be split.

References

[1] D. Berthelot, S. Chaudhuri, and H. Savoj. An Efficient Linear Time Algorithm for Scan Chain Optimization and Repartitioning. In *Proc. Intl. Test Conference*, page 781, 2002.

[2] K. Boese, A. Kahng, and R. Tsay. Scan Chain Optimization: Heuristic and Optimal Solutions.

[3] Y. Bonhomme, P. Girard, C. Landrault, and S. Pravossoudovitch. Power driven chaining of flip-flops in scan architectures. In *Proc. Intl. Test Conference*, 2002.

[4] Y. Bonhomme, P. Girard, C. Landrault, and S. Pravossoudovitch. Efficient scan chain design for power minimization during scan testing under routing constraint. In *Proc. Intl. Test Conference*, pages 448–493, 2003.

[5] C. Chen, K. Lin, and T. Hwang. Layout driven selecting and chaining of partial scan flip-flops. In *Proc. Design Automation Conference*, pages 262–267, 1996.

[6] E. Eichelberger and T. Williams. A Logic Design Structure for LSI Testability. In *J. Design Automat. Fault-Tolerant Comput.*, volume 2, 1978.

[7] L. Guiller, F. Neuveux, S. Duggirala, R. Chandramouli, and R. Kapur. Integrating Dft in the Physical Synthesis Flow. In *Proc. Intl. Test Conference*, page 788, 2002.

[8] P. Gupta, A. B. Kahng, and S. Mantik. Routing-aware Scan Chain Ordering. *ACM Trans. Des. Autom. Electron. Syst.*, 10(3):546–560, 2005.

[9] D. S. Johnson and L. A. McGeoch. *The Traveling Salesman Problem: A Case Study in Local Optimization, Local Search in Combinatorial Optimization*. John Wiley, 1997.

[10] S. Makar. A layout-based approach for ordering scan chain flip-flops. In *Proc. Intl. Test Conference*, page 341, 1998.

[11] B. Paul, R. Mukhopadhyay, and I. Sengupta. Genetic algorithm based scan chain optimization using physical information. In *Proc. IEEE Tencon 2006*, 2006.

[12] K. Rahimi and M. Soma. Layout Driven Synthesis of Multiple Scan Chains. In *IEEE Transactions on Computer Aided Design*, volume 22, 2003.

SESSION B2:
Interconnects

Wiring-Area Efficient Simultaneous Bidirectional Point-to-Point Link for Inter-Block On-Chip Signaling

Charbel J. Akl and Magdy A. Bayoumi
The Center for Advanced Computer Studies (CACS)
University of Louisiana at Lafayette, LA 70504, USA
{cja3455, mab}@cacs.louisiana.edu

Abstract

The continuous semiconductor technology scaling has made on-chip interconnect the major determinant of VLSI design cost and complexity. This necessitates the usage of signaling techniques that reduce the number of long on-chip wires and repeaters. In this paper, we present a point-to-point inter-block on-chip link design that allows simultaneous bidirectional signaling, thus reducing the number of signal lines and repeaters, while achieving high performance. By using accelerating repeaters and inserting a bidirectional latch at the midpoint of the link, high performance simultaneous bidirectional signaling can be achieved with significant reduction in repeater and wire counts. We analyze the switching behavior of the proposed on-chip Simultaneous Bidirectional Link (SBL) and find that it suffers from large switching activity overhead. Therefore, an opposite-polarity transition encoding is also proposed to reduce the power overhead of SBL without affecting its performance.

1. Introduction

Due to the high integration density of deep submicron VLSI chips, the design effort and manufacturing cost are becoming driven by the wiring needs. The large number of global signal lines creates many concerns related to design cost, yield, wiring congestion, repeater placement, as well as routing and clock and power distribution. Methods such as serial transmission [1] and encoding [2] are becoming attractive to deal with the wiring bottleneck. Another approach that can be used at the chip level to provide inter-block data exchange is simultaneous bidirectional signaling where signals are transmitted at the same time from both end of the wire. Traditional unidirectional link requires double the number of wires compared to simultaneous bidirectional link, as shown in Figure 1. For same wiring area, simultaneous bidirectional link achieves double the transmission data rate of unidirectional link. This makes simultaneous bidirectional signaling desirable

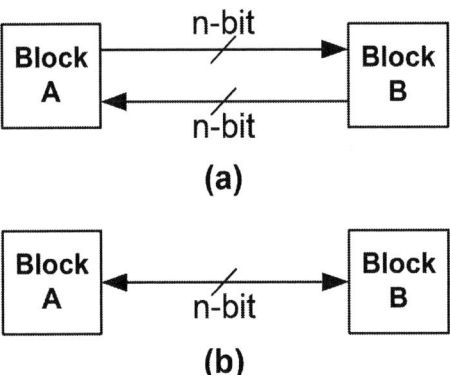

Figure 1. On-chip inter-block communication using (a) unidirectional link, (b) simultaneous bidirectional link.

for cost-effective high-performance designs. Simultaneous bidirectional signaling has been extensively studied and applied on off-chip interconnects to reduce pin count and achieve high data rate [3], [4]. The main idea for off-chip interconnects is that each transceiver subtracts its own transmitted signal from the line voltage to generate the received signal. Applying such approach to a typical on-chip global link may not be possible due to the different characteristics of on-chip and off-chip links. Moreover, the transceivers present large overhead compared to the reduction in on-chip wires, making the off-chip approach not practical for on-chip simultaneous bidirectional links.

Common global on-chip interconnect optimization techniques include shielding which alternates power/ground and signal lines routing to provide well signal integrity and reduced coupling effects between the signal wire and its surrounding [5], and repeater insertion which inserts inverters uniformly throughout a signal line to reduce delay dependence on wirelength from quadratic to linear [6]. One limitation of repeater insertion method is that it does not allow bidirectional signaling since a repeater is unidirectional in nature. Several circuit techniques that overcome this limitation were proposed [7]-[9]. We will refer to those circuits as accelerators. Unlike a repeater, an accelerator attaches to the wire rather than breaking it. Compared to repeaters, accelerators provide higher performance for long wires, less placement sensitivity, and are inserted less frequently than repeaters for optimal performance. The major disadvantage of accelerators is that they require careful tuning and optimization to achieve correct functionality at different design corners, and they can not be used in noisy environments. Accelerators were originally proposed for FPGA's programmable interconnect to reduce the overhead of the unused buffers after programming [7]. Later, they were found advantageous for long on-chip interconnects as well [8], [9]. Bidirectional signaling over a link with accelerators can be done sequentially, but not simultaneously. This introduces a performance penalty since a block at one end of the link can only transmit a signal if the block at the other end is not transmitting. Moreover, this requires a communication protocol or arbitration between the two blocks, similar to the system level bus, which adds significant latency, area and power overhead. Therefore, accelerator links require same number of wires as repeater links to achieve high performance simultaneous bidirectional signaling.

In this paper, we present a design technique that allows high performance simultaneous bidirectional signaling over single wire with accelerators. We study the switching behavior of the proposed Simultaneous Bidirectional Link (SBL), and we show that SBL suffers from high switching activity. An opposite-polarity transition encoding is proposed to reduce the switching activity of SBL without affecting its performance. The remainder of this paper is organized as follows. Section 2 presents the design of on-chip SBL, and discusses its advantages and limitations. Section 3 provides an analysis of SBL switching behavior. An opposite-polarity transition encoded SBL is proposed in Section 4. The simulation results are provided in Section 5, followed by the conclusion in Section 6.

2. On-chip Simultaneous Bidirectional Link (SBL) design

Figure 2 presents the design for 1-bit inter-block simultaneous bidirectional communication using repeater, accelerator and simultaneous bidirectional links. We assume a typical inter-block signaling context where signals are transmitted and received at the rising edge of the clock. Based on the wirelength, the links are optimized via repeater/accelerator tuning, wire width and spacing to achieve a single-cycle transmission. In the case of aggressive clock frequency, a signal may not be able to traverse the link in a single clock cycle even after intensive optimization. Therefore, multi-cycle link should be designed by replicating the link shown in Figure 2.

SBL (Figure 2(c)) uses accelerator to amplify the signal, and midway bidirectional latch and transmit/receive flip-flops to allow simultaneous bidirectional signaling.

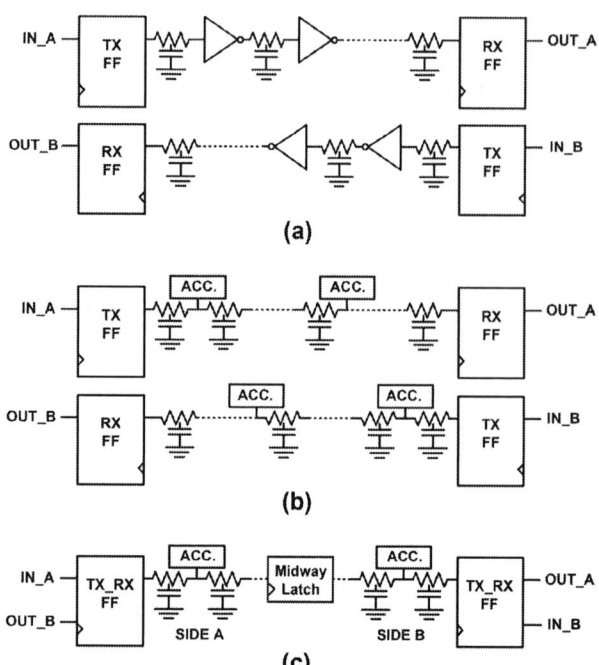

(a)

(b)

(c)

Figure 2. 1-bit simultaneous bidirectional signaling using (a) repeater link, (b) accelerator link, (c) SBL.

Figures 3 and 4 present the transmit/receive (TX_RX) flip-flop and midway bidirectional latch designs, respectively. During the high phase of the clock, the transmit/receive flip-flops transmit signals from both ends of the wire, while the midway latch is in receive mode. After the falling edge of the clock, the midway latch transmits the signal received from side-B over side-A, and the signal received from side-A over side-B. The transmit/receive flip-flops turns into receive mode during the low phase of the clock in order to output the received signals at the rising edge of the clock. As a result, signals from both ends of the wire can be transmitted in a single clock cycle. Sub-sections 2.1 and 2.2 present the performance comparison and the advantages and disadvantages of SBL, respectively.

2.1 Performance comparison

Industrial 90-nm transistor and RC-interconnect models and a supply voltage of 1V are used throughout the simulation. 1-bit simultaneous bidirectional link is considered, and the signal wires are modeled as distributed π-RC segments with shield wires on both sides, as shown in Figure 5, which is a common strategy for performance-critical global wires. The simulated METAL6 wirelength is 8-mm with an aspect ratio of 2.14. We adopt an accelerator circuit design from [7] shown in Figure 6. This accelerator was chosen among the other accelerators due to its high performance and reduced area overhead, which simplifies its tuning and optimization. Guidelines for optimizing this accelerator can be found in [7]. A library of repeater,

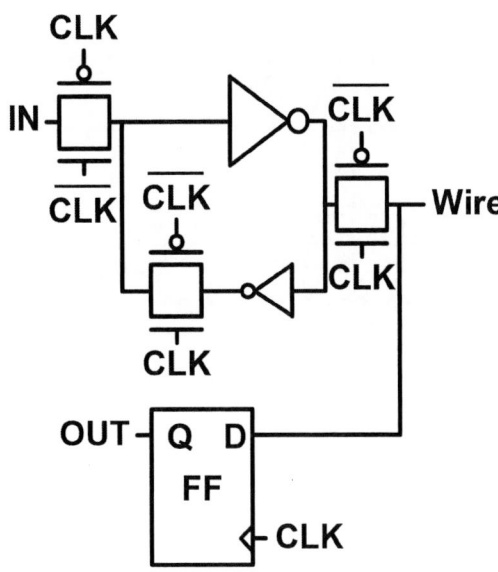

Figure 3.Transmit/receive (TX_RX) flip-flop for SBL.

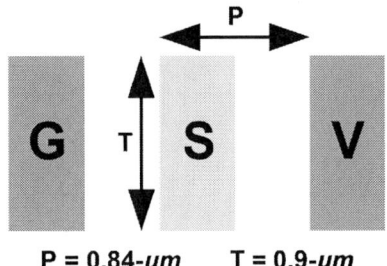

P = 0.84-*um*   T = 0.9-*um*

Figure 5.Global METAL6 wire cross-section.

accelerator, midway latch and flip-flop cells with multiple strengths is considered during interconnect optimization. The maximum drive current of a cell equals 64× the drive current of a minimum size inverter. Using cells with higher driving capability can yield better performance. However, a limitation on the maximum drive current is usually considered to keep the power/ground noise within acceptable range. Figure 7 presents the optimal total latency which includes the wire delay, CLK-to-Q and setup time delays versus number of repeaters/accelerators, and midway latch plus accelerators for SBL. For each number, cell strengths that achieve the minimum latency are considered. Accelerator link provides between 10% to 20% optimal latency improvement compared to repeater link. The number of latch plus accelerators in SBL is always odd due to the midway latch restriction and the similar optimization of side-A and side-B of the link. SBL latency sits between repeater and accelerator link latencies. SBL achieves up to 14% latency improvement compared to repeater link, and has an optimal delay penalty between 7% and 13% compared to accelerator link. When the number of repeaters is more than four, repeater link delay starts to increase since the intrinsic repeater delay adds to the total link delay. Whereas, accelerator link delay keeps

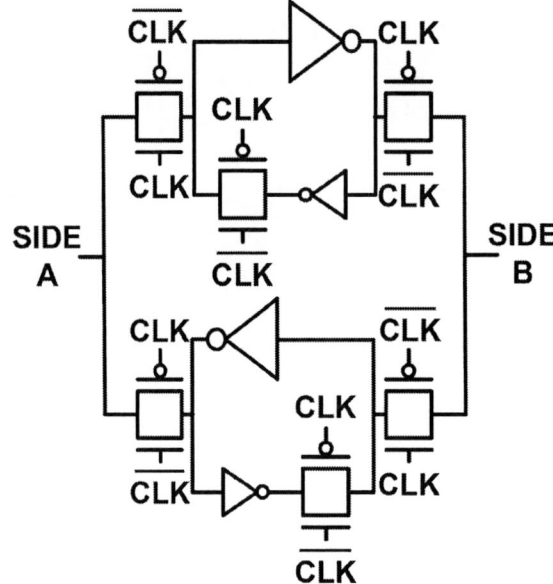

Figure 4.Midway bidirectional latch for SBL.

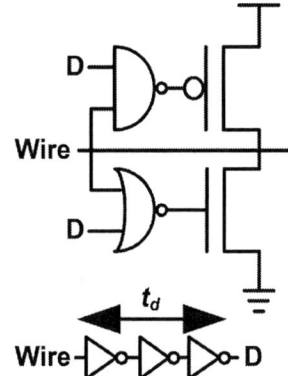

Figure 6.Accelerator design adopted from [7] (Regenerative Feedback Repeater).

decreasing or stays almost constant as more accelerators are added to the line.

2.2 Advantages and disadvantages of SBL

The major advantage of SBL is the efficiency in sharing the wiring resources. Compared to repeater and accelerator links, SBL utilizes half the wiring area for simultaneous bidirectional signaling, while achieving high wire performance. The reduced number of active wires achieves high integration density and reduces design cost and complexity. Routing congestion is minimized, and extra room is available for better global power/ground and clock distributions. The reduced number of active wires leads to reduced accelerator count, which simplifies accelerator and latch placements.

SBL inherits the advantages as well as the disadvantages of accelerators. However, the midway latch makes SBL characteristics closer to a repeater link than a

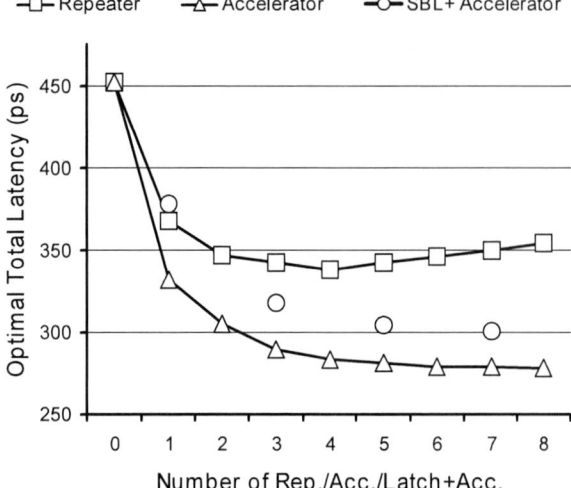

Figure 7. Optimal total latency (including wire delay, CLK-to-Q and setup time) versus number of repeaters, accelerators, midway latch and accelerators, per wire.

Table 1. Number of wire transitions in a conventional 2-wires 1-bit simultaneous bidirectional link, for all possible input transitions.

Input transition		Number of wire transitions
A	B	
-	-	0
↕	-	1
-	↕	1
↕	↕	2

Table 2. Number of wire transitions in 1-bit SBL for all possible current and previous input states.

Previous state		Current state		Number of wire transitions
A	B	A	B	
0	0	0	0	2
		0	1	0.5
		1	0	0.5
		1	1	1
0	1	0	0	1.5
		0	1	0
		1	0	1
		1	1	1.5
1	0	0	0	1.5
		0	1	1
		1	0	0
		1	1	1.5
1	1	0	0	1
		0	1	0.5
		1	0	0.5
		1	1	2

pure accelerator link. SBL has mainly two disadvantages. The first one is related to the midway latch placement and clocking. The placement of the bidirectional latch exactly at the midpoint of the wire may not be possible due to the layout constraints. A small deviation of the midway latch placement can be dealt with by the cycle timing guardband. However, a large placement deviation necessitates different optimizations of side-A and side-B of the link, which increases the tuning and optimization effort. Another issue in the midpoint latch is that 50% or near 50% clock duty cycle is required. A large deviation in the clock duty cycle necessitates the usage of duty-cycle controlling circuit [10], which adds extra power/area overhead to the link. However, for a wide bus (e.g. 32-bit), the controller is shared by all the wires, which make its overhead small compared to the overall power/area of the bus. The second SBL disadvantage is related to power consumption. SBL suffers from increased clock loading, mainly due to the clocked transistors that fall into the critical path of the link. Moreover, the switching activity of SBL is higher than that of a conventional link, which will be analyzed in the next section.

3. SBL switching behavior

In a conventional 2-wires simultaneous bidirectional link, the wire switching activity depends on the input switching activity, as shown in Table 1. When the inputs are stable, no switching occurs over the lines. However, SBL has completely different switching behavior due to the sharing of the wire by both end signals. The major drawback of SBL is the redundant wire switching when the inputs are stable, which make the switching behavior of SBL state-dependant. Table 2 presents the number of wire transitions in SBL for all possible current and previous

input states. When the inputs are not switching and are at same state ("00","11"), 2 wire transitions occur each cycle since each side of the link switches during the high-phase as well as the low-phase of the clock. Using non-inverting latches solves this issue for "00" and "11" states, but it creates the same issue for "10" and "01" states which do not suffer from this problem in the inverting latches SBL. The state-dependence switching behavior adds a significant power overhead to SBL, especially for low switching activity links. When the two inputs A and B switch simultaneously, SBL transmits the signals in a single wire transition on average, which reduces the switching activity compared to a conventional 2-wires link. However, state-dependence switching overhead dominates, even at high input switching activity. Next section presents an encoded SBL to reduce the switching activity overhead.

4. Opposite-polarity transition-encoded SBL

Transition-encoding is a known low-power coding method that is based on the XOR/XNOR operations. Transition signaling was presented in [11] to generate a wire transition when the input state is "1". Therefore, by reducing the number of 1's transmitted over a bus, the

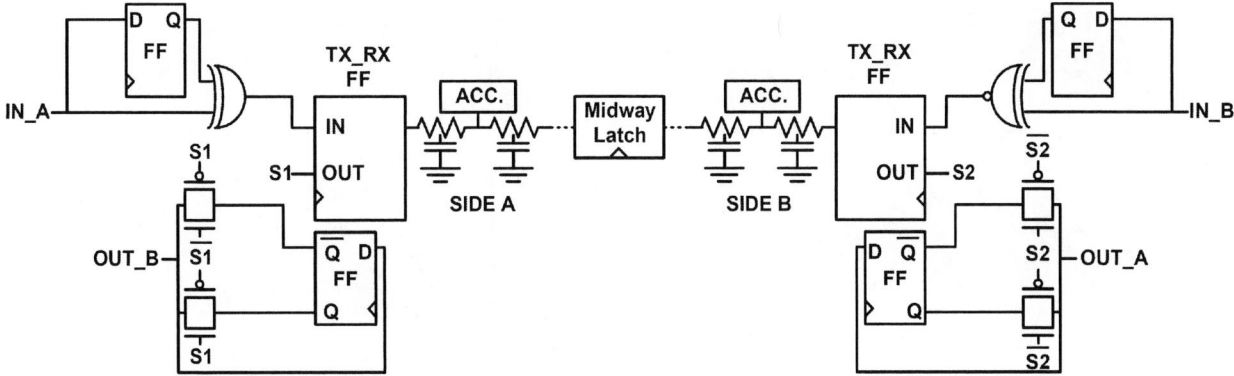

Figure 8.Opposite-polarity transition-encoded SBL (ENC-SBL). Input-A is encoded via an XOR, and Input-B is encoded via an XNOR. Output-A changes its state when the received signal (S2) is high. Output-B changes its state when the received signal (S1) is low. Initialization is required at startup.

switching activity can be reduced. In [12], transition encoding was used to reduce the switching activity of the dynamic bus by evaluating/precharging only when a transition at the input is detected. Transition encoding was also proposed to reduce the switching activity of serialized data-streams [13]. Recently, transition encoding was applied to the inputs as well as to the wire to encode two signals in a single wire transition [2]. In this work, we propose an opposite-polarity transition encoded SBL (ENC-SBL), shown in Figure 8. An XOR gate is used to encode input "A", such that when "A" switches, logic "1" is transmitted over the link. On the other side, An XNOR gate is used to encode input "B", such that when "B" switches, logic "0" is transmitted over the link. The receiver at both sides stores the previous state of the output and uses a multiplexer to select between the previous state and its complement based on the received signal. The latency overhead of the encoder equals to the delay of a single XOR/XNOR gate whose inputs can be reordered to achieve high performance since the previous input state is available earlier in the clock cycle. The encoding flip-flop can be sized to achieve the lowest possible clock load since it is outside of the encoding critical path. The decoder latency overhead equals to a transmission gate delay which is very low.

Encoding SBL greatly changes its switching behavior as shown in Table 3. When the inputs are not switching for two or more consecutive cycles, no wire switching occurs regardless of the inputs state. This reduces the redundant switching problem of SBL. The switching activity saving obtained from ENC-SBL depends on the input switching probability since more savings can be obtained at reduced input activity compared to SBL. Figure 9 shows the transient simulation waveform over 10-ns. Both SBL and ENC-SBL are simulated with same inputs. The outputs in the waveform are the ENC-SBL outputs, which are similar to SBL outputs but slightly delayed due to the decoding latency. It is apparent that the switching activity of ENC-SBL is considerably smaller than SBL, especially when the inputs at both sides of the link have same state.

Table 3.Number of wire transitions in 1-bit ENC-SBL for all possible current and previous input transitions.

Previous transition		Current transition		Number of wire transitions
A	B	A	B	
-	-	-	-	0
		-	↕	1.5
		↕	-	1.5
		↕	↕	1
-	↕	-	-	0.5
		-	↕	2
		↕	-	1
		↕	↕	0.5
↕	-	-	-	0.5
		-	↕	1
		↕	-	2
		↕	↕	0.5
↕	↕	-	-	1
		-	↕	1.5
		↕	-	1.5
		↕	↕	0

5. Simulation results

We consider a METAL6 8-mm 1-bit simultaneous bidirectional link with a target clock frequency of 2.2GHz (clock period around 15FO4s of the technology models in use). 30% of the cycle period is assumed to be reserved for timing guardband to overcome mismatch, variability, clock skew, etc.... The CLK-to-Q, setup time, and wire delays of each link (repeater, accelerator, SBL) are optimized to achieve the target 70% clock period delay with the least number of repeater/accelerator count and size. For SBL and ENC-SBL, the 30% timing guardband is divided equally between Side-A and Side-B of the link (15% each). Table 4 presents the repeater/accelerator count (including the midway latch for SBL and ENC-SBL), and presents the number of wires required for bidirectional signaling as well. Repeater link requires 4 repeaters per line, whereas

199

978-1-4244-3039-0/08 $25.00 © 2008 IEEE

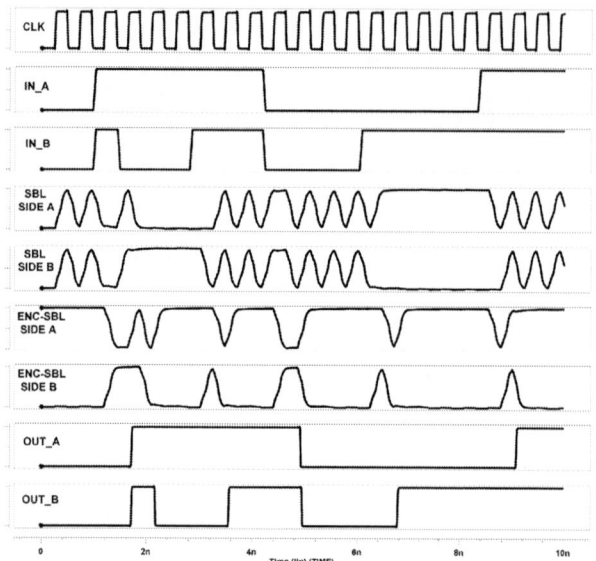

Figure 9.Transient simulation waveform showing the switching activity of SIDE-A and SIDE-B of SBL and ENC-SBL over 10-ns. The wirelength is 8-mm and the clock frequency is 2.2 GHz.

Table 4.Repeater/accelerator count (plus midway latch for SBL) per 1-bit simultaneous bidirectional link over 8–mm wirelength. Also, wire count required for simultaneous bidirectional signaling.

Link	Rep./Acc. count	Wire count
Repeater	8	n
Accelerator	4	n
SBL	3	n/2
ENC-SBL	3	n/2

accelerator link requires 2 accelerators per line. SBL requires 2 accelerators per bidirectional line, one on each side, plus the midway latch. SBL/ENC-SBL reduces wire count by 50%, and repeater/accelerator count by 25% and 62% compared to accelerator and repeater links, respectively. If interconnect area is not of primary concern, the wire count saving can be exploited in several ways [1].

Power measurement was done at different switching activities considering randomly generated, uniformly distributed inputs. The simulation was done over 2000 cycles assuming that both side inputs have same switching activity and their switching time is independent. The clocking power is included in the simulation by feeding the clock signals into buffers, and sizing the clock buffers of each link to achieve similar transition delay of 60ps. Figure 10 presents the average power results versus the input switching probability. Repeater link has the lowest average power, whereas SBL has the highest power that is almost independent of the input switching activity. However, opposite-polarity transition encoding greatly reduces the power of SBL, especially at low switching activity, to a comparable level to repeater and accelerator links.

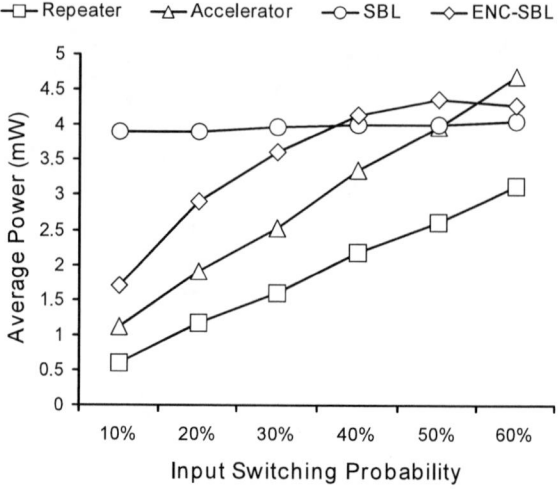

Figure 10.Average power versus switching activity.

6. Conclusion

In this paper, we presented a simultaneous bidirectional link (SBL) design for on-chip inter-block communication to reduce the number of wires and repeaters, while achieving high performance. SBL suffers from large switching activity overhead that can be greatly reduced by using opposite-polarity transition encoding. However, we believe that opposite-polarity transition encoding is not the optimal coding for SBL, and some unique SBL encodings that fully exploit SBL switching behavior are possible, which will be the target of our future work.

7. References

[1] A. J. Joshi, G. G. Lopez, and J. A. Davis, "Design and optimization of on-chip interconnects using wave-pipelined multiplexed routing," IEEE Trans. VLSI, vol. 15, no. 9, pp. 990-1002, Sep. 2007.
[2] C. J. Akl and M. A. Bayoumi, "Transition skew coding: a power and area efficient encoding technique for global on-chip interconnects," in Proc. ASP-DAC, pp. 696-701, Jan. 2007.
[3] R. Mooney, C. Dike, S. Borkar, "A 900 Mb/s bidirectional signaling scheme," IEEE Jour. Solid-State Circuits, vol. 30, no. 12, Dec. 1995.
[4] E. Yeung, and M. A. Horowitz, "A 2.4 Gb/s/pin simultaneous bidirectional parallel link with per-pin skew compensation," IEEE Jour. Solid-State Circuits, vol. 35, no. 11, Nov. 2000.
[5] S. Khatri et al., "A novel VLSI layout fabric for deep sub-micron applications," in Proc. DAC, pp.491-496, June 1999.
[6] H. B. Bakoglu and J. D. Meindl, "Optimal interconnection circuits for VLSI," IEEE Trans. Electron Devices, vol. ED-32, May 1985.
[7] I. Dobbelaere, M. Horowitz, and A. El Gamal, "Regenerative feedback repeaters for programmable interconnections," IEEE Jour. Solid-State Circuits, vol. 30, no. 11, Nov. 1995.
[8] A. Nalamalpu, S. Srinivasan, and W. P. Burleson, "Boosters for driving long onchip interconnects-Design issues, interconnect synthesis, and comparison with repeaters," IEEE Trans. CAD Circuits and Syst., vol. 21, no. 1, pp. 50-62, Jan. 2002.
[9] J. Seo, P. Sing, D. Sylvester, and D. Blaauw, "Self-timed regenerators for high-speed and low-power interconnect," in Proc. ISQED, pp. 621-626, March 2007.
[10] K. Agarwal, R. Montoye., "A duty-cycle correction circuit for high-frequency clocks," in Proc. Symp. VLSI Circuits 2006, pp. 106-107.
[11] M. R. Stan, W. P. Burleson, "Low-power encodings for global communication in CMOS VLSI," IEEE Trans. VLSI, vol. 5, no. 4, pp. 444-455, Dec. 1997.
[12] M. Anders et al., "A transition-encoded dynamic bus technique for high-performance interconnects," IEEE Jour. Solid-State Circuits, vol. 38, no. 5, pp. 709-714, May 2003.
[13] M. Ghoneima, Y. Ismail, M. Khellah, V. De, "Reducing the data switching activity on serial link buses," in Proc. ISQED, March 2006.

21st International Conference on VLSI Design

Energy-Aware Interconnect Optimization for a Coarse Grained Reconfigurable Processor

Andy Lambrechts, *Student Member, IEEE,* Praveen Raghavan, *Student Member, IEEE,* Murali Jayapala, *IEEE Member,* Francky Catthoor, *IEEE Fellow,* and Diederik Verkest, *IEEE Member*

Abstract—**Modern portable embedded devices provide continuously more features and need processors that are of increasingly higher *performance* in order to sustain very demanding multimedia and wireless applications. Larger amounts of *flexibility* need to be built in and the same processor needs to be used for a wide range of evolving products, while very strict *energy constraints* need to be met in order to provide a long battery life. *Coarse Grained Reconfigurable Architectures* (CGRAs) provide a mix of flexible computational resources and large amounts of programmable interconnect. However, this programmable *interconnect* is on average consuming about 50% of the core's energy consumpion for state of the art interconnection topologies. In this work we present an optimized interconnection implementation that selectively activates only the connections that are being used in a certain cycle, in order to reduce the energy spent in the interconnect. Using this optimization, we show the effect on the energy and performance trade-off for the ADRES CGRA. The energy cost of the optimized interconnect topologies that provide a higher performance can be reduced significantly, reducing the total energy consumption of the core with up to 40%. This will enable designers to develop more efficient architectures, tuned to a targeted application domain.**

Index Terms: **Energy-Aware Design, Low Power, Processor Architecture, Interconnect-Aware Design**

I. INTRODUCTION

Our everyday life is filled with objects that contain embedded processors. An increasing amount of these devices are battery operated and provide a functionality which is increasing rapidly. To be able to sustain this trend, processors need to provide an increasing amount of compute power within a very strict energy budget. Moreover, increasing production costs of specialized chips force designers to build flexible processors that can be used for different devices and that support evolving applications, communication standards etc.

Coarse Grained Reconfigurable Architectures (CGRAs) are a class of processors that consist of a high amount of computational resources (Functional Units, or FUs), together with distributed storage (Register Files or RFs), configuration memories and a programmable interconnection topology. The FUs operate on data words (e.g. 32 bit or 16 bit ALU, MUL, MAC etc.) and the mix and the number of units can be optimized for a certain application domain. By storing the required operations and active connections into the configuration memories, applications can be very efficiently mapped

A. Lambrechts, P. Raghavan, F. Catthoor are with IMEC vzw and ESAT, KULeuven, Belgium. E-mail: lambreca@imec.be
M. Jayapala is with IMEC vzw, Belgium.
D. Verkest is with IMEC vzw., KULeuven and VUB, Belgium
This research has been carried out in the context of IMEC's nomadic embedded systems program which is partly sponsored by Samsung and is partly supported by IWT Flanders.

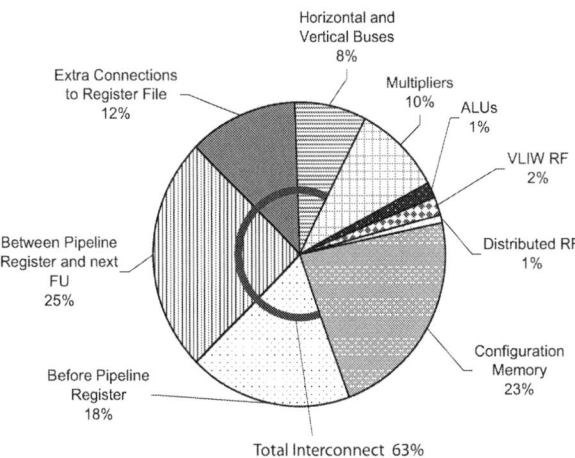

Fig. 1. Detailed Energy Breakdown of *b_neg_nh_rf*, an instance of the ADRES CGRA, for a UMC 130 nm standard cell technology, running a MIMO channel estimation benchmark.

to this type of processor. Performance is boosted through the use of many parallel resources that exploit the high Instruction Level Parallelism, which is commonly found in embedded applications, while energy is kept low through the usage of distributed storage and a lower clock frequency compared to less parallel RISC or VLIW processors.

While a large amount of research has been done to improve the performance of this class of processors, little has been published to improve the energy efficiency. Proposed CGRAs differ in the number and organization of the number and type of their FUs, the number and size of RFs and configuration memories and the flexibility of their interconnection topology. In this paper, we focus on the design of the interconnect topology, which has a large impact on both performance and energy efficiency. As can be seen in Figure 1, the energy consumption of different types of interconnection in a interconnect-rich CGRA instance can go up to 60%. The architecture shown here is typically the result when optimizing for performance, and is commonly used in practice. More energy efficient but less connected instances still spend more than 30% of the core's energy in interconnect, while providing less performance. Over different interconnection topologies, a performance vs. energy efficiency trade-off of about 40% for both energy and performance can be found. A detailed description of the architecture *b_neg_nh_rf* used for the breakdown in Figure 1 can be found in Section V.

201

978-1-4244-3039-0/08 $25.00 © 2008 IEEE

In this paper we provide an energy optimization technique, that selectively activates the connections between different components in the CGRA, only in the cycle when they are used. We insert extra pass gates to be able to activate the required connections, and correctly take the overhead of extra configuration bits into account. This technique significantly reduces the energy consumption that is spent in the interconnect. We investigate the effect of the proposed optimization on the energy vs. performance trade-off, as the selective activation of interconnect leads to a higher reduction for more rich topologies, which in turn reduces the extra cost of moving to an interconnect variant which provides a higher performance. For the experimental results that are shown in this paper, the ADRES CGRA framework has been used [11]. ADRES is a flexible CGRA architecture template. ADRES, as most other CGRAs uses a MUX-based programmable interconnection topology. We propose an optimized implementation, to reduce energy consumption, while keeping functionality the same. We show a reduction of up to **42%** in the energy consumption of the processor core and analyze the impact on the overall energy vs. performance trade-off for a set of representative benchmarks.

The rest of this paper is organized as follows. Section II gives an overview of related work. Section III introduces the ADRES architecture and the MUX-based interconnection implementation. Section IV briefly introduces the used energy estimation frame-work. Section V presents the standard interconnect implementation, introduces the proposed optimization and the effect on energy consumption and the energy vs. performance trade-off. Finally, Section VI concludes this paper.

II. RELATED WORK

The increasing demand for high performance and energy efficient processors has lead to the design of coarse grained reconfigurable processors. Some architectures like [3] are using coarser granularity processing elements and are targeted toward high performance applications and not tuned for embedded systems, because of high energy consumption. A growing number ([11], [12], [14], [16]) are explicitly targeted at low power embedded architectures.

Since CGRA architectures are relatively new and because most architectures do not support a flexible architecture template and automatic retargetable compilation, little work has been done in architectural exploration in this domain. Many earlier approaches, e.g. [9] and most of the commercial architectures, have a relatively fixed base processor and do not allow easy exploration. Others do have a flexible template, but due to the large exploration space no systematic exploration has been published. Wilton et al. [7] explore the various configurations in which the register file can be connected, but their exploration is limited to performance only. Bansal et al. [1] have investigated the impact of different network topologies for mesh-base CGRAs, but their array is homogeneous and the compiler is restricted to only instruction level parallelism.

Interconnect is one of the largest energy consuming parts of the processor [5] and as technology scales the relative energy consumption due to interconnect is expected to grow

as well. Figure 1 shows the energy breakdown for a CGRA architecture. It can be seen that the energy contribution due to interconnect is quite large. Our previous work [8] introduced an energy aware exploration framework for CGRAs. We also showed that exploring interconnect in CGRAs gives a large trade-off between energy consumption and performance for different benchmarks. [6] also presents a trade-off between energy and performance. Their exploration is limited to small kernels like FFT and DCT and they perform power estimation after gate level simulation of the application, which often cannot be done for early architecture exploration. Their approach is not scalable to larger arrays or for larger applications. Also their work does not propose any optimization and analysis of the energy breakdown.

Other work for interconnect energy reduction include utilization of segmented bus for communication [4]. The overhead associated with segmented busses, as they allow concurrent communication on different segments, is larger than the technique proposed here and can only be justified for very large capacitances on the buses that are to be segmented.

To the best of our knowledge there has been no work on directly reducing energy consumption of interconnect in CGRA architectures.

III. ADRES ARCHITECTURE DESCRIPTION

CGRAs offer a large architectural space of exploration, changing any of four categories of parameters: computational resources, data storage, interconnection resources or configuration storage. The number and type of FUs can be changed, the register file architecture and the connectivity between FUs can be centralized or distributed and sizes can be varied, the connection to the data and instruction memory architectures and the topology of the programmable interconnect can explored. In this paper we focus on optimizing the interconnect energy consumption for a given instance, and the effect of this optimization on the energy vs. performance trade-off for the interconnect decision. Finding an *optimal* CGRA instance for a certain application domain and choosing all design parameters accordingly, is outside the scope of this paper.

The energy-aware and interconnect-aware architecture exploration used in this paper is built on the ADRES [11] framework. The ADRES CGRA is a flexible template and many other proposed CGRA architectures can be represented by using this parameterizable template. Additionally, the ADRES architecture has two functional views: in the VLIW-mode the first row of FUs and the shared register file can be used as a normal VLIW processor, while in reconfigurable array mode, loops with a large amount of parallelism, which are often found in wireless communication or multimedia applications, utilize the complete array. When (part of) the application contains little instruction level parallelism, e.g. for irregular control code, it is executed on the VLIW part only.

An example of the detailed CGRA datapath unit is given in Figure 2. This general unit consist of a computational element in the center (FU), that can support a certain instruction set (ADD, MUL, etc.). CGRAs are built of a number of these units, organized in a 1D, 2D array or more irregular and

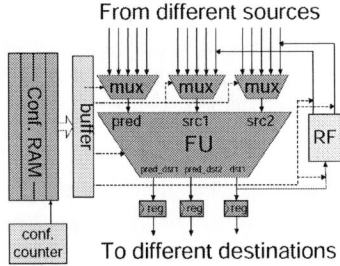

From different sources

To different destinations

Fig. 2. General CGRA datapath unit, consisting of general computation element (FU) and distributed data and configuration storage (RF and Conf. RAM resp.)

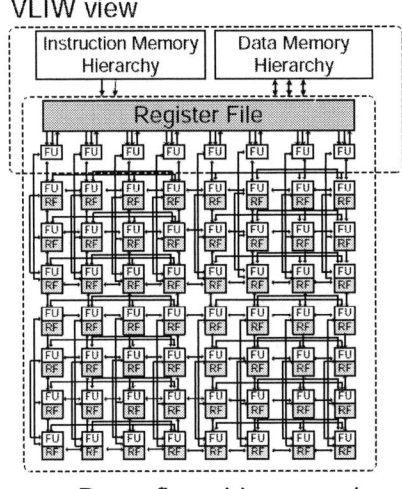

VLIW view

Reconfigurable array view

Fig. 3. 8x8 Instance of the ADRES architecture

sometimes hierarchical structures. Most CGRAs also support distributed storage of data (registers and register files (RF)) and of configuration memories (Conf. RAM). These units can also be shared by more than one FU. The interconnection topology is typically implemented using MUXes. At each input of an FU, a MUX selects the incoming connection that has to be used in a certain cycle. The setting of all MUXes is stored (together with FU instructions and addresses for RFs etc.) in the configuration memory. The output of every FU is connected (via a pipeline register) to the input MUXes of its destination FUs. The produced result can be consumed by any number of those.

Figure 3 shows an example instance of the ADRES template, consisting of a 8x8 array of FUs. It can clearly be seen that the amount of interconnect in this type of processor can be much higher than in traditional RISC or VLIW processors and that the distance over which can be communicated can be much larger. The implementation of the MUX-based communication is discussed in more detail in Section V.

The ADRES architecture template is developed in conjunction with a retargetable simulator and compiler, called DRESC [10]. The ADRES-DRESC framework was chosen because this flexible template and tool-flow allow a designer to vary many of the critical design parameters of a CGRA and quickly

evaluate the effect on performance. The simulator flow was extended to generate more detailed performance results and also include a detailed energy breakdown (see Section IV).

IV. ENERGY MODELING APPROACH

To enable a meaningful optimization of the interconnect architecture for the ADRES CGRA, it is essential to take the energy consumption of all architectural components into account. Therefore a detailed energy breakdown of the core is required. In this section, we introduce the followed energy modeling approach that is used to generate this breakdown.

To support detailed energy estimations, the original ADRES-DRESC framework was extended with additional components, as shown in Figure 4. The different architectural components of the ADRES template were synthesized in Synopsys Physical Compiler, using the UMC130nm Standard Cell Design Ware Library. The extracted layout was back-annotated with switching activity from ModelSim and the results were obtained for the 130nm technology with 1.2V Vdd and a worst case process corner. The energy consumption is computed using Power Compiler.

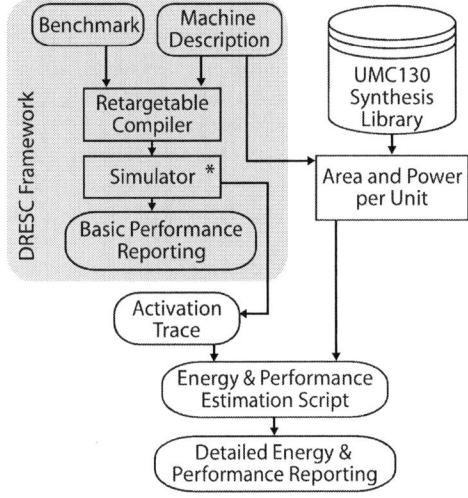

* Extended with Tracing

Fig. 4. Extended Energy-Aware Architecture Exploration Framework

For the register file, the energy consumption has been calculated using the EMPIRE model for register files [15], which gives accurate energy numbers for various instances and is calibrated using the same technology libraries and estimation flow. The configuration memories are modeled using a 130nm, 1.2V SRAM memory macro taken from [2].

Accurate area estimations of different components, taken from the layout, are used to construct a regular high level floor-plan of the ADRES architecture and wire lengths of each individual wire are computed in an automated way. The wire length model for buses and point to point interconnections are taken from the same UMC130nm technology to ensure consistency. To get fast, but fairly accurate estimations on the interconnect energy, without going to a full placement and routing for each complete architecture instance, interconnect

203

978-1-4244-3039-0/08 $25.00 © 2008 IEEE

lengths between the different functional units and connections to register files are computed as the Manhattan distance between these architectural components, based on the area estimates obtained from physical synthesis.

The connections between different FUs are programmable and implemented as a MUX-based interconnection. A specific input of a register file or an FU is selected by setting a MUX at that input, based on the contents of the configuration memory. See also Section V and Figure 6 for a more detailed explanation of the interconnect implementation. This implementation 'broadcasts' the output of every FU to all FUs to which it is connected, where the data can be consumed in the next cycle.

The energy estimation is completed by extending the simulator to keep track of the activation of every component of the architecture, for a given application executing on it. For every operation that produces an output at a certain FU, the list of consumers is traced, together with reads from and writes to the register files. This detailed activation trace is processed and the total energy consumption is computed, per component or per category (e.g. distributed register files or only a certain type of wires) and detailed reports are generated.

V. INTERCONNECT ENERGY OPTIMIZATION

CGRAs consist of a large number of design parameters, of which the programmable interconnect is one of the most important ones. Changing the interconnect topology from a very richly connected to a less connected variant, while keeping all other architectural parameters constant, can have a significant effect on both performance and energy consumption (up to 45% different for both criteria, between the best and the worst architecture, for a certain benchmark, as was demonstrated in [8]). In this paper we focus on optimizing the energy consumption in the interconnect. This section introduces the different interconnect variants, starting from a base architecture for which all other parameters have been fixed. We briefly show the effect of interconnect exploration on both energy and performance for representative benchmarks from both the wireless communication and multimedia application domains. The rest of this section presents the proposed interconnect optimization.

A. Interconnect variant for Energy vs. Performance Trade-off

Different types of interconnections can be added to the architecture. In this work, we have identified four different types of connections that are commonly found in state of the art CGRAs. Using these four types, the flexibility of the resulting interconnect topology is varied from very flexible to restricted. Figure 5 shows four example architectures, illustrating the different interconnect topologies. To denote the different variations of the CGRA interconnect topologies, the following naming conventions are used: 1) **Buses** (horizontal and vertical): *b*: With buses, connecting all FUs in the same row/column; 2) **Interconnect to nearest neighbors**: *_neg*: Connection with nearest neighbors (from the output of an FU to the input of all next neighbors, in both horizontal and vertical direction, but no diagonal connections); 3) **Interconnect to next hop neighbors**: *_nh*: Connection with neighboring FUs

one hop away, both in horizontal and vertical directions, but no diagonal connections; 4) **Interconnect to VLIW register file**: *_rf*: Extra connections from the CGRA to VLIW register file, to facilitate distribution of live-in/live-out variables, to the second and third rows of FUs. Next to the interconnect parameters indicated by *b, neg, nh, rf* all architectures contain connections of the type "before the pipeline register", which connect the FU with the neighboring distributed register files of 4 nearest neighbor FUs in diagonal directions.

E.g. *_b_rf*: the interconnect of this array instance consists of both vertical and horizontal buses and extra connections that connect the second and third row of the array to the VLIW register file.

For a fixed size of the architecture, e.g. 8x8 FUs, these 4 template parameters lead to 16 possible different architecture instances. Excluding the architectures without any interconnect, the one with only the extra connections to the central register file, we end up with 14 valid possibilities. We exclude all architectures that do not have nearest neighbor connections, but do have next hop connections. These two are equivalent from a compiler point of view, but next hop connections are more expensive in the physical implementation. The 10 resulting architectures are compared in the following sections.

B. Standard, Broadcast Interconnect Implementation

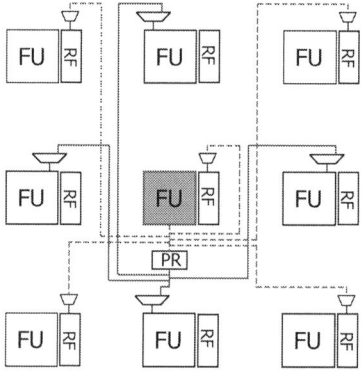

Fig. 6. Detailed Interconnection for one FU in the ADRES array, showing the connections of type "before the pipeline register" in dotted lines and of type "between the pipeline register and the next FU" as solid lines

The way interconnect is implemented in state of the art CGRAs has an influence on the cost of different types of connections. Most CGRAs use MUX-based interconnect. Figure 6 shows a detailed drawing of the interconnection of one FU in the ADRES array. Only connections that are starting from this central FU are shown. We can differentiate between two different types of connections. The first type, called "before pipeline register", connects the output of every FU with the input of distributed register files (RF) of the same FU and of the four neighboring FUs in diagonal directions and connects to the pipeline register (PR). These connections are shown as dashed lines. The second type, "between pipeline register and next FU", connects the output of the PR to MUXes at the inputs of other FUs. In Figure 6 only the connections to the nearest neighbors are shown. Adding connections to

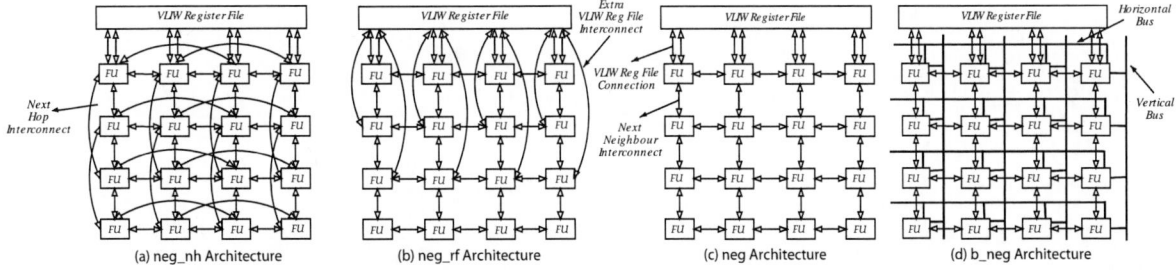

(a) neg_nh Architecture (b) neg_rf Architecture (c) neg Architecture (d) b_neg Architecture

Fig. 5. Examples of Different Interconnection Topologies for the ADRES CGRA: a Nearest Neighbor and Next Hop connections, b Nearest Neighbor connections and extra connections to the central Register File, c Nearest Neighbor and Next Hop connections only, d Nearest Neighbor connections and horizontal and vertical Buses

the next hop neighbors adds even longer wires to the four FUs in both horizontal and vertical directions (not shown). Typically the input of a certain FU that will be used in a certain cycle, is only selected at the MUX located at that input (see Figure 2). The produced result is always "broadcast" to all possible destinations, and this quickly adds to the energy cost of extra interconnection flexibility. A large fanout tree, the cumulative length of the connections to all destination, has to be driven every time an operation is scheduled on a certain FU. The advantage of this approach (and the reason for this design choice), is that the compiler can schedule multiple consumers of the same result on different FUs that are connected to the producer. This compiler-friendly interconnect implementation generates much needed extra freedom, as compilability is one of the biggest problems of many complex Coarse Grained Architectures.

1) Benchmarks: The experiments have been performed on a set of representative benchmarks from the wireless communication and multimedia domains: *MIMO:* MIMO channel estimation kernel with 52 pilots; *Viterbi:* a 189-state Soft Viterbi (SOVA); *AVC_motion:* in-house optimized version of the AVC motion estimation; *AVC_interpolate:* unoptimized version of the AVC half-pixel interpolation filter.

2) Base Architecture: All experiments have been performed using an 8x8 ADRES array of which only the interconnection topology is varied (in our experience, the interconnect study of arrays of similar sizes, e.g. 4x4 to 12x12, leads to the same general conclusions). All FUs are of the ALU-type, except the third and sixth columns, which are multipliers. The base architecture uses configuration memories of 64 entries deep and local register files that can store 16 words, while the central VLIW register file can store 64 words. The exploration of other array parameters is outside of the scope of this paper, and for more information the reader is referred to [8], [11], [13].

3) Energy vs. Performance Trade-off for Interconnect Exploration: Figure 7 shows the simulation results for all four benchmarks. The X-axis shows the normalized number of cycles that is needed to complete the benchmark (inverse of performance), while the Y-axis shows the normalized energy consumption (of the complete core) for that specific task, and this over the ten architecture instances that are evaluated. Diamond shaped markers show the performance and energy results for the un-optimized architectures (*Before Opti*). A line connects all instances that are Pareto optimal, as only these are interesting for a design that considers the energy-performance trade-off. By moving on a line from left to right, top to bottom, we move to more energy efficient architectures, but loose performance. The results show a difference of 35 to 40% between the best and worst architecture variants, both for energy and for performance. For the un-optimized code of the AVC interpolate filter, almost no performance trade-off is found, as the compiler does not succeed in mapping this benchmark efficiently. Standard manual optimizations (e.g. loop merging, loop unrolling, etc.) can improve this, but this was not performed in order to demonstrate the effect on energy consumption of a variation in interconnect topology, even in absence of an effect on performance. The performance variation between the best and the worst architecture shown in Figure 7 is only 2% for this benchmark, while the difference in energy consumption is still 23% when compared to the architecture with the most rich interconnect (*b_neg_nh_rf*).

Because the implementation of this "broadcast"-type interconnection can consume over 60% of the energy consumption of the cores (see Figure 1), we propose an optimization which significantly reduces this cost, without restricting the number of detination FUs. By taking this approach compilation freedom is maintained while still reducing the energy consumption.

C. Optimized, Selective Interconnect Implementation

Figure 8 shows the detailed implementation of the proposed optimization. By adding a pass gate to every outgoing path, and selectively activating the connections depending on which FUs are actually consuming the produced value, the fan-out tree that has to be charged can be significantly reduced. We apply this optimization to all connections *"before the pipeline register"*, *"between the pipeline register and the next FU"* and to the *"extra connections to the VLIW register file"* (not shown in the figure). Each pass gate is controlled by a one bit control signal (Ctrl in Figure 8) which is stored in the configuration memory. These bits can be automatically generated at compile time, together with the control bits of the MUXes at the inputs of the FUs. The cost of storing one bit per extra pass gate is correctly accounted for in our energy estimations, as it as it increases the size of the opcodes stored in the configuration memories. The size of the configuration memory of every FU was increased with 12 bits (for the FUs

205

978-1-4244-3039-0/08 $25.00 © 2008 IEEE

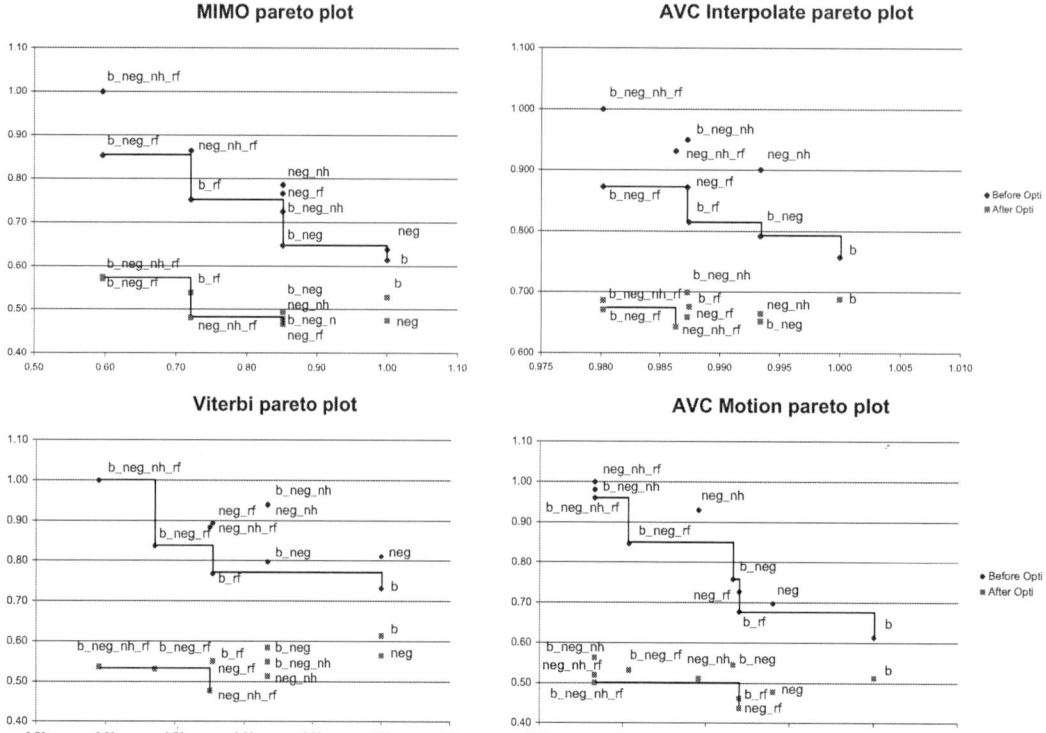

Fig. 7. Pareto plot of interconnect architecture exploration for MIMO, Viterbi, AVC Interpolate and AVC motion estimation benchmarks. The upper line connects un-optimized Pareto points, while the lower points shows the optimized interconnection topologies discussed in Section V-C.

at the side array less bits are needed, as not all connections exist). Extra compression could be used for the encoding of these extra bits in order to reduce the energy consumption of the configuration memory. As the extra area that is needed for the pass gates is less than 1% of the area of the CGRA datapath unit (FU and corresponding distributed data and configuration storage, as shown in Figure 2) this overhead is ignored in the rest of this paper. From a compiler point of view, all information on communication between FUs is available at compile time. Therefore the generation of the control bits for the pass gates can be done at compile time, or even during a simple post-processing step.

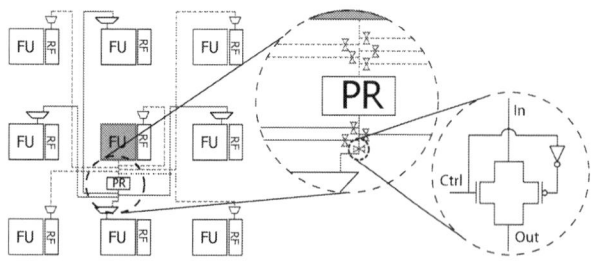

Fig. 8. Detailed Optimized Interconnection Implementation, for One FU in the ADRES array. Pass Gates, controlled by a control signal Ctrl are inserted in order to selectively activate used connections.

1) Result of Energy Optimization: Figure 9 shows a detailed energy breakdown of the *b_neg_nh_rf* architecture running the MIMO channel estimation benchmark, both before

and after the proposed optimization. It can clearly be seen that the energy in the the connections of the type *"before the pipeline register"*, *"between the pipeline register and the next FU"* and *"extra connections to the VLIW register file"* have been heavily reduced: approximately a factor 10 for the connections to the next FU and 33% for the extra connections to the VLIW RF, which lead to an overall reduction of over 40 percent. Given the fact that buses are already activated only when they are used and are preferentially used for longer communications, the expected gains of segmenting the buses are small and they are not optimized further. Adding extra configuration bits (for the Ctrl signal in Figure 8) leads to an increase in the energy consumption of the configuration memory (e.g. 20% more energy for the configuration memory only, as can be seen in Figure 9).

2) Effect on the Energy vs. Performance Tradeoff: Figure 7 shows the effect on the energy vs. performance trade-off, both before and after the proposed optimization, for all interconnect architecture variants. Without the proposed optimization, we observe that adding more connections to the architecture improves performance in some cases, but always leads to a higher energy consumption. In this case the architecture that provides only horizontal and vertical buses has the lowest energy consumption, because adding nearest neighbor or next hop connections quickly leads to large fan-out trees have to be activated. In the un-optimized case, the architectures using the *next hop or nh* connections are therefore almost never Pareto optimal.

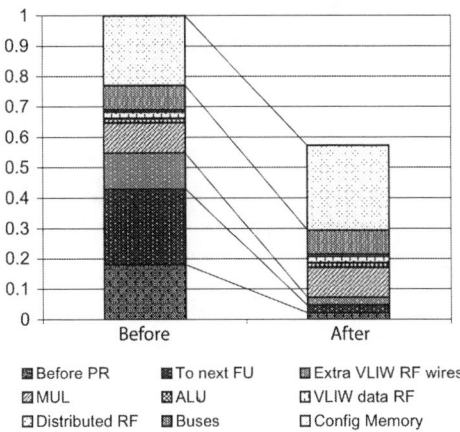

Energy Breakdown for Optimized Interconnect Implementation

Legend:
- Before PR
- To next FU
- Extra VLIW RF wires
- MUL
- ALU
- VLIW data RF
- Distributed RF
- Buses
- Config Memory

Fig. 9. Detailed Energy Breakdown of the b_neg_nh_rf architecture instance, before and after the proposed interconnect optimization.

When using the proposed optimization, the cost of extra connections is heavily reduced, while the performance benefit is maintained. Optimizing the connections of the type *"before the pipeline register"*, which connect the FU to distributed register files of neighboring FUs in the diagonal direction, is performed for all architectures, which e.g. leads to the energy reduction for the *b* architecture. Selectively activating the nearest neighbor, next hop connections and the extra connections to the VLIW RF leads to significant energy cost reductions for the respective architectures, with respect to the cost of using horizontal and vertical buses, which span the complete height and width of the architecture. This results in a relatively bigger energy reduction for the *neg, nh and rf* architectures. Consequently the most connected architecture is Pareto optimal, or very close, for all benchmarks (which was not the case before) and its energy consumption is reduced with over 40% for three out of four benchmarks and for over 30% for the remaining one.

Moving from the optimized high performance architecture *b_neg_nh_rf* to the more energy-efficient but lower performance optimized Pareto points, leads to an energy reduction of about 15% on the complete core for MIMO and about 10% for the other benchmark. For heavily energy constrained embedded applications, this can be very significant still. The performance penalty for the energy consumption reduction trade-off is between 10 and 20% for three out of four benchmarks (the un-optimized AVC Interpolate benchmark is not to be considered here).

VI. CONCLUSIONS

In this paper we have introduced an optimized interconnect implementation and have shown the corresponding result on the energy vs. performance exploration for various interconnect instances of a CGRA architectures. For the given set of benchmarks and the explored architectures, we can conclude that the proposed energy optimization using pass gates significantly reduces the energy consumption in the interconnect

of the Coarse Grained Reconfigurable Architecture. The results show energy reductions compared to the non-optimized variants of over 40% for the most connect architectures. An energy vs. performance trade-off of about 10% for energy for the complete core and between 10 and 20% for performance can still be found between different optimized interconnect topologies. The result of the proposed optimization is that the architecture with the most rich interconnect has become Pareto optimal, and the resulting trade-off range on the energy axis has become much smaller. We can conclude from this experiment that for the used technology, if our proposed energy optimization is used, optimizing the interconnect for performance alone in most cases is sufficient. However, for future scaled technologies it is predicted [5] that interconnect will scale less well than logic, and the impact of interconnect energy consumption on the total energy consumption will grow again. Therefore the energy vs. performance trade-off might again become more significant.

REFERENCES

[1] N. Bansal, S. Gupta, N. Dutt, A. Nicolau, and R. Gupta. Network topology exploration of mesh-based coarse-grain reconfigurable architectures. In *Proc. of Design Automation and Test in Europe (DATE)*, 2004.

[2] L. Benini, D. Bruni, M. Chinosi, C. Silvano, and V. Zaccaria. A power modeling and estimation framework for vliw-based embedded system. *ST Journal of System Research*, 3(1):110–118, April 2002. (Also presented in PATMOS 2001).

[3] D. Burger, S. Keckler, and K. McKinley. Scaling to the end of silicon with edge architectures. In *IEEE Computer*, volume 37(7), pages 44–55, 2004.

[4] J. Y. Chen, W. B. Jone, J. S. Wang, H.-I. Lu, and T. F. Chen. Segmented bus design for low-power systems. *IEEE Transactions on VLSI Systems*, 7(1), March 1999.

[5] H. DeMan. Ambient intelligence: Giga-scale dreams and nano-scale realities. In *Proc of ISSCC, Keynote Speech*, February 2005.

[6] F.Bouwens, M.Berekovic, A.Kanstein, and G.Gaydadjiev. Architectural exploration of the adres coarse-grained reconfigurable array. In *Proceedings of ARC*, 2007.

[7] Z. Kwok and S. Wilton. Register file architecture optimization in a coarse-grained reconfigurable architecture. In *Proc of FCCM*, April 2005.

[8] A. Lambrechts, P. Raghavan, M. Jayapala, D. Verkest, and F. Catthoor. Energy-aware interconnect-exploration of coarse grained reconfigurable processors. In *IEEE Workshop on Application Specific Processors*, September 2005.

[9] G. Lu, H. Singh, M. Lee, N. Bagherzadeh, F. Kurdahi, and E. Filho. The MorphoSys parallel reconfigurable system. In *Proc of Euro-Par*, 1999.

[10] B. Mei, S. Vernalde, D. Verkest, H. D. Man, and R. Lauwereins. DRESC: A retargetable compiler for coarse-grained reconfigurable architectures. In *Proc. of International Conference on Field Programmable Technology*, pages 166–173, 2002.

[11] B. Mei, S. Vernalde, D. Verkest, H. D. Man, and R. Lauwereins. ADRES: An architecture with tightly coupled vliw processor and coarse-grained reconfigurable matrix. In *Proc of FPL*, 2003.

[12] Montium TP Processor, http://www.recoresystems.com. *Montium Tile Processor Reference Manual*, 2005.

[13] D. Novo, B. Bougard, P. Raghavan, H. Souk, and L. V. der Perre. Energy-performance exploration of a cga-based sdr processor. In *Proc of SDR Forum*, 2006.

[14] Philips Research, http://www.siliconhive.com. *Philips SiliconHive Avispa Accelerator.*

[15] P.Raghavan, A.Lambrechts, M.Jayapala, F.Catthoor, and D.Verkest. EM-PIRE: Empirical power/area/timing models for register files. In *International Journal on Embedded Systems (special issue on Media and Stream Processing)*, 2006. To appear.

[16] Sandbridge Technologies, http://www.sandbridgetech.com. *The Sandblaster Architecture.*

21st International Conference on VLSI Design

Integrated TIA-Equalizer for High Speed Optical Link

Saurav Bandyopadhyay
Indian Institute of Technology, Kharagpur
Saurav.Bandyopadhyay@iitkgp.ac.in

Pradip Mandal
Indian Institute of Technology, Kharagpur
pradip@ece.iitkgp.ernet.in

Stephen E. Ralph
Georgia Institute of Technology, Atlanta
stephen.ralph@ece.gatech.edu

Kenneth Pedrotti
University of California, Santa Cruz
pedrotti@ee.ucsc.edu

Abstract

This paper introduces an integrated TIA (Transimpedance Amplifier) Equalizer combination for a low cost Giga-Bit Optical Communication System with perfluorinated Graded Index Plastic Optic Fiber (GI-POF) offering a large optical bandwidth. To realize the potential cost advantages offered by POF based data link, a large area photodiode is used to reduce costly alignment precision, with a TIA-equalizer. Equalization is needed due to bandwidth limitation resulting from the large photodiode. The Photo-Detector capacitance with input impedance of the TIA forms a dominant low frequency pole. Equalizer extends the bandwidth of the whole system by introducing a peak in its frequency response at high frequencies to compensate the loss of bandwidth. The main focus is to use an inexpensive, large area Photo-Detector to increase fiber to detector alignment tolerance simultaneously preserving system speed using the equalizer. This system is for a bit rate varying from 1 to 3Gbps. The TIA-Equalizer combination was implemented with 0.18μm Digital CMOS process using 1.8V Supply.

1. INTRODUCTION

For high-bit-rate long-distance data transport, optical transmission via silica based fiber optics has emerged as the dominant technology. Conversely short distance data transmission, even at high data rates is primarily by copper based interconnects with parallel equalized links used to achieve the highest bitrates-length products. The need for expensive impedance controlled connectors and more costly and complicated cables for copper based solutions contemplated for 1-10Gbps data rates and transmission

distances > 10 m lengths has inspired us to investigate optical solutions that could be cost competitive with electrical interconnects in this regime. The recent development of graded index (GI) multimode perfluorinated plastic optical fiber enables both high bandwidth [1] (40Gbps), due to low modal dispersion, and transparency at 850 nm, a wavelength regime in which relatively lower cost uncooled high bandwidth VCSEL (Vertical Cavity Surface Emitting Laser) emitters are commercially available.

The cost advantages of current POF based applications are chiefly due to the ease of terminating the large core fibers and the use of low-speed inexpensive optical components (LEDs and Si PIN detectors). As 10Gbps data rates are approached it is not readily apparent that the cost advantages of POF based links can continue to be realized. Recently however it has been shown that the mode mixing in the new generation of perfluorinated GI-POF is quite strong [2]. This means that unlike 50 micron core multimode silica fiber the stringent high precision alignment at the transmitter is not required because propagation is fairly independent of launch conditions beyond relatively short distances (~0.1-1m). Unfortunately, however, as 10Gbps data rates are approached the detector size decreases to ~ 50 micron for typical PIN detectors. The optical precision needed for focusing light from 50-120 micron core fibers on to such small detectors threatens to obviate the perceived cost advantages for POF because costs rise nonlinearly as alignment tolerances below ~25 microns are required. Here we investigate the use of a large PIN detector to increase alignment tolerance combined with an equalizer to compensate the bandwidth degradation due to the resulting pole at the input to the TIA. Here we show that by using a cheaper, large diameter Photo-Detector and a large core GI-POF we are able to get the same performance in speed as in a system with a sophisticated high speed, Photo-Detector without requiring a comparably precise alignment.

In this work, a low cost CMOS 0.18μm technology is used. Therefore, the speed is limited by the f_T of the CMOS process. Thus a speed of 1 to 3Gbps is targeted in this paper. Block level schematic of the optical communication system is shown in Fig.1.

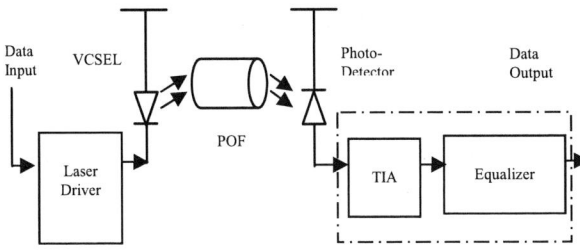

Fig.1. Front End of the Optical Communication System

2. TOPOLOGY

2.1. TIA (Transimpedance Amplifier)

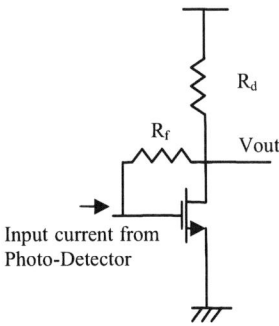

Fig.2. Circuit Diagram of Single Ended TIA

Fig.2.shows the single ended TIA [3]. The Transimpedance by first order approximations is R_f. The input impedance is $R_f /(1+A)$, where A is the voltage gain of the common source amplifier with loading due to the feedback resistor. For Giga bit operation, the pole due to the input impedance and the Photo-Detector capacitance is the dominant pole. This limits the speed of the link. In order to extend the bandwidth, it is required to have low input impedance. This would push the dominant pole to higher frequencies. Lowering the input impedance would mean higher gain and/or lower feedback resistor R_f. As the gain- bandwidth product is limited by the f_T of the

transistor, we cannot increase the gain arbitrarily. On decreasing R_f, the noise current from the TIA feedback resistor R_f, which is the dominant source of noise, increases ($4KT/R_f$). As the TIA is the first block of the receiver, its noise requirements are more stringent than the blocks succeeding it. Decreasing R_f would also decrease the transimpedance of the circuit.

This indicates there is a trade-off between speed and noise performance. Using the equalizer along with the TIA removes the noise-speed tradeoff. The low frequency dominant pole is compensated by the high pass characteristic of the equalizer.

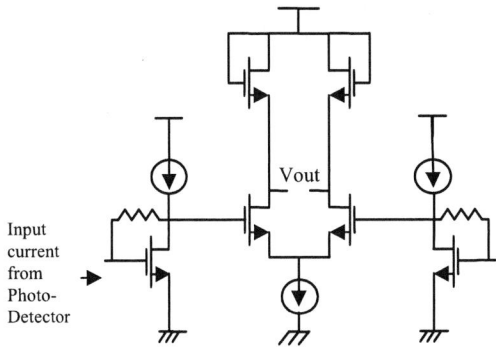

Fig.3. Circuit Diagram of TIA with single ended TIA and Single to Differential Converter

The circuit implemented is a pseudo differential TIA (Fig.3.) which accepts single ended input current from the Photo-Detector and gives a differential output. It is preferred to convert the single ended input current to differential voltage to reduce the effect of noise that may crop up as a common mode signal. The right side dummy TIA is used to get appropriate bias voltage for the second input of the single-to-differential converter. Further, use of the dummy TIA along with an external dummy photodiode helps to treat supply/ground bounce noise as common mode signal and get it rejected by the single-to-differential converter.

2.2. Equalizer

Block Diagram of the Equalizer is shown in Fig.4.

2.2.1. Inductive Peaking Block: The equalizer implemented here boosts the high frequency components of the incoming signal. As shown in Fig.4, the first stage of the equalizer is the inductive peaking [4] stage responsible for the high pass characteristic.

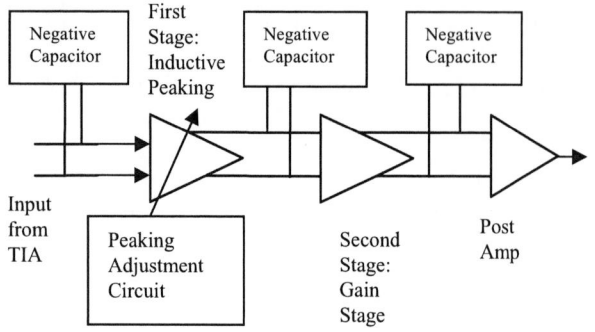

Fig.4. Block Diagram of the Equalizer with Post Amp

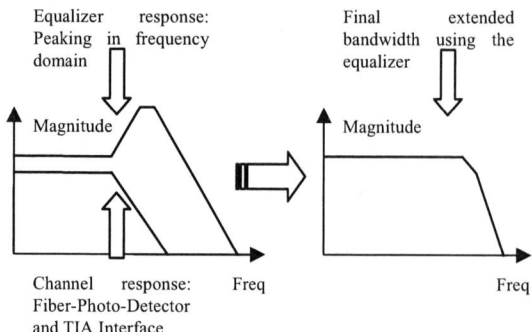

Plot1   Plot2

Fig.5. Plot 1 showing the band limited response before equalization and Plot2 after equalization

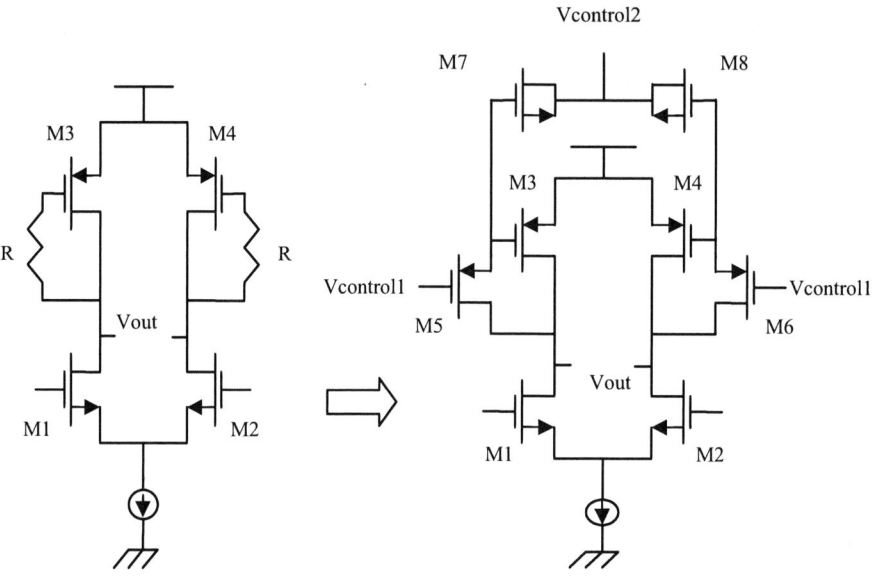

Fig.6. Differential Amplifier with Active Inductor in the first circuit and with Differential Amplifier with Tunable Active Inductor in the second circuit

This stage introduces a peak in the frequency response by using an active inductor. Frequency response, before and after equalization is shown pictorially in Fig.5.

The first circuit in Fig.6 is a general active inductor circuit. The circuit gives a gain of $gm_{1,2}/gm_{3,4}$. But at high frequencies, the gate to source capacitors $C_{gs3,4}$ start acting. Thus, the gain now becomes $gm_{1,2}R$. If R is chosen to be greater than $1/gm_{3,4}$ we get a peaking at high frequencies. By utilizing this circuit, the

tunable version of this circuit was implemented by replacing the R by a PMOS in triode and adding extra MOS capacitance to make the peak location variable. Thus, the location of the peak can be varied by varying the voltages Vcontrol1 and Vcontrol2 as shown in the Fig.6.

This topology of tuning peak location is different from the tuning mechanism used for cable equalizers in [6] and [7]. In this topology, the dc gain of the circuit is fixed and the high frequency gain is tuned.

210

978-1-4244-3039-0/08 $25.00 © 2008 IEEE

But in [6] and [7], the high frequency gain of the peaking block is fixed and the dc gain changes with tuning. That may cause the dc gain to go below 0dB in some cases. DC gain below 0dB in the equalizer would not be desirable as it is the second block in the receiver. The noise requirement may not be as stringent as compared to the TIA which is the first block, but gain below unity will increase input referred noise. Thus, the proposed peaking circuit is superior to [6] and [7].

2.2.2. Negative Capacitor: In this implementation, negative capacitor circuit has been used [5]. As shown in Fig.7, using Miller capacitor between nodes having the same phase, a negative capacitor is realized. The value of the negative capacitor is given by C(1-A) where A is the voltage gain of the differential amplifier in the circuit. For A>1, the effective capacitance is negative. The negative capacitor circuit is used in parallel with the next stage as shown in the block diagram in Fig.4. This helps to decrease the effective capacitive load seen by the previous stage. Even though the circuit is band-limited, it is designed to have impedance inductive (or negatively capacitive) up to a frequency of 3.5GHz. Therefore, for the frequency range of interest, the circuit is inductive.

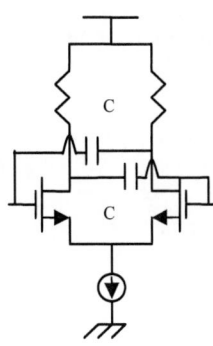

Fig.7. Negative Capacitor Circuit

2.2.3. High Gain Stage with Post Amp: As the single stage will not be able to provide a high gain, a High gain stage was used to provide a gain up to high frequencies. This was followed by a Post Amp to make the signal from low swing analog to high swing digital signal. To achieve this, a traditional comparator circuit was used. The complete schematic is in Fig.8.

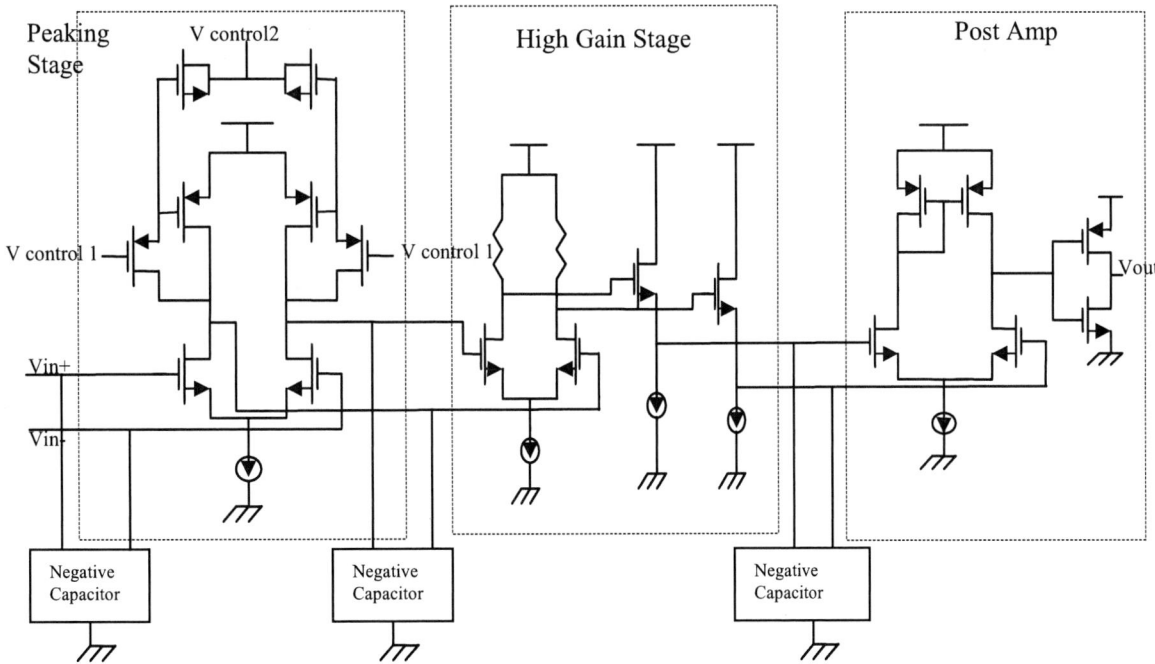

Fig.8. the complete equalizer circuit

978-1-4244-3039-0/08 $25.00 © 2008 IEEE

3. SIMULATED RESULTS

Table.1. shows the extracted simulated performance of the TIA. The bandwidth measured is for 500fF pin capacitance and no external Photo-Detector.

Table.1. Performance of TIA

Parameters	Slow	Typical	Fast	Unit
Bandwidth	1.5	1.9	2.1	GHz
Transimpedance	47.34	45.96	45.33	dB ohm
Input Impedance	39.33	38.1	37.18	dB ohm
Input referred Noise	-	93 at 700MHz	-	pA/√Hz

Table.2. Performance of Equalizer

Photo-Detector Cap	Bandwidth of TIA	Bandwidth of TIA-Equalizer *(Extended)*
5pF	390MHz	1.45GHz
2.5pF	700MHz	1.8GHz
1.5pF	970MHz	2.04GHz

Table.2. shows the effect of adding the Photo-Detector capacitor. The bandwidth of TIA reduces due to large Photo-Detector Capacitance. The Equalizer provides a peaking in frequency domain extending the bandwidth as shown in Fig.9. The plot shows the pole due to the Photo-Detector and TIA is at 400MHz. This is compensated by having a zero and extending the overall bandwidth to about 2GHz.

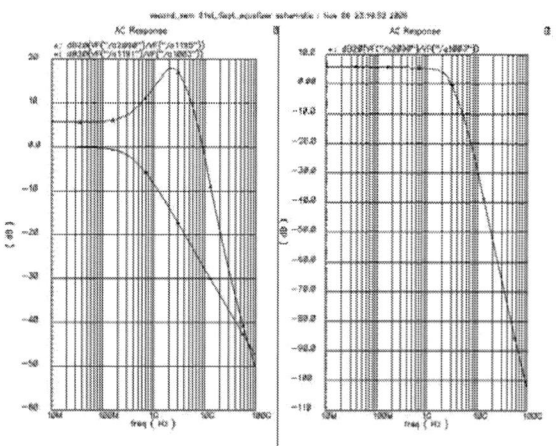

Plot1   **Plot2**

Fig.9. Plot1 showing simulated band limited frequency response before equalization and the equalizer response. Plot 2 shows the response after equalization

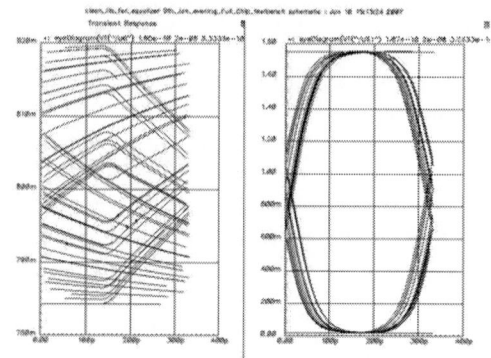

Fig.10. Eye Diagrams before and after equalization at 3Gbps

Fig.10. shows the eye diagrams from post layout transient simulations at 3Gbps. The input eye (before equalization) is closed. The equalizer output eye diagram is open.

4. EXPERIMENTAL RESULTS

The IC was fabricated using National Semiconductor's 0.18μm CMOS9X process.

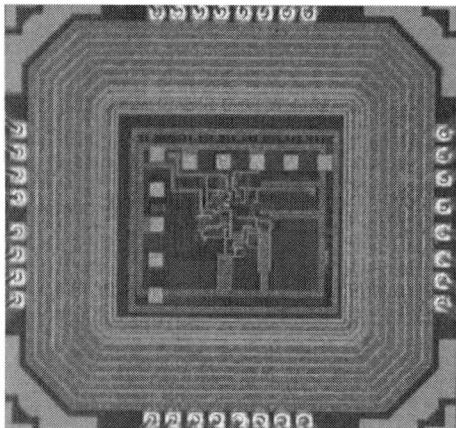

Fig.11. the photograph of the die

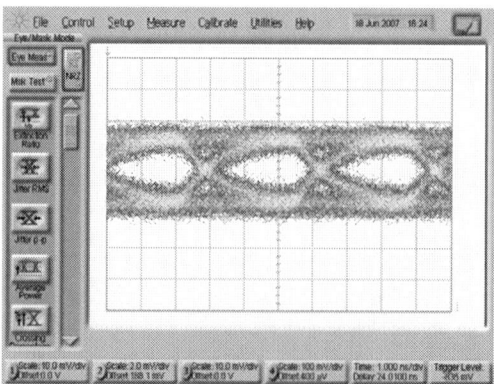

Fig.12. output of the TIA at 300Mbps

The test results were obtained by using a 50 micron plastic optical fiber and photo-detector of effective active area 0.12mm^2. Fig.12. shows the TIA output at 300Mbps. The eye-diagram of the output is open. The response can be approximated to a first order single pole roll off response. When we move to higher speeds like 1.0625Gbps, the eye-diagram completely closes as is shown in Fig.13.

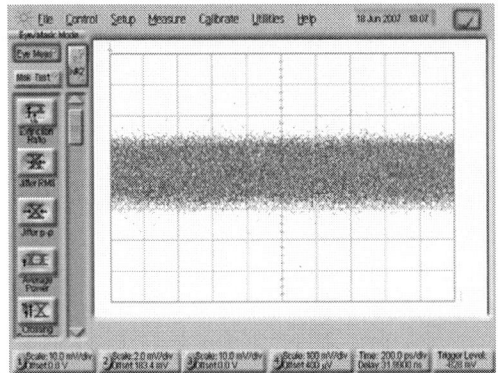

Fig.13 output of TIA at 1.0625Gbps

Fig.14 shows the output of the equalizer at 1.0625Gbps. The TIA output in Fig.13 is the input to the equalizer. The peaking in the equalizer response compensates the loss of bandwidth at the TIA-Equalizer interface. The measured equalizer outputs show a 300mV p-p swing instead of 1.8V swing shown in the simulated Eye-Diagram. This is because of the 50Ω loading due to the measuring instrument.

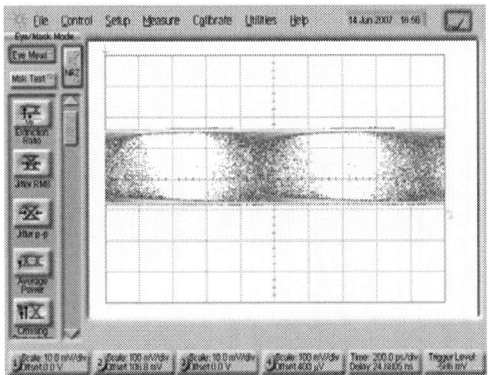

Fig.14 output of the equalizer at 1.0625Gbps

5. CONCLUSIONS

The TIA-Equalizer discussed here is for an optical communication system having POF and large active area Photo-Detector. By using an inexpensive, large area (large Capacitance) Photo-Detector with the Equalizer we are able to get the same performance in speed as with an expensive, small area Photo-Detector (small capacitance). Thus, by using the equalizer, we are able to reduce the overall cost of the system and also increase the fiber to diode alignment tolerance. The concept of such Equalizers is used in Cable or Backplane Equalizers. In cables too, we may have a loss of bandwidth due to skin effect at high frequencies. The speed achieved here can be improved if a high f_T process is chosen. Thus, by using a lower gate length technology CMOS process or a BiCMOS process, 10Gbps speed can be obtained as in [6] and [7].

REFERENCES

[1] A. Polley, R. J. Gandhi, and S. E. Ralph, "40Gbps Links Using Plastic Optical Fiber", Optical Fiber Communication Conference and Exposition and The National Fiber Optic Engineers Conference, OSA Technical Digest Series (CD) (Optical Society of America, 2007), paper OMR5.

[2] S. E. Ralph, A Polley, K. Pedrotti, R. Dahlgren and J. Wysocki, "New Methods for investigating mode coupling in multimode fiber: Impact on High-speed links and channel equalization", SOFM 2006

[3] E. Sackinger, "Broadband Circuits for Optical Fiber Communications," Wiley Publishers, New York, 2005.

[4] M. Maeng, Y. Hur, S. Chandramouli, F. Bien, H. Kim, C. Chun, E. Gebara, J. Laskar, "Fully Integrated 0.18um CMOS Equalizer with an active inductance peaking delay line for 10Gbps data Throughput over 500m Multimode Fiber", Microwave Symposium Digest, 2005 IEEE MTT-S International Publication Date: 12-17 June 2005 On page(s): 4 pp.

[5] B. Shem-Tov, M. Kozak and E. Friedman, "A High Speed CMOS Op-Amp Design Technique using Negative Miller Capacitance", 11th IEEE International Conference on Electronics, Circuits and Systems, ICECS 2004. Proceedings of the 2004, 13-15 Dec. 2004, page(s): 623- 626.

[6] S. Gondi, J. Lee, D. Takeuchi, B. Razavi, "A 10Gb/s CMOS Adaptive Equalizer for Backplane Applications", Digest of Technical Papers ISSCC 2005 IEEE International Publication Date: 6-10 Feb. 2005 On page(s): 328- 601 Vol. 1

[7] G. E. Zhang, M. M. Green, "A 10Gb/s BiCMOS Adaptive Cable Equalizer", IEEE Journal of Solid State Circuits, Vol. 40, No 11, November 2005.

Single Edge Clock (SEC) Distribution for Improved Latency, Skew, and Jitter Performance

Jeff Mueller and Resve Saleh
University of British Columbia
jmueller@ece.ubc.ca and res@ece.ubc.ca

Abstract

Synchronous clock distribution continues to be the dominant timing methodology for VLSI designs. As processes shrink, clock speeds increase, and die sizes grow, more-and-more of the clock period is lost to skew and jitter budgets. We propose to improve clock performance by focusing on the single, critical clock edge while relaxing requirements of the non-critical edge. A novel re-design of the traditional clock buffer is proposed as a drop-in replacement for existing clock distribution networks, yielding timing performance improvements of over 20% in latency and skew and up to 30% in jitter; alternatively, these timing advantages could be traded off to reduce clock buffer area and power by 33% and 12%, respectively.

1. Introduction

Synchronous clock delivery in VLSI has always been a major design challenge. Significant area, power, and metal resources are all required to distribute a high-speed clock across a large die synchronously with minimal skew and jitter. Yet, according to ITRS projections [1], as processes scale further down into the nanometer range, larger percentages of the clock period will be lost to skew and jitter. Therefore, techniques to improve overall clock timing performance are needed.

The vast majority of present day VLSI circuits use synchronous clocks driving single-edge triggered flip-flops (FFs). The design and architecture of many commonly used clock distribution systems are thoroughly described in several papers [2][3]. Standard inverters, commonly referred to as clock buffers, are typically used to drive the clock signal from the root (PLL or DLL source) to the leaves (FFs) of a clock distribution network (e.g., H-tree, grid, hybrid). The buffers have traditionally been designed to produce equal rise and fall times [4]. Such 'symmetric' buffers of the traditional clocking method allocate an equal amount of valuable transistor current drive for each of the two clock edges, propagating a nearly constant clock pulse width throughout the entire clock distribution network, even though only one of those edges is actually *critical* for timing considerations.

Many elaborate design techniques have been proposed and employed to reduce skew and/or jitter: post-silicon tuning (i.e., active deskew) via delay insertion [5][6], dual-Vdd designs [7], and link insertion designs [8], to name a few. However, this paper proposes a new method which is simpler and less disruptive to the ASIC flow, yet significantly increases the clock distribution network's timing performance without changing the interconnect wires and without increasing the clock buffer area or power.

Significant timing gains can be realized by replacing the single chain of equal rise/fall buffers with an alternating pattern of asymmetric rise/fall buffers that focus the majority of their current drive only on the one truly *critical* clock edge – the edge ultimately activating the end FFs. The other *neglected* clock edge will be somewhat degraded; however, depending upon the circuit's clock period and timing requirements, that degradation may present no real liability to the circuit's overall performance. Well-documented methods of clock buffer sizing and insertion along with interconnect wire sizing and routing [9][10] could be carried out to design the distribution network of a traditional clocking (TC) method. Then, in our approach, that TC network could be converted to a single-edge clocking (SEC) method simply by removing the TC clock buffers and dropping-in new SEC clock buffers.

This paper demonstrates how the SEC method makes better use of the clock period. Then, it describes how the SEC clock buffers are designed and provides simulation results showing significant improvements for a typical H-tree clock network. Finally, a few limitations of the SEC method are discussed, and conclusions are drawn.

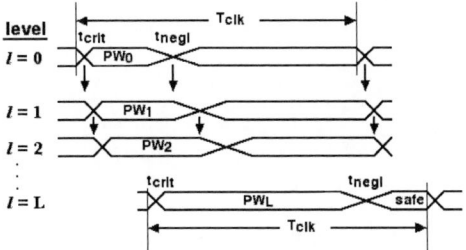

Figure 1. **Optimized clock period utilization**

2. Basics of SEC Distribution

The clock edge which toggles the FF will be referred to as the *critical (or leading) edge*, and the other clock edge will be called the *neglected (or trailing) edge*. By designing the clock tree to focus only on the critical edge, significant timing improvements can be achieved with virtually no penalties in area or power.

Previously, both clock edges were relaxed to reduce switching noise [11], and delay was reduced by using $\beta=1$ instead of the traditional, current balanced $\beta=2.5$ buffers [4]. However, our novel SEC method propagates the clock signal via alternating *strong pull-up* (INVr) and *strong pull-down* (INVf) buffers. Furthermore, since every branch in the clock tree will have the same number and size of buffers, the conditions for minimum skew as derived in [12] are met, even though the buffers themselves are asymmetric.

2.1. Improved Clock Period Utilization

The concept behind SEC clocking is best explained by examining the four regions of the clock period. The leading edge (rise or fall) is followed by the leading level (high or low), followed by the trailing edge (fall or rise), and the trailing level (low or high). In SEC designs only one of these four regions is critical for timing performance – the leading edge should have low skew and low jitter – while the other three regions merely exist to 'setup' for the leading edge of the next clock period. Hence, more focus should be placed on the leading, *critical* edge, even at the expense or neglect of the other three regions.

The clock period diagram of Fig.1 shows the effect of neglecting the trailing edge throughout a typical SEC distribution. The leading edge is propagated quickly and cleanly by each SEC buffer; however, the trailing edge is relaxed, allowing the leading level to grow and the trailing level to shrink. Eventually, the trailing edge would exceed the clock period window if this trend continued, and the clock signal would be lost. Simple equations can be derived to bound the amount

Table 1. **SEC buffer degradation factor choices**

η	PW_0 min	PW_0 max
1	400ps	2400ps
1.5	400ps	1700ps
2	400ps	1000ps
2.3	400ps	580ps
2.5	400ps	300ps

of trailing edge degradation allowable for a given clock period and distribution network. First, assume that t_{crit} and t_{negl} of Fig.1 are constant for all SEC levels. The amount of pulse width (PW) or leading level growth through l SEC buffers (or levels) can be expressed as:

$$PW_l = PW_0 + \frac{1}{2}(t_{negl} - t_{crit})l \quad where \ l = 0, 1, ...L \ (1)$$

Here, if $t_{negl} = t_{crit}$, as in traditional (symmetric) clocking, there is no pulse width growth and the duty cycle remains constant throughout. Next, let T_{clk} = the clock period, let $\eta = (t_{negl}/t_{crit})$ = the degradation factor of each SEC buffer, and let L = the number of SEC levels; so, the initial (root clock) pulse width, PW_0, is bounded by:

$$t_{crit} \leq PW_0 \leq 0.8T_{clk} - t_{crit}(\eta+1) - \frac{1}{2}t_{crit}(\eta-1)L \ (2)$$

A safeguard of 20% of the clock period was chosen to account for generalized assumptions and simulation inaccuracies. The initial clock pulse width, PW_0, must be large enough to propagate through the first buffer, but should be small enough to allow for as much clock period degradation as possible. Choosing a higher degradation factor will result in better critical edge performance, but more of the clock period will be degraded, so a trade-off must be made. The number of SEC levels, L, is limited by the clock distribution network; hence, the design challenge is to minimize PW_0 in order to maximize η.

2.2. Choices for a Typical H-tree

To validate this approach, the H-tree simulation in this paper has a T_{clk} of 4ns, a t_{crit} of 400ps (10% of T_{clk}), and 5 levels of SEC buffers in a typical 90nm process. Plugging these values into Eqn.(2) results in the data shown in Table 1. The table indicates that a degradation factor of 2.5 is not feasible since $PW_0max < PW_0min$, and a factor of 2.3 would be considered the absolute limit. A more suitable factor of 1.5 and an initial clock pulse of 500ps has been selected for later simulations.

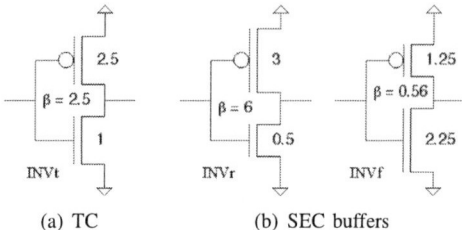

(a) TC   (b) SEC buffers

Figure 2. **Relative xtor sizing for clock buffers**

3. Design of SEC Buffers

3.1. Transistor Sizing

When sizing a chain of buffers with given wire interconnects and fixed end load for minimum delay, the goal is to increase the transistor sizes in order to increase their current drive, but only up to the point at which the transistor capacitances begin to slow themselves down. The two capacitances of the buffers at issue are $C_{self} = C_{diff}(W_P + W_N)$ and $C_{in} = C_{ox}(W_P + W_N)$, where both C_{diff} and C_{ox} are effectively constant for a given process.

When designing SEC buffers if their $(W_P + W_N)$ is kept constant, even as their $\beta = (W_P/W_N)$ is varied, then these new INVr's and INVf's can be used as drop-in replacements for the INVt's; all three are shown in Fig.2. Their capacitive loading on the distribution network and on themselves, and even their total gate area will all remain roughly the same; however, the alternating SEC buffers will deliver the critical edge faster and sharper because they 'overweight' the transistors handling that critical edge and 'underweight' the transistors handling the neglected edge.

3.2. Beta Selection

The timing equations for INVr and INVf can be derived from the Elmore delay formula and a basic circuit model. Furthermore, since $(W_P + W_N)$ is a constant, the two equations, in units of ps, can be expressed in terms of β as follows:

$$t_{rise} = 2.15[20.4(1/\beta) + 81.9] \qquad (3)$$

$$t_{fall} = 2.15[8.5\beta + 70.0] \qquad (4)$$

A fitting factor of 2.15 has been used to account for the difference between the Elmore definition of delay and the simulation transition times, as well as the use of the Slow corner model (SS, 0.9V, 150°C) for worst-case timing versus the Nominal hand calculation

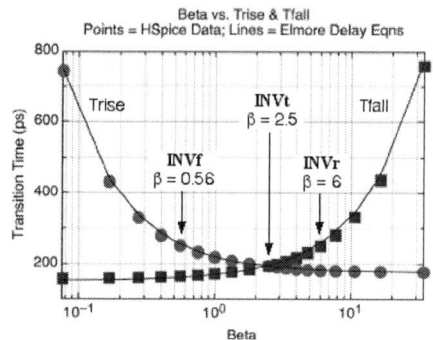

Figure 3. **Trise and Tfall vs. buffer beta**

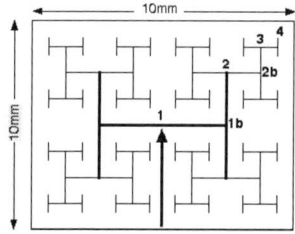

Figure 4. **4-level H-tree (#'s represent buffers)**

data used. The derived Elmore equations match the simulation data extremely well as plotted in Fig.3 and, as such, they can be used to determine the appropriate β's necessary to achieve the degradation factor, $\eta = (t_{negl}/t_{crit})$ as explained in Sec.2.

As indicated in the plot of Fig.3, INVf's have β's less than 2.5 and their critical edge is the falling edge, while INVr's have β's greater than 2.5 and their critical edge is the rising edge.

4. H-tree Simulation Results

We carried out HSPICE simulations to compare the relative performance of the clock buffers of the TC and SEC methods, nominally and over PVT variations, in a realistic 10x10mm^2 chip design. Hence, a standard, balanced fanout-4, 4-level H-tree was constructed. Both methods use exactly the same wire loading and wire process corners.

4.1. H-tree Design Details

4.1.1. H-tree Branch Architecture. The H-tree used in these simulations is shown in Fig.4. It is designed to deliver a supplied clock signal (as represented by the arrow pointing to 1) to 64 leaves covering the entire 10x10mm^2 chip. Buffers are positioned at the center of each H structure (1, 2, and 3), along with extra buffers to handle the long wire loads (1b and 2b), and finally at the leaf ends to drive the FF loads (4).

978-1-4244-3039-0/08 $25.00 © 2008 IEEE

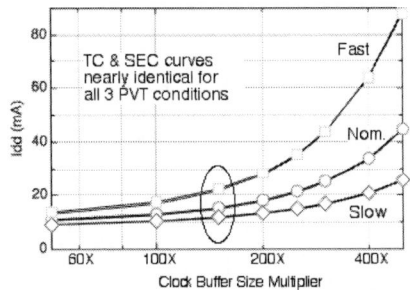

Figure 5. Average current vs. buffer size

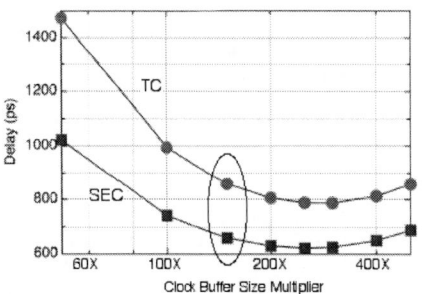

Figure 6. Latency vs. buffer size

The distribution wires from the clock source to the 2nd level buffers are designed to be 2X wide (as indicated by the thick lines), and therefore are modeled as half the resistance per length of the remaining branches. Approximate 90nm process wire parameters of Rwire=300Ω/mm and Cwire=200fF/mm have been used in a single-pi wire model, and the FF loads at each leaf are modeled as a 200fF capacitor.

4.1.2. H-tree Buffer Sizing. Generally, clock buffers are sized as big as necessary to achieve necessary skew performance [9], but not so big as to violate area and power constraints. The average current graphs of Fig.5 show that total clock power begins to grow unmanageably for buffer sizes of 200X and larger; therefore, TC to SEC performance comparisons were made at the 150X size; however, all performance data are plotted over all clock buffer sizes (50X to 500X) to highlight the similar behavior of the two methods.

For TC simulations, all the buffers are INVt as in Fig.2(a). For SEC simulations, the buffers at levels 1, 2, and 3 are INVf and the buffers at levels 1b and 2b are INVr, as in Fig.2(b). The buffers at level 4 are *always* INVt (for both TC and SEC), in order to provide two relatively clean clock edges to the FF loads. The sizing of the buffers is applied identically to *all* buffers of the H-tree for both TC and SEC methods.

4.2. SEC Performance Improvements

4.2.1. Latency Improvements. Latency, or the total path delay from clock root to leaves, is not necessarily as critical for a synchronous system as are skew and jitter. However, when two clock architectures share the same wire H-tree, have the same number of buffers, roughly the same buffer area and power, yet one delivers the critical clock edge significantly faster than the other, the faster method would be preferred.

As seen in Fig.6 at the 150X size, the SEC latency is nearly 200ps or 23% faster than the TC latency (659ps

vs. 858ps). The similar shapes of the curves also demonstrate that capacitive self-loading is behaving as expected for both TC and SEC, as explained in Sec.3.1.

The time-domain graphs of Fig.7 further illustrate the TC and SEC operation in the H-tree, specifically the latency improvement, neglected edges, and pulse width growth of SEC versus TC. Waveforms for the input root clock and all branch buffers (1, 1b, 2, 2b, 3, and 4 [in bold]) are shown for three different clocking methods. The top graph of 7(a) is for the TC method – all buffers are type INVt, all of the leading and trailing edges have similar transition times, and the signal maintains its 50% duty cycle throughout. The middle graph of 7(b) is for a safe SEC method ($\eta = 1.5$) – three type INVf and two INVr are alternated, terminated by a INVt, the leading (critical) edges are sharper and earlier than those of the TC method, but the trailing (neglected) edges are slower for all but the final leaf clock, and therefore the duty cycle grows from the root to the leaf. The bottom graph of 7(c) is for the limit of the SEC method ($\eta = 2.3$) – the INVf and INVr β's are spread further apart increasing the SEC characteristics, but the final leaf clock nearly consumes the entire clock period, increasing the risk of losing the clock signal.

4.2.2. Skew Improvements. Skew, or the spatial variation of clock edges, can be caused by on-die PVT variations of one clock path or branch versus another branch. While the precise level of skew experienced across any clock distribution network requires accurate models of random and systematic intra-die variations [13], a relative comparison of the robustness to skew of two clock distribution methods can be obtained by measuring their worst-case skew. Two simulations are performed on the TC and the SEC H-trees; the first with Slow (SS, 0.9V, 150°C) PVT parameters, and the second with Fast (FF, 1.1V, 0°C) parameters; the difference of their slow and fast latencies (overall path delays from root to leaf), is a measure of their worst-

978-1-4244-3039-0/08 $25.00 © 2008 IEEE

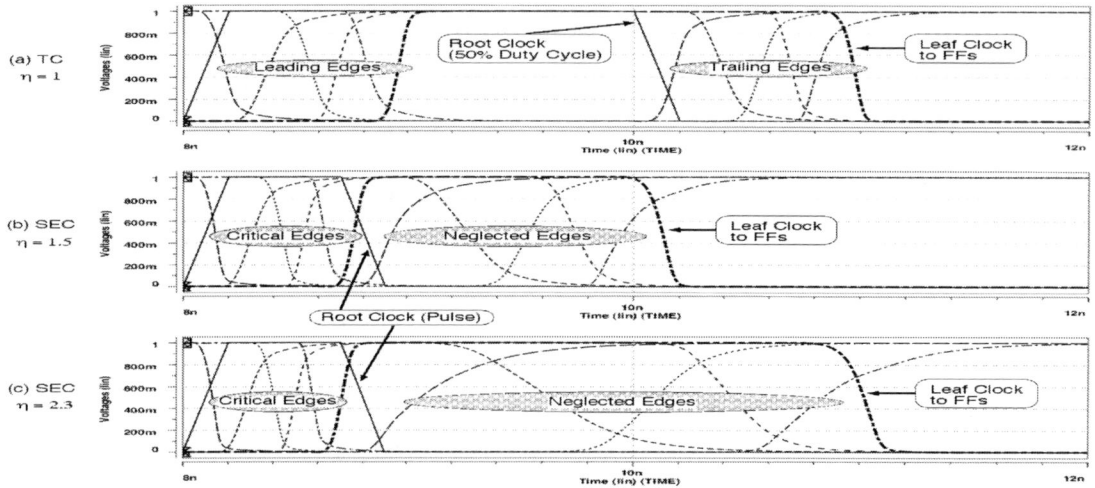

Figure 7. **Timing waveforms for TC and two choices of SEC**

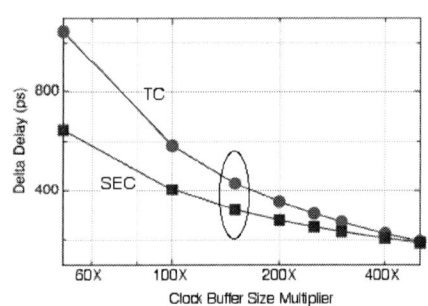

Figure 8. **Worst-case skew vs. buffer size**

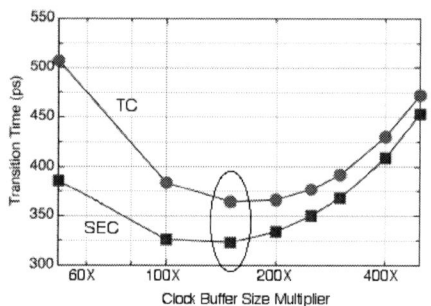

Figure 9. **Average transition time vs. buffer size**

case skew. Wire variation was not modeled, since both methods use the exact same wires, and any wire attributed skew would be similar.

Fig.8 shows that the worst-case skew for SEC is 25% less than that of TC (322ps vs. 431ps) at the 150X size. It would not be accurate to state that SEC has 110ps less skew than TC, since these are unrealistic worst-case numbers, and actual intra-die PVT variations would perhaps be half of these extremes. But, the relative improvement of 25% less skew for SEC versus TC would still hold. So, if for example a TC distribution network has 200ps of skew, then the comparable SEC network could be expected to have only 150ps of skew – a considerable improvement.

4.2.3. Jitter Improvements. Jitter, or the temporal variation of clock edges, is primarily caused by noise impulses on the clock signal, and is therefore very difficult to simulate. However, previous work on jitter, or phase noise, has developed a metric of Impulse Sen-

sitivity Function (ISF) which relates jitter performance to clock edge transition slope [14]. Each transition of the critical edge of the clock leaves it exposed to these noise impulses and therefore vulnerable to jitter; when the clock is not transitioning (at a high or low level) or when the neglected edge is transitioning, then noise impulses are not a jitter concern. Therefore, the transition times of the critical edges (at 1, 1b, 2, 2b, and 3) are averaged to compare the jitter performance of SEC to that of TC.

As seen in Fig.9 the amount of average transition time of the critical edge is 11% less for SEC than TC (323ps vs. 364ps) at the 150X size. Calculating the ISF(rms) across an entire clock period, however, is a more accurate measure of jitter performance; and as derived in [14], that equation becomes a cubed relationship. Hence, the real jitter improvement of SEC over TC would be (0.89^3), which is a full 30% improvement in jitter performance.

218

978-1-4244-3039-0/08 $25.00 © 2008 IEEE

4.2.4. Area and Power Improvements. Alternatively, the better timing performance described above could be traded off to reduce clock buffer area and clock distribution power by essentially 'under-sizing' the clock buffers. As seen in the latency, skew, and jitter performance graphs of Figs.6, 8, and 9, equal or better timing performance can be obtained with the 100X sized SEC buffers as with the 150X sized TC buffers, resulting in a roughly 33% reduction in clock buffer gate area and a 12% reduction in total nominal clock distribution current (i.e., power) – 14.7mA for 150X TC down to 12.9mA for 100X SEC from Fig.5.

5. Limitations

Our novel SEC method improves timing performance by trading off the clock period between the critical and neglected edges. It is not, however, a 1-to-1 trade-off, so this method would only be useful in ASIC circuit designs operating at *less than* the maximum allowable frequency of a given CMOS process.

Also, if both the pull-up and pull-down transistors of a buffer are on simultaneously, short-circuit current will flow from Vdd to Ground. Therefore, the slow transition of the neglected edge in the SEC method must be handled carefully. The smaller devices of these buffers naturally reduce the level of short-circuit current; however, SEC buffers drive a final symmetric TC buffer, so it will have a higher short-circuit current.

Finally, as indicated earlier, at the clock tree leaves a more symmetric clock buffer must be used to clean up the neglected clock edge before driving the FF. The HSPICE simulations performed here use the TC INVt buffer and model the FF load as strictly capacitive. A more accurate model or the use of an actual FF as the load may necessitate a slight redesign or resizing of the final leaf clock buffer and/or the single-phase FF itself.

6. Conclusions

By shifting the clock buffer resources already available from equal rise/fall buffers to alternating fast-rise/fast-fall buffers, significant reductions in latency, skew, and jitter can be achieved with virtually no penalty in clock buffer area or power consumption. HSPICE simulations show that these new INVr and INVf buffers can be simply dropped into an existing TC H-tree buffer distribution network to effectively improve clock timing performance by over 20% or to reduce power by 12%. This new SEC technique could also be utilized in other clock distribution networks (grid, hybrid) with similar results.

Acknowledgements

This work has been funded by the Natural Sciences and Engineering Research Council (NSERC) of Canada, and by PMC-Sierra of Burnaby, B.C.

References

[1] T. K. G. Anthony V. Mule, Elias N. Glytsis and J. D. Meindl, "Electrical and optical clock distribution networks for gigascale microprocessors," *IEEE Trans. on VLSI*, vol. 10, pp. 582–594, Oct. 2002.

[2] E. G. Friedman, "Clock distribution networks in synchronous digital integrated circuits," *Proc. of the IEEE*, vol. 89, pp. 665–692, May 2001.

[3] R. Escovar and R. Suaya, "Optimal design of clock trees for multigigahertz applications," *IEEE Trans. on CAD*, vol. 23, pp. 329–345, Mar. 2004.

[4] N. Hedenstierna and K. O. Jeppson, "Cmos circuit speed and buffer optimization," *IEEE Trans. on CAD*, vol. CAD-6, pp. 270–281, Mar. 1987.

[5] L. Z. Jeng-Liang Tsai and C. C.-P. Chen, "Statistical timing analysis driven post-silicon-tunable clock-tree synthesis," *IEEE/ACM ICCAD*, pp. 575–581, Nov. 2005.

[6] A. K. Yaron Elboim and R. Ginosar, "A clock-tuning circuit for system-on-chip," *IEEE Trans. on VLSI*, vol. 11, pp. 616–626, Aug. 2003.

[7] S. A. Tawfik and V. Kursun, "Dual-vdd clock distribution for low power and minimum temperature fluctuations induced skew," *Proc. of ISQED*, pp. 73–78, Mar. 2007.

[8] B. Y. Makoto Mori, Hongyu Chen and C.-K. Cheng, "A multiple level network approach for clock skew minimization with process variations," *Proc. of ASP-DAC*, pp. 263–268, Jan. 2004.

[9] R. S. Dimitrios Velenis and E. G. Friedman, "Buffer sizing for delay uncertainty induced by process variations," *Proc. of IEEE ICECS*, pp. 415–418, Dec. 2004.

[10] M. H. Shlomo Greenberg, Ido Bloch and A. Maman, "Optimization of chip level clock tree performance by using simultaneous drivers and wire sizing," *Proc. of IEEE ICECS*, pp. 419–423, Dec. 2004.

[11] E. Backenius and M. Vesterbacka, "Reduction of simultaneous switching noise in digital circuits," *IEEE Norchip*, pp. 187–190, 2006.

[12] M. Shoji, "Elimination of process-dependent clock skew in cmos vlsi," *IEEE JSSC*, vol. 21, pp. 875–880, Oct. 1986.

[13] A. Narasimhan and R. Sridhar, "Impact of variability on clock skew in h-tree clock networks," *Proc. of ISQED*, 2007.

[14] S. L. Ali Hajimiri and T. H. Lee, "Jitter and phase noise in ring oscillators," *IEEE JSSC*, vol. 34, pp. 790–804, Jun. 1999.

21st International Conference on VLSI Design

Threshold Voltage Control through Multiple Supply Voltages for Power-efficient FinFET Interconnects

Anish Muttreja, Prateek Mishra and Niraj K. Jha

Dept. of Electrical Engineering, Princeton University,

Princeton, NJ 08544

{muttreja, pmishra, jha}@princeton.edu

Abstract

In modern circuits, interconnect efficiency is a central determinant of circuit efficiency. Moreover, as technology is scaled down, the importance of efficient interconnect design is increasing. In this paper, we explore an option for low-power interconnect synthesis at the 32nm node and beyond, using fin-type field-effect transistors (FinFETs) which are a promising substitute for bulk CMOS at the considered gate lengths. We consider a previously-unexplored mechanism for improving FinFET efficiency, called threshold voltage control through multiple supply voltages (TCMS), which is significantly different from conventional multiple-supply voltage schemes. We develop a circuit design for a FinFET buffer using TCMS. We describe a variation of van Ginneken's classic dynamic programming framework for solving the problem of power-optimal TCMS buffer insertion on a given routing tree. We show that, on an average, TCMS can provide power savings of 50.41% and device area savings of 9.17% compared to a state-of-the-art dual-V_{dd} interconnect synthesis scheme[1].

1 Introduction

Steady miniaturization of transistors with each new generation of bulk CMOS technology has yielded continual improvement in the performance of digital circuits. The scaling of bulk CMOS, however, faces significant challenges in the future due to fundamental material and process technology limits [1]. A prominent manifestation of the above challenges is the growing difficulty VLSI designers experience in meeting power budgets. A dominant portion of the total power consumed in a chip is often consumed in its interconnect network. For instance, it was estimated in [2] that over 50% of the power consumed in modern microprocessors is consumed in repeaters inserted on chip interconnects. In a further challenge to scaling, it is also estimated [3] that 70% of the total cell count at the 32nm node may be due to repeaters, up from roughly 50% in high-performance designs at the 90nm node. Clearly, therefore, interconnect efficiency can be expected to be a central determinant of circuit efficiency. This paper attempts to explore how FinFETs, an emerging transistor technology that might supplement or supplant bulk CMOS, at the 32nm node or beyond, may be used to improve interconnect efficiency.

According to the *2005 International Technology Roadmap for Semiconductors (ITRS)* [4], the scaling of bulk CMOS to sub-32nm gate lengths faces significant obstacles including short-channel effects, high sub-threshold and gate-dielectric leakage and device-to-device variations. It is expected that the use of double-gate field-effect transistors (DG-FETs), which provide better control of short-channel effects, lower leakage and better yield in aggressively-scaled CMOS processes, will be required to overcome these obstacles to scaling [4, 5]. FinFETs are one of the most popular implementation of DG-FETs. It is relatively easy to manufacture FinFETs, which are free of alignment problems which plague many other DG-FET geometries. Also, the use of a lightly-doped channel in FinFETs makes them resistant to random dopant variations.

It was demonstrated [6] that digital logic circuits using FinFETs can be significantly more power-efficient than their counterparts implemented in bulk CMOS at the same gate length. Thus, directly translating bulk CMOS interconnects to FinFETs may also be expected to provide corresponding power savings. However, beyond the obvious technology-driven efficiency benefits, circuits can take advantage of FinFET's double-gate structure to further optimize power and performance. An important FinFET characteristic is dynamic *threshold voltage (V_{th}) controllability*. The V_{th} at each gate of a FinFET can be controlled through the application of a voltage at the other gate. Since the V_{th} governs both transistor power consumption and delay, V_{th} controllability is a powerful tool for circuit optimization. In this paper, a new synthesis style for interconnect circuits with multiple supply voltages, which takes advantage of V_{th} controllability in connected-gate FinFETs, is presented. Traditionally, in multiple supply voltage circuits, power is saved through the use of a lower supply voltage on off-critical paths. For instance, a conventional dual supply voltage (V_{dd}) circuit may use the nominal high-performance V_{dd} for the technology process at hand and, on off-critical paths, a lower V_{dd}, which is typically 60%-70% in magnitude compared to the higher V_{dd}.

In a significant departure from convention, a new multiple-V_{dd} scheme, called TCMS, that does not employ a lower V_{dd} is proposed in this paper. Three supply voltages are used in TCMS: a nominal supply voltage (V_{dd}^L), a slightly *higher* supply voltage (V_{dd}^H), and a slightly negative supply voltage (V_{ss}^H)[2]. The design is based on the principle that in an overdriven FinFET inverter (an inverter which is driven by an input voltage that is higher than its V_{dd}), both leakage and output-current drive (and, thereby, delay) can be controlled simultaneously. As will be seen in Section 3, sub-threshold leakage is reduced because of an increase in the V_{th} of the leaking transistor, while current drive is increased because of the larger gate drive experienced by the active transistor. This allows both sub-threshold leakage and device width, and, thereby, dynamic power consumption, to be reduced.

TCMS is a voltage-level-shifter-free style of circuit de-

---

[1]Acknowledgments: This work was supported by SRC under contract No. 2007-HJ-1602.

---

[2]Superscripts H and L are used to denote high and low magnitudes respectively, without regard to sign, i.e., $|V_{ss}^L| < |V_{ss}^H|$, but $V_{ss}^L > V_{ss}^H$, where V_{ss}^L refers to the nominal ground voltage.

220

978-1-4244-3039-0/08 $25.00 © 2008 IEEE

sign, i.e., inverters tied to high and low V_{dd} are allowed to freely alternate, without the need for dedicated level shifters or level-shifting latches, unlike classic clustered voltage scaling schemes [7]. Instead, level-shifting is built into inverters that require it, through the use of higher V_{th} FinFETs, in a fashion similar to [8]. FinFETs with higher V_{th} are used at the input of buffers that are connected to V_{dd}^H but driven by buffers operating at V_{dd}^L in order to eliminate static leakage current. This allows frequent use of overdriven buffers in TCMS interconnects, thereby increasing the attendant power savings obtainable through the use of overdriven inverters in TCMS.

The remainder of this paper is organized as follows. Related work in the areas of interconnect synthesis and design with FinFETs is reviewed in Section 2. The principle of TCMS, mentioned above, is illustrated in Section 3. Details about TCMS circuit design can be found in Section 4. Power consumption mechanisms in TCMS designs are studied in Section 4.1. The popular textbook example of a uniform infinitely-long interconnect wire is used to explore some general principles regarding TCMS design in Section 4.2. The synthesis of multiple-pin interconnects encountered in real circuits is studied in Section 5, where buffer delay models for use in TCMS synthesis (Section 5.1), and an algorithm for inserting FinFET buffers on a given wiring tree (Section 5.2) are given. TCMS interconnects are compared with conventional dual V_{dd} interconnects in Section 6, where estimated savings in total power consumption (50.41%) are demonstrated. Conclusions are presented in Section 7.

2 Related Work

We review related work in the areas of interconnect synthesis and FinFET-based circuit design. A large and important body of work exists in both areas. Unfortunately, it is impossible to do justice to all related ideas within the constraints of this section. Therefore, only a survey of the most directly related ideas is attempted.

Power consumption in interconnects is not a recent concern. A predictive analysis for interconnect power consumption trends at deeply-scaled technology nodes was performed in [9], where it was shown that sub-threshold leakage power will dominate interconnect power consumption at the 70nm and 50nm technology nodes. Our estimates for the 32nm technology node, in Section 6, are in somewhat alarming agreement with the predictions in [9]. Nevertheless, traditionally, the paramount concern in the design of buffered interconnects has been delay optimization. In practice, such designs often take the form of a problem involving the insertion of buffers on wiring trees. van Ginneken's dynamic-programming-based algorithm, proposed in [10], forms the basis of many subsequent procedures for delay- or power-optimal insertion of buffers on a given wiring tree, including our buffer insertion algorithm in Section 5.2. van Ginneken's algorithm requires $O(n^2)$ operations to insert a single buffer type on a wiring tree with n possible buffer locations. An important extension to this algorithm was proposed in [11] to allow the insertion of $|B|$ buffer types in $O(n^2|B|^2)$ time while minimizing power consumption under a given delay constraint. There has been recent work on improving the complexity of delay- and power-optimal buffer insertion in [12] and [13, 14], respectively. The use of dual power supplies was also considered in [13, 14]. Our algorithm for FinFET buffer insertion in Section 5 is based on the algorithm presented in [14].

FinFET circuit design is a very active area of research. Synthesis of FinFET circuits is, however, a relatively unexplored field. A logical-effort-based algorithm for gate sizing using FinFETs was presented in [6]. A linear-programming-based algorithm for low-power logic design, considering various FinFET logic styles, was presented in [15]. A tool for directly translating bulk CMOS netlists at the physical layer to FinFET netlists was reviewed in [16]. As mentioned in Section 1, the circuit designs proposed in this paper are based on the ability to dynamically control FinFET V_{th}. A number of circuit designs, which take advantage of dynamic V_{th} control in FinFETs and other DG-FETs, have been proposed [15, 17–21] in various domains of digital design. Most of the above designs employed DG-FETs with independently-controllable gates, where a reverse- (forward-) biased voltage was used to control the V_{th} at the front gate in order to reduce sub-threshold leakage [15, 19, 20] (critical-path delay [17]). In [18], reverse- and forward-biased voltages were both employed to reduce leakage power consumption and boost read/write performance in SRAM cells. A clever optimization to exploit gate coupling in connected-gate FinFETs circuits was proposed in [21], where the ratio of gate oxide and fin thickness was optimized to enhance gate coupling.

The circuit designs presented in this paper share a number of features with the ideas espoused in the above works. Nevertheless, the paper makes two important contributions. None of the previous works have addressed interconnect synthesis with FinFETs. Secondly, to the best of our knowledge, the principle of TCMS, or its use in circuit design has also never been explored before.

3 The Principle of TCMS

The V_{th} at each FinFET gate can not only be controlled statically through the control of process parameters, such as channel-dopant concentration or the value of the gate work-function, but also dynamically through the application of a voltage to the other gate (gate-gate coupling). A generalized model for the relationship between the threshold voltage ($V_{th_{g_f}}$) at the front-gate (g_f) of a FinFET and the voltage applied to its back gate (g_b) is derived in [22]. However, for the purpose of this paper, the following approximate relationship suffices.

$$V_{th_{g_f}} \approx \begin{cases} V_{th_{g_f}}^0 - \delta(V_{g_b s} - V_{th_{g_b}}) & \text{if } V_{g_b s} < V_{th_{g_b}}, \\ V_{th_{g_f}}^0 & \text{otherwise.} \end{cases}$$
(1)

where s denotes the source terminal of the FinFET, δ is a positive value determined by the ratio of gate and body capacitances, and $V_{th_{g_f}}^0$ is the minimum observed $V_{th_{g_f}}$. Equation (1) is given for an N-type FinFET, but may also be used for a P-type FinFET with the usual changes in sign. If the FinFET is operated with both its gates tied together, the threshold voltages of both gates respond simultaneously to change in voltage at the other gate. As Equation (1) predicts, gate-gate coupling is observed only in the weak-inversion region of operation. In the region of strong inversion, the presence of inversion charge in the channel shields FinFET gates from each other and no coupling is observed.

As mentioned in Section 1, TCMS is based on the observation that in an overdriven inverter, both sub-threshold leakage and current drive can be controlled simultaneously. We illustrate this using Figure 1, where inverter (a) is a so-called high-V_{dd} inverter, i.e., it is connected to V_{dd}^H and V_{ss}^H.

Inverter (b) is a low-V_{dd} inverter, *i.e.*, it is connected to V_{dd}^L and V_{ss}^L. The values of V_{dd}^H, V_{ss}^H and V_{dd}^L are $1.08V$, $-0.1V$ and $1.0V$, respectively. V_{ss}^L is taken to be tied to ground. Inverter (b) is, thus, overdriven. In the remainder of the paper,

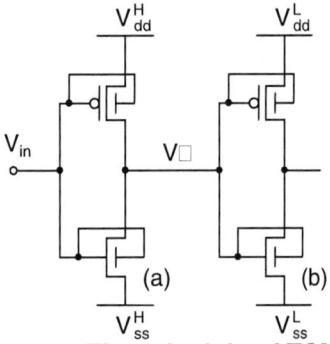

Figure 1. The principle of TCMS

it is assumed that any inverter connected to V_{dd}^H is also connected to V_{ss}^H and similarly for the lower supply voltages.

To illustrate the principle of TCMS, let V_{in} in Figure 1 be held at the logic 0 voltage and $V' = V_{dd}^H$. Then, considering inverter (b) in Figure 1, sub-threshold leakage through the inverter is determined by the leakage through the P-type FinFET, while the inverter's delay is determined, to a large extent, by the current through the N-type FinFET. According to Equation (1), the V_{th} of the P-type FinFET is increased due to the reverse-biased voltage of $0.08V$ observed at its gates, thereby, reducing sub-threshold leakage. The V_{th} of the N-type FinFET, which is operating in the strong-inversion region, is not appreciably altered. Nevertheless, it experiences a forward-biased voltage of $1.08V$, which is higher than the normal gate drive of $1.0V$ at the inverter input. This leads to a somewhat higher drive current. Similarly, the application of a logic 1 voltage at the circuit input leads to a reduction in the leakage of the N-type FinFET and improvement in the drive strength of the P-type FinFET.

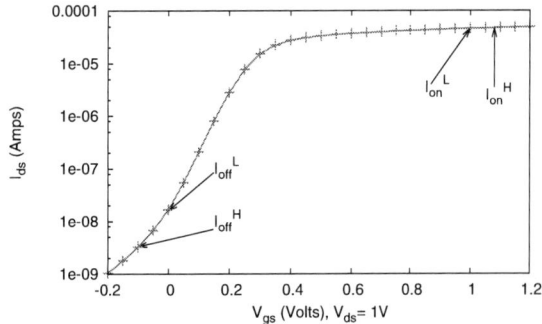

Figure 2. Simulated I_{ds}-V_{gfs} characteristics for an overdriven 32nm N-type FinFET

SPICE-simulated DC transfer characteristics for an overdriven N-type FinFET are shown in Figure 2. A predictive technology model (PTM) for 32nm FinFETs, available from [23], was used for this and all other SPICE experiments reported in this paper. PTM has been validated against manufactured 32nm FinFETs [24] and is widely used in logic simulations [21, 25]. In the simulation, the drain terminal of the N-type FinFET was tied to V_{dd}^L, and the source terminal was tied to ground. The voltage at the gate terminals was swept from V_{ss}^H to V_{dd}^H. On-currents

through the FinFET at normal drive ($V_{gs} = V_{dd}^L$) and overdrive ($V_{gs} = V_{dd}^H$) are indicated by I_{on}^L and $\left(I_{on}^H\right)$ in the figure, respectively. Though it may not be evident on the logarithmic scale of Figure 2, the value of I_{on}^H exceeds the value of I_{on}^L by about 3.4%. The relationship between the corresponding values of off-current I_{off}^H and I_{off}^L, respectively, is evident in Figure 2. I_{off}^H is over $16\times$ smaller in value than I_{off}^L.

4 Circuit Design Considerations

In this section, we first outline the various mechanisms that govern power consumption in TCMS designs (in Section 4.1), following which, a buffer circuit for TCMS links is considered in Section 4.2.

4.1 Power Consumption in TCMS circuits

The mechanisms governing power consumption in TCMS are considerably different from conventional multiple-V_{dd} schemes. Conventionally, the use of a lower supply voltage leads to lower dynamic and leakage power consumption in logic gates tied to it. In TCMS, on the other hand, power savings are observed mainly in overdriven inverters. Three mechanisms that govern the power consumption are observed.

1) Sub-threshold leakage power consumption in overdriven inverters is reduced due to V_{th} control, as discussed in Section 3.

2) The use of a higher V_{dd} leads to higher short-circuit and leakage power consumption in inverters tied to it, as well as higher switching power consumption in capacitances driven by these inverters.

We suggest the following guidelines intended to balance the above opposing mechanisms and minimize power consumption. Values for V_{dd}^H and V_{ss}^H need to be chosen carefully. Higher values lead to larger savings in sub-threshold leakage power consumption through mechanism 1), but also, higher power consumption through mechanism 2). Circuits should be designed such that the parasitic capacitances charged through V_{dd}^H and the widths of high-V_{dd} inverters are minimized. Further, the circuit topology should allow repeated use of overdriven inverters in order to maximize power savings.

4.2 Exploratory Buffer Design for TCMS

A circuit which provides an ideal theoretical substrate to explore TCMS is a long interconnect wire, driven by identical buffers, inserted at regular intervals. The source and sink nodes are also assumed to be driven by identical buffers. The buffers comprise a pair of inverters in series, with the input inverter being smaller in size than the output inverter. Figure 3 depicts a link in this wire, designed

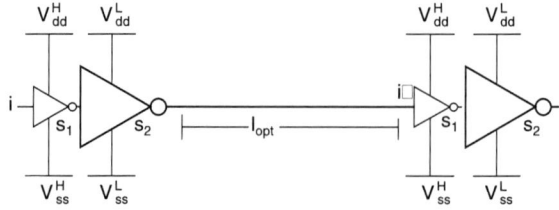

Figure 3. A buffered link on an infinitely long identical interconnect wire

to use TCMS. The size of each inverter (s_1 and s_2) is indicated below it in Figure 3. The length of the link l_{opt},

222

978-1-4244-3039-0/08 $25.00 © 2008 IEEE

is also shown. The smaller inverter in each buffer is tied to V_{dd}^H and the larger buffer is overdriven. The design was chosen because of the following features that might help in maximizing power savings:

1) The size of the high-V_{dd} inverter is kept small.

2) The only parasitic capacitances charged through V_{dd}^H are the input capacitances of the larger inverters, and output capacitances of the smaller inverters. No wire capacitances, which are typically much larger in value than inverter parasitic capacitances, are charged through V_{dd}^H.

3) The larger inverters are overdriven and thus have a much higher I_{on}, presenting a further avenue for saving power: the size of the larger inverters may be reduced, while maintaining delay across the link. Shrinking the larger inverters also reduces the load on the high-V_{dd} inverters feeding them and allows their width to be reduced as well.

4) High- and low-V_{dd} inverters alternate, providing maximum opportunity for power savings.

A significant challenge in the above design is the static leakage that would arise in all the high-V_{dd} buffers since they are driven by voltages lower than their power supply and will not switch off properly. Eliminating static leakage in TCMS designs is the subject of the next section.

4.2.1 Static Leakage Reduction in TCMS Designs

Avoiding static leakage currents in high-V_{dd} buffers requires that the output of a low-V_{dd} buffer be *level-shifted* before being used to drive a high-V_{dd} buffer. As was mentioned in Section 1, existing dual-V_{dd} buffer insertion schemes [13, 14] allow such level conversion to happen at latch boundaries but do not find the use of dedicated level-shifters feasible because the driving strength required of such level-shifters would impose prohibitive area costs. Unfortunately, such a design implies that a net might have at most one overdriven buffer between a pair of latches. This would restrict power savings in TCMS, where large savings are observed in overdriven buffers. TCMS designs, therefore, allow level conversion anywhere by building it within inverters through the use of high-V_{th} FinFETs in high-V_{dd} inverters driven by low-V_{dd} buffers. This obviously imposes a requirement for the FinFET process used to support high-V_{th} FinFETs, which has become possible through the use of gate materials with controllable work functions (typically metals, such as molybdenum [26]). However, FinFET manufacturing processes are still not widely available and the exact range of available V_{th}'s for future FinFETs will be known only as the fabrication process matures. For our study, we manually modified PTM to obtain high V_{th} models by increasing the value of the vth-0 SPICE parameter, by 0.1V.

4.2.2 Design of Average-case circuit

Before considering the design of the TCMS circuit, let us consider the design of a similar link with all buffers tied to V_{dd}^L, which is used as a base case to compare our TCMS design with. Define γ as the ratio of inverter sizes in the buffer *i.e.*, $\gamma = \frac{s_2}{s_1}$. Then, it can be shown that the delay of the link is minimum if the following relationships are true.

$$l_{opt} = \sqrt{\frac{2r_s\left(2c_p + c_o\left(\gamma + \frac{1}{\gamma}\right)\right)}{r_w c_w}}, s_1 = \sqrt{\frac{r_s c_w}{2 r_w c_o}} \quad (2)$$

where l_{opt} and s_1 were defined before and other symbols are defined in Table 1. The proof of Equation (2) can be obtained using a standard delay minimization technique [27] and is, therefore, omitted here. Ratio γ was assumed to be

Table 1. Technology parameters

Wire parameters	
Wire resistance per-unit length (r_w)	$0.73\Omega/\mu m$
Wire capacitance per-unit length (c_w)	$.2fF/\mu m$
Minimum-sized FinFET inverter parameters	
Fin height	$40nm$
N-type FinFET width (W)	$80nm$
P/N ratio (β)	2
Input capacitance (c_o)	$0.255fF$
Output capacitance (c_p)	$0.125fF$
Temperature	$70°C$
Equivalent resistance (r_s)	$3.23k\Omega$

2 in this experiment. A result similar to Equation (2) for $\gamma = 2$ may be found in [28]. Table 1 lists wire, FinFET technology, and inverter design parameters that were used in this and other experiments throughout the paper. The wire parameters in Table 1 are from [29]. The delay of the single-V_{dd} link was measured using SPICE simulations. In order to ensure a properly-shaped input waveform as well as a realistic load, identical link elements were added to the input and output of the link. Design parameters for the link, as well as the measured delay and power, are given in Table 2. Delay was measured between the inputs of the two smaller

Table 2. Link design parameters and measurements

Parameter	Single V_{dd}	TCMS
Link length (l_{opt})	$0.199mm$	$0.199mm$
Inverter widths (s_1, s_2)	42, 84	30, 50
Delay (ps)	12.19	12.27
Power (μW)	1080	647
V_{dd}^H, V_{ss}^H	–	1.08, −0.10

inverters (points i and i' in Figure 3). Rise and fall delays were measured separately, and the maximum is reported here. The width (W) of a FinFET is quantized in terms of its number of fins (n) [16], as $W = 2nH_{fin}$ where H_{fin} is the height of each fin. The quantization makes it difficult to equalize rise and fall delays of inverters exactly. Power and delay were measured in SPICE by applying a 5GHz pulse with a 50ps slope at the input. The power reported is of a somewhat high order because it corresponds to an effective switching probability (α) of 1. Power is reported instead of energy per switch because that measurement would ignore active leakage power consumption which can be a significant component of total power consumption[3]. Measurements in SPICE capture all components of power consumption accurately.

4.2.3 TCMS Circuit Design

Having identified the topology of a TCMS buffered link, we present values for the inverter widths s_1 and s_2 and the high power supply voltages V_{dd}^H and V_{ss}^H in Table 2 to complete the design. Values in Table 2 were obtained by a process of exhaustive SPICE simulations. The length of the link was assumed to be the same as in the average-case circuit and delay was constrained to be less than 101% of the conventional link design. The resultant design met the delay bound

---

[3]Leakage power consumption in 28nm FinFET circuits was estimated to be roughly 30% of total power consumption in [30].

and had 40.1% lower power consumption. Given the large search space of possible link designs, our design is probably far from optimal. It does, however, confirm the feasibility of TCMS and also provides acceptable values for the V_{dd}^H and V_{ss}^H voltages which are used for more detailed evaluation in Section 6. Some other features of the reported design are noteworthy. The sizes of both inverters in each buffer are significantly reduced, leading to a total decrease in device width of 36.51%. The width of the output inverter is reduced in part because of its increased drive strength, owing to overdrive, and in part because of reduced load from the next buffer stage. A comparatively smaller percentage reduction is observed in the size of the input inverter because the use of high-V_{th} FinFETs reduces its drive strength.

5 Interconnect Synthesis

In this section, the problem of TCMS buffer insertion and sizing is considered. We are given a wiring tree, attached to a source and a number of sinks. Before discussing the algorithm in Section 5.2, it is necessary to define models used for delay, power and slew, in the next section.

5.1 Delay, Power and Slew Models

A distributed Elmore delay model is used, as in [10, 11, 13, 14]. The delay $d(l)$ due to a piece of wire of length l is given by

$$d(l) = (\frac{1}{2}c_w l + c_{load})r_w.l \qquad (3)$$

where c_{load} is the capacitive load at the end of the wire and c_w and r_w are as defined in Table 1. Elmore delay times $ln(9)$ is used as a slew metric [31]. As seen in Section 3, the delay of a buffer depends on its threshold voltage, which is controlled by the voltage level at the input of the buffer. The delay of a buffer, $d(b, h^u)$, is assumed to be given by

$$d(b, h^u) = d_{int}(b, h^u) + r_o(b, h^u)c_{load} \qquad (4)$$

where d_{int} is the intrinsic delay of buffer b (delay at no load), h^u indicates the voltage level at the input of buffer b and may be V_{dd}^L or V_{dd}^H and r_o and c_{load} are the output resistance and capacitive loading at the output of buffer b, respectively. d_{int} and r_o are maintained as pairs of values, one for each value of h^u. In fact, different values of d_{int} and r_o are used for falling and rising transitions at the buffer outputs, respectively. However, considering signal polarity, while conceptually simple, complicates the discussion significantly. Signal polarity is, therefore, not considered further in the interest of simplicity. It is observed in [13, 14] that the values of slew in a buffered interconnect are consistently within a few tens of picoseconds of the imposed slew bound (100ps). Therefore, like [13, 14], we conservatively measure buffer delay using a ramp voltage with slope equal to the slew bound. Power consumption in the interconnect is measured as

$$P_w(l, c_{load}) = \alpha f(c_w l + c_{load})\left(h^u\right)^2 \qquad (5)$$

where α is the probability of a $0 \rightarrow 1$ transition and f is the clock frequency. As above, h^u indicates the level of the voltage which drives the wire. Power consumption in a buffer b is given by

$$P(b, h^u) = \alpha f e_d(b, h^u) + p_l(b, h^u) \qquad (6)$$

where $e_d(b, h^u)$ is a function representing the energy consumed by the short-circuit current through buffer b and switching energy consumed at parasitic capacitances at its output and $p_l(b, h^u)$ is the sub-threshold leakage power through the buffer. The power consumed at the input of the buffer is included in the interconnect power in Equation (5).

Algorithm 1: TCMS-BIS

input: T_n

1 **if** n is not a sink terminal **then**
2      $Set(\Phi_n) = \emptyset$
3 **else**
4   **return** $\{(c_n, P_w(0, c_n, V_{dd}^L), 0, \bar{b}, V_{dd}^L),$
         $(c_n, P_w(0, c_n, V_{dd}^H), 0, \bar{b}, V_{dd}^H)\}$
5 **foreach** $b \in set(B)$ **do**
6   **if** $\texttt{voltage}(b) = V_{dd}^H$ and
         $\texttt{threshold}(b) = V_{th}^L$ **then**
7          $Set(\Phi_b) = \{(c(b), P(b, V_{dd}^H), 0, b, V_{dd}^L)\}$
8   **else**
9          $Set(\Phi_b) = \{(c(b), P(b, V_{dd}^L), 0, b, V_{dd}^L),$
             $(c(b), P(b, V_{dd}^H), 0, b, V_{dd}^H)\}$
10   **foreach** child v of n **do**
11         $Set(\Phi_{temp}) = Set(\Phi_b)$
12         $Set(\Phi_{temp}) = \emptyset$
13         $Set(\Phi_v) = $
           $\texttt{3DSample}(\texttt{TCMS-BIS}(T_v), h^d(\Phi_t))$
14   **foreach** $\Phi_t \in Set(\Phi_{temp})$ **do**
15   **foreach** $\Phi_v \in Set(\Phi_v)$ **do**
16                 $\Phi_{new} = \texttt{merge-child}(\Phi_t, \Phi_v)$
17   **if** Φ_{new} meets slew and is not
                   dominated by $\Phi_z \in$
                   $Set(\Phi_n) \cup Set(\Phi_b)$ **then**
18   remove all $\Phi_z \in Set(\Phi_b)$
                       dominated by Φ_{new}
19                     $Set(\Phi_b) = Set(\Phi_b) \cup \{\Phi_{new}\}$
20   **foreach** $\Phi_b \in Set(\Phi_b)$ **do**
21   **if** Φ_b is not dominated by any $\Phi_n \in Set(\Phi_n)$
           then
22   remove all $\Phi_n \in Set(\Phi_n)$ dominated by
               Φ_{new}
23             $Set(\Phi_n) = Set(\Phi_n) \cup \{\Phi_{new}\}$

5.2 Buffer Insertion Algorithm

An algorithm to insert FinFET buffers, and select their sizes and V_{dd}'s on a wiring tree is discussed in this section. It is assumed that the tree is specified as a set of locations (nodes) at which buffers may be inserted, and connections between them. A set of sinks, which are degree-1 nodes, is also given. A loading capacitance (c_{n_s}) and a required arrival time (q_{n_s}) are given at each sink terminal n_s. A unique node, designated as n^{src}, is identified as the source terminal. The direction along the wiring tree, which leads towards (away from) n^{src}, is called the upstream (downstream) direction. Two connected nodes n and v are said to be parent and child if node n is upstream of node v. It is assumed that the source is driven by a high-V_{dd} buffer with output resistance r_{src}. Our TCMS buffer insertion and sizing (TCMS-BIS) algorithms, given in Algorithm 1, is based on the Fast-dBIS algorithm proposed in [14]. However, the use of FinFET buffers, whose delay and power consumption depend on the level of their input voltage, necessitates significant modifications to the Fast-dBIS algorithm. Also both Fast-dBIS and TCMS-BIS are approximate algorithms. Nevertheless, it was shown [14] that the power consumption and delay obtained through Fast-dBIS is accurate to within 3% of exact solutions in a large number of test cases. Moreover, in Section 6, algorithm TCMS-BIS is compared with algorithm `Fast-dBIS` and shown to yield solutions that require 50.41% less power consumption than it at the same delay constraint. Thus, the approximate nature of the algorithm is not a significant drawback in practice.

5.2.1 Preliminaries

The `TCMS-BIS` algorithm is based on a dynamic programming framework. Power-optimal options are constructed at each node using partial solutions (*i.e.*, options) from its subtrees. An option Φ_n at node n is a tuple $\Phi_n = (c, p, q, b, h^u)$ where c, p and q are the upstream capacitance, power consumption and required arrival time, respectively. $b \in Set(B)$ is the buffer inserted at node n under option Φ_n and h^u is as defined earlier in Section 5.1. $Set(B)$ is a set containing all available buffer choices. Buffer $\bar{b} \in B$ is used to indicate that no buffer was used. The required arrival time at n^{src} is used to measure overall delay performance.

Definition 1 *The required arrival time q_n at node n is defined as $q_n = \min_{\forall n_s}(q_{n_s} - d(n, n_s))$ where $d(n, n_s)$ is the delay from node n to sink node n_s.*

As the `TCMS-BIS` algorithm proceeds, a list of options for the subtree rooted at each node is generated by recursively traversing the tree in a bottom-up fashion. At each node n, existing options are iteratively updated by merging them with options from children nodes. Options from parent and children nodes can be merged if the voltage provided downstream by the parent node's option is compatible with the voltage expected upstream by the child node's option. Next, define function $h^d(\Phi)$ as follows.

Definition 2 *Given option $\Phi_n = (c_n, p_n, q_n, b, h_n^u)$, $h^d(\Phi_n)$ returns the V_{dd} value that is supplied downstream of n. If buffer b_n is the null buffer $\bar{b}$, $h^d(\Phi_n) = h_n^u$, otherwise $h^d(\Phi_n)$ is the V_{dd} value to which buffer b_n is tied.*

Definition 3 *Let node v be a child of node n, then options $\Phi_n = (c_n, p_n, q_n, b_n, h_n^u)$ and $\Phi_v = (c_v, p_v, q_v, b_v, h_v^u)$ at node n and node v, respectively, are* voltage-compatible *if $h^d(\Phi_n) = h_v^u$ and may be* merged *to obtain option $\Phi_n' = (c_n', p_n', q_n', b_n', h_n^{u'})$ according to Equation (7).*

$$c_n' = \begin{cases} c_n & \text{if } b \neq \bar{b} \\ c_n + c_v & \text{if } b = \bar{b} \end{cases} \tag{7a}$$

$$p_n' = p_n + p_v + P_w(l(n, v)) \tag{7b}$$

$$q_n' = min(q_n, q_v - d(n, v)) \tag{7c}$$

$$b_n' = b_n \tag{7d}$$

$$h_n^{u'} = h_n^u \tag{7e}$$

where $l(n, v)$ is the length of the wire connecting node n to node v and $d(n, v)$ is the Elmore delay between those nodes.

As new options at node n are generated, not all options need to be kept. An option is redundant if it is dominated by another option and is discarded instead of being propagated upstream.

Definition 4 *Given options $\Phi_n = (c_n, p_n, q_n, b, h_n^u)$ and $\Phi_n' = (c_n', p_n', q_n', b', h_n^{u'})$. Option Φ_n' is said to be dominated by option Φ_n if $h_n^u = h_n^{u'}$, $c_n \leq c_n'$, $p_n \leq p_n'$ and $q_n \geq q_n'$.*

Definition 4 is similar to the definition of dominance used in [11, 14]. However, here, conservatively, only options, which expect the same voltage upstream, are compared. Another property that may be used to prune redundant options, known as b-dominance [12], can be used to detect if a non-redundant option at a node can lead only to redundant options at an upstream node. Such an option can then be pruned without having to be propagated upstream.

Definition 5 *Let $\Phi_n = (c_n, p_n, q_n, b, h_n^u)$ and $\Phi_n' = (c_n', p_n', q_n', b', h_n^{u'})$ be two non-redundant options at node n, where $q_n \leq q_n'$. Suppose $R_{min}(h^u)$ is the minimum resistance for any buffer $b \in B$ such that the V_{dd} of b is h^u. Then, Φ_n b-dominates Φ_n' if Equation (8) holds.*

$$(q_n - R_{min}(h_n^u)c_n) \geq (q_n' - R_{min}(h_n^{u'})c_n') \tag{8a}$$

$$(p_n + c_n h_n^{u^2}) \leq (p_n' + c_n' h_n^{u'^2}) \tag{8b}$$

In Equations (8a) and (8b), respectively, upper bounds on the upstream arrival time and lower bounds on the upstream power consumption, resulting from the use of options Φ_n and Φ_n', are compared. A buffer, which has the minimum resistance, amongst all buffers whose power supply is voltage-compatible with both options is used to calculate the upper bound on the required arrival time. The lower bound on power consumption, calculated above, is independent of the choice of the buffer used upstream. Then, option Φ_n' is pruned if its use leads to a larger lower bound on power consumption but a smaller upper bound on the required arrival time upstream.

5.2.2 The TCMS-BIS Algorithm

We are finally in a position to discuss the `TCMS-BIS` algorithm listed in Algorithm 1. As was mentioned earlier, in Section 5.2.1, this algorithm is part of a dynamic programming framework. It constitutes the bottom-up part of the above framework, which is used to construct all possible options at every node. The other part of the framework, a top-down algorithm used to choose the option with the minimum power at a specified delay bound, is omitted here because of space constraints. However, it is very similar to the top-down algorithm given in [11] to which the reader is referred for details.

The input to the `TCMS-BIS` algorithm is a subtree T_n rooted at node n. If node n is a sink terminal, a set containing two options, one for each expected upstream voltage is returned in line 4. Otherwise, $Set(\Phi_n)$, which is used to hold all options generated at node n, is initialized to an empty set. The algorithm then iteratively merges options from children nodes of node n. This is done for each buffer in buffer library $Set(B)$, including buffer $\bar{b}$. Next, a set, $Set(\Phi_b)$, of options that associate buffer b with node n is initialized. A check is made in line 6 to ensure that no high-V_{dd} buffer is driven by a low V_{dd} input, unless it uses a high V_{th}. This requirement can be encoded in the generated options themselves by only generating options that expect a V_{dd}^H input from upstream.

In the loop starting on line 10, each option in $Set(\Phi_b)$ is iteratively updated with options from each child node v. This operation, if performed exactly, can make the set of options grow exponentially with the depth of tree. To avoid this penalty, we utilize the `3DSample` procedure proposed in [14], which maintains all options in a three-dimensional grid whose axes are capacitance, power and arrival time. Options are sampled by selecting only one option from each

Table 3. Buffer design parameters

Type	Input inverter			Output inverter		
	Width	V_{dd}	V_{th}	Width	V_{dd}	V_{th}
$dvl/32$	$16\times$	$0.7V$	V_{th}^L	$32\times$	V_{dd}^L	V_{th}^L
$dvh/32$	$16\times$	$1.0V$	V_{th}^L	$32\times$	$1.0V$	V_{th}^L
$tcms/32$	$19\times$	V_{dd}^H	V_{th}^H	$32\times$	V_{dd}^L	V_{th}^L

grid cell. In order to maintain optimality, the minimum-power and maximum-arrival time options for each capacitance value are also retained. An enhancement, which was made to the 3DSample procedure in the context of the TCMS-BIS algorithm, is that it is also passed a downstream voltage level. Only solutions that are voltage-compatible with the passed level ($h^d(\Phi_t)$ in line 13) are returned. Internally, this is done simply by maintaining two grids for each value of V_{dd}. It is easy to see that maintaining voltage compatibility, coupled with the check in line 6, is sufficient to ensure that no level conversion problems occur. In the loop beginning in line 15, sampled options from node v are merged with options already in $Set(\Phi_b)$, according to the rules given in Equation (7). In line 17, the newly-generated option, Φ_{new}, is checked for redundancy. Redundancy checks include checks for slew violations at downstream nodes and a search through $Set(\Phi_b)$ and $Set(\Phi_n)$ for options that dominate option Φ_{new} according to Definition 4 or 5. If option Φ_{new} is not found to be redundant, it is added to $Set(\Phi_b)$ and options dominated by it are removed in line 19. Option Φ_{new} cannot yet be added to $Set(\Phi_n)$ because options from all children nodes of node n may not yet have been merged into node Φ_{new}. Options from $Set(\Phi_b)$ are added to $Set(\Phi_n)$ in the loop beginning in line 20 after all children nodes have been considered.

6 Experimental Results

In this section, the performance of TCMS is benchmarked against a conventional dual-V_{dd} scheme. TCMS is implemented as described in Section 5. Fast-dBIS [14] was implemented to provide dual-V_{dd} solutions. As in [12–14], only non-inverting buffers were included in the library, each comprising two inverters in series. The buffer library consisted of three types of buffers in three sizes ($16\times, 32\times, 64\times$) each. Design parameters for the $32\times$ buffers are given in Table 3, which lists the widths, V_{dd} and FinFET V_{th} (V_{th}^L and V_{th}^H are used to denote low and high threshold voltages), for each inverter in the buffer. Two of the buffers, called $dvl/32$ and $dvh/32$, were designed using a conventional dual-V_{dd} style with low and high V_{dd} values set to $0.7V$ and $1.0V$, respectively. The third buffer, $tcms/32$ was designed to use TCMS, according to our experience with the design presented in Section 4.2. The input buffer was connected to V_{dd}^H and the output buffer to V_{dd}^L. High-V_{th} FinFETs were used for the input inverter, which necessitated a slight increase in the size of the input inverter in order to obtain roughly equal rise and fall delays. Sample net instances given in [32] were used to evaluate both multiple-V_{dd} schemes. Since the sample instances were generated for a 180nm process, we scaled capacitance and length values using ITRS [4] trends. The choice of buffers available for each scheme is governed by the supply voltages available to it. Thus, buffers $dvh/16$, $dvh/32$ and $dvh/64$, which are tied to V_{dd}^L, were used in both conventional dual-V_{dd} and TCMS designs. Buffers $dvl/16$, $dvl/32$ and $dvl/64$ were only used in the dual-

V_{dd} scheme. Buffers $tcms/16$, $tcms/32$ and $tcms/64$ were used only for TCMS. In dual-V_{dd} solutions, obtained by using the Fast-dBIS algorithm, no buffer tied to $0.7V$ was allowed to drive a buffer tied to $1.0V$.

The results obtained are summarized in Table 4, which reports the number of fins (a measure of device width), power consumption, and the required arrival time (RAT) constraint at the source node for each benchmark, under both schemes. RAT was defined as $RAT = 1.01 \times RAT^*$, where RAT^* is the minimum possible delay for any benchmark using either scheme. Power was measured using a switching activity of 0.15, as in [9], and with the frequency defined by RAT. It can be seen that on an average, total power consumption is reduced by 50.41% and device area by 9.17%. A dominant fraction of the total power is observed to be leakage power consumption (91.65% and 77.77% for Dual-V_{dd} and TCMS, respectively). Dynamic power consumption increased due to the use of TCMS by 29.27% but leakage power consumption is reduced by 57.88%, on an average. Since leakage power is such a dominant fraction of total power, significant savings in total power consumption are obtained. Leakage power consumption in Table 4 is especially dominant for the larger nets because their clock frequency is limited by the achievable RAT, which is small. Another important consideration for measuring leakage, especially in FinFET circuits, is the expected operating temperature. As mentioned in Table 1, an operating temperature of 70°C was assumed for this study. The assumption is in line with results in [30] where it is shown that temperature in FinFET circuits rises directly with rising switching activity. Also, simulations in [30] for large circuits show that an average temperature above 70°C should be expected at the switching activity (0.15) used here. The dominance of leakage power, while alarming, is actually in line with predictions [9]. While leakage power is controlled somewhat through the use of TCMS, it will clearly need to be managed pro-actively at all levels of interconnect design. On an average, a decrease in device area is observed. However, for smaller nets, an increase is observed. This is probably because TCMS buffers have slightly larger intrinsic delays than buffers connected to V_{dd}^L (1.0V) and more buffers need to be inserted to meet tighter delay constraints.

7 Conclusions

To conclude, we point to some advantages and limitations of TCMS. Clearly, TCMS offers an attractive option for reducing interconnect power consumption. Aside from the interconnect, the use of TCMS in other circuits, such as logic, may also be beneficial. Two major overheads associated with TCMS, other than a modest increase in total device width in some cases, are easily identified. Firstly, the FinFET process used must allow FinFET V_{th} to be adjusted, which is possible with the development of metal-gate FinFETs. Secondly, additional supply voltages for TCMS must be provided. Recent research on the problem of routing multiple supplies [33] might form the basis of a practical solution. In particular, the double-supply double-ground grid [33], which separates the ground lines associated with each V_{dd} may be adapted for use with TCMS by negatively biasing one of the ground lines.

References

[1] E. J. Frank, R. H. Dennard, E. Nowak, P. M. Solomon, Y. Taur, and H.-S. P. Wong. Device scaling limits of Si MOSFETs and their application dependencies. *Proc. IEEE*, 89(3):259–288, Mar. 2001.

Table 4. Interconnect power consumption and delay measurements

Benchmark	#sinks	RAT (ns)	Dual-V_{dd}				TCMS			
			#fins	Power consumption (mW)			#fins	Power consumption (mW)		
				Dynamic	Leakage	Total		Dynamic	Leakage	Total
p1	268	-0.81	36000	3.11	4.15	7.62	41772	1.54	5.40	6.94
r1	267	-1.09	78912	4.36	10.11	14.47	95888	5.62	4.86	10.48
p2	603	-1.29	76824	4.79	9.55	14.34	85338	7.38	3.46	10.84
r2	598	-6.44	202680	1.36	28.62	29.98	228930	1.96	19.53	21.49
r3	862	-7.53	222480	1.35	29.00	30.35	226116	2.03	14.50	16.53
r4	1903	-9.97	443088	1.97	56.36	58.33	365100	2.95	21.53	24.48
r5	3101	-11.02	577800	2.43	69.72	72.15	446472	3.56	18.36	21.92
Total			1637244	19.37	208.07	227.24	1489576	25.04	87.64	112.68

[2] N. Magen, A. Kolodny, U. Weiser, and N. Shamir. Interconnect-power dissipation in a microprocessor. In *Proc. Wkshp. System-Level Interconnect Prediction*, pages 7–13, Feb. 2004.

[3] P. Saxena, N. Menezes, P. Cocchini, and D. A. Kirkpatrick. The scaling challenge: Can correct-by-construction design help? In *Proc. IEEE Intl. Symp. Physical Design*, pages 51–58, Apr. 2003.

[4] 2005 International Technology Roadmap for Semiconductors. http://www.itrs.net/Links/2005ITRS/Home2005.htm.

[5] T.-J. King. FinFETs for nanoscale CMOS digital integrated circuits. In *Proc. Int. Conf. Computer-Aided Design*, pages 207–210, Nov. 2005.

[6] B. Swahn and S. Hassoun. Gate sizing: FinFETs vs 32nm bulk MOSFETs. In *Proc. Design Automation Conf.*, pages 528–531, July 2006.

[7] K. Usami and M. Horowitz. Clustered voltage scaling technique for low-power design. In *Proc. Int. Symp. Low Power Electronics & Design*, pages 3–8, Aug. 1995.

[8] A. U. Dirl, Y. S. Dhillon, A. Chatterjee, and A. D. Singh. Level-shifter free design of low power dual supply voltage CMOS circuits using dual threshold voltages. *IEEE Trans. VLSI Systems*, 13(9):1103–1107, Sept. 2005.

[9] K. Banerjee and A. Mehrotra. Power disspation issues in interconnect performance optimization for sub-180nm designs. In *Proc. VLSI Symp. Technology & Circuits*, pages 12–15, June 2002.

[10] L. P. P. P. van Ginneken. Buffer placement in distributed RC-tree networks for minimal elmore delay. In *Proc. Int. Symp. Circuits & Systems*, pages 865–868, June 1990.

[11] J. Lillils, C. K. Cheng, and T-T. Y. Lin. Optimal wire sizing and buffer insertion for low power and a generalized delay model. *IEEE J. Solid-State Circuits*, 31(3):437–447, 1996.

[12] W. Shi and Z. Li. An O(nlogn) time algorithm for optimal buffer insertion. In *Proc. Design Automation Conf.*, pages 580–585, June 2003.

[13] K. H. Tam and L. He. Power optimal dual-V_{dd} buffered tree considering buffer stations and blockages. In *Proc. Design Automation Conf.*, pages 497–502, June 2005.

[14] Y. Hu, K. H. Tam, T. T. Jing, and L. He. Fast dual-V_{dd} buffering based on interconnect prediction and sampling. In *Proc. Wkshp. System-Level Interconnect Prediction*, pages 95–102, Mar. 2007.

[15] A. Muttreja, N. Agarwal, and N. K. Jha. CMOS logic design with independent gate FinFETs. In *Proc. Int. Conf. Computer Design*, pages 560–567, Oct. 2007.

[16] E. J. Nowak, I. Aller, T. Ludwig, K. Kim, R. V. Joshi, C.-T. Chuang, K. Bernstein, and R. Puri. Turning silicon on its edge. *IEEE Circuits and Devices Magazine*, 20(1):20–31, Jan.-Feb. 2004.

[17] T. Cakici, H. Mahmoodi, S. Mukhopadhyay, and K. Roy. Independent gate skewed logic in double-gate SOI technology. In *Proc. IEEE Int. SOI Conf.*, pages 83–84, Oct. 2005.

[18] R. V. Joshi, K. Kim, R. Q. Williams, E. J. Nowak, and C.-T. Chuang. A high-performance, low leakage, and stable SRAM row-based back-gate biasing scheme in FinFET technology. In *Proc. Int. Conf. VLSI Design*, pages 665–672, Jan. 2007.

[19] L. Wei, Z. Chen, and K. Roy. Double gate dynamic threshold voltage (DGDT) SOI MOSFETs for low power high performance designs. In *Proc. IEEE Int. SOI Conf.*, pages 82–83, Oct. 1997.

[20] P. Beckett. Low-power circuits using dynamic threshold voltage devices. In *Proc. Great Lakes Symp. VLSI*, pages 213–216, Apr. 2005.

[21] T. Sairam, W. Zhao, and Y. Cao. Optimizing FinFET technology for high-speed and low-power design. In *Proc. Great Lakes Symp. VLSI*, pages 73–77, Mar. 2007.

[22] V. P. Trivedi, J. G. Fossum, and W. Zhang. Threshold voltage and bulk inversion effects in nonclassical CMOS devices with undoped ultra-thin bodies. *Solid-State Electronics*, 1(1):170–178, Dec. 2007.

[23] W. Zhao and Y. Cao. New generation of predictive technology model for sub-45nm design exploration. In *Proc. Int. Symp. Quality of Electronic Design*, pages 585–590, May 2006. http://www.eas.asu.edu/~ptm.

[24] W. Zhao and Y. Cao. Predictive technology model for nanoCMOS design exploration. *ACM J. Emerging Technologies in Computing Systems*, 3(1):1–17, Apr. 2007.

[25] F. Wang, Y. Xie, K. Bernstein, and Y. Luo. Dependability analysis of FinFET circuits. In *Proc. IEEE Computer Soc. Symp. Emerging VLSI Technologies and Architectures*, pages 399–404, Mar. 2006.

[26] D. Ha, H. Takeuchi, Y.-K. Choi, and T.-J. King. Molybdenum gate technology for ultrathin-body MOSFETs and FinFETs. *IEEE Trans. Electron Devices*, 51(12):1989–1996, Dec. 2004.

[27] D. A. Hodges, H. G. Jackson, and R. A. Saleh. *Analysis and Design of Digital Integrated Circuits in Deep Submicron Technology*. McGraw Hill, New York, NY, 3rd edition, 2004.

[28] N. Weste and D. Harris. *CMOS VLSI Design*. Addison Wesley, Reading, MA, 3rd edition, 2000.

[29] C. Grecu, P. P. Pande, A. Ivanov, and R. Saleh. A scalable communication-centric SoC interconnect architecture. In *Proc. Int. Symp. Quality of Electronic Design*, pages 343–348, 2004.

[30] J. H. Choi, A. Bansal, M. Meterelliyoz, J. Murthy, and K. Roy. Leakage power dependent temperature estimation to predict thermal runaway in FinFET circuits. In *Proc. Int. Conf. Computer-Aided Design*, pages 583–586, Nov. 2006.

[31] H. B. Bakoglu. *Circuits, Interconnections and Packaging in VLSI*. Addison Wesley, Boston, MA, 1990.

[32] W. Shi and Z. Li. FBI: Fast buffer insertion for interconnect optimization. http://dropzone.tamu.edu/~zhuoli/GSRC/fast_buffer_insertion.html.

[33] M. Popovich, E. G. Friedman, M. Sotman, and A. Kolodny. On-chip power distribution grids with multiple supply voltages for high performance integrated circuits. In *Proc. Great Lakes Symp. VLSI*, pages 2–7, Apr. 2005.

Analysis of Delay Variation in Encoded On-Chip Bus Signaling under Process Variation

Sampo Tuuna   Ethiopia Nigussie   Jouni Isoaho   Hannu Tenhunen
Department of Information Technology
University of Turku
20014 Turku, Finland
{satatu, ethnig, jisoaho, hannu.tenhunen}@utu.fi

Abstract

In this paper, we model on-chip signaling over a bus consisting of encoding, drivers, transmission lines, receivers and decoding. We characterize the signaling circuitry as a function of its load capacitance. The effective load capacitance seen by a driver is derived for the decoupling method and distributed RLC transmission line models. The driver delay and rise time corresponding to the derived effective capacitance are used to derive the far-end voltage of a transmission line bus. The effects of process variation are taken into account in the characterization of the signaling circuitry and in the wire analysis. The overall delay variation of the bus due to device and wire process variation is then calculated. The model is verified by comparing it to HSPICE. We implement regular voltage mode, level-encoded dual-rail and 1-of-4 signaling circuitry and apply the derived model to analyze them. The implementation and analysis are done in 45 nm technology.

1 Introduction

Buses are often used as communication links in SoC designs. Accurate estimation of the bus propagation delay is important in the design process to determine the performance and possible delay violations. The wires in a bus typically run parallel to each other for long distances, making them susceptible to crosstalk effects, i.e. delay variation and noise. The delay of long interconnects can also be significant. To alleviate these problems bus encoding has been proposed [14, 7]. Another approach is to use asynchronous signaling as in globally asynchronous locally synchronous (GALS) design.

Propagation delay is also affected by process variation. The variation affects both devices and interconnects. Devices are affected by e.g. variation in effective channel length, oxide thickness and threshold voltage [2], while interconnects are affected by thickness and width variation due to the copper damascene chemical mechanical polishing process and interference in lithography [8].

The total delay of a bus includes the delay of possible encoding circuitry, drivers, wires, receivers, and possible decoding circuitry. Circuit simulators such as SPICE can be used to analyze the non-linear circuitry while analytical models can be applied to the wires to speed up the analysis for e.g. use in Monte Carlo simulation to analyze the effects of process variation.

The interaction between a driver and an interconnect can be modeled by empirically precharacterizing the driver delays and rise times as a function of load capacitance and input rise time. The interconnect can then be analyzed by modeling the driver as a voltage source with the delay and rise times corresponding to the load. The load to a driver has traditionally been modeled as the total capacitance of an interconnect. Because of the scaling of the interconnect sizes, the actual load seen by a driver is smaller due to resistive and inductive shielding [5]. An RC [11] or RLC [1] driving point load can mapped to an effective capacitance in order to maintain compatibility with the existing efficient empirical driver models.

The driving point models for RLC loads can be constructed using moment matching methods such as AWE [10]. These input admittance models are generated at the same time as the approximate transfer functions. Another common method that is very suitable for for modeling multiple parallel coupled transmission lines, as in a bus, and that does not require the generation of moments, is the decoupling method [9, 6, 3].

In this paper we present a model for analyzing signaling over an on-chip bus consisting of encoding circuitry, drivers, transmission lines, receivers and decoding circuitry. The wires are modeled as capacitively and inductively coupled distributed RLC transmission lines. We derive the driving point effective capacitance for a bus driver in the decoupling method. The delay of the signaling circuitry and rise time of the drivers are characterized as a function of the load capacitance. The equations for the voltage waveforms at the far-end of the bus are derived. The effects of process variation are taken into account in the characterization of the signaling circuitry and in the wire analysis. The overall delay variation of the bus due to process variation is then calculated. The model is verified by comparing it to HSPICE. We implement regular voltage mode, level-encoded dual-rail (LEDR) and 1-of-4 signaling circuitry and apply the derived model to analyze them. The implementation and analysis are done in 45 nm technology.

2 Bus model

The bus structure that we consider in this paper is shown in Fig. 1. The bus consists of a number of input signals that are encoded and sent using voltage-mode signaling over an arbitrary number n wires to the receivers and decoder. In

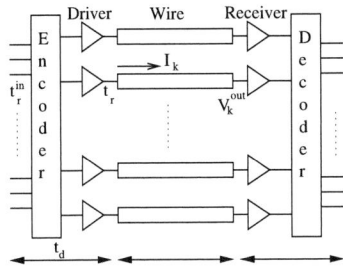

Figure 1. Bus implementation with encoding and decoding.

general, the delay of a driver and the rise time of its output are a function of the load capacitance and the rise time of its input, i.e. $t_d = f(t_r^{in}, C_{load})$ and $t_r = g(t_r^{in}, C_{load})$. We modeled the driver output waveform as a saturated ramp as shown in Fig. 2. The encoder and driver were characterized with HSPICE by determining their rise time t_r and delay t_d from the encoder input to the driver output as a function of load capacitance. In order to simplify the characterization without loss of generality, we assumed that the rise time t_r^{in} of the encoder input signals is constant. In order to connect the encoder and driver characterization with the bus model, we derived the driving point effective capacitance seen by the drivers. The circuit model of the bus is shown in Fig. 3. The bus consists of n wires of length h with inductive and capacitive coupling between them. The wires were modeled as distributed RLC transmission lines. The bus is driven by the saturated ramp voltage source in Fig. 2. C_L is the load capacitance due to the receiver at the end of the wire. The bus can be analyzed by decoupling the transmission

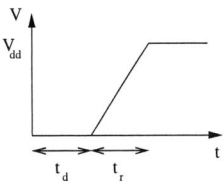

Figure 2. The saturated ramp model of the driver output waveform.

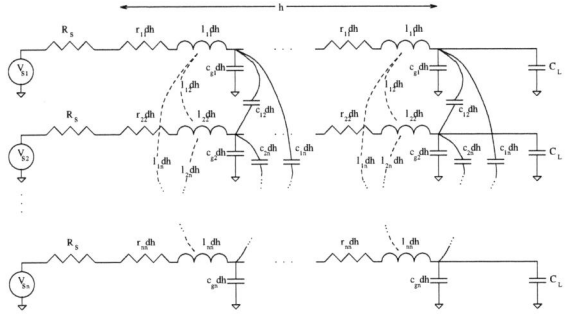

Figure 3. Circuit model of the bus.

line matrices as in [12, 3]. The decoupled system denoted with the superscript '^' is

$$\hat{\mathbf{R}} = \mathbf{R} \qquad (1)$$

$$\hat{\mathbf{C}} = \mathbf{M}^T \mathbf{C} \mathbf{M} \qquad (2)$$

$$\hat{\mathbf{L}} = \mathbf{M}^T \mathbf{L} \mathbf{M} \qquad (3)$$

$$\hat{\mathbf{R}}_S = \mathbf{R} \qquad (4)$$

$$\hat{\mathbf{C}}_L = \mathbf{C}_L \qquad (5)$$

$$\hat{\mathbf{V}}_S = \mathbf{M}^T \mathbf{V}_S \qquad (6)$$

where the transformation matrix $\mathbf{M}$ is the eigenvector matrix of capacitance or inductance matrix. T denotes a transpose. The decoupled mode voltage matrix $\hat{\mathbf{V}}$ and current matrix $\hat{\mathbf{I}}$ are defined as

$$\mathbf{V}(z,t) = \mathbf{M}\hat{\mathbf{V}}(z,t) \qquad (7)$$

$$\mathbf{I}(z,t) = \mathbf{M}\hat{\mathbf{I}}(z,t). \qquad (8)$$

The driver output voltage shown in Fig. 2 is in s-domain

$$V_S(s) = \frac{V_{dd}}{t_r s^2} \left(1 - e^{-t_r s} \right). \qquad (9)$$

For a single interconnect, the current I flowing into it is [13].

$$\frac{I}{V_S} = \frac{Z_L a_{21} + a_{22}}{(a_{11} + R_S a_{21}) Z_L + a_{12} + R_S a_{22}} \qquad (10)$$

229

978-1-4244-3039-0/08 $25.00 © 2008 IEEE

where a are the parameters of a two-port representation of a transmission line. In a manner similar to [13], by combining (9) with (10) and (6), and applying (8), the current I_k flowing into the k-th wire in a bus consisting of n wires can the be derived as

$$I_k(t) = \sum_{i=1}^{n} \mathbf{M}_{ki} \sum_{j=1}^{n} \mathbf{M}_{ij}^{T} I_i(t) \tag{11}$$

where

$$I_i(t) = \sum_{p=1}^{m-1} A_p e^{s_p t} + A_m e^{s_m t} -$$
$$\left(\sum_{p=1}^{m-1} A_p e^{s_p(t-tr)} + A_m e^{s_m(t-tr)} \right) u(t - tr) \tag{12}$$

where m is the order of the terms included in the derivation and $u(t)$ is the unit step function. The current I_i is calculated using the diagonalized resistance, capacitance and inductance values of $\hat{R}_{ii}$, $\hat{C}_{ii}$ and $\hat{L}_{ii}$, respectively. The effective capacitance seen by a driver was calculated by equating the average currents drawn by the interconnect I_k and the effective capacitance I_C over a time interval τ up to the 50% delay point, as proposed in [11]

$$\frac{1}{\tau} \int_0^{\tau} I_k(t) dt = \frac{1}{\tau} \int_0^{\tau} I_C(t) dt. \tag{13}$$

The initial rise time t_r and delay t_d were set according to the total capacitance of the wire, and the effective capacitance seen by the driver was then acquired iteratively using (13). In practice three to five iterations were needed for the effective capacitance to converge. In order to determine the propagation delay of the interconnects, we derived the far-end voltages of the interconnects when they are driven with the voltage source (9). By starting from two-port circuit equations the transfer function for a single interconnect can be derived as

$$\frac{V_{out}}{V_s} = \frac{Z_L}{(a_{11} + R_s a_{21}) Z_L + a_{12} + R_s a_{22}}. \tag{14}$$

By using (6) and (7), the far-end voltage of the k-th wire in an n-bit bus is then obtained as

$$V_k^{out}(t) = \sum_{i=1}^{n} \mathbf{M}_{ki} \sum_{j=1}^{n} \mathbf{M}_{ij}^{T} V_i(t) \tag{15}$$

where V_i is obtained by substituting (9) into (14) and using partial fraction expansion and inverse Laplace transform

$$V_i(t) = \sum_{p=1}^{m-2} B_p e^{s_p t} + B_{m-1} t + B_m - \left(\sum_{p=1}^{m-2} B_p e^{s_p(t-t_r)} + B_{m-1}(t - t_r) + B_m \right) u(t - t_r). \tag{16}$$

Possible non-switching bus drivers are not included in the summations of (15) and (11), while downward switching is included with a minus sign. The RLC transmission line matrices of the interconnects were extracted using analytical equations from [4]. The cross-section view of the wire configuration is shown in Fig. 4. The wires run parallel to each other over a ground plane.

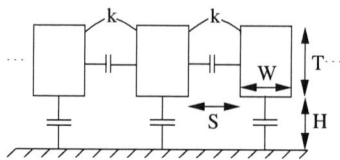

Figure 4. Cross-section view of wire configuration.

3 Signaling techniques

Unlike synchronous design style, which uses a globally distributed clock signal to indicate moments of stability of data, asynchronous circuits exchange information using a handshake to explicitly indicate the validity and acceptance of data. Depending on the type of handshaking, data encoding, channel type, and data-validity schemes, there are a number of alternative communication protocols. There are two types of handshake protocols; four-phase and two-phase schemes. Two-phase signaling is preferred for long on-chip communication since it reduces the required number of transitions by half and avoids the requirement of spacer compared to four-phase signaling. This saves communication time and energy of the system. The most common asynchronous data encoding in GALS design is single-rail (bundled-data) encoding which uses N lines to represent N-bit information and two additional handshake lines indicating data validity and acceptance. Since this encoding has a timing constraint between control (request) and data lines, communication through long on-chip interconnect becomes delay variations sensitive. The delay-insensitive design style in which the data validity is transmitted implicitly operates correctly regardless of the delay in the interconnecting wires. Dual-rail and 1-of-4 encodings are among delay-insensitive data encoding techniques which use two signal wires per data bit.

3.1 Level-encoded two-phase dual-rail encoding

LEDR encoding is a good encoding choice for long on-chip communication because it needs no resetting transitions that consume time and power. It requires only one transition per data bit which makes it energy efficient and faster than bundled and four-phase dual-rail data encodings. The normal two-phase dual-rail protocol has more complex and slower decoding and completion detection circuitry compared to that of LEDR. In normal two-phase dual-rail protocol, if the transmitted data has the value '0' there is a transition on one wire, and if '1' is transmitted there is a transition on the other wire. To detect completion and decode the data, transitions on both wires need to be detected.

978-1-4244-3039-0/08 $25.00 © 2008 IEEE

This makes the circuit more complex and rather slow. In LEDR one information bit is encoded into a 2-bit codeword as follows. A data sequence D(i) of bits is encoded into a sequence DS(i), DP(i) of state and phase bits, respectively. DS(i)=D(i) for all i. Given DP(0), if DS(i+1)=DS(i) then DP(i+1) is the inverse of DP(i), otherwise DP(i+1)=DP(i). The explicit spacer symbol of four-phase dual-rail encoding is implicitly expressed in the LEDR codeword sequencing since each codeword has a phase (even or odd). Fig. 5 shows the four possible codeword organized into overlapping groups of value and phase. The arrows illustrate the allowed transitions between the codeword.

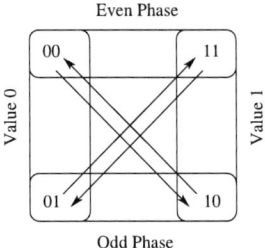

Figure 5. LEDR codeword transitions.

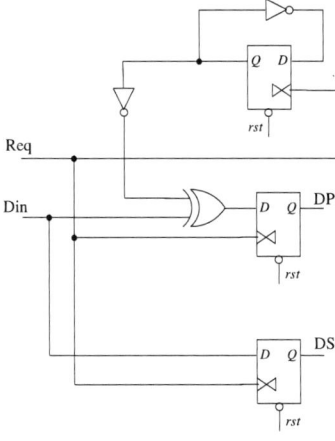

Figure 6. LEDR encoder implementation.

The encoder takes the request and data bit in single-rail encoding and converts into LEDR encoding. It consists of three double-edge triggered flip-flops (DFF), two inverters and one XOR gate as shown in Fig. 6. The DFF1 and DFF2 are used to generate the phase and data signal respectively. The two-input XOR gate detects whether the previous transmitted data is the same as the current data by taking as input the output of the DFF0 and the single-rail input data. An inverter is used as both driver and receiver in this signaling. In LEDR decoding data is decoded out directly from the output of state wire receiver. Completion detection is carried out using N 2-input XOR gates connected to an N-input

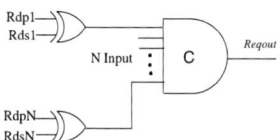

Figure 7. LEDR completion detector implementation.

C-element, where N is the bit-width of the transmitted data. The detector is shown in Fig. 7. The output of the C-element acts as the bundled-data request signal (Reqout) passed to the receiving module. A C-element is a basic building block of asynchronous logic. It is a state-holding element, a special kind of latch. When all of its inputs are 0 or 1 the output is set to 0 or 1, respectively. For other input combinations it preserves its state.

3.2 Two-phase 1-of-4 encoding

In 1-of-4 data encoding, a group of four wires is used to transmit two bits of information per symbol. A symbol is one of the two-bit codes 00, 01, 10, or 11 and it is transmitted through activity on one of the four wires. Besides being delay-insensitive, 1-of-4 encoding is more immune against crosstalk effects as compared to single-rail encoding, because the likelihood of two adjacent wires switching at the same time is much smaller. Furthermore, dynamic power consumption due to wire capacitance is smaller for the 1-of-4 code than for the simpler 1-of-2 (dual-rail) code. This is because the 1-of-4 code conveys two bits of information using only a single transition, while the 1-of-2 code requires two transitions for two bits of information.

The straightforward gate level implementation of the encoder which converts the two-phase single-rail input to the delay-insensitive two-phase 1-of-4 protocol is shown in Fig.8. The encoder consists of NOR gates which generate the select inputs for the multiplexers depending on the two-bit input codes, double-edge triggered flip-flops which are used to sample the symbol value at both edges of the request signal, and multiplexers each of which allows transition on the corresponding flip-flop output only when the appropriate input symbol is present. An inverter is used as both driver and receiver in this signaling. The decoder and completion detector circuit is shown in Fig. 9. It consists of XNOR gates which detect the transitions on the wires, NAND gates and a SR latch to decode the data back into the single-rail form, and a four-input XOR gate together with an N/2-input C-element for detecting completion.

978-1-4244-3039-0/08 $25.00 © 2008 IEEE

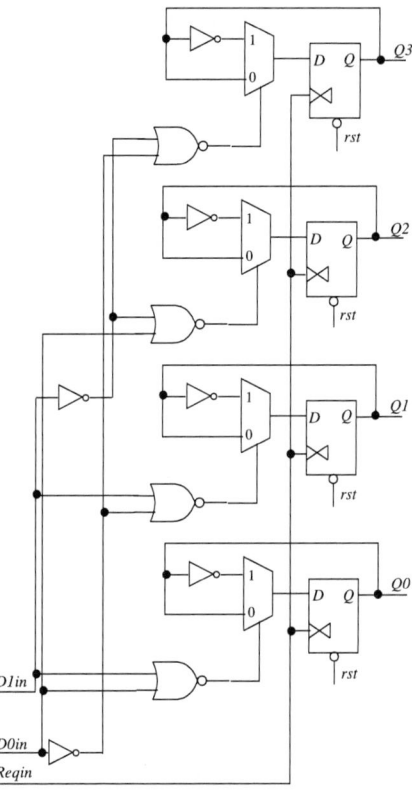

Figure 8. 1-of-4 encoder implementation.

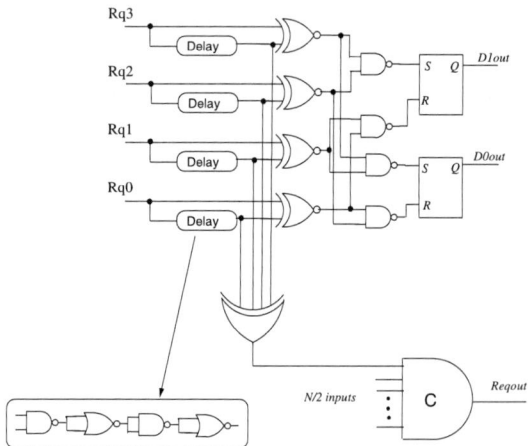

Figure 9. 1-of-4 decoder implementation.

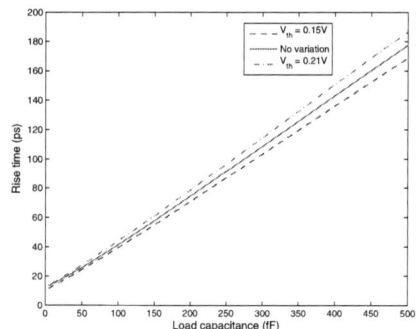

Figure 10. The rise time of the LEDR encoder and driver as a function of load capacitance with different transistor threshold voltages.

4 Verification and case study

We characterized the dependence of the delay and rise time of each signaling circuitry on load capacitance using HSPICE. The HSPICE transistor models for 45 nm technology were obtained using Predictive Technology Model [15]. We focused our transistor process variation analysis on threshold voltage variation. The characterizations were done for each variation in the threshold voltage. In addition to the normal 0.18 V threshold voltage also 0.15 V and 0.21 V were used. Fig. 10 demonstrates the rise time t_r of the LEDR encoder and driver as a function of load capacitance. The 50% delay t_d of the LEDR encoder and driver are shown in Fig. 11. The curve-fitted results were used in the effective capacitance calculations in the model. Regular voltage mode and 1-of-4 signaling circuits were characterized in a similar manner. A change in the physical properties of a wire, e.g. width or thickness due to process variation, also caused the effective capacitance to change. Due to the precharacterization of encoding circuitry as a function of load capacitance, all changes in wire properties could be analyzed analytically instead of using SPICE. The voltage waveforms obtained with the model for driver output and wire far-end are compared to HSPICE in Fig. 12 and in Fig. 13. The comparison was performed for a 16-bit bus.

The wire in the middle of the bus was used in the analysis. The results are shown for regular voltage mode signaling, where the driver was a 200x inverter and the receiver was a 20x inverter. The receiving inverter was modeled in the model as a capacitive load C_L at the end of the wire, and this capacitance was extracted with HSPICE. In the first case, the comparison was performed for a 1 mm long bus where the wire width W and separation distance S were set to 68 nm as approximated in the ITRS roadmap for minimum global wiring pitch in 45 nm technology. The comparison was also performed for a longer 3 mm long bus where wire width and separation distance were 135 nm. The wire thickness T and distance to ground H were in both cases 162 nm. The input signal t_r^{in} to the driver inverter was a falling ramp with a 50 ps fall time as shown in the figures. The model and HSPICE results were close to each other. The propagation delay of the wire was slightly underestimated since the saturated ramp did not accurately capture

232

978-1-4244-3039-0/08 $25.00 © 2008 IEEE

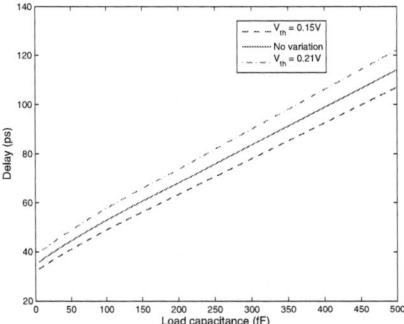

Figure 11. 50% delay of the LEDR encoder and driver as a function of load capacitance with different transistor threshold voltages.

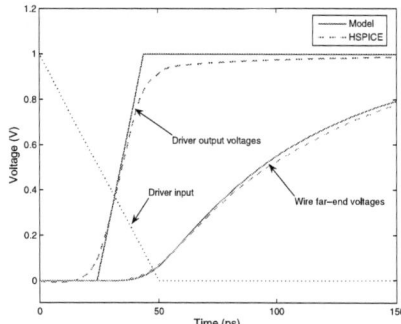

Figure 13. Comparison between HSPICE and the model for driver output voltages and wire far-end voltages in a 16-bit 1 mm bus. Wire width and separation distance are 68 nm.

the exponential tail of the driver output. If needed the exponential tail can be modeled with a resistor as in [11].

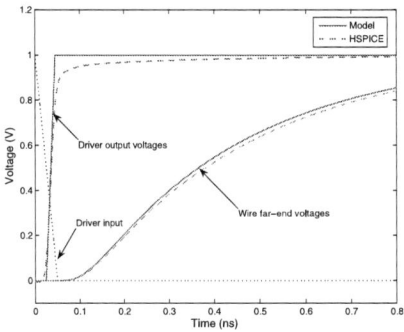

Figure 12. Comparison between HSPICE and the model for driver output voltages and wire far-end voltages in a 16-bit 3 mm bus. Wire width and separation distance are 135 nm.

The total delay variation of the bus was obtained by adding the delay of the receiver and decoder to the 50% delay at the far-end of the wire. The receiver and decoder delay was also characterized with HSPICE for different threshold voltages.

Table 1 shows the amount of delay variation in the 16-bit bus for different signaling techniques. The delay variation was calculated as the difference between the delay acquired in the presence of process variation and the delay acquired without process variation. All signaling techniques had a 200x inverter as a driver and a 20x inverter as the receiver. The length of the bus was 3 mm and the wire width and separation distance were 135 nm. The data to the encoders consisted of all inputs switching. For LEDR and 1-of-4 signaling the 16-bit bus was encoded into 32 wires. The regular voltage mode signaling had no encoding or decoding cir-

cuitry. As shown in the table, the model was further verified by comparing the delay variation obtained with HSPICE and the model. The delay variation was accurately modeled. The first part of the table shows the delay variation due to threshold voltage variation. A lower threshold voltage decreased the delay of the bus while a higher voltage increased it. The second part of the table shows the delay variation when only the wire properties vary. It was assumed that the wire pitch remains constant while the wire width varied. As can be seen, for regular voltage mode signaling the effect of wire variation on bus delay was clearly larger than the effects of device variation. On the other hand, the LEDR and 1-of-4 signaling techniques suffered more delay variation from transistor variation due to their encoding and decoding circuitry. The third part of the table shows the delay variation when both the threshold voltage and wire properties vary simultaneously. The presented analytical model was also applied to demonstrate its use in analyzing the statistical effects of wire variation. Fig. 14 shows the delay variation of the bus for different signaling techniques when the wire width varies. The wire width was varied with a 3-sigma variation of 10% so that the wire pitch remained constant. The Monte Carlo analysis was done with 1000 samples. Although the LEDR and 1-of-4 signaling techniques had in general larger delay variation than regular voltage mode signaling, they are able to operate correctly regardless of the delay due to their delay-insensitivity. The correct operation of regular synchronous voltage mode signaling is however susceptible to delay variation.

5 Conclusion

We developed a model to analyze the effects of process variation on delay in on-chip bus signaling. The model

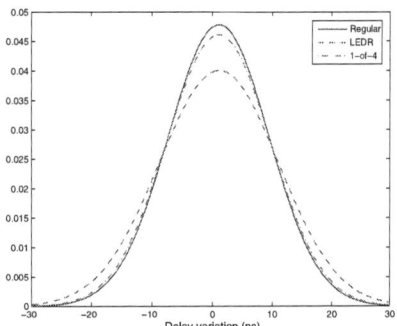

Figure 14. Bus propagation delay variation due to wire width variation.

Table 1. The total delay variation of the bus for different signaling techniques

| Variation source | Regular | | LEDR | 1-of-4 |
	Model	HSPICE	Model	Model
$V_{th} = 0.15V$	-2.2ps	-2.7ps	-28.8ps	-40.2ps
$V_{th} = 0.21V$	+3.4ps	+3.6ps	+34ps	+50.7ps
Wire width -10%	+27.6ps	+27.3ps	+34ps	+31ps
Wire width +10%	-22.7ps	-21.9ps	-26ps	-21ps
Wire thickn. -10%	+54.3ps	+55.4ps	+68ps	+63ps
0.15V, width+10%	-25.1ps	-24.7ps	-53.9ps	-61.2ps
0.21V, thickn.-10%	+57.7ps	+58.9ps	+103ps	+115ps

combined the variation in signaling circuitry and in the wires to calculate the total delay variation of the bus. The wires were modeled as distributed RLC transmission lines including capacitive and inductive coupling between them. The effective load capacitance was derived for the decoupling method.

We characterized the signaling circuitry as a function of its load capacitance. The driver delay and rise time corresponding to the derived effective capacitance were used to calculate the far-end voltage of a transmission line. The effects of process variation were taken into account in the characterization of the signaling circuitry and in the wire analysis. The overall delay variation of the bus due to process variation was then calculated. The model was verified by comparing it to HSPICE. Regular voltage mode, level-encoded dual-rail and 1-of-4 signaling circuitry were implemented and their delay variation was analyzed with the derived model. The implementation and analysis were done in 45 nm technology.

References

[1] R. Arunachalam, F. Dartu, and L. T. Pileggi. Cmos gate delay models for general rlc loading. In *Proc. Int. Conf. on Computer Design*, pages 224–229, 1997.

[2] Y. Cao, P. Gupta, A. Kahng, D. Sylvester, and J. Yang. Design sensitivities to variability: extrapolations and assessments in nanometer vlsi. In *Proc. ASIC/SoC Conference*, pages 411–415, 2002.

[3] J. Chen and L. He. A decoupling method for analysis of coupled rlc interconnects. In *Proc. IEEE/ACM Int. Great Lakes Symposium on VLSI*, pages 41–46, 2002.

[4] N. Delorme, M. Belleville, and J. Chilo. Inductance and capacitance analytic formulas for vlsi interconnects. *IEEE Electronics Letters*, 32(11):996–997, May 1996.

[5] M. A. El-Moursy and E. G. Friedman. Shielding effect of on-chip interconnect inductance. *IEEE Transactions on Very Large Scale Integration (VLSI) Systems*, 13(3):396–400, Mar. 2005.

[6] Y. Eo, S. Shin, W. R. Eisenstadt, and J. Shim. Generalized traveling-wave-based waveform approximation technique for the efficient signal integrity verification of multicoupled transmission line system. *IEEE Transactions on Computer-Aided Design of Integrated Circuits and Systems*, 21(12):1489–1497, Dec. 2002.

[7] M. Lampropoulos, B. Al-Hashimi, and P. Rosinger. Minimization of crosstalk noise, delay and power using a modified bus invert technique. In *Proc. Design, Automation and Test in Europe*, pages 1372–1373, 2004.

[8] V. Mehrotra. *Modeling the effects of systematic process variation on circuit performance*. Ph.D. thesis, Massachusetts institute of technology, 2001.

[9] C. Paul. Decoupling the multiconductor transmission line equations. *IEEE Transactions on Microwave theory and techniques*, 44(8):1429–1440, Aug. 1996.

[10] L. Pillage and R. Rohrer. Asymptotic waveform evaluation for timing analysis. *IEEE Transactions on Computer-Aided Design of Integrated Circuits and Systems*, 9(4):352–366, Apr. 1990.

[11] J. Qian, S. Pullela, and L. Pillage. Modeling the 'effective capacitance' for the rc interconnect of cmos gates. *IEEE Transactions on Computer-Aided Design of Integrated Circuits and Systems*, 13(12):1526–1535, Dec. 1994.

[12] S. Tuuna, J. Isoaho, and H. Tenhunen. Analytical model for crosstalk and intersymbol interference in point-to-point buses. *IEEE Transactions on Computer-Aided Design of Integrated Circuits and Systems*, 25(7):1400–1411, July 2006.

[13] S. Tuuna, L.-R. Zheng, J. Isoaho, and H. Tenhunen. Modeling of on-chip bus switching current and its impact on noise in power supply grid. *IEEE Transactions on Very Large Scale Integration (VLSI) Systems*, In press.

[14] B. Victor and K. Keutzer. Bus encoding to prevent crosstalk delay. In *Proc. Int. Conf. on Computer Aided Design*, pages 57–63, 2001.

[15] W. Zhao and Y. Cao. New generation of predictive technology model for sub-45nm design exploration. In *Proc. Int. Symposium on Quality Electronic Design*, pages 585–590, 2006.

21st International Conference on VLSI Design

Exploiting Variable Cycle Transmission for Energy-Efficient On-Chip Interconnect Design

T. Venkata Kalyan	Madhu Mutyam	P. Vijaya Sankara Rao
IIIT Hyderabad	IIT Madras	IIT Kharagpur
Hyderabad−500032, India	Chennai−600036, India	Kharagpur−721302, India
kalyan_tv@research.iiit.ac.in	madhu@cs.iitm.ernet.in	vijaysankar@ece.iitkgp.ernet.in

Abstract

As on-chip interconnect in deep-submicron designs contribute to the system-wide power consumption, minimization of interconnect power consumption has become one of the important design issues in deep-submicron technologies. As transition activity mainly determines the interconnect power consumption, several bus encoding techniques have been proposed to minimize the activity.

Unlike the existing low-power or energy-efficient bus encoding techniques, in this paper, we propose a scheme which exploits both dynamic voltage scaling and variable cycle transmission mechanisms for minimizing on-chip interconnect energy consumption. We transmit data using variable cycle transmission method and, based on the delay savings achieved through variable cycle transmission method at regular intervals, scale the voltage and frequency to obtain significant energy savings. Using our technique for a 5mm interconnect wire we achieved energy savings of 30% and 45% over the base case in the address bus and data bus, respectively. Our technique also reduces the energy-delay-product by 34% and 52% for address bus and data bus, respectively.

1. Introduction

With the scaling of process technology, system-wide power consumption is increasing. One of the contributors for the system-wide power consumption is the on-chip interconnect. Data transmission on the interconnect causes rail-to-rail voltage swing and charging/discharging of capacitance, which in turn result in dynamic (or switching) power consumption. As switching activity mainly determines the power consumption, several low-power or energy-efficient bus encoding techniques have been proposed in the literature to minimize it.

In coupling-driven bus encoding technique for on-chip buses [7], the data sequences that are closely placed are transformed to minimize the coupling effects, which in turn achieves power savings. In order to ensure the data integrity, the authors proposed to use either extra control lines or extra clock cycles. A technique to minimize power consumption due to coupling transitions is proposed in [15], which modifies the transition profiles to reduce switching energy by 50%. An area and energy efficient coding technique is proposed in [17] to obtain nearly 46% reduction in delay along with 10% reduction in energy, but it requires 48 wires to encode 32-bit data. An adaptive bus encoding technique using weighted code mapping and the delayed bus technique is proposed in [3] to achieve significant reduction in power consumption.

In all the above mentioned techniques, a fixed clock period is considered for data transmission. But data can be transmitted using *variable clock periods*. As data transition patterns determine the necessary delay for data transmission, we can fix necessary delay for each data transition pattern and transmit data with delay corresponding to its worst-case transition pattern. A technique based on this idea called *variable cycle transmission* (VCT) technique [8]. It is shown that VCT technique can achieve significant delay savings.

As system-wide power consumption is one of the critical issues in VLSI community, several techniques are proposed to minimize it. One of the well known techniques for reducing system-wide dynamic power consumption is *dynamic voltage scaling* (DVS) [5]. DVS technique exploits the quadratic dependency between supply voltage and power consumption and the linear relationship between clock frequency and supply voltage, to achieve significant dynamic power savings.

Application of DVS technique for on-chip interconnect is explored in [6]. By considering interconnect designs based on a double sampling latch which detect and correct for timing errors, DVS technique is applied to recover the available slack which results in good power savings. By keeping frequency constant, supply voltage is scaled for

235

978-1-4244-3039-0/08 $25.00 © 2008 IEEE

power reduction. Voltage scaling by keeping frequency as constant can result in timing errors. Errors are detected and corrected using the doubling sampling latch and the voltage scaling is controlled by the error recovery rate.

Motivating from the fact that interconnection network consumes significant portion of system-wide power budget, application of DVS technique for interconnection network links is explored in [13]. Power minimization is achieved by adjusting both frequency and voltage of links.

In this paper, we propose an interconnect design on which data is transmitted using variable clock periods. Delay savings obtained by transmitting data using variable clock periods are exploited for applying DVS technique. As part of the application of DVS technique, we scale both supply voltage and frequency. We validate the effectiveness of our technique by focusing on the L1 cache address/data buses of a microprocessor using the SPEC CPU2000 benchmark suite and show that for a $5mm$ interconnect wire our technique achieves 30% and 45% energy savings over the base case in the address bus and data bus, respectively. In addition to the energy savings, our technique reduces the energy-delay-product by 34% and 52% for address bus and data bus, respectively.

2 Prerequisites

We first review the effects of voltage scaling on dynamic power consumption and delay. Each transition of a digital circuit consumes power because of charging and discharging of the digital circuit's capacitance. The dynamic power consumption ($P_{dynamic}$) is expressed as

$$P_{dynamic} \quad \propto \quad V_{DD}^2 f \qquad (1)$$

where V_{DD} is the supply voltage and f is the clock frequency. It is clear from Equation (1) that the power consumption can be minimized quadratically by reducing V_{DD}, but supply voltage reduction increases the propagation delay. The propagation delay (τ) of a CMOS transistor [12] is expressed as

$$\tau \quad \propto \quad \frac{V_{DD}}{(V_G - V_T)^\alpha} \qquad (2)$$

where V_T is the threshold voltage, V_G is the input gate voltage, and $1 \le \alpha \le 2$. In deep-submicron designs, the value of α is nearly 1.3. The clock frequency is restricted by the propagation delay and it has to be reduced to tolerate the increased propagation delay.

Energy consumption per data transmission for an interconnect, which includes both the self capacitance (C_L) and the coupling capacitance between two adjacent lines (C_I), is given by [10]

Crosstalk Class	Relative delay on the middle wire ($\times C_L R_T$)	Transition patterns
1	0	$x - y$
2	1	$\uparrow\uparrow\uparrow, \downarrow\downarrow\downarrow$
3	$1 + \lambda$	$- \uparrow\uparrow, \uparrow\uparrow -, - \downarrow\downarrow, \downarrow\downarrow -$
4	$1 + 2\lambda$	$- \uparrow -, - \downarrow -, \downarrow\downarrow\uparrow, \uparrow\downarrow\downarrow, \uparrow\uparrow\downarrow, \downarrow\uparrow\uparrow$
5	$1 + 3\lambda$	$- \uparrow\downarrow, - \downarrow\uparrow, \downarrow\uparrow -, \uparrow\downarrow -$
6	$1 + 4\lambda$	$\downarrow\uparrow\downarrow, \uparrow\downarrow\uparrow$

Table 1. Crosstalk classes (here $\lambda = \frac{C_I}{C_L}$ and $x, y : \{a \to b \mid a, b \in \{0, 1\}\}$).

$$E_{bus} \quad = \quad (\alpha_s C_L + \alpha_c C_I) V_{DD}^2 \qquad (3)$$

where α_s and α_c denote the rates at which each capacitance is switched. While α_s represents the self transitions, α_c is related to the coupling transitions on the interconnect. We know from Equation (3) that the voltage scaling can significantly reduces the energy consumption.

We now review an analytical model for propagation delay in deep-submicron buses. By assuming a n-bit parallel bus in a single metal layer, we model a deep sub-micron bus as a distributed RC network with coupling capacitance between adjacent wires. The delay of wire l ($1 < l < n$) of the bus is given by [16]

$$T_l \quad = \quad C_L R_T [(1 + 2\lambda)\Delta_l^2 - \lambda\Delta_l(\Delta_{l-1} + \Delta_{l+1})] \,(4)$$

where R_T is the total resistance, λ is the ratio of the interwire capacitance (C_I) to the wire-to-substrate capacitance (C_L), and Δ_l is the transition occurring on line l. Δ_l is equal to 1 (or $\uparrow$) for 0-to-1 transition, -1 (or $\downarrow$) for 1-to-0 transition, and 0 (or $-$) for no transition.

As data transition patterns determine the propagation delay, they are classified into six different crosstalk classes [8, 16] based on the relative delay of a wire w.r.t. its adjacent wires. Table 1 shows different crosstalk classes.

3 Our Approach

The basic idea of our approach is to transmit data using variable clock periods as proposed in the VCT technique [8] and exploit the delay savings obtained through the VCT technique to apply the DVS technique for significant power savings. As part of the application of DVS technique, we scale both supply voltage and frequency.

Figure 1 shows the basic mechanism used in our approach. The shaded portion in the figure represents the implementation of the VCT technique. In general, data is transmitted using a fixed clock period, which is at least the

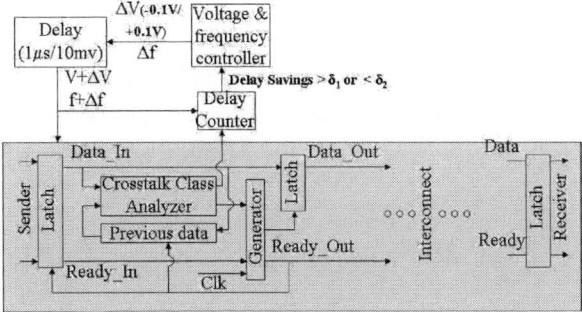

Figure 1. Basic mechanism used in our approach.

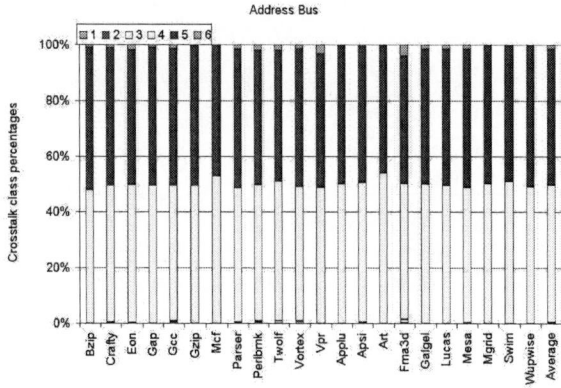

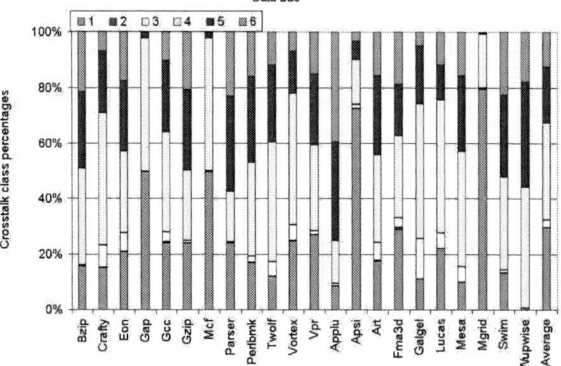

Figure 2. Distribution of transition patterns.

delay of crosstalk class 6 (refer to Table 1). Instead of transmitting data using fixed worst-case crosstalk class delay, we analyze the crosstalk class of a next data w.r.t. the present data and transmit the next data using the necessary delay. In order to determine the crosstalk class of a next data w.r.t. the present data on the bus, we use a *Crosstalk Class Analyzer* [8] (as shown in Figure 1). To support variable clock period for data transmission, we consider x as the unit clock period such that

$$x \geq C_L R_T$$
$$2x \geq C_L R_T(1 + \lambda)$$
$$3x \geq C_L R_T(1 + 2\lambda)$$
$$4x \geq C_L R_T(1 + 3\lambda)$$
$$5x \geq C_L R_T(1 + 4\lambda)$$

Hence, the unit clock period becomes $\frac{C_L R_T(1+4\lambda)}{5}$. Based on the crosstalk class of a next data w.r.t. the present data, we consider delay as an integer multiple of the unit clock period. As variable clock periods are used for data transmission, we use an extra interconnect wire (for *Ready* signal) to indicate the availability of data at the receiver side. The extra wire is separated from the actual data by using a shield wire so that the *Ready* signal takes a single cycle to reach the receiver. The delay due to *Crosstalk Class Analyzer* is overlapped with the propagation delay of the *Ready* signal [8]. Thus, we use two extra wires (i.e., the *Ready* signal and a shield wire) and hence the VCT technique requires 34 wires for a 32-bit bus.

Data transmission using variable clock periods can result in significant delay savings. It is shown in [8] that the VCT technique achieves 31.5% delay savings in the case of L1 cache data bus for on-chip data transmission. Significant delay savings achieved by the VCT technique are due to the fact that on-chip data exhibit high percentage of lower crosstalk class transitions (refer to Table 1) as shown in Figure 2.

As our main focus is on energy minimization rather than

delay minimization, we exploit the delay savings obtained through the VCT mechanism for energy reduction.

In order to apply the DVS technique for energy minimization, the delay savings are measured at regular intervals and based on the delay savings obtained so far we scale the voltage. As voltage scaling alters the propagation delay, we also scale the frequency. As shown in Figure 1, we consider a *Delay Counter* which consists of a *Global Counter* and a *Local Counter*. The *Global Counter* maintains the number of clock cycles that have been saved *so far*. The *Local Counter* maintains the number of clock cycles that have been saved *within the current sampling interval*. We consider the sampling period as 15000 cycles. The *Crosstalk Class Analyzer* provides values 3, 4, 5, and 6 for crosstalk classes 1-3, 4, 5, and 6, respectively, to the *Local Counter*. The *Local Counter*, upon receiving the value i from the *Crosstalk Class Analyzer*, increments its count by $(6-i)$, where $i \in \{3, 4, 5, 6\}$, and resets its value after every 15000 cycles. The *Delay Counter* updates the *Global Counter* value for every 15000 cycles using the following formula:

$$N_g = N_g + N_l - \lceil (\frac{(x' - x)N_l}{x}) \rceil \quad (5)$$

where N_g and N_l are the values of *Global Counter* and *Local Counter*, respectively, x is the original unit clock pe-

B'mark	Address	Data	B'mark	Address	Data
Bzip	104103584	31711265	Crafty	160591861	45039063
Eon	180520910	48659025	Gap	134847111	37827729
Gcc	186621216	56015716	Gzip	181144637	40342073
Mcf	117417966	32797679	Parser	201532006	51105161
Perlbmk	129748403	36285808	Twolf	122384930	31389418
Vortex	114818866	42302045	Vpr	227654678	37423636
Applu	102778392	33394619	Apsi	102578114	28801754
Art	109836092	36995126	Fma3d	131143197	19278215
Galgel	106010236	5657714	Lucas	99952797	18938412
Mesa	130750176	33263659	Mgrid	101415579	32568604
Swim	100232496	24084939	Wupwise	99954963	8159589

Table 2. Number of 32-bit data items considered in different benchmarks.

riod (as defined earlier in the section), and x' is the new unit clock period obtained due to frequency scaling.

If the *Global Counter* value is more than δ_1 and the energy savings due to voltage scaling by $0.1V$ is more than the energy overhead due to voltage regulator (given by Equation (6)), we scale-down the voltage by $0.1V$. On the other hand, if the *Global Counter* value is less than δ_2, we scale-up the voltage by $0.1V$. We consider the upper and lower limits for the supply voltage as $1.2V$ and $0.8V$, respectively, and δ_1 and δ_2 as 4999 and 0, respectively. We adjust the clock frequency in accordance with the supply voltage scaling. The on-chip voltage regulators take few μs to change from one voltage to another voltage [11]. We assume that the original frequency of the bus is $1.5GHz$, the voltage regulator takes $1\mu s$ to adjust the voltage by $10mV$ [6] and t_{bus} (15) cycles are needed for frequency transition [13]. Thus, the voltage regulator takes 15000 cycles to adjust the voltage by $0.1V$. As we use greedy method when applying the DVS technique (i.e., voltage and frequency are scaled based on the clock cycles saved so far), even though the voltage regulator takes 15000 cycles to adjust the voltage by $0.1V$, there is no need of a prediction technique to estimate the bus transition patterns. Although the frequency transitions are much faster compared to voltage transitions, the bus is disabled during a frequency transition in order to avoid timing uncertainty when the receiver is tracking the input clock.

Energy overhead due to transition from initial voltage V_1 to final voltage V_2 is calculated by [18]

$$E_{VR}^{(V_1, V_2)} = (1 - \eta)C|V_2^2 - V_1^2| \qquad (6)$$

where C is the filter capacitance of the power supply regulator and η is the power efficiency. In our experimental setup, we assume C as $5\mu F$ and η to be 94%. It is clear from Equation (6) that $E_{VR}^{(V_1, V_2)} = E_{VR}^{(V_2, V_1)}$. Note that the energy overhead calculated using Equation (6) is included in the calculation of overall energy savings.

Parameter	W (nm)	S (nm)	T (nm)	H (nm)	Dielectric constant
Value	205	205	430.5	398.5	3.3

Table 3. Device parameters for $90nm$ technology nodes based on the ITRS 2004 edition.

Method	# of wires	Codec overhead Area (μm^2)	Energy (pJ)
Base	32	0	0
VCT	34	3603.5	0.458

Table 4. Codec overhead summary.

4 Experimental Validation

We validate our technique by simulating 22 SPEC2000 CPU benchmarks [2] using the Simplescalar 3.0 simulator [4]. For each benchmark, we fast forward 100 million instructions and then simulate next 300 million instructions. Table 2 shows benchmark-wise number of 32-bit data items transmitted on L1 cache address/data bus.

We first discuss the energy, area, and latency overhead due to extra circuitry used to implement our technique. As we transmit data using the VCT mechanism, we design codecs used in the VCT technique [8] in Verilog and synthesize them using the Synopsys Design Compiler with $90nm$ *TSMC* technology library. The Berkeley interconnect model [1] is used to calculate the ground capacitance and coupling capacitance of the interconnect. In our experimental results, we consider metal layer 4 wire parameters as shown in Table 3 and wire length of $5mm$. Energy and area overheads of VCT codecs along with the base case are shown in Table 4. Note that there is no latency overhead due to *Crosstalk Analyzer* used in the VCT technique as its latency is overlapped with the propagation delay of the *Ready* signal [8] and the delay associated with performing activities related to voltage regulations is overlapped with the data transmission delay.

We now discuss the effect of voltage scaling on energy consumption. Energy consumption per data transmission in the base case is given by

$$E_{base} = (\alpha_s C_L + \alpha_c C_I)V_{DD}^2 \qquad (7)$$

where α_s and α_c represent the average self and coupling transitions, respectively.

Energy consumption per data transmission in our technique is given by

$$E_{VCT+DVS} = \frac{(\Sigma_{i=1}^n (\alpha_{s_i} C_L + \alpha_{c_i} C_I)V_i^2)}{n} + E_{OH} \qquad (8)$$

where E_{OH} is the energy overhead due to voltage regulator and the VCT circuitry, n is the number of samples

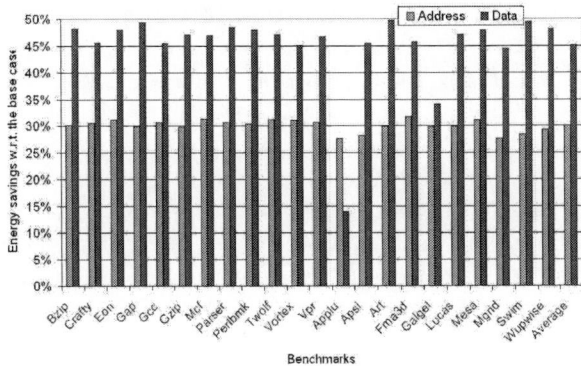

Figure 3. Energy savings of our technique w.r.t. the base case.

considered, α_{s_i} and α_{c_i} represent the average self and coupling transitions, respectively, and V_i is the voltage at the ith sample period. As voltage may change in each sample period depending on the *Global Counter* value, we consider different voltages (V_i) and use the average number of self (α_{s_i}) and coupling (α_{c_i}) transitions occurred during i^{th} sample period to calculate the energy consumption during the sample period. The total number of self (coupling) transitions during the entire execution is equal to the number of self (coupling) transitions at each sample period, i.e., $\alpha_s = \Sigma_{i=1}^{n}\alpha_{s_i}$ and $\alpha_c = \Sigma_{i=1}^{n}\alpha_{c_i}$.

Energy overhead due to voltage regulator and the VCT circuitry is given by

$$E_{OH} = \frac{\Sigma_{i=0.8}^{1.1} n_{(i,i+0.1)} E_{VR}^{(i,i+0.1)}}{m} + t \times E_{Codec} \quad (9)$$

where $n_{i,j}$ is the number of times voltage is changed from i to j, m is the total number of data items transmitted, E_{Codec} is the energy overhead due to the VCT circuitry (as shown in Table 4), and t is a multiplicative factor, which is used to consider energy overhead due to comparator circuit. Note that we use a comparator circuit in our experiments to check whether or not the energy savings due to voltage scaling is more than the energy overhead due to voltage regulation. In our experiments we consider t as 1.1.

Using the above equations, we now give a formula to calculate the percentage of energy savings (E_{save}) obtained through our technique w.r.t. the base case as:

$$E_{save} = (1 - \frac{E_{VCT+DVS}}{E_{base}}) \times 100 \quad (10)$$

Figure 3 shows benchmark-wise energy savings obtained through our technique w.r.t. the base case. Energy savings are almost uniform across different benchmarks in the address bus case and are ranging from 27.53% to 31.66% with

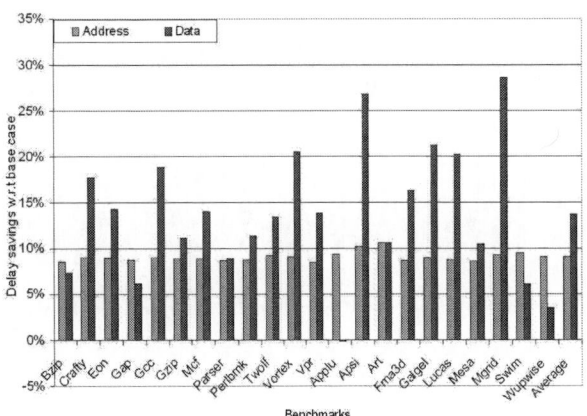

Figure 4. Delay savings of our technique w.r.t. the base case.

an average value of 30%. Variation in the energy savings across different benchmarks is less as data transition pattern behavior is almost uniform across all benchmarks (refer to Figure 2). Energy savings in the data bus case are also almost uniform across different benchmarks except for benchmarks "Applu" and "Galgel". In the case of "Applu" and "Galgel" benchmarks, our technique achieves 14% and 34% energy savings, respectively, while for all other benchmarks, our technique achieves nearly 45% energy savings.

As voltage and frequency scaling can affect data transmission delay, we now discuss the delay savings achieved by our technique w.r.t. the base case. We calculate the percentage of delay savings (d_{save}) obtained through our technique by using the following formula:

$$d_{save} = (1 - \frac{(\Sigma_{i=1}^{n} k_i - \Sigma_{i=0.8}^{1.1} n_{(i,i+0.1)} t_{bus})}{k}) \times 100 \quad (11)$$

where k is the total number of cycles required for data transmission in the base case and k_i is the total number of cycles required during i^{th} sample period for data transmission in our technique. In equation (11), the overhead due to the bus disabling, t_{bus}, is considered for every frequency transition.

Figure 4 shows benchmark-wise delay savings obtained through our technique w.r.t. the base case. Delay savings in the address bus are within the range of 8.50% to 10.64% with an average value of 9.08%, while in the data bus, they range from −0.12% to 28.63% with an average value of 13.69%.

We now consider the normalized energy-delay-product (EDP) of our technique w.r.t. the base case as shown in Figure 5. It is clear from the figure that our technique reduces the EDP by 34% and 52% for address and data buses, respectively. Note that it is easy to show that the EDP of our technique is less than that of the VCT technique. For instance, the VCT technique achieves 31.5% delay savings

239

978-1-4244-3039-0/08 $25.00 © 2008 IEEE

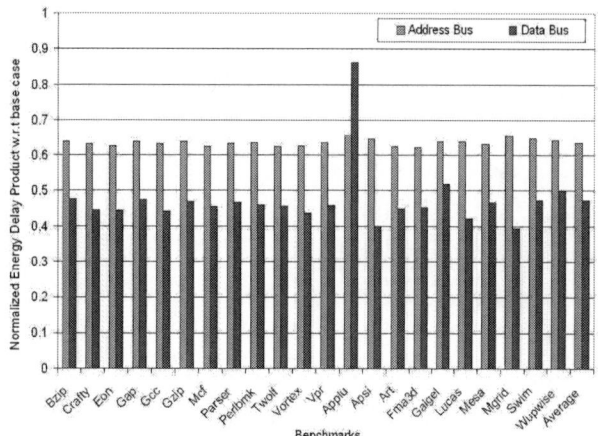

Figure 5. Energy-delay-product of our technique w.r.t. the base case.

in the data bus [8] and incurs some energy penalty (due to codec energy overhead of the VCT technique) as compared to the base case, and hence its normalized EDP is nearly 30%. The sensitivity analysis with wire length of $4mm$ yielded energy savings of 23.68% and 42.70% and EDP reduction of 30.6% and 50.5% for address and data bus, respectively.

5 Conclusion

In this paper, by exploiting variable cycle transmission and DVS mechanisms, we proposed a novel technique for energy-efficient on-chip interconnect design. As part of the application of DVS technique, we scaled both supply voltage and frequency. Delay savings provided by the variable cycle transmission mechanism are exploited while voltage scaling technique is applied so that we obtained significant energy savings as well as delay savings without impacting the throughput. We validated the effectiveness of our technique by focusing on the L1 cache address/data buses of a microprocessor using the SPEC CPU2000 benchmark suite and showed that for a $5mm$ interconnect wire our technique achieves 30% and 45% energy savings over the base case in the address bus and data bus, respectively. We also demonstrated that our technique reduces the energy-delay-product significantly as compared to the base case.

References

[1] Berkeley predictive technology model. http://www-device.eecs.berkeley.edu/~ptm/interconnect.html

[2] SPEC CPU2000 Benchmark. http://www.spec.org

[3] A.R. Brahmbhatt, J. Zhang, Q. Wu, and Q. Qiu. "Low-power bus encoding using an adaptive hybrid algorithm". *DAC*, 2006, pp. 987-990.

[4] D.C. Burger and T.M. Austin. The SimpleScalar tool-set, version 2.0, Technical Report 1342, Department of Computer Science, UW, 1997.

[5] A. Chandrakasan, S. Sheng, and R. Brodersen. "Low-power CMOS digital design". *JSSC*, 27(4), 1992, pp. 473-484.

[6] H. Kaul, D. Sylvester, D. Blauuw, T. Mudge, and T. Austin. "DVS for on-chip bus designs based on timing error correction". *DATE*, 2005, pp. 80-85.

[7] K.-W. Kim, K.-H. Baek, N. Shanbhag, C.L. Liu, and S.-M. Kang. "Coupling-driven signal encoding scheme for low-power interface design". *ICCAD*, 2000, pp. 317-321.

[8] L. Li, N. Vijaykrishnan, M. Kandemir, and M.J. Irwin. "A Crosstalk Aware Interconnect with Variable Cycle Transmission". *DATE*, 2004, pp. 102-107.

[9] D. Liu et al. "Power consumption estimation in CMOS VLSI chips". *IEEE JSSC*, 1994, 26, pp. 663-670.

[10] L. Macchiarulo, E. Macii, M. Poncino. "Low-Energy Encoding for Deep-Submicron Address Buses". *ISLPED*, 2001, pp. 176-181.

[11] M. Meijer, J. Pinede de Gyvez, and R. Otten. "On-Chip Digital Power Supply Control for System-on-Chip Applications". *ISLPED*, 2005, pp. 311-314.

[12] T. Sakurai and A.R. Newton. "Alpha-power law MOSFET model and its applications to CMOS inverter delay and other formulas". *IEEE JSSC*, 1990, 25(2), pp. 584-594.

[13] L. Shang, L.S. Peh, N.K. Jha. "Dynamic Voltage Scaling with Links for Power Optimization of Interconnection Networks". *HPCA*, 2003, pp. 91-102.

[14] P. Sotiriadis and A. Chandrakasan. "Low Power Bus Coding Techniques Considering Inter-wire Capacitances". *CICC*, 2000, pp. 507-510.

[15] P. Sotiriadis and A. Chandrakasan. "Bus energy minimization by transition pattern coding (TPC) in deep submicron technologies". *ICCAD*, 2000, pp. 322-327.

[16] P. Sotiriadis and A. Chandrakasan. "Reducing Bus Delay in Sub-micron Technology using Coding". *ASPDAC*, 2001, pp. 109-114.

[17] S.R. Sridhara, A. Ahmed, and N.R. Shanbhag. "Area and Energy-efficient Crosstalk Avoidance Codes for On-chip Buses". *ICCD*, 2004, pp. 12-17.

[18] A. Stratakos. *High-efficiency low-voltage DC-DC conversion for portable applications*. Ph.D. Thesis, University of California, Berkeley, 1998.

SESSION C2:
Architecture

21st International Conference on VLSI Design

Dynamic Aggregation of Virtual Addresses in TLB using TCAM Cells

Rupak Samanta, Jason Surprise[1] and Rabi Mahapatra[1]

Department of Electrical and Computer Engineering, Texas A&M University, College Station

[1]*Department of Computer Science, Texas A&M University, College Station*

{rupak9, [1]jasonsurprise}@tamu.edu, [1]rabi@cs.tamu.edu

ABSTRACT

In this paper, we propose dynamic aggregation of virtual tags in the Translation Lookaside Buffer (TLB) to increase its storage capacity without increasing the size of the tag array. To support dynamic aggregation, we incorporate a few Ternary-CAM (TCAM) cells into the TLB tag array. The modified TLB architecture demonstrates a compression scheme that increases TLB reach with negligible overhead and no access time penalty. The performance of the proposed TLB architecture is evaluated using SPEC CPU2000 benchmarks. Simulation results indicate a significant reduction in miss ratios, nearly 100% reduction is achieved in several benchmarks, and as much as a 46% increase in IPC (Instructions per cycle) is obtained when compared to a conventional TLB with the same number of tag entries. We also evaluate the performance of our tag compressed TLB against the performance of a conventional TLB that contains an equivalent number of virtual to physical address translations. Our results show that TCAM based compression is able to achieve nearly the same system performance as the large conventional TLB while consuming on average 38% less energy and 42% less area; thus illustrating that tag compression is a more attractive solution for improving TLB performance than simply increasing the size of the TLB.

Keywords

TCAM, TLB, Dynamic aggregation, Miss rate, IPC.

1. INTRODUCTION

Virtual to physical address translation is one of the most critical operations in modern processors. The Translation Lookaside Buffer (TLB) plays a crucial role in expediting the translation process by caching the most recently referenced translations. Previous studies have shown that significant processing time is spent having the kernel handle TLB misses [4, 6, 8]. As instruction level parallelism, processor clock frequency, and application working set size increases, the impact of TLB performance on the total application processing time will continue to grow.

The performance of any TLB design is evaluated based on its access time and miss ratio. Since address translation is in the critical path of instruction fetch and data reference for physically addressed TLB, the TLB look up needs to be as quick as possible. The TLB miss rate should also be minimized because of severe miss penalties, and these penalties will continue to grow with the increasing disparity between processor and memory access speed. Based on the growing disparity between on-chip and off-chip performance predicted in the ITRS report [1], researchers working on cache compression have estimated that main memory latency will grow to hundreds of clock cycles, and they have used main memory access latencies as high as 400 clock cycles to judge the effectiveness of their scheme in next generation processors [3]. Although TLB access time and miss ratio should be kept as low as possible in order to achieve the best performance, improving one of these metrics generally results in degradation of the other due to the inverse relationship among access time and TLB capacity.

In this paper, we present a TLB architecture that is capable of significantly reducing the miss ratio without increasing the access time. Our scheme takes advantage of the fact that most of the TLB access time is spent searching the fully-associative tag array composed of Content Addressable Memory (CAM) cells. We employ the use of Ternary-CAM (TCAM) cells [13] in the second level (L2) Data-TLB tag array to enable dynamic compression (aggregation) of tag entries containing contiguous virtual address values. By compressing TLB tags, a single virtual tag entry is capable of mapping to multiple physical addresses thus increasing the total translation storage capacity (reach) of the TLB. Despite the increased TLB translation capacity, the TLB access time is negligibly affected by TCAM enabled compression. The proposed TCAM based aggregation technique does not incur any decompression overhead when searching the TLB since CAM and TCAM access times have been shown to be equivalent [5]. Also, tag compression is only performed when a TLB miss occurs and can be finished before the missed entry is retrieved from main memory; therefore, the compression latency is completely hidden by the memory access. To the best of our knowledge, this is the first attempt to use TCAM cells to increase the TLB reach.

978-1-4244-3039-0/08 $25.00 © 2008 IEEE

2. TAG AGGREGATION

Tag Aggregation is the process of compressing several tag entries that differ only in their least significant bits into a single entry. The benefits of tag aggregation can be examined under two lights. First, performing tag aggregation frees tag entries, which increases the number of address to data mappings that can be stored thus increasing the effective storage capacity of the cache. In this case, the number of tag entries is the same in both the compressed and uncompressed caches. Although the capacity is increased, the cache access time is unaffected because it is predominantly influenced by the number of tag entries that must be searched. The other benefit of tag aggregation is that it reduces the number of tag entries required to map to a fixed number of data entries. In this case, the number of tag entries is reduced by 2^N, where N is the number of bits considered for aggregation. Although the storage capacity remains the same in both the compressed and uncompressed caches, the power and area consumed by the cache can be drastically decreased due to the reduction in tag entries.

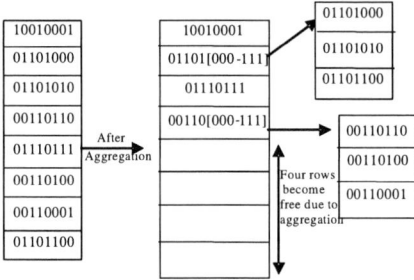

Figure 1- Example of aggregation process

In Figure 1, we illustrate the process of aggregation with an example where we aggregate two entries if they differ in only their three least significant bits. After aggregation, two of the entries each correspond to three tags thus freeing four cache entries.

3. TCAM ENHANCED CACHE DESIGN

TCAM cells are ideally suited for an architecture supporting tag aggregation. In addition to the standard 0 and 1 logic states, TCAM cells have an additional "don't care" state represented with an "X". When a TCAM cell containing the don't care value is searched, a match is returned regardless of whether a 0 or 1 is being requested. Based on this property, two tag entries can be aggregated by setting their least significant differing bits to be don't care values.

The TCAM Enhanced Cache Architecture consists of a tag array that is augmented with a few TCAM cells in order to support dynamic aggregation. The enhanced cache structure can be divided into two components – the CAM/TCAM Cache and the Dynamic Aggregator Module (DAM). Details of each component are given in the following subsections.

3.1 CAM/TCAM Cache

In general, only the W most significant bits of a virtual address are stored in the cache's tag array. We use TCAM cells to store the N least significant bits of the total W bits. The rest of the tag bits are stored using CAM cells. The extent of dynamic aggregation depends on the number of TCAM cells in a single tag array entry. More precisely, a tag entry can map up to a maximum of 2^N cache lines (blocks). We refer to all the blocks that are mapped by a single tag entry as a super-block, and we refer to its tag entry as a super-block tag. Therefore, the number of cache lines in a super-block might be 1, 2, 4, 8, 16....2^N depending on the amount of compression achieved for the tag entry. The CAM/TCAM Cache maintains a single valid and dirty bit for the super-block. The number of cache lines fetched from main-memory on a cache miss is one; however, the numbers of lines that get replaced vary depending on the super-block size. Rather than sequentially storing all the cache lines belonging to a super-block, which would require extra overhead to manage the data array, we propose to use a set of 2^N data banks each storing a cache line for every tag array entry. Figure 2 illustrates the CAM/TCAM design with N, the number of TCAM cells, equal to two. The data bank selector is configured such that Data Bank 0 contains the cache line for a tag value ending in 00, and Data Bank 3 contains the cache line for a tag ending in 11. Based on this design, the appropriate data bank can be activated while the tag array is being searched so that when a matching tag is found only one row is selected from a single data bank. Since new tags are only brought into the cache when a miss occurs, tag compression is only initiated when the cache does not contain the requested data. The steps involved in a CAM/TCAM Cache look up are as follows:

1) W most significant bits of the requested address are searched in the tag array.

2) Bits (N to 0) are passed as input to the Data Bank Selector and output Mux module.

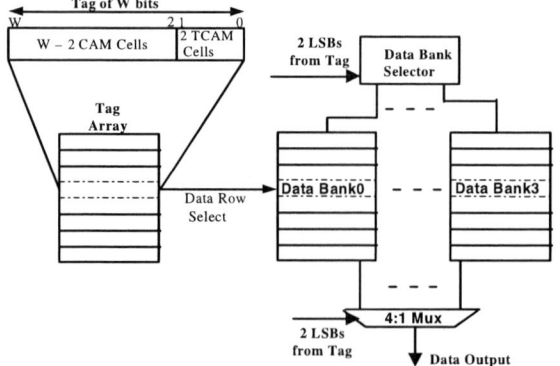

Figure 2 - CAM/TCAM Cache with 2 TCAM cells

3) If there is a match in the tag array, the match line goes high, thus selecting an entire row in the active Data Bank. The appropriate data value is then driven to the cache output.

4) If the requested address is not found in the tag array, the cache requests the cache line containing the data from main memory, and while the data is being fetched, the missed address tag is examined for aggregation with entries already stored in the cache's tag array by the DAM.

3.2 DYNAMIC AGGREGATOR MODULE

Figure 3 gives an overview of the Dynamic Aggregation process carried out by the Dynamic Aggregator Module (DAM). In order to keep aggregation hardware overhead low, we have adopted a sequential tag aggregation process. The figure shown here is based on aggregation of the two LSBs, which requires the use of 2 TCAM cells. With this implementation, a maximum of 4 entries can be merged into a single entry with 2 don't care values stored in the 2 least significant bits. For 2-bit aggregation, at most two rounds of sequential compression are performed depending on how many of the 4 entries are in the cache. Each of the values connected with a square bracket in the figure constitute an aggregation group. If both entries of an aggregation group are present in the cache, then the two entries will be removed and replaced with a single entry having the appropriate number of bits set as "X". Since dynamic aggregation is performed only on a cache miss, only one member of an aggregation group can be present in the cache at a time otherwise the aggregated entry will be present.

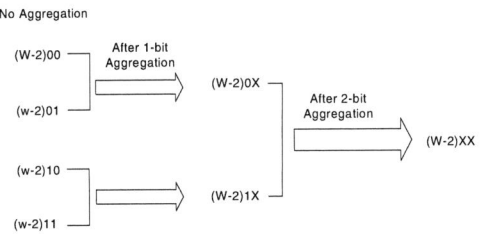

Figure 3 - Sequential TCAM Based Tag Aggregation Process

As previously stated, dynamic aggregation is only performed during a cache miss. On each miss, the address tag that missed is stored in a register, and the requested data is fetched from main memory. While waiting for the data to be retrieved, the DAM is able to use the stored tag to search the cache for aggregation candidates and perform tag aggregation if it is possible. As long as the dynamic aggregation process is completed before the data is returned from main memory, the compression latency will remain hidden. Figure 4 depicts the interconnections among the DAM and CAM/TCAM Cache hardware blocks. Although dynamic aggregation can be implemented with N TCAM cells, we only describe the hardware necessary for 2-bit

aggregation. The DAM consists of seven distinct hardware units. The Temporary Register (TR) stores the address tag when a miss occurs in the cache. Each data bank has its own "Bank Enable Logic" unit and an associated buffer capable of storing a single cache line. The "Bank Enable Logic" unit generates an enable signal for the corresponding bank during aggregation. The "Dynamic Aggregator (DA) Logic" unit generates the DA signal, which initiates the aggregation process when a miss occurs. Each tag array entry has an associated 2-bit Aggregation Counter (AC) unit that stores the current number of bits set as don't care. During a tag array search, Register A (REGA) holds the current Row Address of the matching tag array entry and Register B (REGB) contains the Row Address value previously held by REGA. The "Counter and Comparison Logic" unit maintains state information regarding the aggregation process. The "Update/Replace Logic" unit generates Update and Replace signals for the Memory Management Unit (MMU) to assist storing the data retrieved from main memory due to a cache miss. To perform dynamic aggregation, variable bit length searches must be supported by the CAM/TCAM tag array. To enable searching of (W-1) and (W-2) bit lengths, we add hardware

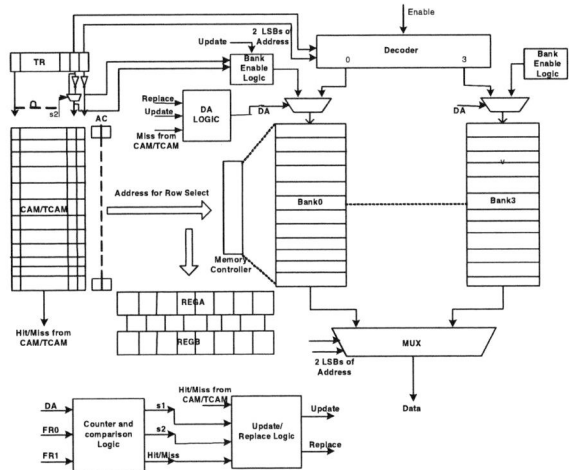

Figure 4 – CAM/TCAM Cache with DAM Hardware

to the TR so that the Counter and Comparison Logic unit can control the values being searched for in the TCAM cells. The hardware added to the TR enables us to match only a single tag entry thus avoiding conflicts and also eliminating the requirement of a complex priority encoder, which would have been required to handle multiple tag array matches.

4. SIMULATION METHODOLOGY

To evaluate the effectiveness of the TCAM Enhanced TLB Architecture, we executed a subset of the SPEC CPU2000 benchmark programs [2] using the SimpleScalar [7] simulator, which was extended to support 3-bit Tag Aggregation in the L2 Data-TLB (DTLB).

978-1-4244-3039-0/08 $25.00 © 2008 IEEE

4.1 Experimental Setup

For our experiments, we configured SimpleScalar to use the Alpha Instruction Set Architecture. In all of the tests, the processor configuration is exactly the same except for the L2 DTLB parameters, thus isolating the effects of the DTLB. Table 1 describes the three L2 DTLB configurations considered for comparison besides the baseline configurations. These test configurations were chosen to evaluate how close the Dynamic Aggregation DTLB comes to its maximum and minimum performance levels. Based on CACTI [13] simulation results, we found the hit latency of a DTLB with 128 tag entries to be 2 clock cycles and the hit latency of a DTLB with 1024 tag entries to be 4 clock cycles.

Table 1 - L2 DTLB Test Configurations

Test Configuration	Parameters
Dynamic Aggregation (3-bit aggregation)	Tag Entries = 128 Replacement = LRU Total Data Entries = 1024 Hit Latency = 2 cycles Miss Latency = 400 cycles
Equivalent Data Size	Tag Entries = 1024 Replacement = LRU Total Data Entries = 1024 Hit Latency = 4 cycles Miss Latency = 400 cycles
Equivalent Tag Size	Tag Entries = 128 Replacement = LRU Total Data Entries = 128 Hit Latency = 2 cycles Miss Latency = 400 cycles

4.2 Workload Selection

We chose twenty three benchmarks from the SPEC CPU2000 suite [2] to evaluate our design – bzip, crafty, eon, gcc, gzip, mcf, parser, perlbmk, twolf, vpr, are Integer applications while ammp, applu, apsi, art, equake, facerec, galgel, lucas, mesa, mgrid, sixtrack, swim and wupwise are floating point programs. For all the experiments, performance metrics were collected by executing 2 billion instructions.

5. EXPERIMENTAL RESULTS

To gauge the performance of the L2 DTLB with 3-bit Dynamic Aggregation, we compare its Miss Rate, Instructions per cycle (IPC), Energy Consumption, and Area overhead to those for a L2 DTLB with Equivalent Data and Tag Sizes.

5.1 Miss Rates

Since Tag Compression is intended to increase the effective L2 DTLB storage capacity, we expect the Dynamic Aggregation DTLB to have significantly fewer misses than the Equivalent Tag Size DTLB. Ideally, if the benchmarks exhibit a considerable amount of spatial locality, then all of the tag entries can be compressed to their maximum capacity resulting in the Dynamic Aggregation DTLB achieving miss rates comparable to the Equivalent Data Size DTLB. Figure 5 presents the miss rates achieved during benchmark simulation. On average, the Dynamic Aggregation DTLB reduces the Equivalent Tag Size miss rate by 39%. Moreover, Dynamic Aggregation on an average achieves 67% of the miss rates of Equivalent Data Size DTLB.

5.2 System Performance

System Performance is typically measured in terms of Instructions per cycle (IPC). Miss rate reductions are usually reflected as an increase in overall system performance. Figure 6 reports the IPC achieved during simulation. On an average, the Dynamic Aggregation achieves 92% of the Equivalent Data IPC. However, excluding three benchmarks (apsi, galgel and mcf) Dynamic Aggregation can achieve 99% of the Equivalent Data IPC. On the other hand, Dynamic Aggregation on an average provides an 8% performance increase over the Equivalent Tag Sized DTLB.

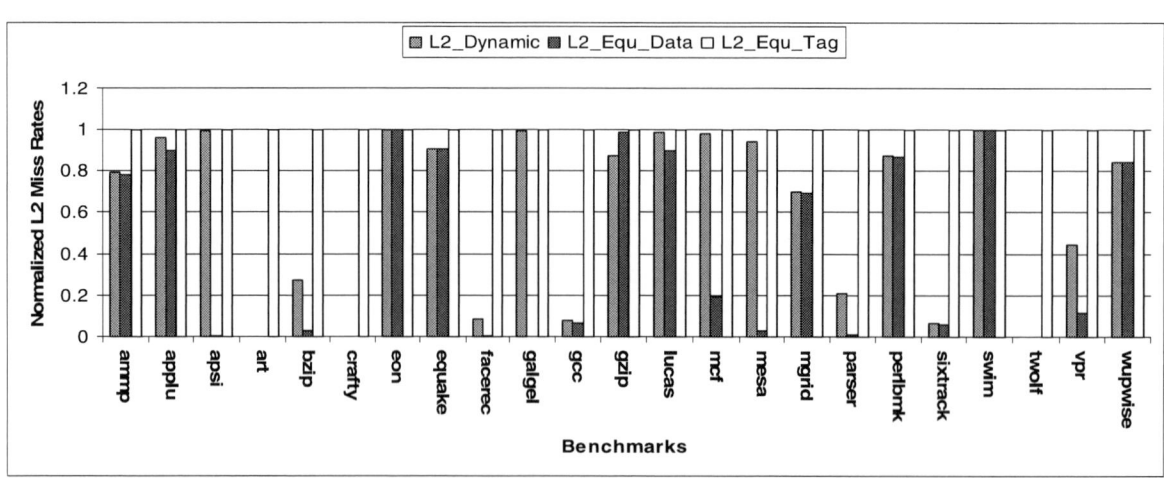

Figure 5 - L2 DTLB Miss Rates Normalized to Equivalent Tag DTLB

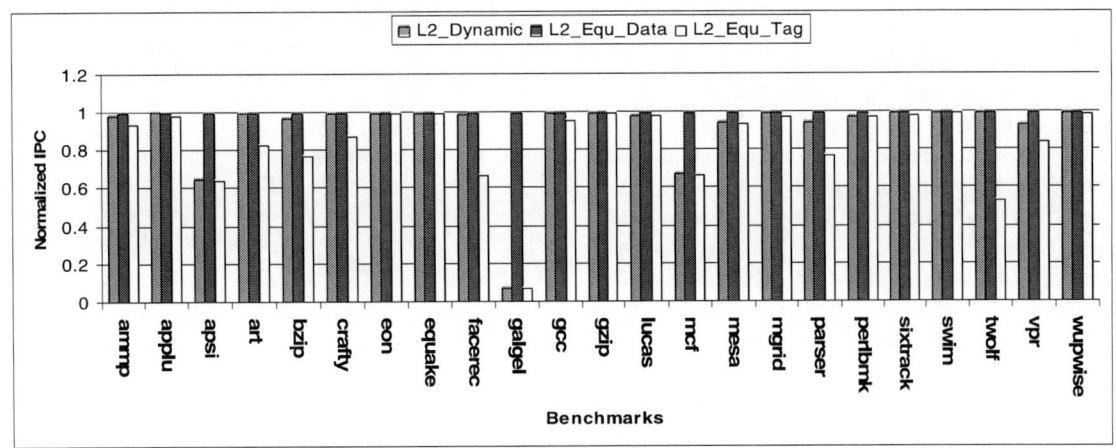

Figure 6 - IPC Normalized to Equivalent Data DTLB

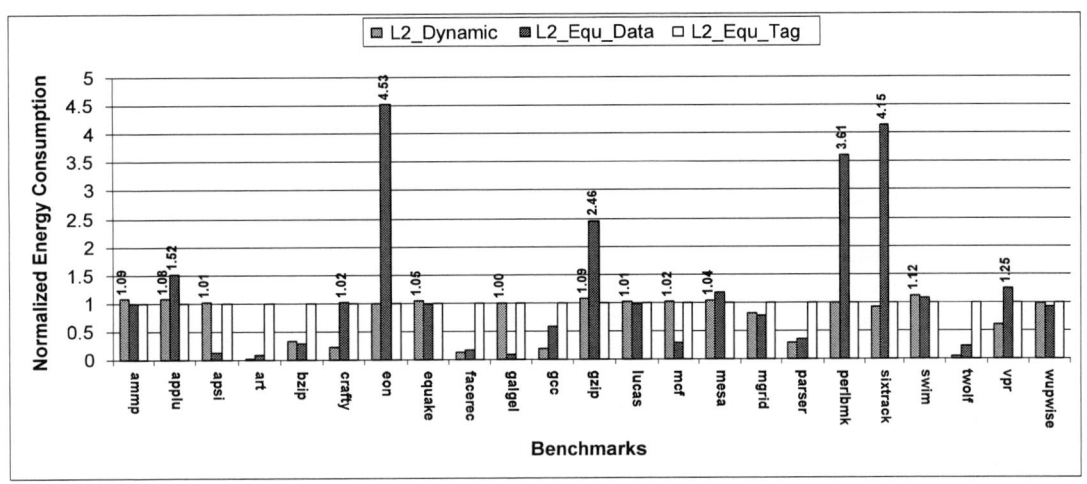

Figure7 - Total Energy Consumption Normalized to Equivalent Tag DTLB

5.3 Energy Consumption

To evaluate the energy efficiency of our architecture, we have calculated the metric "total energy consumption", which is sum of the TLB access energy, aggregation energy and the main memory access energy.

$$E_{Total} = TLB_{Accesses} \times (E_{Tag\_search} + E_{Data\_bank})$$
$$+ E_{Aggregation} \times TLB_{Misses} + E_{Main\_memory} \times TLB_{Misses}$$
$$E_{Aggregation} = 3 \times E_{Tag\_search} + E_{DAM}$$

E_{Total} is the total energy consumed. E_{Tag_search} is the energy consumed during a search in the tag array, and E_{Data_bank} is the energy consumed when accessing the data from the selected row of the data bank. $E_{Aggregation}$ is the worst-case

energy consumption for aggregating all three LSBs using the Dynamic Aggregator Module, and E_{Main_memory} is the energy consumed when accessing the main memory on each TLB miss. Both E_{Tag_search} and E_{Data_bank} are determined using CACTI [14] tool, while E_{DAM} is calculated using SPICE [12] simulation. E_{Main_memory} is calculated by multiplying the number of main memory accesses by 18nJ per access, which was determined based on the DRAM power consumption parameters of 300mW for a 60ns access time specified in [11]. Figure 7 shows the total energy consumption calculated using the formula E_{Total}. On average, the Dynamic Aggregation DTLB consumes 38% and 26% less energy as compared to the

Equivalent Data Size DTLB and the Equivalent Tag Size DTLB.

5.4 Area Comparison

The area comparison was performed based on the number of transistors in the TLB and in the extra hardware needed for 3-bit aggregation. For the area calculations, we used the following transistor relationships, CAM cell = 10 transistors [9], SRAM cell = 6 transistors, and TCAM cell = 17 transistors [10]. The DAM hardware requires approximately 4616 transistors, which is 2% of the total area used by a Dynamic Aggregation TLB storing 128 tag entries. The Dynamic Aggregation TLB requires 84% less tag area than the Equivalent Data TLB, which translates to 42% less total area.

6. CONCLUSIONS

In this paper, we proposed a TCAM Enhanced TLB Architecture that improves the performance of compute-intensive applications by increasing the capacity of a conventional TLB with an equivalent number of tag entries. We use a two level TLB structure, where dynamic tag aggregation is performed only in the L2 TLB. A maximum performance increase of 46% was achieved in our scheme as compared to the conventional Equivalent Tag Sized TLB, while consuming on an average of 26% less energy. Except for a few benchmarks the Equivalent Data Sized TLB was able to achieve on an average 1% performance improvement over the proposed scheme; however, this additional performance costs an average of 38% more energy and 42% more area. The performance to energy and area consumption ratio clearly indicates that Dynamic Tag Aggregation is a more attractive solution to improving system performance rather than simply increasing the size of the TLB.

7. REFERENCES

[1] International Technology Roadmap for Semiconductors. http://public.itrs.net/

[2] S. P. E. Corporation. http://www.spec.org.

[3] A. R. Alameldeen and D. A. Wood, "Adaptive Cache Compression for High-Performance Processors," *Proceedings of International Symposium on Computer Architecture*, 2004, pp. 212-223.

[4] T. E. Anderson, H. M. Levy, B. N. Bershad, and E. D. Lazowska, "The Interaction of Architecture and Operating System Design," *Proceedings of Architectural Support for Programming Languages and Operating Systems*, 1991, pp. 108-120.

[5] I. Arsovski, T. Chandler, and A. Sheikholeslami, "A Ternary Content-Addressable Memory (TCAM) Based on 4T Static Storage and Including a Current-Race Sensing Scheme," *IEEE Journal Of Solid-State Circuits*, vol. 38, no. 1, January 2003, pp. 155-158.

[6] K. Bala, M. F. Kaashoek, and W. E. Weihl, "Software Prefetching and Caching for Translation Lookaside Buffers," In *Proceedings of the Usenix Symposium on Operating Systems Design and Implementation*, Nov 1994.

[7] D. Burger and T. Austin, SimpleScalar Toolset, Version 3.0. http://www.simplescalar.com

[8] J. B. Chen, A. Borg, and N. P. Jouppi, "A simulation based study of TLB performance," *Proceedings of International Symposium on Computer Architecture,* 1992, pp. 114-123.

[9] A. Efthymiou and J. Garside, "An Adaptive Serial-Parallel CAM Architecture for Low-Power Cache Blocks," *Proceedings of International Symposium on Low Power Electronics and Design,* August 2002, pp. 136-141.

[10] B. Gamache, Z. Pfeffer, and S. Khatri, "A Fast Ternary CAM Design for IP Networking Applications," *12th International Conference on Computer Communications and Networks*, October 2003.

[11] N. Kirubanandan, A. Sivasubramaniam, N. Vijaykrishnan, M. Kandemir, and M. J. Irwin, "Memory Energy Characterization and Optimization for the SPEC2000 Benchmarks," In *IEEE International Workshop on Workload Characterization*, December 2001, pp. 193-201.

[12] T. L. Quarles, A. R. Newton, D. O. Pederson, and A. Sangiovanni-Vincentelli, *"SPICE 3 Version 3F5 User's Manual,"* Department of EECS, University of California, Berkeley, March 1994.

[13] J.P. Wade and C.G. Sodini, "A Ternary Content Addressable Search Engine," *IEEE Journal of Solid-State Circuits*, volume. 24, issue. 4, August. 1989, pp. 1003-1013. [22]

[14] S. J. Wilton and N. P. Jouppi, "CACTI: an enhanced cache access and cycle time model," *IEEE Journal of Solid-State Circuits*, volume.31, issue.5, May 1996, pp. 677-688.

21st International Conference on VLSI Design

Continuous Frequency Adjustment Technique Based on Dynamic Workload Prediction[1]

Hwisung Jung and Massoud Pedram
University of Southern California
Department of Electrical Engineering
{hwijung, pedram}@usc.edu

Abstract

Real-time embedded systems increasingly rely on dynamic power management to balance between power and performance goals. In this paper, we present a technique for continuous frequency adjustment (CFA) which enables one to adjust the frequency values of various functional blocks in the system at very low granularity so as to minimize energy while meeting a performance constraint. A key feature of the proposed technique is that the workload characteristics for functional blocks are effectively captured at runtime to generate a frequency value that is continuously adjusted, thereby eliminating the delay and energy penalties incurred by transitions between power-saving modes. The workload prediction is accomplished by solving an initial value problem (IVP). Applying CFA to a real-time system in 65nm CMOS technology, we demonstrate the effectiveness of the proposed technique by reporting 13.6% energy saving under a performance constraint.

1. Introduction

As more power-managed functional blocks (FBs) are being built to realize power-saving opportunities by utilizing dynamic voltage and frequency scaling (DVFS) techniques, the task of integrating multiple power management policies in a single system is becoming ever more challenging. Furthermore, although current CMOS technologies allow an increasing number of different clock and voltage domains to be specified on the same chip, traditional dynamic power management (DPM) methods have not been able to take full advantage of DVFS techniques due to intricate trade-offs between the power-savings and performance constraints. This is because a system-level power manager (PM) has only limited control over power-saving techniques due to additional power and delay costs incurred during power-mode transitions [1]. In addition, the power management routine, most likely residing in the operating system, can itself become a heavy duty since it has to continually monitor the workload of FBs and send DVFS assignment commands to them [2].

DVFS-enabling techniques depend not only on the configuration of the voltage/frequency control circuits, but also on the efficiency of the prediction mechanism used by the PM to set the voltage/frequency levels. As shown in [3]-[8], the problem of determining a power management policy with DVFS

techniques at system-level has received a lot of attention. In [3], the authors present a frequency management method based on variable updated intervals, instead of using a fixed update interval, where a frequency scheduling method is based on effective deadline mechanism. The analytical models for selecting an optimal DVFS under tight performance constraints are presented in [4][5]. Reference [6] presents an online DVFS technique by utilizing interface queues to guide the DVFS control in a multiple clock and voltage domain architecture. A voltage island-based power management technique is proposed in [7] to meet a performance constraint in multi-threshold CMOS technologies. Authors of [8] present an optimization technique for power mode transitions under timing constraints.

Although all of the above techniques perform DVFS based on power management policies, little attention has been paid to handle variable frequency adjustment and prediction techniques by using hardware-control mechanisms, which minimizes the computational efforts by the PM. Furthermore, traditional approaches for DPM, mainly based on the software-control of power-saving techniques, are highly dependent on the speed of operating system, which incur non-negligible overhead. Thus, minimizing the overhead of power-mode transitions in real-time with the hardware-control architecture is an important step to guarantee the quality of DPM techniques. This is precisely the contribution of the present paper.

In this paper, we present a power management framework for dynamic continuous frequency adjustment which provides power-saving opportunities by dynamically and continuously adjusting a variable operating frequency. The basic idea of CFA is to eliminate the power and delay costs incurred by the power-mode transitions which involve clock generators (e.g., PLL). Predicting a workload of tasks is formulated as an initial value problem (IVP) [9], where the frequency is adjusted by the proposed dynamic frequency adapter. Note that the IVP formulation in the paper determines the workload value of future time subsequent to a given time.

The remainder of this paper is organized as follows. Section 2 provides a motivational example, while section 3 describes the proposed architecture. In section 4, we present the details of the workload prediction technique. Experimental results and conclusion are given in section 5 and 6.

---

[1] This research is supported in part by NSF grant no. 0509564.

2. Motivational Example

According to conventional DPM approaches [10], where many FBs in a system are equipped with multiple power-performance states (e.g., sleep, idle, and active modes), a PM sends an command, i.e., DFS (Dynamic Frequency Scaling) values, to each FB. For example, as shown in Figure 1, where we assume each FB has three active states which is controlled by DFS values, the service provider (SP) or the service requestor (SR) can switch between the different speed-levels, where $DFS_1 < DFS_2 < DFS_3$ in terms of frequency values. The PM monitors the current workload of the system by looking into the corresponding service queue (SQ) to adjust the DFS value for each FB. A transition into or out of a power-performance state (i.e., dotted arrows in the figure), however, consumes energy and/or incurs delay penalty that may not be negligible.

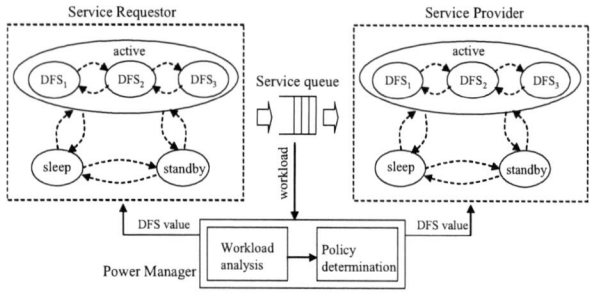

Figure 1. Conventional dynamic power management approach.

Attempting to greedily respond to the workload changes so as to provide an optimal DVFS value can result in significant energy and delay overheads associated with mode transitions. To solve this problem, a software component, which can predict the required future performance level of the system to prevent frequent power mode transitions, has been incorporated into the power managers of [2][3]. Although these prediction methods help reduce the energy overheads, there are some disadvantages because i) a software-oriented prediction algorithm increases the computational overhead of the PM that resides in the driver or the operating system, and ii) when using a PLL (Phase-Lock Loop) to effect a frequency change, the FB may be stalled during the lock time of the PLL. Consequently, use of the PLL to realize the DFS setting commanded by the PM may result in sizeable performance penalty [11]. The main contribution of our work is that we predict the workload level for the next time step and ramp up (or down) the system frequency in a continuous manner until the target frequency value is achieved, where there is never a need for stalling the FB.

3. Power Management Architecture

In this section we present a platform-specific continuous frequency adjustment (CFA) technique. The target system is comprised of various software and hardware modules, which include a power manager (PM), a performance monitoring unit (PMU), and a dynamic frequency adapter (DFA), as depicted in Figure 2. Note that the DFA is implemented as a hardware module to minimize software-oriented computational efforts.

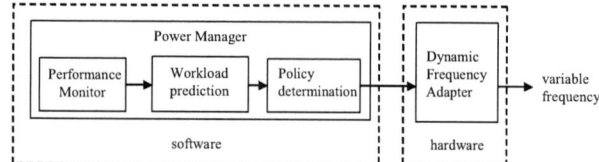

Figure 2. The proposed power management architecture.

3.1 Power Manager

The main goal of the PM is to determine and execute a power management policy (i.e., one that maps workloads to power state transition commands so as to minimize the system energy dissipation), based on the information provided by the PMU. The PM consists of a workload prediction and a policy determination. In this paper, we focus on the selection of optimal frequency value and continuous (gradual) change in system frequency moving from current value to target value.

3.2 Performance monitoring unit

The PMU profiles and analyzes the workload (e.g., the arrival rate of tasks) characteristics by looking into the SQ. In our problem setup, the SQ of each FB is represented by the G/M/1 queuing model, whereby the inter-arrival times are arbitrarily distributed and the service times are exponentially distributed [12]. Note that the service time behavior of each FB is captured in the form of the service time distribution for the FB when it is in the active mode. Similarly, the input request behavior (i.e., workload) for each FB is modeled by the request inter-arrival time distribution at the corresponding input queue. Details of the G/M/1 queuing model are omitted here to save space. Interested reader may refer to [12].

3.3 Dynamic Frequency Adapter

When the workload of an FB changes frequently, the task of deciding what frequency value to assign to the FB becomes increasingly difficult. Furthermore, the conventional PLL-based frequency scaling techniques waste energy and delay when they change the frequency values. To overcome these shortcomings, we present a workload-aware DFA to generate a continuously varying frequency for each FB.

One benefit of using a variable frequency is that the DFA enables each FB to remain functional even when its frequency is being adjusted, while satisfying the required performance. Furthermore, the DFA is able to increase (or decrease) the operating frequency value at a slow or fast rate with the help of the PMU, depending on how slow or fast the workload is changing and what the user-specified preferences are, as depicted in Figure 3.

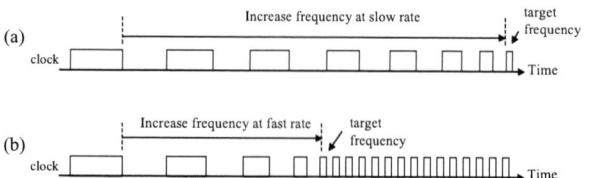

Figure 3. Continuously changing the frequency at a slow pace (a) or fast pace (b).

250

978-1-4244-3039-0/08 $25.00 © 2008 IEEE

The procedure for continuously adjusting the frequency is explained as follows. The PM examines the workload of each FB at decision epoch $n+1$ for the time interval ranging from decision epoch n to $n+1$, and subsequently, sets the frequency value of each FB for the next time ranging from $n+1$ to next decision epoch at time $n+2$ (see the next section for details of the frequency prediction algorithm). Note that each time-based or interrupt-based event occurrence is called a decision epoch. Assume that a mapping table for selecting an optimal operating frequency as a function of the workload has been provided. If the workload change is fast (slow), the interval during the frequency adjustment is performed will be shortened (lengthened) to improve DFA responsiveness. In the proposed framework, determining which frequency level to use in what time interval is implemented in hardware.

Figure 4 shows the block diagram of the proposed DFA which generates a variable frequency by using a pulse width modulation technique. The DFA is implemented inside a chip, where we effectively manage noise and signal integrity problems. Note that we apply this architecture to a specific target system (e.g., high-speed networking controller), where the operating frequency is rather low (e.g., around 200MHz), yet the system provides high throughput.

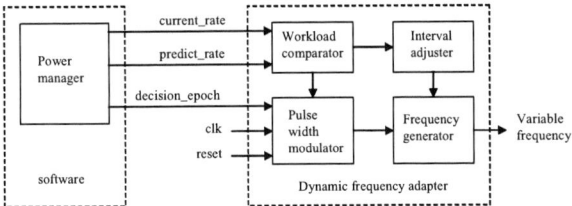

Figure 4. Block diagram of DFA.

4. Workload-Driven Frequency Adjustment

In this section, we present a workload prediction technique based on the initial value problem (IVP) formulation and a procedure of dynamic frequency adjustment.

4.1 IVP-based Workload Prediction

Assume that, by utilizing the PMU, a PM is able to monitor the current workload of the tasks at decision epochs $t_1, ..., t_n$ where $t_{i+1} = t_i + T$. Let $w(t)$ denote the workload (i.e., the arrival rate of tasks) of a target FB at time t and let f be a function providing the operating frequency for the FB in terms of time and workload in every interval $[t_i, t_{i+1}]$. Then, an initial value problem (IVP) may be defined to predict $w(t)$ as follows:

$$\partial w / \partial t = f(t, w), \qquad w(t_i) = w_i \qquad (1)$$

where $t \in [t_i, t_i + T]$, and w_i denotes the workload at the beginning of the current interval. The IVP limits the solution by an initial condition, which determines the value of solution at all future time t in the current interval. Although f can be any general function, in practice, we assume a linear function form: $f = aw(t) + b$ where a and b are appropriately-calculated slope and offset coefficients.

Since the initial workload value, w_i, is provided by the PMU, it is possible to integrate Eqn. (1) to obtain $w(t)$ in the interval $[t_i, t_{i+1}]$. The standard solution method for the IVP is to approximate the solution of the ordinary differential equation (ODE) by calculating the next value of w, i.e., $w(t+h)$ as the summation of the present value $w(t)$ plus the product of the size of a time step h and an estimated slope $w'(t)$ i.e.,

$$w(t + h) = w(t) + hw'(t) \qquad (2)$$

where the smaller this time step h is, the more accurate the results will be. The difference between different ODE solvers is in how they approximate $w'(t)$ and whether and how they adaptively adjust h.

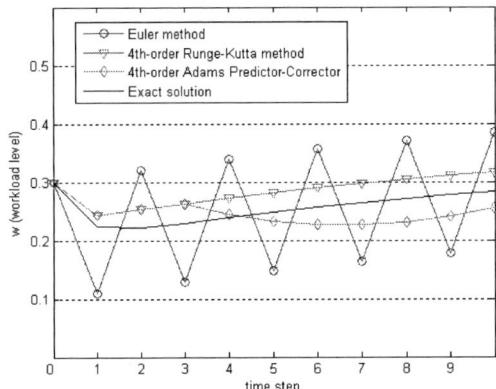

Figure 5. Evaluation of various IVP solutions.

Considering the accuracy and overhead, we have evaluated a number of methods for solving the IVP, which include the Euler's method, the 4th-order Runge-Kutta method, and the 4th-order Adams predictor-corrector method [9]. In Figure 5, we assume that $w(0) = 0.3$ as an initial value. The time step size is defined as $h = T/K$, where the time interval $[t_i, t_i + T]$ is divided into K equal-length segments. It is clearly seen that the Euler method, the simplest approach for solving the IVP, shows low accuracy (i.e., high error) in predicting the workload value, where the error is defined as the difference between the exact values and the computed approximates. However, the 4th-order Runge-Kutta method exhibits low error and consistent stability in predicting the workload value. The 4th-order Adams predictor-corrector is also accurate, but has higher computational complexity.

Figure 6 shows the trade-off between the accuracy and time step h in terms of performance of workload prediction, where time (x-axis) is defined in terms of successive time steps. In this evaluation, the 4th-order Runge-Kutta method is used with an initial value $w(0) = 0.3$. The determination of the time step size is crucial since smaller time step increases the computational overhead in the software (e.g., operating system). We use various values for time step size h (= 2, 5, and 10), where T is fixed, while monitoring the error in predicting the workload values. The time step of size 2 indicates great accuracy, but increases computational efforts by the software (due to more computations in the same interval), whereas step size of 10 exhibits lower computational efforts with lower accuracy. In our problem setup, we have empirically observed that a time step size of 5 provides a reasonable trade-off point.

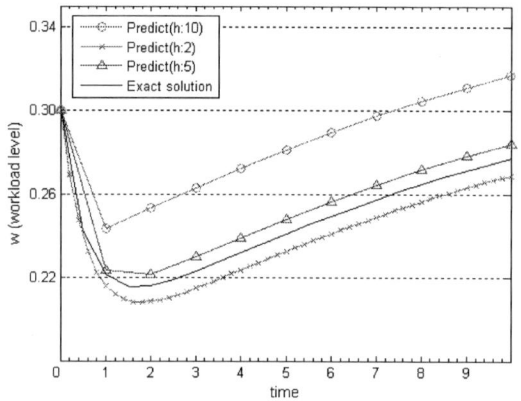

Figure 6. Trade-off between the performance and time step *h*.

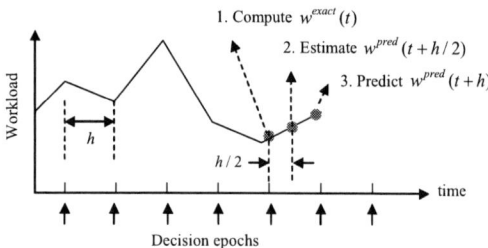

Figure 7. Workload prediction technique based on the midpoint method and IVP.

To make the workload prediction technique more suitable for online implementation, an efficient one-step method known as the *midpoint method* is utilized to solve the IVP. Specifically, at time instance t in $[t_i, t_i + T]$, we predict the workload value for time $t + h$, based on the value at time $t + h/2$, which is obtained by using the midpoint method, as depicted in Figure 7. First, the current workload at time t is monitored by the PMU and a frequency value is read from a pre-characterized workload-frequency mapping table by the power manager. Note that we do not want to use the predicted value for time t, which was previously computed at time $t - h$, because we can achieve the exact frequency value at time t. Next, the workload value at time $t + h/2$ is estimated by using a moving average method, for example, if the window size of the moving average calculator is 2, then, $w^{\text{pred}}(t + h/2) = 1/2 \cdot (w^{\text{exact}}(t) + w^{\text{exact}}(t - h))$. This workload value is subsequently used as the midpoint estimate of the workload in the upcoming period. In particular, it is used along with $w^{\text{exact}}(t)$ to compute $w^{\text{pred}}(t + h)$ by applying the IVP.

The advantage of this prediction method is that we do not attempt to predict $w^{\text{pred}}(t + h)$ directly by using a moving average method only. Instead, we estimate the workload value for a nearer time in the future (which should provide higher accuracy) and use that value to initially estimate the rate of workload change in the upcoming period, followed by finally

computing $w^{\text{pred}}(t + h)$ by solving the IVP.

4.2 Workload-Driven Frequency Adjustment

The decision about the frequency adjustment interval is made based on the difference between $w^{\text{exact}}(t)$ and $w^{\text{pred}}(t + h)$. For example, if $w^{\text{pred}}(t + h) \gg w^{\text{exact}}(t)$, then the DFA will increases the frequency. Clearly, if $w^{\text{pred}}(t + h) \ll w^{\text{exact}}(t)$, then the DFA will decrease the frequency. On the other hand, the DFA increases (decreases) the frequency slowly if $w^{\text{pred}}(t + h)$ is only a little larger (smaller) than $w^{\text{exact}}(t)$.

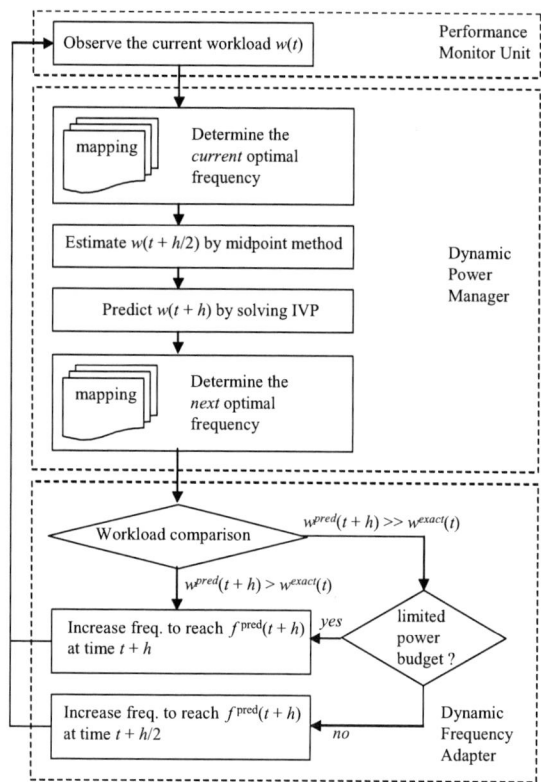

Figure 8. The flow of dynamic frequency adjustment method.

Figure 8 shows the flow of dynamic frequency adjustment technique, where $f^{\text{pred}}(t + h)$ is the frequency value obtained from the predicted workload and the workload-frequency mapping table for the two cases where $w^{\text{pred}}(t + h) >$ or $\gg w^{\text{exact}}(t)$,. In this flow, we have omitted the case of $w^{\text{pred}}(t + h) <$ or $\ll w^{\text{exact}}(t)$, which can be handled in a similar way. Note that when $w^{\text{pred}}(t + h) = w^{\text{exact}}(t)$, the current frequency value is maintained. It is worthwhile to mention that the DFA is capable of handling the throughput and power budget. If there is a target throughput, for example, the DFA will slowly increase the frequency up to a target frequency value that results in just-enough throughput and minimum power dissipation.

4.3 Mapping of Workloads to Optimal Frequency

The entries of the workload-frequency mapping table correspond to various values of workload (i.e., the arrival rate of

252

978-1-4244-3039-0/08 $25.00 © 2008 IEEE

tasks). Figure 9 illustrates the mapping process from workloads to an optimal operating frequency, assuming that $0.1 \leq$ the arrival rate ≤ 0.9. In this figure, the pre-characterized mapping table is achieved through extensive offline simulations during design time, considering performance characteristics of each FB provided by the user or application, in a similar way as [5][14]. For example, when a power manager predicts the workload for the near future, an optimal frequency value for the next decision epoch is selected and provided to the DFA which will continuously change the operating frequency from its present value to the target value. Note that mapping from workload to operating frequencies is achieved by a simple linear function while considering the maximum and minimum operating frequencies that can be applied to the FB in question.

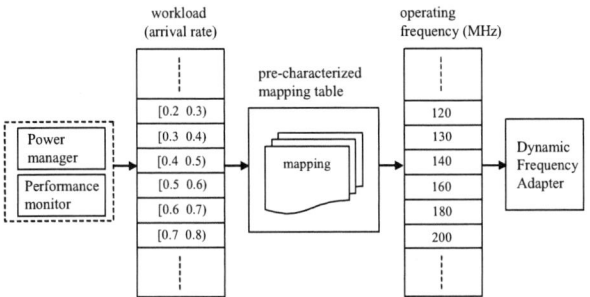

Figure 9. Mapping of workloads to optimal operating frequency values for each FB.

5. Experimental Results

In the experimental setup, we applied the proposed CFA technique to a high-speed network controller (i.e., gigabit Ethernet controller) which includes IEEE 802.3 PHY/MAC blocks, RISC processor, direct memory access (DMA) engine, PCI-E core, etc. as shown in Figure 10. This embedded system is implemented with TSMC 65nmLP library. To capture power-saving opportunities by using the proposed technique, we consider a part of the process of receiving packets inside the system, which involves Ethernet MAC (EMAC) block and control block. Note that it will not hurt the quality of the paper if we concentrate on these blocks (i.e., inside dotted box in Figure 10) to simplify the experimental setup, since they sufficiently exhibit the characteristics of the SR, the SP, and the SQ. Thus, the continuously varying frequency value is applied to the control block (i.e., the SP) by the hardware-implemented DFA, where its frequency is optimally adjusted.

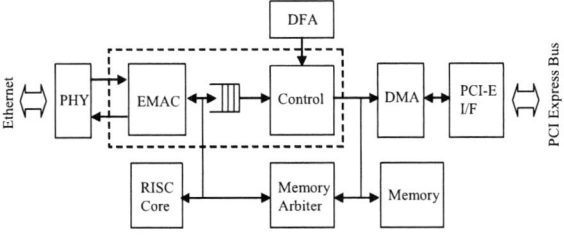

Figure 10. Block diagram of simplified Ethernet controller.

The functions performed by these blocks (i.e., EMAC and control blocks) are explained briefly as follows. The EMAC receives a data stream from the selected physical layer interface and performs address checking, CRC calculation, and CSMA/CD functions [15]. The control block calculates checksum and parses TCP/IP headers and classifies the frames based on a set of matching rules. While processing the control data in the control block, the frame data is temporarily stored in memory buffers before being sent to local interconnect through the PCIE interface.

Table 1. Energy dissipation (normalized) and utilization of the control block.

1000Base-T	Arrival rate of tasks								
	0.1	0.2	0.3	0.4	0.5	0.6	0.7	0.8	0.9
Energy (normalized)	1.96	2.62	4.37	5.04	5.85	6.67	8.89	9.86	10.83
Utilization (%)	9.5	14.2	20.6	32.1	41.8	54.0	63.2	67.5	73.5

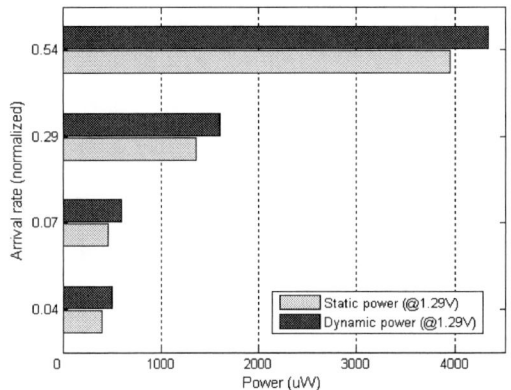

Figure 11. Power consumption of the service queue.

We first achieve the power and energy dissipation of the service provider (i.e., control block) and the service queue (i.e., memory) by using TSMC 65nmLP library, which has 3 optional operating voltages (e.g., 1.08V, 1.20V, and 1.29V). To calculate accurate power values for static and dynamic power consumption, we used SAIF (Switching Activity Interchange File) based on back-annotated RTL simulation of the system with Power compiler [16]. To achieve the energy dissipation of the control block, different workloads (e.g., the arrival rate of traffic) are used to generate the multiple columns in Table 1, where the dynamic and static power values are considered. For simulation setup, we set that the maximum full duplex bandwidth (e.g., 1000Base-T) is achieved. Note that the overhead of designing the DFA block inside the system is negligible due to its small number of gate counts (around 150 standard cells) and power dissipation (around 2uW including dynamic and static power). Figure 11 shows power consumption of the service queue in terms of the normalized arrival rate of the traffic. We set the packet size to 64bytes and the service time to 1 (by using the G/M/1 queuing model) for simplicity. For example, when the arrival rate of the tasks is 0.29 (normalized), the memory size necessary for buffering the incoming data is 5.8 times greater than the case of

253

978-1-4244-3039-0/08 $25.00 © 2008 IEEE

where the arrival rate is 0.04.

Next, we evaluate the effectiveness of the proposed CFA technique. We assume that the workload changes dynamically from 0.1 to 0.9. For comparison purpose, we implemented a couple of power management policies (denoted by PM1 and PM2 and described below) as representatives of the conventional methods, similar to [7][13]. We use three set of frequency values to simplify the experimental setup ($F_1 < F_2 < F_3$ in terms of frequency values).

PM1: Utilize dynamic frequency scaling technique, while accounting for a 100us power-mode transition overhead; the frequency assignment policy is as follows.
  - Use the lowest F_1 value when $0.1 \leq$ the arrival rate ≤ 0.3, i.e., low workload.
  - Likewise, use F_2 and F_3 values when $0.3 <$ the arrival rate ≤ 0.6 and $0.6 <$ the arrival rate ≤ 0.9, respectively.

PM2: The same as PM1 except that a frequency change is avoided when the same frequency changes occurs consecutively. More precisely, let F^i denote the value of frequency at time i.
  - Set $F^{i+1} = F^i$, if the predicted F^{i+1} value is the same as F^{i-1} value.

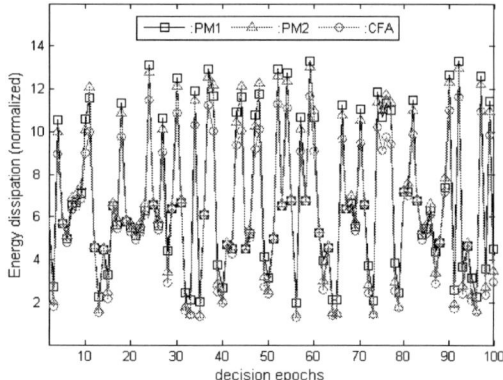

Figure 12. Evaluation of the proposed CFA technique.

Table 2. Power and energy savings of the CFA.

No. of decision epoch	Workload distribution			Average Power (mW)			Power saving over		Energy saving over	
	Low	Mid	High	PM1	PM2	CFA	PM1	PM2	PM1	PM2
100	26	43	31	13.2	11.8	11.4	12.9%	2.3%	11.4%	8.8%
500	143	196	161	13.3	11.7	11.5	13.3%	2.2%	12.0%	9.4%
1000	277	417	306	13.4	11.8	11.5	13.6%	2.1%	12.3%	9.5%
5000	1494	1954	1552	13.4	11.9	11.6	13.1%	2.1%	12.1%	9.6%

Then, we generate dynamic workloads randomly with 100, 500, 1000, and 5000 numbers of power management decision epochs and apply both above-mentioned conventional policies and the proposed power management technique to the control block. The simulation results in Figure 12, which corresponds to the case of 100 decision epochs, show that the proposed CFA technique achieves energy savings compared to the conventional methods. Results in Table 2, which also reports the characteristics of the workload distribution (e.g., Low = $0.1 \leq$ the arrival rate ≤ 0.3), demonstrate that, compared to the PM1

policy, our approach achieves power and energy savings up to 13.6% and 12.3%, respectively.

6. Conclusion

In this paper, we addressed the problem of power management techniques in the context of handling dynamic frequency management, where power-mode transition cost is no longer negligible in the nano-scaled systems. We proposed a continuous frequency adjustment technique based on a workload prediction method, which minimizes the transition cost. Experimental results with a 65nm design show that the proposed technique ensures robust energy savings under dynamic workloads.

References

[1] D. Li, Q. Xie, and P.H. Chou, "Scalable Modeling and Optimization of Mode Transitions Based on Decoupled Power Management Architecture," *Proc. of DAC*, Jun., 2003.

[2] Y-H. Lu, and G. De Micheli, "Comparing System-Level Power Management Policies," *IEEE Design & Test of Computers*, Vol. 18, Issue 2, Mar.-Apr., 2001.

[3] M. Najibi, et al., "Dynamic Voltage and Frequency Management Based on Variable Update Intervals for Frequency Setting," *Proc. of ICCAD*, Nov., 2006.

[4] Y. Cho, N. Chang, C. Chakrabarti, and S. Vrudhula, "High-Level Power Management of Embedded Systems with Application-Specific Energy Cost Functions," *Proc. of DAC*, Jul., 2006.

[5] P. Rong and M. Pedram, "Power-aware Scheduling and Dynamic Voltage Setting for Tasks Running on a Hard Real-time System," *Proc. of ASP-DAC*, Jan., 2006.

[6] A. Iyer, and D. Marculescu, "Power Efficiency of Voltage Scaling in Multiple Clock, Multiple Voltage Cores," *Proc. of ICCAD*, Nov. 2002.

[7] Q. Wu, P. Juang, M. Martonosi, and D.W. Clark, "Voltage and Frequency Control with Adaptive Reaction Time in Multiple-Clock Domain Processors," *Proc. of 11th Symposium on HPCA*, Feb., 2005.

[8] J. Liu, and P.H. Chou, "Optimizing Mode Transition Sequences in Idle Intervals for Component-Level and System-Level Energy Minimization," *Proc. of ICCAD*, Nov., 2004.

[9] J.R. Dormand, *Numerical Methods for Differential Equations: A Computational Approach*, CRC Publisher, Feb., 1996.

[10] L. Benini, and G. De. Micheli, *Dynamic Power Management: Design Techniques and CAD Tools*, Kluwer Academic Publishers, 1998.

[11] R. Zhang, and G. La Rue, "Fast Acquisition Clock and Data Recovery Circuit with Low Jitter," *IEEE Journal of Solid-State Circuits*, Vol. 41, No. 5, May, 2006.

[12] S. M. Ross, *Introduction to Probability Models*, Academic Press, 8th edition, Dec., 2002.

[13] P. Choudhary, et al., "Hardware Based Frequency/Voltage Control of Voltage Frequency Island Systems," *Proc. of CODES*, Oct., 2006.

[14] H. Jung, and M. Pedram, "A Unified Framework for System-level Design: Modeling and Performance Optimization of Scalable Networking System," *Proc. of ISQED*, Mar., 2007.

[15] http://www.ieee802.org/802_tutorials IEEE 802.3 tutorial. Jul., 2005.

[16] http://www.synopsys.com Synopsys Power Compiler Documents.

21st International Conference on VLSI Design

Recursive versus Iterative Algorithms for Solving Combinatorial Search Problems in Hardware

Iouliia Skliarova, Valery Sklyarov

Department of Electronics, Telecommunications and Informatics, IEETA
University of Aveiro, 3810-193 Aveiro, Portugal
iouliia@ua.pt, skl@ua.pt

Abstract

The paper analyses and compares alternative iterative and recursive implementations of combinatorial search algorithms in hardware (in field-programmable gate arrays – FPGA, in particular). The results of experiments and comparisons for three widely used problems from this scope are presented, namely for the Boolean satisfiability, binary matrix covering, and graph coloring. The relevant comparative data have been obtained as a result of synthesis and implementation in FPGAs of the respective circuits from VHDL specifications.

1. Introduction

It is known [1] that combinatorial search algorithms can be implemented using one of two alternative and competitive techniques, which are based on either iterative or recursive specifications. Both of them are very powerful and it is very important to identify some criteria in order to enable the best type of implementation for any particular problem to be chosen. For example, from experience in software development, it is known that recursion is not always appropriate, particularly when a clear efficient iterative solution exists [2]. This is primarily due to the large amount of states that are accumulated during deep recursive calls. Besides, in most high-level programming languages, a function call incurs a bookkeeping overhead. Recursive functions magnify this overhead because a single initial call to the function might generate a large number of recursive invocations of the function. On the other hand, results reported in [3, 4] demonstrate that recursion can be implemented in hardware much more efficiently than in software. This is because any recursive call can be combined with the execution of other operations that are required by the respective algorithm. The same is true when any recursive subsequence is being terminated. The results obtained for

some known methods for implementing recursive calls in hardware, such as the technique [5] based on multi-thread and speculative execution, have shown FPGA circuits to be significantly faster than software programs executing on general-purpose computers. Even in software applications employing recursive algorithms for various kinds of binary (N-ary, N≥2) search (such as that can be executed using binary trees) is considered as a notable exception where a recursive technique might be more efficient [2].

This paper compares iterative and recursive implementations of similar combinatorial algorithms for solving three known problems that are the Boolean satisfiability (SAT), binary matrix covering, and graph coloring. The chosen backtracking search algorithms will be specified in VHDL, synthesized, and implemented in an FPGA of Spartan-3 family of Xilinx. The same algorithms will be described both recursively and iteratively. The resulting circuits will be compared in terms of a) ease of developing, maintaining and changing the specifications; b) the occupied FPGA area; c) the maximum clock frequency; and d) the number of clock cycles needed to solve a particular problem instance.

The remainder of this paper is organized in five sections. Section 2 discusses the considered combinatorial search problems, which can be solved using either iterative or recursive techniques. Section 3 analyses differences between recursive and iterative implementations. Section 4 describes common features of combinatorial search algorithms making it possible to suggest architectures with reusable modules. Section 5 summarizes the results of experiments. The conclusion is in section 6.

2. Combinatorial search problems

There are many combinatorial problems that can be formulated over such mathematical models as graphs, discrete matrices, sets, Boolean equations, etc. The

978-1-4244-3039-0/08 $25.00 © 2008 IEEE

majority of these models are mutually convertible, i.e. any of them might be selected for similar purposes. Many practical problems can be solved by applying various algorithms of combinatorial search over a chosen mathematical model.

Systems for solving combinatorial problems might be implemented in field-programmable devices (FPGA, in particular). The latter have a number of advantages, which have appeared because the considered tasks possess a set of specific features [1]. Firstly, any task involves a huge number of similar operations. As a rule, these operations are not the same for different combinatorial problems. Thus, it is not easy to construct a universal combinatorial processor, i.e. processor's instructions have to be customized for a particular problem that is going to be solved. This can easily be done with the aid of FPGA technology. Secondly, different practical applications might require solving combinatorial tasks with varying complexity. However, optimal results can be achieved in case if the size of processor operands permits any required operation to be performed in one clock cycle. Parameterizable circuits that provide such an opportunity can easily be implemented in FPGA. Thirdly, FPGA enable us to build on the same microchip any desired (customized) interface between a combinatorial accelerator and a general-purpose computational system (or any specialized system that requires an accelerator). Fourthly, the complexity of recent FPGA allows for the construction of a complete system-on-chip and a combinatorial accelerator can be implemented as an application-specific co-processor within this system.

Note that hardware permitting to carry out combinatorial search over various (alternative) models requires different FPGA resources, i.e. the complexity of the respective circuits depends essentially on the adopted mathematical model. Our experience has shown that a discrete (Boolean and ternary) matrix can be seen as a very adequate model for the considered computations. The primary reason for this conclusion is that hardware accelerators that operate over matrices can be reconfigured very easily to solve different problem instances, i.e. in order to solve a new problem instance it is not necessary to synthesize a new circuit; instead it is sufficient to reload matrix data and to configure a limited number of circuit parameters. As a result, the circuit reconfiguration time can be minimized. Examples illustrating how a given combinatorial problem might be formulated over a discrete matrix can be found in [6, 7].

We will consider exact search algorithms that are based on generation and exhaustive examination of all possible solutions until a solution with the desired quality is found [6, 8]. The primary decision to be taken in this approach is how to generate the candidate solutions effectively. A widely accepted answer to this question consists of constructing an N-ary search tree [6, 9], which enables all possible solutions to be generated in a well-structured and efficient way. The root of the tree is considered to be the starting point that corresponds to the initial situation. The other nodes represent various situations that can be reached during the search for results. The arcs of the tree specify steps of the algorithm that have been performed. At the beginning the tree is empty and it is incrementally constructed during the search process.

A distinctive feature of this approach is that at each node of the search tree a similar sequence of algorithmic steps has to be executed. The only thing that is different from node to node is input data. This means that the entire problem reduces to the execution of a large number of repeated operations over a sequentially modified set of data.

Of course, exhaustive checking all possible solutions cannot be used for the majority of practical problems because it requires a very long execution time. That is why it is necessary to apply some optimization techniques that reduce the number of situations that need to be considered. In order to speed up getting the results various tree-pruning techniques can be applied. In general, the pruning process is based on erasing repeated variants and avoiding feasible solutions, which are worse, according to some estimation criteria, than any solution already found. Sometimes it is possible to apply problem-specific methods that allow large segments of the search tree to be pruned by exploiting instance-specific knowledge that can be acquired during the search. A good example of this technique is non-chronological backtracking that is widely used in the state-of-the-art SAT solvers [10].

The other known method of improving the effectiveness of the search is a reduction [9], which permits the current situation to be replaced with some new simpler situation without loosing the possibility to find the solution. As a result, the number of computations that are required for the analysis of situations resulting from the current algorithmic step can be reduced. Returning to the example of the Boolean satisfiability problem, the well-known unit-clause rule is a good illustration of this technique [11].

However, reduction is not possible for all existing situations. In this case another method is used that relies on the divide-and-conquer strategy [8]. This applies to critical situations that have to be divided into N several simpler situations such that each of them has to be examined. The objective is to find the minimum number N. Very often these new situations can be ordered according to some criteria, which essentially reduces the time of computations.

Fig. 1 depicts the basic N-ary backtracking search algorithm, which can be applied to many combinatorial

256

978-1-4244-3039-0/08 $25.00 © 2008 IEEE

problems. The reduction rules permit a given (sub-) problem to be simplified. The selection rules make possible to split the problem into sub-problems and to examine them in such a way that either a solution will be found or the conclusion that the solution does not exist will be done.

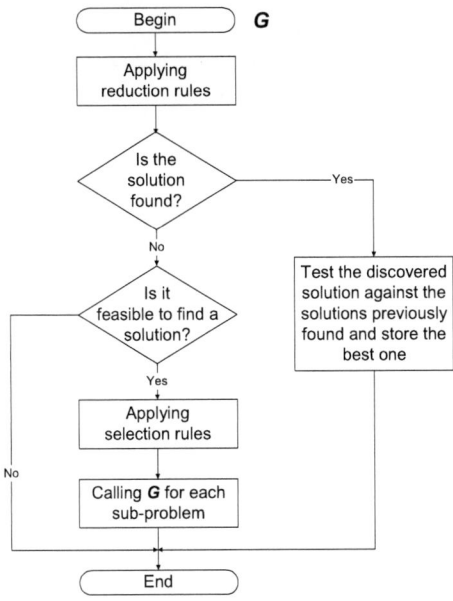

Figure 1. Basic N-ary search algorithm

3. Recursive vs. iterative implementations

The sequence of recursive invocations makes it possible to keep track between the nodes that are examined during forward propagation steps. The execution of recursive algorithms is done in accordance with the general flow shown in Fig. 1, where the node "Calling *G* for each sub-problem" involves recursion.

Iterative algorithms consider the sequence of reduction and selection steps (until the node "Calling *G* for each sub-problem" in Fig. 1) as an individual iteration. If current iteration does not lead to a solution then intermediate results are stored in memory and a new iteration with reduced matrix is executed. The stored intermediate results are needed for backward propagation.

A sub-sequence of returns from recursive procedures easily enables us to encounter any intermediate point on the way from the root of the tree to its leaves. As a result, we could expect that recursive algorithms of this type would be less complicated and would attain better execution time than iterative algorithms, which might require more complex and less understandable sequence of steps. This is because the latter have to provide additional storage and operations for traversing the tree that make it possible to return to any node along the forward propagation path and to recognize from which

branch (for example in case of a binary tree either from the left or from the right branch) the return to the node is being done.

Independently of the implementation (i.e. either in software or in hardware), recursive calls invoke operations over stacks in such a way that the states of the algorithm (where recursive invocations have happened) are saved onto a stack and the stack pointer is incremented to address the storage for a recursively called sub-algorithm. When the recursive sub-algorithm ends, the stack pointer is decremented in order to restore the state of the interrupted algorithm. If we consider an equivalent iterative algorithm, then stack is not required and computations are performed in a loop, which ends as soon as some conditions are satisfied. From the point of view of functional capabilities (including error handling such as stack overflows or endless loops), iterative and recursive techniques are very similar. However, they are not equal in terms of the required resources and the execution time.

Iterative algorithms can be described and implemented using either flat or hierarchical specifications. In the first case, we can recur to traditional finite state machine (FSM) model and employ any suitable language (such as VHDL, Verilog, or Handel-C) for specifying the algorithmic steps.

The hardware model of FSM is shown in Fig. 2. The FSM consists of a combinational circuit (that produces the primary outputs and calculates the next state based on the input values and the present state) and a register that stores the present FSM state. Fig. 2 includes a VHDL language template illustrating how the combinational circuit and the FSM register can be described with the aid of two processes. The presented template is parameterizable and can therefore be used for describing functionality of any FSM with any number of states and using any state encoding technique.

In case of iterative hierarchical specification the algorithm description is decomposed in modules (for example, a module implementing reduction rules, a module for testing the quality of solutions, etc.). The resulting modular descriptions have a number of advantages over traditional FSMs which can be justified by the following. It is well known that the best way to simplify the problem solving process is to divide an initial problem into small, manageable parts. The resulting design will contain modules, which are self-contained circuits. Besides of simplifying the design process, such an approach provides a direct support for reusability since the developed modules might be reused in different parts of the project as well as in other similar projects. Another very important aspect is design's modifiability. Imagine that the initial problem specification is changed after some period of time. When the project is divided in modules, incorporating changes into a single module is a simpler task than changing the implementation of the whole circuit.

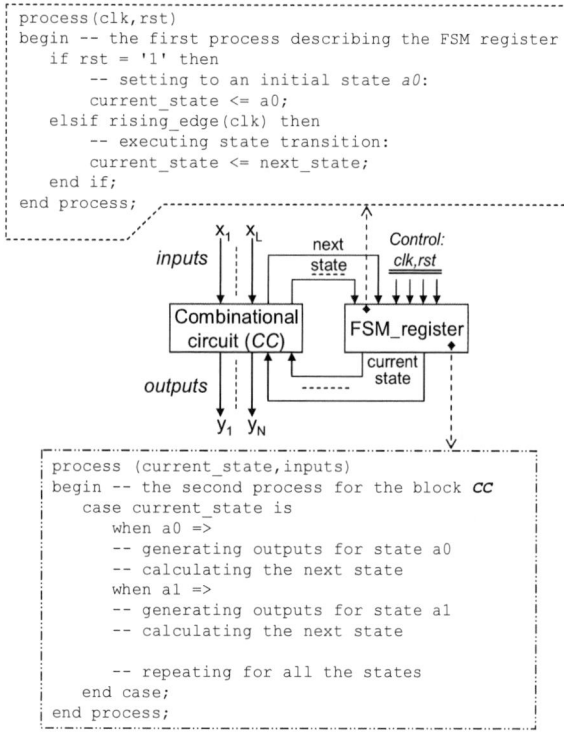

```
process(clk,rst)
begin -- the first process describing the FSM register
   if rst = '1' then
        -- setting to an initial state a0:
        current_state <= a0;
   elsif rising_edge(clk) then
        -- executing state transition:
        current_state <= next_state;
   end if;
end process;
```

```
process (current_state,inputs)
begin -- the second process for the block CC
    case current_state is
        when a0 =>
        -- generating outputs for state a0
        -- calculating the next state
        when a1 =>
        -- generating outputs for state a1
        -- calculating the next state

        -- repeating for all the states
    end case;
end process;
```

Figure 2. The design template for an FSM

Recursive algorithms are also constructed from modules but in this case each module is allowed to invoke itself. This feature is particularly interesting for backtracking search algorithms where in each node exactly the same sequence of actions has to be performed (see section 2).

It is well known that hardware description languages (such as VHDL) and system-level specification languages (such as Handel-C) do not provide direct support for hierarchical algorithms (both iterative and recursive). However, hierarchy can be implemented in a hierarchical finite state machine (HFSM) [12] and the latter can easily be described in hardware and system-level specification languages.

The hardware model of HFSM is depicted in Fig. 3. The HFSM consists of a combinational circuit and two stacks (that keep track of hierarchical module invocations), one for states (*FSM_stack*) and the other for modules (*M_stack*).

The stacks are managed by a combinational circuit that is responsible for new module invocations and state transitions in any active module that is designated by outputs of the *M_stack*. Any non-hierarchical transition is performed through a change of a code only on the top register of the *FSM_stack*. Any hierarchical call alters the states of both stacks in such a way that the *M_stack* will store the code for the new module and two values will be pushed into the *FSM_stack*: first, the code of the next state in the calling module and then the code of the first state in the called module. Any hierarchical return just activates a pop operation without any change in the stacks. As a result, a transition to the state following the state where the terminated module was called will be performed. The stack pointer *stack_ptr* is common to both stacks. If the *End* node (see Fig. 1) is reached when the stack pointer is equal to zero (*stack_ptr*=0), the algorithm terminates execution.

```
process(clk,rst) -- the first process for the blocks
begin             -- M_stack and FSM_stack
   if rst = '1' then
        -- setting to an initial state and initializing
   elsif rising_edge(clk) then
        -- test for possible errors
        -- executing transitions of the following types
        -- a) between states within the same module
        -- b) between states that belong to different modules
   end if;
end process;
```

```
process (current_module,current_state,inputs)
begin     -- the second process for the block CC
    case M_stack(stack_ptr) is
        when z0 =>
            case FSM_stack(stack_ptr) is
                -- state transitions in the module z0
                -- generating outputs for the module z0
            end case;
        when z1 =>
            case FSM_stack(stack_ptr) is
                -- state transitions in the module z1
                -- generating outputs for the module z1
            end case;
        -- repeating for all the modules
end process;
```

Figure 3. The design template for an HFSM

Fig. 3 illustrates an example of VHDL code for HFSM, which makes it possible to describe modular and recursive algorithms [4]. There are two concurrently executing VHDL processes in Fig. 3. The first process describes two stacks (the stack of modules and the stack of states) and the second process describes the combinational circuit, which is able to manage transitions between the FSM modules and FSM states. It is important that the second process can easily be customized for executing any desired hierarchical algorithm.

Although modular design incurs some overhead and therefore occupies more resources, it is not significantly slower than the one that is based on traditional FSMs (as will be evidenced by the results of experiments given in section 5). Moreover, if the stacks are constructed on the

258

978-1-4244-3039-0/08 $25.00 © 2008 IEEE

basis of embedded in FPGA memory blocks, the additional FPGA resources required for stack management are negligible.

4. Common features of combinatorial search algorithms

The considered in section 2 search algorithms have similar characteristics. Their distinctive feature is the execution of problem-specific operations and traversing a search tree starting from the root by involving such procedures as forward search and backtracking. Any branching point can be considered as extracting a sub-tree with a local root. These algorithms possess several common features identified in [1]:
1. They can be formulated either recursively or iteratively.
2. They do not change the initial data (i.e. the initial matrix) because the matrix reduction can be provided by masking some rows/columns and using just the remainder of the matrix.
3. They invoke a very limited number of operations (such as reduction and selection operations), which have to be applied to a huge volume of data.
4. Subsets of the required operations are usually not the same for different combinatorial problems.
5. In order to perform forward and backward propagation we can use a stack memory that stores and restores intermediate results (such as the values of mask registers) in branching points.
6. The algorithms can be decomposed into two levels of control operations. The top-level sequence is the same for different algorithms (see Fig. 1). The bottom level sequence permits the required operations over Boolean and ternary vectors to be executed.

These features make it possible to select a number of reusable functional blocks for constructing application-specific hardware processors that permit combinatorial problems formulated over Boolean and ternary matrices to be solved. The following blocks have been selected on the basis of analysis of different search algorithms and their primary operations [1]:
1. A row/column addressable memory permitting to store Boolean and ternary matrices.
2. Mask registers allowing using the same storage for handling the initial matrix and all the sub-matrices, which have to be constructed during the search for results.
3. Stacks for managing forward and backward propagation steps.
4. General-purpose registers for keeping vectors.
5. A device for computations over discrete vectors.
6. Control circuits, which make possible to execute operations at two levels: operations over vectors; and operations over the entire matrices.

7. Additional auxiliary circuits for testing, debugging and interacting with the hardware processor.

5. Implementation details and the results of experiments

Experiments have been carried out for the following N-ary search problems: P1 – an exact method for solving the Boolean satisfiability problem formulated over ternary matrices (the details of the considered algorithm can be found in [6]); P2 – an exact method [6, 9] for discovering a minimal column cover of a Boolean matrix; P3 – an exact algorithm for graph coloring described in [7].

The synthesis and implementation of circuits from specification in VHDL were done in Xilinx ISE for prototyping system [13] based on FPGA XC3S400 of Spartan-3 family of Xilinx. Note that each circuit includes not only components that are needed for comparison, but also auxiliary blocks for entering input data and visualizing the results on a VGA monitor screen. For each particular problem all the auxiliary components are exactly the same for iterative and recursive implementations. So, the difference in resources and in execution time provides correct data for comparison.

The results of experiments are presented in Table 1. Rows/columns of the table contain the following data: MI – Modular Implementation (otherwise, non-modular implementation); N_s – the number of occupied FPGA slices; F – the maximum attainable clock frequency (in MHz) calculated by the ISE implementation tools; N_{clock} – the number of clock cycles required for solving a given problem instance.

The size of matrices for the satisfiability (P1) and the covering (P2) problems was chosen to be 128 rows × 128 columns. The number of nodes in graphs for coloring was 32. All the projects provide generic parameters for initial data and can therefore be customized.

Table 1. The results of experiments

Problem	N_s/F/N_{clock}	
	Recursive	Iterative
P1		2216/36.0/144320
P1/MI	2036/37.2/135672	2276/36.1/145270
P2		3083/47.8/191162
P2/MI	3056/47.9/183491	3118/47.9/193332
P3		2628/34.1/7554
P3/MI	2574/34.5/0/7311	2661/34.2/7611

For implementation of iterative algorithms two alternative methods were examined: traditional finite state machines and HFSMs. The VHDL projects use identical descriptions of the required operations for the iterative and recursive algorithms.

978-1-4244-3039-0/08 $25.00 © 2008 IEEE

We can summarize the results of experiments and analysis of alternative implementations as follows:
1. Although the results for non-modular iterative implementations are a bit better than for modular iterative implementations, the difference is negligible. On the other hand, we can benefit from reusability of modules in MI and also from the clearness of specifications.
2. Experiments with C++ programs that describe similar iterative and recursive algorithms have shown that benefits in terms of execution time for recursive algorithms take place just in hardware (for all the considered examples iterative algorithms were faster in software than recursive algorithms).
3. In all cases recursive modular algorithms implemented in hardware for solving the considered combinatorial problems are more advantageous than iterative algorithms. It is very important that this result differs from a similar analysis of software implementations where iterative algorithms are, as a rule, more beneficial than recursive algorithms. Nevertheless, it is worthwhile to mention that the improvements are marginal.

Since the presented results are very similar to the results obtained for data sorting [14], we can conclude that for many practical problems hardware recursive algorithms over binary (N-ary, N≥2) trees can be seen a good alternative to iterative algorithms.

6. Conclusion

In this paper alternative recursive and iterative implementations of algorithms for three combinatorial search problems have been analyzed. The main conclusion that can be drawn is that for many practical problems using in hardware recursive algorithms over binary (N-ary, N≥2) trees can be seen as a good alternative to using iterative algorithms. Another important conclusion is that modular implementations based on an HFSM model require almost the same resources as non-modular (FSM-based) implementations but provide support for reusability and more clear specifications. Thus, modular implementations can be strongly recommended.

7. References

[1] I. Skliarova and V. Sklyarov, "Design Methods for FPGA-based Implementation of Combinatorial Search Algorithms", Proc. of the International Workshop on SoC and MCSoC Design - IWSOC'2006, MoMM'2006, Yogyakarta, Indonesia, December 2006, pp. 359-368.

[2] F.M. Carrano, *Data Abstraction and Problem Solving with C++*, The Benjamin/Cumming Publishing Comp., Inc., 1995.

[3] V. Sklyarov, I. Skliarova, and B. Pimentel, "FPGA-based Implementation and Comparison of Recursive and Iterative Algorithms", Proc. of the FPL'2005, Finland, August 2005, pp. 235-240.

[4] V.Sklyarov, "FPGA-based implementation of recursive algorithms", *Microprocessors and Microsystems, Special Issue on FPGAs*, 2004, vol. 28/5-6, pp 197-211.

[5] T. Maruyama, M. Takagi, and T. Hoshino, "Hardware Implementation Techniques for Recursive Calls and Loops", Proc. of the 9th International Workshop on Field-Programmable Logic and Applications, 1999, pp. 450-455.

[6] I. Skliarova, "Reconfigurable Architectures for Problems of Combinatorial Optimization", Ph.D. Thesis, University of Aveiro, Portugal, 2004.

[7] V. Sklyarov, I. Skliarova, and B. Pimentel, "Modeling and FPGA-based implementation of graph coloring algorithms", Proc. of the 3rd Int. Conference on Autonomous Robots and Agents - ICARA'2006, Palmerston North, New Zealand, December 2006, pp. 443-448.

[8] I. Skliarova and A.B. Ferrari, "The Design and Implementation of a Reconfigurable Processor for Problems of Combinatorial Computation", *Journal of Systems Architecture, Special Issue on Reconfigurable Systems*, vol. 49, 2003, pp. 211-226.

[9] A.D. Zakrevski, *Logical Synthesis of Cascade Networks*, Moscow: Science, 1981.

[10] M.W. Moskewicz, C.F. Madigan, Y. Zhao, L. Zhang, and S. Malik, "Chaff: Engineering an Efficient SAT Solver", Proc. 38th Design Automation Conference, June 2001, pp. 530-535.

[11] J. Gu, P.W. Purdom, J. Franco, and B.W. Wah, "Algorithms for the Satisfiability (SAT) Problem: A Survey", DIMACS Series in Discrete Mathematics and Theoretical Computer Science, vol. 35, 1997, pp. 19-151.

[12] V. Sklyarov, "Hierarchical Finite-State Machines and their Use for Digital Control", *IEEE Transactions on VLSI Systems*, vol. 7, n. 2, 1999, pp. 222-228.

[13] M. Almeida, B. Pimentel, V. Sklyarov, and I. Skliarova, "Design Tools for Rapid Prototyping of Embedded Controllers", Proc. ICARA'2006, Palmerston North, New Zealand, December 2006, pp. 683-688.

[14] V. Sklyarov and I. Skliarova, "Recursive and Iterative Algorithms for N-ary Search Problems", International Federation for Information Processing, vol. 218, 2nd IFIP Symp. on Professional Practice in Artificial Intelligence - AISPP'2006, ed. J. Debenham, 19th IFIP World Computer Congress - WCC'2006, Santiago de Chile, Chile, August 2006, pp. 81-90.

21st International Conference on VLSI Design

Exhaustive Enumeration of Legal Custom Instructions for Extensible Processors

Nagaraju Pothineni
nagaraju@cse.iitd.ernet.in

Anshul Kumar
anshul@cse.iitd.ernet.in

Kolin Paul
kolin@cse.iitd.ernet.in

Department of Computer Science & Engineering
Indian Institute of Technology, Delhi
New Delhi, India - 110016

ABSTRACT

Today's customizable processors allow the designer to augment the base processor with custom accelerators. By choosing appropriate set of accelerators, designer can significantly enhance the performance and power of an application. Due to the large number of accelerator choices and their complex trade-offs among reuse, gain and area, manually deciding the optimal combination of accelerators is quite cumbersome and time consuming. This calls for CAD tools that select optimal combination of accelerators by thoroughly searching the entire design space. The term *pattern* is commonly used to represent the computation performed by a custom accelerator. In this paper, we propose an algorithm for rapidly enumerating all the legal patterns taking into account several constraints posed by a typical micro-architecture. The proposed algorithm achieves significant reduction in run-time by a) enumerating the patterns in the increasing order of sizes and b) relating the characteristics of a $(k + 1)$ node pattern with the characteristics of its k node subgraphs. Also, in scenarios where I/O is not a bottleneck, designer can optionally relax the I/O constraint and our algorithm efficiently enumerates all legal I/O unbound legal patterns. The experimental evidence indicate an order of two run-time speedup over state of the art techniques.

1. INTRODUCTION

Today's consumer electronics market is posing increasingly tough challenges on the embedded processor's performance and power. And the device vendors need to meet to these demands of the market, while simultaneously cutting down the cost and time to market. Processor centric system with custom accelerators is amenable to modest evolving standards and also capable of significantly boosting the performance and power of application compared to a general purpose system.

Seeing the potential around this paradigm, in the recent years, several commercial platforms, for facilitating and supporting the high level design of custom processors, were introduced into the market [1, 2]. These platforms quickly provide the designer feedback about the chosen design alternatives and automatically translate the high level design specification into detailed hardware model and also generate the supporting software tool chain. With the availability of these high level design entry tools, now designer can exclusively concentrate his efforts on the optimal custom processor design.

Manually arriving at the optimal mix of patterns is quite cumbersome and time consuming. Practically, any sub operation cluster of the application can be a legal candidate. This means designer has to choose a few among the vastly large number of patterns. Usually, there exists a complex trade-off in performance, reuse and area among the legal patterns. Typically, a small pattern has smaller gain but high reuse factor and smaller area. On the other hand, large patterns would have contrary characteristics. Manually choosing the best set of patterns under the constraints such as area and custom instruction count is practically an infeasible task.

This calls for high level design automation tools for searching and analyzing millions of alternatives. The first and foremost important step is the enumeration of the potential legal candidates. These identified candidates, later, are analyzed and the best set of candidates are selected. Due to the complex trade-offs among the characteristics of patterns, prejudiciously ignoring certain patterns in the first stage itself, just to shorten the design time, may result in the selection of suboptimal candidates. Consequently, it is preferable to enumerate all the legal patterns in the first step and then taking to account different characteristics, judiciously select the optimal combination of patterns.

Legal candidate patterns are the subgraphs of the Control Data Flow Graph (CDFG) that do not violate the constraints imposed by the base processor. Typical constraints include **Load/Store, Input/Output, Convexity**. Among the exponential number of subgraphs, the legal candidates are only polynomial function of the number of operands [3]. This means that for efficient enumeration, an aggressive pruning technique needs to be developed. The existing approaches for legal candidate enumeration are primarily based on evaluation of subgraphs of the CDFG for the above mentioned three properties. And these methods miss to tap certain important inherent relation of characteristics of a graph and its subgraphs. Due to this, the number of subgraphs evaluated is much higher than the legal subgraphs and these existing approaches perform poorly when the allowable operands is large.

On the contrary, our algorithm is a constructive approach and systematically enumerates the legal subgraphs by relating the characteristics of a graph with the characteristics of its subgraphs. The legal candidates are generated and organized in such a way that each legal k node subgraph is linked with its k-1 node legal subgraphs. Since reuse of a pattern is related to its subgraphs, this organization facilitate the

261

978-1-4244-3039-0/08 $25.00 © 2008 IEEE

easy and efficient analysis of reuse of patterns in the later stages. In certain systems such as [4] , micro-architecture of the processor do not impose any limitation on the number of inputs/ outputs. Previous work [5] [6] had shown that selecting I/O unbound patterns is more advantageous than I/O constrained patterns in some benchmarks. In such systems, the enumeration algorithm should be able to generate all legal subgraphs without imposing any constraints on the inputs/outputs. Our algorithm performs significantly better than all the existing approaches while generating the unbound I/O legal patterns.

2. RELATED WORK

The prior work on custom instruction identification can be discussed along the following fronts.

- Starting point
- Scope
- Optimal Vs Heuristic

Starting point refers to the representation of the application from which custom instructions are extracted. Starting point can be the dynamic execution trace [7] or the compiled intermediate representation [8, 9, 10]. Trace Data Flow Graphs(DFGs) or static code transformations such as predication and unrolling are two alternative techniques employed for enlarging the basic block clusters. Handling of traces for custom instruction generation is too complex and to use the generated patterns finally needs some code restructuring. Bonzini et al. [11] performs predication and unrolling with the foresight that custom instructions would be selected on the transformed code. The proposed approach is based on evaluation of each possible design point to arrive at the best combination of predication and unrolling. In our experiments, we manually performed predication/unrolling to bring the entire core computation into a single basic block. Another interesting aspect is the choice of algorithm for the application. Grobschadl et al. [12] notes best implementation on the base processor may not be the best choice for an extended processor. The set of transformations that best magnifies the Instruction Set Extension (ISE) efficiency is still an open ended question.

The scope defines the boundaries for the regions from which custom instructions are extracted. This mainly refer to the capability of the custom instruction generator rather than the topology of custom instructions. Most of the existing work excepting [13] operates within basic block boundaries for custom instruction identification. Yu et al. [13] extracts patterns on most frequently executing paths obtained with the help of Whole Program Paths (WPPs) analysis. Unlike the predication where to speedup one path multiple other paths may be slowed down, this approach speeds up the mostly frequently executing path with little interference on the overlapping paths. However, how the patterns that span multiple basic blocks are utilized is not discussed in the paper.

The heuristic techniques greedily try to build patterns incrementally maximizing certain criteria. Kastner et al. [8] grows patterns along the most frequently occurring edge direction. Clark et al. [14] and Sun et al. [15] grows patterns along a node that maximizes the defined cost function that models appropriately performance gain and penalties

for I/O constraint violation. Biswas et al. [10] also uses similar cost function but gives priority to the nodes on a critical path. Pothineni et al. [6] enumerates all the convex MaxMIMOs as potential candidates for selection. Although, these techniques quickly provide reasonably good solution, but are most likely suboptimal.

The optimal techniques refer to exhaustive generation of legal patterns. Techniques for exhaustive enumeration are proposed in Pozzi et al. [16], Yu et al. [9], and Bonzini et al. [3]. These techniques are mainly based on evaluating different subgraphs for the required properties. Pozzi incorporates pruning by relating the convexity and number of outputs of a graph with a single subgraph of it. Yu et al. [9] builds the connected legal patterns in stages by combining previously built legal patterns. This approach may generate the same pattern multiple times. Also, unlike the Pozzi's approach, in [9] the generation of patterns do not follow any systematic order. Bonzini et al. [3] prunes the search space only based on I/O constraint. Also, the run time of [3] is higher than [16] for many benchmarks.

3. PRELIMINARIES

Let G be an acyclic graph associated with the triple (V, E, L), where V is the set of nodes in G and $E \subset V \times V$ is the set of all directed edges and the function $L : V \rightarrow Z$ is a mapping of nodes to distinct integer labels such that $\forall v \in V, L(v)$ is higher than all its predecessor labels. Since the labels of nodes are unique, we use the label $L(v)$ synonymously with the node v.

Let g^k denote a subgraph of G with k nodes and we use S^k to denote a set of subgraphs with k nodes. A subgraph g of G is said to be convex iff there is no path from a node $u \in g$ to another node $v \in g$ passing through a node $w \notin g$. For $g \subset G$, let mLab(g) denote the maximum label among all the nodes of g, i.e $mLab(g) = \max_{\forall v \in g} L(v)$. Node $v \in g$ is said to be a **primary input** node of g if none of the predecessors of v are in g. Let $PI(g)$ denote the set of all primary input nodes of g. For $g \subset G$, let sources(g) denote the set of nodes in $G - g$ whose successors are in g. Similarly, $results(g)$ denote the set of nodes in g whose successors are in $G - g$. $sources(g), results(g)$ respectively correspond to the set of input and output operands of subgraph g.

Definition : Let g_1^k and g_2^k be two subgraphs of G with exactly k nodes. g_2^k is said to be a **descendant of** g_1^k if

1. $|g_1^k \cap g_2^k| = k - 1$,
2. $L(g_2^k - g_1^k) > mLab(g_1^k)$ and
3. $g_1^k - g_2^k \in PI(g_1^k)$

In other words, the above conditions state that g_2^k, a descendant of g_1^k, has in common $k - 1$ nodes with g_1^k, the label of node in g_2^k but not in g_1^k is higher than $mLab(g_1^k)$ and is obviously $mLab(g_2^k)$ and finally the node in g_1^k but not in g_2^k is a primary input node of g_1^k.

We say g_2^k is a descendant of g_1^k along node v when g_2^k is a descendant of g_1^k and $v \in g_2^k \wedge v \notin g_1^k$. Since v is in g_2^k but not in g_1^k, according to the definition of descendant, $L(v) > mLab(g_1^k)$. Descents of g_1^k along node v can be obtained by deleting a single primary input node of g_1^k and

262

978-1-4244-3039-0/08 $25.00 © 2008 IEEE

including the node v, i.e $\forall u \in PI(g_1^k), (g_1^k - \{u\}) \cup \{v\}$ is a descendant of g_1^k along node v. Let $Desc(g_1^k, v)$ denote the set of all descendant of g_1^k along node v. Again reiterating, $Desc(g_1^k, v)$ is defined only when $L(v) > mLab(g_1^k)$. There would be exactly $|PI(g_1^k)|$ descendants of g_1^k along any node v satisfying the labeling constraint.

Two graphs g_1^k and g_2^k are said to be mergable iff $g_1^k \cup g_2^k$ is a graph with $k+1$ nodes. We call the graph g^{k+1} obtained by merging g_1^k and g_2^k as their child. A set of graphs S^k is said to be mergable iff every pair of graphs in S^k is mergable and the child of every pair is identical, and we call that child as the child of the set S^k, denoted as $child(S^k)$.

Theorem 1 : For $v \in V \wedge L(v) > mLab(g^k)$, $Desc(g^k, v) \cup \{g^k\}$ is a mergable set and its child is $g^k \cup \{v\}$.

PROOF. Proof of this theorem can be found in [17]. $\square$

In the above theorem, the graph $g^k \cup \{v\}$ denote the graph obtained by adding the node v to g^k, implicitly adding all the edges in E connecting node v with any node in g^k. We shall hereafter denote the following mergable set of graphs $Desc(g^k, v) \cup \{g^k\}$ as $MSET(g^k, v)$.

Theorem 2: When $|PI(g^k)| > 1$, for $v \in V \wedge L(v) > mLab(g^k)$, $g^k \cup \{v\}$ is convex if and only if all the graphs in $MSET(g^k, v)$ are convex.

PROOF. Proof can be found in [17]. $\square$

When g^k has exactly one primary input node then for $g^k \cup \{v\}$ to be convex it is not sufficient that all the graphs in $MSET(g^k, v)$ are convex, though this condition is mandatory. The reason is as follows. Let u be the primary input node of g^k. Both the nodes u and v co-exist neither in g^k nor in the descendant. Hence convexity of $g^k \cup \{v\}$ can not be ascertained based on the convexity of g^k and its descendant along node v alone. Once it is asserted that g^k and its descendant along node v are convex, the graph $g^k \cup \{v\}$ can still violate convexity if there is path from the primary input node u of g^k to node v passing through some nodes not in g^k. Let p be such a path from u to v with at least one node not in g^k. The first node of p is u and last node is v. Since g^k is convex, the last but one node of p must be an immediate predecessor of v not in g^k. Hence, we check whether there is a path from node u to any immediate predecessor of v not in g^k and if any such path exists then g^k is not convex otherwise g^k must be convex. In sequel of this paper, for the sake of simplicity of presentation, we don't explicitly repeat this additional check required when the number of primary inputs is one.

The consequence of the above theorem is stated in the following corollary.

Corollary : All the convex subgraphs of size $k+1$ can be derived by suitably merging the convex subgraphs of size k.

4. CANDIDATE ENUMERATION

Instruction Set Extension automation can be divided into two stages, 1. Custom instruction generation and 2. Custom instruction selection. In the first step, the subgraphs of the CDFG of the application that can be implemented on a Custom Functional Unit (CFU) are generated. In the second, out of the generated candidates, few candidates that

yields the highest speedup are selected. The focus of this paper is on fast generation of custom instruction candidates. The reader is referred to [13, 14, 18] for custom instruction selection methods.

Since the CFUs have to be integrated into the execution stage of the processor, the custom instruction generation pass should generate only those candidates that satisfy the constraints imposed by the micro-architecture of the base processor being extended. Typical constraints posed by a standard RISC architecture include the following: *I/O constraint* due to the limited available register ports and instruction encoding restrictions, *convexity constraint* to ensure a legal schedule and *forbidden nodes* such as a load/store and control transfer operations.

Formally, the custom instruction generation is defined as follows:

Problem Statement : Given a Data Flow Graph (DFG) G(V,E), bounds on maximum number of inputs (N_{in}) and outputs (N_{out}) of a legal custom instruction, and a set $P \subset V$ of prohibited nodes, enumerate all subgraphs g of G satisfying the following constraints

1. **Input :** $|sources(g)| \leq N_{in}$

2. **Output :** $|results(g)| \leq N_{out}$

3. **Convexity :** g is convex

4. **Node Type :** $\forall v \in g, v \notin P$

Working with the fourth constraint is trivial, we simply have to find subgraphs which don't include any nodes from P. There are $2^{|V|-|P|}$ subgraphs of G which do not include any node from P. Evaluating each and every one of these exponential number of subgraphs for the three properties *convexity, inputs* and *outputs* is not a practical possibility when $|V| - |P| \geq 20$. Many benchmarks Mediabench [19] suites contain basic blocks with large number of nodes, upto even several hundreds. For such benchmarks, an extremely fast enumerator is a necessity.

Legal candidate enumeration can be practically solved only if the number of legal candidates is not exponential. Bonzini [3] performed theoretical analysis of the problem and proved that number of legal candidates is $O((|V|^{N_{in}+N_{out}})$. This bound suggest that among the exponential number of subgraphs of G most of them violate either convexity or input/output constraint. For fast enumeration of all legal candidates, we must not even generate/evaluate the non-convex or input/output constraint violating subgraphs. Inferring certain subgraphs as illegal without even generating/evaluating them is a challenging task and the remaining part of this section dwell upon our fast legal candidate enumeration algorithm.

The first and foremost requirement of a fast enumerator is that it should not generate the same subgraph more than once. Construction of subgraphs is one of the most crucial steps based on which the other pruning techniques are developed.

In our algorithm, Considering the nodes in a topological order, nodes are uniquely labeled by assigning successively higher numbers starting from 1, i.e $i'th$ node in the topological order is assigned the label i. This labeling ensures that label of a node is higher than all its predecessor labels.

Our algorithm then proceeds enumerating the legal subgraphs in the increasing order of their sizes (no. of nodes). In other words, it enumerates first all the legal subgraphs of size 1, then size 2 etc. In iteration k, all the legal subgraphs of size k are enumerated. The legal subgraphs of size $k + 1$ are generated by considering each subgraph in $g \in CSR[k]$ extending with an appropriate node whose label is higher than $mLab(g)$. The proposed subgraph construction paradigm, i.e adding only the higher labeled nodes to the already generated subgraphs, guarantees that "No subgraph would be generated more than once". It is important to understand how the convexity , input and output properties of the extended graph $g \cup \{v\}$ gets affected compared to g.

P.1 : If g is not convex then $g \cup \{v\}$ is also non-convex.

P.2 : $|sources(g \cup \{v\})| \geq |sources(g)|$, i.e if g violates input constraint then $g \cup \{v\}$ would also violate input constraint

P.3 : $|results(g \cup \{v\})|$ may be smaller than $|results(g)|$.

The implication of P.1 and P.2 is positive and helps to prune several illegal subgraphs. These two properties state that if a graph g violates input or convexity constraint then no matter whatever higher labeled nodes are added to g, that graph would still remain illegal. Once a graph g is known to violate convexity or input constraint then all the subgraphs obtained by adding any number of higher labeled nodes, are not generated/evaluated. In other words, for generating the legal subgraphs of size $k+1$, it is sufficient that only subgraphs satisfying the convexity and input constraint are stored. In our algorithm $CSR[k]$ contains only the subgraphs of size k satisfying the convexity and input constraint.

According to P.3, although a subgraph g^k is violating output constraint, adding higher labeled nodes may reduce the number of outputs and result in a legal subgraph. However, certain outputs of a subgraph can be said to be permanent if they are used outside the basic block or used by nodes whose label is lower than the maximum label of the subgraph. The implication of this property is that, in addition to the legal subgraphs of k, the subgraphs which violate the output constraint but whose number of permanent outputs is below the output limit have to be preserved for generating all the legal subgraphs of size $k + 1$ or higher.

A subgraphs g of G is said to be *useful* iff it satisfies following three properties

1. Convex

2. $|sources(g)| \leq N_{in}$

3. $|perm\_results(g)| \leq N_{out}$

In iteration k, all the useful subgraphs of size k are generated and stored in $CSR[k]$. $CSR[k]$ certainly include all the legal subgraphs and also may include subgraphs which violate output constraint but needed for generating all the legal subgraphs of size k+1. All the illegal graphs in $CSR[k]$ are deleted once all the $(k + 1)$ node useful subgraphs ($CSR[k + 1]$) are generated.

Now, we shall present *the conditions to be satisfied for* $g^k \cup$

$\{v\}$ *to be an useful subgraph*, given that $g^k \in CSR[k]$ and $L(v) > mLab(g^k)$. Based on these constraints, we had to evaluate only the extensions of g^k along the nodes which are likely to be useful subgraphs.

Condition 1: The subgraph $g^k \cup \{v\}$ can be convex only if all the graphs in $Desc(g^k, v)$ are convex.

Condition 2: $perm\_results(g^k \cup \{v\}) \supset \forall_{g \in Desc(g^k, v)} perm\_results(g)$, i.e the graph $g^k \cup \{v\}$ can satisfy output constraint only if each graph in $Desc(g^k, v)$ satisfies the output constraint.

Condition 3:

Let $u_1, u_2, \ldots, u_n$ be the primary input nodes of g^k. Let $g_i^k = (g^k - \{u_i\}) \cup \{v\}$ be a descendant of g^k along node v. The graph g_i^k must obey the input constraint unless $|sources(g^k \cup \{v\})| = N_{in}$, the inputs of u_i are non-exclusive and atleast one successor of u_i is in $g^k \cup \{v\}$.

Putting together the above three observations, the graph $g^k \cup \{v\}$ can be useful only if all the graphs in $Desc(g^k, v)$ which satisfy the input constraints are useful. It is possible that none of the graphs in $Desc(g^k, v)$ are useful but $g^k \cup \{v\}$ is still useful. This happens only when

1. $|sources(g^k \cup \{v\})| = N_{in}$

2. None of the inputs of primary input nodes are exclusive

3. Atleast one successor of each primary input node is in $g^k \cup \{v\}$

We need to find out for each g^k, all the nodes v such that $L(v) > mLab(g^k)$ and $g^k \cup \{v\}$ is useful. Based on the above three conditions established for $g^k \cup \{v\}$ to be useful, we first prune the nodes considered for evaluation. Once, the above three conditions are satisfied then the graph $g^k \cup \{v\}$ is tested for the three properties of an useful subgraph.

For $g^k \cup \{v\}$, the pruning criteria is dealt in one of the following cases based whether atleast one of the graph in $Desc(g^k, v)$ is useful or not.

Case 1: Atleast one descendant is an useful graph, i.e $|Desc(g^k, v) \cap CSR[k]| > 0$

For $g^k \cup \{v\}$ to be useful, all the descendants which satisfy the input constraint must be useful subgraphs. i.e $|Desc(g^k, v) \cap CSR[k]|$ must be equal to $|\{u \in PI(g^k)|sources((g^k - \{u\}) \cup \{v\}) \leq N_{in}\}|$

Case 2: All the descendants are not useful, i.e $|Desc(g^k, v) \cap CSR[k]| = 0$

As stated above, although none of the descendants are useful, the graph $g^k \cup \{v\}$ can still be useful only when

1. $|sources(g^k \cup \{v\})| = N_{in}$

2. None of the inputs of primary input nodes are exclusive

3. Atleast one successor of each primary input node is in $g^k \cup \{v\}$

Let us translate the above three constraints into the graph properties.

Since g^k is an useful graph, its number of inputs is $\leq N_{in}$. Any operation has either one or two inputs and single output. Hence, including node v to g^k can increase its number of inputs by atmost two. Therefore, the above first condition can be true only if $|sources(g^k)| \geq N_{in} - 2$. Moreover, for $g^k \cup \{v\}$ to be convex the node v should be either an immediate successor node of some node in g^k or should be an independent node of g^k. Obviously, when $|sources(g^k)| = N_{in} - 2$ the node v should be an independent node of g^k.

If an input of node v happen to be the exclusive input of a primary input node of g^k then that input would become non-exclusive in $g^k \cup \{v\}$. Since there can be atmost two inputs to any operation, atmost two exclusive inputs of primary input nodes of g^k can become non-exclusive in $g^k \cup \{v\}$. Therefore, number of exclusive inputs of primary input nodes of g^k must be ≤ 2.

Similar to the previous point, the number of primary of input nodes of g^k without any successor in g^k must be ≤ 2.

4.1 Overall Algorithm

Our overall algorithm is shown in Algo. 1. The algorithm accepts the acyclic DDG G, the I/O constraints and the set of prohibited nodes as input and generates the set of all legal subgraphs of G. The set of all k node legal subgraphs are stored in $CSR[k]$.

First, the DDG G is labeled in a topological order. The function $Construct\_Single\_Node\_CSR(G, P)$ constructs the set of all single node legal subgraphs of G and establishes the descendant relationships among them. Then in each iteration of the loop shown in lines 6–27, the useful subgraphs of size $k + 1$ are constructed from the useful subgraphs of size k. As discussed above, for each useful subgraph g^k, finding its useful extensions is dealt in two cases. As per the pruning criteria discussed above, if an extension of g^k along node v, i.e $g^k \cup \{v\}$, passes the pruning criteria then the function $IsUseful(g^k, v)$ tests graph g^k for the three properties input, output and convexity. If it is found useful then it is added to $CSR[k+1]$. The function $DrawEdges(k)$ establishes the descendant relationships among $CSR[k+1]$ based on the descendant relationships among $CSR[k]$. This creates directed edges from each graph to its descendants. Once all the useful $(k + 1)$ node subgraphs and the descendant relationships are derived then all the illegal subgraphs in $CSR[k]$ are deleted.

5. EXPERIMENTAL EVALUATION

5.1 Experiment Setup

The proposed custom instruction generation algorithm was implemented in the trimaran, open source research compiler, framework [4]. Trimaran framework consists of front end (IMPACT) and back end (ELCOR). The front end of the compiler performs classical compiler optimizations such common subexpression elimination, constant folding etc and writes its output in an intermediate representation that is fed as input to the back end. Back end of the compiler uses a REBEL intermediate representation format. ELCOR is a series of architecture dependent transformations where each transformation reads the REBEL file and annotates more low level information and writes its output again in REBEL format to a file. Initially, ELCOR performs global data flow analysis and constructs data dependency graph

---

Algorithm 1 Enumerate all legal subgraphs of G

1: Inputs : G(V,E), N_{in}, N_{out}, $P \subset V$
2: Output : All the legal candidate subgraphs of G
3: Label(G,P)
4: Construct_Single_Node_CSR(G,P)
5: k = 1
6: **while** $|CSR[k]| > 0$ **do**
7:   **for each** $g \in CSR[k]$
8:   **loop**
9:     **Case 1:**
10:     **for each** v such that $g^k \cup \{v\}$ satisfies case 1 conditions
11:     **loop**
12:       **if** $IsUseful(g^k, v)$ **then**
13:         Add $g^k \cup \{v\}$ to CSR[k+1]
14:       **end if**
15:     **end loop**
16:     **Case 2:**
17:     **for each** v such that $g^k \cup \{v\}$ satisfies case 2 conditions
18:     **loop**
19:       **if** $IsUseful(g^k, v)$ **then**
20:         Add $g^k \cup \{v\}$ to CSR[k+1]
21:       **end if**
22:     **end loop**
23:   **end loop**
24:   DrawEdges(k)
25:   DeleteIllegalGraphs(k)
26:   k = k + 1
27: **end while**

---

for each basic block of the application. Immediately after the construction of data dependency graph, we had inserted a call to our custom instruction generation module passing data dependency graphs as the arguments.

The custom instruction generation module accepts the constraints on the number of inputs, outputs and prohibited node types and then proceeds to enumerate all the legal patterns. In our experiments, we had marked all the nodes other than arithmetic/logical, copy and comparison operations as forbidden nodes. The experiments were conducted on a Pentium IV 3.0GHz HT machine with 1GB RAM.

5.2 Experiment Results

We had compared the performance of our algorithm with that of algorithm presented in [16]. The experimental results are presented in Fig. 1. We have tested our approach on a set of five benchmark kernels taken from Mediabench suite [19], which are rich in arithmetic/logical operations. The first column of the table shows the benchmark name and the number of nodes in the kernel basic block. The third column (S_b) and fourth column (S_o) show the number of subgraphs evaluated (for the I/O constraint shown in the second column) by the algorithm in [16] and our algorithm respectively. The fifth column indicate the number of valid subgraphs satisfying the I/O and convexity constraints. Finally, thee last column shows the speedup of our algorithm over [16]. The number of valid candidates do not always increase with the relaxation of I/O constraint. For example, at (4,1) and (5,1) the valid candidates in IDCT are exactly same but different at (3,1). This can be attributed to the fact that most of the operations are binary and number of

Appl.	I/O	S_b	S_o	Valid	Speedup
ADPCM	(3,1)	231035	3252	58	12.8
(56)	(3,2)	689493	5717	592	32.8
	(4,1)	271719	6584	85	17.6
	(4,2)	797280	13409	955	24
	(4,3)	4110801	21203	7760	74
	(5,2)	844450	26832	2472	13
	(5,3)	4333229	45461	9498	37.7
	(5,4)	10124474	63301	55110	60.4
IDCT	(3,1)	73426	6425	319	14.6
(98)	(3,2)	393860	63691	12911	22
	(4,1)	137999	6425	319	37
	(4,2)	959108	63691	16721	63.9
	(4,3)	472637322	6526655	289727	133.1
	(5,2)	1915783	63691	16721	22.8
	(5,3)	695523108	6686327	432451	54.2
	(5,4)	1701279268	96979317	4529367	71.4
DES	(3,1)	454335	9136	129	16.6
(116)	(3,2)	18896895	75438	4695	11.2
	(4,1)	454335	22401	129	24.4
	(4,2)	18896895	107436	7115	31
	(4,3)	579503935	9363127	722831	178.1
	(5,2)	21437820	887914	8830	38.8
	(5,3)	579503935	18059725	722831	83.6
CJPEG	(3,1)	22151	1326	51	21
(128)	(3,2)	272059	22151	1326	37.2
	(4,1)	28736	1452	76	20.4
	(4,2)	27200	22151	1378	48.9
	(4,3)	1270482	72051	22151	94.7
	(5,2)	292182	1502	79	20.6
	(5,3)	182504	48972	1472	42.4
	(5,4)	7105123	96382	38129	96.3
GSM	(3,1)	14631	10231	5325	7.2
(28)	(3,2)	47972	31063	21762	14.9
	(4,1)	16463	10287	5631	6.4
	(4,2)	56282	47922	26827	20.2
	(4,3)	147522	99410	87922	26.1
	(5,2)	66335	49018	28671	4.8
	(5,3)	168511	147977	90111	13.6
	(5,4)	303807	278806	197631	24.5

Table 1: Performance of Our ISE candidate generation algorithm

valid patterns is likely to show much difference at an operand stride of two. In the worst case, for an increase in the allowable operands by one, the number of valid subgraphs may get multiplied by the number of nodes in the graphs. But the experimental results indicate that this is much smaller than what can be possible in the worst case.

The runtime of our algorithm is less than one second for small I/O constraints and at higher I/O constraints, in most of the cases in less than a minute all the valid patterns are identified. The speedup achieved is ranging from order of one to order of two. The speedup is prominently visible as output constraint is relaxed without changing the input constraint. The speedup can be attributed to the reduction in the search space and the fact that the convexity evaluation of subgraphs is practically performed in constant time. Comparing to [16], our algorithm incurs additional time in the new pruning techniques introduced. But, the overall speedup achieved indicate positive improvement over [16].

Although, the number of nodes in the basic blocks is around 100 or less, still the number of legal candidates are in millions at large I/O constraint. This is mainly because we permitted the disconnected patterns also as is done in [16]. The algorithm in [16], although, is able to significantly prune the search space but is upto 1000 times higher than the number of valid patterns. Our algorithm has successfully reduced that search space, sometimes by a factor of 100. In general, the search space reduction factor is in the range of 20 to 100. At relaxed I/O constraints such as (4,2) and (4,3) and (5,3) the reduction is higher but it is not reducing monotonically.

6. CONCLUSION

Exhaustive enumeration of legal patterns is an important task for being able to optimally select the custom instructions. Due to the large gap between the actual number of subgraphs and the number of legal subgraphs, it is important to aggressively prune the search space. In this paper, we had proved certain inherent relations between the characteristics a pattern and its subgraphs. Based on that, we had developed some novel pruning techniques that resulted in significantly speeding up the enumeration process.

7. REFERENCES

[1] R. E. Gonzalez, "Xtensa: A configurable and extensible processor," *IEEE Micro*, vol. 20, no. 2, pp. 60–70, 2000.

[2] T. R. Halfhill, "Mips embraces configurable technology," March 2003.

[3] P. Bonzini and L. Pozzi, "Polynomial-time subgraph enumeration for automated instruction set extension," in *DATE '07*, pp. 1331–1336, 2007.

[4] L. N. Chakrapani, J. C. Gyllenhaal, W. mei W. Hwu, S. A. Mahlke, K. V. Palem, and R. M. Rabbah, "Trimaran: An infrastructure for research in instruction-level parallelism.," in *LCPC*, pp. 32–41, 2004.

[5] L. Pozzi and P. Ienne, "Exploiting pipelining to relax register-file port constraints of instruction-set extensions," in *CASES '05*, pp. 2–10, 2005.

[6] N. Pothineni, A. Kumar, and K. Paul, "Application specific datapath extension with distributed i/o functional units," in *VLSID '07*, pp. 551–558, 2007.

[7] M. Arnold and H. Corporaal, "Designing domain-specific processors," in *CODES '01*, pp. 61–66, 2001.

[8] U. Kastens, D. K. Le, A. Slowik, and M. Thies, "Feedback driven instruction-set extension," in *LCTES '04*, pp. 126–135, 2004.

[9] P. Yu and T. Mitra, "Scalable custom instructions identification for instruction-set extensible processors," in *CASES '04*, pp. 69–78, 2004.

[10] P. Biswas, N. Dutt, P. Ienne, and L. Pozzi, "Automatic identification of application-specific functional units with architecturally visible storage," in *DATE '06*, pp. 212–217, 2006.

[11] P. Bonzini and L. Pozzi, "Code transformation strategies for extensible embedded processors," in *CASES '06*, pp. 242–252, 2006.

[12] J. Grossschadl, P. Ienne, L. Pozzi, S. Tillich, and A. K. Verma, "Combining algorithm exploration with instruction design: a case study in elliptic curve cryptography," in *DATE '06*, pp. 218–223, 2006.

[13] P. Yu and T. Mitra, "Characterizing embedded applications for instruction-set extensible processors," in *DAC '04*, pp. 723–728, 2004.

[14] N. Clark, H. Zhong, and S. Mahlke, "Processor acceleration through automated instruction set customization," in *MICRO 36*, p. 129, IEEE Computer Society, 2003.

[15] F. Sun, S. Ravi, A. Raghunathan, and N. K. Jha, "Synthesis of custom processors based on extensible platforms," in *ICCAD '02*, pp. 641–648, 2002.

[16] L. Pozzi, K. Atasu, and P. Ienne, "Exact and approximate algorithms for the extension of embedded processor instruction sets," *IEEE Transactions on Computer-Aided Design of Integrated Circuits and Systems*, vol. 25, no. 7, pp. 1209–1229, 2006.

[17] N. Pothineni, A. Kumar, and K. Paul, "Exhaustive enumeration of legal custom instructions for extensible processors," Tech. Rep. TR 07/2007, Department of Computer Science, IIT Delhi, July 2007.

[18] K. Atasu, L. Pozzi, and P. Ienne, "Automatic application-specific instruction-set extensions under microarchitectural constraints," *Int. J. Parallel Program.*, vol. 31, no. 6, pp. 411–428, 2003.

[19] C. Lee, M. Potkonjak, and W. H. Mangione-Smith, "Mediabench: A tool for evaluating and synthesizing multimedia and communicatons systems," in *International Symposium on Microarchitecture*, pp. 330–335, 1997.

21st International Conference on VLSI Design

An Optimal Multi-Functional Unit Dynamic Instruction Selection Logic at Submicron Technologies

Terrell Bennett and Rama Sangireddy

Department of Electrical Enginnering

University of Texas at Dallas, Richardson, TX 75080, USA

Abstract

As the technology scales, reduction in transistor size creates many opportunities for increased circuit capabilities in reduced chip area. In modern wide-issue processors, performance of the processor is directly impacted by the time delay complexity of the dynamic scheduling logic. In this paper, we analyze the scaling of time delay of instruction select logic at the submicron technologies, and also present novel designs that provide a single selection tree for two similar functional units. The designs are based on a tree structure using arbiter cells of two and four inputs which can handle one or two functional units. The effects of technology and design decisions are shown based on simulations using four submicron technologies. The delays in the select logic trees are shown to decrease by an average of 60% from 130nm technology to 45nm technology when servicing a single functional unit. The double grant arbiter cells are shown to build a tree that will serve multiple functional units simultaneously with 65% lesser delay as compared to multiple single-grant trees[1].

1. Introduction

The continuous reduction in transistor size is creating many opportunities for increased circuit capabilities in reduced chip area. This trend directly affects the complexity and performance of high performance processors. The decrease in transistor size typically has three goals based on scaling theory as explained by Borkar [1]: (i) reduce gate delay by 30%, resulting in an increase in operating frequency of about 43%; (ii) double transistor density; and (iii) reduce energy per transition by about 65%, saving 50% of power (at a 43% increase in frequency). The circuit delay complexity based on transistor size and logic design in the instruction selection logic, a critical component of wide-issue processor core, will be examined in this paper. Dynamic instruction scheduling in modern processors allows instructions to be executed out of order. The major component that assists in dynamic instruction scheduling is the select logic, which in every cycle selects multiple instructions up to processor's execution bandwidth, from the pool of ready instructions to maximize the instructions per cycle (IPC) throughput. For example, for a 16-instruction window, any number of instructions from 1 to 16 can have their source operands ready in a cycle and so propagate their request signals to a functional unit (FU), which propagates back a grant signal to only one instruction based on preset priority conditions. Besides, when a FU is granted to an instruction, it should be barred from granting to another instruction in the same cycle. Also, a single instruction should be prevented from granted access to multiple similar FUs. These design conditions add complexity to the select logic tree designs

---
[1] The authors can be contacted at rama.sangireddy@utdallas.edu

for the pool of FUs in a dynamically scheduled wide-issue out-of-order processor. The select logic tree is in the critical path of a processor's pipeline, and its circuit delay is desired to fit into one processor clock cycle. In this paper, various designs of select logic are presented and analyzed with the scaling of transistor technology.

The contributions of this paper are as follows. First, we present novel designs for 2-2 arbiter cell and 4-2 arbiter cell, which enable a single select logic tree to serve two similar functional units simultaneously. This is in contrast to the conventional designs, which require select logic trees for multiple functional units to communicate with each other through a masking logic in order to ensure that in a cycle a single instruction is not granted access to multiple similar functional units. Second, we present an extensive analysis of various configurations of select logic tree designs, using both conventional and novel arbiter cell designs, at modern submicron technology sizes ranging from 130nm to 45nm. And, the extensive analysis with results show the circuit delay improvements possible with technology scaling and help to make appropriate decisions for future processor designs.

2. Scheduling Logic Design

2.1 Select Logic Arbiter Cell Design

The basic structure of the selection logic is the arbiter cell. The role of the arbiter cells is to determine which ready instructions are granted access to the functional units. These cells are designed based on a tree selection logic structure and a positional selection policy [2]. There are four arbiter cell designs, which fall into two groups, where the first group includes the single grant arbiter cells that can only handle the grant signal for a single functional unit. The second group includes double grant arbiter cells, where they are able to handle grant signals for two functional units simultaneously. In each of these groups, there is a two input and a four input design. The block diagrams for the arbiter cells are shown in Figure 1. The arbiter cells are referenced using the number of inputs followed by the number of available grant signals. For example, the arbiter 2-2 cell takes two input requests, and can grant access to one or two of the request signals. The gate level and transistor level designs for the 2-1, 2-2, and 4-2 arbiter cells are shown in Figures 2, 3, and 4, respectively. The design for the arbiter 4-1 (not shown) is same as the arbiter cell design presented in [2]. The arbiter 2-1 cell, arbiter 2-2 cell, and arbiter 4-2 cell are designed in this paper. The designs are based on a position based selection policy, which means that instructions in certain positions will always have priority over the other instructions.

Domino logic is used for the outputs that traverse the tree towards the root. Domino logic is faster than standard CMOS

978-1-4244-3039-0/08 \$25.00 © 2008 IEEE

267

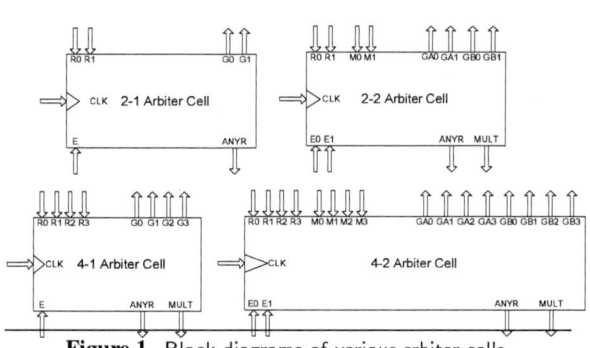

Figure 1. Block diagrams of various arbiter cells.

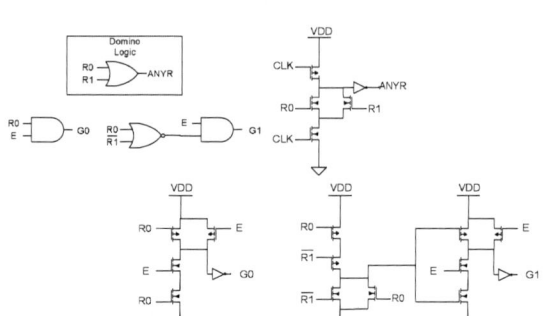

Figure 2. Arbiter 2-1 gate level and transistor level circuits.

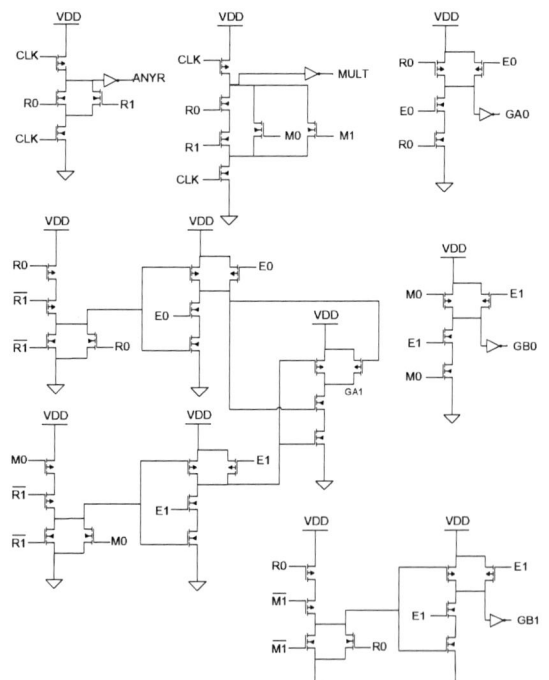

Figure 3. Arbiter 2-2 transistor level circuits.

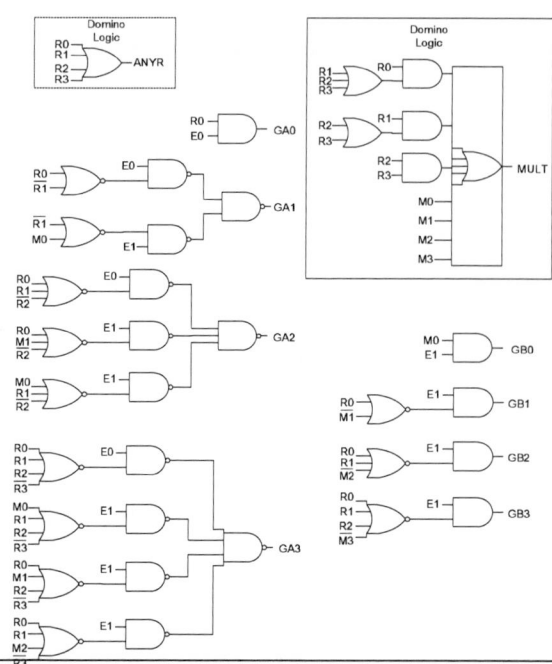

Figure 4. Arbiter 4-2 gate level circuits. The transistor level circuit is not shown here for space paucity.

because it is generally sized to ensure fast transitions. The logic equations used to determine the outputs of the arbiter cells are shown in Table 1. As the equations show, each of the grant outputs uses one or two enable signals. This factor prevented the design in [2] from using the domino logic for signals going up the tree towards the instructions (i.e. grant signals). The arbiters at the top and middle of the tree will not have the enable signal ready for some time as the requests go down the tree to the root node. To take advantage of this extra time, a portion of the grant signal is pre-computed before the enable signal becomes available to make the operation faster.

Table 1. The logic equations for output signals of arbiter cells. The R signals are input requests, E signals are enables, M signals are for multiple input requests, ANYR is output request signal, and MULT is multiple output request signal.

Equations for Single Grant Arbiters
$ANYR = R0 + R1 + R2 + R3$
$G0 = R0 \bullet E$
$G1 = R1 \bullet \overline{R0} \bullet E$
$G2 = R2 \bullet \overline{R0} \bullet \overline{R1} \bullet E$
$G3 = R3 \bullet \overline{R0} \bullet \overline{R1} \bullet \overline{R2} \bullet E$
Equations for Double Grant Arbiters
$ANYR = R0 + R1 + R2 + R3$
$MULT = R0(R1 + R2 + R3) + R1(R2 + R3) + R2 \bullet R3 + M0 + M1 + M2 + M3$
$GA0 = R0 \bullet E0$
$GA1 = R1 \bullet \overline{R0} \bullet E0 + R1 \bullet \overline{M0} \bullet E1$
$GA2 = R2 \bullet \overline{R0} \bullet \overline{R1} \bullet E0 + R2 \bullet \overline{R1} \bullet \overline{M0} \bullet E1 + R2 \bullet \overline{R0} \bullet \overline{M1} \bullet E1$
$GA3 = R3 \bullet \overline{R0} \bullet \overline{R1} \bullet \overline{R2} \bullet E0 + R3 \bullet \overline{R1} \bullet \overline{R2} \bullet \overline{M0} \bullet E1 +$
$\quad R3 \bullet \overline{R0} \bullet \overline{R1} \bullet \overline{M2} \bullet E1 + R3 \bullet \overline{R0} \bullet \overline{R2} \bullet \overline{M1} \bullet E1$
$GB0 = M0 \bullet E1$
$GB1 = M1 \bullet \overline{R0} \bullet E1$
$GB2 = M2 \bullet \overline{R0} \bullet \overline{R1} \bullet E1$
$GB3 = M3 \bullet \overline{R0} \bullet \overline{R1} \bullet \overline{R2} \bullet E1$

In modern wide-issue processors, multiple functional units are available to execute instructions to harness instruction level parallelism (ILP). The double grant arbiter designs proposed were created for this purpose. As shown in Figure 1, the double grant arbiter cell designs basically double the number of inputs and outputs compared to the single grant arbiter cell designs to achieve the added functionality of servicing two

268

978-1-4244-3039-0/08 $25.00 © 2008 IEEE

functional units. Servicing multiple input requests adds some complexity to the grant logic. The double grant designs have two sets of grants. The first set of grants (GA in the equations) is connected to the first set of enable lines (E0 in the equations). Based on the equations, any input request can be serviced by these grant signals. The second set of grants (GB in the equations) is used to grant a second functional unit to a single requesting select tree branch. In other words, GA0 and GB0 going high means that the arbiter at input position 0 will get access to two functional units. This design works with the M input signals to handle multiple requests from a single branch.

2.2 Selection Logic Tree Design

Table 2. Various select logic tree configurations.

The first digit (X) in the tree description CX-Y refers to the basic tree structure. The structures are defined by the number of inputs the arbiters on each level of the tree can accept.			
16 Instructions	C1-Y	2-2-2-2	4 level tree using only 2 input arbiters.
	C2-Y	2-2-4	3 level tree using 2 and 4 input arbiters.
	C3-Y	2-4-2	3 level tree using 2 and 4 input arbiters.
	C4-Y	4-4	2 level tree using only 4 input arbiters.
32 Instructions	C1-Y	2-2-2-2-2	5 level tree using only 2 input arbiters.
	C2-Y	2-2-4-2	4 level tree using 2 and 4 input arbiters.
	C3-Y	2-4-4	3 level tree using 2 and 4 input arbiters.
	C4-Y	4-4-2	3 level tree using 2 and 4 input arbiters.
	C5-Y	4-4-4	3 level tree using only 4 input arbiters.
64 Instructions	C1-Y	2-2-2-2-2-2	6 level tree using only 2 input arbiters.
	C2-Y	2-4-4-2	4 level tree using 2 and 4 input arbiters.
	C3-Y	4-2-2-4	4 level tree using 2 and 4 input arbiters.
	C4-Y	4-4-4	3 level tree using only 4 input arbiters.
The second number (Y) in the tree description CX-Y refers to the tree configuration. The configurations are defined by the types of arbiter cells used and the number of instructions serviced.			
CX-1	These configurations use only single grant arbiters (i.e. 2-1 and 4-1).		
CX-2	These configurations using only double grant arbiters (i.e. 2-2 and 4-2). Two instructions can be serviced.		
CX-3	These configurations use a double grant arbiter for the root, and single grant arbiters for the rest of the tree. Two instructions can be serviced.		

Using the arbiter cells, selection logic trees are designed to simulate the full delay from the initial instruction request and clock signals to an instruction(s) receiving the grant signal(s). The delay of the trees will be based on the depth of the tree and the delay at each level. The structures are designed to test trees ranging from smallest depth (using only 4-input arbiter cells) to largest depth (using only 2-input arbiter cells). To compare which arbiter cells would be most beneficial if only one was available, trees with cells of only one type are included in each group. Table 2.2 shows various tree configurations and explains the naming convention used in the analysis. Four structures each are created for the 16- and 64-instruction trees. An extra structure was created for the 32-instruction tree, since a tree using only 4-input arbiters will have unused inputs at the root level. If only 4-input arbiter cells are available, this is necessary; but the more ideal structure is found in C4-1, C4-2, and C4-3 for the 32-instruction trees. Based on the structure of the tree, the arbiter cell signals that go down the tree will see their worst case delays at the first level. The arbiter cell signals that go up the tree (i.e. grant signals) will see their worst case delays at the root node of the tree. These details factor into the analysis of the delays of the tree structures and configurations. Figures 5 and 6 show the tree structure for the C1-3 and C4-3 configurations, respectively.

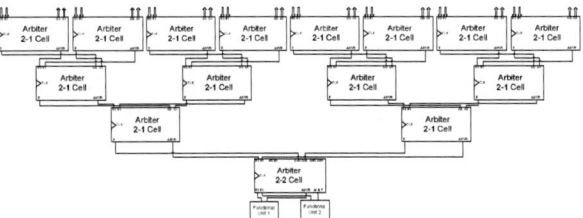

Figure 5. Select logic tree C1-3 configuration for 16 instructions and serving two FUs.

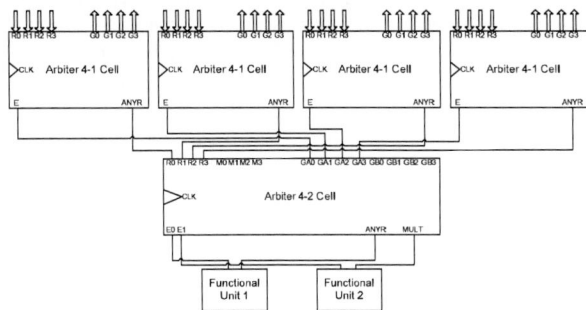

Figure 6. Select logic tree C4-3 configuration for 16 instructions and serving two FUs..

Each tree structure designed is created with three configurations. The first configuration uses only single grant arbiters. The second configuration uses only double grant arbiters. The third configuration uses a double grant arbiter at the root of the tree and single grant arbiters for the rest of the tree. The second and third configurations take advantage of the versatility of the double grant arbiter cells. A tree can be created using only double grant cells to serve two functional units simultaneously (i.e. the second configuration). Two functional units can also be served by using a double grant cell at the root of a tree and with only single grant cells in the other positions (i.e. the third configuration). Delays for designs using the first configuration are only compared with each other as they only allow a single request to be granted. Delays for designs using the second and third configurations are compared with each other as they both allow two requests to be granted. Multiple double grant cells can be used in conjunction with single grant cells to change the position and behavior of the selection policy. For example, second configuration services the first two instructions, while third configuration services the first instruction in each half of the tree when all requests are high. The limitation in this case is that once a double grant is introduced at any level of the tree, all levels toward the root must be double grant cells.

2.3 Sizing

To ensure fair comparisons across designs and technologies, the techniques used to size the transistors have to be consistent. The optimum W/L ratio for one technology might not be the optimum ratio for the other technologies. For each technology, the minimum length of all transistors is made equal to the minimum possible length for the technology. The decision is also made to use the minimum size for the base width of the NMOS transistors. To determine the base width for all PMOS transistors in the different technologies, inverter simulations are done. The PMOS transistors are sized to force the switching threshold (i.e. logical threshold) voltage [3] to equal $V_{DD}/2$ for each technology. With the switching threshold set in this way, the basic inverter does not have a tendency to pull up or down more than the other. In other words, the rise and fall times should be equal in an inverter with this property. This concept is important, because in the circuits presented the pull up and pull down networks should be at equal strength across technologies. The minimum sizes determined by these methods are used whenever possible.

Another sizing decision made is based on the increased resistance of transistors in series. To combat this trend, a sizing factor is used for transistors in series. The number of transistors in series determines the factor applied to the width of the transistor. This sizing adjustment does not exactly match the current in the single transistor case due to the increase in the threshold voltage caused by the non-zero V_{BS}, but is a close approximation. The final sizing decision made is for the pre-charge NMOS and PMOS transistors in the domino logic. In all cases, these are sized much larger than the other transistors. This is not a concern for the PMOS transistor, because the PMOS transistor is off during evaluation and so the size does not affect the delay (ignoring leakage current). The sizing is a concern in the NMOS case, but the decision is made to simplify the sizing of other series components in the domino logic.

3. Results and Analysis

All simulations are done using SPICE 3, and commercial transistor models. The testbenches use input signals with 1ps rise and fall times. It is not likely that the actual inputs to these devices would have such fast rise and fall times. However, this allows all technologies to be compared fairly with inputs that are the same and close to ideal. Output signal delays are measured versus a reference signal (i.e. the signal which causes the change in the output). This measurement is made from $V_{DD}/2$ on the input signals to the same level on the output signals.

The circuit for ANYR signal in 2-1 (Figure 2) and 2-2 (Figure 3) arbiter cells is used as an example here for the best and worst case analysis. The worst case delay for this signal is when only one of the request inputs is high, and the clock signal comes after the request. This is because only one of two pull down paths is used, and the node above the NMOS CLK transistor is pulled high by the PMOS pre-charge transistor. The best case for the ANYR signal is when the clock signal is high before the request signals, and both request signals go high simultaneously. Because the clock signal is high, the node above the NMOS CLK transistor is already low. Therefore, when the two paths conduct, the output will be switched at the fastest possible rate. The other factor considered in the best and worst case delay determination is the load capacitance being driven by the outputs. This primarily affects the ANYR and MULT signals of the arbiter cells, because they can go into different inputs depending on the node position in the tree. As the equations in Table 1 show, lower number inputs (e.g. R0, M0, etc.) are used in more output equations than the higher number inputs (e.g. R3, M3, etc.). Therefore, the worst case for this circuit occurs when driving R0, and the best case occurs when driving R3.

3.1 Arbiter Cell Design Analysis

3.1.1 Arbiter 2-1 Results and Analysis

Figure 7 shows the best and worst case delays for the arbiter 2-1 cell output signals. The signals with the shortest best case delays for the arbiter 2-1 cell are the G0 and G1 signals in all technology sizes. The delays are 17.9ps in the 45nm technology and 42ps in the 130nm technology. In the best case scenario, the grant signals are equally fast. The worst case delay signal is dependent on the technology size. In the 45nm and 65nm technologies, the G1 signal has the worst case delay at 43ps and 48ps, respectively. In the 90nm and 130nm technologies, the ANYR signals worst case delay overtakes that of the G1 signal at 99.8ps and 106.7ps, respectively.

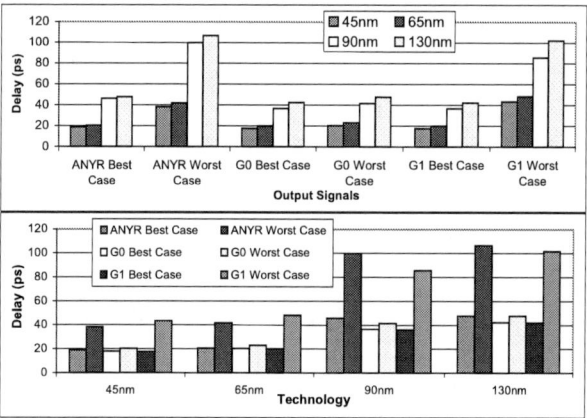

Figure 7. Arbiter 2-1 circuit delay.

A similar analysis for the arbiter 4-1 cell shows that the shortest best case delay in is for the ANYR signal across all technologies. The delay ranges between 14ps in the 45nm technology and 35ps in the 130nm technology. The worst case delay in this cell is the G3 signal in all technology sizes, with 72.9ps in the 45nm technology. The delay for this signal in the 130nm technology is 233% more than that in the 45nm technology. The long worst case delays in the grant signals are due to the fact that in the worst case situation, the pre-compute function discussed earlier does not happen in advance. The difference in the best and worst case delays for the grant signals shows the benefit of the pre-compute circuit.

3.1.2 Arbiter 2-2 Results and Analysis

Figure 8 shows the best and worst case delays for the arbiter 2-2 cell output signals. In the arbiter 2-2 cell, the GA1 and MULT signals have the shortest delay times in the best case. They are within 2ps across all of the technology sizes, between 16ps at 45nm and 39ps at 130nm. In the larger two technologies, the ANYR signal has the largest worst case delay at 114ps at 90nm and 126ps at 130nm. In the two smaller technologies, GA1 has the worst delay; but this delay is not much greater than the ANYR delay, the GB1 delay, and the MULT delay. All of these delays are within 8% of each other in the 45nm technology and 13% in the 65nm technology.

3.1.3 Arbiter 4-2 Results and Analysis

Figure 9 shows the delays for the arbiter 4-2 cell output signals. The MULT signal has the shortest best case delay in

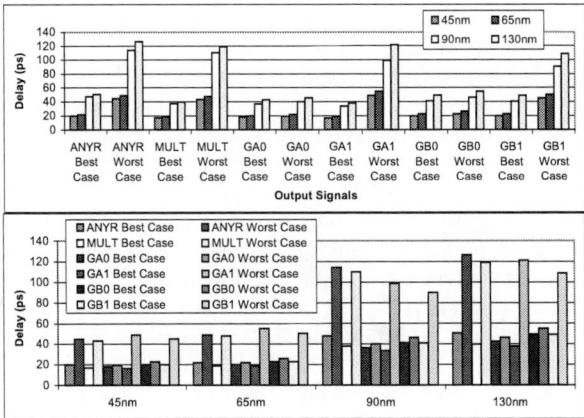

Figure 8. Arbiter 2-2 circuit delay.

the cell across all technology sizes delays of 15.1ps at 45nm, 17.1ps at 65nm, 29.5ps at 90nm, and 32.8ps at 130nm. The GA3 signal is with the worst case delay in the arbiter 4-2 cell across all technologies. The delay increases by 258% going from the 45nm technology to the 130nm technology.

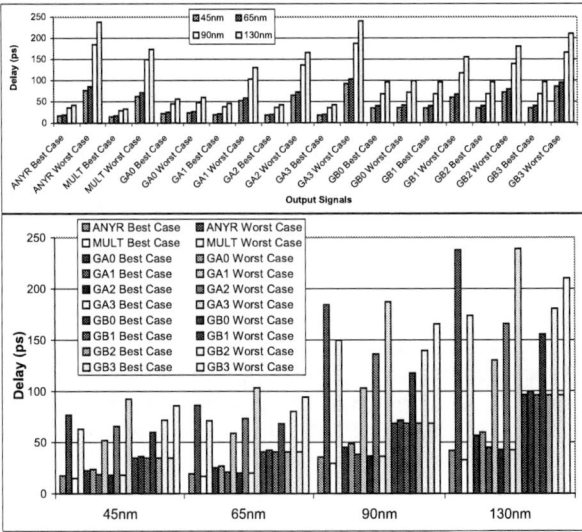

Figure 9. Arbiter 4-2 circuit delay.

3.1.4 Arbiter Cell Summary

In all the arbiter cells, a large increase in delays could be seen between 65nm and 90nm. The best case delays of the 4-input arbiters are smaller for signals that would traverse down the tree towards the root, and larger for signals that would go up the tree relative to the 2-input arbiters best case delays. On the other hand, the worst case delays for these signals are much worse in the 4-input arbiters than the 2-input arbiters. The 2-input cells hold an advantage in best case and worst case delays for signals going up the tree towards the instructions. This is due to the fact that the grant signals have to drive a larger load capacitances in the 4-input arbiters, because the signals are used as inputs to more number of circuits.

3.2 Select Logic Tree Analysis

The tree delays are based on round-trip signal times and are measured from the time the clock goes high to the time the final grant signal is received. This is an important point when comparing the delays for tree configurations that grant one request versus the tree configurations that grant two requests because the configurations granting two requests will be slower, but service multiple functional units simultaneously.

3.2.1 32-Instruction Tree Results and Analysis

Figure 10 shows the delays of the 32-instruction tree. Configuration 4-1 has the shortest delay for the trees that can grant a single instruction with a 142.5ps delay at 45nm, 161.5ps at 65nm, 295.6ps at 90nm, and 354.4ps at 130nm. Again, this is expected since C4-1 and C5-1 are the trees with the shortest depths. The C4-1 configuration is faster as it uses a 2-input arbiter cell at the root instead of a 4-input arbiter cell and the 2-input arbiter cell has much shorter worst case delays for the grant signals. Of the configurations that can grant two instructions, C4-3 has the shortest overall delay. This configuration is faster than C4-2 because the cell delays of the GA1 and GB1 in the arbiter 2-2 cell are relatively close (within 3ps for 45nm) to each other. Continuing up the tree, the best case GA0 delay in arbiter 4-2 is slower than the best case G0 delay in arbiter 4-1. This is why the mixed arbiter configuration provides the best delay for this structure. C3-1 also shows performance very close (within 13ps at 130nm and within 4ps at 45nm) to C4-1 across all technologies. C3-1 uses 2-input cells at the top of the tree while C4-1 uses 4-input cells at the top of the tree. The delay for C3-1 is higher is due to the arbiter cells driving higher load capacitances throughout the structure. The inputs drive the higher capacitance 4-input arbiter cells in case C4-1. With non-ideal inputs to the tree, C3-1 may be just as fast as or slightly faster than C4-1.

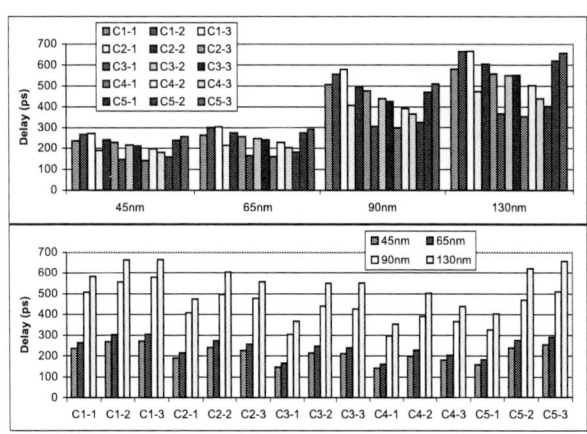

Figure 10. 32-instruction select logic delay.

3.2.2 64-Instruction Tree Results and Analysis

Figure 11 shows the delays of the 64-instruction tree. In the 64-instruction trees, C4-1 has the shortest delay for configurations that can service a single request signal. As the technologies increase in size, the percentage difference in delay between C4-1 and the other configurations remains the same. The delay for C2-1 is 30% larger than the delay for C4-1 in all technologies. In the configurations that can serve two functional units, C4-2 and C4-3 are the fastest as expected. In this case, C4-3 has the overall shortest delay by about 10ps at 45nm and by 19ps at 130nm. This result occurs because of the depth of the tree. Even though the root of C4-3 will

271

978-1-4244-3039-0/08 $25.00 © 2008 IEEE

see a longer delay (see 16-instruction tree analysis above), the shorter delays of the arbiter 4-1 cells going up the tree make enough difference for C4-3 to be faster. C2-3 is only about 20ps slower than C4-3 (at 45nm) since the worst case delays of the 4-input arbiters are never seen. By using 2-input arbiters at the top and bottom of the trees, the 4-input arbiters are assured of having their best case delays throughout. Even with this advantage, the extra depth causes a longer overall delay.

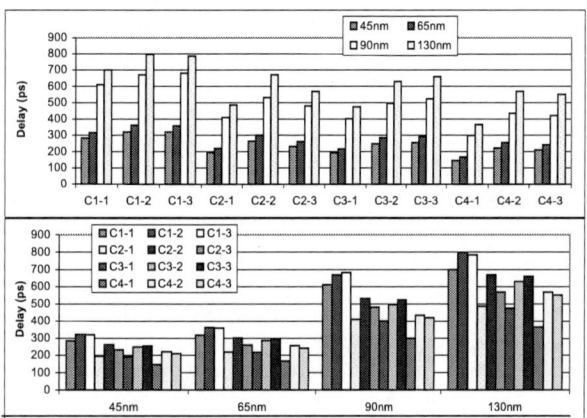

Figure 11. 64-instruction select logic delay.

3.2.3 Select Logic Tree Analysis Summary

While smaller technology sizes generally produce shorter delays, there are some configurations that are slower than other configurations in larger technologies. For example, in the 16-instruction trees, C2-1 at 45nm is more than 30ps slower than C4-1 at 65nm. The results show that the configuration of a tree structure is just as important as the technology available. The depth of the tree seemed to be the most consistent factor in determining which structures would have the shortest delay. In these configurations, the 4-input arbiter cells provided the best tree delays. Larger input cells (e.g. 8-input) seems like they would further decrease tree delays, but it has been found that 4-input arbiter cells are ideal [2].

4. Related Work

The work presented in our paper involves the effects of technology scaling in processors with a specific focus on the selection logic. Palacharla *et al.* [2] studied the effects of scaling on the complexity of processor components, but evaluated the delay with a smallest technology of only 180nm. They also discussed a method to allow access to multiple functional units by masking the request that is granted to the first functional unit into another arbiter tree structure. It was not clear from the work how this would affect delay. If the trees were run serially, the second functional unit would more than double the delay of the system due to the request masking logic and the second tree. In the arbiter cell designs presented in our paper, the additional tree and masking logic are not necessary to serve multiple functional units. Most other techniques only discuss the optimization of the wakeup and select logics in the dynamic scheduling logic. The focus of these techniques has been either to place the wakeup and select logics into two separate stages, or to organize the select logic based on instruction priorities for better IPC. It is noteworthy that most

of these techniques can seamlessly employ the arbiter cell and tree configuration designs presented in our paper. Sato discussed the degradation of IPC when attempting to pipeline wakeup and selection logic [4], showing that performance decreases by 30% in a two stage pipeline and 50% in a three stage pipeline. Stark *et al.* presented the concept of pipelining through speculative wakeup [5], and by scheduling all instructions that wake up and using the selection logic only as a check [6]. In both cases, the pipelined logic is taken out of the critical path to reduce the affect on IPC. Gran *et al.* discuss enhancing pipelining techniques and show performance within 2.6% and 2% of unpipelined schedulers [7]. Buyuktosunoglu *et al.* present an oldest-first selection policy that does not require instruction compaction [8]. The proposed techniques would save energy vs. a position based selection solution with compaction, but there is degradation in performance. Zhou *et al.* present a two-level selection logic structure [9]. The design uses cyclic segmented prefix (CSP) circuits in the first level and a switch network, priority encoders, and filters in the second level. The improvements in delay are based on comparisons to a single CSP select structure, but IPC is still negatively affected in this design. Many of the select logic concepts aim at reducing power by removing the compaction necessary in an oldest first position based selection policy while degrading IPC through the changes. Butler *et al.* show that there is very little difference in performance based on the selection technique [10]. Therefore, compaction is not really necessary when using a position based selection policy.

5. Conclusions

Designs and analysis of selection logic and for superscalar processors were presented in submicron technology sizes. The double grant arbiter cell designs presented add functionality to the selection tree in a variety of ways. The delays in the selection trees were shown to decrease by an average of 60% from 130nm technology to 45nm technology when servicing a single functional unit. The double grant arbiter cells presented allowed access to multiple functional units with 65% lesser delay as compared to multiple single grant trees.

References

[1] S. Borkar, "Design Challenges of Technology Scaling", IEEE Micro, Vol. 19, pp. 23-29, July-August 1999.

[2] S. Palacharla, N. P. Jouppi, and J. E. Smith, "Complexity-effective superscalar processors", Proc. 24th Int. Sym. on Computer Architecture, 1997, pp. 206-218.

[3] J. Rabaey, A. Chandrakasan, and B. Nikolic, "Digital Integrated Circuits", 2nd ed., Upper Saddle River, New Jersey: Pearson Education, 2003.

[4] Toshinori Sato, "A Simulation Study of Pipelining and Decoupling a Dynamic Instruction Scheduling Mechanism," 25th Euromicro Conference (EUROMICRO '99)-Volume 1, 1999.

[5] J. Stark, M. Brown, and Y. Patt, "On pipelining dynamic instruction scheduling logic", Proc. 33rd Int. Sym. on Microarchitecture, December 2000.

[6] M. Brown, J. Stark, and Y. Patt, "Select free instruction scheduling logic", Proc. 34th International Symposium on Microarchitecture, December 2001.

[7] R. Gran, E. Morancho, A. Olive, and J. Llaberia, "An Enhancement for a Scheduling Logic Pipelined Over Two Cycles," Proc. ICCD-2006, October 2006.

[8] A. Buyuktosunoglu, A. El-Moursy, and D. Albonesi, "An oldest-first selection logic implementation for non-compacting issue queues", Proc. 15th IEEE International ASIC/SOC Conference, pp. 31-35, 2002.

[9] J. Zhou and A. Mason, "A two-level hybrid select logic for wide-issue superscalar processors", Proc. IEEE Int. Sym. on Circuits and Systems, 2006, pp. 41-44.

[10] M. Butler and Y. Patt, "An investigation of the performance of various dynamic scheduling techniques,", Proc. 25th International Symposium on Microarchitecture, pp. 1-9, 1992.

21st International Conference on VLSI Design

A 2.1GHz 6.5mW 64-bit Unified PopCount/BitScan Datapath Unit for 65nm High-Performance Microprocessor Execution Cores

Rajaraman Ramanarayanan[2], Sanu Mathew[1], Vasantha Erraguntla[2], Ram Krishnamurthy[1], Shay Gueron[3, 4]

[1] *Circuit Research Laboratory (CRL), Intel Corporation, Hillsboro, OR, USA,*
[2] *Bangalore Design Laboratory, CRL, Intel Corporation, Bangalore, INDIA,*
[3] *Department of Mathematics, University of Haifa, Haifa, ISRAEL,*
[4] *Mobility Group, Intel Corporation, IDC, Haifa, ISRAEL*
rajaraman.ramanarayanan@intel.com

Abstract

This paper describes a unified PopCount/ BitScanForward/BitScanReverse datapath circuit designed for 2.1GHz operation with total power consumption of 6.5mW, targeted for 65nm 64-bit microprocessor execution cores. The unified datapath uses a hybrid 3:2 compressor-based Wallace tree to count the number of '1's in the 64-bit input, along with a novel encoding scheme that enables reuse of the same tree to identify the bit-location of the 1st set bit when scanning the input in the forward and reverse directions. This circuit thus combines the functions of 3 separate units, enabling 26% reduction in total energy and 20% lower area, while achieving single-cycle latency & throughput.

1. Introduction

Present-day high-performance microprocessors use application-targeted hardware accelerators to improve the performance-per-Watt (GOPS/Watt) of general-purpose execution cores. These accelerators introduce special-purpose instructions into the processor ISA, thereby providing performance-optimized, low-latency, lower power instructions for accelerating targeted workloads. The *POPCNT* instruction is an example of an application-targeted instruction that can be effectively used to accelerate search operations involving large data sets [1, 2]. It works by counting the number of set bits in a data object. Applications that benefit from this instruction include genome mining, handwriting recognition, digital health workloads and fast hamming distance counts. Since this instruction is a critical operation in modern-day search engines, an important application in the server market, there is a strong need for energy-efficient

PopCount hardware accelerators with single-cycle latency and throughput in the processor execution core.

BitScanForward (*BSF*) and BitScanReverse (*BSF*) instructions are extensively used in floating-point operations for rounding and normalization of floating-point numbers. These instructions return the bit-position of either the most-significant or least-significant set bit in a 64-bit word. Current implementations of BSF and BSR instructions include separate hardware units for each instruction with counters to sum up the location of the set bit. It will be observed that implementation of the 3 instructions (*POPCNT, BSR and BSR*) involves similar kind of arithmetic circuits. Therefore, there is a potential for sharing hardware between these units, thereby reducing total energy consumption and layout area in the execution core. This paper proposes a novel design and implementation details of a unified datapath that uses a single compressor-tree to perform *PopCount, BSR* and *BSF* operations. The proposed tree uses a hybrid 3:2 compressor circuit-based optimized Wallace tree for fast *PopCount* operations, along with a novel encoding scheme on the inputs, that enable reuse of the same compressor-tree for BSR and BSF operations.

2. Popcount Datapath

The POPCNT instruction returns the total count of set bits in a data word. Hence for a 64-bit word,

$$POPCNT = \sum_{i=0}^{63} b_i$$

Figure 1 shows examples of the popcount instruction for a 64-bit word. The instruction returns a 7-bit count that represents the number of ones in the 64-bit word.

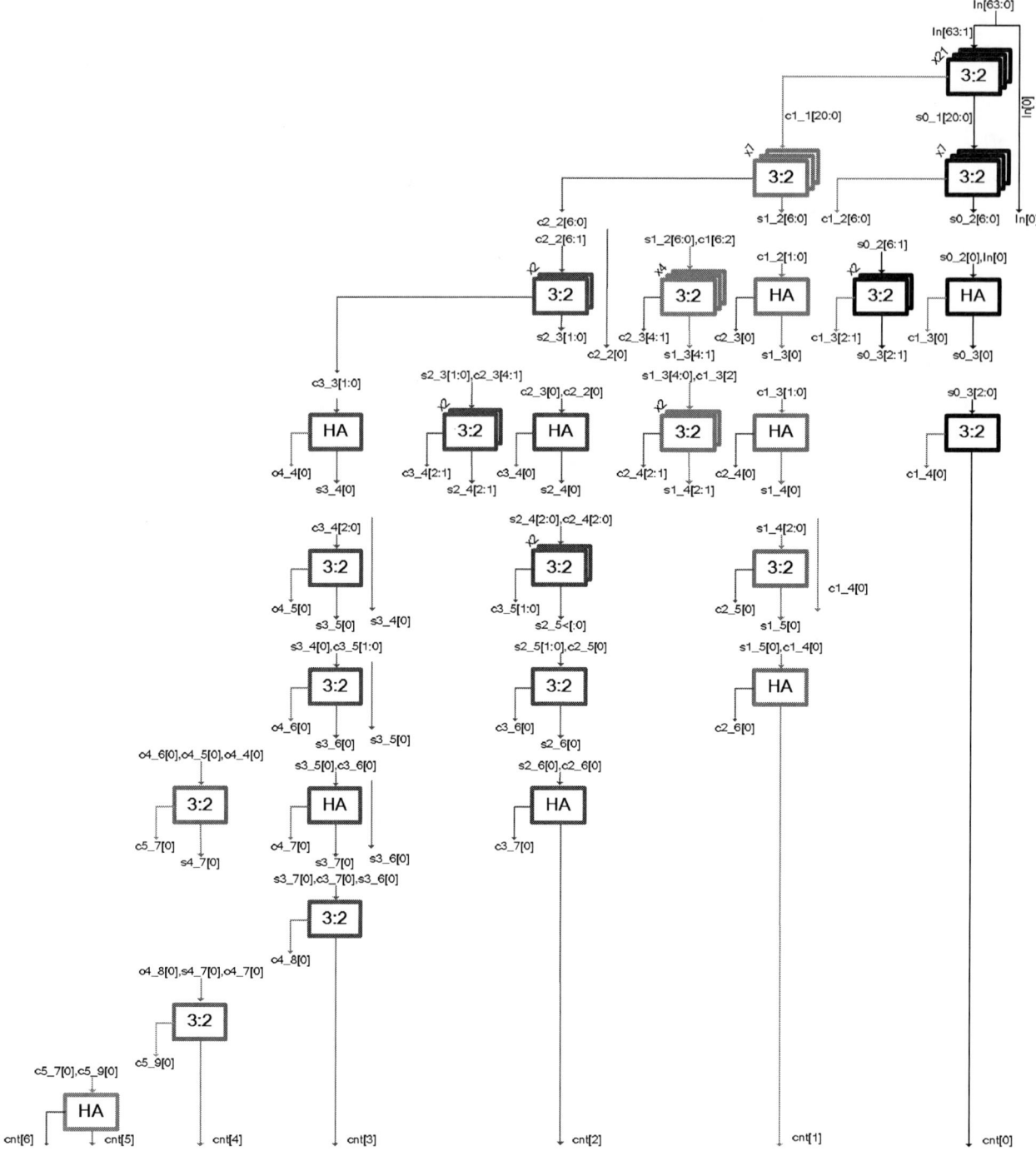

Figure 2: PopCount compressor-tree

The *PopCount* circuit uses a Wallace-tree structure to compress the 64-bit input (In[63:0]) to a 7-bit output (cnt[6:0]). The compressor tree achieves maximum possible compression using 3:2 counters and is composed of 57 3:2 compressors and 8 half-adders (Figure 2). In the first-stage, we have 21 3:2

compressors, each compressor taking in 3 inputs of identical significance, generating a sum-bit of the same significance and 1 carry-bit of higher significance. Compression of input bits In[0], is postponed till the third stage. Subsequent stages of the compressor tree use a combination of 3:2 compressors and half-adders to combine sum & carry bits together to generate the final count outputs. The critical path of the *PopCount* tree is composed of 8 stages of 3:2 compressors and 2 stages of half-adders. We use a hybrid 3:2 compressor (Figure 3) with a static CMOS mirror-gate that generates the carry signal and a transmission gate circuit that implements the 3-way XOR function required for the sum signal. The use of the transmission-gate sum circuits improves the worst-case delay of the 3:2 compressor by 12% compared to the conventional 28-transistor static-mirror compressor circuit [3].

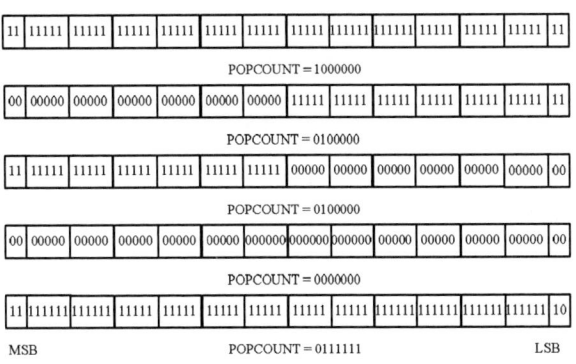

Figure 1 PopCount example

The higher compression efficiency of 3:2 circuits results in 18% shorter critical path (18gates vs. 22gates) in the proposed *PopCount* tree compared to conventional designs that use combinations of 1,2, 3-bit adders along with 4:2 counters and 6-bit adders to achieve the *PopCount* operation.

3. BitScanForward Datapath

The *BitScanForward* circuit computes the location of the 1st set bit when scanning the input from the LSB towards the MSB. Figure 4 illustrates the working of the *BSF* datapath, which takes in a 64-bit input and generates a 6-bit output that represents the bit position of the first '1'. Traditional methods to implement this instruction would split the 64-bit word into 4 bit blocks, generating a 3-bit count that also determines the lower 2 bits of the output. The 3-bit counts are then merged together in a logarithmic tree to generate the remaining output bits. This approach would require a separate datapath of counters, resulting in higher area and power.

In this work, we propose a new method to implement this instruction by reusing the counters in the *PopCount* block, to perform the count required for *BSF*. The merged datapath would require encoding the *BSF* inputs such that the number of 1's in the encoded

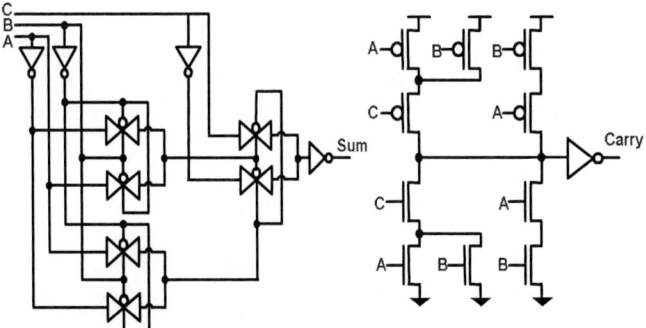

Figure 3: Hybrid 3:2 compressor circuit

output is equal to the bit position of the first set bit in the input. This encoded data can be directly sent into the *POPCOUNT* block to obtain the 6-bit result indicating the bit position of the first set bit.

BSF encoding detects the position of the 1st set bit in each 6-bit section, generating encoded outputs e[5:0]. As shown in Table I, the number of '1's in e[5:0] represents the bit-position of the 1st set bit, scanning b[5:0] from the LSB. The case when the 6 input bits are 0's, is encoded by setting all 6 output bits to 1. This 'zero-detect' condition (e[5]=1) indicates that the set bit is not in the current 6-bit section and therefore must be present in a 6-bit word of higher significance. Ten 6-bit encoders (Figure 5) and a 4-bit version of the same encoder (for the 4 LSB bits) are used to encode the 64-bit input for BSF operation. In addition to encoding the input bits, the compressors in the *PopCount* tree is modified to obtain the correct *BSF* result. When a 'set' bit is detected in a 6-bit word, the counts of all 6-bit blocks of higher significance must be zeroed out.

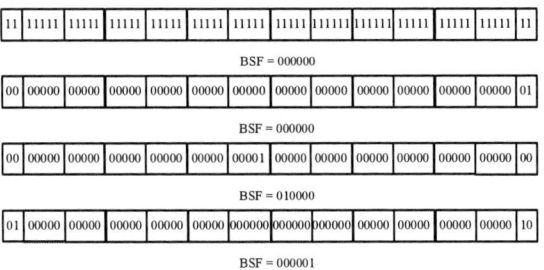

Figure 4: BSF example

The example in Figure 6 illustrates the step by step zeroing out sums/carries of higher significance during BSF encoding. The 64-bit input is first grouped into 10 blocks of 6-bit width and one 4-bit block. This result is shown in the 2nd row in Figure 6. The carry-save outputs of the 1st stage of 3:2 compressors is shown in the 3rd row. A pair of 3:2's compresses each 6-bit encoded word to 2 carry-save results. The 'zero-detect' signal e[5] in each encoded block in the 2nd row is used

978-1-4244-3039-0/08 $25.00 © 2008 IEEE

to conditionally zero-out higher-order carry-save outputs (as indicated by the arrows and the result in row 4). A Zero-Detect Tree (ZDT) is used to merge the zero-detect bits. Zero detection completes in 2 stages and occurs in parallel to the first two 3:2 compressor stages in Figure 2. Thus 3-bit and 2-bit ZDT's are used in the first stage.

Table I: BSF encoding

b[5]	b[4]	b[3]	b[2]	b[1]	b[0]	e[5]	e[4]	e[3]	e[2]	e[1]	e[0]
x	x	x	x	x	1	0	0	0	0	0	0
x	x	x	x	1	0	0	0	0	0	0	1
x	x	x	1	0	0	0	0	0	0	1	1
x	x	1	0	0	0	0	0	0	1	1	1
x	1	0	0	0	0	0	0	1	1	1	1
1	0	0	0	0	0	0	1	1	1	1	1
0	0	0	0	0	0	1	1	1	1	1	1

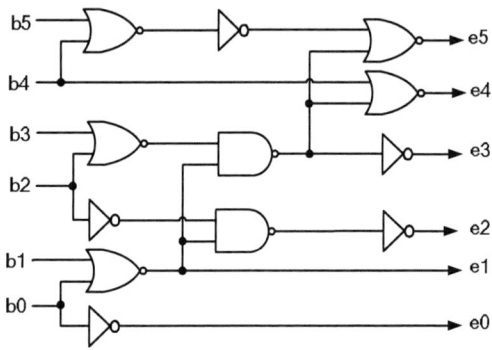

Figure 5: BSF encoder circuit

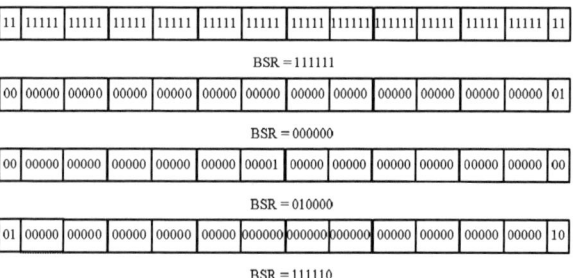

Figure 6: Example to show zeroing out for BSF

Figure 7 shows the 3-bit ZDT circuit. In the first stage of the compressor tree, 3:2 compressor outputs are zeroed out by converting the output inverters in Figure 3 to NOR gates and feeding the other input with one of the ZDT outputs (z00..z02). ZDT computation occurs in parallel to the 1st stage of 3:2 compressors,

thus minimizing performance overheads of zero-detection. A second stage of ZDT is required to extend zero-ing operation from 6, 12 and 18-bit boundaries to the entire 64-bit word. This occurs in the 2nd stage of the compressor tree. Finally stages 3 to 10 of the compressor tree add the bits that do not get zeroed out, resulting in the required result "1100", as shown in the 5th row in Figure 6.

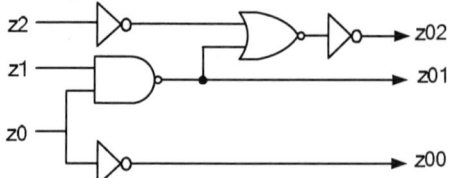

Figure 7: 3-bit Zero Detection Tree (ZDT)

4. BitScanReverse Datapath

The *BitScanReverse* circuit computes the location of the 1st set bit when scanning the input from the MSB towards the LSB. Figure 8 illustrates the working of the *BSR* instruction, which takes in a 64-bit input and generates a 6-bit output that represents the bit position of the first '1'. *BSR* encoding detects the position of the 1st set bit in each 6-bit section, generating encoded outputs e[5:0]. As shown in Table II, the number of '1's in e[5:0] represents the bit-position of the 1st set bit, scanning b[5:0] from the MSB. The case when the 6 input bits are 0's, is encoded by setting the 6 output bits to 0's. Unlike *BSF*, a distinct zero detect signal for *BSR* is generated in *BSR* encoder circuit (Figure 9).

| 11 | 11111 | 11111 | 11111 | 11111 | 11111 | 11111 | 11111 | 111111 | 111111 | 11111 | 11111 | 11111 | 11 |

BSR = 111111

| 00 | 00000 | 00000 | 00000 | 00000 | 00000 | 00000 | 00000 | 00000 | 00000 | 00000 | 00000 | 00000 | 01 |

BSR = 000000

| 00 | 00000 | 00000 | 00000 | 00000 | 00000 | 00001 | 00000 | 00000 | 00000 | 00000 | 00000 | 00000 | 00 |

BSR = 010000

| 01 | 00000 | 00000 | 00000 | 00000 | 00000 | 000000 | 000000 | 000000 | 00000 | 00000 | 00000 | 00000 | 10 |

BSR = 111110

Figure 8: BSR example

Table II: BSR encoding

b[5]	b[4]	b[3]	b[2]	b[1]	b[0]	e[5]	e[4]	e[3]	e[2]	e[1]	e[0]
0	0	0	0	0	1	0	0	0	0	0	0
0	0	0	0	1	x	0	0	0	0	0	1
0	0	0	1	x	x	0	0	0	0	1	1
0	0	1	x	x	x	0	0	0	1	1	1
0	1	x	x	x	x	0	0	1	1	1	1
1	x	x	x	x	x	0	1	1	1	1	1
0	0	0	0	0	0	0	0	0	0	0	0

In the case of *BSR* operation, a set bit in a 6-bit block would require that the outputs of all 6-bit sections of lesser significance to be set to 1. This requires the output inverter of the 3:2 compressors to be converted to a NAND gate driven by ZDT outputs.

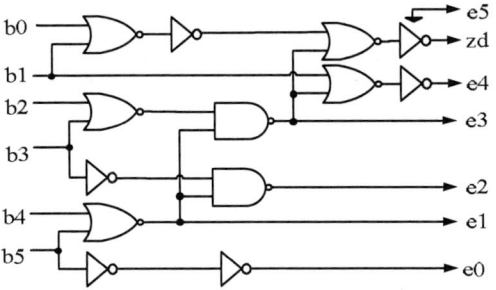

Figure 9: BSR encoder circuit

5. Merged PopCount/BSF/BSR Datapath

The similarity of *BSF* and B*SR* encoder logic leads to the use of a common circuit to achieve both operations. This is done using the merged encoder circuit (Figure 10). This circuit swaps the input bits b0..b5 using 2:1 multiplexers at the inputs of the block. Output XOR gates are used to conditionally invert the e0..e4 outputs during BSF operation. Encoder outputs are sent into the *PopCount* compressor tree during BSF/BSR operations through a 2:1 multiplexer. During *PopCount*, the 64-bit inputs bypass the encoder and directly feed the *PopCount* tree (Fig 11). The output inverter of the 1st 2 stages of 3:2 compressors in the *PopCount* tree is converted to a NOR/NAND gate that enables conditionally resetting/setting the carry and sum outputs during BSF and BSR respectively. During *PopCount* operation the zero-detect signals zdf and zdr are set to 1 and 0 respectively, thus enabling normal inverter operation.

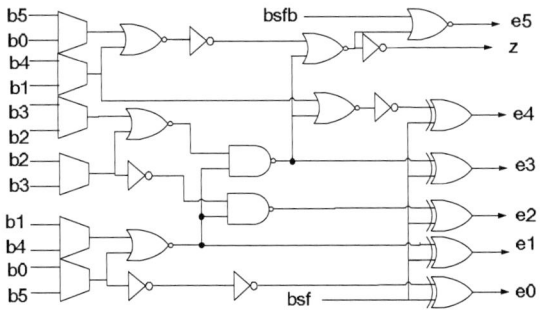

Figure 10: Merged encoder circuit

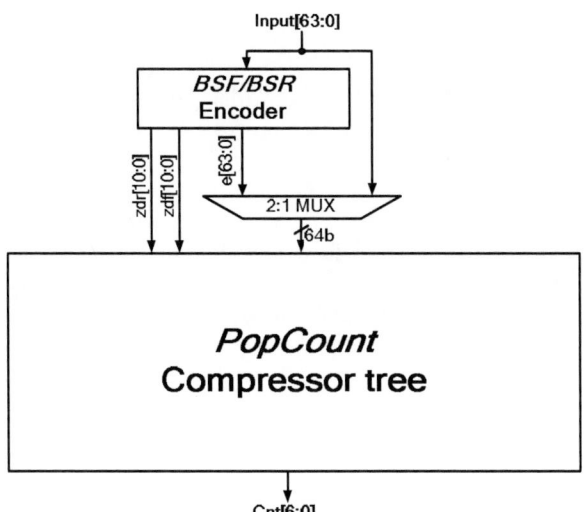

Figure 11: Merged PopCount/BSF/BSR

6. 65nm CMOS Implementation and Simulation Results

In a 1.2V, 65nm CMOS process[4], the merged *PopCount/BitScan* datapath operates at 2.1GHz, with a worst-case total power consumption of 6.25mW and a leakage component of 0.3mW, i.e. 5% of total power at 1.2V, 110°C. Compared to a conventional 65nm microprocessor execution core implementation using separate datapaths for *PopCount, BS*F and *BSR*, the proposed merged datapath achieves 26% lower total energy consumption, 20% lower area while consuming 33% lower leakage energy. At iso-energy consumption of 3.5pJ, the merged design offers 26% higher performance compared to the conventional design.

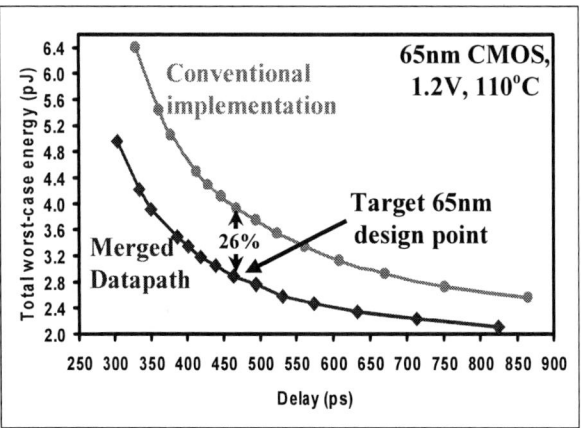

Figure 12: Simulation Results: Energy-delay comparison of PopCount/BSF/BSR datapaths

277

978-1-4244-3039-0/08 $25.00 © 2008 IEEE

7. Conclusions

A unified *PopCount/BSF/BSR* datapath targeted for 64-bit high-performance microprocessor execution cores is implemented in a 65nm CMOS technology. A novel encoding scheme on the 64-bit inputs allows reuse of the *PopCount* compressor tree during *BSF* and *BSR* operations, resulting in a merged implementation that enables 20% lower area and 26% lower energy consumption compared to a conventional separate datapath implementation, while achieving single-cycle throughput and latency at a cycle-time of 2.1GHz. At iso-energy, the unified datapath offers 26% higher performance than the separate datapath implementation.

8. References

[1] R. Ramanathan, "Extending the World's Most Popular Processor Architecture", White Paper, Intel Corporation.
http://download.intel.com/technology/architecture/new-instructions-paper.pdf

[2] E. El-Qawasmeh, Int'l Journal of Info. Tech., vol. 9, 2003, pp1-18

[3] S. Hsu, S. Mathew, M. Anders, B. Zeydel, V. Oklobdzija, R. Krishnamurthy, S. Borkar, "A 110 GOPS/W 16-bit multiplier and reconfigurable PLA loop in 90-nm CMOS", IEEE Journal of Solid-State Circuits, Vol. 41, Issue 1, Jan. 2006, Page(s):256 - 264.

[4] P. Bai et al., A 65nm logic technology featuring 35nm gate lengths, enhanced channel strain, 8 Cu interconnect layers, low-k ILD and 0.57 /spl mu/m/sup 2/ SRAM cell, IEEE International Electron Devices Meeting, 13-15 Dec. 2004, Page(s): 657- 660.

21st International Conference on VLSI Design

Dynamic Error Detection for Dependable Cache Coherency in Multicore Architectures

Hui Wang, Sandeep Baldawa, and Rama Sangireddy

Department of Electrical Engineering
University of Texas at Dallas, Richardson, TX 75080, USA

Abstract

In Chip Multiprocessor (CMP) systems the various effects of technology scaling make the on chip components more susceptible to faults. Most of the earlier schemes that address fault tolerance issues in CMPs adopt redundant-thread techniques. These techniques are mostly effective, except that they fail to detect errors resulting from faults in hardware components on chip that commonly serve multiple cores. The cache coherence controller (CC) logic, which ensures consistency of data shared among multiple threads, is a vital common component in CMPs. A fault in CC logic of any of the processors may lead to errors in the data states in the entire CMP system. It is observed that up to 59.6% of the memory references cause a change in cache state for SPLASH-2 applications. We propose a novel scheme with a verification logic that can dynamically detect errors in the CC logic of multiple cores in a CMP system. The entire verification logic is designed with a negligible area of 0.1372 sq.mm using a TSMC 0.18μ 4-metal layer process technology. Even at highly aggressive fault injection rates, the logic achieves an average error coverage of more than 95% (and almost 100% for some applications)[1].

1. Introduction

The phenomenal growth in transistor budget on chip has ushered newer processor design innovations like chip multiprocessors (CMP) which are one of the most promising and easiest to implement while still offering excellent performance [1]. The industry is adopting CMPs to utilize the high transistor budget resulting from the technology scaling. The technology scaling trends lead to smaller and faster transistors and lower supply voltages. Such technology has helped to increase chip transistor counts and improve performance. However, the low supply voltages also cause noise margins to become narrower. The result is an increased susceptibility to various factors that can affect the value of a transistor. A temporary change of transistor value due to any of those factors is known as a transient fault. The frequency of transient faults is increasing dramatically leading to an increase in number of soft errors that result in the degradation of reliability. Shivakumar *et al.* [2] predicted that the soft error rates of logic circuits will increase nine orders of magnitude from 1992 to 2011. In other words, for every one fault occurring in 1992 there will possibly be one billion faults in 2011. Further, microelectronics industry already has been impacted by the soft errors due to cosmic rays [3]. Hence, fault tolerant systems that are highly robust are imperative to cope with the increasing amount of susceptibility of CMP systems.

Most of the earlier fault tolerance schemes in CMPs adopt spatial redundancy techniques, wherein multiple copies of a

---

[1] The authors can be contacted at rama.sangireddy@utdallas.edu

thread are run on different cores and the results are compared before their finalization [4, 5, 6, 7, 8, 9, 10]. These techniques are mostly effective, except that they cannot detect errors resulting from faults in hardware components on chip that commonly serve multiple cores. Though these schemes can mostly assume that the commonly serving cache and memory components are protected by ECC and other techniques, it cannot be taken for granted that a combinational logic unit that commonly serves multiple cores is not error prone. The cache coherence controller (CC) logic running the coherence protocol is a critical component that ensures consistency of data shared among multiple threads. A fault in CC logic of any of the processors may lead to errors in the data states in the entire CMP system. For example, during a state transition in a cache the next state can be erroneously computed as Shared (*S*) instead of the correct state of Modified (*M*) due to a fault in either the inputs provided or the coherence computation logic. In such a case, the invalidation broadcast that needs to be sent to other copies in the system will not take place and leads to a greater error in the entire system computation. Such faults must be detected to avoid the system failure. Further, our preliminary analysis observes that up to 59.6% of the memory references cause a change in cache state for SPLASH-2 applications in a 4-core CMP system. This implies a considerable size of Architecture Vulnerability Factor (AVF), where AVF is defined as the fraction of faults that become errors [11]. Hence, the robustness of the CC logic becomes a critical factor for dependable computing in a CMP system.

In this paper, we propose a novel scheme with a verification logic that can dynamically detect errors in CC logic of multiple cores in a CMP system at run time. Our base CMP model includes a bus based MESI protocol to ensure the coherency in caches with memory. We design a *verification logic circuit* to detect faults in the entire coherency logic. The verification logic effectively detects any faults in the cache coherence logic under various scenarios, and ensures at run-time the correctness of the various cache state change transactions. The verification logic is evaluated with varied number of checker table entries and at various fault injection rates. The results show that even at highly aggressive fault rates, the verification logic detects more than 95% errors (and almost 100% of errors for some applications). Further, in the proposed scheme the verification logic is scalable with the number of cores.

2. Related Work

For most fault tolerant techniques in CMPs, the key theme has been the the detection of transient faults using thread redundancy. The basis for most of the recently proposed fault tolerance techniques is the Active-Stream/Redundant Stream Simultaneous Multithreading (AR-SMT) [4] technique. In AR-SMT, two redundant copies of the same program run con-

---

279

978-1-4244-3039-0/08 $25.00 © 2008 IEEE

currently on two threads. The active stream (A-stream) runs ahead and the results of each instruction are pushed onto a delay buffer. The redundant instruction stream (R-stream), lags the A-stream by a value not more than the length of the delay buffer. Once instructions is committed from the R-stream, the results are compared with the results in the delay buffer. A fault is detected if there is a difference in the comparison results. Many other fault detection and recovery schemes were proposed later for CMPs and Multiprocessors such as [4, 9, 10, 13, 14, 15] based on the above approach. All of the schemes detect a fault by comparing two threads and recover from a fault by rolling back and re-executing the thread.

In the Chip-level Redundantly Threaded multiprocessor (CRT) scheme [9], proposed by Mukherjee *et al.*, the instructions from the leading thread are committed without checking, except store instructions which are required to be checked for accurate fault detection. Subsequently, Gomaa *et al.* proposed Chip level Redundantly Threaded multiprocessor with Recovery (CRTR) [10], which performs in similar lines to the CRT. Apart from communicating the load and store values, the CRTR sends the register values for fault recovery. Sundaramoorthy *et al.* proposed Slipstream processors [12], that run a shorter but otherwise equivalent version of the original program by removing ineffectual computation and computation related to highly-predictable control flow. The partial-redundancy between programs is leveraged for detecting and recovering from transient hardware faults. Mario Dal Cin *et al.* proposed measures for the hardware fault tolerance in massively parallel multiprocessors [8]. They presented three techniques: self checking, central checking mechanism, and the distributed system-wide checking by mutual tests between nodes. They mention checkpointing as a recovery scheme which helps in resuming from failure by a temporarily failed node. However, the work did not provide results that support the proposed scheme. Sorin *et al.* have proposed dynamic verification of end-to-end, system-wide invariants in shared memory multiprocessors (SMP) [7]. Their scheme proposed two invariant checkers to verify that all nodes in an SMP observe the same total order of broadcast requests.

Banatre *et al.* have proposed a Recoverable Shared Memory (RSM) scheme [16], which provides a hardware supported backward error recovery mechanism which minimizes the propagation of recovery when a processor node fails in shared memory multiprocessor systems. Majzik *et al.* [18] have proposed a watchdog processor (WDP) for checking the faults in the control flow of an application. WDP is a small and simple coprocessor that detects errors by monitoring the behavior of a system. It is shown that a large number of transient faults can be detected by monitoring the control flow and control flow behavior. Morin *et al.* [17] have shown that cache only memory architectures (COMA) are good candidates for building fault-tolerant scalable shared memory multiprocessors (SS-MMs). Cantin *et al.* have proposed a method for improving the fault-tolerance of cache coherent multiprocessors [19]. Sorin *et al.* developed an availability solution called SafetyNet [20], that uses unified, lightweight checkpoint/recovery mechanism to support multiple long latency fault detection schemes, such as the ECC, redundant processors and ALUs, redundant threads and system level state checkers.

Numerous techniques have been proposed in the past, such as [21, 22, 23], for protecting the data in the cache from soft errors. An elaborate discussion on these schemes is out of scope for this paper, as the proposed work in this paper focuses on making the cache coherency controller logic for the entire chip in CMPs tolerant from soft errors.

The salient features of the proposed work in this paper, distinct from the earlier related work, are:

- The proposed scheme specifically deals with the dependability of the cache coherency logic as a whole in CMP systems, as against protecting a localized/private hardware component on the chip. A fault in a private component in a processor can lead to a localized error and can be easily detected/corrected. However, a fault in a global/common logic such as a cache coherence logic can lead to an error that is highly infectious for a major portion of the chip.

- The proposed technique is independent and is run concurrently with the conventional cache coherence protocols, and will lead to any impact on the performance of neither the coherency protocol nor the processor system itself.

- The proposed fault tolerant coherency mechanism for the CMP systems is independent and can be combined with other thread-redundant fault tolerant schemes for a greater dependability of a CMP system.

3. Proposed Dependable CC Logic in CMPs

3.1 System model

In our base CMP model, each processor core has private L1 instruction and data caches, and a unified shared L2 cache. In order to maintain the data coherency among the L1 and L2 caches, we consider MESI, a bus based cache coherence protocol. However, it is essential to note that the proposed mechanism for dependable cache coherency is scalable with number of on-chip cores, and can also be adapted to other cache coherence protocols. The CC logic determines the next state of a cache line based on input parameters of current state, whether the line is accessed by CPU or bus, if the request is read or write, and if the access is a hit or a miss. Figure 1 shows the organization of coherence controller for a two-way set-associative cache. In our base model, the coherence controller logic consists of a hardware representing a state machine of the MESI protocol, where the four states are: *Invalid* (I), *Modified* (M), *Exclusive* (E) and *Shared* (S).

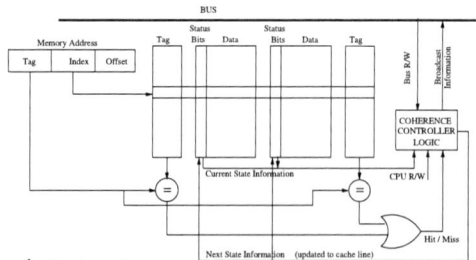

Figure 1: Cache coherence controller logic. The logic determines the next state of cache line based on input information such as the current state, whether read or write request, whether hit or miss, and if request is from CPU or bus.

3.2 Fault Model

From the organization of the CC logic, it can be observed that an error in the logic can be potentially propagated to the other portions on the chip. A fault in a coherence controller logic can be of any of the following types: (i) a faulty current state information input; (ii) a fault in the identification of access as hit in case of a miss or vice versa; (iii) a faulty designation of the request as from CPU instead of bus or vice versa, or as read instead of write or vice versa; (iv) a fault in the computation logic of the coherence controller. We discuss these fault scenarios in detail as below:

- **Fault in current state information:** The current state of a cache line can be one of the M, E, S, and I. Some of the possible scenarios if a faulty current state information is received by the coherence logic are:

  - If the current state information is erroneously received as S instead of M, and if the read request is from bus, then there will be no error in the next state computation as the next state in any case will be S only. Thus, such a fault will not cause an error even if not detected.

  - However, if the current state information is erroneously received as S instead of M, and if the read request is from the CPU, the next state will be computed as S instead of the correct state of M. In such a case, the invalidation broadcast that needs to be sent to other copies in the system will not take place and leads to a greater error in the entire system computation. Such faults must be detected to avoid the system failure.

  - If the current state information is erroneously received as I instead of S, and the request is a CPU read, a read miss request will be placed on the bus to obtain the valid data from another cache causing some performance penalty, unlike when a correct current state information is obtained as S no action is placed on bus. However, such an error may not lead to a system failure, as other copies of the data in the system still reflect S state.

- **Fault in CPU/Bus and R/W information:** Consider that the current state of a cache line is S, and the cache block access is a hit. If the access is determined as write instead of read, the next state will be computed as M instead of S, and an invalidation request will be broadcast to all the copies in the system. This may not lead to a system failure as those invalid blocks in other caches will be treated as misses and the valid data will be fetched again from the appropriate location. However, such a fault will lead to increased memory hierarchy activity in the system and will degrade the system's performance.

- **Fault in coherence computing logic:** A fault in the computation logic of the coherence controller can lead to an error in the determination of the next state of the cache block and may lead to a system failure, or a localized error that may get overwritten with consequent system actions.

It is required to note that not all faults occurring in the logic lead to errors. This is since most of the times when a fault occurs at a particular location, the surrounding logic may not be actively participating in any coherence transactions. On the other hand, faults that lead to various kinds of errors in the CMP coherence logic can be categorized as: faults that lead to perfectly harmless errors (such as leading to same next state even if fault in the logic); or faults that only cause

temporary errors (such as next state as I instead of S) that will be eliminated by natural system action of reading the data again from the lower memory levels and so only cause slight performance degradation; or faults that cause errors that will lead to a potential system failure (such as next state determination of S instead of M). However, at run time it is necessary to detect all errors as it is impossible to dynamically determine the long run impact of the fault and its manifested error. Our proposed scheme is robust to detect any kinds of errors without distinction.

3.3 Proposed Scheme

In order to detect errors in the entire cache coherency logic on chip in a CMP system, we propose a novel and cost effective mechanism of verification logic that is scalable with number of on-chip processors. The verification logic is a centralized structure on the chip, as shown in Figure 2, and it receives inputs from the coherence controllers of all the individual processors. Figure 3 shows the implementation of the verification logic in detail. It consists of:

- A *checker table* that stores the current status of all copies of a recent accessed data item in the system

  which is used to verify the new status changes;

- A *transaction buffer* that holds all the required information for recent state change transactions until verified by the checker table;

- A *coherence computing unit* that redundantly computes the next state for verification.

The verification logic checks the system's actions of data coherency at both local and global levels. For every state change that takes place at a cache line, it verifies the protocol locally for each of the processors. Also for a state change trigger, it verifies the combinations of states between all the cores at a global level. This verification at two levels ensures a greater error coverage. Besides, when number of on chip cores increases, we can simply scale our scheme by inserting corresponding columns in the buffer and table structures.

3.3.1 Local Verification

Whenever a read or write request is made for a cache line, the coherence controller logic (CC) computes the next state of the cache line using the information of current state, and CPU/Bus read/write hit/miss actions. In the proposed scheme, upon computing the next state the CC logic sends the transaction consisting of the address, current state, next state, and all the other input signals to the verification logic. The verification logic stores the inputs from the CC in the *transaction buffer*. Each entry for a transaction address in the transaction buffer

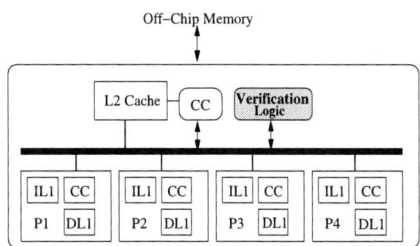

Figure 2: Integration of the verification logic in a CMP system

978-1-4244-3039-0/08 $25.00 © 2008 IEEE

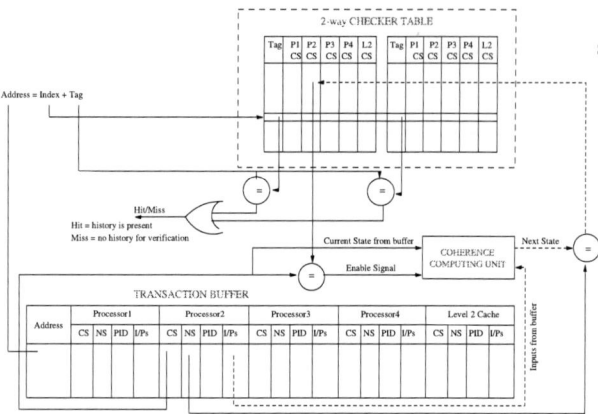

Figure 3: Detailed Verification Logic Implementation

holds the current state (CS), next state (NS), processor ID (PID), and other inputs for each of the L1 caches and the L2. Here, we define a transaction as a complete state change actions at all caches for an initiated at a cache for a memory address. For a new transaction with a unique address that is initiated by a state change at one of the caches a new entry is made in the buffer, and all immediate state changes at other caches are registered in the same entry at appropriate fields. Thus, the buffer holds the transactions of all the CC units corresponding to all the L1 caches and L2 (for the base model) in a FIFO fashion. Another new transaction initiated at the same cache for the same address will automatically be made as a new entry if it finds an already existing entry in the buffer for earlier transaction. An entry at the top of the buffer is taken for verification, and the entry is deleted upon completion of the verification. As the entries in the transaction buffer are verified at run time in parallel with the cache state transactions, our simulation studies have shown the requirement of at most eight transaction buffer entries. Although we have never encountered the problem of buffer overflow during our extensive evaluation, we have designed the buffer to just drop the verification request, for design simplicity, in case the buffer is full.

The checker table is a 2-way set associative lookup table that contains the most recent state information of the cache line. Each entry in the checker table stores the tag of the transaction address and current state information of the corresponding cache line in each of the L1 caches and L2. The checker table is looked up with the index portion of the address, and the tag obtained is matched with that from the buffer entry. A match of the address tag indicates the existence of a prior recording of the state information for that particular address and further verification is continued. However, the tag mismatch indicates the absence of state information for the address, and the checker table entry is updated with state information by replacing an existing entry for the conflicting address in the 2-way table using least recently used (LRU) algorithm. As the number of checker table entries implies the amount of state information stored, our simulation study has conducted a detailed evaluation of the impact on the error detection capabilities for a varied number of checker table entries.

When the tag for a lookup matches, two steps of comparison can be made to detect errors.

- Current state (CS) information in the table for the particular address is compared against the CS information stored in the buffer. If there is a mismatch, an error is detected.

- If there is a match is CS comparison, the *coherence computing unit* recomputes the next state (NS) based on the transaction information in the buffer. The computed NS is compared against the NS in the buffer. If the result mismatches, an error is detected. Otherwise, an enable signal is sent to checker table which updates the new next state value into the current state field. Thus, the updated state information will become the current state for verification of next transaction for the address.

when any error is detected in above two steps, a recovery using check-pointing or other relevant mechanisms can be triggered which is out of the scope of this paper. These two steps are repeated for all the fields (four processors and L2 for base model) in an entry of the buffer corresponding to a transaction address. This will ensure that all the processors have completed the local checking process.

3.3.2 Global Verification

Due to the underlying redundant computing mechanism, local checking sometimes might fail in error detection due to the same inputs which may already be faulty. The global verification is proposed to reinforce local checking by double checking if any illegal combinations of states exist for a cache line among multiple cores. Table 1 shows the valid state combinations among cores and L2 cache that should exist for an address after a transaction.

Processor 1 State	Processor 2 State				L2 Cache State			
	M	E	S	I	M	E	S	I
M	X	X	X	V	X	X	X	V
E	X	X	X	V	X	V	X	X
S	X	X	V	V	X	X	V	X
I	V	V	V	V	V	V	V	V

Table 1: Example valid state combinations for MESI protocols (X: incorrect combination; V: valid combination).

The next states at all the caches corresponding to an address are sent as inputs to the global checking unit, which verifies for any variation in the state combinations. For example, if a cache line at processor 1 is in M state, all the other copies in the system should reflect an I state. Also, if a cache line at processor 1 is in S state, the corresponding cache lines at processor 2 and L2 cache can only be in a S state. And, if a cache line at processor 1 is in I state, the corresponding cache lines at processor 2 and L2 cache can be in combination of M and I or E and E, respectively. The global checking unit signals an error if it finds any deviation from allowed combinations.

4. Evaluation Study

4.1 Experimental Setup

For our study, we use CMP-SIM tool [24] which is a multicore micro architectural simulation environment with a detailed cycle-accurate model. CMP-SIM extends the Simplescalar [25] out-of-order simulator to model a CMP environment. We ran simulations for a CMP system with four cores. Each core has its own private L1 data and instruction caches where as the L2 cache is shared among the cores. Table 2 shows the configuration parameters used for each core except for the L2 cache and memory parameters. We have evaluated the proposed scheme with SPLASH-2 [26] suite of parallel applications. The benchmarks are run with input data sets as listed in Table 3. The applications are run until completion to obtain the results.

Parameter	Value
L1 Instruction cache	16KB, 2-way, 32B line 1 cycle latency
L1 Data cache	16KB, 4-way, 32B line 1 cycle latency
Shared L2 Unified cache	1MB, 8-way, 64B line 11 cycle latency
Processor Pipeline	4-wide o-o-o issue
Instruction window	128 entries
Load/store queue	64 entries
Functional units	4 Int ALU, 2 Int Mult 4 FP ALU, 2 FP Mult
Memory	400'4 cycles, 8B

Table 2: Processor configuration parameters.

Benchmark	Input Parameter
BARNES	16K particles
FFT	256K integers
FMM	16K particles
RADIX	1024K keys
OCEAN	258X258
LU-Cont	512x512 matrix
LU-NonCont	512x512 matrix

Table 3: Benchmarks and Input Data Sets.

4.2 Results and Analysis

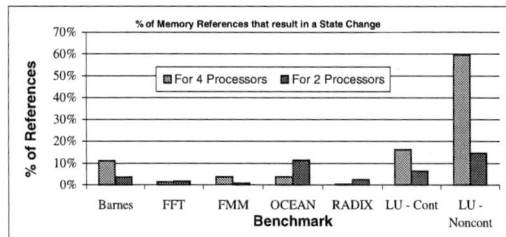

Figure 4: Memory References that result in a state change

The fault detection scheme runs in parallel and without any modification in the underlying cache coherence mechanism, and hence the proposed verification logic implementation will not impact the base system's performance. In order to determine how often the verification logic has to verify a coherence transaction, the percentage of memory references (loads and stores) that result in a state change is examined (Results shown in Fig 4). This can be in one way construed as the basis for the measurement of the AVF. In other words, if a large portion of memory references lead to state changing transaction in a CMP, a fault in a single coherence control logic unit can lead to a potential global error. From Figure 4, the decrease in the percentage of memory references in case of a 2-processor CMP system can be attributed to reduced memory traffic resulting from lesser coherence misses and other related aspects. Thus, we can infer that as the number of processors on a chip increase, the cache coherency transactions increase thus demanding a robust fault tolerant mechanism with highest possible error coverage. We define the error coverage to be the percentage of errors detected using the verification logic.

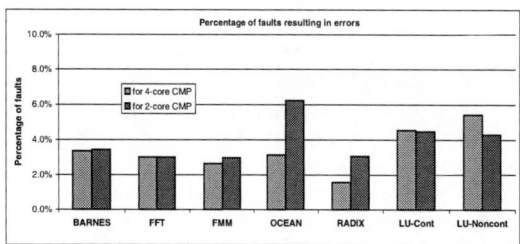

Figure 5: Percentage of faults that lead to errors.

In our simulation set up, we inject fault randomly in any part of the coherence logic on the chip at regular processor clock intervals of 100 cycles, 1000 cycles, 10,000 cycles, and 100,000 cycles in order to conduct the evaluation at varied levels of soft error rates. The smaller the time interval, the higher is the number of faults injected in a given time period. Hence, we restrict our fault injection interval to 100,000 cycles to show the performance evaluation of the proposed at aggressive error rates. Our analysis showed that, in the base 4-processor CMP system up to 5.7% of the injected faults (as shown in Figure 5 at an injection interval of 10,000 cycles) lead to errors. This pattern is observed to be similar across all fault injection intervals, and is also in line with the pattern of memory references leading to state changes as shown in Figure 4. As discussed in the fault model, some of the faults do not lead to errors, or the errors manifested may not lead to the failure of the system. However, some faults will be propagated as errors and infect the entire chip leading to a system failure, which have to be detected without fail.

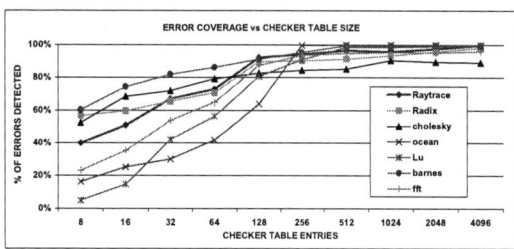

Figure 6: Error coverage with fault injection interval of 1000 cycles

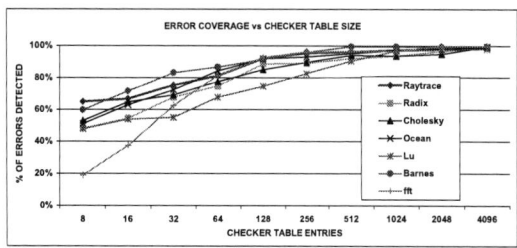

Figure 7: Error coverage with fault injection interval of 10,000 cycles

Figure 6 shows for different benchmarks the percentage of errors detected by the verification logic with varied number of checker table entries, when the faults are injected at an interval of 1000 cycles. The error coverage increases almost linearly with increase in number of entries in the checker table till 256 entries. As expected, as the fault injection interval increases the number of errors decreases and so lesser number

283

978-1-4244-3039-0/08 $25.00 © 2008 IEEE

of checker table entries will be able to detect a significant portion of the errors. A checker table with 128 entries detects a significant portion of errors for most applications, and almost 100% of errors are covered for most applications with 256 entries. Thus, the simulations suggest that a 128-256 entry checker table is sufficient to cover majority of errors even for a highly aggressive fault injection interval of 1000 cycles. Similar observations are made when the faults are injected at an interval of 100 cycles.

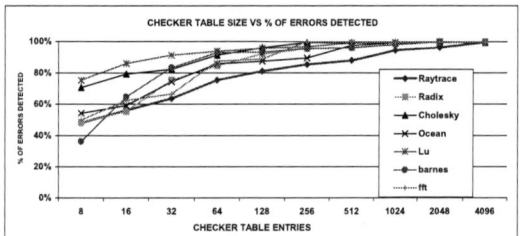

Figure 8: Error coverage with increase in checker table size and fault injection interval of 100,000 cycles

Figure 7 shows the percentage of errors detected when the faults are injected at an interval of 10,000 cycles. The error coverage performance behavior of the verification logic almost saturates at 256 entries of checker table. Similarly, Figure 8 shows the error coverage when the faults are injected at an interval of 100,000 cycles.

In order to estimate the area overhead of the verification logic, we have done a hardware implementation. A commonly used hardware size estimate is the number of 2-input NAND (4 transistors) equivalent gates. We have implemented the entire verification logic in Verilog, synthesized it to gates, and divided the total transistor count by 4 to obtain 2-input NAND equivalent gate count. The verification logic implemented included a checker table of 512 entries, a transaction buffer of eight entries, and the coherence controller logic. We have implemented each of the block and obtained the 2-input NAND equivalent gate count. The checker table is implemented in the form of an SRAM, where SRAM design is done by the SRAM compiler which gives an area estimate. The input buffer logic, the coherence controller, and the comparators in the verification logic altogether consume a 2-input NAND gate equivalent of 4250 gates. The entire verification logic consumes an area of 0.1372 sq.mm using a TSMC 0.18μ 4-metal layer process technology, which is negligible when compared to the total CMP system chip area.

5. Conclusions

The CC logic which is a common component in CMPs is vital to the dependability of the entire system. Our preliminary analysis observed that up to 59.6% of the memory references cause a change in cache state for SPLASH-2 applications. In this paper, we proposed a novel, cost-effective, and scalable technique for dependable data coherency in CMP systems. The proposed verification logic can dynamically detect errors in CC logic of multiple cores in a CMP system. The entire cache consistency verification logic was designed with a negligible area of 0.1372 sq.mm using a TSMC 0.18μ 4-metal layer process technology. Even at highly aggressive fault injection rates, the verification logic was shown to achieve an average error coverage of more than 95%.

References

[1] L. Hammond, B. A. Nayfeh, and K. Olukotam, "A Single-Chip Multiprocessor", *IEEE Computer Society*, Sept. 1997, pp. 79-85.

[2] P. Shivakumar, M. Kistler, S. W. Keckler, D. Burger, and L. Alvisi, "Modeling the effect of technology trends on the soft error rate of combinational logic", *Proc. the 2002 Int. Conf. on Dependable Systems and Networks*, June 2002, pp. 389 - 398.

[3] R. Baumann, "Soft Errors in Commercial Semiconductor Technology: Overview and Scaling Trends", *Proc. IEEEReliability Physics Tutorial Notes, Reliability Fundamentals*, April 2002, pp. 121.01.1-121.01.14.

[4] Eric Rotenberg, "AR-SMT: A Microarchitectural Approach to Fault Tolerance in Microprocessors", *Proc. of Fault Tolerant Computing Systems*, 1999.

[5] V. Sieh, A. Pataricza, B. Sallay, W. Hohl, J. Honig and B. Benyo, "Fault Injection Based Validation of Fault-Tolerant Multiprocessors", *Proc. of the eighth Symposium on Microcomputer and Microprocessor Applications*, Vol. 1, 1994.

[6] H. K. Ku and J. P. Hayes, "Systematic Design of Fault-Tolerant Multiprocessors with Shared Buses", *IEEE Trans. on Computers*, Vol. 46, No. 4, April 1997.

[7] D. J. Sorin, M. D. Hill, and D. A. Wood, "Dynamic Verification of End-End Multiprocessor Invariants", *Proc. Int. Conf. on Dependable Systems and Networks*, June 2003, pp. 281-290.

[8] M. D. Cin, W. Hohl, and V. Sieh, "Hardware-Supported Fault Tolerance for Multiprocessors", *Proc. Architecture of Computing Systems*, ARCS'97, pp. 13-22.

[9] S. S. Mukherjee, M. Kontz and S. K. Reinhardt, "Detailed design and evaluation of redundant multithreading alternatives", *Proc. 29th Intl. Symp. on Computer Architecture*, May 2002.

[10] M. Gomaa, C. Scarbrough, T. N. Vijaykumar and I. Pomeranz, "Transient-fault recovery for Chip Multiprocessors", *Proc. 30th Int. Symposium on Computer Architecture*, June 2003, pp.98-109.

[11] S. S. Mukherjee, C. Weaver, J. Emer, S. K. Reinhardt, and T. Austin, "A Systematic Methodology to Compute the Architectural Vulnerability Factors for a High-Performance Microprocessor",*Proc. International Symposium on Microarchitecture*, MICRO-36, 2003, pp. 29-40.

[12] K. Sundaramoorthy, Z. Purser, and E. Rotenberg, "Slipstream Processors: improving both performance and fault tolerance", *Proc. 9th Int. Sym. on Architectural Support for Programming Languages and Operating Systems*, Nov. 2000, pp. 257-268.

[13] S. K. Reinhardt and S. S. Mukherjee, "Transient-fault detection via simultaneous multithreading", *Proc. of the 27th Annual International Symposium on Computer Architecture*, June 2000, pp. 25-36.

[14] T. N. Vijaykumar, I. Pomeranz and K. Cheng, "Transient-fault recovery using simultaneous multithreading", *Proc. 29th Intl. Symposium on Computer Architecture*, May 2002, pp. 87-98.

[15] Wenbin Yao, Dongsheng Wang and Weimin Zheng, "A Fault-Tolerant Single-Chip Multiprocessor", *Asia-Pacific Computer Systems Architecture Conference*, Sep. 2004, pp. 137-145.

[16] M. Banatre, A. Gefflaut, P. Joubert, C. Morin and P. A. Lee, "An Architecture for Tolerating Processor Failures in Shared Memory Multiprocessors", *IEEE Transactions on Computers*, Vol. 45, Oct 1996, pp. 1101-1115.

[17] C. Morin, A. Gefflaut, M. Banatre and A. Kermarrec, "COMA: an Opportunity for Building Fault-tolerant Scalable Shared-Memory Multiprocessors", *Proc. of 23rd Annual Intl. Symp. on Computer Architecture*, 1996, pp. 56-65.

[18] I. Majzik, W. Hohl, A. Pataricza and V. Sieh, "Multiprocessor Checking Using Watchdog Processors", *Int. Journal of Computer Systems Science and Engineering*, Vol. 5, 1996, pp. 301-310.

[19] J. F. Cantin, M. H. Lipasti and J. E. Smith, "Dynamic Verification of Cache Coherence Protocols", *Workshop on Memory Performance Issues*, June 2001.

[20] D. J. Sorin, M. K. Martin, M. D. Hill and D. A. Wood, "SafetyNet: Improving the Availability of Shared Memory Multiprocessors with Global Checkpoint and Recovery", in *Proc. 29th Intl. Symp. on Computer Architecture*, May 2002.

[21] S. Kim and A. K. Somani, "Area Efficient Architectures for Information Integrity in Cache Memories", in *Proc. International Symposium on Computer Architecture*, May 1999, pp. 246-255.

[22] W. Zhang, S. Gurumurthi, M. Kandemir, and A. Sivasubramaniam, "ICR: In-Cache Replication for Enhancing Data Cache Reliability", in *Proc. International Conference on Dependable Systems and Networks (DSN)*, June 2003, pp. 291-300.

[23] V. Sridharan, H. Asadi, M.B. Tahoori, D. Kaeli, "reducing data cache suceptibility to soft errors", *IEEE Transactions on Dependable and Secure Computing*, October-December 2006, pp. 353-364.

[24] S. Baldawa and R. Sangireddy, "CMP-SIM: An Environment for Simulating Chip Multiprocessor Architectures", University of Texas at Dallas, http://www.utdallas.edu/~rama.sangireddy/CMP-SIM, October 2006.

[25] D. Burger and T. M. Austin, "The SimpleScalar Tool Set, Version 2.0", Technical Report No. 1342, C.S Dept., Uniersity of Wisconsin-Madison, June 1997.

[26] S. C. Woo, M. Ohara, E. Torrie, J. P. Singh, and A. Gupta, "The SPLASH-2 Programs: Chracterization and Methodological Considerations", *Proc. of the 22nd Annual International Symposium on Computer Architecture*, June 1995, pp. 24-36.

SESSION D2:
Analog

21st International Conference on VLSI Design

Mismatch Aware Analog Performance Macromodeling using Spline Center and Range Regression on Adaptive Samples *

Shubhankar Basu, Balaji Kommineni and Ranga Vemuri
Electrical and Computer Engineering,
University of Cincinnati, Cincinnati, OH 45221
{basusr,komminb,ranga}@ece.uc.edu

Abstract

Analog design traditionally relies on designer's knowledge and expertise. Numerous automated synthesis methods have been proposed over the years; they reduce time complexity and explore wider design space. Manufacturing induced defects in the process parameters, render device characteristics inconsistent with their prediced behavior. Device mismatch casues significant variation in analog circuit performance. Monte-carlo simulation is known to be the most accurate method of measuring performance under random variation. But monte-carlo simulation is prohivitively expensive during synthesis process. In this work we present a novel Spline Center and Range Regression (SCRR) technique on adaptive samples to model performance in the presence of process variation. Mismatch aware macromodels can provide considerable speedup during synthesis with minimal loss in accuracy. Experimental results demonstrate the accuracy of the macromodels on an independent validation set using 180nm and 65nm technologies.

1 Introduction

Conventional analog design had suffered from longer design times and overdesigned circuits. Analog design automation had tried to address both these issues through extensive research [4, 6, 9, 1]. This had led to the inclusion of synthesis in the analog design flow [2].

To measure randomly varying performance in the presence of mismatches due to process variation (dopant concentration, T_{ox}, V_{th_0} etc), monte-carlo simualtion is needed to replace nominal simulation. However monte-carlo simulation is even more expensive when used in a synthesis loop. Consequently an alternative to multiple simulations have been proposed in the form of fast evaluating performance macromodels. Most of the reported works in performance macromodeling however do not consider the impact of process variation on circuit performance. This makes the models insufficient for use in variation tolerant synthesis.

Varying data may be represented as intervals [13] as long as they are continous in the bounds. The use of in-

---

*This research is supported in part by the National Science Foundation under grant numbers CNS-0421092 and CCF-0429717

terval data types for computation and optimization however exhibit limitations like expansion in width of results and non-monotonic intervals. Neto et al. [8] proposed an alternative to perform regression modeling on interval valued data using the center and the range information. However the authors' use of linear regression technique fail to capture the non-linearity of performance data for analog circuits accurately.

In this work we present a novel methodology to build mismatch aware performance macromodels using Spline Center and Range Regression (SCRR) techniques on adaptively sampled design space. SCRR makes use of pseudo-cubic splines on center and range of interval valued data. The SCRR toolbox can be used across multiple designs and technology nodes. Experiments performed on two benchmark circuits using 180nm and 65nm predictive technology model demonstrate the effectiveness of the models generated. They correlate above 90% with the HSpice montecarlo simulation results. The macromodels also provide around 5000X reduction in computation time over spice monte-carlo simulations.

The remainder of the paper is organized as follows. Section 2 illustrate the motivation behind our work. In Section 3 we present the theory and algorithm used in SCRR toolbox. In Section 4 we describe the experimental setup and present the results. Finally Section 5 presents the conclusions and future work.

2 Motivation

Several researchers [1, 5, 10, 12] have proposed performance macromodeling techniques using neural networks, support vector machines, spline etc. However none of these works consider the effect of process variation causing deviation in device model parameters, in their performance macromodeling techniques. In [14], the authors present a variation tolerant synthesis methodology using geometric programming. The overhead of manual posynomial model creation is tackled in the work using a combination of implicit power iteration and rank-one projection algorithms. While the technique seems promising, the use of static sampling in the work can lead to unnecessary sample points during modeling, while failing to capture the necessary non-linear regions of the sampled space. Also the rank-one projection methodology used in this work, may be insufficient for use in sub-100nm technologies,

978-1-4244-3039-0/08 $25.00 © 2008 IEEE

Table 1. SEO Performance in 180nm

Parameter	Synth.	Variation
Gain	50.55db	[40.78, 57.95]
BW	6.29e5 Hz	[2.25e5, 23.2e5]
UGF	2.0e8 Hz	[1.27e8, 8.3e8]
PM	67.98 Deg	[19.15, 74.31]

Table 2. SEO Performance in 65nm

Parameter	Nominal	Variation
Gain	52.06db	[-7.63, 63.6]
BW	2.22e7 Hz	[5.11e6, 4.76e9]
UGF	6.94e9 Hz	[1.24e8, 1.18e10]
PM	55.69 Deg	[-2.15, 100.396]

where the random variation in each process parameter is significantly high.

To illustrate the impact of process variation on circuit performance, we use a sized SEO circuit in 180nm and subject it to variation in the device parameters as presented in [15] for 500 sample cases. Due to the impedance mismatch occuring out of such random variation in device parameters (T_{ox}, L_{eff}, W_{eff}, V_{t0}), the performance function (gain) of the opamp shows significant deviation from normal. The experiment shows that for open loop gain, around 11% of the single ended opamps with same sizes but varying parameters fall below the specifications. Table.1 presents the variation in the different performance parameters for the synthesized SEO due to process variations. In Table. 2, the results demonstrate how the performance degrades even further (around 70%) due to process variation on the SEO circuit's nominal performance due to the scaling of technologies to 65nm.

The above situation prompts the need for the development of accurate macromodels that computes the bounds on performance parameters under the effect of such variations. In the next section we present an automated toolbox which can generate accurate mismatch aware performance macromodels.

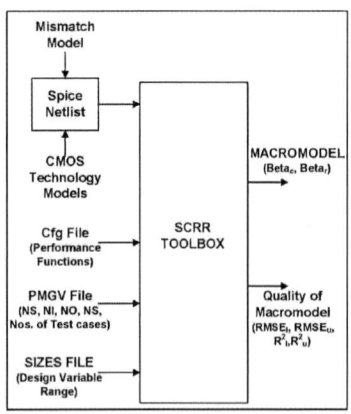

Figure 1. SCRR Toolbox

3 SCRR Toolbox

In this section, we describe the SCRR toolbox, used to build mismatch aware performance macromodels for analog circuits. The toolbox is essentially a black box system. Fig. 1 shows the box diagram for the top level toolbox. The input to the toolbox are the circuit, technology models, mismatch models and the performance functions to be modeled. The output of the toolbox is a validated macromodel for the performance functions of the circuit. The toolbox is capable of working with different CMOS technologies. It has been implemented in C++ and $MATLAB^{\circledR}$. Algorithm. 1 presents the pseudo-code employed in SCRR. The SCRR toolbox comprises of three major components:

- Training Data Generation

- Macromodel Generation

- Validation Data Generation and Verification

The following sub-sections explain the components in detail.

3.1 Training Data Generation

The accuracy of any statistical modeling process is dependent on the sampled raw data, here termed as the training data for the system. In high dimensional design space, the problem of sparse data samples is an active area of research [11]. The training data generation process in our work involves two steps. They are adaptive sampling and simplified performance analysis using spice simulation. Fig. 2 shows the block diagram for the training data generation process.

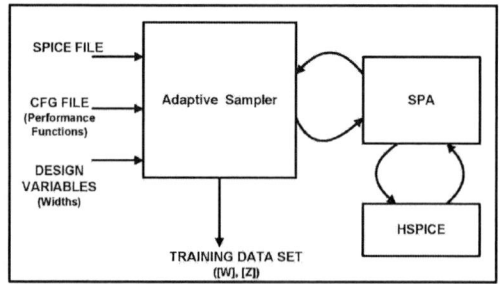

Figure 2. Training Data Generation

3.1.1 Sampling

Random sampling techniques like pseudo-random and quasi-random sampling are capable of sampling design spaces reliably with no prior knowledge of the function. However static sampling often leads to useless samples because the sampling technique has no knowledge of the

Algorithm 1: SCRR Macromodeling

procedure SCRRModelgen (Spice file, Cfg file, Sizes file, NS, NI, NO, NC)

Input: Spice file (ckt. netlist, model parameters, mismatch models), Cfg file (performance parameters), Sizes file (range of design variables), NS (nos. of total samples), NI (nos. of internal montecarlo samples), NO (nos.of initial samples), NC (nos. of adaptive candidate samples)

Output: Final performance macromodels

/* Training Data Generation */

for $i \leftarrow 0$ **to** NS **do**
  N_p = NO
  N_d = Nos. of design variables (widths)
  Halton_Matrix = Halton_seq(N_p, N_d)
  X_0 = null, Z_0 = null
  /* Adaptive Sampling */
  for $Height(X) <$ NS **do**
    /* Initial static samples */
    for $i \leftarrow 0$ **to** N_p **do**
      w_i = Halton(w)
      /* Halton generates matrix of widths based on range of widths and Halton_Matrix */
      /* Functions to model are obtained from Cfg File */
      [params] = $[w_i], [V_{th}], [t_{ox}], [l_{eff}]$
      [functions] = Gain, PM, BW, UGF
      /* SPA: Simplified Performance Analysis */
      pr_metrics = SPA(sim_data, functions)
      sim_data = HSPICE (spice netlist, $params_j$, monte ($j \leftarrow 0$ **to** NI))
      $params_j$ = uniform($w_{ij}, v_{th_j}, t_{ox_j}, l_{eff_j}$)
      $X_i = X_{i-1} \cup w_{ij}$
      $Z_i = Z_{i-1} \cup$ pr_metrics
    endfor
    /* Adaptive sampling for remaining samples */
    $X = X_i, \quad Z = Z_i$
    $X_c \leftarrow NC$ candidate sample widths
    X_{new} = Adaptive_sampler(X_c)
    $pr\_metrics_{new}$ = SPA(sim_data_{new}, functions)
    sim_data_{new} = HSPICE (spice netlist, $param_j$, monte($j \leftarrow 0$ **to** NI))
    $params_j$ = uniform($X_{new}, v_{th_j}, t_{ox_j}, l_{eff_j}$)
    $Z_{new} = pr\_metrics_{new}$
    $X = X \cup X_{new}$
    $Z = Z \cup Z_{new}$
  endfor
endfor

/* Performance macromodel generation */

$[Z]$ = Interval_transform (Z)
$[Z_c, Z_r]$ = CR_transform ($[Z]$)

/* SCRR: Spline based Center and Range Regression */

β_c = SCRR(X, Z_c)
β_r = SCRR(X, Z_r)
Macromodel = $\{\beta_c, \beta_r\}$
$[a] = (a_l, a_u)$

---

function at hand and fails to capture the relative non-linearity. Analog building blocks such as operational amplifiers exhibit weakly non-linear behavior when subjected to process variation.

Adaptive samplers are capable of placing higher density of sample points in regions where the function is most non-linear while sparsely sampling the linear regions. In our work we use an adaptive sampling scheme as presented in [5].

Let us assume that we need to adaptively sample points to explore a real valued function 'f' that defines a hyper-surface of dimension $d+1$ with d independent real variables. Suppose X_0 is an initial set of sample points for which the function 'f' is evaluated. Then for $x \in \Re^d$, we have:

$$X_0 = x_1, x_2, ..., x_n \qquad (1)$$
$$Y_0 = f(x_1), f(x_2), ..., f(x_n) \qquad (2)$$

In our case, X_0 is chosen as quasi-random uniform samples using Halton sequence generator. The function domain Ω is a hyper-cube of dimension d. During each iteration, n_c candidate points are considered from the domain Ω for sampling. For each candidate point x in X_c, the algorithm generates k-nearest neighbor points X_k in X_0 based on Euclidean Norm. For non-linear regions of the fucntion, the nearest neighbor points will not precisely intersect the least squares plane, providing the estimate of non-linearity (ξ). The sample points which have the largest measure of ξ are added to the lists X_0, Y_0. The sampling iteration continues until the stopping criterion (maximum number of samples etc.) is met.

Table.3 compares the accuracy of macromodels generated using statically sampled and adaptively sampled design space for single-ended opamp using 180nm technology. As evident from the table, except for 'gain', all other performance functions are much superior modeled using adaptive samples compared to same number of static samples..

Table 3. SEO sampling accuracy comparison

Function	Static Correlation		Adaptive Correlation	
	r_l^2	r_u^2	r_l^2	r_u^2
Gain	95%	94%	95%	93%
log(BW)	83%	87%	97%	99%
PM	85%	77%	94%	91%
log(UGF)	90%	86%	95%	90%

3.1.2 Simplified Performance Analysis

Simplified Performance Analyzer (SPA) is implemented in C++ and contains the definition of each of the performance parameters. HSpice is called by SPA in the loop to run the monte-carlo simulation using random value of process and model parameters (L_{eff}, T_{ox}, W_{eff} and V_{t0}) as provided by the specific technology model and its corresponding variation. The corresponding performance values are fed back to adaptive sampler for selection of new samples from candidate samples.

The output from the training data generation process is a pair of matrices ([X],[Z]). Here [X] represents the design variables (nominal widths of transistors) and [Z] represents the performance values.

3.2 Macromodel Generation

The objective of our work is to generate an interval valued performance macromodel. The lower and upper bounds of the intervals represent the least and the most values that a performance may vary due to device mismatches. If the intervals faithfully describe the behavior of a system, then it can provide a multi-fold advantage in computation time over a numerical system operating on scalar valued data. Fig. 3 shows the box diagram for the modeling process.

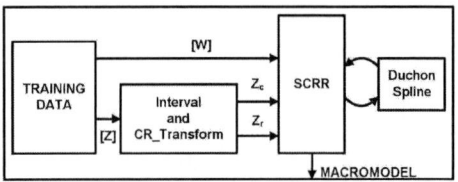

Figure 3. Macromodeling

We present a novel technique for performance macromodel generation using Duchon pseudo-cubic Spine [7] based Center and Range Regression (**SCRR**). SCRR works on an initial interval data set by transforming them into real valued center and range for each of the intervals. The main advantage of the use of interval data type is to group the scalar data into meaningful statistical 3σ 'min'/'max' data which are assumed to be continuous in the interval. This is performed using interval transformation, which converts the vector of scalar data ([Z]) for each iteration obtained during montecarlo simulation into corresponding $[Z_l, Z_u]$ intervals.

Given an interval valued data system as an input, the transformation into 'center' and 'range' is done as a secondary step to prepare data for regression or other curve fitting techniques. The transformation [8] is obtained as follows:

$$Input : X = [X_l, X_u] \tag{3}$$
$$Z = [Z_l, Z_u] \tag{4}$$
$$Output : X_c = (X_l + X_u)/2 \tag{5}$$
$$X_r = (X_u - X_l) \tag{6}$$
$$Z_c = (Z_l + Z_u)/2 \tag{7}$$
$$Z_r = (Z_u - Z_l) \tag{8}$$

With the CR_transformed data sets, the task is to find the model coefficients (β_c, β_r) that leads to the least error measured as the root mean squared error (RMSE) or the highest square of the correlation co-efficient (r^2) between the modeled and actual performance values.

3.2.1 Duchon pseudo-cubic spline

The design space for analog performance function is nonlinear in the presence of process variation. This makes common techniques like linear regression with one or higher order polynomials an inaccurate technique. Spline is found to be a better fitting technique in a non-linear and unstructured bounded space than other regression techniques such as neural networks or support vector machines. While neural networks or SVM provides an approximation of the data space, spline interpolation provides exact values by solving a system of linear equations. In [3], the authors compare the accuracy of different regression techniques with spline. However due to the limitations of generating gridded samples, conventional multivariate spline techniques like tensor products cannot be used for modeling analog performance functions.

In this work, we use the Duchon pseudo-cubic splines [7] that work well on scattered data.

Duchon pseudo-cubic spline can be mathematically expressed as:

$$Z_i = \sum_{j=1}^{k} W_j \phi(x_i - x_j) + P^m(x_i) \tag{9}$$
$$\phi = \|x\|_2^3 \tag{10}$$

The height (z_i) of the N-dimensional point to interpolate (x_i) is a weighted (W_j) summation of basis functions (ϕ) applied to the difference between the unknown point and all 'k' number of sampled nearest neighbor points (x_j) currently defining the spline, plus a polynomial of degree $m = 1$. The basis function is the Euclidean norm cubed between the points x_i and x_j. The coefficients together with the location of the sampled data points completely define the spline interpolate.

Mathematically, the two SCRR regression models are expressed by Equation. (11) and Equation. (12).

$$z_i^c = \sum_{j=1}^{k} W_j^c \phi(x_i^c - x_j^c) + P^m(x_i^c) \tag{11}$$

$$z_i^r = \sum_{j=1}^{k} W_j^r \phi(x_i^r - x_j^r) + P^m(x_i^r) \tag{12}$$

3.3 Model Verification

Qualifying the performance macromodels is the last step in the SCRR toolbox. This verification process is performed on an independent validation set comprising of randomly generated test cases. Spice has been the defacto standard to measure accuracy of circuit simulation. In our work we compare the performance measures using SCRR toolbox and HSpice monte-carlo simulation. The quality of the models are expressed using standard statistical goodness-of-fit measures like $RMSE$ and R^2. Fig. 4 shows our verification strategy for the performance macro models. The quality measures are performed for each of the lower bound and upper bound for each test cases.

290

978-1-4244-3039-0/08 $25.00 © 2008 IEEE

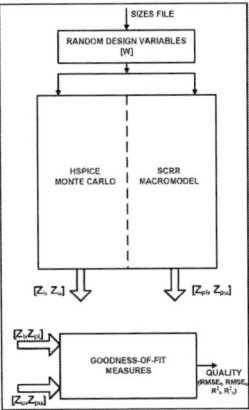

Figure 4. Model Verification

4 Experimental Results

In this section we demonstrate the effectiveness of the spline center and range regression generated macromodels in accurate estimation of analog performance. We also prove through the experiments that the methodology is capable of working across several technology generations and with varying degrees of mismatches.

Experiments are performed using Synopsys $HSPICE^®$ for Windows as the circuit simulator. For this work, all transistors are considered to be MOS transistors using TSMC 180nm and 65nm predictive technology models. Experiments are run on Windows PC with $Intel^® CORE^{TM}2$ Duo processor.

We perform our experiments on two operational amplifiers:

- Two-Stage Opamp

- Single-Ended Opamp

The performance metrics for the experiments are generic to the class and are independent of the specific opamp circuit topology. The performance metrics measured for this work are:

- Open Loop Gain (Gain)

- 3dB Bandwidth (BW)

- Unity Gain Frequency (UGF)

- Phase Margin (PM)

For our experiments we select 2000 samples. The lengths of the transistors are kept at their minimum sizes for this work. The corresponding percentage variations for the model and device parameters are chosen from [15] for the 180nm and 65nm technologies.

For each of our experiments we generate 100 random and independent test cases using a combination of nominal widths of the transistors. Through our results, we present the verification outputs which qualifies the macromodels

generated using SCRR toolbox. For convenience we use **MM** to imply the macromodels and **MC** to mean spice monte-carlo simulation.

The goodness-of-fit measures as explained in the previous sections are **root mean square error** and R^2. Though the value of RMSE is closely tied to the values of the functions, R^2 is expressed as a percentage. A higher percentage for R^2 implies higher correlation between the macromodel (MM) generated predicted results and monte-carlo (MC) evaluated golden results. Thus the higher accuracy of SCRR macromodel would be validated by higher R^2 values.

Alike any other regression modeling techniques, the accuracy of the macromodels improve with the increase in the number of samples. However the number of samples also increase the time complexity of the modeling process. While, we do not intend to minimze the simulation time, we try to improve the modeling time and accuracy of the macromodels.

Fig. 5 plots the modeling time required by SCRR based on the number of samples. Fig. 6 compares the accuracy of the macromodel generation using the correlation metric for different number of samples.

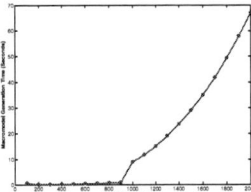

Figure 5. SCRR Modeling Time Comparison

Fig. 7, plot the MM (predicted) and MC (actual) values of the four performance parameters for the single-ended opamp circuit using 20 random test cases from the validation data set. As is eveident from the plot that most of predicted function values closely follow the monte-carlo generated results. This proves the accuracy of our models.

Table. 4(a) presents the lower and upper bounds on the performance functions for the 100 test cases verified. Re-

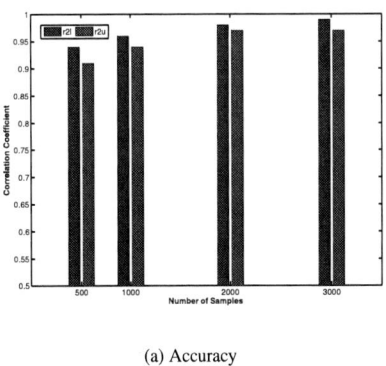

(a) Accuracy

Figure 6. Accuracy Comparison

Table 4. Summary of Experimental Results

Ckt.	Perf.	MC $[MC_l, MC_u]$	MM $[MM_l, MM_u]$	$rmse_l$	r_l^2 (%)	$rmse_u$	r_u^2 (%)	$Time_{MC}$	$Time_{MM}$
TSO	Gain	[-35.7 , 55.4]	[-34.7 , 59.0]	2.9	98	5.1	95	3124.2sec	0.589sec
	log(BW)	[15.9 , 21.8]	[15.5 , 21.7]	0.27	96	0.09	99		
	log(UGF)	[18.4 , 24.1]	[16.7 , 24.1]	0.4	91	0.17	96		
	PM	[-37.5 , 148.1]	[-38.4 , 144.8]	4.26	98	6.2	95		
SEO	Gain	[-46.3 , 60.5]	[-45.3 , 61.0]	6.2	95	4.7	94	3551.4sec	0.5343sec
	log(BW)	[10.2 , 17.9]	[11.0 , 17.9]	0.44	97	0.28	99		
	log(UGF)	[16.5 , 20.7]	[15.7 , 20.4]	0.22	95	0.31	90		
	PM	[36.6 , 101.8]	[28.5 , 97.8]	3.8	94	3.6	91		

(a) Comparison of SCRR Macromodel and Monte-Carlo (180nm)

Ckt.	Perf.	MC $[MC_l, MC_u]$	MM $[MM_l, MM_u]$	$rmse_l$	r_l^2 (%)	$rmse_u$	r_u^2 (%)	$Time_{MC}$	$Time_{MM}$
TSO	Gain	[-55.5 , 67.1]	[-52.4 , 67.8]	5.2	93	6.7	91	3888.4sec	0.53461sec
	log(BW)	[16.1583 , 22.5]	[14.18 , 22.4]	0.53	89	0.14	96		
	log(UGF)	[20.4 , 24.2]	[20.3 , 24.5]	0.23	94	0.13	96		
	PM	[-77 , 147.9]	[-84.2 , 143.4]	4.3	99	6.46	97		

(b) Comparison of SCRR Macromodel and Monte-Carlo (65nm)

Ckt.	Function	Lower Bound Intervals		Upper Bound Intervals	
		MC	MM	MC	MM
TSO	Gain	[-35.7, 44.5]	[-34.7, 41.7]	[-23.4, 55.4]	[-19.2, 59.0]
	log(BW)	[15.9, 20.3]	[15.5, 20.3]	[17.7, 21.8]	[17.6, 21.7]
	PM	[-37.6, 104.6]	[-38.5, 100.1]	[-11.3, 148.1]	[-17.3, 144.8]
	log(UGF)	[18.4, 23.2]	[16.7, 23.2]	[20.1, 24.1]	[19.9, 24.1]
SEO	Gain	[-46.3, 42.3]	[-45.3, 42.1]	[-18.2, 60.5]	[-15.9, 61.0]
	log(BW)	[10.2, 16.4]	[11.0, 16.5]	[14.1, 17.9]	[13.6, 17.9]
	PM	[36.6, 90.4]	[28.5, 87.6]	[56.9, 101.8]	[51.1, 97.8]
	log(UGF)	[16.5, 19.7]	[15.7, 19.7]	[17.4, 20.7]	[17.3, 20.4]

(c) Lower and Upper Intervals using Macromodel and Monte-Carlo (180nm)

sults show that for the 100 independent test cases, the MM generated results have a R^2 value greater than 90% with the MC results. Table. 4(b) compares the similar results for 65nm for the TSO circuit.

In Table. 4(a) and Table. 4(b) we also compare the time required to evaluate the performance functions for 100 test cases using MM and MC techniques. The results show around 5000X improvement in computation time of the macromodels over the monte-carlo simulation. During the verification process, each performance function is computed for each of the test cases and their lower and upper bounds are reported as an interval. Table. 4(c) presents the lower and upper bound intervals for each of the performance functions as evaluated using MM and MC for the two benchmark circuits for 100 testcases in 180nm technology.

The high correlation of the results together with the fast evaluation time demonstrate the effectiveness of the macromodels for repeated use in the synthesis loop in the presence of process variations. Also the robustness of the modeling technique is proven thorough its use with multiple technologies with difference range of variations.

5 Conclusion

In this paper we presented the Spline Center and Range Regression technique on adaptively sampled space to generate mismatch aware performance macromodels for analog circuits. These macromodels may find wide use in fast and variation-tolerant synthesis of analog circuits with minimal loss in accuracy as compared to the expensive monte-carlo simulation-in-the-loop measurements. For future work, the authors intend to improve the accuracy of SCRR macro modeling by considering layout induced mismatches in the circuit.

978-1-4244-3039-0/08 $25.00 © 2008 IEEE

References

[1] A. Agarwal and R. Vemuri. Hierarchical Performance Macromodels of Feasible Regions for Synthesis of Analog and RF Circuits. In *IEEE ICCAD:pages 430-436*, 2005.

[2] Cadence Design Systems. *Virtuoso Neocircuit*.

[3] D. Hurley, J. Hussey et al. An evaluation of splines in linear regression.

[4] Debyser, G.; Gielen, G. Efficient analog circuit synthesis with simultaneous yield and robustness optimization. In *ICCAD*, pages 308–311, 1998.

[5] G. Wolfe and R. Vemuri. Adaptive sampling and modeling of analog circuit performance parameters with pseudo-cubic splines. In *IEEE ICCAD: pages 931-938*, 2004.

[6] Harjani, R.; Rutenbar, R.A.; Carley, L.R. Oasys: A framework for analog circuit synthesis. *IEEE Transaction on Computer Aided Design*, 8(12):1247–1266, December 1989.

[7] J. Duchon. *Constructive Theory of Functions of Several Variables, Lecture Notes in Mathematics*. Springer-Verlag, 1977.

[8] L. Neto, F. Carvalho et al. Linear regression methods to predict interval-valued data. In *IEEE Ninth Brazilian Symposium on Neural Networks*, 2006.

[9] Lohn et al. *Automated Analog Circuit Synthesis using Linear Representation, Lecture Notes in Computer Science*. Springer Berling/Heidelberg, Berlin, 2004.

[10] M. Ding and R. Vemuri. A combined feasibility and performance macromodel for analog circuits. In *IEEE Design Automation Conference: pages 63-68*, 2005.

[11] P. L. Pingli. A sketch-based sampling algorithm on sparse data.

[12] R. Harjani and J. Shao. Feasibility and performance region modeling of analog and digital circuits. *Analog Integrated Circuits and Signal Processing*, 10(1-2):23–43, December 1996.

[13] Ramon E. Moore. *Interval Analysis*. Prentice-Hall, 1966.

[14] Xin Li; Gopalakrishnan, P.; Yang Xu; Pileggi, L.T. Robust analog/rf circuit design with projection-based posynomial modeling. In *ICCAD*, pages 855–862, 2004.

[15] Xin Li; Jiayong Le; Gopalakrishnan, P.; Pileggi, L.T. Asymptotic probability extraction for non-normal distribution of circuit performance. In *IEEE ICCAD: pages 2-9*, 2004.

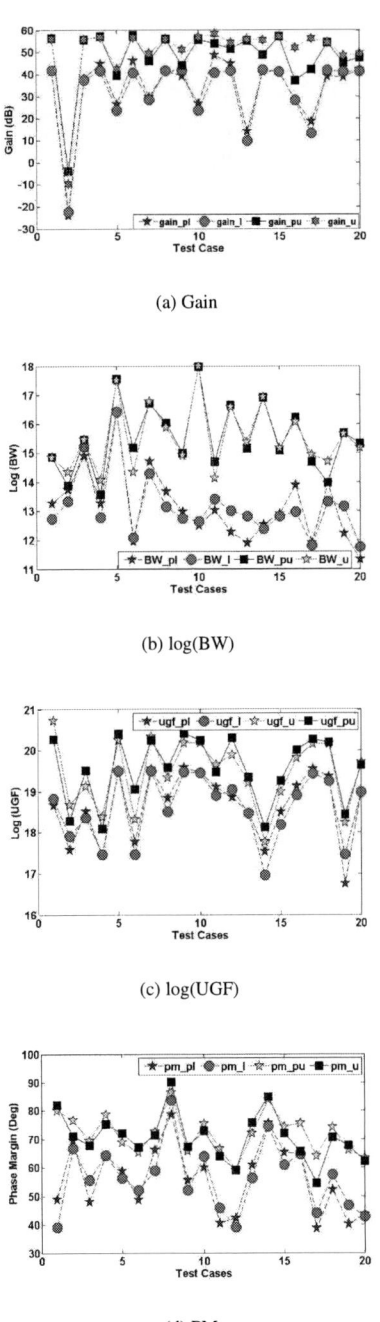

(a) Gain

(b) log(BW)

(c) log(UGF)

(d) PM

Figure 7. Comparing Performance for SEO

21st International Conference on VLSI Design

An Input Stage for the Implementation of Low-Voltage Rail to Rail Offset Compensated CMOS Comparators

Jaime Ramirez-Angulo, Lalitha Mohana Kalyani-Garimella, Annajirao Garimella, Sri Raga
Sudha Garimella, Antonio Lopez-Martin and Ramon Gonzalez Carvajal
New Mexico State University, Las Cruces, NM 88003 USA
jramirez@nmsu.edu, lalitha.garimella@intel.com, garimella@ieee.org, sudha@nmsu.edu,
antonio.lopez@unavarra.es, carvajal@gte.esi.us.es

Abstract

A rail-to-rail differential input stage with programmable threshold levels and offset compensation is introduced. Applications for the implementation of differential and double differential comparators are discussed. Experimental results obtained from a MOSIS 0.5μm CMOS technology test chip are shown that validate rail-to-rail operation with a 1.5V supply voltage.

1. Introduction

The comparator is the basic analog to digital interface element [1]-[2]. It generates digital output signals as a result of the comparison of two analog voltages V_{in} and V_{ref} respectively. It generates a high output voltage (close to V_{DD}) if $V_{in} > V_{ref} + \Delta V - V_{os}$ and a low output voltage (close to ground or V_{SS}) for $V_{in} < V_{ref} - \Delta V + V_{os}$, where V_{os} is the offset voltage, $\Delta V = V_{DD}/A$ is the minimum overdrive voltage or comparator resolution (if offset compensation is used) and A is the gain of the comparator. Technological and power dissipation constraints require modern deep submicron VLSI circuits to operate with continuously decreasing supply voltages V_{DD}, which will take sub-volt ($<1V$) values within the next few years. Many applications require comparators where both V_{in} and V_{ref} can take rail-to-rail (R2R) values. A typical example is in flash A/D converters that require a large number of very compact R2R comparators with low power consumption. Double differential R2R comparators with two signal inputs (V_{inP} and V_{inN}) and two reference inputs (V_{refP} and V_{refN}) are also commonly required in two-step A/D flash converters. These generate a high output voltage if $V_{inP} - V_{refP} > V_{inN} - V_{refN} + \Delta V - V_{os}$ and a low output voltage otherwise.

2. Traditional approaches to implement rail to rail comparators

2.1. Switched capacitor approach

R2R comparators have been implemented using switched capacitor (SC) techniques by means of what some authors denote "common mode jump circuits" (Fig. 1a) [3]. These use an input differential stage with two capacitors C that have one of their terminal connected to the gate of the differential pair (DP). During a pre-charge phase (ϕ_1) the gate terminal of each capacitor is connected to a common mode biasing voltage V_{CMbias} (with a value close to the upper rail for *NMOS* DPs) while the other capacitor's terminal is connected to a common mode input voltage V_{CMinp}. This causes both capacitors to charge to a voltage $V_{bat} = V_{CMinp} - V_{CMbias}$. During the evaluation phase (ϕ_2) both capacitors are first disconnected from V_{CMbias} (on the gate terminal) which leaves the gate terminals floating. The other terminal of each capacitor is then connected to V_{in} and V_{ref} respectively. Given that the capacitors retain their charge (and voltage), they act as floating batteries with value V_{bat} and transfer to the gate terminal changes on the other (control) terminal. During ϕ_1 the gate terminals have voltages $V_{g1} = V_{g2} = V_{CMbias}$. During ϕ_2 they take values $V_{g1} = V_{CMbias} + V_{in} - V_{CMinp}$ and $V_{g2} = V_{CMbias} + V_{ref} - V_{CMinp}$. During ϕ_2 their difference takes a value $V_d = V_{in} - V_{ref}$. Selection of values $V_{CMbias} = V_{DD}$ and midsupply $V_{CMinp} = V_{DD}/2$ leads to gate voltages $V_{g1}, V_{g2} > V_{DD}/2$ for V_{in} and V_{ref} taking R2R values when $V_{DD}/2 > V_{GS} + V_{DSsat}$. The DP transistors remain ON with R2R voltages V_{in}, V_{ref}. This technique allows R2R operation if $V_{DD} > 2(V_{GS} + V_{DSsat})$. A disadvantage of this approach besides the inherent speed limitation of discrete time operation is that for $V_{in}, V_{ref} = V_{DD}$ the gate voltages can take during ϕ_2 values higher than the supply rail

294

978-1-4244-3039-0/08 $25.00 © 2008 IEEE

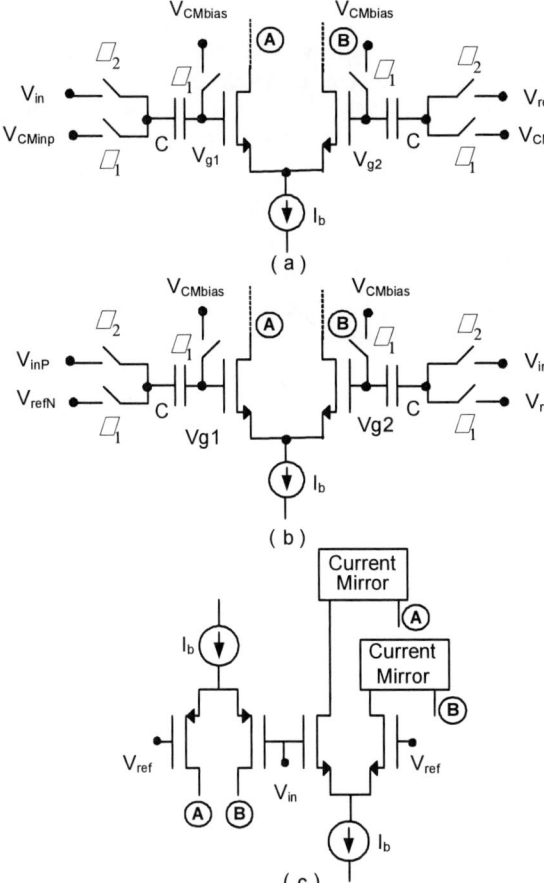

Figure 1. Traditional R2R comparator input stages: (a) Switched capacitor single ended input stage (b) SC double differential input stage (c) Complementary input stage.

$V_{g1}^{MAX} = V_{g2}^{MAX} = 3V_{DD}/2$. This can lead to reliability problems (like oxide breakdown) in deep submicron technologies.

2.2. Complementary differential pairs approach

Another common approach to implement R2R input stages makes use of a complementary differential pair (CDP) input stage consisting of a *PMOS* and an *NMOS* DP (Fig. 1c) [4]-[5]. This requires a current folding circuit to add the currents from the two DPs and a bias control circuit (see Fig. 1c). This technique has been used to implement R2R op-amps [6]. Complementary input stages have following disadvantages: 1) they are relatively complex and have increased power consumption, noise and offset. 2) they can also be subject to severe CMRR and PSRR degradation over some regions of the input range where the tail biasing transistors operate in triode mode. Both approaches described above require a relatively large minimum supply voltage $V_{DD}=2(V_{GS}+V_{DSsat})$. Some authors have

combined the two techniques discussed above in order to achieve R2R low-voltage operation.

2.3. Contributions of the present work

In this paper, application of a simple R2R differential input stage to implement comparators that operate in continuous time with a single supply voltage that can be as low as $V_{DD}=V_{GS}+V_{DSsat}$ is discussed. It is based on the floating gate (FG) technique reported in [7],[8]. Offset compensation and implementation of double and multiple differential comparators is also discussed. The rest of the paper is organized as follows. The proposed R2R input stage is described in Section 3. The architecture and stability of the comparator are discussed in Section 4. Simulation and experimental results are disclosed in Section 5 with conclusions in Section 6.

3. Proposed rail to rail input stage

The proposed R2R input stage is shown in Fig. 2a. As mentioned before, it operates in continuous time and is based on the low- voltage FG basing technique reported in [7],[8].

3.1. Analysis of rail to rail operation

By using charge conservation and assuming zero charge on the floating gates of M_1 and M_2 (a condition easily achieved by utilization of the layout technique reported in [9]) it can be shown that the FG voltages are given by

$$V_{fg1}=V_{bias}(C_{bias}/C_{total})+V_{in}(C/C_{total}) \qquad (1)$$
$$V_{fg2}=V_{bias}(C_{bias}/C_{total})+V_{ref}(C/C_{total}) \qquad (2)$$

where $C_{total}=C+C_{bias}$.

By selection of $C_{bias}=C=C_{total}/2$ and $V_{bias}=V_{DD}$ the FG voltages V_{fg1}, V_{fg2} take values ranging from $V_{DD}/2$ to V_{DD} for V_{in} and V_{ref} taking R2R voltages (from 0V to V_{DD}). The differential input voltage of the differential pair is given by $V_d = V_{fg1}-V_{fg2}= k(V_{in}-V_{ref})$. The effective gain of the input stage is reduced by the attenuation factor $k=(C/C_{total})=0.5$. This selection for C_{bias} and V_{bias} allows operation with a minimum supply voltage $V_{DD}^{MIN}=2(V_{GS}+V_{DSsat})$ but in this case the gate voltages do not go over the supply rail and unlike in the CDP approach both the DP transistors and the tail biasing source remain ON in saturation mode over the complete R2R common mode input range. For this reason no CMRR or PSRR degradation takes place and no additional power dissipation or complex circuitry is required as in the CDP technique. Selection of a value $C_{bias}>>C$ leads to voltages V_{fg1} and V_{fg2} that remain close to V_{DD} with R2R signals V_{in} and V_{ref}. In this case it is possible to operate the circuit of Fig. 2a with a

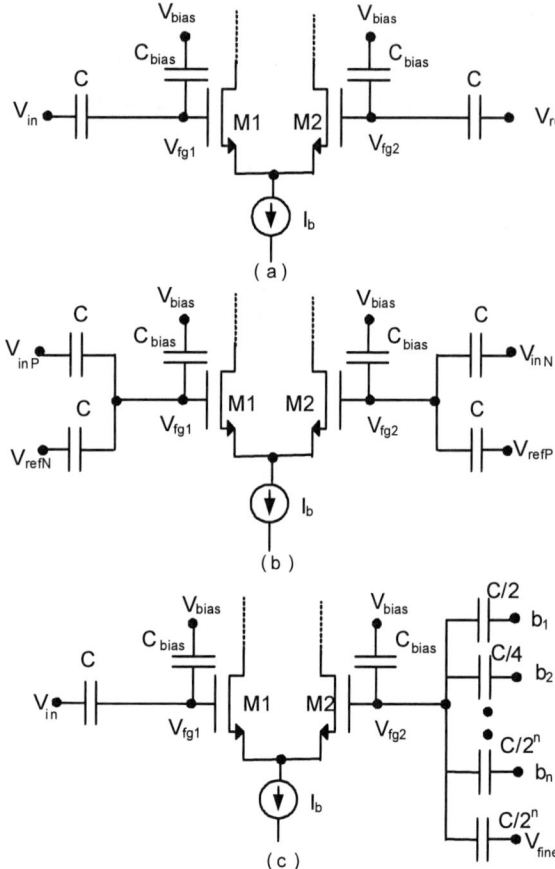

Figure 2. FG approach for R2R operation. (a) Differential input stage (b) Double differential input stage (c) Differential input stage with R2R programmable reference voltage.

single supply voltage $V_{DD}^{MIN} \approx V_{GS} + V_{DSsat}$. The tradeoff is a larger attenuation factor $k = C/C_{total}$ that leads to a smaller effective comparator gain and increased effective input offset and noise (by the factor $1/k$).

3.2. Implementation of double and multiple differential pairs

Fig. 1b shows the implementation of a double differential input using the SC approach and Fig. 2b using FG biasing technique. In Fig. 2b two capacitors with value C are connected to the FG of each DP transistor. The control terminal of the two capacitors on the side of M_1 is connected to voltages V_{inP}, V_{refN} and on the side of M_2 to V_{inN}, V_{refP} respectively. In this case FG voltages are given by

$$V_{fg1} = V_{bias}(C_{bias}/C_{total}) + V_{inP}(C/C_{total}) + V_{refN}(C/C_{total})$$
$$V_{fg2} = V_{bias}(C_{bias}/C_{total}) + V_{inN}(C/C_{total}) + V_{refP}(C/C_{total})$$

$$(3), (4)$$

with $C_{total} = 2C + C_{bias}$. The differential voltage is given by

$$V_d = k((V_{inP} - V_{refP}) - (V_{inN} - V_{refN})) \qquad (5)$$

Selection of biasing capacitors with values $C_{bias} = 2C$ lead to an attenuation factor $k = 0.5$ and similar conditions as discussed above for V_{DD}^{MIN} with R2R input voltages $V_{inN}, V_{inP}, V_{refN}$ and V_{refP}. This technique can be easily extended to triple, quadruple or more differential pairs with larger number of differential inputs which might be required for some applications.

3.3. Coarse and fine adjustment of threshold voltage using the floating gate technique

Fig. 2c shows a scheme to achieve digital coarse adjustment and analog fine adjustment of the reference voltage V_{ref}. In this case a binary weighted capacitor array (BWCA) is connected to the floating gate of M_2. Binary control voltages $b_1, b_2, ..b_n$ (with values V_{DD} or *ground*) are applied to the control terminal of the capacitors in the BWCA and if desired an analog voltage is applied to a termination capacitor with value $C/2^n$. Another BWCA with all control terminals in parallel connected to V_{in} is used on the side of M_1. BWCAs have a total capacitance with value C. This leads to a differential voltage $V_d = k(V_{in} - V_{ref})$ with a reference voltage given by

$$V_{ref} = V_{DD}(b_1/2 + b_2/4 + .. + b_n/2^n) + V_{fine}/2^n \qquad (6)$$

It can be seen that V_{ref} can take R2R values. Coarse adjustment can take place with the digital control word $b_1, b_2, ..b_n$, while fine adjustment (if required) can take place can take place with voltage V_{fine}.

3.4. Offset compensation scheme

The resolution of the comparator is limited by the offset voltage V_{os} which is random in nature. For this reason high accuracy applications require offset compensation [10]. This is done in the proposed input stage by including two small valued capacitors $C_{os} \ll C$ connected to M_1 and M_2 as shown in Fig. 3a. Assuming the total capacitance (excluding C_{bias}) connected to the gate of each transistor has a value $C_{os} \sim C/10$. During the offset measurement phase ϕ_1 (Fig. 3b) C_{os} is connected to the comparator output terminal while the remaining capacitance $C - C_{os}$ is connected to V_{CMinp}. This leads to an output voltage $V_{osamp} = V_{os}(1 + (C - C_{os})/C_{os})$ which is an amplified version of the input offset voltage by the factor $A_{os} = (1 + (C - C_{os})/C_{os})$ (~10). This voltage is stored in a capacitor C_{hold} that serves also as load capacitor (and as compensation capacitor as discussed later) during this phase. This voltage is applied to the negative input during the evaluation phase ϕ_2 though a voltage divider formed by C_{os} and $C - C_{os}$ (Fig. 3c). This leads to a voltage on the negative input terminal with value: $V_{i-} = V_{os}$ that compensates the offset V_{os} on the positive input terminal. Given that this scheme uses an amplified version of V_{os} for offset compensation

296

978-1-4244-3039-0/08 $25.00 © 2008 IEEE

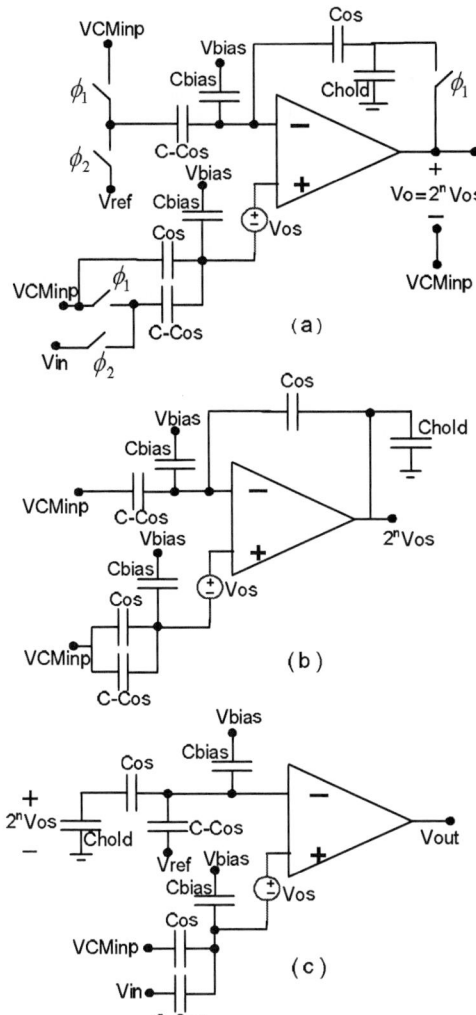

Figure 3. Offset compensation scheme (a) Scheme showing both phases and switches (b) Circuit connections in ϕ_1 c) circuit connections in ϕ_2.

it is highly insensitive to charge injection, clock feedthrough and leakage. Utilization of the comparator as an amplifier with gain A_{os} has the added advantage that no frequency compensation elements are required to prevent possible instability during ϕ_1 in which the comparator is used with negative feedback. This is discussed in detail in section 4.2. Offset can be compensated on a periodic basis [10].

4. Comparator architecture and stability considerations

4.1. Comparator architecture

In order to test the proposed R2R input stage a two (three) stage comparator was designed and fabricated in 0.5μm CMOS technology with nominal *NMOS* and

PMOS threshold voltages $V_{TN}=0.73V$ and $V_{TP}=-0.95V$ respectively as shown Fig. 4. It consists of a cascade of a R2R differential input stage (without hysteresis) and two CMOS inverters Following transistor dimensions were used M_1,M_2: W/L=6/0.6; M_3,M_4,M_B: W/L=12/0.6; M_5: W/L=4.5/0.6; M_6: W/L=3/0.6. Outputs V_{out} and V_{out}' from the two CMOS inverter output stages were available externally for testing purposes. The circuit was tested with load capacitance $C_L=75pF$, $V_{DD}=1.5V$ and $I_{bias}=30\mu A$. The total area of the fabricated circuit was only $140 \times 52\mu m^2$. Eight unit capacitors with value $C_u=50fF$ (area~$25\mu m^2$) and a biasing capacitor $C_{bias}=400fF$ were connected to the floating gate of M_1 and M_2. A unit capacitor on each side was used for offset compensation ($C_{os}=C_u$). The terminals of the remaining unit capacitors were available as individual input terminals that could be grouped (if desired) in a binary weighted array with values C_u, $2C_u$ and $4C_u$, in parallel or in any other test arrangement.

4.2. Stability considerations

An advantage of the proposed offset compensation scheme, together with the fact that minimum length L values were used for all transistors, is that no Miller compensation capacitor is required when the comparator is used with negative feedback during the offset measurement phase. This is due to the fact that the output pole is the dominant pole and the load capacitor serves as a hold and compensation element at the same time and that the loop gain is reduced by the feedback factor as explained next:

1) In the offset compensation phase the open loop gain is reduced by the feedback factor $\beta=C_{os}/(C-C_{os})$

2) Utilization of minimum L values leads to relatively low output resistance r_0 and very small parasitic capacitances C_x at the internal node X. For this reason this node has a high frequency pole of approximately $f_{px}=40MHz$.

3) For typical load capacitances $C_L=C_{hold}>1pF$ at V_{out} the output pole f_{pout} (or f_{pout}' if output V_{out}' is used instead) is a dominant pole that satisfies easily the stability condition on the unity gain frequency f_{unity} of the negative feedback loop: $f_{unity}=A\beta f_{pout}<f_{px}/2$. This requires approximately $C_L>40C_x$ for values $A=200$, $\beta=0.1$ with $f_{pout}=1/(2\pi r_0 C_L)$ and $f_{px}=1/(2\pi r_0 C_x)$. This results in values $C_L>0.4pF$ for the estimated value $C_x=0.01pF$. In this case the comparator has the capability to resolve signals of approximately $\Delta V =1.5/200=7.5mV$. If the high gain output V_{out}' is used there are two internal high frequency poles with approximately equal values $f_{px}=f_{py}=40MHz$ at nodes x and y (also labeled V_{out}). In this case the stability

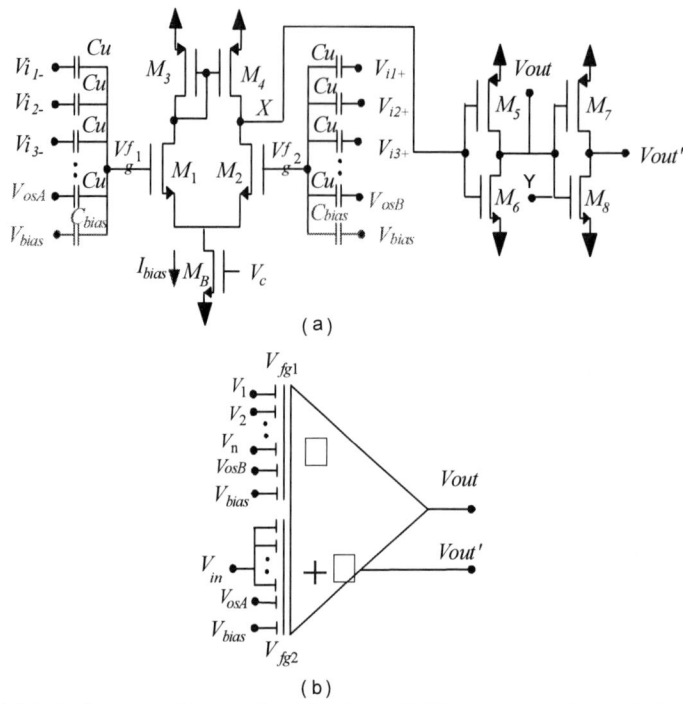

Figure 4 (a) Scheme of two (three) stage R2R comparator (b) Symbol.

condition is given by $A'\beta f_{pout} < f_{px}/4$ (equivalently $C_L > 1600C_x$ or $C_L > 16pF$) and the resolution with offset compensation is $\Delta V = 1.5/4000 = 0.375mV$. These considerations determine the minimum value of C_{hold} which operates both as hold and compensation capacitance during the offset measurement phase in which negative feedback is applied to the comparator. Besides providing stable behavior a relatively large value of C_{hold} is of advantage to minimize charge injection errors in offset compensation. C_{hold} determines the rise and fall time in phase 1. This is given by $t_{rise} = 2.2(2\pi/(A\beta f_{pout}))$ and has an approximate value $t_{rise} = 650ns$. This value is in good agreement with simulations and determines the maximum operating speed of the comparator when offset compensation is used.

5. Simulation and experimental results

5.1. Simulation results

As stated above given that minimum L was used for all transistors their output resistance r_0 is relatively low. This causes the gain of each stage to be also relatively low. From simulations in Cadence DFII design environment the first stage was determined to have a gain of 10 (due to a factor 2 attenuation of the capacitive divider formed by C and C_{bias}), the second and third stages have gains of approximately 20. The internal pole had a frequency $f_{px} = 40MHz$. The

comparator's gain at output V_{out} was $A = 200$. This gain was in very good agreement with experimental results. With the comparator connected in unity gain voltage follower configuration the gain bandwidth was $GB = 62MHz$ with $C_L = 1pF$ and 63° phase margin. The gain for output V_{out}' is $A' = 4000V/V$. The supply voltage used to test the comparator was $V_{DD} = 1.5V$. The bias current was $I_b = 30\mu A$.

5.2. Experimental results

Given that both outputs were available externally and loaded with relatively large breadboard capacitances the comparator could be only characterized experimentally at relatively low frequency on the first (low gain) output V_{out}. The load capacitance (breadboard, test probes and wiring) was approximately $C_L = 75pF$. Fig. 5a, 5b and 5c show experimental input and output waveforms V_{in}, V_{out} (and V_{ref}) upon application of a $10kHz$ R2R ($1.5V_{pp}$) triangular input signal for $V_{ref} = 100mV$, $500mV$ and $1.4V$. This validates R2R operation of the comparator. Rise and fall times are in this case limited by slew rate which is given by $SR = I_o^{MAX}/C_L$. Positive and negative slew rates have values $SR+ = 1.5V/\mu s$ and $SR- = 1.9V/\mu s$ which are determined by the maximum positive and a negative output currents $I_o^{pos} = 110\mu A$, $I_o^{neg} = 190\mu A$ and the $C_L = 75pF$. The micrograph of the circuit is shown in Fig. 6.

6. Conclusions

A low-voltage R2R differential input stage with offset compensation was introduced. Its application for the implementation of single differential and double differential comparators was discussed. An offset compensation technique with reduced sensitivity to gain errors was used in the proposed stage. Experimental results of a test chip fabricated in $0.5\mu m$ CMOS technology validated R2R operation of the comparator with a single supply $V_{DD}=1.5V$ and with close to R2R reference voltages.

10. References

[1] R. Gregorian, *Introduction to CMOS Op-amps and Comparators*, John Wiley and Sons, Inc., ISBN: 0-471-31778-0, 1999.

[2] B. Razavi and B.A. Wooley, "Design Techniques for High-speed, High-Resolution Comparators", *IEEE J. of Solid State Circuits*, vol. 27, no. 12, Dec. 1992, pp. 1916-1926.

[3] C. Toumazou, G. Moschytz and B. Gilbert (Ed), *Trade-Offs in Analog Circuit Design, The designers Companion*, Kluwer Academic Publishers, Boston, Chapter 14 pp. 407-439, 2002.

[4] Y. Hung and B. Liu, "1V CMOS Comparator for Programmable Analog Rank-Order Extractor, *IEEE Trans. on Cir. and Sys. I*, vol. 50, no.5, May 2003, pp. 673-677.

[5] W. Redman-White, "A High Bandwidth Constant gm and Slew-Rate Rail-to-Rail CMOS Input Circuit and its Application to Analog Cells for Low Voltage VLSI Systems," *IEEE J. of Solid State Circuits*, vol. 32, no. 5, May 1997, pp. 701-712.

[6] S. Yan, J. Hu, T. Song and E. Sánchez-Sinencio, "Constant-gm techniques for rail-to-rail CMOS input stages: A comparative study", in *Proc. IEEE International Symp. on Circuits and Systems*, pp. 2571-2574, May 23-26, 2005.

[7] J. Ramírez-Angulo, S.C. Choi and G. Gonzalez-Altamirano, "Low-voltage circuits building blocks using multiple input floating gate transistors", *IEEE Trans. on Circuits and Systems I*, vol. 42, no. 11, Nov. 1995, pp.971-974.

[8] J. Ramírez-Angulo, R.G. Carvajal, J. Tombs and A. Torralba, "Low-voltage CMOS Op-amp with rail-to-rail input and output signal swing for continuous-time signal processing using multiple-input floating-gate transistors", *IEEE Trans. on Circuits and Systems II*, vol. 48, no. 1, Jan. 2001, pp. 111-116.

[9] E. Rodriguez-Villegas and H. Barnes, "Solution to trapped charge in FGMOS transistors", *Electronics Letters*, vol. 39, no. 19, 18 Sep. 2003, pp.1416-1417.

[10] C.C. Enz and G.C. Temes, "Circuit Techniques for Reducing the Effects of Op-Amp Imperfections: Autozeroing, correlated double sampling, and Chopper Stabilization," *Proceedings of the IEEE*, vol. 84, no. 11, Nov. 1996 pp. 1584-1614.

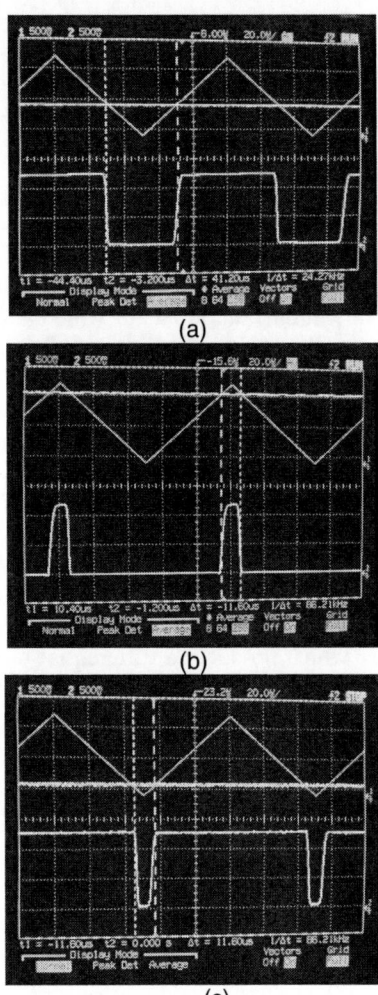

Figure 5. Experimental R2R pulse output (bottom trace) and triangular input, reference (top trace) waveforms (a) Vref=500mV (b) Vref=1400mV, (c) Vref=100mV.

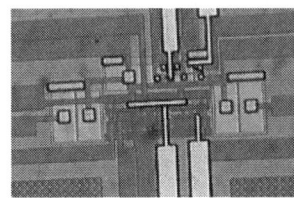

Figure 6. Micrograph of the comparator circuitry. (Area 140μm x 52μm).

21st International Conference on VLSI Design

Highly Linear Wide Dynamic Swing CMOS Transconductance Multiplier using Source-degeneration V-I Converters

Sri Raga Sudha Garimella
New Mexico State University, Las Cruces, NM 88003 USA
sudha@nmsu.edu

Abstract

A novel compact four quadrant CMOS transconductance analog multiplier with wide dynamic swing and wide gain bandwidth product using source-degeneration V-I converters is proposed. The design consists of two stages. First stage is a voltage adder and utilizes two V-I converters with diode connected load and source-degeneration resistor which can provide high bandwidth. The second stage consists of two cross connected differential pairs with source-degeneration resistor which act as current steering elements performing V to I conversion with wide dynamic swing and continuous adjustable gain. Unlike conventional multipliers, in the proposed scheme all the significant intermediate terms generated are linear reducing the non-linear term cancellation, making the circuit power efficient. SPICE simulation results in 0.5μm CMOS AMI technology are presented which validate the proposed work.

1. Introduction

Analog multipliers find potential applications in frequency doubling, frequency shifting, phase angle detection, real power computation, signal multiplication, division and squaring. CMOS analog multipliers have wide application range in analog signal processing, real time systems, communication systems such as filters, mixers, modulators etc. Trade-offs between linearity, gain, dynamic range, bandwidth and low-voltage operation are the challenging aspects for CMOS multiplier designers. [1] presents a comprehensive survey on the CMOS transconductance multiplier architectures. A four-quadrant CMOS multiplier which makes use of voltage adder is described in [2]. This scheme uses complex circuitry with duplicated V–I converters to achieve four quadrant multiplication. The same configuration is reported in [3]–[5] using a capacitive adder. Most of the multipliers make use of differential pairs or op-

amps in their circuit implementation [6]-[7] resulting in generation of non-linear terms which causes power dissipation. [7] is a two quadrant multiplier using a cascade operational amplifier for wide bandwidth with 16MHz gain-bandwidth product. [8]-[9] uses current mirrors with transistors operating in triode mode. In this case gain adjustment is very sensitive to the gain control voltage. The Gilbert Cell [10] has been used to implement four quadrant multipliers in bipolar and CMOS technology [11]-[12]. These approaches require large supply voltages and complex circuitry.

1.1. Contributions of the present work

The proposed scheme is a transconductance multiplier which produces a differential output current that is proportional to the product of two independent input voltages. Fig. 1 shows the block diagram of the proposed scheme. The scheme consists of two stages, two V–I converters with diode connected load (acting as voltage adder) in first stage and cross connected differential pairs in second stage. Each V-I converter in first stage acts as voltage adder. The diode connected load results in wide bandwidth. Both the stages utilize source-degeneration resistor leading to low power dissipation and wide dynamic swing. Also the proposed scheme represents a very compact circuitry compared to other multipliers which belong to this family

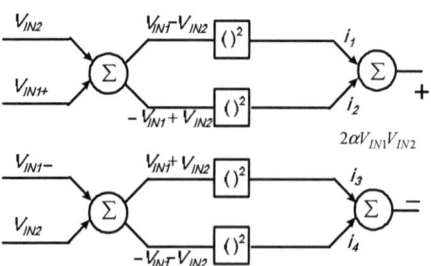

Figure 1. Block diagram of the proposed Multiplier.

300

978-1-4244-3039-0/08 $25.00 © 2008 IEEE

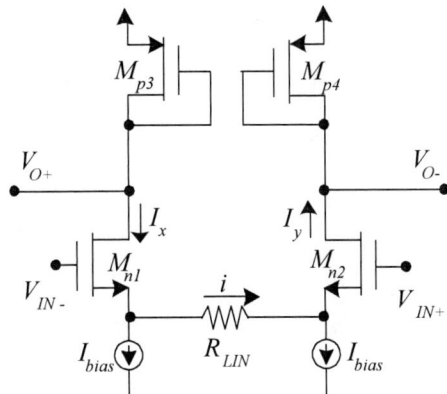

Figure 2. A Fully Differential Source-degeneration V–I Converter with diode connected load.

and shows high performance characteristics compared to other CMOS multiplier topologies reported in the literature. Section 2 describes the source-degeneration V–I converter and diode connected load. In section 3, the multiplier circuit realization, analysis and characteristics are presented. The simulation results are discussed in section 4. Section 5 concludes the work.

2. V–I converter with Diode connected load and Source-degeneration resistor

Fig. 2 shows the fully differential V–I converter cell with diode connected load and source-degeneration resistor. It has an NMOS differential pair formed by transistors M_{n1}, and M_{n2} with diode connected PMOS formed by transistors M_{p3}, M_{p4}. Two tail current sources I_{bias} are connected at the sources of the differential pair. R_{LIN} is the source–degeneration resistor for the V to I conversion. Differential input signals V_{IN-}, V_{IN+} are applied to the gates of M_{n1} and M_{n2}. The source-degeneration resistor R_{LIN} allows a current i to pass through it from transistor M_{n1} to transistor M_{n2}. Due to current steering current through M_{n1} is $I_x = I_{d1}^Q - i$ and current through M_{n2} is $I_y = I_{d2}^Q + i$, where I_{d1}^Q and I_{d2}^Q are the drain currents of M_{n1} and M_{n2}. The differential outputs are given by $v_{O+} = A_v(v_{IN+} - v_{IN-})/2$ and $v_{O-} = A_v(v_{IN-} - v_{IN+})/2$ where A_V is the gain of the V–I converter. Hence the output is the addition of the inputs. The gain of the V–I converter is given by $A_V = 1/((R_{LIN} + 2/g_m)(g_m)) \approx 1/(R_{LIN}.g_m)$. Neglecting $1/g_m$ and body effect of the differential pair transistors M_{n1}, M_{n2} the output of this V–I converter is a linear term. The impedance of the

diode connected load is $1/g_m$ which is typically a low value and hence bandwidth is high.

3. Proposed CMOS Multiplier

Fig 3. shows the proposed four quadrant CMOS transconductance multiplier scheme. The multiplier is implemented using two stages as described below.

3.1. The First Stage – Voltage addition using V–I Converters

The two identical V–I converters X_1 and X_2 are formed by differential pairs M_1, M_2 and M_5, M_6 as shown in Fig. 3. Each V–I converter utilizes the circuit shown in Fig. 2. M_3, M_4 and M_7, M_8 are the PMOS diode connected loads. R_{LIN1} and R_{LIN2} are the source–degeneration resistors which allow currents i_a and i_b to pass through them. V_{IN1} and V_{IN2} are the two independent inputs to the multiplier. V_{IN1+}, V_{IN1-} are the differential signals generated from V_{IN1}. Signals V_{IN2} and V_{IN1+} are applied to X_1 and V_{IN1-}, V_{IN2} are applied to X_2. The differential output voltages are

$$V_{01+} = A_{VX1}(V_{IN1} - V_{IN2}) \tag{1}$$

$$V_{01-} = A_{VX1}(-V_{IN1} + V_{IN2}) \tag{2}$$

$$V_{02+} = A_{VX2}(V_{IN1} + V_{IN2}) \tag{3}$$

$$V_{02-} = A_{VX2}(-V_{IN1} - V_{IN2}) \tag{4}$$

where A_{VX1} is the gain of the X_1 and A_{VX2} is the gain of the X_2. This (first) stage acts as a voltage adder providing a fully differential output of the two inputs and is fed to the second stage.

3.2. The Second Stage – Cross Connected Differential pairs

The second stage of the CMOS multiplier has two cross connected identical differential pairs with source–degeneration resistor and two resistive loads. M_9, M_{11} forms one of the differential pair whereas M_{10}, M_{12} forms the other. A source–degeneration resistor R_{LIN3} is connected to sources of the above two differential pairs. I_B is the tail current of the differential pairs. The signals V_{O1+}, V_{O1-}, V_{O2+}, V_{O2-} generated from the first stage are given as inputs to the differential pairs in the second stage. The gain of the second stage $A_{vII} = g_{meff}.R_L = R_l/(R+2/g_m) \approx R_l/R_{LIN3}$, where R_L is the load resistance of the multiplier. The cross connection of the differential pairs can add the two drain currents of the transistors from differential pairs. i_c is the current through the source–degeneration resistor R_{LIN3}.

301

978-1-4244-3039-0/08 $25.00 © 2008 IEEE

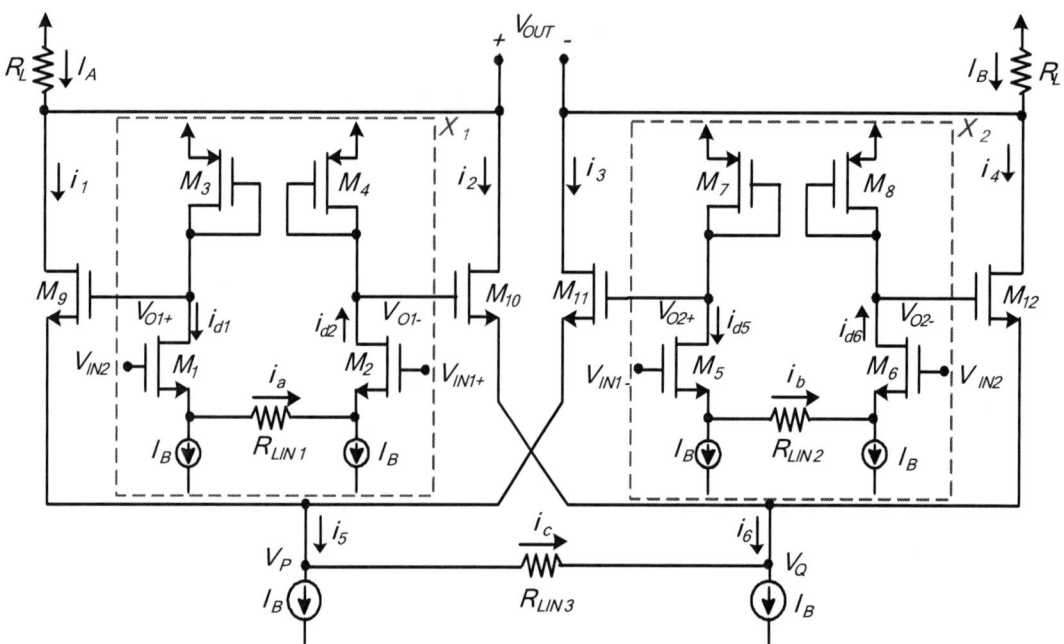

Figure 3. Proposed two stage four quadrant CMOS transconductance multiplier using two source–degeneration V–I converters X_1, X_2 in first stage and cross coupled differential pair in the second stage.

3.3. Analysis

The second stage is a fully differential multiplying stage which is based on square law characteristics of the MOS transistors. All the MOS transistors in the circuit operate in the saturation region.

The signal currents of transistors M_1, M_2, M_5 and M_6 are given by

$$i_{d1} = (1/R_{LIN1})\,(v_{IN2}\text{-}v_{IN1}/2) \tag{5}$$
$$i_{d2} = \text{-}(1/R_{LIN1})(v_{IN2}\text{-}v_{IN1}/2) \tag{6}$$
$$i_{d5} = \text{-}(1/R_{LIN2})(\,v_{IN2}\text{+}v_{IN1}/2) \tag{7}$$
$$i_{d6} = (1/R_{LIN2})(\,v_{IN2}\text{+}v_{IN1}/2) \tag{8}$$

The signal currents of transistors M_9, M_{10}, M_{11} and M_{12} are given by

$$i_1 = \sqrt{(\beta i_5)}.\,(v_{O1}/2\text{-}v_{O2}/2) \tag{9}$$
$$i_2 = \text{-}\sqrt{(\beta i_6)}.\,(v_{O1}/2\text{-}v_{O2}/2) \tag{10}$$
$$i_3 = \text{-}\sqrt{(\beta i_5)}.\,(v_{O1}/2\text{-}v_{O2}/2) \tag{11}$$
$$i_4 = \sqrt{(\beta i_6)}.\,(v_{O1}/2\text{-}v_{O2}/2) \tag{12}$$

β is the transconductance factor of NMOS given by

$$\beta = \mu_n C_{ox}(W/L)_n \tag{13}$$

where the transistor parameters have their usual meanings.

i_5 and i_6 are the signal tail currents of the cross connected differential pairs. Differential output currents through the load resistor R_L are given by

$$I_A = i_1 + i_2 \tag{14}$$
$$I_B = i_3 + i_4 \tag{15}$$

$$I_{OUT} = I_A - I_B \tag{16}$$

From (14) and (15)

$$I_{OUT} = 2\sqrt{(\beta(i_5\text{-}i_6))}.\,(v_{O1}/2\text{-}v_{O2}/2) \tag{17}$$

where

$$v_{O1}/2 = 1/g_{m3}i_{d1} \tag{18}$$
$$v_{O2}/2 = 1/g_{m7}\,i_{d5} \tag{19}$$
$$i_5 = (v_P\text{-}v_Q)\,/R_{LIN3} \tag{20}$$
$$i_6 = \text{-}(v_P\text{-}v_Q)/R_{LIN3} \tag{21}$$

where

$$v_P\text{-}v_Q = (\,v_{O1}+v_{O2})/2 \tag{22}$$

From the above equations, I_{OUT} can be written as

$$I_{OUT} = 2\alpha V_{IN1}V_{IN2} \tag{23}$$

where α is the effective gain of the multiplier. Hence for the proposed transconductance multiplier the output current is proportional to product of the two input voltages V_{IN1} and V_{IN2}.

3.4. Characteristics of the Multiplier

The use of V–I converter with source-degeneration reduces the dependence of the gain of the circuit upon the input level i.e. making the gain relatively independent of the transistor bias current. The headroom needed for the second stage is $2V_{DS,SAT}$: headroom for differential pair transistors ($M_9, M_{10}, M_{11}, M_{12}$) and the headroom for tail current I_B (implemented by current mirror). This less headroom requirement can leverage wide dynamic swing. The scheme can accept high resistive loads R_{L1} and R_{L2} and hence gain can be

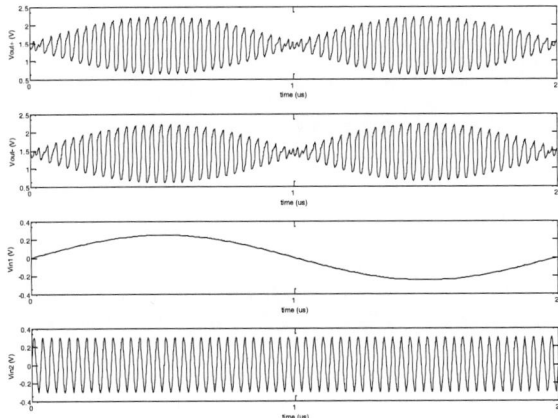

Figure 4. Transient response of the Multiplier (top to bottom): Differential outputs V_{OUT+}, V_{OUT-}, 500KHz - 0.5V_{PP} modulating sinusoidal wave V_{IN1}, 30MHz - 0.6V_{PP} carrier sinusoidal wave V_{IN2} (Response for 2µs)

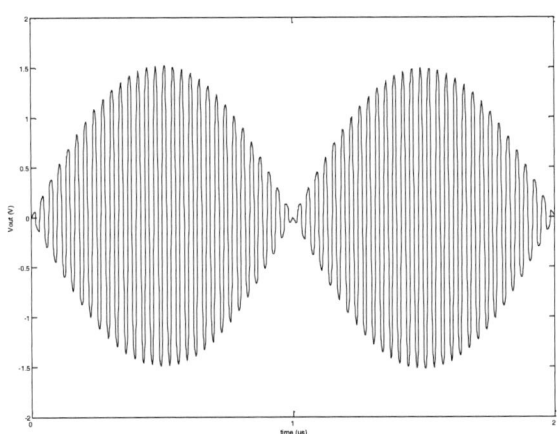

Figure 5. Output V_{OUT} of the proposed multiplier: Difference of the differential outputs V_{OUT+} and V_{OUT-} of Fig. 4. (Response for 2µs)

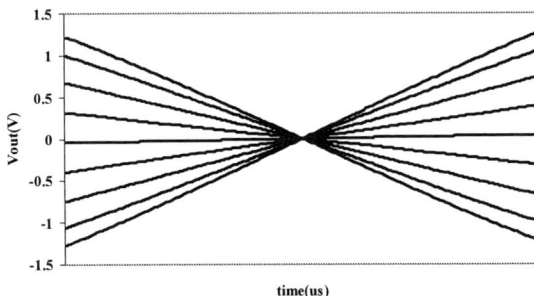

Figure 6. DC characteristics of Output. Input V_{IN1} is a 500KHz - 0.5V_{PP} triangular wave and V_{IN2} is DC voltage stepped from -240mV to 240mV in 9 steps. (Response for 1µs).

quite high. The source–degeneration resistors R_{LIN1}, R_{LIN2}, R_{LIN3} provide a very linear gain. Since there are no significant non-linear intermediate terms generated, their cancellation is obviated. Any higher order non-linear terms (caused by body effect and other factors) get cancelled easily because of the addition of two currents using cross connection. Hence power dissipation is quite reduced and the output of the multiplier is linear. Because of the diode connected load the output node of first stage is a low impedance node. Since there are no significant high impedance nodes in the signal path wide bandwidth is obtained. The voltage gain of the multiplier is $A_{V,MUL} = A_{V1}.A_{V2} = R_L/(R_{LIN3}.R_{LIN2}.g_m)$ where g_m is the transconductance of the diode connected transistor. As shown this is a very compact scheme. Since the circuit is fully differential it can give wide output swing.

TABLE I.   TRANSISTOR AND DEVICE DIMENSIONS

Transistor Dimensions		W/L (µm/µm)
Fig. 2. V–I Converter	M_{n1}, M_{n2}	10/1
	M_{p3}, M_{p4}	5/1
	Current Mirror Transistors (NMOS) for Ibias	50/1
	R_{LIN}	1kΩ
Fig. 3. Multiplier	M_1, M_2, M_5, M_6, M_9, M_{10}, M_{11}, M_{12}	20/1
	M_3, M_4, M_7, M_8	5/1
	Current Mirror Transistors (NMOS) for Ibias = 100µA	50/1
	R_{LIN1} , R_{LIN2}	1kΩ
	R_{LIN3}	2kΩ
	R_L	10 kΩ

TABLE II.   CHARACTERISTICS OF THE MULTIPLIER.

Type	Two-Stage Transconductance
Quadrant	IV
Technology	CMOS 0.5µm AMI
Gain Bandwidth Product	555 MHz
Input Dynamic Swing for V_{IN1}	500 mV_{PP}
Input Dynamic Swing for V_{IN2}	600 mV_{PP}
Output Dynamic Range	3.0 V_{PP}
THD at 30MHz	0.98 %
Power Rails	±2.5 V

303

978-1-4244-3039-0/08 $25.00 © 2008 IEEE

4. Simulation results

The proposed four quadrant transconductance multiplier circuit shown in Fig. 3 is simulated using 0.5µm CMOS AMI technology with V_{DD}=2.5V, V_{SS}= -2.5V and NMOS and PMOS threshold voltages $V_{TH,NMOS}$ =0.73V and $V_{TH,PMOS}$ =−0.95V respectively. The bias current I_B is 100µA. All current sources I_B in the schematic are implemented using NMOS current mirrors with size W/L= 50/1. Rest of the NMOS transistors have a size W/L= 20/1. Table I. lists all the transistor and resistor dimensions.

Fig. 4 shows transient response of the inputs and two differential outputs V_{OUT+} and V_{OUT-}. Fig. 5 shows the transient response of the output $V_{OUT} = (V_{OUT+} - V_{OUT-})$ of the proposed scheme with sinusoidal wave of 0.5V_{PP}, 500KHz at V_{IN1} and 0.6V_{PP}, 30MHz sinusoidal wave at V_{IN2}. The output exhibits high output dynamic swing of 3V_{PP}. Fig. 6 shows the DC characteristics with triangular input of 0.5V_{PP}, 500KHz to V_{IN1} and DC voltage to V_{IN2} stepped from -240mV to 240mV in 9 steps. The characteristics exhibit high linearity for a wide range of input values. The gain bandwidth product is 555MHz. The given scheme is stable within the bandwidth limit since it has no feedback. Though the scheme has two stages it has only one dominant pole at the output of second stage. The pole at the output of first stage is at very high frequency because of the low impedance node at the diode connected load. Table II. enumerates the characteristics of the multiplier.

5. Conclusion

A novel compact high performance four quadrant CMOS transconductance multiplier having linearity, wide dynamic swing and high gain bandwidth product is introduced and validated with simulation results. This is also a power efficient scheme with wide gain adjustment range used in current mode systems like OTAs, mixers and modulators.

6. References

[1] G. Han and E. Sanchez-Sinencio, "CMOS Transconductance Multipliers: A Tutorial", *IEEE Trans. on Circuits and Systems II:* vol. 45, no. 12, Dec. 1998, pp. 1550–1563.

[2] K. Bult and H. Wallinga, "A CMOS four-quadrant analog multiplier," *IEEE J. of Solid-State Circuits*, vol. 21, NO. 3, Jun. 1986, pp. 430–435.

[3] Z. Hong and H. Melchior, "Four-quadrant CMOS analog multiplier," *Electronic Letters*, vol. 20, Nov. 1984, pp.1015–1016.

[4] H. R. Mehrvarz and C. Y. Kwok, "A large-input-dynamic-range multi input floating gate MOS four-quadrant analog multiplier," *IEEE International Solid-State Circuits Conference* pp. 60–61, 15-17 Feb. 1995.

[5] J. F. Schoeman and T. H. Joubert, "Four quadrant analogue CMOS multiplier using capacitively coupled dual gate transistor," *Electronic Letters*, vol. 32, no. 3, 1 Feb. 1996, pp. 209–210.

[6] J. Ramirez-Angulo, S.R.S. Garimella, A.J. López-Martn, R.G. Carvajal, "Gain Programmable current mirrors based on current steering", *Electronics Letters*, vol. 42, no. 10, 11 May 2006, pp. 559-560.

[7] G.A. Hadgis and P.R. Mukund, "A novel CMOS monolithic analog multiplier with wide input dynamic range," *8th International Conference on VLSI Design*, 4-7 Jan. 1995, pp.310-314.

[8] M. S. Sawant, J. Ramirez-Angulo, R.G. Carvajal, A. Lopez-Martin "Wide gm adjustment range highly linear OTA with programmable mirrors operating in triode mode", *IEEE 48th Midwest Symposium on Circuits and Systems*, vol. 1, pp.21-23, 7-10 Aug. 2005.

[9] G. Palmisano and S. Pennisi, "New CMOS tunable Transconductor for Filtering Applications", Proc. *IEEE International Symp. on Circuits and Systems*, vol. 1, 6-9 May 2001, pp.196-199.

[10] B. Gilbert, "A precise four-quadrant multiplier with subnanosecond response", *IEEE J. of Solid State Circuits*, vol. 3, no. 4, Dec. 1968, pp.365-373.

[11] B.S. Bong, "CMOS RF circuits for data communication applications", *IEEE J. of Solid State Circuits*, vol. 21, no. 2, Apr. 1986, pp.310-317.

[12] S.C. Quin and R.L. Geiger, "A +-5v CMOS analog multiplier", *IEEE J. of Solid State Circuits*, vol. 22, no. 6, Dec. 1987, pp.1143-1146.

21st International Conference on VLSI Design

Chaos-modulated ramp IC for EMI reduction in PWM buck converters- design and analysis of critical issues

Rupam Mukherjee, Amit Patra and Soumitro Banerjee
Department of Electrical Engineering
Indian Institute of Technology, Kharagpur, India
Email: rupam@ee.iitkgp.ernet.in

Abstract

Various non-conventional methods have been employed in the past, to reduce the cost and weight of traditional conducted EMI filters and radiation screens for EMI suppression in switching power electronic converters. This paper points out various shortcomings of these methods which are mainly frequency modulation based, and describes the design of a ramp-generator IC based on a modified modulation scheme. This IC can be used on any voltage mode controlled converter and has a feature that enables the user to tune the same converter to various EMC norms. Test results from a prototype showing significant reduction in harmonic power level have been presented. Moreover, this paper discusses a theoretical formulation for calculating the output capacitor size to maintain ripple specifications, when operating under chaotic modulation.

1. Introduction

One of the most important problems faced by today's electronic hardware manufacturers is the EMI generated by switching power converters. They interfere with communication signals and degrade power supply quality.

Conventional methods to tackle the EMI problems include LC filters for blocking conducted EMI and metal shielding for radiated EMI. These solutions come with their own problems like increased cost, size and weight. Moreover, such a design is not portable across different EMC norms because each time the application domain changes, filters and screens have to be redesigned even if the converter specifications themselves are unchanged. This incurs additional expenses and increased design time.

Some unconventional methods to counter the above problems involve spectral modification at source. They aim at reducing EMI power at harmonic frequencies rather than screening the generated EMI. The feasibility of utilizing the broadband nature of chaotic waveform for EMI reduction was explored in [3] and [7]. Furthermore, it was explained in [2, 10], that all switching converters are highly non-linear in nature and have chaotic operation in certain parameter range. In [3], converter operation under chaotic mode was reported to be simple in terms of quantity of hardware used. However, [7] reported some inherent problems of chaotic operation like increased ripple and higher emissions floor. A common shortcoming of all these methods is that they are not portable i.e., the converter once designed for a particular EMC norm cannot be tuned to satisfy some other norm, and hence have to be over-designed.

To counteract the above problems, modulating the converter oscillator was considered. In [1], a pseudo-random modulation of the switching signal or clock was explored. In [8] chaotic modulation was shown to outperform pseudo random modulation. In [6, 5], chaotic modulation of the clock/ramp was suggested as a method of spectral modification. There are two basic schemes for achieving this target, i.e., *frequency modulation* and *amplitude modulation*. In [5], chaotic modulation through FM was described.

However the basic problem with FM is that actually the frequency keeps shifting and the full power remains concentrated at the instantaneous frequencies. Hence, though in a wide sense, power level reduction is achieved, effective reduction is much less in small time intervals. This is precisely the reason why existing schemes like linear frequency sweep are also inadequate.

In [6], a chaos-based scheme similar to AM was suggested, in which there exists no time interval in which the full power is concentrated at any particular frequency. The present IC has been designed based on this scheme. Additionally, the portability issue has been addressed through a feature which enables on-board variation of the magnitude of an internal chaotic signal in order to vary the amount of EMI reduction.

305

978-1-4244-3039-0/08 $25.00 © 2008 IEEE

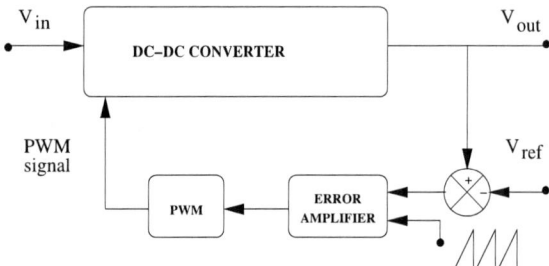

Figure 1. Block diagram voltage mode control

2. Modulation scheme

In brief, Fig. 1 describes the block diagram of voltage-mode controlled buck converter. The falling edge of the ramp in Fig. 1 is made to vary chaotically, creating a jitter as in Fig. 2. This causes the power at harmonic frequencies to be spread within a certain band about the mean frequency and consequently, the height of the fundamental component goes down. The extent of this modulation and thus, the width of the spread is tunable. The IC we have developed, gives as output, this modulated ramp and the amount of modulation is adjusted with a variable resistor on the test board, which is connected to an internal gain block.

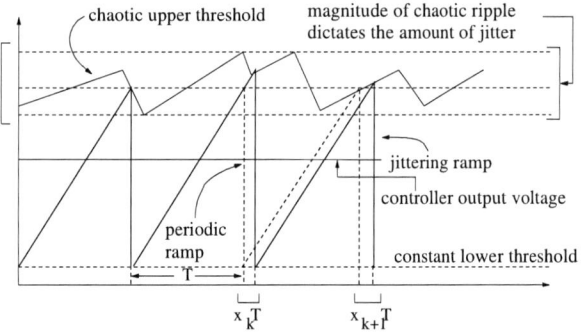

Figure 2. Nature of jittering output ramp

This modulation is achieved using the scheme described in Fig. 3. It uses the chaos generator circuit described in [9]. It has been seen that such a circuit generates chaotic i.e. bounded aperiodic signal under certain magnitudes of circuit elements. An experiment involving a discrete PCB realization of this scheme and giving significant power-level reduction was reported in [6]. A buck converter was used, but in general, this hardware may be used with other switching regulators also.

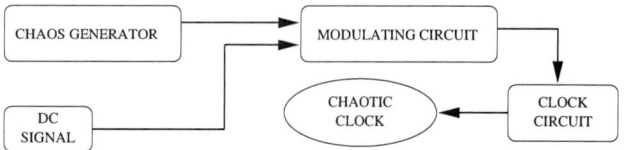

Figure 3. Chaotic modulation scheme

3. Chaos generator

The chaos generator used in this hardware is a very simple circuit having the schematic of Fig. 4. It does not involve inductors like other popular chaos generator architectures like Chua's circuits and hence, particularly compatible to fully analog on-chip realization.

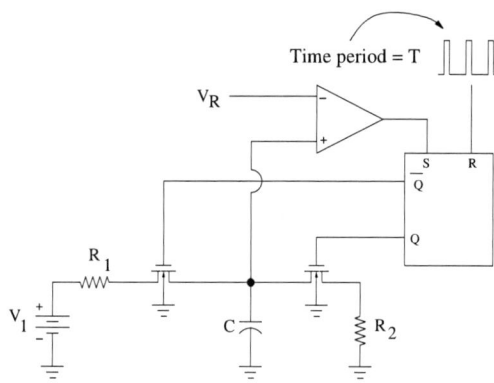

Figure 4. Chaos generator schematic

It was first introduced in [9]. Considering the capacitor voltage $v(t)$ as the output, it has been shown in [9] that it will be chaotic for certain combination of parameters $T_1 = \frac{T}{R_1 C}$, $T_2 = \frac{T}{R_2 C}$ and $\beta = \frac{V_R}{V_1}$

4. Chaotic ramp hardware

The chaotic ramp hardware designed, has the scheme described in Fig. 3. The chaos generator is designed as above, with the parameters $T_1 = 0.1$, $T_2 = 0.45$, $\beta = 0.56$. A separate DC level is added to the chaotic waveform to set the frequency externally. The clock is generated using the standard scheme shown in Fig. 5. A capacitor charges and discharges between upper and lower thresholds and thus, the nature of the output ramp follows Fig. 2.

The gain block amplifies the difference of the chaotic signal and its average value. Thus the mean frequency remains same while the amount of gain decides the amount of chaos that is injected into the system and hence, the amount of power spectrum reduction. The gain block structure is depicted in Fig. 6. The average value is extracted by an internal RC filter.

978-1-4244-3039-0/08 $25.00 © 2008 IEEE

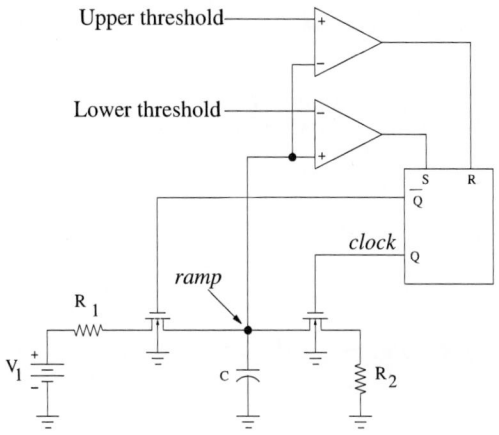

Figure 5. Schematic of clock circuit

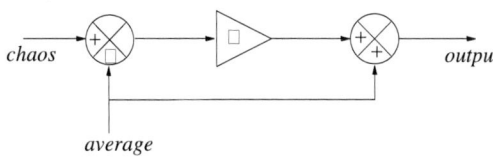

Figure 6. Structure of gain block

5. Experimental results

We have fabricated the chaotically modulated clock/ramp IC in a National Semiconductors 0.5μ Bi CMOS technology. Its salient features are shown in Table 1. The chaos generator and gain block are included within the IC. The chaotic waveform and the modulated ramp are present at the output. A variable resistor on the test board does the job of adjusting the amount of jitter to get different levels of power spectrum reduction. Fig. 7 shows a typical chaotic waveform generated in the IC.

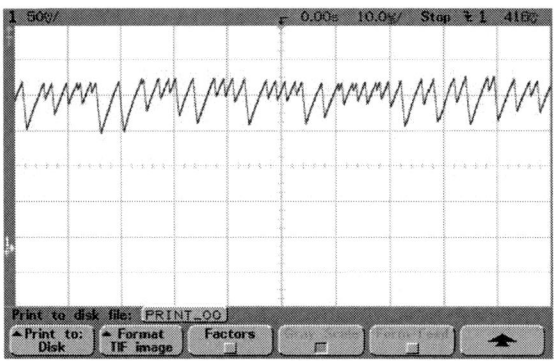

Figure 7. Output of chaos generator

Figs. 8- 9 show the ramp output under different magnitudes of jitter. It is to be noted that the ramp is dual sloped because it is designed with a futuristic target of driving

higher frequency PWM converters and in such situations, comparator delay makes a sawtooth waveform infeasible.

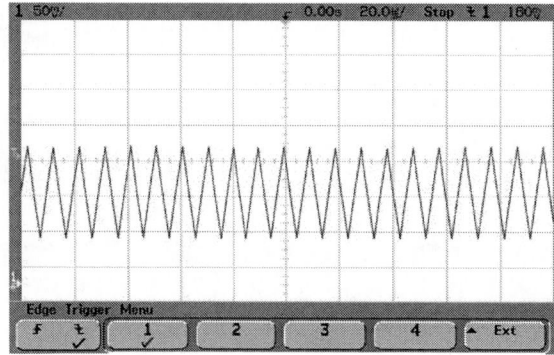

Figure 8. Ramp output with minimum amount of jitter

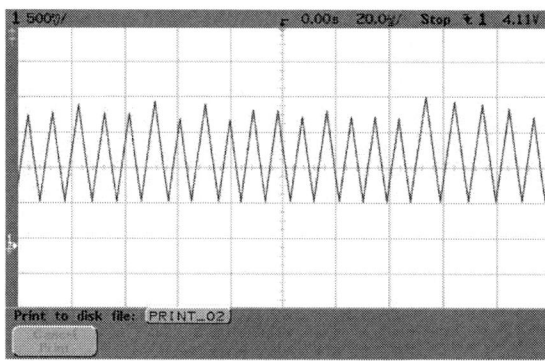

Figure 9. Ramp output with a greater amount of jitter

The ramp output from this IC was applied to a voltage-mode controlled buck converter having the specifications as in Table 2.

Features	Specifications
Supply voltage	5 V
Power consumption	50 mW
Die area (without pad)	$1950\mu \times 1850\mu$
Die area (with pad)	$2562\mu \times 2307\mu$

Table 1. IC specifications

The inductor current waveform was acquired and its FFT calculated under different amounts of jitter. Figs. 10-11 show the inductor current and its FFT under different amounts of jitter.

A significant reduction of close to 20 dB is achieved in the magnitude of the fundamental harmonic component.

978-1-4244-3039-0/08 $25.00 © 2008 IEEE

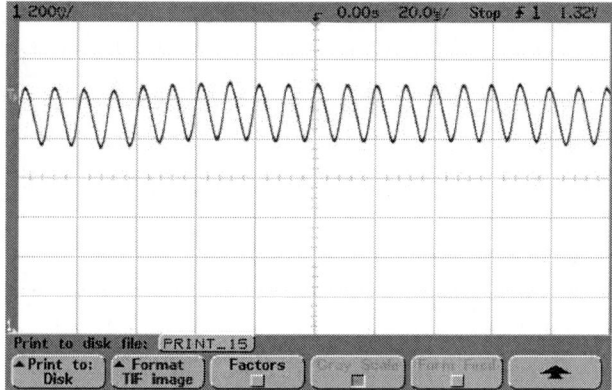

(a) Inductor current waveform with minimum amount of jitter

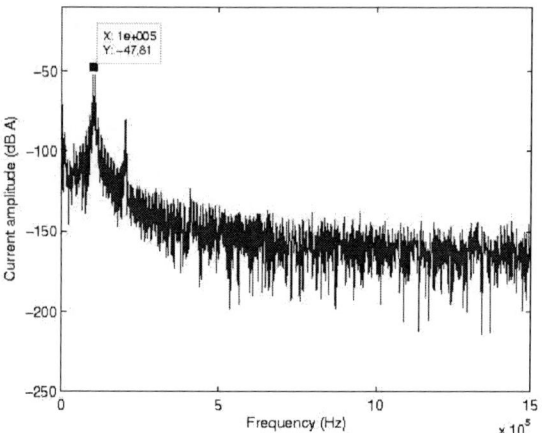

(b) FFT of inductor current waveform with minimum amount of jitter

Figure 10. Nature of inductor current with minimum amount of jitter

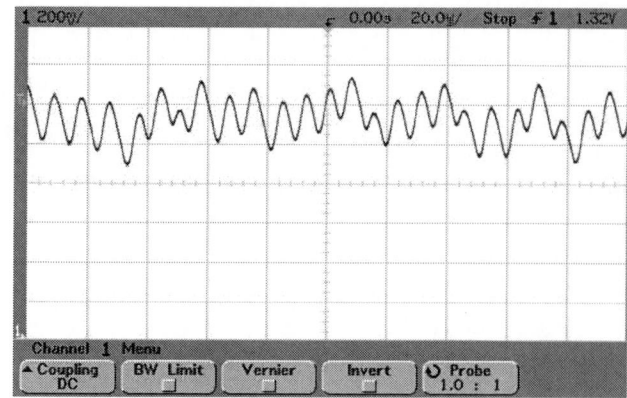

(a) Inductor current waveform with a greater amount of jitter

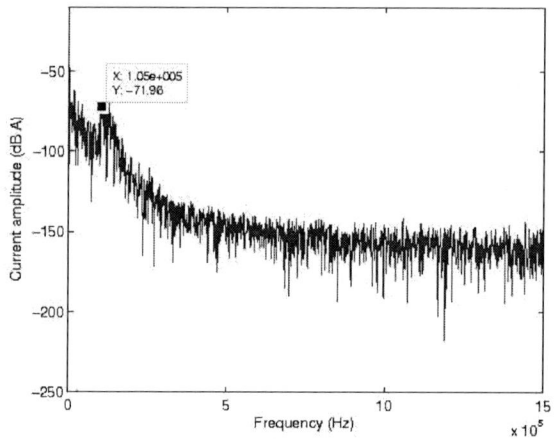

(b) FFT of inductor current waveform with a greater amount of jitter

Figure 11. Nature of inductor current with a greater amount of jitter

Simply adjusting the variable resistor on the test board achieved this reduction from Fig. 10 to Fig. 11. The maximum jitter of the ramp waveform was about 0.4 times T, T being the time period.

6. Effect of modulation on output voltage ripple

Modulation techniques like the ones referred to and discussed in this paper tend to increase the output voltage ripple. However, ripple specifications are tightly dictated by the application and any increase of ripple must lead to increase of output capacitor to maintain specifications. This will, in turn put additional real estate burden on the system board, a detailed analysis of which is essential. In Appendix A, such an analysis has been presented and a relation between capacitance increase and maximum amount of jitter

has been arrived at, in eqn. 17.

We now consider a typical input filter design example. A converter with specifications as in the experiment is considered. A typical single stage input filter with configuration shown in Fig. 12 has been designed once for attenuation 90 dB and again for 70 dB, assuming that our chaotic ramp hardware makes up for the remaining attenuation. Design equations presented in [4] have been used for this purpose. Keeping L_f constant, we have calculated C_b values in both cases as $2200 \mu F$ and $396 \mu F$ respectively.

Eqn. 17 gives us, assuming duty cycle of 0.25 and maximum jitter of $+/- 0.4$, the additional output capacitance required, to maintain ripple specifications, is 3.4 times the original output capacitance. Thus, the increase of output capacitance is less than the reduction of filter capacitance.

978-1-4244-3039-0/08 $25.00 © 2008 IEEE

Input	5 V
Output	2 V, 1 A
Frequency	100 kHz
Control	Voltage-mode (proportional control)

Table 2. Test-converter specifications

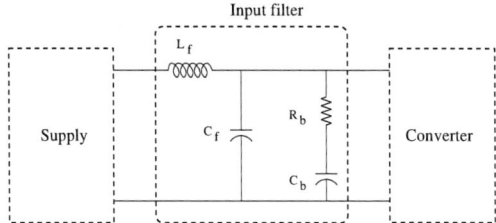

Figure 12. Configuration of single stage input EMI filter for buck converter

7. Conclusion

This paper presents the design and test results of a ramp generator IC for chaos-based EMI reduction in voltage mode controlled converters. It has a feature by which the user may tune his converter hardware to match various EMC standards. A prototype was tested on a voltage-mode controlled buck converter and results have shown up to 20 db power reduction. Also, a theoretical discussion for the effect of modulation on output capacitor requirement, has been given.

8. Acknowledgement

The authors are thankful to National Semiconductors Corp. USA for the fabrication facillity and Indian Space Research Organisation for technical mentorship of the project.

References

[1] A.M.Stankovic, G.C.Verghese, and D.J.Perreault. Analysis and synthesis of randomized modulation schemes for power converters. *IEEE Transactions on Power Electronics*, 10(6):680–693, 1995.

[2] J.H.B.Deane and D.C.Hamill. Instability, subharmonics, and chaos in power electronic systems. *IEEE Transactions on Power Electronics*, 5(3):260–268, July 1990.

[3] J.H.B.Deane and D.C.Hamill. Improvement of power supply EMC by chaos. *Electronic Letters*, 32(12):1045, June 1996.

[4] D. Maximovic and R. Ericksson. *Fundamentals of Power Electronics*. Springer Verlag.

[5] M.Balestra, M.Lazzarini, G.Setti, and R.Rovatti. Experimental performance evaluation of a low-EMI chaos-based current-programmed dc/dc boost converter. In *Proc. IEEE International Symposium on Circuits and Systems (IS-CAS'05)*, pages 1489–1492, 2005.

[6] R.Mukherjee, S.Nandi, and S.Banerjee. Reduction in spectral peaks of dc-dc converters using chaos-modulated clock. In *Proc. IEEE International Symposium on Circuits and Systems (ISCAS'05)*, pages 3367–3370, Kobe, Japan, May 2005.

[7] S.Banerjee, A.L.Baranovski, J.L.R.Marrero, and O.Woywode. Minimizing electromagnetic interference problems with chaos. *IEICE Transactions on Fundamentals*, E87-A(8):2100, 2004.

[8] S.Callegari, R.Rovatti, and G.Setti. Chaotic modulations can outperform random ones in electromagnetic interference reduction tasks. *IEE Electronics Letters*, 38(12):543–544, June 2002.

[9] S.Mandal and S.Banerjee. Analysis and CMOS implementation of a chaos-based communication system. *IEEE Transactions on Circuits and Systems-I*, 51(9), 2004.

[10] G. C. Verghese and S. Banerjee, editors. *Non-linear Phenomena in Power Electronics*. IEEE Press., New York, 1992.

Appendices

A. Treatment of voltage ripple

In this appendix, we present the theoretical analysis required for determining the increase in output capacitance due to chaotic modulation.

A.1. Problem formulation

Let $x_k T$ be the deviation of the kth falling edge from kT. Fig. 2 describes the jitter of the ramp.

The inductor current(I_L) and capacitor voltage(V_C) is shown in Fig. 13 Lets' assume the following variables:

$$
\begin{aligned}
s_1 &= \text{charging slope of inductor current in Fig. 13} \\
s_2 &= \text{discharging slope of inductor current in Fig. 13} \\
V &= \text{controller output voltage in Fig. 2} \\
m &= \text{slope of the ramp in Fig. 2} \quad (1)
\end{aligned}
$$

A.2. Calculation of I_L maxima and minima

$$
\begin{aligned}
y_{k+1} &= y_k + s_1 \frac{V}{m} - s_2 \left[(k + 1 + x_{k+1})T \right. \\
&\quad \left. - (k + x_k)T - \frac{V}{m} \right] \\
\Rightarrow y_{k+1} &= y_k + (s_1 + s_2)\frac{V}{m} - s_2 T \left[1 + (x_{k+1} - x_k) \right]
\end{aligned}
$$

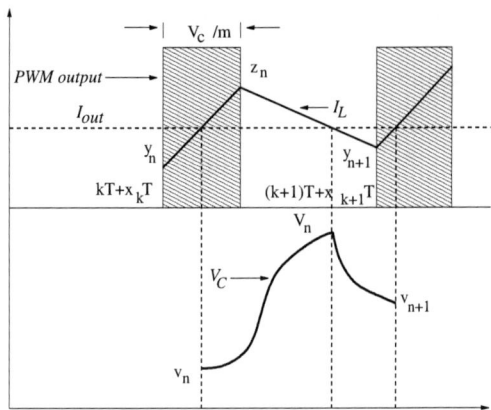

Figure 13. I_L and V_C **waveforms**

$$\Rightarrow y_{k+2} = y_{k+1} + (s_1 + s_2)\frac{V}{m}$$
$$- s_2 T \left[1 + (x_{k+2} - x_{k+1}) \right] \quad (2)$$

Summing up to $k + n$ and noticing, $\frac{V}{m} = DT$ and $s_1 DT = s_2(1 - D)T = \Delta I$, we have

$$y_{k+n} = y_k - \frac{\Delta I}{1 - D}(x_{k+n} - x_k) \quad (3)$$

Also,

$$z_n = y_n + \Delta I \quad (4)$$

Therefore,

$$z_n = y_0 - \frac{\Delta I}{1 - D}(x_n - x_0) + \Delta I \quad (5)$$

A.3. Calculation of capacitor voltage maxima and minima

Referring Fig. 13, we have

$$C(V_n - v_n) = \frac{1}{2C}\left(\frac{1}{s_1} + \frac{1}{s_2}\right)(z_n - I_{out})^2 \quad (6)$$

$$C(V_n - v_{n+1}) = \frac{1}{2C}\left(\frac{1}{s_1} + \frac{1}{s_2}\right)(I_{out} - y_{n+1})^2 \quad (7)$$

$$\Rightarrow v_{n+1} - v_n = \frac{1}{2C}\left(\frac{1}{s_1} + \frac{1}{s_2}\right)(z_n - y_{n+1}) \times$$
$$(z_n + y_{n+1} - 2I_{out}) \quad (8)$$

Now, referring to equations 5 and 3 and making the following choice of initial conditions:

$$y_0 = I_{out} - \frac{\Delta I}{2}$$
$$x_0 = 0 \quad (9)$$

we can write

$$z_n - y_{n+1} = A(x_{n+1} - x_n) + B$$
$$z_n + y_{n+1} - 2I_{out} = -A(x_n + x_{n+1}) \quad (10)$$

Please note, that we have substituted $A = \frac{\Delta I}{1-D}$ and $B = \Delta I$ to reach the above relations. Hence, referring eqn. 8 and substituting $\alpha = \frac{1}{2C}\left(\frac{1}{s_1} + \frac{1}{s_2}\right)$ we get

$$v_{n+1} - v_n = -\alpha A^2 \left(x_{n+1}^2 - x_n^2\right) - \alpha AB(x_n + x_{n+1})$$
$$\Rightarrow v_{n+2} - v_{n+1} = -\alpha A^2 \left(x_{n+2}^2 - x_{n+1}^2\right)$$
$$- \alpha AB(x_{n+1} + x_{n+2})$$
$$\Rightarrow v_{n+3} - v_{n+2} = -\alpha A^2 \left(x_{n+3}^2 - x_{n+2}^2\right)$$
$$- \alpha AB(x_{n+2} + x_{n+3}) \quad (11)$$

Adding up the series, we get

$$v_{n+k} - v_n = -\alpha A^2 \left(x_{n+k}^2 - x_n^2\right) + \alpha AB(x_{n+k} - x_n)$$
$$- 2\alpha ABkE[x_k] \quad (12)$$

Now, assuming large value of k and $E[x_k] = 0$, we have

$$v_{n+k} - v_n = \alpha \left[\frac{\Delta I}{1 - D}\right]^2 \left[(1 - D)(x_{n+k} - x_n) - \left(x_{n+k}^2 - x_n^2\right)\right] \quad (13)$$

Referring to eqn. 8 and considering the initial conditions already mentioned,

$$V_n - v_n = \alpha \left[\frac{\Delta I}{1 - D}\right]^2 \left[\frac{1 - D}{2} - x_n\right]^2 \quad (14)$$

Therefore under jittery conditions

$$V_n - v_{n+k} = \alpha \left[\frac{\Delta I}{1 - D}\right]^2 \left[\left(\frac{1 - D}{2} - x_n\right)^2 - (1 - D)\right]$$
$$(x_{n+k} - x_n) + \left(x_{n+k}^2 - x_n^2\right)] \quad (15)$$

Under nominal conditions, when there is no jitter i.e., $x_n = x_k = 0$,

$$V_n - v_{n+k} = \alpha_1 \left[\frac{\Delta I}{1 - D}\right]^2 \left(\frac{1 - D}{2}\right)^2 \quad (16)$$

However, the maximum deviation must remain unchanged as it is dictated by the ripple requirements of the converter. Hence, equating the RHS of equations 15 and 16, we get

$$\frac{C}{C_1} = \left[\frac{2}{1 - D}\right]^2 \left[\left(\frac{1 - D}{2} - x_n\right)^2\right.$$
$$+ (x_{n+k} - x_n)(x_{n+k} + x_n - (1 - D))]$$
$$\left(\frac{C}{C_1}\right)_{max} = \left[\frac{2}{1 - D}\right]^2 \left[\left(\frac{1 - D}{2}\right)^2 + (|x_n|_{max})^2\right.$$
$$- (1 - D)x_{n(min)} + |x_{n+k} - x_n|_{max} \times$$
$$\left(2x_{n(max)} - (1 - D)\right)] \quad (17)$$

This is the expression for the output capacitance value as a function of the maximum jitter of the clock about the mean position.

21st International Conference on VLSI Design

A Fast settling 100dB OPAMP in 180nm CMOS process with compensation based optimisation.

Amal Kumar Kundu, Student, E and ECE Dept. IIT Kharagpur,
Subho Chatterjee, Student, Electrical Dept. IIT Kharagpur and
Tarun Kanti Bhattacharyya, Faculty,E and ECE Dept. IIT Kharagpur.

Abstract

A two-stage gain-boosted OPAMP with 100dB DC gain, 807MHz unity gain bandwidth (UGB) and rail to rail output swing in 180nm digital CMOS process is presented. A compensation based optimisation methodology for fast settling response and closed loop stability is also described for this topology. Optimised settling response provides 0.001% settling time equal to 9.7ns. This OPAMP is designed to be used as a current to voltage converter for 15-bit 100MSamples/s DAC application. It could drive an off-chip capacitive load of 10pF in parallel with a 500ohm resistor and provides -94dB THD at 100KHz with a closed loop gain of 20dB and 600mV output swing.

Gain boosting, Fast settling, High gain-bandwidth, off-chip capacitive load, Improved THD, pole-zero doublet, global optimisation.

1 Introduction

since OPAMP is the most important component for many analog circuits, its performance makes significant impact on the analog system. For example, resolution of data converters are limited by the DC gain, settling accuracy, THD and SNR of the OPAMP. Where as, sampling rate is limited by settling time and bandwidth. Continuous research work is going on to improve its critical parameters like, DC gain, Unity gain-bandwidth, Settling time and accuracy, THD etc. Due to short channel effect, DC gain of a simple differential pair is restricted to 20dB only. On the other hand, UGB is limited by the parasitic poles for multi-stage implementation and settling response depends on relative pole positions.

High-speed and high gain OPAMP use only single stage implementation to get rid off parasitics. To achieve high gain and band width telescopic or folded cascode are vastly used[1]. Gain-boosting technique shown in [2] is used to further increase the DC gain. However, cascoding of transistors limits output swing and creates voltage headroom

problem in low voltage application. Again single stage regulated cascode is not sufficient to fulfil the gain requirement due to high current density in today's transistors. Further, telescopic cascode stage could not be used as unity gain buffer and folded cascode stage suffers from lower DC gain, higher current consumption and low frequency pole. OPAMP presented in [4] has been designed for sample/hold application with 88dB DC gain, 725MHz UGB and 5V power supply.

This OPAMP has been designed with higher DC gain, improved UGB, better linearity and faster settling response. Here a two stage implementation is used to overcome output swing limitation in low voltage application. Again, single stage telescopic cascode could not drive off-chip capacitive load. Class-A output stage in this two stage implementation, helps to achieve rail to rail output swing and drive large capacitive load without affecting gain-bandwidth. It also enables us to configure the OPAMP as unity gain buffer. Gain boosted input stage helps to increase DC gain without affecting the unity gain bandwidth. A brief description of the architecture is given in section II. section III describes Pole-zero analysis for closed loop stability and fast settling. Optimisation methodology for fast settling response followed by simulation results, conclusion and references are given in section IV, V, VI and VII respectively.

2 Architecture

This two stage OPAMP consists of three amplifiers: Telescopic cascode input stage, Folded cascode gain boosting amplifier (GBamp) and class-AB output stage. Forward signal path including both stages and GBamp is shown in Fig.1. DC gain of the input stage would be

$$A_{DC(in)} \approx g_{m1} \times r_5 \qquad (1)$$

$$where \; r_5 \approx (r_{d8}g_{m6}r_{d6})||(r_{d2}g_{m4}r_{d4}A_{DCgb}) \qquad (2)$$

311

978-1-4244-3039-0/08 $25.00 © 2008 IEEE

Where A_{DCgb} is the low frequency gain of gain-boosting amplifier. Output amplifier has been implemented with class-AB stage to have more DC gain and rail to rail output swing which makes the design power efficient compared to class-A stage. DC gain of the output stage is given by

$$A_{DC(out)} \approx \frac{g_{m9} + g_{m10}}{g_{ds9} + g_{ds10}} \qquad (3)$$

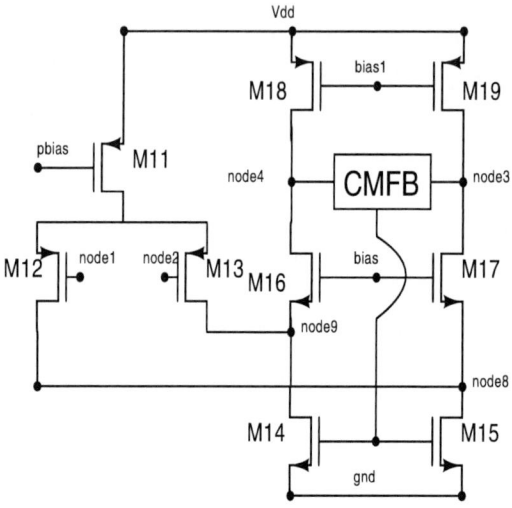

Figure 2. CMFB Circuit.

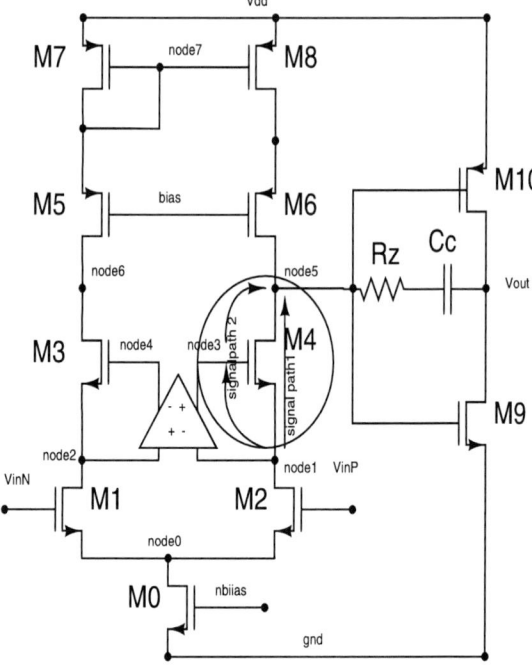

Figure 1. OPAMP circuit diagram

Overall DC gain is $A_{DC} = A_{DC(in)} \times A_{DC(out)}$. Since parasitic poles due to gate capacitance of transistors $M_5 - M_8$ does not affect high frequency response of the OPAMP, these transistors are implemented with larger length to avoid gain boosting at the load side of the input stage. Transistors $M_1 - M_4$ are implemented with lower length to push non dominant poles and zeroes to higher frequency. Reduction in output impedance is compensated by GBamp. GBamp has been implemented with folded cascode stage as described in Fig.2. DC gain of this stage is given by

$$A_{DC(gb)} \approx g_{m12}r_3 \qquad (4)$$

$$where \quad r_3 \approx r_{d19}||((r_{d15}||r_{d12})g_{m17}r_{d17}) \qquad (5)$$

Output impedance of input stage would be multiplied by the gain of GBamp. Thus GBamp helps to increase overall

DC gain without affecting high frequency response of the OPAMP. However existence of GBamp creates two parallel signal path at the input stage as shown in Fig.1. This parallel path creates a pole-zero doublet around the UGB of the GBamp. This pole-zero doublet makes settling response slower than that expected for a two pole system. Pushing this doublet towards high frequency may lead to closed loop instability. Thus it is essential to find out the optimum position of this pole-zero doublet to get fast settling as well as closed loop stability. Gain bandwidth product for this OPAMP is given by

$$\begin{aligned} Gain\ bandwidth &= P_1 \times A_{DC} \\ &= \frac{1}{r_5 A_{DCout}C_c} \times g_{m1}r_5 A_{DCout} = \frac{g_{m1}}{C_c} \end{aligned}$$

where P_1 is the first dominant pole.

Again, 15-bit DAC application needs around -90dB THD at the signal frequency. Negative feedback helps to reduce THD, if open loop gain is large enough at high frequency. Since at high frequency open loop gain reduces, THD reduction needs to be done at the source i.e. linearity of the OPAMP needs to improve. To improve the linearity of the OPAMP, tail current source(M_0) of the input stage is made as a matched device with the input transistors($M_1 and M_2$). This matched current source minimise odd harmonics as shown in [8].

3   Pole Zero analysis for closed loop stability and fast settling:

This two stage implementation contributes two dominant poles to the system transfer function. Conventional

Miller compensation technique has been employed to separate these two dominant poles. Location of the poles and zeroes are as follows:

1. The first dominant pole appears at the output of input stage(node5 of Fig.1.) and is given by $P_1 = \frac{1}{R_1 \times A_{DCout} \times Cc}$.

2. Second dominant pole is due to output node (Vout of Fig.1) of overall OPAMP and defined by $P_2 = \frac{g_{m9}+g_{m10}}{C_L+Cc}$.

3. One non-dominant pole is due to gaiain boosting at node3 (Fig.1) located at the -3dB bandwidth of GBamp.

4. One non-dominant pole appear at cascoding point (node1 Fig.1).

5. Another non-dominant pole appear at cascoding point of GBamp (node8 Fig.2).

6. One zero is due to the feed forward path around M4 in Fig.1.

7. Another LHP zero is due to nulling resistor Rz placed just beyond the UGB of overall OPAMP.

Miller capacitor Cc (in Fig.1) separates the dominant poles $P_1 and P_2$ to assure closed loop stability in unity gain mode. Thus, circuit is optimised for DC-gain, UGB, phase margin and power consumption. Settling response depends on the relative position of dominant poles $P_1 and P_2$ (i.e separation factor $\frac{P_1}{P_2}$), phase margin of over all amplifier, and position of the pole-zero doublet.

Hence an optimisation procedure is used to find out the global optimum solution in terms of current consumption and performance parameters like DC-gain, UGB, settling time. Complete design cycle consists of three distinct phases:

1. Selection of proper circuit topology which can achieve required specification. Based on the specification calculation of bias voltages for various transistors.

2. Identification of the design variables and determining their range of variation based on circuit knowledge and first order analytical calculations.

3. Exploration of the feasible design space within the specified boundary limits and finding the global optimum solution for settling response, DC-gain, UGB and current consumption.

4 Optimisation Methodology

The constraints imposed on the GBamp in terms of the high end specifications call for an efficient exploration of its design space. For efficient exploration of the design space and achievement of the design objectives, we use a hybrid optimisation technique with the Differential Evaluation algorithm based optimisation as the global optimisation algorithm. The good convergence properties and suitability

for parallelisation coupled with the conceptual simplicity and ease of use make DE an efficient method for optimising real valued objective functions. DE also has the added advantage of having a few control variables for the process which remain constant throughout the procedure. The key idea behind the differential evolution algorithm is to add the weighted difference vector between two population members to a third member of the population. If the resulting vector has the value of the objective function for it to be less than any predetermined existing population members, the offs-pring goes on to replace that particular member in the population. Thus, if V represents a vector of the Ith generation and R represents the resulting vector in the $(I+1)$th generation then:

$$R_{I+1} = V_{r1,I} + F(V_{r2,I} - V_{r3,I}) \tag{6}$$

The real and constant factor $F \epsilon [0, 2]$controls the differential amplification. We have used the scheme "DE/rand-to-best/1/exp" for the optimisation routine as described in the following section.

4.1 DE/rand-to-best/1/exp

This scheme places the perturbation between a randomly chosen member of the population and the best population member. The relationship between the resulting vector and the parent vectors is established by:

$$R_{I+1} = V_{r1,I} + \lambda(V_{best,I} - V_{r1,I}) + F(V_{r2,I} - V_{r3,I}) \tag{7}$$

where λ is the controlling factor for the greediness of scheme. For simplification purposes, λ is taken to be equal to F. The exponential cross over mechanism increases the diversity of the population as shown in [7].

4.2 The Choice of Cost Function and Control Variables

The cost function or the objective function formation is one of the most important aspects of the optimisation procedure. The proper formation of the objective function and setting of control variables is crucial to the fast convergence of the algorithm. In this application, the circuit parameters(transistor widths, resistor values etc) are chosen as the optimisation variables. Furthermore, if the circuit has "m" optimisation objectives(maximisation or minimisation) and "n" specification constraints to be achieved, the cost function is represented as:

$$Cost = \sum_{i=1}^{m} W_{obj_i} P_{sim_i} + \sum_{j=1}^{n} W_{con_j} \frac{P_{obs_j} - P_{sim_j}}{P_{obs_j}} \tag{8}$$

where P_{sim} and P_{obs} are the simulated and observed specifications for the circuits. The weights are usually adjusted after an initial first cut run.In low dimensional problems, high crossover probability values work better to ensure the diversity of the population. Hence for our case (dimension < 10) we keep the value of Crossover factor(CR) as in [7] almost equal to 1. The value of F should not be too small to avoid local minimum convergence, and not too large as in that case it will explore a wider search space at the cost of an increased number of simulations and consequently simulation time. Typically to strike a balance, the value of F is chosen to be equal to 0.85. The objective of the GBamp circuit we decided was to achieve a minimum settling time(objective) meeting the demands for constraints on the gain, UGF and phase margin. Thus the weights are adjusted as 10 and 100 for the objectives and constraints respectively. The optimisation procedure yields a design point in the design space of the GBamp satisfying our objectives. To ensure convergence to a global minimum, the process is made to run for a large (> 500) number of generations. If the evolutionary algorithm converges to the minimum even before that the cost variance is a good metric for its detection.

4.3 Choice of optimisation variables

The choice of the appropriate parameters for the optimisation is a crucial step. We initially try to map the relevant specifications on to the circuit parameters such as the $\frac{W}{L}$ ratios of transistors, resistances and capacitances. The settling time is determined by the ratio of the two dominant poles as well as the nulling zero and the pole zero doublet arising from the gain boosting stage. The pole ratios are controlled by $\frac{W}{L}$ of the transistors M_0, M_{10}, M_9. The nulling zero is controlled by the nulling resistor R_z and compensation capacitor C_c. The dB amplifier cutoff frequency is the influencing factor for the pole zero doublet and in turn is attributed to the $\frac{W}{L}$ ratios of the transistors $M_{12}, M_{13}, M_{11}, M_{14}$ and M_{15}. The gain for the expression has been deduced at an earlier stage and is given by the aspect ratios of the transistors M_{10}, M_1 and M_2. The current source transistor M_5 also has a controlled influence also. The GBW is also dependent on aspect ratios of transistors M_0, M_1, M_2 and the compensation capacitor C_c. Thus we begin the procedure of optimisation with the identified circuit parameters as the optimisation variables and choose the initial parameter values as those corresponding to a valid operating point in the design space.

5 Results

The optimiser has been tuned with eight design variables: $(\frac{W}{L})_0$, $(\frac{W}{L})_{1-2}$ $(\frac{W}{L})_9$, $(\frac{W}{L})_{10}$, $(\frac{W}{L})_{11}$, $(\frac{W}{L})_{12-13}$, R_z, C_c. The objective function is defined by three constraints (DC-gain, UGB and Phase margin) and the objec-

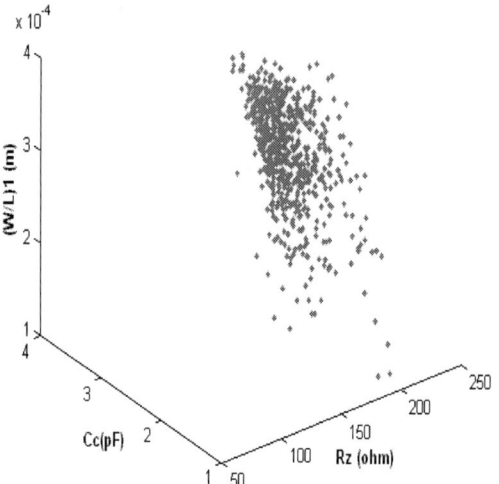

Figure 3. Design space.

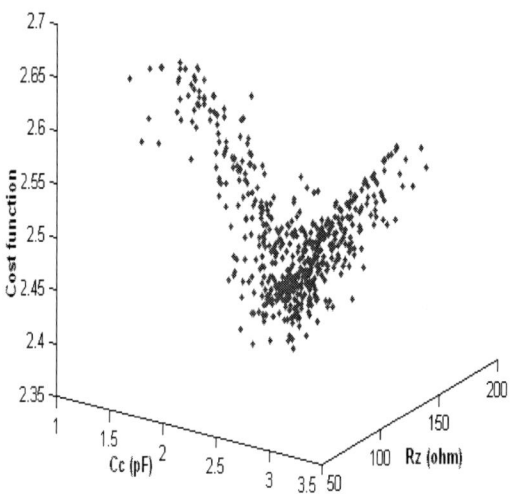

Figure 4. Cost function variation.

tive is defined to be minimisation of settling time. The maximum number of generations for the optimisation procedure is initially set to 100. The optimiser converges to a global minimum after running for 41 generations for the DE/rand-to-best/1/exp procedure with CR=1 and F=0.85(typ). The design specifications corresponding to the global minima after 41 generation with the results are given in the table-1. Optimiser also proves an estimate of the feasible design space under the constraint conditions. Fig.3 shows the feasible design space for three design variables: $(\frac{W}{L})_{12}, R_z, C_c$. Fig.4 represents the distribution of cost function as a function of R_z and C_c and the concentration of points in the zone around $R_z$in the zone 100 - 150ohm and C_c in the zone $1-1.6pF$ is indicative of the convergence to the global minimum of the cost function in this zone.

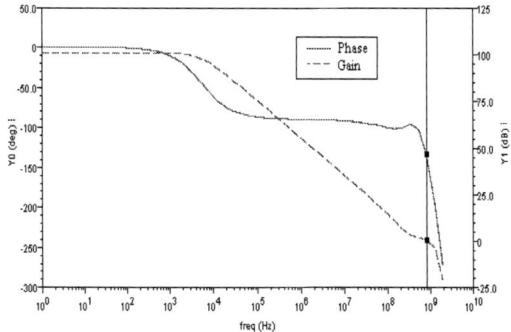

Figure 5. AC response.

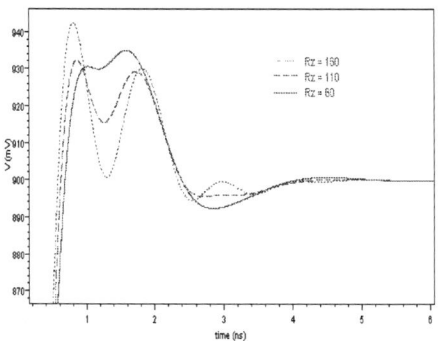

Figure 6. Settling response for different value of nulling resistor(R_z)

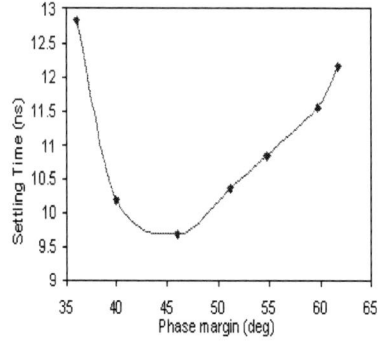

Figure 7. Variation of settling time with phase margin.

Fig.5 represents the AC gain and phase plot. It indicates the location of P_1 and P_2 at 4KHz and 1GHz respectively. Step response in unity gain feedback mode is given in Fig.6 for different value of nulling resistor R_z. It indicates that, optimumally damped settling response occurs

for R=110ohm. Again, Fig.7 represents variation of settling time with phase margin indicating the minima at 45degree. Summary of the simulation result is given in table-I.

Table 1. Simulation result summary

PARAMETER	RESULT
Supply	1.8V
Process	0.18μm epi-CMOS
DC Gain	101.43dB
Unity gain bandwidth	807MHz
Phase margin	46 degree
Settling time (0.001%)	9.7 nS
Power consumption	20mA
CMRR	-60dB
PSRR+	-41dB
PSRR-	-38dB
Input offset	300μV
THD (closed loop gain = 10, 100KHz, output swing = 600mV)	-94dB

A comparative study of the critical performance parameters are shown in Table-II. It shows that

Table 2. Comparison of specification

OPAMP	[3]	[4]	This work
Process	.6u	.6u	.18u
Supply(V)	3	5	1.8
DC gain(dB)	101	88	101
Load (pF)	1.5	2	10
UGB(MHz)	901	725	807
power(mW)	9.3	26	36
Output swing	0.75	3	rail to rail
Settling Time(ns)	8	5@.1%	9.7@.001%
Resolution (Bit)	14	12	15
THD(dB)	–	–	-94@100KHz

6 Conclusion

This 100dB OPAMP has been designed for 15-Bit 100MSPS DAC application. Settling accuracy of 0.001% has been achieved at 9.7ns. Unlike the all previous work, it could drive external load of 10pF and provide rail to rail output swing making the design suitable for low voltage application. Implementation of this OPAMP is going on in CMOS 180nm technology. All the silicon results and detailed analysis for linearity analysis will be publised as soon as data is available.

References

[1] D. Johns and K. Martin, "Analog integrated circuit design," in *John Wiley & Sons, New York*, 1997.

[2] K. Bult and G. Geelen, "A fast-settling cmos op amp for sc circuits with 90-db dc gain," in *IEEE J. Solid-State Circuit, vol. 25*, 1990, pp. 1379–1384.

[3] M. Das, "Improved design criteria of gain-boosted cmos ota with high-speed optimizations," in *IEEE Trans. on Circuits and Systems II, vol. 49*, 2002, pp. 294–297.

[4] Nabil Farhat Jie Yuan, "A compensation-based optimization methodology for gain-boosted opamp," in *Circuits and Systems, ISCAS '94 IEEE International Symposium*, 1994, vol. 5, pp. 517–522.

[5] DAVID J. ALLSTOT HOWARD C. YANG, "Considerations for fast settling operational amplifiers," in *IEEE TRANSACTIONS ON CIRCUITS AND SYSTEMS*, 1990, vol. 37, NO 3.

[6] Jan Van der Spigel Jie Yuan, Nabil Farhat, "Gbocad a synthesis tool for high performance gain-boosted opamp design," in *IEEE transactions on circuits and systems*, 2005, vol. 52, NO 8, pp. 1535–1543.

[7] R Storn, "On the usage of differential evolution for function optimization," in *In NAFIPS*, 1996, pp. 519–523.

[8] Morimoto M; Hadidi K; Futami K; Matsumoto T, "A novel design technique for input differential pairs in single-ended operational amplifiers," in *Electronics, Circuits and Systems, 1998 IEEE International Conference*, 1998, vol. 3, pp. 365–368.

21st International Conference on VLSI Design

VLSI Implementation of a Digitally Tunable G_m-C Filter with Double CMOS Pair

S.Ramasamy, B.Venkataramani, K.Anbugeetha

Department of Electronics & Communication Engineering,
National Institute of Technology, Tiruchirappalli ,India
phone: +91 431 2503301   e-mail : bvenki@nitt.edu

Abstract

This paper proposes a modified, inverter based transconductor using double CMOS pair for implementation of G_m-C filters . The advantage of this scheme is that, instead of varying the power supply, the bias voltages at high impedance nodes are varied for frequency (F) tuning. A current steering DAC is proposed for controlling these bias voltages. Another major contribution of this paper is the use of switchable transconductance cell for Q-tuning. This dispenses with the need for two separate biasing circuits (for F and Q tuning). To study the performance of proposed schemes, a bandpass filter is implemented on TSMC-0.18µm CMOS process using G_m/I_d design methodology. The simulation results show a good centre frequency (10MHz–120MHz) and pass band (10MHz–80MHz) tuning. The proposed approach guarantees the upper bound on THD to be -40dB for 1 V_{pp} signal swing. The use of inverters with double CMOS pair results in 21dB higher PSRR compared to those using push pull inverter.

1. Introduction

Transconductors have a wide range of applications in the area of analog signal processing [1], [2]. Continuous time filters implemented with transconductance amplifiers and capacitors are known as G_m-C or OTA-C filters and are quite popular for a host of applications such as IF filters, hard disk drive linear phase filters, LC-oscillators and RF filters.

A number of architectures have been proposed in the literature for implementing the transconductor. Push-pull inverters are proposed in [3] for realizing the transconductor. This does not have any internal node and results in large bandwidth. However, for realizing programmable filters, this scheme requires the power supply voltage to be varied. This is not suitable for low voltage applications and it results in poor power supply rejection ratio (PSRR). To solve this problem, floating battery implementation is proposed in [4]. But this scheme requires a large value of biasing resistance,

which introduces an additional pole in the region of interest.

The transconductor using double CMOS pair is proposed in [5]. In this paper, a double CMOS pair is proposed for the push pull inverter based transconductors reported in [3] and the resulting filter is studied in detail. The centre frequency (f_c) of the filter is varied by tuning the gate bias voltage. Symmetrical bias generator built by current steering DAC provides the bias voltage for the filter.

This paper is organized as follows. Section 2 explains the structure of a transconductor block using double CMOS pair. Section 3 presents the structure of the proposed digitally tunable second order band pass filter. The F-tuning and Q-tuning schemes used are also discussed. The filter design using G_m/I_d method is illustrated in Section 4. The simulation results are given in Section 5 followed by the conclusions in Section 6.

2. CMOS Pair based transconductor

In this section, the linear V-I conversion of double CMOS pair is described, followed by the common mode control and dc gain enhancement of the transconductor. The two-transistor circuit shown in Fig.1 is referred to as the CMOS pair. It may be considered to be a single transistor.

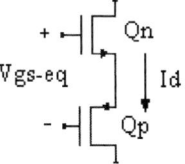

Fig 1. CMOS pair

Assuming that both transistors remain in the saturation region, the current I_d can be written as [2],

$$I_d = K_{eff} (V_{GS-eq} - V_{T-eq})^2 \qquad (1)$$

where,

$$\frac{1}{\sqrt{K_{eff}}} = \frac{1}{\sqrt{K_n}} + \frac{1}{\sqrt{K_p}}$$

317

978-1-4244-3039-0/08 $25.00 © 2008 IEEE

and $\quad V_{T-eq} = V_{Tn} + |V_{Tp}|$

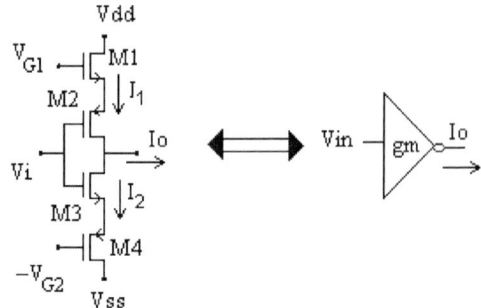

Figure 2. Double CMOS pair

The circuit in Fig.2 shows the double CMOS pair [5], which acts as a transconductance cell. Assuming that all the MOS devices are operated in the saturation region, using the simplified square-law model for devices and neglecting the channel length modulation, the output current can be expressed as [5],

$$
\begin{aligned}
I_o = I_1 - I_2 &= -2 K_{eff}[V_{G1}+V_{G4}-\textstyle\sum V_T] V_i \\
&+ \Delta V_T K_{eff}[V_{G1}+V_{G4}-\textstyle\sum V_T]
\end{aligned}
\tag{2}
$$

where,

$$
\textstyle\sum V_T = V_{Tn1} + V_{Tn3} + |V_{Tp2}| + |V_{Tp4}|
$$
$$
\Delta V_T = (V_{Tn3} - V_{Tn1}) + (|V_{Tp4}| - |V_{Tp2}|) + (V_{G1} - V_{G4})
$$

Assuming that the twin tub process is used, the transistors can have their bulks connected to their own source terminals and this eliminates the body effect. Hence, the threshold voltages are almost constant, given by,

$$
V_{Tn1} = V_{Tn3} = V_{Tn0} \tag{3a}
$$
$$
V_{Tp2} = V_{Tp4} = V_{Tp0} \tag{3b}
$$

where V_{Tn0} and V_{Tp0} are the zero-bias threshold voltages of the "N" and "P" MOS transistors. Now applying $V_{G1}=V_{G4} = V_G$, and by (3), ΔV_T becomes zero and the circuit behaves as an ideal g_m cell. i.e. from (2),

$$
I_0 = g_m V_i
$$

where,

$$
g_m = -4 K_{eff}(V_G - V_{Tn0} + |V_{Tp0}|) \tag{4}
$$

Thus the transconductance value can be varied by changing the bias voltage V_G.

In order to ensure that all transistors in g_m cell remain in the saturation region, the input voltage V_i and bias voltage V_G must satisfy the inequalities [5] given in (5) and (6).

$$
V_{G4} + V_{Tn3} + |V_{Tp4}| \le V_i \le V_{G1} - V_{Tn1} - |V_{Tp2}| \tag{5}
$$

$$
V_{Tn} + |V_{Tp}| \le V_G \le V_{dd} + V_{Tn} \tag{6}
$$

2.1. Common mode control and DC gain enhancement

To operate filters at very high frequencies, we need a transconductor with high DC gain of at least 40 dB and the parasitic poles should be located far from the cutoff frequency [3]. To increase the DC gain, the negative resistance loading is proposed in [3]. Let us consider two transconductance cells which are connected in a cross coupled manner as shown in Fig. 3.

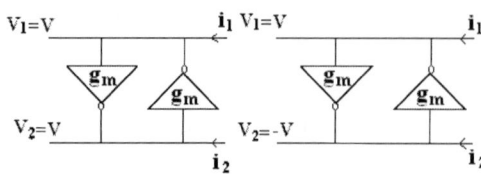

Figure 3. Cross coupled g_m cell

Applying a common mode voltage V at both nodes results in output currents $i_1 = i_2 = g_m V$, yielding common mode resistance $R_{ocm} = 2V/(i_1+i_2) = 1/g_m$. Application of differential voltage at the nodes results in output currents $i_1 = -g_m V$, $i_2 = g_m V$. The differential mode resistance is $R_{odm} = 2V/(i_1-i_2) = -1/g_m$. Thus the differential signal sees negative load resistance.

The structure of a complete transconductor block (G_m Block) using CMOS pair (g_m cell) with differential input and output is shown in Fig. 4. This is similar to Nauta's transconductor [3], but in this design, each g_m cell is constructed with a double CMOS pair and g_m tuning is done by varying the bias voltages (4). The g_m cell 1 and g_m cell 2 act as the main transconductor, g_m cells 3-6 provide common mode control and dc gain enhancement.

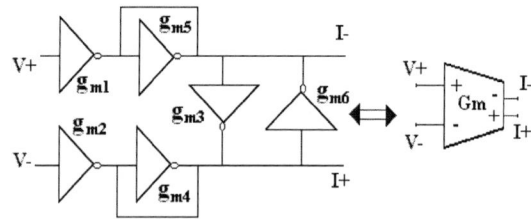

Figure 4. CMOS Pair based balanced Gm block

This transconductor has a differential architecture made of two identical sub circuits: g_m cells 1,5,6 and g_m cells 2,4,3 respectively. Thus all equations which correspond to the first set of g_m cells hold good for the second set as well, with the indexes in the given order. The common-mode output resistance at the node I- is,

$$
R_{ocm} \cong 1/(g_{ds1}+g_{ds5}+g_{ds6}+g_{m5}+g_{m6}) \tag{7}
$$

while the differential-mode output resistance at the node I- is,

$$
R_{odm} \cong 1/(g_{ds1}+g_{ds5}+g_{ds6}+g_{m5}-g_{m6}) \tag{8}
$$

where, $g_{mi} = 2(g_{mni} * g_{mpi}) / (g_{mni} + g_{mpi})$
$g_{dsi} = 2g_{dspi} + 2g_{dsni}$

g_{mn}, g_{mp} denote the g_m of NMOS and PMOS transistor respectively. Similarly g_{dsp}, g_{dsn} denote the output transconductance of NMOS and PMOS transistors respectively.

Assuming that for all the transconductance cells, $g_{dsi} = g_d$ and $g_{mi} = g_m$, the common-mode dc gain (A_{cm}) and differential-mode dc gain (A_d) at the output nodes are computed to be [3],

$$A_{cm} = \frac{g_m}{3g_d + 2g_m} \qquad (9)$$

$$A_d = \frac{g_m}{3g_d} \qquad (10)$$

Since A_{cm} is less than unity, common-mode stability is maintained. The differential mode gain can be boosted by choosing $g_{m5} \cong g_{m6} - (g_{ds1} + g_{ds5} + g_{ds6})$.

3. Digitally tunable second order band pass G_m-C filter

In programmable continuous time filters, the center frequency and the quality factor of the filter can be tuned by varying G_m, which in turn is controlled by changing either the bias current or the device dimensions. We propose a symmetrical bias generator, to vary the bias current of the transconductor to achieve F-tuning. Output switchable transconductance technique is also proposed to vary the Q-factor of the filter by combining the outputs of multiple transconductors.

The differential G_m-C realization of second order band pass filter structure based on double CMOS pair, with digitally assisted centre frequency tuning (F-tuning) and quality factor tuning (Q-tuning) is shown in Fig. 5.

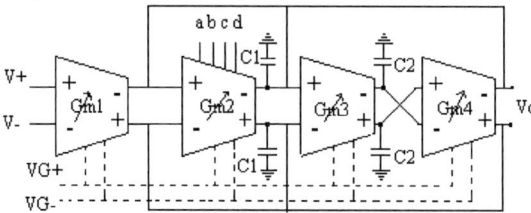

Figure. 5. Schematic diagram of tunable second order band pass filter

In this parallel resonance circuit, G_{m1} is the V-I converter, the resistor is realized by G_{m2}, the inductor is simulated by the Gyrator (G_{m3}, G_{m4} and C_2) and C_1 is the capacitor of resonant circuit. The transfer function of the above biquad structure [2] is given by,

$$H(s) = sC_2G_{m1}/(s^2C_1C_2 + sC_2G_{m2} + G_{m3}G_{m4}) \qquad (11)$$

From (11), the center frequency and the quality factor, Q, are given by,

$$\omega_0 = \frac{\sqrt{G_{m3}G_{m4}}}{\sqrt{C_1C_2}} \qquad (12)$$

$$Q = \left(\frac{1}{G_{m2}}\right)\frac{\sqrt{G_{m3}G_{m4}C_1}}{\sqrt{C_2}} \qquad (13)$$

From (12), the center frequency of the filter can be varied either by the constant-C or constant-G_m method. In constant-C technique, the load capacitance is maintained constant and the value of G_m is changed to alter the centre frequency of the filter. In constant-G_m technique, G_m is kept constant and the value of load capacitance is changed to alter the centre frequency of the filter. Detailed analysis of these two approaches is carried out in [9] based on noise, total capacitance required and power dissipation. It suggests that constant-C approach is suitable for tunable filter realizations. We follow the constant-C approach for both F-tuning and Q-tuning. In this approach, the value of transconductance is varied by changing the bias current and combining the outputs of multiple transconductors. F-tuning is achieved by controlling the bias current, whereas Q-tuning is achieved by switched g_m cells. The tuning schemes are discussed in the subsequent sections.

3.1. F-tuning

The bias current for the transconductor is varied by changing the bias voltage (V_{G+}) to the NMOS transistor in the top CMOS pair and by changing the bias voltage (V_{G-}) to the PMOS transistor in the bottom CMOS pair of the g_m cells of G_m block in Fig. 5. The transconductor is powered by $\pm$ 0.9V. From (6), it may be noted that the bias voltage has to be varied from +0.9V to +1.4V for V_{G+}, and -0.9V to -1.4V for V_{G-} in order to keep the transconductor in saturation. To produce the above bias voltage, a symmetrical bias voltage generator circuit as shown in Fig. 6 is used. The bias generator circuit uses $\pm$ 1.5V supply. Two 4-bit current steering DAC [10], namely positive DAC (P-DAC) and negative DAC (N-DAC) produce the necessary bias voltages V_{G+} and V_{G-} required for F-tuning respectively. The control word D[3:0] of the DAC determines the frequency tuning range. For the control word of "0000", a bias voltage of $\pm$ 0.9V is generated, yielding a centre frequency of 10MHz. For the control word of "1111", a bias voltage of $\pm$ 1.4V is generated, yielding a centre frequency of 120MHz. The bias voltage generated by the circuit is applied to all the G_m blocks. Transconductance is varied from 250μS to 2500μS using this scheme.

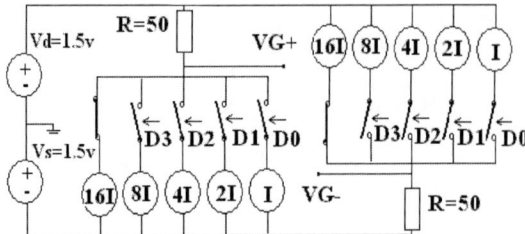

Figure 6. Symmetrical bias generator circuit

3.2. Q-tuning

The pass band of the band pass filter can be varied by varying the quality factor. From (13), if the product $G_{m3}G_{m4}C_1 / C_2$ is fixed, then the Q-factor is controlled by G_{m2} alone. To increase the Q-factor, the transconductance of the $G_{m2 \ block}$ has to be reduced. Since bias current to all the G_m block is controlled by the symmetrical bias generator circuit (F-tuning), switched Gm technique is applied to G_{m2} block to achieve Q-tuning. However, if switchable g_m cells are used to implement the programmable G_m, then each time such cells are switched in and out, the total value of the parasitic capacitance at each node changes, which in turn changes the centre frequency of the filter. Hence, constant capacitance scaling [9] can be used. But using the approach suggested in [9] will limit the signal swing because of the use of pass transistor between V_{dd} and V_{ss} for switching. Hence, we propose output switchable transconductance cell to switch the G_m cells.

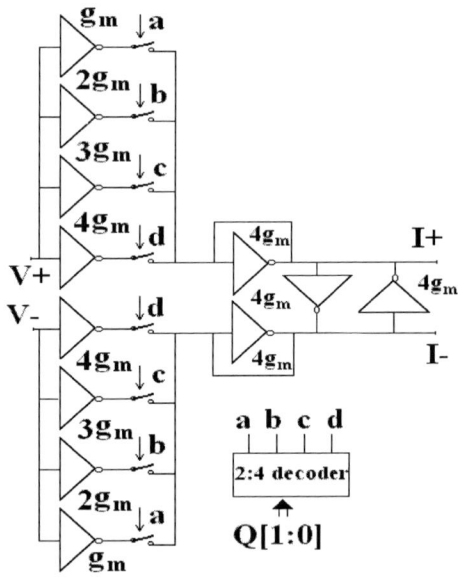

Figure 7. Output switchable Gm cell for Q-tuning

The proposed structure of switchable g_m cell, suitable for lower supply voltages is shown in Fig. 7. It uses a pass transistor at the output of the main g_m

cell in the G_{m2} block so as to vary the Q factor. The control word Q[1:0], which is the input of the 2:4 decoder, switches the output pass transistor of the switchable g_m cell to control the Q-factor. Table 1 summarizes the Q-tuning procedure for f_c=100 MHz. If C_i be the input capacitance of a unit g_m cell, then $10C_i$ will be the input capacitance of the G_{m2} block irrespective of Q-tuning. Hence, the capacitance at the input node is maintained constant, thus maintaining a constant capacitance scaling.

Table.1. Q-Tuning Summary

Sl.No	Ctrl.Bits Q[1:0]	Active Switch	G_m (μS)	Q
1	00	d	2400	1.2
2	01	c	1800	4.6
3	10	b	1200	9.1
4	11	a	600	12.8

The small signal equivalent of the output switchable g_m cell, when it is switched on and off is shown in Fig. 8(a) and 8(b) respectively. C_0 and g_0 are the output capacitance and conductance of the gm cell, r_{on} is the ON state resistance of the output pass transistor and C_{os} is its output capacitance. From Fig. 8, it is seen that the capacitance at the output node is C_{os} for both on and off conditions and hence the output parasitic capacitance remains constant.

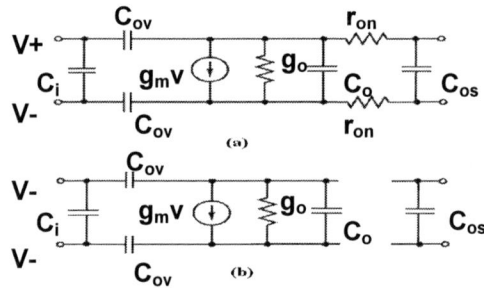

Figure 8. Small signal model of the output switched g_m cell

4. Filter design

Majority of methods for analytical synthesis of analog circuits assume that the MOS transistors are either in strong inversion or weak inversion region. The design methodology, based on the G_m/I_d characteristics [6],[7], allows a unified synthesis technique which is valid in all regions of operation of the MOS transistor. Detailed explanations on G_m/I_d method are given in [6], [7], [8].

The extracted G_m/I_d versus $I_d/(I_0 W/L)$ curves for both NMOS and PMOS transistors of TSMC 0.18μm process are shown in Fig. 9(a) and 9(b). I_0 is the specific current [6] given by,

978-1-4244-3039-0/08 $25.00 © 2008 IEEE

$$I_O = 2nKU_T^2 \qquad (14)$$

where K is the process parameter(μC_{OX}), U_T is the thermal voltage and n is the slope factor[6].

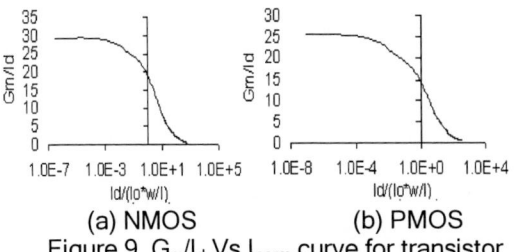

(a) NMOS (b) PMOS

Figure 9. G_m/I_d Vs I_{norm} curve for transistor

Transistors operating in strong inversion occupy less area but dissipate more power where as, transistors operating in weak inversion occupy more area but dissipate less power. Transistors operating in moderate inversion provide a compromise between these two cases in terms of both power and area. The Operational transconductance amplifier (OTA) structure (G_m block) with the double CMOS pair (g_m cell) is shown in Fig.4. Here g_m cell 1, g_m cell 2 act as differential pair, gm cell 3-6 provide common mode control and dc gain enhancement.

The following steps are used in the design of the OTA .

1. Assume the unity gain frequency f_T and the load capacitance C_L. The required G_m can be calculated using $G_m = f_T 2\pi C_L$.
2. Assume G_m/I_d based on the region of operation of the transistor and calculate the value of drain current as $G_m / (G_m/I_d)$.
3. Find the value of normalized drain current $I_d/(I_OW/L)$ from the G_m/I_d curve corresponding to the assumed G_m/I_d.
4. Calculate the W/L corresponding to this normalized current.
5. Once the W/L values are determined, their lengths are chosen based on both gain and area requirement and then the corresponding width is found.

Similarly, the dimensions of current source transistor in DAC are obtained by G_m/I_d methodology [10]. The dimensions of the various transistors used in the band pass filter are shown in Table 2.

5. Simulation results

This section presents pre-layout simulation results of the band pass filter, obtained using Mentor Graphics tools, Eldo and ezwave for the TSMC 0.18µm CMOS technology model. Simulated results are summarized in Table 3. THD of -41dB is obtained for a signal swing of 1V_{pp} with PSRR of 31 dB at 100MHz. The DC transfer characteristics (I_{od} versus V_{id}) for the various bias voltages are shown in Fig.10. From this figure, it can be observed that the output current is

linearly proportional to differential voltage of upto ±500mV . Simulated transconductance values versus differential input voltage (V_{id}) for various bias voltages are shown in Fig.11.

Table 2. Transistor dimensions for BP filter

Transistors in	G_m/I_d (1/V)	Region of operation.	W/L	
			N	P
G_m block	6	Strong	100	220
Unit Current source of DAC	5	Strong	0.25	--
Switch in G_m block	9	Moderate	10	--
Switch in DAC	5	Strong	2	--

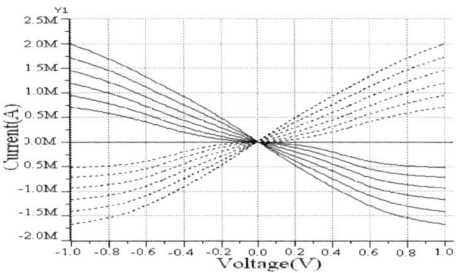

Figure 10. DC Transfer characteristics for V_G =±(0.9V-1.4V)

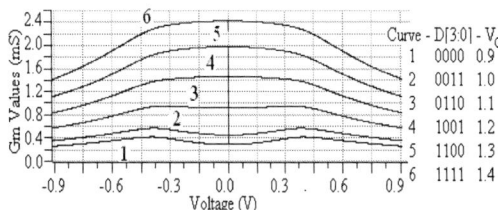

Figure 11. Transconductance values for F-Tuning

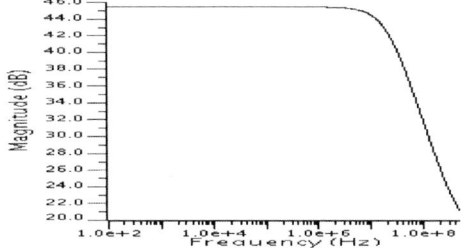

Figure 12. PSRR characteristics (± 0.9V)

978-1-4244-3039-0/08 $25.00 © 2008 IEEE

PSRR characteristics for the supply voltage of ± 0.9 V is shown in Fig. 12. Fig. 13 shows the simulated frequency response of the band pass filter as the centre frequency varies from 10 MHz to 120 MHz for various bias voltages. The quality factor tuning of the filter for various settings of Q control word is shown in Fig. 14 for a centre frequency of 100 MHz.

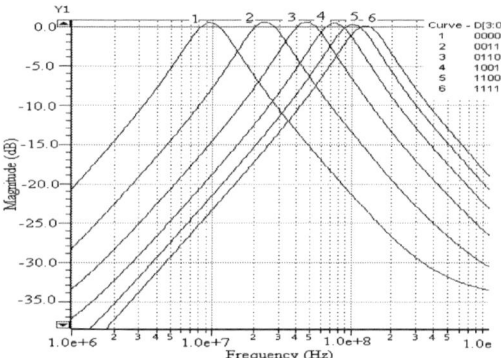

Figure 13. BP filter response for F-tuning

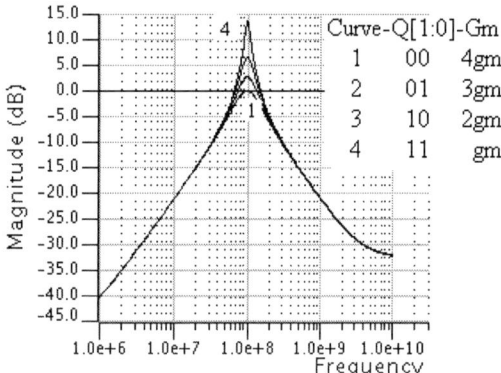

Figure 14. BP filter response for Q-tuning

Table 3. Simulation results summary

Technology	TSMC 0.18 µm CMOS process
Supply voltage	± 0.9V for Filter ± 1.5V for Bias circuitry
Frequency tuning	10MHz -120MHz with bias ± 0.9V ≤ V_G ≤ ± 1.4V
Q -tuning	1.2 – 12.8 with G_m 2400 μ S to 600 μ S
THD @ 1V_{pp}	-41.2dB
Group delay	<2ns
PSRR (± 0.9V) at 100MHz	31.2dB
Load capacitance	1pF

6. Conclusions

A tunable second order G_m-C band pass filter based on double CMOS pair is implemented in TSMC 0.18 µm digital CMOS process. The designed band pass filter features a good center frequency tuning range between 10MHz and 120MHz and uses the current steering DAC based symmetrical bias voltage generator. Q- factor ranging between 1.2 and 12 is obtained by output switchable G_m cell. The programmability of the filter requires ± 1.5V supply, even though the filter operates at ± 0.9V. This filter is proposed to be used for software defined FM radio.

7. References

[1] M.Ismail, and T.Fiez, *Analog VLSI Signal and Information Processing*, McGraw-Hill, New York, 1994.
[2] David A.Johns, and Ken Martin, *Analog Integrated Circuit Design*, Wiley & Sons Inc, 1997.
[3] B. Nauta, "A CMOS transconductance-C filter technique for very high frequencies", *IEEE J. Solid-State Circuits*, vol. 27, pp. 142–153, Feb.1992.
[4] F. Munoz, A. Torralba, R. G. Carvajal, and J. Ramirez-Angulo, "Two new VHF tunable CMOS low- voltage linear transconductors and their application to HF gm-C filter design", in *Proc. ISCAS*, May 2000, pp.V-173–176.
[5] C.S. Park, and R.Schaumann, "A High-Frequency CMOS Linear Transconductance Element", *IEEE Transactions on Circuits and Systems*, 33, pp. 1132- 1137, 1986.
[6] Daniel Foty, David Binkley, and Mathias Bucher, "Starting Over:G_m/Id -Based MOSFET Modeling as a Basis for Modernized Analog Design Methodologies", *Nanotech 2002*, Vol.1, Chapter 13, pp. 682 – 685.
[7] F. Silveira, D. Flandre, and P. G. A. Jespers, "A G_m/I_D Based methodology for the Design of CMOS Analog Circuits and Its Application to the Synthesis of a Silicon-on-Insulator Micro power OTA", *IEEE Journal of Solid-State Circuits*, Vol.31, no. 9, September 1996.
[8] F.P. Cortes, E. Fabris and S.Bampi ,"Applying the G_m/I_D Method in the analysis and design of Miller amplifier, comparator and G_m-C filter", *Proceedings of IFIP VLSI-SoC2003, Germany,* December 2003.
[9] Shanthi Pavan, Yannis.P. Tsividis, and Krishnaswamy Nagaraj, "Widely programmable high frequency continuous time filters in digital cmos technology",*IEEE Journal of Solid-State Circuits*,Vol-35,April 2000.
[10] S.Ramasamy, B.Venkataramani, and Sreekanthbabu Nukaraju, "Design and implementation of a 14-bit 200 MSPS Current Steering DAC using G_m/I_d method", *Proceedings of VLSI Design and Test symposium*, pp. 105-113, India, August 2007

21st International Conference on VLSI Design

A 9 bit 400 MHz CMOS double-sampled Sample-and-Hold Amplifier

Sounak Roy
Electronics and Electrical Communication Engineering Dept.
Indian Institute of Technology
Kharagpur 721302
Email: rajroy04@gmail.com

Prof. Swapna Banerjee
Electronics and Electrical Communication Engineering Dept.
Indian Institute of Technology
Kharagpur 721302
Email: swapna@ece.iitkgp.ernet.in

Abstract— **A fully differential CMOS sample and hold amplifier(SHA) is described here. The circuit is designed as a front end sampler of a low-power,high-speed analog to digital converter. The SHA uses double-sampling technique to achieve high speed with reasonably low power consumption. Using 0.18∞ CMOS technology,a resolution of 9 bit has been achieved at a sampling rate of 400MHz. Also,to acquire superior linearity, boot-strapping technique has been used while implementing the switches and to reduce clock feed through, concept of bottom plate sampling has been utilized. Using a supply voltage of 1.8 V and a signal swing of $0.6V_{pp}$ the circuit consumes approximately 10 mW of power.**

I. INTRODUCTION

In any traditional sample and hold amplifier,an open loop architecture can achieve high sampling rate at the cost of resolution. Operational amplifiers or operational transconductance amplifiers(OTAs),when operated in closed loop topology,improve the resolution of an SHA ,but in order to achieve high sampling rate in closed loop,the concept of double sampling must be adopted [1].

Other than sacrificing the accuracy,an open-loop sample-and-hold amplifier also suffers from clock feedthrough generated by the switch induced charge injection. Applying op-amps in the negative feed-back loop can remove this feed-through to some extent. To further remove this error,techniques such as bottom plate sampling can be used. Bottom-plate sampling generates a constant charge injection from the switches,which can be effectively removed by using differential architecture. Also, by using boot-strapping,switch resistances can be made independent of the signal variations thus improving the linearity of the circuit.

The SHA described in this paper uses both boot-strapped switches and bottom-plate sampling while working differentially. An input signal of 600 mV swing and signal frequency of 40 MHz has been sampled at 400 MHz. With a load capacitor of 1 pF, total harmonic distortion (THD) has been measured at -58.51 dB which in terms of ENOB (effective number of bits) turns out to be 9.43 bits.

In Section A,the functionality of the SHA is detailed. The high-speed OTA is described in Section B. Finally,simulated results of the SHA are shown in Section C. Conclusions are drawn in Section II.

A. Functional description:SHA

A closed loop SHA architecture, commonly used in switched capacitor (SC) circuits and referred to as flip-around SHA [2] [3] [4] is shown in Figure 1. Instead of using an op-amp in the negative feedback loop,this circuit uses passive circuits in the feedback,thus making faster acquisition possible. This differential circuit operates in two phases. In acquisition phase ($\square$1) switches S1 and S3 are closed and the capacitor C_S is charged to (Vin-)-VCM. In the amplification phase($\square$2),swithes S2 and S7 are closed. In this phase op-amp works in closed loop and the input signal gets reflected to the output.

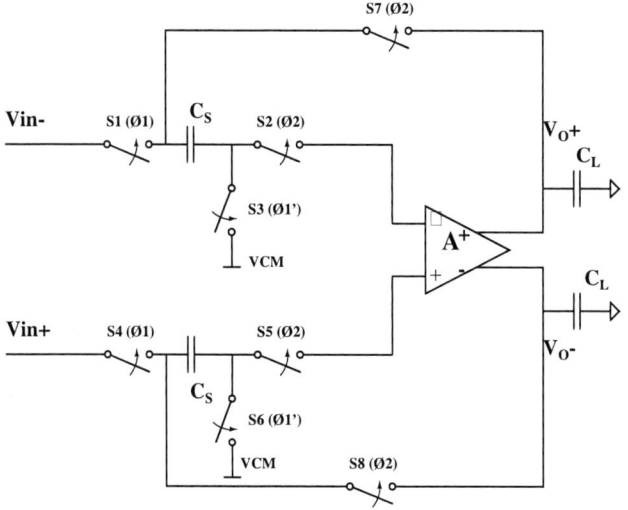

Fig. 1. A flip around differential SHA

In CMOS technology,switches are generally implemented using MOS transistors. A conducting MOS switch has a finite amount of mobile charges and when it turns off,charges distribute themselves through the drain,source and bulk terminals of the device. There are various methods to reduce these leakge charge error which is popularly known as charge injection error. Applying dummy switches working in opposite phase is a primitive but useful idea. The errors can be further reduced by applying transmission gates. But if the errors can be made

978-1-4244-3039-0/08 $25.00 © 2008 IEEE

independent of the signal variations, superior linearity can be achieved by using differential architecture.Bottom-plate sampling is one such technique [5] [7].

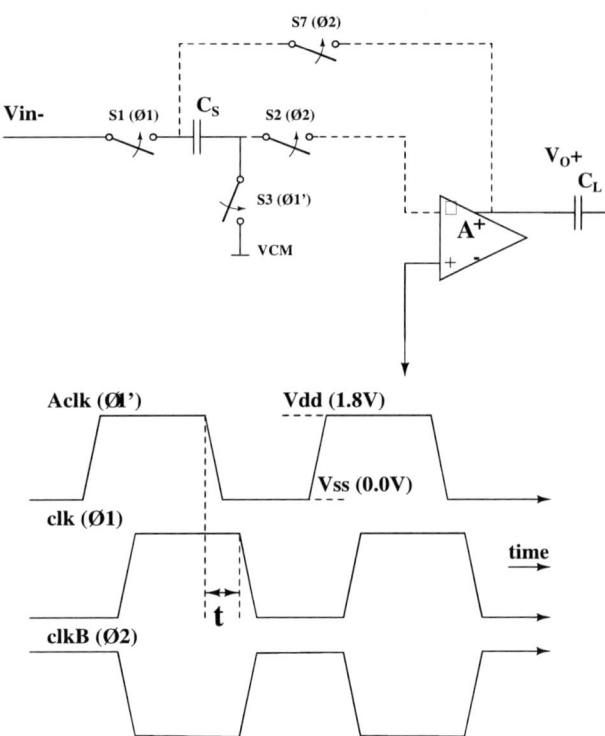

Fig. 2. Bottom plate sampling

As shown in Figure 2 bottom plate sampling actually uses 3 clock phases.Phases ☐1 and ☐2 are of opposite polarity.Whereas clock Aclk (phase ☐1′) is t secs advanced than clock clk.At t secs prior to the termination of phase ☐1 ,clock Aclk forces switch S3 to turn off.As one terminal of this switch is permanently connected to VCM,charge injected by this switch in the off period is constant.So, the bottom plate of the capacitance C_s is at a fixed erroneous potential thanks to switch S3.After t secs,switch S1 turns off and charge induced by this switch gets no path to discharge as bottom plate of C_s is left floating. In differential architecture, this error gets cancelled as it is same for both halves of the circuit.

In SC circuits while implementing the switches NMOS transistors are preferred over PMOS ones because of higher trans-conductance.While switched on,they operate in triode region and show a resistance governed by the following equation:

$$R_{on} = \frac{1}{\propto_n C_{OX}\frac{W}{L}(V_{DD}-V_{in}-V_{TH})} \quad (1)$$

Where V_{DD} is applied at the gate terminal and V_{in} is applied at the source of the device.

From the the above equation it is evident that resistance of the transistor is dependent on its gate-to-source voltage

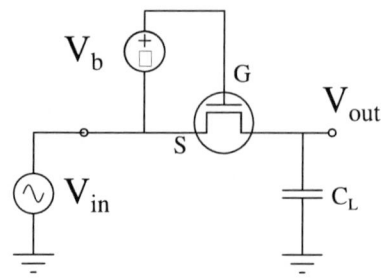

Fig. 3. Boot-strapping technique

V_{GS}.Non-idealities arising due to this signal dependency can be removed if V_{GS} can be made constant.Boot-strapping is an efficient technique where V_{GS} of an NMOS switch is made constant as shown in Figure 3.Here V_{GS} of the NMOS is boosted by a fixed voltage V_b.

Over the last few years various implementations of boot-strap switches have been made [1] [5] [6].In deep sub-micron technologies high voltage levels cause reliabilty problems while designing boot-strap switches.The nominal supply voltage of a deep sub-micron technology is typically set as high as the reliability permits.Generally, for devices to work effectively,the terminal voltages of a transistor may not be much higher than the supply voltage.

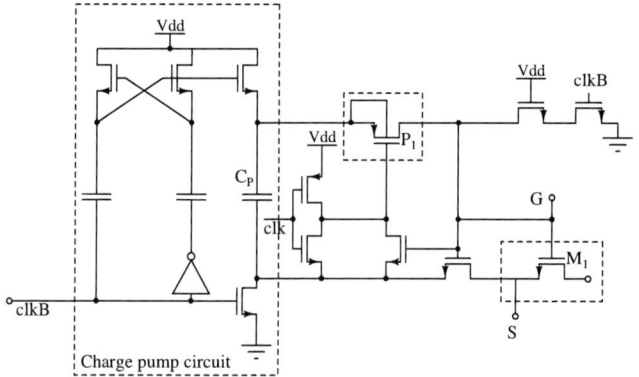

Fig. 4. Boot-strapped switch

The boot-strapped switch designed in this paper is shown in figure 4.Here,transistor M_1 acts as the switch which has its gate-to-source voltage boosted by the boot-strap circuit.The charge pump circuit shown works as a battery when the switch M_1 is off and charges the capacitor C_p to V_{dd}.When M_1 gets on,the voltage of the capacitor C_p is switched to its the gate-to-source voltage.As a result,V_{GS} of M_1 becomes relatively independent of the input signal.As the present design has been done using n-well CMOS process,generally bulk node of a PMOS transistor is attached to supply voltage V_{DD}.But in the boot-strap circuit,switch P1 will have its source terminal reach as high as $2V_{DD}$.Thus to avoid latch-up, bulk of this PMOS is shorted to the source terminal.

In the SHA presented here, both the techniques to improve

linearity have been applied and is shown in figure 5.Here, only the switches associated with the input signal (switches S1 and S1a) are modified using the boot-strap cicuit.Reason being these circuits are complicated and uses large die area compared to other swithes.Rest of the switches in this SHA are designed with transmission gates.

In the differential SHA described earlier (Figure 1) during the acquisition phase switches S1 and S3 are closed while the opamp A gets isolated.Thus for half the time period,the op-amp remains idle.To fully utilize the op-amp double sampling technique comes into play.As shown in Figure 5,when switches S1 and S3 do the job of acquisition,S2a and S4a put the op-amp into negative feed-back configuration.

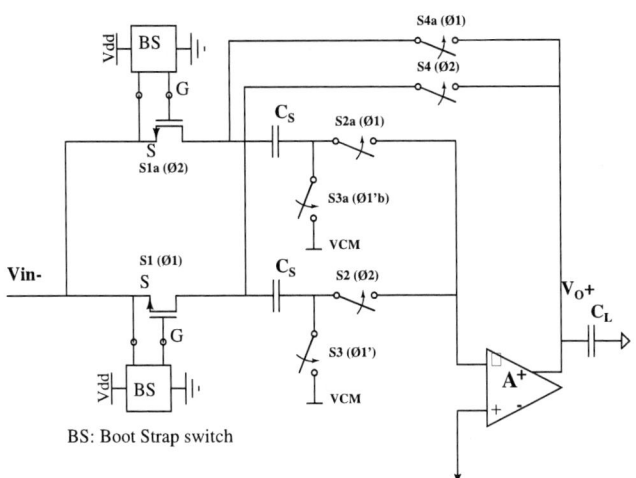

Fig. 5.   Double sampled SHA

As the power consumption of an SHA is primarily defined by the op-amp or the OTA,this double sampling technique does not increase the power consumption by two fold.In the present design,the OTA uses 6.56 mW of power out of the 9.6 mW of total power consumed by the SHA circuit.

B. Design of the OTA

To desgn a high speed closed loop SHA, it is imperative to have a OTA of high gain-band-width (GBW) and fast settling behaviour.Also, to achieve the required accuracy the OTA must show sufficient dc gain.Keeping in mind the above require-ments,a folded cascode architecture has been selected with a few modifications.It uses complementary input transistors and regulated cascode load for fast settling as well as high dc gain.

The design of the OTA is shown in Figure 6.To extend the input range of the OTA i.e. to increase the ICMR (Input common mode range),both NMOS and PMOS differential pairs are used so that when input range has forced one of the pairs to cut-off,other pair can keep the circuit alive.As input common mode (CM) level reaches the ground po-tential,transconductance (g_{mn}) of the NMOS pair drops,but PMOS pair remains active.Conversely,when CM level reaches

V_{DD},g_{mp} becomes dominant.Variation of the G_{mTOT} with input CM level is shown in Figure 7.Figure 7 suggests that if the OTA can be biased at $V_{DD}/2$ then effective transconductance of the OTA can be increased by a factor of almost 2.

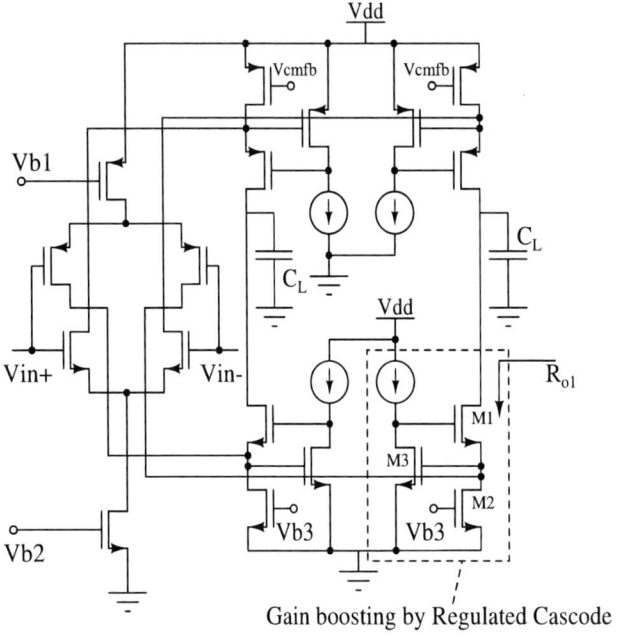

Fig. 6.   Folded Cascode OTA

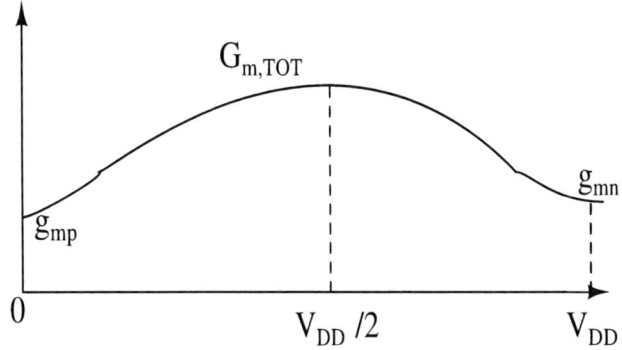

Fig. 7.   Variation of transconductance with input CM voltage

The OTA designed here has used tranistors of minimum feature length to get good stabilty.

Gain of this OTA is enhanced by using regulated cascode loads.As DC gain of the OTA follows the equation

$$A_v = G_{mTOT}R_{out} \qquad (2)$$

increasing the out put resistance does the job of gain improvement.In Figure 6 the dashed box shows how regulated cascode increases the output resistance.Here, the transistor M3 is put in a local feed back loop and the out put resistance as

325

978-1-4244-3039-0/08 $25.00 © 2008 IEEE

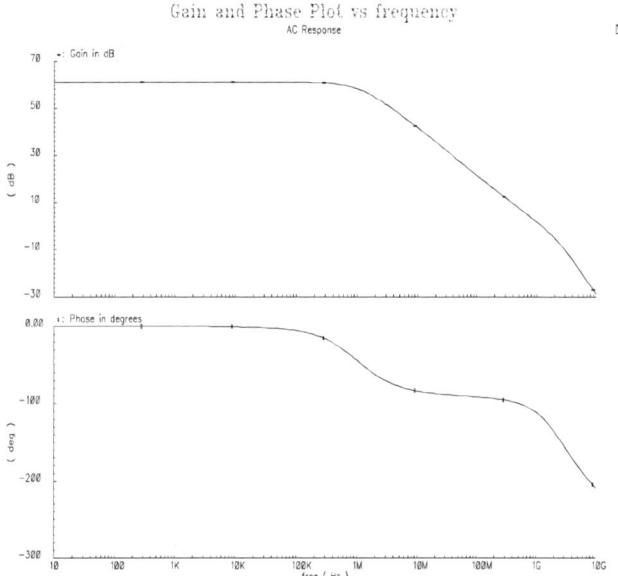

Fig. 8. Frequency response of the OTA

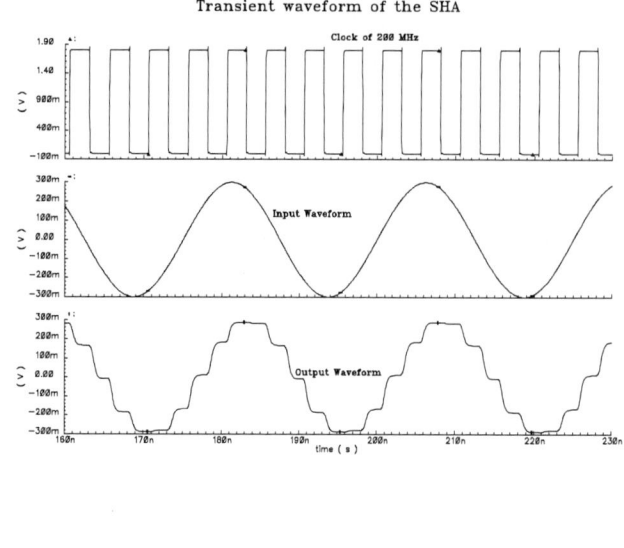

Fig. 9. Transient response of the SHA

seen from the drain of the transistor M1 to ground is governed by the following equation.

$$R_{o1} = (g_{m1}r_{o1}r_{o2})(g_{m3}r_{o3}) \qquad (3)$$

Thus, the output resistance of the OTA increases by a factor of $g_m r_o$. Simulated waveforms of the OTA is shown in Figure 8. With 0.5 pF load the OTA exhibits a dc gain of 61.22 dB, a GBW of 1.21 GHz, phase margin of 63.2^0, settling time of 1.2 nS and consuming 6.56 mW of power.

C. Simulation Results

The SHA described in this paper has been simulated at an input signal frequency of 40 MHz. Input swing is 600 mV differential. Using a clock of 200 MHz, a sampling rate of 400 MHz has been achieved. The transient behaviour of the circuit is shown in Figure 9. The frequency domain analysis is performed using DFT of the transient signal and is presented in Figure 10.

In Figure 10, the DFT of a 40 MHz signal sampled at 400 MHz has been shown. As the clock frequency is 200 MHz, at frequencies 160 MHz and 240 MHz intermodulation of the clock frequency with signal frequency ocuurs and 4th and 6th harmonic become very dominant. While calculating the total harmonic distortion, these even harmonics are neglected and due to differential nature of the output and only odd harmonics

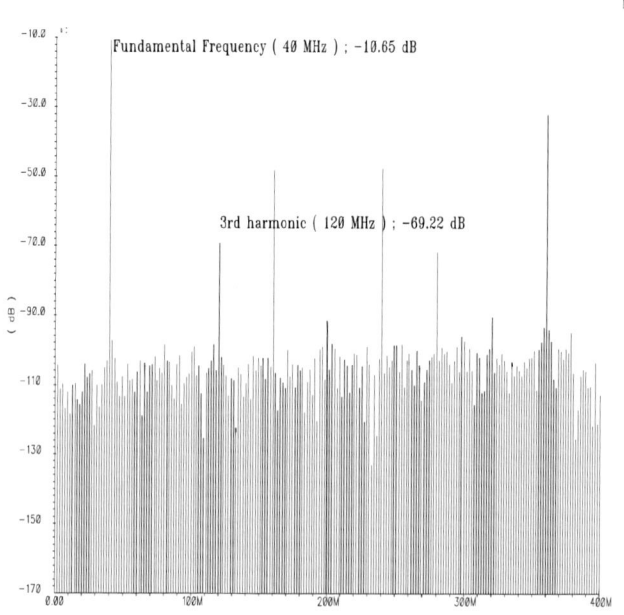

Fig. 10. Dynamic behaviour of the SHA

978-1-4244-3039-0/08 $25.00 © 2008 IEEE

are taken into consideration.From Figure 10 total harmonic (which is approximately equal to 3rd harmonic) distortion is found at -58.51 dB.

II. CONCLUSION

A CMOS sample-and-hold amplifer capable of sampling at a frequency of 400 MHz has been described.A 0.18∞ CMOS technology was used to design and simulate the circuit.Removal of non-idealities and sampling rate enhancement have been the key features of the presented design.Selection of boot-strap switches have been carefully made to optimize between linearity and power consumption.Bottom plate sampling has helped further to reduce the non-linearity.A complementary-input regulated-folded-cascode OTA has been designed to fulfil the requirements of the high operating speed and fast settling.The circuit is capable of working as a front-end sample-and-hold amplifier of a high speed pipeline A/D converter.

REFERENCES

[1] Mikko Waltari,Kari Halonen; A 10-bit 220-Msamples/s cmos sample-and-hold circuit ,*Proceedings of IEEE international symposium on circuits and systems,*ISCAS'98.

[2] T. Mukherjee,R.L. Carley;High-speed low-power integrating CMOS sample-and-hold amplifierarchitecture,*Custom Integrated Circuits Conference, 1995., Proceedings of the IEEE 1995*

[3] A.Bascirotto, A low-voltage sample-and-hold circuit in standard CMOS technologyoperating at 40 ms/s,*IEEE Transactions on Circuits and Systems II: Analog and Digital Signal Processing*

[4] B. Razavi, Design of sample-and-hold amplifiers for high-speed low-voltage A/Dconverters,*Custom Integrated Circuits Conference, 1997., Proceedings of the IEEE 1997*

[5] Mikko Waltari,Kari Halonen; Circuit techniques for low voltage and high speed A/D converters,*Kluwer Academic Publishers*

[6] A.M. Abo,P.R. Gray; A 1.5-V, 10-bit, 14.3-MS/s CMOS pipeline analog-to-digitalconverter,*IEEE Journal of Solid-State Circuits*

[7] D. Haigh and B. Singh,A switching scheme for switched capacitor filters which reduces the effect of parasitic capacitances with switch control terminals,*Proc. IEEE 1983 ISCAS, May. 1983, pp. 586-589.*

SESSION A3:
Physical Design/CAD

21st International Conference on VLSI Design

A New Approach for Estimation of On-Resistance and Current Distribution in Power Array Layouts

Jyotirmoy Ghosh[1], Siddhartha Mukhopadhyay[1], Amit Patra[1], Barry Culpepper[2] and Tawen Mei[2]
[1]*Department of Electrical Engineering*
Indian Institute of Technology, Kharagpur, India
[2]*National Semiconductor Corp., Santa Clara, USA*
Email: jyotirmoy@vlsi.iitkgp.ernet.in

Abstract

This paper presents an accurate and fast technique for the estimation of on-resistance ($R_{DS(on)}$) of large lateral power MOSFET switch layouts in on-chip DC-DC converter and determination of the current distribution pattern in the switch layouts. In the proposed approach an extracted netlist is created which consists of the lumped parasitic resistances formed in the metal interconnects and the MOS devices present in the layout. The extracted resistance values are computed from the metal geometry using models that relate resistance values to the geometric patterns in the layout. This approach exploits the highly symmetric and repetitive pattern of power MOSFET layouts to generate the resistance netlist efficiently. Similarly the modeling of very high W/L MOS finger channels is also described in this paper. Results from the numerical experiments show that the extracted resistances are within 2.6% of results obtained from standard FEM solver tool ANSYS.

1 Introduction

On chip DC-DC switching converters have a wide range of application in power management domain starting from less than one watt to several tens of watts. These on-chip DC-DC converters use power MOSFET switches which have large W/L ratio typically in the order of 10^5 to 10^7 and can carry currents up to a several amperes. The layouts of the lateral MOS switches are made by breaking the MOS device into a number of fingers and arranging the fingers to form a matrix like structure popularly known as 'power array'. The power arrays occupy 60% to 70% of full chip area. In a power array the MOS fingers are connected to interleaved source and drain metallizations which are in turn connected to higher level of metals by contacts and vias up to the pad level. Thus a complex metal interconnect pattern,

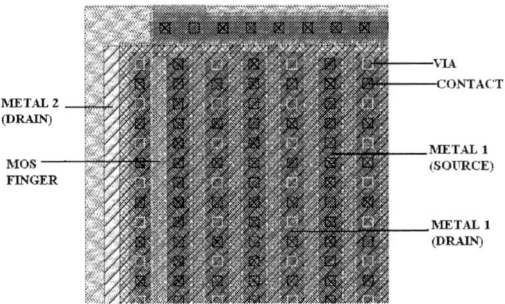

Figure 1. Layout of Power Array

described in [4], is formed in power array. Fig. 1 shows sectional view of a power array layout with two metal layers.

A major challenge for the designers of on chip DC-DC converters is to reduce the losses of the power MOSFET switches. For the power MOSFET switches operating near full load, the conduction loss is a significant part of total losses in the switches and this loss is directly proportional to $R_{DS(on)}$ of the switches [3]. Hence accurate modeling of the conduction loss is dependent on accurate estimation of $R_{DS(on)}$ at a specific gate voltage V_{gs} and gate junction temperature. Also determination of the optimum value of W of a power MOSFET switch is dependent on the conduction loss model. Due to the presence of parasitic resistances in the complex metal interconnect pattern of a power array, the actual $R_{DS(on)}$ deviates much from the value calculated without consideration of the metallization effect.

In layout design for power array it is also necessary to identify probable locations of high current concentration zones which may lead to electromigration and formation 'Hot Spots' during operation of the device. Thus, along with the accurate estimation of $R_{DS(on)}$ including the layout parasitic resistances, profiling of current distribution pattern is also important.

Many resistance extraction methods and programs have

978-1-4244-3039-0/08 $25.00 © 2008 IEEE

been reported in the literature. These include approaches based on Conformal Transformation [12], and the Finite Difference Method [7]. In Finite Element Method described in [2, 6] the resistance can be computed by numerically solving Laplace's Equation ($\bigtriangledown^2 \phi = 0$) with boundary conditions for both body and contact regions. The advantage of this process is its high level of accuracy and applicability for varied layout geometries. But this process is computationally highly expensive both in terms of run time and memory overhead for large and complex power array layouts. An improvement in FEM is presented in [9]. Two dimensional and three dimensional Boundary Element Methods, described in [8, 11, 10], solve the Laplace's Equation using discretized equations only for domain boundary resulting in a reduction in volume of data input and required memory space. However, BEM requires a fairly complicated mathematical treatment for irregular shapes, large number of holes or nodes in the system and singular points. Recent trends in resistance extraction methods have been discussed in [5]. Information related to ANSYS can be found in [1].

The advantages of available RC extraction tools are that they are much more generic in nature since they can be run on digital, analog and mixed-signal circuit layouts mainly for RC delay estimation and full chip performance verification. These tools are fast and can handle large databases. However, the accuracy level desired by power array designers to determine $R_{DS(on)}$ for large and complex power array, is difficult to achieve from these tools due to the limitations in fracturing of metal layers (which is one-dimensional, non uniform and dependent on aspect ratio only) and extraction of very wide MOS channel as a single device.

2   The Proposed Approach

From a study made on the resistance extraction methods and available tools, it is evident that there are two trends in extraction. In the first one, the results are accurate, but memory/time requirement is high. On the contrary, in the second trend, the extraction is fast with a loss in accuracy level. In our approach an attempt is made to use the advantages of both the trends exploiting the symmetry of power array layouts and by building computationally efficient behavioral models for the common layout structures. For metal layers with a cluster of N closely spaced via/contact holes in the 'Basic Cell' (the basic structure which on its row and column wise repetition can generate the whole power array), an N port network of lumped resistances has been proposed. The value of each lumped resistance model in this network is determined by modeling relationships connecting typical geometrical layout parameters to resistance values of this network. The

data used to form the models are generated for layouts having metal layers along with via arrangement patterns, using accurate FEM simulations. Since the metal layer structures and the via/contact arrangement patterns are repetitive in nature, the N port network is needed to be formed only for the Basic Cell. The wide MOS fingers are modified as a number of MOS devices of smaller width connected in parallel so that the variation of current in channel can be captured. Finally, an extracted netlist is formed by combination of netlists of the Basic Cells for simulation in SPICE or SPECTRE.

3   Modeling of Metal layers

3.1   Modeling of Distributed Metal Layers

It has been observed from the experimental results that the value of lumped inter via resistance model in a metal layer, which is actually distributed in nature, depends on a) sheet resistivity of the metal layer, b) inter via separation, c) distance of a via system from the boundary edge, d) the dimensions of vias, and e) via position in a set or cluster of vias. Each of these parameters have been modeled individually to obtain the resistance value between two vias placed at two arbitrary positions on the metal layer. The graphs in the following sections are generated from 2D FEM analysis of metal layers considering the via-holes and show the effect of each of these parameters in detail.

3.1.1   Sheet Resistivity of Metal Layer

The lumped resistance value between any two vias on a metal layer is directly proportional to the sheet resistivity of the metal layer. The experiment to measure the value of lumped resistance had been carried out with setting one via hole at 1V and the other via hole at 0V. The magnitude of the resistance is the inverse of the current flowing between the two via holes. Fig. 2 shows the dependency of resistance value on metal sheet resistivity.

3.1.2   Inter Via Separation

The resistance value between two vias does not vary linearly with the separation between the two via positions. The nonlinear characteristic becomes prominent when the two vias are closely spaced as shown in Fig. 3.

3.1.3   Distance of Via system from Boundary

When a cluster of vias is shifted towards the boundary in Y direction, the mutual resistance between two member vias (located at (x_1, y_1) and (x_2, y_1)) increases following the

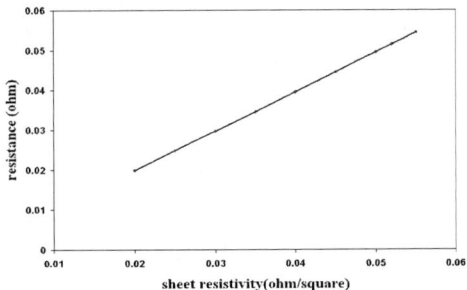

Figure 2. Resistance value vs. Metal Sheet Resistivity

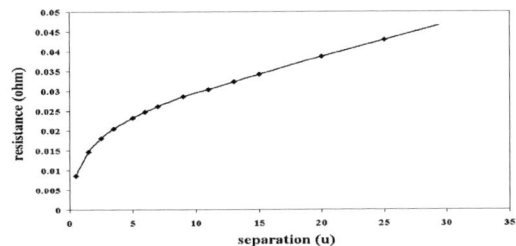

Figure 3. Resistance value vs. Separation between Via/Contact

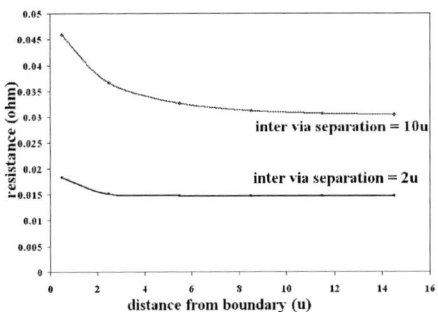

Figure 4. Resistance value vs. Distance from Boundary

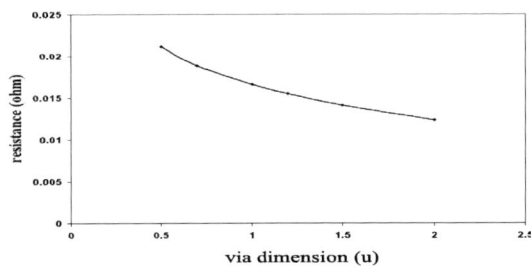

Figure 5. Resistance value vs. Via/Contact Dimension

characteristics as shown in Fig. 4. The boundary effect is negligible when separation of the vias from the boundary in Y direction is greater than the separation between the vias. If the cluster of vias is shifted towards boundary in X direction, the boundary has almost no effect on the resistance between the two vias located at (x_1, y_1) and (x_2, y_1).

3.1.4 Via/Contact Dimensions

The resistance between vias/contacts increases with decrease in cross-sectional area of the via/contact. The graph in Fig. 5 shows the change in resistance with the width of via/contact with square cross section.

3.1.5 Via Position in a Cluster of Vias

The position of a via in a set or cluster of vias is very important to identify and model its connectivity with other vias in the system. Except the four corner vias in a set of vias, others are connected to only the first ring of vias encircling that via with the lumped resistances. To establish this fact, the via arrangement as shown in Fig. 6 had been considered, and 1V was applied to the central via hole. The other 0V via holes were placed one at a time in the given order of number in the first via-circle around the central via and then in the

subsequent via-circles. The total incremental current drawn from central 1V via was measured with the placing of each 0V via. As evident from the Fig. 7, the vias of second and third via rings draw negligible current (less than 10%) from the central via. Hence, lumped resistance models between a via to its neighbouring vias of first circle only, give a fairly accurate result. However, for the vias placed at four corners, around 80% of its total current is drawn by first neighbouring via circle. For these vias lumped resistance model with the second neighbouring circle is also required.

It has been also observed that the lumped resistance net-

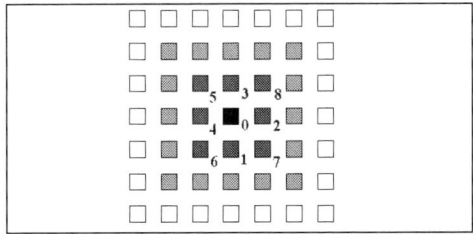

Figure 6. Cluster of Vias

work for a via system obey the *superposition principle*, i.e.

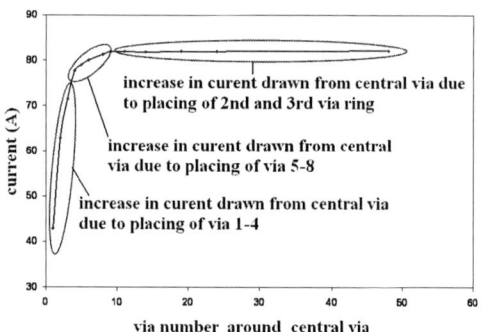

Figure 7. Increment in Current Drawn from Central Via

doubling the applied voltage on a via hole increases the current to twice the previous value between that via hole and a 0V via hole. This implies that the resistance values between the vias are constant and independent of the potential differences between the vias. Fig. 8 establishes this fact.

From the analysis performed in this section, the lumped

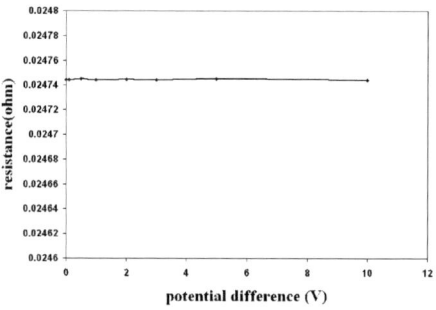

Figure 8. Resistance Value vs. Potential Difference between Vias

resistance model developed for a cluster of vias is shown in Fig. 9. All the clusters are separated from each other using 'cut lines' which are assumed to be equipotential. Thus we may assume parallel flow of currents through the cut lines from one via cluster to another and a model of parallel lumped resistances connected between the via clusters to form the complete resistive network for a distributed metal layer.

3.2 Modeling of Metal-1 Layer

Since metal-1 layers have narrow and long strip like structure, the boundary effect is very prominent in the resistance model for metal-1 layer. The effect of boundary is being shown in Fig. 10. Due to close proximity of bound-

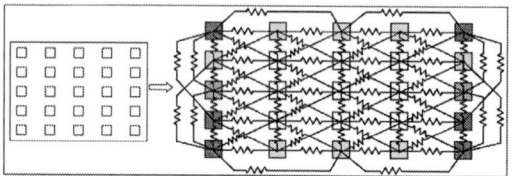

Figure 9. Lumped Resistance Network Model for a Cluster of vias in Distributed Metal Layer

ary edge, the current from one via/contact can only travel up to the adjacent via/contact. Thus lumped resistance model is developed between the adjacent vias in metal-1 (Fig. 11). The lumped resistance value for any inter via separation and metal width can be determined accurately using first order or second order interpolation methods.

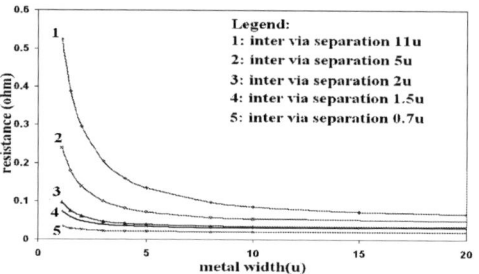

Figure 10. Boundary Effect on Resistance in Metal Strip

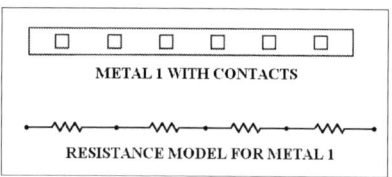

Figure 11. Lumped Resistance Model for Metal-1

4 Modeling of MOS Fingers

In power array layouts, the width of MOS finger typically varies from 50μ to 200μ depending on the MOS channel length. Hence, modeling each finger as a single device cannot capture the variation of current distribution in the MOS channel. Also, for modeling one MOS channel as a single device, the extractor has to merge and short all the contacts connected to the source and drain diffusion regions into one source and one drain contact respectively. A significant error creeps in due to this reason since the contact potentials are not equal for large value of W. A more accurate approach is to break the MOS finger into parallel MOS models depending on the number of contacts (Fig. 12). Simulation of this modified MOS finger can capture the variation in current in the device channel. However, the Threshold voltage V_T, which is dependent on W and L of MOS is being wrongly calculated in this case. Standard simulators will simulate the MOS models with V_T value of the broken parts, while the actual V_T should be calculated from the actual channel width.

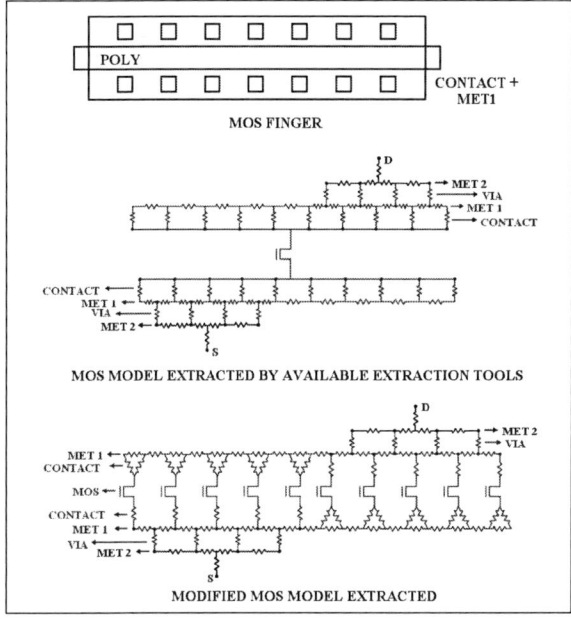

Figure 12. Modeling of MOS Finger

5 The Overall Process Flow

The overall process flow in this approach is being described elaborately in Fig. 13. The final extracted netlist is created from the combination of the netlists of the Basic Cells. The netlist of the Basic Cell is created in two parallel paths. The left path creates the netlist for the metal interconnects for the Basic Cell, while the right path creates the modified MOS model by breaking up the MOS channels. Proper geometrical position of each of the vias/contacts and metal polygons are recorded in order to get one to one correspondence between the actual layout and the extracted netlist. The vias and contacts (along with their contact resistances) have been modeled as lumped resistances in this approach with the resistance values are taken form Electrical Rule file of the process.

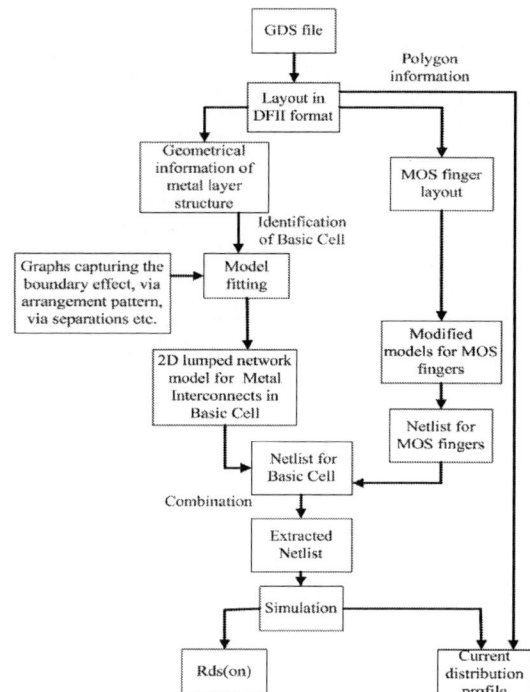

Figure 13. The Process Flow Diagram

The current distribution profile is created from the current values through different resistances obtained from simulation and the information of geometrical position of each of these resistances in the layout polygons. As the current density in vias and contacts are much higher than that in the metal layers and this method can determine the via and contact currents, the probable zones with high current density can be predicted fairly accurately in this method.

6 Results

For comparison, resistance estimates by the proposed technique, ANSYS, and a commercially available extraction tool for test structures in 0.5μ BiCMOS Process with two metal layers, are presented below.

6.1 Extracted $R_{DS(on)}$

Table I presents the comparison between the $R_{DS(on)}$ extracted by ANSYS, available extraction tool and the proposed method. The W/L ratio of test power array structure is given in column-1. $R_{DS(on)}$ (in Ω) extracted by ANSYS is given in columns 2. Columns 3 and 4 show the extracted $R_{DS(on)}$ (in Ω) by available extraction tool and the proposed method respectively along with percentage deviation from ANSYS results.

With the increase of W/L ratio of the power array, the metal interconnect structure becomes more complex and

335

978-1-4244-3039-0/08 $25.00 © 2008 IEEE

W/L ratio	ANSYS	Available Tool	Proposed Method
150/0.5	17.463	17.410/0.30%	17.431/0.18%
250/0.5	8.710	8.981/3.11%	8.677/0.38%
500/0.5	4.454	4.719/5.94%	4.396/1.37%
1000/0.5	2.246	2.501/11.33%	2.188/2.58%

Table 1. Comparison of Performance

the interconnect resistances become significant part of the extracted $R_{DS(on)}$. Consequently, the deviation of results from available tool is more prominent than the result obtained from ANSYS with the increase of W/L ratio.

In the proposed approach, for N number of nodes on metal layer, the number of elements in resistance network is of the order of N since each node is connected to neighbouring nodes only. In standard FEM/BEM based extraction process the number of elements in the network is of the order of N^2, due to formation of complete graph. Thus, in this approach the number of elements in the extracted netlist reduces significantly. This makes the simulation much faster and reduces problems related to convergence of the simulator, while ensuring a high accuracy level.

While the results from ANSYS are considered to have maximum accuracy, for power arrays with W/L ratio more than $2000\mu/0.5\mu$, ANSYS could not mesh the domains due to memory constraint on a stand alone machine.

6.2 Current Distribution

Fig. 14a shows the current distribution profile and high current density zones obtained from 3D simulation in ANSYS. Fig. 14b shows probable high current density zone as identified in the proposed method. Note that identical regions have been shown to have the highest current densities. Thus from the comparison of the two figures, we may conclude that the proposed method can locate the probable high current density zone accurately.

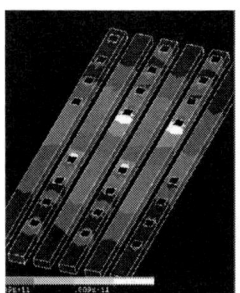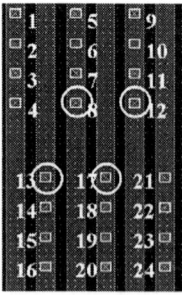

Figure 14. ANSYS 3d Simulation Result (a) and Result from Proposed Method (b)

7 Conclusion

A new approach for estimation of on-resistance and current distribution of on-chip lateral power MOS switches has been presented in this paper. This approach is expected to be computationally fast since it uses simple but accurate behavioral models to replace FEM. Secondly, it attempts to use the inherent 'Basic Cell' concept that makes the extraction process efficient. Accurate results have been obtained on test structures of relatively smaller sizes. Further validation of the approach is required on larger power arrays with more complex structure using bench test measurements.

Acknowledgement

Useful discussion with William Meier, National Semiconductor Corp., Santa Clara, USA is acknowledged.

References

[1] Ansys inc.- corporate homepage. [online] available: http://www.ansys.com.
[2] E. Barke. Resistance calculation from mask artwork data by finite element method. *in Proc. 22nd Design Automation Conference*, pages 305–311, 1985.
[3] R. W. Erickson and D. Maksimovic. *Fundamentals of Power Electronics*. Springer International Edition, 2006.
[4] A. Hastings. *The Art of Analog Layout*. Prentice-Hall, 2001.
[5] W. H.Kao, C. Y. Lo, M. Basel, and R.Singh. Parasitic extraction:current state of the art and future trends. *in Proceedings of the IEEE*, 89(5):729–738, May 2001.
[6] T. Mitsuhashi and K. Yoshida. A resistance calculation algorithm and its application to circuit extraction. *IEEE Transaction on Computer-Aided Design*, CAD-6(3):337–345, May 1987.
[7] S. P.McCormick. Excl: A circuit extractor for ic design. *in Proc. 21st Design Automation Conference*, pages 616–619, 1984.
[8] X. Qinfang, L. Shiyu, and R. Meizhi. Resistance calculation from geometric layout data by boundary element method. *International Conference on Circuits and Systems*, pages 863–866, June 1991.
[9] A. van Genderen and N. van Mejis. Using articulation nodes to improve the efficiency of finite-element based resistance extraction. *in Proc. 33rd Design Automation Conference*, pages 758–763, 1996.
[10] X. Wang, W. Yu, D. Liu, and Z. Wang. Fast extraction of 3-d interconnect resistance: Numerical-analytical coupling method. *in Proc. 5th International Conference on ASIC*, 1:315–318, Oct. 2003.
[11] Z. Wang and Q. Wu. A two-dimensional resistance simulator using the boundary element method. *IEEE Transaction on Computer-Aided Design*, 11(4):497–504, Apr. 1992.
[12] Y.Okamura, Y.Muraishi, T.Sato, and Y.Ikemoto. Las: Layout pattern analysis system with new approach. *in Proc. ICCC*, pages 308–311, 1982.

An Elitist Non-Dominated Sorting based Genetic Algorithm for Simultaneous Area and Wirelength Minimization in VLSI Floorplanning

Pradeep Fernando and Srinivas Katkoori

Department of Computer Science and Engineering, University of South Florida
{prfernan, katkoori}@cse.usf.edu

Abstract

VLSI floorplanning in the gigascale era must deal with multiple objectives including wiring congestion, performance and reliability. Genetic Algorithms lend themselves naturally to multi-objective optimization. In this paper, a multi-objective genetic algorithm is proposed for floorplanning that simultaneously minimizes area and total wirelength. The proposed genetic floorplanner is the first to use non-domination concepts to rank solutions. Two novel crossover operators are presented that build floorplans using good sub-floorplans. The efficiency of the proposed approach is illustrated by the 18% wirelength savings and 4.6% area savings obtained for the GSRC benchmarks and 26% wirelength savings for the MCNC benchmarks for a marginal 1.3% increase in area when compared to previous floorplanners that perform simultaneous area and wirelength minimization.

1. Introduction

The complexity of current VLSI systems has increased drastically due to rapidly improving integration technology. This has lead to an increased use of IP modules and hierarchical design methods making floorplanning a critical step in the VLSI physical design cycle. Traditional floorplanning consists of the problem of finding the relative module positions that result in a minimal bounding box area for the entire design without any module overlaps. But due to the interconnect and reliability issues of the current technologies acting as a bottleneck to realizing their full potential, current floorplanners must use multiple metrics including total wirelength and heat dissipation in the objective function. Genetic Algorithms (GA) [1, 2] are a natural optimization engine for multi-objective optimization as they maintain a population of individuals [3]. In this paper,

a multi-objective genetic algorithm is proposed for the VLSI floorplanning problem that simultaneously optimizes area and total wirelength. The main contributions of this paper are:

- Application of non-domination concepts to rank floorplans in a GA population
- A novel heuristic crossover operator (HOOX) that promotes the multiplication of good sub-floorplans
- A new crossover operator (MTOX) that is a novel combination of the classical two point crossover operator and order crossover operator.

The efficiency of the proposed method is shown by the 26.1% and 18.1% average savings in total wirelength with only 1.3% increase and 4.6% average savings in area for the MCNC and GSRC benchmarks respectively.

Of the numerous floorplanning works in the literature, very few floorplanners work explicitly on simultaneous area and wirelength optimization. Kim and Kim [4] proposed a linear programming approach coupled with simulated annealing. Sheqin et al [5] proposed a Quadratic Floorplanner based on Less Flexibility First Principles that reports the best results on the MCNC benchmarks. Other floorplanners [6, 7] and all the previous genetic floorplanners [8] use the weighted normalized single objective methodology to perform multi-objective optimization by assigning equal weights to all the objectives. This weighted sum approach has many limitations which will be discussed in section 2.1.

The rest of the paper is organized as follows: Section 2 gives a brief introduction to GAs and multi-objective optimization. Section 3 describes the proposed genetic floorplanner in detail. Section 4 summarizes the results obtained for the floorplanning experiments on the MCNC and GSRC benchmarks. Section 5 concludes with a summary of the proposed work.

2. Genetic Algorithms

A genetic algorithm is a robust, stochastic search technique that simulates the process of natural evolution [2]. Genetic algorithms are attractive because unlike simulated annealing that considers a single solution at a time, they maintain a population of solutions or individuals that are evolved throughout the optimization process. An *elitist* genetic algorithm preserves some of the current best solutions into the next generation.

2.1 Multi-objective Optimization and Genetic Algorithms

Multi-objective optimization is the process of finding a solution that is optimal in terms of multiple objectives. Traditional techniques transform the multi-objective optimization problem into a single objective optimization problem using a weighted sum of the normalized objectives. But normalization of the objectives and finding the correct weight vector to be used may not be a trivial process. In such a case, the preference given to the individual objectives by the user assigned weight vector might result in an undesirable bias towards a particular objective.

A more direct approach to multi-objective optimization involves finding pareto-optimal solutions. A solution is said to be *pareto-optimal* if no other solution to the multi-objective optimization problem is better than it in terms of all the objectives under consideration. The set of all such solutions is called the pareto-optimal set. All these solutions are said to lie in the non-domination level zero with a non-domination rank of zero as they are not inferior to any other solution in the entire solution space. For example in Figure 1, the solutions numbered 1 and 2 are better than all the other solutions in terms of both area and wirelength. But solution 1 cannot be deemed better than solution 2 as solution 1 is better than solution 2 only in terms of area but not wirelength. Hence solutions 1 and 2 belong to the global pareto-optimal set and are said to have a non-domination rank of zero. If solutions 1 and 2 are not considered, then the solutions numbered 3 and 4 are better than solutions 5 and 6 in terms of both objectives. Hence solutions 3 and 4 form the next *local* pareto-optimal front and are assigned a non-domination rank of 1.

The weighted single objective method can yield only one of the multiple pareto-optimal solutions and will also bias search engines such as Simulated Annealing towards a particular solution of the pareto-optimal set. This might lead to rejection of nearby solutions closer to the global pareto-optimal set and cause increased difficulties in finding a good solution.

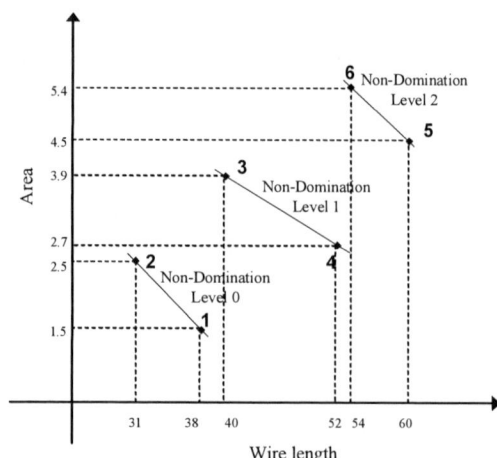

Figure 1 Non-domination levels in multi-objective optimization

Multi-objective genetic algorithms can avoid all these pitfalls as they can maintain a subset of the pareto-optimal front in their population. The user can then choose any one of these optimal solutions.

Many different templates have been proposed for multi-objective optimization using genetic algorithms [3]. In this paper, a Non-Dominated Sorting based GA template with elitism (NSGA-II) [9] is chosen as it can be implemented efficiently in terms of runtime.

3. The Proposed Approach

The proposed floorplanner is an Elitist Non-Dominated Sorting based Genetic Algorithm that uses two novel crossover operators and a set of mutation operators. The overall flow of the proposed genetic floorplanner is shown in Figure 2.

The genetic floorplanner generates an initial population of randomly generated solutions encoded using the sequence pair representation [10]. Sequence Pairs can represent both slicing and non-slicing floorplans using two permutations (Γ+, Γ-) of the module indices. In this work, all modules are assumed to have fixed widths and heights and are allowed to be rotated 90 degrees which is represented using a Boolean orientation vector (θ). Thus an individual's chromosomal encoding consists of two sequences namely, X sequence (Γ+) and Y sequence (Γ-), and an orientation vector (θ). The population size used by the genetic floorplanner is determined based on the problem size as shown in Table 1.

At the start of each generation, the non-domination ranks of all the individuals are computed and the population is sorted in non-decreasing order of the individuals' non-domination ranks using the procedure described in [3]. An individual in the population dominates a second individual if it is better than the second individual in terms of both area and wirelength. All the individuals in the current population are assigned a non-domination rank starting with zero with the fittest individuals having the lowest non-domination ranks. All the individuals with a non-domination rank of zero are marked as elite.

```
Proposed Multi-Objective Genetic Floorplanner
{
    Generate random Initial Population(population);
    for gen in N_generations do
    {
        EliteSet ← Non-Dominated Sort(population, Area, Wire);
        MatePool←Select Mating Pool(population);
        for i in 1 to cx_rate do
        {
            Select Parents(MatePool, P1, P2);
            Crossover(P1, P2, Off1, Off2);
            Update Population(population, Off1, Off2);
        }
        for i in 1 to mut_rate do
        {
            mutInd ← Select Non-Elite Individual(population);
            Mutate(mutInd);
            Update Population(population, mutInd);
        }
    }
    Output Best Individual;
}
```

Figure 2 Overall flow of the proposed elitist NSGA-based floorplanner

A mating pool of individuals is then selected to serve as parents for all the crossover operations in the current generation. The size of this pool is set to half the population size. If the number of elite individuals is greater than this size or if only a subset of individuals from a particular non-domination level have to be selected, then the individuals are picked randomly to form the mating pool.

Table 1 Problem-size dependent parameters

Number of modules (N)	Population Size	Tournament Size
N<100	200	4
100≤N≤200	400	8
N=300	600	16

Crossover is then performed according to the specified crossover rate by selecting parents only from the mating pool. The proposed GA uses tournament selection to select the two parents for crossover from the mating pool of individuals. In tournament selection, a group of parent candidates are selected randomly from the mating pool and the fittest two individuals will serve as parents for the crossover

operation. The tournament size depends on the population size as shown in Table 1.

Crossover. The proposed genetic floorplanner uses two novel crossover operators - namely Modified Two-Point Order-based Crossover Operator (MTOX) and Heuristic One-Point Order-based Crossover Operator (HOOX). While the MTOX operator is an unbiased operator that searches the entire solution space for a good solution by randomly combining segments from the parents to form the offspring, the Heuristic One-point Order Crossover operator (HOOX) biases the search towards promising regions of the solution space by promoting the transfer of good sub-floorplans from the parents to the offspring. In the first half of the genetic search, the unbiased operator is preferred (selection probability=0.75) while the heuristic operator is preferred (selection probability=0.6) in the latter phase of the genetic search. The selection probabilities values were derived empirically from numerous experiments.

Modified Two-Point Order-based Crossover (MTOX). The MTOX operator is a novel crossover operator proposed specifically to work on Sequence Pairs. This operator is a combination of two classical crossover operators used in several genetic algorithms, namely the Two-Point crossover operator and the Order crossover operator.

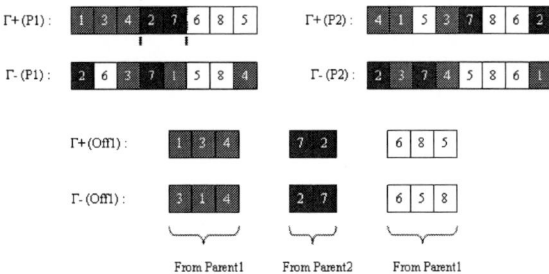

Figure 3 Generation of offspring using the MTOX operator

The original two point crossover operator [11] was proposed for use with individuals that were encoded as binary strings. The two point crossover operator randomly chooses two cut-points that split both the parents into three parts each. The offspring is formed by concatenating the first part of the first parent, the second part of the second parent, and the third part of the first parent together. If this traditional method is used with sequence pairs it will result in invalid Γ+ and Γ- sequences due to module duplication and deletion. To eliminate such violations, the traditional two point crossover operator is combined with the order crossover operator to yield the Modified Two-point Order Crossover (MTOX). The original order crossover operator [2] was proposed to work on individuals encoded as a single permutation. It copies over a part of the first parent's string into the offspring and copies over the missing genes in the order of their occurrence in the second parent. The MTOX operator is not a simple extension of the two-point crossover to handle the two permutations of the sequence pair. The MTOX operator couples the crossover of the Γ-sequence with the crossover of the $\Gamma+$ sequence to ensure that the relative positions of the modules contributed by the same parent are maintained in the offspring. The orientation of a module is contributed by the parent that dictates the position of the module ensuring that a good configuration within a parent is preserved in the offspring. Figure 3 illustrates the MTOX operator with an example.

The MTOX crossover operation is described in the procedure below:

1. Generate 2 random cut-points, c1 and c2, on $\Gamma+$ sequence of the first parent individual ($P1$) to obtain the three segments x_1^{p1}, x_2^{p1}, and x_3^{p1}.

2. Generate segment y_1^{p1} using the order of the modules in segment x_1^{p1} in the first parent's ($P1$) Γ-sequence. Similarly, generate segments y_2^{p1} and y_3^{p1} using $P1$'s Γ- sequence ordering of the modules in the segments x_2^{p1} and x_3^{p1}.

3. Generate segment x_1^{p2} using the order of the modules in segment x_1^{p1} in the second parent's ($P2$) $\Gamma+$ sequence. Similarly, generate segments x_2^{p2} and x_3^{p2} using P2's $\Gamma+$ sequence ordering of the modules in the segments x_2^{p1} and x_3^{p1}.

4. Generate segment y_1^{p2} using the order of the modules in segment x_1^{p1} in the second parent's ($P2$) Γ-sequence. Similarly, generate segments y_2^{p2} and y_3^{p2} using $P2$'s Γ- sequence ordering of the modules in the segments x_2^{p1} and x_3^{p1}.

5. Concatenation of the sequences x_1^{p1}, x_2^{p2}, and x_3^{p1} forms the $\Gamma+$ sequence of the first offspring (Off1).

6. Concatenation of the sequences y_1^{p1}, y_2^{p2}, and y_3^{p1} forms the Γ- sequence of the first offspring (Off1).

7. The orientation of a module in the offspring is copied over from the respective parent that contributes the position of the module.

Heuristic One-Point Order-based Crossover (HOOX). The HOOX operator is a heuristic crossover operator that identifies good sub-floorplans in the parents and preserves them in the offspring. The HOOX operator uses the traditional One-point Crossover operator [2] to partition the parents into two sub-floorplans as shown in Figure 4. The order crossover operator is used to eliminate any violations

in the sequence pairs for the two sub-floorplans. The two sub-floorplans with the better area usage are then combined to form the offspring as shown in Figure 5.

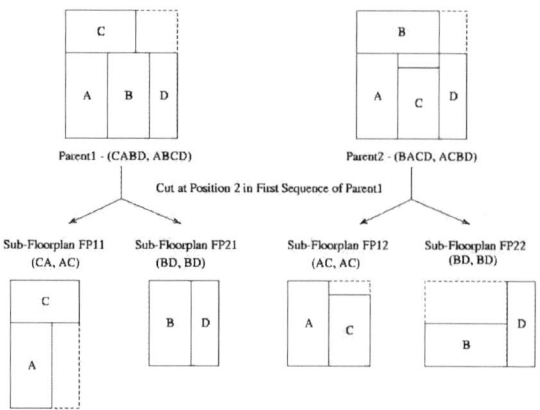

Figure 4 Generation of sub-floorplan alternatives in the HOOX operator

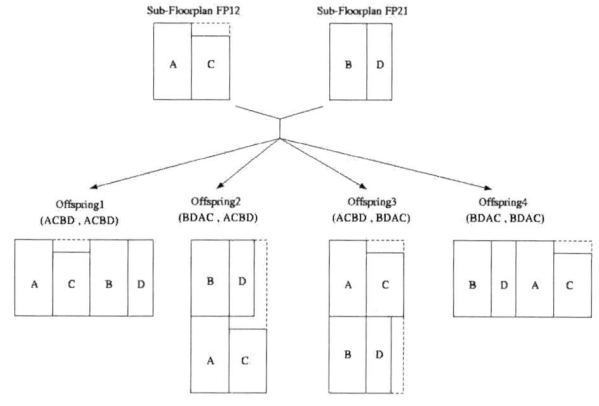

Figure 5 Offspring configurations from sub-floorplan alternatives in the HOOX operator

The HOOX crossover operation is described in the procedure below:

1. Split the $\Gamma+$ sequence of the first parent, $P1$ to obtain the two segments x_1^{p1} and x_2^{p1}.

2. Generate segment y_1^{p1} using the order of the modules in segment x_1^{p1} in the first parent's ($P1$) Γ-sequence. Similarly, generate segment y_2^{p1} using $P1$'s Γ- sequence ordering of the modules in the segment x_2^{p1}. The sequence pairs (x_1^{p1}, y_1^{p1}) and (x_2^{p1}, y_2^{p1}) correspond to two sub-floorplans, FP11 and FP21.

3. Generate segment x_1^{p2} using the order of the modules in segment x_1^{p1} in the second parent's ($P2$) $\Gamma+$ sequence. Similarly, generate segment x_2^{p2} using P2's $\Gamma+$ sequence ordering of the modules in the segment x_2^{p1}. Also generate segments y_1^{p2} and y_2^{p2} using $P2$'s Γ- sequence ordering of the modules in the segments

340

978-1-4244-3039-0/08 $25.00 © 2008 IEEE

x_1^{p1} and x_2^{p1}. Sequence pairs (x_1^{p2}, y_1^{p2}) and (x_2^{p2}, y_2^{p2}) correspond to two sub-floorplans, FP12 and FP22.

4. Sub-floorplans FP11 and FP12 are two alternatives for building a sub-floorplan using the modules in the segment x_1^{p1}. Similarly, sub-floorplans FP21 and FP22 are two alternatives for building a sub-floorplan using the modules in the segment x_2^{p1}. The sub-floorplan alternatives with the better area usage are picked to be transferred to the offspring floorplan.

5. Four different offspring configurations are possible using the two sub-floorplan alternatives. The best configuration is chosen as the final offspring.

The module orientations in the parents are preserved in the sub-floorplans and the offspring.

Mutation. Three mutation operators are used in the proposed genetic floorplanner. A non-elite individual is randomly chosen from the population and one of the three mutation operators is selected uniformly at random and applied to the individual.

- **Mutation Operator 1** – It picks 2 random modules and exchanges their positions in both the sequences, $\Gamma+$ and $\Gamma-$, of the chosen individual.

- **Mutation Operator 2** – It picks 2 random modules and exchanges their positions only in the first sequence, $\Gamma+$, of the chosen individual.

- **Mutation Operator 3** – It picks a random module, b_i, and changes its orientation (θ_i) by exchanging its width and the height.

After crossover and mutation have generated the offspring pool, the new population is formed by combining the offspring with all the elite individuals from the current generation. The above process repeats for the specified number of generations. The individual in the pareto-optimal front of the final population with the lowest weighted sum fitness (equal weights to area and wirelength) is reported as the best solution.

4. Experimental Results

The proposed floorplanner was implemented using C++/STL and run on a 200MHz SUN SPARCv9 workstation with 256MB memory. A crossover rate of 0.5 and a mutation rate of 0.1 were empirically found to give the best results for a fixed period of 1500 generations. Floorplan area is computed using the longest common subsequence method [12] proposed for floorplans represented by sequence pairs. Half-Perimeter wirelengths (HPWL), pin-to-pin for MCNC

and center-to-center for GSRC, are computed for all the nets to estimate the total wiring required.

4.1 Performance on the MCNC benchmarks

The proposed genetic floorplanner is compared with another GA-based floorplanner [13] (referred to as AdaptGA here on) that reports the best results for a genetic floorplanner using the sequence pair representation. The proposed genetic floorplanner is also compared with Sheqin et al [5] (referred to as QP-LFF here on) which reports the best results for the MCNC benchmarks among floorplanners that simultaneously optimize both area and wirelength. The proposed floorplanner is also compared to a SA-based publicly available floorplanning tool, Parquet, whose hard block outline-free floorplanning results using the sequence pair representation are reported in Chan et al [14]. It is to be noted that the area and wirelength reported for the proposed genetic floorplanner belong to a single best individual that is found as described in Section 3.

Table 2 summarizes the area and wirelength comparison for the two MCNC benchmarks for which QP-LFF and AdaptGA report results. The proposed genetic floorplanner outperforms all the other methods in terms of wirelength for the ami*XX* benchmarks. The marginal increase in area for some cases is justified given the high wirelength savings obtained. The proposed GA was faster than AdaptGA but the runtimes are not directly comparable as AdaptGA was run on an UltraComp model60 workstation. The runtime of the proposed genetic floorplanner was 640 seconds for the ami33 benchmark and 1384 seconds for ami49 averaged over five runs.

4.2 Performance on the GSRC benchmarks

We compare our floorplanning results against a recent Simulated Annealing (SA) based floorplanner [15] that uses the weighted sum approach for optimizing area and wirelength.

The proposed genetic floorplanner compares favorably against the SA-based floorplanner producing better results in total wirelength for all the benchmarks and in area for 12 of the 16 benchmarks as shown in Table 3. The proposed floorplanner obtained a savings of 4.6% and 18.14% in area and total wirelength respectively averaged over all the 16 benchmark circuits. The average run-times of the proposed genetic floorplanner ranged from 181 seconds for the n10 benchmarks to 4789 seconds for the biggest benchmark (n300).

Table 2 Area and wirelength comparisons with AdaptGA [13], QP-LFF [5], and Parquet [14] floorplanners. (A=Area (mm^2); W=Wirelength(mm); %S=Percentage Savings)

	Proposed Floorplanner		AdaptGA Floorplanner				QP-LFF Floorplanner				SA (Parquet) Floorplanner			
	A	W	A	%S	W	%S	A	%S	W	%S	A	%S	W	%S
ami33	1.20	31.33	1.22	1.64	39.37	20.4	1.177	-1.95	45.3	30.84	1.32	9.1	75	58.2
ami49	37.81	677.9	37.16	-1.75	971.3	30.2	36.6	-3.3	879.9	22.96	40.69	7.1	849	20.2
				-0.05		25.3		-2.63		26.9		8.1		39.2

Table 3 Area and wirelength comparisons with SA-based floorplanner [15]

Circuit	Proposed Genetic Floorplanner		SA-based Floorplanner		Percent Savings	
	Area	Wire	Area	Wire	Area	Wire
N10	233352	13572	258152	18164	9.61	25.28
N10b	232162	13320	251178	15128	7.57	11.95
N10c	237965	16129	268865	19880	11.49	18.87
N30	224910	42164	245115	54586	8.24	22.76
N30b	210090	37789	234574	45931	10.44	17.73
N30c	235304	40398	233867	55979	-0.61	27.83
N50	217000	85327	231431	104395	6.24	18.27
N50b	217765	86633	237266	94790	8.22	8.61
N50c	217434	93928	234567	106562	7.30	11.86
N100	205758	133497	210378	180413	2.20	26.00
N100b	186095	135469	185868	169767	-0.12	20.20
N100c	199320	148445	208616	185215	4.46	19.85
N200	215232	339324	214349	393644	-0.41	13.80
N200b	212800	294663	208960	336236	-1.84	12.36
N200c	201056	331065	206954	394358	2.85	16.05
N300	336396	534157	329589	658162	-2.07	18.84
Average Savings Percentage					4.60	18.14

5. Conclusion

A GA-based multi-objective floorplanner was presented that uses the concept of non-dominated sorting to simultaneously reduce area and total wirelength of a 2D floorplan. Two novel crossover operators are proposed that produce very good results on both the MCNC and GSRC benchmark circuits. The proposed floorplanner can be easily modified to accommodate more number of objectives. Thus with efficient operators, genetic algorithms can be used effectively for multi-objective floorplanning.

6. References

1. Holland, J.H., *Adaptation in Natural and Artificial Systems.* 1975: University of Michigan Press, Ann Arbor.
2. Goldberg, D.E., *Genetic Algorithms in Search, Optimization, and Machine Learning.* 1989: Addison-Wesley Publishing Company, Inc.
3. Deb, K., *Multi-Objective Optimization using Evolutionary Algorithms.* 1st ed. 2001: John Wiley & Sons, Ltd.
4. Jae-Gon, K. and K. Yeong-Dae, *A linear programming-based algorithm for floorplanning in VLSI design.* Computer-Aided Design of Integrated Circuits and Systems, IEEE Transactions on, 2003. **22**(5): p. 584-592.
5. Sheqin, D., et al., *Module placement based on quadratic programming and rectangle packing using less flexibility first principle*, in *International Symposium on Circuits and Systems.* 2004. p. V-61-V-64 Vol.5.
6. Lin, J.-M. and Y.-W. Chang, *TCG: A Transitive Closure Graph-Based Representation for Non-Slicing Floorplans*, in *Design Automation Conference.* 2001.
7. Pei-Ning, G., C. Chung-Kuan, and T. Yoshimura. *An O-tree representation of non-slicing floorplan and its applications.* in *Design Automation Conference.* 1999.
8. Cohoon, J., J. Karro, and J. Lienig, *Evolutionary Algorithms for the Physical Design of VLSI Circuits.* Natural Computing Series, 2003: p. 683-711.
9. Deb, K. and T. Goel, *Controlled elitist non-dominating sorting genetic algorithms for better convergence*, in *International Conference on Evolutionary Multi-Criterion Optimization.* 2001.
10. Murata, H., et al., *VLSI module placement based on rectangle-packing by the sequence-pair.* Computer-Aided Design of Integrated Circuits and Systems, IEEE Transactions on, 1996. **15**(12): p. 1518-1524.
11. Eshelman, L.J., R.A. Caruana, and J.D. Schaffer, *Biases in the crossover landscape.* Proceedings of the third international conference on Genetic algorithms, 1989: p. 10-19.
12. Xiaoping, T., T. Ruiqi, and D.F. Wong, *Fast evaluation of sequence pair in block placement by longest common subsequence computation.* Computer-Aided Design of Integrated Circuits and Systems, IEEE Transactions on, 2001. **20**(12): p. 1406-1413.
13. Nakaya, S., T. Koide, and S. Wakabayashi, *An adaptive genetic algorithm for VLSI floorplanning based on sequence-pair*, in *IEEE International Symposium on Circuits and Systems.* 2000. p. 65-68 vol.3.
14. Chan, H.H., S.N. Adya, and I.L. Markov, *Are floorplan representations important in digital design?* Proceedings of the 2005 international symposium on physical design, 2005: p. 129-136.
15. Shiu, P.H., et al., *Multi-layer floorplanning for reliable system-on-package*, in *International Symposium on Circuits and Systems.* 2004. p. V-69-V-72 Vol.5.

Fast Congestion Aware Routing for Pin Assignment

Shashank Prasad (shasha@cadence.com)

Cadence Design Systems

Macroblock (aka partition) pin assignment and routing are important tasks in typical top-down hierarchical physical design. Routers use pin locations as connection points to route the design with a goal of minimizing congestion. However, determining suitable pin locations it self depends on availability of congestion free routing topology as a seed input. This results in a catch-22 situation. In this paper, we present an approach, during prototyping phase, to generate fast-and-dirty congestion free routing topology, in top channels. This is real chip routing topology, in the sense that, the routing topology of every net adheres to physical hierarchy, as would happen during hierarchical implementation. This is passed as seed to pin assignment engine, which thus, results in congestion-free pin locations. The novelty of this approach lies in efficient detection of those inter-partition nets whose routing topology have little or no bearing to top channel congestion. These nets are then either not routed or routed in a fast hierarchy unaware manner. We will show that this routing topology is good enough (less than 10% error margin) to establish suitable cross points at partition boundaries, while the speed up achieved is around 6X compared to routing all nets in hierarchy aware manner. Experimental results demonstrate its efficiency and effectiveness. Furthermore, it can also be effectively used as seed input for decisions like channel sizing between partitions, and budgeting timing constraints to partitions.

1. INTRODUCTION

Due to the enormous complexity of very large scale integration design, a hierarchical approach is needed for the placement and routing of millions of standard cells in order to reduce runtime and improve solution quality. Pin assignment and routing for macroblocks (aka partitions) are important steps in a typical top-down hierarchical approach.

In hierarchical design methodology, pin assignment is usually the last and a crucial step in defining physical interface of partitions. The challenge here is to come up with a permutation of pin locations on partition boundaries, which when becomes the physical interface of the partitions, results in minimal congestion, and improves routability in top channels. State-of-the-art designs today are top channel starved which makes it very critical to reduce congestion in top channels. Generally, pin assignment algorithms depend heavily on the routes generated by Steiner routers (Steiner tree based router; also called non partition aware (**NPA**) router) to find routing

cross-points with partition boundaries. These routing cross-points eventually become seed to find pin locations, along with other pin constraints, like pin spacing, improve pin alignment, no pin overlaps and drc rules.

However, NPA routing, in hierarchical design, is not indicative of real chip routing, as it does not adhere to physical hierarchy. Pin assignment based on this routing, leads to channel congestion, which are very much visible in case of channel-starved designs (i.e., designs having thin channels between partitions).

Alternative method is to instead use partition aware (PA) router, which is generally indicative of real chip routing topology. But, PA routing is an extremely time consuming operation. In state-of-the-art designs having 6 to 8 partitions, it could easily take up to 6 to 10 hours to do PA routing, while NPA routing finishes in less than 1 hour. Such large run time, coupled with prototyping iterations, is simply unacceptable to the designers. So, they often resort to NPA routing – multiply faster but prone to congestion – to generate seed routing for pin assignment engine. While a lot of work has been done ([1], [2], [3], [4]) to do simultaneous global routing (which is mostly PA) and pin assignment, however, there is no attempt made to increase the performance of PA routing, which becomes a major bottleneck, especially in today's multi-million SoCs.

In this paper, we describe an algorithm by which it is possible to do PA routing of most of the inter-partition nets (>90%) in multiply faster run time than current algorithms. In the course of this run time speed up, it does bring in some inaccuracies in routing topology, but, for purposes like pin assignment, they are all too small and well within limits, as will be illustrated by quantitative data in this paper. As a result, the decisions on pin locations based on these fast-and-dirty PA routes are far more superior to those based on NPA routes. To our best knowledge, this is the first use of PA routes in the manner described here. In the experimental results presented in this paper, we compare congestion reports after pin assignment is done based on routing topology of a NPA router, verses our proposed fast-and-dirty PA router.

The organization of the rest of the paper is as follows - -Section 2 presents basic definitions including PA routing. Section 3 describes existing hierarchical flow for pin assignment, and why PA routing takes enormous time to finish. Section 4 describes the underlying philosophy we have used to speed up PA routing at the cost of a little loss to accuracy. Section 5 presents experimental results wrt congestion when pin assignment is done based on this new fast-and-dirty routing technique. And finally, Section 6 concludes with hints and ideas on how fast-and-dirty PA routes can be applied to other floor planning applications like channel estimator, master floor planer, and some ideas for future work.

2. BACKGROUND

2.1 Terminology: *Partition, Intra-Partition nets, Inter-Partition nets, top nets, top channel, channel starved design*

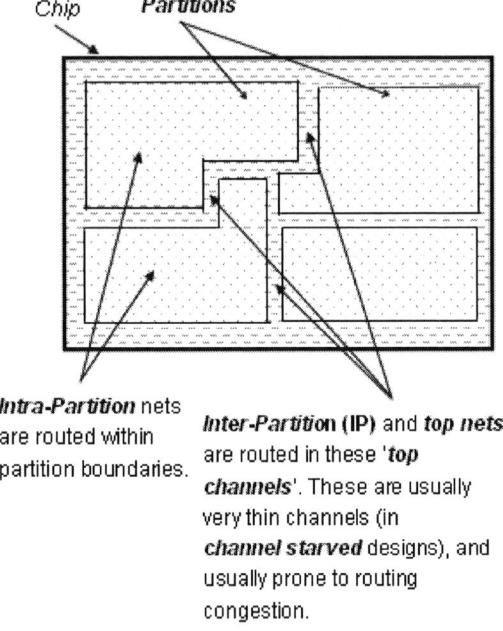

Intra-Partition nets are routed within partition boundaries.

Inter-Partition (**IP**) and **top nets** are routed in these 'top channels'. These are usually very thin channels (in **channel starved** designs), and usually prone to routing congestion.

Figure 1: Basic terminology

2.2 NPA routing, PA routing

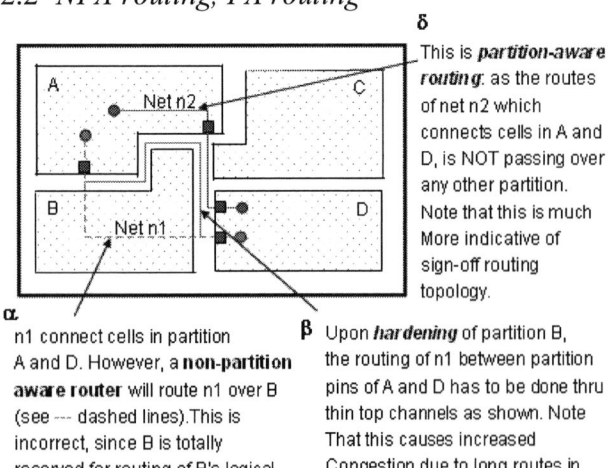

α.
n1 connect cells in partition A and D. However, a **non-partition aware router** will route n1 over B (see --- dashed lines).This is incorrect, since B is totally reserved for routing of B's logical nets ONLY.

β Upon *hardening* of partition B, the routing of n1 between partition pins of A and D has to be done thru thin top channels as shown. Note That this causes increased Congestion due to long routes in thin top channels.

δ
This is **partition-aware routing**: as the routes of net n2 which connects cells in A and D, is NOT passing over any other partition. Note that this is much More indicative of sign-off routing topology.

Figure 2: Partition aware (PA) and non-partition aware (NPA) routing

The default behavior of NPA router is to generate non-partition aware routes (like Steiner routes), as explained by α in Figure 2 above for net n1, by the dashed lines ---. The routes of n1 are going through B, although they are not connected to any standard cell in B. Thus, the routes of n1 violate physical hierarchy. Note that, in hierarchical design, such routing topology in top channels is not indicative of the real routing topology of the chip. Upon hardening of partitions, since the subsequent routing of inter-partition nets can occur only in top channels, it often results in long routes in top channels, as indicated by β in Figure 2; see solid _____ lines. Hence, pin assignment based on such NPA routing often leads to very bad congestion in most parts of thin top channels. Also, note that creation of multiple physical ports for a single pin is not an option, as different delay values due to multiple physical ports of a pin can not be represented today in timing delay formats (SPEF, SDF). Also, feed thru insertion is often rejected as it makes hierarchical logical equivalence checking complicated.

The above problems could have been avoided if the routes were based on a PA router (as shown by δ in Figure 2). The key difference, wrt to NPA routing, is that the route of n2 – which connects standard cells in A and D - is not going thru either B or C. In other words, the routes of n2 adhere to physical hierarchy. PA routing is indicative of real routing which happens in hierarchical physical implementation. However, today's PA routers are 5-10X slower than NPA routers. This run time degradation is simply unacceptable for floor planning applications, which usually need several routing iterations.

3. WHY PA ROUTER IS SO SLOW?
To understand this sluggishness of PA routers, please refer to Figure 3 below:

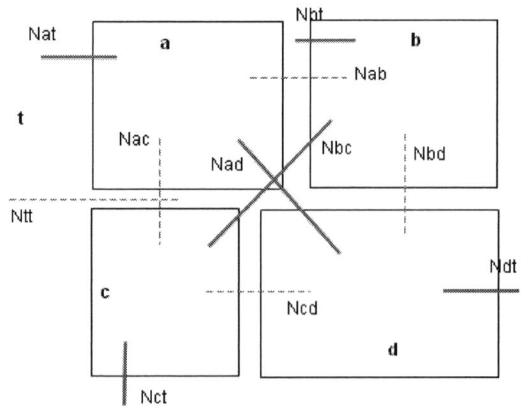

Figure 3: Various net groups illustrated

The above figure has:

978-1-4244-3039-0/08 $25.00 © 2008 IEEE

- 4 partitions: **a, b, c, d**
- Default top chip, occupying rest of the chip space: **t**
- Several sets of nets (henceforth called *netGroups*), categorized according to their connectivity to various partitions. For example, *Nab* netGroup denotes the set of those inter-partition nets which connect instances in partition **a** and **b** only. *Nat* netGroup denotes the set of those inter-partition nets which connect instances in partitions **a** and **t** only.

In order to generate PA routing topology for all netGroups, most PA routers would do several iterations of routing runs: separate NPA routing run to route each netGroup as shown above. This is because, to route netGroup *Nab*, it will have to put routing blockages over **c** and **d**, and call NPA router on Nab nets only – this will avoid routing any *Nab* net over **c** and **d**, thus emulating PA routing topology for these nets. Once, it has finished routing all nets of *Nab* netGroup, in order to route nets of another netGroup *Ncd*, it will now remove routing blockages over **c** and **d**, and instead put them over other partitions, namely **a** and **b** here. And so on. Thus, in order to obtain PA routes, the router does multiple runs of NPA routing, each time over selected set of nets, which makes its run time multiply more than a NPA router. In a typical customer design having 6-8 partitions, usually the number of netGroups formed is around 40. This makes PA routing around 10-20X slower than NPA routing.

4. FAST-AND-DIRTY PA ROUTING ALGORITHM AND SUITABLE PIN ASSIGNMENT

4.1 *Design configurations*

Typically, customer chip designs show the following statistical trend for inter-partition nets:

- Roughly, around **0.1%** of all nets are inter-partition nets.
- **99%** of these inter-partition nets are either *1P* or *2P* nets. By *1P (netGroup connecting a partition and top chip)*, we mean netGroups of the type *Nat, Nbt, Nct, Ndt*. By *2P (netGroup connecting two partitions)*, we mean any of *Nab, Nad, Nad, Nbc, Nbd, Ncd* types.
- Rest **1%** of the inter-partition nets is of the type *3P, 4P* and so on combined. (Usually they are clock nets, reset, enable, global nets).

The above statistics is true for most of the customer designs. This is understandable as chip net list is designed in a manner to minimize communication nets between partitions. Also, most of the communication is between groups of two partitions each. Only for global nets - like clock, enable, and reset – the communication goes across multiple partitions.

Also, typically, the number of nets in these netGroups shows the following statistical trend:

- When netGroups are sorted in descending order wrt to number of nets contained in each, top **30%** of the netGroups contain around **90%** of the total number of inter-partition nets.

The fast-and-dirty PA routing algorithm heavily exploits the above statistics to come up with its routing topology.

4.2 *Fast-and-dirty PA routing*

4.2.1 *Reduce number of NPA runs during PA routing*

As noted above, since top 30% of the netGroups contain more than 90% of inter-partition nets, then it is possible to do PA routing of 90% of these inter-partition nets within just $1/3^{rd}$ of the total number of NPA runs. So, our first enhancement to the algorithm is to:

- Submit top 30% of the netGroups to NPA runs. This will route 90% of inter-partition nets in PA manner.
- Then, all remaining nets are routed using only one run of NPA router. This introduces NPA-ness in these routes, but since they are small in number, it has little affect on the overall PA-ness of all inter-partition nets, while at the same time, it brings an overall speed up of 3X wrt a complete PA run time (i.e. if all nets are routed in PA manner).
- In the top 30% netGroups, some more netGroups - like *Nac, Nab, Nbd, and Ncd* - can be further discarded from PA routing runs. These are the ones indicated by dotted lines ---- in Figure 4 below. This is because since they contain 2P nets which belong to two completely facing partitions, such nets are likely to be routed in a similar manner by either NPA or PA router. So, this further reduces the number of netGroups to be submitted to PA routing runs thereby reducing overall run time.

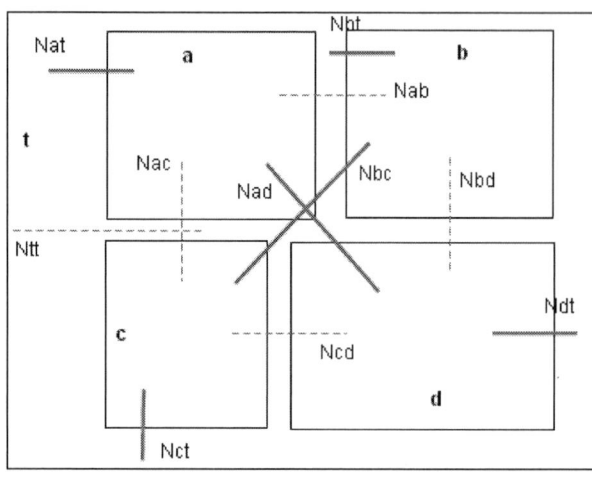

Denotes the InterPartition nets which are subjected to partition aware routing runs

Denotes the InterPartition nets which are subjected to non-partition aware (shortest-distance-based) runs

Figure 4: Net groups submitted to partition aware routing runs, while others to non-partition-aware routing runs

4.2.2 Reduce run time of each PA routing run

In order to speed up each single run of PA router, following changes are done:

- Grid cell size (used by routers) is increased by 4X2. This does bring in some accuracy loss in global router, but this is again acceptable, as for pin assignment purposes, we are interested in a rough estimation of routing cross-point only.

- It is observed that a router spends a lot of time in creating routing obstacles for standard cells. Since most of these obstacles are useful only during routing of intra-partition nets, and have close to no significance for inter-partition nets, we discarded creation of these obstacles.

- Since we are interested in the routing of only inter-partition nets, we hide all other nets (like intra-partition nets) from the router.

The above enhancements bring up a total speed up of around 2X in each routing run. Coupled with reduction in the total number of runs, we get a total speed up of around 3X * 2X = 6X.

5. EXPERIMENTAL RESULTS

We tried the new algorithm on two latest state-of-the-art customer tape out designs. Both the pin assignment flows are illustrated below:

Flow A: NPA routing, pin assignment by enforcing the pins at routing cross-points; measure top channel congestion, via count and net length.

Flow B: PA routing (using the new fast-and-dirty algorithm proposed in this paper), pin assignment by enforcing the pins are routing cross-points; measure top channel congestion, via count and net length.

The result is illustrated below for a representative customer design. The other design also shows similar results:

Table 1: Comparison between Flow A and B

Criteria	Flow A	Flow B	Comparison
Run time	**5500 sec***	620 sec	9X speed up
Top channel congestion	21000 congested cells	16000 congested cells	20% improvement
Top channel via count	53024 vias	46918 vias	13% reduction
Top channel net length	1.892e+07u m	1.747e+07u m	8.3% reduction

* **5500 sec** is the run time of our partition-aware router.

The following diagrams show visual illustration of improvement in congestion hotspots via the new algorithm.

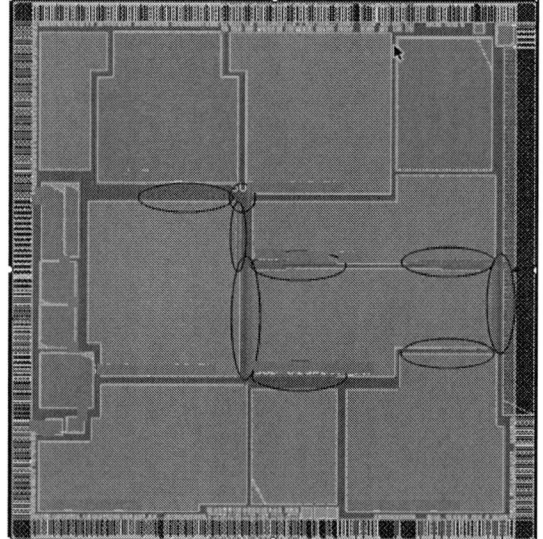

Figure 5: Flow A congestion hotspots indicated by red congestion markers (and also by black ovals)

Flow B shows significant improvement in congestion:

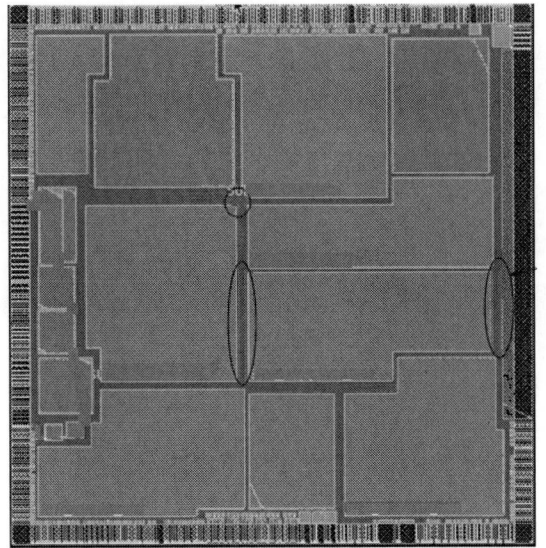

Figure 6: Flow B shows significantly reduced number of congestion hotspots

6. FUTURE WORK AND APPLICATIONS

PA routing is the heart of hierarchical design methodology. But because of its run time overhead, most of the prototyping applications today use alternate means to estimate congestion, net length etc; these are not based on PA routes, are heuristic based and highly inaccurate. But, with the speed up achieved by our algorithm (with acceptable loss to accuracy), we believe it has become possible to use the enormous power of PA router for critical and accurate decision making. As for pin assignment, it can be equally well used for other applications like partition channel size estimation, automatic floor planner.

Since we are interested in the accuracy of congestion estimates primarily in top channels only, we are planning to invest some efforts on how can we make prototyping router to give extra focus on top channels during its global routing phase. One idea is to have refined global cell size in top channels and have coarser ones inside partitions. Another idea is to do multiple runs on router by utilizing grid map of previous run, rather than re-creating it during every run. Prototypes are yet to be made to validate these ideas.

7. References

[1] S. G. Choi and C. M. Kyung, "Three-step pin assignment algorithm for building block layout," Electron. Lett, vol. 28, no. 20, pp. 1882–1884, 1992.

[2] J. Cong, "Pin assignment with global routing for general cell designs," IEEE Trans. Computer-Aided Design, vol. 10, pp. 1401–1412, Nov. 1991.

[3] T. Koide, S. Wakabayashi, and N. Yoshida, "An integrated approach to pin assignment and global routing for VLSI building-block layout," in Proc. Eur. Conf. Design AutomationWith Eur. Event ASIC Design, 1993, pp. 24–28.

[4] L. E. Liu and C. Sechen, "Multilayer pin assignment for macro cell circuits," IEEE Trans. Computer-Aided Design, vol. 18, pp. 1452–1461, Oct. 1999.

21st International Conference on VLSI Design

A Novel Approach to Compute Spatial Reuse in the Design of Custom Instructions

Nagaraju Pothineni
nagaraju@cse.iitd.ernet.in

Anshul Kumar
anshul@cse.iitd.ernet.in

Kolin Paul
kolin@cse.iitd.ernet.in

Department of Computer Science & Engineering
Indian Institute of Technology, Delhi
New Delhi, INDIA - 110016

ABSTRACT

In the automatic design of custom instruction set processors, there can be a very large set of potential custom instructions, from which a few instructions are required to be chosen, taking into account their spatial as well as temporal reuse and cost. Using the existing pattern matching techniques, finding complete reuse of every identified pattern in the entire application would be very slow and may even be computationally infeasible. Due to this, the existing selection methods employ pattern matching techniques at a very later stage of selection process on a small set of patterns, compromising the quality of selected candidates. In this paper, we propose a method by which each pattern's reuse information can be derived at an early stage of selection process even when there are very large number of potential patterns. The novel contributions of this paper include a simple and efficient algorithm for finding all the isomorphic convex subgraphs (termed as Recurring Pattern Information(RPI)) of the given application's Control Data Flow Graph (CDFG). The proposed technique is integrated into the estimation phase of the Instruction Set Extension(ISE) automation. Experimental results show the efficiency of the proposed algorithm and demonstrate its utility in generating high quality custom instructions.

1. INTRODUCTION

In recent years, several commercial extensible processor platforms have become available in the market. Examples of these platforms include Tensilica Xtensa [1], Altera NIOS [2], MIPS CorExtend [3] etc. Using these platforms, application specific functionality can be introduced into the base processor to speedup the critical parts of applications. The designer specifies the application specific functionality in a high level language and the platforms automate the integration of custom hardware units into the processor pipeline and generate the software tool kit for the custom processor. Over the last decade, several heuristic techniques [4] [5] [6] [7] were proposed for automating the identification of best possible set of Custom Functional Units (CFUs). The approach followed is to identify a set of legal computation patterns and then select a few of the most beneficial patterns based on cost-benefit analysis.

The gain of a pattern mainly depends on a) efficiency of its hardware implementation and b) reuse. Reuse of a pattern may be due to its multiple instances in different parts of the application and due to the repeated execution of same pattern instance at run time. We call the former *spatial*

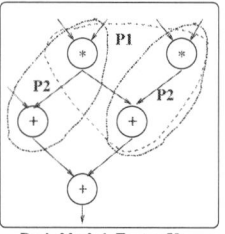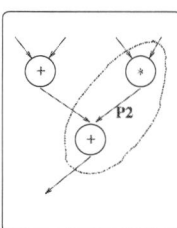

| | Basic block 1, Freq = 50 | | Basic block 2, Freq = 80 | |

Pattern	Gain	Complete Reuse	Area(um^2)	Total Gain
P1	2	50	19	100
P2	1	50 + 50 + 80 =180	12	180

Figure 1: Motivating example for considering complete reuse

reuse and the latter *temporal reuse*.

To ensure good results, the area and performance estimates should be as accurate as possible. This implies that, while computing the gain of a pattern, its complete reuse should be taken into account. However, the current techniques that exhaustively identify all legal patterns, do not take into account complete spatial reuse while selecting the patterns. Due to this, the estimated gain is often an underestimate. This has two main disadvantages 1). Underestimates would influence the pattern selection and thus the final quality of solution 2). The true speedup potential of patterns remains hidden. Our experimental results (in section 5) clearly show that these two problems indeed occur in real benchmarks.

Consider the two patterns P1 and P2 shown in Example 1. Pattern P1 occurs exactly once, whereas the pattern P2 occurs at three different places. If spatial reuse is not taken into account, then P1 would incorrectly appear to have higher performance benefit. But when spatial reuse is taken, the true benefits of patterns P1 (100), P2(180) would be unveiled and P2 would be preferred over P1. Selecting P2 instead of P1 is beneficial not only from the performance point of view but also P2 has smaller area than P1.

In this paper, we present an efficient method for computing the spatial reuse of each identified pattern. The custom instruction selection phase utilizes the spatial reuse information for judging the usefulness of each potential candidate. Experimental results indicate upto 25% improvement in the speedup achieved over state of the art techniques.

978-1-4244-3039-0/08 $25.00 © 2008 IEEE

2. RELATED WORK

Instruction extension methods can be classified into two categories. In the first [8], new instructions are created to encode frequently occurring parallel micro-ops or micro-op dependency chains. This kind of ISE requires little or no modification to the data path of the processor but microprogram is augmented for the new instruction. In the second category [4] [5] [6] [7] [9], a Custom Functional Unit (CFU) is tightly integrated into the data path of a base processor of RISC/VLIW class.

Automatic Instruction Set Extension (ISE) method can be divided into three stages: Identification, Cost-Benefit analysis and Selection.

Broadly, custom instruction identification methods can be grouped into two categories: Exhaustive and Non-Exhaustive heuristic techniques. Atasu et al. [7] and Yu et al. [10] enumerates exhaustively all the patterns satisfying the microarchitectural constraints imposed by the base processor. Exhaustive identification has the advantage of not missing any potential candidate. But the number of legal candidates could be very large and the pattern evaluation and selection should be very efficient to handle those large number of patterns. In contrast to the exhaustive technique, the techniques proposed in [5,6,9,11,12] identify a few patterns through incremental growing and/or combination of generated patterns guided by a heuristic. Due to the inherent greedy nature of these heuristics, the patterns identified in this manner are likely to be suboptimal.

The scope of identification is usually a basic block. However, Arnold et al. [11] identify patterns from trace, hence the identified patterns may span multiple basic blocks. Using such patterns may require a lot of bookkeeping code to preserve the semantics. But the paper does not discuss how actually such patterns are used.

In [5,6], where identification pass generated few patterns, graph matching based techniques are used for obtaining the spatial reuse. But the approach [7] generating millions of patterns does not consider the reuse factor while evaluating the patterns. Standard graph matching techniques [13] are quite slow, because each identified pattern is treated independently even if there is a strong correlation among them. Our approach calculates the spatial reuse of each pattern very efficiently through the technique of recurring pattern computation and can be used in combination with exhaustive pattern identification.

Finally, few patterns that maximize/minimize the objective function under the given constraints such as area, power etc, are selected. The selection problem is shown to be NP Complete even for the simple formulation of selecting N best patterns [14] and is even so when more constraints are included. The non-exhaustive identification techniques [5, 6] formulate an ILP for custom instruction selection. But ILP based methods are not scalable to large number of patterns. The approach in [7] is essentially a greedy approach that iteratively selects the best pattern without taking account reuse factor. Our selection strategy judiciously selects the patterns taking into account the reuse of patterns.

3. ISE METHODOLOGY

Our automatic application specific ISE methodology, shown in Fig. 2, consists of three steps. 1. Custom instruction identification, 2. Cost-Benefit analysis and 3. Custom instruction selection.

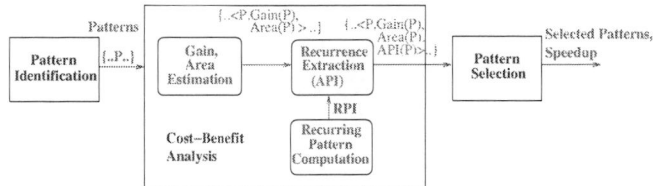

Figure 2: Our ISE methodology

In the first step, all subgraphs of the given application's CDFG that satisfy the micro-architectural constraints imposed by a typical RISC processor are identified. We employed an algorithm similar to [7] for enumerating all the legal candidates. In the second step, the isomorphic instances of each identified pattern are found with the aid of the Recurring Pattern Compuation (RPC) algorithm. RPC preprocesses the application and classifies the convex subgraphs into isomorphic equivalence classes. In addition to the spatial reuse, the latency and area of hardware implementation are found in the second step. Latency of hardware implementation can be obtained by dividing the critical path delay by the base processor's clock period. And area can be obtained by summing the areas of individual operations.

In the final step, a set of N custom instructions that maximize the performance gain are selected. When an instance of a pattern p_1 overlaps with an instance of another pattern p_2 then only one of p_1 and p_2 can be utilized in that portion of the code. The problem at hand is more complex than the code covering problem, well known in the compiler theory. Because, we not only have to decide the instances each pattern would cover but also decide the best set of patterns. The pattern selection problem is shown to be NP Complete [14]. We employ a greedy heuristic that iteratively selects the best of the remaining patterns. After selecting a pattern p, the instances of the remaining patterns that overlap with p are discarded.

Let $P1, P2, \ldots, Pk$ be the patterns finally selected. The speedup of application A is calculated as follows.

$$Speedup = \frac{SW(A)}{(SW(A) - \sum_{1 \le i \le k} Gain(P_i))} \quad (1)$$

Where $SW(A)$ is the execution cycles of application A on the base RISC processor.

4. RECURRING PATTERN COMPUTATION

Let $G_1, G_2, \ldots, G_N$ be the Data Flow Graphs (DFGs) corresponding to the N basic blocks of the application. Each Graph G_i is associated with the triple (V_i, E_i, t_i) where V_i is the set of operation nodes, E_i is the set of directed data flow edges and t_i specifies the operation class of each node.

Let S_i denote the set of all convex subgraphs of G_i and S be the set of all convex subgraphs of the given graphs, i.e $S = S_1 \cup S_2 \ldots \cup S_N$. Let $S^i \subset S$ denote the set of all convex subgraphs having i nodes.

Problem definition : Find all the isomorphic equivalence classes of S (where S is the set of convex subgraphs of $G_1, G_2, \ldots, G_N$) which consists of atleast two elements.

Let $\Upsilon(S^k)$ denote the set of all isomorphic equivalence classes of S^k and let $\Psi(g)$ denote the equivalence class containing g. We are interested in finding all equivalence classes

$H \in \Upsilon(S)$ such that $|H| > 1$.

4.1 Our Algorithm

We partition $S^1, S^2, \ldots$ in that order, and while partitioning S^k, the equivalence class information that is already computed for S^{k-1} is effectively utilized. Our overall algorithm is depicted in Algorithm 1.

Algorithm 1 Recurring Pattern Computation

1: Let $G_1, G_2, \ldots, G_N$ be the N given DDGs
2: Partition S^1
3: $R^1 = \{ g \in S^1 | \ |\Psi[g]| > 1 \}$
4: k = 2
5: **while** $|R^{k-1}| > 1$ **do**
6:    **Step 1 :**
7:    PR^k = Potential-Recurring-Subgraphs(R^{k-1})
8:    **Step 2 :**
9:    Compute-Equivalence-Classes($PR^k, \Upsilon(R^{k-1})$)
10:    $R^k = \{ g \in PR^k | \ |\Psi[g]| > 1 \}$
11:    k = k + 1
12: **end while**

To start the process, S^1 is partitioned as follows: All the nodes with same type would have to be placed into the same equivalence class. Hence, each possible operation type defines an equivalence class. We scan each node of the given DDGs and simultaneously place it into the equivalence class defined by its type. After placing each node into their corresponding equivalence classes, the set of single node recurring subgraphs, R^1, is initialized to the set of nodes whose equivalence class size is higher than one.

After partitioning S^1, in each iteration of the loop (lines 5-10) S^k is partitioned, for k = 2, 3, \ldots in that order. In our approach, $\Upsilon(R^k)$ is efficiently calculated based on $\Upsilon(R^{k-1})$.

Our algorithm has two major steps for finding $\Upsilon(R^k)$. In the first step, the function *Potential-Recurring-Subgraphs*, instead of generating/evaluating all the k-node graphs, avoids the generation/evaluation of non-convex and non-recurring subgraphs based on R^{k-1}. The inference mechanism employed in pruning those two categories is described in detail in the following subsection.

In the second step, the function *Compute-Equivalence-Classes* finds the equivalence classes of generated potential recurring subgraphs. It does not compare every pair of generated potential recurring subgraphs. Rather, it compares two graphs g_1 and g_2 only if a $(k-1)$ node convex subgraph of g_1 is isomorphic to a $(k-1)$ node convex subgraph of g_2. Detailed description of which graphs are compared is presented in section 4.1.2.

After finding the equivalence classes of PR^K, the actual recurring subgraphs, R^k, whose equivalence class size is more than one, are preserved and the non-recurring subgraphs in PR^k are discarded. Since, the set of potential recurring subgraphs PR^k are computed based on R^{k-1}, the loop terminates once there are no more recurring subgraphs of size $(k-1)$.

In the following subsections, we describe in detail the key ideas of partitioning S^k using the partitioning information about S^{k-1}.

4.1.1 Potential Recurring Subgraphs(PR^k)

Let us classify all the k-node graphs denoted as A^k into three categories, **non-convex**, **convex recurring** (R^k) and

convex non-recurring (NR^k). From the practical examples, we observed that majority of the graphs in A^k are non-convex and among the convex, majority of them are non-recurring. Since we are not interested in the non-convex and non-recurring subgraphs, in our approach, we use a heuristic which prunes most of these kinds of graphs based on the computed equivalence class information of S^{k-1}. The graphs generated after pruning the above mentioned two categories from A^k, are termed as *potential recurring subgraphs* (PR^K). PR^k includes all the recurring subgraphs (i.e $R^k \subset PR^k$) and may include some non-recurring subgraphs as well, which would be eliminated after partitioning PR^k into equivalence classes.

A graph $g^k \in A^k$ can be inferred as non-recurring if it contains a non-recurring convex subgraph. This criteria can, in principle, be verified by checking whether every $(k-1)$ node convex subgraph of g^k is part of R^{k-1}. We are interested in diligently enumerating all the graphs in S^k that satisfy this criteria. Any potential recurring subgraph $g^k \in PR^k$ can be seen as a single node extension of recurring subgraph in R^{k-1}. Our potential recurring subgraph generation method would be to consider a recurring subgraph in R^{k-1} and extend it with appropriate node such that the resulting graph qualifies our definition of *potential recurring subgraph*. Let $g^{k-1} \in R^{k-1}$ be a subgraph of the DDG G_i. Single node extension of g^{k-1} with a node $v \in G_i \wedge v \notin g^{k-1}$ is denoted as $g^{k-1} \diamond v$.

Before proceeding further, we need to be cautious that the proposed method doesn't generate the same graph more than once. This can be achieved by assigning unique labels to each node and each recurring subgraph is extended only with nodes whose label is higher than the labels of all its nodes. However, arbitrarily assigning the labels may necessitate the generation of non-convex subgraphs on the way to generation of convex subgraphs later. We assign the labels in such a way that the label of a node is higher than the labels of all its predecessors. This labeling is feasible since each DDG G_i is acyclic. This labeling strategy guarantees that no non-convex graph need to be generated on the way to generation of convex subgraph later [7].

Consider a graph $g^k = g^{k-1} \diamond v, v \notin g^{k-1}$ such that $Lab(v) > max_{\forall u \in g^{k-1}} Lab(u)$, i.e g^k is an extension of g^{k-1} along node v. We define all the graphs obtained by deleting a single primary input node (a node without a single predecessor) of g^k from g^k as the descendants of g^{k-1} along the node v. Comparing g^{k-1} with its descendants along node v, we have the following two important observations.

1. Node v is present in each descendant but is absent in g^{k-1}.

2. Exactly one primary input node of g^{k-1} is absent in each descendant, and the descendants differ based on what primary input node of g^{k-1} is absent.

Let $Desc(g^{k-1}, v)$ denote the set of descendants of g^{k-1} along node v. Let $PI(g)$ denote the set of primary input nodes of graph g. It can be observed that, for any node node v satisfying labeling constraint, there would be exactly $|PI(g^{k-1})|$ descendants of g^{k-1} along node v.

Convexity of g^k :

We provide the following theorem for checking whether g^k is convex or not. This theorem mainly relates the convexity of g^k with convexity of some subgraphs of it.

Theorem 1 : Let $g^k = g^{k-1} \diamond v$, where $g^{k-1} \in S^{k-1}$ and $Lab(v) > max_{\forall u \in g^{k-1}} Lab(u)$. If $|PI(g^{k-1})| > 1$ then $g^k \in S^k$ if and only if $Desc(g^{k-1}, v) \subset S^{k-1}$.

Proof : Proof of this theorem in outlined in [15]

Considering the criteria for a potential recurring subgraph, the following corollary is a straight forward consequence of the above theorem.

Corollary : Let $g^k = g^{k-1} \diamond v$, where $g^{k-1} \in R^{k-1}$ and $Lab(v) > max_{\forall u \in g^{k-1}} Lab(u)$. $g^k \in R^k$ only if $Desc(g^{k-1}, v) \subset R^{k-1}$.

Two graphs $g_1^k, g_2^k \in S^k$ are said to be mergable if $g_1^k \cup g_2^k \in S^{k+1}$ and we call $g_1^k \cup g_2^k$ as the *child* of g_1^k and g_2^k.. A set of graphs is said to be mergable if every pair of graphs in the set are mergable and the child of each pair is identical. It can be proved that the set containing g^{k-1} and all the descendants of g^{k-1} along node v is a *mergable set* and $g^k = g^{k-1} \diamond v$ is its *child*. In other words, a graph g and all its descendants along a node v forms a mergable set, we denote that mergable set as $MSET(g, v)$. Graph g^k is a potential recurring subgraph only if $MSET(g^{k-1}, v) \subset R^{k-1}$. It can be observed that any two graphs in $MSET(g^{k-1}, v)$ have $k-2$ nodes in common and merging any two graphs would give g^k. Another point to note is that a graph may be a member of several mergable sets and although the size of $MSET(g^{k-1}, v)$ may be more than two, still given any two members of $MSET(g^{k-1}, v)$ the other members could be derived. This sets an upper bound on the number of potential recurring subgraphs PR^k to be atmost $|R^{k-1}|^2$.

The potential recurring subgraph generation can be summarized as follows:
For each graph $g^{k-1} \in R^{k-1}$, its extension along node v (i.e $g^k = g^{k-1} \diamond v$) such that $Lab(v) > max_{\forall u \in g^{k-1}} Lab(u)$ is potentially recurring iff all descendants of g^{k-1} along node v are in R^{k-1}.

Algorithm 2 Potential-Recurring-Subgraphs(R^{k-1})

1: $PR^k = \phi$
2: **for each** $g^{k-1} \in R^{k-1}$
3: **loop**
4:    Let $g^{k-1} \subset G_i$
5:    **for each** $v \in G_i - g^{k-1} \mid Lab(v) > Lab(u), \forall u \in g^{k-1}$
6:    **loop**
7:      **if** $Desc(g^{k-1}, v) \subset R^{k-1}$ **then**
8:        $PR^k = PR^k \cup \{g^{k-1} \diamond v\}$
9:      **end if**
10:    **end loop**
11: **end loop**

Algorithm 2 show the steps for the generation of potential recurring subgraphs PR^k. The algorithm finds all different mergable sets in R^{k-1} and adds the child of each mergable set to the potentially recurring subgraphs set PR^k.

In the new algorithm, explicit convexity check is necessary only when g^{k-1} has exactly one primary input node. In this case, it is just sufficient to check whether there is a path from the primary input node of g^{k-1} to an immediate predecessor of v not in g^{k-1}, and if it exists then $g^{k-1} \diamond v$ is not convex, otherwise it is convex. To simplify this task, we, apriori, compute the path matrix for each of the given DDGs

and since each node has atmost two predecessors, checking for convexity requires constant number of operations as opposed to general convexity checking time of $O(|V_i| + |E_i|)$. Hence, our algorithm almost completely saves the time spent in convexity check.

4.1.2 Incremental isomorphism checking

The function *Compute-Equivalence-Classes* computes the equivalence classes of the generated potential recurring subgraphs, PR^k. In our algorithm , the isomorphism checking of graphs of size k is performed using the knowledge of isomorphism between subgraphs of size k-1. Consider a graph $g_1^k = g_1^{k-1} \diamond v$. Graph g_1^k can only be isomorphic to a graph containing an isomorphic image of g_1^{k-1}. Let g_2^{k-1} be an isomorphic graph of g_1^{k-1} and graph $g_2^k = g_2^{k-1} \cup \{v'\} \in PR^k$ is isomorphic to g_1^k if $type(v) = type(v')$ and there exists an isomorphic mapping (as there can be many) between g_1^{k-1} and g_2^{k-1} such that the predecessors of v and v' match. The matching criteria is defined based on whether type of node v is commutative or not. If v is commutative then no distinction is made between its predecessors and it is sufficient if all the predecessors of v are mapped onto all the predecessors of v'. Incase v is not a commutative operation, then left node of v must necessarily map onto the left node of v' and similarly right node.

In the search for isomorphic graphs of g_1^k if we could find one such isomorphic graph g_2^k, the potential subgraph g_1^k is added to the equivalence class in which g_2^k part of. Since isomorphism is an equivalence relation g_1^k is guaranteed to be isomorphic to every subgraph in the equivalence class in which g_2^k is part of. In case no subgraph among the generated potential subgraphs is found to be isomorphic to g_1^k then a new equivalence class is generated which contains only the subgraph g_1^k.

5. EXPERIMENTS & RESULTS

The proposed ISE methodology was implemented in Trimaran retargetable compiler framework. The front end of the compiler performs several classical optimizations on the given application program. The ISE pass is inserted in the backend of the compiler immediately at the point where the control and data flow analysis is complete. We have taken the same approach as [7] for estimation of pattern's software and hardware latencies. The latency of the patterns is derived for the UMC 0.18 micron CMOS process using the Synopsys Design Compiler. The latency is normalized with the latency of 16-bit MAC unit to convert into the processor cycles. All the timings reported are collected on a PIV 3.0GHz machine with 1GB RAM.

The proposed ISE methodology was applied to a total of four benchmark applications taken from Mediabench [16] and Cryptography [17] suites. The selected benchmarks are representative of applications having a rich set of fixed point arithmetic/logical operations. The kernels of these benchmarks consists of basic blocks having more than 50 operations that can be safely implemented on a custom functional unit. For each benchmark, number of patterns extracted in the identification step with the I/O constrain (4,3) and the time taken for identification are shown in Table. 3(a). Number of identified patterns is of the order 10^6.

The summary of performance analysis of our recurring pattern computation is shown in Table. 3(b) and detailed analysis of the performance of our algorithm is plotted in

Fig. 3(c)-(g). For each benchmark, the Table. 3(b) shows a) No. of recurring subgraphs (N_r), b) No. of equivalence classes (N_e), c) Average equivalence class size $EC\_SIZE = \frac{N_r}{N_e}$, d) No. of potential recurring subgraphs (N_p), e) Prediction accuracy $P_a = N_r/N_p$, f) No. of equivalence checks (N_{ec}), g) Average equivalence checks $EC_{avg} = \frac{N_{ec}}{N_p}$ and h) Time taken for recurring pattern computation

The prediction accuracy P_a indicates the non-recurring subgraph pruning efficiency. Prediction accuracy of 1 indicate every non-recurring subgraph is pruned. The values prediction accuracy, average no. of equivalence checks, recurring pattern distribution and average size of equivalence class for all pattern sizes are plotted in Fig. 3(c)-(g). The most notable observations from those plotted results are a) Prediction accuracy P_a is above 0.95 for all the studied benchmarks, b) the number of equivalence checks of k-node patterns is less then the average equivalence class size of k-node and c) the recurring pattern distribution follows a bell shaped curve and recurrence factor is steeply grows and falls and then reaches a saturation level. The recurrence factor touches peak at around 2 to 4 pattern size range.

Interestingly, the time taken for recurring pattern computation is much smaller (around 2 orders) than pattern identification time. This can be attributed to the fact that number of recurring patterns are smaller than the identified patterns and the pattern identification algorithm is a brute force technique with pruning based only the number of inputs/outputs.

The impact of incorporating spatial reuse inform into the design space exploration is shown in Fig. 3(h)-(l). For the speedup studies, we finally selected four patterns that gives highest possible performance gain. For different I/O constraints((m,n) indicates m read and n write register ports), the speedups achieved for the two cases, a) RPI is used and b) RPI is not used, are shown in Fig. 3(h)-(l). Introduction of spatial reuse information into the gain estimation has altered the fate of finally selected patterns. When spatial reuse information is not used, due to the greedy selection technique, the larger patterns used get selected. But when spatial reuse information is used, average size patterns with high recurrence factor got selected. For instance, in the IDCT benchmark, a pattern with 3-inputs and 2-outputs has reuse factor of 12. The gain of the single instance of the pattern is 3. When reuse information is not used, this pattern doesn't get selected. With the inclusion of its spatial reuse, the gain of this pattern dominated the gain of other patterns and gets finally selected. The impact of selecting that pattern resulted in improving the speedup by around 15% at (4,2) I/O constraint. The average speedup improvement using RPI is around 10% and it is upto 25% for some I/O constraints.

6. CONCLUSION

The paper presents a simple and fast algorithm for identifying all the convex isomorphic subgraphs in the entire application. An efficient method of organizing the recurring patterns is also presented. Both of these enabled the application of recurrence information at an early stage of selection process when there are very large number of identified candidates, thus achieving better quality solutions as observed from the results. The proposed method can be extended to group patterns that approximately match. Approximately matched patterns can be synthesized into a ciruit whose area

is smaller than the sum of the individual areas of the patterns, through resource sharing. We would like to perform area conscious exploration that takes into account the possibility of resource sharing as suggested by Brisk et al. [18] and several others.

7. REFERENCES

[1] R. E. Gonzalez, "Xtensa: A configurable and extensible processor," *IEEE Micro*, vol. 20, no. 2, pp. 60–70, 2000.

[2] D. Lau, O. Pritchard, and P. Molson, "Automated generation of hardware accelerators with direct memory access from ansi/iso standard c functions," in *FCCM '06: Proceedings of the 14th Annual IEEE Symposium on Field-Programmable Custom Computing Machines (FCCM'06)*, pp. 45–56, 2006.

[3] T. R. Halfhill, "Mips embraces configurable technology," March 2003.

[4] D. Goodwin and D. Petkov, "Automatic generation of application specific processors," in *CASES '03: Proceedings of the 2003 international conference on Compilers, architecture and synthesis for embedded systems*, pp. 137–147, 2003.

[5] F. Sun, S. Ravi, A. Raghunathan, and N. K. Jha, "Synthesis of custom processors based on extensible platforms," in *ICCAD '02: Proceedings of the 2002 IEEE/ACM international conference on Computer-aided design*, pp. 641–648, 2002.

[6] N. Clark, H. Zhong, and S. Mahlke, "Processor acceleration through automated instruction set customization," in *MICRO 36: Proceedings of the 36th annual IEEE/ACM International Symposium on Microarchitecture*, p. 129, 2003.

[7] K. Atasu, L. Pozzi, and P. Ienne, "Automatic application-specific instruction-set extensions under microarchitectural constraints," *Int. J. Parallel Program.*, vol. 31, no. 6, pp. 411–428, 2003.

[8] H. Choi, S. H. Hwang, C.-M. Kyung, and I.-C. Park, "Synthesis of application specific instructions for embedded dsp software," in *ICCAD '98: Proceedings of the 1998 IEEE/ACM international conference on Computer-aided design*, pp. 665–671, 1998.

[9] R. Kastner, A. Kaplan, S. O. Memik, and E. Bozorgzadeh, "Instruction generation for hybrid reconfigurable systems," *ACM Trans. Des. Autom. Electron. Syst.*, vol. 7, no. 4, pp. 605–627, 2002.

[10] P. Yu and T. Mitra, "Scalable custom instructions identification for instruction-set extensible processors," in *CASES '04: Proceedings of the 2004 international conference on Compilers, architecture, and synthesis for embedded systems*, pp. 69–78, 2004.

[11] M. Arnold and H. Corporaal, "Designing domain-specific processors," in *CODES '01: Proceedings of the ninth international symposium on Hardware/software codesign*, pp. 61–66, 2001.

[12] N. Pothineni, A. Kumar, and K. Paul, "Application specific datapath extension with distributed i/o functional units," in *VLSI Design*, pp. 551–558, 2007.

[13] L. P. Cordella, P. Foggia, C. Sansone, and M. Vento, "A (sub)graph isomorphism algorithm for matching large graphs," *Pattern Analysis and Machine Intelligence, IEEE Transactions on*, vol. 26, no. 10, pp. 1367–1372, 2004.

[14] Y. Guo, G. J. Smit, H. Broersma, and P. M. Heysters, "A graph covering algorithm for a coarse grain reconfigurable system," in *LCTES '03: Proceedings of the 2003 ACM SIGPLAN conference on Language, compiler, and tool for embedded systems*, pp. 199–208, 2003.

[15] N. Pothineni, A. Kumar, and K. Paul, "Exhaustive enumeration of legal custom instructions for extensible processors," Tech. Rep. TR 07/2007, Department of Computer Science, IIT Delhi, July 2007.

[16] C. Lee, M. Potkonjak, and W. H. Mangione-Smith, "Mediabench: A tool for evaluating and synthesizing multimedia and communicatons systems," in *International Symposium on Microarchitecture*, pp. 330–335, 1997.

[17] T. R. Halfhill, "Eembc releases first benchmarks," May 2000.

[18] P. Brisk, A. Kaplan, and M. Sarrafzadeh, "Area-efficient instruction set synthesis for reconfigurable system-on-chip designs," in *DAC '04: Proceedings of the 41st annual conference on Design automation*, pp. 395–400, 2004.

Benchmark	No. of Patterns
ADPCM	7760
DES	722831
COMPRESS	76222
IDCT	289727
SHA	1266867

(a) Pattern identification step

Benchmark	N_r	N_e	EC_SIZE	N_p	P_a	N_{ec}	EC_{avg}	Time
Adpcm	7015	1044	7	7375	0.95	79755	11	0.33s
DES	28288	706	40	28353	0.97	135716	5	1.57s
COMPRESS	16032	1018	16	16095	0.99	221763	14	0.84s
IDCT	97613	2968	33	97828	0.99	1047373	11	7.50s
SHA	827232	6787	121	816931	0.98	21628741	26	89.2s

(b) Performance of our RPI algorithm

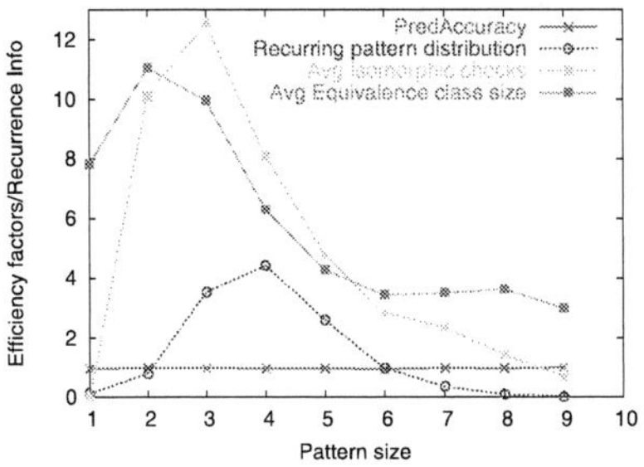

(c) ADPCM

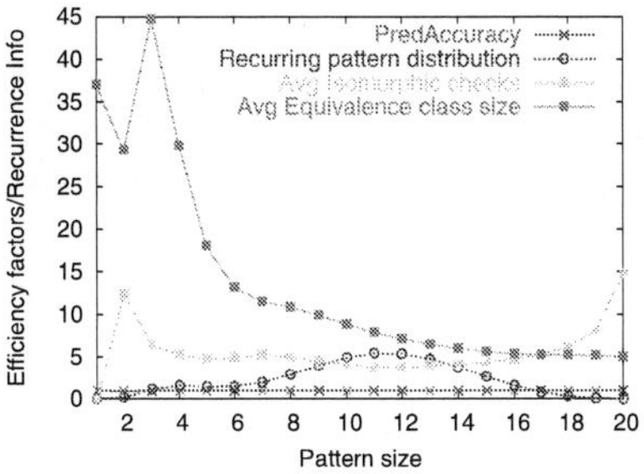

(d) DES

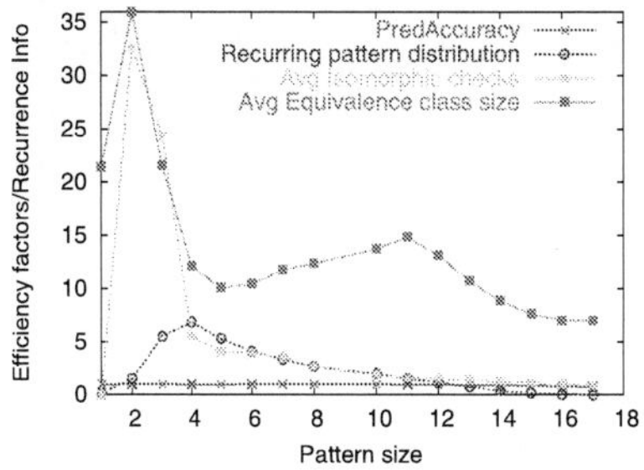

(e) COMPRESS

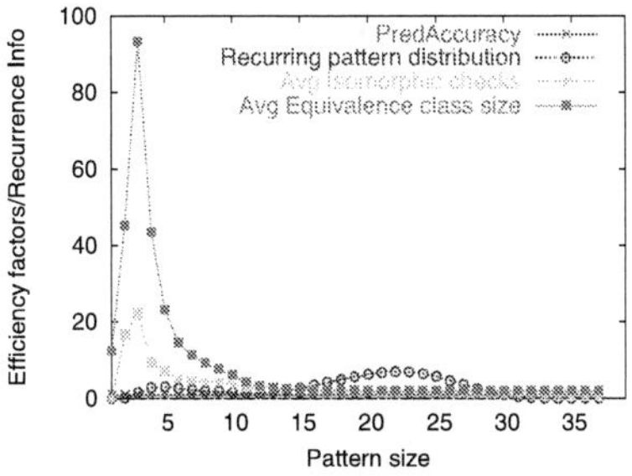

(f) IDCT

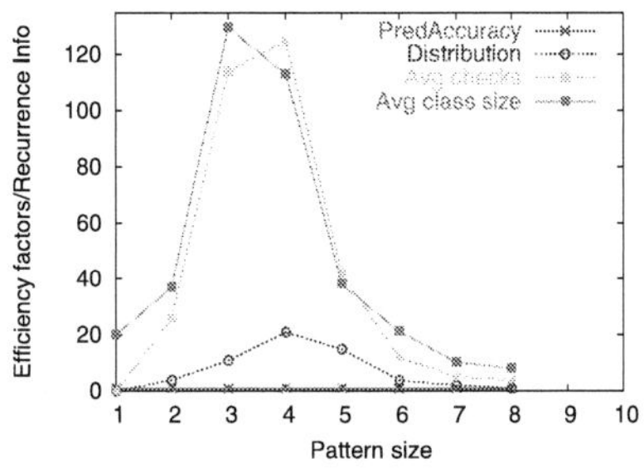

(g) SHA

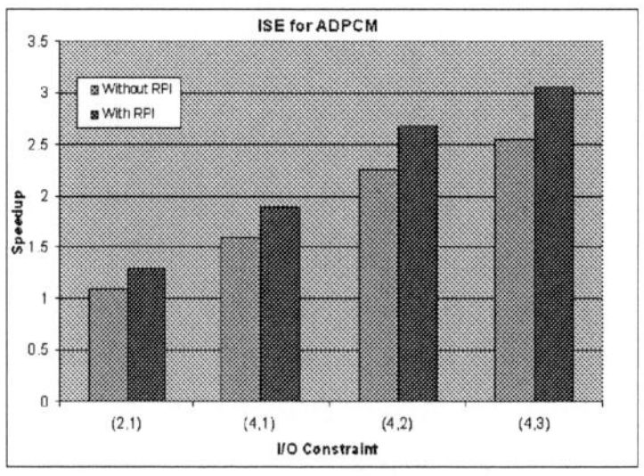

(h) ADPCM

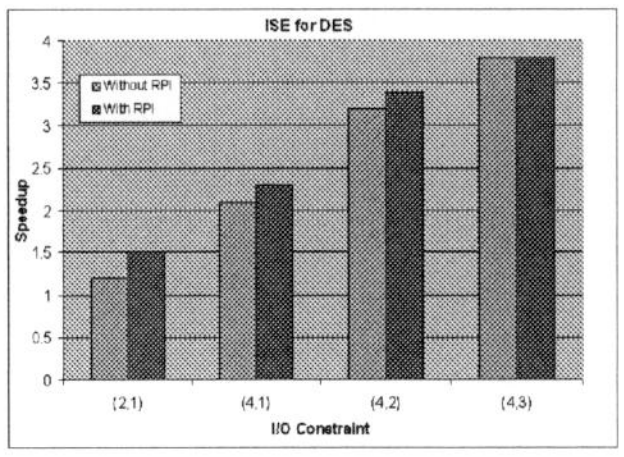

(i) DES

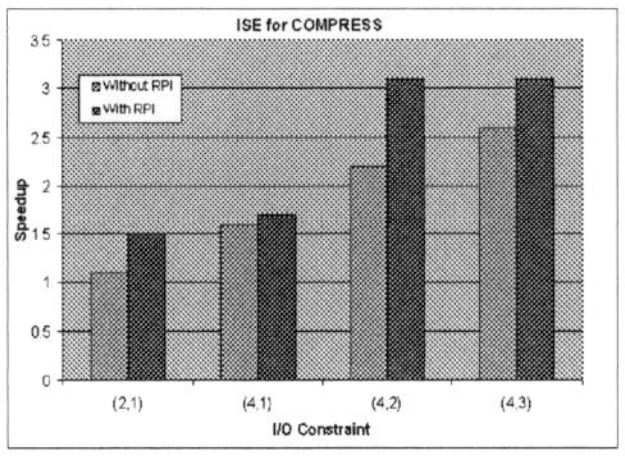

(j) COMPRESS

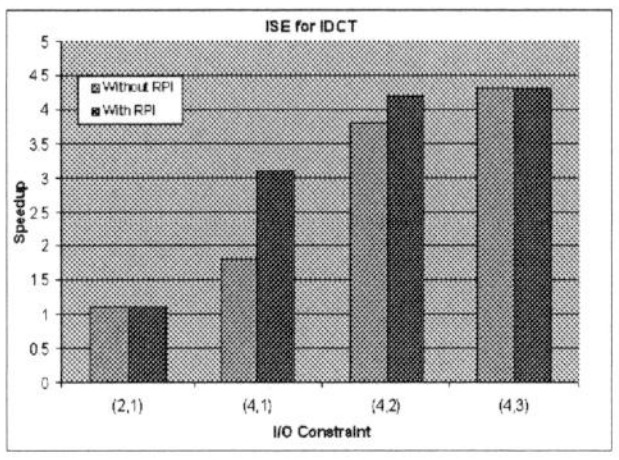

(k) IDCT

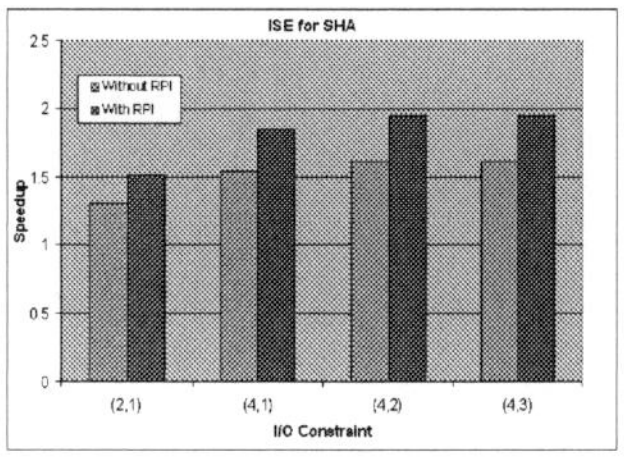

(l) SHA

Figure 3: RPI generation algorithm's performance and its impact on ISE

21st International Conference on VLSI Design

Addressing the Challenges of Synchronization/Communication and Debugging Support in Hardware/Software Cosimulation

Banit Agrawal   Timothy Sherwood
Department of Computer Science
University of California, Santa Barbara
Email: {banit, sherwood}@cs.ucsb.edu

Chulho Shin   Simon Yoon
ARM Inc., Irvine, CA
Email: {chulho.shin, simon.yoon}@arm.com

Abstract

With increasing adoption of Electronic System Level (ESL) tools, effective design and validation time has reduced to a considerable extent. Cosimulation is found to be a principal component of ESL tools to simulate the hardware designs and software models concurrently. It helps in providing an integrated system-on-chip design platform to get rid of most of the design errors in the early stage. To nail down these design errors early, a better debugging support of RTL memory on the software side is extremely useful. We present a just-in-time shadow memory technique that can allow debugging of RTL memory from a software perspective.

While cosimulation is fast compared to a complete hardware based simulation, the communication and synchronization overhead between the hardware and software simulators can be very significant. Since the two simulators have to communicate almost every cycle, a good communication platform is necessary to reduce this extra overhead. To evaluate this overhead, we implement and evaluate three communication primitives for a real system design with ARM926EJ-S processor and RTL memory. We find that a message-queue based communication backplane can alleviate the communication overhead to a considerable extent compared to other alternatives.

1. Introduction

Recently, software simulation tools (or ESL tools) are found to be very useful tools for the hardware designers to quickly explore a set of architectures with relaxed but acceptable timing accuracy [4]. The relaxed accuracy is acceptable because in the beginning stage of design, exploration speed is more important than accuracy to its traditional users. In hardware simulation, design is expressed in a hardware description language (HDL) and

simulated in logic simulators (or HDL simulators). As this design is closer to hardware, it can provide more accurate results. But the overall design time and simulation speed makes embedded software development and validation impractical considering the ever-increasing scale of System-on-chip (SoC) designs. Hardware-software cosimulation is a hybrid approach where some IP models are expressed in a hardware description language and others are modeled in software. Cosimulation exploits benefits of both hardware simulation and software simulation to provide better flexibility and a fast simulation platform. It is found to be a viable solution to achieve fast and efficient architecture exploration. As HDL simulators require the models to be described in HDL, early-stage software models cannot be directly simulated with other hardware models. Similarly, software simulators require the models to be described in high-level languages like SystemC or C++, it is not straightforward to integrate with hardware models. Hence, cosimulation tools bridge the gap between hardware and software simulation and provide an architecture exploration platform for complete SoC design and in addition a verification platform where software models and hardware models can communicate. In ESL domain, developing an ESL model that accurately models the behavior of an existing hardware model takes significant effort. Though once the model is available, faster design exploration becomes possible. Transaction-level models (TLMs) can be integrated with already-available RTL IP models through cosimulation for validating a design, not requiring development efforts needed for building TLMs which in turn allows fast designer exploration.

Along with these advantages, there are some unique challenges associated with cosimulation such as how to provide better debugging support on the software side or how to achieve fast simulation speed by reducing the communication and synchronization overhead. Debugging

354

978-1-4244-3039-0/08 $25.00 © 2008 IEEE

IEEE
computer
society

on the software side can be extremely difficult if there is no debugging support for RTL memory components on the software side. Hence, a good RTL memory view capability is required. To this end, we present a just-in-time shadow memory technique (*patent pending*) that can provide better debugging capabilities for RTL memory on the software side. To address the communication overhead, we evaluate three communication primitives using a real example with ARM926EJ-S processor with RTL memory. We find that message queue based implementation can provide much faster cosimulation compared to other alternatives. Overall, our contributions in this paper are as follows:

o A just-in-time shadow memory technique is presented, which aids cosimulation by providing better debugging capabilities (memory view), fast cosimulation, disassembly support, and generating memory traces. (Section 3.1)

o To achieve fast simulation, we need to minimize the communication and synchronization overhead between hardware and software simulators. We implement and evaluate three communication primitives on a real system design to find the better communication backplane in different scenarios. (Sections 3.2 and 4)

In the next section, we discuss some of the related works. In Section 3, we present just-in-time shadow memory techniques for better debugging support and the implementation details of communication and synchronization in our cosimulation platform. In Section 4, we present the communication backplane evaluation to find the best available communication primitive. We conclude in Section 5.

2. Related Works

Cosimulation has been very attractive among the system designers and researchers since the last decade. The main challenges associated with hardware-software cosimulation are achieving faster simulation, better synchronization in heterogeneous cosimulation environments, visibility of internal state for debugging, getting better timing accuracy and availability/type of software models. The strength and weakness of various cosimulation techniques based on these challenges are compared in [8]. Most of these techniques are categorized based on the models used on the software and hardware side. Similarly, in [14] several key issues have been presented to combine the capabilities for software simulation and hardware simulation in a best possible manner.

Improving the speed of cosimulation with better timing accuracy is one of the most important challenges of cosimulation. Most of the previous works have focused on improving the speed of cosimulation [1,2,5,7,9,11,12,13], whereas only a few of them have looked into other different challenges [3,10]. Passerone et al. [7] proposed a technique to do fast cosimulation using constrained software synthesis that utilizes the run time estimation of a target processor. The estimation accuracy is limited by the caches, pipelined architectures, and communication cost. In [10], an integrated cosimulation environment is presented which allows the designers to model the entire system in one language C/C++. It does not interface with any HDL simulators. This technique limits the use of any third-party hardware core in the overall system design and verification process. A compiled simulation technique is presented in [13] that can generate bit-, cycle- and pin-accurate cosimulation engines which are much faster than the interpretive simulators. But any design changes require the recompilation of the system design. The techniques mentioned above are well suited in homogenous environments where only one integrated environment is used.

In heterogeneous environments, different simulators interact with each other that present a different set of problems compared to homogeneous environments. Kim et al. [1] presented an integrated and heterogeneous cosimulation environment that provides an automatic interface generation. Becker et al. [6] used distributed processes for cosimulation and tested their cosimulation environment by designing a network interface unit. The communication timing was less predictable in this approach due to communication between multiple distributed processes. Bishop et al. [2] present another heterogeneous cosimulation environment where time management and synchronization issues are addressed. In this scheme, as always, synchronization plays a big role in the overall cosimulation performance.

In [9], a trace-driven HW/SW cosimulation technique is proposed. In this paper, synchronization is addressed as major performance bottleneck to the system cosimulation. They alleviate the effect of the synchronization issues by using a virtual synchronization technique that makes use of the execution traces and the timing management of the execution traces. But in this scheme, the accuracy of cosimulation is dependent on the OS and channel model. Sung et al. [12] present a backplane approach for hardware-software cosimulation. This approach tries to minimize the communication of control data between the simulators.

A completely different approach is presented by Lee et al. [5] where architecture simulators inherit circuit modeling capabilities and react to circuit characteristics such as latency, energy on a per-cycle basis, and still provides a considerable throughput. This technique can be applied to existing cosimulation tools to get delay and energy estimation along with performance estimation. In a similar direction, a cosimulation based power estimation method is presented by Lajolo et al. [3]. Power estimation for different components of system-on-chip is done using concurrent and synchronized execution of multiple power estimators.

We use an ESL tool [15] that allows us to quickly evaluate the designs. It uses a SystemC core on the software side to provide fast cycle-based simulation and attempts to minimize the flow of information between the software simulator and RTL simulator. Since most of the time, the memory components are being simulated on the HDL side, it gets extremely slow and inflexible to debug the memory on the software side. To provide better debugging support of memory components, we present the just-in-shadow memory technique. Similarly, to address the issues of communication and synchronization overhead, we implement three communication primitives available on the Linux platform and in the next section we present its implementation details. Then, we use the ESL tool [15] to evaluate the overhead of a real system design with ARM926EJ-S processor.

3. Addressing Cosimulation Challenges

In this section, we address some of the key challenges in cosimulation by presenting the idea of just-in-time shadow memory that provides better debugging capabilities and also present how it is implemented in our synchronization and communication paradigm.

3.1 RTL Memory Debugging

One of the most common practices of cosimulation is that the core models run on an ESL simulator while memory subsystems (including the bus subsystems in most cases) are simulated on an RTL simulator. The reason being is that off-the-shelf ESL core models are more increasingly available in a mature state while the memory subsystem is likely to be the major target of a system design. In such cases, debugging gets difficult on the system simulation side because it doesn't know anything about the memory on the RTL side and it cannot provide any debug accesses to memory directly. Many times, protected IP cores are simulated on the RTL side and debug accesses to the RTL memory or RTL peripherals with memory-mapped registers gets extremely difficult. A just-in-time shadow memory[1] technique is presented, which provides a shadow memory on the ESL simulator side and it captures all debug writes to the RTL side and services all the debug reads. This technique is described in detail in the next subsection.

3.1.1 Just-in-time shadow memory

Just-in-time shadow memory technique[1] helps in providing a uniform memory view, debug read/write access to RTL memory, disassembly support. In this technique, an identical copy of the RTL memory is maintained on the ESL simulator. All the debug read requests from the ESL simulator are served from shadow memory instead of RTL memory. The debug writes from the ESL simulator has to be written to both shadow memory and RTL memory.

The key difficulties in maintaining the identical copy are debug writes from ESL simulator, normal writes from the ESL simulator, and writes on the RTL side. All of these

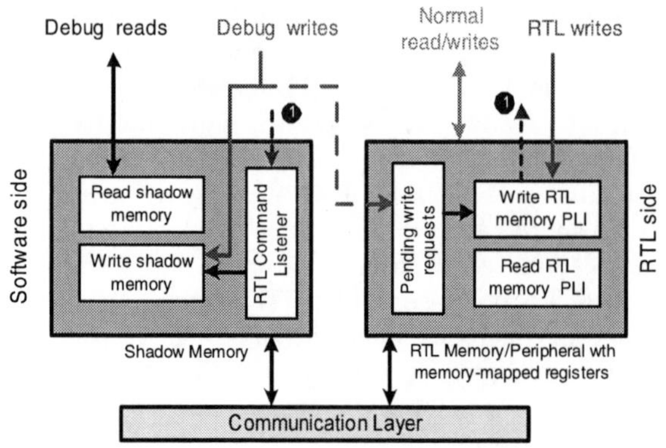

Figure 1: Just-in-time shadow memory block diagram. All the debug reads are read from shadow memory. Debug writes are written to shadow memory and pending writes are queued on RTL side. All the normal RTL writes are written back to shadow memory using a running thread on the ESL simulator side.

different types of write accesses are updated on both sides as shown in Figure 1.

The debug writes from ESL simulator are sent to the RTL side as memory write command using file-based sockets. Just before the starting of the next cycle, any pending write accesses are updated on the RTL side by the use of programming language interface (PLI) of the RTL simulator. In this case, it becomes necessary to know the signal name of memory object in RTL design to do PLI/FLI/VPI/VHPI accesses. We provide a cosimulation configuration interface where user can specify the signal name of the memory and type of object. The wrapper library on the RTL side looks for the memory signal names when it is loaded for the first time in the RTL simulator.

Normal writes through signals should also be updated on the shadow memory side. This requires little changes to the RTL memory code to do a PLI/FLI/VPI/VHPI access. This access is responsible to send a memory write command on the ESL simulator. A separate thread on ESL simulator side is listening for any write commands from the RTL side. On receiving the memory write command from the RTL side, shadow memory is updated instantaneously. All RTL writes are also communicated to shadow memory in a similar fashion. A very minimal change is required on ESL simulator to make available the just-in-time shadow memory support, which can be quickly done with the help of provided examples.

All the write updates are communicated to either side to maintain an identical copy on both sides. Hence, the shadow memory and RTL memory are kept cycle-wise consistent. This considerably reduces the time to view the memory, to enable debug read/writes, and helps providing better debugging support on the ESL simulator through a well-defined debug interface.

---

[1] Just-in-time shadow memory technique is filed for patent.

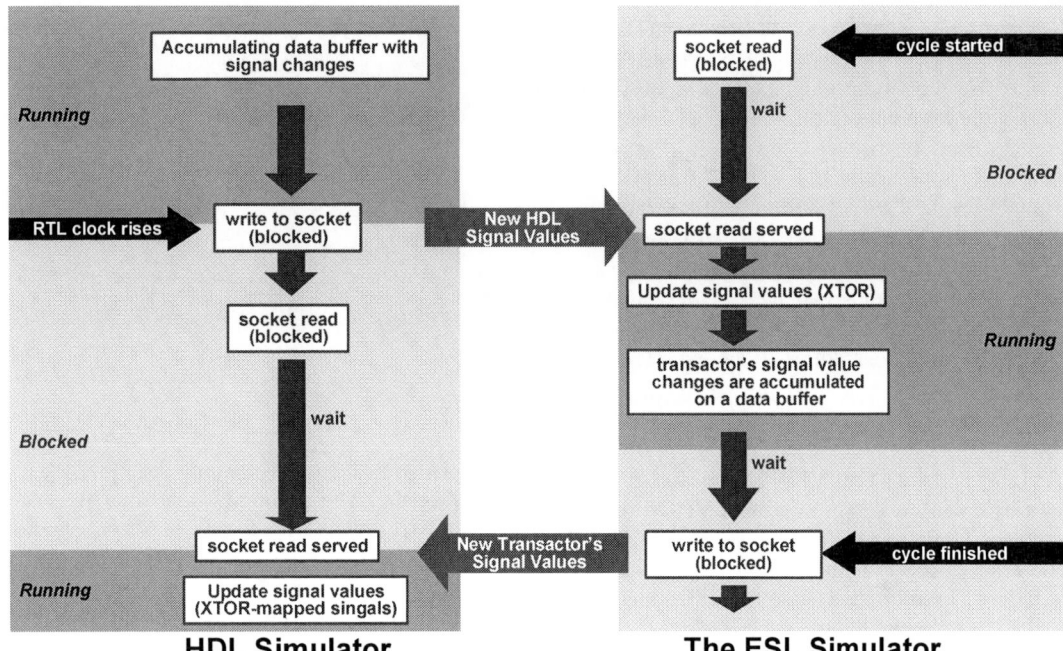

Figure 2: Synchronization between the ESL simulator and logic simulators. The light grey part shows when the simulator is blocked for a read/write operation, whereas the dark grey region shows the running status of the simulators.

3.2 Synchronization/Communication

In this subsection, we talk about the synchronization and communication details.

3.2.1 Synchronization

In our implementation, all events taking place between two clock cycle edges are abstracted to one point during the cycle. At this point, two simulation interface functions, *communicate* and *update* are called [15]. In *communicate* function, all inter-component communication is done as the name implies. In *update* function, shared resources are updated while no communication takes place. The main scheduler calls these two functions of each component that comprises a system while emulating concurrency of the hardware components of the system. For cosimulation, we have chosen the rising edge of a clock cycle for synchronization between the ESL simulator and the logic simulator. At the rising edge, the two functions, *communicate* and *update* are both called before returning control to the logic simulator.

Figure 2 illustrates how the ESL simulator and a logic simulator are synchronized while communicating for changes in signal values. There are four important events involved with the synchronization:

1) *Beginning of the ESL simulator cycle*
The ESL simulator waits on a blocked read. The logic simulator executes and upon next rising edge of the clock it will send data via socket. The data arriving at the socket releases the blocked read on the ESL simulator side. All updates on signals in HDL simulation are sampled by the ESL simulator.

2) & 3) *Communicate and Update*
The ESL simulator's cycle-based computation is done in all models existing in the simulator. Changes in all the signals are accumulated in a data buffer while the HDL simulator is waiting on a blocking read.

4) *End of the ESL simulator cycle*
Within the callback for this event, changes in all the signals are sent to the HDL simulator via a socket. (This releases the HDL simulator's blocking read).

The HDL simulator samples the changes on signal values sent by the ESL simulator and applies them to the corresponding signal objects. Once the HDL simulator is free, it computes the changes on the signals connected to the transactor and accumulates them in a buffer for next event. The ESL simulator is also let go free and the callback for the event of the beginning of ESL simulator cycle will be called incurring the next cycle's wait on blocking read. The same entire process is continued for every cycle for both simulators.

3.2.2 Communication

As described in the previous subsection, at the synchronization points we need to transfer the changes in all exported signals using a communication primitive. Communication layer between the ESL simulator and RTL simulator plays a key role in overall performance and cycle-accurate synchronization. In our implementation, we generate a proxy module (in HDL) which specifies all the exported signals from the software side to the hardware

357

978-1-4244-3039-0/08 $25.00 © 2008 IEEE

side. All the master signals are driven from the ESL simulator's side and all the slave signals are driven from the RTL side. The communication data is in the form of driving these signals on either side.

All communications between the system components on ESL simulator side are in the form of transactions or signals. But RTL side does not understand the transaction-level modeling. Hence, all the transactions to the RTL side must be converted into signals. A transactor that converts a transaction to a set of signals is placed in the ESL system design to export required signals. The transactor is responsible for driving and reading all the exported signals accurately. The implementation of the converter is based on the protocol of the transaction and should be implemented by the system designer.

The RTL layer library uses VPI/PLI accesses in case of Verilog and FLI/VHPI accesses in case of VHDL to access all the exported signals in RTL simulation. We find that our cosimulation speed is mainly limited by the speed of RTL simulation. Therefore, we try to minimize the number of PLI/VPI/FLI accesses to the RTL code as much as possible. There is also some delay overhead associated for communicating the data from RTL side to our side and *vice versa* on each cycle. We try to minimize the amount of data to be communicated by sending only the value changes in signals rather than the values of all the exported signals in each cycle.

All the data exchange between the simulator and logic simulators is done using file-based socket in Unix-based systems. While in window platform, a TCP/IP socket is used to transfer the data. Although TCP/IP based socket is a solution for remote cosimulation, but remote cosimulation is generally less preferred. For local cosimulation, file-based socket communication provides good performance on Unix-based systems because it does not incur the TCP/IP protocol overhead. System V IPC in Unix-based systems provides communication primitives such as shared memory, semaphores, and message queue. We compare the performance of cosimulation by implementing these communication primitives, which we discuss next.

4. Evaluating Communication Primitives

In this section, we present evaluation of communication backplane between the hardware and software simulators. Since the communication and synchronization overhead affects the overall speed of simulation, there is a pressing need to realistically quantify the cosimulation performance considering different communication primitives. We implement three communication primitives available in Linux system: 1) shared memory 2) message queue and 3) file-based socket. Since shared memory does not come with its own synchronization, we had to implement the synchronization part of the shared memory using semaphores. Due to this extra overhead, we found that cosimulation performance is almost more than hundred

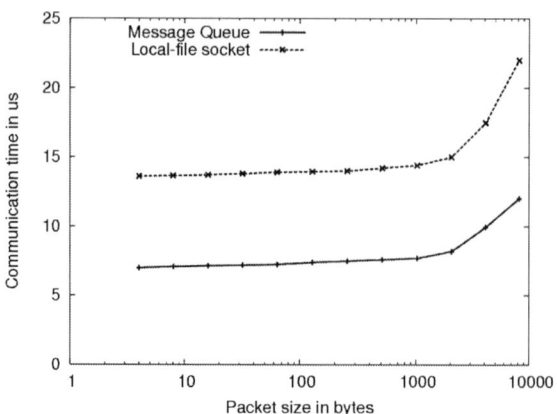

Figure 3: Evaluating communication primitives by varying packet size. The communication time overhead is shown on y-axis and packet size is shown on the x-axis in log scale.

times worse than message queue or file-based socket implementation. So we need to find better synchronization mechanism for shared memory and we concentrate on message queue and file-based socket for the rest of our analysis.

We implement message queue communication routines and use the built-in features for synchronization. The main limitation in message queue is that the maximum size of the packet supported is 8192 bytes. Hence, to send larger size packets, it has divided into many chunks before communicating. We implement the file-based socket in the same way as message queue and the maximum size of the packet supported is 64 Kbytes.

4.1 Changing Message Size

We first evaluate the message queue and file-based socket primitives using a standalone environment where we change the size of the message. We communicate a million packets using both primitives and record the time of execution and using this time, we calculate the time of communicating a message of particular size for both primitives. The communication time for different sizes of messages (for both primitives) is shown in Figure 3. The message size is shown on the x-axis in log scale, whereas y-axis shows the communication time in μs. As we can see that increase in communication time is less than 7% when we increase the message size up to 1024 bytes. But as we increase the message size from 1024 to 8192 bytes, we see a sharp increase in the communication overhead. In the case of message queue, the communication time increases about 20% when we increase the message size from 4kbytes to 8Kbytes. For the same size increase in file-based socket, the percentage increase in communication time is found to be 25%. When we compare both message queue and local socket, we find that message queue provides a significant lower communication overhead for all the message sizes. For example, to communicate a message of

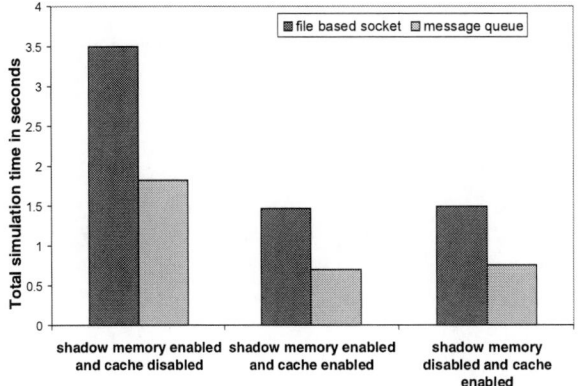

Figure 4: Communication/Synchronization overhead for a real system with ARM926EJ-S processor with RTL memory. Three configurations are compared as shown. Message queue based implementation is shown to perform better in all the three cases,

4kbytes, local socket based implementation requires more than two times of the communication overhead in message queue based implementation.

4.2 Evaluating ARM926EJ-S Based System

We evaluate both the communication primitives using an ARM926EJ-S system with RTL memory. We consider three scenarios: 1) When shadow memory is enabled and cache is disabled 2) Shadow memory is enabled and cache is enabled 3) shadow memory is disabled and cache is disabled. We measure the cosimulation performance in these three different scenarios for both message queue and file-based socket and the results are presented in Figure 4. When cache is disabled, the memory traffic is much higher and it involves much higher communication overhead as shown in the figure. We find that message queue implementation provides almost 50% more performance compared to file-based socket implementation for communication backplane. This is true for all three scenarios. From Figure 4, we can also see that we get slight increase in cosimulation performance when shadow memory is enabled (comparing scenario 2 and 3). For all the three scenarios, communication overhead is two times less in message queue based communication platform than that compared to local file based socket. Hence, message queue based communication backplane can provide significant performance advantage compared to a local file based socket implementation.

5. Conclusion

Hardware-software cosimulation is becoming more popular because of simulation speed and its usefulness in co-verification space. We addressed some of the key challenges associated with cosimulation including debugging support and communication overhead. We presented the just-in-time shadow memory technique that provides RTL memory view, debug read/write accesses to RTL memory and better debugging capabilities on the ESL simulator side. The synchronization issues along with our implementation details between an RTL simulator and an ESL simulator was also described. We also presented a study of communication backplane implemented using different communication primitives. We find that message queue is a better solution for communication backplane instead of file-based sockets when cosimulation performance is evaluated for a real system design. In the future, we plan to investigate the communication primitives for cosimulation using a multi-core system design.

References

[1] Kyuseok Kim, Yongjoo Kim, Youngsoo Shin and Kiyoung Choi, "An integrated hardware-software cosimulation environment with automated interface generation", *Seventh IEEE International Workshop on Rapid System Prototyping*, pp. 66 – 71, June 1996.

[2] William D. Bishop and Wayne M. Loucks, "A Heterogeneous Environment for Hardware / Software Cosimulation," *in the Proceedings of the 30th Annual Simulation Symposium*, pp. 14-22, Atlanta, Georgia, April 1997.

[3] M. Lajolo, A. Raghunathan, S. Dey, L. Lavagno, "Cosimulation-Based Power Estimation for System-on-Chip Design", *IEEE Transactions on VLSI Systems*, Vol. 10, No. 3, pp. 253-266, June 2002.

[4] C. Lennard and D. Mista. "Taking Design to the System Level", *ARM White paper*, April 2005.

[5] S. Lee, S. Das, V. Bertacco, T. Austin, D. Blaauw, and T. Mudge, .Circuit-Aware Architectural Simulation,. *in the 41st Design Automation Conference (DAC-2004)*, June 2004.

[6] D. Becker, R. K. Singh, and S. G. Tell. "An engineering environment for hardware/software co-simulation". *In Readings in Hardware/Software Co-Design*, Kluwer Academic Publishers, Norwell, MA, 550-555.

[7] C. Passerone, L. Lavagno, M. Chiodo, and A. Sangiovanni-Vincentelli, "Hardware/Software Co-Simulation for Virtual Prototyping and Trade-off Analysis", in *Proceedings of the 34th Design Automation Conference (DAC'97)*, Anaheim, California, USA, June 9-13, 1997, pp. 389-394.

[8] J. Rowson. "Hardware/Software Co-Simulation," *Design Automation Conference Proceedings*, June 1994, pg 439.

[9] D. Kim, Y. Yi and S. Ha. "Trace-Driven HW/SW Cosimulation Using Virtual Synchronization Technique", *Design Automation Conference Proceedings* June 13-17 2005

[10] L. Séméria and A. Ghosh. "Methodology for hardware/software co-verification in C/C++". *In ASP-DAC* 2000: pages 405-408.

[11] C. Liem, F. Naçabal, C. A. Valderrama, P. G. Paulin, and A. A. Jerraya. "System-on-a-Chip Cosimulation and Compilation". *IEEE Design & Test of Computers* 14(2), pages 16-25, 1997.

[12] W. Sung and S. Ha. "Optimized Timed Hardware Software Cosimulation without Roll-back". *In DATE* 1998, pages 945-946.

[13] V. zivojnovic and H. Meyr. "Compiled HW/SW co-simulation". *In Proceedings of the 33rd Annual Conference on Design Automation*, Las Vegas, Nevada, United States, June 03 - 07, 1996.

[14] B. Bailey, R. Klein, S. leef. "Hardware/software Co-Simulation Strategies for the future", White paper.

[15] ARM Ltd., "Cycle-Accurate Simulation Interface (CASI) – RealView ESL API" (http://www.arm.com/products/DevTools/RealViewESLAPIs.htm

SESSION B3:
Low Power - I

21st International Conference on VLSI Design

Incorporating PVT Variations in System-level Power Exploration of On-Chip Communication Architectures

Sudeep Pasricha, Young-Hwan Park, Fadi J Kurdahi, Nikil Dutt

Center for Embedded Computer Systems
University of California, Irvine, CA
{spasrich, younghwp, kurdahi, dutt}@uci.edu

Abstract

With the shift towards deep sub-micron (DSM) technologies, the increase in leakage power and the adoption of power-aware design methodologies have resulted in potentially significant variations in power consumption under different process, voltage and temperature (PVT) corners. In this paper, we first investigate the impact of PVT corners on power consumption at the System-on-Chip (SoC) level, especially for the on-chip communication infrastructure. Given a target technology library, we then show how it is possible to "scale up" and abstract the PVT variability at the system level, allowing characterization of the PVT-aware design space early in the design flow. We conducted several experiments to estimate power for PVT corner cases, at the gate-level, as well as at the higher system-level. Our preliminary results are very interesting and indicate that: (i) there are significant variations in power consumption across PVT corners, and (ii) the PVT-aware power estimation problem may be amenable to a reasonably simple abstraction at the system-level.

1. Introduction

With the advent of the deep submicron (DSM) era, more and more System-on-Chip (SoC) designs are being fabricated in sub-100nm technologies. Unfortunately, *process, voltage and temperature (PVT) variability* makes it hard to achieve 'safe' designs in such nanometer technologies. This is because PVT variability causes fluctuation in timing as well as power for SoC designs [1]. Consequently, timing and power estimates derived early in the design flow are no longer valid, and considerable redesign effort is needed to account for these variability-induced fluctuations. Recently, many research efforts have focused on statistical timing analysis [2-3] to address variability in timing. However, till now very few efforts [4] have looked at addressing the effect of PVT variability on system-level power estimation. Since reducing power consumption is increasingly becoming the most important goal for SoC designs, especially for portable battery-driven embedded systems [5], it becomes essential to address the issue of reliable power estimation for these designs, in the face of PVT variability.

In modern IP-based design, the communication architecture

backbone has become a significant factor in influencing overall system power, performance, cost and time-to-market [6]. In particular, it has been shown that for some SoC designs, on-chip communication architectures (wires and bus logic) can consume anywhere between 20-50% of overall system power [7]. The amount of power consumed in the various bus logic components is also steadily increasing with design complexity, and has been shown to be as high as 80% of the total on-chip communication power [7][17]. Furthermore, a significant portion of the on-chip communication architecture power consumption is converted into heat, which has been shown to not only increase interconnect delay (reducing performance), but also increase electro-migration (EM), which significantly increases device failure rate [8]. These observations motivate the need for system-level estimation of on-chip communication architecture power consumption early in the design flow, where design decisions have a much greater impact on power consumption than at lower levels.

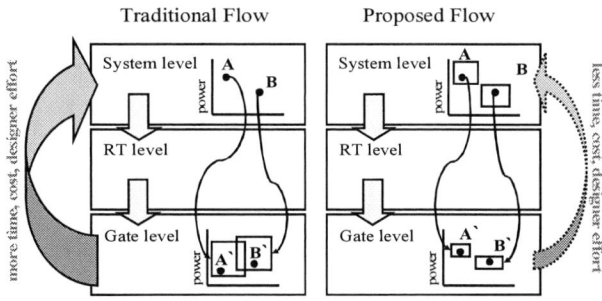

Figure 1. Traditional approach compared with proposed PVT variation-aware system-level exploration approach

While in the past, leakage was negligible and dynamic power did not vary much between technology corners, today the increase in leakage power and the adoption of power-aware design methodologies (such as voltage islands and DVS/DFS) has resulted in considerable variations in power consumption under different process, voltage and temperature (PVT) technology corners. In this paper, we first explore the impact of PVT corners on power consumption at the system level, especially for the on-chip communication architecture. We then show how the variability due to different PVT corners can be abstracted up to the system-level for the on-

363

978-1-4244-3039-0/08 $25.00 © 2008 IEEE

chip communication architecture, where the corners can be explored early in the design flow. To the best of our knowledge, this is the first piece of work to incorporate PVT variations at the system-level during power exploration of the on-chip communication architecture.

Figure 1 illustrates the difference between the traditional SoC design approach, and the approach proposed in this paper. In the *traditional approach*, designers explore the power space of the design at the system-level, and select the configuration with the least power consumption. In the figure, points A and B represent design configurations, and a designer would select configuration B, with the lower power consumption at the system-level. Later in the design flow, at the gate-level, designers encounter process, voltage and temperature (PVT) variations that alter the power characteristics and behavior of the synthesized design. Each of the configuration points becomes a large region of uncertainty (representing possible power consumption under different PVT conditions), and it is no longer clear whether configuration A or B is the superior one in terms of lower power consumption. It is possible that an instance of the design configuration A (shown as A` in the figure) is found to be superior to the best instance of design configuration B (shown as B` in the figure). As a result, designers end up spending considerable time and effort to explore design configurations at the gate-level. It is also important to ensure that PVT variations do not cause a violation of design constraints for the selected design configuration. Design reiterations might be required if violations are detected (requiring changes in the design at the system-level), which can severely influence design cost and time-to-market. In contrast, in our *proposed approach*, we attempt to "scale up" and abstract the PVT variability at the system-level, to provide a more realistic characterization of the design space, early in the design flow. The designer can then select a design configuration with greater confidence, after analyzing its behavior under PVT variations. This significantly reduces the exploration and redesign effort later in the design flow. We conducted several experiments to estimate power for PVT corners of DSM technology libraries, at the gate-level as well as at the higher system-level, especially for on-chip communication architectures. Our preliminary results are very interesting and indicate that: *(i)* there are significant variations in power consumption across PVT corners, and *(ii)* the PVT-aware power estimation problem may be amenable to a reasonably simple abstraction at the system level.

2. Related Work

System-level power estimation approaches typically create power models for heterogeneous system components (e.g. buses, memories, caches, processors) and integrate them to get overall power estimates [10-12]. Several approaches have proposed power estimation techniques for bus-based communication architectures [13-17]. While early work focused mainly on power estimation for bus wires [13-14], more recent work has shown the importance of considering bus logic components as well [7]. System-level power estimation approaches for communication architectures that consider the contribution of both bus logic and wires, have been proposed for the AMBA hierarchical shared bus [15], STBus interconnection network [16] and the AMBA bus matrix [17]. None of the abovementioned power estimation approaches have studied the effects of PVT variability on power consumption at the system-level. To the best of our knowledge, our work is the first to try and understand how PVT variability affects power consumption, especially for on-chip communication architectures, and then attempt to abstract up this variability to the system-level, for early power exploration of the true design space.

Table 1. PVT Corners in UDSM Technologies

Nominal V_{dd}	Corner	Process	Temp	V_{dd}
1.0V	MaxPerf	F-F	0	1.1
	TypPerf	T-T	25	1
	WorstPerf	S-S	125	0.9
	WorstLeakage	F-F	125	1.1
	TypLeakage	T-T	125	1
0.7V	MaxPerfLowV	F-F	0	0.77
	TypPerfLowV	T-T	25	0.7
	WorstPerfLowV	S-S	125	0.7
1.2V	MaxPerfHighV	F-F	0	1.32
	TypPerfHighV	T-T	25	1.2
	WorstPerfHighV	S-S	125	1.08
	WorstLeakageHighV	F-F	125	1.32
	TypLeakageHighV	T-T	125	1.2

3. PVT Corners in Ultra-Deep Submicron (UDSM) Technologies

Traditionally, the most important means by which a foundry communicates process, voltage and temperature variations to designers is through library characterization at design corners, known as PVT corners, relating cell metrics (timing, power) to Process, Voltage and Temperature variations. Up until the 130nm technology library node, design tools relied on three corners: *Typical*, *Worst*, and *Best* corners. The adjectives associated with these corners relate mainly to timing. The *Worst* corner combines high temperature, low V_{dd} (nominal-10%), and a Slow-Slow (S-S) process that leads to worst case timing. The *Best* corner goes the opposite way, combining low temperature, high V_{dd} (nominal+10%), and Fast-Fast (F-F) process to achieve maximum performance. The *Typical* performance corner lies between these two extreme corners. Synthesis tools currently use the *Worst* and *Best* corners during synthesis, guaranteeing that all the resulting functional chips would meet timing (by using *Worst* case corner to avoid setup time violations at register inputs, and using *Best* case corner to guarantee that no hold time violations occur). The *Typical* corner is usually used to characterize power consumption under nominal conditions. Since leakage was negligible up until the 130nm technology library node, the only factor affecting power consumption (mostly dynamic) was V_{dd}, and variations of about $\pm$ 20% were expected between the three corners.

With ultra-deep submicron (UDSM) technologies under 100nm gearing into production, some changes became necessary with corner characterization. This is due to a variety of factors, the most important of which is the drastic increase in the device leakage power. Other factors include IR drop as well as power management strategies such as DVS/DFS and voltage islands. Today, IPs, especially cell libraries, I/O and memories are thus characterized at many more PVT corners, shown in Table 1. These corners become necessary for a variety of reasons: A *TypicalPerf* corner, for example, does not provide a realistic assessment of leakage power under typical conditions because when the application is running, a die would heat up to well above 25°C. With temperature being an exponential factor in leakage, a more realistic *TypicalLeakage* corner must be considered with Typical-Typical (T-T) process, nominal V_{dd} and 125°C. A *WorstLeakage* corner is used to assess the absolute maximum leakage under Fast-Fast (F-F) process (i.e. low V_t) and high V_{dd} (Nominal+10%).

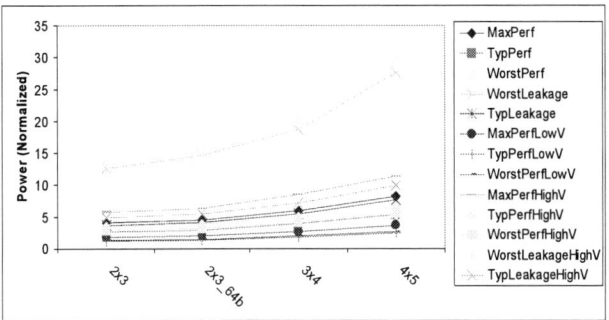

Figure 2. Normalized Power for Bus Matrix Configurations at 90nm

Power management strategies such as voltage islands and discrete voltage scaling (DVS) cannot be validated at the chip level unless IPs are characterized under several low V_{dd} conditions. This requires another set of corners. IPs such as the Metro libraries from Artisan (ARM) [19] are characterized for V_{dd} increments of 100mV for a range of possible V_{dd} values. *MaxPerfLowV*, *TypPerfLowV* and *WorstPerfLowV* are needed for each V_{dd}. For 90nm, the lowest safe V_{dd} is 0.7V. On the other hand, and under certain condition, more performance may be needed. For that case, some cell libraries are characterized for higher than normal operating conditions. In the case of 90nm, IPs can operate up to 1.2V nominal V_{dd}. Thus, another set of corners are needed including *MaxPerfHighV*, *TypPerfHighV*, and *WorstPerfHighV*. Since leakage can be significantly higher under those conditions, additional *TypicalLeakageHighV* and *WorstLeakageHighV* are sometimes available in order to assess typical and worst case leakage under high V_{dd} conditions. Note that the PVT corners shown in Table 1 do not constitute a maximal set. Many more corners can be added during library characterization, to support more elaborate design methodologies and possible operating environments and conditions. Alternatively, some technology libraries may not support *HighV* corners for

reliability reasons. The existence of a multitude of these corners motivates the need to mitigate the complexity and achieve more reliable designs by understanding and incorporating PVT effects early in a design flow, at the system-level.

In the case of timing analysis, one may argue that corner characterization is of limited use. However, we note that while timing analysis concentrates on critical path characterization, power characterization introduces significantly more degrees of freedom to the analysis, such as data dependence, power management, etc., all of which are very hard to incorporate into an amenable statistical analysis, especially at the system level. Thus, employing corner based analysis with a larger and more realistic corner set helps reduce the complexity of the designers' task in exploring the design space, albeit at the cost of perhaps slightly more pessimistic assumptions.

4. Impact of PVT Variability on Power Consumption

We will now present some experimental results to show how PVT variability affects power consumption in ultra deep submicron technology nodes, for the on-chip communication architecture (Section 4.1), the entire SoC design (Section 4.2) and a multi-Vt design flow (Section 4.3). Unlike the technology nodes upto the 130nm node, the ultra deep submicron technology nodes such as the 90nm and 65nm have many more PVT corners that need to be explored during the design phase, to ensure that power constraints in a design are satisfied. The experiments in this section explore power consumption characteristics for these different PVT corners, for different ultra deep submicron technology nodes, designs and operating frequencies.

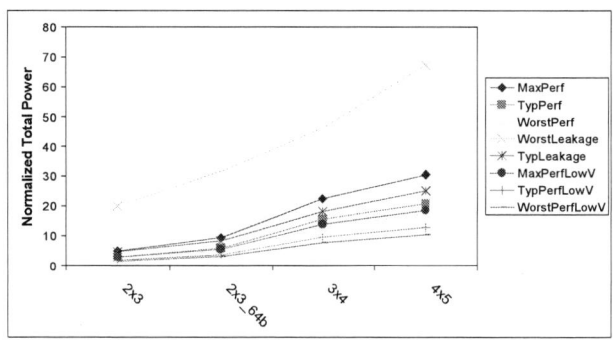

Figure 3. Normalized Power for Bus Matrix Configurations at 65nm

4.1. Impact on the On-Chip Communication Architecture

Our first set of experiments was conducted with the aim of understanding the impact of PVT variation on power consumption of the on-chip communication architecture. In the first experiment, we selected four SoC designs of varying

365

978-1-4244-3039-0/08 $25.00 © 2008 IEEE

complexity. Each of the designs had an AMBA AHB bus matrix [18] communication architecture configuration with a different structure and traffic characteristics: *(i)* a 2 master, 3 slave bus matrix with 32 bit data bus width (2x3), *(ii)* a 3 master, 4 slave bus matrix with 32 bit data bus width (3x4), *(iii)* a 4 master, 5 slave bus matrix with 32 bit data bus width (4x5) and *(iv)* a 2 master, 3 slave bus matrix with 64 bit data width (2x3_64b). We targeted the 90nm and 65nm general purpose technology libraries, synthesized these designs for a 100 MHz bus clock frequency and estimated power for the different PVT corners shown in Table 1 using Synopsys PrimePower [20] at the gate-level.

The normalized power of the bus matrix communication architecture, for the different PVT corners is shown for all four bus matrix configurations, in Figure 2 (90nm) and Figure 3 (65nm). It can be seen from the figures that there is significant variability in estimated power for 90nm and 65nm libraries, especially between the *WorstLeakageHighV* and *TypPerfLowV* corners (more than a 10× difference). As mentioned earlier, just considering the traditionally used corners (e.g. *TypPerf* vs. *TypLeakage*) is not realistic because there is a large variation in power consumption for sub-100nm libraries due to DSM effects (Section 3) which can only be captured by additional corners. In order to meet power goals, designers thus need to consider multiple PVT corners to understand the power characteristics of a design. *It can be concluded from our experimental results that there is a significant (as much as 10×) variation in power consumption across PVT corners, for on-chip communication architectures.*

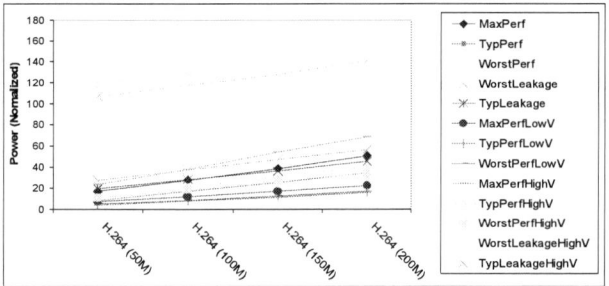

Figure 4. Normalized Power for H.264 SoC Subsystem

4.2. Impact on SoC Design

In our next set of experiment, we were interested in investigating the impact of PVT corners on power consumption for not just the communication architecture, but for an entire SoC design. We were also interested in estimating the impact of clock frequency on power consumption for the corners. For the purpose of this experiment, we implemented a complex SoC subsystem for the H.264/AVC codec [21] at the RTL level, consisting of the chroma inter-prediction IP, buffers and several memory blocks interconnected together using the AMBA AHB bus matrix [18]. Figure 4 illustrates normalized total power of the synthesized H.264 subsystem, obtained after detailed gate-

level power analysis [20], for a 90nm technology library implementation, with clock frequency ranging from 50 MHz to 200 MHz. A near linear increase in power is observed with frequency, which implies that power increases by an approximately constant ratio due to clock scaling. The impact of leakage power can be observed by considering the difference in power (almost **2×**) between *TypPerf* and *TypLeakage* corners under the same V_{dd}. *From our experimental result, it can be concluded that there is significant variation (as much as 60×) in power consumption across PVT corners, for SoC designs.*

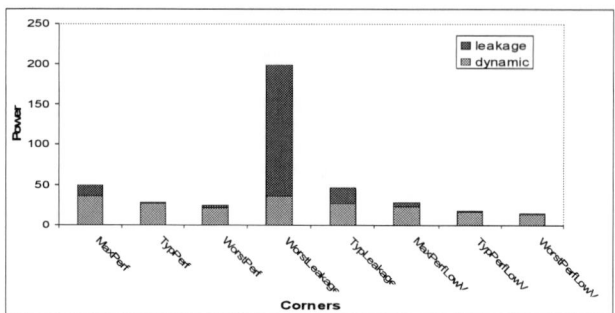

Figure 5. Decomposed Power for Bus Matrix Configurations at 65nm

4.3. Impact on Multi-Vt Design Flow

The multi-Vt technique is an effective way to reduce sub-threshold leakage current without sacrificing performance. High-Vt libraries can be used to reduce leakage current while low-Vt libraries can be used to get high performance on critical paths. We conducted several experiments for the 90nm and 65nm libraries, where we performed multi-Vt synthesis for different configurations of the AMBA AHB bus matrix communication architecture, running at 100 MHz bus clock frequency. The bus matrix power consumption across corners for the Multi-Vt case was found to be similar to the results shown in Figures 2 and 3, for the single-Vt case. Synthesis of the bus matrix configurations across different frequencies ranging from 100-300 MHz showed some interesting results. Unlike the near linear increase in power with frequency that was noticed in Figure 4, some distortions in the power consumption were noticeable at high frequencies. This was because additional low-Vt libraries were used to meet the more stringent timing requirements at higher frequencies in multi-Vt synthesis, unlike for the single-Vt synthesis case. *Our experimental results indicated significant variations in power consumption across PVT corners (as much as 6×) for multi-Vt implementations of the on-chip communication architecture.* Detailed experimental results obtained with multi-Vt synthesis can be found in our technical report [26].

5. Scaling Relation for Power Estimation Across PVT Corners

From the results of the experiments presented in the

previous section, we found that the power consumption for a PVT corner scales almost linearly with frequency. However, a much more interesting and important observation is that the power consumption numbers obtained for the different PVT corners show an almost constant ratio relative to each other. Thus if the power consumption of the bus matrix on-chip communication architecture for an implementation with PVT corner $C1$ is expressed as:

$$P_{C1} = P_{L1} + P_{D1} \times f \qquad \dots (1)$$

where P_{C1} gives the base level total power for a corner $C1$ that has base level leakage power P_{L1} and base level dynamic power P_{D1}, at frequency f; then the power consumption for an implementation under any other PVT corner $C2$ can be expressed as:

$$P_{C2} = \alpha_{1\text{-}2} \times P_{L1} + \beta_{1\text{-}2} \times P_{D1} \times f \qquad \dots (2)$$

where the total power P_{C2} for another corner $C2$ can be linearly scaled from the base level power relation in Eq. (1) by using scaling factors $\alpha_{1\text{-}2}$ and $\beta_{1\text{-}2}$ for leakage and dynamic power respectively. The scaling factors can be easily obtained by decomposing the total power into dynamic and leakage components (as shown in Figure 5, for the 65nm case shown in Figure 3), and averaging the ratio values. *Thus, knowing the leakage and dynamic power for one PVT corner can enable us to obtain the power for other PVT corners using a simple linear model with good accuracy, which speeds up power exploration across corners considerably.*

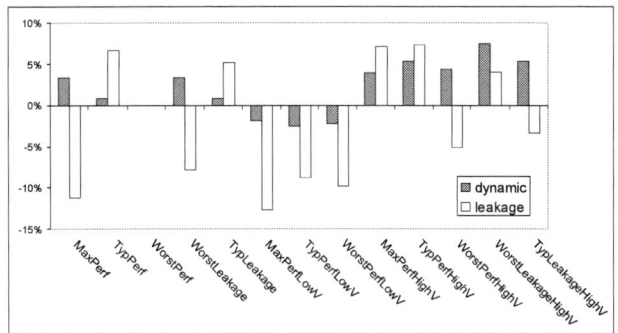

Figure 6. Normalized Power Estimation Error (90nm)

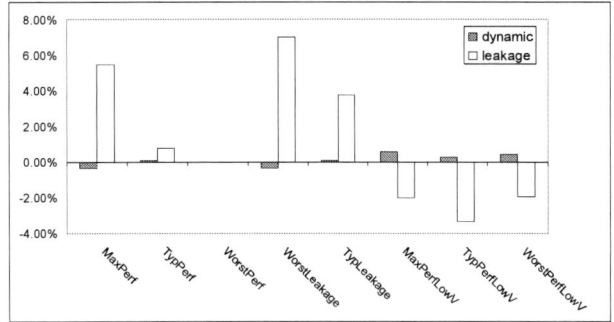

Figure 7. Normalized Power Estimation Error (65nm)

It is important to differentiate the proposed linear corner-to-

corner scaling proposed in (2) from the power (specifically leakage) dependence on P, V, and T. While it is well known that a super-linear relation exists between leakage and each of those P, V, and T parameters [27], what (2) reflects is an observation that, outside of frequency, which affects dynamic power, the ratio of power (dynamic+leakage) shows little variations *across design instances.* In other words, once a design implementation is characterized for power across the PVT corners (say through detailed gate level simulation), the corner ratios obtained can be re-used to "scale" another design instance without having to do another full characterization run for that instance.

Figures 6 and 7 show the maximum estimation error for the leakage and dynamic power for the 90nm and 65nm libraries respectively, when the scaling factors are used to estimate power consumption for different PVT corners, for different design configurations: the AMBA AHB bus matrix communication architecture, the H.264 SoC subsystem and a register file, operating at different clock frequencies. It can be seen from the figures that using the scaling technique we propose, it is possible to estimate power consumption for different PVT corners with extremely good accuracy for dynamic power (< 5% in most cases), and fairly good accuracy for leakage power (< 10% in most cases). *This amenability to scaling for PVT corners is an extremely important result, and has been obtained for cell libraries characterized with industrial strength numbers.*

6. PVT-Aware System Level Power Exploration

In this section, we will use the results of our observations from the previous sections, and show how to abstract up the power exploration across PVT corners early in the design flow, up to the system level, specifically for the AMBA AHB bus matrix communication architecture. The results of all the power exploration experiments conducted at the system-level have been verified by detailed gate-level simulation.

6.1. Incorporating PVT Corners in a System-level Power Estimation Methodology

To estimate power for the bus matrix communication architecture early in the design flow, at the system-level, we make use of the power estimation methodology for the bus matrix proposed by us in [17]. The approach makes use of energy macro-models to determine power consumption for the bus matrix logic components such as the input buffer stages, decoders, arbiters and output stages. A high level simulated annealing floorplanner [22] is used for early core placement and Manhattan routing is used to determine wire lengths. The wire lengths are subsequently used to determine wire energy, in formulations proposed in [23], which we extended to incorporate power for delay-optimally inserted repeaters. The power estimates for the bus matrix communication architecture were shown to be within 5% accuracy of gate-level power estimates in [17]. Creating the energy macro-

367

978-1-4244-3039-0/08 $25.00 © 2008 IEEE

models for the bus matrix however requires a one-time effort to identify macro-model variables and coefficients using multiple linear regression analysis [25], which can take several hours. Since a different energy macro-model is required for every PVT corner, the task can take a long time.

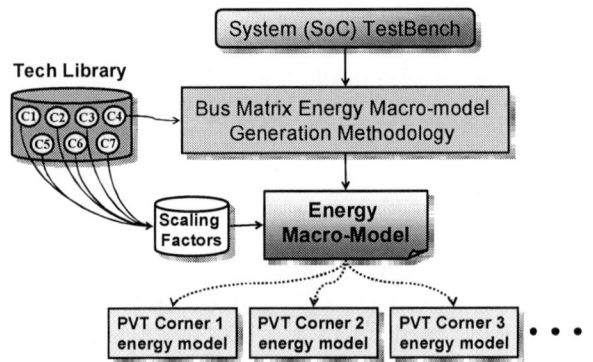

Figure 8. System-level Energy Macro-model Generation for PVT Corners

Figure 8 shows our proposed methodology to speed up bus matrix energy macro-model creation for different PVT corners of a technology library. This approach extends the energy macro-model based power estimation methodology from [17]. Initially, a system testbench consisting of a diverse set of bus matrix-based SoC designs is used to generate an energy macro-model for one of the PVT corners of a technology library, according to [17]. Subsequently, scaling relations presented in Section 5 are used for the other PVT corners to modify the base energy macro-model and create macro-models for each of the other PVT corners. This enables a considerable saving in time because only one macro-model generation iteration needs to be performed in order to obtain the energy models for all the corners in the selected technology library. *Since a single macro-model iteration can take in the order of hours, this approach can save us in the order of days to estimate power for the various PVT corners in ultra deep submicron technology library nodes.* In the next section, we present experiments to show how accurately these PVT corner macro-models can estimate power at the system-level for the bus matrix communication architecture.

6.2. System-level PVT-Aware Power Estimation for Bus Matrix Communication Architectures

To verify if the scaling factor-based PVT corner power estimation approach can be used to accurately explore power consumption across different corners for the bus matrix communication architecture, we performed an experimental study at the system-level. We selected an industrial strength multi-processor networking SoC application used for data packet processing and forwarding, with 4 ARM processors and more than 25 master, slave and memory IP blocks interconnected using the AMBA AHB bus matrix. We modeled the SoC application in SystemC [9] at the

Transaction-based Bus Cycle Accurate (T-BCA) [24][28] abstraction. The goal of our experiment was to estimate power for the PVT corners at the system-level using two methods: first, with the traditional approach of creating energy macro-models for each PVT corner separately, and second, with the proposed scaling based approach.

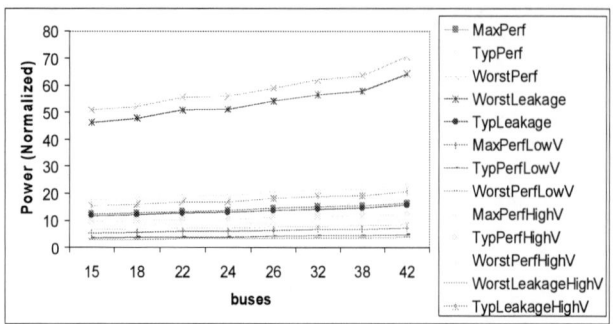

Figure 9. Normalized Power for networking SoC

In the first experiment, we used our communication architecture synthesis framework from [17] to generate a set of bus matrix solutions (each solution having a different number of buses) that satisfies the performance constraints of the networking SoC application. Next, we created bus matrix energy macro-models for each of the different PVT corners of the 90nm general purpose technology library. We plugged the energy macro-models into a T-BCA simulation model of the networking SoC, and simulated the design for each of the solutions in the solution set to get power consumption for the application. Figure 9 and 10 show the normalized power and energy obtained after simulating the design for all the PVT corners, for each of the solutions. As mentioned earlier, these numbers are within 5% accuracy of gate-level power estimates. The X-axis shows different bus matrix solutions, each having a different number of buses (and consequently different number of logic components such as arbiters, decoders etc.).

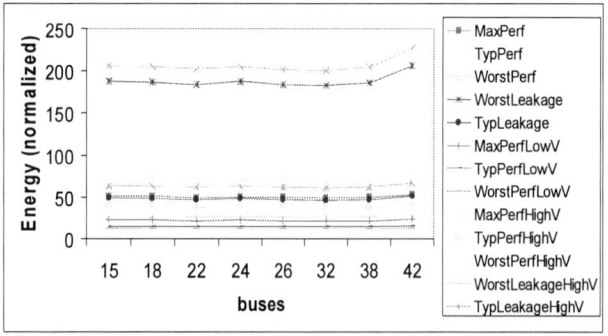

Figure 10. Normalized Energy for networking SoC

From the figures, it can again be seen that there is a significant variation in power and energy consumption across different PVT corners, and the power and energy ratios for different corners are fairly constant. Table 2 shows the percentage change in performance for the different solutions

368

978-1-4244-3039-0/08 $25.00 © 2008 IEEE

having fewer buses, compared to the solution with 42 buses. It can be seen that solutions having fewer number of buses have lower performance. This is because there are more delays due to traffic conflict when data streams originating from different masters must share fewer buses. *Note that the performance numbers in Table 2 represent a lower bound on performance that can be achieved for all the PVT corners.*

Table 2. % performance variation for bus matrix configs

No. of buses	15	18	22	24	26	32	38	42
% perf. variation	-26.5	-21.9	-12.6	-14	-6.1	-0.7	-0.2	0

Creating energy macro-models for each PVT corner case turns out to be very time consuming, requiring a few days in designer effort. There is a need to speed this process up. Clearly, one solution is to use the scaling relations from Section 5, and the methodology presented in Figure 8. The question is, *can this approach be used at the system-level to accurately estimate power for PVT corners?* To answer this, we selected one of the PVT corners (*WorstPerf*) as a base reference and scaled its power results to create energy-macro models for the other PVT corners. We then plugged these energy macro-models into our T-BCA simulation model, and simulated the design for each solution in the bus matrix solution set, to obtain power numbers.

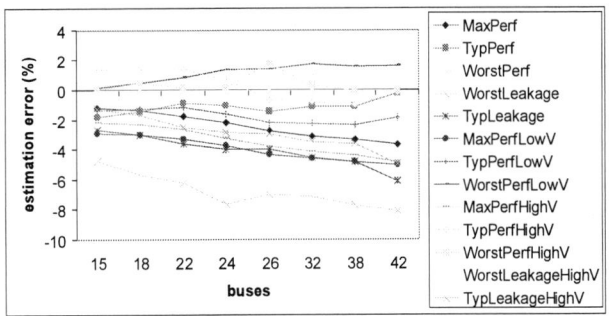

Figure 11. % power estimation error for corner cases when Eq, (2) is used to obtain power at PVT corners

Figure 11 shows the error in power estimation for the solutions, when we used the scaling technique to estimate power at PVT corners, compared to the traditional technique of creating energy macro-models separately for the corners (as was done in the previous experiment). It can be seen that the maximum absolute error compared to macro-model estimates is less than 9%. *The maximum absolute error compared to gate-level estimates is 14%* (since the macro-models are within 5% accuracy of gate-level estimates [17]), *which is an extremely good accuracy for PVT corner power estimation at the system-level.* The results imply that we only need to create an energy macro-model for one of the PVT corners, and use scaling factors to quickly obtain power for other PVT corners at the system-level. The overall time taken for creating energy macro-models for all the PVT corners from the base reference macro-model in this case is in the order of a few seconds, and

several orders of magnitude less than the case where macro-models have to be separately created for every PVT corner.

Figures 12 and 13 show the percentage change in power and energy for the solutions in the bus matrix solution set, compared to the solution with the most number of buses (42). It can be seen that solutions with fewer number of buses have lower power and energy dissipation, at the cost of performance (Table 2). There is a large variation in energy and power consumption across the different PVT corners, for the solutions.

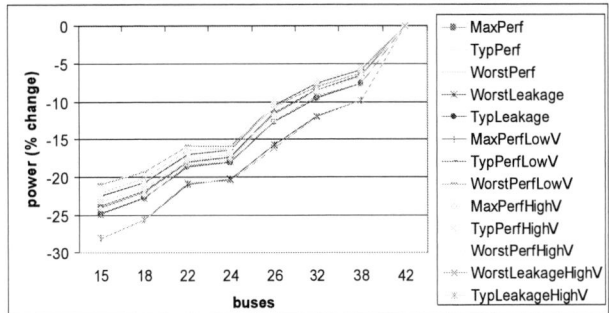

Figure 12. % change in power for networking SoC

6.3. Discussion of Experimental Results

From the experiments, it is clear that it is extremely important for designers to consider different PVT corners during early power and energy exploration at the system-level. Consider, for instance, the case where the average power dissipation constraint for the design is 80 mW. Assume that average power dissipation for the solution with 42 buses is 100 mW. In such a case, from Figure 12, it can be seen that the solution with 15 buses meets the constraint for all corners (i.e. % power reduction compared to 42 bus case > 20% for all corners). The solution with 18 buses meets the constraint for all corners except the *TypPerf* and *WorstPerfLowV* corners. The solutions with 22 and 24 buses meet the constraint only for the *WorstLeakage* and *WorstLeakageHighV* corners. If the designer used traditional techniques, he would have only a single power number for a technology library to rely upon during early exploration; now the designer can select a solution based on a more comprehensive characterization of the technology library across various PVT corners. The designer in this case can select the 15 bus solution conservatively, or go for the better performing solution with 18 buses (refer to Table 2), with a reasonable degree of confidence that the solution will not violate constraints under most PVT conditions.

Consider another case where a solution with minimum energy dissipation needs to be selected. From Figure 13, it can be seen that under maximum leakage conditions, the lowest energy is obtained for the solution with 15 buses. However, if such maximum leakage conditions will not be encountered, then the lowest energy dissipation (for the other corners) is obtained for the solution with 22 buses. Such PVT-aware

369

978-1-4244-3039-0/08 $25.00 © 2008 IEEE

exploration information can aid the designer in selecting the appropriate solution with greater confidence and more accuracy, than the traditionally used approach of considering only a single corner for a technology library, during early system-level power exploration.

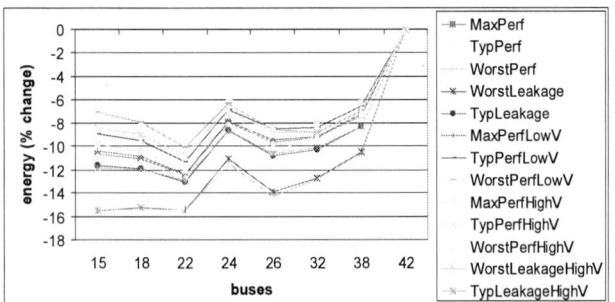

Figure 13. % change in energy for networking SoC

7. Conclusion

In this paper, we investigated the impact of PVT corners on power consumption at the system level, especially for the bus matrix on-chip communication architecture. We first conducted several experiments to show how there are significant variations in power consumption across different corners for a given technology library. Next we showed how it is possible to "scale up" and abstract the PVT variability to characterize the true design space early on in the design flow, at the system level. We used scaling relations to quickly create power models for the different PVT corners, to estimate the power consumption of the bus matrix at the system level. The scaled power models took several orders of magnitude less time to create than the traditional macro-modeling approach, with a maximum absolute estimation error of 14%, which is extremely good for early power estimation at the system-level. Finally, we experimentally established the importance of considering PVT corners during system-level power exploration. While we currently do not consider intra-die PVT variations, we believe that the simplicity of the abstractions described in this paper will make it feasible to incorporate such variability into future work.

Acknowledgement

This research was partially supported by grants from SRC (2005-HJ-1330).

References

[1] S. Borkar et al., "Design and reliability challenges in nanometer technologies", *DAC 2004*

[2] V. Khandelwal, A. Srivastava, "A General Framework for Accurate Statistical Timing Analysis Considering Correlations", *DAC 2005*

[3] J. Le, X. Li, L. Pileggi. "STAC: Statistical Timing Analysis with Correlation", *DAC 2004*

[4] A. Papanikolaou et al. "A System Level Methodology for Fully Compensating Process Variability Impact of Memory Organizations in Periodic Applications", *CODES+ISSS 2005*

[5] K. Lahiri et al., "Battery driven system design: A new frontier in low power design", *ASPDAC 2002*

[6] R. Ho, K. W. Mai, M. A. Horowitz, "The Future of Wires", *IEEE, vol. 89, April 2001*

[7] K. Lahiri, A. Raghunathan, "Power Analysis of system-level on-chip communication architectures", *CODES+ISSS 2004*

[8] K. Banerjee et al., "Coupled Analysis of Electromigration Reliability and Performance in ULSI Signal Nets", *ICCAD 2001*

[9] SystemC initiative, www.systemc.org

[10] M. Lajolo et al, "Cosimulation-based power estimation for system-on-chip design," *IEEE TVLSI, vol. 10, no. 3, 2002*

[11] N. Dhanwada, I-C. Lin, V Narayanan, "A Power Estimation Methodology for SystemC Transaction Level Models", *CODES+ISSS 2005*

[12] I. Lee et al. "PowerViP: SoC Power Estimation Framework at Transaction Level", *ASPDAC 2006*

[13] P. P. Sotiriadis, A. P. Chandrakasan, "A Bus Energy Model for Deep Submicron Technology," *IEEE TVLSI June 2002*

[14] L. Benini et al. "Architectures and Synthesis Algorithms for Power-efficient Bus Interfaces," *IEEE TCAD Sept. 2000*

[15] M. Caldari et al. "System-level power analysis methodology applied to the AMBA AHB bus", *DATE 2003*

[16] A. Bona et al., "System level power modeling and simulation of high-end industrial network-on-chip", *DATE 2004*

[17] S. Pasricha, Y. Park, F. Kurdahi, N. Dutt, "System-Level Power-Performance Trade-Offs in Bus Matrix Communication Architecture Synthesis", *CODES+ISSS 2006*

[18] AMBA AHB Interconnection Matrix, www.synopsys.com/ products/designware/amba_solutions.html

[19] http://www.arm.com/products/physicalip/metro.html

[20] Synopsys CoreTools, PrimePower www.synopsys.com

[21] Join Video Team (JVT) ISO/IEC MPEG, ITU-T VCEG: Final draft international standard of joint video specification (ITU-T Rec. H.264/ISO/IEC 14496-10 AVC). *JVT-G050, 2003*

[22] S. N. Adya, I. L. Markov, "Fixed-outline Floorplanning: Enabling Hierarchical Design", *IEEE TVLSI, Dec. 2003*

[23] C. Kretzschmar, et al., "Why transition coding for power minimization of on-chip buses does not work", *DATE 2004*

[24] S. Pasricha, N. Dutt, M. Ben-Romdhane, "Extending the Transaction Level Modeling Approach for Fast Communication Architecture Exploration", *DAC 2004*

[25] J.J. Faraway, "Linear Models with R", *CRC Press, 2004*

[26] S. Pasricha, Y. Park, F. Kurdahi, N. Dutt, "PVT Variation Aware On-Chip Communication Architecture Synthesis", CECS Technical Report, *Jan 2008*

[27] J. Rabaey, A. Chandrakasan, B. Nikolic, "Digital Integrated Circuits: A Design Perscpective", *2nd Ed. Prentice-Hall, 2003.*

[28] S. Pasricha, "Transaction Level Modeling of SoC with SystemC 2.0", *SNUG 2002*

978-1-4244-3039-0/08 $25.00 © 2008 IEEE

21st International Conference on VLSI Design

Energy – Efficient, High Performance Circuits for Arithmetic Units

Sundeepkumar Agarwal	Pavankumar V K	Yokesh R
Final year BE ECE	*Final year ME VLSI*	*Final year BE ECE*
PSG College of Technology	*PSG College of Technology*	*PSG College of Technology*
sun_ece_psg@yahoo.com	*kalangi.pavan@gmail.com*	*r_yokesh@yahoo.com*

Abstract

Adders and multipliers are the most important arithmetic units in a general microprocessor and the major source of power dissipation. Various architecture styles exist to implement these units, each having their own merits and demerits. However, due to continuing integrating intensity and growing needs of portable devices, low power design is of prime importance. In addition, much power is dissipated due to a large number of spurious transitions on internal nodes in power hungry multiplier structures. We present a new full adder structure based on complementary pass transistor logic (CPL) which is faster and more energy efficient than the existing structures. We also propose a new technique of implementing multiplier circuit using decomposition logic which improves speed and reduces power consumption by reducing the spurious transitions on internal nodes. Combined with the new adder structure and the decomposition logic, there is substantial improvement in the performance of the multiplier structures. With the help of these state of the art designs, it would be possible to design highly power efficient processors, especially digital signal processors. We have used TSPICE for simulation in the TSMC 180nm technology.

1. Introduction

Low power design is the need of today's integrated systems. In portable electronic devices, it is important to prolong the battery life as much as possible. Adder is the core component of an arithmetic unit. The efficiency of the adder determines the efficiency of the arithmetic unit. Various structures have evolved trying to improve the performance of the adder in terms of area, power and speed. Low power design with high speed of operation is more essential. We present a new adder design which is faster as well as more energy efficient than the existing adder structures.

The multiplier circuit is a core component of most of the present day digital signal processors. To improve the performance of multiplier structures, we propose a new technique using the decomposition logic. The new structure not only improves speed of operation, but also reduces power consumption.

2. Full adder structures

Various adder structures have been presented in [1], [2], [3], [4], [5]. The adders proposed in [6] and [7] are reported to have better performance than the other adder structures. We therefore confine to these adder structures for comparing the proposed adder structure. Figure 1 shows the adder structure reported in [7] along with size of transistors. The numbers depict the aspect ratio of the transistors keeping the length fixed at 180nm. For comparison purpose, we name this adder as Hybrid adder. The adder proposed in [6] is shown in figure 2 and is named as Hybrid Exor adder. The numbers in figure 2 depict the actual width of the transistors in μm as used in [6]. Sizes given in [6] were chosen because of better performance than the sizes used in [7]. The CMOS mirror adder [8] shown in figure 3 was also used for comparison due its robustness to scaling and stability at low voltages. The CMOS mirror adder uses 'carryout' signal to generate the 'sum' signal. So, the structure is unbalanced and causes generation of glitches due to skewness in arrival of signals in tree structured arithmetic circuits.

3. Proposed full adder structure

The proposed full adder has been implemented using the CPL logic. The schematic of the proposed adder is shown in figure 4 along with a logic diagram for better understanding. For reference, we use the name CPL for this adder. The aspect ratio of the transistors is shown in figure 4 and the length of the transistors is kept at 180nm. The adder is mainly composed of NMOS transistors with pull – up PMOS

371

978-1-4244-3039-0/08 $25.00 © 2008 IEEE

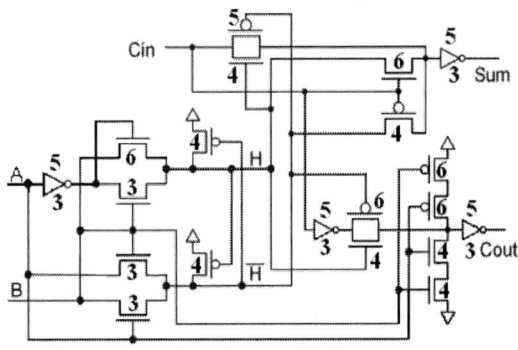

Figure 1. Hybrid adder

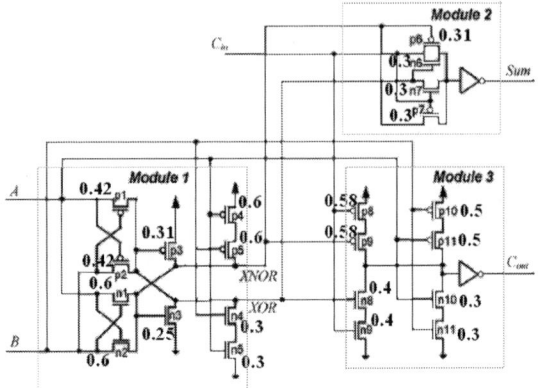

Figure 2. Hybrid Exor adder

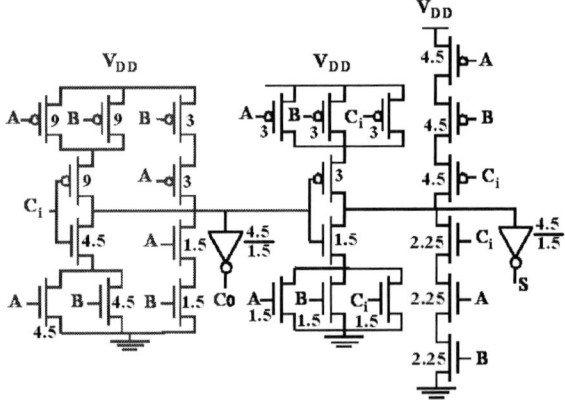

Figure 3. CMOS mirror adder

transistors to obtain full swing output voltage. Due to positive feedback and use of NMOS transistors, the circuit is inherently fast. This property is utilized to reduce the width of the transistors to reduce power consumption without much speed degradation. The proposed adder has a balanced structure with respect to generation of 'sum' and 'carryout' signals. This helps in simultaneous arrival of signals in tree structured circuits like the Dadda multiplier and thus reduces generation of unwanted glitches.

The number of transistors used in this design is more compared to other designs. This is due requirement of seven inverters to generate complement signals. However, when this adder is used in designs such as the multiplier, the input complementary signals can be derived from previous stage outputs. This reduces the number of transistors. Also, the drivability of this adder is fairly good even without the use of inverters. This is due to use of pull – up PMOS transistors. Hence, the output inverters can be used in alternate stages of the design. As an example, we consider the design of a 4 bit ripple carry adder which is shown in figure 5. Here, the full adders marked with an asterisk '*' do not use output inverters for carry generation. By doing so, one inverter delay is eliminated for every two full adders in the adder chain and four transistors are reduced. Similarly, in complex designs like the multiplier, the output inverters for generating sum and carry can be used in alternative stages, thereby improving speed and reducing area.

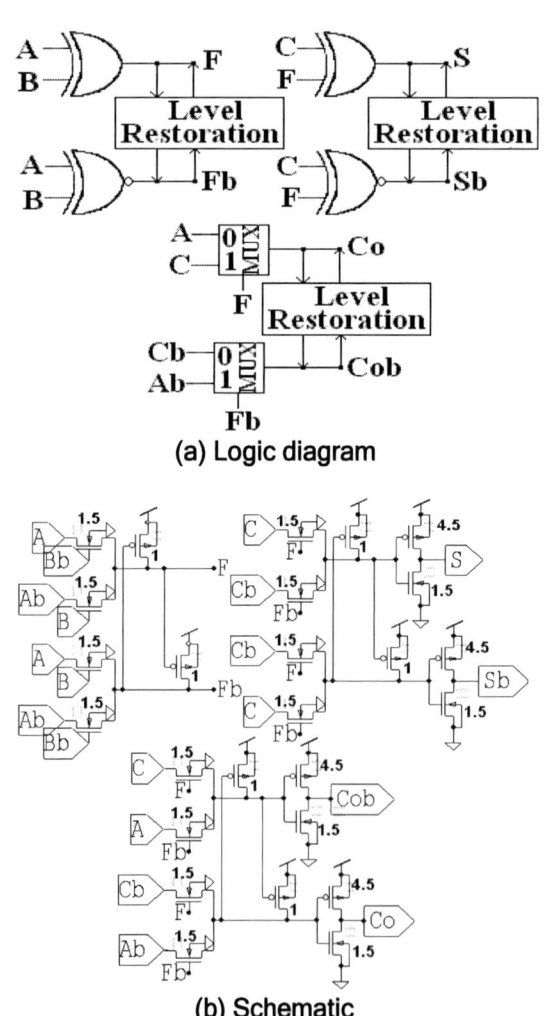

(a) Logic diagram

(b) Schematic
Figure 4. Proposed adder

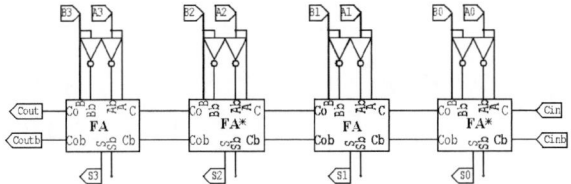

Figure 5. Ripple carry adder - 4 bit

4. Multiplier design

The two well-known fast multipliers are those presented by Wallace and Dadda. Both consist of three stages. In the first stage, the partial product matrix is formed. In the second stage, this partial product matrix is reduced to a height of two. In the final stage, these two rows are combined using a carry propagating adder. In the Wallace method, the partial products are reduced as soon as possible. In contrast, Dadda's method does the minimum reduction necessary at each level to perform the reduction in the same number of levels as required by a Wallace multiplier. We have used an 8x8 Dadda multiplier for comparison. The Dadda multiplier was chosen because it uses less number of full adders and half adders than the Wallace structure. The dot diagram representation of the Dadda multiplier is shown in figure 6.

We propose a new technique to implement digital multipliers using the decomposition logic. Here, the multiplication process is split into smaller sub – units (smaller multipliers) and their outputs are combined to get final results. By doing so, parallel processing is also introduced in addition to the benefits from tree structured implementation of the multiplier.

Figure 7 shows an 8x8 multiplier implemented using the decomposition logic. In the first stage we use four 4x4 multipliers to combine all the partial products. The outputs from these 4x4 multipliers are then combined to get final results. For a 4x4 multiplier, the Wallace tree and Dadda tree structures are almost similar. We have used the Wallace tree structure [8] shown in figure 8.

The decomposition logic requires extra circuitry to perform final addition of outputs obtained from the 4x4 multipliers. However, due to parallel processing of the 4x4 multipliers, significant improvement in speed is achieved. Since the inputs to the final adder circuitry arrive in parallel, glitches are reduced resulting in lower dissipation of power.

The decomposition process can further be extended to implementing a 4x4 multiplier using two 2x4 multipliers or four 2x2 multipliers. However, the benefits derived from parallel processing of data are outweighed by degradation due extra circuitry.

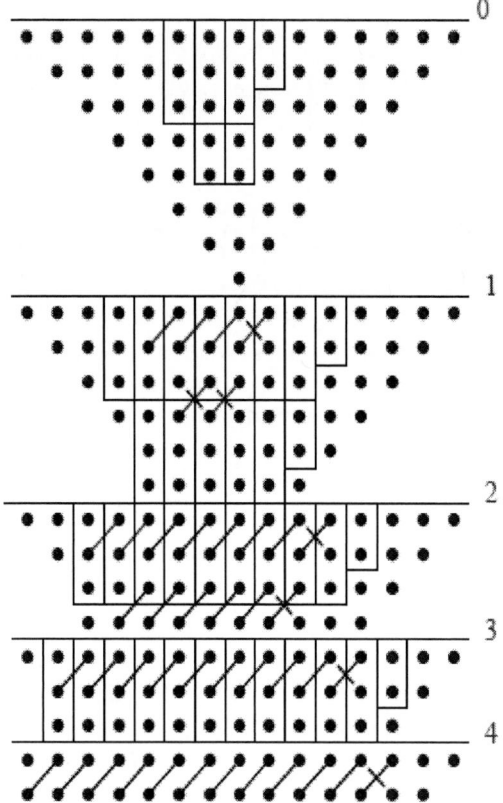

Figure 6. Dot diagram representation of 8x8 Dadda multiplier

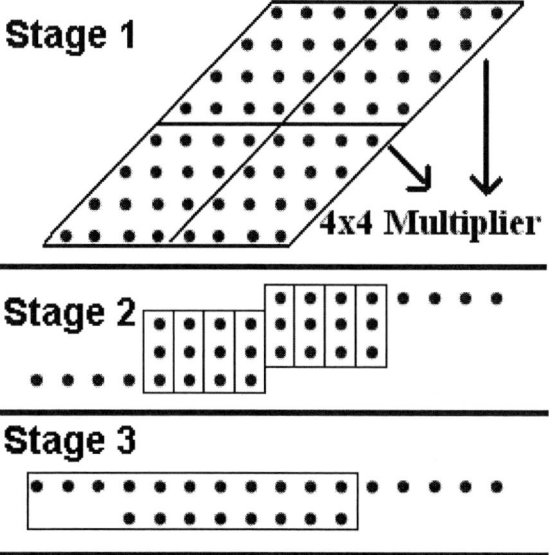

Figure 7. 8x8 multiplier using decomposition logic

978-1-4244-3039-0/08 $25.00 © 2008 IEEE

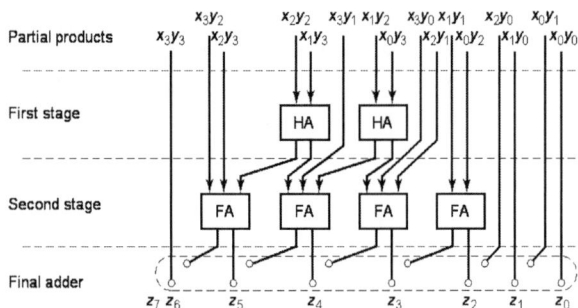

Figure 8. 4x4 Wallace multiplier

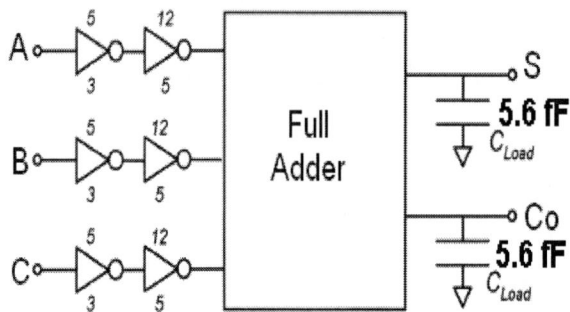

Figure 9. Full adder test bench given in [7]

5. Simulation environment

Simulation was done using TSPICE in the TSMC 0.18μm (Level 49) technology. The threshold voltages of NMOS and PMOS transistors are around 0.39V and -0.41V respectively.

To evaluate the performance of the full adders, test benches suggested in [6] and [7] were used. Figure 9 shows the test bench used in [7]. This set up is useful to test the performance of the adders as a single unit. However, for tree structured applications, such as the multiplier, it is important to test the driving capability of the adder. Figure 10 shows the simulation set up used in [6] to test the drivability of adders.

As mentioned earlier, the complementary signals required for the proposed adder can be derived from previous stages. So, for the tree structured simulation set up shown in figure 10, only the first stage requires inverters for input signals. Apart from these two test benches, we also designed a 32 – bit ripple carry adder to test the delay in propagation of carry signals.

6. Simulation results for adder structures

The circuits were tested at different supply voltages from 0.7V to 1.8V. The input frequency was kept at 200 MHz for 1.8V and suitably reduced for lower supply voltages. The parameters considered for comparison are power consumption, worst case delay and energy – delay product. Delay was calculated from 50% of input voltage level to 50% of output voltage level. Energy – delay product was chosen to emphasize more on speed performance of the circuit. The simulation results are tabulated in table 1.

When compared to other adders, the proposed CPL adder showed significant improvement in speed at all supply voltages. This is due to use of positive feedback PMOS pull – up transistors. However, the CMOS mirror adder shows lesser delay for the 32 – bit ripple carry adder. This is due to faster 'carryout' signal generation. The Hybrid Exor adder dissipated the least

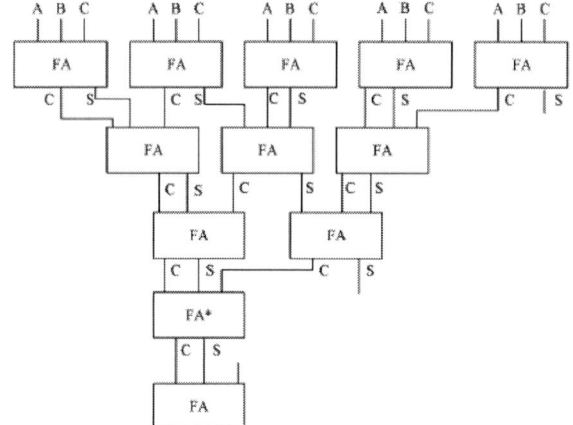

Figure 10. Test bench given in [6]

power among the adders. The power dissipation of proposed CPL adder was slightly more than Hybrid Exor adder due to more number of transistors in the design. The performance of CMOS mirror adder is more stabilized as the supply voltage is reduced when compared to other adders. The CPL adder is next better in performance at low voltages.

Energy – delay product was used as the metric to compare the performance of the adders. Figure 11 shows the energy – delay product of the all adder structures for different supply voltages.

7. Simulation results for multipliers

The 8x8 Dadda multiplier was implemented using the proposed CPL adder, Hybrid Exor adder, Hybrid adder and the CMOS mirror adder. These multiplier structures were compared with the 8x8 multiplier designed using decomposition logic. As mentioned earlier, the output inverters in the CPL adder can be used in alternate stages. Table 2 summarizes the simulated results for all the multipliers. The energy – delay product for the multipliers is shown in figure 11. Table 3 shows a comparison of number of transistors used in different circuits.

374

978-1-4244-3039-0/08 $25.00 © 2008 IEEE

Table 1. Simulation results for adder circuits

Full adder test bench								
Supply voltage (V)	Power in micro watt				Delay in nano second			
	CPL	Hybrid Exor	Hybrid	CMOS	CPL	Hybrid Exor	Hybrid	CMOS
1.8	25.2	20.8	27.0	21.8	0.28	0.35	0.38	0.31
1.5	16.9	13.9	18.3	14.7	0.37	0.46	0.51	0.42
1.2	10.6	8.77	11.5	9.29	0.57	0.64	0.67	0.63
1.0	7.33	6.09	7.93	6.41	0.82	1.00	1.02	0.91
0.7	3.57	2.90	3.69	3.01	3.30	3.45	3.82	3.37
Full adder tree structure test bench								
Supply voltage (V)	Power in micro watt				Delay in nano second			
	CPL	Hybrid Exor	Hybrid	CMOS	CPL	Hybrid Exor	Hybrid	CMOS
1.8	114	110	159	117	0.75	1.23	2.07	1.06
1.5	74.3	73.4	106	78.9	1.10	1.63	2.77	1.40
1.0	28.1	28.7	40.9	31.3	3.20	4.08	7.30	3.10
0.7	3.34	3.40	4.80	3.60	11.7	15.2	26.1	11.1
32 bit ripple carry adder								
Supply voltage (V)	Power in micro watt				Delay in nano second			
	CPL	Hybrid Exor	Hybrid	CMOS	CPL	Hybrid Exor	Hybrid	CMOS
1.8	93.7	83.1	160	89.3	3.62	6.06	6.22	3.31
1.5	59.7	55.9	108	59.6	5.00	7.87	8.03	3.38
1.0	17.8	17.3	35.2	19.3	12.3	18.2	18.3	10.0
0.7	2.13	2.05	4.25	2.27	45.0	69.3	72.3	40.3

Table 2. Simulation results for 8x8 multiplier

8x8 Multiplier										
Supply voltage (V)	Power in micro watt					Delay in nano second				
	Decom-position	CPL	Hybrid Exor	Hybrid	CMOS	Decom-position	CPL	Hybrid Exor	Hybrid	CMOS
1.8	567	569	854	1300	851	1.12	1.51	1.83	2.28	1.81
1.5	184	189	282	433	283	1.45	1.92	2.28	3.17	2.32
1.2	112	117	175	273	175	2.51	3.23	3.77	4.96	3.60
1.0	76.7	80.8	119	188	120	4.00	5.19	5.85	8.23	5.38
0.8	24.4	25.9	37.3	59.8	37.6	8.38	11.1	11.8	17.8	10.8

Table 3. Comparison of transistor count

Circuit	CPL	Hybrid Exor	Hybrid	CMOS
Full adder test bench	44	38	36	40
Full adder tree structure	342	312	288	336
32 bit ripple carry adder	832	832	768	864
8x8 multiplier	Decomposition	CPL	Hybrid Exor	Hybrid	CMOS
	1648	1476	1742	1644	1840

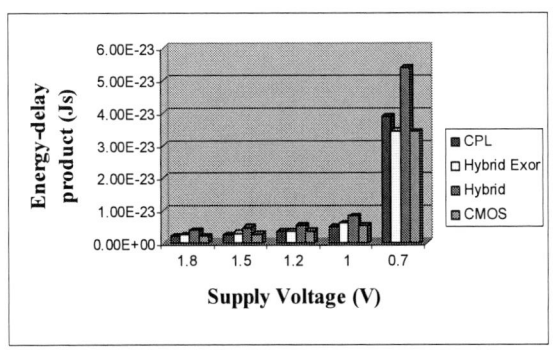

(a) Full adder test bench

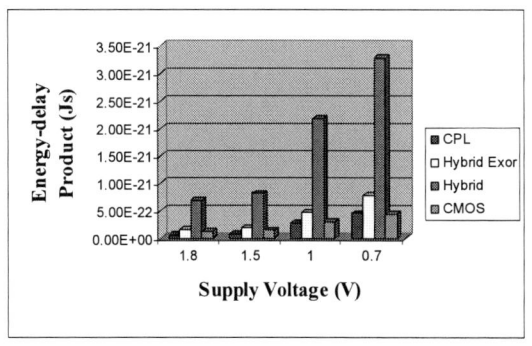

(b) Full adder tree structure

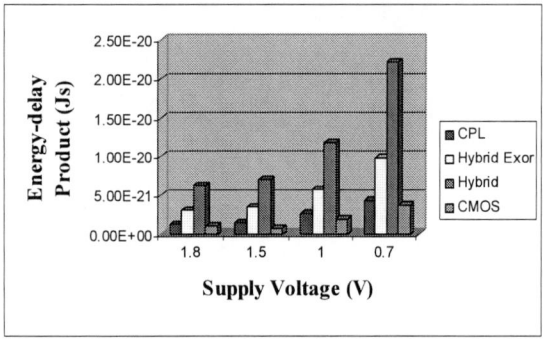

(c) 32 - bit ripple carry adder

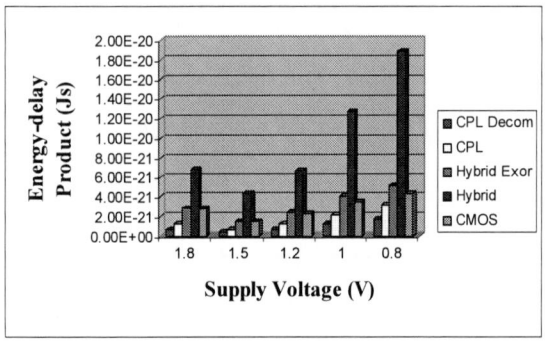

(d) 8x8 multipliers

Figure 11. Comparison of energy - delay product for various circuits

8. Conclusion

A new design for adder was presented. The proposed adder was compared with latest adder structures and tested in different test benches to analyze various performance factors. Simulation results show that the proposed adder is faster and more energy efficient than the existing adders.

Also, a new technique of implementing digital multipliers using decomposition logic was presented. When compared to Dadda multiplier, the proposed multiplier was faster and dissipated lesser power in spite of extra logic circuitry. The

decomposition logic can be extended to other multiplier designs also.

9. References

[1] N. Zhuang and H. Wu, "A new design of the CMOS full adder," *IEEE J. Solid – State Circuits*, vol. 27, no. 5, pp. 840 – 844, May 1992.

[2] M. Vesterbacka, "A 14 – transistor CMOS full adder with full voltage – swing nodes," in *Proc. IEEE Workshop Signal Processing Systems*, pp. 713 – 722, Oct. 1999.

[3] D. Radhakrishnan, "Low – voltage low – power CMOS full adder," *IEEE Proc. Circuits Devices Syst.*, vol. 148, no. 1, pp. 19 – 24, Feb. 2001.

[4] H. A. Mahmoud and M. A. Bayoumi, "A 10 – transistor low – power high – speed full adder cell," in *Proc. Int. Symp. Circuits Syst,*, pp. I-43 – 46, 1999.

[5] H. T. Bui, Y. Wang, and Y. Jiang, "Design and analysis of low – power 10 – transistor full adders using novel XOR – XNOR gates," *IEEE Trans. Circuit Syst. II, Analog Digit. Signal Process.*, vol. 49, no. 1, pp. 25 – 30, Jan. 2002.

[6] C. H. Chang, J. Gu, and M. Zhang, "A review of 0.18μm full adder performances for tree structured arithmetic circuits," *IEEE Trans. Very Large Scale Integr. (VLSI) Syst.*, vol. 13, no. 6, pp. 686 – 695, Jun. 2005.

[7] S. Goel, A. Kumar, and M. A. Bayoumi, "Design of robust, energy – efficient full adders for deep – submicrometer design using hybrid – CMOS logic style," *IEEE Trans. Very Large Scale Integr. (VLSI) Syst.*, vol. 14, no. 12, pp. 1309 – 1321, Dec. 2006.

[8] Jan. M. Rabaey, Anantha Chandrakasan and Borivoje Nikolic, "*Digital Integrated Circuits A Design Perspective*", Prentice-Hall of India Pvt Ltd, New Delhi, 2004.

Delay and Energy Efficient Design of On-Chip Encoded Bus with Repeaters

Qingli Zhang, Jinxiang Wang, Yizheng Ye

Microelectronics Center, Harbin Institute of Technology, 150001, Harbin, China
qinglee@hit.edu.cn

Abstract

In this paper, we propose a new spatial and temporal encoding approach for generic on-chip global buses with repeaters that enables higher performance while reducing peak energy and average energy. The proposed encoding approach exploits the benefits of temporal encoding circuit and spatial bus-invert coding techniques to simultaneously eliminate opposite transitions on adjacent wires and reduce the number of self-transitions and coupling-transitions. In the design process of applying encoding techniques for reduced bus delay and energy, we present a repeater insertion design methodology to determine the repeater size and inter-repeater bus length which minimizes the total bus energy dissipation while satisfying target delay and slew-rate constraints. This methodology can be employed to obtain optimal energy vs. delay trade-offs under slew-rate constraint for various encoding techniques.

1. Introduction

With shrinking of feature sizes, increasing chip area, and increasing interconnect density, on-chip global buses are suffer from large propagation delay due to capacitive crosstalk and high power consumption due to both parasitic and coupling capacitance [1]. Many encoding techniques using spatial redundancy have been presented to alleviate power or delay problems. For example, the bus-invert method [3] can limit the maximum number of self-transitions and coupling-transitions to 50% and result in potential gains of 50% in peak energy; various encoding schemes [2, 4] are applied to minimize both average self-transition and coupling-transition activity for bus average power reduction; and crosstalk avoidance codes [5] have shown to provide both power efficiency and elimination of the worst-case crosstalk delay simultaneously. These encoding techniques have a large routing area overhead due to the need for additional bus wires. We refer to such an encoding scheme as *spatial encoding*. There are also some encoding techniques [6, 7] using temporal redundancy to minimize both delay and power. These encoding techniques have no routing area overhead compared with the spatial encoding techniques. We refer to such an encoding scheme as *temporal encoding*.

However, all the aforementioned work only focus on the effects of encoding schemes on bus delay and power, but do not

consider the effects of repeater insertion. Some previous work can be found in the literature, which attempt to address the issue of optimizing the repeater insertion in the design process of applying bus encoding techniques for reduced delay and power [7, 8]. However, these papers do not provide any closed-form expressions for their repeater optimization methods; instead, they performed the optimization by exhaustively sweeping the number and size of repeaters at the expense of considerable SPICE simulation time. Therefore, their design methods are not very suitable for integration in a CAD tool flow.

In the previous work [9], we presented a delay and energy efficient temporal encoding circuit. In this paper, we propose a spatial and temporal encoding technique to further improve energy reduction by combining the temporal encoding circuit technique with the spatial bus-invert coding [3]. In addition, we further present a repeater insertion design methodology to determine the repeater size and inter-repeater bus length which minimizes the total bus energy dissipation while satisfying the target delay and slew-rate constraints. This methodology can be employed to obtain energy vs. delay trade-offs under slew-rate constraint for the various encoding techniques.

2. Spatial and temporal encoding method

2.1. Temporal encoding circuit technique [9]

The topologies of the original uncoded and temporally encoded buses with repeaters are shown in Fig. 1. We are interested in the optimal energy-delay trade-off for transmitting data from the node *bus_in* to the node *bus_out*. For the uncoded (UNC) bus, the worst-case coupling-transitions between adjacent wires are '↑↓' and '↓↑'. The temporal encoding (TE) technique, which eliminates the worst-case coupling-transitions without the cost of additional wire area, is based on the following property: for any given *n*-bit input data stream, if some *n*-bit shield signals (e.g. all 0' or all 1's) is inserted in the data stream every other data value, it is observed that there exist no '↑↓' and '↓↑' transitions in the newly generated data stream. The key idea of the technique is that the TE circuit (as shown in Fig. 2) can dynamically build shield signals depending on the results of the logic *AND* operation of the current and previous state of input data signals instead of inserting some fixed shield signals (e.g. all 0's). Thus, on the one hand, the TE-coded bus not only achieves bus switching activity dependent on input

377

978-1-4244-3039-0/08 $25.00 © 2008 IEEE

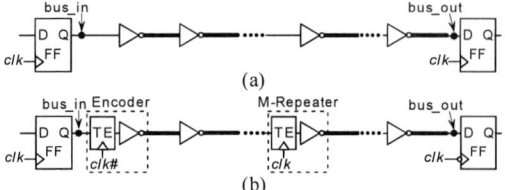

Fig. 1. Topology of (a) original uncoded and (b) temporally encoded bus with repeaters.

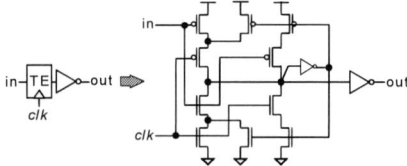

Fig. 2. Temporal encoding circuit.

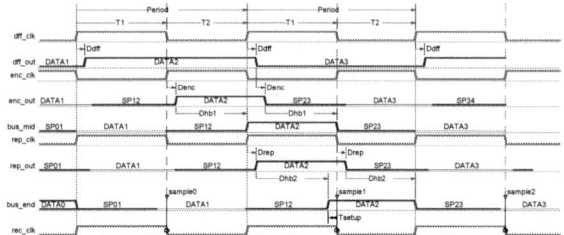

Fig. 3. Timing plan of the TE-coded bus.

switching behavior, but also switches only once in a cycle when static input data switches; that is, the TE technique has the self transition profile of an uncoded bus. On the other hand, since the TE technique eliminates opposite transitions on adjacent wires completely, the technique can reduce both the maximum number of coupling-transitions in any given cycle and the average number of coupling-transitions per bus cycle.

The M-Repeater near the middle of the TE bus topology has the data latching property, which allows data signal to traverse the second half of the bus while the first half of the bus transmits shield signal. This ensures full throughput of useful data transmission. This can be illustrated by the timing plan of the TE-coded bus in Fig. 3. From the timing plan, we can observe that the T_1 and T_2 phase of clock are determined by data signal arrival instead of shield signal arrival; that is, $T_1 \geq (D_{mrep}^{TE} + D_{hb2} + T_{setup})$ and $T_2 \geq (D_{enc}^{TE} + D_{hb1})$, respectively. Therefore, the bus delay of the TE technique is given by

$$
\begin{aligned}
T_d^{TE} &= T_1 + T_2 - D_{dff} - T_{setup} \\
&= D_{hb1} + D_{hb2} + D_{codec}^{TE} - D_{dff}
\end{aligned}
\tag{1}
$$

where D_{hb1} and D_{hb2} are the delays of the first and second half of the bus respectively, D_{codec}^{TE} is the sum of D_{enc}^{TE} and D_{rep}^{TE} which are the delays of the TE circuits (Encoder and M-Repeater, respectively), D_{dff} is the clock-to-out delay of the master flip-flop, T_{setup} is the setup time of the slave flip-flop.

2.2. Combination of TE and Bus-Invert coding

In order to further improve the power efficiency, we combine a spatial bus-invert (BI) coding with the TE technique. This is because the BI coding is a simple but effective low-power

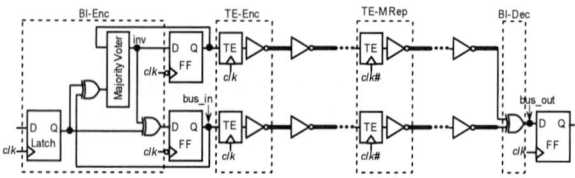

Fig. 4. Bus topology of the BI(g)TE joint coding.

coding scheme through self transition activity reduction. Because the TE technique has the self transition profile of an uncoded bus, the BI encoder can be followed by TE encoding without destroying the effectiveness of self transition activity reduction of BI coding. The bus topology of the joint coding based on the combination of TE and BI coding is shown in Fig. 4. The original input data is first encoded through the BI encoder which computes the Hamming distance Hd between the next data value and the present bus value (including the invert bit). The data value is inverted for transmission and the invert bit is set to high level if $Hd > n/2$ for n-bit bus; otherwise, the data value is unchanged and the invert bit is set to low level. Then, the newly generated data and invert bit are encoded through the TE technique to eliminate opposite transitions on adjacent wires completely. In sum, the joint coding exploits the benefits of both coding techniques. It should be pointed out that the effectiveness of BI coding decreases with the increase in the bus width. Therefore, for wide buses, the whole bus is partitioned into several sub-buses each with its own invert bit to improve switching activity reduction [10]. We refer to the joint coding as the BI(g)TE method, where g is the number of sub-buses. By modifying (1), we can get the bus delay of BI(g)TE as follows

$$
T_d^{BITE} = D_{hb1} + D_{hb2} + D_{codec}^{TE} + D_{dec}^{BI} - D_{dff}
\tag{2}
$$

where D_{dec}^{BI} is the delay of BI decoder (XOR gate).

3. Delay and energy models of bus with repeaters

3.1. Delay and transition time model of uncoded bus

As shown in Fig. 5, a global bus of length L, resistance r per unit length, self capacitance c_s per unit length, and coupling capacitance c_c per unit length is evenly divided into k segments of length l by identical repeaters. For a repeater of size s, the total output resistance $R_{tr} = r_s / s$, the total output capacitance $C_p = s c_p$ and the total input capacitance is $C_g = s c_g$, where, r_s, c_p and c_g are the output resistance, output capacitance and input capacitance of a minimum-sized repeater. The 50% delay t_{ds} and transition time t_{rs} of a wire segment in the bus can be obtained by modifying the expressions in [11, 13] as follows

$$
t_{ds}(p, t_{rin}, C_L) = \ln 2 \left[\frac{r_s'}{s} \left(s c_p + (c_s + p c_c)l + C_L \right) + r l C_L \right] \\
+ (0.1 + 0.4 \ln 2) r (c_s + p c_c) l^2 + \gamma t_{rin}
\tag{3}
$$

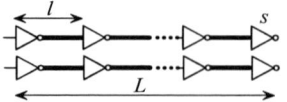

Fig. 5. Repeater insertion in a global bus.

$$t_{rs}(p, C_L) = \frac{t_{90\%} - t_{10\%}}{0.8}$$
$$= 2.5 \ln 3 \left[\frac{r_s''}{s} \left(sc_p + (c_s + pc_c)l + C_L \right) \right. \tag{4}$$
$$\left. + rlC_L + 0.4r(c_s + pc_c)l^2 \right].$$

where, p is referred to as the coupling factor, which takes values $p = 0, 1, 2, 3$ and 4 depending on the transitions occurring on adjacent wires [1]; t_{rin} is the input transition time of the driving repeater of the wire segment; C_L is the load at the end of the segment, and $C_L = C_g$ if the load is the driving repeater of the next segment; r_s' and r_s'' are the different version of the output resistances r_s used to calculate the delay t_{ds} and transition time t_{rs}, respectively; γ is the coefficient representing the contribution of t_{rin} to the repeater delay, which is determined as [11]

$$\gamma = \frac{1}{2} \left(1 - \frac{1 - v_{tn}}{1 + \alpha_n} - \frac{1 - v_{tp}}{1 + \alpha_p} \right) \tag{5}$$

where $v_{tn} = V_{tn} / V_{dd}$, $v_{tp} = V_{tp} / V_{dd}$, and α is the velocity saturation index. The total delay of a wire in the bus is given by

$$T_{wire}(p, L, t_{rdi}, C_L) = (\frac{L}{l} - 1)t_{ds}(p, t_{rri}, C_g)$$
$$+ t_{ds}(p, t_{rdi}, C_g) + t_{ds}(p, t_{rri}, C_L) \tag{6}$$

where t_{rdi} is the input transition time of the first driver of the wire, t_{rri} is the input transition time of the repeaters in the middle of the wire, thereby, $t_{rri} = t_{rs}(p, C_g)$.

When coding is not employed, p takes the worst-case value of 4. The worst-case delay and transition time of an uncoded bus with repeaters are, respectively, given by

$$T_d^{UNC} = T_{wire}\left(4, L, t_{rdi}, C_{in}^{dff}\right) \tag{7}$$
$$T_r^{UNC} = t_{rs}\left(4, \max(C_g, C_{in}^{dff})\right) \tag{8}$$

where C_{in}^{dff} is the input capacitance of the slave flip-flop.

3.2. Energy dissipation model of uncoded bus

The energy dissipation of an uncoded bus with repeaters is composed of dynamic, short-circuit, and leakage energy dissipation in the repeaters and dynamic energy dissipation in the interconnects. Assume that the number of self-transitions and coupling-transitions on the bus are N_s and N_c in a given clock cycle, respectively. Therefore, the dynamic energy E_{ds}, leakage energy E_{ls} and short-circuit energy E_{ss} dissipated in a bus segment during that clock cycle are obtained by modifying the expressions in [12] as follows

$$E_{ds}(N_s, N_c, C_L) = \frac{1}{2} V_{dd}^2 \left(N_s (lc_s + sc_p + C_L) + N_c lc_c \right) \tag{9}$$

$$E_{ls}(B) = B \frac{1}{2} s \left(I_{offn} W_{nmin} + I_{offp} W_{pmin} \right) \frac{V_{dd}}{f_{clk}} \tag{10}$$

$$E_{ss}(N_s, t_{rin}) = \frac{N_s}{2} \left(t_{1 - |v_{tp}|} - t_{v_{tn}} \right) V_{dd} W_{nmin} s I_{sc}$$
$$= \frac{N_s}{5 \ln 3} \ln \left(\frac{1 - v_{tn}}{|v_{tp}|} \right) t_{rin} V_{dd} W_{nmin} s I_{sc} \tag{11}$$

where V_{dd} is the power supply, f_{clk} is the clock frequency, I_{offn} (I_{offp}) is the leakage current for NMOS (PMOS) per unit

width, W_{nmin} (W_{pmin}) is the width of the NMOS (PMOS) transistor in minimized sized repeater, I_{sc} is the per unit width short-circuit current, and B is the bit width of bus. Therefore, the total energy dissipation in the whole bus is given by

$$E_{bus}(N_s, N_c, B, L, t_{rdi}, C_L)$$
$$= (\frac{L}{l} - 1)\left(E_{ds}(N_s, N_c, C_g) + E_{ss}(N_s, t_{rri}) \right) + \frac{L}{l} E_{ls}(B) \tag{12}$$
$$+ E_{ds}(N_s, N_c, C_L) + E_{ss}(N_s, t_{rdi}).$$

About the number of both self-transitions and coupling-transitions, we have the following computing principle. Both charging and discharging transitions on the self capacitance are counted as self-transitions for energy dissipation. For the transitions on the coupling capacitance, there are three possible cases: charging, discharging and toggling. Toggling is defined as the case where adjacent lines switch simultaneously in opposite direction. A toggling case has four times more energy dissipation than a charging or discharging case [2]. Thus, if the number of coupling-transitions is 1 for a charging or discharging case, a toggling event is equivalent to 4 coupling-transitions. Assume that $P_s(X_i)$ is the probability of self capacitance i (of line i) with X_i self-transitions per bus cycle and $P_c(X_i)$ is the probability of coupling capacitance i (between lines i and $i+1$) with X_i coupling-transitions per bus cycle. Then, the expected number of self-transition $\tilde{N}_s$ and coupling-transitions $\tilde{N}_c$ per bus cycle on B-bit bus with both outer bus wires having a grounded (shield) wire as a neighbor are, respectively, given by

$$\tilde{N}_s = \sum_{i=1}^{B} P_s(X_i = 1) \tag{13}$$

$$\tilde{N}_c = \sum_{i=1}^{B-1} \sum_{m=1}^{4} m P_c(X_i = m) + P_s(X_1 = 1) + P_s(X_n = 1). \tag{14}$$

Hereafter, we assume that the original uncoded input data are spatially and temporally independent and uniformly distributed random n-bit pattern stream. Then, for an uncoded bus (i.e. $B^{UNC} = n$), we have $P_s(X_i = 1) = 0.5$, $P_c(X_i = 1) = 0.5$, $P_c(X_i = 2) = P_c(X_i = 3) = 0$, and $P_c(X_i = 4) = 0.125$ for any i [10]. Consequently, the expected number of self-transitions $\tilde{N}_s^{UNC}$ and coupling-transitions $\tilde{N}_c^{UNC}$ per bus cycle on an uncoded bus are

$$\tilde{N}_s^{UNC} = n/2 \qquad \tilde{N}_c^{UNC} = n. \tag{15}$$

Also it's easy to prove that the maximum number of self-transitions $\hat{N}_s^{UNC}$ and coupling-transitions $\hat{N}_c^{UNC}$ in any given cycle are

$$\hat{N}_s^{UNC} = n \qquad \hat{N}_c^{UNC} = 4n - 2. \tag{16}$$

Therefore, the peak energy and average energy of the uncoded bus with repeaters are, respectively, given by

$$E_{peak}^{UNC} = E_{bus}\left(\hat{N}_s^{UNC}, \hat{N}_c^{UNC}, B^{UNC}, L, t_{rdi}, C_{in}^{dff} \right) \tag{17}$$

$$E_{avg}^{UNC} = E_{bus}\left(\tilde{N}_s^{UNC}, \tilde{N}_c^{UNC}, B^{UNC}, L, t_{rdi}, C_{in}^{dff} \right). \tag{18}$$

3.3. Effects of bus encoding techniques on delay and transition time

As described in the section 2, the TE and BI(g)TE bus

encoding techniques can eliminates opposite transitions on adjacent wires completely, that is, reducing the maximum coupling factor from $p = 4$ to $p = 2$, thereby, achieving the worst-case delay reduction of the bus. However, it is noted that the delay overhead of encoding circuits has a negative effect on the bus delay reduction. From (1), (2) and (6), we get the worst-case delays of the TE and BI(g)TE coded buses with repeaters, respectively, given by

$$T_d^{TE} = (\frac{L}{l} - 2)t_{ds}\left(2, t_{rri}, C_g\right) + t_{ds}\left(2, t_{r_{out}}^{enc}, C_g\right) + t_{ds}\left(2, t_{r_{out}}^{mrep}, C_g\right) \\ + t_{ds}(2, t_{rri}, C_{in}^{mrep}) + t_{ds}\left(2, t_{rri}, C_{in}^{dff}\right) + D_{codec}^{TE} - D_{dff} \quad (19)$$

$$T_d^{BITE} = (\frac{L}{l} - 2)t_{ds}\left(2, t_{rri}, C_g\right) + t_{ds}\left(2, t_{r_{out}}^{enc}, C_g\right) + t_{ds}\left(2, t_{r_{out}}^{mrep}, C_g\right) \\ + t_{ds}(2, t_{rri}, C_{in}^{mrep}) + t_{ds}\left(2, t_{rri}, C_{in}^{dec}\right) + D_{codec}^{TE} + D_{dec}^{BI} - D_{dff} \quad (20)$$

where C_{in}^{mrep} is the input capacitance of M-Repeater, C_{in}^{dec} is the input capacitance of BI decoder (*XOR* gate), $t_{r_{out}}^{enc}$ and $t_{r_{out}}^{mrep}$ are the output transition times of TE circuits (Encoder and M-Repeater, respectively), L_{hb1} and L_{hb2} are the length of the first and second half of the bus, respectively. The maximum transition times of the both buses are, respectively, given by

$$T_r^{TE} = t_{rs}\left(2, \max(C_g, C_{in}^{mrep}, C_{in}^{dff})\right) \quad (21)$$

$$T_r^{BITE} = t_{rs}\left(2, \max(C_g, C_{in}^{mrep}, C_{in}^{dec})\right). \quad (22)$$

3.4. Effect of bus encoding techniques on energy

For the self-transitions on each line, since the TE and BI(g)TE bus techniques have the switching characteristics of an uncoded bus and BI-coded bus, respectively, some results [10] can be applied here.

For TE coding: $P_s(X_i = 1) = 0.5$.

For BI(g)TE coding: $P_s(X_i = 1) = \dfrac{1}{2} - 2^{-(n+1)}C\left(n, \dfrac{n}{2}\right)$.

For the coupling-transitions between adjacent wires, the TE and BI(g)TE techniques transform a toggling event into a discharging followed by a charging event (or vice-versa), reducing the number of coupling-transitions from 4 to 2. Thus, in various switching scenarios on a coupling capacitance, there are only three possible values: 0, 1 and 2 for the number of coupling-transitions. Similar to the Markov-based approach employed in [10], by modeling the TE and BI(1)TE coding processes as Markov chains, we can obtain $P_c(X_i=1)$ and $P_c(X_i=2)$ for TE and BI(1)TE techniques, respectively, as follows

For TE coding: $P_c(X_i = 1) = 0.5$ and $P_c(X_i = 2) = 0.125$.

For BI(1)TE coding:

$$P_c(X_i = 1) = 2\sum_{h=0}^{n/2-1} C(n-1, h)2^{-n} = 0.5$$

$$P_c(X_i = 2) = 2^{-3} - 2^{-n-1}C\left(n-1, \dfrac{n}{2} - 1\right).$$

Also we have $B^{TE} = n$, and $B^{BI(1)TE} = n+1$. Consequently, the expected number of self-transitions and coupling- transition per bus cycle for the both coding techniques are, respectively, given by

$$\tilde{N}_s^{TE} = \frac{n}{2}; \quad \tilde{N}_c^{TE} = \frac{3(n-1)}{4} \quad (23)$$

$$\tilde{N}_s^{BI(1)TE} = (n+1)\left(\frac{1}{2} - 2^{-(n+1)}C\left(n, \frac{n}{2}\right)\right) \quad (24)$$

$$\tilde{N}_c^{BI(1)TE}(n) = 1 + \frac{3}{4}n - (n+2)2^{-n}C\left(n-1, \frac{n}{2} - 1\right) \quad (25)$$

Now, we consider the coupling-transitions for BI(g)TE buses with $g \geq 2$. Suppose the bus lines are partitioned into g equal-sized groups, each of which has $m=n/g$ lines excluding invert lines. Then we have $B^{BI(g)TE} = n + g$. Using the similar approach in [10], we can obtain the expected number $\tilde{N}_c^p$ of coupling-transitions between the invert line of group j and the first bus line of group $j+1$ as follows

$$\tilde{N}_c^p = P_s(X_i = 1)\left(2 - P_s(X_i = 1)\right) \\ = \left(\frac{1}{2} - 2^{-(n+1)}C\left(n, \frac{n}{2}\right)\right)\left(\frac{3}{2} + 2^{-(n+1)}C\left(n, \frac{n}{2}\right)\right) \quad (26)$$

Therefore, the expected number of coupling-transitions per bus cycle on an BI(g)TE bus is

$$\tilde{N}_c^{BI(g)TE} = g\tilde{N}_c^{BI(1)TE}(m) + (g-1)\tilde{N}_c^p \quad (27)$$

It's easy to prove that the maximum number of self-transitions and coupling-transitions for the both coding techniques in any given cycle are

$$\hat{N}_s^{TE} = n \qquad \hat{N}_c^{TE} = 2n \quad (28)$$

$$\hat{N}_s^{BI(g)TE} = n/2 \qquad \hat{N}_c^{BI(g)TE} = n \quad (29)$$

From (12), (28) and (29), we can get the peak energy dissipation of the TE and BI(g)TE buses with repeaters, respectively, given by

$$E_{peak}^{TE} = (\frac{L}{l} - 2)\left(E_{ds}(\hat{N}_s^{TE}, \hat{N}_c^{TE}, C_g) + E_{ss}(\hat{N}_s^{TE}, t_{rri})\right) \\ + E_{ds}(\hat{N}_s^{TE}, \hat{N}_c^{TE}, C_{in}^{mrep}) + E_{ds}(\hat{N}_s^{TE}, \hat{N}_c^{TE}, C_{in}^{dff}) \\ + E_{ss}(\hat{N}_s^{TE}, t_{r_{out}}^{enc}) + E_{ss}(\hat{N}_s^{TE}, t_{r_{out}}^{mrep}) \\ + \frac{L}{l}E_{ls}(B^{TE}) + \hat{E}_{codec}^{TE} \quad (30)$$

$$E_{peak}^{BITE} = (\frac{L}{l} - 2)\left(E_{ds}(\hat{N}_s^{BITE}, \hat{N}_c^{BITE}, C_g) + E_{ss}(\hat{N}_s^{BITE}, t_{rri})\right) \\ + E_{ds}(\hat{N}_s^{BITE}, \hat{N}_c^{BITE}, C_{in}^{mrep}) + E_{ds}(\hat{N}_s^{BITE}, \hat{N}_c^{BITE}, C_{in}^{dec}) \\ + E_{ss}(\hat{N}_s^{BITE}, t_{r_{out}}^{enc}) + E_{ss}(\hat{N}_s^{BITE}, t_{r_{out}}^{mrep}) \\ + \frac{L}{l}E_{ls}(B^{BITE}) + \hat{E}_{codec}^{BI} + \hat{E}_{codec}^{TE} \quad (31)$$

where $\hat{E}_{codec}^{TE}$ is the peak energy dissipation of the TE codec circuits; $\hat{E}_{codec}^{TE}$ is the peak energy dissipation of the BI codec circuits. The average energy dissipation per bus cycle of the both buses can be obtained in a similar way, but they are not shown here due to limited space.

4. Repeater insertion optimization method

Herein, we assume that a semiconductor technology (r_s, c_g, c_p, V_{dd}, V_{tn}, V_{tp}, α_n, α_p, I_{off_n}, I_{off_p}, I_{sc}), a global interconnect design specification (r, c_s, c_c, f_{clk}, n, p, N_s, N_c, B), codec circuit overheads and other related design parameters are given. Based on this assumption, we have the fact that the worst-case

380

978-1-4244-3039-0/08 $25.00 © 2008 IEEE

delay, maximum transition time and energy dissipation of bus are function of repeater size s and segment length l (or the number of segments $k=L/l$). There exits the delay optimal point in the design space of l and s. However, this delay-minimal repeater design methodology is not necessarily an appropriate strategy in practical circuits. This is because the delay is not sensitive to the size of the repeaters near the optimal point; therefore, significant power and area are wasted to achieve only a small improvement in speed when approaching the optimal point for minimum delay. So, in many instances, such as non-critical global buses, a target delay is desired rather than minimal delay to reduce energy dissipation. Here, we present a repeater insertion design methodology for achieving the minimum bus energy dissipation at each target delay point while satisfying a maximum slew-rate constraint in the design process of applying bus encoding techniques for reduced delay and energy.

Here, we investigate the repeaters insertion of an uncoded bus to illustrate the optimization method. Assume that the desired delay is $T_{d\,target}$ and the maximum transition time (slew-rate) constraint is $T_{r\,max}$, then we set

$$T_d^{UNC} \leq T_{d_{target}} \qquad (32)$$

$$T_r^{UNC} \leq T_{r_{max}} \qquad (33)$$

If $T_{d\,target}$ is greater than the optimal delay, then there exist many combinations of l and s that satisfy (32). For each s, an optimal l exists to achieve the minimum peak energy (17) or average energy (18). If l is too large, the signal transition time will be large. Also there are many combinations of l and s that satisfy (33). The minimum energy dissipation satisfying the slew-rate constraint can be achieved with minimum-sized repeaters. For minimum-sized repeaters, the corresponding number of segments and bus delay, however, are impractically large. Hence, in order to produce an effective repeater insertion, the delay and slew-rate constraint should be considered simultaneously. We apply the genetic algorithm (GA) and sequential quadratic programming (SQP) method in the Matlab toolbox to solve the non-linear constrained optimization problem with objective function being equation (17) or (18), and constrained functions being equation (32) and (33). First, we use the GA to find a good starting point (l, s) of global solution. Next, since k ($=L/l$) obtained is usually not an integer, the nearest two integers are used to determine the minimum energy dissipation whereas the SQP based solver is applied to further refine the value of s.

5. Experimental results

The methodology described in Section 4 is employed to obtain energy vs. delay trade-offs under slew-rate constraint for uncoded and encoded buses with repeaters. In all simulations, we use an industrial 0.13μm CMOS technology. We consider a 16-bit bus in the metal-6 layer while both outer bus wires having a shield wire as a neighbor. If all uncoded and encoded buses are assumed to be routed at minimum pitch, there is an area penalty for BI(g)TE (i.e., a 12.5% increase in routing area for g=2, and 25% increase for g=4). For a fair comparison, the wire width

and spacing of the uncoded and TE-coded buses are re-optimized for minimum energy within the increased routing area. This re-optimization always resulted in increased spacing since the coupling capacitance decreases rapidly, reducing both energy and delay of the bus. The capacitances per unit length for the different bus pitches were extracted using Synopsys's Raphael and listed in Table 1. Device parameters were extracted using SPICE simulation similar to [11, 12]. The relevant technology parameters are shown in Table 2. The minimized-size repeater is defined to have $W_n = 2L_{drawn} = 0.26\mu m$ and $W_p = 2.4W_n$. The codec circuits are sized so as to operate at the "knee" of their respective energy-delay curves. The knee points typically resulted in energies that were 10%-20% higher than the minimum energy of codec circuits and yielded nearly constant delay over a range of load capacitances. Hence for simplicity we assume that the codec circuits have fixed configurations, delay and energy overheads for all delay targets. These values measured by SPICE simulation are shown in Table 3. In this paper, buses of length 9mm are optimized, and the maximum transition time constraint is set to 240ps (~3X the transition time at the output of an inverter driving a FO4 load).

The peak and average energy versus worst-case delay trade-off curves for the uncoded and encoded buses with repeaters are shown in Fig. 6 and Fig. 7. The left most point of each curves represents the delay-optimized solution and hence consumes the

Table 1. Wire RC parasitic for different bus pitches

Bus pitch	r (Ω/mm)	c_s (fF/mm)	c_c (fF/mm)
1X min. pitch (W/S=0.2μm/0.2μm)	303	76.29	91.55
1.125X min. pitch (W/S=0.2μm/0.25μm)	303	84.06	72.65
1.25X min. pitch (W/S=0.2μm/0.3μm)	303	91.59	59.30

Table 2. Device parameters for an industrial 0.13μm CMOS technology

Parameters	Value	Parameters	Value
r_s' (kΩ)	7.44	r_s'' (kΩ)	3.54
c_g (fF)	1.35	c_p (fF)	2.59
V_{tn} (V)	0.42	V_{tp} (V)	-0.36
α_n	1.23	α_p	1.45
I_{off_n} (mA/m)	4.15	I_{off_p} (mA/m)	6.97
I_{sc} (μA/μm)	50	V_{dd} (V)	1.2

Table 3. Overhead of codec circuits and other related design parameters

Overheads	Value	Parameters	Value
D_{enc}^{TE} (ps)	53	D_{dff} (ps)	120
D_{rep}^{TE} (ps)	67	t_{rdi} (ps)	100
D_{dec}^{BI} (ps)	15	t_{rout}^{enc} (ps)	240
$\hat{E}_{codec}^{TE}$ (pJ)	1.93~1.66	t_{rout}^{enc} (ps)	240
$\tilde{E}_{codec}^{TE}$ (pJ)	1.43~1.55	C_{in}^{mrep} (fF)	16
$\hat{E}_{codec}^{BI}$ (pJ)	0.79	C_{in}^{dec} (fF)	4
$\tilde{E}_{codec}^{BI}$ (pJ)	0.65	C_{in}^{dff} (fF)	4

highest peak(average) energy; while the right most point of each curves represents the peak(average) energy-optimized solution which is due to the slew rate constraints. For the routing area constraint of 1.125× minimum pitch (Fig. 6a and Fig. 7a), we observe that TE and BI(2)TE achieve peak energy gains of 59.2% and 74.4%, respectively, and average energy gains of 55.1% and 52.4% over the uncoded bus, respectively, at the minimum achievable delay points of the uncoded bus. Also, TE and BI(2)TE allow more aggressive delay targets to be met (23.6% and 17.2% faster, respectively), while still dissipating less peak energy (38% and 62.1%, respectively) and average energy (29.7% and 28.9%, respectively). In many instances, such as no-critical global buses, a target delay is larger than the delay number corresponding to the minimum energy point, and hence reducing energy is the primary goal. In such cases, TE and BI(2)TE can result in 32% and 58.9% reductions in peak energy, respectively, and 10.8% and 9.1% reductions in average energy, respectively. In a word, TE and BI(2)TE can provide peak(average) energy savings over the uncoded bus at all target delays. Similar analyses were carried out for the routing area constraint of 1.25× minimum pitch (Fig. 6b and Fig. 7b). The relevant data points for all cases have been tabulated in Table 4.

From Table 4, it can be seen that the proposed BI(g)TE technique yields much better results in peak energy reduction compared with the TE technique, but trivial (or even negative) improvements in average energy reduction. This is because the energy gains that result from a weak reduction in average number of self- and coupling-transitions are counteracted, due to the fact that the TE-coded bus energy can also be reduced further under the increased area penalty incurred by BI(g)TE. With respect to the efficiency of BI(g)TE with various g values to reduce peak/average energy, we see that BI(2)TE results in larger gains in peak and average energy than BI(4)TE at all target delay values (though BI(2)TE should ideally result in the same peak energy gains with BI(4)TE and less average energy gains) since, under the same area constraint, the bus pitch for BI(2)TE is more relaxed than BI(4)TE due to fewer control signals.

6. Conclusions

A new spatial and temporal encoding approach has been proposed for global on-chip buses in DSM SoC design to allow higher performance than an uncoded repeater tuning strategy while achieving peak and average energy reduction. Compared with the temporal encoding technique presented in our previous work, the proposed spatial and temporal encoding approach further improves peak energy reduction. In the design process of applying bus encoding techniques for reduced bus delay and energy, we have presented a repeater insertion design methodology for achieving the minimum bus energy with delay and slew-rate constraints, which can be employed to obtain energy vs. delay trade-off curves under slew-rate constraint for buses with repeaters, thereby, providing convenience for comparisons of the efficiency of various encoding techniques.

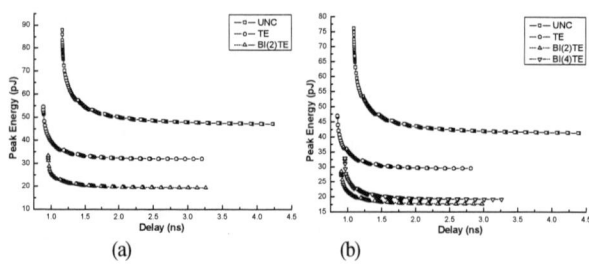

Fig. 6. Peak energy vs. delay curves for a routing area constraint of (a) 1.125X and (b) 1.25X minimum pitch.

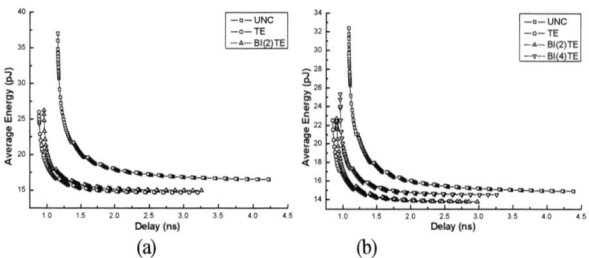

Fig. 7. Average energy vs. delay curves for a routing area constraint of (a) 1.125X and (b) 1.25X minimum pitch.

Table 4. Gains achieved by TE, BI(2)TE and BI(4)TE over uncoded bus with repeaters for 9mm bus

Metrics	Coding	E_{peak}		E_{avg}		Delay	
		X min. Pitch		X min. Pitch		X min. Pitch	
		1.125	1.25	1.125	1.25	1.125	1.25
Max. gains (%)	TE	59.2	56.3	55.1	51.5	23.6	21.7
	BI(2)TE	74.4	73.3	52.4	50.5	17.2	16.4
	BI(4)TE	-	69.8	-	45.8	-	11.6
Gains at min. energy point (%)	TE	32	28.4	10.8	7.3	24.2	36
	BI(2)TE	58.9	57.4	9.1	7.5	22.9	32.2
	BI(4)TE	-	53.7	-	2.5	-	25.7

7. References

[1] P. P. Sotiriadis, "Interconnect modeling and optimization in deep submicron technologies," Ph.D. dissertation, Massachusetts Inst. Technol., Cambridge, 2002.

[2] Y. Zhang, J. Lach, et al, "Odd/Even bus invert with two-phase transfer for buses with coupling," in Proc. *ISLPED*, pp.. 80-83, 2002.

[3] M. Stan, and W. Burleson, "Bus-invert coding for low-power I/O," *IEEE Trans. VLSI System*, Vol. 3. pp. 49-58, Mar 1995.

[4] P. P. Sotiriadis and A. Chandrakasan, "Low power bus coding techniques considering inter-wire capacitances," in Proc. *CICC*, pp. 507–510, 2000.

[5] S. R. Sridhara, A. Ahmed, and N. R. Shanbhag, "Area and energy-efficient crosstalk avoidance codes for on-chip buses," in Proc. *ICCD*, pp. 12-17, 2004.

[6] M. Mutyam, M. Eze, et al, "Delay and energy efficient data transmission for on-chip buses," Proc. ISVLSI, pp. 6-11, 2006.

[7] M. Anders, et al, "A transition-encoded dynamic bus technique for high performance interconnection," *IEEE J. Solid-State Circuit*, Vol. 38, pp. 709-714, 2003.

[8] H. Kaul, D. Sylvester, et al, "Design and analysis of spatial encoding circuits for peak power reduction in on-chip buses," *IEEE Trans. VLSI Systems*, Vol. 13, pp. 1225-1238, 2005.

[9] Q. L. Zhang, J. X. Wang, et al, "An energy-efficient temporal encoding circuit technique for on-chip high performance buses," in Proc. *GLSVLSI*, pp. 273-279, 2006.

[10] R. B. Lin, "Coupling reduction analysis of bus-invert coding," in Proc. *ISCAS*, pp. 5862-5866, 2005.

[11] G. Q. Chen, and E. G. Friedman, "Low-power repeaters driving RC and RLC interconnects with delay and bandwidth constraints," *IEEE Trans. VLSI System*, Vol. 14, pp. 161-172, 2006.

[12] K. Banerjee, and A. Mehrotra, "A power-optimal repeater insertion methodology for global interconnects in nanometer designs," *IEEE Trans. Electron Devices*, Vol. 49, pp. 2001-2007, 2002.

[13] T. Sakurai, "Closed-form expressions for interconnection delay, coupling,and crosstalk in VLSI's," *IEEE Trans. Electron Devices*, Vol. 40, pp. 118-124, 1993.

21st International Conference on VLSI Design

A Robust Level-Shifter Design for Adaptive Voltage Scaling

Ankur Gupta, Rajat Chauhan, Vinod Menezes, Vikas Narang, Roopashree H.M.
Texas Instruments (India) Pvt. Ltd
{ankur.gupta, rajat.chauhan, v-menezes, vikasnarang, roopa} @ti.com

Abstract

Voltage scaling is one of the knobs that is used today to control both static and the active power for SoCs. The SoC core supply voltage is scaled adaptively based on the performance needs. But it is also required to maintain the external electrical chip interface protocol, which may run at a different voltage level. The chip interfaces need to operate reliably under adaptively scaling core voltage and fixed IO supply voltage. Within the IO circuits, Voltage Level shifters are used to communicate between two voltage domains. This paper examines the performance of a conventional voltage level shifter and describes a novel high performance level shifter that is more robust under adapting voltage scaling.

I. Introduction

The need for the high performance and low power budgets is a requirement for both battery operated mobile devices and the large data and compute centers. While the traditional power management techniques like clock-gating [3] and frequency scaling [1] will continue, voltage scaling is one of the knobs that is used today to control both static and the active power for SoCs. It is possible to adaptively scale the core voltage [2][4][5] based on the performance needs as required by a scheduling algorithm (fig.-1) [7]. The process monitor senses the device strength such as delays and supplies information to the performance scheduler. The performance scheduler tries to match the sensed delay with the performance requirement. If the sensed delay is less than the maximum allowed and there is a margin to increase, the performance scheduler reduces the supply voltage in order to save power.

Both the mobile device and the compute center applications use adaptive core voltage scaling to meet

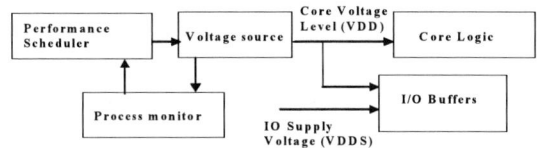

Fig. 1. Block Diagram for Adaptive Voltage Scaling on SoCs

the external electrical chip interface protocol, which may run at a different voltage level. The chip interface need to operate reliably under adaptively scaling core voltage and fixed IO supply voltage. Circuit techniques are required to level shift the signals communicating between scaled and un-scaled voltage domains. Within the IO circuits Voltage Level shifters are used to communicate between the two voltage domains [8]. The level shifters need to be fast, and also need to be robust in their operation when the difference between the two voltages is high. Under such a scenario, suboptimal designs or conventional circuit topologies may result in performance as well as functional failure.

This poses a significant challenge when a device using a scaled core supply (VDD) communicates to the off-chip devices using the IO voltage (VDDS) for signaling.

This paper is organized into five sections. Section II describes the simulation setup and the metric developed to quantify voltage-performance correlation of a circuit. Section III examines the performance of a conventional level-shifter under AVS (Adaptive Voltage Scaling), using the setup described in Section II. Section IV describes the novel robust level-shifter for best operation under AVS. Conclusions are provided in Section V.

II. Performance Variation with Voltage Scaling

To reduce the power consumption, during the low performance need, supply voltage level is scaled down.

383

978-1-4244-3039-0/08 $25.00 © 2008 IEEE

Supply level is set based on the performance need and therefore each supply level, set by performance scheduler (fig-1), is associated with a specific performance goal [7].

A consistent metric is required to estimate the performance of a circuit with voltage scaling. This section describes the simulation setup and the metric developed to correlate the performance of a circuit with scaling core voltage.

A. Simulation Setup

The setup is designed to capture the performance Vs VDD behavior of the circuit, keeping VDDS fixed. The results are then compared with the performance goal criteria set for a VDD level.

Consider the block shown in fig.-2. The block can be a small CMOS circuit, like a level-shifter, supplied by the voltage VDD. The simulations and the analysis are performed through following steps:

- The DC voltage level of VDD is scaled down in steps. This increases the difference between VDD and VDDS, hence impacting the performance. Table-1 defines the various VDD voltage levels.
- For each VDD voltage level, the cell is run through multiple process points. The transistor process points are picked randomly, using the statistical device models. [9]
- For each process point, the cell delay from input IN to output OUT is measured.

B. Performance measure

The cell delay is used as a measure for performance. The lowest delay (maximum performance) is achieved at maximum VDD level. Assuming this as the 100% performance level, performance at each VDD level is measured relative to this. Table 1 lists the VDD levels and corresponding minimum required performance across multiple process points. VDD1 is highest VDD level and VDD9 is the lowest voltage level. 'P0' is a parameter defining maximum acceptable performance loss (maximum delay permitted) normalized to 100% performance level.

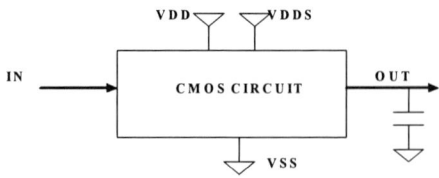

Fig. 2. Simulation setup

Table 1. Vdd levels Vs performance requirement

VDD Level	VDD Voltage Normalized to VDD1	Minimum Required Performance	P0
VDD1	1	100%	1.0
VDD2	0.84	50 %	2.0
VDD3	0.82	45%	2.2
VDD4	0.80	40%	2.5
VDD5	0.79	35%	2.9
VDD6	0.77	30%	3.3
VDD7	0.76	25%	4.0
VDD8	0.74	20%	5.0
VDD9	0.72	15%	6.7

Table-1 is obtained empirically from the silicon characterized data for a given technology node and is used as a specification for this analysis and design.

Performance loss of the cell is measured as a ratio of cell delay Td at given VDD level to cell delay Td1 at VDD1.

$$P = \frac{Td}{Td1} \qquad ...(1)$$

A cell performance loss 'P' higher than P0 at a given VDD level is considered as failure.

III. Conventional Cross-Coupled Level-Shifter

This section describes the operation and performance of one of the conventional voltage level-shifter with scaling VDD levels.

Fig. 3 shows a conventional cross-coupled level-shifter (Topology 1) circuit. Level-shifter uses thin oxide, low voltage MOSFETs (MN1, MN3, MN5 and MP5) and the thick oxide, high voltage MOSFETs (MN2, MN4, MP2 and MP4). The thin oxide transistors operate at a lower core voltage supply VDD and the thick oxide transistors operate at a higher I/O voltage supply VDDS. The VSS is ground supply for the circuit. The node S2 is connected to VDDS level. MN2 and MN4 protect the thin oxide transistors MN1, MN3 from the high VDDS level at their drains. For the further details on this topology [8] can be referred.

The topology-1 gives good results when the voltage difference between VDD and VDDS is small, but it starts malfunctioning and even fails functionally when the difference becomes large.

The reason for the failure of the circuit is the cross-coupled gates of PMOSs MP4 and MP2 using regenerative feedback for switching. In the topology, switching is initiated by the low voltage input signal

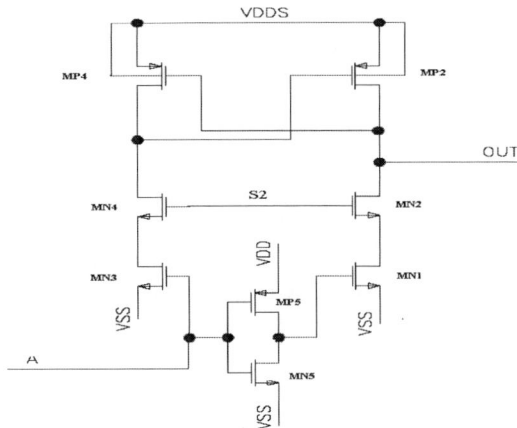

Fig. 3. Conventional cross-coupled voltage level-shifter (Topology 1).

'A', going to the gates of thin oxide NMOS MN1, MN3 and finally controlled and concluded by regenerative feedback. Switching initialization by the input 'A' has to ensure that some threshold voltage is reached at the node OUT before switching is handed over to regenerative feedback. Pull-down path transistors MN1, MN2 have to fight with ON PMOS MP2 to pull-down the node OUT to threshold voltage level. If the initialization process (pulling down) is weak, which is the case at lower VDD levels, then the topology 1 switch late or may not switch at all, and the output OUT become either distorted or is held at an intermediate voltage.

The equation (4) correlates the minimum drive strength of MN1 to MP2, required for the level-shifter to function. To get the equation, it is assumed that when the gate of MN1 switches from VSS to VDD, the node OUT must go low to at least 'VDDS-V_{tp}' (2) in order to turn-on MP4.

$$VOUT < VDDS - Vt_{MP4} \qquad ...(2)$$

For the OUT node at 'VDDS-V_{tp}', MP2 operates in the linear region and can be assumed a resistor, while MN1 operates in the saturation region and can be assumed a current sink. The voltage at node OUT can be written as

$$VOUT = VDDS - Ids_{MN1} \times Ron_{MP2} \quad ...(3)$$

Using (2), (3) and transistor current equations [10], we get the following correlation.

$$\frac{\beta_{MN1}}{\beta_{MP2}} > \frac{V_{Tp}\left(V_{DDS} - V_{Tp}\right)}{\left(V_{DD} - V_{Tn}\right)^2} \qquad ...(4)$$

Equation (4) shows that the size of MN1 (β_{MN1}) should be increased as the differences between VDDS and VDD increases. It also shows that for bigger MP2 (β_{MP2}), bigger MN1 is required.

IV. Feedback based Level-Shifter

Fig.4 shows the new feedback based level-shifter topology (Topology 2). The Level-shifter circuit uses thin oxide, low voltage MOSFETs (MN3, MN4, MN7 and MP6) and high voltage MOSFETs (MN5, MN6, MP3, MP1 and MP5). The thin oxide transistors operate at a lower core voltage supply VDD and the thick oxide transistors operate at a higher IO voltage supply VDDS. The VSS is ground supply for the circuit. Nodes S2 and S1 are connected to the VDDS and VSS level respectively. Similar to topology-1, the transistors MN5 and MN6 protect the thin oxide transistors MN3, MN4 from high VDDS level at their drains. The level-shifter get a low voltage level input at A and level shifts it to high voltage level output at OUT. A feedback path is formed in the circuit by connecting the output OUT to the gates of MN5 and MP5.

For the better understanding, the circuit operation is explained separately for high-to-low and low-to-High transition.

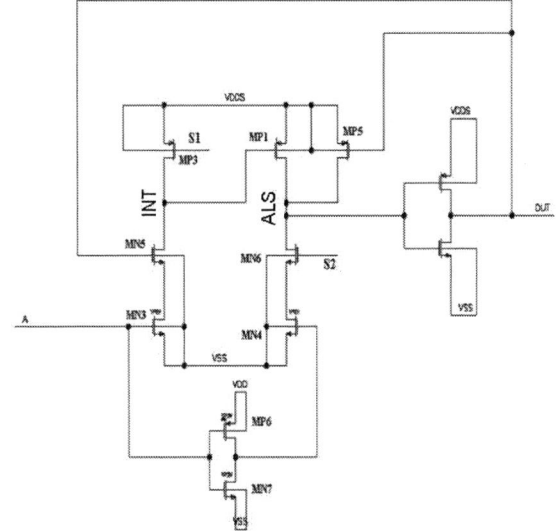

Fig. 4. New Level-shifter architecture (Topology 2).

978-1-4244-3039-0/08 $25.00 © 2008 IEEE

A. High-to-Low transition at input

For steady state logic high at the input A, the output node OUT is at logic low. The nodes ALS and INT are held at VDDS level through the 'ON' PMOSs MP5 and MP3 respectively. When the input transition from logic High–to-Low, transistor MN3 turns-off and MN4 turns-on, pulling down the node ALS through ON NMOS MN6. This in turn switches OUT to logic high, turning-off MP5 and turning-on MN5.

While pulling the ALS node down to VSS, the cascade of MN4 and MN6 easily overcomes the very small PMOS MP5. The MP5 is kept minimum sized as it is required there to just hold the logic state and not for any signal transition.

B. Low-to-High transition at input

The circuit uses feedback mechanism to shutoff the static current path which is created during low-to-high transition. As the input transitions from VSS to VDD, MN3 turns-on and MN4 turns-off. Turning-on of MN3 pulls down the node INT through the already ON NMOS MN5. A current path "MP3->MN5->MN3" is created between VDDS and VSS.

The low going node INT turns-on MP1, which pulls the node ALS to VDDS and eventually switching the output OUT to low. OUT going low turns-off MN5, shutting off the current path and pulling the node INT back to VDDS. MP1 turns-off and logic high at node ALS is held by MP5.

The pulse depth in Fig.5 is adjusted by the size of MP3 and the duration can be adjusted by adjusting the delay of the feedback path. By adjusting the pulse dimensions and the size of MP1, the output timings can be achieved.

It is clear from the circuit operation explanation that MN4 does not have to fight with the big PMOS MP1 while pulling down the node ALS. Referring to

fig(4), since the size of MP5 (β_{MP5}) is very small, the size of MN4 (β_{MP4}) can be kept small even for the operation at lower VDD. Notably this was not the case with topology-1 and was the main cause of its operational problem.

As the output transition is fast, topology-2 draws less switching current compared the conventional level-shifter (Topology 1).

Fig.7 shows the waveform at the node ALS for VDD level VDD6. It shows the overlapped waveforms for multiple process points and worst case VDDS and temperature level. The waveforms at node ALS is shown to get a better comparison with topology-1. Waveform W2 shows the behavior of topology-2 at VDD1 level. The smooth waveform edges imply that even at lower VDD level VDD6, the circuit gives good performance across process.

The curve P2 in Fig-8 plots the performance loss for new level-shifter (topology 2). It can be seen that curve P2 crosses over the curve P0 at VDD8 level. This implies that topology-2 meets the performance need up to VDD8 level.

V. Results

Fig.6 shows waveforms at the output OUT of topology 1 at VDD level VDD6. It shows overlapped waveforms for multiple process points at worst case VDDS and temperature level. Waveform W1 shows the behavior of the same circuit at VDD1 level. The smooth waveform W1 implies that, across process, the NMOSs MN1, MN3 have sufficient drive strength to pull the output node OUT down. While the distorted waveforms imply that at lower

VDD level VDD6, at some of the process points, the NMOSs do not have sufficient strength to fight the 'ON' PMOSs and pull-down the output node quickly. As a result the output is held at an intermediate voltage which causes large delays and crow-bar

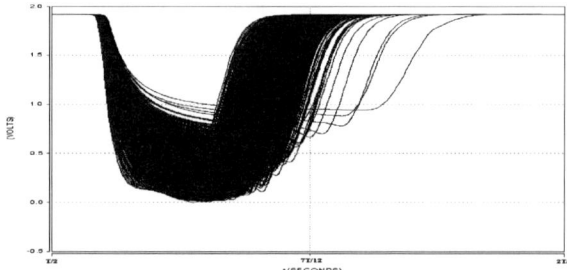

Fig. 5. Low going pulse waveforms at node INT. MP1 is tuned-on for this short period of time to produce low-to-high transition at ALS.

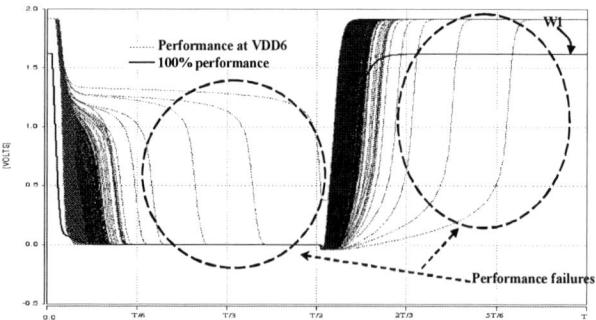

Fig. 6. Output of conventional cross-coupled level-shifter at VDD6 level.

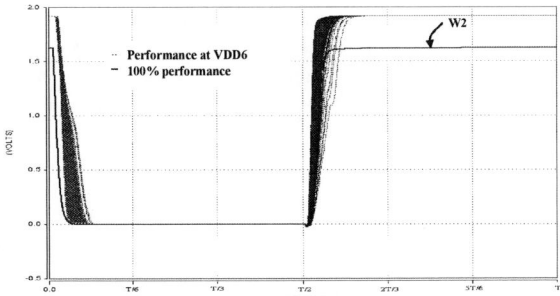

Fig. 7. Output of new level-shifter (topology-2) at core voltage level VDD6.

current in the next CMOS stage.

Performance loss of conventional cross-coupled level-shifter at different VDD levels is plotted in fig. 8. Fig. 8 plots the performance loss 'P', as in expression 1, against VDD voltage levels. The curve 'P0' is the maximum allowed performance loss, as defined in table-1. The curve P1 is the performance loss for conventional level-shifter (topology 1).

In the graphs it can be seen that curve P1 crosses and goes higher than curve P0 at VDD5 level. This implies that topology-1 fails the performance need beyond VDD5 level.

Beyond VDD6 level, topology-1 fails functionally and no delay data is produced, therefore P1 is not plotted beyond VDD6. Curve P2 is for Topology-2, which is already explained earlier in the text.

Improvement can be achieved in topology 1 but at the cost of area by highly over sizing the NMOSs MN1, MN3. It also involves power cost because increase in thin oxide transistor's GIDL leakage [3][6] which is directly proportional to the transistor dimensions.

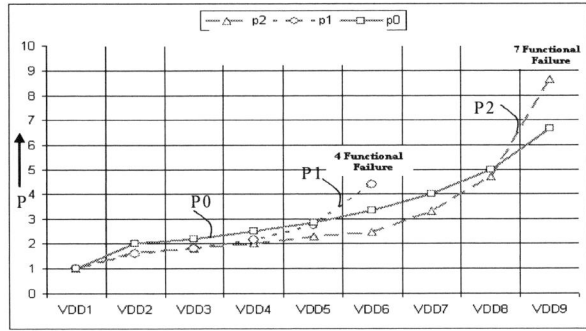

Fig. 8. Performance with respect to VDD voltage level scaling.

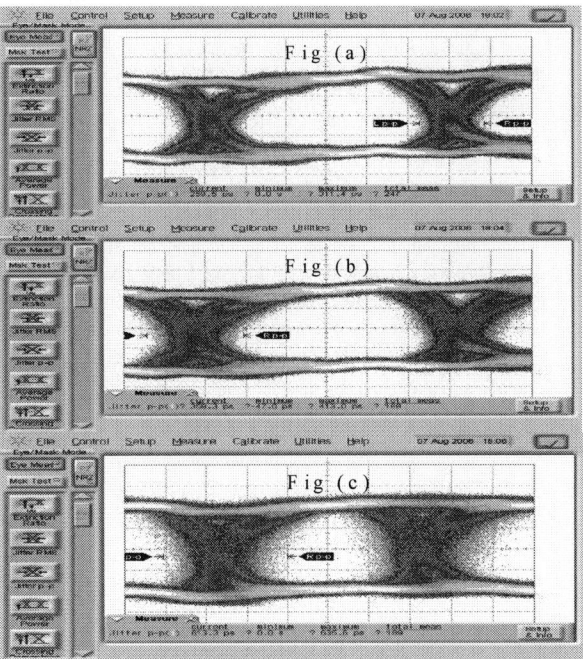

Fig. 9. Silicon validation result of an output buffer using proposed level-shifter topology-2, for three different VDD levels. Eye plotted for output driving 4 inches, 50 Ohms trace and 5 pF load, running at 500 MHz, 125C, VDDS= 1.98V. Fig (a) = VDD1, Fig (b) =VDD4, Fig (c) = VDD8.

V. Conclusion

The new level-shifter topology shows more robust operation under scaled core voltage supply levels compared to conventional cross-coupled topology. At VDD6 level, the new level-shifter gives double the performance than topology-1.

Implementation of this proposal is used in an IO buffer in a 65nm bulk CMOS process. In silicon validation the IO buffer shows good performance at lower supply voltages without imposing any area or power penalty.

References

[1] Hai Li, et. al., "Combined circuit and architectural level variable supply-voltage scaling for low power," in IEEE Trans. on VLSI Systems, vol.13,no.5, pp.564-576, May.2005.

[2] R. Brodersen, et. al., "Design issues for dynamic voltage scaling," in Proc. 2000 Int. Symp. Low Power Electronics and Design (ISLPED00), Jul.2000, pp.9-14.

[3] K.Roy, et. al., "Deterministic clock gating for microprocessor power reduction," in Proc. 9th Int. Symp. On High Performance Computer Architecture (HPCA), Feb.2003, pp.113-122.

[4] D. Sylvester, et. al., "The limit of dynamic voltage scaling and insomniac dynamic voltage scaling," in IEEE Trans. on VLSI Systems, vol.13, no.11, pp. 1239-1252, Nov. 2005.

[5] B. Zhai, et al., "Theoretical and practical limits of dynamic voltage scaling," in DAC, 2004, pp.868-873.

[6] Kaushik Roy et. al, "Leakage current mechanisms and leakage reduction techniques in deep sub-micrometer CMOS circuits." In Proceedings of the IEEE, vol.91, no.2, Feb. 2003.

[7] Kihwan Choi et al, "Fine-grained dynamic voltage and frequency scaling for precise energy and performance trade-off based on the ratio of off-chip access to oc-chip computation Times". In IEEE Proceedings of the Design, Automation and Test in Europe (DATE'04)

[8] K. Usami et al, "Automated low-power technique exploiting multiple supply voltages applied to a media processor," IEEE J. Solid State Circuits, pp463-472, Mar.1998.

[9] Liou, J.J. et. al., "Statistical modeling of MOS devices based on parametric test data for improved IC manufacturing," Electron Devices Meeting, 2001. Proceedings. 2001 IEEE Hong Kong

[10] Behzad Razavi, Design of analog CMOS integrated circuits, chap.2. 2nd Edition, McGraw-Hill Publication.

21st International Conference on VLSI Design

Low Power Hardware Architecture for VBSME using Pixel Truncation

Asral Bahari, Tughrul Arslan and Ahmet T. Erdogan
School of Engineering and Electronics, University of Edinburgh
United Kingdom
Email: {a.bahari, tughrul.arslan, ahmet.erdogan}@ed.ac.uk

Abstract

This paper presents an efficient architecture to implement low power variable block size motion estimation (VBSME) using full search. Power reduction is achieved by performing the search in two steps: low pixel resolution and full pixel resolution. We analysed the computation and memory units needed to support these two search modes. The proposed architecture reduces the total energy consumption by 50% with 6% additional area compared to the conventional architecture.

1. Introduction

Video compression plays an important role in today's wireless communications. It allows raw video data to be compressed before it is sent through a wireless channel. However, video compression is computing intensive and dissipates a significant amount of power. This is a major limitation in today's portable devices. Existing multimedia devices can only play video applications for a short time before the battery is depleted.

Motion estimation (ME) has been identified as the main source of power consumption in video encoders. It consumes more than 40% of the total power used in video compression [1]. This results from the high computational load needed to predict the current frame. Full-search motion estimation predicts the current macroblock by finding the candidate that gives the minimum sum of absolute difference (SAD), as follows:

$$SAD(i,j) = \sum_{k=0}^{M-1} \sum_{l=0}^{N-1} |C(k,l) - R(i+k,j+l)| \quad (1)$$

where $C(k,l)$ is the current macroblock, $R(i+k,j+l)$ is the candidate macroblock located in the search window within the previously encoded frame. From Eq. 1,

the power consumption in motion estimation is affected by the number of candidates and the total computation to calculate the matching cost. Thus, the power in ME can be minimised by reducing these two parameters.

To maximise the available battery energy, the computational power should be adapted to the supply power, picture characteristics and available bandwidth. Because these parameters change over time, the ME computation should be adaptable to different scenarios without degrading the picture quality.

[2] used pixel truncation to reduce the computational load for the matching calculation unit. The authors in [3] implemented adaptive pixel truncation on the least significant bits (LSB). The number of truncated pixels is modified according to the quantisation parameter (QP). Truncating the pixel's most significant bits (MSB) was discussed in [4].

Truncating pixels at a 16 x 16 block size results in acceptable performance, as shown in [3]. However, for smaller block partition, as defined in the H.264 standard, pixel truncation increases the truncation error. To solve this problem we have proposed a two-step method, as in [5], for truncating pixels in VBSME. The aim of this technique is twofold: to reduce the computational load and to minimise the memory access.

In this paper we analyse the hardware implementation for this technique in terms of area, power and energy efficiency. The aim of this architecture is to include low power capability on top of the existing full search motion estimation. The rest of the paper is organised as follows. Section II reviews our proposed technique and Section III discusses the proposed architecture. The results are discussed in Section IV. Finally, Section V concludes the paper.

2. Pixel truncation for VBSME

In [5], we analysed and proposed pixel truncation approaches for VBSME. In this method we perform a two-step methodology for VBSME. In the first step we

Table 1. Memory access vs second search range

Memory access

	Full Search	Proposed 2-Step
1st search	$256 \times (2p)^2 \times 8bit$	$256 \times (2p)^2 \times 2bit$
2nd search	-	$256 \times p^2 \times 8bit$
Total	$8192p^2$	$4096p^2$

Computation cost

	Full Search	Proposed 2-Step
1st search	$256 \times (2p)^2 \times 1$	$256 \times (2p)^2 \times 0.25$
2nd search	-	$256 \times p^2 \times 1$
Total	$1024p^2$	$512p$

perform a full search using the first two MSBs with difference-pixel-count (DPC) [6] as the matching cost. The best match for 8 x 8 blocks found in the first step is used to guide the next step.

In the second step, we perform full-pixel resolution to refine the result obtained from the first search. To ensure that the overall computation cost does not exceed the conventional full search computation, the second search is done at a quarter of the size of the first search area. At this stage, all 41 block partitions are evaluated.

The results in [5] show that our method can reduce the computation load and memory access without significantly degrading the picture quality. Table 1 shows this method's estimated memory bandwidth and computational load.

3. Hardware design

In this section we first review the conventional ME architecture that is used in our analysis. Next, we discuss the architectures needed to support the proposed two-step method. The area and power overhead for the computation and memory unit are also investigated. Based on these analyses, we propose three low power ME architectures with different areas and power efficiencies.

In this analysis we implement the ME architecture based on 2D ME as discussed in [7]. We choose 2D ME because it can cope with the high computational needs of the real-time requirement of H.264 using a lower clock frequency than 1D architecture.

3.1. Computation unit

Figure 1 shows the functional units in the conventional 2D ME (me_sad) [7]. The ME consists of search area (SA) memory, a processing array which contains 256 processing elements (PEs), an adder tree, a comparator and a decision unit. The SA memory consists of 16 memory banks where each bank stores 8-bit pixels in a $2 \times H \times \frac{W}{M}$ total word, where H and W are the search area windows' height and width, respectively, and M is the macroblock's (MB) width. During motion prediction, 16 pixels are read from the 16 memory banks simultaneously. The data in the memory are stored in a ladder-like manner to avoid delay during the snake scan [8].

At each initial search, the current and the first candidate MB are loaded into the processing array's registers. Then, it calculates the matching cost for one candidate per clock cycle. The 256 absolute-different from the PEs are summed by the adder tree, and outputs the sum of absolute-different (SAD) for 41 block partitions. The adder tree reuses the SAD for 4x4 blocks to calculate a larger block partition. In total, the adder tree calculates 41 partitions per clock cycle.

Throughout the scanning process, the comparator updates the minimum SAD and the respective candidate location for each 41-block partition. Once the scanning is complete, the decision unit outputs the best MB partition and its motion vectors. The ME requires 256 clock cycles for each MB search.

Figure 2 (a) shows the PE for me_sad. The input and output for the PE are 8bits wide. The input for the adder tree is 8 bits wide, and the SAD output is 12 to 16 bits wide, depending on the partition size. These data are then input into the comparator, together with the current search location information.

Compared to me_sad, DPC-based ME (me_dpc) requires 2 bits input for the current and reference MB as shown in Fig. 2 (b). The matching cost is calculated using XOR and OR. The input into the adder tree for me_dpc is one bit. Thus the adder tree requires 5 to 9 bitwidths of output. A similar bitwidth is applied to the comparator's input.

Table 2 shows the total area (mm^2) and power consumption (μW) for me_sad and me_dpc. The table shows that me_sad's area is dominated by the 256 PE (78%). Thus, with the significantly smaller area for 256 PE, the me_dpc will require less area than the me_sad. The overall me_dpc requires 20% of the me_sad area.

Based on the above analysis, we propose two type of architecture for the ME computation unit that can perform both low resolution and full resolution searches, as follows:

1. me_split

2. me_combine

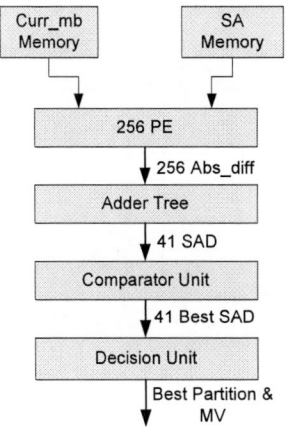

Figure 1. Me_sad block diagram

Table 2. Me_sad and me_dpc area and power

	me_sad		me_dpc	
	Area	Power	Area	Power
256 PE	0.841	22.437	0.086	1.237
sad_tree	0.090	5.131	0.022	0.987
comparator	0.073	0.838	0.050	0.512
best_part	0.071	0.487	0.044	0.453
Total	1.075	28.893	0.202	3.189

Me_split implements both me_sad and me_dpc as two separate modules, as shown in Fig. 3. During the low resolution search, me_sad is switched off while the me_dpc is used to perform the low resolution search. The second step uses the me_sad and the me_dpc is switched off. This architecture allows only the necessary bit size to be used during different search modes. While potential power savings is possible, this architecture requires additional area for the adder tree, comparator and decision unit to support the low resolution search.

Because the functions of the adder tree, comparator and decision are similar for both me_sad and me_dpc, me_combine shares these units during low resolution search and full pixel resolution (Fig. 4). This architecture results in a much smaller area compared to me_split. However, higher power consumption is expected during low resolution search because the adder tree, comparator and decision unit operate at higher bit size then needed.

3.2. Memory architecture

Conventional ME architecture implements the SA using single port SRAM with a pixel (8 bits) per word. To implement the two-step search, we need to access the

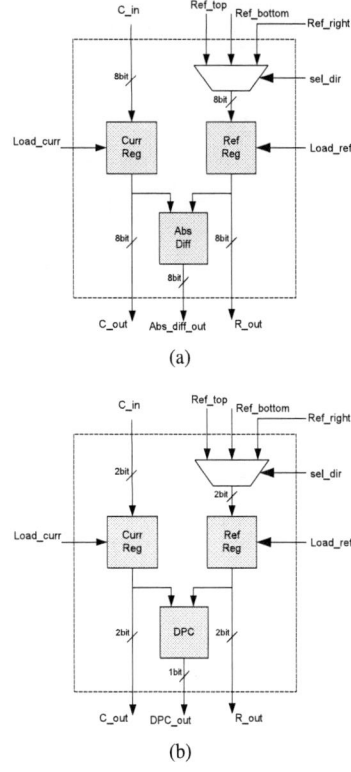

(a)

(b)

Figure 2. PE to support (a) SAD and (b) DPC

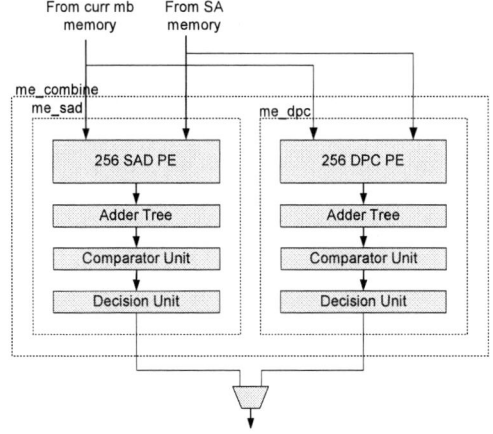

Figure 3. me_split

978-1-4244-3039-0/08 $25.00 © 2008 IEEE

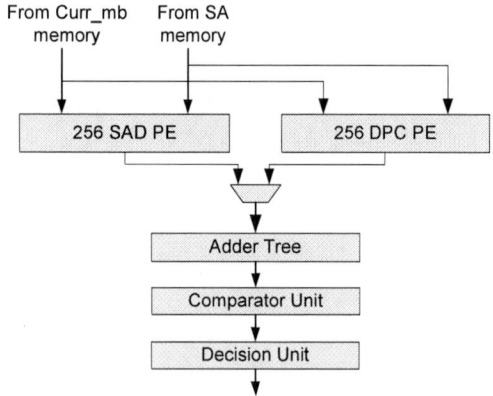

Figure 4. me_combine

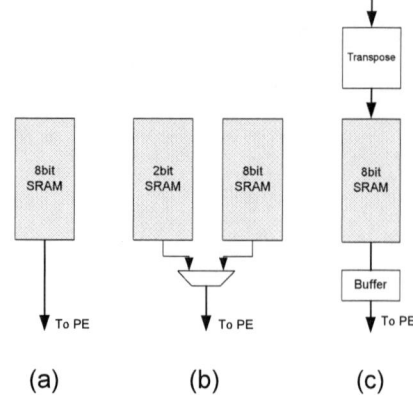

(a)   (b)   (c)

Figure 5. SA memory arrangement (a) mem8 (b) mem28, (c) mem8reg

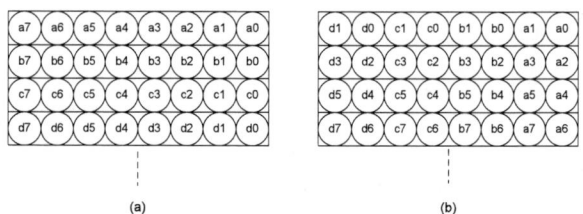

Figure 6. Storing 8bit pixels in 8bit memory (a) conventional arrangment (b) mem8reg

first two MSBs for each pixel during the first search, and 8 bits in the second stage. Thus, the pixels need to be stored to allow two reading modes. For this, three types of memory architecture are proposed as follows (refer Fig 5):

1. 8bit memory (mem8)

2. 2bit and 8bit memory (mem28)

3. 8bit memory with prearranged data and transposed register (mem8reg)

Mem8 stores the data in the same way as in conventional ME. We access 8-bit data during both low resolution and the refinement stage. However, during the low resolution search, the lower 6 bits are not used by the PE. Because the memory is accessed during both low resolution and the refinement stage, it results in higher memory bandwidth than the conventional ME architecture.

To overcome the problem in mem8, mem28 use two type of memory: 2-bit and 8-bit. The 2-bit memory stores the first two MSBs of each datum, and the 8-bit memory stores the complete full pixel bitwidth. During the low resolution search, the data from the 2-bit memory are accessed. This allows only the required bits to be accessed without wasting any power during low resolution. In the refinement stage, the 8-bit memory is read into the PEs. Although this architecture can potentially reduce memory bandwidth and power consumption, it needs an additional area for the 2-bit memory.

Mem8reg prearranges the data before storing them in 8-bit memory. Four pixels are grouped together, and then transposed according to their bit position, as shown in Fig. 6. During the low resolution search, we read only the memory locations that store the first two MSBs of the original pixels. Thus, the total memory accessed

during the low resolution is $\frac{1}{4}$ of the conventional full pixel access.

In full resolution search, we read four memory locations that contain the first upto eighth bits in four clock cycles. Delay buffers realigns these words to match the original 8-bit pixel, as shown in Fig. 5 (c). By prearranging the pixels this way, we can use the same memory size as in the conventional full search while retaining the ability to access the first two MSBs, as well as the full bit resolution. The drawback of this approach is it needs additional circuitry to transpose and re-align the pixels during the motion prediction.

3.3. Overall architecture

From the above discussion, we propose three different architectures that can perform both low resolution and full resolution searches. By combining different computation and memory units, we propose the following architectures:

1. me_split+mem8reg (ms_m8r)

2. me_combine+mem8 (mc_m8)

3. me_combine+mem28 (mc_m28)

Table 3. me_split and me_combine area and power

	me_split			me_combine		
	Area	Power		Area	Power	
		low res	high res		low res	high res
pe_16x16	0.928	1.237	22.437	0.928	1.210	22.437
sad_tree	0.112	0.987	5.131	0.091	1.180	5.131
comparator	0.123	0.512	0.838	0.073	0.510	0.838
best_part	0.115	0.453	0.487	0.060	0.470	0.487
Total	1.277	3.189	28.893	1.152	3.370	28.893

The performance of each of these is discussed in the next section.

4. Results

This section presents the synthesised results of our analysis. First we discuss the synthesised results for the computation and memory architectures; then we discuss the results for the overall ME architectures that can provide efficient area and power consumption. We synthesised our design using the UMC 0.13 CMOS library. We used Verilog-XL for functional simulation, and Design Compiler for synthesis and performing gate-level power analysis.

Table 3 shows the results for me_split and me_combine. Two modes of power consumption are shown to indicate the power consumption during low resolution and full resolution searches. Note that the low resolution search requires 256 clock cycles, and the full pixel resolution requires 64 clock cycles.

From the table, the me_split consumes 5% less power during the low resolution search than the me_combine. However, this comes at the cost of an additional 19% of area on top of existing me_sad. On the other hand, because me_combine shares some modules, it requires only 7% additional area compared to me_sad.

Tables 4 show the area and power comparison for the proposed memory arrangement. The SRAM model is generated using a UMC 0.13 memory compiler with estimated area and power provided by the datasheet. The table shows that the mem8reg provided the lowest bandwidth and power compared to the other memory configurations. However, it requires extra area for the additional circuits needed to arrange the pixels. On the other hand, mem8 gives the minimum area compared to the other configurations. Because the full pixel bitwidth is accessed in both low and full resolution, it needs higher memory bandwidth and uses more power than the others.

Table 4. Memory architecture comparison

	Area	Memory Bandwidth		Power	
		Low Res	High Res	Low Res	High Res
mem8	0.23	$MWH \times 8bit$	$M\frac{WH}{4} \times 8bit$	4.7	4.7
mem28	0.39	$MWH \times 2bit$	$M\frac{WH}{4} \times 8bit$	3.3	4.7
mem8reg	0.42	$MWH \times 2bit$	$M\frac{WH}{4} \times 8bit$	2.2	4.7

Table 5. ME architecture power comparison

	Area	Power(mW)					
		Low Res			Full Res		
		Logic	Mem	Total	Logic	Mem	Total
SAD + mem8	1.39	NA	NA	NA	28.89	4.7	33.59
ms_m8r	1.78	4.239	1.2	5.41	31.69	4.7	36.39
mc_m8	1.48	3.370	4.7	8.07	28.89	4.7	33.59
mc_m28	1.64	3.370	3.3	6.67	28.89	4.7	33.59

Tables 5 show the area, power and normalised energy comparison for the overall ME architecture. The normalised energy is the total energy used to perform both low and full resolution searches. Architecture ms_m8r consumes the least energy compared to the others, and saves 56% compared to conventional ME. However, because it requires additional area for the adder tree, comparator, decision unit and buffer to arrange the pixels, 30% area overhead is required to implement this method.

Compared to other architectures, mc_m8 gives the best results in terms of both area and energy efficiency. By using the two-step method, this architecture can save 50% of the energy compared to conventional ME with only 6% additional area required to implement a low resolution search.

5. Conclusion

This paper discusses the hardware architecture to support 2-step search method as proposed in [5]. The objective of this analysis is to enhance the conventional ME architecture so that it can perform both low resolution search and full resolution search. Based on our

Table 6. ME architecture comparison (Normalised)

	Area	Energy	Area*Energy
me_sad	1.00	1.00	1.00
ms_m8r	1.28	0.43	0.55
mc_m8	1.06	0.49	0.52
mc_m28	1.18	0.45	0.53

analysis on the conventional architecture, we proposed three architecture that is suitable for 2-step method namely ms_m8r, mc_m8 and mc_m28. From the results, the ms_m8r give the highest energy saving (56%) and lower memory bandwidth compared to the conventional architecture. On the other hand, mc_m8 is more energy and area efficient compared to other architecture where it save 50% energy with only 6% additional area overhead compared to the conventional architecture.

References

[1] Z. He, Y. Liang, and I. Ahmad, "Power-rate-distortion analysis for wireless video communication under energyconstraint," vol. 5308, (San Jose, CA, United States), pp. 57 – 68, 2004.

[2] Y. Chan and S. Kung, "Multi-level pixel difference classification methods," *IEEE International Conference on Image Processing*, vol. 3, pp. 252 – 255, 1995.

[3] Z.-L. He, C.-Y. Tsui, K.-K. Chan, and M. L. Liou, "Low-power vlsi design for motion estimation using adaptive pixel truncation," *IEEE Transactions on Circuits and Systems for Video Technology*, vol. 10, pp. 669 – 678, 2000.

[4] V. G. Moshnyaga, "Msb truncation scheme for low-power video processors," *Proceedings - IEEE International Symposium on Circuits and Systems*, vol. 4, pp. 291–294 –, 1999.

[5] A. Bahari, T. Arslan, and A. T. Erdogan, "Low computation and memory access for variable block size motion estimationusing pixel truncation," *IEEE 2007 Workshop on Signal Processing Systems (SiPS 2007)*, 2007.

[6] S. Lee, J.-M. Kim, and S.-I. Chae, "New motion estimation algorithm using adaptively quantized low bit-resolutionimage and its vlsi architecture for mpeg2 video encoding," *IEEE Transactions on Circuits and Systems for Video Technology*, vol. 8, no. 6, pp. 734 – 44, Oct 1998.

[7] C.-Y. Chen, S.-Y. Chien, Y.-W. Huang, T.-C. Chen, Tung-Chienand Wang, and L.-G. Chen, "Analysis and architecture design of variable block-size motion estimationfor h.264/avc," *IEEE Transactions on Circuits and Systems I: Regular Papers*, vol. 53, no. 3, pp. 578 – 593, 2006.

[8] T.-C. Chen, Y.-H. Chen, S.-F. Tsai, and S. Y. C. andLiang Gee Chen, "Fast algorithm and architecture design of low-power integer motion estimationfor h.264/avc," *IEEE Trans. Circuits Syst. Video Technol. (USA)*, vol. 17, no. 5, pp. 568 – 77, 2007.

SESSION C3:
NoC/SoC

21st International Conference on VLSI Design

MoCSYS: A Multi-Clock Hybrid Two-Layer Router Architecture and Integrated Topology Synthesis Framework for System-Level Design of FPGA based On-Chip Networks

Arun Janarthanan, Karen A. Tomko
Department of Electrical & Computer Engineering and Computer Science
University of Cincinnati, Cincinnati OH 45221, USA
{janarta, ktomko}@ececs.uc.edu

Abstract

Complex System-on-Chip designs targeted for FPGAs merit sophisticated communication architectures to support a host of high performance applications. In this research we implement a hybrid two-layer router architecture for FPGA based NoCs and quantify its area and performance trade-offs by characterizing a network component library (Mo-Clib). Results from the VHDL and SystemC models of the advanced router architecture show an average improvement of 20.4% in NoC bandwidth (maximum of 24% compared to a traditional NoC). As a part of the CAD flow, we develop an algorithm that utilizes the above NoC framework and includes bandwidth capacity and area as a cost during an automatic NoC topology synthesis phase. Experimental results for a set of real applications and synthetic benchmarks show an average reduction of 21.6% in FPGA area (maximum of 26%) for equivalent bandwidth constraints when compared with a baseline approach.

1 Introduction

Present device capabilities enable FPGAs to replace ASICs in a variety of high performance applications. However, with increasing FPGA size, it is a challenge to preserve clock signal integrity between Intellectual Property (IP) cores that are farther apart. Network-on-Chip is a recent methodology, primarily adopted for ASICs and later extended to the FPGA domain to increase the performance of multi-core System-on-Chip (SoC) designs.

An FPGA based on-chip network has a unique set of design goals that includes satisfying an application-suitable bandwidth requirement with a minimum (limited) resource availability. Choice of appropriate network aspects have a significant impact on the performance and area of the NoC. The mechanism by which the data moves across the network is termed as switching. Packet-switching performs on-line scheduling by dynamically negotiating communication between the cores. Serialization/Packetization overhead and unpredictable latencies are some of the downsides that this mechanism sustains. An alternate approach, namely circuit-switching offers high throughput dedicated connections to overcome these drawbacks. However, connecting

a large number of modules by this mechanism reduces the overall performance due to path reservation, thereby limiting the time-slice available in the resource. Further, this network requires preserving synchronous operation between communicating cores and is therefore not appropriate for FPGAs due to their large device sizes.

To overcome these drawbacks, we propose a modified router architecture which interfaces multiple IP cores to the router and supports packet-switching for inter router transfers and time-multiplexed circuit-switching for IP cores connected to the same router. This technique also eliminates the latency in $req/grant$ protocol, serialization and control overheads for data transfers between cores placed close to each other in FPGAs and mapped to the same router.

Motivation: Interconnects in FPGAs are available at a premium and contribute significantly to the total circuit delay and power. It is therefore appropriate to share the communication resources using overlaid NoC architectures. We propose a novel two-layer hybrid router architecture that offers high bandwidth with minimum area and develop a library of network components (MoClib). By varying the library component design parameters, we characterize the router for area and performance in order to rapidly explore the NoC design space.

Topology Synthesis and IP Mapping: Utilizing the above library of network components and mapping the IP cores onto such a complex NoC presents a great challenge. We develop an algorithm and an application-generic design flow that includes required bandwidth and area in the cost function to synthesize the NoC topology for FPGAs. For a given task communication graph, the algorithm effectively automates the design cycle by synthesizing a minimum area topology that satisfies all bandwidth requirements.

To summarize, the contributions of the paper are,

1. Presentation of a novel hybrid two-layer router architecture that supports packet-switching for inter router transfers and time-multiplexed circuit-switching for IP cores connected to the same router.

2. Development of an integrated design flow that automatically performs topology synthesis to optimize for area while satisfying required bandwidth by importing a characterized library of network components.

978-1-4244-3039-0/08 $25.00 © 2008 IEEE

397

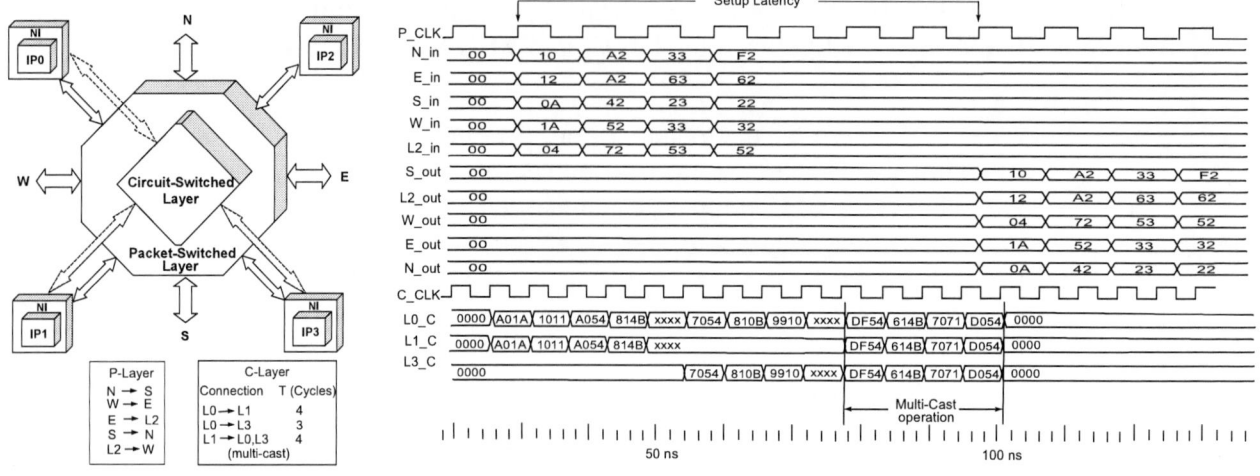

(a) Hybrid Router Architecture

(b) SystemC Functional Simulation

Figure 1. Two-Layer Hybrid Router

The rest of the paper is organized as follows: Section 2 presents related work. Section 3 describes the router architecture, its merits and design issues. The synthesis algorithm proposed for our architecture and its design flow are presented in Section 4. Experimental results for a set of benchmarks are analyzed in Section 5. Section 6 concludes the paper and presents future work.

2 Related Work

We target our proposed NoC framework for reconfigurable computing platforms and therefore we restrict our discussions in this section primarily to existing FPGA based NoCs. NoCs were introduced into the FPGA domain mainly to simplify tile-based reconfiguration [1] [2], and its potential as an effective communication architecture is largely unexplored [3]. Research in [4] [5] address the capabilities of FPGAs to support NoC based multi-processor applications. Hilton et al. [6] incorporate flexibility into their design for FPGA based circuit-switched NoCs. However, their strictly circuit-switched router suffers from signal integrity and path reservation issues which we overcome in our design. SoCBUS [7] proposes a circuit-switched router with a packet based setup, which is different from our two-layer approach. Research in [8] [6] [9] also present FPGA based NoCs. The above designs ignore implementation level area-performance trade-offs while proposing the architecture, thereby limiting to a system-level performance analysis.

In MoCReS [10], we present an area efficient multi-clock packet-switched router with virtual cut-through flow control, supporting a high data rate with minimum area overhead. Further, the NoC framework enables the routers to function at independent operating frequencies that are dictated by the placement and routing constraints in FP-GAs. We obtain this router and construct the hybrid two-layer router that is used in our NoC framework.

While ASIC implementations have a well developed CAD flow for NoC design [11] [12], there is no automated methodology for FPGAs that takes into account the features of the underlying architecture. Moreover, the limitation in resources, higher power consumption, and increasing heterogenity of the FPGA device complicates the design flow. Research in [2] is the first work to address automated design for FPGAs. Their underlying NoC model enables fast performance verification and is not customized for FPGA based NoCs as opposed to the characterized *MoClib* NoC library used in this research.

To the best of our knowledge, this is the first work to propose an FPGA-suitable hybrid router architecture integrated with an automatic topology synthesis framework that satisfies the bandwidth requirements of an application while optimizing its area overhead.

3 Architecture Description

In this section, we first present the modified router micro-architecture, followed by its architectural advantages and design issues involved.

3.1 Router Micro-Architecture

Network Topology: Mesh networks have minimum area overhead (reduced long lines) [4] [10], low power consumption and map well to the underlying routing structure of FP-GAs. Hence, we choose a mesh topology to optimize logic and routing in FPGAs, and to provide sufficient resources for the IP cores.

Flow Control: Our router supports multi-clock virtual cut-through flow control with a deadlock-free XY routing. The router complexity involved in the above choice is more suitable for a light-weight implementation [10].

398

978-1-4244-3039-0/08 $25.00 © 2008 IEEE

Architecture Modifications The modified router comprises of two layers of operation: a high throughput time-multiplexed circuit-switched layer (C-layer) and a multi-clock packet-switched layer (P-layer). Variable number of IP cores connected to the router participate in the C-layer, thereby achieving guaranteed throughput (more predictable latencies) between IP cores placed close to each other in the FPGA. Figure 1a presents the modified router architecture that has three IPs communicating over the C-layer. It is to be noted that all the four IPs can dynamically negotiate packets between neighboring routers through the P-layer.

1. Cross-Point Matrix: Traditionally the switch matrix supports all connections between directional (two to four in number) ports and the local ports. We prune the mux-based connections for each of the inter-local ports from the P-layer. This translates into gain in area, which we utilize to support an additional C-layer cross-point with increased bus width (32 bits, correspoinding to Microblaze [13] data width). Figure 1b presents the functional simulation of the router. For the sake of clarity, the C-layer channel width is set to 16 bits.

2. Central Arbiter: The central arbiter is responsible for configuring simultaneous connections by setting the Cross-Point. We perform state reduction in the FSMs corresponding to inter-local port connections. The simplicity of round-robin arbitration coupled with the above state reduction translates into significant area savings.

3. NI Design: The network interface (NI) transitions the mode of communication between the two layers. Further, in the P-layer, the NI is also responsible for encoding the packet header with its size (as a fraction of bRAM depth), and co-ordinates (X, Y) of the destination IP. Supporting variable packet sizes reduces the average latency by increasing the buffer utilization.

4. Design Parameters: We parameterize our router for total number of ports, channel width, virtual channels/port and number of ports participating in the C-Layer. To explore the NoC design space rapidly, we develop a component library (MoClib) and characterize it for operating frequency and area (Figure 2,3).

3.2 Architectural Advantages

Bandwidth Increase: Bandwidth available in a router is the product of number of ports, operating frequency and channel width. The C-layer has minimum logic overhead with no buffering and can operate at a clock rate significantly higher than the P-layer. Further, increasing the number of ports also scales the available bandwidth in a router. Moreover, absense of control/serialization overheads (req/grant) also increases the throughput.

Power Savings: The amount of logic required reduces with router count, thereby saving static power. Further, with increasing number of ports within a router, the average packet latency reduces [8]. Therefore dynamic power reduces considerably with the reduction of router hops.

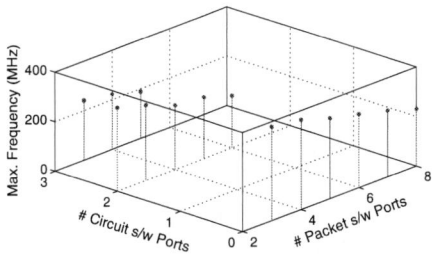

Figure 2. Design Parameters Vs Frequency

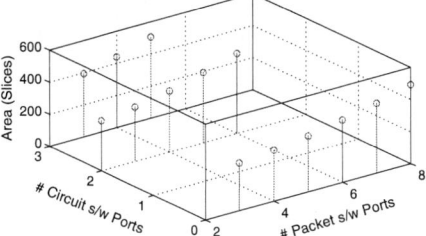

Figure 3. Design Parameters Vs Area

Guaranteed Throughput: The time-multiplexed nature of scheduling provides good Quality of Service (QoS) to the application, particularly, between cores placed close to each other. Otherwise, these cores would require area expensive QoS protocols to ensure the required bandwidth.

Inherent Multi-Cast Capability: The cross-point in the C-layer can be configured simultaneously for a multi-cast (one to many destinations) operation without any penalty in performance. Further, this capability also optimizes the area required for storing the schedules (with fewer bits required to encode the configuration data of the circuit-switched network).

3.3 Design Issues

Even though it appears intuitively that an increase in number of ports in the C-layer gives performance benefits with low area overhead, there are certain design issues that can potentially limit the performance due to increase in router complexity.

Operating frequency Vs Router Complexity: There is a depletion of critical resources associated with an increase in router complexity (number of ports, bus width). As a result, the operating frequency of the router degrades which in turn affects the bandwidth offered by the router. For the NoC paradigm to efficiently be an alternative to the bus-based architecture, the performance bottleneck needs to be eliminated by operating the routers at the highest possible frequency.

Router Power vs Link Power: By increasing the number of ports, we can reduce the average hop count [8], i.e we minimize the routers and links. This translates into a reduction in power consumed by the links, but an increase in power consumed by the switches. Beyond a cut-off, the increase in router power can potentially overshadow the gain in link power, thereby it can increase the power/flit to a high value.

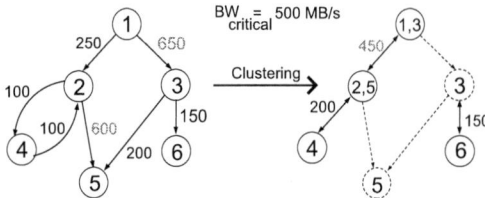

Figure 4. Clustering Phase of Topology Synthesis

Explosion of Schedule Memory: With increasing number of C-layer ports, the schedule memory also scales linearly. The schedule memory, expressed in number of LUTs is a function of number of schedule cycles and C-layer ports present.

Algorithm 1 Topology Synthesis Design Flow

Input: Core Task Graph G(V,E), with $|V|$ Cores, $|E|$ Edges/Links and Edge E_{ij} associated with a bandwidth weight function $b_{ij} : E_{ij} \rightarrow \mathbf{R}$.

Input: $A_{p/c-ovrhd(i,j)}$ Area overhead in replacing a packet/circuit-switched router with $i \rightarrow j$ ports (*MoClib* Library), Critical Bandwidth b_c, A_{av} Available FPGA area in Slices for NoC, Maximum C-layer ports $C_{critical}$.

Output: Low Area FPGA based NoC Topology, Satisfying Bandwidth

1:   **foreach** Edges $e_{ij} \in$ E, G(V,E) **do**
2:       **if** Required bandwidth for edge $e_{ij} : b_{ij} > b_c$ **then**
3:           *Clustering*: Group Vertices i, j, update edges, and add no. circuit-switched ports by one, thereby removing the bandwidth violation
4:       **end if**
5:   **end for**
6:   Output Modified Core Task Graph G'(V,E) such that $|V| = U'$, the upper bound on the no. of router instances
7:   γ: generate all mesh topologies possible with U' network components
8:   **foreach** Mesh Topology in γ **do**
9:       δ: permute all IP mapping combinations of G'(V,E) $\rightarrow$ C(V,E)
10:      **foreach** Topology mapping in δ **do**
11:          *Candidate Topology Selection*: foreach source, destination pair $(i, j) \in |V|$ of G'(V,E), estimate required link capacities in C(V,E) by applying XY routing
12:          *MoClib* Library: Estimate the bandwidth available in the edge E $\in$ C(V,E), Link BW = Channel Width (b)$\times$ Router Operating Frequency (f)
13:          Determine one valid topology instance such that available BW meets required BW $b_{i,j} \forall e_{ij} \in E$
14:      **end for**
15:      **repeat**
16:          $U' = U' - 1$ and $C_{max} = C_{max} + 1$
17:          Area gain: sum the area of the routers for every instance
18:          ensure $\sum_{i=0}^{N} A_i < A_{av}$
19:          $best.topology \leftarrow current.topology$
20:      **until** $\{U' = 1$ .or. $C_{max} > C_{critical}\}$
21:   **end for**
22:   Output $best.topology$ to ISE design flow

Clock Signal Integrity: Operation of the C-layer ports require the participating IP cores to be synchronous, as there is no buffering done, as opposed to packet-switch where multi-clock FIFOs separate the clock domains. Therefore increasing the number of C-layer ports could potentially increase the distance between the connected IP cores. In this case, the signal integrity acts as a limitation to the number of C-layer ports.

It can be seen that all of the above factors limit the amount of performance gain that can be achieved using our hybrid approach. This trade-off between performance, area and port count merits a balance and requires an application-

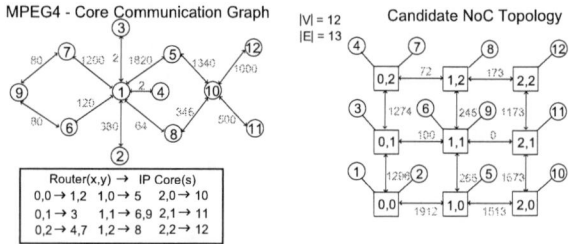

Figure 5. Bandwidth Estimation

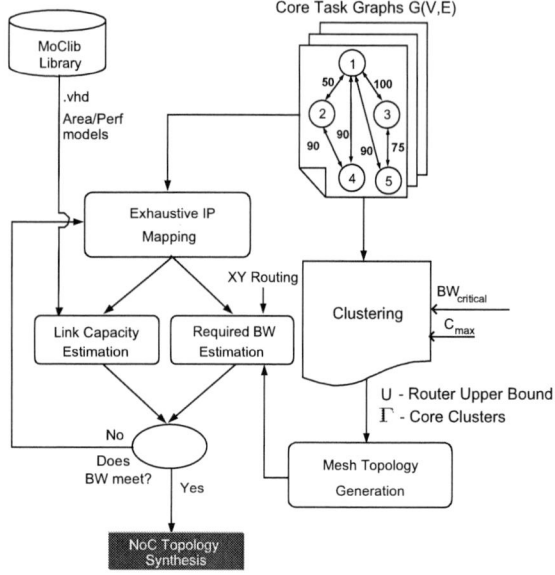

Figure 6. Topology Synthesis Framework

suitable tuning of the NoC topology. We present an algorithm along with a CAD flow in Section 4 to automate topology synthesis for FPGA based NoCs and address the above trade-offs.

4 FPGA Topology Synthesis Flow

In this section, we present the synthesis algorithm and a description of the phases it supports, its functioning and complexity. Further, this tool is a part of our larger design flow called, MoCSYS which is also presented here.

The input to the algorithm consists of a task communication graph (G), annotated with bandwidth requirement between modules. Further, the design space parameters, namely, the maximum bandwidth supported in a packet-switched link (critical bandwidth b_c), available FPGA area (A_{av}) along with area and performance models of our router architecture are also provided as input. Our algorithm (Algorithm 1) operates in multiple phases. During the *clustering phase*, the edges in the task graph whose required bandwidth violates the critical bandwidth are identified. The cores requiring these bandwidths that exceed the available inter-router capacity are grouped to form clusters of multiple IPs connected via the C-layer of a single router. Fig-

400

978-1-4244-3039-0/08 $25.00 © 2008 IEEE

ure 4 demonstrates clustering, where by incrementing the C-layer ports for two routers, we eliminate the bandwidth violations. Upon completion, this phase outputs the clustered core graph (Γ) along with the upper bound (U') on the number of routers. During the second phase, we generate all mesh topologies with U' routers. Due to its suitability for FPGAs, we consider only mesh based topologies in this research. Figure 6 shows our topology synthesis framework. Router models from the *MoClib* library ($\Re$) are input during link capacity estimation.

Core to Router Mapping: Mapping of IPs to a mesh, based on an unconventional hybrid router architecture presents a great challenge. During this phase, we iterate through a large search space by permuting the possible IP mappings exhaustively. The complexity of this phase dominates that of the algorithm. We select a candidate topology based on whether a mapping satisfies the required bandwidth and optimize it for area during the last phase.

Link Bandwidth Estimation: Selecting a candidate topology requires verifying if the bandwidth required for all (source,destination) pairs in the clustered core graph meets the available topology bandwidth. Our choice of XY routing simplifies this phase in addition to reducing the router complexity due to its simple logic. The cumulative bandwidth requirement on each edge (contributed by each source to destination route) establishes the required link capacity constraint. For a given MPEG4 application as a task graph, Figure 5 presents a candidate topology and its cumulative link bandwidth requirement (cost) in MB/s. As a result of Task 1 mapped to router (0,0), the link connecting the router to (1,0) requires a 1912 MB/s bandwidth (equal to the sum of bandwidths between Task 1 and rest of the tasks).

Problem Formulation: *Given* a task graph $G(V, E)$, where each $v_i \in V$ represents an IP core, and directed edge $e_{ij} = \{v_i, v_j\} \in E$ denotes a communication edge with a bandwidth weight function $b_{ij} : E_{ij} \rightarrow \mathbf{R}$, *Find* a mapping $G(V, E) \rightarrow C(V, E)$, where C represents a mesh topology graph, such that, $\forall i, j \in V$ the available topology bandwidth meets the required bandwidth $\sum_{\forall e_{ij} \in E} b_{ij}$ with minimum NoC area, $\sum_{\forall i \in C} \Re_i$.

5 Experimental Results and Analysis

Experimental Platform: Our target FPGA device is XC4VLX100 [13], on a Nallatech BenDATA^{TM} [14] development board. Functional verification of the hybrid router is performed using the VCD (value change dump) output obtained from simulating the SystemC (2.0.1) model of the framework. Further, this transaction level model can be used as an abstraction to quickly perform system-level simulation by easily varying the architectural parameters. To characterize our library of routers accurately for area and performance, we model them in structural VHDL. We use Xilinx ISE 8.2i [13] to follow the FPGA design flow for the router models. Xilinx FIFO Generator v2.3 [13] produces

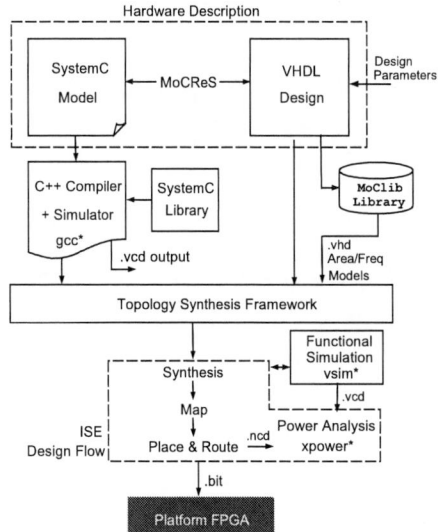

Figure 7. Experimental Flow for NoC Topology Synthesis

independent clock FIFO buffers for seperating the router's clock domain (multi-clock operation).

Algorithm 1 is implemented in C++ using Standard Template Library (STL) data structures and is supported with Perl for benchmark processing. Both benchmark execution and SystemC simulation were run on a AMD Opteron Processor with Linux, operating at 2.4 GHz and having $3GB$ RAM. For our chosen benchmarks, the average execution time was around 11 minutes. Figure 7 presents the experimental design flow used in this research.

Analysis of Results: We compare our results in this research with a traditional NoC used as a baseline approach [10]. We discuss the results below:

Area vs Average Bandwidth / Port: The bandwidth increase associated with the hybrid router architecture is compared in this section with the baseline approach. For equivalent area overheads (in slices) on a similar FPGA, Figure 8 presents the bandwidth capacity (in MB/s) of the NoC for both approaches. For the area window utilized in our library of routers, there is an average 20.4% gain in bandwidth (maximum of 24%) offered by our NoC. This gain in performance is due to supporting a high throughput circuit-switched layer with a marginal area overhead.

Benchmarks: Directed Communication Task Graphs (CTG) can successfully model multi-processor SoC applications. Vertices in the graph represent IP cores and edges represent precedence and bandwidth requirement. We apply our technique on four widely used application benchmarks, (FFT, MPEG4, VOPD and MWD) [15], represented as task graphs. These classes of application permit traffic (bandwidth) characterization early in the design cycle which is utilized in this research to fine-tune the NoC topology. In addition to the application benchmarks, we apply our technique to a set of synthetic benchmarks [8] which represent a variety of communication patterns that are frequently en-

Table 1. Application and Synthetic Benchmarks

	Benchmark	|V|	|E|	In-degree	Out-degree	Bandwidth MB/s	
						Max	Min
Application	MPEG4	12	26	7	7	910	1
	FFT	15	22	2	2	473	26
	VOPD	12	14	2	2	500	16
	MWD	12	13	2	2	128	64
Synthetic	LU Decomposition	9	11	2	3	510	76
	Laplace Solver	9	12	2	2	378	68
	Basic -1	6	8	1	4	196	34
	Parallel -1	8	14	3	4	225	47
	Packed -1	6	16	3	5	334	59
	Packed -2	9	15	3	4	412	106

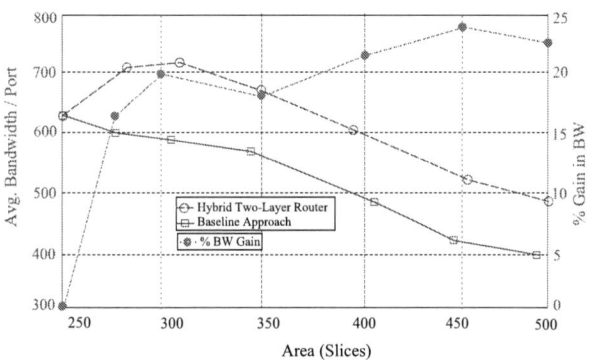

Figure 8. Area (Slices) Vs Avg. Bandwidth / Port

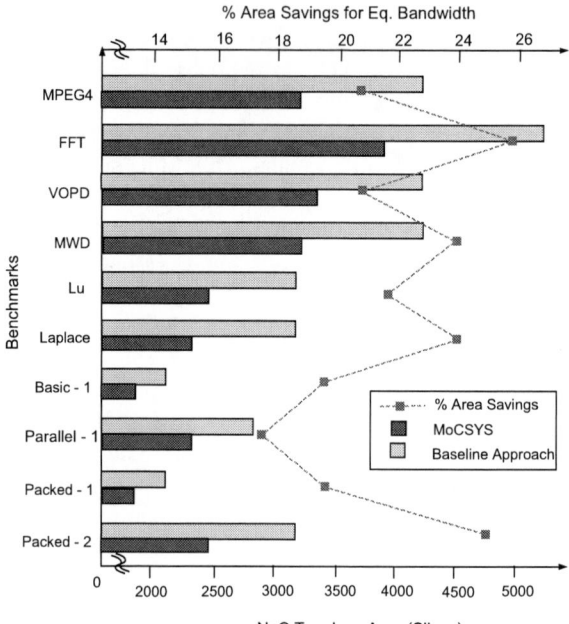

Figure 9. NoC Benchmark Results

countered in multi-processor designs. Due to space limitations, we abstain from a detailed description of the benchmarks and concentrate on a discussion on the results.

Using our hybrid architecture and integrated design flow, results were obtained for various benchmarks. For similar bandwidth constraints applied through task graph edges, Figure 9 compares the synthesized topology area between the proposed and baseline approaches. With the number of cores in the benchmarks varying between 6 and 15, it can be seen that there is an average reduction of 21.6% (maximum of 26%) in the NoC area which can be used for efficient implementation of application logic. The bandwidth constraints were translated into the original design and estimation of area was performed in slices. It is to be noted that the CAD tool does not optimize the design for execution time. However, ensuring that the required bandwidth is satisfied is the primary goal. For all the applications, our approach was able to obtain alternate topologies utilizing our hybrid router library with fewer FPGA resources.

6 Conclusion and Future Work

In this paper, we present a multi-clock hybrid two-layer router architecture suitable for FPGAs. We analyze the merits and issues involved with the architecture and characterize a library of network components for area and performance. For equivalent area overhead, our proposed architecture supports 20.4% increased bandwidth when compared with a baseline approach. We effectively automate the NoC design cycle by integrating the router with an algorithm that optimizes for FPGA area while satisfying the required bandwidth. We validate our approach using a set of

application and synthetic benchmarks and report the results.

As a future work, we are incorporating power trade-offs to the MoCSYS design flow based on the activity data we obtained through SystemC and VHDL simulations.

Acknowledgements

We would like to acknowledge Xilinx and Mentor Graphics for the CAD tools that we have utilized in this research.

References

[1] T. Marescaux et al. Interconnection Networks Enable Fine-Grain Multi-Tasking on FPGAs. In *FPL'2002*, pages 795–805, 2002.

[2] A. Kumar et al. An FPGA design flow for Reconfigurable Network-Based Multi-Processor Systems-on-Chip. In *DATE'07*, 2007.

[3] T.S.T Mak et al. On-FPGA Communication Architectures and Design Factors. In *FPL*, 2006.

[4] M. Saldaña et al. The Routability of Multiprocessor Network Topologies in FPGAs. In *SLIP'06*, pages 49–56, 2006.

[5] T.A Bartic et al. Topology Adaptive NoC Design and Implementation. In *Computer and Digital Techniques, IEE Proceedings*, 2005.

[6] C. Hilton and B. Nelson. PNoC: a flexible circuit-switched NoC for FPGA based systems. In *Computer and Digital Techniques*, 2006.

[7] D. Wiklund and D. Liu. SoCBUS: Switched Network for Hard Real Time Embedded Systems. In *IPDPS'03*, 2003.

[8] B. Sethuraman and R.Vemuri. optiMap: A Tool for Automated Generation of NoC Architectures using Multi-Port Routers for FPGAs. In *DATE'06*, 2006.

[9] N. Kapre et al. Packet-Switched vs. Time-Multiplexed FPGA Overlay Networks. In *FCCM'06*, 2006.

[10] A. Janarthanan et al. MoCReS: an Area-Efficient Multi-Clock On-Chip Network for Reconfigurable Systems. In *IEEE Annual Symposium on VLSI*, 2007.

[11] D. Bertozzi et al. NoC Synthesis Fow for Customized Domain Specific Multiprocessor Systems-on-Chip. In *IEEE Transaction on Parallel and Distributed Systems*, 2005.

[12] K. Srinivasan and K.S Chatha. A Low Complexity Heuristic for Design of Custom Network-on-Chip Architectures. In *DATE'06*, 2006.

[13] Xilinx Inc. http://www.xilinx.com.

[14] Nallatech Inc. http://www.nallatech.com.

[15] A. Jalabert et al. xpipescompiler: A tool for instantiating application specific networks on chip. In *DATE'04*, 2004.

21st International Conference on VLSI Design

MPSoC Communication Architecture Exploration Using an Abstraction Refinement Method

Hao Shen   Frédéric Pétrot

System Level Synthesis Group, TIMA Laboratory, INP Grenoble
43, Avenue Félix Viallet, 38031, Grenoble, France
{hao.shen@imag.fr   frederic.petrot@imag.fr}

Abstract

*The complexity of today's Multi-Processors System-on-Chip (MPSoC) requires new design methodologies to solve time-to-market and design cost problems. In SoC for which several subsystems are connected together, we notice that lots of design time is wasted on solving the inter-subsystem (global) communication problem. In this paper, we propose a novel communication exploration method based on a multi-abstraction levels exploration. With this work, the inter-subsystem communication structure can be optimized at the beginning of the design process by using simulation models at three different abstraction levels. The simulation at the higher abstraction level allows designers to explore parameters of the interconnection model at the more detailed abstraction level. Some design loop cases can be avoided by using this exploration method. With the **Motion-JPEG** case study, we illustrate the whole communication exploration process step by step. From experimental results, we show that compared with the cycle accurate simulation, the inter-subsystem communication can be well optimized and evaluated at higher abstraction levels.*

1 Introduction

With the popularity of complex video compression algorithms and high performance network communication protocols, the original single MCU/DSP core based SoC solution can not satisfy the computation requirement with the low power consumption constraint. The industry and academy put more and more emphasis on Heterogeneous *Multi-Processors System-on-Chip* (**MPSoC**) solutions. Generally, such system includes several heterogeneous processors, coprocessors and hardware acceleration blocks. Compared with single core SoC solutions, system architectures of MPSoC are more complex. Designers should fix all parameters of the system architecture to meet constraints of timing, chip area and power consumption. Traditional design space exploration methods normally cause design loops which waste much design time and become the bottleneck of the whole design process. A loop-less design space exploration method is required to solve upcoming MPSoC design challenges.

In one MPSoC, with the increasing number of IP modules, the communication exploration becomes the bottleneck of the whole MPSoC design flow. There are lots of papers talking about communication optimization methods, but most of them need very detailed architecture informations which are only fixed and available at the end of the whole design process [9][10]. Some others have restricted requirements which are hard to meet in general design projects [6][16][15].

The method provided by this paper is more generic and can be well integrated into existed design processes. By using the abstraction level modeling approach, some detailed parameters of the system can be ignored at the beginning. Meanwhile, at high abstraction levels, the simulation speed is much higher than that at the traditional ISS/RTL level. These two advantages provide the ability of fast communication architecture exploration at the beginning of the design process. For some complex MPSoC design projects, this ability can significantly shorten the whole chip design schedule.

The input of our communication exploration flow is the generic application source code which is written in standard C language. After the application is transfered into the well known *Kahn Process Networks* (**KPN**) model [7], we can easily convert them into the system level simulation model. Though the timing information is not available at this level, it provides enough information to extract transfer numbers and the data length of each transfer. This information can be used for the basic inter-subsystem communication optimization.

As the timeless model is not enough for the detailed exploration, we create a new abstraction level which can provide accurate inter-subsystem communication informations

403

978-1-4244-3039-0/08 $25.00 © 2008 IEEE

with relatively fast simulation speed. This level is called the *Virtual Architecture* (**VA**) level in this paper. At VA level, the whole MPSoC is decomposed into several subsystems and one inter-subsystem communication module. All subsystems are described with the timeless model and several tasks are mapped on each of them. However, we use the *Cycle Accurate* inter-subsystem communication model instead of the original timeless system level model to support the detailed exploration.

Finally, we introduce the *Virtual Prototype* (**VP**) level model. At this level, the software is executed by *Instruct Set Simulators* (**ISS**) and hardware is modeled with *cycle accurate* SystemC. The VP level model is used as the golden performance reference model in this paper. With the *Motion-JPEG* case study, we show the inter-subsystem communication exploration process and compare the system simulation speed and communication performance evaluation results at all these three levels. At the end, we get the conclusion that our multi-abstraction levels inter-subsystem communication exploration method is effective for general MPSoC design projects.

The paper is organized as follows. Section 2 provides the background on the communication exploration problem. Section 3 discusses the communication exploration at system level. Section 4 shows the communication exploration at VA level. The VP level is introduced in section 5. Section 6 presents the Motion-JPEG case study and experimental results. Summary and future work are given in section 7.

2 Related Work

Because the communication exploration is an important step of the SoC design flow, many researchers focus on it since more than ten years ago. Some works [6][16][15] try to solve this problem by automatically synthesizing bus-based communication architectures. These works depend on the accurate timing information of each task. Because the accurate execution time of each task depends on inputs and is hard to predict, some methods [9][10] try to eliminate this problem by using cycle-accurate trace files to extract the exact execution time. Generally, all of these methods require static scheduling and fixed task execution times which are strong assumptions at the beginning of design process for normal industrial projects.

There are also some design platforms which support the multi-abstraction levels heterogeneous MPSoC simulation. These platforms can help designers explore communication architectures. Ptolemy [5] mainly focuses on application modeling and simulation. StepNP [12] provides an ISS based system level simulation environment. MetroPolis [8] uses formal methods for the architecture exploration, while Artemis [14] takes the Y chart and trace file to do this work. Compared with these platforms, our method emphasis more

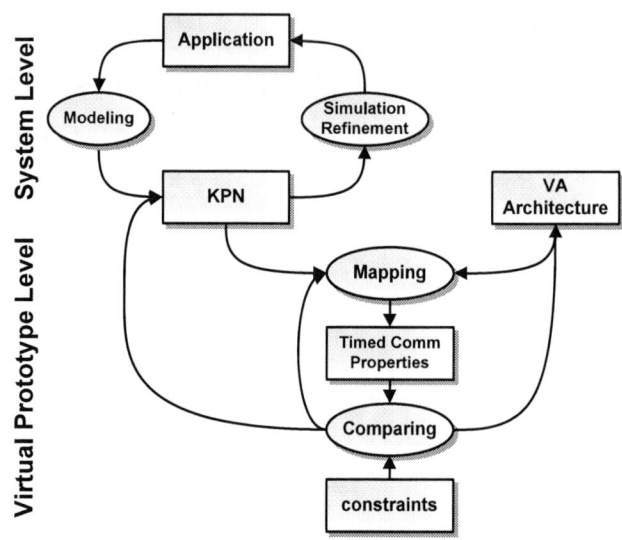

Figure 1. The System Level and Virtual Architecture Level Communication Exploration Flow

on how to use predefined abstraction levels to accelerate the communication exploration.

There is work [11] that aims to use different abstraction level bus models to speed up the simulation. As we found, the application tasks execution is the real speed bottleneck of the simulation. Our paper focuses more on the abstraction of all application tasks instead of only the bus communication.

3 System Level Communication Exploration

In this paper, the *System Level* is the highest abstraction level in our design flow. Because the key point in the MPSoC design is the application parallelism, this abstraction level is designed to provide a parallel execution environment. In our system level, the application is described as a set of communicating processes exchanging data exclusively through blocking lossless point-to-point *First In First Out* (**FIFO**) channels, known as *KPN*. At system level, there is no difference between software and hardware. From Table.1, the application is executed natively and no timing information is given for communications.

There are several advantages of exploring the communication architecture at this level. Though compared with more detailed abstraction levels, application tasks at system level are timeless, the functional simulation can give out the accurate number of transfer and the data size of each transfer. The information can be used to optimize the communication behavior and estimate the required bandwidth for real time requirements. We should notice that the sim-

404

978-1-4244-3039-0/08 $25.00 © 2008 IEEE

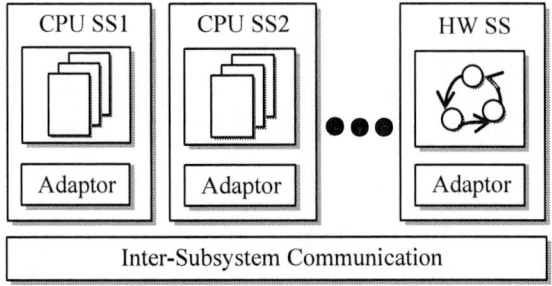

Figure 2. System Architecture at VA Level

	System Level	VA Level	VP Level
Software	executed natively	executed natively	executed by ISS
Hardware	executed natively	executed natively	cycle accurate model
Communication	no timing information	cycle accurate / timed TLM	cycle accurate

Table 1. Difference between Three Abstraction Levels

ulation speed at system level is much higher than that of more detailed levels. This advantage is really useful for the communication exploration of complex applications.

The system level exploration flow is shown in Fig.1. Square blocks represent exploration statuses while ellipses represent actions. From this exploration flow, we first model the original application into a KPN. The system level simulation model is build based on KPN and using SystemC. For communication exploration requirements, we design a specific channel module which captures the number of transfers and transfer sizes. After each channel in the KPN model is implemented, we obtain the executable system with the detailed transfer burst size and transfer numbers. The statistic communication information can be shown in figures which designers can use for bus usage optimization and latency minimization. By modifying the application source code and the KPN modelling, communications can be optimized at this level. The data shown in the experimental section gives detailed information of the system level exploration. Because the simulation speed is really high, design loops at this level are relatively low-cost.

4 VA level Communication Exploration

To continue refining the MPSoC, information provided at system level is not enough. In this section, we introduce a more detailed abstraction level: the *Virtual Architecture* (**VA**) level. By using the VA level, we can continue optimizing inter-subsystem communications. Detailed experimental results will be given at the experimental section.

4.1 VA Level Introduction

The VA level is a more detailed abstraction level compared with the system level. The most important difference is that the VA level includes architecture information. At this level, the architecture is coarse grain and abstracts the real system into *Subsystems* and *Inter-subsystem Communications*. Fig.2 presents a general VA architecture. In this figure, the system is composed with several subsys-

tems which can be divided into *CPU Subsystems* (**CPUSS**) and *Hardware Subsystems* (**HWSS**). All these subsystems are connected with each other by using the inter-subsystem communication module.

The aim of the VA level is to provide a execution model which can give more detailed inter-subsystem information but with relative high simulation speed. From Table.1, we find all the subsystems are timeless models which can significant speed up the system simulation, while the inter-subsystem communication can be cycle accurate or timed TLM to provide detailed timing information for communication exploration.

With all these requirements, we build the VA level platform based on SystemC also. We replace the original timeless communication module at system level with the timed communication module, and we add adaptors to connect subsystems to the communication module with the *Virtual Component Interface* (**VCI**)[3].

4.2 Exploration Process

After communications are optimized at system level, we minimize the communication transfer number and utilize the communication burst mode. At VA level, we continue exploring communications by fixing the type and protocol of the inter-subsystem communication. For communication types, we can choose one type from {bus, crossbar and Network-On-Chip}. After we fix the communication type as bus, we may continue to select one specific protocol from several kinds of buses {APB, PI and so on}.

These exploration works are divided into two steps. The first step is to map application tasks into subsystems. Then we can simulate the whole system at VA level and get inter-subsystem communication performance results. If results are not compatible with system constraints, designers should change the tasks mapping based on inter-subsystem communication performance results and choose another communication type and protocol. The exploration flow is shown in Fig.1 and we give details of these two methods

405

978-1-4244-3039-0/08 $25.00 © 2008 IEEE

separately.

4.2.1 Communication Based Task Mapping

Because computation and communication parameters are related, task mapping becomes really complex. At RTL/binary level, it is a time consuming process to simulate and find a suitable mapping solution. At VA level, there are no computation information because all CPUSS and HWSS are timeless. The designer needs only take into account of communication data during the mapping process. With a specific number of subsystems, we map application tasks from KPN to these subsystems based on the inter-subsystem communication. So a good mapping solution is the one uses all subsystems and requires less inter-subsystem communication. This is the guideline for the VA level communication mapping.

Concerning computation requirements, we satisfy them at the next abstraction level. The performance of one CPUSS can be improved by increasing the number of CPUs, changing the CPU instruction set and adding co-processors. These computation refinement approaches are another topic and not concerned in this paper.

4.2.2 Constraints Based Refinement

Many requirements have to be met in SoC designs. The most important ones are speed, power and chip size requirements. In our exploration method, we decompose these requirements into detailed one for each subsystem and inter-subsystem communication.

With the MJPEG example, the first requirement we meet is the speed. Decoding a stream at real time requires at most 40 ms for one frame. For each subsystem, it should provide enough computation power for all tasks mapped on it. Meanwhile, global communication system should also support enough transfer ability to deliver all inter-subsystem communication requests of one frame in 40 ms. The requirement is decomposed and given to detailed modules. Only all decomposed requirements are well processed at VA level, designers can continue exploring lower abstraction levels.

5 VP Level Introduction

The *Virtual Prototype* (**VP**) level is the abstraction level which is defined to be a cycle accurate simulation platform. From Table.1, we know that at this level, all software tasks are compiled into binary codes and executed by ISS. Both hardware tasks and inter-subsystem communication blocks should be implemented as cycle-accurate SystemC models to provide very accurate simulation results.

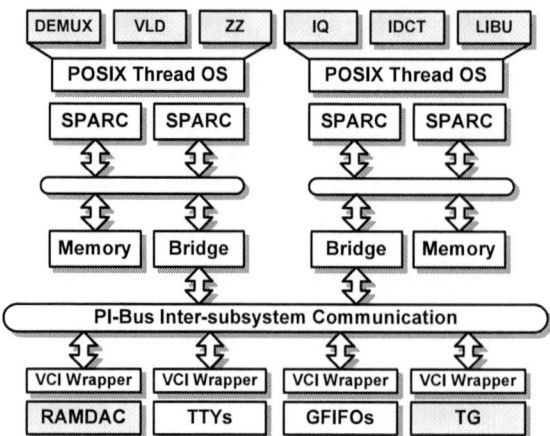

Figure 3. VP Level System Architecture of the Motion-JPEG Case Study

One VP level system architecture example is shown in Fig.3. This system is composed with two CPUSS and two HWSS. All these subsystems are connected with each other by using the inter-subsystem communication module. Each CPUSS includes two SPARC CPUs, one local memory and one local bus. Besides the hardware architecture, we use the *POSIX Thread Operating System* [13] to provide the multi-tasks execution and communication environment for 6 software tasks of the MJPEG case study. For each HWSS, we create a specific *VCI wrapper* which realizes the VCI protocol to connect with the inter-subsystem communication. By adding bridge modules, we separate intra-subsystem communications with inter-subsystem communications. The intra-subsystem communication includes internal memory accesses and all information send/receive between tasks in the same subsystem. The inter-subsystem communication only focuses on data transfers between tasks in different subsystems. All data given at this level is referenced as a golden model for following experiments.

6 Experimental Results

In this section, we present experimental results of the whole Motion-JPEG case study. At first, we show the Motion-JPEG case study. Then we give out how to evaluate and optimize communication properties at system level, VA level and VP level separately. At the end, we show the simulation speed difference between three abstraction levels which is the key advantage of the high level communication exploration.

978-1-4244-3039-0/08 $25.00 © 2008 IEEE

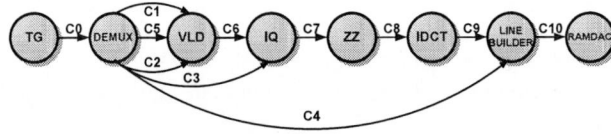

Figure 4. Functional Model of the Motion-JPEG Case Study

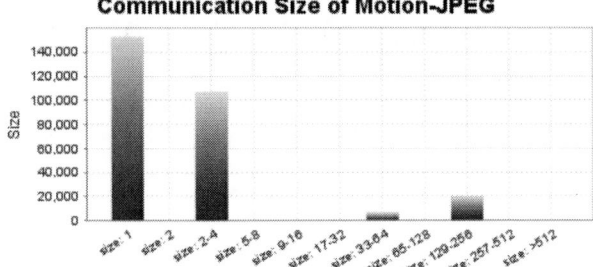

Figure 5. Communication Property before Optimization

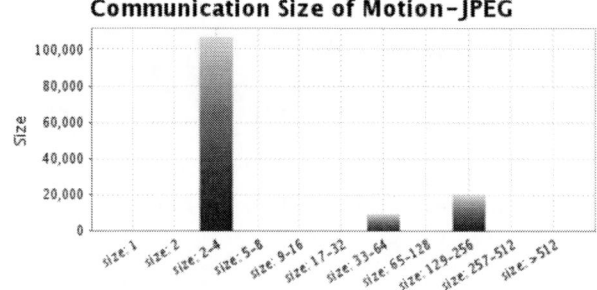

Figure 6. Communication Property after Optimization

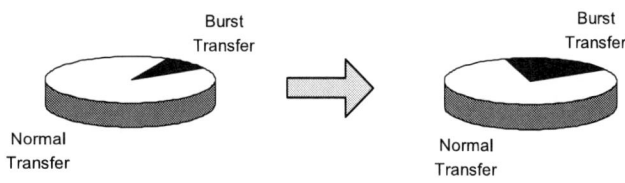

Figure 7. Burst Mode Percentage Change with the System Level Optimization

6.1 Motion-JPEG Case Study

The *Motion-JPEG* is a multimedia format in which a video sequence is separately compressed as JPEG images. This format is often used in mobile applications such as digital cameras. In this case study, we realize the MJPEG decoder application with eight initial tasks: TG, DEMUX, VLD, IQ, ZZ, IDCT, LINE BUILDER and RAMDAC [4]. It works by reading a stream of JPEG images with the Traffic Generator (TG) task and writing decoded pixels into the Random Access Memory Digital-to-Analog Convert (RAMDAC). After some modifications, we get the KPN model of this case study. The model is shown in Fig.4.

Because this paper is aimed at the communication architecture exploration, we focus on 11 channels (C0 to C10) which transfer data and control commands between each task. In following parts, we will explore and optimize these channels by using the system level platform and the VA level platform.

6.2 Optimization at System Level

Because the bus is the most popular and well used interconnection in SoC designs, in this MJPEG case study, we choose bus protocols as our inter-subsystem communication. One of the most interesting property of bus communication models is the burst mode. Most of the modern bus protocols such as ARM AHB [1] and PI-Bus [2] support the burst mode. Though they may have different burst prop-

erties, the transfer size is always the key point. The burst mode is effective only when the transfer size is larger than a specific number.

Fig.5 shows the transfer size property of the MJPEG application before optimizations. From this diagram, we find that there are lots of 1 byte transfers. Because inter-subsystem communication is normally uncached to avoid the cache coherency expense, these small transfers are low efficiency. Our optimization method is to merge these small data transfers together and transfer them by using the burst mode provided by bus protocols.

The transfer size property after optimization is shown in Fig.6. We can clearly find that all single byte data transfers are eliminated. To evaluate the effectiveness of this optimization, we take the *PI-Bus* models as the reference protocol. Fig.7 shows the percentage of burst mode used under the PI-Bus protocol. From this diagram, we notice the burst ratio is significantly improved. The effectiveness of the system level communication exploration is obvious.

6.3 Optimization at VA Level

From the VA level communication exploration flow given at Fig.1, task mapping is extremely important for the whole system performance. We design a VA architecture which includes two CPU subsystems and two HW subsystems. The two HW subsystems are used by TG and RAMDAC tasks while two CPU subsystems are free for mapping

407

978-1-4244-3039-0/08 $25.00 © 2008 IEEE

CPUSS1 Tasks	CPUSS2 Tasks	Comm Size
DEMUX,VLD,IQ	ZZ,IDCT,LIBU	2.2MB
DEMUX,LIBU	VLD,IQ,ZZ,IDCT	1.0MB
DEMUX,VLD,ZZ	IQ,IDCT,LIBU	5.6MB

Table 2. Communication Size with Different Mapping Solutions (25 Frames MJPEG)

	Sim Time(s)	Speed Improvement
System Level	0.08	5913x
VA Level(CA)	0.9	526x
VP Level	473	1

Table 3. Simulation Time Difference

6 software tasks. In Table.2, we show three different mapping solution and the final inter-subsystem communication size. After 25 frames Motion-JPEG decoder simulation, the 3rd solution needs 5.6 MB communication size which is more than five times compared with that of the 2nd solution. From this table, we can easily understand advantages of task mapping at the beginning of the whole design process.

6.4 Simulation Speed at Different Abstraction Levels

The speed is the key advantage of high abstraction level simulation models. We build the simulation model at system level, VA level and VP level. Table.3 gives out the simulation speed at these three different levels.

From this table, we clearly find that the higher abstraction level requires less simulation time. Compared with the most detailed VP level, the system level simulation model can finish one frames MJPEG decoding with 5913 times faster. The simulation speed at the VA level with cycle accurate communication is also much higher than that of the VP level , so this level can provide the possible to test many different mapping solutions to meet all those constraints.

7 Conclusions and Future Work

In this paper, we introduce a new inter-subsystem communication exploration method. With this method, we can optimize the communication fast and avoid much design loops by using two relatively high abstraction levels. With the Motion-JPEG case study, we find that the simulation at the high abstraction level is much faster than at the traditional cycle accurate VP level. Our inter-subsystem communication exploration method is proven to be useful for future applications.

In the near future, we will finish the intra-subsystem communication exploration method and the computation exploration method to build the whole design space exploration flow.

References

[1] AHB 3 Specification. [Online]. Available: http://www.arm.com/products/solutions/AMBA3AXI.html.

[2] Draft Standard omi 324: PI-Bus, rev. 0.3d, open microprocessor systems initiative. Siemens AC. Munich., 1994. [Online]. Available: http://www.cordis.lu/esprit/src/omihome.htm.

[3] On-chip bus development working group. virtual component interface standard version 2. VSI Alliance, April 2001. OCB 2 2.x. [Online]. Available:http://www.vsia.org.

[4] I. Augé, F. Pétrot, F. Donnet, and P. Gomez. Platform-based design from parallel c specifications. *IEEE Trans. on CAD of Integrated Circuits and Systems*, 24(12):1811–1826, 2005.

[5] J. Buck, S. Ha, E. A. Lee, and D. G. Messerschmitt. Ptolemy: A framework for simulating and prototyping heterogenous systems. *Int. Journal in Computer Simulation*, 4(2):0–, 1994.

[6] M. Gasteier and M. Glesner. Bus-based communication synthesis on system-level. In *ISSS*, pages 65–70, 1996.

[7] G. Kahn. The semantics of simple language for parallel programming. In *IFIP Congress*, pages 471–475, 1974.

[8] S. Kakita, Y. Watanabe, D. Densmore, A. Davare, and A. L. Sangiovanni-Vincentelli. Functional model exploration for multimedia applications via algebraic operators. In *ACSD*, pages 229–238. IEEE Computer Society, 2006.

[9] K. Lahiri, A. Raghunathan, and S. Dey. Efficient exploration of the soc communication architecture design space. In E. Sentovich, editor, *ICCAD*, pages 424–430. IEEE, 2000.

[10] P. Lieverse, P. V. D. Wolf, K. Vissers, and E. Deprettere. A methodology for architecture exploration of heterogeneous signal processing systems. *J. VLSI Signal Process. Syst.*, 29(3):197–207, 2001.

[11] S. Pasricha and M. Ben-Romdhane. Using tlm for exploring bus-based SoC communication architectures. In *ASAP*, pages 79–85. IEEE Computer Society, 2005.

[12] P. G. Paulin, C. Pilkington, and E. Bensoudane. StepNP: A system-level exploration platform for network processors. *IEEE Design & Test of Computers*, 19(6):17–26, 2002.

[13] F. Pétrot and P. Gomez. Lightweight implementation of the posix threads api for an on-chip mips multiprocessor with vci interconnect. In *DATE*, pages 20051–20056. IEEE Computer Society, 2003.

[14] A. D. Pimentel, L. O. Hertzberger, P. Lieverse, P. van der Wolf, and E. F. Deprettere. Exploring embedded-systems architectures with artemis. *IEEE Computer*, 34(11):57–63, 2001.

[15] A. Pinto, L. P. Carloni, and A. L. Sangiovanni-Vincentelli. Constraint-driven communication synthesis. In *DAC*, pages 783–788. ACM, 2002.

[16] T.-Y. Yen and W. Wolf. Communication synthesis for distributed embedded systems. In R. L. Rudell, editor, *ICCAD*, pages 288–294. IEEE Computer Society, 1995.

978-1-4244-3039-0/08 $25.00 © 2008 IEEE

An NoC Test Strategy Based on Flooding with Power, Test Time and Coverage Considerations

[1,2]Mahshid Sedghi, [2]Elnaz Koopahi, [2]Armin Alaghi, [1]Mahmood Fathy and [2]Zainalabedin Navabi

[1]*Computer Engineering Department*
Iran University of Science and Technology
16846-13114 Tehran, Iran

[2]*Electrical and Computer Engineering Department*
Faculty of Engineering, Campus #2
University of Tehran, 14399 Tehran, IRAN

[1] *ma_sedghi@comp.iust.ac.ir; mahfathy@iust.ac.ir*
[2] *{koopahi, armin, navabi}@cad.ece.ut.ac.ir;*

Abstract

A test strategy for testing NoC switches based on flooding is presented in this paper. This test strategy tests all switch ports and network routes, while it avoids sending a test packet arriving at a switch in every direction. This test strategy is referred to as pseudo-exhaustive, versus the exhaustive testing that sends an incoming test packet of a switch in every direction. As compared with the exhaustive strategy, the pseudo-exhaustive testing consumes lower power consumption, has a lower test time and still has 100% switch port fault coverage. This paper discusses our test strategy, test mode switch hardware requirements, and evaluates test power, time, and coverage.

1. Introduction

Global on-chip interconnects for the billion transistor System-on-Chips (SoCs) are becoming a bottleneck in fast transmission of data between various SoC processing elements [1, 2]. A possible solution for this performance degradation is the use of on-chip packet-based communication, giving rise to what is generally referred to as Network-on-Chips (NoCs).

A typical NoC consists of switches to route the data packets, interfaces that connect each core to a switch in the NoC, and interconnections among the switches. Switches are connected together in a regular (e.g., mesh-based) or irregular topology [3]. NoC switches consist of FIFO buffers and the routing logic.

For the NoCs to be effectively used as a means of implementation of complex digital systems, an efficient test strategy must be devised for this technology. DfT methods including scan chains and BIST architectures have been proposed for NoC structural testing.

While scan testing NoC switches targets stuck-at faults, higher level testing using data packets targets higher level routing and switching faults. Analysis of test methods targeting stuck-at faults requires exact hardware model of the circuit under test, while higher level fault analysis requires hardware models that can be injected by such switch level or routing faults [4].

The work described in this paper focuses on high level testing of NoC switches. Some of our previously proposed high-level fault models and a high-level test platform will be used in this work. The primary goal of this test platform is evaluation of test packet requirements and test packet routing methods into an NoC-under-test. This test method, to which we refer to as a test strategy, is evaluated by our platform for test time, high level fault coverage, and power consumption. A pseudo-exhaustive test strategy for NoC switches will be presented which is a modification of the previously presented test method named exhaustive test strategy [4]. While keeping the full fault coverage of the exhaustive test strategy, we will show that pseudo-exhaustive test strategy is more efficient both for test time and total power consumption.

The rest of this paper is organized as follows. Some related works on NoC switch test strategies will be reviewed in Section 2. Section 3 presents the description of the NoC architecture used in this paper, the test platform used for our simulations, and our proposed high level fault model. Section 4 gives a brief description of the *exhaustive* test strategy. Disadvantages of this strategy will also be mentioned here. Description of the *pseudo-exhaustive* test strategy along with its advantages over the *exhaustive* strategy will be presented in Section 5. Finally, the results of comparison of test time, total number of the packets, power consumption and switch test hardware overhead necessary for these two test strategies will be given in Section 6. This section is followed by conclusions in Section 7.

2. Related Works

A wide variety of standard Design-For-Test (DFT) techniques has been proposed for testing of NoC based designs [5]. Hosseinabady et al. [6] propose a scan-based concurrent testing method for NoC switches. Serial test vectors are generated by a test source and then are broadcasted through the scan chains of all of the switches. Fault detection is carried out through comparison of output responses of switches with each other. All switches in the network are assumed to be identical. However, in [3] it is shown that this test mechanism can be extended to cover different switch structures in an NoC.

In [7], different test methodologies have been proposed for the FIFO and routing logic of an NoC switch block; they have inserted scans as a DFT strategy for routing logic blocks, while a distributed BIST methodology is used for FIFO buffers. Their reason for using a distributed BIST methodology for FIFOs is that they believe that using a dedicated BIST mechanism per FIFO will raise the silicon area overhead unacceptably.

The works described above take the structural test approach, while the works done in [8, 9] test NoC interconnect network in a functional-like mode. In [8], Petersen et al. propose a scalable BIST strategy for testing 2D-mesh NoCs which achieves almost full fault coverage for logic-level stuck-at faults. Their proposed methodology is able to detect faults as well as specifying their locations. Eventually, they use the information attained from fault detection and location to reconfigure the NoC architecture to keep the full functionality of the NoC. Another functional logic BIST test strategy for 2-D mesh NoCs is presented in [9]. The timing of this BIST does not depend on the NoC size. The authors have shown that the only test logic inserted into each switch is a number of 2-to-1 multiplexers which has a very less hardware overhead compared to conventional scan approaches.

The works reviewed so far deal with logic-level stuck-at fault model, while some research has been done recently on targeting faults modeled at higher levels of abstraction. For example, Raik et al. [10] propose an scalable test method for NoC switches which is based on functional fault models. The proposed functional fault model targets single stuck-at faults in multiplexers and registers of the switches.

A non-scan method for testing NoC switches is presented in [11] which targets specific high level direction faults that are modeled by *stuck-at port* fault model. In case of an *stuck-at port* fault, all packets in the input ports of a switch are sent to a specific output port regardless of their destination. Different methods of test packet generation are also introduced and their test times are compared. [12, 4] use the same fault model: in [12] NoC switches are tested in an online fashion and fault diagnosis is carried out. Several methods of fault detection are introduced and their area overhead and fault coverage are compared with each other. [4] proposes a test strategy for NoC switches which detects 100% of *stuck-at port* faults. Because of this coverage, this test method is referred to as *exhaustive* test strategy.

3. Preliminaries

This section briefly describes the NoC architecture we have used, *stuck-at port* fault model, and our test platform.

3.1. Our NoC Architecture

The NoC we will use has a 2-D mesh architecture and uses XY routing algorithm, where a packet is first routed in the X direction and then in the Y direction before reaching its destination. We also assume that there are only two externally accessible switches in the mesh that we refer to as Test Access Switches (TASs). One of the TASs is located at the lower left of the mesh (TAS1) and the other is at the upper right (TAS2). Processing elements connected to TASs can act as test packet generators and output packet analyzers in addition to their specific operations. In other words, they are test sources and sinks of the circuit. Switching is based on packet switching approach where a packet consists of several fields like source address, destination address, and data payload.

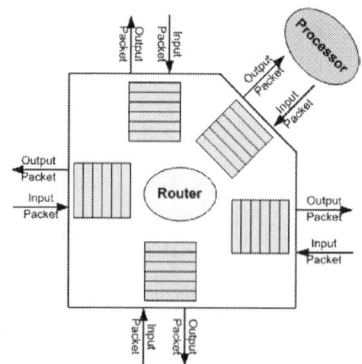

Figure 1. NoC switch architecture

Figure 1 shows the structure of our NoC switch. Each switch has five I/O ports, four ports for connecting each switch to adjacent switches and one port to connect the switch to its related processor.

3.2. Stuck-at port Fault Model

In general, NoC switch faults can be divided into two categories: data faults and control faults. The data faults have to do with data contents of the packets (payload), while the control faults deal with faults in routing data in switches and between processing elements. The *stuck-at port* fault model proposed in [11] focuses on this kind of faults in NoC switches. In case of an *stuck-at port* fault in a switch, all packets in the input ports are sent to a specific output port regardless of their destination. Accordingly, 5 *stuck-at port* faults are introduced:

1. Stuck-at East, *SaE*
2. Stuck-at South, *SaS*
3. Stuck-at West, *SaW*
4. Stuck-at North, *SaN*
5. Stuck-at Processor, *SaP*

In [11] it is also mentioned that other switch control faults such as packet dropping, lost-destination and mis-routing can also be modeled with these faults.

3.3. Test Platform

We will use a high level VHDL-based platform that we developed in [4] for fault simulation, fault coverage calculation, test time measurement, and test packet number calculation. Our test platform consists of models of actual NoC switches and processing elements in the test mode. The platform is modified in order for it to support *pseudo-exhaustive* test strategy.

4. Exhaustive Test Strategy

In this section, we first explain the main motivation of *exhaustive* test strategy and then give a brief description of this test method.

4.1. Flooding Routing Technique

The main idea of *exhaustive* test strategy is taken from a routing technique used in packet-switched networks called flooding and works as follow: A packet is sent by a source node to every one of its neighbors. At each node, an incoming packet is retransmitted on all outgoing links except for the link on which it arrived. Unless something is done to stop the incessant retransmission of packets, the number of packets in circulation just from a single source packet grows without bound. A simple technique to keep track of the traffic is to include a hop count field with each packet. The count can originally be set to some maximum value, e.g. length of the longest minimum-hop path through the network.

Each time a packet passes through a switch, the switch decrements its hop count by one. When the count reaches zero, the packet is discarded.

4.2. Flooding for Fault Tolerant NoCs

Flooding is an effective fault tolerant technique; therefore, some variations of it have been proposed as fault tolerant communication algorithms. Dumitras et al. [13] propose a probabilistic flooding scheme based on the well known gossip techniques as a possible fault tolerant solution [14, 15]. In this algorithm, every time a message is generated, it will be passed to all of its neighbors with probability p, and will be dropped with probability $1 - p$. Parameter p must be carefully chosen to obtain near flooding performance without sending as many redundant copies of the message [16].

The probabilistic flooding algorithm is destination ignorant in that packets get routed irrespective of where they are in the network. This results in significant network resources being used to transmit the packet to areas of the chip where it will not be needed. The directed flooding algorithm makes use of an NoC's highly regular structure to direct a flood towards the destination. In this algorithm, the probability p of passing a message to each outgoing link is not fixed and varies according to the destination of the packet [16].

Although flood-based communication algorithms are highly fault tolerant, it has been shown that they have an exceedingly high communication overhead [16]. Since this level of fault tolerance is usually not necessary, we assume that the NoC we are designing routes packets in the XY fashion in its normal mode, and does flooding only in its test mode.

4.3. Test Mode Operation

For our *exhaustive* test strategy, we are using the concept of *Manhattan* distance. The *Manhattan* distance between the point P_1 with coordinates (x_1, y_1) and the point P_2 at (x_2, y_2) is $|x_1 - x_2| + |y_1 - y_2|$. A *Manhattan* path between two points is an alternating sequence of horizontal and vertical segments and its length is equal to their *Manhattan* distance. Figure 2 shows several *Manhattan* paths between the two TASs.

Exhaustive test strategy works as follows: the processing element connected to TAS1 generates a packet and sets its destination field value to TAS2 address and sets its hop count field value to *the Manhattan* distance between the two TASs and then floods it as discussed in Subsection 4.1. Simultaneously the same sequence of actions will take place in TAS2. For TAS2, the generated packet will have a destination field value equal to the address of TAS1. Every switch in the mesh decre-

411

978-1-4244-3039-0/08 $25.00 © 2008 IEEE

ments the hop count field value of its incoming packets by 1 unit and floods them to its output ports. Packets arrived at each switch with hop count filed value of zero will be discarded. Also, packets arrived at each TAS with a destination filed other than its address will be discarded.

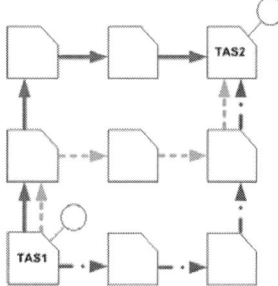

Figure 2. Manhattan paths between the TASs

According to the combinational analysis, the total number of the *Manhattan* paths between the two TASs can be calculated by the following expression:

$$K(m , n) = \frac{(m - 1 + n - 1)!}{(m - 1)! \, (n - 1)!}$$

where *m, n* are the number of the rows and columns of the mesh, respectively [17].

If no *stuck-at port* fault exists in the network, the total number of the packets arrived at each TAS will be equal to the total number of *Manhattan* paths, i.e. *K(m,n)*; otherwise, the total number of the packets received at each TAS will be less than *K(m,n)*.

It should be mentioned that a test-mode controller is responsible for the operation of the network in test mode. The controller configures the switches so that they will be able to flood, and also sets a counter in the processing elements connected to the TASs to count the total number of the packets received at each TAS.

The main advantage of the *exhaustive* test strategy is its simple implementation, and low hardware overhead. A very small amount of hardware must be added to switches for their test mode operations. In addition, this method has 100% *stuck-at port* fault coverage. Another advantage is that the packets are routed along all paths from source to destination. This detects some faults many times and therefore N-detect coverage is achieved which strengthens the method further.

4.4. Disadvantages of Exhaustive Strategy

The main disadvantage of *exhaustive* test strategy is the high traffic load that it generates, and its high power consumption. The main problem with this strategy is that some of the packets generated while flooding do not travel toward the test sink and gets discarded in the

midway. The second disadvantage is that using this test strategy, the routing logic will not be tested. This is because the test-mode controller will manage routing of the packets in the network regardless of the routing decided by the routing logic. For overcoming these problems, we have proposed a *pseudo-exhaustive* test strategy which will be described in the next section.

5. Pseudo-exhaustive Test Strategy

Pseudo-exhaustive test strategy is similar to the *directed flooding* algorithm described in Subsection 4.2. In this strategy, every switch in the network sends its incoming packets in both XY and YX fashion simultaneously, thus generating only two copies of each incoming packet. This strategy yields lower power consumption and lower test time compared to the *exhaustive* test strategy. At the same time full fault coverage for *stuck-at port* faults is still retained.

Therefore, if we design our switches so that they can route packets in both XY and YX fashion at the same time, we will be able to test the routing logic embedded in switches. Adding the capability of routing packets in YX fashion in the normal mode as well as test-mode will also make the NoC design more fault tolerant Figure 3 shows the operation of the network in test mode applying *pseudo-exhaustive* test strategy.

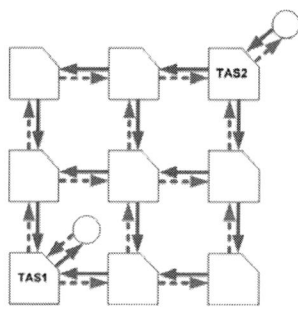

Figure 3. *Pseudo-exhaustive* test strategy

Some of the advantages of *pseudo-exhaustive* strategy over *exhaustive* strategy include a simpler test mode operation that simplifies the analysis of the test time, and power consumption. Test time and power consumption become analytically predictable. Furthermore, there will be fewer number of packets thus lowering the power consumed in the NoC during test session. Finally, the *pseudo-exhaustive* method makes it possible to test the routing logic of the switches.

6. Experimental Results

We evaluated previously proposed *exhaustive* test strategy as well as *pseudo-exhaustive* test strategy for different NoC sizes of up to 10x10. These strategies were simulated in our high level VHDL- based platform. Test time, total number of test packets and energy consumption and hardware overhead for these test strategies will be discussed in the following subsections and a comparison between their results will be done to illustrate the efficiency of *pseudo-exhaustive* over *exhaustive* test strategy. As we mentioned in the previous section, analysis of *pseudo-exhaustive* is simpler than the *exhaustive* strategy. Therefore, we have been able to extract formulas for calculating test time and power consumption for *pseudo-exhaustive* strategy, while the corresponding values for *exhaustive* strategy has been obtained by simulation.

6.1. Test Time

Test time of the *pseudo-exhaustive* strategy begins when a test packet is sent from a TAS and ends when the last test packet arrives at the other TAS. This test time is calculated as follows:

$$T(m , n) = K(m, n) + (m + n - 2)$$

Figure 4 shows the test time of the *exhaustive* and *pseudo-exhaustive* test strategies on logarithmic scale for different NoC sizes.

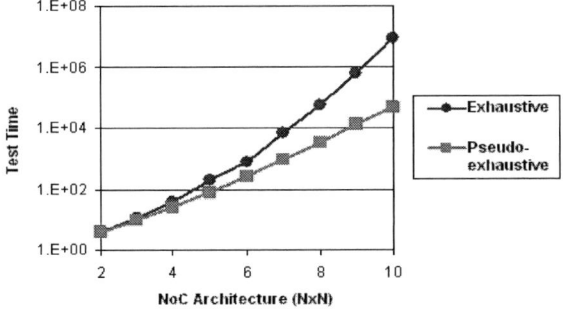

Figure 4. Test time comparison

As shown, the *pseudo-exhaustive* strategy has a significantly lower test time than the *exhaustive* strategy. This difference becomes more significant as the size of the NoC increases.

6.2. Total Number of Test Packets

As discussed before, the processing element connected to TAS1 generates a packet and sets its destination field value to TAS2 address. Every switch in the network sends its incoming packets in both XY and YX fashion simultaneously, generating two copies of each incoming packet. The packet generating path looks like an incomplete binary tree with height of *m+n-1*. A complete tree with *m+n-1* height has $2^{m+n-1} - 1$ nodes.

nodes. The number of packets that reaches a switch is equal to $k(i,j)$ in which i and j represent the location of the switch. Boundary switches, which do not have any adjacent switch on their east or north, generate only one copy of incoming packets. These boundary switches are located in column n or row m.

The same sequence of actions takes place in TAS2, but this time the generated packet has a destination field value equal to the address of TAS1. Therefore, the total number of packets can be obtained using this formula:

$$P(m , n) = 2 \times [2^{m+n-1} - 1 - \sum_{j=1}^{n-1} k(m , j) \times (2^{n-j} - 1)$$

$$- \sum_{i=1}^{m-1} k(i , n) \times (2^{m-i} - 1)]$$

The graph shown in Figure 5 represents the total number of test packets of *Pseudo-exhaustive* and *exhaustive* test strategies for different NoC sizes.

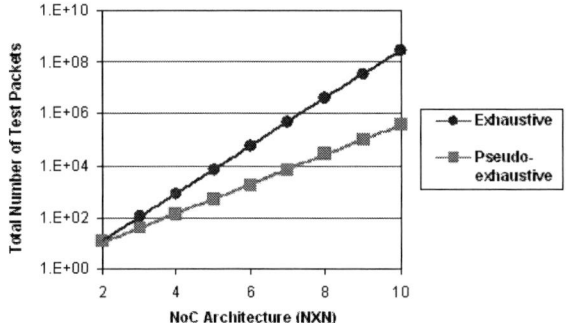

Figure 5. Comparing number of test packets

6.3. Energy Consumption

In order to compare different test strategies, a method for estimating energy consumption is needed. Typically, a design's energy consumption is evaluated using a register transfer level or gate level power simulator with a lengthy run-time. A fast estimation model can provide useful insights into different test strategies.

Hu et al. [18] propose an energy model for NoCs. They show that for a 2-D mesh network with routing whose total communication energy is minimal, the average energy consumption of sending one bit of data from a node to another node in the network is determined by the *Manhattan* distance between them.

Energy consumption in the NoC in a high- level model can be estimated by the number of switched packets and the length of the path they pass through. Therefore, we can compare the energy consumption of different test strategies implemented on a specific NoC architecture by comparing sum of the lengths of the paths through which test packets travel.

In the *pseudo-exhaustive* strategy, almost all packets travel only one link between switches. This excludes those that are generated in the processing elements connected to TAS1 or TAS2 and go only to that TAS. Therefore, the total energy consumption in this strategy can be estimated as follows:

$$E(m,n) = [P(m,n) - 2] \times e$$

where e is the average amount of energy required for sending a packet from a node to its adjacent node.

6.4. Hardware Overhead

The proposed method has a small amount of hardware overhead. In the normal mode, the routing algorithm has to check a received packet's destination and make a decision based on the XY routing algorithm, and assigns a proper output multiplexer to it. In the test mode, in addition to the decision made based on XY routing, another output port must also be selected based on the YX routing scheme. So, upon receiving a packet, the two multiplexers identified as such, duplicate the incoming packet to enable multicasting. The amount of the hardware overhead remains the same for larger NoCs, thus making it a less significant overhead.

7. Conclusions

This paper discussed a method of NoC switch testing that we referred to as the *pseudo-exhaustive* test strategy that is based on the network switching algorithms. The proposed method is an alternative of the previously proposed test method called *exhaustive* test strategy. The only requirement for this method is that switches must be enabled to route packets both in XY and YX fashion simultaneously; a relatively small hardware overhead is added to each switch for this test strategy. The packets propagate through the entire NoC and a fault is detected by counting the number of packets received in the TASs.

We simulated these two test strategies in a high level VHDL-based test platform for NoCs of different sizes using a high level fault model. The simulation results showed that the *pseudo-exhaustive* test strategy consumes less power and its test time is smaller as compared with the *exhaustive* test strategy. The fault coverage of the *pseudo-exhaustive* test strategy is the same as that of *exhaustive* test strategy.

8. References

[1] P.P. Pande, C. Grecu, A. Ivanov, R. Saleh, and G. De Micheli, "Design, synthesis, and test of Networks on Chips", *IEEE Design & Test of Computers*, Vol. 22, No. 5, September–October 2005, pp. 404-413.

[2] C. Grecu, A. Ivanov, R. Saleh, E.S. Sogomonyan, and P.P. Pande, "On-line Fault Detection and Location for NoC Interconnects", Proc. *IEEE International Symposium on On-Line Testing (IOLTS)*, Lombardy, Italy, July 2006, pp. 145-150.

[3] M. Hosseinabady, A. Dalirsani, and Z. Navabi, "Using the Inter- and Intra-Switch Regularity in NoC Switch Testing", Proc. *Design Automation and Test in Europe (DATE)*, Nice, France, April 2007, pp. 361-366.

[4] M. Sedghi, E. Koopahi, A. Alaghi, M. Fathy, and Z. Navabi, "An Exhaustive Test Strategy Based on Flooding routing for NoC Switch Testing", Proc. *IEEE East-West Design and Test Symposium (EWDTS)*, Yerevan, Armenia, September 2007, pp. 262-267.

[5] R. Ubar, and J. Raik, "Testing Strategies for Network on chip", *Networks on Chip*, Jantsch, A., and H. Tenhunen (ed.), Kluwer Academic Publisher, January 2003, pp. 131-152.

[6] M. Hosseinabady, A. Banaiyan, M.N. Bojnordi, and Z. Navabi, "A Concurrent Testing Method for NoC Switches", Proc. *Design Automation and Test in Europe(DATE)*, MESSE Munich, Germany, March 2006, pp. 1171-1176.

[7] C. Grecu, P.P. Pande, B. Wang, A. Ivanov, and R. Saleh, "Methodologies and Algorithms for Testing Switch-Based NoC Interconnects", Proc. *IEEE International Symposium on Defect and Fault Tolerance in VLSI Systems(DFT)*, California, USA, October 2005, pp. 238-246.

[8] K. Petersen, and J. Oberg, "Toward a Scalable Test Methodology for 2D-mesh Network-on-Chips", Proc. *Design Automation and Test in Europe (DATE)*, Nice, France, April 2007, pp. 367-372.

[9] K. Petersen, J. berg, and B. Magnhagen, "Towards an Almost C-Testable NoC Test Strategy", Proc. *IEEE East-West Design and Test Symposium (EWDTS)*, Yerevan, Armenia, September 2007, pp. 126-133.

[10] J. Raik, V. Govind, and R. Ubar, "An External Test Approach for Network-on-a-Chip Switches", Proc. *Asian Test Symposium (ATS)*, Fukuoka, Japan, November 2006, pp. 437-442.

[11] M. Sedghi, A. Alaghi, E. Koopahi, and Z. Navabi, "An HDL-Based Platform for High Level NoC Switch Testing", Proc. *Asian Test Symposium (ATS)*, Beijing, China, October 2007, pp. 453-458.

[12] A. Alaghi, N. Karimi, M. Sedghi, and Z. Navabi, "Online NoC Switch Fault Detection and Diagnosis Using a High Level Fault Model", Proc. *IEEE International Symposium on Defect and Fault Tolerance in VLSI Systems (DFT)*, Rome, Italy, September 2007, pp. 21-29.

[13] T. Dumitras, S. Kerner, and R. Marculescu, "Towards on-chip fault-tolerant communication", Proc. *Asia and South Pacific Design Automation Conference (ASP-DAC)*, Kitakyushu, Japan, January 2003, pp. 225-232.

[14] S.M. Hedetniemi, T. Hedetniemi, and A.L. Liestman, "A survey of gossiping and broadcasting in communication networks", *NETWORKS*, Vol. 18, No. 4, December 1988, pp. 319-349.

[15] D.W. Krumme, G. Cybenko, and K.N. Venkataraman, "Gossiping in minimal time", *SIAM Journal on Computing*, Vol. 21, No. 1, February 1992, pp. 111-139.

[16] M. Pirretti, G.M. Link, R.R. Brooks, N. Vijaykrishnan, M. Kandemir, and M.J. Irwin, "Fault tolerant algorithms for network-on-chip interconnect", Proc. *IEEE International Symposium on Very Large Scale Integration (ISVLSI)*, Louisiana, USA, February 2004, pp. 46-51.

[17] Grimaldi, R.P., *Discrete and Combinatorial Mathematics: An Applied Introduction*, Addison Wesley, October 1998.

[18] J. Hu, and R. Marculescu, "Energy- and Performance-Aware Mapping for Regular NoC Architectures", *IEEE Trans. on Computer-Aided Design of ICs and Systems*, Vol. 24, No. 4, April 2005, pp. 551-562.

21st International Conference on VLSI Design

High-Level Modeling Approach for Analyzing the Effects of Traffic Models on Power and Throughput in Mesh-based NoCs

S. Koohi[+], M. Mirza-Aghatabar[+], S. Hessabi[*], M. Pedram[†]

[+*]Sharif University of Technology, Tehran, Iran, [†]University of Southern California, CA, USA
[+]{Koohi,Aghatabar }@ce.sharif.edu, [*]Hessabi@sharif.edu, [†]Pedram@usc.edu

Abstract

Traffic models exert different message flows in a network and have a considerable effect on power consumption through different applications. So a good power analysis should consider traffic models. In this paper we present power and throughput models in terms of traffic rate parameters for the most popular traffic models, i.e. Uniform, Local, HotSpot and First Matrix Transpose (FMT) as a permutational traffic model. We also select Mesh topology as the most prominent NoC topology and validate the presented models by comparing our results against simulation results from Synopsys Power Compiler and Modelsim. From the comparison, we show that our modeling approach leads to average error of 2% for power and 2.8% for throughput modeling.

1. Introduction

NoC is an efficient on-chip communication architecture for SoC architectures that is structured, reusable, scalable, and has high performance [1][2]. Nowadays, power issues are becoming of primary design constraints for even very high-end microprocessors. Different applications impose different traffic patterns into the network, which have a direct effect on behavior of power consumption. So, modeling this behavior in terms of traffic patterns leads to power consumption changes under traffic variation.

While network performance analysis due to different message flow (which stems from different traffic pattern) based on queuing and markov models, has been studied rigorously in the past [3][4][5], but network power analysis has not been explored.

N.Eisley proposed a framework in [6] that takes as input, message flows, and derives a power profile of the network fabric. Although this framework provides accurate power estimation, it is not always applicable. For instance when the generated traffic by application is not specified in terms of message flows, this method is not applicable. After partitioning and mapping stages, which lead to process allocation, we can specify nearly accurate traffic load between different nodes which can be summarized in terms of some popular traffic models. For example, if we have a memory node which acts as a global memory for all other nodes, we can predict

that the traffic behavior is a certain percentage of the hotspot traffic model. In all, specifying traffic patterns in terms of some predefined traffic models is easier than determining message flow, which enforces us to implement, or at least simulate the whole application.

None of these works has modeled power consumption and throughput in *high level of abstraction,* and therefore, enforces designers to provide a detailed knowledge on generated traffic. In this paper, we present high level power and throughput models in terms of traffic parameters for the most popular traffic models, i.e. Uniform, Local, HotSpot and FMT, in Mesh topology as the most prominent NoC topology. In Section 2, fundamental concepts such as traffic models and evaluation metrics are presented. Section 3 presents experimental results, and finally in Section 4, we conclude our work and give the summary.

2. Fundamental Concepts

In this section, we briefly explain power and throughput as our modeled performance criteria, and also present selected traffic models.

2. 1. Power and Throughput

We have modeled power and throughput in the mesh architectures under different traffic models with XY routing algorithm.

Definition 2.1: In message passing systems, throughput (TP) may be defined as average number of flits arrived at their destination node for each node in each cycle [7].

Power consumption in NoCs consists of two components, the power consumed in routers, and the one associated with links. We calculated router s power consumption using Synopsys Power Compiler. Link s power consumption is determined as follows:

$$P_{link} = \alpha C_{wire} V_{DD}^{2} f \qquad (2.1)$$

We have calculated the power dissipated in links of each router, and used UMC18 [8]. The switching activity of each link is extracted from simulation. The length of metal wires is estimated as 2mm for the mesh topology. Based on the above data and technology s parameters [9], we have $C_{wire} = 0.7 \, pf.$

2.2. Traffic Models

Based on address distribution method [10], we define our traffic models as follows:

Uniform: each node sends its messages to any other node with equal probability.

HotSpot: each node sends specific portion of its generated messages to the hot node.

Local: each node sends specific portion of its generated messages to its neighbors within a predefined distance, called neighborhood radius.

FMT (First Matrix Transpose): In this traffic model the destination address for each node is fixed [11]. Each node with address of (i,j) in an $M \times N$ topology sends its message to a node with address of $(M-1-j, N-1-i)$. This traffic is generated by different applications such as Fast Fourier Transform (FFT).

Usually, generated traffic is composed of multiple address distribution methods, and therefore it is necessary to consider their composite effect while modeling power and throughput of the NoC.

3. Power and Throughput Analysis

The goal of power and throughput modeling under different traffic models is to predict power and throughput at high level of abstraction and their behavior with traffic pattern changes. Our presented power and throughput models are based on described traffic models in Section 2. Therefore, generated traffic patterns in our mesh-based NoC consist of these traffic models with l (local), h (hotspt), f (FMT), and u (uniform) percentages which vary in the range of $(0,1)$ and satisfy the relation $l+h+f+u = 1$.

Definition 3.1: *Pure* traffics are defined as traffics with one nonzero parameter among f, l, and h. For example, in a pure local traffic with $l=0.8$, in each node, 80% of packets are sent locally and 20% of them are sent uniformly.

Definition 3.2: Power and throughput models presented in this paper are defined as *Power* $=P (f, l, h)$, *Throughput* $= T (f, h, l)$, where $f, h, l \in (0,1)$.

Definition 3.3: *Mixed* traffic is defined as a traffic pattern in which more than one parameter among f, l, and h is nonzero.

So, we present our models in a 3D space with l, f and h axes. The key steps of our modeling strategy are as follows:

1. First, power and throughput on three l, f and h axes (pure traffics) are extracted through simulation.
2. Mutual effects between different traffic models are analyzed.
3. Finally, power and throughput models are obtained using extracted results and analysed effects presented in the previous steps, respectively.

Now, we go through the detailed description of each step of our modeling strategy.

3.1. Power and throughput for pure traffics

Figure 1 shows throughput and power consumption of a 6! 6 mesh topology under pure traffic models with XY routing algorithm. These values are extracted by simulation at the brink of saturation, because these points show maximum throughput that an NoC can deliver and maximum power that should tolerate.

Figure 1(a) shows power consumption under pure traffics versus traffic model percentage in the range of $(0,1)$. Usually, with percentage increment, blocking probability in the network increases, which leads to reduction of switching activity, and so the power consumption drops as shown in Figure 1(a). But under different traffic models, this reduction occurs with a different behavior. We base our power model on these behaviors which are described below.

Since our topology is a 2D Mesh, when traffic percentage increases under local traffic models, the number of messages sent to neighbors of each node will increase and so the average distance between source and destination will decrease, which leads to reduction of switching activity in both X and Y dimensions. This reduction in two dimensions results in quadratic power reduction in terms of traffic percentage. Under FMT traffic model, the destination address for each node is fixed, so the XY routing algorithm as a deterministic routing always selects a specific path between each two nodes. It means that when FMT percentage increases in the network (which leads to uniform traffic reduction), the traffic load between each two nodes increases in only one path (as opposed to two dimensions which was the case for local traffic model). Therefore, we can conclude that traffic load increases linearly in terms of traffic percentage, which leads to linear increase in blocking time and decrease in power consumption. When percentage increases under hotspot traffic model, traffic load around hot node increases in both X and Y dimensions. On the other hand, when a flit is blocked in a buffer (e.g. the buffer size is equal to one flit), the blocking easily propagates in the network due to the wormhole switching nature and results in a hot area in the network. Therefore, when the probability of blocking occurrence in this traffic model increases with hot percentage increment, hot area s diameter grows linearly. Quadratic increase of traffic load due to traffic load increment in two dimensions around hot node, together with linear increase of hot area, lead to cubic increase in blocking time and so cubic decrease in switching activity, which leads to cubic power reduction in terms of traffic percentage. Based on these behaviors, following power models are proposed for pure traffics:

$$P_h = \alpha_1 h^3 + \alpha_2 h^2 + \alpha_3 h + \alpha_4 \qquad \alpha_1 < 0 \qquad (3.1)$$

$$P_r = \beta_1 l^2 + \beta_2 l + \beta_3 \qquad\qquad \beta_1 < 0 \qquad (3.2)$$

$$P_f = \gamma_1 f + \gamma_2 \qquad\qquad\qquad \gamma_1 < 0 \qquad (3.3)$$

where $f, h, l \in (0,1)$ are traffic percentages and P_h, P_l and P_f are power associated with pure Hotspot, Local and FMT traffic models, coefficients α_4, β_3 and γ_2 represent power consumed when $h=l=f=0$. Assuming power consumption under uniform traffic as P_u:

$$\alpha_4 = \beta_3 = \gamma_2 = P_u \qquad\qquad (3.4)$$

Coefficients of (3.1), (3.2) and (3.3) are calculated using MATLAB s regression functions, and are based on real values extracted for power consumption under pure traffic models using Synopsys Power Compiler.

Figure 1(b) shows throughput of NoC under pure traffics versus traffic model percentage in the range of *(0,1)*. To model throughput versus traffic percentage, first we should analyse its behavior under different traffic models in terms of traffic percentage.

Under local traffic model, when locality increases, the number of messages sent to neighbors of each node will increase and so the distance between source and destination will decrease which leads to smaller delay and greater throughput. Average distance between source and destination nodes decreases linearly with traffic percentage increment which leads to linearly increasing throughput. Let us consider d as an average distance for uniform messages and 1 (one) for local traffic. Then, the average distance in an N! N network (*AvgD*) is:

$$AvgD = l \times 1 + (1 - l) \times d = (1 - d) \times l + d \qquad (3.5)$$

and d is extracted as following:

$$d = \frac{\displaystyle\sum_{x=0}^{N-1}\sum_{y=0}^{N-1}\left(\frac{\displaystyle\sum_{i=0}^{N-1}\sum_{j=0}^{N-1}\left(|x-i|+|y-j|\right)}{N^2-1}\right)}{N^2} \qquad (3.6)$$

In a 6! 6 mesh topology, *d* is 4. It is clear from (3.5) that

average distance increases linearly with traffic percentage increment. On the other hand, latency has a direct relation with average distance, while throughput from (2.1) has a reverse relation with latency:

$$Throughput \;\propto\; \frac{1}{AvgD} = \frac{1}{(1-d) \times l + d} \qquad (3.7)$$

We can write (3.7) as follows:

$$Throughput \;\propto\; \frac{1}{1-k} \quad; k = \frac{(d-1)}{d} \times l \qquad (3.8)$$

In (3.8), we know that $l < 1$ and $(d-1)/d$ is lower than one, therefore, $k < 1$ and using Tailor series from (3.9), we can extend (3.10) from (3.7) :

$$\frac{1}{1-x} = 1 + x + x^2 + x^3 + x^4 + ... = \sum_{n=0}^{\infty} x^n : |x| < 1 \qquad (3.9)$$

$$Throughput \;\propto\; 1 + k + k^2 + k^3 + k^4 + ... \qquad (3.10)$$

We showed that $k < 1$, so we can ignore the terms with high orders of k in (3.10) and rewrite this equation as follows:

$$Throughput \;\propto\; (1 + k = 1 + \frac{(d-1) \times l}{d}) \propto 1 + 0.75\,l \qquad (3.11)$$

Relation (3.11) shows that throughput has a linear relation with traffic percentage which is clear from Figure 1(a), too.

On the other hand, in FMT and Hotspot traffic models, traffic percentage increment leads to more blocking time and so switching activity reduction, which results in more delay and smaller throughput. As mentioned before, under FMT traffic model, blocking time increases linearly in terms of traffic percentage which results in linear decrease in throughput. Similarly, under FMT traffic model, we can show that throughput has a linear relation with traffic percentage as follows. Let us consider d and r as average distances for uniform and FMT traffic models, respectively. Then we have:

$$AvgD = f \times r + (1 - f) \times d = (r - d) \times l + d \qquad (3.12)$$

The value of d is 4 as calculated in (3.6). For calculating value of r, we can extract (3.13) from the definition of FMT traffic model as follows:

$$r = \frac{\displaystyle\sum_{i=0}^{N-1}\sum_{j=0}^{N-1}\left(2|N-i-j+1|\right)}{N^2} = 4.22 \qquad (3.13)$$

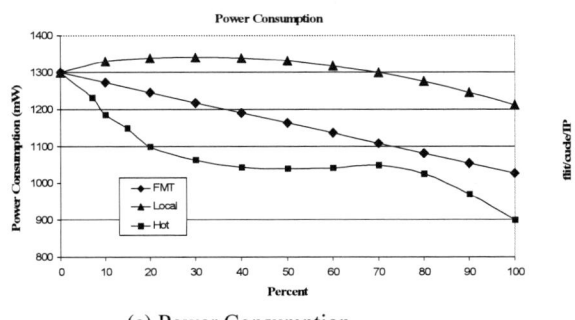

(a) Power Consumption

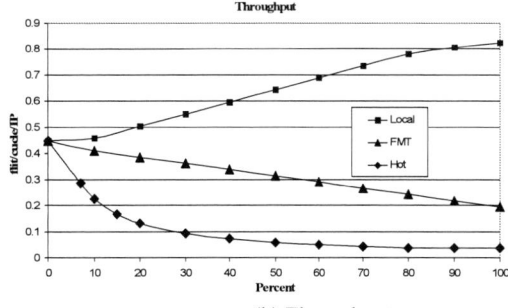

(b) Throughput

Figure 1: Power consumption and Throughput for mesh topology under pure traffic models versus traffic percentages

From (3.8) we can conclude that throughput is related to r and d as follows:

$$\text{Throughput} \propto \frac{1}{1-k} \quad ; k = \frac{(d-r)}{d} \times f \qquad (3.14)$$

The value of $|k|$ is lower than one too, so we can use (3.9) again to write throughput as (3.10) and, due to small value of k we can ignore k^2, k^3, etc and therefore (3.15) is derived as follows:

$$\text{Throughput} \propto (1+k = 1+\frac{(d-r)\times f}{d}) \propto 1-0.05\ f \quad (3.15)$$

Under hotspot traffic model, percentage increment leads to more blocking time, which results in throughput reduction like FMT. But unlike FMT, in hotspot traffic model, relative change of throughput versus percentage is not constant; i.e., with percentage increment, slope of throughput versus traffic percentage decreases. Because when percentage traffic increases from low to high, number of blocking increases noticeably. In traffic percentage closer to 100%, due to high number of blocking around hot node, blocking get closers to its maximum value and therefore slope of throughput reduces to its minimum value. This leads to slight throughput variation. So, we infer that throughput has an inverse relation with traffic percentage.

Based on these behaviors, following throughput models are proposed for pure traffics:

$$T_h = \frac{1}{\alpha_5 h + \alpha_6} \qquad \alpha_5 > 0 \qquad (3.16)$$

$$T_l = \beta_4 l + \beta_5 \qquad \beta_4 > 0 \qquad (3.17)$$

$$T_f = \gamma_3 f + \gamma_4 \qquad \gamma_3 < 0 \qquad (3.18)$$

where $f, h, l \in (0,1)$ are traffic percentages and T_h, T_l and T_f are throughputs associated with pure Hotspot, Local, and FMT traffic models, respectively. When $h= l= f=0$, the values of $1/\alpha_6$, β_5 and γ_4 are equal to throughput of the network under uniform traffic (T_u):

$$\frac{1}{\alpha_6} = \beta_5 = \gamma_4 = T_u \qquad (3.19)$$

Values of all coefficients presented in (3.16), (3.17) and (3.18) are calculated using MATLAB s regression functions, based on real values extracted for throughput under pure traffic models using Modelsim.

3.2. Analyzing mutual effects between different traffic models on power and throughput

As mentioned before, generated traffic is usually a mixture of multiple address distribution methods (mixed traffic). Therefore, before power and throughput modeling, we should investigate **composite effects** of various traffic models in a mixed traffic. As we know:

Superposition principle is not applicable to power and throughput of a traffic pattern; i.e., power consumption (throughput) of mixed traffics is not the sum of power consumed (throughput) of each traffic model./

In step 1, we have extracted power and throughput formulas for pure traffic models, but as stated, we cannot simply sum up these formulas to extract the final power and throughput models for the mixed traffics. Hence, we should analyze the mutual effects of traffic models on each other.

We know that switching activity, which determines the value of power consumption, depends on the average number of blocking occurred in the network. In presence of all the mentioned traffic models; i.e., Local, FMT, Hotspot and Uniform models, number of blocking depends on traffic load generated by all these traffics. As mentioned before, number of blocking will significantly increases with hotspot percentage increment (step1), while this effect does not appear under local traffic, and is weak under FMT traffic (compared to hotspot traffic model). Therefore, presence of hotspot traffic model compared to any other traffic in a mixed traffic will increases the number of blocking and so decreases switching activity, which leads to power reduction. We explain this effect with an example.

Example1: Assume that we have two different traffic models T1 and T2. T1 is a pure local traffic model with *l=0.2 (u=0.8)* and T2 is a mixture of local and hot traffic models with *l=0.2* and *h=0.4 (u=0.4)*. Power consumed by 20% messages sent locally in both traffic patterns, strongly depends on the blocking situations and traffic load formed by other messages in the network. Since the hotspot percentage in T2 is 0.4, we have increased the number of blocking, which leads to message blocking even for locally sent messages. So, the power consumed by packets sent locally decreases (while local percentage is fixed). The same statement holds for simultaneous presence of FMT and hotspot traffic models. On the other hand, the messages sent locally do not increase the number of blocking in the network, and so, do not increase blocking probability for messages sent under FMT traffic mode, which leads to no power change for FMT messages, and vice versa.

From the above explanation, it is obvious that there is a mutual effect between hotspot and other traffic models. Therefore, in presence of hotspot traffic model, we cannot apply superposition principle to power, and similarly throughput values. We will present our final power and throughput formulas base on this mutual effect.

3.3. Final power and throughput models

In previous steps, we have described basic power and throughput models for pure traffics and their mutual effects in a mixed traffic. In this section, we will present our final power and throughput models using these parts.

As mentioned in step2, presence of hotspot traffic model reduces the slope of power (throughput) variation versus traffic percentage in Local and FMT traffic models. With hot percentage close to 1, this effect is stronger and power (throughput) of messages sent under other traffic models slightly changes with percentage increment. This effect stems from high blocking occurrence for high hot percentage. In the other words, as the simplest form, we can model the effect of hot percentage increment on the slope change of power, and similarly throughput values, versus traffic percentage under FMT or local traffic models, as follows:

$$\frac{dP}{dt} = k_1(1-\sqrt{h}) \; ; \qquad h, t \in (0,1) \quad (3.20)$$

$$\frac{dT}{dt} = k_2(1-\sqrt{h}) \; ; \qquad h, t \in (0,1) \quad (3.21)$$

where P and T are power and throughput values, respectively, for local or FMT traffic models. h is the hot percentage and t is FMT or Local traffic percentage in the mixed traffic model. k_1 and k_2 are constants in term of h. From (3.1)-(3.4) and (3.20), we conclude that the formula for power consumption under mixed traffic model is:

$$P_{total} = P'_h + P'_l + P'_f$$

$$= \alpha_1 h^3 + \alpha_2 h^2 + \alpha_3 h + \underbrace{(\beta_1 l^2 + \beta_2 l)(1-\sqrt{h})}_{k_1} \quad (3.22)$$

$$+ \underbrace{\gamma_1 f(1-\sqrt{h})}_{k_1} + P_u$$

where P_{total} is the total power consumed under mixed traffic, P'_h, P'_l and P'_f are power consumed by messages sent under hotspot, local and FMT traffic models considering mutual effects between different traffics. Coefficients α_i $i = 1..3$, β_j $j = 1,2$ and γ_1 are defined in (3.1)-(3.3), h, l and f are hotspot, local and FMT traffic percentages, respectively. P_u is the power consumption under uniform traffic model (when $f=l=h=0$). Similarly, from (3.16)-(3.19) and (3.21), we conclude the following formula for throughput under mixed traffic model:

$$T_{total} = T'_h + T'_l + T'_f$$

$$= \frac{1}{\alpha_5 h + \alpha_6} - \frac{1}{\alpha_6} + \underbrace{\beta_4 l(1-\sqrt{h})}_{k_2} \quad (3.23)$$

$$+ \underbrace{\gamma_3 f(1-\sqrt{h})}_{k_2} + T_u$$

where T_{total} is the total throughput under mixed traffic, T'_h, T'_l and T'_f are throughput values for messages sent under Hotspot, Local and FMT traffic models considering mutual effect between different traffics. Coefficients α_i $i = 5,6$, β_4 and γ_3 are defined in (3.16), (3.17) and (3.18), h, l and f are Hotspot, Local and FMT traffic percentages, respectively. T_u is the throughput value under uniform traffic model (when $f=l=h=0$). From (3.19) we know that $1/\alpha_6 = T_u$, therefore we can derive (3.24) from (3.23) as follows:

$$T_{total} = \frac{1}{\alpha_5 h + \alpha_6} + \beta_4 l(1-\sqrt{h}) + \gamma_3 f(1-\sqrt{h}) \quad (3.24)$$

(3.22) and (3.24) present our final power and throughput models. The main important point in these models is their simplicity, which leads to rapid power and throughput calculation under mixed traffic, while calculating these values using simulation needs considerable time.

Beside power and throughput as two performance criteria for evaluating NoC architecture, energy consumption as the product of these factors plays an important role for NoC evaluation. Therefore, as a direct conclusion from power and throughput modeling, we can obtain the model for power per throughput ratio as follows:

$$Energy \propto \frac{P}{T} = \frac{\alpha_1 h^3 + \alpha_2 h^2 + \alpha_3 h + (\beta_1 l^2 + \beta_2 l)(1-\sqrt{h})}{\frac{1}{\alpha_5 h + \alpha_6} + \beta_4 l(1-\sqrt{h}) + \gamma_3 f(1-\sqrt{h})} \quad (3.25)$$

3.4. Validation and Results

In this part, we validate our power and throughput models presented in (3.22) and (3.24) by comparing against simulation results from Power Compiler and Modelsim.

As definition 3.2 shows, our power and throughput models are 4D functions whose domains are 3D space (f,l,h). According to the range of traffic parameters, this 3D space is a unit cube. Validating (3.22) and (3.24) by computing all points in this volume and then comparing them with true values obtained from simulation, is impractical. Therefore, we selected a subset from this volume and validated the model upon it. In the 1x1x1 3D space, power consumption and throughput values associated to 12 lines and a 3D cube (consisting of about 110 points with 0.1 steps along each axis) are calculated by proposed models and compared against true values from simulation. Selected cube is associated to 3D subset $f \in [0,0.3]$, $l \in [0,0.3]$ and $h \in [0,0.3]$.

Simulation values for 12 lines (on three planes, including Hot_FMT, FMT_Local and Local_Hot) are

Table 1- Comparing modeling values against real values in 3D cube

Traffic Percentage			Power (mW)			Throughput		
h	*l*	*f*	Sim.	Mod.	Err	Sim.	Mod.	Err
0.1	0.1	0.3	1157	1175	1.5	0.252	0.249	1.1
0.1	0.2	0.2	1140	1177	3.2	0.240	0.245	2.2
0.1	0.3	0.2	1126	1107	1.7	0.239	0.231	3.2
0.2	0.1	0.2	1118	1149	2.8	0.162	0.165	1.8
0.2	0.2	0.1	1112	1105	0.6	0.157	0.161	2.6
0.2	0.3	0.2	1108	1110	0.2	0.157	0.152	3.5
0.3	0.1	0.1	1092	1118	2.4	0.118	0.111	6.0
0.3	0.1	0.2	1090	1110	1.8	0.122	0.117	4.0

compared against their power and throughput values extracted from proposed models. But because of their huge volume, only resulted average error for each line is shows in Figure 2. As seen from the figure, maximum average error obtained is about 3.04% for throughput and 3.74% for power. Also, the maximum value for error s standard deviation is equal to 1.98. Computing average error for all 12 selected lines leads to total average value of 2% for power and 2.8%. Since 4D functions cannot be drawn, we were enforced to present the extracted results for the simulated cube in Table 1 (for brevity results are shown for some selected points). In this table, columns named as *Mod* and *Sim* refer to data obtained from modeling and simulation and column named *Err* refers to modeling error, for each triple *(l,f,h)*. As seen from this table, modeling in 3D cube results in accurate values with average error of 1.8% for power and 2.8% for throughput. Another interesting conclusion can be drawn by comparing modeling errors from Figure 1 for 2D planes and those form Table 1 for 3D cube which show 2% against 1.8% error for power, and 2.8% error in both cases for throughput. From this comparison, we conclude that our modeling approach remains accurate by dimension increment from 2D to 3D parameter space.

4. Conclusions

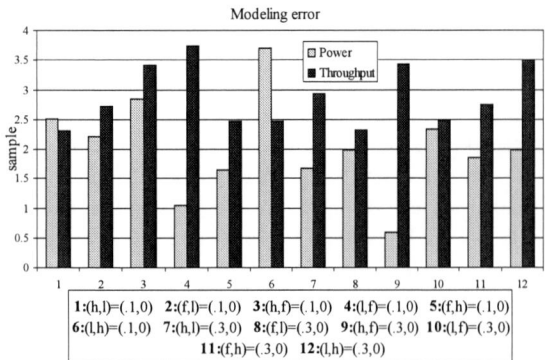

Figure 2: Modeling error, (h,f) = (0.1,0) means percent of Hotspot and FMT is 10%, and 0% respectively, and percent of local is varying from 0% to 90%.

In this paper, we proposed a power and throughput modeling approach that takes as input the traffic parameters of the application, and outputs the values of power and throughput of the network. Although our proposed models are based on mesh topology and some popular traffic models, this modeling approach can be generalized to other topologies with variety of traffic models.

We stated that different traffic models have mutual effects on each other. This fact prevents us from applying superposition principle to power (throughput) of pure traffics for calculating mixed traffic s power (throughput). Based on this observation, and extracted power and throughput values for pure traffics, we presented our power and throughput models for mixed traffics. In the proposed models, mutual effects between different traffics are taken into account.

We validated the proposed models by comparing our results against simulation results from Synopsys Power Compiler and Modelsim. From the comparison, we showed that our modeling approach leads to average errors of 2% for power and 2.8% for throughput values. Low error rates verify our modeling accuracy. Using this modeling approach, we can estimate NoC s behavior under different traffic patterns.

5. References

[1] L. Benini and G. de Micheli, "Networks-on-Chip: A new Paradigm for System on Chip Design,# *Design Automation and Test in Europe*, IEEE computer, vol. 35, no. 1, pp. 70-78, January 2002.

[2] V. Tiwari et al. Reduction power in high-performance micro processors. In *35th Design Automation Conference*, 1998.

[3] V. S. Adve, M. K. Vernon, "Performance Analysis of Mesh Interconnection Networks with Deterministic Routing,# *IEEE Trans. on Par. and Dist. Syst.*, vol. 5, no. 3, pp. 225-246, March 1994.

[4] A. Agarwal, "Limits on Interconnection Network Performance,# *IEEE Trans.on Par. and Dist. Syst.*, vol. 2, no. 4, pp 398-412, October, 1991.

[5] W. J. Dally, "Performance Analysis of *k*-ary *n*-cube Interconnection Networks,# *IEEE Trans. on Computers*, vol. 39, no. 6, pp. 775-785, June 1990.

[6] N. Eisley and L. Peh. High-level power analysis for on-chip networks. In *CASES*, Sept. 2004.

[7] Partha Pratim Pande, Cristian Grecu, Michael Jones, André Ivanov, Resve A. Saleh, "Performance Evaluation and Design Trade-Offs for Network-on-Chip Interconnect Architectures,# *IEEE Trans. Computers, vol. 54, no. 8, pp* 1025-1040, 2005.

[8] United Microelectronics Corporation (UMC), http://www.umc.com/, Dec, 2006.

[9] International Technology Roadmap for Semiconductor (ITRS), http://www.itrs.net/, Dec, 2006.

[10] W. Hsh, #Performance issues in wire-limited hierarchical networks,# PhD Thesis University of Illinois-Urbana Champaign, 1992.

[11] K. Hwang, Advanced computer architecture: parallelism, scalability and programmability, McGraw-Hill (Ed.), 1993.

21st International Conference on VLSI Design

PTSMT: A Tool for Cross-Level Power, Performance, and Thermal Exploration of SMT Processors

Deepa Kannan†, Aseem Gupta‡, Aviral Shrivastava†, Nikil D. Dutt‡, Fadi J. Kurdahi‡

† Compiler and Microarchitecture Laboratory, Arizona State University, Tempe, AZ 85281 USA
{deepa.kannan, aviral.shrivastava}@asu.edu
‡Center for Embedded Computer Systems, University of California Irvine, CA 92697 USA
{aseemg,dutt,kurdahi}@uci.edu

Abstract

Simultaneous Multi-Threading (SMT) processors are becoming popular because they exploit both instruction-level and thread-level parallelism by issuing instructions from different threads in the same cycle. However, the issues of power and thermal management hinder SMT processors fabricated in nano-scale technologies. Power and thermal issues in SMT processors not only limit the achievable performance, but also have a direct impact on the cost and viability of these processors. While several performance simulation tools to explore the performance aspect of SMT processors early in their design phase exist, there is a lack of early power and performance evaluation tools for SMT processors. To this end, we have developed PTSMT: a tightly coupled power, performance and thermal exploration tool for SMT processors. In this paper, we demonstrate that PTSMT can automatically and effectively accomplish power, performance and thermal exploration of SMT processors at various levels of design hierarchy, at the application level, microarchitecture level, and physical level. Our experimental results show that: at the application level, number of contexts into which an application is divided could affect performance by 2.2x, energy by 52%, and peak temperature by 35°C; and at the microarchitecture level, context swapping during run time could reduce energy by 9% and improve performance by 8%. These observations indicate the size of the design space which can be explored using PTSMT.

1. Introduction

Simultaneous multithreading [1] is a processor design that combines hardware multithreading with superscalar processor technology to allow multiple threads to issue instructions each cycle. Unlike conventional superscalar processors, which suffer from a lack of per-thread instruction-level parallelism, simultaneous multithreading uses multiple threads to compensate for low single-thread ILP. The performance consequence is significantly higher instruction throughput and program speedups.

While SMT processors have become viable only due to the level of integration provided by technology scaling, technology scaling has resulted in high power density (power per unit area), and therefore high operating temperatures, both of which have a profound impact on SMT processors. While high power density leads to an increase in the cooling cost, [2], high operat-

ing temperatures results in increased probability of timing violations because of higher signal propagation delay and switching time, reduced lifetime because of phenomena such as electromigration [3]. High power densities lead to higher temperatures which in turn increase the leakage power. This close coupling can lead to thermal runaway and therefore requires a joint power, and thermal simulation approach to detect and avoid such situations. These thermal effects in modern sub-nanometer fabrication technologies necessitate a closely coupled power performance and thermal modeling of SMT processors. Furthermore, temperature-aware performance modeling [4, 5] becomes critical because the temperatures of individual blocks in the processor play a very important role in estimating the reliability and performance of the whole processor design. For example, the temperature of any block in a chip is not only dependent on its own power density but also on the power densities of adjacent blocks, due to thermal diffusion (i.e. the flow of heat from hot blocks to cold blocks).

While there exist several tools for performance exploration of SMT processors [6, 7, 8], there is a lack of closely coupled power, performance and thermal simulation tools, which can effectively detect thermal emergencies, and evaluate the impact of various thermal policies.

In this paper, we present **PTSMT**: A tool for closely coupled Power, Performance and Thermal simulation of SMT processors. We present experimental results to demonstrate the effectiveness of PTSMT in early exploration of SMT processor designs at various levels of design hierarchy: at the **Application level**, where decisions such as the number of contexts into which an application should be divided are made; **Microarchitectural level** at which optimal selection of hardware resources, instruction fetch policies from different threads is done; **Physical level**, where different floorplans and other technology parameters are examined. PTSMT empowers the designer with the capability to quantitatively evaluate various design alternatives and power and thermal management techniques at each of the three design levels, very early in the SMT processor design process.

2. Related Work

2.1. PPT Exploration

Temperature estimation of VLSI chips is a well defined area of research. Industrial tools (e.g., Flotherm [9]) and Firebolt [10]

are based on finite element analysis methods and have been in use for a long time. Such tools were primarily used by package and cooling system designers. A thermal model can be converted to an equivalent electrical model dual [5]. Temperature estimation tools [5], [11] based on this model are often used in academic settings. In our work we have used Hotspot [5].

2.2. SMT - Performance Exploration

Many previous works have shown the importance of SMT architectures, analyzing and evaluating their performance characteristics. Early work by Goncales et al. [7] demonstrated the need for an efficient SMT simulator to model the multi-threaded architecture including the memory hierarchy. They presented an SMT simulator that has been developed on top of the Simplescalar toolset [12]. Tullsen et al. presented the SMT-SIM [6]-an instruction-level simulator for a detailed simulation of a pipelined SMT processor with all sources of latency modeled. Madon et al. [8] developed the SSMT simulator, to evaluate the performance aspects of SMT processors. The SSMT is an SMT simulator based on SimpleScalar and augmented to simulate an SMT pipeline.

2.3. SMT - Power and Thermal Exploration

The power consumption of SMT processors has been examined in [13]. They used PowerTimer [14] to estimate the power of the underlying hardware components of the SMT processor. [15] have carried out investigations on thermal management of SMT processors. They have proposed increasing the area of the blocks with high power densities by adding redundant devices in an effort to reduce the power densities and hence reduce temperatures. A taxonomy of different thermal management schemes for SMT processors is given in [16]. [17] presents an adaptive fetch algorithm where new instructions are fetched depending on history based profiling of threads as integer or floating-point intensive.

In all of the above works, SMT processors were simulated and different power estimators such as [14], [18] were used to estimate power. Then temperatures were estimated using tools like [5] or in many cases power densities were used as a indicative representation of temperatures. Such an approach ignores the effects of **thermal diffusion** among blocks with different temperatures. It has been observed that because of thermal diffusion, within a chip the temperature of a block with lower power density can be much higher than the temperature of another block with higher power density [19]. The floorplan of a processor significantly affects thermal diffusion and hence the temperature profile [20]. Our work is different from all the above in that we **tightly couple performance, power, and temperature estimation** for SMT processors. The reason for tight coupling is that *we should be able to examine the effect of thermal management policies on performance and power during run time*. For example, if a thermal emergency occurs and a thermal management technique kicks in at the Nth cycle, we should be able to observe the resulting changes (because of the thermal management technique) in the performance, power, and temperature traces from $(N+1)^{th}$ cycle onwards.

Another motivation behind coupling power and temperature es-

timation is **leakage power**. Leakage power of a transistor increases with temperature in a super-linear manner [21]. Increased leakage power raises the temperature further. This positive interdependency continues to raise both temperature and leakage power until the total power consumed is dissipated to the environment by the package and steady temperatures are reached. Leakage power and temperature estimations can be off by as much as 20% if this positive feedback between temperature and leakage power are ignored [19]. Existing SMT simulators do not take into account the power and thermal considerations during their performance simulation. In our work we have coupled performance, power, and temperature estimation, and presented a platform for concurrent PPT exploration of SMT processors across different levels in a design.

3. PTSMT: Power, Performance & Temperature Simulator for SMT Processors

Power and temperature are fast becoming the bottleneck in increasing the performance of SMT processors. Consequently, both power and thermal management techniques must be investigated in a tightly coupled manner for these technique to be effective. To this end we have developed a **P**ower-**T**emperature **S**imultaneous **M**ulti **T**hreading simulator called PTSMT.

3.1. Performance Model

At the heart of the PTSMT performance model is the Simplescalar [12] engine. The SSMT simulator used to model the performance of SMT processors supports out-of-order issue and execution. PTSMT models a very detailed out-of-order issue SMT processor with a two-level memory system and speculative execution support. This simulator is able to execute multiple contexts simultaneously i.e. dispatch instructions provided by multiple contexts to functional units.

3.2. Power Model

The power models are based on WATTCH [18], but also include the leakage energy models, similar to those in the PTScalar tool setPTSMT models both the dynamic power, and the leakage power, including the temperature dependence of the leakage power. Dynamic energy is independent of temperature, but has a quadratic dependence on supply voltage. In our power model, dynamic energy for a circuit block in each cycle is calculated as: $P_d = N_{gate} \cdot C \cdot V_{dd}^2$, where P_d represents the dynamic energy of the block per cycle, N_{gate} is the number of gates in the circuit block, C represents the switching capacitances and V_{dd} represents the supply voltage. Leakage power mainly consists of subthreshold and gate leakage power. In our power model, we consider both subthreshold and gate leakage for logic circuits and all memory-based units using Equation (1):

$$P_{leak} = N_{gate} \cdot I_{avg}(T, V_{dd}) \cdot V_{dd} \qquad (1)$$

where, P_{leak} is the total leakage power, $I_{avg}(T, V_{dd})$ is the total leakage current per gate at a given temperature T and supply voltage V_{dd}.

422

978-1-4244-3039-0/08 $25.00 © 2008 IEEE

3.3. Thermal Model

The thermal model in PTSMT is based on the HotSpot [5] models. Temperature is modeled by equivalent RC thermal circuits, based on the duality between heat transfer and electrical phenomena where two parameters: thermal resistance R_t and thermal capacitance C_t are used to characterize thermal behavior. The equivalent RC thermal circuit consists of three layers: heatsink, heat spreader and chip die. Provided the average power within a time period computed using the power model, the transient temperature is calculated by solving the differential equations for the RC circuit with a fourth-order Runge-Kutta method.

3.4. Capabilities

PTSMT provides the designer with a fast, cycle-accurate simulation of SMT processors for a coupled evaluation of performance, power, and temperature. The SMT architecture is parameterizable and many of these features were inherited from the SimpleScalar and SSMT simulators. Some of the *parameterizable options* include: instruction and data caches configurations, number of integer and floating point ALUs, branch predictor configurations, size of Register Update Unit (RUU), latency for memory accesses, Translation Lookaside Buffer (TLB) configurations, and other execution parameters such as the maximum number of instructions to simulate etc. The output of PTSMT includes many *performance parameters* such as: Total number of instructions committed from each context; number of loads, stores, and branches; simulation statistics such as total number of cycles, instructions per cycle, time; branch predictor statistics on lookups, hits, misses, updates; cache behavior parameters such as accesses, hits, misses, replacements etc. for instruction and data L1 caches, L2 cache, and instruction and data TLBs; access counts, power, and temperature values at each cycle for all the SMT architectural components.

3.5. Construction

PTSMT is derived from the SSMT simulator. We have supplemented SSMT with a power estimator which is similar to WATTCH [18] and temperature estimator which is similar to HotSpot [5]. Figure 3.5 shows the basic construction of PTSMT. PTSMT performs a fast, cycle-accurate, power and temperature coupled simulation of SMT architecture. However, the architectural configuration can be easily changed which allows the tool to be used for architectural exploration.

The tool takes binaries compiled for the SimpleScalar architecture, floorplan of the processor, configuration files with parameters for the thermal and power models, package and other technology dependent libraries. PTSMT also inherited an advanced cache simulator [26] from Simplescalar which models behavior like cache misses etc. Outputs of PTSMT simulation include performance statistics, power (leakage + dynamic), and temperature profiles for all the components in the architecture at any cycle intervals.

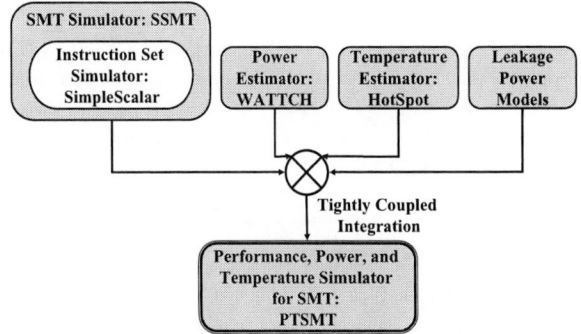

Figure 1. Construction of PTSMT

4. Cross-level PPT Exploration with PTSMT

4.1. Experimental Setup

We have used a SMT version of Alpha-21464 as the floorplan to conduct our experiments and demonstrate the need and usefulness of PTSMT. Alpha-21464 is not a SMT, therefore, we have added the required additional hardware to enhance the architecture for SMT. Then we used Parquet [22] to generate a floorplan for this variant. We used the benchmarks **benchmarks** from the MiBench [23] suite. We use 45nm technology, clock speed of 3 GHz and a relatively large L2 cache, and other parameters scaled accordingly. For our SMT, we allow complete sharing of the Instruction and Data L1 caches, functional units and all other blocks except the Integer and Floating point register files and the corresponding register alias tables and the RUU.

In our results, we have used a **base case** for comparison, which we define as follows: To do a fair comparison we assume that the SMT processor in base case has a thermal control mechanism. The temperatures of all the blocks in the processor are continuously monitored during the execution of the program. If the temperature of any block goes beyond a defined thermal threshold of 100^oC (can be different for other designs or technologies), we stop fetching instructions till the temperature wanes (i.e., stall the complete processor pipeline) in order to avoid a thermal runaway.

4.2. Physical level Exploration

The phenomenon of power and temperature are very closely related to physical design choices such as selection of cell libraries, floorplan etc. The floorplan of the processor affects the temperature profile because it changes the thermal diffusion among neighboring components. As a result, different floorplans have different temperature profiles. Since the leakage power of a transistor has a super-linear dependency on temperature, the floorplan also has an affect on the leakage power dissipation.

Effect of Floorplan

We simulated our benchmarks using five different SMT floorplans with PTSMT. Figure 2 shows the energy consumption (normalized to the base case) and peak temperatures of the processor for different floorplans. The energy consumption values are normalized to the energy for the first floorplan. It was ob-

423

978-1-4244-3039-0/08 $25.00 © 2008 IEEE

served that there was a 13% variation in the overall energy consumption of the processor and a 21°C variation in the maximum temperature of the chip across different floorplans. This variation in the peak temperature has a foremost effect on the cost of the package and the cooling mechanism which increases significantly with temperature.

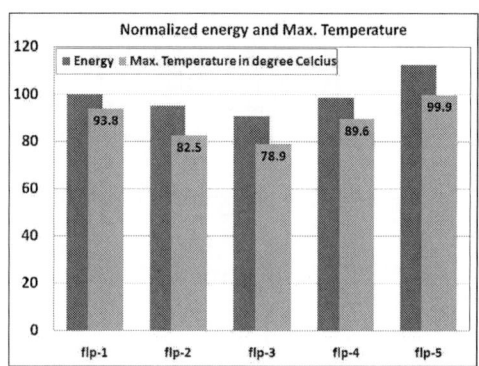

Figure 2. Energy and maximum temperatures for different floorplans of an SMT Processor

If the cooling mechanism includes a fan, a lower peak temperature will warrant less use of the fan and thus save power consumed by fans (which can range from 4W for hand-held devices to 30W for fixed devices). Other benefits of a lower peak temperature include: improved reliability, reduced leakage current, less delay for transistors and wires, etc. Different floorplans have different thermal profiles because the placement of the blocks affects thermal diffusion. This experiment motivates the need for a coupled power and temperature estimation tool which takes into account the effect of floorplans on temperature and hence on the power consumption. If we consider only those floorplans which meet the timing constraints of the processor, there is a design space with multiple floorplans having different power dissipation and temperature profiles.

4.3. Microarchitecture level Exploration

SMT processors need replication of certain hardware so that multiple threads can issue instructions in the same cycle. There are multiple ALUs of integer and floating point type so that multiple instructions can execute in one cycle. The extra hardware offers increased performance but at a cost of increased area and power. Thus, the choice of microarchitecture and the analysis of the accompanying tradeoffs must be done very early in the design cycle of SMT processors, because they have a direct impact on the PPT metrics.

Effect of Context Swapping

If the temperature of any of the resources corresponding to one of the contexts rises above a defined threshold temperature, we swap the execution of this context with another context. The context responsible for the increased register file temperature now uses another register file which was at a lower temperature, thereby allowing the original register file to cool down. This runtime swapping is repeated again when needed. We simulate the benchmarks and compare the runtime, energy, and temperature of the processor against the case when no context swapping was

employed. From Figure 3 we observe that on an average there is a 9% reduction in energy consumption with a 8% reduction in runtime (performance improvement) because of this swapping of contexts compared to the base case of stalling the processor.

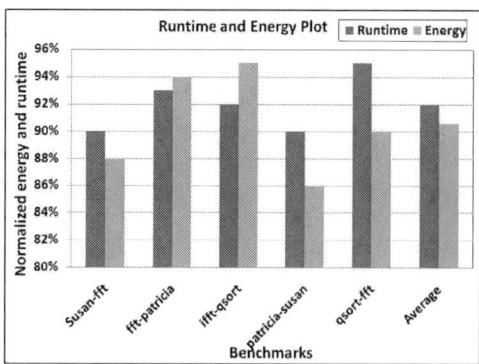

Figure 3. Energy and performance for context swapping to do thermal management

Effect of Architectural Choices

The number of functional units, Integer ALUs (IALUs) and Floating Point ALUs (FALUs), is one of the microarchitectural parameters that can be used to optimize power and performance. We simulate our benchmarks using PTSMT with varying number of IALUs and analyze the energy and runtime in each case. Figure 4 plots the processor's total energy consumption and runtime for the representative susan corners benchmark as the number of IALUs in the microarchitecture vary from 1 to 8.

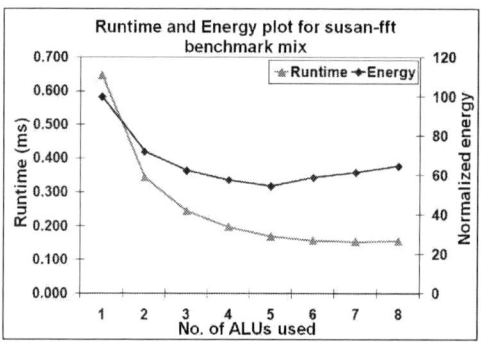

Figure 4. Energy and performance for a benchmark at different number of IALUs in the SMT architecture

It can be seen from the curve that the runtime decreases steeply as the number of IALUs increase from 1 to 4. This is because of higher the number of IALUs. But as the number of IALUs increases to 8, runtime decreases slightly before stabilizing because there are not too many integer operations in the benchmark. Functional units such as IALUs, being regions of very high activity, are one of the major contributors to the processors total energy. Consequently, we observe a significant difference in the total energy with the different number of IALUs used. The energy consumption curve shows a decrease in energy as the number of IALUs increases from 1 to 5 because: the instructions are distributed among all the IALUs, thereby decreasing

424

978-1-4244-3039-0/08 $25.00 © 2008 IEEE

the dynamic energy consumed by each of them. The energy consumption increases as the number of IALUs increases from 5 to 8 because: though there are more number of IALUs, there are lower number of instructions committed to them, which is also implicit from a corresponding halt in the decrease of runtime. Hence, though the additional IALUs consume only small dynamic power, they consume a significant standby power. Thus different number of IALUs provide a tradeoff between energy consumption and runtime.

Effect of Fetch Policies

PTSMT allows the exploration of different policies for fetching instructions from the ready queues of hardware contexts, for SMT processors. The 'fetch policy' governs the way in which multiple instructions are fetched from contexts which are ready with executable instructions. The fetch policy has a direct impact on the PPT metrics trio. We implement two such fetch policies and compare the power and performance of each of them with that of the base case.

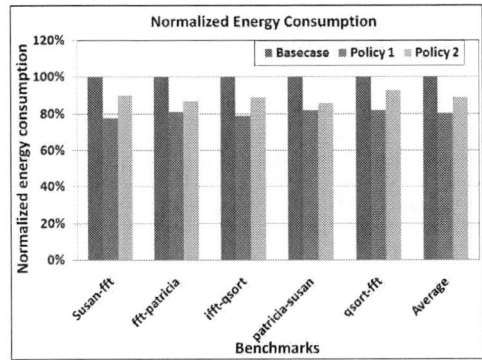

Figure 5. Energy for different fetch policies to regulate maximum temperature

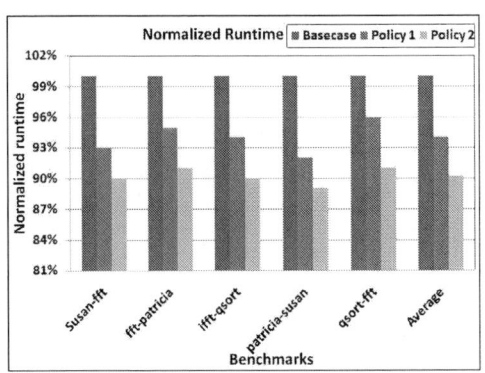

Figure 6. Performance for different fetch policies to regulate maximum temperature

Policy 1: The temperature of any of the resource corresponding to each context (such as the Register Files and the Register Update Unit) corresponding to all contexts is monitored. If the temperature of any of them goes beyond the defined threshold, instructions corresponding to that context are stopped from being fetched, until the temperature of the hottest block belonging to that context drops below the lower threshold temperature.

Policy 2: In this policy, similar to policy 1, the temperature of any of the resource corresponding to each context is monitored. If the temperature of any of them goes beyond the defined threshold the number of instructions being fetched from the hottest context in each cycle is reduced.

Figure 5 and Figure 6 show the normalized (to base case) energy consumption and run time for the two policies respectively. On an average there is up to 20% and 11% energy reduction for fetch policies 1 and 2 respectively compared to the base case. Thus there is an average 6% and 10% performance gain respectively for the two fetch policies compared to the base case. Using PTSMT, the designer can choose to implement several other power and thermal management techniques for SMT processors at the micro architecture level, for a combined power, performance and temperature exploration.

4.4. Application level Exploration

Since SMT processors offer the capability to execute instructions from multiple threads in a single cycle, the top level application is divided into multiple threads or *contexts*. Each context has its own set of registers and program counters to maintain its state. While it is desirable to partition an application into the maximum number of threads, it is not always beneficial to do so. The maximum number of threads that an application can be broken into is limited by the hardware resources available on the SMT processor. The number of threads running on the SMT processor will not only affect the performance of the application but they also alter the power consumption and thermal profiles.

Selection of Number of SW Contexts for an Application

The division of an application into a suitable number of contexts may be done explicitly at functional level by the designer or it may be done using special compiler solutions.

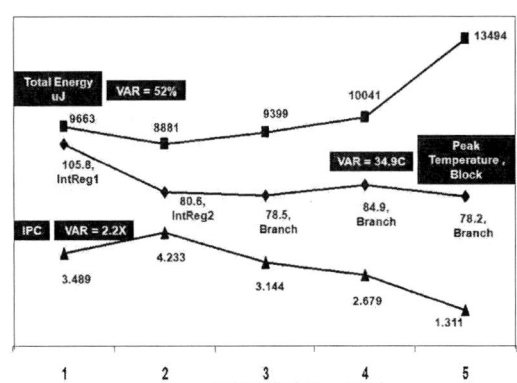

Figure 7. Variation in Performance, Energy & Temperature metrics when an application is run as different number of contexts on SMT

We have taken a large application based on MiBench benchmarks. We divided this application into a different number of sub-applications ranging from 2 to 5 for which the results are in Figure 7. We observe that maximum IPC is obtained when the application is divided into 2 contexts because the architecture has two hardware replications for context registers, program counters etc. When the application is divided into more contexts the performance goes down because there is a penalty of swapping contexts. We observe large variations (VAR) in energy, peak temperature, and performance. Thus, there is an exploration space which needs to be traversed in order to identify the right number of threads for an application in order to meet PPT budgets.

4.5. Overhead of PTSMT

In order to allow designers to take full advantage of PTSMT to do an early evaluation of design choices at the three levels it is necessary to have a low simulation time for PTSMT. On an average, PTSMT takes 71 seconds to simulate 5 million instructions. On the other hand, the SSMT simulator takes 29 seconds to simulate the same 5 million instructions. However, SSMT simulator only does performance simulation and not a tightly coupled power and temperature estimation as done by PTSMT. This execution overhead of PTSMT is justified by the additional power and temperature estimation for SMT processors at each cycle. The overhead can be reduced greatly if the temperature estimation is done after every tens of thousands of cycles instead of each cycle. This extends the utility of PTSMT beyond research on performance enhancement techniques to optimization of the PPT metrics.

5. Conclusion

There has been extensive research on analyzing the performance characteristics of SMT processors and evaluating their tradeoffs. However, current generation of technologies have imposed additional constraints of power dissipation and temperature on designs. There is a need for simultaneous evaluation of Performance, Power, and Temperature (PPT) metrics for SMT processors because of inter-relation between these metrics. In this paper we attempted to bridge this gap using a platform which does a tightly coupled performance, power, and temperature estimation for SMT processors and also allows the designer to evaluate choices at three different levels of design: Application, Microarchitecture, and Physical. We demonstrated that such a cross-level exploration will give the designer an opportunity to explore different choices at all the above levels to do early trade-off analysis for PPT.

Acknowledgment

This work was partially supported by NSF grant: CCF-0702797. The authors would like to thank Kaiyu Chen from Princeton University and Prof. Sarma Vrudhula and Vipin Mohan from Arizona State University.

References

[1] D. Tullsen et al., " Simultaneous Multithreading: Maximizing On-Chip Parallelism," *Proceedings of International Symposium on Computer Architecture*, 1995.

[2] S. Velusamy et al., "Monitoring Temperature in FPGA based SoCs," *Proceedings of International Conference on Computer Design*, 2005.

[3] K. Banerjee, A. Sangiovanni-Vincentelli et al., "On Thermal Effects in Deep Sub-Micron VLSI Interconnects," *Proceedings of Design Automation Conference*, 1999.

[4] W. Liao et al.,"Microarchitecture level power and thermal simulation considering temperature dependent leakage model,"*Proceedings of International Symposium of Low Power Electronic Devices*, 2003.

[5] W. Huang et al.,"HotSpot: Thermal Modeling for CMOS VLSI Systems,"*IEEE Transactions on Component Packaging and Manufacturing Technology*, 2005.

[6] D. Tullsen et al., "The SMTSIM Multithreading simulator", 1996.

[7] Ronaldo Goncalves et al., "SS SMT - A Simulator for SMT Architectures", 1998

[8] Dominik Madon et al., "The SSMT Simulator", 1999

[9] Flotherm, http://www.flomerics.com/flotherm/

[10] Firebolt,http://www.gradient-da.com/tech/firebolt.htm

[11] Y. Yang et al.,"Adaptive Chip-Package Thermal Analysis for Synthesis and Design,"*Proceedings of Design, Automation, and Test in Europe*, 2006.

[12] T. Austin et al., "The SimpleScalar Tool Set, Version 2.0," *University of Wisconsin-Madison Computer Sciences Department Technical Report*, June 1997.

[13] Y. Li et al., "Understanding the Energy Efficiency of Simultaneous Multithreading," *Proceedings of International Symposium on Low Power Electronics and Design*, 2004.

[14] D. Brooks et al., "Microarchitectural-Level Power-Performance Analysis: The PowerTimer Approach," *IBM Journal of Research and Development*, Oct/Nov, 2003.

[15] J. Donald et al.,"Temperature-Aware Design Issues for SMT and CMP Architectures,"*Proceedings of the Wkshop on Complexity-Effective Design*, 2004.

[16] J. Donald et al.,"Techniques for Multicore Thermal Management: Classification and New Exploration,"*Proceedings of International Symposium on Computer Architecture*, 2006.

[17] J. Donald et al.,"Leveraging Simultaneous Multithreading for Adaptive Thermal Control,"*Workshop on Temperature-Aware Computing Systems*, 2005.

[18] D. Brooks et al.,"Wattch: A Framework for Architectural-Level Power Analysis and Optimizations,"*Proceedings of International Symposium on Computer Architecture* , 2000.

[19] Aseem Gupta et al.,"STEFAL: A System Level Temperature- and Floorplan-Aware Leakage Power Estimator for SoCs," *Proceedings of VLSI Design Conference*, 2007.

[20] I. Koren et al.,"Temperature Aware Floorplanning," *Proceedings of Workshop on Temperature Aware Computer Systems,*, 2005.

[21] A. Agarwal, K. Roy et al., "Leakage in nano-scale technologies: mechanisms, impact and design considerations," *Proceedings of Design Automation Conference*, 2004.

[22] S. Adya et al., "Fixed-outline Floorplanning: Enabling Hierarchical DesignEfficient," *IEEE Trans. on VLSI Systems*, 2003.

[23] M. Guthaus et al., "MiBench: A free, commercially representative embedded benchmark suite," *IEEE Workshop in workload characterization*, 2001.

[24] D. Tullsen et al., " Simultaneous Multithreading: A Platform for Next Generation Processors," *Proceedings of International Symposium on Computer Architecture*, 1995.

[25] T. N. Vijaykumar et al.,"Heat Stroke: Power-Density-Based Denial of Service in SMT,"*Proceedings of International Symposium on High-Performance Computer Architecture* , 2005.

[26] R. Sugumar et al., "Efficient Simulation of Caches under Optimal Replacement with Applications to Miss Characterization," *Proceedings of ACM Sigmetrics Conference on Measurements and Modeling of Computer Systems*, 1993.

[27] Joachim Clabes et al., "Design and implementation of the POWER5 microprocessor", *Proceedings of the 41st annual conference on Design automation*, 2004.

[28] Shailender Chaudhry et al., "High-Performance Throughput Computing", *IEEE Micro*, 2005.

[29] Intel Hyper-threading technology, http://www.intel.com/technology/platformtechnology/hyper-threading/index.htm

[30] M. Onouchi et al., "A System-level Power-estimation Methodology based on IP-level Modeling, Power-level Adjustment, and Power Accumulation," *Proceedings of Asia South Pacific Design Automation*, 2006.

[31] J.Zejda et al., "Gate-Level Power Estimation Using Transition Analysis," *Workshop on Design Methodologies for Microelectronics*, 1995.

978-1-4244-3039-0/08 $25.00 © 2008 IEEE

SESSION D3:
Nano

Single Event Upset: An Embedded Tutorial

Fan Wang and Vishwani D. Agrawal

Dept. of Electrical and Computer Engineering, Auburn University, Auburn, AL, 36849, USA

Email: *wangfan@auburn.edu, vagrawal@eng.auburn.edu*

Abstract— With the continuous downscaling of CMOS technologies, the reliability has become a major bottleneck in the evolution of the next generation systems. Technology trends such as transistor down-sizing, use of new materials, and system on chip architectures continue to increase the sensitivity of systems to soft errors. These errors are random and not related to permanent hardware faults. Their causes may be internal (e.g., interconnect coupling) or external (e.g., cosmic radiation). To meet the system reliability requirements it is necessary for both the circuit designers and test engineers to get the basic knowledge of the soft errors. We present a tutorial study of the radiation-induced single event upset phenomenon caused by external radiation, which is a major source of soft errors. We summarize basic radiation mechanisms and the resulting soft errors in silicon. Soft error mitigation techniques with time and space redundancy are illustrated. An industrial design example, the IBM z990 system, shows how the industry is dealing with soft errors these days.

I. INTRODUCTION

From the beginning of the recorded history, man has believed in the influence of heavenly bodies on the life on Earth. Machines, electronics included, are considered scientific objects whose fate is controlled by man. So, in spite of the knowledge of the exact date and time of its manufacture, we do not draft a horoscope for a machine. Lately, however, we have started noticing certain behaviors in the state of the art electronic circuits whose causes are traced to be external and to the celestial bodies outside our Earth. The *Single Even Upset* (SEU) phenomenon, as this non-permanent (i.e., random or soft) error behavior is termed, in digital systems affects the modern nanotechnology electronic devices. We believe SEU will assume greater importance in the future [12]. Sifting through the literature of the last half a century, we have collected the necessary material for a starter. Our aim is not to cram up these six pages with most information, but to provide the essentials that can be assimilated conveniently to help a reader to become an effective contributor. We begin with the definition.

"**Single Event Upset (SEU):** Radiation-induced errors in microelectronic circuits caused when charged particles (usually from the radiation belts or from cosmic rays) lose energy by ionizing the medium through which they pass, leaving behind a wake of electron-hole pairs". ⋯ *NASA Thesaurus*

The objective of this tutorial is to familiarize the reader with the SEU in digital electronics – definitions and terms, causes (mostly experimental), measurement and estimation, reliability standards, and the related design methods. You should expect to get almost complete, but not comprehensive, information. Looking over the Appendix on the last page will improve the comprehension as you read through this article.

We will present an up-to-date understanding of the SEU phenomena. Following the historical note of the following section, we summarize the concept of basic radiation mechanisms and explain how a soft error occurs in silicon in Section III. Examples of soft error mitigation techniques are presented in Section IV. In Section V, a case study of soft error detection and tolerance in IBM z990 system is given.

II. HISTORICAL NOTES

Soft errors have been studied by electrical, aerospace, nuclear and radiation engineers for almost half a century. In the period 1954 through 1957 failures in digital electronics were reported during the above-ground nuclear bomb tests. These were treated as electronic anomalies in the monitoring equipment because they were random and their cause could not be traced to any hardware fault [27]. Perhaps the first paper concerning the role of cosmic rays on electronics is by Wallmark and Marcus [24]. As quoted in the recent literature [16], these authors predicted that cosmic rays would start upsetting microcircuits due to heavy ionized particle strikes and cosmic ray reactions when feature sizes become small enough. Through 1970s and early 1980s, the effects of radiation received attention and more researchers examined the physics of these phenomena. Also from 1950s, theories of fault tolerance and self-repairing computing were being developed due to the increased reliability requirement of critical applications like the space-mission [23].

May and Woods of Intel Corporation [13] determined that these errors were caused by the alpha particles emitted in the radioactive decay of uranium and thorium present just in few parts-per-million levels in package materials. Their paper represented the first public account of radiation-induced upsets in electronic devices at sea level and these errors were referred to as "soft errors". The term soft error was used to differentiate from the repeatable errors traceable to permanent hardware faults. Guenzer and Wolicki [10] reported that the error causing particles came not only from uranium and thorium but that nuclear reactions generated high energy neutrons and protons, which could also cause upsets in circuits. Because the title of their paper was "Single Event Upset of Dynamic RAMs by Neutrons and Protons", the term "SEU" has been in use ever since [10] (refer to [16]). In 1979, Ziegler and Lanford from IBM [28] predicted that cosmic rays could

TABLE I

PROJECTED FAILURE RATE ON SRAM-BASED FPGAS APPLICATIONS DUE TO NEUTRON EFFECTS (ACTEL)

Application Examples	Altitude (feet)	Neutron Flux (relative)	FPGAs/ System	#upsets/1M-gate FPGA/day(.13μ)	MTBF[1] (hours)		FIT[1] (in million)	
					0.13μ	0.09μ	0.13μ	0.09μ
(1) Ground-based Communication Network	5000	1	512	4.19E-4	112	58	8.92	17.24
(2) Civilian Avionics System	30,000	~40	4	1.85E-2	324	162	3.09	6.17
(3) Military Avionics System	60,000	>160	16	8.33E-2	18	9	55.56	111.11

result in the same upset phenomenon in electronics (not only memories) even at sea level.

Recent *Soft Error Rate* (SER) testing result of SRAM-based FPGAs from Actel [1] shows a significant and growing risk of functional failures due to the corruption of configuration data, especially when the system has higher densities. Table I shows the failure rate projection for different applications without using any protection. The number of upsets per 1 million gates per day increases for cases (1) through (3) because of the altitude dependent increase in neutron flux density. The table includes projected failure rates for the 90nm process. It is expected that neutron-induced soft errors get worse by a factor of two as we move from 0.13μ to 0.09μ technology. Note that this table ignores alpha particle effects, which are also expected to be significant for nanometer technologies and will further increase the system failure rate.

The radiation induced soft errors have become one of the most important and challenging failure mechanisms in modern electronic devices. SER of commercial chips is controlled to within 100~1000 FITs[1]. Compared to most hard failure mechanisms that produce failure rates on the order of 1~100 FIT, the SER of a low-voltage embedded SRAM can easily be 1000 FIT/Mbit. Therefore, a four-phase approach to deal with them is in progress [21]:

1) Methods to protect chips from soft errors (prevention).
2) Methods to detect soft errors (testing).
3) Methods estimate the impact of soft errors (assessment).
4) Methods to recover from soft errors (recovery).

III. WHAT IS SOFT ERROR?

A. Soft Error Categories

An electronic circuit, that bears no permanent hardware fault, may witness unexplained events resulting in single bit changes spontaneously in the system, and there is no way to repeat such failures. Within the computer industry such phenomenon is known as a "soft fail", to differentiate from the "hard or permanent fail", which may be repairable [28]. After observing a soft error, there is no implication that the system hardware is any less reliable than before because the soft fail is completely random. These soft fails may be caused by the well-known electronic noise sources such as a noisy power supply, lighting, and electrostatic discharge (ESD), or the thermal radiation from the galaxy, such as from radiation-emitting stars and atmospheric gases. A soft or non-permanent fault is a non-destructive fault and falls into two categories [22]:

1) Transient faults, caused by environmental conditions like temperature, humidity, pressure, voltage, power

[1]See Appendix

supply, vibrations, fluctuations, electromagnetic interference, ground loops, cosmic rays and alpha particles.

2) Intermittent faults caused by non-environmental conditions like loose connections, aging components, critical timing, power supply noise, resistive or capacitive variations or couplings, and noise in the system.

With advances in the design and manufacturing technology, non-environmental conditions may not affect the sub-micron semiconductor reliability. However, the errors caused by cosmic rays and alpha particles remain the dominant factors causing errors in electronic systems.

B. Radiation Mechanisms in Semiconductors

Three principal radiation sources cause soft errors in advanced semiconductor devices [5]:

1) Alpha particles are emitted when the nucleus of an unstable isotope decays to a lower energy state. They contain kinetic energy in the range of 4 to 9 MeV. There are many radioactive isotopes, however, uranium and thorium have the highest activity among naturally occurring materials. In the terrestrial environment, major sources of alpha particles are radioactive impurities such as lead-based isotopes in solder bumps of the flip-chip technology, gold used for bonding wires and lid plating, aluminum in ceramic packages, lead-frame alloys and interconnect metalization [8].

2) High-energy (> 1 MeV) neutrons from cosmic radiation can induce soft errors in semiconductor devices via secondary ions produced by the neutron reaction with silicon nuclei. Cosmic rays that are of galactic origin react with the Earth's atmosphere to produce complex cascades of secondary particles. Less than 1% of the primary flux reaches ground level and the predominant particles include muons, neutrons, protons, and pions. Because pions and muons are short-lived and proton and electrons are attenuated by Coulombic interaction with the atmosphere, neutrons are the most likely cosmic radiation sources to cause SEU in deep-submicron semiconductors at terrestrial altitude. The neutron flux is dependent on the altitude above the sea level, the density of the neutron flux increases with altitude.

3) The third significant source of ionizing particles in electronic devices is the secondary radiation induced from the interaction of cosmic ray neutrons and boron. It is the radiation induced by low-energy cosmic neutron interactions with the isotope *boron-10* (^{10}B is commonly used as *p*-type dopant for junction formation

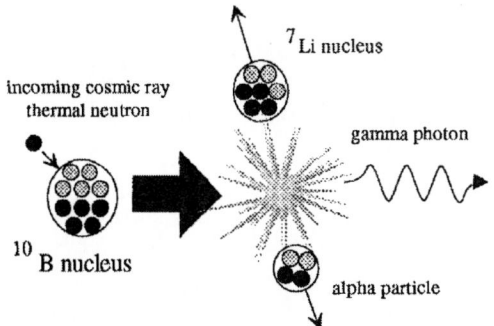

Fig. 1. Fission of ^{10}B induced by the capture of a neutron (commonly happened in SRAMs) [3].

in IC package). Specifically, BPSG (*Borophosphosilicate glass*) dielectric layer is commonly used to form insulator layers in IC manufacturing. Boron has two isotopes: ^{10}B and ^{11}B of which ^{10}B is unstable. The reaction scheme is shown in Figure 1. In the $^{10}B(n, \alpha)$ Li reaction the lithium nucleus is emitted with a kinetic energy of 0.84 MeV 94% of the time and 1.014 MeV 6% of the time. The gamma photon has energy of 478 KeV, while the alpha particle is emitted with an energy of 1.47 MeV [3]. This mechanism has recently been found to be the dominant source of soft errors in 0.25 and 0.18μ SRAM fabricated with BPSG. Modern microprocessors use highly purified package materials and this radiation mechanism is greatly reduced, making the high-energy cosmic rays the major reason for soft errors.

C. Sensitive Regions in Silicon

A *single event transient* (SET) is caused by the generation of charge due to a single particle (proton or heavy ion) passing through a sensitive node in the circuit. SETs in linear devices differ significantly from other types of *single event effects* (SEE) like SEU in a memory. Each SET has its unique characteristics like polarity, waveform, amplitude, duration, etc. These characteristics depend on particle impact location, particle energy, device technology, device supply voltage and output load. In CMOS circuits, the "off" transistors struck by a heavy ion in the junction area are most sensitive to SEU by particles with high enough LET (*linear energy transfer*; see Appendix) of around 20 MeV-cm²/mg. When these particles hit the silicon bulk, the minority carriers are created and if collected by the source/drain diffusion regions, the change of the voltage value of those nodes occurs [20]. A particle can induce SEU when it strikes at the channel region of an off NMOS transistor or the drain region of an off PMOS transistor. The ionization can induce a current pulse in a p-n junction. Conceptually, when the charge injected by the current pulse at a sensitive node exceeds the critical charge (Q_{crit}), a SET is generated at the affected junction.

D. Single Event Transient

In Figure 2, an SET is produced after an energetic ionizing particle has been brought to the silicon near sensitive device

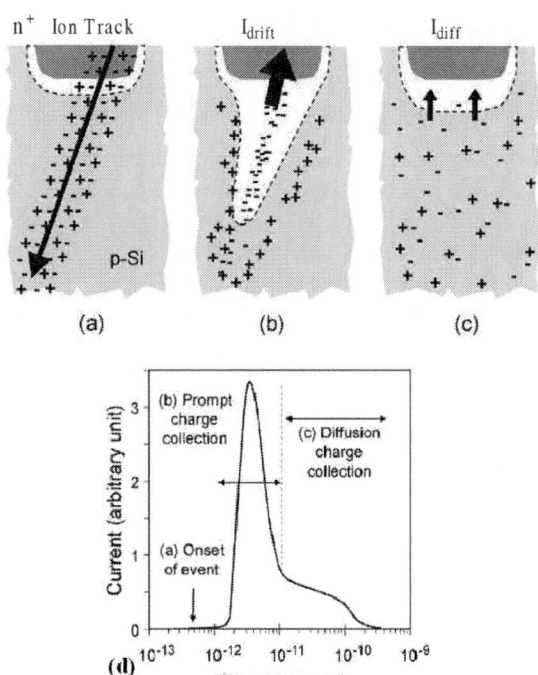

Fig. 2. Schematic representation of charge collection in a silicon junction immediately after (a) an ion strike, (b) prompt (drift) collection , (c) diffusion collection, (d) the junction current induced as a function of time [4].

nodes [4]. Along the traversed path, the particle produces a dense radial distribution of electron-hole pairs as illustrated in Figure 2(a). If the resultant ionization track traverses the depletion region, carriers are rapidly collected by the electric field, thus compensating the charge stored in the junction. Outside the depletion region the non-equilibrium charge distribution induces a temporary funnel-shaped potential distortion along the trajectory of the event, thus further enhancing charge collection by drift (Figure 2(b)). A "prompt" collection phase typically follows for tens of picoseconds and as the funnel collapses, diffusion then dominates the collection process (Figure 2(c)) until all excess carriers have been collected, recombined, or diffused away from the junction area (about nanoseconds). The transient charge collected from the radiation event produces a current pulse at the junction as illustrated in Figure 2(d) [4]. The current transient typically lasts 200 picoseconds with the bulk of the charge collection occurring within 2~3 microns of the junction region for modern submicron CMOS technologies. The time constants depend strongly on the type of ion, its initial energy and the properties of the specific technology [4]. If enough charge is collected by a node the data state may change. The collected charge (Q_{coll}) is a function of the ionizing particle's energy and trajectory, silicon substrate structure and doping, and the local electric field [4].

A commonly used approximate analytical model for the induced transient current waveform for ion track charge collection has a double-exponential form [15] with a rapid rise

431

978-1-4244-3039-0/08 $25.00 © 2008 IEEE

time and a gradual fall time:

$$\begin{cases} I(t) = \frac{Q_{coll}}{\tau_\alpha - \tau_\beta}\left(e^{-\frac{t}{\tau_\alpha}} - e^{-\frac{t}{\tau_\beta}}\right) & \text{(a)} \\ Q_{coll} = 10.8 \times L \times LET & \text{(b)} \end{cases} \quad (1)$$

where Q_{coll} is the collected charge (in femto coulomb) in the sensitive region, τ_α is a process-dependent collection time constant of the junction, and τ_β is the ion-track establishment time constant, which is relatively independent of the technology. Typical values are approximately $1.64 \times 10^{-10} sec$ for τ_α and $5 \times 10^{-11} sec$ for τ_β [7]. In bulk silicon, a typical *charge collection depth* (L in micron) is 2 for every linear energy tranfer (LET) of 1 MeV-cm^2/mg, and an ionizing particle deposits about $10.8 fC$ charge along each micron of its track.

The induced transient voltage pulse may propagate through several levels of logic gates. Because a particle can induce an SEU when it strikes either the channel region of an off NMOS transistor or the drain region of an off PMOS transistor, we will consider the strike at an off PMOS drain area as an illustrative example. The critical charge depends on the total charge collected at the sensitive node as well as on the temporal shape of the current pulse and the device supply voltage. So, a parameter called "switching time (t_{th})" or "feedback time" is defined as the interval after the particle strikes at which the affected node voltage exceeds the threshold voltage. The charge on the output capacitor equals Q_{crit} at that time. Q_{crit} can be calculated by integrating the current that flows at the sensitive node after the strike [9]. The condition for the SEE to propagate is that output node voltage follows Equation 2.

$$V \geq \frac{Q_{crit}}{C} = \frac{1}{C}\int_0^{t_{th}} I_{drain}(t)dt \quad (2)$$

The pulse width of the voltage pulse depends on the value of the capacitance and the RC time constant of the discharging path. For example, in ami12 technology, when the output load capacitance is 100fF and the cumulative collected charge is 0.65pC, the amplitude of the voltage pulse is $0.65pC/100fF = 0.65 \times 10^{-12}C/100 \times 10^{-15}F = 0.65V$. We observe that for the same charge collected in the sensitive area a smaller load capacitance will have a larger amplitude of the SEE-induced voltage pulse. The discharge process can be modeled by a simple RC-circuit. So, the voltage as a function of time is $v(t) = v(0)^{\frac{-t}{RC}}$. Thus, smaller the RC value, faster is the discharge process. A schematic view of how the SEE-induced current pulse translates into an SEE-induced voltage pulse is given in Figure 3.

IV. SOFT ERROR MITIGATION TECHNIQUES

Soft error tolerant techniques can be classified into two types: prevention and recovery. The methods to protect microchips from soft-errors are the prevention methods. They are used during the chip design and development. The recovery methods include on-line recovery mechanisms from soft-errors in order to achieve the chip robustness requirement. These include fault tolerant computing, ECC/parity, online-testing and redundancy. One should note that soft error is not the only reason why computer systems need to resort to a

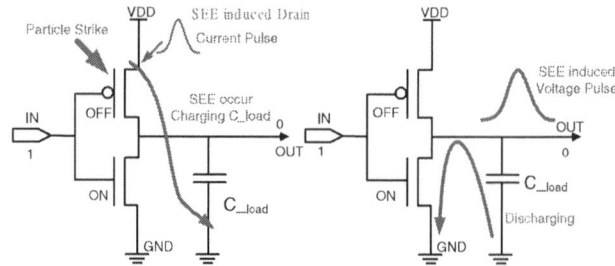

Fig. 3. An schematic view of how SEE-induced current pulse translates into a voltage pulse in a CMOS inverter.

recovery procedure. Random errors due to noise, unreliable components, and coupling effects may also require recovery mechanisms [21]. The need for a recovery mechanism stems from the fact that prevention techniques may not be enough for contemporary microchips, because the supply voltage keeps reducing, feature size keeps shrinking, and the clock frequency keeps increasing. Also, the cost of prevention techniques for a fault tolerant design may be too high. Because the error-tolerant computing is a broad area, here we only give a few examples of techniques used for soft error mitigation. In addition, a built-in soft error resilience (BISER) technique for correcting radiation-induced soft errors in latches and flip-flops may be found [25]. In that work, the error-correcting latch and flip-flop designs are power efficient and can correct both flip-flop errors and combinational logic errors, and employ reuse of on-chip scan design-for-testability in cell-level SER.

A. Prevention Techniques

1) Purify the Fabrication Material: A significant improvement in the SER performance of microelectronics can be achieved by eliminating or reducing the sources of radiation. To reduce the alpha particle emission in the final packaged IC, high purity materials and processes are employed. Uranium and thorium impurities have been reduced below one hundred parts per trillion for high reliability. Going from the conventional IC packaging to an ultra-low alpha packaging materials the alpha emission is reduced from 5~10 alphas/cm^2-hr to less than 0.001 alphas/cm^2-hr. To reduce the SER induced by the ^{10}B activation by low energy neutrons, BPSG is replaced by other insulators that do not contain boron. In addition, any processes using boron precursors is carefully checked for ^{10}B content before introducing them to manufacturing process [4]. When these measures are employed the SER of the IC is reduced dramatically, but the SER caused by the cosmic high energy neutron interactions cannot be easily shielded.

2) Radiation Hardened Process Technologies: SER performance can be greatly improved by adapting the process technology either to reduce the collected charge (Q_{coll}) or increase the critical charge (Q_{crit}) [26]. One approach is to use additional well isolation (triple-well or guard-ring structure) to reduce the amount of charge collected by creating potential barriers, which can limit the efficiency of the funneling effect and reduce the likelihood of parasitic bipolar collection paths [6].

Another approach replaces bulk silicon well-isolation with silicon-on-insulator (SOI) substrate material. The direct charge collection is significantly reduced in SOI devices because the active device volume is greatly reduced (due to thin silicon device layer on the oxide layer) [18]. Recent work shows a 10X reduction in SER achieved over conventional bulk devices when a fully depleted SOI substrate is used. Unfortunately, SOI substrates are more expensive than conventional bulk substrates and phenomena like parasitic bipolar action limit further reduction of SER [4]. Circuit-level solutions such as the addition of cross-coupled resistors and capacitors to decrease bit-line float time are also employed.

B. Recovery Techniques

Fault-tolerant computing methods have existed in the literature for quite some time [23] but have seen renewed interest due to the SEU phenomenon. On-line testing techniques are frequently used as recovery solutions for soft error mitigation. Specific techniques includes self-checking design [19], concurrent error detection for finite state machines (FSM) by signature monitoring, error detection and correction (EDAC) codes, and redundancy.

1) Redundancy: The basic idea of redundancy in design is to gain higher system reliability by sacrificing the minimality of time or space, or both. The classic triple modular redundancy (TMR) with a majority voter [2] continues to be widely used.

Mitra *et al.* [17] combine a self-checking design with time redundancy based on the C-element gate to compare two samples of the outputs signal from a combinational circuit at times t_0 and $t_0 + d$. The C-element has the ability to eliminate glitches at combinational outputs. Their error correction structure is illustrated in Figure 4. The space redundancy and time redundancy are often combined together to meet high fault-tolerance requirement with reduced hardware overhead, such as duplication and comparison instead of TMR.

2) ECC and Parity: Memories have a significant role in modern systems. Because of very high density of storage cells, a large memory is more sensitive to ionized particles than the logic. A simple solution for protecting a memory is to add parity bits to each memory word. During each write operation, a parity generator computes the parity bits of the data to be written with the data in the memory. If a particle strike alters the state of a single bit of a memory word, the error can be discovered by checking the parity code during the read operation. Depending on the number of parity bits used, this scheme may only detect an error, or correct it as well. Such schemes are often combined with system-level approachs for error recovery [19]. In most situations, however, the error recovery in a memory is more complex so protection of the memory by means of codes, like error correcting code (ECC), is preferable. Table II summarizes sample EDAC methods for memory, data and systems [11].

V. A CASE STUDY

The IBM eServer z990 system is designed to detect and recover from both soft and permanent errors [14]. The z990

TABLE II
SAMPLE EDAC METHODS FOR MEMORY OR DATA DEVICES [11]

EDAC Method	EDAC Capability
Parity	Single Bit Error Detect
Hamming Code	Single Bit Error Correct, double bit detect
RS Code	Correct consecutive and multiple bytes in error
Conventional Encoding	Corrects isolated burst noise in a communication stream
Overlying Protocol	Specific to each system implementation

contains up to four pluggable nodes connected through a planar board in a daisy chain interconnect structure. Each node contains up to 64 GB physical memory and 32 MB L2 cache for a system capacity of 256 GB memory and 126 MB L2 cache. In IBM z990 system, microarchitecture-level SEU mitigation features include: extensive use of ECC and parity with retry on data and controls; full SRAM ECC and parity protection; operational retries; microprocessor mirroring, checkpointing and rollback, and some hardware derating techniques. These approaches may be useful for future mainframe, general purpose, and application-specific computing systems.

VI. CONCLUSION

Soft error rate in logic and and memory chips will continue to increase as devices become more sensitive to soft errors even at sea level. The logic FIT rate is expected to increase faster due to internal phenomena such as cross coupling, ground bounce and delay faults, becoming comparable to the prevailing FIT rate of memory. The IBM z990 system provides an illustration of how the soft error issue might be handled in the industry. Open soft error issues are in the areas of EDA tools, radiation tests and measurement, analysis of newer radiation mechanisms, device hardening, soft error rate analysis, and error mitigation methods, on which research is being conducted. We hope we have given a running start to our reader.

APPENDIX
Definitions and Terminology[2]

Collected Charge (Q_{coll}): The charge collected by a particular device node during the passage of a particle. The collected charge is dependent on the geometry and doping of the node, the particle property like mass, energy and trajectory, and the density and type of material in the volume being penetrated by the incident radiation.

Cross Section (σ): The device SEE response to ionizing radiation. Normally, the units for cross section are $cm^2/device$ or cm^2/bit.

Critical Charge (Q_{crit}): The minimum amount of charge that when collected at any sensitive node will cause the node to change state. The critical charge is usually generated by incident radiation and, it is dependent on the linear energy transfer effective which is usually a function of the angle of incident particle radiation.

LET: Linear Energy Transfer. LET is a measure of the energy transferred to the device per unit length as an ionizing particle travels through a material. The common unit is MeV-cm²/mg of material (Si for MOS devices).

LET_{th}: LET threshold (LET_{th}) is the minimum LET to cause an effect at a given particle fluence.

SEE: Single Event Effect. Any measurable or observable change in state or performance of a microelectronic device, component, subsystem or system resulting from a single energetic particle strike. SEE includes **SEU** (Single

---

[2]These miscellaneous definitions and terms are collected from JEDEC standard and relevant papers.

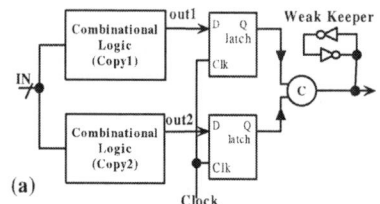 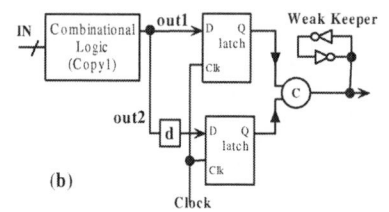 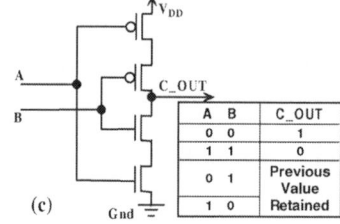

Fig. 4. Error correction using duplication, (a) space redundancy structure, (b) time redundancy structure, and (c) C-element [17].

Event Upset), **SEL** (Single Event Latchup), **SEB** (Single Event Burnout), **SEFI** (Single Event Functional Interrupt), and **SET** (Single Event Transient).

Sensitive Volume: A region, or multiple regions affected by SEE-induced radiation. The sensitive volume is determined by the angle of the incident radiation, the mass and energy of the incident particles and the density, type of the material in the volume being penetrated by the incident radiation. Is is not easy to know the geometry of the sensitive volume of the device but some information can be gained from the test cross section data.

Units and Conversion Factors

Energy Unit: Electron Volt (eV) One eV is the energy gained by one electron in accelerating through a potential difference of 1 volt. Energy in radiation is usually in unit of MeV (10^6eV) or KeV (10^3eV). 1eV = 1.6×10^{-19} J, 1MeV = 1.6×10^{-13} J.

FIT: Failure in Time; the number of failures per 10^9 device hours. 1 year MTTF (Mean Time To Failure) = $10^9/(24 \times 365)$ FIT = 114,155 FIT.

REFERENCES

[1] Actel, "Effects of Neutrons on Programmable Logic.–a white paper," Technical report, Actel corporation, Dec., 2002.

[2] A. Avizienis, "Faulty-Tolerant Computing: An Overview," *Computers, IEEE Trans. Computers*, vol. 4, no. 1, pp. 5–8, 1971.

[3] R. Baumann, "Soft Errors in Advanced Semiconductor Devices-Part I: The Three Radiation Sources," *IEEE Trans. Device and Materials Reliability*, vol. 1, no. 1, pp. 17–22, 2001.

[4] R. Baumann, "Soft Errors In Commercial Integration Integrated Circuits," *International Jour. High Speed Electronics and Systems*, vol. 14, no. 2, pp. 299–309, 2004.

[5] R. Baumann, "Soft Errors in Advanced Computer Systems," *IEEE Design & Test of Computers*, vol. 22, no. 3, pp. 258–266, 2005.

[6] D. Burnett, C. Lage, and A. Bormann, "Soft-Error-Rate Improvement in Advanced BiCMOS SRAMs," in *Proc. 31st Annual IEEE Reliability Physics Symp.*, Mar. 1993, pp. 156–160.

[7] V. Carreno, G. Choi, and R. K. Iyer, "Analog-digital simulation of transient-induced logic errors and upset susceptibility of an advanced control system," in *NASA Technical Memo 4241*, 1990.

[8] C. L. Claeys and E. Simoen, *Radiation Effects in Advanced Semiconductor Materials and Devices*. Springer, 2002.

[9] C. Detcheverry, C. Dachs, E. Lorfevre, C. Sudre, G. Bruguier, J. M. Palau, J. Gasiot, and R. Ecoffet, "SEU Critical Charge and Sensitive Area in A Submicron CMOS Technology," *IEEE Trans. Nuclear Science*, vol. 44, no. 6, pp. 2266–2273, 1997.

[10] C. S. Guenzer, E. A. Wolicki, and R. G. Allas, "Single Event Upset of Dynamic RAMs by Neutrons and Protons," *IEEE Trans. Nuclear Science*, vol. 26, pp. 5048–5052, Dec. 1979.

[11] K. L. LaBel, P. W. Marshall, J. L. Barth, E. Stassinopoulos, C. Seidleck, and C. Dale, "Commercial Microelectronics Technologies for Applications in the Satellite Radiation Environment," in *Proc. 1996 IEEE Aerospace Applications*, (New York), 1996, pp. 375–390.

[12] J. Maiz and N. Seifert, "Introduction to the Special Issue on Soft Errors and Data Integrity in Terrestrial Computer Systems," *IEEE Trans. Device and Materials Reliability*, vol. 5, no. 3, pp. 303–304, Sept. 2005.

[13] T. C. May and M. H. Woods, "A New Physical Mechanism for Soft Errors in Dynamic Memories," in *Proc. 16th Annual Reliability Physics Symp.*, 1978, pp. 33–40.

[14] P. J. Meaney, S. B. Swaney, P. N. Sanda, and L. Spainhower, "IBM z990 Soft Error Detection and Recovery," *IEEE Trans. Device and Materials Reliability*, vol. 5, no. 3, pp. 419–427, 2005.

[15] G. C. Messenger, "Collection of Charge on Junction Nodes from Ion Tracks," *IEEE Trans. Nuclear Science*, vol. 29, no. 6, pp. 2024–2031, 1982.

[16] G. C. Messenger and M. Ash, *Single Event Phenomena*. Chapman & Hall, 1997.

[17] S. Mitra, Z. Ming, S. Waqas, N. Seifert, B. Gill, and K. S. Kim, "Combinational Logic Soft Error Correction," in *Proc. International Test Conference*, 2006, pp. 1–9.

[18] O. Musseau, "Single-Event Effect in SOI Technologies and Devices," *IEEE Trans. Nuclear Science*, vol. 43, no. 2, pp. 603–613, 1996.

[19] M. Nicolaidis, "Design for Soft Error Mitigation," *IEEE Transactions on Device and Materials Reliability*, vol. 5, no. 3, pp. 405–418, 2005.

[20] M. Omana, G. Papasso, D. Rossi, and C. Metra, "A Model for Transient Fault Propagation in Combinatorial Logic," in *Proc. 9th IEEE On-Line Testing Symp.*, 2003, pp. 111–115.

[21] K. Roy, S. Kundu, R. Galivanche, V. Narayanan, R. Raina, and P. N. Sanda, "Is the Concern for Soft-Error Overblown?," in *Proc. International Test Conf. (Panel Discussion)*, 2005.

[22] A. J. van de Goor, *Testing Semiconductor Memories: Theory and Practice*. Wiley, 1991.

[23] J. von Neumann, "Probabilistic Logics and the Synthesis of Reliable Organisms from Unreliable Components (1959)," in A. H. Taub, editor, *John von Neumann: Collected Works, Volume V: Design of Computers, Theory of Automata and Numerical Analysis*, Oxford University Press, 1963, pp. 329–378.

[24] J. T. Wallmark and S. M. Marcus, "Minimum Size and Maximum Packing Density of Non-Redundant Semiconductor Devices," *Proc. IRE*, vol. 50, pp. 286–298, Mar. 1962.

[25] M. Zhang, S. Mitra, T. M. Mak, N. Seifert, N. J. Wang, Q. Shi, K. S. Kim, N. R. Shanbhag, and S. J. Patel, "Sequential Element Design With Built-In Soft Error Resilience," *Very Large Scale Integration (VLSI) Systems, IEEE Transactions on*, vol. 14, no. 12, pp. 1368–1378, 2006. 1063–8210.

[26] Q. Zhou and K. Mohanram, "Gate Sizing to Radiation Harden Combinational Logic," *IEEE Trans. CAD*, vol. 25, no. 1, pp. 155–166, 2006.

[27] J. F. Ziegler, "IBM Experience in Soft Fails in Computer Electronics (1978-1994)," *IBM Jour. Res. and Dev.*, vol. 40, no. 1, pp. 3–18, 1996.

[28] J. F. Ziegler and W. A. Lanford, "Effect of Cosmic Rays on Computer Memories," *Science*, vol. 206, no. 4420, pp. 776–788, Nov. 1979.

21st International Conference on VLSI Design

Fault-Tolerant Computing Using a Hybrid Nano-CMOS Architecture

Muzaffer O. Simsir
Dept. of Electrical Engineering
Princeton University,
Princeton, NJ 08544
msimsir@princeton.edu

Srihari Cadambi
Franjo Ivancic
Martin Roetteler
NEC Labs America,
Princeton, NJ 08540
{cadambi,ivancic,mroetteler}
@nec-labs.com

Niraj K. Jha
Dept. of Electrical Engineering
Princeton University,
Princeton, NJ 08544
jha@princeton.edu

ABSTRACT

Architectures based on nanoscale molecular devices are attracting attention for replacing CMOS architectures at the end of the semiconductor roadmap. The two most promising nanotechnologies, according to ITRS, are silicon nanowires and carbon nanotubes. Although they offer unmatched densities for building logic, interconnect and memory, they suffer from very defect-prone manufacturing processes. This is further exacerbated by testing complexities where it is nearly impossible to detect all defects in a large nanoscale chip. Furthermore, the small structures in nanoscale architectures are susceptible to transient faults which can produce arbitrary soft errors. As a result, fault tolerance is necessary to make nanoscale architectures practical and realistic. We propose an architecture that can tolerate a large number of undetected manufacturing faults as well as a large rate of transient faults. Our architecture is characterized by multiple levels of redundancy and majority voting to correct errors caused by such faults. A key aspect of the architecture is that it contains a judicious balance of both nanoscale and traditional CMOS components. A companion to the architecture is a compiler with heuristics tailored to quickly and compactly map logic onto partially defective components. Experimental results demonstrate the efficacy of the architecture.

1. INTRODUCTION

As CMOS technology nears the end of the semiconductor roadmap, nanoscale molecular devices constructed from silicon nanowires and carbon nanotubes are emerging as alternatives. ITRS 2005 [1] rates these devices as among the most promising of all the new technologies under investigation.

Nanowires and carbon nanotubes offer unmatched densities for building logic, interconnect and memory. However, a major drawback is that their manufacturing processes are defect-prone; the defect rate is expected to be as high as 10% [2]. Furthermore, testing chips built from nanoscale devices is difficult because of their extremely high densities, and low controllability and observability. Most testing methods proposed in the literature [3–6] are probabilistic, i.e., they isolate regions of the chip in which a fault likely exists, but do not identify or localize faults with certainty. It is unlikely that such methods will scale to be able to test a large chip with reasonable speed and accuracy. The complexity of testing, coupled with high defect levels, imply that we must build fault-tolerant architectures if nanoscale molecular devices are to become practical.

One aspect that permits fault tolerance is configurability. Diode-based logic [2], which uses intersecting horizontal and vertical nanowires, lends itself to easy reconfiguration. When a junction is configured by applying a voltage, it is held together effectively until the application of another voltage. Thus, the junction implicitly stores a configuration bit. Configurability allows us to use defective chips by "compiling around" the defects, i.e., the compiler can choose not to use defective com-

ponents. Nevertheless, this is only useful for defects that have been discovered and localized during testing. Given the complexity of testing, however, fault tolerance by configurability is neither feasible nor practical.

Another problem associated with nanoscale molecular devices are transient faults, which are temporary glitches that cause erroneous behavior. Transient faults can occur due to cosmic rays, heat and noise [7–10]. Small structures, such as nanoscale circuits, are particularly susceptible to them.

While fault tolerance is an important issue and is being addressed in the nanotechnology community, a composite architecture with the ability to handle undetected permanent faults as well as transient faults has not been well investigated. In this paper, we present such an architecture and its accompanying compiler.

Our work is characterized by the following key contributions:

- We propose an architecture with two levels of redundancy: first, each processing element is based on nanowire crossbars that contains extra nanowires and second, the processing elements themselves are replicated.

- We "hybridize" the architecture by using CMOS logic to vote on the outputs of the redundant nanoelectronic processing elements. The redundancy and voting mechanisms resolve transient faults as well as undetected permanent faults.

- We propose fast compilation heuristics so that compilation time is largely unaffected by the defect rate.

- Our compiler has distinct defect-aware and defect-unaware phases, and is retargetable across defect models. A key idea that helps achieve this is that all defect information is completely subsumed within the cost function of the simulated annealing-based placer.

The rest of the paper is organized as follows. In Section 2, we provide background information about nanowire-based design and CMOS-nanowire interface. In Section 3, we present our proposed hybrid fault-tolerant architecture. In Section 4, we present the compilation framework and algorithms. In Section 5, we present experimental results and conclude in Section 6.

2. BACKGROUND

We next present some background material.

2.1 Nanowire-based Architectures

Nanowires are wires that have diameters on the order of nanometers and lengths up to several hundreds of micrometers. By using metallic, p-type and n-type nanowires, basic computation blocks, such as AND, OR and NOR gates [11], can be implemented.

Furthermore, nanowires can be aligned to construct the so-called crossbar architectures, in which a layer of nanowires is placed orthogonal to another layer. In such architectures, the

435

978-1-4244-3039-0/08 $25.00 © 2008 IEEE

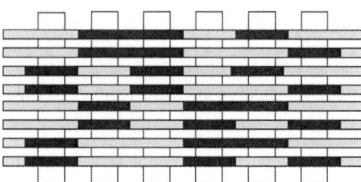

Figure 1: CMOS-nanowire interface using nanowire decoder.

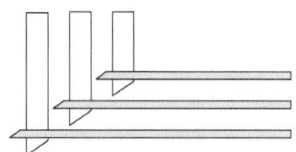

Figure 2: CMOS-nanowire interface using one-to-one mapping.

region where two wires in different layers overlap is called a junction. When activated by a voltage, each junction behaves like a diode, FET or resistor, depending on the electrical properties of the nanowires and the layer between them. A function can be realized by configuring selected junctions in the crossbar.

Among all nanowire architectures, crossbar architectures are the most dominant since nanoscale manufacturing processes yield regular structures which make them easier to build. These architectures also permit easy incorporation of redundancy, which makes them suitable in the presence of high manufacturing defect rates.

A crossbar architecture proposed by DeHon [2] uses programmable diode junctions to construct OR planes, and nanowire FET-based buffers and inverters, to make up NOR and OR planes. Another crossbar architecture from HP Labs [7] has junctions that can be configured as FETs or diodes, depending on the material between the nanowire layers. n-type and p-type FETs are used to implement logic, while diode junctions are used for routing. In these two proposals, known defects are avoided by routing around them. Another architecture, which is a CMOS-nanowire hybrid chip, proposed in [12], uses an island-style FPGA architecture. The logic blocks are implemented using nanowires, and the interconnect using CMOS. However, it is assumed that nanowire-based logic blocks are defect-free.

2.2 CMOS-Nanowire Interface

In order to implement logic on nanowire crossbars, each nanowire must be accessed individually to be able to program selected junctions. Since it is not possible to have direct access to nanoscale wires, a CMOS-nanowire interface should be provided. Furthermore, this interface can not only be used to program junctions, but can also serve as a signal input/output interface between nanowire and CMOS components for hybrid designs.

In the literature, there are three main CMOS-nanowire interface proposals. In the first proposal, a nanowire decoder is used to activate exactly one nanowire for each signal assignment on the micro-scale wires [2] (Figure 1). A nanowire decoder can be built by using nanowires that have unique addresses encoded in their axial profile.

The second proposal [13] uses masking methods to provide one-to-one mapping between micro-scale wires and nanowires (Figure 2). In this method, a diagonal mask is used to match micro-scale wire pitch to nanowire pitch.

In [14], yet another alternative for achieving a CMOS-nanowire interface has been described. In this proposal, a nanowire crossbar is placed on top of a CMOS plane. The interconnect between these two planes is provided by special pin/pad

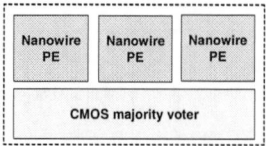

Figure 3: Hybrid nano-CMOS computing element example.

junctions which are distributed uniformly across the CMOS layer. The nanowire crossbar is slightly rotated so that each nanowire connects to only one pin.

3. HYBRID ARCHITECTURE

In this section, we describe our hybrid architecture consisting of both nanowire and CMOS logic elements. The primary goals of the architecture are tolerating high manufacturing defect rates, reducing the general unreliability of computations caused by transient faults, and facilitating fast compilation. Before we introduce the architecture in detail, we identify three basic problems germane to all nanoscale computing systems.

3.1 Problems in Nanoscale Circuits

Circuit design using devices at structure sizes in the range of few nanometers faces several fundamental problems. We classify them into three categories: *detected permanent faults, undetected permanent faults* and *transient faults*.

Detected permanent faults are permanent manufacturing defects in the chip that have been detected during testing. They may be wire defects in the form of broken nanowires or unprogrammable junctions due to stuck-open (stuck-closed) faults in which the junctions are permanently open (closed). **Undetected permanent faults** are permanent faults in the chip which are not discovered during testing. Since testing methods for nanoarchitectures are probabilistic in nature [2,3], it is reasonable to expect that, due to the sheer scale of the chip, a significant fraction of defective junctions and broken nanowires will remain undetected and this can cause the chip to malfunction. **Transient faults** are dynamic faults that occur at runtime even if the device is free of permanent manufacturing defects. Transient faults, which lead to soft errors, can be caused by perturbations to the substrate, e.g., by sub-atomic particles causing charge fluctuations, noise or heat [7, 9]. In CMOS logic circuits, transient faults have been addressed [8, 10], but do not pose serious problems yet. Nanoscale circuits, on the other hand, are expected to be prone to transient faults due to their smaller structures [7].

3.2 Architectural Details

The proposed architecture addresses all the above types of faults. The two major ideas involved are: (i) using redundancy in a "hierarchical" manner, and (ii) using majority voting with both nanowire and CMOS components to build a hybrid, fault-tolerant architecture. Other key contributions relate to the compiler and will be discussed in Section 4.

The basic unit of our architecture is a nanowire-based crossbar which is formed by intersecting h horizontal and v vertical nanowires. We refer to this structure as a *nanowire processing element (NPE)*. A natural outcome of using NPEs is the easy realization of Boolean look-up tables (LUTs), where a k-input LUT is a logical unit capable of implementing any single-output Boolean function of up to k inputs. We refer to a k-input LUT as a k-LUT for convenience. Although we restrict ourselves to k-LUTs with one output, the framework can easily be extended to allow multi-output LUTs or any other functional representation.

The first level of hierarchical redundancy is obtained by replicating each NPE several times. The outputs of all NPEs within such a redundant set are fed to a majority voter constructed using CMOS. Building this structure is possible by us-

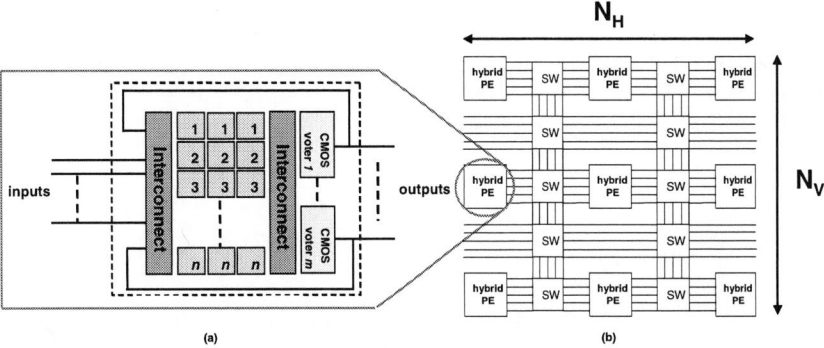

Figure 4: (a) An $(n, 3, m)$ hybrid cluster, and (b) the overall hybrid architecture.

ing one of the CMOS-nanowire interface methods explained in Section 2.2. The CMOS voter corrects errors due to transient and undetected permanent faults. We assume that the voter is fault-free (this is a reasonable assumption since CMOS processing is much more reliable than nanowire processing).

Although using the CMOS voter has a high area overhead, it is essential to correct transient faults. This implies that keeping the CMOS logic elements simple will decrease the area of the overall structure.

The second level of hierarchical redundancy is obtained by providing additional nanowires *within* each NPE. This is different from replicating NPEs and is applied on the level of individual nanowires. Recall from Section 2 that our nanowire architecture implements diode-based logic. The input signals to the LUTs are provided via h rows. While realizing a function, v columns are used to implement product terms. Therefore, the maximum number of inputs to a LUT is determined by the number of rows as well as the number of columns. For example, to implement a k-LUT, we need at least $2k + 1$ rows, since one wire for the output and two wires per input (one for the signal and another for its complement) are required. The number of columns should be at least 2^{k-1} because this is the maximum number of product terms required to implement any function with k inputs. Having spare rows and columns improves fault tolerance since certain nanowires may be defective. In addition, if a function does not require all 2^{k-1} product terms, the remaining nanowires can also enhance fault tolerance.

Figure 3 shows three NPEs and a CMOS majority voter. A generalized version of Figure 3 is the basic building block of the hybrid architecture. The architecture consists of a two-dimensional array of *hybrid clusters*. Each hybrid cluster is characterized by three parameters n, r and m. An (n, r, m) hybrid cluster consists of n fully interconnected NPEs, each replicated r times, and m CMOS voters. m and n allow us to trade-off fault tolerance for area: if m is large (m cannot be greater than n since we need at most one CMOS voter for one redundant set of NPEs), we achieve good fault tolerance at the expense of a comparatively large CMOS area. If m is small, a CMOS voter votes on the output of several sets of interconnected NPEs, rather than on each set of NPEs individually. In other words, not all NPE outputs are voted on in each cycle.

Figure 4(a) shows an $(n, 3, m)$ hybrid cluster and Figure 4(b) the entire hybrid architecture. Each hybrid cluster accepts inputs from other clusters, and provides outputs to other clusters. In addition, feedback wires, which are micro-scale, provide fast routing between closely-connected NPEs. Using NPEs within clusters makes it possible to use redundancy schemes that are considered very area-expensive fault-tolerant techniques in the CMOS world. However, it is possible to reduce the area overhead of CMOS elements by organizing computations

in the form of a directed acyclic graph (DAG) of NPEs with voting only performed at the output of the DAG. The complete DAG can then be replicated, and outputs of these replications can be checked for faults using CMOS voters.

Since nanoscale manufacturing processes lead to highly regular structures, we assume that all NPEs within a hybrid cluster are identical. A given function is mapped onto each copy separately (if such a mapping is possible) by routing around the defects, and the outputs are fed to a CMOS majority voter.

The overall nano-CMOS architecture is a 2D array of hybrid clusters with N_V rows and N_H columns (see Figure 4(b)). Since errors caused by faults in nanowire interconnects between clusters are difficult to deal with, we assume that the inter-cluster interconnect is realized using metal. However, as mentioned above, nanowires are used to implement logic. By using NPEs, we can apply redundancy schemes without significant area overhead; whereas, if CMOS logic elements were used, it would be very costly in terms of area.

4. HYBRID ARCHITECTURE COMPILER

The function of the hybrid architecture compiler is to map circuits to the nano-CMOS hybrid architecture. In this section, we present details of the compilation flow, including the algorithms used. The compiler has a technology-independent portion that is unaware of manufacturing defects, and a technology-dependent portion that uses an optional defect map. The defect map specifies defects that were discovered during testing. Note that, the defect map does not need to be complete (defects not specified in the defect map will be treated as undiscovered permanent defects).

As stated in Section 1, it is difficult to obtain an exact defect map for manufactured chips. This problem has been addressed in [15]. By looking at the expected defect levels of the manufacturing process, a method has been suggested to identify defect-free subsets within the original partially-defective crossbars. However, the assumption that functions should be implemented on defect-free subsets is an overkill since it is possible to map functions onto partially defective crossbars.

Figure 5 shows the major parts of the compiler. The three main inputs to the compiler are: (i) a circuit description in the Berkeley Logic Interchange Format (BLIF), (ii) a set of architectural parameters (the important ones are shown in Figure 5), and (iii) an optional defect map. We explain each part of the compiler next.

4.1 Logic Synthesis and Clustering

The logic netlist is first synthesized and mapped to k-LUTs. We use SIS [16] for logic synthesis, and the built-in PLD package for LUT mapping. The synthesized and LUT-mapped circuit, which is a network of k-LUTs, is then grouped into hybrid clusters using T-VPack [17]. The clustering process tries

437

978-1-4244-3039-0/08 $25.00 © 2008 IEEE

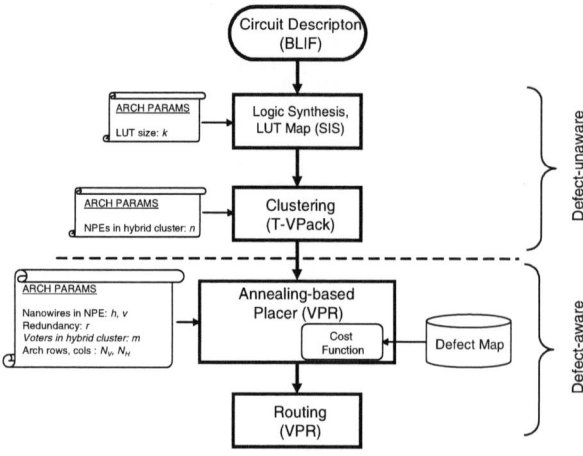

Figure 5: Hybrid architecture compiler.

to pack as many k-LUTs as possible into a single (n, r, m) hybrid cluster. In particular, T-VPack tries to maximize intra-cluster communication, thereby minimizing inter-cluster routing. This is crucial because significant delay and power dissipation are incurred in inter-cluster routing.

In the compiler, both logic synthesis and clustering are defect-unaware. We defer the use of defect knowledge to the subsequent placement phase, as described next. In that sense, our compiler is easily retargetable to different defect models.

4.2 Defect-aware Placement Strategy

The output of the clustering phase is a network of k-LUT clusters where each cluster has no more than n LUTs (recall n is the number of NPEs in an (n, r, m) hybrid cluster). Other inputs to the placer, which is based on VPR [17], are relevant architectural parameters – such as the number of CMOS voters in a hybrid cluster (see Figure 5 for a complete list) – and the optional defect map. The output of the placer is a netlist in which each LUT cluster in the output of the clustering phase is mapped to a physical hybrid cluster.

The defect map contains information about broken nanowires and defective junctions. For a particular chip, this information can be collected using one or more of the testing methodologies proposed for nanowire-based circuits [3–6]. If a defect map is not provided, placement onto a defect-free physical chip is assumed.

The simulated annealing-based placement method proceeds as follows. First, all clusters are randomly "assigned" to specific hybrid clusters on the chip. At each temperature, random swaps are performed, and the cost of the mapping is calculated. The cost function itself is a key contribution using which we make the placer defect-aware. The cost function C of a placement P is given by:
$$C(P) = C_{VPR} + num\_invalid\_mappings * N_{large} \quad (1)$$
where C_{VPR} is the timing and area cost function used by VPR [17], and $num\_invalid\_mappings$ the number of infeasible assignments to a physical hybrid cluster. An assignment is declared infeasible if the LUT cluster cannot be mapped to a physical cluster due to defects. N_{large} is a large positive number to drive the annealer towards a feasible solution.

Checking the feasibility of mapping a cluster of LUTs to a physical, possibly defective, hybrid cluster can be broken down into two parts: (i) mapping a k-LUT to a defective NPE, and (ii) mapping a cluster of k-LUTs to a defective (n, r, m) hybrid cluster. We describe each of these steps next.

4.2.1 Mapping a k-LUT to an NPE

Using diode junctions, it is possible to construct AND-OR planes, as shown in Figure 6(a). Input signals are fed to hori-

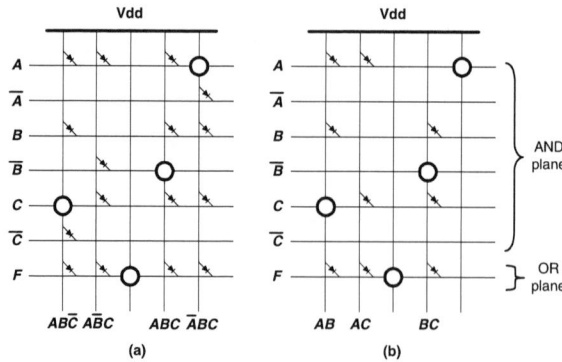

Figure 6: Two realizations of a three-input majority function.

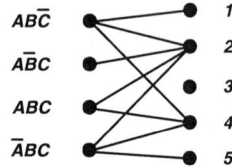

Figure 7: Bipartite graph for Figure 6(a).

zontal nanowires, while the vertical nanowires are configured to implement product terms by activating appropriate junctions. Junctions on the output horizontal wire are activated to implement the OR-plane.

Since we provide all input signals in true and complemented form, it is possible to implement any k-input Boolean function using an NPE. Figure 6(a) shows an unoptimized implementation of a three-input majority gate that uses every minterm, and Figure 6(b) shows the same function after minimizing the product terms. This example illustrates the use of "functional redundancy," which we describe in the next subsection.

In practice, it may not be possible to map any k-input Boolean function to an NPE due to defects. For instance, a broken nanowire cannot be used at all, neither can nanowire junctions with stuck-closed defects. On the other hand, the more common nanowire junction stuck-open defects can be tolerated by using other wires that do not have stuck-open defects on desired junctions. For example, in Figure 6, the function is mapped in such a way that it avoids all stuck-open defects (shown by circles), and uses an extra vertical nanowire.

We use bipartite matching to decide if a k-LUT can be mapped to a defective NPE. A bipartite graph is an undirected graph whose vertices can be partitioned into two sets such that no edge connects vertices in the same set (Figure 7). While creating the bipartite graph, we add a node on the "left side" for each product term, and a node on the "right side" for each vertical wire.

We now pick an assignment of input signals to horizontal wires (pin assignment). We do not attempt to find an optimal pin assignment since that is provably NP-hard. Instead, we assign pins based on a heuristic that we will soon explain. Given a pin assignment, an edge between a product term p and a vertical nanowire w indicates that p can be realized by w. Figure 7 shows the bipartite graph corresponding to Figure 6(a). The existence of a matching in the bipartite graph of a function implies that the function can be realized. On the other hand, if there are unmatched product term nodes, then bipartite graph construction and matching is repeated with another pin assignment. After a reasonable number of pin assignments, we deem the function as not mappable, and proceed to attempt mapping on a different hybrid cluster.

Table 1: Experimental results for 3-LUT networks

Circuit	#LUTs	#clusters	chip size (N_H, N_V)	$P_{so} = 1\%$			$P_{so} = 5\%$			$P_{so} = 10\%$		
				delay (ns)	chip size (N_H, N_V)	compile time (s)	delay (ns)	chip size (N_H, N_V)	compile time (s)	delay (ns)	chip size (N_H, N_V)	compile time (s)
c3540	575	144	12, 12	3.900	12, 12	46.9	3.903	12, 12	46.9	3.903	12, 12	46.8
c5315	650	163	13, 13	2.570	13, 13	67.6	2.570	13, 13	66.9	2.570	13, 13	67.4
c6288	965	242	16, 16	7.630	16, 16	81.3	7.630	16, 16	81.0	7.630	16, 16	81.0
c7552	653	164	13, 13	3.390	13, 13	75.1	3.390	13, 13	74.3	3.390	13, 13	74.3
dalu	602	151	13, 12	5.084	13, 12	48.1	5.080	13, 12	48.2	5.080	13, 12	47.9
des	1631	408	21, 20	2.560	21, 20	201.9	2.560	21, 20	201.8	2.560	21, 20	203.8
i8	533	134	12, 12	1.710	12, 12	53.5	1.710	12, 12	53.6	1.710	12, 12	53.7
i10	976	244	16, 16	4.480	16, 16	118.4	4.480	16, 16	118.6	4.480	16, 16	118.2
too_large	212	53	8, 7	1.837	8, 7	27.7	1.840	8, 7	27.8	1.840	8, 7	27.7

Table 2: Experimental results for 4-LUT networks

Circuit	#LUTs	#clusters	chip size (N_H, N_V)	$P_{so} = 1\%$			$P_{so} = 5\%$			$P_{so} = 10\%$		
				delay (ns)	chip size (N_H, N_V)	compile time (s)	delay (ns)	chip size (N_H, N_V)	compile time (s)	delay (ns)	chip size (N_H, N_V)	compile time (s)
c3540	389	98	10, 10	2.609	10, 10	44.4	2.617	10,10	49.7	n/a	n/a	n/a
c5315	517	130	12, 11	2.201	12, 11	74.4	2.196	12, 11	96.2	n/a	n/a	n/a
c6288	919	230	16, 15	7.330	16, 15	97.2	7.330	16,15	96.3	7.395	16, 15	96.3
c7552	516	129	12, 11	2.997	12, 11	85.2	3.006	12, 11	143.9	3.022	14, 14	136.2
dalu	379	95	10, 10	4.290	10, 10	40.8	4.290	10, 10	40.9	4.292	10, 10	40.5
des	1155	289	17, 17	2.250	17, 17	206.2	2.250	17, 17	207.1	2.242	17, 17	208.5
i8	364	91	10, 10	1.510	10, 10	49.8	1.510	10, 10	49.4	1.510	10, 10	48.8
i10	721	181	14, 13	3.735	14, 13	130.2	3.763	14, 13	145.7	3.751	16, 15	124.3
too_large	152	38	7, 6	1.520	7, 6	28.7	1.520	7, 6	28.7	1.518	7, 6	28.7

During simulated annealing, functions will be mapped to defective NPEs several times. Hence, the above process must be fast. Our fast heuristic begins with pin assignment. To choose the pin assignment, we assign signals most frequently used in the product terms to the nanowires with the smallest number of defects. Next, instead of constructing the complete bipartite graph, which is time-consuming, we start by finding a single edge for each product term. Vertical nanowire nodes that have defects on the junction of the output nanowire are skipped. The other vertical nanowires are checked to see if they can realize the product term, starting with the ones with the least number of edges. If the first vertical nanowire node that can realize the product term has other edges, it implies that there are other product terms using this nanowire. Those edges are marked. If a product term has no unmarked edges, we try to add more edges to that product term to see if it can be mapped to a free vertical nanowire. If no vertical nanowire can implement a product term, we restart the search with a different pin assignment. Although logic mapping heuristics for nanowire crossbars exists [18], they are not applicable to our scenario. We propose our own heuristics to map LUTs to a hybrid cluster.

4.2.2 Mapping LUTs to a Hybrid Cluster

Our goal here is to map a network of k-LUTs (the output of the clustering phase) to a physical hybrid cluster. To do this, we use the heuristic of Section 4.2.1 to determine if each k-LUT can be mapped to a candidate NPE within the cluster. We construct another bipartite graph with a node for each k-LUT on the left side, and a node for each set of redundant NPEs on the right side. The fast heuristic for checking a valid cluster mapping is similar to the heuristic used in Section 4.2.1. When a matching is found, a valid cluster mapping exists. On the other hand, if any function node is left unmatched after all trials, the algorithm terminates unsuccessfully.

To improve reliability, each function is mapped to redundant NPEs in different ways. If we use the same realization for all copies, then for a given input vector, if one copy produces an incorrect output, the probability of other copies having incorrect outputs will be higher; this is known as common-mode or common-cause failure [19]. By choosing different realizations, we aim to reduce the effect of common-mode failures.

Since we use metal for inter-cluster routing, we were able to use an unmodified third-party router (VPR) together with our placer.

5. EXPERIMENTAL RESULTS

In this section, we evaluate the hybrid architecture using our compiler. We vary the architectural parameters of Figure 5 to elicit useful information about the performance of the architecture under different defect rates. In particular, we show the effect of defect rates on critical path delay, chip size and compilation time by using $(4, 3, 2)$ hybrid clusters (recall from Section 3.2 that these hybrid clusters have two CMOS voters and four NPEs with triple modular redundancy). For the hybrid clusters, a minimum number of rows and columns are selected in order to analyze the effect of redundancy within an NPE.

Since manufacturing processes for nanowire chips are not yet mature, for the experiments, we create a random defect map based on defect probabilities that have been observed so far [2]. We vary the probability of stuck-open defects (P_{so}) between 1% and 10% and assume probabilities of broken nanowire (P_{bnw}) and stuck-closed (P_{sc}) defects to be zero. Furthermore, to statistically decide the minimum chip size required to implement a circuit, we generate several random defect maps for each defect rate. If 5% of the compilations for a specific defect rate fail, we increase the number of rows and columns of hybrid clusters on the chip and repeat the procedure.

We compiled several MCNC and ISCAS'85 benchmarks to the hybrid architecture to obtain critical path delays, chip sizes and compilation times. Tables 1, 2 and 3 show experimental results for $(4, 3, 2)$ hybrid clusters using 3-, 4- and 5-LUT NPEs, respectively. In these tables, the second column shows the number of LUTs required to realize each circuit. The third column denotes the number of hybrid clusters used for implementation. The chip size, in terms of the number of rows and columns of hybrid clusters, is given in the fourth column. The following columns denote critical path delay, chip size and compilation time for 1%, 5% and 10% P_{so} values. The delay values are calculated by using the nanowire models given in [2], and $32nm$ CMOS models provided by the predictive technology model [20].

The experiments indicate that the critical path delay typically increases slightly with an increasing defect rate. This result is expected as it may not be possible to place clusters optimally due to invalid mappings caused by defects. However, it is useful to note that the defect rate does not drastically affect critical path delay.

The chip size, on the other hand, does not increase with in-

Table 3: Experimental results for 5-LUT networks

Circuit	#LUTs	#clusters	chip size (N_H, N_V)	$P_{so} = 1\%$ delay (ns)	$P_{so} = 1\%$ chip size (N_H, N_V)	$P_{so} = 1\%$ compile time (s)	$P_{so} = 5\%$ delay (ns)	$P_{so} = 5\%$ chip size (N_H, N_V)	$P_{so} = 5\%$ compile time (s)	$P_{so} = 10\%$ delay (ns)	$P_{so} = 10\%$ chip size (N_H, N_V)	$P_{so} = 10\%$ compile time (s)
c3540	306	77	9, 9	2.482	9, 9	47.7	2.493	9, 9	59.1	n/a	n/a	n/a
c5315	397	100	10, 10	1.745	10, 10	95.8	1.742	10, 11	144.6	n/a	n/a	n/a
c6288	482	121	11, 11	4.860	11, 11	82.9	4.860	11, 11	82.3	4.860	11, 11	81.1
c7552	413	104	11, 10	2.372	11, 10	123.0	2.384	13, 12	250.3	n/a	n/a	n/a
dalu	323	81	9, 9	2.790	9, 9	48.1	2.790	9, 9	48.1	2.790	9, 9	47.4
des	962	241	16, 16	2.010	16, 16	228.7	2.010	16, 16	227.7	2.010	16, 16	223.8
i8	302	76	9, 9	1.400	9, 9	61.2	1.400	9, 9	60.7	1.400	9, 9	60.0
i10	610	153	13, 12	3.205	13, 12	163.7	3.191	13, 13	192,9	n/a	n/a	n/a
too_large	131	33	6, 6	1.380	6, 6	31.2	1.380	6, 6	31.1	1.380	6, 6	30.9

creasing defect levels for 3-LUT NPEs. However, for 4- and 5-LUT NPEs, some of the circuits require more NPEs than a defect-free realization would require. This implies that redundancy in terms of extra hybrid clusters was enough to successfully map to defective chips. Nevertheless, it is not possible to realize some circuits under 5% and 10% defect rates at all. This implies that NPEs are fully utilized and no extra nanowire is left for redundancy suggesting that redundant nanowires within NPEs are also important for defect tolerance. Furthermore, as nanowire crossbars become larger, they will contain more defective junctions. Therefore, full utilization of larger nanowire crossbars will require more redundant nanowires.

The average compilation time is not affected by the defect level if it is possible to find mappings to the same chip size for different defect levels. On the other hand, if extra hybrid clusters are used for redundancy, then compilation time typically increases. This is expected, because the placer will have more candidates at each iteration of simulated annealing and it will take longer to try those options. All compilation times are on a 2.2GHz PC with 1GB DRAM running under Linux kernel version 2.4.21-20.

6. CONCLUSION

In this paper, we proposed and evaluated a fault-tolerant architecture based on nanowire crossbar-style processing elements. The aim of this architecture was to target errors caused by not only permanent defects, which may or may not be detected during testing, but also transient faults that occur at runtime. This goal has been achieved by (i) using a hierarchical, multi-level redundancy based on replicating nanowires within an NPE, as well as replicating the NPEs themselves, and (ii) employing CMOS voting logic to correct errors produced by the NPEs.

A compiler companion to the architecture was also described. Two other key contributions relate to this. First, we proposed fast heuristics to map logic to defective, redundant components. Second, we proposed an annealing-based placer that uses defect information within its cost function. By restricting defect-awareness to the back-end, we increase the retargetability of the compiler across different defect models.

REFERENCES

[1] "International Technology Roadmap for Semiconductors." http://public.itrs.net.

[2] A. DeHon, "Nanowire-based programmable architectures," *ACM J. Emerging Technologies in Computing Systems*, pp. 109–162, July 2005.

[3] M. Mishra and S. C. Goldstein, "Scalable defect tolerance for molecular electronics," in *Proc. First Wkshp. Non-Silicon Computation*, Feb. 2002.

[4] A. DeHon and H. Naeimi, "Seven strategies for tolerating highly defective fabrication," *IEEE Design and Test of Computers*, pp. 306–315, July 2005.

[5] J. G. Brown and R. D. Blanton, "CAEN-BIST: Testing the NanoFabric," in *Proc. Int. Test Conf.*, pp. 462–471, June 2004.

[6] M. Jacome, C. He, G. de Veciana, and S. Bijansky, "Defect tolerant probabilistic design paradigm for nanotechnologies," in *Proc. Design Automation Conf.*, pp. 596–601, June 2004.

[7] G. Snider, P. Kuekes, T. Hogg, and R. S. Williams, "Nanoelectronic architectures," *Appl. Phys. A*, vol. 80, pp. 1183–1195, 2005.

[8] P. Shivakumar, M. Kistler, S. Keckler, D. Burger, and L. Alvisi, "Modeling the effect of technology trends on the soft error rate of combinational logic," in *Proc. Int. Conf. Dependable Systems and Networks*, pp. 389–398, June 2002.

[9] T. Rejimon and S. Bhanja, "Probabilistic error model for unreliable nano-logic gates," in *Proc. IEEE Conf. Nanotech.*, pp. 717–722, June 2006.

[10] S. S. Mukherjee, J. Emer, and S. K. Reinhardt, "The soft error problem: An architectural perspective," in *Proc. Int. Symp. High-Performance Computer Architecture*, pp. 243–247, Feb. 2005.

[11] Y. Huang, X. Duan, Y. Cui, L. J. Lauhon, K.-H. Kim, and C. M. Lieber, "Logic gates and computation from assembled nanowire building blocks," *Science*, vol. 294, pp. 1313–1317, Nov. 2001.

[12] R. M. P. Rad and M. Tehranipoor, "A new hybrid FPGA with nanoscale clusters and CMOS routing," in *Proc. Design Automation Conf.*, pp. 727–730, July 2006.

[13] M. Ziegler and M. Stan, " The CMOS/nano interface from a circuits perspective," in *Proc. Int. Symp. Circuits and Systems*, pp. 904–907, May 2003.

[14] D. B. Strukov , and K. K. Likharev, "CMOL FPGA: a reconfigurable architecture for hybrid digital circuits with two-terminal," *Nanotechnology*, vol. 16, pp. 888-900, June 2005

[15] M. B. Tahoori, "A mapping algorithm for defect-tolerance of reconfigurable nano-architectures," in *Proc. Int. Symp. Computer-Aided Design*, pp. 668–672, Nov. 2005.

[16] E. M. Sentovich *et al.*, "SIS: A system for sequential circuit synthesis," Tech. Rep., UC Berkeley, 1992. citeseer.ist.psu.edu/sentovich92sis.html.

[17] V. Betz and J. Rose, "VPR: A new mapping, placement and routing tool for FPGA research," in *Proc. Wkshp. Field Programmable Logic and Applications*, pp. 213–222, 1997.

[18] W. Rao, A. O"railoglu, and R. Karri, "Topology aware mapping of logic functions onto nanowire-based crossbar architectures," in *Proc. Design Automation Conf.*, pp. 723–726, July 2006.

[19] S. Mitra, N. R. Saxena, and E. J. McCluskey, "Common-mode failures in redundant VLSI systems: A survey," *IEEE Trans. Reliability*, vol. 49, pp. 285–295, Sept. 2000.

[20] "Predictive technology model." http://www.eas.asu.edu/~ptm/.

978-1-4244-3039-0/08 $25.00 © 2008 IEEE

21st International Conference on VLSI Design

Analysis and Robust Design of Diode-Resistor Based Nanoscale Crossbar PLA Circuits

Rajat Subhra Chakraborty, Somnath Paul and Swarup Bhunia

Dept. of Electrical Engineering and Computer Science, Case Western Reserve University,
Cleveland, OH-44106, USA
e-mail: {rsc22,sxp190,skb21}@case.edu

Abstract—Logic circuit design with future nanoscale devices using dense and regular fabrics such as crossbar is promising in terms of integration density, performance and power dissipation. Among the emerging alternatives to CMOS, molecular electronics based "diode-resistor logic" has generated considerable interest in recent times. However, some major challenges associated with circuit design using molecular switches are: 1) high defect rate; 2) lack of voltage gain of these switches that prevent logic cascading; and 3) large output voltage level degradation that affect robustness of operation. In this paper, we analyze the issue of input-dependent logic level degradation in diode-resistor style molecular crossbar and develop a simple analytical model for fast and accurate estimation of logic level degradation in a circuit. We also propose a voltage level-aware circuit design technique that limits the worst-case output level degradation. We verify the model by SPICE simulation which shows an average absolute error of less than 2%. Moreover, the proposed design technique improves the logic degradation level from 27% to 7% on an average compared to conventional design.

Keywords*: Diode-resistor logic, logic-level degradation, nano-crossbar circuit, robust circuit design.*

I. INTRODUCTION

The advancement of *Nanoelectronics* in recent years has opened up the possibility of designing circuits with integration densities which are not realizable by present day photo-lithography based traditional IC fabrication techniques. As has been mentioned in [1], integration densities in the order of 10^{10} gate equivalents per cm^2 are also possible. However, having an integration density of this order would mean a significant change in design philosophy [2]. In place of the traditional lithography based "top-down" design methodology, "bottom-up" self-assembly based design approach appear plausible for future nano devices .

Also, there is widespread agreement about the fact that these "self-directed, self-assembly" techniques would be unable to replicate the arbitrarily complex structures achievable by the conventional top-down design methodology [2, 3]. Instead, the bottom-up molecular circuits would be restricted to either randomly assembled structures or regular periodic structures. Although building functional circuit blocks from randomly assembled molecular structures have been reported [4], a more attractive

approach is to go for new circuit architectures that rely on a regular and periodic fabric, which can then be configured to map any target application. The "nanoscale crossbar" fabric [5-10, 13] is particularly suitable in this regard, where we have a two-dimensional array of devices arranged in rows and columns. The devices that form the nanoscale crossbar structure can be modeled as resistors, diodes or Field Effect Transistors (FETs) depending on their electrical characteristics.

Nanoscale crossbar fabric realized with two-terminal diode-like devices (such as molecular monolayer of chemicals such as rotaxanes sandwiched between metal nanowires [11]) has emerged as a promising platform for circuit implementation due to its simplicity and regularity. The nanoscale circuits implemented in this platform fall under the "Diode-Resistor Logic" family [2, 11]. These fabrics are favorable for realization of large and complex functionalities within a small area either in the form of Programmable Logic Array (PLA) or as Look up Table (LUT) [2]. However, this circuit style suffers from several major shortcomings. The devices suffer from high intrinsic defect rate. Besides, the diode-like molecular switches are unable to provide any gain, which necessitates level-restoring circuitry (typically implemented with another technology like CMOS) at the interfaces to make cascading of stages possible [2]. Another deficiency of this circuit style is that the output logic level is not well-defined (unlike static CMOS logic), and can vary appreciably depending on the input pattern. This, in turn, implies that the noise margin of this logic family is input-dependent, which makes the design of CMOS-nano interface circuitry highly challenging. In this paper, we focus on the level degradation issue in nanoscale crossbar circuits. In particular:

1. We analyze the impact of input-dependent voltage level degradation in diode-resistor crossbar circuits.
2. We develop a simple and accurate analytical model to have fast pre-simulation estimate output voltage levels of diode-resistor logic circuits for a given input combination.
3. We propose a design methodology to address the problem of level degradation while minimizing design overhead. We introduce a new design optimization parameter to limit the worst-case output voltage level degradation. Although previous investigations have reported the effects of crossbar size and wire resistivity over the logic levels [2], to the best of our knowledge, this is the first work

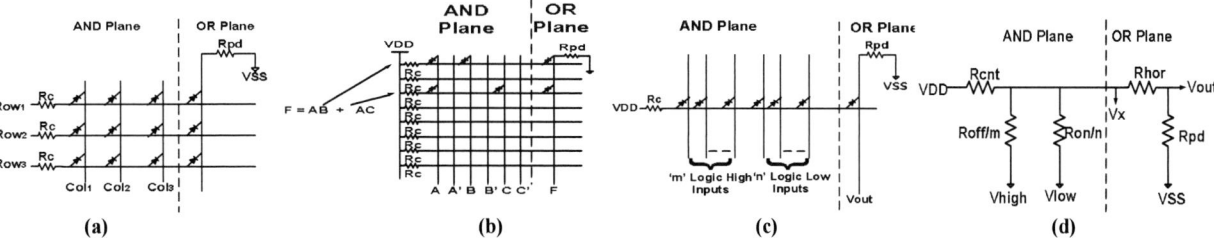

Fig. 1: Analysis of the Diode-resistor Logic Crossbar Circuit: (a) General Structure (b) PLA Implementation of the Logic Function *F(A,B,C)=AB+AC*
(c) Single Horizontal Line with Multiple Logic Inputs, and (d) Simplified Equivalent Circuit

441

978-1-4244-3039-0/08 $25.00 © 2008 IEEE

targeting analysis and modeling of input-dependent level degradation in nano-crossbar based diode-resistor logic. Furthermore, we develop a design method for robust (with respect to voltage level) diode-resistor circuits using nanoscale devices.

The rest of the paper is organized as follows. Section II presents the analysis of diode-resistor logic with respect to output level degradation. Section III focuses on modeling of diode resistance, while Section IV considers several cases of input-dependent noise margin in diode-resistor crossbar. Section V describes a design technique to alleviate logic level degradation. Section VI provides simulation results and Section VII concludes the paper.

II. ANALYSIS OF DIODE-RESISTOR LOGIC

The electrical structure of a Boolean function implemented in the diode-resistor logic style has been shown in Fig 1(a). The "diode-like" devices at the junctions are configurable, and the devices at the selected junctions can be independently activated and deactivated. The "row" and "column" terminals are in different planes. A device is turned "off" if a positive voltage is applied at the column terminal corresponding to it, because then the diode does not experience any forward bias. Each row in the crossbar structure generates a unique minterm of a logic function. The contact resistance is usually modeled as a lumped resistance attached to each row (R_c). An explicit pull-down resistance (R_{pd}) is added to each column corresponding to an output term. Any Boolean logic function can be realized by this circuit architecture primarily by two different approaches [2]: 1) the Look-up Table (LUT) approach and 2) the Programmable Logic Array (PLA) approach. In the LUT approach, the address decoder determines the output corresponding to the minterm "address" that has been decoded, while in the PLA realization, only primes of the minimized function are mapped. The PLA implementation is usually more economical in terms of area, and the savings in area increases exponentially as the number of inputs increases. Fig. 1(b) shows the PLA implementation of the simple logic function $F(A,B,C) = AB + AC$. Details of their implementation methodology can be found in [2]. The basic working principle of this logic-style is to control the amount of current flowing through the "pull-down" output resistor by the voltage applied at the input terminals. We concentrate on the analysis of the PLA style of implementation in this paper. The merit of the proposed approach lies in the fact that this saves the designer the effort of analyzing different scenarios (using circuit simulations) with respect to output voltage level corresponding to the different input conditions.

The following assumptions have been made in this analysis:

a) The nano-devices at the junctions can be modeled as semi-conducting diodes for purposes of circuit analysis. This assumption is justified by published results like [6], and analyses like [2].

b) Within a small range of variation in the terminal voltage, the resistance of these "diode-like" devices show a piece-wise linear or polynomial variation, which is a consequence of the exponential I-V characteristics of diodes. These lumped equivalent resistance values need to be estimated once for each device family either by parameter extraction from measured data or by an accurate circuit simulation. The proposed approach of output-level analysis will, however, remain valid for any device showing "diode-like"

characteristics.

c) The voltage of each horizontal line can be analyzed separately, and their effects can be "superimposed" to calculate the overall voltage of the logic function output. This allows the application of the voltage and current laws (KVL and KCL) and considerably simplifies the analysis.

Fig. 1(c) shows a situation where "m" logic-1 inputs have been connected to a horizontal line, and "n" logic zero inputs have been connected to the horizontal line. The electrical equivalent circuit of this structure (following assumptions (a) and (b)) has been shown in Fig. 1(d). Here, R_{on} represents the resistance of a "On" diode which has a logic-0 as an input at their corresponding column terminal, and "n" such resistances in parallel result in an equivalent resistance of $\dfrac{R_{on}}{n}$. Similarly, R_{off} represents the resistance of an "Off" diode which has a logic-1 as an input at its corresponding column terminal, and "m" such devices in parallel result in an equivalent resistance of $\dfrac{R_{off}}{m}$. R_{hor} represents the resistance of the diode connected between the horizontal line and the vertical output line.

Applying KCL at node V_x:

$$\frac{V_{DD}-V_x}{R_{cnt}} = \frac{V_x - V_{low}}{\dfrac{R_{on}}{n}} + \frac{V_x - V_{high}}{\dfrac{R_{off}}{m}} + \frac{V_x - V_{SS}}{R_{hor}+R_{pd}} \qquad (1)$$

where in our case we have chosen $V_{DD}=3.0$ volt, $V_{SS}=-1.5$ volt, $V_{high}=1.5$ volt and $V_{low}=0.0$ volt.

Solving for V_x yields:

$$V_x = \frac{(R_{hor}+R_{pd})(R_{on}R_{off}V_{DD}+mR_{cnt}R_{on}V_{high}+nR_{cnt}R_{off}V_{low})+R_{cnt}R_{on}R_{off}V_{ss}}{(R_{hor}+R_{pd})(R_{on}R_{off}+nR_{cnt}R_{off}+mR_{cnt}R_{on})+R_{on}R_{off}R_{cnt}} \qquad (2)$$

Two special cases arise:

a) m=0, i.e., all the input signals connected with the horizontal line are at logic-0

Then, keeping in mind that $V_{low}=0$,

$$V_x = \frac{(R_{hor}+R_{pd})R_{on}V_{DD}+R_{on}R_{cnt}V_{SS}}{(R_{hor}+R_{pd})(R_{on}+nR_{cnt})+R_{on}R_{cnt}} \qquad (3)$$

a) n=0, i.e., all the input signals connected with the horizontal line are at logic-1

Then, as before,

$$V_x = \frac{(R_{hor}+R_{pd})(R_{off}V_{DD}+mR_{cnt}V_{high})+R_{off}R_{cnt}V_{SS}}{(R_{hor}+R_{pd})(R_{off}+mR_{cnt})+R_{off}R_{cnt}} \qquad (4)$$

Once all the voltages on the horizontal lines have been calculated, we can proceed to calculate the voltage on the output line of the logic function (V_{out}), by "superposition" of all the horizontal line voltages. Fig. 2 shows the schematic where we have "n" horizontal lines, each with voltage V_i, i = 1, 2,.........n.

Applying KCL at node V_{out}, we obtain:

$$\sum_{i=1}^{n} \frac{V_i - V_{out}}{Rhor_i} = \frac{V_{out}-V_{SS}}{R_{pd}} \qquad (5)$$

On solving equation (5) for V_{out}, one obtains:

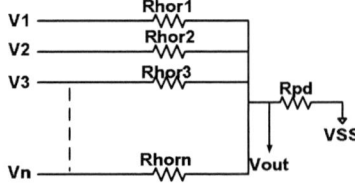

Fig. 2: Superposition of the horizontal line (minterm) voltages to calculate the output voltage of a logic function.

978-1-4244-3039-0/08 $25.00 © 2008 IEEE

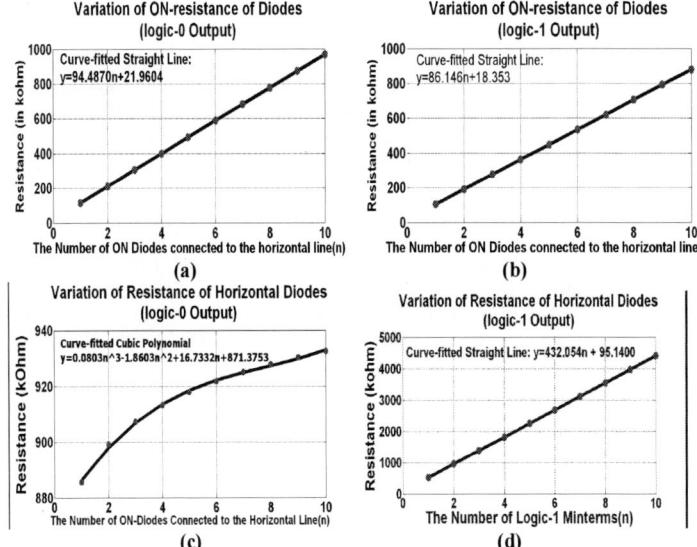

(a)   **(b)**

(c)   **(d)**

Fig. 3: (a) Variation of R_{on} with n for logic-0 output (b) Variation of R_{on} with n for logic-1 output (c) Variation of R_{hor} with n for logic-0 output (d) Variation of R_{hor} with the number of logic-1 minterms for logic-1 output.

$$V_x = \frac{V_{SS} + \sum_{i=1}^{n} \frac{V_i}{Rhor_i}}{1 + R_{pd} \sum_{i=1}^{n} \frac{1}{Rhor_i}} \qquad (6)$$

III. MODELING THE DIODE RESISTANCE

As mentioned before, all the diode resistances are functions of their terminal voltages. So, the values of R_{on}, R_{off}, and R_{hor} used in equations (1)-(4) are not constants, and depend on the voltages at the cathode and anode terminals of the diode. Four distinct situations arise that determine the diode resistance values:

CASE 1: The function is at logic-0, and all the input signals are applied to the horizontal line are at logic-0.

CASE 2: The function is at logic-0, but not all input signals attached to the horizontal line are at logic-0.

CASE 3: The function is at logic-1 and the prime is at logic-1, i.e., all the input signals attached to this horizontal line are at logic-1.

CASE 4: The function is at logic-1, but the prime is at logic-0, i.e., at least one of the input signals attached to the horizontal line is at logic-0.

Extensive SPICE simulations were carried out for these four different situations with a model of the diode derived from the data of [6], and following the approach of [2]. The following trends were observed:

a) In case (1), the resistance of the on-diodes (R_{on}) was found to be a linear function of the number of on-diodes connected to the horizontal line. The resistance of the horizontal diodes (R_{hor}) was a cubic polynomial function of the number of on-diodes connected to

the horizontal line.

b) In case (2), the resistance of the on-diodes (R_{on}) and the resistance of the horizontal diodes (R_{hor}) followed more or less the same relationship as for case (1). The resistance of the off-diodes was more or less constant.

c) In case (3), the resistance of the horizontal diodes (R_{hor}) is a linear function of the number of logic-1 minterms connected to the output line. The resistance of the off-diodes (R_{off}) was nearly constant.

d) In case (4), the resistance of the on-diodes (R_{on}) was a linear function of the number of on-diodes connected to the horizontal line. The resistance of the horizontal diodes (R_{hor}) and the resistance of the off-diodes (R_{off}) were nearly constant.

Fig. 3 shows the plots of the diode on-resistance and the resistance of the horizontal diodes for the four cases discussed, along with the curves that were fitted to the data (using MATLAB). It should be noted that due to process variation, the resistances of all the nano-devices in a given fabric might not follow the same trend. However, the trends observed might be expected to be the same, and the discussion in the following sections would be still applicable.

IV. INPUT DEPENDENT NOISE MARGIN

The equations derived in Section II clearly show the dependence of the logic voltage levels on the number of input signals, and their states. For example, from equation (6), it is clear that the output voltage level corresponding to logic-1 increases as the number of minterms at logic-1 increases. Fig. 4(a) shows the variation of the output logic level of a 10-input OR gate, as the number of inputs at logic-1 vary from one to ten. The worst case logic-1 value (corresponding to the situation when there is only one input at logic-1) is 1.29V, while the best case voltage level is 1.54V. Note that the voltage level tends to saturate as the number of inputs at logic-1 increases. This is because the value of R_{hor} increases with the number of minterms at logic-1. In Fig. 4(b), we show the variation in the output voltage of a 10-input AND gate, as the number of inputs at logic-0 of the AND gate varies from one to ten. Here, the worst case logic-0 (corresponding to only one input at logic-0) is 0.081V and the best logic-low voltage level is 0.001V (corresponding to all ten inputs at logic-0)). These two plots show the general trends in variation of logic-0 and logic-1 values. Thus, we can say that the noise margin of the diode-resistor logic circuits depends on the number of inputs and the logic values at the inputs. In general there are four issues related to noise-margin and delay variation that arise in connection with the diode-resistor logic circuits. Two of them are associated with the AND plane and two others are associated with the OR plane. The issues associated with the AND plane are:

a) Variation in logic-0 voltage and delay with the number of inputs at logic-0, and

b) Variation in delay due to the difference in the number of minterms sharing the same input signal.

The issues associated with the OR plane are:

c) Effect of multiple outputs sharing the same minterm(s) over each other, and

d) Variation in logic-1 voltage and delay with the number of inputs at logic-1.

Next we consider these issues in detail.

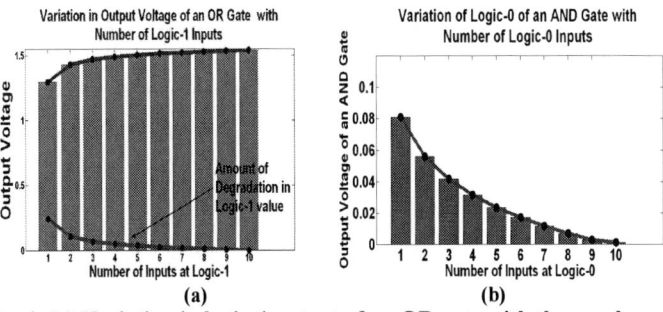

(a)   **(b)**

Fig. 4: (a) Variation in logic-1 output of an OR gate with the number of inputs at logic-1; (b) Variation in logoc-0 output of an AND gate with the number of inputs at logic-0.

443

978-1-4244-3039-0/08 $25.00 © 2008 IEEE

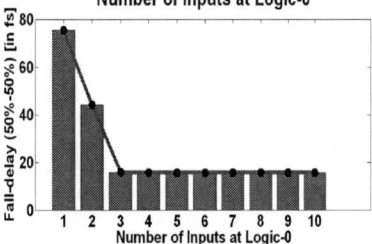

Fig. 5: Variation of Fall Delay (50%-50%) of an AND Gate with the Number of Inputs at Logic-0.

a) Variation in logic-0 voltage and delay with the number of inputs at logic-0

From Fig. 4(b) and similar other simulations carried out, we found that the variation in logic-0 level for different input combinations was always small and within tolerable limits. For example, in case of the 10-input AND gate considered, this was around 70mV, which is about 4.67% of the "rail-to-rail" swing (=1.5V). So, this effect can be considered to be a minor design concern.

The variation in delay might also be a concern in this case, because of the difference in the *total* pull-down strength of the diodes. Fig. 5 shows the variation in the fall delay with the number of inputs at logic-0. With the increase in this number, initially the delay decreases significantly. However, then it saturates because the resistance of the pull-down diodes increases with the number of inputs at logic-0, as can be seen from the plot in Fig. 3(a). Although the variation is significant for small values of *n*, the variation is less than 0.1ps in all cases, and hence is not is a major cause of concern. The reason for such small delay is that the capacitance associated with these nano-diodes is very small ($\sim 10^{-18}$ F [2]). However, note that for a design methodology targeting both logic-0 improvement and delay improvement are the same, because both of them improve with the number of logic-0 inputs. So, designing for the improvement of one automatically ensures the improvement of the other.

(b) Variation in delay due to the difference in the number of horizontal lines sharing the same input

The fall delay increases with the number of horizontal lines connected to the same input signal. This is because the same pull-down diode now has to sink more current. Fig 6 is a plot of the output fall-delay as the number of horizontal lines connected to the input logic-0 line increases. The trend matches the expected one. The designer may not have too much control over this issue because the logic function might be such that multiple horizontal lines might share the same input signal. Moreover, the current-carrying capability of the diodes is limited by the technology, and for the nano-scale diodes we are considering, this can be extremely small.

c) Effect of multiple outputs sharing the same horizontal line(s)

There might be two different cases: i) the logic functions are at different logic levels and ii) the logic functions are at the same logic level. In the first case, no discernable effect of different logic functions at different logic levels sharing the same minterm was observed from the simulations. A close look at the structure of the diode-resistor logic circuits makes this observation intuitively obvious. If a horizontal line (corresponding to a prime) is connected to two logic function outputs at different logic levels, it is always connected through two *different* diodes corresponding to the two *different* functions – one of which is forward-biased and the other is reverse-biased. The reverse-biased diode has a very high resistance and essentially isolates the horizontal line from the corresponding output. The forward-biased diode on the other hand has a well-

defined resistance according to the relationships depicted in Figs. 3(c) and 3(d). Thus, the voltages at the two output lines become effectively de-coupled. The analytical treatment of Section II remains valid as before and it lets us calculate the voltage at the individual horizontal lines (primes), which in turn lets us estimate the voltage at the output(s) according to equation (6).

In the second case, two separate conditions might be considered. 1) We have a horizontal line at logic-1 connected to multiple outputs at logic-1. From simulations, the variation in rise-delay with the number of functions sharing the same logic-1 horizontal line was found to be practically negligible. 2) We have a horizontal line at logic-0 connected to multiple outputs at logic-0. Here also the variation in delay with the change in the number of functions sharing the same logic-0 minterm was found to be negligible. Thus, they present minor design challenge.

d) Variation in logic-1 voltage and delay with the number of horizontal lines at logic-1

This effect was found to be significant enough to deserve more attention. From Fig. 4(a), it can be observed that the logic-1 output of a 10-input OR gate varies by 0.25V, which is 16.67% of the "rail-to-rail" swing (=1.5V). This is a significant variation, and makes the design of the interface circuitry in noisy environment very challenging. Fig. 7 shows the variation in rise delay in a 10-input OR gate with the number of inputs at logic-1 (all switching to logic-1 simultaneously). The observed improvement in delay is expected from the analysis of Section II. The delay and level improvements happen simultaneously with the increase in the number of minterms at logic-1. Hence, we concentrate on this aspect and in the next section develop a design technique to overcome this challenge, while simultaneously minimizing the hardware overhead.

V. ROBUST NANO-PLA DESIGN

As discussed in the previous section, the logic-1 level of a nano-PLA output can vary by as much as 16.67% depending on the state of the input. This degradation can be more severe with diodes of less drive strength. We also observe that with two logic-1 primes, the logic-1 output of the function increases to 1.44V, which reduces the degradation to just 4% (for a nominal logic-1 value of 1.5V). This means that adding a single prime at logic-1 can improve the logic level by over 12%. This figure varies with the strength of the device used, e.g. if the drive strength of the nano-device is 100 times smaller, the improvement might be more drastic. Fig. 8 shows the variation (and degradation) in the logic-level of a 10-input OR gate for such a device as the number of inputs at logic-1 varies from one to ten. Note that the improvement due to addition of a single prime at logic-1 in this case is 20.34% of rail-to-rail swing. This is the main motivating factor for our design methodology aimed at improving the logic-1 level, and thus in turn to avoid the worst-case level degradation in nano-PLA. *Therefore, a robust (with respect to output voltage level) nano-PLA design requires considering an additional design constraint N_{min} during design space exploration, which indicates the minimum number of primes evaluating to logic-1 for any input combination, during design space exploration.*

We assume that less than 10% degradation at output voltage

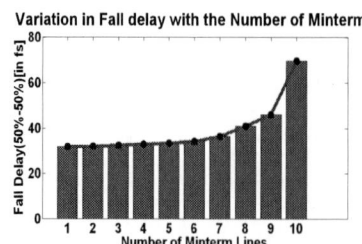

Fig. 6: Variation of Fall Delay (50%-50%) with the Number of Minterms Connected to the Same Input.

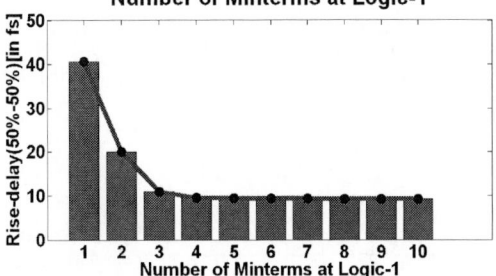

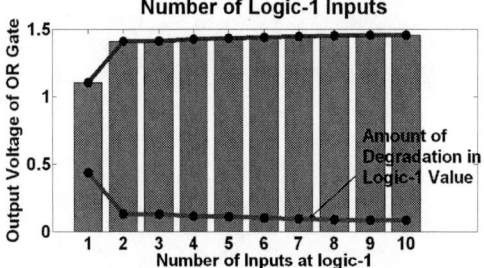

Fig. 7: **Variation of Rise Delay (50%-50%) of an OR Gate with the Number of Inputs at Logic-1.**

provides acceptable noise margin. This translates to the fact that we need at least two primes to evaluate to logic-1 for any input combination. Hence, the proposed design technique involves duplicating primes to ensure that the worst case logic-1 voltage level remains within acceptable limits. For example, the logic function $F(A,B,C) = A + BC$ has the worst-case logic-1 voltage value either when $A=1, B=0, C=0$ or when $A=0, B=1, C=1$, i.e. for the cases where only one prime is at logic-1. Using the Boolean identity $A = A + A$, the prime can be duplicated and the expression for $F(A,B,C)$ can be modified as:

$$F(A,B,C) = A + BC = A + A + BC + BC \qquad (7)$$

This modification ensures that at least two primes would be at logic-1 whenever $F(A,B,C)$ would be at logic-1, and thus the logic-1 value would not be degraded the worst case, where only one prime evaluates to logic-1. In general, depending on the characteristics of the diode and the technology used, there is a threshold of the number of primes simultaneously at logic-1 which would ensure a proper logic-1 value at the output in the worst case. Hence, this design methodology might be extended to any technology.

(a) Design Overhead Considerations:

The size of a function implemented in diode-resistor logic crossbar is given by [2]:

$$A = c(2N + f)P_{wire}^2 \qquad (8)$$

where N is the number of literals in all the functions implemented by the crossbar, f is the number of functions, P_{wire} is the pitch of the nanowires, and c is the number of unique two-level minimized product terms in all the functions. If the design methodology described earlier for logic-1 level enhancement is followed, the number of two-level minimized terms increases by a factor of two, thus making the area double of the original implementation.

However, there are special situations in which the overhead can

Fig. 8: Variation in logic-1 output of an OR gate (with low device current) with the number of inputs at logic-1.

be improved. In case of functions which can share primes, the duplication of primes can be *partially avoided*. For example, consider the logic function $F(A,B,C) = AB + BC + CA$ which can be identified to be the *CARRY* of a full-adder. Consider a situation where this function is to be implemented along with three other logic functions: $F_1 = A'BC$, $F_2 = AB'C$ and $F_3 = ABC'$. We can identify that the non-minimized representation for the full-adder carry function F is $F(A,B,C) = A'BC + AB'C + ABC' + ABC$. If we express F as $F(A,B,C) = F_1 + F_2 + F_3 + ABC + ABC$, and duplicate each minterm of F_1 through F_3 (i.e. represent F_1 as $F_1 = A'BC + A'BC$, etc.), we can implement all the four functions with a combined minterm count of 8 (6 in total for F_1 through F_3 and 2 for the duplicated ABC) and also solving the logic degradation problem. If each function is considered separately, 12 minterms will be required (6 for F and 6 in total for F_1 through F_3) to ensure proper logic-1 level for all input combinations. Note that in this modified implementation F has been realized in a non-minimized form. Hence, a choice must be made to realize the proper forms. A similar approach can be adopted for the LUT based implementation of logic functions.

(b) Testing Challenges:

The increase in the number of devices (caused by duplication of horizontal lines) in a system with high device defect density might cause additional complications in testing these circuits. However, with the development of better nano-fabric testing techniques and innovative "defect-aware" circuit configuration techniques [14], these shortcomings might be expected to be overcome.

Table I: Comparison of the Simulated and Modeled Output Logic Values for Diode-Resistor nano-PLA

Circuit	Input Condition	V_{out} (Simulation)	V_{out} (Modeled)	% Absolute Error (with respect to rail-to-rail voltage=1.5V)
AND(2 i/p)	A=1,B=1	1.2966	1.2640	2.17
AND(2 i/p)	A=0,B=1	0.0808	0.1050	1.61
AND(2 i/p)	A=0,B=0	0.0561	0.0594	0.22
OR (3 i/p)	A=0, B=0, C=0	0.1237	0.1740	3.35
OR (3 i/p)	A=0, B=1, C=1	1.4323	1.3928	2.63
OR (3 i/p)	A=1, B=1, C=1	1.4719	1.4500	1.46
Full-adder Sum	A=0, B=0, C=0	0.1279	0.1560	1.87
Full-adder Sum	A=0, B=0, C=1	1.2956	1.2632	2.16
Full-adder Sum	A=1, B=1, C=1	1.2956	1.2632	2.16
Full-adder Carry	A=0, B=0, C=0	0.0990	0.1353	2.42
Full-adder Carry	A=0, B=1, C=1	1.2966	1.2640	2.17
Full-adder Carry	A=1, B=1, C=1	1.4559	1.4491	0.45

978-1-4244-3039-0/08 $25.00 © 2008 IEEE

Table II: Area Savings by Modified Implementation of Logic Circuits ($N_{min} = 2$)

* Requires an inverter at the interface for implementation

Circuit	Orig. Minterm Count	Orig. Area (μm^2) [2]	Modified Area to Avoid Logic Degradation (μm^2)	Modified Minterm Count with Sharing	Modified Area with Minterm Sharing (μm^2)	% Improvement
Full-adder Carry and F_1-F_3	6	0.294	0.588	8	0.392	33.3
Full-adder*	7	0.343	0.686	10	0.463	32.5
c17*	5	0.245	0.490	8	0.392	20.0

VI. RESULTS

In this section, we provide simulation results for the proposed analysis as well as for the output level-aware design approach. To verify the accuracy of the proposed analysis (as described in Section II), we simulated several test circuits for different input conditions using HSPICE. The SPICE level-1 diode model was derived from the experimental data presented in [6], following the approach in [2]. Table I shows the simulated and estimated values of output logic voltage levels (following the analysis and modeling presented in Section II and III). The average absolute estimation error is 1.89% of rail-to-rail voltage, which shows quite close agreement between the analytical and simulation results. This table validates the analytical approach to be a fast and simple design-time estimate of the output logic level for diode-resistor logic circuits.

To show the effectiveness of the proposed scheme of minterm sharing to improve logic-1 level (as described in Section V), we consider two different circuits. The first of them is a full-adder. The Boolean expression suitable for implementation in diode-resistor logic for the *sum* output of a full-adder is: $S(A,B,C) = A'B'C + A'BC' + AB'C' + ABC$, while the non-minimized expression for the carry of a full-adder is: $Ca(A,B,C) = A'BC + AB'C + ABC' + ABC$ (the minimized expression being $Ca(A,B,C) = AB + BC + CA$). If we implement the non-minimized form of the complement of the carry, i.e., if we implement $Ca' = A'B'C' + A'B'C + A'BC' + AB'C'$, we would have 3 minterms in common with the sum, and thus can implement the total full-adder with only 10 minterms (6 for duplicating the 3 common minterms, plus 4 for the two unshared minterms ABC and $A'B'C'$). Thus, we save 4 minterms compared to the straightforward minimized implementation, which would have required 14 minterms (6 for the carry, 8 for the sum). We can retrieve the actual carry by using the inverting logic restoring circuitry at the interface, which are required in the diode-resistor logic style [2].

Another example that we consider is the ISCAS85 benchmark circuit c17 [12]. This circuit has 5 inputs and 2 outputs. The outputs are linked by the relation $O_1 = O_2$ for 5 input combinations and are "Don't Care" for the others. Hence, as before we can implement a single output O_1 and implement O_2 by inverting it. The Boolean expression for O_1 is: $O_1(a,b,c,d,e) = ac' + ad' + bc' + bd'$. In this way, we can implement both functions using just 8 minterms, instead of 10 that would have been required for a straightforward implementation. Table II shows the improvement in PLA area by the proposed minterm sharing approach over minterm duplication, following equation (8) with P_{wire} assumed to be 70nm. As discussed earlier, we consider representation of a function in terms of either its ON-set (true form) or OFF-set (complementary form) in order to maximize minterm sharing. The average improvement in area for the three circuits considered in Table II is 28.6% over simple duplication of minterms to ensure acceptable logic level.

VII. CONCLUSIONS

We have addressed the issue of output voltage level degradation in diode-resistor based nanoscale crossbar fabric. To analyze the level-degradation effect, we have developed a simple analytical model for the accurate estimation of output voltage levels in diode-resistor based nano-PLA for any input combination. We have shown the input-dependent variation in logic level and associated impact on circuit delay for some test circuits and have identified that logic-1 degradation is more severe than logic-0 degradation in nano-PLA. We have also proposed a design approach which alleviates this problem of input-dependant logic-1 level degradation at the cost of moderate hardware overhead. The analysis and the design technique presented here for nano-PLA can be easily extended to LUT-based logic implementation using nanoscale crossbar.

REFERENCES

[1] S.C. Goldstein and M. Budiu: "NanoFabrics: Spatial Computing Using Molecular Electronics," *Proc. Intl. Symp. On Computer Architecture*, pp. 178-189, 2001.

[2] M.M. Zieglar and M.R. Stan: "CMOS/Nano Co-Design for Crossbar-Based Molecular Electronic Systems," *IEEE Trans. on Nanotechnology*, pp. 217-229, vol. 2, no. 4, December 2003.

[3] J. Huang, *et al*: "On the Defect Tolerance of Nano-scale Two-Dimensional Crossbars", *DFT*, 2004.

[4] J. Tour, *et al*: "Nanocell logic gates for molecular computing," *IEEE Trans. on Nanotechnology*, pp. 100-109, vol. 1, 2002.

[5] C.P. Collier *et al*: "Electronically Configurable Molecular-Based Logic Gates," *Science*, vol. 285, pp. 391-394, July 1999.

[6] X. Li *et al*: "Nanoscale molecular-switch devices fabricated by imprint lithography," *App. Phys. Letters,* vol. 82, no.10, pp. 1610-1612, March 2003.

[7] T. Rueckes *et al*: "Carbon Nanotube-Based Nonvolatile Random Access Memory for Molecular Computing," *Science*, vol. 289, pp. 94-97, July 2001.

[8] Y. Huang *et al*: "Logic Gates and Computation from Assembled Nanowire Building Blocks," *Science*, 2001.

[9] X. Duan *et al*: "Indium Phosphide nanowires as building blocks for nanoscale electronic and optoelectronic devices," *Nature,* vol. 409, pp 66-69. January 2001.

[10] T. Rueckes *et al*: "Carbon Nanotube-Based Nonvolatile Random Access Memory for Molecular Computing," *Science*, vol. 289, pp. 94-97, July 2001.

[11] G. Snider *et al*: "Nanoelctronic Architectures," *Applied Physics A,* vol. 80, issue 6, pp. 1183-1195, March 2005.

[12] ISCAS85 Circuits: http://www.fm.vslib.cz/~kes/asic/iscas/

[13] A. DeHon: "Array-Based Architecture for FET-Based, Nanoscale Electronics", *IEEE Transactions on Nanotechnology*, vol. 2, no.1, pp. 23-32, March 2003.

[14] S. Paul, R.S. Chakraborty and S. Bhunia: "Defect-Aware Configurable Computing in Nanoscale Crossbar for Improved Yield", *Proc. IOLTS*, pp. 29-36, 2007.

21st International Conference on VLSI Design

A New Threshold Voltage Model for Omega Gate Cylindrical Nanowire Transistor

Biswajit Ray
Nano Scale Device Research Lab,
CEDT, Indian Institute of Science
Bangalore-560012
rbiswajit@cedt.iisc.ernet.in

Santanu Mahapatra
Nano Scale Device Research Lab,
CEDT, Indian Institute of Science
Bangalore-560012
santanu@cedt.iisc.ernet.in

Abstract

In this work, for the first time, we present a physically based analytical threshold voltage model for omega gate silicon nanowire transistor. This model is developed for long channel cylindrical body structure. The potential distribution at each and every point of the of the wire is derived with a closed form solution of two dimensional Poisson's equation, which is then used to model the threshold voltage. Proposed model can be treated as a generalized model, which is valid for both surround gate and semi-surround gate cylindrical transistors. The accuracy of proposed model is verified for different device geometry against the results obtained from three dimensional numerical device simulators and close agreement is observed.

1. Introduction

To continue the CMOS (Complementary Metal Oxide Semiconductor) scaling trend for the next decade several non-classical MOSFET (Metal Oxide Semiconductor Field Effect Transistor) architecture have been proposed which offer better electrostatic integrity, higher transconductance and lower sub-threshold swing than the traditional MOSFETs [1]. Double-gate Silicon on insulator (SOI) devices, multiple gate FinFETs and surrounding gate nanowire transistors are few examples of such attractive device structures. Although the surrounding gate devices offer the best immunity to Short Channel Effect (SCE) and Drain Induced Barrier Lowering (DIBL) [2], these structures requires complex processing steps that are difficult to make compatible with conventional CMOS fabrication processes. Considering the issues of mass production, semi-surrounded omega shaped gate structure has been proposed [3] to make a compromise between the device performance and manufacturability. Hence it is very important to

formulate simple compact models for these omega gate nanowire FETs for its successful implementation in future VLSI.

Most of the previous works on the cylindrical nanowire structure are based on three dimensional numerical simulations [2, 4]. However, an analytical model for the threshold voltage and potential distribution was presented in [5] for fully surrounding gate nanowire transistor, which is the special case of more general omega gate MOSFET geometry. In this paper we propose an analytical model for calculating the potential $\Psi(r,\theta)$ at any point of the wire cross section of omega gate transistor based on the closed form solution of Poisson's equation. Using the analytical expression of $\Psi(r,\theta)$ we have modeled the inversion charge and threshold voltage for the long channel undoped omega gate device. The models are then verified with the results obtained from professional three dimensional numerical device simulator [6].

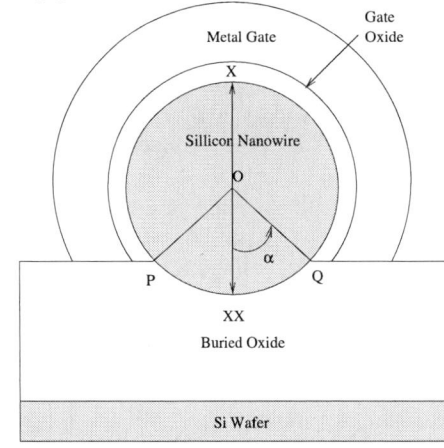

Figure 1. Cross sectional view of the omega gate cylindrical Si nanowire transistor

447

978-1-4244-3039-0/08 $25.00 © 2008 IEEE

2. Derivation of Potential model

The cross section of a omega gate transistor for which we have developed our model is shown in Fig 1 The assumptions used for the derivation of the potential model are as follows:

i) The channel is assumed to be undoped (intrinsic), as in these ultra thin body devices, SCE and DIBL are controlled by the device geometry rather than channel doping. Also the addition of dopants would lead to detrimental statistical fluctuations of threshold voltage.

ii) The model is derived for long channel nMOS devices and the gradual channel approximation [7] has been used in solving the Poisson's equation.

iii) Effects of energy quantization in the inversion layer are not considered in the present work.

With the above assumptions the 2D-Poisson's equation along a vertical cut (X-XX) can be written in terms of polar coordinates as follows [5]:

$$\nabla^2 \psi(r,\theta) = \frac{q}{\varepsilon_{Si}} n_i e^{\frac{q(\psi - V)}{kT}}$$

$$= V_T \delta e^{\frac{\psi}{V_T}} \qquad (1)$$

with boundary condition : $\Psi(R, \theta) = \Psi_S + g(\theta)$ (2)

where $\delta = \frac{q^2 n_i}{kT\varepsilon_{Si}} e^{\frac{-qV}{kT}}$, $V_T = \frac{kT}{q}$, q being the electronic charge, n_i the intrinsic carrier concentration , ε_{Si} the permittivity of silicon, V the electron quasi-Fermi potential, Ψ_S is the surface potential with $g(\theta)$ as a correction term which takes care of potential of the area which is not covered by the gate. It should be noted that the RHS of the Poisson's equation contains only the mobile charge term (assumption (i)) and also valid for the gate voltage higher than flat band voltage (V_{FB}), where the hole concentration could be neglected.

Now for the surrounding gate nanowire transistor, due to symmetry, equation (1) reduces to one-dimensional ordinary differential equation which can be solved analytically [5]. But in the case of omega gate FET, as the channel is not fully surrounded, we have to solve the two dimensional non linear partial differential equation to obtain potential profile $\Psi(r,\theta)$. To solve the equation (1) we split the total potential profile as follows:

$$\psi(r,\theta) = \psi_1(r) + \psi_2(r,\theta) \qquad (3)$$

Now the Eq. (1) can be divided into two following equations

$$\nabla^2 \psi_1(r) = V_T \delta e^{\frac{\psi_1}{V_T}} \qquad (4)$$

with boundary condition: $\psi_1(R) = \psi_S$ (5) and

$$\nabla^2 \psi_2(r,\theta) = V_T \delta e^{\frac{\psi_1}{V_T}} (e^{\frac{\psi_2}{V_T}} - 1) \qquad (6)$$

with boundary condition: $\psi_2(R,\theta) = g(\theta)$ (7)

Equation (4) can be analytically solved as in the case of surrounding gate nanowire FET [5] with solution

$$\psi_1(r) = V_T \log \frac{8B}{(1 - Br^2)^2 \delta} \qquad (8)$$

where the constant B is evaluated from the boundary condition (5) and hence related to Ψ_S as below:

$$B = \frac{1}{R^2}\left(1 + \frac{A}{R^2}\left(1 - \sqrt{1 + \frac{2R^2}{A}}\right)\right), \quad A = \frac{4}{\delta} e^{-\frac{\psi_S}{V_T}} \quad (9)$$

It can be shown that for all practical values of Ψ_S the following relation holds good which makes the logarithmic term in Eq (8) real.

$$0 < BR^2 < 1 \qquad (10)$$

Now, to solve the Eq. (6) we use the series expansion of the exponential term in braces and take first two terms only. This approximation is found to be an excellent one for points near the central region of the channel but deviates slightly at the points near boundary where the gate is not covered. With this approximation Eq (6) reduces to

$$\nabla^2 \psi_2(r,\theta) = \delta e^{\frac{\psi_1}{V_T}} \psi_2 \qquad (11)$$

Or

$$\frac{\partial^2}{\partial r^2}\psi_2 + \frac{1}{r}\frac{\partial}{\partial r}\psi_2 + \frac{1}{r^2}\frac{\partial^2}{\partial \theta^2}\psi_2 = \frac{8B}{(1 - Br^2)^2}\psi_2$$

$$(12)$$

One needs to determine the exact form of boundary condition $g(\theta)$ in order to solve this non linear partial differential equation. From the simulation we perceive that $g(\theta)$ can be approximated by the following empirical expression

$$g(\theta) = C\cos\left(\frac{\pi\theta}{2\alpha}\right) \quad \text{for } -\alpha \le \theta \le \alpha$$

$$= 0 \quad \text{otherwise.} \qquad (13)$$

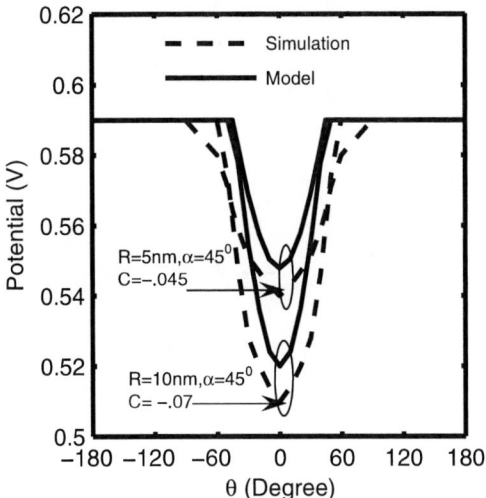

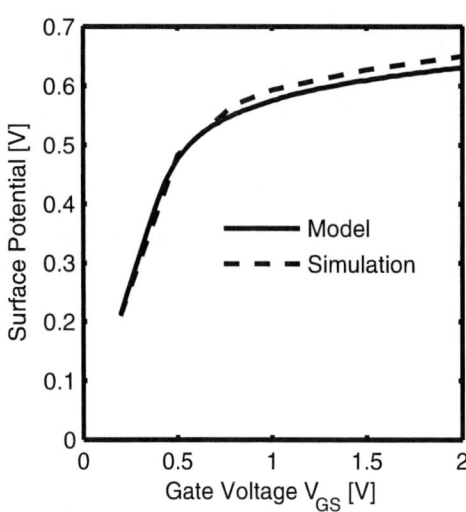

Figure 2. Potential distribution at the boundary of the nanowire((R,)) as a function of , plotted for two different radius values with the corresponding fitting parameter values mentioned in the figure

Figure 4. Surface potential variation as a function of Gate voltage. R=10nm. $\alpha = 45^0$, mid gap metal is used as gate and T_{ox} = 2nm.

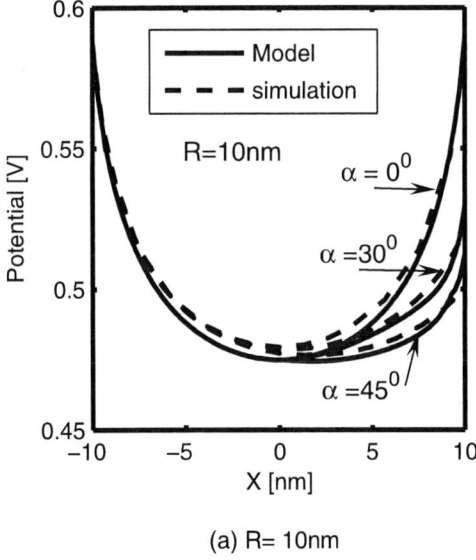

(a) R= 10nm

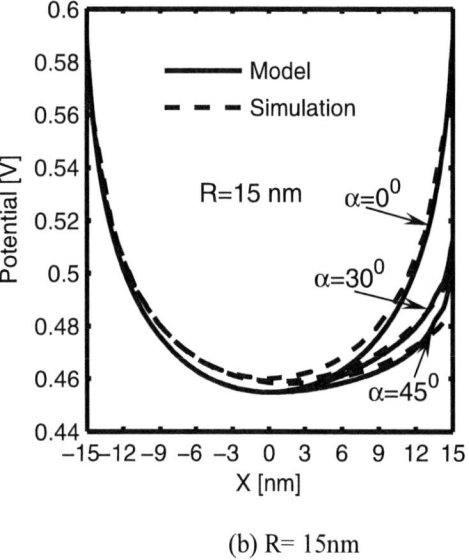

(b) R= 15nm

Figure 3. Potential distribution for different α along the vertical diameter (X-XX) (shown in the fig. 1) for two different radii. For both the cases V_{GS} = 1V, mid gap metal is used as gate and T_{ox} = 2nm.

where C is a fitting parameter . With this approximation the potential profile at the boundary $\Psi(R,\theta)$ is plotted along with simulation result in Fig. 2. We also observe from the simulation studies that the contribution from Ψ_2 to the total potential distribution Ψ is significant only for the uncovered region of the body ($-\alpha \le \theta \le \alpha$; that is the sector OPQ in Fig 1).

Hence we solve Ψ_2 only for that region and as derived in the Appendix the solution for Ψ_2 can be written as (for any practical values of α)

$$\psi_2(r,\theta) = \lambda\, g(\theta)\, r^{2k} e^{-2Br^2} M(k+1, 2k+1, 4Br^2) \quad (14)$$

where M is the Kummer's function [8] of first kind, λ is a constant determined from the normalizing condition discussed in Appendix and k is a constant depending on α as

$$k = \frac{\pi}{4\alpha} \qquad (15)$$

Now, since it has been reported [4] that for the 75% gate coverage ($\alpha = 45^0$) the performance of omega gate nanowire transistor becomes very close to surrounding gate transistor, hence we present a simpler expression for potential distribution for the special case of $\alpha = 45^0$ for which $k=1$ $k=1$.

Now, for $k=1$ the expression of ψ_2 simplifies to

$$\psi_2(r,\theta) = \psi_2(z = 2Br^2, \theta) = \frac{\lambda}{2B} g(\theta)(e^z - \frac{\sinh(z)}{z}) \qquad (16)$$

Hence the overall Potential model for the omega gate nanowire transistor (for $\alpha = 45^0$) reduces to

$$\psi(r,\theta) = V_T \log \frac{8B}{\delta(1 - \frac{1}{2}z)^2} + \frac{\lambda}{2B} g(\theta)(e^z - \frac{\sinh(z)}{z})$$

where $z = 2Br^2$. $\qquad (17)$

Using the general potential model the potential distribution has been plotted in Fig 3 along the vertical diameter (X-XX in Fig 1) for different α values. A closed agreement between model and numerical simulation (obtained from Sentaurus Device Simulator [6]) has been observed except at the oxide box boundary. This is because at the boundary some of the approximations used for explicit solution are not very accurate. But this minor deviation at the boundary will be smoothened out while we integrate this expression over θ for the calculation of total inversion charge and the threshold voltage.

3. Inversion charge and threshold voltage

Once the potential distribution at every point of the cross section of the cylindrical channel is known we can calculate the inversion charge density (Cm^{-2}) by using Gauss law over the surface area of the channel. The total charge inside the cylindrical nanowire channel is given by (using Gauss law)

$$Q_T = \varepsilon_{Si} \oint_S E.dS$$

$$= \varepsilon_{Si} \int_0^{2\pi} E.R(d\theta)L \qquad (18)$$

Hence the surface charge density is

$$Q = \frac{Q_T}{2\pi RL} = \frac{\varepsilon_{Si}}{2\pi} \int_0^{2\pi} E.d\theta \qquad (19)$$

where L is the length of the channel and E is the electric field given by

$$E = \frac{d\psi}{dr}\Big|_{r=R} \qquad (20)$$

Now expression for Q can be analytically evaluated for any value of α, as shown at the bottom of the page (Eq. (21)).But a much simpler expression can be obtained for the special case of 75% gate coverage that is $\alpha = 45^0$ which is written below:

$$Q = \frac{2\varepsilon_{Si} V_T}{R}\left[\frac{Z}{(1-.5Z)} + \frac{\lambda CR^2}{2\pi Z}(Z e^Z - \cosh(Z) + \frac{\sinh(Z)}{Z})\right]$$

where $Z = 2BR^2$ $\qquad (22)$

Now the variation of inversion charge with gate voltage is plotted in Fig 5 using Eq (21). We note that for a given gate voltage inversion charge decreases as we increase α which is expected because for higher α channel is less covered by gate.

The definition of threshold voltage that is used for bulk MOSFET (gate voltage at which $\Psi_s = 2\phi_B$; whereϕ_B is the separation between extrinsic and intrinsic Fermi level) does not apply for undoped body devices. In this work, we define the threshold voltage for the omega gate nanowire transistor as the gate voltage for which the $\Psi_s = \frac{1}{2}E_G$ (E_G is the band gap for Si). With this definition, the threshold voltage of the device can be written as shown in the following Eq:

$$Q = \varepsilon_{Si}\left[\frac{4V_T BR}{1 - BR^2} + \frac{2C\lambda\alpha}{\pi^2}e^{-2BR^2}\left\{\begin{array}{l}(2kR^{2k-1} - 4BR^{2k+1})M(k+1, 2k+1, 4BR^2) \\ + 4BR^{2k+1}\frac{2k+2}{2k+1}M(k+2, 2k+2, 4BR^2)\end{array}\right\}\right] \qquad (21)$$

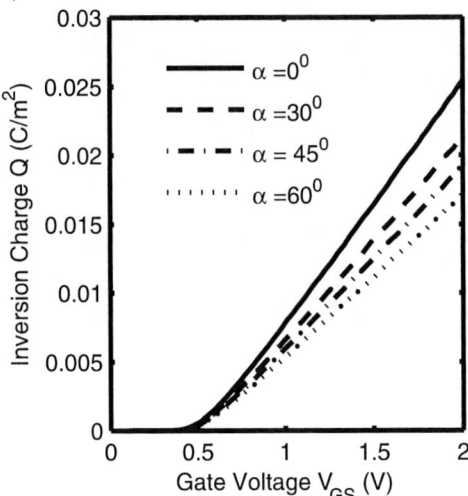

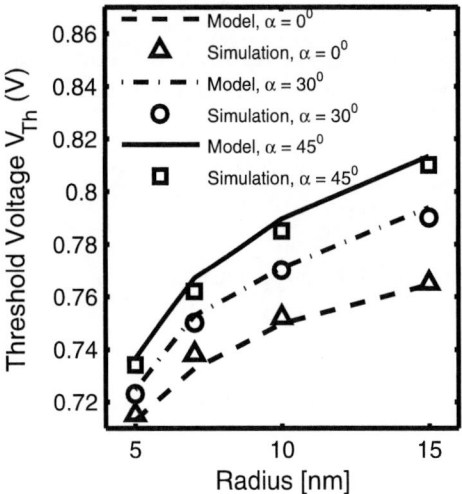

Fig. 5 Inversion charge (Cm⁻²) variation as a function of gate voltage using equation (7), for different . R=10nm.

Fig. 6. Threshold voltage variation for different gate coverage (or) as a function of different nanowire radii values. Mid gap metal is used as gate and T_{ox} = 2nm.

$$V_{TH} = V_{FB} + \frac{E_G}{2} + \frac{Q}{C_{ox}} \qquad (23)$$

where V_{FB} is the flat band voltage, Q is defined in the equation (21) and C_{ox} can be calculated by the same procedure used as in the case of co-axial gate geometry [9] .The only modification we do here is that we assume the field in the uncovered region of the nanowire to be zero and neglect the fringing fields at the two ends of the omega gate (which are indeed good approximations as verified by simulation). With this modification the gate capacitance for omega gate structure can be given by

$$C_{ox} = \frac{\varepsilon_{ox}(\pi - \alpha)}{\pi R \log(1 + T_{ox}/R)} \qquad (24)$$

where ε_{ox} is the permittivity of oxide and T_{ox} is the oxide thickness. Hence the final expression for threshold voltage for 75% gate coverage can be expressed as

$$V_{TH} = V_{FB} + \frac{E_G}{2} + \frac{2\varepsilon_{Si}V_T}{C_{OX}R}\left[\frac{Z}{\left(1 - \frac{1}{2}Z\right)} + \frac{\lambda C R^2}{2\pi Z}\left(Z\,e^Z - \cosh(Z) + \frac{\sinh(Z)}{Z}\right)\right]$$

$$(25)$$

where the various terms are defined as in earlier Eqs.

Although this expression is derived for 75% gate coverage, the error is found to be very small (less than 5%) if the same model is used for (± 10% extra) other gate coverage. However if we need the exact value we can use the general expression for Q in Eq. (23) for threshold voltage calculation. In Fig 6 the variation of threshold voltage has been plotted for various set of values of R & α by considering general expression for Q and verified with simulation results.

4. Conclusion

In this paper an explicit potential distribution model is derived for the cross sectional area of omega gate cylindrical nanowire transistor by solving the 2-D Poisson's equation. Analytical expressions for the inversion charge and threshold voltage are also been presented. A simple compact model for potential distribution and threshold voltage has also been presented for the special case of 75% gate coverage (α = 45⁰).The model is verified with 3D numerical simulation result and a good agreement has been found between the model we have introduced and the classical 3-D numerical simulation results.

5. Appendix

To solve this partial differential equation (12) in the interval $-\alpha \leq \theta \leq \alpha$ we use variable separation technique and re-expressed $\Psi_2(r, \theta)$ as

$$\psi_2(r,\theta) = \phi_1(r)\phi_2(\theta) \qquad (26)$$

With this representation the partial differential Eq (12) can be separated into two ordinary differential eq as shown below:

$$r^2\frac{\phi_1''}{\phi_1} + r\frac{\phi_1'}{\phi_1} - \frac{8Br^2}{(1-Br^2)^2} = -\frac{\phi_2''}{\phi_2} = \omega^2 \qquad (27)$$

where ω is the separating constant.

The angular part is easy to solve which gives the general solution

$$\phi_2(\theta) = C_1\cos(\omega\theta) + C_2\sin(\omega\theta) \qquad (28)$$

But the boundary condition (equation(7)) implies that

$$\phi_1(R)\phi_2(\theta) = g(\theta) \qquad (29)$$

Hence we let $\phi_1(R)=1$ and solve to get

$$\phi_2(\theta) = g(\theta) = C\cos(\omega\theta) \;\; ; \;\; \omega = \frac{\pi}{2\alpha} \qquad (30)$$

To solve the radial part ϕ_1 we approximate the $(1-Br^2)^{-2}$ term as $(1+2Br^2)$ as $Br^2 \ll 1$. So the differential eq for the radial part becomes

$$r^2\frac{d^2\phi_1}{dr^2} + r\frac{d\phi_1}{dr} - \left[8Br^2(1+2Br^2) + \omega^2\right]\phi_1 = 0$$
$$(31)$$

Now this differential equation can be transformed into the well known Hypergeometric differential eq[8] by the substitution $x=4Br^2$ and $y=e^{2Br^2}r^{-2k}\phi_1$ where k=ω/2. The transformed equation is written below :

$$xy'' + (2k+1-x)y' - (k+1)y = 0 \qquad (32)$$

Now this differential eq has solution in terms of Kummar's functions [8] as written below

$$y = \lambda_1 M(k+1, 2k+1, x) + \lambda_2 U(k+1, 2k+1, x) \quad (33)$$

where M and U are Kummar's function of first and second kind respectively.

For x very close to zero the U term becomes infinite which is not feasible as potential at all points of cross

section is finite. Hence we let $\lambda_2=0$. The other constant λ_1 can be calculated from the boundary condition $\phi_1(R)=1$ or

$$\lambda\, r^{2k}e^{-2Br^2}M(k+1, 2k+1, 4Br^2)\,\big|_{r=R} = 1 \qquad (34)$$

So , the solution can be written as

$$\psi_2(r,\theta) = \phi_1(r)\phi_2(\theta)$$
$$\text{or}$$
$$\psi_2(r,\theta) = \lambda\, r^{2k}e^{-2Br^2}M(k+1, 2k+1, 4Br^2)\,g(\theta) \quad (35)$$

6. Acknowledgement

This work has been supported by Department of Science and Technology (DST), India.

7. References

[1] J.-P. Colinge, "Multiple gate SOI MOSFETs," *Solid State Electron.*, vol. 48, no. 6, pp. 897-905, June 2004.

[2] J.Wang, E. Polizzi, and M. Lundstrom, "A computational study of ballistic silicon nanowire transistors," in *IEDM Tech. Dig.*, Dec. 8–10, 2003, pp. 695-698.

[3] T. Park, et al., "Fabrication of body-tied FinFET's (omega MOSFET's) using bulk Si wafers," in *VLSI Technology Tech. Symp. Dig.*, Jun. 10–12, 2003, pp. 135-136.

[4] Yiming Li; Hung-Mu Chou; Jam-Wem Lee, "Investigation of electrical characteristics on Surrounding-gate and omega-shaped-gate nanowire FinFETs" *IEEE Trans. Nanotechnology,* vol. 4 pp 510-516, Sep 2005.

[5] Hamdy Abd El Hamid; Benjamin Iniguez; Jaume Roig Guitart "Analytical Model of the Threshold Voltage and Subthreshold Swing of Undoped Cylindrical Gate-All-Around-Based MOSFETs" *EEE Trans. Electron Devices,* vol. 54, pp. 572-579, March 2007.

[6] Sentaurus Device Version 2006.06 , Synopsys Inc

[7] Y. Taur and T. H. Ning, *Fundamentals of Modern VLSI Devices* Cambrigde, U.K.: Cambridge Univ. Press, 1998.

[8] Abramowitz and Stegun, *Handbook of Mathematical Functions,* Dover Publications, Inc.,New York.

[9] R.Plonsey and R.E. Collin, *Principles and Applications of Electromagnetic Fields*. New York: McGraw-Hill,1961.

Design of Reversible Finite Field Arithmetic Circuits with Error Detection

Jimson Mathew, Hafizur Rahaman, Babita R. Jose*, Dhiraj K. Pradhan
Department of Computer Science, University of Bristol, Bristol, BS8 1UB, UK.
*Cochin University of Science and Technology, India.

Abstract— **Motivated by the potential of reversible computing, we present a systematic method for the designing reversible arithmetic circuits for finite field or Galois fields of form GF(2^m). It is shown that an adder over GF(2^m) can be designed with m garbage bits and that of a PB multiplier with 2m garbage bits. To tackle the problem of errors in computation, we also extend the circuit with error detection feature. Gate count and technology oriented cost metrics are used for evaluation. The expression for the upper bound for gate size is also derived for special primitive polynomials. Our technique, when compared with existing CAD tool gives the same gate size and quantum cost.**

I. INTRODUCTION

The design of reversible networks has received much attention in recent years[1],[2]. The primary motivation of this work is huge power consumption in irreversible circuit. As observed by Landauer [1], use of traditional (irreversible) logic gates results in information loss and causes inherent energy dissipation in a circuit, regardless of its realization. A system is said to be reversible if it is information lossless. Bennett [4] showed that zero-energy dissipation would be possible only if the network consists of reversible gates. Thus, reversibility may play a significant role in future circuit design. Improved process technologies, higher levels of integration, and low-power design methods and tools have significantly reduced the energy loss for irreversible gates over the last decades. Although the heat generation due to the information loss in modern CMOS is still small, the recent work by Zhirnov et al. [5] shows the potentially prohibitive difficulty of heat removal with the increasing density of CMOS. The second motivator is that all quantum gates are reversible [6].

Hence, the interest in reversible logic is motivated by its applications in quantum computing, low power CMOS, nanotechnology, and optical computing. Several works have been reported for reversible synthesis of practical circuits in[2][3]. This paper presents a systematic technique for reversible synthesis of adder and PB multiplier circuits over finite fields of form GF(2^m). The Galois field multiplier design is one of the most well-researched and widely investigated topics, having great impact on the solvability of large class of design problems in cryptography, coding theory, Galois switching theory and digital signal processing. In the Galois field, there are two basic arithmetic operations: addition and multiplication. Over the years, many solutions have been proposed for efficient multiplier design, such as bit-serial, digit-serial and bit parallel. The hardware implementation of

addition operation in the finite field over GF(2^m) is relatively straightforward and requires at most m XOR gates. On other hand, implementation of the multiplication is much more complicated. The other operations like exponentiation, division, and inversion in GF(2^m) field can be performed by repeated multiplications [8]. For error detection in finite field multipliers, a number of schemes have been proposed in the literature [17]. One way to detect errors in a finite field multiplier is to use parity prediction technique[16]. A second approach is to scale the inputs of the multiplier by a factor and at the end of the multiplication, the correctness of the result is checked by one or two divisions. The main techniques that can be used for single error correction are (1) error detection and retry and (2) error masking. Error detection and retry involves using concurrent error detection (CED) circuitry that monitors the outputs of a circuit for the occurrence of an error. If an error is detected, the system recovers through rollback and retry thereby preventing a failure. Error masking involves using circuitry that masks (*i.e.*, *corrects*) errors using schemes such as triple modular redundancy (TMR).

In this paper, a systematic method for reversible synthesis of adder and PB multiplier over GF(2^m) circuits is proposed. In [16] the authors have considered detection of single stuck at fault in polynomial basis Galois field multipliers. They use simple parity prediction technique for error detection. Our technique fundamentally differs from this technique in two critical issues. Firstly, our technique addresses the problem of reversible circuit design. Secondly, more importantly, due to quantum mechanical restrictions, the design of reversible logic is done with no feedback and no fanout. Fanouts and feedbacks are also not allowed for some other technologies that use reversible gates. This leaves the cascade structure as the only network topology satisfying those conditions. Therefore, the parity prediction technique described in [16] can not be used reversible paradigm.

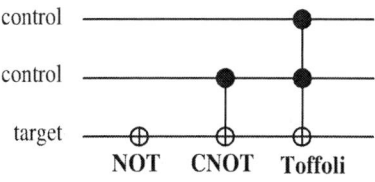

Fig. 1. NOT, CNOT , and TOFOLLI gates

A. Contributions to this Paper

To date, design of reversible Galois field circuits has not explored. The contribution of this paper is mainly a systematic design technique finite field arithmetic circuits. Closed form expression for the overall gate count is derived for special cases. Efficiency of reversible computation realized in a circuit form depends on many parameters including number of gates, number of auxiliary bits and the number of logic levels. In this paper we also report the cost of adder and multiplier circuits assuming appropriate cost for the basic gates. Furthermore, for the first time error detection feature is incorporated into the design, using parity prediction technique. The overhead for the proposed error detection technique is also analyzed.

The paper is structured as follows. Section II presents the background on Galois filed and reversible logic basics necessary for this paper. Section III presents a basic design steps and closed form expressions are derived for primitive polynomials of type trinomial and All One Polynomials (AOP). The technique is illustrated by appropriate example. Experimental results are presented in Sections IV. The paper concludes with some observations and suggestions for further research in Section V.

II. PRELIMINARIES

An n-input, m-output Boolean function F is said to be reversible if and only if m = n, and F is one-to-one. In other words, a reversible function, to produce outputs, simply induces a permutation on the set of input vectors. Therefore, given an output vector, its corresponding input vector can be uniquely restored. The function $\{x, y \rightarrow xy\}$ denoting AND operation is not reversible. By adding one extra input and two extra outputs, a modified but reversible function can be constructed. By garbage we meant the number of extra outputs required to realize the given function.

In reversible logic, one can restore the input pattern to a circuit, given its output. In conventional non-reversible logic design, such derivation may not always be possible. In practice, not all of the possible reversible functions can be realized as a single reversible gate. Several reversible gates have been proposed. Toffoli gates and Fredkin gates are the best known and most widely studied [6]. The basic three gates are illustrated in Fig.1. Gates with more controls are drawn similarly. We will only consider CNOT and Toffoli gates in this paper. Design of reversible logic is often considered in conjunction with quantum technologies. Due to quantum mechanical restrictions, the design of reversible logic is done with no feedback and no fanout. Fanouts and feedbacks are also not allowed for some other technologies that use reversible gates. This leaves the cascade structure as the only network topology satisfying those conditions. In this paper, our design uses cascades of CNOT and Toffoli gates. Further, we assume that the signal propagated from left to right. The reversible cost (or simply, cost) of a function implementation is defined as the number of gates in the network realizing it. Next, some basics of Galois field is explained.

Let GF(N) denote a set of N elements, where N is a power of a prime number, with two special elements 0 and 1 representing the additive and multiplicative identities respectively, and two operators addition '+' and multiplication '.'. GF(N) defines a finite field, if it forms a commutative ring with identity over these two operators in which every element has a multiplicative inverse.

Finite fields can be generated with irreducible polynomials of the form $P(x) = x^{m-1} + \sum_{i=0}^{m-2} c_i x^i$, where $c_i \in$ GF(2) [15]. It is conventional to represent the elements of GF(2^m) as a power of the primitive element α where α is the root of p(x), i.e. $p(\alpha)$. The set $\{1, \alpha, \alpha^2 \ldots, \alpha^{m-1}\}$ is referred to as the polynomial basis. Each element A$\in$ GF(2^m) can be expressed with respect to the PB as a polynomial of degree m over GF(2), i.e. where ai$\in$ GF(2). Given , the PB multiplication over GF(2^m) can be defined as A(x) = $\sum_{i=0}^{m-1} a_i x^i$. Given A(x), B(x), the polynomial representation of elements A and B, the PB multiplication over GF(2^m) can be defined as W(x)=A(x)B(x) mod P(x) . Details can be found in [10], [12].

Mastrovito has proposed an algorithm along with its hardware architecture for PB multiplication [12] popularly known as the Mastrovito algorithm/multiplier. In [10], based on Mastrovito algorithm, a new formulation for PB multiplication and generalized bit-parallel hardware architecture for special reduction polynomials, namely: trinomials, equally spaced polynomials (ESPs), and two classes of pentanomials has been presented. Their formulation is described below.

Consider a multiplier with A and B inputs where A = [a_0, a_1, a_2, ...,a_{m-1}]and B =[b_0,b_1,b_2,...,b_{m-1}]. The ai, and bi are the coordinates of A and B respectively where $0 \leq$ i $\leq$m-1. The formulation is based on three matrices: (i) an m – 1 by m reduction matrix Q, (ii) L matrix and the (iii) U matrix. The L and U matrices are formed for implementation of this multiplication scheme. L is an lower triangular matrix and U is an upper triangular matrix. The outputs (d's and e's) of the IP-network are defined by the following two vectors which are functions of A and B.

$$d = L.b \tag{1}$$

$$e = U.b \tag{2}$$

where b= [b_0, b_1, b_2,...,b_{m-1}]^T.

$$L = \begin{bmatrix} a_0 & 0 & 0 & \cdots & 0 \\ a_1 & a_0 & 0 & \cdots & 0 \\ a_2 & a_1 & a_0 & \cdots & 0 \\ \vdots & \vdots & \ddots & \vdots & \vdots \\ a_{m-2} & a_{m-3} & \cdots & a_0 & 0 \\ a_{m-1} & a_{m-2} & \cdots & a_1 & a_0 \end{bmatrix}$$

$$U = \begin{bmatrix} 0 & a_{m-1} & a_{m-2} & \cdots & 0 & a_1 \\ 0 & 0 & a_{m-1} & \cdots & 0 & a_2 \\ \vdots & \vdots & \vdots & \ddots & \vdots & \vdots \\ 0 & 0 & \cdots & 0 & a_{m-1} & a_{m-2} \\ 0 & 0 & \cdots & 0 & 0 & a_{m-1} \end{bmatrix}$$

The multiplication outputs are given by the equation:

$$c = d + Q^T e \qquad (3)$$

Where Q matrix can be derived from [10].

III. PROPOSED REVERSIBLE GALOIS CIRCUITS

In this section, we present the basic steps for designing Galois circuits with minimum garbage.

A. Reversible adder

First, the reversible Galois adder is designed and its design is straight forward. It is basically bit wise exclusive operation of the input operands A and B . Figure 2a shows an example of $GF(2^4)$ reversible adder. In Figure 2a, the Gs outputs are the garbage outputs and Si's are the functional outputs. It is clear that an m bit finite field adder over GF(2) requires m garbage outputs.

B. Error detection in reversible adder

The output bit Si of of an adder is defined as $S_i = a_i + b_i$, where '+' denotes XOR operation. Therefore the output parity is given by,

$$P_s = \sum_{i=0}^{m-1} (a_i + b_i) \qquad (4)$$

$$P_s = \sum_{i=0}^{m-1} (a_i) + \sum_{i=0}^{m-1} (b_i) \qquad (5)$$

Equation 5 can be simplified as $P_s = P_a + P_b$, where P_a and P_b respectively are the parity of the input operand A and B. Figure 2b shows the implementation of the reversible adder with error detection, here we assumed that parities of the inputs are known. In the experimental section we analyze the overhead for error detection.

C. Reversible Multiplier

Galois multiplier design is bit more complex. The following steps must be followed for designing reversible GF multiplier.

Step 1. Using equations1, 2 and 3 derive the expressions for d, e and c.

Step 2. Implement the common sub expressions e_i's in respective outputs.

Step 3. Implement the common e_i's in the output expression.

Step 4. Complete the GF circuit by implementing remaining d_i, where $0 \leq i \leq m-1$. We explain the above steps using example 1.

Example 1: : Reversible design of $GF(2^4)$ multiplier with P(x) = x^4+ x+1.

Step 1: The Boolean expressions as derived from using equations are given below:

The d and e outputs of the IP-network are given below. '+' sign represents XOR operation.

$d_0 = a_0 b_0$
$d_1 = a_1 b_0 + a_0 b_1$
$d_2 = a_2 b_0 + a_1 b_1 + a_0 b_2$
$d_3 = a_3 b_0 + a_2 b_1 + a_1 b_2 + a_0 b_3$
$e_0 = a_3 b_1 + a_2 b_2 + a_1 b_3$
$e_1 = a_3 b_2 + a_2 b_3$
$e_2 = a_3 b_3$

we have, $Q = \begin{bmatrix} 1 & 1 & 0 & 0 \\ 0 & 1 & 1 & 0 \\ 0 & 0 & 1 & 1 \end{bmatrix}$

The multiplier outputs ci, where $(0 \leq i \leq 3)$ are derived as follows.

$c_0 = d_0 + e_0$; $c_1 = d_1 + e_1 + e_0$; $c_2 = d_2 + e_1 + e_2$; $c_3 = d_3 + e_2$,

Step 2: Here e_0, e_1, and e_2 are implemented first because these are the common sub-expression.

Step 3: The e_0, e_1, and e_2s in output expressions are implemented next shown in Figure 3.

Step 4: Final outputs are completed by implementing d_i's associated with outputs c_i's.

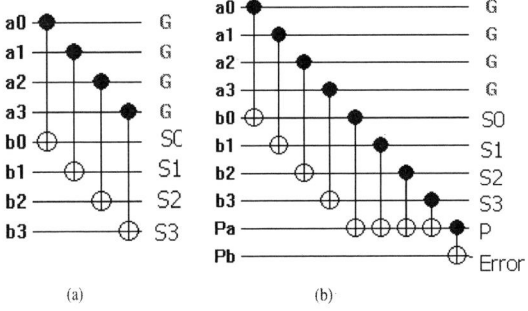

(a)　　　　(b)

Fig. 2.   GF(16) Adders (a) without Error Dection (b) with Error Dection

1) Multipliers Using Special Polynomials: The methodology described above can be applied to primitive polynomials of any type, however for special cases closed form expression for the total gate size can be derived. Two special type of primitive polynomials are considered next and it's closed for expression for the gate size is derived. Table 1 gives expression for these cases.

Lemma 1: A reversible Galois Multiplier with a trinomial P(x) can be designed with m^2+m-1 reversible gates.

Proof: Follows from the equations 1, 2 and 3, we have

$$d_0 = a_0 b_0 \qquad (6)$$

$$d_1 = a_1 b_0 + a_0 b_1 \qquad (7)$$

$$d_2 = a_2 b_0 + a_1 b_1 + a_0 b_2 \qquad (8)$$

455

978-1-4244-3039-0/08 $25.00 © 2008 IEEE

TABLE I
TOTAL GATE SIZE FOR DIFFERENT TYPES OF POLYNOMIALS

Types of Polynomials	Number of Gates
Trinomials (of the form $x^m + x + 1$)	$m^2 + m - 1$
Equally spaced polynomials(AOP) (of the form $P(x) = x^m + x^{m-1} + \ldots + x^2 + x + 1$	$m^2 + m - 1$

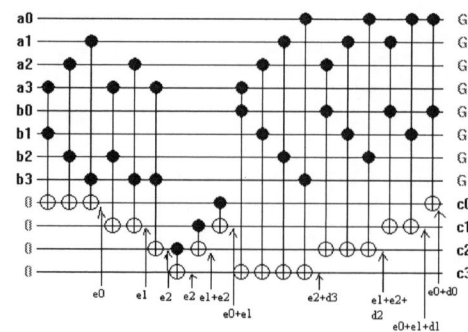

Fig. 3.   Reversible $GF(2^4)$ Multiplier with $P(x) = x^4 + x + 1$

$$d_{m-2} = a_{m-2}b_0 + a_{m-2}b_1 + \ldots + a_1b_{m-3} + a_0b_{m-2} \quad (9)$$

$$d_{m-1} = a_{m-1}b_0 + a_{m-1}b_1 + \ldots + a_1b_{m-1} + a_0b_{m-1} \quad (10)$$

$$e_0 = a_{m-1}b_1 + a_{m-2}b_2 + \ldots + a_1b_{m-1} \quad (11)$$

$$e_1 = a_{m-1}b_2 + a_{m-1}b_3 + \ldots + a_2b_{m-1} \quad (12)$$

$$e_{m-1} = a_{m-1}b_{m-1} \quad (13)$$

$$c_0 = d_0 + e_0; \quad (14)$$

$$c_1 = d_1 + e_1 + e_0 \quad (15)$$

$$\ldots$$

$$c_i = d_i + e_i + e_{i-1} \quad (16)$$

$$c_{m-1} = d_{m-1} + e_{m-2} \quad (17)$$

where c_is are the multiplication outputs ($0 \leq i \leq$ m-1).

Let T be the total number required to implement a reversible Galois multiplier with a trinomial P(x), therefore

T = Gates required to implement Eqns. 14-17, which is equivalent to implementing Eqns.6-13 and common subexpressions in Eqns. 14-17. Furthermore, assuming that each product term is implemented using a single gate, we have

$$T = \sum_{i=1}^{m-1} i + \sum_{i=1}^{m-2} (i) + m - 1 \quad (18)$$

$T = \frac{m(m+1)}{2} + \frac{m(m-1)}{2} + m - 1$
Simplifying, $T = m^2 + m - 1$ QED.

Lemma 2: A reversible Galois Multiplier with an All One Polynomial (AOP) P(x) can be also be designed with $m^2 + m - 1$ reversible gates.

Proof: It can be derived in the same as way as in Lemma 1.

D. Reversible GF multiplier with error detection

In this section, error detection of the above reversible multiplier is considered. For error detection, we use the well known parity prediction principle. We consider, multipliers with primitive polynomial of the form $P(x) = x^n + x + 1$

The predicted output parity is given by

$$P_c = \sum_{i=0}^{m-1} (c_i) \quad (19)$$

where P_c is the predicted parity of the output and c_is are the multiplier output bits. Substituting for equations 14-17

$$P_c = d_0 + e_0 + d_1 + e_1 + e_0 + \ldots + d_i + e_i + e_{i-1} + \ldots + d_{m-1} + e_{m-2}; \quad (20)$$

Simplifying, we get

$$P_c = d_0 + d_1 \ldots + d_i + \ldots d_{m-1}; \quad (21)$$

Equation 21 is used to predict the parity from the input operands A and B, and it is compared against output parity. If there is a mismatch, an error is reported. Figure 4 shows the modified architecture of Figure 3 with error detection. In this new design the additional gates required to implement the d_is which is approximately about 50 percent of the overall classical Galois field multiplier. Our experimental results shows that on an average about 53 percent extra gates required to implement error detection feature for multipliers with primitive polynomial of the form $P(x) = x^n + x + 1$. In this case we have a total of 2m garbage outputs.

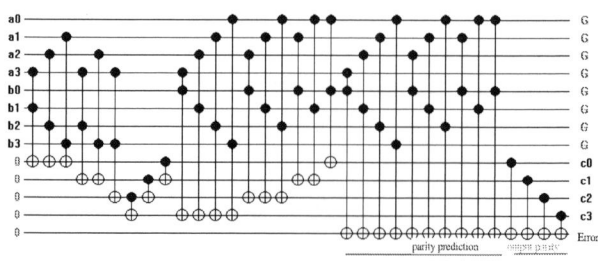

Fig. 4.   Error Decting Reversible $GF(2^4)$ Multiplier with $P(x) = x^4 + x + 1$

978-1-4244-3039-0/08 $25.00 © 2008 IEEE

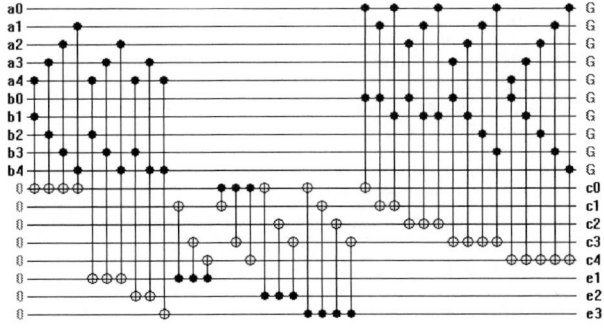

Fig. 5. Reversible GF(2^5) Multiplier with P(x) = $x^5 + x^4 + x^3 + x + 1$

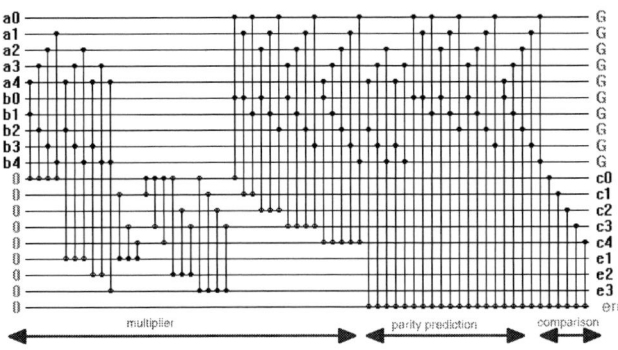

Fig. 6. Reversible GF(2^5) Multiplier with P(x) = $x^5 + x^4 + x^3 + x + 1$

IV. EXPERIMENTAL RESULTS

It is a common practice in reversible logic design to synthesize a network using multiple control Toffoli gates and report its cost as the number of gates. However, from the point of view of technological realization, multiple control Toffoli gates are not simple transformations. Rather they are composite gates themselves and Toffoli gates with a large set of controls can be quite expensive. Considering these fact, the cost of quantum implementation, we optimized the direct realization using the technique proposed in [2]. Table II shows performance analysis of different PB multipliers, with respect to gate counts and Quantum Cost (QC). It is assumed that cost of CNOT is unity and that of Tofoli is 5 unit [2].Our derived upper bound of these circuits exactly matched with the gate size and quantum cost of the synthesized multiplier circuits using the popular reversible CAD tool [2]. However, the tool can not handle variables more than 20 input variables, (maximum Galois multiplier size of GF(2^{10})) whereas the proposed method does not have any limitation on the value of m. The '-' columns in Table 2 indicate that the tools fails to converge. Table III shows comparison of the proposed architecture without error detection and with error detection feature.

Figure 7 shows overhead analysis for the proposed adder architectures, with different primitive polynomials. The x-axis shows the decimal representation of the primitive polynomial, for example primitive polynomial P(x) = $x^7 + x^5 + x^4 + x^3 + x^2 + x + 1$, represented as $(10111111)b$ and its decimal equivalent is 191. This decimal notation is shown in figures. For error detection, there is an increase of approximately 113 percent in gate depth. Figure 8 shows overhead analysis for the proposed multiplier architectures. Due to special properties of the trinomials the average overhead is only about 53 percent.

The above design procedure is the same for other generic polynomials as well. The following example illustrates a pentanomial P(x).

Example 2: : Consider a GF(2^5) multiplier with P(x) = $x^5 + x^4 + x^3 + x + 1$. The Boolean expressions derived using equations 1, 2 and 3 are given below:

$c_0 = d_0 + e_0 + e_1 + e_3$; $c_1 = d_1 + e_0 + e_2 + e_3$; $c_2 = d_1 + e_1 + e_3$; $c_3 = d_3 + e_0 + e_1 + e_2 + e_3$; $c_4 = d_4 + e0 + e_2$, where $c_i's$ are the multiplication outputs ($0 \leq i \leq 4$) and we have,

$d_0 = a_0 b_0$

$d_1 = a_1 b_0 + a_0 b_1$

$d_2 = a_2 b_0 + a_1 b_1 + a_0 b_2$

$d_3 = a_3 b_0 + a_2 b_1 + a_1 b_2 + a_0 b_3$

$d_3 = a_4 b_0 + a_3 b_1 + a_2 b_2 + a_1 b_3 + a_0 b_4$

$e_0 = a_4 b_1 + a_3 b_2 + a_2 b_3 + a_1 b_4$

$e_0 = a_4 b_2 + a_3 b_3 + a_2 b_4$

$e_1 = a_4 b_3 + a_3 b_4$

$e_2 = a_4 b_4$

Figure 5 shows the complete design of the above multiplier and the same mulitiplier design with error detection feature added is shown in Figure 6.

TABLE II

PERFORMANCE ANALYSIS

GF	Poly.	[2]		proposed	
Mult.		# gates	QC	# gates	QC
GF(2^2)	$x^2 + x + 1$	5	21	5	21
GF(2^3)	$x^3 + x + 1$	11	47	11	47
GF(2^4)	$x^4 + x + 1$	19	83	19	83
GF(2^6)	$X^6 + x + 1$	41	186	41	186
GF(2^7)	$X^7 + x +$	55	245	55	245
GF(2^9)	$X^9 + x + 1$	89	408	89	408
GF(2^{10})	$X^{10} + x + 1$	-	-	109	509
GF(2^{13})	$X^{13} + x + 1$	-	-	181	857

TABLE III

PERFORMANCE ANALYSIS WITH AND WITHOUT ERROR DETECTION(ED)

GF	Poly.	Without ED		with ED	
Mult.		# gates	QC	# gates	QC
GF(2^2)	$x^2 + x + 1$	5	21	10	38
GF(2^3)	$x^3 + x + 1$	11	47	20	80
GF(2^4)	$x^4 + x + 1$	19	83	33	137
GF(2^6)	$X^6 + x + 1$	41	186	68	297
GF(2^7)	$X^7 + x +$	55	245	90	392
GF(2^9)	$X^9 + x + 1$	89	408	143	642
GF(2^{10})	$X^{10} + x + 1$	109	509	174	794
GF(2^{13})	$X^{13} + x + 1$	181	857	285	1325

978-1-4244-3039-0/08 $25.00 © 2008 IEEE

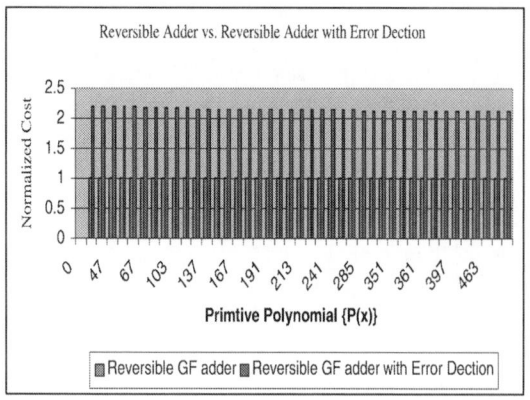

Fig. 7.   Overhead Analysis GF(2^m) Adder

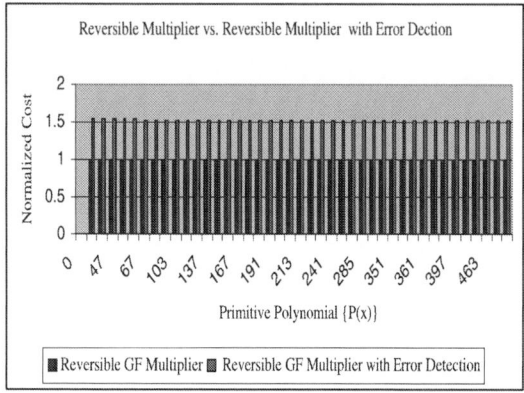

Fig. 8.   Overhead Analysis GF(2^m) Multiplier

V. CONCLUSION

In this paper, a methodology for designing reversible adder and PB multiplier circuits over GF(2^m) with error detection is presented. Our goal here is to design a reversible Galois circuit with minimum garbage bits. Gate count and technology oriented cost metrics are used for evaluation. It is shown that Galois adders with m and multipliers with 2m auxiliary bits. Since the proposed method is generic, there is no limitation in number of input bits whereas CAD tool fails to converge for value of m greater than 10. Design examples show that our method achieves error detection with an average gate overhead of 53 percentage.

REFERENCES

[1] R. Landauer "Irreversibility and heat generation in the computing process". IBM J. of R&D, 5, 1961, pp. 183–191.

[2] D. Maslov and G. W. Dueck. Reversible cascades with minimal garbage. IEEE TCAD, 23(11), 2004, pp. 1497–1509.

[3] V. V. Shende, A. K. Prasad, I. L. Markov, and J. P. Hayes. Synthesis of reversible logic circuits. IEEE TCAD, 22(6), June 2003, pp. 710–722.

[4] C. H. Bennett, Logical reversibility of computation. IBM J. of R&D, 17, November 1973, pp. 525–532.

[5] V.V Zhinov, R. K. Kavin, J. A. Hutchby and G.I.Bourianoff. Limits to binary logic switch scaling - a Gedanken model. Proc. IEEE 91(11), 2003, pp. 1934–1939.

[6] M. Nielsen and I. Chuang, Quantum Computation and Quantum Information. Cambridge University Press, 2000.

[7] P.A. Scott, S.J. Simmons, S.E. Tavares, and L.E. Peppard, "Architectures for Exponentiation in GF(2m)," IEEE J. Selected Areas in Comm., vol. 6, no. 3, pp. 578-586, Apr. 1988.

[8] H. Wu and M.A. Hasan, "Efficient Exponentiation of a Primitive Root in GF(2m)," IEEE Trans. Computers, vol. 46, no. 2, pp. 162-172, Feb. 1997.

[9] J.H. Guo and C.L. Wang, "Systolic Array Implementation of Euclid's Algorithm for Inversion and Division in GF(2m)," IEEE Trans. Computers, vol. 47, no. 10, pp. 1161-1167, Oct. 1998.

[10] A. Reyhani-Masoleh, and M. Anwar Hasan, "Low Complexity Bit Parallel Architectures for Polynomial Basis Multiplication over GF(2m)", IEEE Transactions on Computers, vol.53, no.8, pp.945-959, August 2004.

[11] M. Nicoliadis, "Carry Checking/Parity Prediction Adders and ALUs, " IEEE Trans. VLSI Systems, vol. 11, no. 1, Feb. 2003.

[12] E.D. Mastrovito, "VLSI Architectures for Computation in Galois Fields," PhD thesis, Linkoping Univ., Sweden, 1991.

[13] B. Sunar and C.K. Koc, "Mastrovito Multiplier for All Trinomials," IEEE Trans. Computers, vol. 48, no. 5, pp. 522-527, May 1999.

[14] A. Halbutogullari and C.K. Koc, "Mastrovito Multiplier for General Irreducible Polynomials," IEEE Trans. Computers, vol. 49, no. 5, pp. 503-518, May 2000.

[15] D. K. Pradhan, "A Theory of Galois Switching Functions", IEEE Trans. Computers, vol. 27, no. 3, pp.239-248, Mar. 1978.

[16] A. Reyhani-Masoleh, and M. A. Hasan, "Fault Detection Architectures for Field Multiplication Using Polynomial Bases ", IEEE Trans. Computers, vol. 55, no. 9, pp.1089-1103, Sept. 2006.

[17] S. Fenn, M Gossel, M Benaissa, and D. Taylor, "On-line Error Dection for Bit-seial Multipliers in GF(2m) ", J. Electronic Testing: Theory and Applications, vol. 13, pp.29-40, 1998.

SESSION A4:
Verification

Exploiting Circuit Reconvergence through Static Learning in CNF SAT Solvers

Yinlei Yu   Cameron Brien   Sharad Malik

Dept. of Electrical Engineering, Princeton University

E-mail: {yyu, cbrien, sharad}@princeton.edu

Abstract

Most contemporary SAT solvers use a Conjunctive-Normal-Form (CNF) representation for logic functions due to the availability of efficient algorithms for this form, such as deduction through unit propagation and conflict driven learning using clause resolution. The use of CNF generally entails transformation to this form from other representations such as logic circuits [23]. However, this transformation results in loss of information such as direction of signal flow and observability of signals at circuit outputs [12][13]. This has prompted the development of various circuit-based solvers [2][14][17][22], hybrid CNF+circuit-based solvers[13], as well as augmented CNF solvers [12]. Having the circuit available provides for additional capabilities at a cost, and thus requires careful analysis to determine the viability of each approach.

This paper highlights one specific capability provided by a circuit: the ability to consider reconvergent paths in unit propagation. Unit propagation is the workhorse of contemporary SAT solvers, thus any improvement to this has significant practical potential. We first demonstrate that the Tseitin circuit-to-CNF transformation limits backward unit propagation and how additional implications can be derived when unit propagation across multiple paths is considered. Next, we show how these implications can be exploited by statically learning clauses during circuit pre-processing. The results of the practical implementation of these algorithms show that the static learning can provide significant speed-up on several classes of benchmark circuits. Finally, we discuss how this work compares with other circuit-based approaches, especially those arising from the automatic-test-pattern-generation (ATPG) community (e.g. recursive learning) and circuit and non-circuit based pre-processors.

1. Introduction

Boolean Satisfiability (SAT) solvers are widely used in electronic design automation. Most modern SAT solvers

(e.g. [11][21]) use the Conjunctive-Normal-Form (CNF) representation for logic functions. In CNF, a formula f is described as

$$f = C_1 \wedge C_2 \wedge \ldots \wedge C_n, C_i = l_1 \vee l_2 \vee \ldots \vee l_m$$

where C_i is a clause and l_i is a literal, *i.e.*, a variable in positive or negative phase. $\wedge$ and $\vee$ are the AND and OR operators. This representation has been shown to be efficient for implementing the Davis-Putnam-Logemann-Loveland (DPLL) algorithm [7]. In particular, the three main steps of modern DPLL solvers – decision-making, unit propagation/Boolean Constraint Propagation (BCP), and conflict driven learning using resolution, have all been shown to be efficient with this representation. Further, conversion to this form from other general representations (e.g. logic circuits) has been shown to be straightforward. By introducing a new variable for each gate output in the circuit, the Tseitin Tautology Transformation [23] converts a logic circuit into a CNF in time linear in the circuit size. While this transformation is efficient, it is known to be "lossy" in the sense that some of the circuit information is not captured in the resulting CNF. Specifically, information regarding the direction of signals, as well as observability of signals at the circuit outputs is lost. Various circuit-based SAT solvers [14][17], hybrid circuit+CNF solvers [13] and augmented CNF solvers [12] attempt to overcome these limitations with mixed success.

In this paper, we highlight an additional capability that is lost/difficult to exploit in CNF solvers. We show that while unit propagation in the forward direction through the circuit is adequately captured by the CNF resulting from the Tseitin transformation, backward unit propagation (justification in logic circuits) is only limited to single paths. The introduction of the intermediate variables in the transformation blocks the reasoning from considering reconvergent paths. We then provide a calculus for capturing and propagating this information during DPLL search. Due to the overhead of implementing the calculus dynamically during the search, we propose to use a static learning step to encode this information into additional clauses to be followed by a CNF SAT solver. This approach has been evaluated

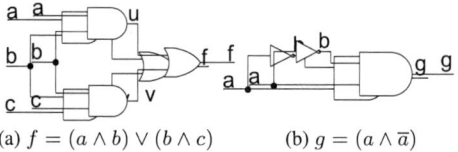

(a) $f = (a \wedge b) \vee (b \wedge c)$ (b) $g = (a \wedge \overline{a})$

Figure 1. Partial circuits with reconvergence

on a variety of benchmark circuits, mostly from the formal verification area. Moderate to significant speed up has been shown for several categories of benchmarks. A statistic-based heuristic is employed to identify instances that are most likely to benefit from our static reasoning algorithm. The heuristic also determines the best compromise between the performance gain because of the added unit propagation capability and the loss due to the additional pre-processing time and overhead due to a larger clause database.

This paper is organized as follows. In Section 2, we highlight the limitations of unit propagation in CNF solvers through some motivating examples. In Section 3, we describe the calculus of *unit sets* and show how this is used in reconvergent unit propagation. Section 4 describes the static pre-processing step for learning clauses from circuit reconvergence. In Section 5, we discuss how this work fits in the spectrum of efforts in circuit and circuit+CNF-based solvers including classic automatic-test-pattern-generation techniques (ATPG) such as Recursive Learning. In Section 6, we provide our experimental results on various benchmarks. Finally, we draw some conclusions in Section 7.

2. Motivating Examples

The DPLL algorithm determines satisfiability through a search and backtrack approach [7]. A unit literal is identified or created through a decision and unit propagation used to determine the implications (forced assignments). When a conflict is encountered, learned clauses are generated and the algorithm backtracks.

Our approach is motivated by how the transformation from circuit description to CNF may negatively impact unit propagation in the DPLL algorithm. This is best illustrated by two distinct cases. To facilitate discussion, we assume every gate g has a single output and denote g as its output signal as well.

Case 1: Consider the partial circuit $f = (a \wedge b) \vee (b \wedge c)$ illustrated in Figure 1(a). Note that a and c have single paths to f while b diverges and then reconverges at f.

By the Tseitin Tautology Transformation [23] we obtain a satisfiability equivalent expression in CNF, *i.e.*, an assignment is consistent with the circuit *iff* it satisfies the following CNF: $T = (a \vee \overline{u}) \wedge (b \vee \overline{u}) \wedge (\overline{a} \vee \overline{b} \vee u) \wedge (b \vee \overline{v}) \wedge (c \vee \overline{v}) \wedge (\overline{b} \vee \overline{c} \vee v) \wedge (\overline{u} \vee f) \wedge (\overline{v} \vee f) \wedge (u \vee v \vee \overline{f})$

Consider forward implications generated by reconvergent paths with the partial assignment $b = 0$. With this assignment, we generate forward implications $u, v, f = 0$. Now T is reduced to $T = 1$ as all variables are assigned consistently.

If consider justifications along reconvergent paths. Let $f = 1$, we get: $T = (a \vee \overline{u}) \wedge (b \vee \overline{u}) \wedge (\overline{a} \vee \overline{b} \vee u) \wedge (b \vee \overline{v}) \wedge (c \vee \overline{v}) \wedge (\overline{b} \vee \overline{c} \vee v) \wedge (u \vee v)$

This assignment results in no implications from unit propagation. However, if we step back and examine the circuit itself we see that for $f = (a \wedge b) \vee (b \wedge c)$ clearly $f \Rightarrow b$. Here, the introduction of intermediate variables u and v in the CNF has masked the relationship between f and reconvergent b. This case highlights the limitation of the CNF in generating unit literals corresponding to backward unit propagation.

Case 2: Let us now examine the partial circuit $g = (a \wedge \overline{a})$ depicted in Figure 1(b).

Again, we use Tseitin to determine a satisfiability equivalent CNF expression $T = (\overline{a} \vee \overline{b}) \wedge (a \vee b) \wedge (a \vee \overline{g}) \wedge (b \vee \overline{g}) \wedge (\overline{a} \vee \overline{b} \vee g)$

If we make the partial assignment $g = 1$, we have $T = 0$ through unit propagations, which indicates a conflict. However, notice that this conflict is obvious – clearly $g = (a \wedge \overline{a}) \equiv 0$. Yet, if g were not assigned, the fact that g is constant 0 would not be revealed. In this case Tseitin transformation hid the relationship between a and b.

These two cases are important because though simple, they illustrate the power of exploiting reconvergence in determining satisfiability. When assignments are made in a CNF representation of a circuit, the use of intermediate variables masks the reconvergence of input variables and unit propagation suffers as a result. In some instances, as in Case 2, reconvergence goes unnoticed until an assignment is made. Even worse, in instances like Case 1, reconvergence is unexploited all-together. It is precisely these gaps that the technique proposed in this paper fills.

3. Unit Propagation across Reconvergent Paths

To capture the reconvergence highlighted in the previous section, we introduce the concept of a *unit set* for each gate output, which similar to the *extended backward propagation* in [2]. We show the propagation rule of unit sets. We consider circuit representations consisting exclusively of 2-input gates and invertors.

3.1. Unit set definition

We define two sets of literals for every gate g in the circuit: a *unit-one* set ($u1_g$) and a *unit-zero* ($u0_g$) set. If a

462

978-1-4244-3039-0/08 $25.00 © 2008 IEEE

literal $l \in u1_g$, then $g \Rightarrow l$; if $l \in u0_g$, then $\overline{g} \Rightarrow l$. For example, if $u1_g = \{a, b\}$, then if g is assigned to 1, a, b must also be assigned to 1. Note that if $u0_g$ contains both a and $\overline{a}$, then g should never be assigned 0, so $g = 1$.

3.2. Calculating unit sets

Here, we present the rules for calculating the unit sets for an arbitrary gate g in a circuit. If g is a 2-input gate, let it have inputs l and r. A gate may have a constant value 0/1 or unknown value x. If g is a NOT gate, let it have input i with value v_i. Let g have unit-one set $u1_g$ and unit-zero set $u0_g$. Similarly, let l, r and i have respective unit-one and unit-zero sets $u1_l, u1_r, u1_i$ and $u0_l, u0_r, u0_i$. If g is a primary input, we define $u1_g = \{g\}$ and $u0_g = \{\overline{g}\}$.

The value of g can be obtained by calculating the corresponding logic function of its inputs. For g with constant value 1 or 0, the unit set calculations are simple: If $g = 0$, $u1_g = \mathbb{U}$ and $u0_g = \varnothing$. If $g = 1$, $u1_g = \varnothing$ and $u0_g = \mathbb{U}$. Here $\mathbb{U}$ is the universal set and $\varnothing$ is the empty set.

If v_g is not constant, we must calculate g using its inputs' unit sets using the rules summarized in Table 1.

Gate	$u1_g$	$u0_g$
AND	$(u1_l \cup u1_r) \cup \{g\}$	$(u0_l \cap u0_r) \cup \{\overline{g}\}$
OR	$(u1_l \cap u1_r) \cup \{g\}$	$(u0_l \cup u0_r) \cup \{\overline{g}\}$
NOT	$(u0_i) \cup \{g\}$	$(u1_i) \cup \{\overline{g}\}$

Table 1. Unit set calculations when g is not constant

3.3. Application of unit sets

With these unit sets computed at every gate, we are able to exploit reconvergence in the two circumstances where CNF-based solvers cannot. Recall Case 1 and Case 2 from Section 2.

In Case 1, the CNF expression's intermediate variables prevented justification of a reconvergent literal. Let us revisit this example and make use of unit sets. Consider the partial circuit in Figure 1(a). Here, employing the calculation rules outlined earlier, we get $u1_u = \{a, b, u\}$, $u1_v = \{b, c, v\}$ and $u1_f = \{b\}$. Hence, if we make the partial assignment $f = 1$ as we did in Section 2, we now observe that b must also be assigned 1. Using unit sets, we can now generate unit propagation (justification) along reconvergent paths.

In Case 2, CNF expression did not reveal that g is constant 0 until attempting to make an assignment on g. However, as presented in Figure 1(b), unit sets reveal this information immediately. Here, $u1_g = \{\overline{a}, b, a, g\}$ and given that $a, \overline{a} \in u1_g$ we know that g must be constant 0 without making any assignment.

Therefore, in calculating unit sets we are able to obtain implications along reconvergent paths in addition to all of the implications generated when using the CNF. This can be applied repeatedly until a fixed point is reached. However, A practical implementation of the dynamic unit set updates resulted in a 10-100x slowdown because of the overhead incurred, which makes a dynamic solution impractical.

4. Static Unit Set Driven Learning

In this section, we present a new static algorithm to exploit these unit set implications without incurring the unit set computation overhead during the running of the SAT solver. We do this by a *pre-processing* step of static learning which is used to generate a set of clauses based on unit sets *prior to* the running of the DPLL algorithm.

These new clauses are then added to the CNF clause database and followed by a traditional CNF SAT solver. In a single pass, generating these unit set clauses is linear in the size of the circuit times depth of the circuit. With the exception of very easy-to-solve instances, the pre-processing time is typically negligible. Once these clauses have been added to the clause database, no additional dynamic updates are needed beyond the normal unit propagation during the DPLL search.

4.1. Basic idea

The main idea of this algorithm is to statically examine the reconvergent circuit structures and generate the clauses that will result in reconvergent unit propagation. Given a pair of paths in a circuit that share the same two ends, but disjoint in the rest of the paths, a *reconvergent section* is the set of all the gates in the two paths. In Figure 2(a), there are a few pairs of such reconvergent paths, such as $(d \rightarrow e \rightarrow g)$ and $(d \rightarrow f \rightarrow g)$.

We now need to learn the conditions that will result in the appearance of the starting gate literal of a pair reconvergent path in the unit set of the reconverging gate. For example, we would like to find out the condition for d to appear in $u0_g$. The condition when the literal of a starting gate such as d may propagate through the intermediate unit sets such as $u0_e$ and $u0_f$ depends on the type of intermediate gates. As e is an NAND gate, d will always be in $u0_e$, whereas f is an NOR gate, c must be 0 for d to be in $u0_f$.

Once d is in both $u0_e$ and $u0_f$, it is in $u0_g$. Thus, we obtain the result that when $c = 0$, $\overline{g} \Rightarrow d$. This corresponds to a new learned clause $(c \vee g \vee d)$, which can be added to the clause database.

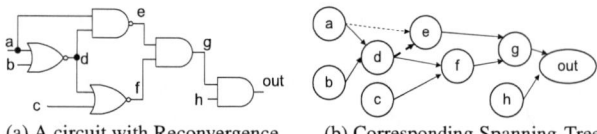

(a) A circuit with Reconvergence   (b) Corresponding Spanning-Tree

Figure 2. Circuit with reconvergence and bypass and its spanning tree

4.2. Determining reconvergent sections

The previous paragraph outlines the algorithm to generate a learned clause corresponding to a reconvergent section. We now need to select such reconvergent sections. However, the number of all reconvergent sections is often exponential in the depth of the circuit. Generating clauses for the reconvergent sections may not be desirable as too many additional clauses may slow down SAT solving by introducing overhead in unit propagation. Thus, we will focus on using a potentially useful subset of reconvergent sections.

Spanning tree algorithm We use a spanning-tree for selecting reconvergent sections. For a DAG (*directed acyclic graph*) G with a spanning tree T, each edge e in G that is not in T forms a reconvergent section with T [6]. Using the small set of non-tree edges to generate the reconvergent sections provides us a reasonable compromise between completeness and the number of unit set clauses.

Without loss of generality, we assume: 1) Each gate has multiple inputs and one output. 2) There are no dead gates, *i.e.* gates with no path to the output of the circuit. 3) There is only a single output in the circuit.

Consider the circuit as a DAG $G(V, E)$, in which vertex set V is the set of all gates and an edge (s, t) represents gate t is a fanout gate of gate s (edge direction is from input to output). So the vertex *out* representing the output of the circuit has a path from every other vertex.

Consider the circuit in Figure 2(a). The circuit is transformed to a directed graph as shown in Figure 2(b). The solid edges represent the edges traversed in a Breath First Search (BFS) from the output, forming a spanning-tree.

To find reconvergent sections in G, we perform a breadth first search from the output gate *out* to build a directed spanning tree $T(V, E^*)$, in which $E^* \subset E$. Consider a link edge (edge that is not in T) $e \in E - E^*$ such as the edge (d, e) in Figure 2(b). Consider the root of the smallest subtree in the spanning tree T that contains both d and e, (g in this case). We have the two reconvergent paths from d to g, one is the path from d to g in T; the other is the link edge (d, e) and the path from e to g in T. These two paths are disjoint except at d and g as T is a tree, which is considered as a *reconvergent section*.

Complexity Note that this method has complexity $O(|E| + (|E| - |V|) \cdot d)$ in which d is the depth of the circuit (the longest path from an input to the output). The Breadth First Search itself requires $O(|E|)$, whereas there are $|E| - |V|$ reconvergent edges, each may cause a backward traversal of no more than d steps. Note that if the circuit is a Binary And-Invertor Graph (BAIG) [15], $|E| = 2|V| - |I| \leqslant 2|V|$, in which $|I|$ is the number of primary inputs. Therefore, the complexity can be simplified to $O(|V| \cdot d)$.

4.3. Unit set clause generation

The detailed algorithm to generate a unit set clause based on a pair of reconvergent paths is as follows:

The reconvergent paths are denoted as $\mathcal{R} = (p_1, p_2)|_{(s,r)}$, in which the fanouts of gate s reconverges at gate r along two paths p_1 and p_2.

The first step is checking the parity of the two paths to identify the type of reconvergence. If p_1 and p_2 have same parity (the number of inversions of p_1 and p_2 are both even or both odd), we denote this reconvergence as *positive*, otherwise, *negative reconvergence*. Case 1 in Section 2 is a positive reconvergence, in which both paths from b to f have no invertors. The reconvergence from a to e in Figure 2(a) is negative, as one path $(a \to e)$ has one invertor while the other $(a \to d \to e)$ has two invertors.

In the algorithm, we need to find the condition under which the source gate of reconvergence s or its inverse (if the number of inversions in both paths are odd) is propagated to the unit-zero set of the reconvergence gate r on both paths. This requires the assignments on the side inputs of the paths that determining g_s to be propagated along both paths. The required assignments are determined as in Section 4.1. The new clause contains the inversion of all required side assignment literals as well as r and $\bar{s}$ (or s if the number of inversions is odd).

For negative reconvergence, both s and $\bar{s}$ occur in $u1_r$ if the side inputs are set to allow such unit set propagation. This means r must be 0 (for AND and NOR gates) or 1 (for OR and NAND gates) if the propagation conditions are met. The static learning clause should include $\bar{r}$ or r respectively as well as the inversion of all the required side assignment literals.

Algorithm 1 shows the framework of the proposed static reconvergence pre-processing algorithm. Building spanning tree by BFS is standard [6]. $path(T, i, r)$ is the path from i to r in tree T. For each link edge (p, q) in link edge set L_T, the common subtree root $r = \text{SubTreeRoot}(T, (i, j))$ in spanning-tree T is found and the reconvergent section $\mathcal{R}$ is built up.

In some reconvergent sections, the fanin of an interim gate of one path may be a gate on the other path. For ex-

Algorithm 1 Static Unit Set based Learning
___
1: $T, L_T \leftarrow \text{SpanningTree}(G(V, E))$
2: **for all** $(i, j) \in L_T$ **do**
3:     $r \leftarrow \text{SubTreeRoot}(T, (i, j))$
4:     $\mathcal{R} \leftarrow \text{path}(T, i, r) \cup \text{path}(T, j, r) \cup (i, j)$
5:     **repeat**
6:         **if** $\exists (p, q) \in L_T$, $p \in \text{path}(T, i/j, r)$, $q \in \text{path}(T, j/i, r)$, $p \neq i, q \neq r$ **then**
7:             $\mathcal{R} \leftarrow \text{path}(T, i, p/q) \cup \text{path}(T, j, q/p) \cup (i, j) \cup (p, q)$
8:         **end if**
9:     **until** (p, q) does not exists any more
10:    $C \leftarrow$ the reconvergent implication clause from $\mathcal{R}$
11:    $\text{AddClause}(C)$
12: **end for**
___

ample, the circuit in Figure 2(a), the negative reconvergent sections from a to g has a "bypass" link edge from d to e (as shown in a thick dotted line in Figure 2(b)). This is less desirable as allowing unit set literal a to propagate through e requires d to be some specific value (here it requires $d = 1$ if d is in $u1_g$), which will eliminate a from the unit sets of d. As a to e has two reconvergent paths, they are a reconvergent section by themselves. We just use this reconvergent section instead of the section formed by a to g. After checking bypass in line 5-9 in Algorithm 1, a reconvergent section is found and its corresponding static learned clause is added to the clause database.

4.4. Limits and parameters of the static algorithm

Using clauses generated by reconvergent pairs cannot completely capture all possible unit set implications as recovergence. Implications from more than two paths cannot be found using the given algorithm. The algorithm may not generate as many implications as a dynamic algorithm with unit set propagation. However, the overhead is significantly lower than a dynamic implementation. By calculating the unit set clauses just once, we do not incur the large overhead of the dynamic case.

To further reduce the overhead introduced by adding these static learned clauses, we introduce two parameters for clause generation. Since reconvergence from long paths is likely to result in long learned clauses, they may be not as useful as reconvergence from relatively short paths. A longer learned clause is also easier to satisfy and more time-consuming in unit propagation. Therefore, we use two parameters to limit the path-length as well as the size of the new clause.

The first parameter k is the maximum length allowed in a reconvergent path. The second one, l, is the maximum number of literals in a statically learned clause. Our exper-

imental results show that k and l have significant impact on running time as well as the quality of the result; therefore they must be chosen properly.

5. Previous Work

Several researchers – in particular, Ganai [13], Lu [17], Thiffault [22], Fu [12] and Jain [14] – have investigated circuit-driven SAT solvers. While our paper is in the same space as their work, none of the above use unit sets for determining implications across reconvergence.

Ganai's work shows a practical speed-up over zChaff by using circuit-driven unit propagation in place of zChaff's original two-literal-watching scheme for CNF implications [13]. However, this does not consider implications across reconvergence. Lu *et al.* discusses using circuit analysis and random simulation in the decision heuristic and when choosing between potential candidates for branching [17]. Thiffault improves Ganai's work [13] by implementing various circuit-specific optimizations, including direct gate-watching and the use of *don't cares* [22]. Fu implements a concise method of exploiting *don't cares* within a CNF solver framework [12], while Jain's work details a novel implication chain algorithm utilizing AND-OR trees [14].

Arora's work [2], however, uses a similar approach as the one we proposed here, referred to as "Extended Backward Implications". In this work, the authors have shown reconvergence implications similar to what we proposed. They generate static learning clauses from reconvergence. However, their approach uses an implication engine to probe possible reconvergent implication candidates. Each variable is assigned and the circuit is simulated. As a result, only direct implications, that is implications requiring no other assignments, may be found through the process. Therefore, their static processing approach can only generate two-literal clauses whereas our approach can find reconvergence clauses of any size.

Eén's work [10] and Manolios' work [19] are also related to our approach. Instead of converting the circuit to CNF directly, these approaches optimize the circuit before make the conversion. Eén's work uses Look-Up-Table based resynthesis whereas Manolios' work uses *NICE (Negation, Ite, Conjunction and Equivalence) DAGs*. These approach are largely orthogonal to our approach as we can apply static learning after these optimizations.

HyperBinRes+eq [3] is another SAT pre-processor that generates new learned clauses based on *HyperResolution* of clauses. In HyperResolution, multi-literal clauses resolve with two-literal clauses to generate new multi-literal clauses. This is similar to our algorithm but works directly on CNF. But they are different in that by limiting multi-literal clauses to resolve with two literal clauses only; reconvergent clauses generated by our static algorithm with side input requirements cannot be generated by HyperResolutions. For example, the clause we generated in Section 4.1 cannot be found by HyperResolution since generating it by resolution requires resolving two three-literal clauses $(\overline{g} \vee e \vee f)$ and $(\overline{e} \vee \overline{a} \vee \overline{d})$.

The unit set propagation and its static learning algorithm is also related to Recursive Learning (RL) used primarily in Automatic Test Pattern Generation (ATPG) as outlined in Kunz's book [16]. RL is used to derive additional implications in a circuit by allowing justifications across reconvergent paths. Under RL, one first chooses an unassigned gate g on a frontier (a boundary between assigned and unassigned sets of gates). g is assigned 1 and then assigned 0 and the backward implications generated in both cases are compared. If an identical implication is found, this implication holds and the implied gate h is set to a constant. A one-step RL will just choose one unassigned gate while a n-step RL need requires 2^n assignments. Multi-step RL is rarely used due to its high overhead. A one-step RL has stronger implication-finding capacity than our static learning clauses, but it requires significant effort to calculate it dynamically whereas our approach only requires one step.

6. Experimental Results

We have evaluated our proposed static pre-processor by testing it in conjunction with the MiniSat 2.0 solver [11] on nearly 500 circuit benchmarks from a variety of academic and industrial sources and have obtained some promising results. MiniSat 2.0 has been chosen because it is a leading modern SAT solver and any improvements obtained by the coupling with our static pre-processing algorithm represent advances over today's best SAT solving capacities. MiniSat 2.0 has a built-in pre-processor derived from SatELite [9]. Our experimental results are shown with MiniSat pre-processing enabled because we believe that their pre-processing is orthogonal to ours.

Our experimentation begins with the conversion of all circuit benchmarks to Binary-And-Inverter-Graph (BAIG) representation [15]. The Tseitin Tautology Transformation [23] is then used to create equivalent CNF instances for MiniSat to solve. Next, we run the same BAIG instances through our pre-processor to create CNF instances that contain extra clauses representative of the circuits' reconvergence and again run MiniSat. We compare the MiniSat execution times of the original instances against the combined MiniSat execution and pre-processing times for pre-processed instances. All experiments were performed on a Pentium 4 2.8GHz processor with 1GB memory running Linux with gcc-4.0.

6.1. Benchmarks

Our test benchmark suite contains 480 instances from a number of different classes, including verification, bounded model checking, software modeling and algebraic problems. The suite is comprised of all of the circuits available to us and contains both satisfiable (indicated by a *) and unsatisfiable (indicated by a †) instances. The benchmarks include aiger*† [5], alloy*† [8], fvp-unsat-1.0†, fvp-unsat-2.0†, sss-sat-1.0* [24], iscas89_comp† [4], necla-sat-ckt*† [18], oablif† [1], parity† and qg6*† [20]. Details on individual circuits can be obtained from these cited sources.

6.2. Optimal results

As discussed in Section 4, our reconvergence-finding algorithm is parameterized by the maximum search depth k from each node and the maximum generated clause length l. To this end, we have iterated through a broad range of (k, l) pairs for each benchmark to determine which pairs were responsible for our pre-processor generating the 'best' set of additional clauses. Here, the 'best' set of clauses were those that led to the lowest MiniSat solving time. We have recorded the set of optimal (k^*, l^*) pairs for each instance with (k, l) typically ranging from $(5, 5)$ to $(20, 35)$. In some instances $(k^*, l^*) = (0, 0)$ is optimal, indicating that these cases are solved fastest without the pre-processor.

In 35% of the cases, we observe reductions in the MiniSat solving time upon the addition of these best clauses and we find an average speed-up of 2.5x across the test suite as a whole. We have separated our suite into an 'easy' and a 'difficult' group – those instances that are and are not solvable by MiniSat alone in less than 1s. We present the observed results for these two groups in Tables 2 and 3. For the easy cases, we note that there is not much difference in the absolute solving times. However, we observe a speed-up of more than 2x in more than 25% of the difficult instances. Even more, the pre-processor allows MiniSat to successfully solve 2 instances which, when not pre-processed, timeout after 2000s. While these results are specific to our particular choice of reconvergence-finding algorithm outlined in Section 4.3, they point to the potential for speed-ups across a variety of benchmarks when circuit reconvergence is considered during CNF generation.

6.3. Pre-processor parameterization

Of course, in practice we cannot simply choose the optimal (k^*, l^*) for an unseen benchmark since the optimal pairs presented above vary significantly across the benchmark suite. Moreover, we have observed that choosing a

Benchmark Set	#	Speed-up factor									
		1	1-1.5	1.5-2	2-3	3-5	5-7	7-10	10-15	≥15	∞
`aiger`	6	5				1					
`alloy`	5		1	1						2	1
`fvp-unsat-1.0`	2	1			1						
`fvp-unsat-2.0`	7		2	1	1	1	1				1
`iscas89_comp`	3		1	2							
`necla-sat-ckt`	47	39	4	2	1					1	
`oablif`											
`parity`	42		1	12	14	12	1	1	1		
`qg6`	30	3	14	5	6					2	
`sss-sat-1.0`	95	66	2	14	10	3					
Total	**237**	**114**	**25**	**37**	**33**	**17**	**2**	**1**	**1**	**5**	**2**

Table 3. Speed-up factors using optimal (k^*, l^*)

Benchmark Set	#	Base time (s)		Time w/ pp. (s)	
		<0.5	0.5-1	<0.5	0.5-1
`aiger`	2		2	1	1
`alloy`					
`fvp-unsat-1.0`	3	3		3	
`fvp-unsat-2.0`					
`iscas89_comp`	6	5	1	6	
`necla-sat-ckt`					
`oablif`	22	22		22	
`parity`	57	19	38	55	2
`qg6`	148	145	3	145	3
`sss-sat-1.0`	5	2	3	2	3
Total	**243**	**196**	**47**	**234**	**9**

Table 2. Solver times for easy instances using optimal (k^*, l^*)

fixed (k, l) results in significant slow-downs in many cases. As such, we are left with the task of optimizing our parameter choices for specific instances based on their characteristics. For each circuit, we track the number of inputs and gates as well as the circuit's minimum depth and width and then formulate a correlation between these circuit parameters and (k, l) using non-linear regression analysis. All four of these metrics are easily obtained in one-pass of the circuit and their measurement incurs a negligible overhead. The offline regression analysis is also of negligible cost. With a correlation between (k, l) and the four circuit properties in hand, we are able to use our pre-processor with the best-possible regressed (k_r, l_r) pair for each specific instance.

For instances where $k_r < 3$, the benchmark is simply transformed to CNF and then solved by MiniSat since a reconvergent cycle must have at least depth 3. In all other cases, we pre-process the benchmark using the parameters (k_r, l_r) and then run MiniSat. As one would expect, regression-derived (k_r, l_r) do not produce as significant speed-ups as the optimal (k^*, l^*). Nonetheless, using a 4th-order regression of the optimal parameters, we obtain a formulation for (k_r, l_r) that performs quite well. In Table 4, we provide the results of static pre-processing with these regressed parameters for the easy instances. The results for the difficult instances are given in Table 5 and we have indicated the number of benchmarks of each class for which the pre-processor was run (i.e. those benchmarks for which $k_r \geq 3$).

Using this approach, we only run our pre-processor with 41% of the suite – in general, we successfully avoid running the pre-processor on benchmarks for which we obtain no speed-up. For the easy instances, there is not much change in the absolute running times. Only 1 instance is extended from the <1s range to the 1-1.5 range. With the difficult instances, speed-ups occur for nearly 70% of the instances in which the pre-processor is used. Only 6.7% of the instances experience significant slow-downs and the 2 most difficult instances are still brought out of the time-out limit.

Benchmark	#	Base time (s)		Time w/ preproc. (s)		
		<0.5	0.5-1	<0.5	0.5-1	1-1.5
`aiger`	2		2	2		
`alloy`						
`fvp-unsat-1.0`	3	3		3		
`fvp-unsat-2.0`						
`iscas89_comp`	6	5	1	5	1	
`necla-sat-ckt`						
`oablif`	22	22		22		
`parity`	57	19	38	50	7	
`qg6`	148	145	3	141	6	1
`sss-sat-1.0`	5	2	3	2	3	
Total	**243**	**196**	**47**	**223**	**19**	**1**

Table 4. Solver times for easy instances using regression-optimized (k_r, l_r)

6.4. General findings

In experimenting with our pre-processor, we have developed an automated methodology for using a static reconvergence algorithm alongside a modern SAT solver. While here we have only had the opportunity to experiment with one pre-processing algorithm and one method of parameter-optimization, the speed-ups in the optimal case and the promising results from regression analysis suggest that reconvergence-based static learning is a useful new technique in CNF SAT solving. We also note that certain classes (*e.g.* `alloy`, `parity`) show strong benefit while others (e.g. `oablif`) show limited benefit. This warrants additional investigation into the characteristics of these benchmarks as well as the recovergence selection. For a new class of circuit SAT problems, test runs and regressions may be employed to determine the optimal k and l's for that class of circuit SAT problems.

978-1-4244-3039-0/08 $25.00 © 2008 IEEE

Benchmark	#	# preproc.	Speed-up factor										
			<0.5	0.5-0.8	0.8-1	1-1.5	1.5-2	2-3	3-5	5-7	7-10	10-15	∞
`aiger`	6	2	1				1						
`alloy`	5	5	2	1							1		1
`fvp-unsat-1.0`	2	1				1							
`fvp-unsat-2.0`	7	6	3	1			1						1
`iscas89_comp`	3	3			1	1	1						
`necla-sat-ckt`	47	11		3	3	4		1					
`oablif`													
`parity`	42	42			1	4	10	13	12		1	1	
`qg6`	30	18		5	6	7							
`sss-sat-1.0`	95	2		1		1							
Total	**237**	**90**	**6**	**11**	**11**	**18**	**13**	**14**	**12**		**2**	**1**	**2**

Table 5. Speed-up factors using regression-optimized (k_r, l_r)

7. Conclusions

In this paper we first demonstrate the limitations of traditional CNF unit propagation in determining backward implications across reconvergent paths in circuits. We then describe a calculus for determining these implications using the concept of unit sets. The dynamic updates to these unit sets is too slow in the inner decision/implication loop of modern SAT solvers. Instead, we propose the use of an alternative method using a static learning pre-processing step. Each reconvergence in a circuit contributes one learned clause. These learned clauses can then be used by a CNF SAT solver. Given that the number of paths and thus reconvergences may be exponential in the circuit depth, we propose a heuristic algorithm that limits the number of learned clauses. We illustrate the practical benefits of using this using a pre-processor based on this algorithm followed by a state-of-the-art SAT solver (MiniSat) on several classes of circuit benchmarks. The experiments show significant speed-up in several benchmark classes.

References

[1] http://openaccess.si2.org.

[2] NECLA verification benchmarks: necla-sat-ckt. http://www.nec-labs.com/~fsoft.

[3] R. Arora and M. S. Hsiao. Enhancing SAT-based equivalence checking with static logic implications. In *Proc. HLDVT'03*, pages 63–68.

[4] F. Bacchus and J. Winter. Effective preprocessing with hyper-resolution and equality reduction. In *Proc. SAT'03*, volume 2919 of *LNCS*, pages 341–355, 2003.

[5] F. Brglez, D. Bryan, and K. Kozminski. Combinational profiles of sequential benchmark circuits. In *Proc. ISCAS'89*, pages 1929–1934, 1989.

[6] R. Brummayer and A. Biere. Local two-level and-inverter graph minimization without blowup. In *Proc. MEMICS'06*, 2006.

[7] T. H. Cormen, C. E. Leiserson, R. L. Rivest, and C. Stein. *Introduction to Algorithms*. MIT Press, 2nd edition, 2001.

[8] M. Davis, G. Logemann, and D. Loveland. A machine program for theorem proving. *Comm. of ACM*, 5(7):394–397, July 1962.

[9] G. Dennis, F. S.-H. Chan, and D. Jackson. Modular verification of code with SAT. In *Proc. ISSTA'06*, 2006.

[10] N. Eén and A. Biere. Effective preprocessing in SAT through variable and clause elimination. In *Proc. SAT'05*, 2005.

[11] N. Eén, A. Mishchenko, and N. Sörensson. Applying logic synthesis for speeding up SAT. In *Proc. SAT'07*, 2007.

[12] N. Eén and N. Sörensson. An extensible SAT-solver. In *Proc. SAT'03*, volume 2919 of *LNCS*, pages 502–518, 2003.

[13] Z. Fu, Y. Yu, and S. Malik. Considering circuit observability don't cares in CNF satisfiability. In *Proc. DATE'05*, 2005.

[14] M. K. Ganai, L. Zhang, P. Ashar, A. Gupta, and S. Malik. Combining strengths of circuit-based and CNF-based algorithms for a high-performance SAT solver. In *Proc. DAC'02*, pages 747–750, June 2002.

[15] H. Jain, C. Bartzis, and E. Clarke. Satisfiability checking of non-clausal formulae using general matings. In *Proc. SAT'06*, volume 4121 of *LNCS*, pages 75–89, 2006.

[16] A. Kuehlmann, V. Paruthi, F. Krohm, and M. K. Ganai. Robust Boolean reasoning for equivalence checking and functional property verification. *IEEE Trans. on CAD of Integrated Circuits and Syst.*, 21(12):1377–1394, Dec. 2002.

[17] W. Kunz and D. Stoffel. *Reasoning in Boolean Networks – Logic Synthesis and Verification Using Testing Techniques*. Kluwer Academic Publishers, 1997.

[18] F. Lu, L.-C. Wang, K.-T. Cheng, and R. C.-Y. Huang. A circuit SAT solver with signal correlation guided learning. In *Proc. DATE'03*, 2003.

[19] P. Manolios and D. Vroon. Efficient circuit to CNF conversion. In *Proc. SAT'07*.

[20] A. Meier and V. Sorge. A new set of algebraic benchmark problems for SAT solvers. In *Proc. SAT'05*, volume 3569 of *LNCS*, pages 321–326, 2005.

[21] M. W. Moskewicz, C. F. Madigan, Y. Zhao, L. Zhang, and S. Malik. Chaff: Engineering an efficient SAT solver. In *Proc. DAC'01*, pages 530–535, 2001.

[22] C. Thiffault, F. Bacchus, and T. Walsh. Solving non-clausal formulas with DPLL search. In *Proc. CP'04*, volume 3258 of *LNCS*, pages 663–678, 2004.

[23] G. C. Tseitin. On complexity of derivations in propositional calculus. In *Studies in Constructive Mathematics and Mathematical Logic, part II*, pages 466–483, 1970.

[24] M. N. Velev and R. E. Bryant. Superscalar processor verification using efficient reductions of the logic of equality with uninterpreted functions to propositional logic. In *Proc. CHARME'99*, volume 1703 of *LNCS*, pages 37–53, 1999.

21st International Conference on VLSI Design

Efficient Linear Macromodeling via Discrete-Time Time-Domain Vector Fitting

Chi-Un Lei and Ngai Wong
Department of Electrical and Electronic Engineering
The University of Hong Kong, Pokfulam Road, Hong Kong
Email: {culei, nwong}@eee.hku.hk

Abstract

We present a discrete-time time-domain vector fitting algorithm, called TD-VFz, for rational function macromodeling of port-to-port responses with discrete time-sampled data. The core routine involves a two-step pole refinement process based on a linear least-squares solve and an eigenvalue problem. Applications in the macromodeling of practical circuits demonstrate that TD-VFz exhibits fast computation, excellent accuracy, and robustness against noisy data. We also utilize an quasi-error bound unique to the discrete-time setting to facilitate the determination of approximant model order.

1 Introduction

VLSI signal integrity analysis constantly requires efficient modeling and simulation of passive structures such as packages and interconnect networks [1] for efficient simulation. The ever-rising operating frequencies have, however, posed serious challenges to simulators due to the emergence of high-frequency effects such as interconnect dispersion and mutual couplings [2]. A full-wave electromagnetic (EM) analysis over the global system is impractical. A continuous-time frequency-domain system identification technique, called vector fitting (VF) [3], is recently utilized for package and interconnect macromodeling [2]. However, frequency-domain macromodeling requires spectral conversion in conventional VF-based signal integrity analysis which involves complicated measurement and relatively long data sequences to be captured. Time-domain vector fitting (TD-VF) [4], similar to other time-domain approximation techniques [5], is developed to directly admit measured or calculated (and usually truncated) time-sampled data. However, TD-VF does not fully exploit the advantages of time-domain approximation, and requires discretization during approximation. The discretization is application-specific and introduces extra arithmetics and approximation errors.

On the other hand, vector fitting has recently been extended to the digital frequency domain, called discrete-time vector fitting (VFz), for digital filter design [6]. It has been shown that VFz exhibits high computational efficiency and produces highly accurate approximants at low model orders. However, the deployment of VFz in the digital time domain has not been explored.

Since physical measurements or electronic simulations typically result in discrete time-sampled data rather than frequency-domain responses, it is advantageous to perform broadband approximation directly in the time domain. In this paper, we reformulate VFz to its discrete-time counterpart, called TD-VFz. TD-VFz respects the discrete nature of input/output sequences and works exclusively in the discrete-time domain, and is particularly suitable for generating macromodels based on truncated finite-difference time-domain (FDTD) solution or transient response analysis. Unlike TD-VF, TD-VFz treats the system as a digital black box and skips the continuous-time interpolation or reconstruction, thereby rendering faster computation and generally comparable or more accurate fitting. Though the fitting is performed in the (digital) time domain, the response is identified with underlying (digital) frequency-domain poles such that global accuracy is achieved further to local matching. The algorithm robustness and convergence are discussed. An error bound is presented for model order selection. Application examples then confirm the remarkable efficiency and accuracy of TD-VFz to generate better macromodel.

2 Vector Fitting

The original continuous-time frequency-domain VF [3] attempts to fit the rational function

$$\left(\sum_{n=1}^{N} \frac{c_n}{s - a_n} \right) + d + se \tag{1}$$

to the desired response $f(s)$ at a set of calculated/sampled data points at frequencies $\{s_k\}$, $k = 1, 2, \cdots, N_s$. The

978-1-4244-3039-0/08 $25.00 © 2008 IEEE

poles a_n and residues c_n are either real or in complex conjugate pairs, and d and e are real. Starting with a set of N prescribed or approximated poles $\{\alpha_n^{(0)}\}$, $n = 1, 2, \cdots, N$, and a scaling function $\sigma(s)$, the problem is linearized into a separable denominator calculation for the ith iteration, namely,

$$\underbrace{\left(\sum_{n=1}^{N} \frac{c_n}{s - \alpha_n^{(i)}}\right) + d + se}_{(\sigma f)(s)} \approx \underbrace{\left(\left(\sum_{n=1}^{N} \frac{\gamma_n}{s - \alpha_n^{(i)}}\right) + 1\right)}_{\sigma(s)} f(s), \quad (2)$$

$i = 0, 1, \cdots, N_T$, where N_T denotes the number of iterations when convergence is attained or when the upper limit is reached. The unknowns, c_n, d, e and γ_n, are solved through an overdetermined linear equation formed by evaluating (2) at the N_s sampled frequency points. The poles are refined iteratively. Any unstable pole is flipped about the imaginary axis to the open left half plane for stability. Upon convergence, the update in $\alpha_n^{(i)}$ diminishes and $\sigma(s) \approx 1$. Recently, it is shown that VF is a reformulation of the Sanathanan-Koerner (SK) iteration [7], an iterative continuous-time frequency-domain system identification technique [14].

There exists two problems in VF: 1) The convergence is affected by the choice of initial poles in (2). Orthogonal vector fitting (OVF) has been proposed to alleviate this degradation [7]. 2) The algorithm is not robust for noisy data. Different techniques have been proposed to alleviate this problem [8,9]. Furthermore, a modified VF (MVF) [10] has recently been proposed for model reduction with estimated time delay (the propagation of the main pulse) to significantly reduce the number of poles.

3 Discrete-Time Time-Domain Vector Fitting

We start by formulating the discrete-time frequency-domain (i.e., z-domain) VF, called VFz [6], and subsequently transform it to the discrete-time time-domain counterpart, called TD-VFz. As in VF, VFz uses partial fractions to seek a rational approximation, $\hat{F}(z)$, to the desired z-domain response $F(z)$, namely,

$$\hat{F}(z) = \left(\sum_{n=1}^{N} \frac{c_n}{1 - z^{-1}a_n}\right) + d \approx F(z), \quad (3)$$

where we assume $\overline{F(e^{j\Omega})} = F(e^{-j\Omega})$, $\forall \Omega \in [-\pi, \pi)$, such that $F(z)$ corresponds to a *real* time-domain sequence. Subsequently, c_n and a_n are either real or in complex conjugate pairs. To ensure stability, the set of poles $\{a_n\}$ in (3) must be within the unit circle or $|a_n| < 1$. As in (2), suppose an initial set of poles $\{\alpha_n^{(0)}\}$, $|\alpha_n^{(0)}| < 1$, are specified, we build

$$\underbrace{\left(\sum_{n=1}^{N} \frac{c_n}{1 - z^{-1}\alpha_n^{(i)}}\right) + d}_{(\sigma F)(z)} \approx \underbrace{\left(\left(\sum_{n=1}^{N} \frac{\gamma_n}{1 - z^{-1}\alpha_n^{(i)}}\right) + 1\right)}_{\sigma(z)} F(z),$$

$$(4)$$

$i = 0, 1, \cdots, N_T$, where $\sigma(z)$ is matched to unity as z approaches the origin. It can be observed that (4) constrains $(\sigma F)(z)$ and $\sigma(z)F(z)$ to share the same set of poles, which in turn implies that the original poles of $F(z)$ are canceled by the zeros of $\sigma(z)$. Therefore, (4) can be re-expressed as

$$\underbrace{\frac{\prod_{n=1}^{N+1} \left(1 - z^{-1}\beta_n\right)}{\prod_{n=1}^{N} \left(1 - z^{-1}\alpha_n\right)}}_{(\sigma F)(z)} \approx \underbrace{\frac{\prod_{n=1}^{N+1} \left(1 - z^{-1}\widetilde{\beta}_n\right)}{\prod_{n=1}^{N} \left(1 - z^{-1}\alpha_n\right)}}_{\sigma(z)} F(z), \quad (5)$$

$$\Rightarrow F(z) \approx \frac{(\sigma F)(z)}{\sigma(z)} = \frac{\prod_{n=1}^{N+1} \left(1 - z^{-1}\beta_n\right)}{\prod_{n=1}^{N+1} \left(1 - z^{-1}\widetilde{\beta}_n\right)}. \quad (6)$$

Subsequently, solving the zeros of $\sigma(z)$ results in, in the least-squares (LS) sense, an approximation to the poles of $F(z)$, i.e., $\{\alpha_n^{(i+1)}\} := \{\widetilde{\beta}_n^{(i)}\}$, which are then fed back to (2) as the new set of known poles for better fitting in the next iteration. Next, we apply an input $X(z)$ to $F(z)$ and let $Y(z) = F(z)X(z)$ be the output. The time-domain relationship is then given by the inverse z-transform,

$$y[k] \approx dx[k] + \sum_{n=1}^{N} c_n x_n[k] - \sum_{n=1}^{N} \gamma_n y_n[k],$$

$$y_n[k] = \left((\alpha_n^{(i)})^k u[k]\right) * y[k], \, x_n[k] = \left((\alpha_n^{(i)})^k u[k]\right) * x[k],$$

$$(7)$$

where $*$ denotes convolution and $u[k]$ is the Heaviside unit step sequence.

3.1 Computing the Real Poles

Continuing from (7), suppose N_s samples of the input and output sequences, $x[k]$ and $y[k]$, are captured, the following overdetermined linear equation is set up,

$$Av = b,$$

$$A = \begin{bmatrix} x_1[0] & \cdots & x_N[0] & x[0] \\ x_1[1] & \cdots & x_N[1] & x[1] \\ \vdots & & \vdots & \vdots \\ x_1[N_s - 1] & \cdots & x_N[N_s - 1] & x[N_s - 1] \\ -y_1[0] & \cdots & -y_N[0] \\ -y_1[1] & \cdots & -y_N[1] \\ \vdots & & \vdots \\ -y_1[N_s - 1] & \cdots & -y_N[N_s - 1] \end{bmatrix},$$

$$v = \begin{bmatrix} c_1 & \cdots & c_N & d & \gamma_1 & \cdots & \gamma_N \end{bmatrix}^T,$$

$$b = \begin{bmatrix} y[0] & y[1] & \cdots & y[N_s - 1] \end{bmatrix}^T, \quad (8)$$

with $N_s \geq 2N + 1$. Using the last N elements of the LS solve of v, i.e., γ_1 to γ_N, $\sigma(z)$ of (4) can be reconstructed, whose zeros, denoted by $\{\alpha_n^{(i+1)}\}$, then form the new set of poles in the next TD-VFz iteration. Similar to the formulation in VF [3], the zeros of $\sigma(z)$ are implicitly obtained as the eigenvalues of

$$
\Psi = \left(\begin{bmatrix} 1/\alpha_1^{(i)} & & \\ & \ddots & \\ & & 1/\alpha_N^{(i)} \end{bmatrix} + \begin{bmatrix} 1/\alpha_1^{(i)} \\ \vdots \\ 1/\alpha_N^{(i)} \end{bmatrix} \begin{bmatrix} \gamma_1 & \cdots & \gamma_N \end{bmatrix} \right)^{-1}
$$

$$
= \begin{bmatrix} \alpha_1^{(i)} & & \\ & \ddots & \\ & & \alpha_N^{(i)} \end{bmatrix} - \begin{bmatrix} 1 \\ \vdots \\ 1 \end{bmatrix} R^{-1} \begin{bmatrix} \alpha_1^{(i)}\gamma_1 & \cdots & \alpha_N^{(i)}\gamma_N \end{bmatrix},
$$

$$(9)$$

where $R = 1 + \sum_{i=1}^{N} \gamma_i$. When only real poles are present, Ψ is a real matrix. To ensure stability, it is required that every $|\alpha_n^{(i+1)}| < 1$. Otherwise, its reciprocal is taken, viz. $\alpha_n^{(i+1)} := 1/\alpha_n^{(i+1)}$, such that the pole is flipped back inside the unit circle. Here a real $\alpha_n^{(i+1)}$ is assumed but the flipping of conjugate poles follows exactly by multiplying two conjugate reciprocals. Till now, only real poles are considered. Special attention must be paid to complex conjugate poles as will be discussed below.

3.2 Modification for Complex Poles

The transfer function $F(z)$ in (3) may, and in fact more often than not, contain complex conjugate poles and residues whose time-domain transforms are also complex conjugate, thus conforming to a real response. Apart from accelerating convergence, allowing complex poles in the (real) TD-VFz arithmetics is critical, if not necessary, as practical digital systems generally have poles distributed in certain sectors in the complex plane. Without loss of generality, we assume $\alpha_2^{(i)} = \overline{\alpha_1^{(i)}}$ and $c_2 = \overline{c_1}$ ($\gamma_2 = \overline{\gamma_1}$), so $x_2[k] = \overline{x_1[k]}$ ($y_2[k] = \overline{y_1[k]}$). However, if these complex quantities are directly used in (8), finite-precision arithmetics would almost always result in inexact cancellation of imaginary parts and lead to erroneous time-domain responses. Consequently, (8) should be modified accordingly to restrict all quantities to be real. To achieve this, (8) is equivalently cast as

$$Av = b,$$

$$
A = \begin{bmatrix} 2\Re(x_1[0]) & -2\Im(x_1[0]) & \cdots & x[0] \\ 2\Re(x_1[1]) & -2\Im(x_1[1]) & \cdots & x[1] \\ \vdots & \vdots & & \vdots \\ 2\Re(x_1[N_s-1]) & -2\Im(x_1[N_s-1]) & \cdots & x[N_s-1] \end{bmatrix}
$$

$$
\begin{bmatrix} -2\Re(y_1[0]) & 2\Im(y_1[0]) & \cdots \\ -2\Re(y_1[1]) & 2\Im(y_1[1]) & \cdots \\ \vdots & \vdots & \\ -2\Re(y_1[N_s-1]) & 2\Im(y_1[N_s-1]) & \cdots \end{bmatrix},
$$

$$v = \begin{bmatrix} \Re(c_1) & \Im(c_1) & \cdots & d & \Re(\gamma_1) & \Im(\gamma_1) & \cdots \end{bmatrix}^T,$$

$$b = \begin{bmatrix} y[0] & y[1] & \cdots & y[N_s-1] \end{bmatrix}^T, \qquad (10)$$

where $\Re(\circ)$ and $\Im(\circ)$ denote the real and imaginary parts, respectively. Modification for other complex conjugate

poles and residues follows similarly. The setup of (10) is eased by the fact that

$$\Re(x_n[k]) = \left(|\alpha_n^{(i)}|^k \cos(\theta_n^{(i)}k) \right) u[k] * x[k], \qquad (11a)$$

$$\Im(x_n[k]) = \left(|\alpha_n^{(i)}|^k \sin(\theta_n^{(i)}k) \right) u[k] * x[k], \qquad (11b)$$

where $\theta_n^{(i)} = \arg(\alpha_n^{(i)})$. $\Re(y_n[k])$ and $\Im(y_n[k])$ are obtained in the same manner by replacing $x[k]$ with $y[k]$ in (11). To compute the zeros of $\sigma(z)$ which now contains complex poles, we apply similarity transform to (9) to bring it back to a real matrix. Each pair of conjugate poles now manifest as a 2×2 diagonal block in Ψ. For example, when $\alpha_2^{(i)} = \overline{\alpha_1^{(i)}}$, Ψ takes the form

$$
\Psi = \begin{bmatrix} \Re(\alpha_1^{(i)}) & \Im(\alpha_1^{(i)}) & \\ -\Im(\alpha_1^{(i)}) & \Re(\alpha_1^{(i)}) & \\ & & \ddots \end{bmatrix} -
$$

$$
\begin{bmatrix} 2 \\ 0 \\ \vdots \end{bmatrix} R^{-1} \begin{bmatrix} \Re(\alpha_1^{(i)}\gamma_1) & \Im(\alpha_1^{(i)}\gamma_1) & \cdots \end{bmatrix}, \quad (12)
$$

and R is obviously real. The entries in the solution of v in (10) can be reused in forming the row matrix in (12), namely,

$$\Re(\alpha_1^{(i)}\gamma_1) = \Re(\alpha_1^{(i)})\Re(\gamma_1) - \Im(\alpha_1^{(i)})\Im(\gamma_1), \qquad (13a)$$

$$\Im(\alpha_1^{(i)}\gamma_1) = \Re(\alpha_1^{(i)})\Im(\gamma_1) + \Im(\alpha_1^{(i)})\Re(\gamma_1). \qquad (13b)$$

3.3 Reconstructing the Rational Function

Suppose a converged set of poles $\{\alpha_n^{(N_T)}\}$ are obtained, the final step is to reconstruct the rational function $\hat{F}(z)$ in (3). Referring to (4) and (7), we should now have $\sigma(z) \approx 1$ and

$$y[k] \approx dx[k] + \sum_{n=1}^{N} c_n x_n[k], \, k = 1, 2, \cdots, N_s. \qquad (14)$$

The residues c_n of $\hat{F}(z)$ are computed in the same manner as in (8) or (10) by LS fitting, except that the last N columns in A and the last N elements in v are omitted, leaving the unknowns c_1 to c_N and d. Besides reconstructing the rational function, the parameters can be easily converted into causal parameters and used for recursive convolution for fast simulation [11].

3.4 Algorithmic Convergence and Model Order Selection

Similar to the equivalence between VF and SK iteration [7], VFz can be regarded as a reformulation of the rational function fitting procedure called Steiglitz-McBride (SM) iteration [15]. The convergence and error bound properties of VFz, as will be shown below, then carry over to

471

978-1-4244-3039-0/08 $25.00 © 2008 IEEE

TD-VFz as they are equivalent representations in different domains. First, given a transfer function or response $F(z)$, SM iteration replaces the nonlinear LS approximation objective $\widehat{G}_{L_2} = \Sigma_{k=1}^{N_s} |F(z_k) - \frac{P(z_k)}{Q(z_k)}|^2$ with a linearized $\widehat{G}_{SM}$ where

$$\widehat{G}_{SM} = \sum_{k=1}^{N_s} \frac{1}{\left|Q^{(i-1)}(z_k)\right|^2} \left| Q^{(i)}(z_k)F(z_k) - P^{(i)}(z_k)\right|^2. \quad (15)$$

Here $P^{(i)}$ and $Q^{(i)}$ are respectively the numerator and denominator determined during the ith SM iteration (thus $Q^{(i-1)}$ is assumed predetermined). Although $\widehat{G}_{SM}$ is not equivalent to $\widehat{G}_{L_2}$, by using the triangle inequality, if we approximate $F(z)$ by an Nth-order system, we get $\|\widehat{G}_{L_2} - \widehat{G}_{SM}\|_2 \leq 2\sigma_{N+1}$. Here σ_i, which denotes the ith Hankel singular value (HSV) of a Hankel-form matrix constructed by the time-domain impulse coefficients h_n's of $F(z)$, measures the significance of the ith approximant order [13]. In general, SM iteration converges to a near-global-optimal approximant in the LS sense for noise-free data, with an *a priori* error bound for an Nth-order approximant,

$$\min_{\deg(\frac{P}{Q}=N)} \left(\frac{1}{2\pi} \int_{-\pi}^{\pi} \left| F(e^{j\omega}) - \frac{P^{(i)}(e^{j\omega})}{Q^{(i)}(e^{j\omega})} \right|^2 d\omega \right)^{1/2} \leq \sigma_{N+1}. \quad (16)$$

Such error bound is important as it provides a certificate for the approximation accuracy and can be used to select the approximant order. That is, the order N of $\hat{F}(z)$ should be selected such that $\sigma_{N+1} \approx 0$. This also constitutes an analytical and non-heuristic way to determine the model order which, as far as we know, is not available in VF or TD-VF.

Moreover, unlike point-matching (curve-fitting) algorithms, a crucial feature in TD-VFz is that it incorporates the (digital) frequency-domain system poles in its (digital) time-domain fitting machinery. A consequence is that further than *local accuracy* in the context of the *finite* captured data set, which forms the objective function of most direct point-matching schemes, TD-VFz achieves *global accuracy* by simultaneously identifying the intrinsic system poles.

3.5 Noise Robustness

Noise robustness and convergence of an algorithm can be attributed to two factors: 1) numerical conditioning of the algorithm itself in finite-precision arithmetics, e.g., the solve of the overdetermined equation in (8) or (10); 2) sensitivity of the algorithm against noise-corrupted input data.

Although OVF has been proposed to improve the conditioning of the LS solve in traditional VF [7], it is known that the VF convergence will still be hampered even by a small spectral noise, e.g, when SNR=30dB [8], due to the normalization of the unity basis [the '+1' term in (2)]. This can be alleviated by relaxing that unity basis to a free variable and inserting a relaxed constraint [9]. However, extra processing and computation are required in these schemes.

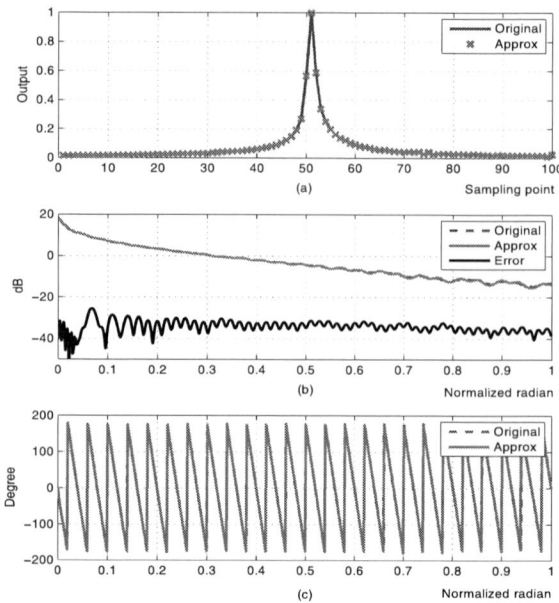

Figure 1. Backplane: (a) time response for first 0.67ns; (b) & (c) magnitude and phase responses in frequency domain.

In (TD-)VFz, instead of approximating in the whole s-domain wherein the poles can take on largely dynamic magnitudes, the z-domain poles are constrained inside the unity circle which lead to significantly improved numerical conditionings. Moreover, the scaling function $\sigma(z)$ in (4), which is sensitive to noisy data in VF, is mapped by the inverse z-transform to $\sigma[k] = \left(\sum_{n=1}^{N} \gamma_n (\alpha_n^{(i)})^k u[k] \right) + \delta[k]$ in TD-VFz. Insensitivity of $\sigma[k]$ against noisy data ($x[k]$ and $y[k]$) is obviously but the filtering action of the convolution operations in (7). These robust features of TD-VFz are verified in the examples that follow.

4 Numerical Examples

The proposed TD-VFz algorithm is coded in Matlab m-script (text) files and run in the Matlab 7.4 environment on a 1GB-RAM 3.4GHz PC. The test example arises from modeling a 40.5-inch differential transmission channel on a full mesh ATCA backplane [10]. The time-domain response is obtained by an excitation from 50 MHz to 15GHz. The signal is normalized and fitted using TD-VFz with a 62-pole approximant. Time samples are taken at 6.7ps intervals for the first 425 points. It takes 8 seconds and 21 iterations for TD-VFz to reach convergence. Fig. 1(a) plots the fitted

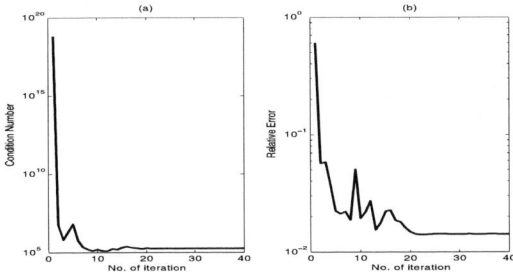

Figure 2. (a) Condition number of overdetermined system matrix and (b) relative error in LS solution.

Table 1. Backplane: comparison between TD-VFz and other algorithms.

	TD-VFz	VF	MVF	TD-VF
Number of poles	62	204	66	64
Relative error (dB)	-37	-39	-38	-36
CPU time (sec.)	8	44	33	9

time response and shows the excellent accuracy of TD-VFz. This is further verified in Fig. 1(b) & (c) where the fitted results are transformed to the frequency-domain wherein the error is seen to be negligible. This demonstrates the power of TD-VFz in achieving multi-domain analyses simultaneously (viz., the time, frequency, and pole-plane domains) with fast computation and small models. In contrast, obtaining the time-domain response from frequency-domain data as in VF is nontrivial and may require complicated Hilbert transform due to causality consideration [12].

We also test the same example with VF [3], MVF [10] (which considers time delay estimation to reduce approximant size), and TD-VF [4]. The results are shown in Table 1. It is seen that TD-VFz is the fastest, and produces the smallest model with similar accuracy to other algorithms. Fig. 2 shows the condition number of the system matrix in (10) and the relative error in this LS solve in each TD-VFz iteration. The fast pole convergence of TD-VFz results in the rapid drop of the condition number after the first iteration to the order of 10^5, which is much more robust than VF whose condition number remains in the order of 10^{15} in the first few iterations [7].

Next, we study the robustness of TD-VFz under noisy data. We repeat the example but with the output sequence $y[k]$ corrupted by white noise under an SNR of 35dB. In this case, TD-VFz converges in 30 runs in 9 CPU seconds with 500 sampling points and a 62-pole approximant, ending up at a relative error of -24dB. The magnitude of the scaling function $\sigma[k]$ in the first TD-VFz iteration is shown

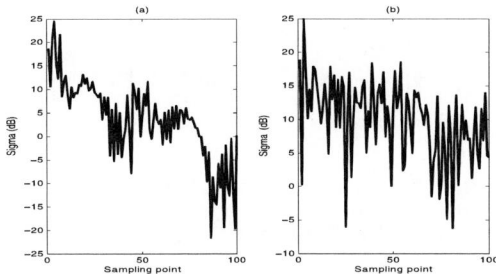

Figure 3. Magnitude of the scaling function $\sigma[k]$: (a) noise-free; (b) with noisy data (SNR=35dB).

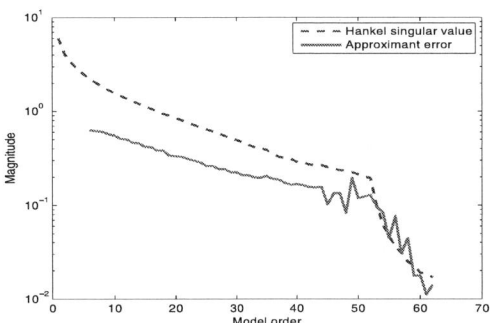

Figure 4. Hankel singular values and relative error of approximant.

in Fig. 3, for both the noise-free and noisy cases. We note that the ideal $\sigma[k]$ should be a delta function which corresponds to $\sigma(z) = 1$ under TD-VFz convergence. Apparently, the level of noise thus introduced in $\sigma[k]$, due to the filtering action discussed in Section 3.5, is not influential on the delta-function envelope of $\sigma[k]$ and has not caused too much impact to the TD-VFz convergence.

Finally, we investigate the use of the HSV in guiding the model order selection. Fig. 4 shows the HSVs found by the impulse response of the backplane system (cf. Section 3.4), as well as the relative error of different approximant orders. An evidential correlation can be seen between these two parameters, with similar slope and tipping point. We can therefore choose the number of poles, i.e., the approximant order, to be bigger than or equal to the corner cut-off point. Therefore, we choose 62-pole approximant in the numerical example. Such HSV design guideline, as far as we are aware, is unavailable in the continuous-time VF and its variants like MVF, OVF, TD-VF etc.

The TD-VFz-fitted rational macromodels can then be easily integrated into popular mixed-signal simulation environments like Verilog-A, Synopsys Saber, Matlab etc. For

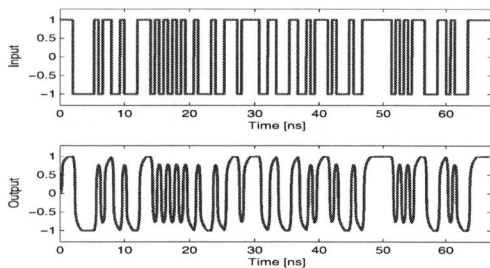

Figure 5. Backplane: random input and the corresponding output.

example, an output signal response of a high frequency random input signal to the backplane example is shown in Fig. 5 using Matlab.

5 Conclusions

Some remarks are in order:

1. TD-VFz is the fastest macromodeling algorithm among various VF variants studied here. In our implementation, TD-VFz is always about 8% faster than TD-VF as there is no need to discretize the continuous-time convolution integrals.

2. Perhaps more importantly, the convergence error bound which translates into an analytical model order selection guideline, as far as we are aware, is unique to the discrete-time domain.

3. As discussed and exemplified, restriction of poles inside the unit circle renders much better numerical conditioning of (TD-)VFz over the conventional (TD-)VF which works with poles in the infinite open left half plane.

4. Time-domain fitting, due to its inherent convolutional filtering, is much more robust against noisy data than frequency-domain fitting.

In summary, this paper has generalized the vector fitting (VF) algorithm to its discrete-time time-domain counterpart, called TD-VFz, for the fast and robust linear macromodeling of port-to-port responses in time-sampled data. The partial fraction basis in digital time domain improves the numerical conditioning, convergence and robustness under noisy data. A quasi-error bound unique to the discrete-time formulation allows deterministic choice of approximant order. Application examples have confirmed that TD-VFz exhibits efficient computation and produces highly accurate approximants, in terms of both time and frequency responses.

Acknowledgment

This work was supported in part by the Hong Kong Research Grants Council under Project HKU 717407E, by the University Research Committee of The University of Hong Kong.

References

[1] L. P. M. Celik and A. Obadasioglu. *IC Interconnect Analysis*. Norwell, MA: Kluwer, 2002.

[2] D. Ioan, G. Ciuprina, M. Radulescu, and E. Seebacher. Compact modeling and fast simulation of on-chip interconnect lines. *IEEE Trans. Magn.*, 42(4):547–550, Apr. 2006.

[3] B. Gustavsen and A. Semlyen. Rational approximation of frequency domain responses by vector fitting. *IEEE Trans. Power Delivery*, 14(3):1052–1061, July 1999.

[4] S. Grivet-Talocia. Package macromodeling via time-domain vector fitting. *IEEE Microwave Wireless Compon. Lett.*, 13(11):472–474, Nov. 2003.

[5] T. Wu, C. Kuo, H. Chang, and J. Hsieh. A novel systematic approach for equivalent model extraction of embedded high-speed interconnects in time domain. *IEEE Trans. Electromagn. Compat.*, 45(3):493–501, Aug. 2003.

[6] N. Wong and C.-U. Lei. IIR approximation of FIR filters via discrete-time vector fitting. *IEEE Trans. Signal Processing*, accepted.

[7] D. Deschrijver, B. Haegeman, and T. Dhaene. Orthonormal vector fitting: A robust macromodeling tool for rational approximation of frequency domain responses. *IEEE Trans. Adv. Packag.*, 30(2):216–225, May 2007.

[8] S. Grivet-Talocia and M. Bandinu. Improving the convergence of vector fitting for equivalent circuit extraction from noisy frequency responses. *IEEE Trans. Electromagn. Compat.*, 48(1):104–120, Feb. 2006.

[9] B. Gustavsen. Improving the pole relocating properties of vector fitting. *IEEE Trans. Power Delivery*, 21(3):1587–1592, July 2006.

[10] R. Zeng and J. Sinsky. Modified rational function modeling technique for high speed circuits. In *IEEE MTT-S Int. Microwave Symp. Digest*, June 2006.

[11] S. Luo and Z. Chen. Extraction of causal time-domain network parameters from their band-limited frequency-domain counterparts using rational functions. *IEEE Trans. Circuits Syst. I*, 52(6):1205–1210, June 2005.

[12] R. Mandrekar and M. Swaminathan. Delay extraction from frequency domain data for causal macro-modeling of passive networks. In *Proc. Int. Symp. Circuits and Systems*, volume 6, pages 5758–5761, May 2005.

[13] P. Regalia and M. Mboup. Undermodeled adaptive filtering: an a priori error bound for the Steiglitz-McBride method. *IEEE Trans. Circuits Syst. II*, 43(2):105–116, Feb. 1996.

[14] C. Sanathanan and J. Koerner. Transfer function synthesis as a ratio of two complex polynomials. *IEEE Trans. Automat. Contr.*, 8(1):56–58, Jan. 1963.

[15] K. Steiglitza and L. McBride. A technique for the identification of linear systems. *IEEE Trans. Automat. Contr.*, 10(4):461–464, Oct. 1965.

978-1-4244-3039-0/08 $25.00 © 2008 IEEE

21st International Conference on VLSI Design

Formal verification of a public-domain DDR2 controller design

Abhishek Datta
Oski Technology
New Delhi, India
abhishek@oskitech.com

Vigyan Singhal
Oski Technology
Fremont, CA, USA
vigyan@oskitech.com

Abstract

This paper demonstrates a formal verification-planning process and presents associated verification strategy that we believe is an essential (yet often neglected) step in an ASIC or SoC functional formal verification flow. Our contribution is to present a way to apply the verification planning process and a set of abstraction techniques on a non-trivial open-source example (the Sun OpenSPARC™ DDR2 controller). The process and verification strategy can be applied to DDR2 controllers in particular and generalized for other designs.

1. Introduction

Model checking (referred to as Formal verification elsewhere in this paper) usage has been increasing in recent times. This is partly due to the increase in performance and capacity of formal engines and partly due to the standardization of property specification languages. However, formal verification can still not lay claim to mainstream acceptance as a necessary part of verification sign-off. This can be attributed to a dearth of published material on three key aspects of effective formal verification - structured verification *planning*, re-usable *verification IP* and good verification *strategy*.

Recently, there have been several attempts made to address the verification-planning aspects (Foster et al [1]). Formal verification IP re-use has also been reasonably well-documented [6]. However, there remains a need to disseminate effective verification strategies and methods to simplify complexity without necessarily compromising the completeness of the desired proof. Without good strategy most formal verification projects will be overwhelmed by complexity issues [7]. This paper is intended to address this perceived need.

We use the open-source OpenSPARC T1 DDR2 controller design [2] as a context to demonstrate formal verification planning and implementation. The design is of comparable complexity to many modern memory controllers which, it is hoped, renders the presented ideas immediately applicable to the verification of similar designs.

We discuss the re-use of formal verification IP to shorten schedules, use of interesting design abstractions and explain how cut-points can be a powerful compositional reasoning technique. The verification strategy employed to handle the design complexity is described in detail. Most memory controllers have common design strategies, even though they have been designed independently. This is partly because common design techniques permeate the industry over time and partly because of the requirements imposed by the standard specifications [4]. Therefore, we believe that the complexity issues described in this paper are pertinent to most DDR controller implementations. By extension, the techniques used to tackle the complexity are applicable to the whole class of memory controller designs and can be leveraged for other design types.

The design and verification material discussed in this paper has been made freely available and published on a public website. Readers can download the material for further experimentation and jump-starting their own verification efforts.

2. Review of the OpenSPARC design

The OpenSPARC-T1 processor is a highly integrated processor that implements the 64-bit SPARC V9 architecture. The processor, codenamed "Niagara" has been made publicly available [2] by Sun to promote participation in SPARC processor architecture and application design. The design is representative of cutting edge processor architecture with multi-threading, multi-CPU and other advanced features.

Our focus area comprises the four on-chip dynamic random access memory (DRAM) controllers directly

475

978-1-4244-3039-0/08 $25.00 © 2008 IEEE

interfacing to the double data rate-synchronous DRAM (DDR2 SDRAM) modules. The controllers implement 144-bit interface per channel with 25 GB/sec peak total bandwidth. The downloadable package for the chip contains the Verilog RTL source, documentation and a comprehensive simulation-based regression environment.

3. OpenSPARC DDR2 Controller

The functional block diagram of the DDR2 controller is shown below.

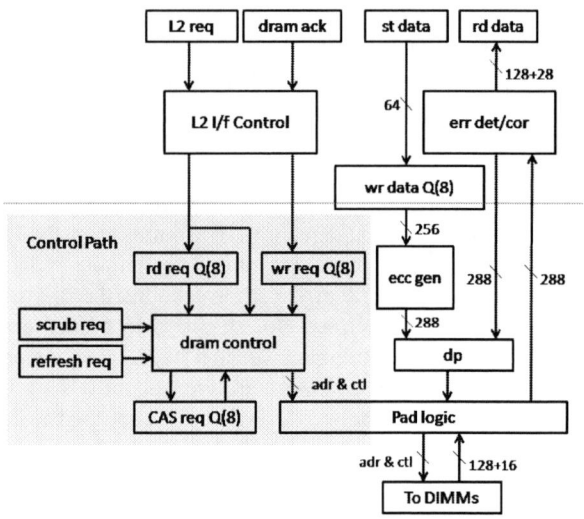

Figure 1. DDR2 Controller Functional Diagram

The shaded region in the figure (sourced from [3]) above shows the address and control logic block inside the controller. There exists high concurrency in the control path due to the multiple independent sources of command words to the DDR2 DIMMs. In order of arbitration priority, these sources are –

o Periodic Refresh requests.
o Pending CAS Requests.
o Scrub Row-Address-Strobe (RAS) requests.
o Incoming Read RAS requests.
o Incoming Write RAS requests.

The read-write priority can change depending upon per-bank write-starvation counters and handling of write-before-read hazards.

This concurrency in the design makes the control logic in the shaded box a good candidate for formal verification. Also, the control logic is encapsulated as a separate verilog module which means that we need make no effort to isolate the target logic.

The data-path contains elements like ECC-based error correction which due to the arithmetic operations involving large data-paths is not ideal for formal

analysis. Suitable abstractions, such as black-boxing the ECC module and applying data-tagging patterns would make the data-path more amenable to analysis. However, for the purposes of this paper, we shall restrict the discussion to the control logic in the shaded area in *Fig. 1*.

4. Formal Verification-planning Review

The initial planning stages of a 7-step verification plan [1] are shown in *Fig. 2*.

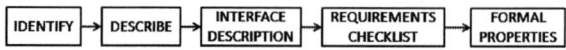

Figure 2. Initial stages of a structured plan

The intuitive steps impose a structure upon the planning effort. We briefly outline the preliminary steps for the DDR2 controller verification.

Identification involves determination of design components that are in the formal verification "sweet-spot". Blocks that have concurrency with limited sequential depth are good candidates. We have already identified the DDR2 control path logic as our target for the reasons mentioned in Section 3.

Description involves a high-level functional specification of the target block. This can be derived from internal implementation specifications as well as the standard protocols that the target implements. In our case, the JEDEC DDR2 protocol specification [4] serves as the interface description.

Interface description refers to the documentation of the inputs and outputs of the target block. The DRAM section of the OpenSPARC micro-architecture document [3] describes the interface in detail. The point of the interface listing is to identify missing constraints (are all my inputs constrained appropriately?) and missing assertions (do I have assertions on all my outputs, wherever applicable?).

Generating *requirements* refers to the creation of a set of English language properties derived from the high-level design description. Typically, these are classified as interface and end-to-end requirements. We shall restrict our verification scope to the DDR2 interface. A set of English requirements for the control logic might read as follows –

1. *Activate cannot be issued to a non-Idle bank.*
2. *Read and write commands can only be issued to active banks.*
3. *Check that the minimum Write to Read turnaround time is (CL-1) + (BL/2) + T_WTR ...*

978-1-4244-3039-0/08 $25.00 © 2008 IEEE

The next step is to translate the requirements to a *formal description* i.e. a formal property specification language. We use a pre-verified set of DDR2 properties, packaged as verification IP, in this effort.

5. Review of DDR2 Formal VIP

The DDR2 protocol is an industry standard specification for SDRAM memory modules [4]. Compliant interface implementations must obey the protocol rules in the specification. A set of DDR2 properties must therefore be re-usable in all DDR2 verification. The re-use of proven, canned sets of properties is an important element of formal verification planning and re-use. It can significantly shorten verification schedules with better ROI since less time is spent implementing properties and verifying correctness.

The DDR2 Formal VIP is articulated into module-level and per-bank properties with minimum auxiliary code sharing between different properties. There are two approaches to implementing property sets for non-trivial protocols that require supporting RTL logic to track design states. One approach is to code one or more monolithic FSMs and then writing properties that are derived from the states (or state transitions) in the large FSMs. The other approach is to minimize the shared logic between the different properties. The former approach is considered better for implementing *constraints* and the latter for *assertions*.

Since the DDR2 control path has only outputs, all our properties are assertions. The approach was to come up with a set of properties that each target distinct design attributes, but together cover all of the protocol. This minimizes code sharing and duplication. It keeps the logic cone of the assertions pared down to the minimum relevant state elements.

6. Verification Overview

Some points need to be made about the verification effort.

We had no access to the designers of the block and an element of guesswork was involved in constraining the internal interfaces. Further, we had to spend more time looking through the RTL than normally required. The functional specification of the block [4] and the simulation-based regression infrastructure were well-documented. We restricted our efforts to the control path in the controller block. Also, we focused solely on the DDR2 *interface* properties of the block.

The total effort comprises about *three weeks* of single person activity. The shortened schedule owes a

great deal to the availability of DDR2 verification IP and reasonably good documentation.

7. Verification Strategy

In this section we present a step-by-step discussion of the verification strategy. Note that while some of the elements, like initialization and exploiting symmetry, were anticipated *a priori*, the other implementation-specific abstractions were refined as we gained familiarity with the target design. This is fairly typical of the verification process.

7.1. CSR Value Selection and Initialization

The DDR2 protocol specifies a large set of timing parameters. These parameters are typically stored in Controller Core CSRs accessible through a register programming interface. These (and other) CSRs need to be initialized by software before normal operation can commence. The programming of the registers is well-defined and typically done during boot-up. This highly sequential operation of significant depth is best done in simulation. We use the OpenSPARC simulation environment to take the design through initialization to an initial "normal" operation state from which point the formal engines take over.

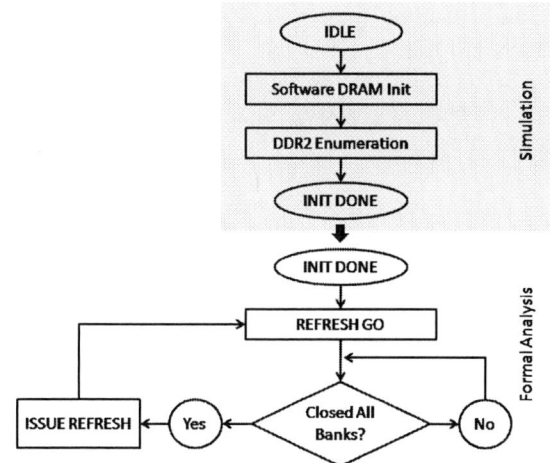

Figure 3. Simulation-based Design Initialization

Once initialized, it is important to block off the register write interface so that the CSRs do not change during formal analysis. Suitable constraints are written on the design inputs to disable register writes.

Further, the DDR2 protocol defines an enumeration sequence of command words to initialize the DDR2 SDRAM module. A DDR2 controller cycles through this sequence after boot-up. Once again, this sequential

477

978-1-4244-3039-0/08 $25.00 © 2008 IEEE

"hump" needs to be crossed before we can begin formal analysis of the controller design. We again turn to simulation to take the design over the enumeration sequence. The initial state for our formal analysis is the state of the design after CSR initialization and DDR2 enumeration sequence is complete.

7.2. Identifying Formal Sweet-spots

Fig. 4 shows the DDR2 protocol state-diagram. The states and transitions outside the shaded area involve significant sequential depth that can compromise formal analysis. The design inputs were constrained such that these state transitions are avoided.

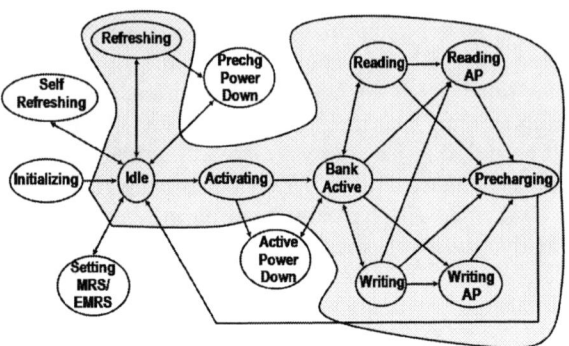

Figure 4. DDR2 Formal Sweet-spots

Simulation-based techniques are very good at exercising well-defined state transitions with high sequential depth that are avoided in formal analysis. This demonstrates how the two verification techniques are complimentary to each other.

7.3. Exploiting Design Symmetry

Design symmetry can be used to reduce analysis complexity. For instance, a 32-bit data-path can be reduced to a single bit if the data-path is manipulated atomically (no bit-slice manipulation). In this case, the symmetry of the data-path allows us to consider a single bit in lieu of the 32 bits.

The DDR2 controller is capable of accessing four DDR2 modules concurrently with a 4-bit chip-select. The symmetry of the implementation means that if we verify the correctness for a *single* module (one bit of chip-select) while allowing the other modules to be accessed (but not monitored) then we have reasonable confidence in the completeness of the proof.

Further, the DDR2 controller implements eight banks. Similar to the chip-selects, there is symmetry in the per-bank queue management logic. Verifying the operation for a single bank would be equivalent to verifying each bank separately. Note that to exploit this symmetry the formal VIP must have corresponding symmetry in its implementation of per-bank properties.

7.4. Abstracting Large Counters

Counters can add significant sequential depth to formal analysis. The DDR2 controller has a surfeit of counters. Most of these are small but the following are large enough to significantly impact performance –

Refresh Interval – The DDR2 controller issues periodic Refresh commands to the DDR2 modules. The period is established by a 13-bit counter.

Scrub Interval – The controller issues internal scrub commands at periodic intervals. The period is established by a 12-bit counter (see elaboration below).

Write Starvation Counters – The controller gives higher priority to Read as opposed to Write. To prevent write starvation, a 6-bit counter begins counting after the Write Queue is full. A pending Write has priority over Read when the counter is full.

CAS Latency Counters – Each bank has an 8-bit CAS latency counter that blocks further RAS picks if a CAS has been pending for more than 64 clock cycles.

These counters can add substantial amount of sequential depth (diameter) to the analysis. As a result, the analysis might not converge. We show how cut-points can serve as equivalent abstractions.

Consider the scrub-interval decision logic –

```
assign que_scrb_read_en = que_scrb_time &
    ~(que_scrb_write_valid | que_prev_scrb_wr_pending);

assign que_scrb_time =
    (que_scrb_cnt[11:0] >= chip_config_reg[20:9]) &
    que_data_scrub_enabled & ~que_hw_selfrsh;
```

The listing shows how the decision about the scrub timing interval is dependent on the value of the "*que_scrb_time*" net. This net is driven by an equation involving the comparison of the interval counter with a programmed interval in a CSR. If we place a *cut-point* on this net i.e. we direct the formal engines to treat this net as if it were a primary input, then we remove the interval counter from the logic cone of (relevant) assertions on the controller outputs. This increases the likelihood of analysis convergence.

We have replaced the counter-driven logic by a simpler abstraction (an arbitrary un-driven net) that eliminates the sequential depth imposed by the counter. An analysis failure might arise due to the cut-point, which is easy to debug. On the other hand, a proof obtained with the cut-point is applicable to the original design as well. This is because we have replaced the specific counter logic that drives (constrains) the net, with an abstraction that under-constrains the analysis.

For each of the counters listed earlier, a similar cut-pointing was done to eliminate them from the analysis.

7.5. Handling Address Comparisons

The controller assigns higher priority to Read over Write unless there is a Read after Write (RAW) hazard. It handles RAW hazards by comparing an incoming Read address with all the write addresses currently pending in the write queue. The corresponding decision logic is shown below –

```
assign que_b0_wr_index_pend = (
    ( que_rd_addr_picked[31:0] == {writeqbank0addr0[35],
        writeqbank0addr0[33], writeqbank0addr0[31:5],
        writeqbank0addr0[2:0]}) & writeqbank0vld0_arb
    | ...
    | (que_rd_addr_picked[31:0] == {writeqbank0addr7[35],
        writeqbank0addr7[33], writeqbank0addr7[31:5],
        writeqbank0addr7[2:0]}) & writeqbank0vld7_arb)
    & que_wr_addr_picked_vld & que_rd_addr_picked_vld ;
```

The net "*que_b0_wr_index_pend*" is asserted when a RAW hazard is detected. The 32-bit address comparison pulls in all the 32-bits of the address space, not to mention the recorded address bits in the Write queues. Consider the modified equation -

```
wire free;  // Undriven net considered primary input
assign que_b0_wr_index_pend = (
    writeqbank0vld0_arb | writeqbank0vld1_arb | ...
    writeqbank0vld6_arb | writeqbank0vld7_arb)
    & que_wr_addr_picked_vld & que_rd_addr_picked_vld
    & free;
```

We have removed the address comparison terms from the equation driving the net. Instead we introduce a *free* variable that can be arbitrarily asserted by the formal analysis to introduce a hazard condition. This simpler abstraction does not rule out any hazard condition that is admitted by the original equation. Therefore, a proof with the simpler equation is applicable to the original design. We have accomplished two things with this abstraction –

First, we have eliminated the 32-bit address words stored in the write queue from the logic cone of relevant assertions. Second, we have reduced the need to monitor the 32-bit incoming read address. The *free* net removes the address logic from the simplified hazard equation.

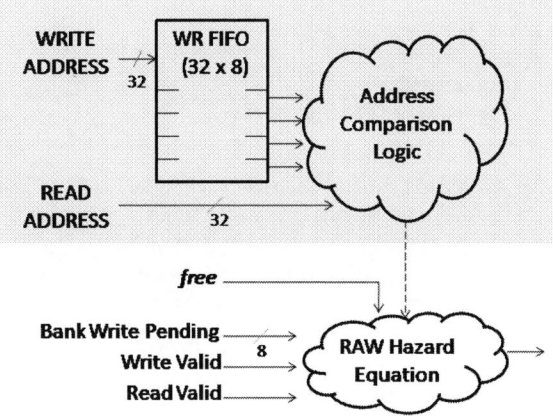

Figure 5. Eliminating Address-bits from COI

Address comparisons of this type are common. For instance, code that checks whether an address falls within an acceptable range. This technique can be judiciously used in these situations to eliminate complexity without compromising the quality of the proof.

7.6. Applying Verification Patterns

Patterns can be defined as generalized solutions in a given engineering domain that find recurring application to the problems of that domain. We illustrate the use of one such verification pattern, the *Floating Pulse* [5], which was used to implement a DDR2 SVA property that uncovered a potential bug in the control-path. The pattern expresses intent to –

*Specify that a single bit value can be asserted for only **one** cycle in **any** cycle of an infinite sequence.*

We use the pattern to implement the following DDR2 property –

No more than 4 ACTIVATE commands may be issued to the DDR2 SDRAM within a window of T_FAW clock cycles.

We can use the Floating Pulse pattern here to fix the lower bound of our sampling window of width *T_FAW* clock cycles. Then a counter that counts off the *T_FAW* cycles is used to define the sampling window within which we check if more than 4 ACTIVATE were issued.

We first setup a *pulse* element through the following assumption. This is an SVA implementation of the *Floating Pulse* pattern.

```
wire pulse;
assume_floating_pulse_pattern :
  assume property (
    @(posedge clk)
    pulse ##0 (1'b1)[*0:$] |=> !pulse
);
```

We *bind* the *pulse* to (any and all) occurrence of an ACTIVATE command. The *pulse* assertion marks the start of our sampling window. We then load a counter at the *start* of the window and begin counting down.

```
always @ (posedge CK or negedge RST_N) begin
  if (!RST_N)
    num_banks_activated <= 4'd0;
  else if(pulse && activate_issued)
    num_banks_activated <= 4'd1;
  else if (activate_issued && !num_banks_activated )
    num_banks_activated <= num_banks_activated + 4'd1;
end

always @ (posedge CK or negedge RST_N) begin
  if (!RST_N) tfaw_counter <= 4'd0;
  else if(tfaw_counter != 4'd0)
    tfaw_counter <= tfaw_counter - 4'd1;
  else if(pulse && activate_issued) // start of window
    tfaw_counter <= T_FAW - 4'b1;
end
```

The following assertion ties everything together.

```
property p_tFAW_rolling_window;
  @(posedge CK) disable iff (!RST_N)
  !((tfaw_counter != 0) && (num_banks_activated > 4'd4));
endproperty
```

This property failed for the OpenSPARC DDR2 controller design. The controller issued five ACTIVATE commands within a T_FAW window violating the protocol. It is possible that some element outside the design controls the ACTIVATE commands but purely in the context of the controller design this can be construed as an error of omission. *Fig. 6* shows the property violation on the last clock cycle.

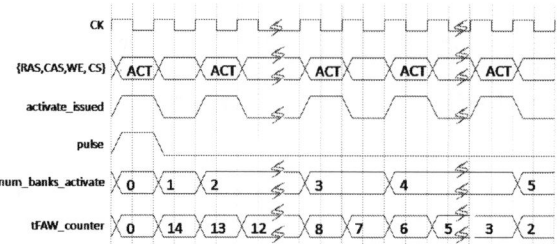

Figure 6. T_FAW Window Violation

8. Conclusion

In this paper we presented elements of verification strategy for a DDR2 controller design. We showed how simulation-assisted initialization, exploiting design symmetry, using cut-points, design abstractions and verification patterns can be used to reduce complexity. The techniques and abstractions presented have been found to be equally effective in the verification of other classes of designs. The work presented is in the public domain. The design and verification material can be downloaded from the Formal Verification Patterns repository [5].

9. References

[1] Harry Foster, Lawrence Loh, Bahman Rabii, Vigyan Singhal, "Guidelines for creating a formal verification testplan", *DVCON*, 2006.

[2] OpenSparc Initiative - *http://www.opensparc.net*

[3] Sun Microsystems - OpenSPARC T1 Micro-Architecture Specification Document.

[4] JEDEC DDR2 SDRAM Specification – JESD79-2B, *www.jedec.org/download/search/JESD79-2B.pdf*, Jan 2005.

[5] The Formal Verification Pattern Repository (Wiki) - *http://oskitech.com/wiki*

[6] Andrea Fedeli, Matteo Moriotti, Umberto Rossi, Franco Toto, "Addressing IP Reuse with Formal Verification and Assertion Based Verification", *D&R* Article - *http://www.us.design-reuse.com/articles/article9511.html*

[7] Adnan Aziz, Vigyan Singhal, Robert K. Brayton, "Verifying Interacting Finite State Machines: Complexity Issues", *Technical Report UCB/ERL M93/52*. Electronics Research Lab, Univ. California, Berkeley, CA 94720.

21st International Conference on VLSI Design

Enhanced TED: A New Data Structure for RTL Verification

P. Lotfi-Kamran, M. Massoumi, M. Mirzaei*, and Z. Navabi
School of Electrical and Computer Engineering, University of Tehran
* *School of Electrical Engineering, Sharif University of Technology*
plotfi@computer.org, massoumi@sbcglobal.net,
mohammadmirzaei@ee.sharif.edu, and navabi@ece.neu.edu

Abstract

This work provides a canonical representation for manipulation of RTL designs. Work has already been done on a canonical and graph-based representation called Taylor Expansion Diagram (TED). Although TED can effectively be used to represent arithmetic expressions at the word-level, it is not memory efficient in representing bit-level logic expressions. In addition, TED cannot represent Boolean expressions at the word-level (vector-level). In this paper, we present modifications to TED that will improve its ability for bit-level logic representation while enhancing its robustness to represent word-level Boolean expressions. It will be shown that for bit-level logic expressions, the Enhanced TED (ETED) performs the same as the BDD representation.

1. Introduction

Increasing size and complexity of digital designs has made it essential to address critical verification issues at the early stages of design cycle. This requires robust, automated verification tools at the higher (RT or behavioral) levels of abstraction.

There already are tools for formal verification of designs implemented at the lower levels of abstraction, such as gate-level and circuit/layout level. However tools that can verify a high-level specification have not yet reached the desired level of automation and maturity. This is partly due to the lack of a good graph-based representation for the RT-level descriptions. While for the gate-level designs, BDD [1] and its variations [3] [4] offer good solutions for their representation, most high-level verification tools synthesize such descriptions to be able to take advantage of BDDs that are at the gate-level.

The aim of this paper is to provide a high-level graph-based representation for manipulation of RT-level descriptions. To achieve this, we have used Taylor Expansion Diagrams (TEDs) [9]. TED has superiority over other graph-based representations in manipulation of RTL designs, but it has some limitations. We enhance TED to overcome its shortcomings to achieve Enhanced TED (ETED). Our experimental results show that ETED is suitable for manipulation of RT-level designs.

The rest of paper is organized as follows: Section 2 presents a brief overview of the previous works in this area. In Section 3, the Taylor Expansion Diagram is briefly introduced. In Section 4, RT-level issues are introduced. In Section 5, our proposed solution will be discussed.

Experimental results are presented in Section 6 and conclusions appear in the last section.

2. Previous Works

Boolean functions are often represented and manipulated by Decision Diagrams (DDs). Ordered Binary Decision Diagrams (OBDDs) [1] are the most commonly used form of decision diagrams in EDA applications [2]. OBDDs and their variations [3] [4] have been successfully used in manipulating gate-level designs, but have limitations in representing arithmetic circuits.

For representing arithmetic circuits, Word-Level Decision Diagrams (WLDDs) are proposed. MTBDDs [5], EVBDDs [6], *BMDs [7], k*BMDs [8] are examples of WLDDs. They are graph-based representations of functions with Boolean domain and integer range; therefore an arithmetic function should be broken down into its bit-level format in order to be represented by a WLDD.
With increasing complexity of digital systems, the need for higher level abstraction becomes more evident. TED [9] is proposed as an answer to this need. TED can be used for representing functions with integer domain and integer range. Therefore, in contrast to WLDDs, an arithmetic function should not be broken down into bit-level in order to be represented. In response to these abilities, in many researches, TED is used as a basis for the contribution to be built upon it (e.g., Reference [10]).

Although TED has advantages over WLDDs, it has some limitations in representing RT-level designs. Some work is done to address TED shortcomings [11][12][13], but they are not considered to be sufficient. In this paper, we eliminate some limitations of TED in order to achieve a better representation for RT-level manipulation.

3. Taylor Expansion Diagram

TED is a graph-based representation, which uses the Taylor series as its decomposition method [9]. The Taylor series of a real differentiable function $f(x)$ around $x=0$ are:

$$f(x) = f(0) + xf'(0) + \frac{1}{2!}x^2 f''(0) + \frac{1}{3!}x^3 f^{(3)}(0) + \cdots \quad (1)$$

where $f'(0)$, $f''(0)$ and $f^{(3)}(0)$ are first, second and third derivations of function f around $x=0$ respectively. The decomposition will be performed recursively using Equation 1.

481

978-1-4244-3039-0/08 $25.00 © 2008 IEEE

TED has three kinds of nodes: vector nodes, scalar (bit) nodes, and terminal nodes. Vector and scalar nodes demonstrate variables, while terminal nodes demonstrate constants. The difference between vector and scalar is that $A * A = A^2$, if A is a vector while $a * a = a$, if a is a scalar. Every node of a TED representation has a label, which demonstrates its associated variable. As in most canonical decision diagrams, e.g., OBDD, the variables of TED are ordered. The function of a node is determined by the Taylor series expansions, according to Equation 1. The out-degree of a variable node depends on the order of the associated variable of that node. The out-degree of a terminal node is 0.

Figure 1 shows TED decomposition of function f at variable x. In this paper, we refer to the k-th derivative of a function rooted at a node as k-child of that node: $f(x=0)$ is the 0-child, $f'(x=0)$ is the 1-child, $f''(x=0)$ is the 2-child, etc. We also refer to the corresponding edges as 0-edge (dotted), 1-edge (solid), 2-edge (double), etc. From the Taylor expansion, it is evident that each edge has an implicit multiplicative factor: x^0 for the 0-edge, x^1 for the 1-edge, $x^2/2$ for the 2-edge, etc. In addition, each edge in a TED has a multiplicative weight, which is computed from the Taylor expansions.

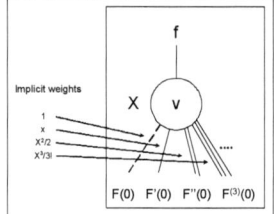

Figure 1. Decomposition in TED

TED can also represent functions containing both algebraic and bit-level Boolean expressions. To represent Boolean expressions, the following formulas should be used (x and y are scalar variables) [9]:

$$\text{NOT}(x) = \bar{x} = 1 - x \tag{2}$$

$$\text{AND}(x, y) = x \wedge y = x * y \tag{3}$$

$$\text{OR}(x, y) = x \vee y = x + y - x * y \tag{4}$$

4. Proper RT-Level Representation

A uniform RT-level representation is required for efficient manipulation of RTL designs. This section discusses peculiarities of RT-level structures.

4.1. RT-Level Issues

RT-level issues are considered from three different perspectives. These are dealing with bit and vector RT-level variables, proper handling of Boolean and arithmetic RTL operators, and efficient bit-level Boolean representation. These issues will be elaborated here.

4.1.1. RTL Variables

In most RT-Level designs, vectors and scalars (bits) co-exist. Consider an 8 bit adder with 2 input vectors A and B, and a carry in (ci). The RT-level description of this adder is A

+ B + ci. However many decision diagrams cannot represent and manipulate vectors, and hence they break vectors into their bit elements. In this example, for representing the addition, a typical decision diagram breaks vector variables A and B into 8 one-bit variables. This makes this representation inconsistent with RT-level description and more inclined towards a bit-level representation. For an RT-level representation, it is essential to represent vectors as a whole and not in terms of their bits.

In addition to the large volume of data structure needed for bit representation, in many cases vector and bit representations are fundamentally different. For example, if A is a vector, $A * A = A^2$ while if a is a bit, $a * a = a$. This means that present bit-level representations cannot be used for efficient representation of RTL variables. Also it is essential to represent mixed vectors and bits simultaneously (as our adder example suggests).

4.1.2. RTL Operators

Many different operators are used in RT-level designs. They include Arithmetic operators, and Boolean operators. These operators are shown in Table 1. Representation and manipulation of these operators is essential for a data structure used for an RT-level verification.

Table 1. RT-Level Operators

Arithmetic Operators	+ - *
Boolean Operators	& | ! ∧ ~ ∧

These operators apply to bits and vectors, but have different and opposing natures. Starting with *Arithmetic Operators* (Table 1), these operations are arithmetic and apply to vector and bit variables as a whole, whereas, the next group of operations (*Boolean Operators*) apply to vector on bit-by-bit basis. These mixed bags of properties should be handled in a uniform data structure.

For the bit-level representation, handling only *Arithmetic Operators* or *Boolean Operators* is sufficient, because *Arithmetic Operators* can be constructed from *Boolean Operators* and vice versa. For example, a BDD can handle *Boolean Operators* and a BMD can handle *Arithmetic Operators* at the bit-level, and either format can handle all operators at the bit-level. On the other hand, for the word-level representation, a data structure that handles only *Arithmetic Operators* or *Boolean Operators* is not sufficient. This is due to the fact that one cannot construct *Arithmetic Operators* form *Boolean Operators* without having to break vectors into bits and vice versa. An example for this discussion is the *multiplication* operator (*) that is the same as the "*and*" operator (&) at the bit-level. But, at the vector-level, not only one cannot replace the other, but also no relation exists between the two without breaking them into their bits.

A data structure corresponding to RT-level descriptions should have a proper representation for bit and vector-level variables and operators. Bit-level variables and operators can easily be handled by either an arithmetic or Boolean representation. However, vector-level Boolean operations require a Boolean representation and vector-level arithmetic operations require an arithmetic representation.

482

978-1-4244-3039-0/08 $25.00 © 2008 IEEE

4.1.3. Efficient Bit-level Representation

In addition to the above issues, many RT-level designs have a large bit-level Boolean part. For example, many microprocessors have a datapath and one or several controllers. The controller has bit-level signals for controlling the datapath operations. Controllers are usually described by bit-level Boolean expression. Therefore having a good bit-level Boolean representation is essential for efficient RT-level manipulation.

For handling issues discussed in Subsections 4.1.1, 4.1.2, and 4.1.3 finding a uniform and efficient data structure is crucial.

4.2. Existing RT-Level Data Structures

This subsection presents several data structures that are used for representing RT-level designs.

4.2.1. BDD Solution

While BDD has a good bit-level representation, it cannot represent vectors. All vectors should be broken into bits in order for them to be represented. Also all operators are represented through bit-level Boolean operators. For example, multiplication should be represented through many Boolean operators. Because of lack of vectors and arithmetic operators, BDD is not a good solution for RT-level manipulation.

4.2.2. WLDD Solution

Word-Level Decision Diagrams (WLDDs) represent several data structures such as MTBDD, EVBDD, *BMD, and k*BMD. Like BDDs, these WLDDs represent bits and not vectors. Distinction of WLDDs and BDDs is in handling arithmetic operators at the bit-level, whereas BDDs translate arithmetic operators to their Boolean equivalent for representing them. Many WLDDs have weights in their edges for better sharing of sub-graphs. These weights separate constants from arithmetic expressions, making the remaining part of these expressions more sharable. Bit-level representation of arithmetic operators and WLDD weights make these representations more compact than BDDs for RT-level representation. However, because of lack of proper handling of vectors, they are not still suitable for RT-level manipulation.

4.2.3. TED Solution

Unlike BDDs and WLDDs, a TED can represent both vectors and bits. However TED has some limitations. As we saw in Section 3, TED only represents arithmetic operators at bit or vector-level. Because Boolean operation at the bit-level can be expressed in terms of arithmetic operations (as shown in Equation 2 to 4), TED can easily represent Boolean operations. However, because of lack of the relation between Arithmetic and Boolean operators at the vector-level, TED cannot represent vector-level Boolean expressions. The only possibility for representing vectored Boolean operations is to break them into their bits and apply Equation 2 to 4 to their individual bits. Obviously, vectored-Boolean operations occur very frequently in RTL descriptions. In addition, when an arithmetic operator and a Boolean operator are applied to the same vector, the vector should be broken into its bits, and TED capabilities cannot be utilized. For example, consider an 8 bit arithmetic unit of a simple processor that has two functions + and |. The circuit's select input (*sel*) selects between the + and | operation. For this circuit, the following expression should be represented by a TED.

$$sel' \times (A+B) + sel \times (A \mid B) \qquad (5)$$

Since TED cannot represent $A \mid B$ at the vector-level, A and B should be broken down into their individual bits. This forces the representation of the $A + B$ operation at the bit-level, which degrades the TED's performance down to BMD's. Therefore, for many RT-level designs, TED cannot do better than BMD. Table 2 summarizes the ability of BDDs, WLDDs, and TEDs in bit and vector-level representation.

Table 2. Comparing BDDs, WLDDs, and TEDs

	Arithmetic		Boolean	
	Bit	Vector	Bit	Vector
BDDs	Y	N	Y	N
WLDDs	Y	N	Y	N
TEDs	Y	Y	Y	N

Another limitation of a TED is its weak bit-level Boolean function representation; usually a Boolean function can more efficiently be represented by a BDD than a TED. Figure 2 shows this disadvantage of TED. As shown, TED representation of a simple OR function has 3 nodes while BDD representation of this function has only 2. This situation can be worse for larger Boolean expressions. As we discussed in Subsection 4.1.3, the ability of efficient bit-level Boolean representation is needed for RT-level manipulation.

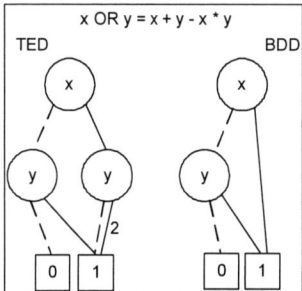

Figure 2. TED and BDD representation of x OR y

In spite of the fact that TED is an improvement over BDDs and WLDDs for representing RT-level representation, it has shortcomings for manipulating RTL descriptions. It is essential to improve TED for efficient RT-level manipulation.

5. Our Proposed Solution

In this part, we modify TED to address the above shortcomings.

5.1. Efficient Bit Level Boolean Representation

Representing bit-level Boolean functions is one of the problems of TED. This means that the TED representation of a Boolean function is larger than the function's BDD representation.

ETED that is discussed below capitalizes on the fact that, in some cases, representation of TED, when its decomposing variable has degree 1, can be simplified. This greatly simplifies the presentation of bit-level Boolean functions. As discussed in the previous section, in each level, TED decomposes a function into its Taylor series expansions, i.e., Equation 1. When the degree of the decomposing variable is 1, the Taylor series becomes:

$$f(x) = f(0) + xf'(0) \qquad (6)$$

in which the following equation holds.

$$f'(0) = f(1) - f(0) \qquad (7)$$

Using Equation 7 in 6 and rearranging the variables, Equation 8 results.

$$f(x) = (1 - x)f(0) + xf(1) \qquad (8)$$

As shown, Equation 8 is similar to the Shannon decomposition, which is used as the decomposing method in Binary Decision Diagrams. To improve bit-level logic representation of TED, in bit-level nodes with two outgoing edges (i.e., nodes with associated variable of degree 1), we use Equation 8 instead of Equation 6 for our decomposing method. By doing this and using Equations 2, 3 and 4 as NOT, AND, and OR respectively, ETED will behave like BDDs for bit-level Boolean representation. Also because of the global restriction of all bit-level nodes to Shannon decomposition, the ETED still remains canonical.

It should be noted that the Shannon expansion in used only in bit-level nodes while in a previously published paper (Reference [12]), Shannon decomposition is used for all nodes that have associated variable of degree 1. By our restriction, the manipulation of ETED is greatly simplified.
In theory the following can be stated about TED and BDD Boolean representation [14].

a) There exist Boolean functions that have a BDD representation of polynomial size but only a TED representation of exponential size.

b) There exist Boolean functions that have a TED representation of polynomial size but only a BDD representation of exponential size.

But in practice, our experiences show that BDD is more compact than TED in representing bit-level Boolean parts of many RT-level designs. We will show this in our experimental results.

5.2. RTL Logical Operators

As discussed in the previous section, TED cannot represent Boolean operators at the vector-level. In this section, we show modifications that will enable TED to represent such operators.

For representing vector-level Boolean operators, as we did in the bit-level Boolean operators, we try to convert them to arithmetic operators. For this, the following formulae should be used (A and B are vectors).

$$NOT(A) = -1 - A \qquad (9)$$

$$AND(A, B) = A \& B \qquad (10)$$

$$OR(A, B) = A + B - A \& B \qquad (11)$$

In contrast to bit-level Boolean operators, vector-level Boolean operators cannot completely be represented through arithmetic operators. However, the resulting expressions have simpler vector-level Boolean operators. With the help of the formulae 9-11, vector-level Boolean operators are converted to arithmetic operators and the vector-level "&" operation. The arithmetic part of these equations can be represented by a TED. In the rest of this subsection, we first show how to represent the vector-level "&" operations. In the next step, we will show how to represent expressions consisting of both arithmetic and vector-level "&" operators in a single data structure (ETED).

We use the generalized Shannon decomposition for representing the vector-level "&" operation. The generalized Shannon decomposition can be used to represent the vector-level "&" operation, where an all-zero value for a variable in a node constitutes a 0-child while for constructing a 1-child, an all-one value is used for a variable in a node. Some examples of the vector-level Boolean "&" representations are shown in Figure 3. In this figure, A and B are 8 bit variables. In each case, in the terminal nodes, the results are converted into a bit pattern. All numbers are treated as 2's complement numbers. For example, the all-one bit pattern (11111111) becomes a -1 as shown on an edge of Figure 3. Small solid circles in the nodes of this figure show that these nodes represent the vector-level Boolean "&" expressions, and not the traditional arithmetic expressions.

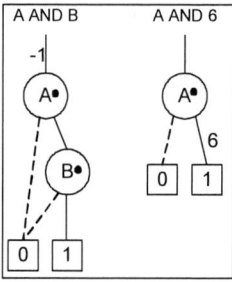

Figure 3. Vector-level Boolean & representation of ETED

The above discussion and the material of the previous section enable us to express the vector Boolean "&" and arithmetic operators. The discussion that follows focuses on simultaneous handling of mixed Boolean and arithmetic operators. For this, we first convert vector-level Boolean operators to arithmetic operators and "&" operations. In the next step, we break a mixed expression into a minimal number of expressions so that each expression is completely arithmetic or the "&" operation. For example, using Equation 9-11 in 5, results in Equation 12 which has simpler vector-level Boolean operators (only the "&" operator).

$$sel' \times (A + B) + sel \times (A + B - A \& B) \qquad (12)$$

With this separation, arithmetic parts of a mixed expression are expressed separately by the standard TED and "&" expressions are expressed as discussed earlier. A special edge is used for connecting "&" expressions and arithmetic graphs of a mixed expression. We will use an arrowed edge for showing this edge. To connect two such expressions, we

484

978-1-4244-3039-0/08 $25.00 © 2008 IEEE

encapsulate the graph of one in a special node using an intermediate variable and use that node in the other graph. For example, Equation 12 can be broken down into the following expressions.

$$sel' \times (A + B) + sel \times (A + B - G)$$
$$G = A \& B \tag{13}$$

The first expression is an arithmetic expression, and the second is a Boolean vector "&" operation. ETED representation of these expressions is shown in Figure 4. Because *sel* is a one-bit variable, as mentioned in Subsection 5.1, we use Shannon decomposition for representing it. *A*, *B* and *G* are vector variables, so Taylor expansion is used for their decomposition. An arrowed edge coming out of *G* indicates that *G* is an intermediate variable representing *A & B* in the first arithmetic expression. Advantage of this intermediate variable becomes clearer when mixing of operations occurs at several levels. As an example consider the *(A & B) + 1* expression instead of *(A & B)* in the above expression. In this case, *(A & B)* must be encapsulated in a special node for representing the whole expression.

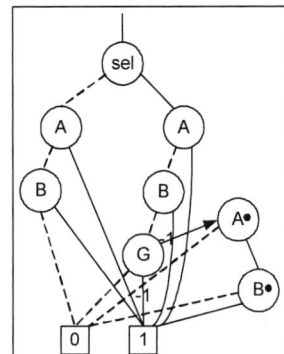

Figure 4. ETED representation of Equation 5

When two Boolean vector "&" expressions are combined to form a single representation (through arithmetic operators), care should be taken if the two expressions are the result of *anding* by a constant value. There are cases that the result should be a simpler Boolean vector "&" expression. A simple check can identify these cases to ensure the uniqueness of the resulting diagram.

The only remaining issue that should be considered is the De Morgan's law. By De Morgan's law, the following vector-level Boolean expressions are equivalent.

$$NOT(A) \ AND \ NOT(B) == NOT(A \ OR \ B) \tag{14}$$

Using Equations 9 to 11 in 14 and rearranging the variables, Equation 15 results.

$$A \& B == 1 + A + B + (-A - 1) \& (-B - 1) \tag{15}$$

Equation 15 says that there are two ways for representing a vector-level "&" operation. ETED should handle "&" operators in such a way that the resulting diagram is canonical. For this, ETED looks at multiplicative weights of the two expressions that should be *anded*. It represents the "&" operation based on right or left hand sides of Equation 15 which results in operands with positive multiplicative weight. If the multiplicative weights of "&" operands are negative, the "&" operation is represented according to right hand side of

Equation 15. If the multiplicative weights of "&" operands are positive, the "&" operation is represented according to left hand side of Equation 15. Finally, if one operand has a negative and the other has a positive multiplicative weight, ETED represents the "&" operation based on the multiplicative weight of the node with greater *id*. This *id* is a unique integer value that is assigned to each node in our ETED implementation.

6. Experimental Results

In this section, we describe experimental results that have been carried out on a PC Pentium 4 with 1 GBytes memory. All runtimes are given in CPU milliseconds. *BMD, TED and ETED packages are implemented by the authors in Visual C++ v6.

In the first series of experiments, we compare bit-level Boolean representation of TED and ETED. The RT-level benchmarks of this experiment are selected so that only arithmetic operators exist in the datapath. Therefore, the difference of these representations originates from representing bit-level Boolean parts. Table 3 provides a summary of the results obtained for these benchmark circuits. Of the five benchmarks, *Paulin* is a differential equation solver and is described in detail in [15]. *Chain_mult* is an ASIC based on the circuit given in [16]. The *SimpleCPU* and *SimpleRTL* are described in [17]. All benchmarks of this experiment have a word-length of 8. All diagrams are built based on the same variable orderings.

Table 3. Comparing *BMD, TED, and ETED on some simple RTL Benchmarks

	*BMD		TED		ETED	
	Nodes	Time	Nodes	Time	Nodes	Time
SimpleRTL	774	16	159	6	94	5
Paulin	1017	109	380	16	208	9
Chain_mult	282	33	114	12	89	7
SimpleCPU	687	46	407	15	249	10
Avenhaus	1093	50	372	15	211	9

As Table 3 shows, ETED is better than TED in terms of the number of nodes and time of conversion. The vector-level representation of ETED and TED is the same for arithmetic operators. Therefore in this experiment, these advantages are due to the fact that bit-level logic representation of ETED is the same as BDD, which is more compact than TED. Because of the lower number of nodes, the manipulation of ETED takes less CPU time. Also, TED is better than *BMD in terms of the number of nodes and time of conversion. This is due to the vector-level representation of TED, because bit-level representation of TED and *BMD are the same.

In the second series of experiments, we compare total ability of ETED with TED and *BMD in representing general RTL designs. Table 4 provides a summary of the results obtained for these benchmark circuits. *Tseng* is described in detail in [18]. *ASPP4* is taken from Reference [19] and can emulate the behavior of either *Paulin* or *Tseng*. These two benchmarks have world-length of 8. *Parwan* is an 8-bit and *Sayeh* is a 16-bit benchmark processors.

978-1-4244-3039-0/08 $25.00 © 2008 IEEE

Because of the Boolean vector-level representation, ETED is better than TED and *BMD. *Parwan*, which is an 8-bit processor, is translated to *BMD, TED, and ETED within reasonable CPU time. On the other hand, *BMD and TED representations of *Sayeh* cannot be constructed within 60000 milliseconds, but its ETED is constructed almost the same timing as *Parwan*. Note that *Sayeh* and *Parwan* have almost the same datapath complexity with different word-lengths.

The number of nodes and therefore construction time of ETED are independent of the word-length but TED's and *BMD's are directly related to the word-length of benchmarks. In many typical datapaths, TED has no significant advantage over *BMD. On the other hand, ETED presents significant improvements over *BMD and TED.

Table 4. Comparing *BMD, TED, and ETED on some realistic RTL Benchmarks

	*BMD		TED		ETED	
	Nodes	Time	Nodes	Time	Nodes	Time
Tseng	771	30	771	31	170	1
ASPP4	3297	1130	3297	1132	612	37
Parwan	5321	1500	5321	1501	724	40
Sayeh	X	X	X	X	731	42

7. Conclusions

In this paper Enhanced TED (ETED) was proposed. ETED is the same as BDD in representing bit-level Boolean expressions. It is also the same as TED in representing word-level arithmetic operators. On the other hand, ETED is able to represent vector-level Boolean operators. Therefore, there is no need to transform these operations to their bit-level equivalents.

ETED provides an efficient representation for RTL designs. It can represent a number of RT-level designs in a compact form, and can also represent bit-level Boolean functions with complexity equivalent to a BDD.

8. References

[1] R. E. Bryant, "Graph-based algorithms for Boolean function manipulation," IEEE Transactions on Computers, Vol. 35, August 1986, pp. 677-691.

[2] R. E. Bryant, "Symbolic Boolean manipulation with ordered binary decision diagrams," ACM Computing Surveys, Vol. 24, September 1992, pp. 293-318.

[3] U. Kebschull, E. Schubert, and W. Rosenstiel, "Multilevel logic synthesis based on functional decision diagrams," European Design Automation Conference, March 1992, pp. 43-47.

[4] S.-i. Minato, "Zero-suppressed BDDs for set manipulation in combinatorial problems," Design Automation Conference, June 1993, pp. 272-277.

[5] R. I. Bahar, E. A. Frohm, C. M. Gaona, G. D. Hachtel, E. Macii, A. Pardo, and F. Somenzi, "Algebraic Decision Diagrams and their Applications," International Conference on Computer Aided Design, November 1993, pp. 188-191.

[6] Y.-T. Lai, M. Pedram, and S. B. K. Vrudhula, "Formal Verification Using Edge-Valued Binary Decision Diagrams," IEEE Transactions on Computers, Vol. 45, February 1996, pp. 247-255.

[7] R. E. Bryant, and Y-A. Chen, "Verification of Arithmetic Circuits with Binary Moment Diagrams," Design Automation Conference, June 1995, pp. 535-541.

[8] R. Drechsler, B. Becker, and S. Ruppertz, "The K*BMD: A Verification Data Structure," IEEE Design & Test of Computers, Vol. 14, April-June 1997, pp. 51-59.

[9] M. Ciesielski, P. Kalla, S. Askar, "Taylor Expansion Diagrams: A Canonical Representation for Verification of Data Flow Designs," IEEE Transactions on Computers, Vol. 55, No. 9, September 2006, pp. 1188-1201.

[10] M. R. Kakoee, and P. Faradji, "A Feasible Approach to Check Extended SAT Using TED Equations," The 20th Conference on Design of Circuits and Integrated Systems, November 2005, pp. 217-222.

[11] A. Hooshmand, S. Shamshiri, M. Alisafaee, B. Alizadeh, P. Lotfi-Kamran, M. Naderi, and Z. Navabi, "Binary Taylor Diagrams: An Efficient Implementation of Taylor Expansion Diagrams," International Symposium on Circuits and Systems, Vol. 1, Kobe, Japan, May 2005, pp. 424-427.

[12] P. Lotfi-Kamran, M. Hosseinabady, H. Shojaei, M. Massoumi, and Z. Navabi, "TED+: A data structure for microprocessor verification," Asia and South Pacific Design Automation Conference, January 2005, pp. 567-572.

[13] P. Lotfi-Kamran, and Z. Navabi, "Improving Logic-Level Representation of Taylor Expansion Diagram Using Attributed Edges" Iranian Journal of Science and Technology, Transaction B, Engineering, Vol. 30, No. B6, December 2006, pp. 735-748.

[14] B. Becker, R. Drechsler, and R. Enders, "On the representational power of bit-level and word-level decision diagrams," Asia and South Pacific Design Automation Conference, January 1997, pp. 461-467.

[15] I. Ghosh, A. Raghunathan, and N. K. Jha, "A design-for-testability technique for register-transfer level circuits using control/data flow extraction," IEEE Transactions on Computer Aided Design of Integrated Circuits and Systems, Vol. 17, August 1998, pp. 706-723.

[16] S. Ravi, I. Ghosh, R.K. Roy, S. Dey, "Controller resynthesis for testability enhancement of RTL controller/data path circuits," International Conference on VLSI Design, January 1998, pp. 193-198.

[17] S. Ravi, G. Lakshminarayana, N.K. Jha, "TAO: regular expression-based register-transfer level testability analysis and optimization," IEEE Transactions on Very Large Scale Integration Systems, Vol. 9, December 2001, pp. 824-832.

[18] I. Ghosh, N.K. Jha, S. Bhawmik, "A BIST scheme for RTL controller-data paths based on symbolic testability analysis," Design Automation Conference, June 1998, pp. 554-559.

[19] I. Ghosh, A. Raghunathan, N.K. Jha, "Hierarchical test generation and design for testability methods for ASPPs and ASIPs," IEEE Transactions on Computer Aided Design of Integrated Circuits and Systems, Vol. 18, March 1999, pp. 357-370.

Simulation Acceleration with HW Re-Compilation Avoidance [1]

Kyuho Shim, Kesava Talupuru*, Maciej Ciesielski*, and Seiyang Yang
Dept. of Computer Enineering., Pusan National University, Korea
<syyang@pusan.ac.kr>
* Logic-Mill Technology, LLC, USA
<mciesielski@logic-mill.com>

Abstract

This work is based on a premise that in traditional, simulation-based RTL functional verification reducing total debugging turnaround time (which includes both the simulation execution time and the compilation time) is much more desirable than simply increasing the simulation speed. This is due to the repeated nature of the debugging process, which includes a large number of simulation and compilation steps. While the HDL compilation process is fast, pure HDL simulation suffers from extremely long simulation execution time. On the other hand, HW-assisted simulation acceleration is characterized by fast execution, but suffers from a long HW re-compilation time, required whenever the design is modified for debugging. This paper proposes an efficient HW-assisted simulation acceleration method based on HW re-compilation avoidance, which can significantly reduce the debugging turnaround time, while maintaining its high execution speed.

1. Introduction

In order to handle the ever-increasing design and verification complexity, engineers have explored several verification methods: Simulation, Simulation-Acceleration (SA), Emulation, Prototyping, and Formal Verification techniques. Simulation is the most widely used technique for functional verification and, because of its ease of use and the availability of inexpensive general computing hardware (e.g. Linux machines) and software, it will remain such for a foreseeable future. However, simulation suffers from very low performance dictated by its sequential nature:

---

[1] This research was supported in part by SBIR grant DMI-0339399 from the National Science Foundation.

---

a numerical simulation of the software (SW) model of the design, specified using a Hardware Description Language (HDL). In this situation, hardware-assisted verification (such as simulation acceleration, emulation, or prototyping) becomes a necessity. In simulation acceleration mode, the DUV (design under verification) is emulated in hardware (HW), while the stimulus is applied from the HDL simulator, which offers high visibility of the design signals. Software stimulus drives the actual hardware, achieving (at least in principle) very high speeds, because of the inherent parallel nature of the hardware. However, HW-assisted simulation acceleration does not solve the verification problems for several reasons:

- It is characterized by disappointingly low speedup compared to pure HDL simulation (at best 10× for complex designs), which is caused by two major factors: a) a large <u>communication overhead</u> between the design emulated in HW and a non-synthesizable testbench (SW stimulus which cannot be mapped onto HW and as such must reside in software); and b) the <u>testbench overhead</u>.
- It suffers from long hardware re-compilation time: every time the design is found to contain a bug, it must be re-compiled onto the hardware emulation platform.
- HW-assisted verification tools are characterized by high price, mostly due to the cost of the hardware platforms; and
- They are difficult to use, requiring complicated setup and configuration.

Recently, by utilizing high-capacity FPGAs the cost of HW-assisted simulation acceleration has been

decreased and transaction-based acceleration has been used to successfully boost the performance of HW-assisted simulation acceleration [1,2]. However, a major deficiency of HW-assisted simulation is a much longer compilation time due to the logic and physical synthesis (P&R) processes (for FPGA-based acceleration); or the compilation/synthesis process (for processor-based acceleration), compared to a pure HDL simulation. This very long compilation time is a major bottleneck in RTL functional verification because every debugging step requires at least one HW re-compilation and there are very many such debugging steps occurring repeatedly during the functional verification process.

This paper presents an efficient simulation acceleration technique based on HW re-compilation avoidance. It is based on the observation that not every debugging step, which results in a subsequent design modification, automatically requires a HW re-compilation. Instead, we propose a technique, in which the modified part of the design is simulated in the HDL simulator, and re-compilation of hardware is deferred. In Chapter 2, we describe a new simulation acceleration method with such defined HW re-compilation avoidance. An experimental results and analysis are presented in Chapter 3, followed by the conclusion.

2. HW Re-compilation Avoidance

2.1 Motivation

Recently, transactional-level modeling (TLM) methodology has been adopted to cope with the ever-increasing complexity of functional verification of SOCs [3, 4]. HW-assisted simulation acceleration is also gaining the ground to address the need of RTL simulation which is too slow for verifying complex SOCs with a meaningful number of simulation cycles. Therefore, there is a growing demand for HW-assisted simulation accelerator to become a hub of verification platforms by replacing pure HDL simulators, even at the early stage of functional verification [4].

However, the long compilation time of HW-assisted simulation accelerators is a serious obstacle to eficient verification, in addition to its high cost. This is because of the large number of hardware compilations required during the entire functional verification process, since every design modification resulting from debugging requires hardware re-compilation. Therefore, the total number of HW re-compilations must be reduced to complete the functional verification as early as possible.

2.2 Basic Concept

The basic idea is to replace the portion of the design that either contains the bug or is the root cause of the bug with the corresponding simulation model, and to co-simulate that portion of the design modified in the HDL simulator with the rest of the design in a HW platform unmodified. To make this possible, the bug-free portion of the design in a HW platform should be logically decoupled from the portion of the design to be modified; furthermore, a communication with the portion of the modified design modeled in the HDL simulator should be provided.

This process is illustrated in Fig. 1, where module B(M1) has a root cause of a bug. As part of the debugging process, module B(M1) has been modified in order to remove the bug. Then, instead of performing FPGA re-compilation, the modified B(M1) is migrated to the HDL simulator and compiled for simulation. To do this, additional communication channel is established between the HDL simulator and FPGA(s), as shown in Fig. 1(b). It is expected that this additional communication channel may decrease the simulation acceleration speed to some extent. However, a transaction-based communication using a transactor can effectively minimize the effective performance degradation caused by this modification. To provide this additional communication channel, additional logic needs to be instrumented to a design, as explained in more detail in Section 2.3.

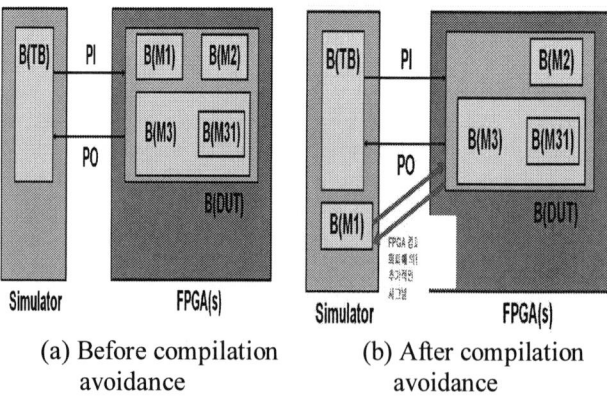

(a) Before compilation   (b) After compilation
     avoidance   avoidance

Fig. 1 The concept of HW re-compilation avoidance

In this way, time-consuming HW re-compilation is avoided and replaced by fast compilation for HDL simulation. Therefore, a new simulation acceleration run can proceed right after the design has been partially modified, which results in a much shorter debugging turnaround time.

However, the major difficulty in performing such a module migration comes from the fact that it is not know in advance which portion of the design is going to be a root cause of a bug. Therefore, the design should be decomposed into a set of modules so that each module could be easily migrated into the HDL simulator once it is modified. This requires adding instrumentation logic at the boundary of each of these modules. The role of the instrumentation logic is to provide a logical communication channel between the (previously buggy) migrated modified module in a HDL simulator and the rest of designs that is still residing in a HW platform.

The size of those modules should be carefully determined to trade off the instrumentation logic overhead and the simulation acceleration performance. If the size of a module is too small, the overhead of the instrumentation logic can be too high. On the other hand, if the modules are too big, the simulation acceleration performance could degrade significantly with this HW re-compilation avoidance technique, because they have to be simulated by the HDL simulator. In deciding on the decomposition of the design into modules, one can take advantage of the hierarchy present in most practical designs, and assign module boundaries in accordance with such a hierarchy. For example, in SOC design, each IP block or a newly designed component can be a module for our HW re-compilation avoidance technique.

2.3 Instrumentation Logic for HW Re-compilation Avoidance

Once the boundaries of the modules are determined, an additional logic should be instrumented at their boundaries, prior to the first HW compilation, to facilitate the communication of the modified module in a HDL simulator with the rest of unmodified design.

The instrumented logic for HW re-compilation avoidance can be added automatically to the original RTL code or to a synthesized netlist by the instrumentation software. The structure of such logic is similar to the IEEE 1149.1 boundary scan, except that its data channel bandwidth is 32 bit wide to support fast interface [5]. The instrumentation logic acts as *virtual I/O* for the modified module, once it is migrated from HW platform to HDL simulator. The output values of the modified module drive the corresponding inputs of the rest of unmodified modules and some output values of the unmodified modules drive the corresponding inputs of the modified sub-design.

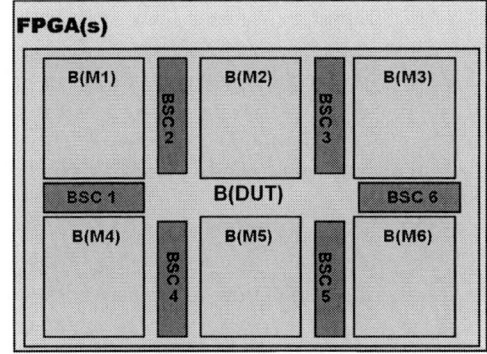

(a) Instrumentation logic for HW re-compilation avoidance for six modules

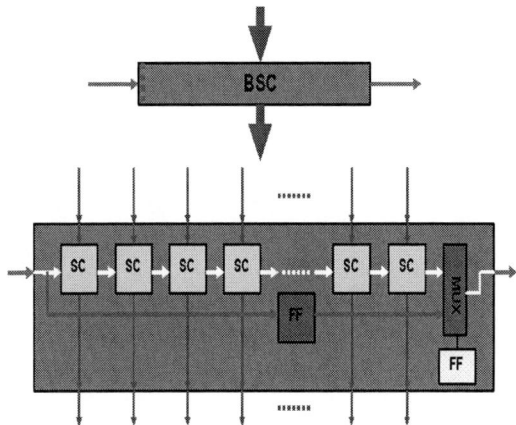

(b) The structure of virtual I/Os for a module

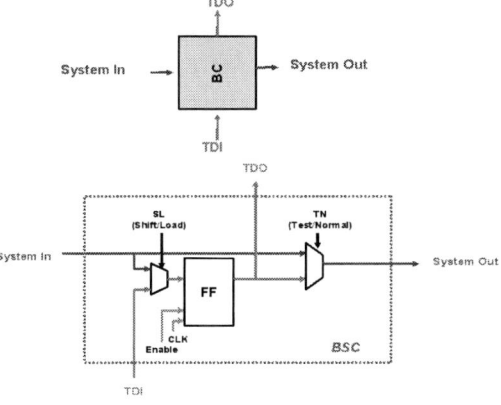

(c) A scan cell used in a virtual I/Os

Fig. 2 Instrumentation logic for HW re-compilation avoidance

2.4 Overall Simulation Acceleration Architecture

Fig. 3 shows the overall architecture of our simulation acceleration with HW re-compilation avoidance. The DUV is implemented with the instrumentation logic on a HW platform, e.g. FPGA-based, for HW re-compilation avoidance.

HW re-compilation can be performed in background mode, while HDL simulation takes place. As soon as the background HW re-compilation is completed, the modified module can be brought back to the HW platform to maximize the simulation acceleration speed.

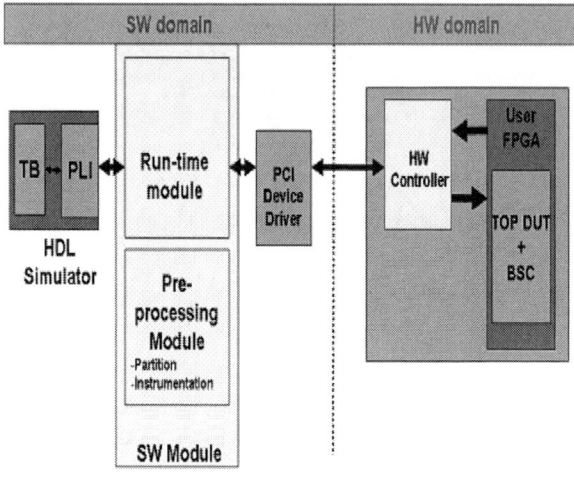

Fig. 3 The overall architecture of simulation acceleration with HW re-compilation avoidance

3. Experimental Results

A prototype of the simulation acceleration with FPGA re-compilation avoidance, presented in this paper, has been built using an FPGA development board from Xilinx. The board has one SPARTAN2 200 FPGA chip and a 32-bit PCI slot [6]. The PCI core is provided by Xilinx to facilitate the communication between hardware and software. The board has an expansion slot to accommodate bigger designs. A single daughter board with a Xilinx XC2V800 FPGA chip has been used for the PICx100 design.

For the purpose of this experiment we used our own design of the Peripheral Interface Controller (PIC) micro-controller from MicroChips. The VHDL specification of the design was analyzed, and the

instrumentation logic was added manually to implement HW re-compilation avoidance. The PIC has an 8-bit microcontroller unit with a 12-bit instruction set, an ALU, a 9-level stack, and an SRAM. Since the HDL simulation time of the basic PIC controller is too short to demonstrate the advantages of our technology, the design size was increased by replicating the basic design 100 times. This makes the speed of pure simulation at the range of a few hundred cycles per second at most, which is reasonable and conservative enough. It is also observed that the speed of traditional simulation acceleration is at the range of a few thousand cycles per second, which is also reasonable.

Table I shows the simulation results (speed) for our PICx100 design. Mentor Graphics ModelSim was used in our experiments as the HDL simulator [7].

Table I. Experimentation Results

	Simulation time for 1M cycles (sec) [cycles/sec]	Re-compilation time (sec)	Total debug turn-around time(sec)
Pure HDL simulation	4,090 [244]	53	4,143
Traditional simulation acceleration	630 [1,587]	2,720	3,350
Simulation acceleration with HW-recompilation avoidance	670 [1,493]	9	679

After fixing one module and migrating it to the software side, the effective simulation speed is degraded by only 6% compared to a conventional simulation acceleration. This is due to the additional simulation time needed for the migrated module by an HDL simulator and to the additional communication overhead between the HDL simulator and an FPGA emulation board. However, the simulation speed is still much higher than that of a pure HDL simulation.

A drastic decrease in the debugging turn-around time (the sum of simulation time and re-compile time) has been observed in this experiment. The debugging turn-around time was 679 sec per bug for 1M cycles of simulation; whereas, the debugging turnaround time of traditional simulation acceleration and pure simulation was 3,350 and 4,143 sec, respectively. It proves that our HW re-compilation is very effective for reducing the debugging turnaround time, which has closer

978-1-4244-3039-0/08 $25.00 © 2008 IEEE

correlation to the total functional verification time than the simulation run time. The area overhead for logic instrumentation for HW re-compilation avoidance was less than 10% of the total design area.

Another interesting point is that the re-compilation time in our approach is much shorter than the re-compilation time in pure HDL simulation because only the modified module needs to be re-compiled by the HDL simulator with our HW-recompilation avoidance technique. This ultra fast re-compilation could have a significant performance advantage even over a pure HDL simulation, because the compilation time of the HDL simulation tends to be very long if the size of design is large, as observed in the contemporary complex SOCs.

In this experiment, the communication was performed at a signal level, not at a transaction level. If the transaction-level communication based on SCE-MI is used, this communication overhead can be reduced further, and the performance degradation can be minimized [8].

4. Conclusions

We proposed a novel technique for HW re-compilation avoidance in simulation acceleration to reduce the debugging turnaround time and implemented the prototype system demonstrating the advantages of our technology over the standard simulation acceleration.

We believe that the proposed approach is equally applicable to FPGA-based simulation acceleration and processor-based simulation acceleration, which suffer from much longer re-compilation time than pure HDL simulation. The proposed HW re-compilation avoidance technique is especially attractive when HW-assisted simulation acceleration is deployed even in the early stage of function verification, where the bug density is relatively high.

5. References

[1] "The Target Platform Methodology for HW/SW Debugging Before Silicon", White paper, Mentor Graphic.(http://www.mentor.com), 2007

[2] "Accelerated Hardware/Software Co-verification", White paper, Cadence Design Systems Inc.(http://www.cadence.com), 2005

[3] Murali Kudlugi, Soha Hassoun, Charles Selvidge, Duaine Pryor, "A Transaction-Based Unified Simulation/Emulation Architecture for Functional Verification", in Proc. of 38th DAC, pp.623-628, June 2001

[4] Stuart Swan, "SystemC Transaction Level Models and RTL Verification", in Proc. of 43rd DAC, pp. 90-92, July 2006

[5] JTAG Boundary Scan, IEEE Std. 1149.1

[6] Spartan-II 200[TM] PCI Development Board User's Guide V2.0, Memec (http://www.memec.com), 2002

[7] ModelSim SE User's Manual, Mentor Graphics (http://www.model.com), 2006

[8] Draft SCE-MI 2.0 Standard, Accellera, 2006

21st International Conference on VLSI Design

A Module Checking based Converter Synthesis Approach for SoCs

Roopak Sinha, Partha S Roop and Samik Basu

Abstract— Protocol conversion involves the use of a converter to control communication between two or more protocols such that desired system-level specifications can be satisfied. We investigate this problem in a formal setting and propose, for the first time, a temporal logic based automatic solution to *convertibility verification and synthesis*. At its core, our technique is based on local module checking and determines the existence of the converter and if a converter exists, it is automatically generated. A number of key features of our technique distinguishes it from all existing formal and/or informal approaches. Firstly, we handle both data and control mismatches using a single unifying module checking based solution. Secondly, the proposed approach uses temporal logic for the specification of correct behaviors (unlike earlier automaton based specifications) which is both elegant and natural to express event ordering and data-matching requirements. Finally, we have experimented extensively with the examples available in existing literature to evaluate the applicability of our technique in a wide range of applications.

Index Terms— protocol mismatches, protocol conversion, module checking.

I. INTRODUCTION

A SYSTEM-on-a-chip (SoC) contains individual processing and peripheral components (called intellectual property or IP blocks) connected together using a common bus [8]. Components in a SoC are usually developed in isolation and may follow independent communication protocols. Therefore, when several components are interconnected, it is possible that their may suffer from *protocol mismatches* [11]. Mismatches occur when the exchange of control signals and/or data between components is not consistent with the intended behaviour of their interaction [8], [15], [18] (leading to *control* and/or *data* mismatches).

In order to resolve mismatches, it is required that mismatched components be redesigned to achieve desired system-level behaviour. This is usually a very expensive process. Due to this overhead, *protocol conversion*, a term broadly used to refer to techniques that resolve mismatches without requiring manual modification of components, has been studied extensively for over two decades [3], [8], [15], [16], [19]. Protocol conversion typically involves the automatic generation of extra glue-logic, called a *converter*, to control the communication between components in order to satisfy system-level behaviour. Consider the example of a SoC that uses the AMBA high-performance bus (AHB) [8] to connect two masters - a producer processor and a consumer processor. These two processors in the SoC [20] communicate using the slave RAM block to read/write shared data, as shown in

Roopak Sinha is a PhD student at the Department of Electrical and Computer Engineering, University of Auckland, *rsin077@ec.auckland.ac.nz*

Partha S Roop is a Senior Lecturer at the Department of Electrical and Computer Engineering, University of Auckland, New Zealand. *p.roop@auckland.ac.nz*

Samik Basu is an Assistant Professor at Department of Computer Science, the Iowa State University. *sbasu@cs.iastate.edu*

Fig. 1. However, the masters and slave have inherent control and data mismatches (explained in later sections), which prevent their integration into the AHB system system. Protocol conversion, in this case, will look at creating a converter for each master (shown in Fig. 1), such that mismatches can be eliminated.

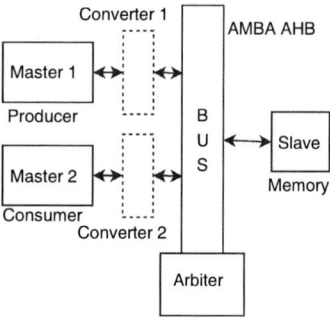

Fig. 1. Protocol conversion overview

A formal protocol conversion technique concerns itself with a range of issues. Firstly, participating protocols and their interaction, and specifications must be formally described. A protocol conversion technique must also be able to detect mismatches (*mismatch detection*) and have an algorithm to automatically generate converters if mismatches exist (*converter generation*). Additional issues include determining *scope*–the range of mismatches that can be handled, *converter existence*–which checks whether a converter to resolve mismatches exists, and *converter correctness*–which checks whether a given converter indeed bridges mismatches. The answers to the above questions differ between individual protocol conversion techniques.

Related Work. Existing protocol conversion techniques can be broadly categorized as informal or formal. Informal approaches like [2]–[4] and [17] lack mathematical rigor, have very restricted scope and focus mainly on converter generation without addressing the questions of converter correctness and existence. Formal approaches, like [8], [10] and [19], on the other hand are based on mathematical techniques and proofs and solve protocol conversion within well-defined but restricted scopes, that differ for each technique. [19] present a game-theoretic formulation to resolve control mismatches between protocols with only unidirectional communication and do not address data mismatches. [8]- [9] provide synchronous protocol automata to precisely model protocols, and their solution, based on checking for a compatibility relation between protocols, can only handle a restricted set of data mismatches along with control mismatches. Additionally, model-checking based verification for proving the correctness of the synthesized converter is performed as an additional step. In [10], a hybrid simulation/verification approach to protocol conversion in SoC designs is proposed, where both simulation and formal

492

978-1-4244-3039-0/08 $25.00 © 2008 IEEE

verification is combined. However, the questions of converter correctness and generation are not addressed comprehensively.

Protocol conversion seems superficially similar to the problem of controller synthesis in discrete event systems [1], [12], [22]. In this setting, controllers perform selective *disabling* of controllable events to control a given plant. However, in protocol conversion, converters may use additional techniques like event buffering and forwarding, and generation of extra control signals (described in later sections) in addition to disabling. Also, data mismatches require additional effort and cannot be addressed in the discrete-event setting.

Given the above summary of available protocol mismatches techniques, it is evident that no single unifying approach to automatically resolve control and data mismatches in a formal setting exists. With the increasing use of SoCs, the lack of such a technique significantly increases design effort.

Our solution. This paper provides the first protocol conversion technique to handle control and data mismatches under a unifying framework. Its key features and contributions are as follows:

- *Protocol representation, interaction and detection of mismatches*: We use *Synchronous Kripke Structures* (SKS) to formally represent protocols and their interaction. SKS are finite state descriptions that precisely model of control and data behaviour of SoC protocols. SKS description also leads to straightforward mismatch detection.
- *Specifications*: Temporal logic CTL is used to describe the intended control and data behaviour of the interaction between protocols. Temporal logics provide a natural way of writing such requirements succinctly and effectively. Although this paper presents CTL-based conversion, our approach can be extended to other temporal logics like LTL and CTL*.
- *Converter generation*: We develop a module checking [14] based formulation for converter generation. Model checking is not used as it can only verify whether protocols satisfy given CTL properties. However, using module checking, we can construct an environment (converter) which can control protocols to satisfy given properties. This is in fact the first known practical application of module checking.
- *Converter representation and control*: In our approach, converters are described using SKS. A converter controls protocols by using techniques such as disabling, buffering and event forwarding, and generation of extra control signals to bridge mismatches. Converters generated using our approach can, for the first time in the protocol conversion setting, deal with arbitrary data-widths between two protocols.
- *Converter existence and correctness*: Converters generated by our algorithm are guaranteed to bridge protocol mismatches and satisfy given specifications. Also, given a protocol pair and a set of constraints on conversion, if a converter cannot be generated by our technique, mismatches cannot be resolved by any converter having the same features as converters in our setting.

The rest of this paper is organized as follows. Section II provides the description of protocols and specifications. Section III describes how converters operate. Section IV describes how converters are automatically synthesized in our setting. Finally, section V provides results obtained from a wide range of protocol mismatch problems, with concluding remarks in section VI.

II. PROTOCOLS AND SPECIFICATIONS

A. Protocol description and interaction

We define Synchronous Kripke Structures for protocol description as follows.

Definition 1 (SKS): A Synchronous Kripke structure (SKS) is a *finite state machine* represented as a tuple $\langle AP, S, s_0, I, O, R, L, clk \rangle$ where AP is a set of propositions; S is a finite set of states with $s_0 \in S$ being the initial state; I is a finite and non-empty set of inputs and O is a finite non-empty set of outputs. $R \subseteq S \times \{t\} \times B(I) \times 2^O \times S$ is the transition relation where $\mathbf{B(I)}$ represents the set of all boolean formulas over I; and $L : S \rightarrow 2^{AP}$ is the state labelling function. Finally, the event t represents ticking of the clock clk.

All transitions trigger with respect to the *ticks* of the clock clk and a boolean combination of inputs. A transition can therefore trigger with respect to the presence or absence of a single input a (represented as a or $\neg a$ respectively), a combination of one or more inputs (e.g. $\neg a \wedge b \vee \neg c$). In our current setting, the same clk is used to drive all protocols and converters. Hence we remove references to clk and its ticks from SKS transitions described henceforth. Also, for transitions of the type $(s, t, b, o, s') \in R$, we use the shorthand $s \xrightarrow{b/o} s'$.

States of a SKS are labelled using atomic propositions. In addition, we use some atomic propositions that have an integer suffix to indicate data input or output over ports of specific widths. These are subsequently used by our algorithm to address data-width mismatches (illustrated in section IV).

Fig. 2 presents the SKS description of the various parts of the AMBA AHB-based SoC system presented in Fig. 1. The SKS P_S for master 1 (producer) consists of 7 states with s_0 as its initial state (Fig. 2(a)). In s_0, the master keeps requesting bus access by emitting the HBUSREQ1 signal every tick. When it receives the grant signal HGRANT1, it moves to state s_1 from where it either chooses to perform a single write operation or an incrementing burst operation. A burst transfer (transition to s_2) is selected when the signal *int*, denoting an internal choice made by the protocol, is present whilst a single write (s_3) is performed when *int* is absent. To perform a burst operation, the protocol emits relevant control signals to indicate an incrementing burst operation (and keeps requesting further bus access using HBUSREQ1 output) and reaches state s_2 where it writes a 16-bit data packet (denoted by the label $DOut_{16}$) onto the AHB's data bus. It then waits for the HREADY signal, which signifies a successful read by the slave, while persistently requesting bus access. If HREADY is received before HGRANT1, implying that the master no longer has bus access, the protocol resets back to s_0. However, if HGRANT1 is available when HREADY is read, the protocol moves to state s_6 to write another 16-bit packet. When the protocol chooses to perform single write, the protocol writes 16-bit data in state s_3 and then moves back to its initial state upon confirmation that data has been received (HREADY).

Fig. 2(b) shows the SKS P_T for the bus-arbiter. In its initial state s_0, the arbiter awaits bus request signals HBUSREQ1 and HBUSREQ2 from masters 1 and 2 respectively, and grants access to the first requester. Once access has been granted, the arbiter waits for the completion of a transfer by awaiting the HREADY signal, and then moves back to its initial state.

The SKS for the slave memory block P_U is shown in Fig. 2(c). In its initial state u_0, the slave waits for signal HSELECT1 (its

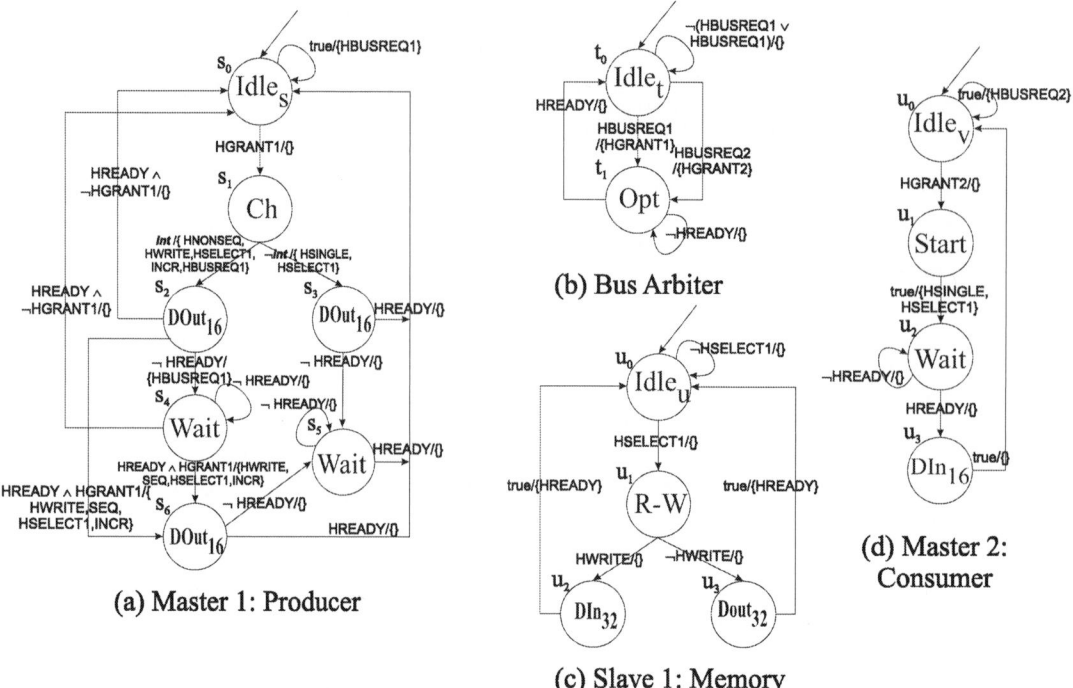

(a) Master 1: Producer

(b) Bus Arbiter

(c) Slave 1: Memory

(d) Master 2: Consumer

Fig. 2.   SKS representation for various parts of a SoC.

enable signal) to activate read/write options in state u_1. In u_1, if the write signal HWRITE has been activated by a bus-master, the memory reads 32-bit data from the AHB's data bus (u_2). Otherwise, it writes 32-bit data to the AHB's data bus (u_3).

Finally, the SKS P_V for master 2 (consumer) is shown in Fig. 2(d). From its initial state v_0, it keeps requesting bus access by emitting the HBUSREQ2 signal. When the grant signal HGRANT2 is received, it moves to state v_1 where it emits control signals HADDR and HSINGLE (requesting a read from the slave) and moves to state v_2. It then awaits HREADY before moving to state v_3 where it reads a 16-bit packet from the AHB's data bus. More 16-bit packets can be read if the signal MORE is present, otherwise the protocol resets back to its initial state.

We now define the interaction between protocols, called their *parallel composition*.

Definition 2 (Parallel Composition): Given two SKS $P_1 = \langle AP_1, S_1, s_{0_1}, I_1, O_1, R_1, L_1, clk \rangle$ and $P_2 = \langle AP_2, S_2, s_{0_2}, I_2, O_2, R_2, L_2, clk \rangle$, their parallel composition is the SKS $P_1 || P_2 = \langle AP_{1||2}, S_{1||2}, s_{0_{1||2}}, I_{1||2}, O_{1||2}, R_{1||2}, L_{1||2}, clk \rangle$ where $AP_{1||2} = AP_1 \cup AP_2$; $S_{1||2} = S_1 \times S_2$; $s_{0_{1||2}} = (s_{0_1}, s_{0_2})$; and $I_{1||2} \subseteq I_1 \cup I_2$; $O_{1||2} = O_1 \cup O_2$, and $L_{1||2}((s_1, s_2)) = L_1(s_1) \cup L_2(s_2)$.

Finally, the transition relation $R_{1||2} \subseteq S_{1||2} \times \{t\} \times B(I_{1||2}) \times 2^{O_{1||2}} \times S_{1||2}$ such that:

$$\frac{(s_1 \xrightarrow{b_1/o_1} s_1') \wedge (s_2 \xrightarrow{b_2/o_2} s_2')}{((s_1, s_2) \xrightarrow{b_1 \wedge b_2/o_1 \cup o_2} ((s_1', s_2'))}$$

Intuitively, the parallel composition contains all possible states and transitions that can be reached by making simultaneous transitions from each protocol. For example, given the bus-arbiter P_T and the memory slave P_U, the parallel composition $P_T || P_U$ has the initial state (t_0, u_0) which will have transitions to states

(t_0, u_0), (t_0, u_1), (t_1, u_0) and (t_1, u_1). Each transition of the combined state (t_0, u_0) combines a transition of t_0 (which can reach t_0 and t_1) and another transition of u_0 (which can reach u_0 and u_1).

Given the SKS descriptions of the various parts of a SoC system, conversion can be performed between various components, such as the bus-arbiter and a master, a slave and a master, or a master and the rest of the system. In this paper, we perform conversion between the consumer processor P_V and the combined bus-arbiter and slave memory chip represented by $P_T || P_U$ (Fig. 2). The two SKS have certain inherent mismatches. The consumer master reads 16-bit data whilst the memory writes 32-bit data, leading to a data-width mismatch. Similarly, the sequence of control signals exchanged between them may result in executions contrary to intended behaviour.

B. Specifications in CTL

In our setting, the temporal logic CTL is used to describe desired control and data behaviour of the interaction between mismatched protocols. CTL is defined over a set of propositions using temporal and boolean operators as follows:

$$\phi \rightarrow p \mid \neg p \mid t\!t \mid f\!f \mid \phi \wedge \phi \mid \phi \vee \phi \mid AX\phi \mid EX\phi \mid A(\phi \, U \, \phi) \mid E(\phi \, U \, \phi) \mid AG\phi \mid EG\phi$$

Semantics of a CTL formula, φ denoted by $[\![\varphi]\!]_M$ are given in terms of a set of states in a SKS M, which satisfies the formula. A state $s \in S_M$ is said to satisfy a CTL formula φ, denoted by $M, s \models \varphi$, if $s \in [\![\varphi]\!]_M$. Typically, we omit the M in $[\![\ ]\!]_M$ if the context is clear. We also say that $M \models \varphi$ to indicate $M, s_0 \models \varphi$. In this paper, we restrict ourselves to formulas where negations are applied to propositions only.

Control constraints. For the consumer processor P_V and the combined arbiter-memory protocols represented by $P_T \| P_U$, the following CTL control-constraints can be used:

[φ_1] $\texttt{AGEF}Start$[1]: The consumer must always eventually receive bus access in order to read data from the memory.

[φ_2] $\texttt{AGEF}DOut_{32}$: The combined system should also allow the memory slave write data.

[φ_3] $\texttt{AG}((Idle_v \wedge Idle_t) \Rightarrow \texttt{A}(\neg Start \, \texttt{U} \, Opt))$: The master cannot move further from its initial state before the arbiter grants it bus access.

[φ_4] $\texttt{AG}((Idle_u \wedge Idle_t) \Rightarrow \texttt{A}(\neg DOut_{32} \, \texttt{U} \, Wait))$: The memory cannot write data unless requested by the master.

Data constraints. In addition to control constraints, it is also essential to describe the correct data communication behaviour of participating protocols. To restrict protocols such that no data *underflow* or an *overflow* happen, we introduce *data counters* as follows.

Given the data-widths N and M (which are integers) of the outputs and inputs respectively, we compute the minimum width needed for the communication medium (usually a buffer) between the two protocols. If $N < M$, then the minimum capacity must be $N \times f$ such that f is the smallest integer for which $N \times f \geq M$; otherwise the minimum capacity is N. This assumption ensures that there are enough preceding outputs before any input. While the minimum bound of communication medium buffer can be computed as above, the maximum bound can be any value greater than the minimum bound. In our setting, we assume that the maximum bound of the communication medium buffer is $\texttt{LCM}(N, M)$. Given a capacity K of the communication medium between these bounds, the maximum number of outputs possible when the medium is empty is $x = \lfloor K/N \rfloor$; while the maximum number of inputs possible when the medium is full is $y = \lfloor K/M \rfloor$. We use an auxiliary counter for every input/output pair such that the counter is *incremented* by y for every output and *decremented* by x for every input. We then verify that the counter always remains between 0 and $x \times y$ using the CTL property $\texttt{AG}(0 \leq counter \leq (x \times y))$.

Example. Consider the AMBA-based SoC example in Figure 2. Data outputs from the arbiter-memory protocol $P_T \| P_U$ are 32-bits while while data inputs by the consumer-master are 16-bits. Hence, $N = 32$ and $M = 16$. As $N > M$, the communication medium capacity needs to be at least 32-bits. As the bus connecting these protocols is the AMBA AHB, which has a data-bus of size 32, the protocols can be possibly handled in our setting (if the data-bus's width was less than 32-bits, our technique would fail). Given this 32-bit capacity of the communication medium, the maximum number of write operations possible when the medium is empty is $x = \lfloor K/N \rfloor = 1$. Similarly, the maximum number of read operations is $y = \lfloor K/M \rfloor = 2$. Given these values for x and y, we introduce a counter variable $counter$ which is incremented by 2 (y) for each $\texttt{DOut}_{32}$ and decremented by 1 (x) for each $\texttt{DIn}_{16}$. To verify that the counter always remains between 0 and $x \times y = 2$, we use the following property $\varphi_d \equiv \texttt{AG}(0 \leq counter \leq 2)$.

In addition to overflow/underflow prevention using the above CTL formula, stronger restrictions can be described. For example, the formula $\texttt{AG}((counter = (x \times y)) \Rightarrow \texttt{A}(\neg DOut \, \texttt{U} \, (counter = 0)))$ requires that once the communication medium is completely full, all data on it is read completely before more data is

added. Such constraints cannot be handled by existing protocol conversion techniques. Furthermore, multiple counters can be used, allowing for conversion for protocols with multiple data-communication channels.

III. CONVERTERS: DESCRIPTION AND CONTROL

In the presented approach, converters are described using SKS.

Definition 3 (Converter): Given two protocols $P_1 = \langle AP_1, S_1, s_{0_1}, I_1, O_1, R_1, L_1, clk \rangle$ and $P_2 = \langle AP_2, S_2, s_{0_2}, I_2, O_2, R_2, L_2, clk \rangle$ with their protocol composition $P_1 \| P_2 = \langle AP_{1\|2}, S_{1\|2}, s_{0_{1\|2}}, I_{1\|2}, O_{1\|2}, R_{1\|2}, L_{1\|2}, clk \rangle$, a converter $\mathcal{C}$ for P_1 and P_2 is a SKS: $\langle AP_\mathcal{C}, S_\mathcal{C}, s_{\mathcal{C}0}, I_\mathcal{C}, O_\mathcal{C}, R_\mathcal{C}, L_\mathcal{C} \rangle$ where $AP_\mathcal{C} = \emptyset$, each $c \in S_\mathcal{C}$ corresponds to some $s \in S_{1\|2}$ with the initial state c_0 corresponding to $s_{0_{1\|2}}$; $I_\mathcal{C} \subseteq (O_1 \cup O_2)$ and $O_\mathcal{C} \subseteq (I_1 \cup I_2)$.

The transition relation $R_\mathcal{C} \subseteq S_\mathcal{C} \times \{t\} \times B(I_\mathcal{C}) \times 2^{O_\mathcal{C}} \times S_\mathcal{C}$ such that for any $c \in S_\mathcal{C}$ corresponding to $s \in S_{1\|2}$, if $s \xrightarrow{b/o} s'$, then c can have a transition $c \xrightarrow{b'/o'} c'$ where c' corresponds to s', $b' = \bigwedge o_i$ ($i = 1 \ldots |o|, o_i \in o$) is the conjunction of all elements of o, and o' is a set that satisfies the boolean formula b.

Converter states are not required to satisfy any temporal property ($AP = \emptyset$) as all desired propositional properties must be satisfied by the participating protocols. A converter acts as an intermediary between the participating protocols. Converters read outputs generated from participating protocols and produce outputs that form the inputs to the protocols. This basic strategy helps converters control protocols in the following ways[2]:

- *Disabling*: Converters may prevent signals emitted by one protocol from being visible to another protocol. This helps to disable transitions that may lead to faulty states and paths.
- *Buffering and event forwarding*: Converters contain buffers which can store signals emitted by one protocol that can be forwarded to another protocol at a later stage.
- *Generation of missing control signals*: A converter can generate signals which are required by a protocol but are not generated by any other protocols that it controls.

Converters exercise lock-step control over protocols. Each converter state c corresponds to a single state s in the protocol composition of the participating protocols. A transition $s \xrightarrow{b/o} s'$ of s is allowed only when c has a matching transition of the form $c \xrightarrow{b'/o'} c'$, and both transitions are made simultaneously (relating c' to s'). Of course, the inputs required by a transition might be present as outputs of the same transition, forwarded from the converter's buffers, and/or be generated by the converter itself.

Fig. 3 shows the converter for the consumer master and arbiter-slave protocols. In its initial state, it lets the consumer processor P_2 continue to request for a grant until a grant HGRANT2 is received from the arbiter. The grant is immediately provided to P_2, enabling it to make a transition from its initial state v_0 to v_1 and the converter makes a transition from c_0 to c_1. The converter then receives the control signals from P_2 which include the slave-memory enable signal HSELECT1, which is immediately conveyed to the slave. In the next tick, both the slave-memory and consumer processor make transitions without requiring or emitting any signals. At this stage, the slave-memory makes a

---

[1] EFp is an abbreviation for $\texttt{E}(tt \, \texttt{U} \, p)$.

[2] Note that controllers in discrete-event systems [13] can only perform disabling.

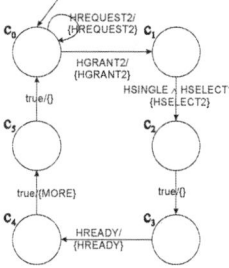

Fig. 3. Converter for the arbiter-slave and consumer processor protocols.

transition to u_3 allowing it to write a 32-bit packet to the data-bus. After writing, the slave-memory emits HREADY marking the end of the transaction. The converter passes this signal to the master allowing it to read a 16-bit packet from the data-bus. The converter then generates the missing input MORE to enable the master to read another 16-bit data, ensuring that the 32-bits written on the data-bus are read completely. In the next tick, the converter allows the master to reset back to its initial state.

IV. CONVERTER GENERATION ALGORITHM

The algorithm to automatically generate converters in our setting is based on module checking [14], also known as model checking for *open systems*. Model checking can merely check whether given protocols satisfy given CTL properties. Module checking, on the other hand, is used to construct an environment (converter) under which the given protocols satisfy given CTL formulas. We demonstrate the working of the algorithm by showing how the converter presented in Fig. 3 is obtained for the AMBA AHB based SoC example provided in Fig. 2.

Intialization. The main inputs to the algorithm are the SKS descriptions of two participating protocols (each one can be a composition of multiple protocols), a set of counters (one for each matching data input/output pair), and a set of CTL formulas to be satisfied. Additionally, each input/output of the protocols must be identified as either *controllable* or *uncontrollable*. Controllable protocol inputs can be disabled by a converter as opposed to *uncontrollable* inputs that can never be disabled. Uncontrollable inputs model signals that are generated internally by protocols or by other IPs that the converter has no control over. Furthermore, each controllable signal must be further selected as either *buffered* or *non-buffered*. Buffered signals are those that can only be presented to a protocol if they have previously been generated by another protocol (and hence buffered by the converter). Non-buffered signals, on the other hand, may be generated by the converter without reading them from other protocols. Finally, the parallel composition of the participating protocols is computed and used during the tableau generation phase.

To carry out converter generation between the consumer master protocol P_V and the arbiter-memory slave pair $P_T || P_U$, we provide the following inputs to the algorithm. P_V and $P_T || P_U$ form the SKS descriptions of participating protocols. We introduce one counter for the data-communication between the slave and the master and the input specification set contains the formulas φ_1, φ_2 and φ_d (see section II-B). Furthermore, signals like HBUSREQ2, HGRANT2, HREADY, HSELECT1 etc that are emitted by one protocol and read by the other are marked as controllable and buffered. The input MORE to the master P_V is marked controllable

and non-buffered because it is not emitted in $P_T || P_U$. The input HBUSREQ1 to the arbiter which is not presented by P_V is similarly marked controllable. Finally, the parallel composition $P_V || (P_T || P_U)$ is computed.

Note that properties φ_3 and φ_4, which describe the correct control signal sequencing between the master and the arbiter-slave pair, are not included in the specification set. This is so because these properties are handled *implicitly* by our algorithm. Consider for example, the property φ_1 (AG$((Idle_v \wedge Idle_t) \Rightarrow$ A$(\neg Start \,\text{U}\, Opt)))$ that requires the master to move to state labelled by $Start$ (v_1) only after the arbiter has granted access by moving to state t_1. Now, the master can only move to v_1 when the signal HGRANT2 is provided to it by the arbiter. Furthermore, the arbiter only emits this signal when it makes a transition to t_1. Now because HGRANT2 is marked as a *buffered* signal, any transition in the master using HGRANT2 is disabled by the converter before the signal is read (and buffered) from the arbiter-slave pair. This achieves the behaviour intended by φ_3, making its addition the specification set redundant. φ_4 is also be handled in a similar manner.

Tableau Generation. Given the parallel composition of participating protocols, the algorithm attempts to build a successful *tableau* for the parallel composition given the control and data constraints to be satisfied. A tableau is a graph that contains a finite number of nodes and edges. Each node NODE relates to a state in the parallel composition of participating protocols, a set FS of CTL formulas, a set I of counter valuations, a set H of already visited nodes to ensure termination, and a set of events E that have been buffered at the node. The tableau generation starts by creating a *root* node which corresponds to the initial state of the parallel composition, the original set of CTL properties to be satisfied, a set of counter valuations where each counter is set to 0 and empty H and E sets. The above root node is passed to the main recursive function. Given such a NODE, its corresponding state s is checked against formulas in FS. The function either returns the NODE back indicating that a converter for the given node can be achieved or returns a false-node indicating failure. If the recursive function returns successfully for the root node, the algorithm returns tt, implying that a converter can be automatically constructed.

Given the parallel composition $P_V || (P_T || P_U)$, each tableau node relates to a state (v, t, u) in the composite system, a valuation for *counter*, and a set of CTL formulas. The root node corresponds to the initial state (v_0, t_0, u_0) of $P_V || (P_T || P_U)$, the formula set $\{\varphi_1, \varphi_2, \varphi_d\}$, the set I where *counter* is set to 0 and empty H and E sets.

The above root node is passed to the main function which proceeds as follows. Each formula in FS ($\{\varphi_1, \varphi_2, \varphi_d\}$) is broken down into present and next-state commitments. Present-state commitments are those which must be satisfied by the state s corresponding to the current node. Next-state commitments are formulas of the type AX or EX which must be satisfied by the successors of s. The steps involved in breaking down the formula set FS for the root node of the consumer master and arbiter-slave example is shown are shown in Tab. II. Consider how the formula φ_1 = AGEF$Start$ is processed (Tab. II, steps 1-3). The formula is initially broken down into the conjunction EF$Start \wedge$ AXAGEF$Start$ and the conjuncts are added back to FS. The formula EF$Start$ is a present-state commitment whereas AXAGEF$Start$ is a next-state commitment. EF$Start$ is processed

496

978-1-4244-3039-0/08 $25.00 © 2008 IEEE

Inputs/Outputs	State	Type
¬(HBUSREQ1 ∧ HBUSREQ2) ∧ ¬HSELECT1/{HBUSREQ2}	(v_0, t_0, u_0)	**Enabled**
¬(HBUSREQ1 ∧ HBUSREQ2) ∧ HSELECT1/{HBUSREQ2}	(v_0, t_0, u_1)	Disabled (buffering)
¬(HBUSREQ1 ∧ HBUSREQ2) ∧ HSELECT1 ∧ HGRANT2/{ }	(v_1, t_0, u_1)	Disabled (buffering)
¬(HBUSREQ1 ∧ HBUSREQ2) ∧ ¬HSELECT1 ∧ HGRANT2/{ }	(v_1, t_0, u_0)	Disabled (buffering)
HBUSREQ1 ∧ ¬HSELECT1/{HGRANT1, HBUSREQ2}	(v_0, t_1, u_0)	Disabled (buffering)
HBUSREQ1 ∧ HSELECT1/{HGRANT1, HBUSREQ2}	(v_0, t_1, u_1)	Disabled (buffering)
HBUSREQ1 ∧ HSELECT1 ∧ HGRANT2/{HGRANT1}	(v_1, t_1, u_1)	Disabled (buffering)
HBUSREQ2 ∧ ¬HSELECT1 ∧ HGRANT2/{HGRANT1}	(v_1, t_1, u_0)	Disabled (buffering)
HBUSREQ2 ∧ ¬HSELECT1/{HGRANT2, HBUSREQ2}	(v_0, t_1, u_0)	**Enabled**
HBUSREQ2 ∧ HSELECT1/{HGRANT2, HBUSREQ2}	(v_0, t_1, u_1)	Disabled (buffering)
HBUSREQ2 ∧ HSELECT1 ∧ HGRANT2/{HGRANT2}	(v_1, t_1, u_1)	Disabled (buffering)
HBUSREQ2 ∧ ¬HSELECT1 ∧ HGRANT2/{HGRANT2}	(v_1, t_1, u_0)	**Enabled**

TABLE I

SUCCESSORS OF (v_0, t_0, u_0).

further to form the disjunction $Start \lor \text{EXEF} Start$. The first disjunct $Start$ is a proposition and is checked against the labels of (v_0, t_0, u_0). As the state is not labelled with $Start$, the other disjunct $\text{EXEF} Start$ must be satisfied by s. However, this formula is not broken down further because it is a next-state commitment. The other formulas are processed in a similar manner till FS only contains next-state commitments (see step 8, Tab. II).

Step	FS
1	AGEF$Start, \varphi_2, \varphi_d$}
2	{EF$Start$, AXAGEF$Start, \varphi_2, \varphi_d$}
3	{$Start \lor$ EXEF$Start$, AXAGEF$Start, \varphi_2, \varphi_d$}
.	...
.	...
8	{EXEF$Start$, EXEF$DOut_{32}$, AXAGEF$Start$, AXAGEF$DOut_{32}$, AXAG$(0 \leq counter \leq 2)$}

TABLE II

PROCESSING THE ROOT NODE

Once only next-state commitments remain in FS, the algorithm passes these commitments to the successors of s in the following manner. Firstly, all AX-formulas in FS are aggregated in the set FS_AX whilst the set FS_EX contains the remaining EX formulas of FS. Next, a *conforming* subset of the set of successors of s is computed. Given the state s with n successors, there are $2^n - 1$ possible non-empty successor subsets. Some of these successors must always be enabled by the converter as the events triggering transitions from s to such successors are *uncontrollable* by the converter. Hence a conforming subset must contain all such successors. Similarly, some successors of s cannot be reached because the signals required to trigger transitions from s to such successors are not buffered in E. Hence, a conforming subset must not contain any such successors.

Take for example the root node for the consumer master and arbiter-slave pair. After all present-state commitments are checked, FS_AX and FS_EX are computed. FS_AX is the formula-set {AGEF$Start$, AGEF$DOut_{32}$, AG$(0 \leq counter \leq 2)$} whilst FS_EX is {EF$Start$, EF$DOut_{32}$} (see FS in Step 8, Tab. II).

Next, a conforming subset $Succ$ is selected. Although the state (v_0, t_0, u_0) has a number of successors, only three can possibly be enabled (see Tab. I). This is because signals like HREADY, HBUSREQ2 and HGRANT2 are buffered signals and transitions involving the presence of such signals can only be triggered if the signals are present in E (which is empty for the root node) or are present as outputs in the corresponding transition (emitted at the same time as when they are required). Hence a conforming

subset of (v_0, t_0, u_0) must not contain any disabled transitions and contain at least one enabled transition.

Then, for each state s' in the conforming subset $Succ$, a node NODE' is formed (with NODE as its parent). All AX commitments contained in FS_AX are passed to every NODE'. Each formula in FS_EX is distributed as a commitment to any one of the newly created nodes. I is updated to I' by checking the labels of s' (some labels may increment/decrement a counter), H' is updated to include the node NODE (the parent of NODE') and E' contains all signals emitted in the transition from s to s' along with any remaining signals in E (some buffered signals may have been used for the transition from s to s'). Then, the same recursive process described above is used to check if each NODE' satisfies all commitments (all AX commitments and some EX commitments) passed to it. If any NODE' returns a failure, we move to select a different distribution of the formulas in FS_EX for the children nodes. If a distribution that allows the satisfaction of all future commitments is found, the node NODE' is returned (signifying success). On the other hand, if no distribution of the formulas in FS_EX returns success, another conforming subset is chosen. If no conforming subset that satisfies the future commitments of NODE under any possible distribution can be found, we return failure.

If a successful tableau is generated by the above procedure, a converter is automatically extracted. A tableau node refers to a state s in the parallel composition of participating protocols, a specific valuation I of all counter variables, a set of CTL formulas FS and a set of buffered events E. The children of each successful tableau node represent the transitions that can be enabled by the converter in s (given specific I, FS and E). This information is stored in the converter which can then guide each state s of the protocol parallel composition such that given constraints are met.

For the consumer-master and arbiter-slave example, a successful tableau can be generated. The converter shown in Fig. 3 is generated automatically by processing the successful tableau in the manner described above.

Complexity. The complexity of the algorithm can be obtained from the number of recursive calls. It is of the order $O(|\text{I}| \times 2^{|S|} \times 2^{|\text{FS}_i|} \times 2^{|\text{E}|})$ where $|\text{I}|$ is the size of the counter set, $|S|$ is the size of the state space of the parallel composition of participating protocols, $|\text{FS}_i|$ is the size of the set of formulas to be satisfied by the initial state of the protocol composition, and $|\text{E}|$ is the maximum size of the buffered signal set contained in the converter.

Converter existence, correctness and algorithm termination.

No.	Name	No. of states in $P_1 \| P_2$	Properties	Result
1	Handshake-serial [19]	4	No read before corr. write (φ_{1_1}). Outputs a and b alternate along every path (φ_{1_2}).	Success
1.1	Handshake-serial (data-mismatch)	12	$\varphi_{1_1}, \varphi_{1_2}$ Data is eventually consumed before more is written.	Success
1.2	2-way communication	12	$\varphi_{1_1}, \varphi_{1_2}$	Success
2	ABP sender (8-bit data)- NS receiver (8-bit data) [5]	18	Each output is eventually read (φ_{2_1}). Another output allowed only after an input (φ_{2_2}).	Success
2.1	16-bit ABP, 8-bit NS	18	$\varphi_{2_1}, \varphi_{2_2}$	Success
2.2	8-bit ABP, 16-bit NS	18	$\varphi_{2_1}, \varphi_{2_2}$	Success
3	ABP receiver (8-bit data)- NS sender (8-bit data) [5]	24	Each output is eventually read (φ_{3_1}). Another output allowed only after an input (φ_{3_2}).	Success
3.1	16-bit ABP, 8-bit NS	24	$\varphi_{3_1}, \varphi_{3_2}$	Success
3.2	8-bit ABP, 16-bit NS NS sender (16-bit data)	24	$\varphi_{3_1}, \varphi_{3_2}$	Success
4	Poll-End receiver (8-bit data)- Ack-Nack sender (8-bit data) [21]	6	No overflow or underflow during data communication (φ_{4_1}).	Success
4.1	8-bit poll-end, 16-bit Ack-Nack	9	$\Rightarrow \varphi_{4_1}$	Success
4.2	16-bit poll-end, 8-bit Ack-Nack	9	$\Rightarrow \varphi_{4_1}$	Failure
5	16-bit Master, 8-bit slave	9	Correct handshaking sequence. Each output is consumed before another output.	Success
6	Mutex	16	Mutual exclusion always achieved	Success
7	8-bit Reader, 24-bit writer	12	Correct handshaking sequence (φ_{7_1}) No data overflows or underflows (φ_{7_2}). Both protocols reset after each transaction (φ_{7_3}).	Success
7.1	5-bit Reader, 4-bit writer	12	($\varphi_{7_1}, \varphi_{7_2}, \varphi_{7_3}$)	Success
7.2	2-bit Reader, 9-bit writer	12	($\varphi_{7_1}, \varphi_{7_2}, \varphi_{7_3}$)	Success
8	MCP missionaries-	30	All missionary-cannibal pairs transported without loss	Success
9	4-bit ABP sender-modified receiver	166432	Sender can always eventually read data.	Success
10	SoC Master-Slave	15	Each protocol resets after one transaction.	Success
11	Pipeline-Handshake [7]	24	Pipeline can always read data in a transaction.	Success
12	AMBA AHB-master and BVCI Target	20	Data is eventually read.	Success
13	16-bit Producer and AMBA **AHB**, 32-bit consumer,arbiter, 32-bit mem	224	Correct control signal sequencing. No data loss and fairness.	Success
14	Producer, consumer, arbiter, memory on AMBA **ASB**	224	Correct control signal sequencing. No data mismatches and fairness.	Success
15	Producer, consumer, arbiter, memory on AMBA **APB**	224	Correct control signal sequencing. No data mismatches and fairness.	Failure
16	Modified producer, consumer, arbiter , memory on AMBA **APB**	128	Correct control signal sequencing. No data mismatches and fairness.	Success

TABLE III

BENCHMARKING RESULTS

The following theorem theorem proves that our approach comprehensively handles the questions of converter correctness and existence. It can also be proved that the given algorithm always terminates even in the presence of bounded counters and recursive CTL formulas.

Theorem 1 (Sound and Complete): Given the protocol composition SKS $P_1 \| P_2 = \langle AP_{1\|2}, S_{1\|2}, s_{0_{1\|2}}, \Sigma_{I_{1\|2}}, \Sigma_{O_{1\|2}}, R_{1\|2}, L_{1\|2}, clk \rangle$ of two protocols P_1 and P_2, an initial set of counter valuations I, a set E describing bufferable of signals toevent buffering rules E and a set of CTL formulas FS, a converter that can control the protocol composition to satisfy all properties in FS exists *iff* the tableau generation algorithm returns *tt*.

V. RESULTS

The conversion algorithm has been developed using C/C++ and by extending the NuSMV model checker [6]. Tab. III provides conversion results obtained from a number of mismatched protocol pairs that have been chosen to showcase the capabilities of the proposed approach. The first two columns describe the mismatched protocols and the third column contains the number of states in their parallel composition. The description of CTL properties used for conversion is provided in the next column. The final column states the result of conversion (success/failure).

Problems 1–4 are classical protocol conversion examples explored in earlier works on protocol conversion [5], [19], [21]. Our approach is able to handle these examples in a similar manner to earlier works *and* can handle variations that extend these examples. These variations, which involve bidirectional communication between protocols and data-width mismatches, cannot be handled by earlier works which explored these problems. Problems 5–9 are synthetically created benchmarks which model commonly encountered protocol mismatches such as control sequencing. Problems 8 and 9 are well-known NuSMV examples that were modified to create control sequencing mismatches. For these examples, our approach was generate converters to ensure proper control sequencing and mutual exclusion. Furthermore, problem 7 (and variants 7.1 and 7.2) demonstrate the ability of our algorithm to handle arbitrary data-width mismatches. No other existing technique can handle arbitrary data-widths.

Finally, some SoC protocols are presented in problems 10–16. These problems model control and data mismatches that cannot be handled by other protocol conversion techniques. Problem 13 looks at matching the producer-master P_S with the rest of the SoC system presented in this paper (Fig. 2). Problems 14–16 look at the same problem but use different AMBA buses, namely ASB (Advanced System Bus) and APB (Advances Peripheral Bus).

In some cases our algorithm fails to generate a converter. Problem 4.2 fails because the 8-bit Ack-Nack sender is unable to write more than 8-bit data per-transaction while the poll-end receiver reads 16-bits, leading to underflows. Problem 15 fails because of

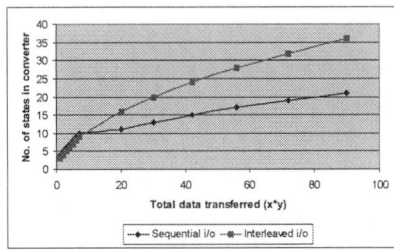

Fig. 4. Reader-writer protocol pair

the inability of the AMBA APB to allow burst transfers. When failures occur, mismatches between protocols cannot be bridge by any converter (that exercises the same control over protocols as in our setting) under the given inputs (protocols, properties, buffering and counter rules–see theorem 1) and therefore, the inputs need to be modified manually. Although our technique fails to generate a converter for some problems, it provides traces (counter-examples) that may help during component modification. Faulty traces returned by our algorithm can pinpoint which CTL and protocol state caused a failure. Failures caused by empty signal buffers or uncontrollable actions can also be revealed.

Fig. 4 shows how the number of states in the generated converter changed when the total number of reads and writes per-transaction in the reader-writer protocol pair (problem 7) was varied. In the first sequence (sequential i/o), the data communication behaviour was constrained such that in each transition, the writer must first write data to completely fill the communication medium (fixed at $x \times y$ where x and y are the weights for reads and writes respectively) and the reader must then completely read all data. For this purpose, CTL properties of the form $AG(TransactionBegin \Rightarrow A(\neg Read \, U \, counter = (x \times y))$) and $AG(counter = (x \times y) \Rightarrow A(\neg Write \, U \, (counter = 0 \wedge TransactionEnd))$) were used. In the second case (interleaved), the protocols were allowed to arbitrarily read and write data as long as counter boundaries were not breached.

It was observed that as the total size of data to be exchanged was increased, the number of states in the converter also increased in an almost linear fashion. The number of converter states in the sequential i/o case never exceeded the interleaved case because of the stronger data constraints used. Converters were generated even when the data sizes were not integer multiples of each other.

These results demonstrate that we have developed, the first formal technique for SoC protocol conversion problem that can handle control mismatches, data-width mismatches and additional specification verification (CTL constraints) using a single unifying solution. We have also demonstrated through a series of experiments a practical SoC application based on the AMBA bus that the proposed approach can be used in real SoC designs. Another advantage of this technique is the feedback obtained in case of a failure to match protocols.

VI. CONCLUSIONS

In this paper, we present a unifying approach towards performing conversion for protocols with control and data mismatches. Earlier works could only handle control or data mismatches separately or handle both in a restricted manner. Our fully automated approach is based on CTL module checking, allowing temporal logic specifications for both control and data constraints.

Converters synthesized in our setting are capable of buffering signals and can handle several protocol mismatch problems. We also present comprehensive experimental results to show the practical applicability of our approach.

We are currently working on extending the proposed approach to deal with clock mismatches. We claim that the same algorithm can be extended with little effort to handle extensions to the logic while clock mismatches can be represented and addressed by using multi-clock Synchronous Kripke Structures [20].

REFERENCES

[1] M Antoniotti. *Synthesis and verification of discrete controllers for robotics and manufacturing devices with temporal logic and the Control-D system*. PhD thesis, New York University, New York, 1995.

[2] G V Bochmann. Deriving protocol converters for communication gateways. *IEEE Transactions on Communications*, 38(9):1298–1300, September 1990.

[3] G Borriello and R H Katz. Synthesis and optimization of interface transducer logic. In *Digest of Technical Papers of the IEEE International Conference on Computer-Aided Design*, pages 274–277, 1987.

[4] F M Burg and N D Iorio. Networking of networks: Interworking according to osi. *IEEE Journal on Selected Areas in Communications*, 7(7):1131–1142, September 1989.

[5] K L Calvert and S S Lam. Formal methods for protocol conversion. *IEEE Journal on Selected Areas in Communication*, 8(1):127–142, 1990.

[6] R Cavada, A Cimatti, G Keighren, E Olivetti, M Pistore, and M Roveri. *NuSMV 2.1 User Manual*, 2006.

[7] V D'Silva, S Ramesh, and A Sowmya. Bridge over troubled wrappers: Automated interface synthesis. In *VLSID'04*, 2004.

[8] V D'Silva, S Ramesh, and A Sowmya. Synchronous protocol automata : A framework for modelling and verification of soc communication architectures. In *DATE*, pages 390–395, 2004.

[9] V d'Silva, S Ramesh, and A Sowmya. Synchronous protocol automata: a framework for modelling and verification of soc communication architectures. *IEE Proc. Computers & Digital Techniques*, 152(1):20–27, 2005.

[10] S Gorai, S Biswas, L Bhatia, P Tiwari, and R S Mishra. Directed-simulation assisted formal verification of serial protocol and bridge. In *Proceedings of the 43rd annual conference on Design automation DAC '06*, pages 731 – 736, 2006.

[11] P Green. Protocol conversion. *IEEE Transactions on Communications*, 34(3):257–268, March 1986.

[12] S Jiang and R Kumar. Supervisory control of discrete event systems with ctl* temporal logic specifications. *SIAM Journal on Control and Optimization*, 44(6):2079–2103, 2006.

[13] R Kumar and S S Nelvagal. Protocol conversion using supervisory control techniques. In *IEEE International Symposium on Computer-Aided Control System Design*, pages 32–37, 1996.

[14] O Kupferman, M Y Vardi, and P Wolper. Module checking. *Information and Computation*, 164:322–344, 2001.

[15] S Lam. Protocol conversion. *IEEE Transactions on Software Engineering*, 14(3):353–362, 1988.

[16] F Maraninchi and Y Remond. Argos: an automaton-based synchronous language. *Computer Languages*, 27:61–92, 2001.

[17] S Narayan and D Gajski. Interfacing incompatible protocols using interface process generation. In *Design Automation Conference*, pages 468–473, 1995.

[18] K Okumura. A formal protocol conversion method. In *ACM SIGCOMM 86 Symposium*, pages 30–37, 1986.

[19] R Passerone, L de Alfaro, T A Henzinger, and A L Sangiovanni-Vincentelli. Convertibility verification and converter synthesis: Two faces of the same coin. In *International Conference on Computer Aided Design ICCAD*, 2002.

[20] I Radojevic, Z Salcic, and P S Roop. Mccharts and multiclock fsms for modelling large scale systems. In *Fifth ACM-IEEE International Conference on Formal Methods and Models for Codesign (MEMOCODE'2007)*, 2007. Accepted for publication.

[21] M Rajagopal and R E Miller. Synthesizing a protocol converter from executable protocol traces. *IEEE Transactions on Computers*, 40(4):487–499, 1991.

[22] P J G Ramadge and W M Wonham. The control of discrete event systems. *Proceedings of the IEEE*, 77(1):81–98, January 1989.

978-1-4244-3039-0/08 $25.00 © 2008 IEEE

SESSION B4:
Low Power - II

21st International Conference on VLSI Design

Energy Reduction in SRAM using Dynamic Voltage and Frequency Management

Mohammed Shareef I, Pradeep Nair, Bharadwaj Amrutur

Abstract

This paper describes a dynamic voltage frequency control scheme for a 256X64 SRAM block for reducing the energy in active mode and stand-by mode. The DVFM control system monitors the external clock and changes the supply voltage and the body bias so as to achieve a significant reduction in energy. The behavioral model of the proposed DVFM control system algorithm is described and simulated in HDL using delay and energy parameters obtained through SPICE simulation. The frequency range dictated by an external controller is 100MHz to 1GHz. The supply voltage of the complete memory system is varied in steps of 50mV over the range of 500mV to 1V. The threshold voltage range of operation is ±100mV around the nominal value, achieving 83.4% energy reduction in the active mode and 86.7% in the stand-by mode. This paper also proposes a energy replica that is used in the energy monitor subsystem of the DVFM system.

***Index Terms*—Delay Monitor, DVFM, Energy reduction, Energy monitor, Pareto optimal curve, Replica circuits, SRAM.**

1. Introduction

Dynamic Voltage Frequency Management has been a very effective technique for optimizing the energy consumption of logic circuits [1], [2]. The same technique can be extended to memories to gain power benefits as memory accounts for a large part of the system power. In this work, we have designed a 256×64 SRAM which adjusts its operating voltage and device threshold voltage in response to a change in the operating frequency. The algorithm used to arrive at the operating point is tested using VHDL to ensure stable operation. Finally, the energy saving with our scheme is found by simulating the SRAM with the operating values obtained by VHDL simulations.

Several techniques for reducing power in SRAMs have been discussed in [3]. However at sub-micron technologies, less than 65nm, the leakage energy is a major component of the total energy, even in the active mode. The DVFM control scheme promises to optimize the total energy consumption by suitably controlling the supply voltage and the threshold voltage of the transistors in response to a change in the operating frequency.

Section II explains the SRAM design and its timing models. Section III explains the proposed DVFM algorithm and its subsystems like the delay monitor and the proposed energy replica. Section IV reports the simulation results and Section V summarizes the work.

2. SRAM design

2.1. SRAM cell sizing

The Static Noise Margin (SNM) of 6T cell depends on supply voltage, threshold voltage and trans-conductance (β) ratios $r=\beta_d/\beta_a$ and $q/r= \beta_p/\beta_d$ where β_a, β_d, β_p are the trans-conductance factors of the access, driver and pull up transistors [3]. The worst case read SNM is reached at the lowest supply voltage of 500mV and a body bias corresponding to the lowest device threshold voltage ($V_{Th (nominal)} - 100\text{mV}$). The cell is sized to give a reasonable SNM for the worst case. The aspect ratios of the NMOS driver, NMOS access and the PMOS pull-up transistors are $8\lambda/2\lambda$, $4\lambda/2\lambda$ and $3\lambda/2\lambda$ respectively. This sizing yields an acceptable read SNM of 85mV at 500mV and a threshold voltage of ($V_{Th (nominal)} - 100\text{mV}$).

The write SNM at the worst case is 200mV. To enable writing at low voltages, power-line-floating write technique is used, which also reduces power while writing [4]. The V_{DD} lines of all the cells in a row are connected through a PMOS transistor. During write the V_{DD} line of the selected row is made to float, this makes the cells in that row unstable thus easing the write operation.

2.2. Timing Signal Generation

The timing signals such as write, precharge and

503

978-1-4244-3039-0/08 $25.00 © 2008 IEEE

sense clock are generated internally based on the chip select and read/write signal from the external controller. The bit-line swing should be restricted in order to avoid switching of high capacitance associated with the bit lines and large delays in reading and precharging. The bit line swing of a SRAM cell is replicated using a dummy cell, with capacitance ratio method [5], which tracks the process variations closely. The sense clock timing generation using dummy cell is shown in Fig.1. The dummy cell is designed such that its bit-line falls ten times more rapidly than the bit-line of a SRAM column. So a latch type sense amplifier is used which can amplify a differential of one-tenth of the supply voltage.

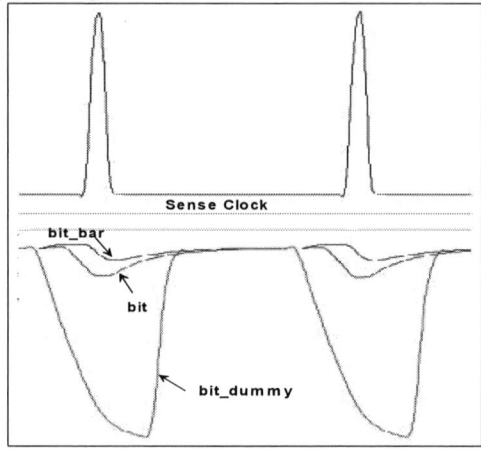

Fig. 1. Sense Clock Generation

The word line has to reset when the sense clock is activated to avoid wastage of power. This is achieved by the word-line reset circuit shown in Fig.2. So there is a dummy column in the SRAM memory array with 256 dummy cells, one for each word line.

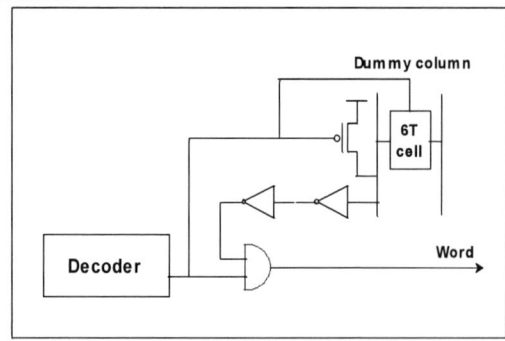

Fig 2. Word resetting circuitry

2.3. Decoder Design

Decoder design issues for Fast Low Power operation have been explained in [6]. There are a few architectures which have advantage in terms of performance and power (like DWL [7] and SCPA

[8]). However, these have a greater advantage when applied to bigger SRAM blocks.

There are many different circuit styles implementing n-input AND function [6]. However, the conventional NAND implementation has the advantage of lower power as compared to the NOR implementation if the performance constraints are not tight.

The layout of a unit 6T cell, $17\lambda \times 38\lambda$, was considered in the modeling of the wire resistance and capacitance. The simulations were done on a 65nm PTM model files.

3. Dynamic voltage and frequency Management (DVFM)

3.1. DVFM Algorithm

The basic philosophy behind DVFM control as explained in [1] is to first find the minimum voltage at which the chip can operate, for a given frequency, and then to find the optimum body bias for which the switching and leakage currents are in a certain ratio.

The active energy consists of switching component and a leakage component.

$$E_{ACTIVE} = E_{SW} + E_{LEAK}$$

The switching and leakage energies are a strong decreasing function of V_{DD} and V_T respectively as shown in Fig. 3 and Fig. 4.

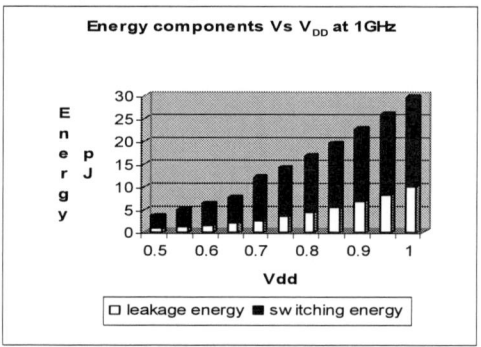

Fig.3. Energy variation w.r.t Vdd

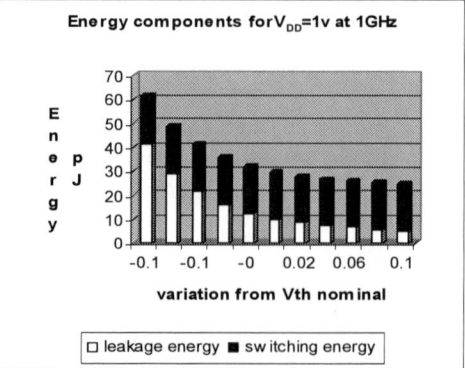

Fig.4. Energy variation Vs Vth

Our proposed DVFM algorithm for a memory system is shown in Fig. 5. The sequence of events is explained below.

1. The controller senses a change in frequency. A DVFM control cycle is initiated.

2. The supply voltage is varied in steps of 50mV until the lower bound of delay (discussed in Section 3.3) is met.

3. The body voltage is varied to change the device thresholds in steps of 100mV until the upper bound of delay (discussed in Section 3.3) is met.

4. The supply voltage is adjusted for the last time to ensure that the delay is above the upper bound by a convenient margin.

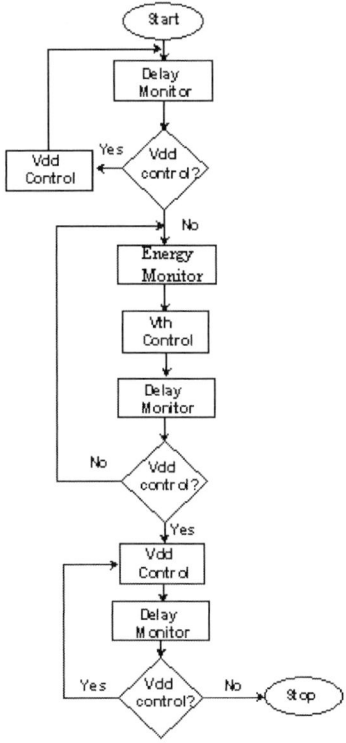

Fig.5. DVFM Algorithm for Active mode

3.2. DVFM Block diagram

The DVFM system has two main subsystems. The Delay monitor to check the delay by varying the supply voltage and the Energy monitor which monitors the energy consumed by the energy model at that operating point. Both the blocks are discussed in detail in the following section.

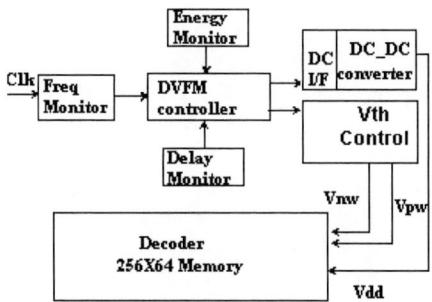

Fig.6. Block Diagram of Proposed DVFM

3.3. Delay monitor and Delay Synthesizer

The delay monitor circuit [1] [2] has a critical path delay model and buffers with 32 latches. The delay of each buffer is tailored to reflect a change in the delay monitor circuit by a Vdd step of 50mV. Each buffer in the delay line gauge has a delay of 20ps. Out of the 32 bits the first 9 bits, 180ps of delay margin, is reserved for process variations. The 21st bit and the 9th bit are, respectively, the lower and upper bounds of delay used by the DVFM algorithm.

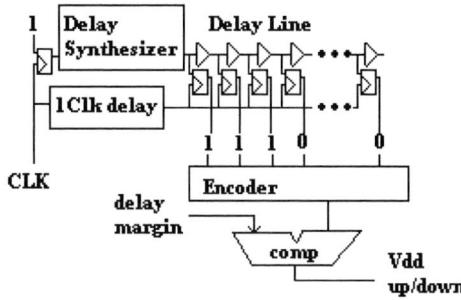

Fig.7. Delay monitor circuit

The delay synthesizer mimics the read delay from the time of arrival of address bits on to the decoder to the time data is put on to the data bus via the sense amplifiers. The circuit of delay synthesizer is shown in Fig. 8. The sense amplifier timing replica is not shown, which is very similar to the word reset replica as already discussed.

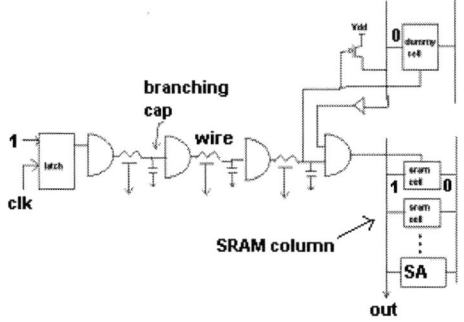

Fig.8. Delay Synthesizer.

505

978-1-4244-3039-0/08 $25.00 © 2008 IEEE

3.4. Energy Monitor and energy replica circuit

The energy monitor circuit proposed in [1] is inaccurate because the entire circuit is mimicked by a single capacitor in the switching current monitor. Our energy monitor circuit works on the following principle. For a constant Vdd, the first step is incrementing the Vth by a step and comparing the new energy with the previous one. If the new energy is lesser than the previous then Vth is taken towards higher values. Else, it is decremented. When it reaches its nearest minima the loop settles and Vdd control is done once before exiting the DVFM loop. For every Vth step the delay is monitored and if the delay crosses the 9th bit then Vth control is stopped.

The energy monitor circuit is shown in Fig. 9. The energy monitor circuit works as follows. A capacitor C1 is charged to the supply voltage. C1 is then connected to the energy replica as its supply and the replica is operated for a particular number of read cycles. The expression $0.5 \ast C1 \ast (V_{INITIAL}^2 -- V_{FINAL}^2)$ gives the true energy consumed by the energy replica. Now this voltage on C1 is transferred to the other capacitor C2 through a unity gain buffer [10] which is shown in Fig. 10. Now the Vth is taken to the next step and a similar read process is done on the energy model. The voltage left on the capacitor C1 is compared to the previous voltage, which is now in C2. If voltage on C1 is higher then the VTH_UP signal is generated else VTH_DOWN.

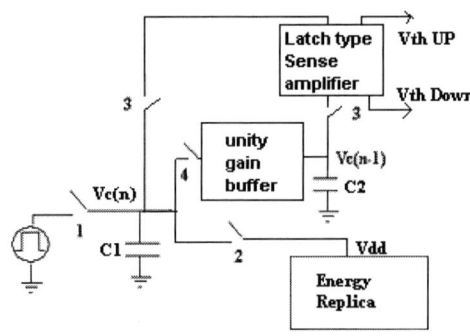

Fig.9. Proposed energy monitor circuit

A latch type sense amplifier is used to compare the voltages on the two capacitors. The capacitors are of a known value and they don't have any relation with the chip switching capacitance. The switches are closed as per the shown sequence.

TABLE I
ENERGY OVERHEAD OF DVFM BLOCKS

Entity	Energy consumed at 100MHz
Delay Monitor	51.8pJ for 11 iterations in worst case
Energy Monitor	615pJ for an average of ten steps of each ten read cycles from the energy replica.

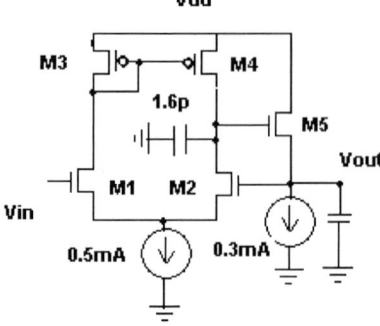

Fig.10. Unity gain buffer.

The energy replica is designed to mimic the true energy of the memory array along with the decoder faithfully. The energy replica consumes an energy $1/30^{th}$ to that of the original system. Two dummy memory columns are employed in the energy replica to replicate 64 columns. Here it is to be noted that along with the energy scaling of nearly $1/30^{th}$ both the switching and the leakage energies are also scaled proportionately, thus giving true measure at all operating voltages and frequencies. Similarly the decoder is also separated into switching and leaking gates appropriately and down-sized to get the same ratio as shown in Fig. 11. So the energy measured by the energy replica will be a scaled down version of the true energy. The area occupied by the energy replica is also in the same ratio which is a very small percentage (3%) of the whole memory area.

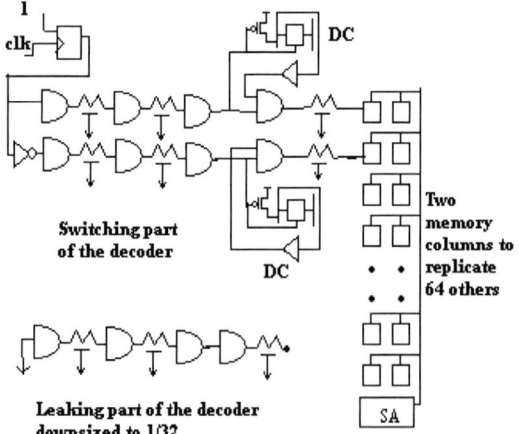

Fig.11. Proposed Energy Replica

The validation of the energy replica can be seen from Fig. 12. It shows the percentage of replica energy w.r.t the actual energy across supply voltages and across different threshold voltages. It is confined in the range 3.15% to 3.8%. This proves that the energy replica has both the switching and leakage components in the same proportion as the actual system.

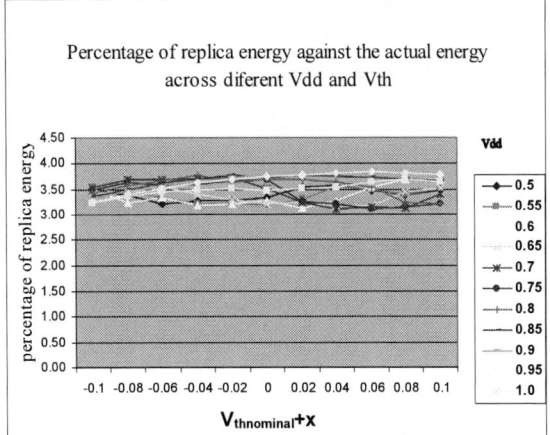

Fig.12. Plot showing the validation of replica energy across different threshold voltages and supply voltages.

3.5. Standby energy reduction

The SRAM block supply voltage is scaled down to the minimum possible voltage which is decided by the data retention voltage of the 6T cell. The SRAM cell has been sized to give acceptable noise margin at Vdd of 500mV and hold data. Then energy monitor scheme as described for the active mode is used to
vary Vth by comparing the energies consumed in the present step with the previous step. The one difference in standby mode using the energy model is to monitor the leakage energy for a particular interval of time.

3.6. DVFM system overhead

Overhead of the DVFM blocks in terms of energy is shown in Table I. The DVFM system comes into active mode of operation only when there is a frequency change indication to the memory controller. Once it arrives at the optimum operating points for that particular frequency it shuts off, till the next frequency change. The Delay monitor energy is measured assuming it works for 11 V_{DD} iterations in the worst case. The Energy Monitor similarly is assumed to work for an average of ten iterations of threshold step till the Vth control terminates. With these values the overhead of DVFM system can be calculated with an assumption that change of frequency occurs once in 100 clock cycles, which is reasonable. This energy overhead is 18% of the total energy saved through DVFM.

3.7. Integration of DVFM sub-systems with Memory

The layout of a 6T SRAM cell is shown in Fig. 13. In order to enable the implementation of power-line-floating write the V_{DD} lines of cells in the same row have to be grouped and separated from the V_{DD} lines of the other column. Fig. 14 gives an idea about the placement of DVFM sub-blocks like the delay monitor and more importantly the energy replica which has two replica memory columns. The column to the immediate left of the decoder is the dummy column used for resetting the word-line.

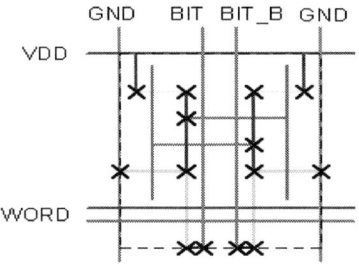

Fig.13 Layout of a 6T SRAM cell with horizontally running VDD line

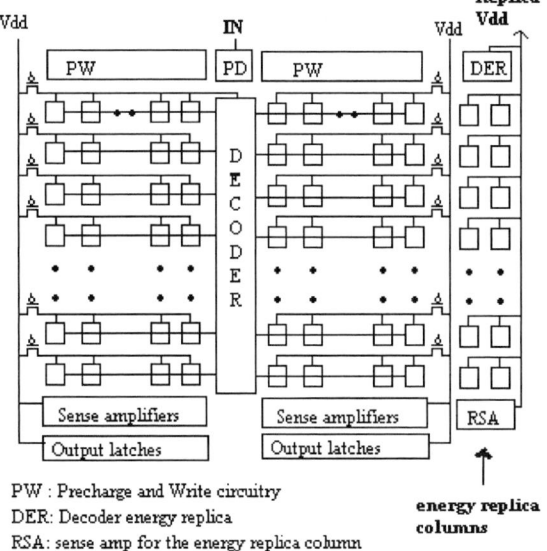

PW : Precharge and Write circuitry
DER: Decoder energy replica
RSA: sense amp for the energy replica column
PD: pre-decoder

Fig.14 Placement of energy replica columns along side memory array

The supply lines of the energy replica memory columns and the decoder energy replica are connected together and given to the capacitor C1 in Fig.9. It is to be noted that the voltages of the entire memory and its peripherals are not changed until the optimum values are not arrived at and the memory is out of access for the external world. With the optimum operating values, the DVFM shuts off and these values are applied to the whole system till the next frequency change.

507

978-1-4244-3039-0/08 $25.00 © 2008 IEEE

4. Simulation results

Fig.15 gives the energy Vs performance plot in which the pareto-optimal curve (the lower boundary) is seen, where the sensitivities of Vdd and Vth are same. The DVFM system tries to choose the optimum Vdd and Vth points on this curve. This curve was obtained by plotting energy values Vs the corresponding inverse delay values that is a representation of performance, which were obtained from exhaustive SPICE simulations of the entire SRAM cell array.

The two light gray point encircled white (lower left side among the points) on the Fig. 15 were the operating point chosen by the DVFM algorithm for 100MHz, which are found to be on the Pareto optimal curve. This shows that the algorithm converges to an optimum value.

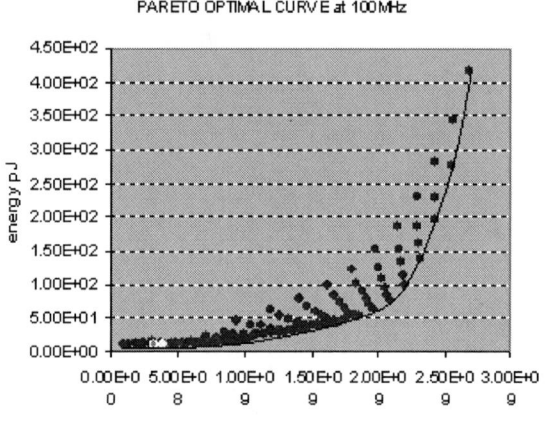

Fig.15. Pareto Optimal curve

At 100MHz operation the energy consumption at nominal Vdd and Vth is 4.35nJ for 100 cycles of read operation. With the simulated DVFM control loop in HDL with the help of SPICE values, the optimum Vdd and Vth arrived at, are 0.5V and Vthnom+40mV respectively and the energy consumed at this operating point is 0.734nJ for 100 cycles of operation. Therefore, the energy saving at 100MHz operation is 83.4% which includes the energy overhead also.

In the standby mode 86.7% energy saving is achieved with data retention at minimum operating voltage of 500mV and at a threshold voltage of Vthnom+20mV.

5. Conclusion

We have described a DVFM scheme for a 256 X 64 SRAM block for active and stand-by energy reduction. The scheme includes an iterative

monitoring of delay and energy to achieve the reduction in energy. The results showed a saving of 83.4% active energy at 100 MHz and 86.7% stand-by energy. The maximum latency of the controller is found to be 46 clock cycles during the active mode control. The operating point obtained from the DVFM simulation, when plotted on a energy Vs performance scale is found to be on the Pareto optimal curve, which reiterates that the values arrived at are optimum.

6. References

[1] M. Nomura et al., "Delay and Power Monitoring Schemes for Minimizing Power Consumption by means of Supply and Threshold Voltage Control in Active and Standby Modes", IEEE Journal of Solid State Circuits, Vol. 41, No. 4, pp.805-814, April 2006.

[2] M. Nakai et al., "Dynamic Voltage and Frequency Management for Embedded Microprocessor", IEEE Journal of Solid State Circuits, Vol. 40, No. 1, pp 28-35, Jan 2005.

[3] Evert Seevinck, F. J. List and J. Lohstroh, "Static-Noise Margin Analysis of MOS SRAM Cells" IEEE Journal of Solid State Circuits, Vol. SC-22, No. 5, pp. 748-754, October 1987.

[4] Masanao Yamaoka et al., "90-nm Process-Variation Adaptive Embedded SRAM Modules With Power-Line-Floating Write Technique", IEEE Journal of Solid State Circuits, Vol. 41, No.35, pp.705-711, March 2006.

[5] Bharadwaj S. Amrutur and Mark A. Horowitz, "A Replica Technique for Word line and Sense Control in Low-Power SRAM's", IEEE Journal of Solid State Circuits, Vol. 33, No. 8, pp. 1208-1219, August 1998.

[6] Martin Margala "Low Power SRAM Circuit Design", Records of the 1999 IEEE International Workshop on Memory technology, Design and Testing, pp.115 – 122, August 1999.

[7] Bharadwaj S. Amrutur and Mark A. Horowitz "Fast Low Power Decoders for SRAMs", IEEE Journal of Solid State Circuits, Vol.36, No.10, October 2001, pp. 1506-1515.

[8] K. Itoh et al., "Trends in Low-Power RAM Circuit Technologies", Proceedings of the IEEE, Vol. 83, No. 4, pp.524-543, April 1995.

[9] M. Ukita et. al., "A Single Bitline Cross-Point Cell Activation (SCPA) Architecture for Ultra Low Power SRAMs", IEEE Journal of Solid State Circuits, Vol. 28, No. 11, pp. 1114-1118, November 1993.

[10] Behzad Razavi "Design of Analog CMOS Integrated Circuits", Tata McGraw Hill Publications.

21st International Conference on VLSI Design

Unified V_{dd} - V_{th} optimization based DVFM controller for a Logic block

Kannan S A
akannan@cedt.iisc.ernet.in

Sreeram N S
nssreeram@cedt.iisc.ernet.in

Bharadwaj S Amrutur
amrutur@ece.iisc.ernet.in

Department of ECE, Indian Institute of Science, Bangalore, India - 560012

Abstract

In this paper analytical expressions for optimal V_{dd} and V_{th} to minimize energy for a given speed constraint are derived. These expressions are based on the EKV model for transistors and are valid in both strong inversion and sub threshold regions. The effect of gate leakage on the optimal V_{dd} and V_{th} is analyzed. A new gradient based algorithm for controlling V_{dd} and V_{th} based on delay and power monitoring results is proposed. A V_{dd}-V_{th} controller which uses the algorithm to dynamically control the supply and threshold voltage of a representative logic block (Sum of Absolute Difference computation of an MPEG Decoder) is designed. Simulation results using 65 nm predictive Technology models are given.

1. Introduction

Dynamically varying frequency of a logic block based on load requirements is effective in minimizing the energy consumptions. It has been shown that optimal supply and threshold voltages exist which minimizes the power consumed while meeting the speed constraint [1] [2] [3]. Most of the earlier attempts to derive the optimal V_{dd}, V_{th} focus either on the strong inversion region or the sub threshold region of operation. But in a generic case where the frequency specifications of a circuit has a wide dynamic range operations in both regions may be necessary. In this paper delay and power models based on the EKV model [4] for the transistors are developed from which we obtain the optimal V_{dd} and V_{th} solutions which are valid in both the regions.

In [5] delay and power monitoring schemes for dynamically controlling V_{dd}, V_{th} is proposed. The algorithm used in [5] varied V_{dd} based on the delay monitoring results and V_{th} based on power monitoring results. This is not satisfactory as the delay depends on V_{th} also and power has a V_{dd} dependency also. A control algorithm which varies V_{dd} and V_{th} in a combined fashion based on the delay and power will give a faster convergence and more accurate results. So a combined V_{dd}-V_{th} control algorithm based on

the sensitivities of delay and power to V_{dd} and V_{th} is discussed in the paper. These sensitivities are obtained from the unified models developed.

The paper is organized as follows. Section 2 describes the analytical solution for optimal V_{dd}, V_{th} based on the EKV model. Section 3 describes the proposed V_{dd}-V_{th} control algorithm, delay and power monitoring circuits and the DVFM controller simulation results. Section 4 concludes the paper.

2 Unified V_{dd} - V_{th} optimization based on EKV model

2.1 Proposed EKV based current model

The V_{dd}-V_{th} optimization for a given target frequency has been done in [1] for above threshold operation and in [2] [3] for sub threshold operation. The transistor current models in these work were based on the classical alpha power law model. Fig.1 shows the comparison of the SPICE simulated I_d values and the alpha power law model fitted values. There are two issues in using this model for V_{dd}-V_{th} optimization.

(i) It uses two different equations to model the current. A power law model for strong inversion region and an exponential model for subthreshold region. So V_{dd}-V_{th} optimizations has to be done separately for these two regions.

(ii) The model does not fit well for regions close to V_{th}.

To overcome these difficulties we chose to use a transistor current model based on the EKV model. The proposed model is given by

$$I_d = I_0[ln(1 + e^{\frac{V_{gs}-V_{th}}{\alpha N_s}})]^\alpha \qquad (1)$$

For high V_{gs} and low V_{gs} the proposed model reduces to the alpha power law model. But for V_{gs} values close to V_{th} it uses an interpolation function to predict the transistor current. This single equation closely follows the measured

509

978-1-4244-3039-0/08 $25.00 © 2008 IEEE

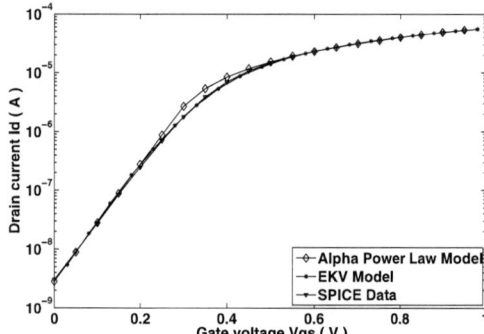

Figure 1. I_d - V_{gs} plot for 65nm transistor of w/l = 65nm/65nm at V_{ds} = 1V.

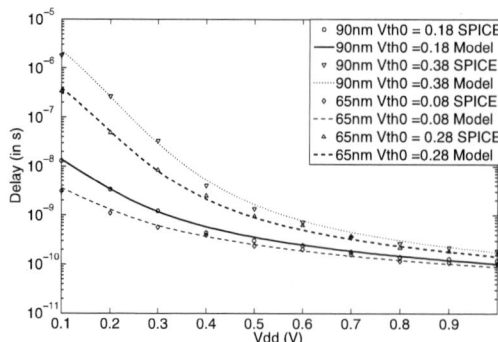

Figure 2. Comparisons of proposed model and SPICE data for 90nm and 65nm processes. Delays shown are that of a ring oscillator in the test circuit.

values in all regions with an maximum error of around 15%. Fig.1 shows the comparison between the spice simulated transistor currents and that given by (1).

2.2 Test circuit

To evaluate the validity of the proposed models a test circuit given in [8] is used. The circuit has a ring oscillator made of 11 NAND gates feeding chains of NAND gates. All the SPICE results given in this section have been obtained by the simulations of this test circuit.

2.3 Delay model

A simple delay model for a gate is given by

$$t_{gate} = \frac{KC_L V_{dd}\theta}{I_0[ln(1 + e^{\frac{V_{dd}-V_{th0max}+\eta V_{dd}}{\alpha N_s}})]^\alpha} \quad (2)$$

where N_s is the subthreshold slope, V_{thmax} is the maximum V_{th} across temperature and process corners, C_L is the load of the gate and K is a delay fitting parameter. θ is a degradation factor to account for increase in delay due to finite rise/fall times. θ is a function of V_{dd}, V_{th}. Through simulations the degradation due to slope was found and modelled as a function V_{dd}, V_{th} as

$$\theta = \frac{1}{[ln(1 + e^{\frac{V_{dd}-V_{th0max}+\eta V_{dd}}{\alpha N_s}})]^\beta} \quad (3)$$

β is a fitting parameter. Fig.2 shows the accuracy of the model in predicting the delay across a wide range of V_{dd} for 65nm(PTM) [9] and 90nm(UMC) processes. Maximum discrepancies of around 17% were seen across a wide range of V_{dd} and V_{th}.

2.4 Power model

Dynamic power is given by

$$P_{dyn} = aC_{Leff}V_{dd}^2 f \quad (4)$$

Where C_{Leff} is the effective switching capacitance and a is the activity factor. In this work the contribution of short circuit power dissipation has been neglected for simplicity as it is a small fraction of total power [1]. Leakage current in deep submicron devices is predominantly due to subthreshold leakage and gate leakage. In this optimization the gate leakage current is neglected. However in a subsequent section the validity of this assumption and its limitations are discussed. Fig.3 shows the comparison of leakage current of the test circuit using 65nm PTM process with the corresponding models.

$$I_{leak} = I_0 W_{eff} e^{\frac{-(V_{th0min}-\eta V_{dd})}{N_s}} \quad (5)$$

$$P_{leak} = I_0 W_{eff} e^{\frac{-(V_{th0min}-\eta V_{dd})}{N_s}} V_{dd} \quad (6)$$

$$V_{th0min} = V_{th0max} - \Delta V_{th0} \quad (7)$$

ΔV_{th0} is peak-to-peak V_{th0} variation across process corners and temperature. V_{th0min} is lowest V_{th0} in operation temperature and process variation range [1]. It is to be noted that delay is calculated at V_{th0max} and leakage power at V_{th0min} as in [1]. The total power is given by

$$P_t = P_{leak} + P_{dyn} \quad (8)$$

2.5 V_{dd} - V_{th} optimization

Using above models optimization for minimum energy is done. solving (2) for V_{thmin} in terms of V_{dd}

$$V_{th0min} = (1+\eta)V_{dd} - \alpha N_s ln(e^{\chi V_{dd}^{\frac{1}{\gamma}}} - 1) - \Delta V_{th0} \quad (9)$$

510

978-1-4244-3039-0/08 $25.00 © 2008 IEEE

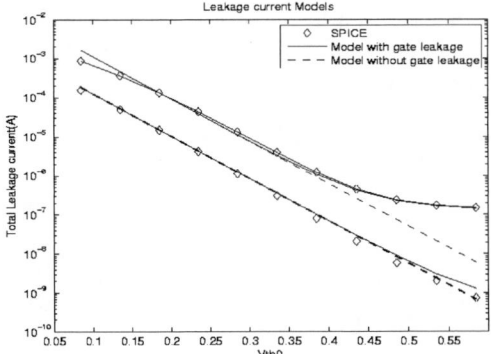

Figure 3. Leakage current of the test circuit. Comparison of SPICE results, equation(5) and equation(14).

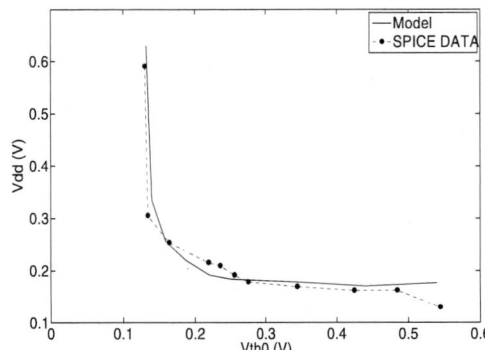

Figure 4. Optimal energy points for various frequencies on the V_{dd}-V_{th} plane for 65nm PTM process. Model and SPICE Data comparison.

Where $\gamma = \alpha + \beta$. Using (6), (9) leakage power is given by

$$P_{leak} = I_0 W_{effe} e^{\frac{-((1+\eta)V_{dd} - \Delta V_{th0})}{N_s}} (e^{\chi V_{dd}^{\frac{1}{\gamma}}} - 1)^{\alpha} V_{dd} \tag{10}$$

where

$$\chi = \left(\frac{KC_L f L_d}{I_0}\right)^{\frac{1}{\gamma}} \tag{11}$$

the optimum values are given by solving

$$\frac{dP_t}{dV_{dd}} = 2aC_{Leff}V_{dd}f +$$

$$P_{leak}\left\{\frac{1}{V_{dd}} - \frac{1+\eta}{N_s} + \frac{\alpha \chi V_{dd}^{\frac{1}{\gamma}-1} e^{\chi V_{dd}^{\frac{1}{\gamma}}}}{\gamma(e^{\chi V_{dd}^{\frac{1}{\gamma}}} - 1)}\right\} = 0 \tag{12}$$

V_{ddopt} is obtained by solving (12) for V_{dd}. Equation (12) is transcendental and is solved numerically. V_{thopt} obtained by substituting V_{ddopt} in (9).

Fig.4 compares the optimal V_{dd}/V_{th} values obtained using the models with those obtained from exhaustive SPICE simulation. The model and SPICE values are very close, further validating the models.

2.6 Discussion on the optimization results

(i) In the strong inversion region V_{thopt} remains almost constant with frequency whereas V_{dd} varies to achieve the necessary speed whereas in the subthreshold region the V_{ddopt} remains almost constant and V_{thopt} varies with frequency [3] [2] [1].

(ii) The optimal energy points occur where the leakage energy and the dynamic energy are comparable. [1] report that optimal energy point occurs when the ratio of

dynamic to leakage power remains constant. This is true both in the strong inversion and the sub threshold regions but the optimality occurs for different ratios in these two regions. If a single ratio is used for the whole range of frequencies, the power level for some range of frequencies will be sub optimal. In fig.5 ratio of energy along the optimal curve to that along the curve Edyn/Eleak = constant is plotted for different delays. The sub optimality is within 10% as shown in fig.5.

2.7 Effect of gate leakage on the optimal curve

We model Gate leakage of a circuit empirically as

$$I_{gateleak} = I_{0gate} * V_{dd}^{k1} * e^{-k2*V_{th0min}} \tag{13}$$

$$I_{leak} = I_{gateleak} + I_{subleak} \tag{14}$$

Where k1, k2 are fitting parameters.

Fig.3 shows the accuracy of the empirical fit for 65nm process. Gate leakage is strongly dependent on V_{dd} and increases with V_{dd}. However in the strong inversion region, optimal V_{th} is very low and hence subthreshold leakage dominates. In the sub-threshold regions V_{dd} is very low and hence gate leakage is again negligible. Hence gate leakage term affects the optimal curve very little. When the frequency is extremely low and optimal $V_{th}0$ is very high the subthreshold leakage falls even below the gate leakage. Only for such extremely low frequencies gate leakage tends to affect the optimal V_{dd} - V_{th} curve.

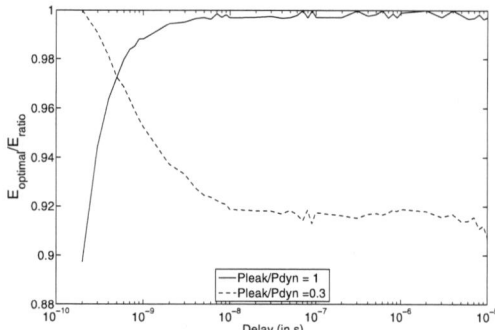

Figure 5. Suboptimality in choosing constant Eleak/Edyn ratio as the minimum energy criteria.

3 DVFM controller

A DVFM controller based on delay and power measurement results is shown in fig.6(a). It adjusts the supply and body bias to achieve minimum power for a frequency target. The control algorithm uses the measured values of delay and power to adjust V_{dd} and body bias and is described next.

3.1 Gradient based V_{dd} - V_{th} control algorithm

In [5] V_{dd} is controlled based on the results of the delay measurement and V_{th} is varied based on the results of power measurements. V_{dd} and V_{th} affects both delay and power and hence a combined control of V_{dd} and V_{th} based on the power and delay measurements will be more effective. We propose a new control algorithm below.

The difference between targeted delay(Dtarget) and the actual circuit delay(Dactual) is obtained. Also the ratio of Leakage Energy to Dynamic Energy(Eleakratio) is obtained and its difference from the optimal Energy ratio (Eratioptimal) is computed.

$$delD = Dtarget - Dactual \quad (15)$$

$$delEratio = Eratiotarget - Eratioactual \quad (16)$$

The Sensitivity matrix is defined as

$$S = [\quad \frac{\partial D}{\partial V_{dd}} \quad \frac{\partial D}{\partial V_{th}}$$

$$\frac{\partial Eratio}{\partial V_{dd}} \quad \frac{\partial Eratio}{\partial V_{th}}] \quad (17)$$

We can write

$$[delD \ delEratio]^T = S * [delvdd \ delvth]^T \quad (18)$$

Then the direction in which we should move in the V_{dd}-V_{th} space to reach the optimal point is computed by inverting the above equation.

$$[delvdd \ delvth]^T = R * [delD \ delEratio]^T \quad (19)$$

where R is related to the inverse of the sensitivity matrix S.

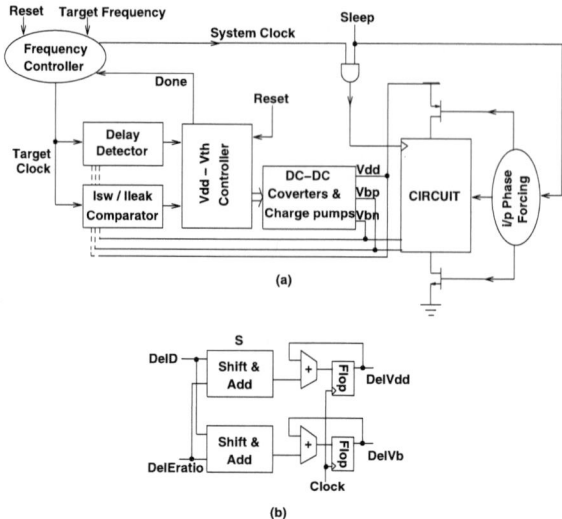

Figure 6. (a)Block Diagram of DVFM controller. (b)Proposed V_{dd} - V_{th} controller.

The matrix multiplication in (19) is implemented in a shift and add fashion as shown in fig.6(b). Choosing the elements of R is critical as they control the dynamics and stability of the loop. The following heuristic steps are developed to avoid oscillations in the proposed algorithm.

(i) Given a range of frequencies the optimal V_{dd} - V_{th} points are estimated using the delay and power models given in equations (2) and (8). The range of V_{dd} and V_{th} is thus known.

(ii) The sensitivity matrix S is evaluated using (17) at each point in this operational range. Then S^{-1} is computed. R(i,j) is then computed as $minS^{-1}(i,j)$ over the operational range.

The performance of the proposed algorithm was evaluated using MATLAB simulations. The operation of the algorithm to control the V_{dd} and V_{th} of test circuit for a frequency range on 100M - 1G was simulated. The initial values of V_{dd} and V_{th} was chosen to be the optimal V_{dd} - V_{th} for a frequency of 2GHz. A step decrease in frequency is then given and the number of iterations to converge was found for the proposed algorithm and the algorithm in [5]. The proposed algorithm is faster by more than 50% than the previous work as shown in fig.7. In a DVFM

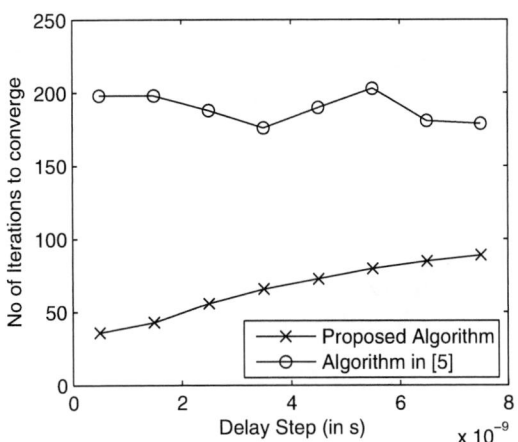

Figure 7. Performance comparison of the proposed algorithm and algorithm in [5].

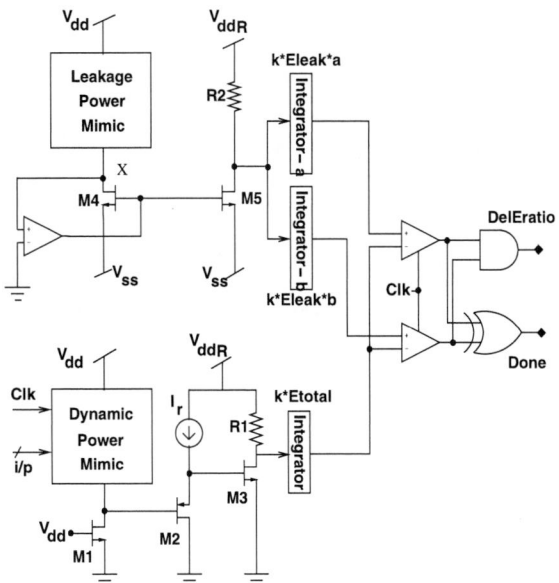

Figure 8. $I_{leak} - I_{sw}$ **comparator.**

controller each iteration must be followed by a change in control to the power controller(to change V_{dd} and V_{th}). In algorithm proposed in [5] in each iteration either V_{dd} or V_{th} is changed. So the convergence time is

Tc = No.of V_{dd} iterations*time taken per V_{dd} step + No.of V_{th} iterations*time taken per V_{th} step

In the proposed algorithm V_{dd} and V_{th} are changed simultaneously. So the convergence time is

Tc= No.of iterations * Max(time taken per V_{dd} step, time taken per V_{th} step)

Time taken per V_{dd} and V_{th} step depends on the response time of the power regulators, it can be in the order of microseconds. Hence in [5] algorithm even if equal no.of V_{dd} and V_{th} steps are assumed, the proposed algorithm performs better.

3.2 Architecture of the Logic block

The logic block to compute the sum of sixteen 16 bit numbers is similar to the SAD computation architecture used in [6]. It uses 6 stages of carry save adders and the final carry propagation is done using a Hans Carlson adder. A pipelining of 12 FO4 delay per stage found to be optimal [7] is used.

3.3 Delay and Power monitoring Circuits

The Delay monitoring circuit given in [5] is used. The monitoring circuit to measure the total energy consumed by the circuit is given in fig.8. The SAD computing structure is highly regular and is composed of adders, flops and wires only. Due to the regularity of the structure we believe that the power mimic made of adders, flops and wires in a proportionate ratio should mimic the total circuit power across

wide range of V_{dd} and V_{th} satisfactorily. The mimic used is shown in fig.9. The voltage drop across the sleep transistor is proportional to the current flowing through it and hence can serve as a measure of the instantaneous current flow. The total energy measurement is done by switching the inputs of the power mimic using a counter and integrating the drop across the sleep transistor over a few cycles. Since the voltage across the sleep transistor is close to ground a PMOS level shifter is used to shift the voltage up by V_{th}. The shifted voltage is amplified and integrated to obtain a voltage proportional to the total energy consumed. The leakage power measurement circuit proposed in [5] is used. The Op-Amp shown maintains the node x at ground potential. The leakage current of the mimic flows through M4 which is mirrored to M5 producing a proportional voltage across R2. This voltage is then integrated and compared with the total energy using a clocked comparator to generate *delEratio*. A small window for comparison is provided with scaling factors aand b. The circuit proposed uses a negative voltage V_{ss} of -0.75V. The linearity of the total energy and leakage energy are shown in fig.10(a) and fig.10(b).

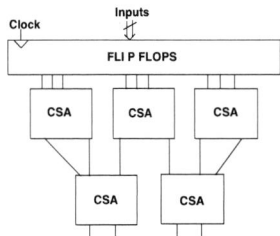

Figure 9. Power Mimic.

513

978-1-4244-3039-0/08 $25.00 © 2008 IEEE

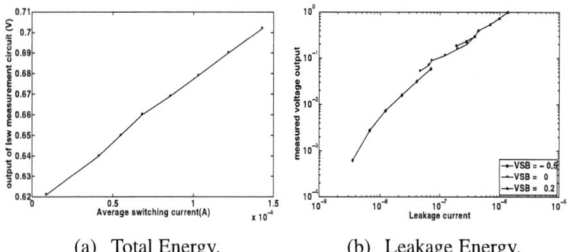

| (a) Total Energy. | (b) Leakage Energy. |

Figure 10. Linearity of Measurements

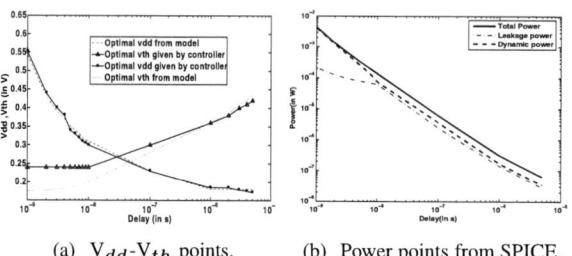

| (a) V_{dd}-V_{th} points. | (b) Power points from SPICE. |

Figure 11. Optimal Results.

3.4 DVFM Controller Simulation results

The DVFM controller with the measurement circuits was simulated with behavioral models for all the digital components of the controller. 65nm PTM model is used to build the logic block and the measurement circuits. An ideal DC-DC converter with voltage resolution of around 10mV was assumed in the behavioral model. The target frequencies Vs the optimal V_{dd}-V_{th} points are given in fig.11(a) and the power consumed by the block at different frequencies is given in fig.11(b). In the super threshold region the optimal V_{th} turns out to be lesser than the V_{th} obtainable by forward body bias. So the controller settles down the minimum obtainable V_{th} and V_{dd} to maintain the delay constraint. That is the controller settles to the reachable optimal point in the V_{dd}-V_{th} plane. The power consumed by the circuit for various frequencies at the optimal V_{dd}-V_{th} given by the controller is given in fig.11(b). The leakage power and the dynamic power are comparable indicating that the optimal points given by the controller are indeed close to the minimal power points.

4 Conclusions

Analytical expressions for optimal V_{dd}-V_{th} valid in both strong inversion and sub threshold regions has been derived. Gate leakage is found to affect the optimal V_{dd}-V_{th} curve very little except for very low frequencies. Minimum energy operation for a given frequency occurs when the leakage energy and the dynamic energy are comparable to each other and this criteria holds quite well for a wide range of frequencies. V_{dd}-V_{th} controller based on a new control algorithm has been designed and simulated using behavioral models. The proposed algorithm performs better than the previously reported algorithms in terms of convergence speed.

References

[1] K. Nose and T. Sakurai. Optimization of V_{dd} and V_{th} for Low-power and high-speed applications.*Proc. Asia*

South Pacific Design Automation Conference, pp. 469-474, Jan.2000.

[2] A. Wang, A.P. Chandrakasan and S.V. Kosonocky. Optimal supply and threshold scaling for subthreshold CMOS circuits. *Proc. IEEE Computer Society Annual Symposium on VLSI* 2002, pp. 5-9.

[3] Benton H. Calhoun and Anantha Chandrakasan. Characterizing and Modeling Minimum Energy Operation for Subthreshold Circuits. *Proceedings of the 2004 International Symposium on Low Power Electronics and Design* 2004, pp. 90-95.

[4] Christian.C.Enz, Francois Krummeenacher and Eric A.Vittoz. An analytical MOS transistor model valid in all regions of operation and dedicated to low voltage and low current applications. *Analog Integrated circuits and signal processing* 1995, pp. 81-114.

[5] Masahiro Nomura et al. Delay and power monitoring schemes for minimizing power consumption by means of supply and threshold voltage control in active and standby modes. *IEEE Journal of solid-state circuits*,41(4):805-814, April 2006.

[6] J. Vanne, E. Aho, T.D. Hamalanen and K. Kuusilinna. A High-Performance Sum of Absolute Difference Implementation for Motion Estimation. *IEEE Transactions on Circuits and Systems for Video Technology*, 16(7):876-883, July 2006.

[7] Kim et al. Total power-optimal pipelining and parallel processing under process variations in nanometer technology *IEEE/ACM International Conference on Computer-Aided Design*, November 2005, pp. 535-540.

[8] Dejan Markovic, Mark Horowitz et al. Methods for true energy performance optimization. *IEEE Journal of solid-state circuits*, 39(8):1282-1293, August 2004.

[9] PTM. *http://www.eas.asu.edu/ptm/*.

978-1-4244-3039-0/08 $25.00 © 2008 IEEE

Temperature and Process Variations aware Power Gating of Functional Units

Deepa Kannan†, Aviral Shrivastava†, Vipin Mohan‡, Sarvesh Bhardwaj‡, Sarma Vrudhula‡

† Compiler and Microarchitecture Laboratory,
School of Computing and Informatics, Arizona State University, Tempe, AZ 85281 USA
{ deepa.kannan, aviral.shrivastava}@asu.edu
‡VLSI Electronic Design Automation Laboratory
School of Computing and Informatics, Arizona State University, Tempe, AZ 85281
{vipin.mohan, sarvesh.bhardwaj, vrudhula}@asu.edu

Abstract

Technology scaling has resulted in an exponential increase in the leakage power as well as the variations in leakage power of fabricated chips. Functional units (FUs), like Integer ALUs are regions of high power density and significantly contribute to the variation in the whole processor power consumption. Hence, it is important to reduce both the power consumption and the variation in power consumption of the FUs. Among existing FU power reduction techniques, power gating (PG) has been most effective. In this paper, we introduce a leakage sensor inside the FUs and propose a temperature and process variation aware power gating scheme, Leakage Aware Power Gating (LA-PG). Our experimental results demonstrate that LA-PG results in 22% reduction in mean and a 25% reduction in standard deviation of the ALU energy consumption when compared to existing power gating techniques, without significant performance penalty.

1. Introduction

Ever increasing performance demand of electronic devices has been the primary driving force behind aggressive technology scaling. Two important consequences of technology scaling are, the increase in leakage power, and increase in variation in the characteristics of manufactured devices. Leakage power is projected to contribute more than 40% of total power budget in processors fabricated in 65 nm technology and beyond [18]. Unlike dynamic power, leakage power is highly sensitive to variations in gate dimensions as well as the operational temperature.

High variation in the power consumption results in significant overestimation of the specification, leading to in-

creased design time/effort and results in significant loss of parameterized yield [4, 3]. *Hence, reducing both the total power, and the variation in the power consumption of FUs is an important problem.*

Leakage power is a very important concern for functional units (FUs) such as Integer ALUs, Floating point ALUs and Multipliers which are significant contributors to the total energy consumption of the processor [7]. In addition, FUs being regions of high activity, are among the hottest regions on the chip. Therefore reducing both the leakage power, and the variation in leakage power of FUs is an important research problem.

Among the existing techniques to reduce the leakage power of FUs, power gating is one of the most promising approaches. [5]. Power gating is a technique which reduces leakage by shutting off the power supply to a unit during periods of inactivity. However existing power gating mechanisms [10, 17] do not consider dependence of leakage on temperature and process variations.

In this paper, we propose to introduce a leakage sensor in FUs, and develop a temperature and process variations aware power gating technique. We present a power gating technique based on the IPC and propose Leakage Aware Power Gating (LA-PG) scheme, which is both temperature and process variations aware, to decide on which FUs are to be power gated. Our technique, LA-PG results in 22% reduction in the average, and 25% reduction in the standard deviation of the total ALU energy consumption, without any performance loss, as compared to existing power gating techniques.

2. Experimental Setup

Microarchitecture Model: Our simulation framework is depicted in Figure 1A. We perform our experiments on

515

978-1-4244-3039-0/08 $25.00 © 2008 IEEE

the ALPHA DEC 21364 processor. This is a 4-wide super-scalar processor, whose floorplan is shown in Figure 1B.

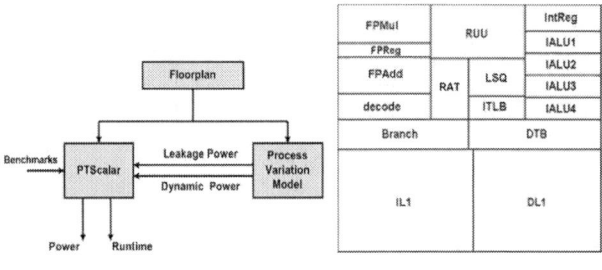

A. Simulation Framework

B. Floorplan of ALPHA DEC 21364 Processor

Figure 1. Simulation Setup

The power, performance and temperature modeling of alpha processor is done using a modified version of sim-outorder of the PTScalar toolset. PTScalar is a coupled power and thermal simulator built over SimpleScalar [6]. We execute several benchmarks from the MiBench [9], and Spec 2000 [1] suite.

Process Variation model: We model the variations in device features (gate length and threshold voltage) using the stochastic process corresponding to gate length using the Karhunen-Loève Expansion proposed in [3]. The process variation model generates the dynamic power and leakage power values for the ALUs in the processor corresponding to one die. We generate 1000 such die samples, which are fed into PTScalar for power, performance, and temperature modeling. The power numbers are scaled to correspond to 45nm technology.

3. Motivation

In this section, we perform experiments on the representative susan-corners benchmark from the MiBench suite [9], to demonstrate the need to reduce the FU energy consumption, as well as the variation in the FU energy consumption in the presence of temperature and process variations.

3.1. FUs are regions of high energy density

Figure 2 shows the total energy consumption (dynamic + leakage) of all the units sorted by their energy consumption for all the 1000 die samples. It can be observed from the plot that the total energy consumed by the Integer ALUs is 11.2% of the total processor energy.

In addition, FUs are one of the most active units in a processor, and therefore have very high energy density. This is exacerbated by the exponential relation of leakage on temperature. Figure 3 shows that ALUs have second highest energy density among all the units, only next to the IntReg-File.

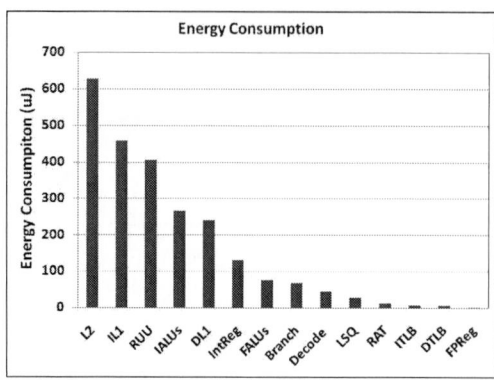

Figure 2. Energy consumption of all units in the alpha processor

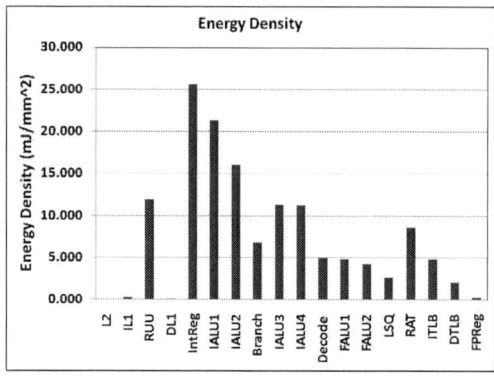

Figure 3. Energy density of all units in the alpha processor

3.2. FUs contribute significantly to variation in processor energy

Figure 4 plots the standard deviation of the energy consumption of each unit, across the 1000 die samples. The plot shows that ALUs have the highest variation in energy consumption. This is also due to the strong exponential dependence of leakage on temperature.

4. Related Work

Butts and Sohi [5] demonstrated that due to the exponential dependence of leakage on temperature, combinational logic has an order of magnitude larger leakage current relative to cache RAM transistors. Since Functional Units (FUs) are regions of high power density in the processor, techniques to reduce the leakage power of FUs were explored. Of various FU leakage reduction techniques, power gating [8, 10, 17] has proven to be the most effective for FU leakage reduction. These techniques address the question of how to implement power gating.

978-1-4244-3039-0/08 $25.00 © 2008 IEEE

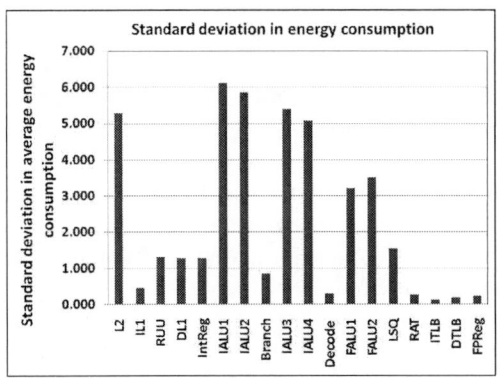

Figure 4. Standard deviation of the energy consumption of all units in the alpha processor

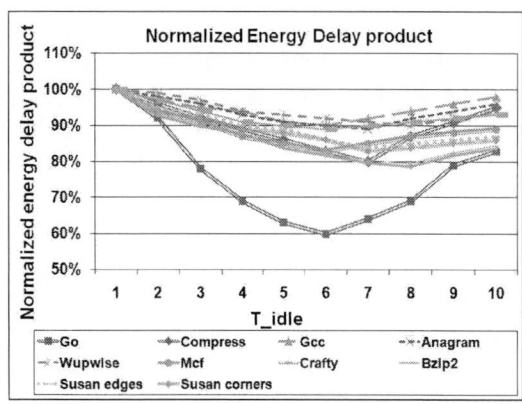

Figure 5. Energy delay product variation with t_{idle}

When to do power gating, has been approached from two directions, i) compiler-based solutions, and ii) hardware solutions. Compiler-based FU leakage reduction techniques were studied in [17]. But this technique requires that the entire code be examined off-line to identify suitable regions for turning the functional units off. Hardware based techniques for identifying the idle regions consume additional power throughout the execution.

Previous techniques for FU power gating in superscalar processors are idle-time based [8]. Whenever an FU is predicted to be idle for more than break-even time, the FU is power gated.

Previous works have attempted to design effective and accurate leakage current sensors [15, 12]. Our power gating scheme, reads the reading of the leakage sensor and power gates functional units in order to reduce both the power consumption, as well as the variation in the power consumption of the FUs. To the best of our knowledge, this is the first work in this direction.

5. Previous Approach: Idle Time-based Power Gating (IT-PG)

In the idle-time based power gating technique (we call it IT-PG), t_{idle} is the key parameter. The activity of FU is monitored, and if the FU is idle for more than t_{idle} cycles, the power supply to the FU is gated off. Once in a power gated state, the FU will be woken up (power gating is disabled) when an operation is issued to it. The parameter t_{idle} can be varied to obtain a tradeoff between performance and leakage savings.

Figure 5 plots the normalized energy delay product of all our benchmarks for varying values of t_{idle}. The average of energy delay product over all our benchmarks is the least for $t_{idle} = 7$. This is consistent with previously published results [10], who found the optimal value of t_{idle} as

between 6-9 cycles, and therefore we choose $t_{idle} = 7$ for our comparison experiments.

6. Our Approach: Leakage Aware Power Gating (LA-PG)

Our first observation is that temperature increases are gradual, and like a plethora of previous works, we assume that appreciable temperature changes occur only at $10,000$ cycle granularity, and therefore it is reasonable to implement temperature dependent policies at this granularity [11, 16]. Our power gating mechanism is a two step process: we use the current IPC information to find out how many FUs to power gate, and then we use leakage sensor values to determine which FUs to power gate.

6.1. How many FUs to Power Gate?

At each decision moment (i.e., every 10,000 cycles), we compute the *average IPC*, or the average number of instructions that are *ready to be issued* every cycle. Note that this is different from the regular definition of IPC or Instructions Per Cycle, which is the number of instructions issued each cycle. However, due to it's close similarity to IPC, and since we do not use IPC otherwise in this paper, we call our approach as IPC based technique. The number of FUs to power gate is determined by comparing our computed average with a *threshold*. For a n FU configuration, we have $n - 1$ thresholds.

The *average IPC* is computed as an average of IPCs of the last *history* number of cycles. The value of *history* determines the accuracy of our power gating technique. Therefore, the history and the thresholds are the two key parameters of our IPC threshold-based power gating technique. Designers can vary these parameters to trade off power, performance, and architectural complexity. To determine suitable values of history and thresholds, we simulated all the 10 benchmarks with IPC threshold-based power gating technique for several values of history and thresholds. We vary the history from 10 to 1000 cycles, and the threshold value for the case when a single ALU is in the active mode from 1.0 to 1.20 in steps of 0.01. The corresponding values for two and three ALUs to be in the active mode are varied from 2.0 to 2.20 and 3.0 to 3.20.

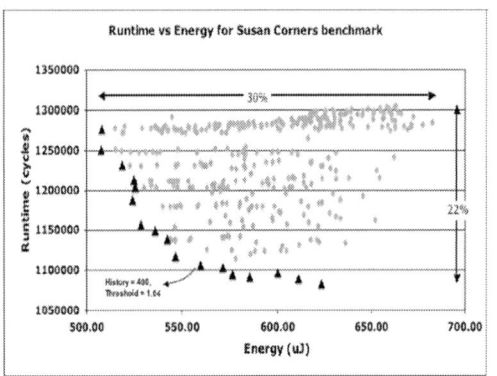

Figure 6. Runtime vs Energy showing the pareto optimal points for Susan Corners benchmark

Figure 6 shows the runtime vs energy plot for all history-threshold configurations for the representative susan-corners benchmark. The figure shows that a variation of 30% in the energy of the ALUs and a variation of 22% in the runtime is possible by power gating. We have identified and marked the pareto-optimal points by the dark triangles in the figure. A configuration is pareto optimal if it is not worse than any other configuration in both power and performance. Designers can choose any of these pareto-optimal design points to trade-off power and performance.

We varied the values of history and thresholds for all 10 benchmarks, and computed the energy-delay product for each history-threshold configuration. We then compute the summation of the energy-delay product for all benchmarks for each history-threshold configuration. The optimal values of history and thresholds came out to be 400, 1.04, 2.04, and 3.04 respectively, and we use these values in estimating the effectiveness of our approach.

6.2. Which FUs to Power Gate?

In order to reduce the leakage, we want to power gate the FUs which have the highest leakage. An FU may have high leakage either because of process variations, or because it's temperature is high. Thus LA-PG is both temperature and process variation aware. Power gating the FU with the highest leakage, minimizes the FU power consumption; in addition it also reduces the variation in the leakages of FUs.

Introducing Leakage Sensors: We propose to introduce the leakage sensor proposed by Kim et al. [15] inside each FU and continuously measure the FU leakage during the chip operation. A single channel leakage sensor is shown in Figure 7A. M2 is the only transistor that is sensitive to the variation in leakage of the ALU due to the impact of temperature and process variations. Therefore, the accuracy of leakage sensor itself is not affected by process and temperature variations.

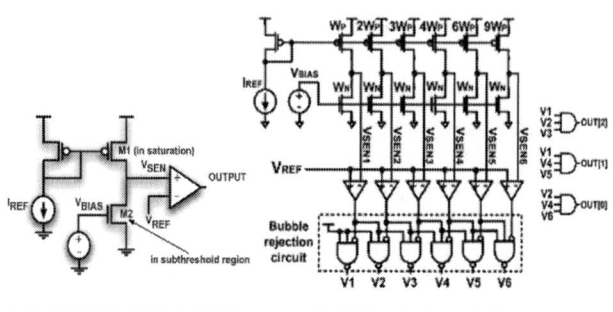

Figure 7. Six Channel Leakage sensors [15]

We explicitly model the area, power and inaccuracy introduced when converting the leakage sensor in our experimental setup. The overhead of using leakage sensors accounts to around $3-4\%$ reduction in the total power savings obtained using our LA-PG.

Leakage Sensor Placement: To find a good location for the leakage sensor, we compared the leakage of a device located at various locations (x_i, y_i) inside the ALU, and the average leakage of the ALU ($I_{av} = I_{S,T}/N$). We found that mean of the percentage difference between the average ALU leakage and the leakage of a device located at the center of the ALU for a sample of 1000 dies to be less than 1%. The maximum percentage error over the same set of samples was 7%. Thus a single leakage sensor located at the center of the ALU can provide accurate estimation of the leakage power of the entire ALU.

Microarchitectural Overheads: Figure 8 shows the implementation of the circuit required for our technique. A naive implementation could have high power and performance overheads. Therefore, we introduce several optimizations in the implementation. (i) We limit the range of IPC to be only from 0 to n_{issue}, instead of 0 to $n_{reorder}$, for a $n - issue$ superscalar processor with a re-order buffer size of $n_{reorder}$. In other words, if the number of instructions that are ready is more than n_{issue}, the IPC saturates at n_{issue}. This reduces the size of the microarchitectural overhead tremendously. (ii) Instead of adding 400 values, we add the IPC every 4th cycle for a period of 512 cycles. This results in 128 samples of IPC over 512 cycles. On a 4-issue superscalar, the maximum value of the sum of the IPC over the entire sampling period will not exceed 512. Hence, a

518

978-1-4244-3039-0/08 $25.00 © 2008 IEEE

9 bit adder is sufficient for this purpose and it can be implemented as very low-power *ripple carry adder*, and still meet the timing constraint. This further reduces the power consumption of the architectural overhead. The logic circuit required is a small combinational logic block that determines how many ALUs to power gate based on the IPC sum and the threshold values and which ALUs to power gate based on the 3 threshold values which are scaled to 133, 233 and 333, and the 3 bit output of the leakage sensor placed in each of the 4 ALUs.

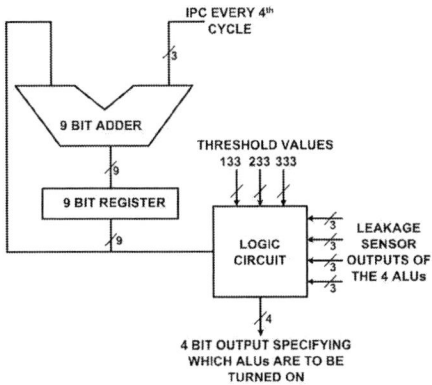

Figure 8. Microarchitectural enhancements for the IPC Threshold based power gating technique

We synthesized this logic using Synopsys Design Compiler and implemented it in Cadence Spectre toolset (Virtuoso Schematic editor) using TSMC 0.25um CMOS deep submicron process, and scaled the numbers to 45 nm. We also synthesized the logic for the IT-PG technique for comparison purposes. This logic has an area overhead of less than 3% and energy overhead of $< 0.15\%$, as compared to the architectural overhead of idle-time based technique. We include the energy overhead due to our logic in the power computations using PTscalar in all our simulations.

7. Experiments

7.1. LA-PG reduces ALU energy consumption

Figure 9 plots the mean of the ALU energy consumption computed over 1000 sample dies, for LA-PG, normalized to IT-PG, for all the 10 benchmarks. The 11th bar to the extreme right denote the average energy reduction achieved over all the benchmarks.

The figure shows that LA-PG decreases the average energy consumption by 22% as compared to the IT-PG. The performance penalty of applying our IPC threshold-based

power gating techniques is less than 2%. This performance loss is lesser than the performance loss of IT-PG, which is 2.2%, as compared to the case with no power gating. Another important observation from the graph is that the energy reductions are quite uniform across the benchmarks. Hence, the effectiveness of our technique is consistent through the benchmark spectrum.

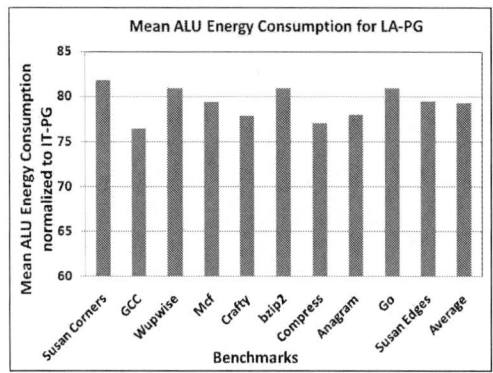

Figure 9. Mean ALU energy consumption by IPC threshold based techniques

7.2. LA-PG mitigates process variation

In the presence of process variation, the power gating priorities assigned for one die may not be the best for the other dies. For the same priorities, the variation in the total ALU energy consumption in different dies may be quite significant. We simulate all the 10 benchmarks with IT-PG and LA-PG techniques, for 1000 die samples.

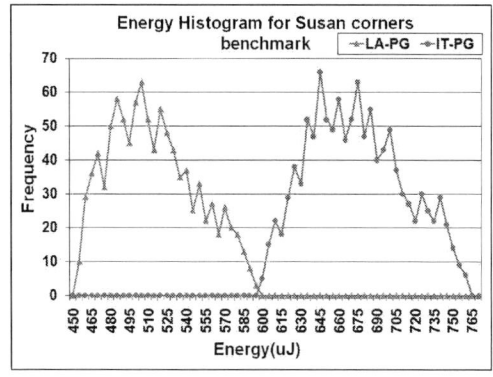

Figure 10. Energy histogram for various power gating techniques for susan corners benchmark

Figure 10 plots the energy histogram for IT-PG and LA-PG techniques, for susan corners benchmark for 1000 die samples. The rightmost curve (lines connected by circles)

519

978-1-4244-3039-0/08 $25.00 © 2008 IEEE

corresponds to the energy distribution for IT-PG. The mean and standard deviation of the energy distribution are 675.18 μJ and 33.76 μJ. For processors that will incorporate leakage energy sensors, LA-PG technique is very effective. It reduces the mean and standard deviation to 521.98 μJ and 23.2 μJ respectively, as shown by its energy histogram depicted by the leftmost curve (lines joined by triangles). As compared to IT-PG, LA-PG reduces the energy consumption by 22% and reduces the standard deviation by 25%.

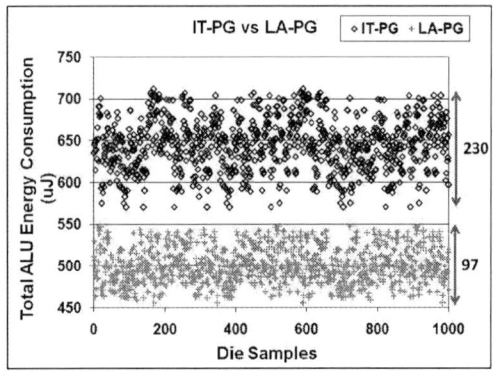

Figure 11. IT-PG vs LA-PG

Figure 11 plots another view of comparison between the ALU power consumption for IT-PG and LA-PG in 1000 die samples for susan corners benchmark. The same two observations can be made from this graph: (i) all the LA-PG points are lower than the IT-PG points, and (ii) the width of the vertical band in which points of LA-PG lie is lesser than the width of the band in which the points of IT-PG lie. The difference between the lowest and highest energy dies is around 230 μJ in the case of IT-PG when compared to 97 μJ in the case of LA-PG.

8. Summary

The exponential dependence of leakage on the temperature and device dimensions has made leakage an increasingly important concern in the nano-design era. In this paper, we presented an IPC threshold based power gating technique that reduces the energy consumption and the variation in the total ALU energy consumption across dies. Our LA-PG is both temperature and process variation aware, through a leakage sensor in the FUs. LA-PG reduces the mean ALU energy consumption by 22% and reduces the standard deviation in the ALU energy consumption by 25%, without any performance penalty, as compared to existing techniques.

Acknowledgement

We would like to thank Microsoft Corporation for their generous support.

References

[1] SPEC2000 Benchmarks, www.spec.org/benchmarks/html, 2000.

[2] A. Abdollahi, F. Fallah, and M. Pedram. Leakage current reduction in cmos vlsi circuits by input vector control, 2004.

[3] S. Bhardwaj, S. Vrudhula, P. Ghanta, and Y. Cao. Modeling of intra-die process variations for accurate analysis and optimization of nano-scale circuits. In *Proc. of IEEE/ACM Design Automation Conference*, 2006.

[4] D. Boning and S. Nassif. Models of process variations in device and interconnect, 2000.

[5] J. A. Butts and G. S. Sohi. A static power model for architects. In *Micro33*, pages 191–201, 2000.

[6] D.Burger and T.Austin. The simplescalar tool set version 3.0, 1997.

[7] J. Deeney. Reducing power in high-performance microprocessors. In *International Symposium on Microelectronics*, 2002.

[8] D. Duarte, Y.-F. Tsai, N. Vijaykrishnan, and M. J. Irwin. Evaluating run-time techniques for leakage power reduction. In *VLSI Design*, pages 31–38, 2002.

[9] M. R. Guthaus, J. S. Ringenberg, D. Ernst, T. M. Austin, T. Mudge, and R. B. Brown. MiBench: A free, commercially representative embedded benchmark suite. In *IEEE Workshop in workload characterization*, 2001.

[10] Z. Hu, A. Buyuktosunoglu, V. Srinivasan, V. Zyuban, H. Jacobson, and P. Bose. Microarchitectural techniques for power gating of execution units. In *Proc. of ISLPED*, pages 32–37, New York, NY, USA, 2004. ACM Press.

[11] W. Huang, S. Ghosh, S. Velusamy, K. Sankaranarayanan, K. Skadron, and M. R. Stan. Hotspot: A compact thermal modeling methodology for early stage vlsi design. *IEEE Transactions on Component Packaging and Manufacturing Technology*, 14(5), 2006.

[12] J. Hurst and A. Singh. A differential built-in current sensor design for high speed iddq testing. *vlsid*, 00:419, 1995.

[13] J. Kao and A. Chandrakasan. Dual-threshold voltage techniques for low-power digital circuits, 2000.

[14] S. Kaxiras, Z. Hu, and M. Martonosi. Cache decay: Exploiting generational behavior to reduce cache leakage power. In *Proc. of ISLPED*, pages 240–251, 2001.

[15] C. H. Kim, K. Roy, S. Hsu, R. Krishnamurthy, and S. Borkar. A Process Variation Compensating Technique with an On-Die Leakage Current Sensor for nanometer Scale Dynamic Circuits. *IEEE Transactions on VLSI*, 14(6):646–649, 2006.

[16] W. Liao, L. He, and K. Lepak. Ptscalar version 1.0, 2004.

[17] S. Rele, S. Pande, S. Onder, and R. Gupta. Optimizing static power dissipation by functional units in superscalar processors. In *Computational Complexity*, pages 261–275, 2002.

[18] J. W. Tschanz, S. G. Narendra, Y. Ye, B. A. Bloechel, S. Borkar, and V. De. Dynamic sleep transistor and body bias for active leakage power control of microprocessors. *IEEE Journal of Solid State Circuits*, 38, Nov 2003.

978-1-4244-3039-0/08 $25.00 © 2008 IEEE

21st International Conference on VLSI Design

A Robust Top-down Dynamic Power Estimation Methodology for Delay Constrained Register Transfer Level Sequential Circuits

Sriram Sambamurthy*, Jacob A. Abraham* and Raghuram S. Tupuri[†]

*Electrical and Computer Engineering,
University of Texas at Austin, U.S.A.
sambamur, jaa@cerc.utexas.edu
[†]Advanced Micro Devices Inc., U.S.A.
raghuram.tupuri@amd.com

Abstract— We present a top-down dynamic power estimation methodology for delay constrained sequential circuits. The methodology works at the register transfer level (RT-Level), and applies to both structural and behavioral descriptions of circuits. The average power consumption of a circuit varies with the worst case cycle-time or frequency of operation. As the cycle-time is reduced, the increase in the capacitance of the circuit due to technology mapping and optimization is captured by our technique at the RT-Level using the principles of logical effort. Switching activity is obtained at the RT-Level visible nodes through RT-Level functional simulation. This information is utilized to approximate the activities at the remaining nodes of the circuit and combined with capacitance to estimate dynamic power. Power estimation results for RT-Level sequential circuits indicate good accuracy (average error<10%) with respect to the reference values obtained by detailed gate-level power analysis. The power consumed by a circuit varies with the target library and technology. Our methodology is parameterizable and the results obtained for different target libraries at 0.18um TSMC and 0.13um UMC technologies are consistent, indicating the robustness of our technique. The applicability of our methodology in design frameworks consisting of bottom-up techniques is also discussed.

I. INTRODUCTION

Power estimation has become an integral part of the design flow of both high-performance and low-power VLSI chips. There is an increasing interest to develop techniques for higher (early) levels of design hierarchy, namely RT-Level and architectural level, to better optimize the design with respect to the specifications. With the increasing usage of RT-Level IPs in the world of embedded chip-design, different tradeoffs have to be considered for different applications. A design may be used in a high performance application or modified slightly for usage in a low-power application, as part of design reuse. In such cases, it is essential to analyze the power-delay tradeoffs of the design [1]. An RT-Level power estimation methodology that has the provision to estimate power over a range of target delays is preferred in this situation.

Various bottom-up methods have been developed to estimate power at the RT-Level [2], [3]. These bottom-up methods typically use macro-modeling to establish a database of power

This research was partly supported by Advanced Micro Devices Inc.

estimates for different constraints. The process is constraint specific and the macro-modeling has to be performed for each set of constraints through characterization. This can become quickly explosive for building an extensive database. For instance a macro-model has to be built for every module at all n delay or frequency targets. Moreover, these bottom-up techniques cannot be used when the design is being built. On the other hand, a top-down technique can quickly analyze the design, even in its rudimentary stage to give guidelines that can be used to build the design in a better manner, for it does not require macro-modeling.

Target technology and library are important factors that need to be taken into account, for they strongly drive the power-delay characteristics of any circuit. A top-down technique is not constrained to a single library, as against the bottom-up techniques. Therefore, given a target library consisting of any m gates and n constraints (e.g. n cycle-time points), it provides an accurate power estimate. The top-down power estimation methodology should also be independent of the type of macros used during implementation [4], as it improves its range of applicability.

In this work, we present a novel top-down methodology that estimates the dynamic power of a generic synthesizable sequential RT-Level design over different delay targets. We abstract the principles of logical effort [5] to the RT-Level and use them to estimate the variation in capacitance over the different target frequencies. The wire-load model is used to approximate the interconnect capacitance. We combine the estimated capacitance with activity factor obtained through RT-Level simulation to estimate the dynamic power of the given design. Specifically, we observe the activity values at the RT-Level visible nodes (wires, registers) and use interpolation to estimate the activities at the nodes that are not visible at the RT-Level.

The underlying algorithms of our technique are evaluated using different target libraries consisting of different gates, at two technology points (0.18um and 0.13um). The results obtained on opencores [6] verilog designs (behavioral) indicate a average accuracy of over 90% with respect to gate-level

521

978-1-4244-3039-0/08 $25.00 © 2008 IEEE

reference values, over the delay curve. We estimate the ratio between the estimated power at the delay point (cycle-time) under consideration and at the maximum cycle-time, for each point on the delay curve. These ratios are used to estimate power accurately at the corresponding delay points. The robustness with respect to different technologies and libraries is a huge advantage of our methodology. To our knowledge, the top-down technique presented for the power estimation over various target frequencies is the first of its kind for RT-Level sequential circuits.

We also evaluate our technique at the structural RT-Level, and the results obtained for the ISCAS sequential benchmarks show that our methodology is consistent. In addition, we present results for varying switching activities which were obtained by using different input vectors during simulation. For demonstrating the smooth integration of our technique with existing frameworks, we also discuss how our technique can co-exist with bottom-up methods.

The related research in this field is discussed in the next section. In Section 3, we introduce our methodology and present the power estimation algorithm for sequential circuits. In Section 4, we evaluate the power estimates obtained by our technique against those obtained through gate-level analysis and discuss the applications of our work. We conclude in Section 5.

II. RELATED WORK

Extensive research has been performed in the field of power estimation at the RT-Level [7]. The techniques can be classified into top-down and bottom-up methods. The top-down method [8] attempts to first do a quick mapping of the design to a set of primitives to get an estimate of capacitance, and then uses RT-level simulation to get activity factor. However, the method in [8] does not estimate power for various target frequencies and it applies only to combinational circuits. The only other top-down method [9] works for combinational circuits, but there is no top-down method which estimates power for delay constrained sequential circuits. A bottom-up method first characterizes a large number of macros for different target constraints based on a number of parameters like input signal probability, input activity, output transition density etc. [2], [3]. Then based on the parameters and the constraints, power is looked-up from a table during estimation. Another type of power estimation extracts parameters from the RT-level code, builds a model and characterizes it for a particular technology [10], [11], [12]. All the above bottom-up methods are library and technology dependent. Thus the model has to be built and the characterization has to be performed for each constraint, library and technology.

At the architectural level, we have well known tools to estimate power [13], [14]. They are complementary to our current work. For example, Wattch [13] uses pre-characterized power values for individual units and operates at the processor level. Our current method can be used in conjunction with Wattch to estimate power for the individual local units, dynamically estimating the local switching factor and the capacitance for various target frequencies, provided we have the RT-Level code for those units.

III. POWER ESTIMATION METHODOLOGY

In this section, we present our methodology in detail and discuss the underlying algorithms. Average power is the average amount of energy consumed per unit time. The dynamic component of power is the sum of active and short circuit power components.

$$P = (V_{dd}^2 * f * \sum_{i=1}^{n} C_i * sf_i) + (I_{sc} * V_{dd}) \qquad (1)$$

where C_i denotes the capacitance at node i, sf_i denotes the switching factor at node i, f denotes the frequency of the clock, I_{sc} denotes the direct-path short-circuit current and V_{dd} denotes the supply voltage. Active power is further subdivided into zero-delay (single switch per cycle) power and glitch power. In this methodology, the active zero-delay dynamic power is obtained at various frequencies by estimating the capacitance and switching activities separately, in a top-down manner. Next, we describe our power estimation algorithm in detail.

A. Power estimation algorithm

The algorithm to estimate the average dynamic power for RT-Level circuits is given in Figure 1. The first step involves estimating the minimum number of logic stages (N) necessary to implement the function. The RT-Level code is converted to a Control flow Data Flow Graph (CDFG) format where each node (RT-Level visible) represents a RT-Level expression. Based on the maximum input size of the gates available in the target library, N is obtained for every node. For example, if the node represents "$y = (a\&b)|(b\&c)|(c\&a)$", for a library consisting of two-input gates, the value of N will be 3. If our target library consists of three input gates, then the parameter N will be 2. The maximum size of the gates available in the target library is the only information utilized here. For example, the previous expression can be written as "$x1 = (a\&b)$, $x2 = (b\&c)$, $x3 = (c\&a)$, $x4 = x1|x2$, $y = x4|x3$", giving N as 3, for both two-input and three-input target libraries. We can observe from this example that N depends on the style of the RT-Level programmer, a bigger expression suitably decoded better for different libraries. We use *booldnf* of the Boolstuff package [15] to convert the expressions into disjunctive normal form before estimating the number of stages as above.

The second step involves estimating the local switching factors. The global switching factors at the IOs of the modules can be obtained either by RT-level simulation, architectural-level simulation or through static propagation of probabilities and activities. Local switching factor is the average toggle rate at every node in the gate-level implementation of the circuit. Since we do not know how the design is implemented at the RT-Level, we approximate it using the observation given in [16] for the logic levels of the design. The switching factor at a particular logic depth is represented by

$$sf_i = (sf_{in} - sf_o) \cdot (1 - \frac{i}{N})^2 + sf_o \qquad (2)$$

522

978-1-4244-3039-0/08 $25.00 © 2008 IEEE

where sf_{in} denotes the input switching factor, sf_o denotes the output switching factor, i denotes the ith level and N denotes the total number of levels (logic depth) in the circuit. This is an empirical observation which states that the switching factor varies quadratically with respect to the logic depth of the circuit. This equation does not hold good for all circuits. For example, consider the same RT-Level expression "$w1 = (a\&b)$, $w2 = (b\&c)$, $w3 = (c\&a)$, $y = w1|w2|w3$". If the signal probability and activity [17] are 0.5 for all inputs, then the output activity for signal "y" is 0.2. When implemented with a three-input target library, the number of stages is approximated as 2. Using Equation 2, the activity at the levels are 0.275 and 0.2 for levels 1 (input of y) and 2 (output at y) respectively. But the actual activity at level 1 obtained is 0.2. The error in this case is 37.5%. To mitigate this error, we considered the activity at every RT-Level visible node (w1, w2, w3) and adjusted the values at the other nodes obtained using Equation 2. Considering the same example, the activity at level 1 is now 0.2. Instead of a single logic depth expression, if we have the expression without any RT-Level visible net (like w1, w2 and w3), we use Equation 2 to approximate the activity at the intermediate levels. We can observe from this example that the more the number of RT-Level visible intermediate nodes, the better will be the switching activity estimate from step 3.

The fourth step of the algorithm involves estimating the capacitance for a given delay target for the output functions of the given design using the algorithm in Figure 2 for sequential circuits. The fifth step combines the capacitance at the nodes with activity factor to give the average dynamic power. The summation of the average dynamic power over the output functions gives the total average dynamic power consumed by the RT-Level circuit, provided there is no any shared logic between them. In the sixth step, the power consumed by the logic of each shared node is decremented from the total power, for the number of times it is used by the output functions, to adjust for shared logic. This process is repeated for every delay point D_i, to get the corresponding power values.

We now explain the capacitance estimation algorithm in detail. Logical Effort is a widely used technique to evaluate different gate level designs for minimal delay. We use it at the RT-Level to estimate the capacitance variation with respect to different delay targets. The delay of a path built using logic gates is given by

$$\frac{D}{\tau} = N \times F^{(\frac{1}{N})} + P \qquad (3)$$

where N denotes the number of stages in the path, τ denotes the technology constant, P denotes the sum of the parasitic efforts of the gates in the path and F denotes the product stage effort of the gates in the path.

At the RT-Level, we do not have knowledge of the gate-level implementation. Therefore, we model each RT-Level function (node in the CDFG) as a tree of blackboxes, each blackbox being associated with its own logical and parasitic efforts. We rewrite Equation 3 to obtain the relationship between the delay

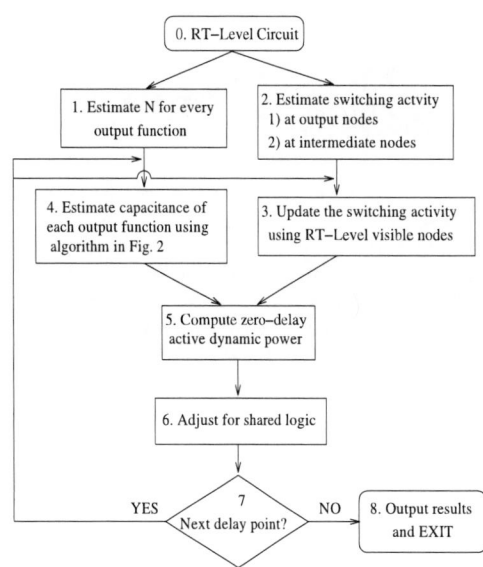

Fig. 1. Algorithm to estimate the average dynamic power consumption of RT-Level circuits

target and the stage effort,

$$f = F^{\frac{1}{N}} = \frac{(\frac{D}{\tau}) - P}{N} \qquad (4)$$

This individual stage effort (f) is used along with the logical effort g to size the blackboxes, during the capacitance estimation step.

1) Capacitance estimation: Sequential circuits are composed of memory elements whose type and drive strength have to be determined for power estimation. The algorithm for the capacitance estimation of sequential circuits is shown in Figure 2.

The logical efforts of the gates in the target technology library are pre-computed as per the directions given in [5]. The inputs to the algorithm are the sequential circuit and the delay target. The first step involves breaking the circuit at the sequential element (flip-flop) boundaries to identify the pseudo-primary inputs and outputs. The second step is performed to take care of the dependencies in the given circuit. Sizing is performed in the third step. The cycle time is first adjusted according to whether we have a flop at the output and(or) at one of the inputs. If we have a flop at the output, average setup time is reduced from the cycle time. The Clk-to-Q time is reduced from the cycle-time, if the critical path at the node starts at a pseudo-primary input. Then, the combinational part of the circuit is sized according to the adjusted cycle time.

For computing the capacitance of the combinational portion, a non-uniform tree is constructed with the output node as the root node. For the worst case logic depth, the average input size (in_{avg}) of the blackbox nodes in the local tree (RT-level expression) is calculated. According to (in_{avg}), the average values of logical efforts (g_{avg}), parasitic efforts (p_{avg}), mimimum and maximum capacitances are computed. Starting with the CDFG representation of the function, each node is processed, starting with the output node in topological order.

Input:	Sequential RT-Level circuit, worst case delay target D
Output:	Capacitance of IO nets, type and drive-strengths of the flip-flop

Algorithm

(1) Mark the inputs of the flip-flops as pseudo-primary outputs and the outputs of the flip-flops as pseudo-primary inputs

(2) Sort or order the list of output and pseudo-primary outputs in such a way that the dependent nodes are considered after the independent nodes

(3) For the delay point D, for every primary and pseudo-primary output function

  (A) Adjust cycle time for

    (i) Setup time: if it is a pseudo-primary output

    (ii) Clk-to-Q: if the critical path starts at a pseudo-primary input

  (B) Compute the capacitance of the combinational part

    (i) Calculate the average input size (in_avg) for the worst number of stages

    (ii) Calculate the average logical effort (g), average parasitic effort(p), corresponding to in_avg

    (iii)Form a non-uniform tree with RT-Level visible nodes as intermediate nodes

    (iv) For each node in topographical order from output to input,

      (a) Form a uniform tree at the current node with its inputs and number of logic stages

      (b) Calculate local in_avg

      (c) Calculate the stage effort (f) for the delay D using Equation 4

      (d) Size the local tree using the sizing ratio (g/f)

  (C) Account for the flip-flops

    Case 1: The output is pseudo-primary and no input is pseudo-primary

    Case 2: One of the inputs is pseudo-primary and the output may or may not end at a flop

  (D) Repeat steps A, B and C for different flip-flop configuration, if target cycle-time is not satisfied

Fig. 2. Algorithm to estimate the capacitance of sequential circuits

For each node, the logic depth from its immediate inputs is known from Step 1 of Figure 1. A uniform tree is formed with these inputs and the logic depth at each node. The stage effort for this local tree is computed and the sizing of every node is performed as per the sizing ratio (g_{avg}/f). The wire capacitance was taken into account by using the wire-load model, during the sizing of each node. The minimum and maximum capacitances according to the target library are used as bounds during the sizing process. This process is recursively repeated for the immediate inputs of the RT-level expression of the current node, till the primary inputs are reached. It should be noted here that since we process the expression at every intermediate node instead of approximating the output expression as a whole as in [9], the blackbox nodes are not over-sized.

TABLE I

LIBRARY CHARACTERISTICS

Library	Gate input size	Gates	Flip-flop
lib1	one, two	inv, nand, nor, and, or	Dff
lib2	one, two, three	inv, nand, nor, and, or	Dff
lib3	one, two, three, four	inv, nand, nor, and, or	Dff

In step C, flip-flops are taken into consideration. Two cases are considered, the first one is when the output ends in a flip-flop, i.e. it is pseudo-primary and all inputs are primary inputs. In this case, the target cycle-time is checked with the timing of the sized combinational portion and the setup time of the output flip-flop. Case2 deals with the situation where atleast one of the inputs of the output function is a pseudo-primary input. Here, depending on the number of stages of the pseudo-primary input in the critical path to the output, the flip-flop is selected. If the total cycle-time is not satisfied, steps A, B and C are repeated for different flip-flop configurations. In the next section, we evaluate our algorithms for different libraries, activity factors, technologies and discuss the results.

IV. RESULTS

To evaluate the accuracy of our power estimation method, we estimated the dynamic power of RT-Level cicuits over their delay curve of operation and compared them with the reference values. The reference values were obtained by synthesizing the RT-Level circuits to gates and performing power estimation. We used Synopsys Design Compiler for synthesis and Power Compiler to obtain the reference power values. We used the 0.18um Artisan TSMC and 0.13um Virtual Silicon UMC libraries for our experiments. Three target gate libraries (drive strengths x1, x2 and x4) were extracted from the libraries and are listed in Table I.

For evaluating the algorithm on behavioral RT-Level circuits, we considered modules from the OR1200 processor and the exceptions module from the Floating point co-processor [6]. The OR1200 programmable interrupt controller ($OR1200\_pic$), instruction fetch ($OR1200\_fetch$) and control ($OR1200\_ctrl$) modules were considered for our experiments. The circuit sizes approximately ranged over (24 to 149) flip-flops and (170 to 1200) gates. The ISCAS sequential benchmarks (s298, s344, s386, s526, s641, s820, s1196, s1423, s5378, s15850 and s38584) were considered for evaluation at the the structural RT-Level. The circuit sizes ranged over (5 to 1200) flip-flops and (100 to 10000) gates. We generated input vectors for simulation using the method given in [18]. The input vectors were generated by considering different signal probabilities, transition densities and spatial correlations among the input signals, for the circuit under consideration. The modules were simulated using Synopsys VCS at the RT-Level and the switching factors were computed. The switching factors were taken as inputs by our algorithm, and the corresponding switching activity interchange format (SAIF) file was used to generate the reference power values. Due to lack of space, we have presented the results only for $OR1200\_ctrl$ and s38584 circuits which are the largest of the circuits under consideration, corresponding to *lib1* and *lib2* (0.18um) respectively, in Table II. Column-2 of the table lists

524

978-1-4244-3039-0/08 $25.00 © 2008 IEEE

TABLE II

POWER ESTIMATION RESULTS FOR $s38584$ AND $OR1200\_ctrl$

Circuit	Cycle-time in ns	R_{Pest}	R_{Pref}	Err%
$OR1200\_ctrl$	2	1.00	1.00	-NA-
	1.75	1.31	1.19	10.4
	1.5	1.63	1.53	6.6
	1.25	2.14	2.18	1.9
$s38584$	8	1.00	1.00	-NA-
	6	1.37	1.45	5.1
	4	2.12	2.24	5.3
	2	5.05	4.99	1.2

the target cycle-time and columns 3 and 4 list the power ratios. We now discuss the metrics that we have used in the table.

The metric Err denotes the deviation between the relative ratios of the estimated and the reference power. It is defined by

$$Err = \frac{abs(R_{Pref} - R_{Pest})}{R_{Pref}} \qquad (5)$$

where $Pref$ is the reference power and $Pest$ is the corresponding estimated power for a fixed point on the delay curve. A low value of Err implies that the accuracy of the method in tracking the variation in power over the delay curve is high.

The ratios R_{Pref} and R_{Pest} are

$$R_{Pref} = \frac{Pref}{Pmaxdel\_ref}, R_{Pest} = \frac{Pest}{Pmaxdel\_est} \qquad (6)$$

where $Pmaxdel$ is the power of the circuit at its maximum delay point of operation. The power ratios for a particular delay point denote how far the power value is with respect to the power at the maximum delay target or minimum frequency. For the circuit $OR1200\_ctrl$, the maximum delay target is 2ns and minimum is 1.25ns. The power ratios were obtained with respect to the maximum delay point (2ns). We now explain how the power ratios obtained from our estimated values can be used to get accurate power values, by considering the $OR1200\_ctrl$ module (Table III). First, we obtain the golden value of the power estimate at the maximum cycle-time using either a bottom-up power estimation method or by using synthesis. Here at 2ns, the golden value is found to be 2.52 mW. The estimated power ratios are listed in column 3. The product of the golden value and the power ratios are given in column 4, these being the estimated power values at the delay points. The power values in column 5 (Pref) list the actual reference values obtained from using synthesis. The error between the estimated and reference values are shown in Column 6, which matches the $Err\%$ of $OR1200\_ctrl$ from Table II. In this way, our method can make use of a golden value obtained from macro-modeling and provide power estimates at other delay points, co-existing as part of a power estimation framework.

The results for the rest of the circuits are summarized in Table IV and plotted in Figure IV for the 0.18um library. The data in Figure IV were collected from all the circuits under consideration over all three libraries of 0.18um, covering 3 to 5 points on the delay curve of each circuit. The overall estimation accuracy is consistently high across libraries, as observed from the table. The data in Column-3 of Table

TABLE III

POWER ESTIMATION RESULTS FOR $OR1200\_ctrl(lib1)$

Circuit	Cycle-time in ns	R_{Pest}	Pest in mW	Pref in mW	Error%
$OR1200\_ctrl$	2	1.00	2.52	2.52	-NA-
	1.75	1.31	3.30	2.99	10.4
	1.5	1.63	4.11	3.86	6.6
	1.25	2.14	5.39	5.49	1.9

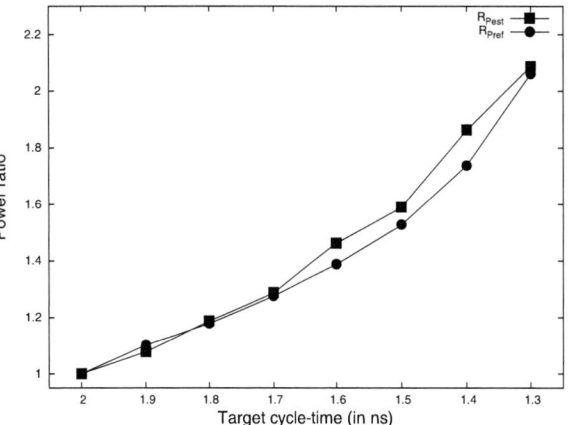

Fig. 3. Estimated vs. Reference power ratios ($OR1200\_ctrl(lib2)$)

IV were obtained with respect to a set of input vectors (af1) with input activity 0.15. For different sets of input vectors, the results are shown in Column-4,5 of Table IV. As it is known that the nominal activity factor is around 0.15 (af1) to 0.25 (af2), we used those values for input vector generation. To analyze how the prediction accuracy scales with high activity, we also applied a sequence of vectors with 0.5 (af3) as the activity at the inputs. It can be observed that the dynamic power estimation is consistent, the average estimation accuracy close to 90%. This is quite encouraging as our method is top-down and generic. The ISCAS sequential benchmarks were not evaluated with three and four input libraries, as there were not enough three or four input expressions in their structural verilog code. The tracking of the power ratios for $OR1200\_ctrl$ with respect to *lib2* is shown in Figure 3. As the cycle-time decreases, we can observe from the graph that the estimated power ratio tracks the reference ratio by increasing proportionately. All the above experiments were repeated for 0.13um, and Table V compares the results obtained using af3 with corresponding results from 0.18um. It can be observed that the average error in prediction does not vary much between the technologies, proving the robustness of our methodology.

The outlier points in the graph of Figure IV are explained as

TABLE IV

SUMMARY OF POWER ESTIMATION RESULTS FOR 0.18UM

Circuits	Library	Avg. Err% af1 (0.15)	Avg. Err% af2 (0.25)	Avg. Err% af3 (0.5)
Behavioral 4 modules	lib1	6.57	7.07	6.80
	lib2	5.89	7.15	7.89
	lib3	4.45	5.65	6.79
Structural 11 ISCAS	lib1	8.14	7.98	10.19

525

978-1-4244-3039-0/08 $25.00 © 2008 IEEE

TABLE V
COMPARISON OF RESULTS BETWEEN 0.18UM AND 0.13UM

Circuits	Library	Avg. Err%_0.13	Avg. Err%_0.18
Behavioral	lib1	6.48	6.80
	lib2	5.75	7.89
4 modules	lib3	5.47	6.79
Structural 11 ISCAS	lib1	8.26	10.19

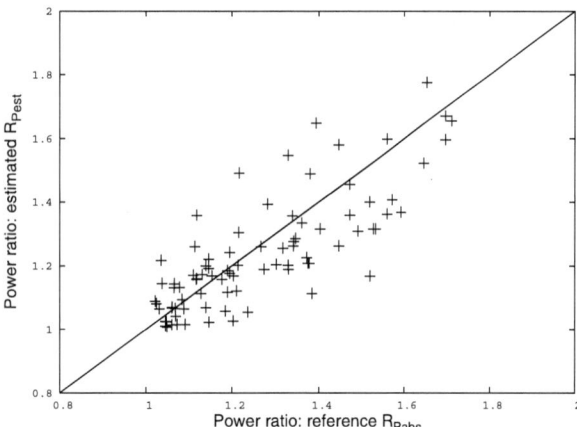

Fig. 4. Estimated ratio vs. Reference ratio (af2, 0.18um)

follows. One of the important factors influencing the accuracy is that the input RT-Level verilog code was not written in a balanced manner, using early and late arriving signal optimization. Since our method does not perform functional analysis, we were not able to modify the structure to account for timing. Moreover, there must be enough RT-level visible nodes to account for both activity factor approximation, as well as for detecting shared logic. The greater the number of intermediate points, the better will be the accuracy. The behavioral RT-level code was modified to introduce RT-Level visible nodes before it was given as input to our technique. The visible nodes were added only to detect shared logic between the primary outputs, and so there were not enough points to estimate the switching activity accurately. Analyzing and modifying the algorithms to account for these factors is very much part of our future work. In addition, we do not analyze the code functionally for evaluation with complicated libraries (like AOI, OAI etc.), as this study is about how much accuracy can be achieved by not including functional analysis. The average estimation error is less than 10%, which is acceptable for this kind of analysis [19]. We plan to extend this study to designs that involve macros like adders, comparators and evaluate it on bigger behavioral designs. The prototype tool was implemented in *perl* and the total running times for $OR1200\_ctrl$ and $s38584$ over all four delay points were 1.5 and 3.5 minutes respectively. For balanced circuits that are written using early and late arriving signal optimizations, we plan to include an intelligent expression analyzer to achieve better results. We plan to implement the prototype in C++ to compare its running times with other power estimation methodologies and also include wire capacitance due to routing, as part of our future work.

V. CONCLUSION

We have demonstrated the usefulness of our technique in estimating the zero-delay dynamic power for different target frequencies. Our method can be employed in early stage power estimation at both behavioral and structural RT-Level. It is robust across technology libraries and applies to generic circuits and can co-exist with bottom-up techniques.

REFERENCES

[1] C. Chen and C. Tsui, "Towards the Capability of Providing Power-Area-Delay Trade-off at the Register Transfer Level," in *Proceedings of the International Symposium on Low Power Electronics and Design*, pp. 24–29, 1998.

[2] S. Gupta and F. N. Najm, "Power Macromodeling for High Level Power Estimation," in *DAC '97: Proceedings of the 34th annual conference on Design automation*, pp. 365–370, 1997.

[3] G. Bernacchia and M. C. Papaefthymiou, "Analytical Macromodeling for High-Level Power Estimation," in *Proceedings of the International Conference on Computer-Aided Design*, pp. 280–283, 1999.

[4] M. Bruno, A. Macii, and M. Poncino, "A Statistical Power Model for Non-synthetic RTL Operators," in *Proceedings of the International Workshop on Power and Timing Modeling, Optimization and Simulation*, pp. 208–218, 2003.

[5] I. Sutherland, R. Sproull, and D. Harris, *Logical Effort: Designing Fast CMOS Circuits*. San Francisco, CA: Morgan Kaufmann, 1999.

[6] "OR1200 processor and Floating-point co-processor," *http://www.opencores.org/*.

[7] E. Macii, M. Pedram, and F. Somenzi, "High-Level Power Modeling, Estimation and Optimization," *IEEE Transactions on Computer-Aided-Design of Integrated Circuits and Systems*, vol. 17, no. 11, pp. 1061–1079, 1998.

[8] R. P. Llopis and K. Goossens, "The Petrol Approach to High-Level Power Estimation," in *Proceedings of the International Symposium on Low Power Electronics and Design*, pp. 130–132, 1998.

[9] S. Sambamurthy, J. A. Abraham, and R. S. Tupuri, "Delay Constrained Register Transfer Level Dynamic Power Estimation," in *Proceedings of the International Workshop on Power and Timing Modeling, Optimization and Simulation*, pp. 36–46, 2006.

[10] R. Zafalon, M. Rossello, E. Macii, and M. Poncino, "Power Macromodeling for a High Quality RT-level Power Estimation," in *Proceedings of the International Symposium on Quality of Electronic Design*, p. 59, 2000.

[11] S. Ravi, A. Raghunathan, and S. Chakradhar, "Efficient RTL Power Estimation for Large Designs," in *Proceedings of the International Conference on VLSI Design*, pp. 431–439, 2003.

[12] K. M. Buyuksahin and F. N. Najm, "Early Power Estimation for VLSI Circuits," *IEEE Transactions on Computer-Aided-Design of Integrated Circuits and Systems*, vol. 24, no. 7, pp. 1076–1088, 2005.

[13] D. Brooks, V. Tiwari, and M. Martonosi, "Wattch: A Framework for Architectural-level Power Analysis and Optimizations," in *Proceedings of the 27th annual international symposium on Computer architecture*, pp. 83–94, 2000.

[14] W. Ye, N. Vijaykrishnan, M. Kandemir, and M. J. Irwin, "The Design and Use of Simplepower: A Cycle-Accurate Energy Estimation Tool," in *Proceedings of the 37th conference on Design automation*, pp. 340–345, 2000.

[15] "Boolstuff," *http://perso.b2b2c.ca/sarrazip/dev/boolstuff.html*.

[16] M. Nemani and F. N. Najm, "Towards a High-Level Power Estimation Capability," *IEEE Transactions on Computer-Aided-Design of Integrated Circuits and Systems*, vol. 15, no. 6, pp. 588–598, 1996.

[17] F. N. Najm, "A Survey of Power Estimation Techniques in VLSI Circuits," *IEEE Transactions on Very Large Scale Integrated Systems*, vol. 2, no. 4, pp. 446–455, 1994.

[18] X. Liu and M. C. Papaefthymiou, "A Markov Chain Sequence Generator for Power Macromodeling," in *Proceedings of the 2002 IEEE/ACM international conference on Computer-aided design*, pp. 404–411, 2002.

[19] S. Gupta and F. N. Najm, "Analytical Models for RTL Power Estimation of Combinational and Sequential Circuits," *IEEE Transactions on Computer Aided Design of Integrated Circuits and Systems*, vol. 19, no. 7, pp. 808–814, 2000.

21st International Conference on VLSI Design

Total Power Minimization in Glitch-Free CMOS Circuits Considering Process Variation

Yuanlin Lu[*]
Intel Corporation
Folsom, CA 95630, USA
yuanlin.lu@intel.com

Vishwani D. Agrawal
Auburn University
Auburn, AL 36849, USA
vagrawal@eng.auburn.edu

Abstract

Compared to subthreshold leakage, dynamic power is normally much less sensitive to the process variation due to its approximately linear relation to the process parameters. However, the average dynamic power of a circuit optimized by deterministic glitch elimination (using hazard filtering and path balancing) increases because glitches randomly start reappearing under the influence of process variation. Combining existing techniques, we propose a new statistical mixed integer linear programming (MILP) formulation, which combines glitch elimination and dual-threshold design to statistically minimize the total power in a glitch-free circuit under process variation.

1. Introduction

With the continuous increase of the density and performance of integrated circuits due to the scaling down of the CMOS technology, reducing power dissipation becomes a serious problem that every circuit designer has to face. At the same time, the increase in variability of several key process parameters can significantly affect the design and optimization of low power circuits in the nanometer regime [1-3]. Due to the exponential relation of leakage current with some process parameters, such as the effective gate length, oxide thickness and doping concentration, process variations can cause a significant increase in the leakage current. To minimize the effect of process variation, some techniques [1-3] statistically optimize the leakage power and circuit performance by dual-V_{th} assignment. Leakage current and delay are treated as random variables. A dynamic programming approach for leakage optimization by dual-V_{th} assignment has been proposed [2] using two pruning criteria that stochastically identify pareto-optimal solutions and prune the sub-optimal ones. Another approach [1] solves the statistical leakage minimization problem by a theoretically rigorous formulation for dual-V_{th} assignment and gate sizing.

Glitches are unnecessary signal transitions that account for 20%-70% of the dynamic switching power [4]. To eliminate glitches, we combine the techniques of

*Formerly with Department of Electrical and Computer Engineering, Auburn University, Auburn, AL 36849, USA.

hazard filtering [5, 8-12] and path balancing [6, 8, 11], referred to in this paper as glitch elimination. Compared to leakage power, dynamic power is normally much less sensitive to the process variation because of its approximately linear dependency on the process parameters. However, any deterministic glitch elimination technique becomes less effective under process variation, since the perfect hazard filtering conditions can be easily corrupted even with a small variation in some process parameters. Hu and Agrawal [13-14] proposed a technique to eliminate glitches under process variation. However, performance is sacrificed to obtain a process-variation-resistant circuit, and the effect of process variation on leakage power is not considered.

Our work is motivated by the above research. To minimize the leakage power, we use a mixed integer linear programming (MILP) model to determine the optimal assignment of V_{th} while controlling any reduction in performance. To eliminate the glitch power, additional MILP constraints determine the positions and values of the delay elements to be inserted to balance path delays. Statistical delay and leakage models are further adopted to reduce the total power in glitch-free circuits considering process variation.

2. Background

Lu and Agrawal [17] propose a statistical MILP formulation to minimize the impact of process variation on the subthreshold leakage. In this section, we extend that discussion to study the impact of process variation on dynamic power. Dynamic power comprises of two parts, logic switching power and glitch power:

$$P_{dyn} = \frac{1}{2}C_L V^2 \cdot A \cdot F$$
$$= \text{Logic switching power} + \text{Glitch power} \qquad (1)$$

where A is switching activity and F is the circuit operating frequency.

Logic switching power is directly proportional to the loading capacitances, C_L, which linearly depends upon gate sizes, gate width and gate length. Local (intra-die) process variation causes gate sizes to vary randomly and hence does not affect logic switching power too much. Global (inter-die) process variation changes gate sizes in similar ways and does vary the logic switching power. This

978-1-4244-3039-0/08 $25.00 © 2008 IEEE

also does not affect the solution of our MILP formulation, since gate delays and gate sizes in the MILP constraints either increase or decrease by the same percentage when global process variation is considered, and T_{max} (critical path delay that affects the circuit performance) is assumed to change accordingly [15].

The impact of process variation on glitch power is different and more complicated. Glitches are generated if the glitch filtering condition (2) [6] is not satisfied for cell i. Since inertial gate delays d_i vary with process variations, inequality (2) may not be satisfied.

$$d_i > T_i - t_i \qquad (2)$$

Where $T_i - t_i$ is the differential path delay at gate i. We consider the impact of global process variation and local process variation on glitch power, separately.

- **Impact of global process variation on glitches**

For every gate i, the timing window $T_i - t_i$ is actually determined by two timing paths, the fastest path (*FPath*) and the slowest path (*SPath*) from primary inputs to gate i. T_i is the cumulative inertial gate delay along the slowest path, and t_i is the cumulative inertial gate delay along that fastest path. Thus,

$$T_i - t_i = \sum_{m \in SPath} d_m - \sum_{n \in FPath} d_n \qquad (3)$$

Assuming that there is $r \cdot 100\%$ (r: 0~1) of global variation applied to the circuit, glitch filtering condition (2) for gate i remains unchanged since both timing window, $T_i - t_i$, and gate delay vary by $r \cdot 100\%$. Therefore, the technique of glitch elimination is resistant to the global process variation.

- **Impact of local process variation on glitches**

Let us consider the impact of local process variation on glitch elimination. When local variations occur in a circuit, T_i and t_i are the sum of gate delays, which vary randomly, along the slowest and fastest paths from primary inputs to cell i's inputs, so, $T_i - t_i$ is not very sensitive to process variations, while d_i does change with the process variation. Therefore, it is very possible that the original glitch filtering conditions (2) can not be satisfied in the presence of local process variation.

As shown in Figure 1, there are three possible glitch filtering conditions. Both Figures 1(b) and 1(c) are glitch free while Figure 1(a) has a glitch. In an un-optimized circuit (with glitches), Figures 1(a) or 1(b) is represents a much more common condition for a gate. Although the condition of Figure 1(c) is still possible it has lower possibility. On the contrary, in an optimized glitch-free circuit, Figure 1(c) applies to many gates because Figure 1(a) is always forced to become Figure 1(c) by path balancing for glitch elimination.

With local process variation, Figures 2(a) and 2(b) show that the original condition is not so easily corrupted if only the variation of the timing window or the gate delay falls into the shaded areas, while Figure 2(c) is extremely sensitive to the local process variation, since a slight increase in the timing window or decrease in the gate delay can simply let an original glitch-free gate generate glitches at its output.

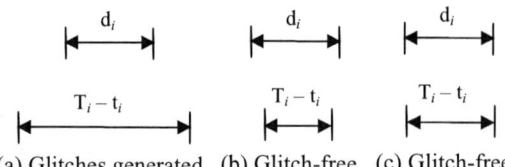

(a) Glitches generated (b) Glitch-free (c) Glitch-free

Figure 1. Three possible glitch elimination conditions.

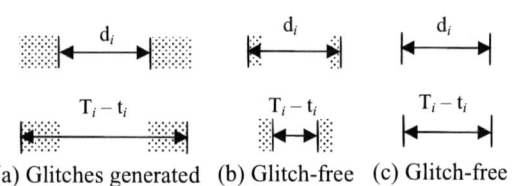

(a) Glitches generated (b) Glitch-free (c) Glitch-free

Figure 2. Three glitch elimination conditions under local process variation.

This explains why the dynamic power of an un-optimized circuit is much more resistant to local process variation than that of an optimized glitch-free circuit. The glitch elimination condition shown in Figure 1(c) cannot be really satisfied even with quite small process variation.

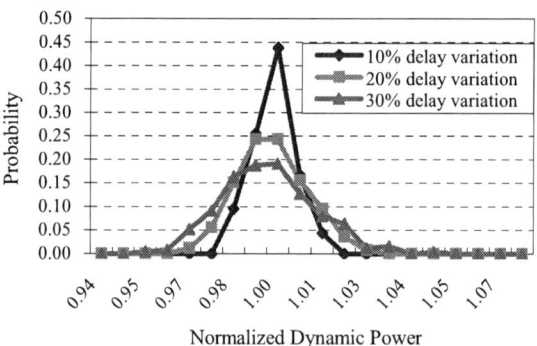

Figure 3. Normalized dynamic power distribution of un-optimized (with -glitches) C432 under local delay variation.

Figure 3 demonstrates the resistance of un-optimized circuits to the local process variation. We applied 10%, 20% and 30% local delay variations, as may be caused by variations in gate-length-independent V_{th}, to an un-optimized (with-glitch) version of circuit C432. The largest percentage of the mean value deviated from the nominal value is 0.22% and the maximum spread, 3×standard deviation / mean, is only 4.5%.

The sensitivity of optimized glitch-free circuits to local process variation is illustrated in Figure 4. Both mean value and standard deviation of dynamic power distribution increase significantly with the increase of the local process variation. When 30% local variation was applied to the optimized glitch-free C432, its average dynamic power increased by 32% and almost became equal to the normalized dynamic power (1.34) of the un-optimized C432. In Figure 4, some samples of optimized C432 have dynamic power even larger than 1.34. We also note that every sample in Figure 4 consumes more than the nominal value, 1, which is the expected minimum-normalized-dynamic-power of the optimized glitch-free C432. Process variation causes some glitches to be generated in the glitch-free circuit and hence increases the dynamic power.

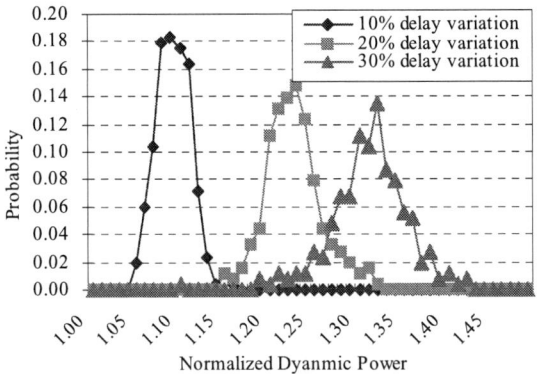

Figure 4. Normalized dynamic power distribution of optimized (glitch-free) C432 under local delay variation.

It is remarkable that the advantage of glitch elimination is totally lost due to the local process variation. Hence, the deterministic approach of glitch elimination is not useful for power optimization with process variation. In the following section, we combine the MILP formulations introduced in [15-17], and thus a new statistical MILP formulation is proposed to optimize total power under process variation and to fully utilize the advantage of the glitch elimination procedure.

The deterministic MILP [15-16] using glitch elimination and dual-V_{th} assignment to reduce the total power consumption is a prerequisite procedure, which is modified to consider process variation.

3. Statistical MILP for Total Power Optimization with Process Variation

In the statistical MILP formulation, we treat all gate delays and timing window variables as random variables with normal distribution whose standard deviation is σ_r.

3.1 Variables

- **Integer variables:**

In our cell library, each standard cell has two possible threshold voltages and three alternative sizes (1X, 2X and 4X). Therefore, this MILP has six integer variables to allow alternative choices. The variables are denoted as,

$X1L[i]$, $X2L[i]$, $X4L[i]$, $X1H[i]$, $X2H[i]$, $X4H[i]$

- **Continuous Variables:**

$\delta[i]$ - relaxed variable for the glitch filtering constraint of cell i. It will be discussed in Section 3.3.

$Size[i]$ - size of cell i.

$I_{leak}[i]$ - nominal value of leakage of cell i.

$u\_D[i]$ - mean of inertial gate delay of cell i.

$s\_D[i]$ - standard deviation of inertial gate delay.

$u\_T[i]$ - mean of $T[i]$.

$s\_T[i]$ - standard deviation of $T[i]$.

$u\_t[i]$ - mean of $t[i]$.

$s\_t[i]$ - standard deviation of $t[i]$.

$u\_\Delta d[i,j]$ - mean of $\Delta d[i,j]$ (the delay of the inserted delay element).

$s\_\Delta d[i,j]$ - standard deviation of $\Delta d[i,j]$.

3.2 Constants

σ_r - standard deviation of the process parameter variations.

T_{max} - the maximum expected circuit performance.

$S_{X2}[i]$ - gate size of cell i with 2X driving strength.

W_1, W_2, W_3 - weight factors.

$I_{X2L}[i]$, $I_{X2H}[i]$ - nominal values of the subthreshold leakage of cell i with 2X driving strength.

$D_{X1L}[i]$, $D_{X2L}[i]$, $D_{X4L}[i]$, $D_{X1H}[i]$, $D_{X2H}[i]$, $D_{X4H}[i]$ - nominal values of the inertial gate delay of cell i at all six corners.

3.3 Constraints

- **Basic constraints**

Let LP solver choose one and only one optimal version for cell i.

$$X1L[i] + X2L[i] + X4L[i] + X1H[i] + X2H[i] + X4H[i] = 1 \quad (4)$$

Nominal value of the subthreshold leakage of cell i:

$$
\begin{aligned}
I_{leak}[i] = &(0.5 \cdot X1L[i] + X2L[i] + 2 \cdot X4L[i]) \cdot I_{X2L}[i] + \\
&(0.5 \cdot X1H[i] + X2H[i] + 2 \cdot X4H[i]) \cdot I_{X2H}[i]
\end{aligned} \quad (5)
$$

Mean and standard deviation of the gate delay of cell i:

529

978-1-4244-3039-0/08 $25.00 © 2008 IEEE

$$u\_D[i] = D_{X1L}[i] \cdot X1L[i] + D_{X2L}[i] \cdot X2L[i] + D_{X4L}[i] \cdot X4L[i] + D_{X2L}[i] \cdot X1H[i] + D_{X2L}[i] \cdot X2H[i] + D_{X4L}[i] \cdot X4H[i]$$

$$(6)$$

$$s\_D[i] = \sigma_r \cdot u\_D[i] \qquad (7)$$

The size of cell i:

$$Size[i] = \begin{cases} 0.5 \cdot (X1L[i] + X1H[i]) + (X2L[i] + X2H[i]) + \\ 2 \cdot (X4L[i] + X4H[i]) \end{cases} \cdot S_{X2}[i]$$

$$(8)$$

- **For glitch elimination**

Instead of using inequality (2), in the statistical method, we adopt the following glitch filtering constraint:

$$u\_D[i] - 3 \times s\_D[i] \geq (u\_T[i] + 3 \times s\_T[i]) - (u\_t[i] - 3 \times s\_t[i])$$

$$(9)$$

This constraint can leave certain margin for process variation in advance as shown in Figure 2(b) instead of Figure 2(c). However, normally the above worst case constraint is too tight to make CPLEX LP solver find a feasible solution. So, we add one nonnegative relaxed variable $\delta[i]$ to each glitch filtering constraint (9).

$$\delta[i] + (u\_D[i] - 3 \times s\_D[i]) \geq (u\_T[i] + 3 \times s\_T[i]) - (u\_t[i] - 3 \times s\_t[i])$$

$$(10)$$

In the objective function, by minimizing $\Sigma\delta[i]$, CPLEX LP solver will try to find one optimal solution to make as many of the constraints (10) satisfied as possible with a zero $\delta[i]$, which means the glitches of corresponding cells can be truly eliminated even in the worst case condition of process variation. Those constraints only being satisfied with the help of a positive $\delta[i]$ quite likely fail to filter glitches.

- **For maximal performance**

To keep the maximal performance, at every primary output k, let,

$$u\_T[k] + 3 \times s\_T[k] \leq T_{max}. \qquad (11)$$

3.4 Objective function

The objective function minimizes the impact of process variation on the total power consumption:

Min {the impact of process variation on the total power consumption}

= Min {mean and standard deviation of leakage power + mean and standard deviation of dynamic power}

$$= Min \begin{cases} W_1 \cdot C_1 \sum_i I_{leak}[i] + \\ W_2 \cdot \left(C_2 \sum_i size[i] + C_3 \sum_i \sum_j \mu\_\Delta d[i,j] \right) + \\ W_3 \cdot \sum_i \delta[i] \end{cases}$$

$$(12)$$

C_1, C_2 and C_3 are fitting parameters to let three terms ($C_1\Sigma I_{leak}[i]$, $C_2\Sigma size[i]$ and $C_3\Sigma\Sigma u\_\Delta d[i,j]$) have the same units (μW).

The impact of process variation on both mean and standard deviation of the power consumption should be considered. For leakage, a smaller mean value automatically implies a narrower spread of leakage power distribution since more gates are assigned high V_{th}. Min($C_1\Sigma I_{leak}[i]$) should be enough to minimize the impact of process variation on the total subthreshold leakage. For the dynamic power, standard deviation of the dynamic power distribution is determined by $\Sigma\delta[i]$, and ($C_2\Sigma size[i] + C_3\Sigma\Sigma u\_\Delta d[i,j]$) affects the average dynamic power. Therefore, we should minimize ($C_2\Sigma size[i] + C_3\Sigma\Sigma u\_\Delta d[i,j]$) and $\Sigma\delta[i]$, simultaneously.

The objective function (12) is composed of three parts (three single objectives), namely, minimize the average leakage power, minimize the average dynamic power and minimize the standard deviation of the dynamic power. It is a multi-objective function in which individual objectives conflict. For instance, minimization of $\Sigma\delta[i]$ results in an increase of $\Sigma\Sigma u\_\Delta d[i,j]$, and optimization of $\Sigma I_{leak}[i]$ leads to a larger $\Sigma size[i]$, *etc.* It is not easy to get one optimum value for every single objective. What we can do instinctively is to carefully select weight factors, W_1, W_2 and W_3 to make a tradeoff among the three objectives.

It should be noticed that the solution provided by a deterministic MILP [15-16] gives us a rough idea of which one is the dominant component between leakage and dynamic power. We also get their exact optimal values (power consumption) for the optimized circuit. Based on that information, we can choose weight factors and add some empirical constraints on the largest allowable minimal leakage or dynamic power in the statistical MILP formulation.

The choice of minimizing the impact of process variation either on leakage or on dynamic power depends on which one is the dominant power consumer, and the circuit applications as well. For a circuit optimized by the deterministic MILP, we consider:

- Case 1 - if the optimal leakage is much less than the optimal dynamic power and its large spread due to process variation (for example, 5X difference under 30% global process variation) can still be ignored, we need put much more emphasis on dynamic power changes being resistant to process variation;

- Case 2 - if the optimal leakage is comparable to the optimal dynamic power, and most of the time the circuit remains in standby mode(for example, circuits of cell phones) the impact of process variation on the optimal leakage should be minimized with priority since leakage is much more sensitive to the process variation;

- Case 3 - if the optimal leakage is comparable to the optimal dynamic power, and most of the time the circuit is in the active mode (for example, circuits of portable GPS, portable game machines, *etc.*) both the mean and standard deviation of the dynamic power distribution should be optimized.

3.5 Minimizing impact of process variation on leakage

In case 1 and case 3, dynamic power is the dominant component of the total power consumption. Its standard deviation is determined by the number of glitch filtering constraints (10) whose $\delta[i]$ have positive values. So, in the MILP objective function (13), we first let $W3$ be infinitely large to put the highest priority on minimizing $\Sigma\delta[i]$:

$$Min\left\{W1\cdot\sum_i I_{leak}[i]+W2\cdot\left(\sum_i size[i]+\sum_i\sum_j\Delta d[i,j]\right)+\underset{W3\to\infty}{W3}\sum_i\delta[i]\right\}$$
(13)

Although MILP tries to minimize $\Sigma\delta[i]$, $\delta[i]$ for some gates may still be positive since the constraint (9) is too tight to be satisfied without the help of a positive $\delta[i]$. Every positive $\delta[i]$ possibly results in the glitch generation at gate i's output. From Figure 4, we also see that the average dynamic power almost linearly increases with the process variation. This increase is contributed by the glitches caused by the process variatio. To counteract the increase in the average dynamic power due to those glitches, or to let the really average dynamic power in process variation condition still be close to that achieved by the deterministic MILP formulation, we sacrifice some leakage power and get a smaller logic switching power. This can be achieved by letting $W1$ and $W2$ both equal to 1 in the MILP objective function (14) and adding a new constraint (15) to the statistical MILP formation.

$$Min\left\{C_1\sum_i I_{leak}[i]+\left(C_2\sum_i size[i]+C_3\sum_i\sum_j\Delta d[i,j]\right)+\underset{W\to\infty}{W3}\sum_i\delta[i]\right\}$$
(14)

$$C_2\sum_i size[i]+C_3\sum_i\sum_j\Delta d[i,j]<P_{dyn\_opt}/\rho \quad (\rho>1) \quad (15)$$

P_{dyn_opt} is the optimal dynamic power obtained by the deterministic MILP [15-16] and ρ is a constant determined by the process variation. By letting ρ larger than 1, the statistical MILP formulation can give an optimal circuit with less dynamic power.

3.6 Minimizing impact of process variation on leakage

In case 2, leakage almost equals or is even larger than the dynamic power. Since leakage is so sensitive to the process variation that we cannot minimize the effect of process variation on the dynamic power by sacrificing leakage any more. The technique of eliminating glitches has to be discarded since the increase in the average dynamic power under process variation may be close to or even larger than the glitch power saved. To make the

leakage of optimized circuits resistant to the process variation, we can still use the MILP proposed in [17] except every gate has six possible choices instead of two choices.

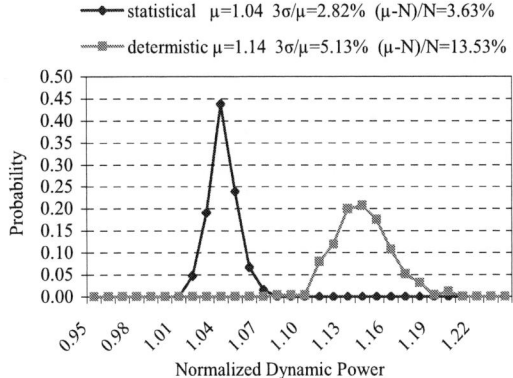

Figure5. Comparison of the impacts of 15% local process variation on the **dynamic power** in C432 which is optimized by the statistical MILP with the emphasis on the resistance of dynamic power to process variation, or by the deterministic MILP [15-16]. (Dynamic power = 1 is the expected normalized minimum dynamic power in the optimized glitch-free C432).

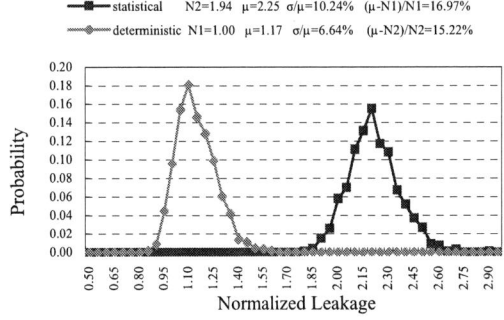

Figure 6. Comparison of the impacts of 15% local L_{eff} process variation on the **leakage power** in C432 which are optimized by the statistical MILP with the emphasis on the resistance of dynamic power to process variation, or the deterministic MILP [15-16]. (N1 and N2 are the normalized nominal leakage power in the optimized glitch-free C432).

4. Results

In C432 optimized by the deterministic MILP formulation [15-16], the optimized total power comprises 59.3μW dynamic power and 5.54μW leakage power. With 15% local process variation, the average dynamic power increases 13.53% with 5.13% standard deviation. To reduce the impact of process variation on the dynamic power, the objective function (14) and constraint (15) (let P_{dyn_opt}=59.3μW and ρ=1.10) are adopted in the statistical MILP formulation. The two curves in Figure 5 show that the average dynamic power only increases 3.63% instead of 13.53%, and standard deviation is also reduced to 2.82% from 5.13% when 15% local process variation is applied to the optimized glitch-free C432, although at a cost of 94% average leakage power increase (from 1.0 to 1.94) and a little bit wider spread of leakage power distribution, which is shown in Figure 6.

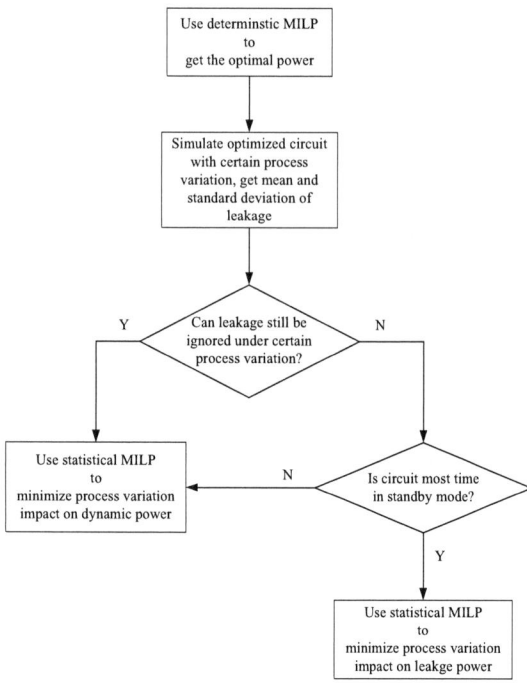

Figure 7. An algorithm to determine whether leakage or dynamic power should be optimized with process variation.

5. Summary

In this paper, the impact of process variation on dynamic power is analyzed, and a statistical MILP formulation is presented to minimize the total (dynamic and leakage) power in glitch-free circuits considering process variation. The impact of process variation on dynamic power can be minimized by giving up some leakage if the dynamic power is still the dominant power component under process variation. Figure 7 gives a flowchart of how to make a decision about which one, leakage or dynamic power, should be optimized with process variation.

6. References

[1] M. Mani, A. Devgan, and M. Orshansky, "An Efficient Algorithm for Statistical Minimization of Total Power Under Timing Yield Constraints," *Proc. Design Automation Conference*, 2005, pp. 309-314.

[2] A. Davoodi and A. Srivastava, "Probabilistic Dual-V_{th} Optimization Under Variability," *Proc. ISLPED,* 2005, pp. 143-147.

[3] A. Srivastava, D. Sylvester, D. Blaauw, "Statistical Optimization of Leakage Power Considering Process Variations Using Dual-V_{th} and Sizing," *Proc. Design Automation Conf.*, 2004, pp. 773-778.

[4] A. P. Chandrakasan and R. W. Brodersen, *Low Power Digital CMOS Design.* Boston: Springer, 1995.

[5] V. D. Agrawal, "Low Power Design by Hazard Filtering," *Proc. 10th Int. Conf. VLSI Design*, 1997, pp. 193-197.

[6] V. D. Agrawal, M. L. Bushnell, G. Parthasarathy, and R. Ramadoss, "Digital Circuit Design for Minimum Transient Energy and a Linear Programming Method," *Proc. 12th International Conf. VLSI Design*, 1999, pp. 434-439.

[7] E. Jacobs and M. Berkelaar, "Using Gate Sizing to Reduce Glitch Power," *Proc. PRORISC/IEEE Workshop on Circuits, Systems and Signal Processing*, 1996, pp. 183-188.

[8] S. Kim, J. Kim, and S. Y. Hwang, "New Path Balancing Algorithm for Glitch Power Reduction," *IEE Proc. Circuits, Devices and Systems,* vol. 148, no. 3, pp. 151-156, 2001.

[9] C. V. Schimpfle, A. Wroblewski, and J. A. Nossek, "Transistor Sizing for Switching Activity Reduction in Digital Circuits," *Proc. European Conference on Theory and Design*, 1999, pp. 114-117.

[10] A. Wroblewski, C. V. Schimpfle, and J. A. Nossek, "Automated Transistor Sizing Algorithm for Minimizing Spurious Switching Activities in CMOS Circuits," *Proc. IEEE International Symposium on Circuits and Systems*, 2000, pp. 291-294.

[11] T. Raja, V. D. Agrawal, and M. L. Bushnell, "Minimum Dynamic Power CMOS Circuit Design by a Reduced Constraint Set Linear Program," *Proc. 16th International Conf. VLSI Design*, 2003, pp. 527-532.

[12] T. Raja, V. D. Agrawal, and M. L. Bushnell, "Variable Input Delay CMOS Logic for Low Power Design," *Proc. 18th International Conf. VLSI Design*, 2005, pp. 596-603.

[13] F. Hu and V. D. Agrawal, "Input-Specific Dynamic Power Optimization for VLSI Circuits," *Proc. Int. Symp. Low Power Electronics and Design*, 2006, pp. 232-237.

[14] F. Hu, "Process-Variation-Resistant Dynamic Power Optimization for VLSI Circuits," PhD Thesis, Auburn, Alabama: Auburn University, May 2006.

[15] Y. Lu and V. D. Agrawal, "CMOS Leakage and Glitch Power Minimization for Power-Performance Tradeoff," *Journal of Low Power Electronics,* vol. 2, no. 3, pp. 378-387, Dec. 2006.

[16] Y. Lu and V. D. Agrawal, "Leakage and Dynamic Glitch Power Minimization Using Integer Linear Programming for Vth Assignment and Path Balancing," *Proc. of the International Workshop on Power and Timing Modeling, Optimization and Simulation*, 2005, pp. 217–226.

[17] Y. Lu and V. D. Agrawal, "Statistical Leakage and Timing Optimization for Submicron Process Variation," *Proc. 20th International Conf. VLSI Design*, 2007, pp. 439-444.

978-1-4244-3039-0/08 $25.00 © 2008 IEEE

Power Reduction of Functional Units considering Temperature and Process Variations

Deepa Kannan, Aviral Shrivastava

Compiler and Microarchitecture Laboratory
School of Computing and Informatics
Arizona State University, Tempe, AZ 85281
{deepa.kannan, aviral.shrivastava}@asu.edu

Sarvesh Bhardwaj, Sarma Vrudhula

VLSI Electronic Design Automation Laboratory
School of Computing and Informatics
Arizona State University, Tempe, AZ 85281
{sarvesh.bhardwaj, vrudhula}@asu.edu

Abstract

Continuous technology scaling has resulted in an increase in both, the power density as well as the variation in device dimensions (process variations) of the manufactured processors. Both power density and process variations have a significant impact on the leakage power. Therefore, power optimization techniques should be sensitive to the variation in leakage power due to both temperature as well as process variations. Operation to Functional Units Binding Mechanism (OFBM) is the mechanism to dynamically issue operations to Functional Units (FUs) in superscalar processors. We propose a Leakage-Aware OFBM (LA-OFBM), which is both temperature and process variation aware. Our experimental results demostrate that LA-OFBM reduces the mean and standard deviation of the total energy consumption of ALUs by 18%, and 46% respectively, as compared to the traditional OFBM, without any performance penalty.

1. Introduction

The ever increasing performance demands from high-end microprocessors have been one of the most important forces behind continuous technology scaling for the past four decades. As a consequence of incessant technology scaling, leakage power has become a major component of the total power budget of the processors developed in nano-scale CMOS technologies. In fact, according to [19], leakage power in 65 nm technology amounts to almost 40% of modern microprocessors' total power budget.

Leakage power, in contrast to dynamic power is highly sensitive to variations in the operational temperature. In-fact, leakage of a CMOS gate increases exponentially with increase in temperature. According to [15] a 10^oC rise in temperature at 35^oC will result in leakage currents going up by 126%.

Another important consequence of technology scaling is a significant loss of control in the lithography as well as channel doping steps during the manufacturing, resulting in large variations in the characteristics of the manufactured devices (ITRS 2003) [7]. This phenomenon, called process variation has significant impact in terms of power consumption, yield, reliability, and design processes. Since leakage power has an exponential dependence on device characteristics, even small variations in the device characteristics result in large variations in the leakage power. Leakage power being a major contributor to the total power consumption in present day microprocessors, the variations in leakage power results in a significant variation in the processor power consumption across dies. The impact of process variations was demonstrated recently in [1] wherein the authors demonstrated 20X variation in leakage power for a 1.3X performance variation in Intel processors.

Given the significant and increasing impact of process variations and temperature variations on the power consumption of processors, it is important for power optimization techniques to consider both these factors, and aim at reducing both the power consumption, and the variation in the power consumption across dies.

Functional units (FUs) such as ALUs and multipliers are significant contributors to the total energy consumption of the processor [4, 3]. In addition, owing to high energy densities of FUs, the effect of process variations on the leakage of FUs is amplified by the exponential dependence of leakage on temperature. *Consequently, reducing both the total power, and the variation in the total power consumption of FUs is an important problem.*

OFBM (Operation to FU Binding Mechanism) is the mechanism by which the ready operations are issued to the FUs in a superscalar processor. Traditional OFBMs statically bind the operations to FUs, without considering the process and temperature variations. They may therefore issue operations to FUs that will leak more, due to either of

533

978-1-4244-3039-0/08 $25.00 © 2008 IEEE

the two effects. Traditional OFBMs therefore result in high FU power, as well as high variation in FU power.

In this paper, we propose to introduce leakage sensors in each FU, and develop LA-OFBM (Leakage-Aware Operation to FU Binding Mechanism) that is cognizant of both process and temperature variations through the leakage sensor. Our experimental results show that LA-OFBM reduces: (i) the average (mean) ALU energy consumption by 18% and (ii) the variation in the total ALU energy consumption (standard deviation) across 1000 die samples by 46%, as compared to previous OFBMs, without any performance overhead.

2. Experimental Setup

Our simulation framework is depicted in Fig. 1A. We perform our experiments on the ALPHA DEC 21364 processor. This is a 4-wide superscalar processor, whose floorplan is shown in Fig. 1B.

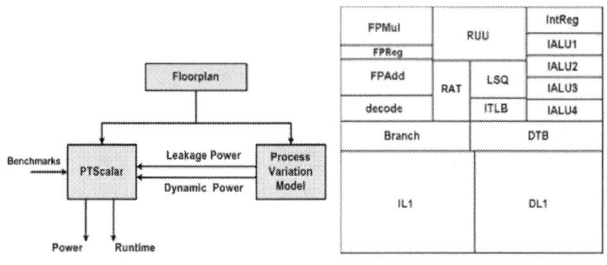

A. Simulation Framework

B. Floorplan of ALPHA DEC 21364 Processor

Figure 1. Simulation Setup

The power, performance and temperature modeling of the alpha processor is done using a modified version of sim-outorder of the PTScalar toolset. PTScalar is a coupled power and thermal simulator built over SimpleScalar [2]. The floorplan of the processor, leakage and dynamic powers of all units in the processor are given as input to PTscalar, which then simulates the benchmark to estimate the power, performance and temperature of all the units. We execute several benchmarks from the MiBench [8], and Spec 2000 [16] suite.

Process Variation Model: We model the variations in device features (gate length and threshold voltage) using the stochastic process corresponding to gate length using the Karhunen-Loève Expansion proposed in [5]. The floorplan of the processor is given as an input to the process variation model to accurately model the spatial correlation in the device parameters. The process variation model generates the dynamic power and leakage power values for all the units in the processor corresponding to one die. We generate 1000 such die samples, which are fed into the PTScalar for

power, performance, and temperature modeling. The power numbers are scaled to correspond to a 45nm technology.

3. Related Work

The impact of process variations and temperature on leakage, has been extensively researched, and the importance of leakage reduction in FUs has been recognized for long. However, to the best of our knowledge, there are no prior OFBMs, which are aware of temperature and process variations.

3.1. Leakage Reduction of Functional Units

Due to their dominant transistor budget in a processor, earlier research focused on leakage reduction of storage structures [9, 12, 6]. However, recognizing that FUs can be the spots of highest leakage density, recent efforts have focused on mitigating leakage in the FUs. Power gating has been the most researched technique for reducing the leakage of FUs. [10] proposed a mechanism to reduce leakage through power-gating of FUs. [14] detects the idle intervals of FUs dynamically to power gate the FUs and thereby reduce leakage. Talli et al [18] use the profile information to identify the idle periods of the functional units and use the compiler to issue corresponding on/off instructions.

3.2. Operation-to-FUs Binding Mechanisms

Several thermal-aware scheduling approaches to balance temperature distribution across the functional units of a VLIW architecture are proposed in [13]. This is one of the prior works that is closely related to ours. Our approach is different from theirs in the sense that ours is a microarchitecture level binding mechanism unlike their compiler-based approaches. The approaches proposed in [13], though are simple, require recompiling the application, which may not be desirable/possible. The applicability of those techniques are limited by the fact that they insist on the availability of the source code for analysis and recompilation. Another significant difference in our approach is that the technique we propose are for superscalar processors which has the scalability to be extended to other processors, as compared to their techniques which are only for VLIW processors. Also, their techniques do not take into account the process variations. There exists several other such scheduling schemes proposed for functional units. However, no prior work on operation to FU binding mechanisms considers the dependency of leakage power on both temperature and process variations.

978-1-4244-3039-0/08 $25.00 © 2008 IEEE

4. Prior OFBMs

4.1. Fixed Priority OFBM (FP-OFBM)

Operation of FU binding mechanism (OFBM) is the technique to issue the ready operations to FUs. In the absence of any process or temperature variations, all OFBMs are the same. Consequently a direct mapping or FP-OFBM is usually employed in all existing processors. In FP-OBM, FUs are assigned static priorities at design time, and the priorities do not change over time. In every cycle, ready instructions are distributed to the FUs in order of their priority. A lower priority FU will get an operation iff FUs with higher priority also get an operation. FP-OFBM results in a very high activity in the highest priority FU, increasing it's temperature and leakage.

4.2. Load Balancing OFBM (LB-OFBM)

Recognizing the impact of temperature variations on the leakage of the FUs, [13] observed that FP-OFBM causes a skew in the activity of ALUs, i.e., higher priority ALU gets much more operations than a lower priority one. As a result the ALU with the highest priority leaks a lot. Consequently, authors in [13] proposed to reduce leakage by distributing the activity equally among the ALUS, and proposed Load Balancing OFBM (LB-OFBM). In LB-OFBM, operations are issued to FUs in a round robin fashion and since all FUs are equally active, the leakage of any one FU does not rise significantly more than the other FUs.

5. Our Approach: Leakage Aware OFBM (LA-OFBM)

We propose LA-OFBM to reduce the leakage energy and therefore the total energy of processors. To achieve this, we introduce a leakage sensor in each FU, and issue operations to the FUs based on the leakage information of the FUs. In the LA-OFBM, the sensors within the ALUs are used to accurately detect leakage and set the ALU priorities dynamically. The leakage sensor values are continuously read and the FU priorities are updated to be in the decreasing order of increasing leakage. Since the temperature of the ALUs vary over time, the priorities assigned to the ALUs will also change dynamically. LA-OFBM is therefore both process and temperature variations aware OFBM.

5.1. Introducing Leakage Sensor in FUs

We propose to introduce the leakage sensor proposed by Kim et al. [11] inside each FU and continuously measure the FU leakage during the chip operation. A single channel leakage sensor is shown in Fig. 2A. The bias circuits

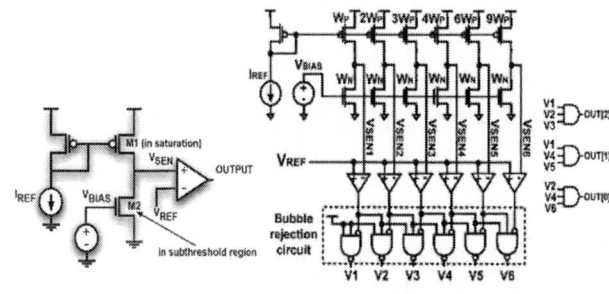

Figure 2. Single and six Channel Leakage sensors [11]

for generating I_{REF} and V_{BIAS} are process variation insensitive. M2 is the only transistor that is sensitive to the variation in leakage of the ALU due to the impact of temperature and process variations. Therefore, the accuracy of the leakage sensor itself is not affected by process and temperature variations. We explicitly model the area, power and inaccuracy introduced when converting the leakage sensor in our experimental setup. The overhead of using leakage sensors accounts to around $3-4\%$ reduction in the total power savings obtained using our LA-OFBM.

5.2. Leakage Sensor Placement

If the leakage measured by the sensor is not a good estimate of the total leakage of the ALU, we might not get the correct ordering of the ALUs in terms of their leakage power. This could potentially result in instructions being bound to a higher leakage ALU instead to the lowest leakage ALU, thus eliminating the power savings obtained using our approach. To find a good location for the leakage sensor, we compared the leakage of a device located at various locations (x_i, y_i) inside the ALU, and the average leakage of the ALU ($I_{av} = I_{S,T}/N$). We found that mean of the percentage difference between the average ALU leakage and the leakage of a device located at the center of the ALU for a sample of 1000 dies to be less than 1%. The maximum percentage error over the same set of samples was 7%. Thus a single leakage sensor located at the center of the ALU can provide accurate estimation of the leakage power of the entire ALU.

5.3. Architecture Model

In the LA-OFBM, whose architecture model is shown in Fig. 3, operations are issued to the FUs based on their leakage information. The sensors within the ALUs are used to accurately detect leakage and set the ALU priorities dynamically in the FU Priority Updater. The leakage sensor values

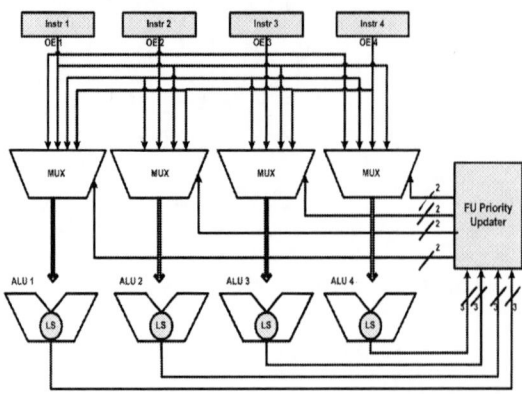

Figure 3. OFBM in 4-issue superscalar processor

are continuously read and the FU priorities are updated after every 10,000 cycles, to be in the decreasing order of increasing leakage. We introduce four 4-to-1 line multiplexors in the operation issue path, to select the ALUs to which the incoming instructions are to be issued. Since the temperature of the ALUs vary over time, the priorities assigned to the ALUs will also change dynamically. LA-OFBM is therefore both process and temperature variations aware OFBM.

Microarchitectural Overheads: We accurately model the impact of microarchitectural enhancements structurally in PTScalar. The leakage sensor is very small, only a few gates, and is not in the critical path of the ALU. Therefore there is no performance impact of the sensor. The multiplexors lie in the critical path of execution, they might cause some extra delay. However this is very small, and in our experiments, we observe that it can be accommodated in the cycle time slack. We synthesized the multiplexors and the FU Priority Update logic using Synopsys design compiler [17]. The energy overhead of multiplexors and the FU priority Updater is less than 0.75 μJ which is very small as compared to the 500 μJ energy of all the 4 ALUs. But we included both their leakage and dynamic powers in the power computation using PTscalar in all our simulations.

6. Experimental Results

6.1. FP-OFBM has a higher total energy consumption as well as variation in total energy

Fig. 4 plots the total energy consumption of all the ALUs in each of the 1000 die samples for FP-OFBM, for the representative susan corners benchmark. It can be observed that due to process variations, there can be upto 25% difference in the total ALU energy consumption between the lowest

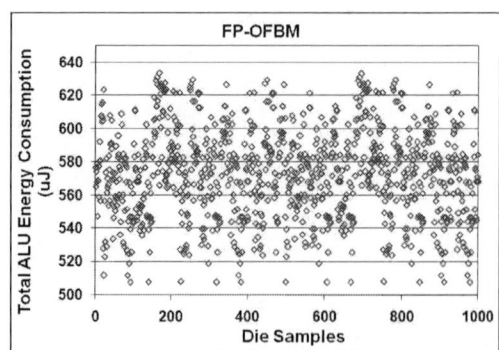

Figure 4. Total ALU Energy consumption for 1000 die samples using FP-OFBM

and the highest power dies.

6.2. LB-OFBM increases total energy consumption but reduces variation in total energy

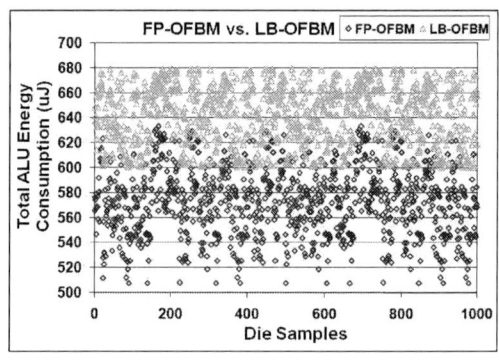

Figure 5. Total ALU Energy consumption for 1000 die samples using LB-OFBM and FP-OFBM

Fig. 5 plots the total energy consumption of all the ALUs in each of the 1000 die samples, for FP-OFBM and LB-OFBM , for susan corners benchmark. LB-OFBM shows 13% increase in the mean leakage, but a reduction of 15% in the variation in the total energy of the ALUs, as compared to FP-OFBM.

Fig. 6 plots the mean of the ALU energy consumption computed over 1000 sample dies, normalized to FP-OFBM, for all the OFBMs, for all the 10 benchmarks. The last set of bars plot the average ALU power reduction over all the benchmarks. This plot shows that the difference in mean energy consumptions between all the OFBMs is consistent over benchmarks.

The increase in the total energy consumption by the LB-OFBM is an important and counter-intuitive result. This is

536

978-1-4244-3039-0/08 $25.00 © 2008 IEEE

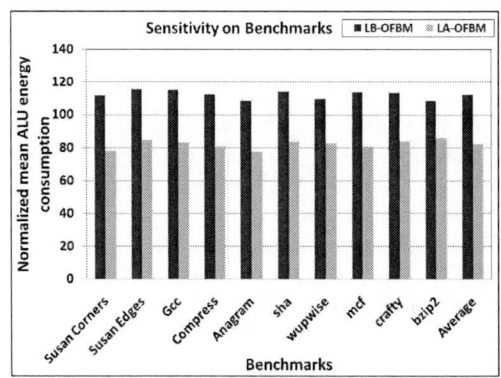

Figure 6. Total ALU energy reduction is consistent across benchmarks

because, it could be argued that FP-OFBM will concentrate the ALU activity on the lowest priority ALU, and increase the temperature of that ALU, which in turn would result in higher energy dissipation. However this is not the observation. On closer investigation, we found that the cooling efficiency improves with the increase in temperature. Therefore after some time it becomes very difficult to increase the temperature. Thus issuing more instructions to the same FU may not increase the temperature and therefore leakage much.

6.3. LA-OFBM reduces total energy consumption

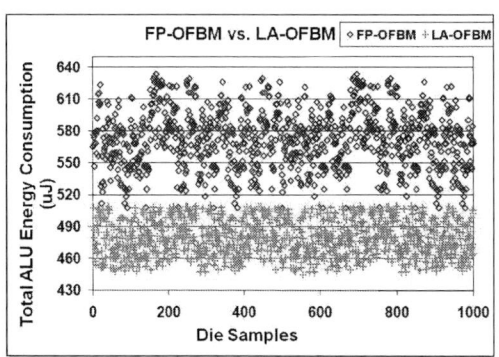

Figure 7. Total ALU Energy Consumption for FP-OFBM and LA-OFBM

Fig. 7 plots the total energy consumption of the ALUs for the FP-OFBM (baseline) and LA-OFBM techniques for 1000 die samples, for susan corners benchmark. The first observation we make from this figure is that all the LA-OFBM points are lower than the FP-OFBM points. As compared to the FP-OFBM, LA-OFBM reduces the total energy consumption of the ALUs by 18%. In terms of leakage energy alone, the LA-OFBM decreases the total leakage en-

ergy of all the ALUs by 44% as compared to FP-OFBM. Fig. 6 further bolsters our claim by demonstrating that LA-OFBM consistently reduces the energy consumption across the benchmark spectrum.

6.4. LA-OFBM reduces variation in total energy

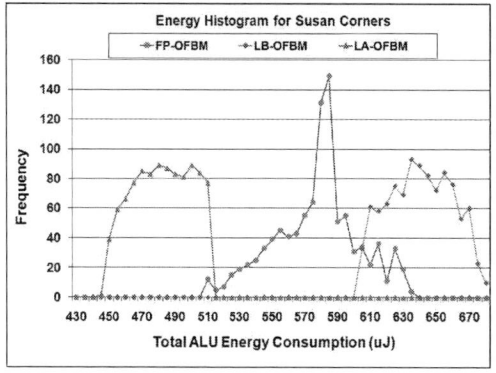

Figure 8. Variation in total energy consumption

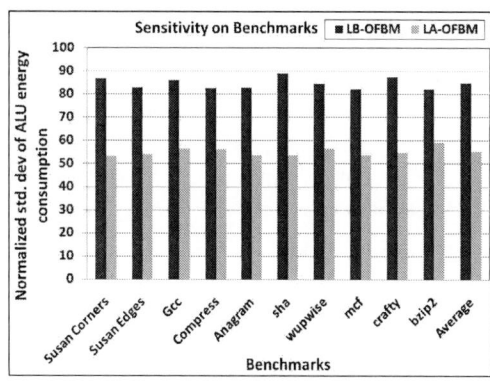

Figure 9. Variation in total ALU energy consumption is consistent across benchmarks

Another observation that we make from Fig. 7 is that the width of the vertical band in which points of LA-OFBM lie is lesser than the width of the band in which the points of FP-OFBM lie. In fact, the standard deviation of the LA-OFBM is just 15 μJ, which is 46% lesser than the standard deviation in the FP-OFBM case. Fig. 9 plots the standard deviations of the ALU energy consumption computed over 1000 sample dies, normalized to FP-OFBM, for all the OFBMs, for all the 10 benchmarks. This plot shows that the reduction in standard deviation in the energy consumptions in the OFBMs is consistent over the benchmarks.

Fig. 8 plots another view of the same data, It plots the energy histogram for each of the OFBMs for susan corners benchmark for 1000 die samples. The second curve from the right (lines connected by circles) corresponds to the energy distribution for FP-OFBM. The rightmost curve (lines connected by diamonds) is the energy histogram of LB-OFBM, which increases the mean energy consumption by 13% but reduces the standard deviation in total energy by 17%. Finally LA-OFBM, as compared to FP-OFBM, reduces the mean and standard deviation in energy consumption by 18% and 46% respectively, as shown by its energy histogram depicted by the leftmost curve (lines joined by triangles). Thus our LA-OFBM reduces the total energy as well as the variation in total energy in the presence of both temperature and process variations.

7. Summary and Future Work

Continuous technology scaling for the last four decades, has lead us to the point, where the leakage energy has become a significant portion of the total energy budget of the processor. Leakage energy is highly sensitive to process and temperature variations. However existing Operation to Functional Unit Binding Mechanisms (OFBMs) do not take the process and temperature variations into consideration. In this paper, we propose to introduce leakage sensor in the ALU and propose a Leakage-Aware OFBM (LA-OFBM), which is sensitive on both the process and temperature variations, and is therefore able to effectively reduce both the total ALU power consumption, as well as the variation in the total ALU power consumption. Our experiments on a 0.45nm, 4-wide superscalar processor demonstrates that LA-OFBM reduces the total ALU energy consumption of the ALUs by 18%, and the variation in the total ALU energy consumption by 46%.

Acknowledgement
We would like to thank Microsoft Corporation for their generous support.

References

[1] S. Borkar, T. Karnik, S. Narendra, J. Tschanz, A. Keshavarzi, and V. De. Parameter variations and impact on circuits and microarchitecture. *dac*, 00:338, 2003.

[2] D.Burger and T.Austin. The simplescalar tool set version 3.0, 1997.

[3] J. Deeney. Reducing power in high-performance microprocessors. In *International Symposium on Microelectronics*, 2002.

[4] S. Dropsho, V. Kursun, D. Albonesi, S. Dwarkadas, and E. Friedman. Managing static leakage energy in microprocessor functional units, 2002.

[5] S. B. et al. Modeling of intra-die process variations for accurate analysis and optimization of nano-scale circuits. In *Proc. of IEEE/ACM Design Automation Conference*, 2006.

[6] K. Flautner, N. S. Kim, S. Martin, D. Blaauw, and T. Mudge. Drowsy caches: simple techniques for reducing leakage power. In *Proc. of ISCA*, pages 148–157, Washington, DC, USA, 2002. IEEE Computer Society.

[7] P. Friedberg, Y. Cao, J. Cain, R. Wang, J. Rabaey, and C. Spanos. Modeling within-die spatial correlation effects for process-design co-optimization. In *Proc. of ISQED*, 2005.

[8] M. R. Guthaus, J. S. Ringenberg, D. Ernst, T. M. Austin, T. Mudge, and R. B. Brown. MiBench: A free, commercially representative embedded benchmark suite. In *IEEE Workshop in workload characterization*, 2001.

[9] H. Hanson. Static energy reduction techniques for microprocessor caches, in proc. iccd 2001, 2001.

[10] Z. Hu, A. Buyuktosunoglu, V. Srinivasan, V. Zyuban, H. Jacobson, and P. Bose. Microarchitectural techniques for power gating of execution units. In *Proc. of ISLPED*, pages 32–37, New York, NY, USA, 2004. ACM Press.

[11] C. H. Kim, K. Roy, S. Hsu, R. Krishnamurthy, and S. Borkar. A Process Variation Compensating Technique with an On-Die Leakage Current Sensor for nanometer Scale Dynamic Circuits. *IEEE Transactions on VLSI*, 14(6):646–649, 2006.

[12] P. Li, Y. Deng, and L. T. Pileggi. Temperature-dependent optimization of cache leakage power dissipation. In *Proc. of ICCD*, pages 7–12, Washington, DC, USA, 2005. IEEE Computer Society.

[13] M. Mutyam, F. Li, V. Narayanan, M. Kandemir, and M. J. Irwin. Compiler-directed thermal management for vliw functional units. In *Proc. of LCTES*, pages 163–172, New York, NY, USA, 2006. ACM Press.

[14] S. Rele, S. Pande, S. Onder, and R. Gupta. Optimizing static power dissipation by functional units in superscalar processors. In *Computational Complexity*, pages 261–275, 2002.

[15] M. Santarini. Thermal integrity: A must for low-power integrated circuit digital design, 2005.

[16] SPEC2000 Benchmarks, www.spec.org/benchmarks/html, 2000.

[17] Synopsys Design Compiler, www.synopsys.com/products/logic/design compiler.html.

[18] S. Talli, R. Srinivasan, and J. Cook. Compiler-directed funcitonal unit shutdown for microarchitecture power optimization. In *Proc. of IPCCC*, NewOrleans, LA, USA, 2007.

[19] J. W. Tschanz, S. G. Narendra, Y. Ye, B. A. Bloechel, S. Borkar, and V. De. Dynamic sleep transistor and body bias for active leakage power control of microprocessors. *IEEE Journal of Solid State Circuits*, 38, Nov 2003.

SESSION C4:
Architecture/Arithmetic

21st International Conference on VLSI Design

Stall Power Reduction in Pipelined Architecture Processors

Pejman Lotfi-Kamran, Amir-Mohammad Rahmani, Ali-Asghar Salehpour,
Ali Afzali-Kusha, and Zainalabedin Navabi
Nanoelectronics Center of Excellence, School of Electrical and
Computer Engineering, University of Tehran
plotfi@computer.org, am.rahmani@ece.ut.ac.ir, a.salehpour@ece.ut.ac.ir,
afzali@ut.ac.ir, and navabi@ece.neu.edu

Abstract

This paper proposes a technique for dynamic power reduction of pipelined processors. Pipelined processors frequently insert NOP instruction to the pipe for generating delay or resolving dependency. Our study shows that the percentage of power consumed by NOP instructions in a pipelined processor is significant. This article studies the detail behavior of NOP instruction and proposes a technique for eliminating unnecessary transitions that are generated during execution of NOP instructions. Initial results demonstrate up to 10% reduction in power consumption for some benchmarks at a cost of negligible performance (almost zero) and area overhead (below 0.1%).

1. Introduction

Computer scientists have always tried to improve the performance of processors. Today's processors are much faster and far more versatile than their predecessors [1]. These chips are still somewhat below the power and power density limits afforded by the package/cooling solution of choice in server markets targeted by such processors. In designing future processors, however, energy efficiency is known to have become one of the primary design constraints [2] [3].

There are many microprocessor applications, typically battery-powered embedded applications, where energy consumption is the most critical design constraint. In these applications, where performance is less of a concern, relatively simple RISC like pipelines are often used [4] [5]. In the current CMOS technology, the most energy consumption occurs when transistor switching or memory access activity takes place [6] [7]. Among the instructions that a pipelined processor executes, the NOP instruction is one that

does not contribute to any useful work. Therefore, the power consumed for its execution is wasted.

In the pipelined architectures, the NOP instruction is inserted for hazard elimination in addition to delay generation. There are three types of hazards; structural, data and control. The structural hazard may occur when there are not enough hardware resources for execution of combination of instructions [8] [9].

A data hazard occurs when an instruction needs the result of a prior instruction that is still in the pipeline and there is not enough latency between these instructions. Two instructions are data dependent when the second instruction requires the result of the first one to begin its execution. A technique for preventing data hazard is to use a forwarding unit. The forwarding unit detects dependencies and forwards the required data from the running instruction to the dependent instructions. In some cases, it is impossible to forward the result because it may not be ready. In these situations, using a NOP instruction is inevitable [8] [9].

The last type of hazard is control hazard that occurs when a branch prediction is mistaken or in general, when the system has no mechanism for branch prediction. There are two mechanisms for handling the miss-prediction. The first mechanism is flushing the pipe after the miss-prediction. Generally, flush mechanisms are not cost effective. A better solution is to fill the pipe after the jump instruction with specific number of NOPs [8] [9].

NOP insertion eliminates hazards but also degrades the performance of the processor. Many solutions are presented for stall reduction (e.g., Forwarding [8], Branch Prediction [8], Speculative Execution [8], etc.) but a significant number of stalls still remain.

The aim of this paper is to optimize dynamic power consumption of a pipelined processor by eliminating useless transitions that are generated in the pipeline when a stall happens. This article shows that in pipelined architectures a number of useless transitions

541

978-1-4244-3039-0/08 $25.00 © 2008 IEEE

is generated when a NOP passes through pipe stages. We slightly modify the architecture of RISC processors to reduce the useless transitions generated when a stall happens. Our experimental results show that, with a negligible hardware overhead, a dynamic power reduction of up to 10 percent is achievable.

The rest of paper is organized as follows. The next section underlines related works and their properties. Section 3 presents a simple example that illustrates inserted NOP instructions in pipelined architectures contribute to unnecessary transitions. In Section 4 our proposed technique for reducing the unnecessary transitions is presented. Experimental results are presented in Section 5 and conclusions come in the last section.

2. Related Work

Hartstein and Pusak [10] explored the impact of pipeline length on both the power and performance of a microprocessor by theory and by simulation. Their results show that the more important power metric is, the shorter the optimum pipeline length that results.

In another article, authors present a bipartition dual-encoding architecture for low-power pipelined circuits [11]. They exploit the bipartition approach as well as encoding techniques to reduce power dissipation not only of combinational logic blocks but also of the pipeline registers. Based on Shannon expansion, they partition a given circuit into two sub-circuits such that the number of different outputs of both sub-circuits are reduced, and then encode the output of both sub-circuits to minimize the Hamming distance for transitions with a high switching probability.

The main goal of another article in low power design is to introduce a dynamic branch prediction scheme suitable for energy-aware VLIW (Very Long Instruction Word) processors. The proposed technique is based on a compiler hint mechanism to filter the accesses to the branch predictor blocks [12].

In the same way, another group proposes a low complexity and low power Re-Order Buffer (ROB) design [13] that exploits the fact that the bulk of the source operand values is obtained through data forwarding to the issue queue or through direct reads of the committed register values. Their ROB design uses an organization that completely eliminates the read ports needed to read out operand values for instruction issue.

It is widely known that branch prediction has enabled micro-processors to increase instruction level parallelism (ILP) by allowing programs to speculatively execute beyond control boundaries. In a related work, authors present an innovative method for power reduction which, unlike the previous work that

sacrificed flexibility for performance, reduces power in high-performance microprocessors without impacting performance [14]. In particular; they introduce a hardware mechanism called pipeline gating to control rampant speculation in the pipeline. They present inexpensive mechanisms for determining when a branch is likely to mispredict, and for stopping wrong-path instructions from entering the pipeline.

There are many works that target power optimization of pipelined processor, but almost all of them neglect the redundant transitions of NOP instructions and their useless power consumption. In this article by studying the behavior of dynamic power consumed when a NOP instruction executes, an efficient mechanism for power reduction of this instruction is proposed.

3. A Simple Scenario

As discussed, NOP instructions are inserted into the pipeline by many pipelined processors to eliminate hazards. In this section, through a simple example, we demonstrate how these inserted NOP instructions contribute to the overall dynamic power of a pipelined processor by generating a number of unnecessary transitions. For the sake of simplicity and clarity, we discuss the scenario in the usual 5 stage MIPS pipeline [8], but the problem can easily be generalized to any pipelined architectures.

In pipelined architectures and in the DECODE stage, each instruction is analyzed and control signals for running that instruction are generated. In the later stages of pipeline, the generated control signals are used to control the flow of data. If the control unit determines that the current instruction depends on the former instructions and the forwarding cannot resolve the dependency, the control unit inserts a NOP instruction by deactivating some critical control signals to be used in the later stages of pipeline including control signals for writing to memory and register file. We demonstrate through a simple example, the inserted NOP contributes to unnecessary transitions.

```
LOAD $1, 100($2)
ADD  $3, S1, $3
```

Figure 1. A simple program

A simple program is shown in Figure 1. The first instruction is a load from memory and the second instruction is an ADD instruction that uses the loaded data. Because of the dependency between these two instructions, after load instruction, a NOP instruction should be inserted into the pipeline. During the execution of the simple program of Figure 1, when the LOAD instruction is in the DECODE stage, the control signals and the required data corresponding to this instruction are generated/extracted. On the rising edge of the clock the generated/extracted control/data are latched into the DE/EXE pipeline register. In the next clock cycle, the ADD instruction is in the DECODE stage and the control unit determines that a NOP instruction should be inserted into the pipeline. Therefore, critical control signals are deactivated and these deactivated control signals along with the other control signals and the required data of the ADD instruction (current instruction in the DECODE stage) are latched on the rising edge of the clock. Generally, the data parts of the current and previous instructions are different. It means that data part of NOP is different form the former instruction (i.e., LOAD). Therefore, passing the NOP instruction in the pipe generates a number of transitions. In the third clock cycle, the ADD instruction should be passed to the pipeline. Therefore, control signals corresponding to ADD are generated and are getting latched along with its required data. The expectation is that the data and some control signals of NOP and ADD are the same, so the number of transitions of passing ADD in the pipeline stages is negligible, but it is not the case. At least parts of the data of the ADD instruction are different from those of NOP because those data were not available when they were extracted in the DECODE stage (in fact the lack of availability of those data is the reason of NOP insertion). It means that a number of useless transitions is generated because of the change of data part of the ADD instruction relative to that of the NOP instruction. These unnecessary transitions contribute to the overall dynamic power but do not contribute to any useful work. The aim of this paper is to eliminate (minimize) these transitions, thus optimizing the dynamic power consumption of a pipelined processor.

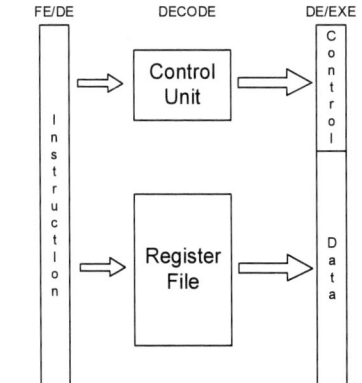

Figure 2. Decode stage of a simple processor

4. Our Proposed Solutions

As discussed, the data part of an inserted NOP instruction is not the same as that of its preceding or subsequent instruction. It means that passing a NOP instruction in the pipe generates a number of transitions. In addition, passing the pending instruction after NOP generates still a number of other transitions. A NOP instruction does not perform any useful work; therefore, the component of dynamic power used for running it is wasted. The technique that is proposed here tries to minimize this component of dynamic power dissipation.

For the NOP instruction to generate as few transitions as possible, its data part should be the same as that of its preceding or subsequent instruction. As discussed, because of the unavailability of certain data of the pending instruction (the instruction passing the pipe after NOP), the data part of the NOP instruction cannot be the same as its subsequent instruction. Therefore the best choice for the power reduction is to use the data part of the instruction preceding NOP as data part of the NOP instruction. In this way, as a NOP instruction passes through a pipe, relative to the previous cycle, the same operations are performed on the same data in all stages of the pipeline; therefore only a small number of transitions is generated as a result of the NOP insertion and propagation.

For this to be implemented, it is sufficient to add a load enable signal to the data and non-critical control parts of DE/EXE pipe register (i.e., only critical control signals [e.g., write to memory and register file signals] that should be loaded in each clock cycle are not controlled by the added load enable). When a NOP is decided to be inserted into the pipe, the controller should deactivate the load enable signal. This way, the content of data and non-critical control part of the inserted NOP instruction are not changed relative to those of its preceding instruction.

4.1. Propagation Boundary Limitation

The technique proposed in the previous section decreases the number of unnecessary transitions generated when a NOP is inserted into the pipe. When data part of the instruction preceding NOP is valid when it is extracted in the DECODE stage, the proposed technique guarantees no useless transitions is generated as a NOP instruction passes the pipe. However, if parts of the data of the instruction preceding NOP are not valid when they are extracted in the DECODE stage, for the correct execution, valid data are prepared by the forwarding unit. In order to minimize the number of transitions generated during execution of NOP, the same data should be prepared for the NOP instruction. If valid data of the instruction preceding NOP are still in some pipe registers when the NOP instruction needs them, the forwarding unit prepares the data for the NOP as well. In this case, a few number of transitions is generated during execution of the NOP instruction. On the other hand, if the valid data are not available in any pipe register when the NOP instruction needs them (because the instruction that generates those data is finished and goes out of the pipe), different data are loaded into some operators, therefore a number of useless transitions is generated. This causes the inputs of some other operators to change and this cycle continues until transitions reach to the last stage of the pipeline. The technique that we propose here limits the propagation boundary of transitions to a single stage, i.e., the transitions do not propagate to all stages of the pipeline.

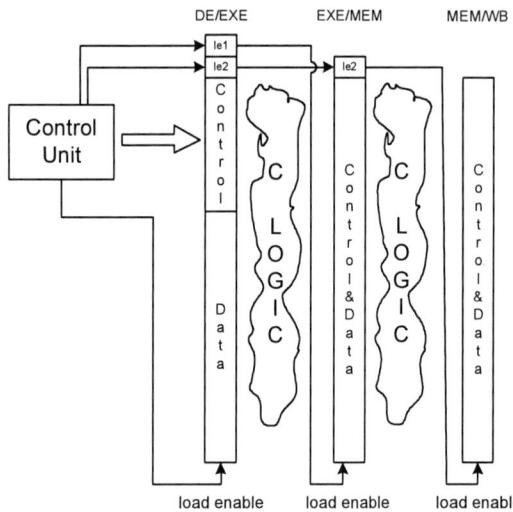

Figure 3. Load enable propagation in a pipeline

The ultimate goal is that the NOP instruction produces the same results as those of its preceding instruction in all pipe stages. In this condition, the value that is loaded in each pipe register when NOP and its preceding instruction execute are the same except for a few critical control signals (e.g., write to memory or register file). Therefore, loading to pipe registers can almost be deactivated during execution of NOP instructions. For this purpose, a load enable is added to each pipe register. This control signal is only applied to data and non-critical control parts of the pipe registers. When this signal is activated, the pipe register performs its usual operation. When this signal is deactivated, only critical control signals are loaded into the pipe register and the value of data and non-critical control signals do not change. By deactivating a pipe register's load enable when NOP results are

written to it, only critical control signals of that pipe register are changed and its other parts remain unchanged. If data of a NOP instruction are not valid (i.e., NOP data defer from those of instruction preceding NOP), in some pipe stages a number of transitions is generated. These transitions are propagated until they reach a pipe register. They do not propagate any further.

These load enables are generated by the controller in the DECODE stage and are propagated through pipe registers like other control signals to the desired destination (i.e., specific pipe register). Figure 3 illustrates the mechanism of propagating load enable control signals in the pipe registers.

5. Experimental Results

In this section, we analyze and report the power reduction, area overhead, and timing penalty of our proposed power reduction technique. The described techniques have been implemented in three general processors: MIPS [8], DLX [8], and PAYEH [9].

MIPS is a 5 stage pipelined processor and its architecture is RISC with fixed-width 32-bit instructions.

DLX is a text book example of a RISC processor with a 5 stage pipeline using forwarding to avoid data hazards.

PAYEH is a pipelined version of SAYEH [15] with a similar instruction set and has five pipe stages. SAYEH is a multi-cycle RISC processor with 16-bit data and 16-bit address buses. PAYEH architecture uses a forwarding unit. This forwarding unit can resolve all dependencies by forwarding the required data from the next pipe stages to the previous ones.

In the first series of experiments, we evaluated the effectiveness of our proposed technique when an ASIC is targeted. The modified and original MIPS, DLX, and PAYEH processors are synthesized using the CUB library and the area usage and the maximum clock cycle of each of them are extracted. Table 1 compares the area usage and maximum clock cycle of the original and modified processors. As Table 1 indicates, the performance penalty of the modified processors is approximately 0%. Also the area overhead of the proposed approach is about 0.1% of the original processors.

Four benchmark programs are used to evaluate the effectiveness of our proposed technique. The Factorial benchmark reads a number and calculates its factorial. Fibonacci reads a number and calculates the Fibonacci series. Power reads two numbers, a and b, and calculates a to power b (i.e., a^b), and Vector Addition reads two vectors and calculates their addition element by element.

978-1-4244-3039-0/08 $25.00 © 2008 IEEE

Table 1. Area and frequency characteristics of original and modified processors

Processor	Area Characteristic			Frequency Characteristic		
	Original Area (mil^2)	Modified Area (mil^2)	Overhead (%)	Original Frequency (MHz)	Modified Frequency (MHz)	Overhead (%)
MIPS	13457	13470	0.097	21.2	21.2	≈ 0
DLX	8400	8411	0.13	39.5	39.5	≈ 0
PAYEH	5974	5981	0.12	46.7	46.7	≈ 0

Table 2. Dynamic power characteristics of original and modified MIPS processor (ASIC)

Benchmark	Power Characteristic		
	# Transitions (Original)	# Transitions (Modified)	Improvement (%)
Factorial	1381280	1247710	9.67
Fibonacci	1317690	1203841	8.64
Power	1385450	1298028	6.31
Vector Addition	1057050	993309	6.03

Table 3. Dynamic power characteristics of original and modified DLX processor (ASIC)

Benchmark	Power Characteristic		
	# Transitions (Original)	# Transitions (Modified)	Improvement (%)
Factorial	1763100	1653430	6.22
Fibonacci	1660220	1560440	6.01
Power	1864770	1760530	5.59
Vector Addition	1507060	1432460	4.95

Table 4. Dynamic power characteristics of original and modified PAYEH processor (ASIC)

Benchmark	Power Characteristic		
	# Transitions (Original)	# Transitions (Modified)	Improvement (%)
Factorial	2763960	2519070	8.86
Fibonacci	2565480	2352030	8.32
Power	2663630	2456400	7.78
Vector Addition	2202560	2043540	7.22

These benchmark programs are applied to the original and modified synthesized processors and dynamic power consumption of each processor is estimated by counting the number of transitions that are generated when the benchmark is running. Tables 2, 3, and 4 show the results obtained for MIPS, DLX, and PAYEH respectively.

As Table 2 indicates, for the MIPS processor, a maximum dynamic power reduction of 9.67% is achieved. The average power reduction of the proposed approach is about 7.66% for this processor. Almost the same results are achieved for DLX and PAYEH processors. For the DLX processor, a maximum and average power reduction of 6.22% and 5.69% are achieved. For PAYEH, 8.86%, and 8.04% are the percentage of maximum and average amount of power saving that is achieved by the proposed technique. The results of Table 1 to 4 indicate that with a 0.13% area overhead, an average dynamic power reduction of 5 to 8 percent is realizable.

In the second series of experiments, we evaluated the effectiveness of our proposed technique when the synthesis target is an FPGA. The modified and original MIPS, DLX, and PAYEH are synthesized. Four benchmark programs are applied to the original and modified synthesized processors and dynamic power consumption of each processor for execution of the benchmarks are estimated. Tables 5, 6, and 7 show the results obtained for MIPS, DLX, and PAYEH respectively.

Table 5. Dynamic power characteristics of original and modified MIPS processor (FPGA)

Benchmark	Power Characteristic		
	Power (Original)	Power (Modified)	Improvement (%)
Factorial	191.06 mw	176.07 mw	7.85
Fibonacci	182.26 mw	169.51 mw	7
Power	191.63 mw	181.56 mw	5.25
Vector Addition	146.21 mw	138.89 mw	5

As Table 5 indicates, for the MIPS processor, a maximum dynamic power reduction of 7.85% is achieved. The average power reduction of proposed approach is about 6.28% for this processor. For DLX, a maximum and average power reduction of 5.5%, and 5.23% is achieved. For the PAYEH processor, 8.7% and 7.82% are the percentage of maximum and average amount of power savings respectively. The results of Table 5 to 7 indicate that an average power reduction of 5 to 8 percent is realizable by using the proposed approach.

Table 6. Dynamic power characteristics of original and modified DLX processor (FPGA)

Benchmark	Power Characteristic		
	Power (Original)	Power (Modified)	Improvement (%)
Factorial	243.87 mw	230.45 mw	5.5
Fibonacci	229.51 mw	217.34 mw	5.3
Power	256.43 mw	243.27 mw	5.13
Vector Addition	207.44 mw	197.13 mw	4.97

Table 7. Dynamic power characteristics of original and modified PAYEH processor (FPGA)

Benchmark	Power Characteristic		
	Power (Original)	Power (Modified)	Improvement (%)
Factorial	350.4 mw	319.9 mw	8.7
Fibonacci	343.2 mw	315.3 mw	8.11
Power	344.5 mw	319.5 mw	7.24
Vector Addition	305.2 mw	283.1 mw	7.23

6. Conclusions

In this paper, a technique is proposed for eliminating unnecessary transitions that are generated when a NOP instruction is inserted into the pipe of a pipeline processor. The proposed technique is applicable to many pipelined architectures. While hardware overhead and timing penalty of the proposed approach is negligible, the dynamic power reduction of up to 10% on some pipelined processor and benchmark programs is achieved.

7. References

[1] Vasanth, Venkatachalam, and M Franz, "Power Reduction Techniques for Microprocessor Systems", ACM Computing Surveys, Vol. 37, No. 3, September 2005, pp. 195–237.

[2] D. Brooks et al., "Power-aware Microarchitecture: Design and Modeling Challenges for the next-generation microprocessors", *IEEE Micro*, Nov./Dec. 2000, pp. 26-44.

[3] M. J. Flynn, P. Hung, and K. Rudd, "Deep-Submicron Microprocessor Design Issues", *IEEE Micro*, July/Aug. 1999, pp. 11-22.

[5] J. Montanaro and et al. "A 160-MHz, 32-b, 0.5 W CMOS RISC Microprocessor", Digital Tech. J'rnal, Vol. 9, Dec. 1997, pp. 49 - 62.

[5] "PowerPC 405CR User Manual", IBM/Motorola, 6/2000.

[6] G. Cai and C. H. Lim, "Architectural Level Power/ Performance Optimization and Dynamic Power Estimation", Cool Chips tutorial in conjunction with MICRO 32, November 1999, Vol. 17, Issue 11, pp. 1061-1079.

[7] Ramon Canal, Antonio González and James E. Smith, "Very Low Power Pipelines using Significance Compression", 33rd Annual IEEE/ACM International Symposium on Microarchitecture, 2000, pp. 181-190.

[8] D. A. Patterson, and J. L. Hennessy, 2003. *Computer Architecture: A Quantitative Approach, 3rd Edition*. Morgan-Kaufmann, San Francisco, CA.

[9] S. Shamshiri, H. Esmaeilzadeh, and Z. Navabi, "Instruction-Level Test Methodology for CPU Core Self-Testing", ACM Transactions on Design Automation of Electronic Systems, Vol. 10, No. 4, October 2005, pp. 673–689.

[10] A. Hartstein, and T. R. Puzak, "The Optimum Pipeline Depth Considering Both Power and Performance", ACM Transactions on Architecture and Code Optimization, Vol. 1, No. 4, December 2004, pp. 369-388.

[11] S.-J. Ruan, K.-L. Tsai, E. Naroska, and F. Lai, "Bipartitioning and Encoding in Low-Power Pipelined Circuits", ACM Transactions on Design Automation of Electronic Systems, Vol. 10, No. 1, January 2005, pp. 24–32.

[12] M. Monchiero, G. Palermo, M. Sami, C. Silvano, V. Zaccaria, R. Zafalon, "Power-Aware Branch Prediction Techniques: A Compiler-Hints Based Approach for VLIW Processors", *GLSVLSI'04*, April 26–28, 2004, Boston, Massachusetts, USA, pp. 440-443.

[13] G. Kucuk, D. Ponomarev, K. Ghose, "Low–Complexity Reorder Buffer Architecture", *ICS'02*, June 22–26, 2002, New York, USA, pp. 57 - 66.

[14] S. Manne, A. Klauser, D. Grunwald, "Pipeline Gating: Speculation Control for Energy Reduction", IEEE Transactions on Computer-Aided Design of Integrated Circuits and Systems, Vol. 17, Issue 11, Nov 1998, pp. 1061-1079.

[15] Z. Navabi, 2004, *Digital Design and Implementation with Field Programmable Devices*. Kluwer Academic Publisher.

21st International Conference on VLSI Design

A Novel Carry-look ahead approach to an Unified BCD and Binary Adder/Subtractor

Sreehari Veeramachaneni, Kirthi Krishna M, Prateek G V, Subroto S, Bharat S, M.B. Srinivas

Centre for VLSI and Embedded System Technologies (CVEST)
International Institute of Information Technology (IIIT)-Hyderabad
Gachibowli, Hyderabad, 500032, India.
Email: srihari@research.iiit.ac.in, {kirthikrishna, pgvijay, sankhlecha, subrotosen}
@students.iiit.ac.in, srinivas@iiit.ac.in.

Abstract

Increasing prominence of commercial, financial and internet-based applications, which process decimal data, there is an increasing interest in providing hardware support for such data. In this paper, new architecture for efficient binary and Binary Coded Decimal (BCD) adder/subtractor is presented. This employs a new method of subtraction unlike the existing designs which mostly use 10's complements, to obtain a much lower latency. Though there is a necessity of correction in some cases, the delay overhead is minimal. A complete discussion about such cases and the required logic to process is presented. The architecture is run-time reconfigurable to facilitate both BCD and binary operations, including signed and unsigned numbers. The proposed circuits are compared (both qualitatively as well as quantitatively) with the existing circuits in literature and are shown to perform better. Simulation results show that the proposed architecture is at least 11% faster than the existing designs.

1. Introduction

Due to growing importance of decimal arithmetic in commercial, financial and internet-based applications, which cannot tolerate errors from converting between binary and decimal formats, hardware support for decimal arithmetic is receiving an increased attention. Recently, specifications for decimal floating point arithmetic have been added to the draft revision of the IEEE-P754 Standard for Floating Point Arithmetic [1]. Despite the widespread use of binary arithmetic, decimal computation remains essential for many applications. Not only it is required whenever numbers are presented for human inspection, but it is also often a necessity when fractions are involved. Decimal fractions are pervasive in human endeavors, yet most cannot be represented by binary fractions. The value 0.1, for example, requires an infinitely recurring binary number. If a binary approximation is used instead of an exact decimal fraction, results can be incorrect even if subsequent arithmetic is correct. [2]

It is anticipated that once the IEEE-P754 Standard is finally approved, hardware support for decimal floating point arithmetic will be incorporated on processors for various applications. Still, the major consideration while implementing Binary Coded Decimal (BCD) arithmetic will be to enhance its speed as much as possible which is being addressed in this paper.

But to facilitate even binary applications on the same hardware a reconfigurable approach needs to be adopted. This paper deals with the design of an architecture that can perform both binary and BCD addition/subtraction. It also supports both signed and unsigned operations.

Most of the existing architectures use 10's or 9's complement to implement subtraction in BCD. But this has been found to have a very higher latency. Hence a new approach has been proposed to overcome this problem.

The architecture has been designed to have maximum hardware utilization. The rest of the paper is organized as follow: Section 2 provides a brief mathematical background of BCD while section 3 gives a brief explanation about the existing architectures. The proposed algorithm for the unified BCD and binary adder/subtractor is given in section 4. In section 5, the proposed architecture is presented. Simulation results for the proposed and existing circuits are given in section 6 and comparisons are carried out. Finally a conclusion is presented in section 7.

547

978-1-4244-3039-0/08 $25.00 © 2008 IEEE

2. Mathematical Background for BCD

BCD is a decimal representation of a number directly coded in binary, digit by digit. For example the number $(9527)_{10} = (1001\ 0101\ 0010\ 0111)_{BCD}$. It can be seen that each digit of the decimal number is coded in binary and then concatenated, to form the BCD representation of the decimal number.

To use this representation all the arithmetic and logical operations need to be defined. As the decimal number system contains 10 digits, at least 4 bits are needed to represent a BCD digit. Consider a BCD digit A. The BCD representation of A is $A_4A_3A_2A_1$ where all $A_k \in (0,1)$. The only point of note is that the maximum value that can be represented by a BCD digit is 9. The representation of $(10)_{10}$ in BCD is $(0001\ 0000)$.

Addition in BCD can be explained by considering two decimal digits A and B with BCD representations as $A_4A_3A_2A_1$ and $B_4B_3B_2B_1$ respectively. In the conventional algorithm, these two numbers are added using a 4-bit binary adder. It is possible that the resultant sum can exceed 9 which results in overflow. If the sum is greater than 9, the binary equivalent of 6 is added to the resultant sum to obtain the exact BCD representation. This can be illustrated with the following example

A	0110	(6)
B	<u>0101</u>	(5)
Sum	1011	(11)
Add	<u>0110</u>	(6)
BCD	1 0001	(11 *in BCD*)
Answer =	(0001 0001)	

3. Related Work

There is a wide range of literature available in field of BCD arithmetic. Some of the first contributions were made by Schmookler et. al.[10] and Adiletta et. al. [14]. An approach towards to architecture dealing with both BCD and binary was shown by Levine et. al. and Anderson, while one of the first BCD sign-magnitude adder/subtractor architecture was presented by Grupe [17]. An area efficient sign-magnitude adder was later developed by Hwang [13]. In his approach two additional conversions were introduced before and after the binary addition. Area

occupied by this design was least amongst all the previous designs.

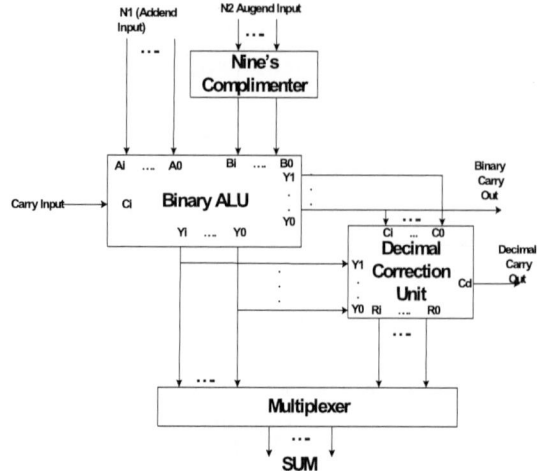

Fig 1. Hwang's proposal [13]

Flora [11] presents an adder similar to the carry-select adder. This employs the use of duplicate hardware to compute the output in the presence of a carry and in its absence. It then selects the appropriate one as the carry is computed. This was improvised by Fischer et. al. where only a single adder was employed to reduce the area over-head. But there was a higher latency due to the additional correction block employed.

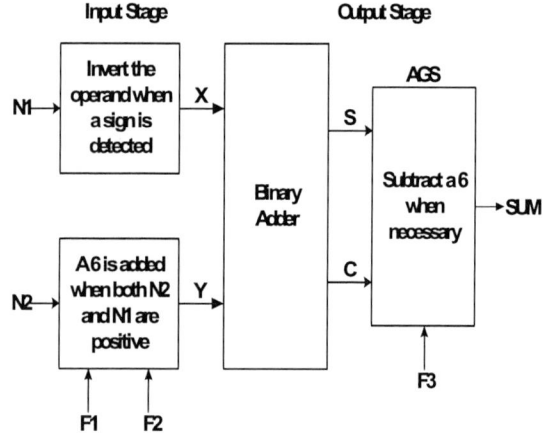

Fig 2. Fischer's proposal [11]

The approach to construct BCD architectures in many IBM processors is based on the work presented by Haller et. al. in [12]. This architecture shown in Fig. 3 operated in a single cycle, though requiring corrections in some cases. In the case of subtraction there is a need for the computation of the complement after the subtraction to obtain the correct difference, hence increasing the latency. Another improvement in the same architecture was the

978-1-4244-3039-0/08 $25.00 © 2008 IEEE

optimization of the carry chain resulting in a slight delay improvement with an increased area of the unit.

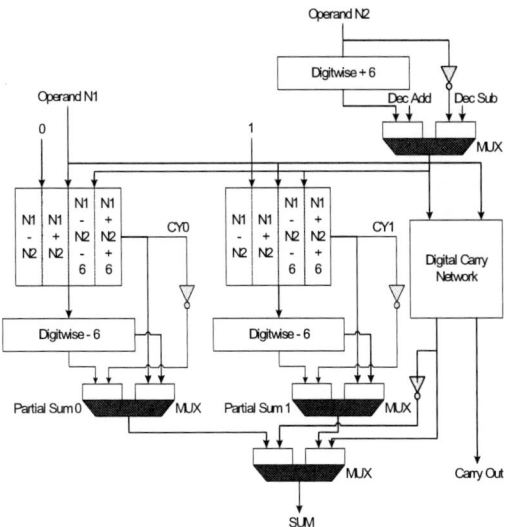

Fig 3. Haller's proposal [12]

The Universal adder micro architecture design (Fig. 4) proposed by Humberto et. al. [18] uses effective addition/subtraction operations on unsigned, sign-magnitude, and various complement representations. This design also overcomes the limitations of previously reported approaches that produce some of the results in complement representation when operating on sign-magnitude numbers. This design proposed that the major disadvantage of the previous designs i.e. having the subtrahend the smaller number in magnitude, was eliminated by their approach. One of the major points to note is that all these proposals make use of complement addition to perform subtraction. This paper proposed a new method without the essential use of 10's complement to perform BCD subtraction.

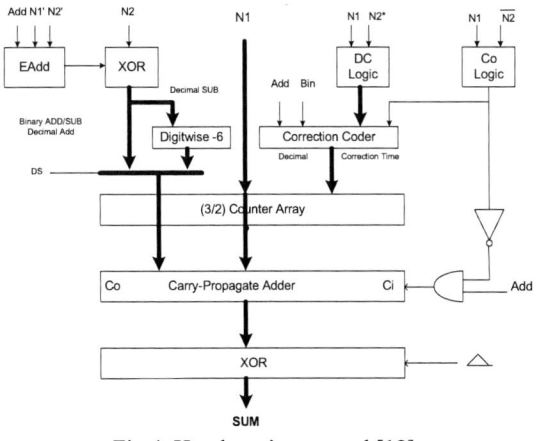

Fig 4. Humberto's proposal [18]

4. Proposed Algorithm

The proposed algorithm aims at performing both BCD and binary addition/subtraction. The major concern is to avoid 10's complement to perform subtraction which is the reason for the high latency in the existing architectures. The proposed design can be divided into three major parts, the pre-computation stage, the prefix network [20] and the post-computation stage.

The pre-computation stage generates control signals named propagate (P) and generate (G). These control the operation of the prefix network. These control signals are generated for every significant stage in the N-digit number i.e. there $2*N$ propagate and generate signals. These denote whether the kth stage propagates the carry/borrow signal or generates it respectively. Essentially there is only one tree that is maintained but the inputs to that tree are determined by the operation i.e. addition/subtraction.

The concept of propagate and generate in both binary and BCD is illustrated below with equations and examples. In the case of addition of binary numbers the following equations denote propagate and generate for A and B (two bits at the kth stage):

$$P = A \oplus B$$
$$G = A \bullet B$$

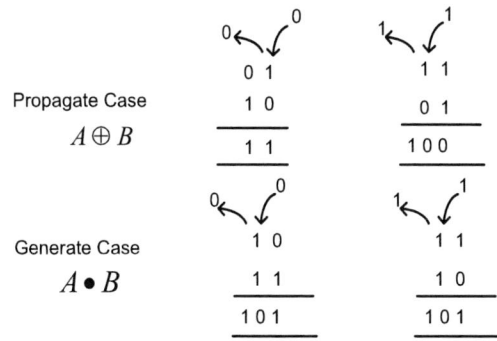

Fig 5. Examples of Binary addition illustrating the concept of propagate and generate

In the case of addition of BCD digits A and B (two digits at the kth stage) the following equations denote propagate and generate:

P => A+B=9

G => A+B>9

This operation can be better illustrated from the following example:

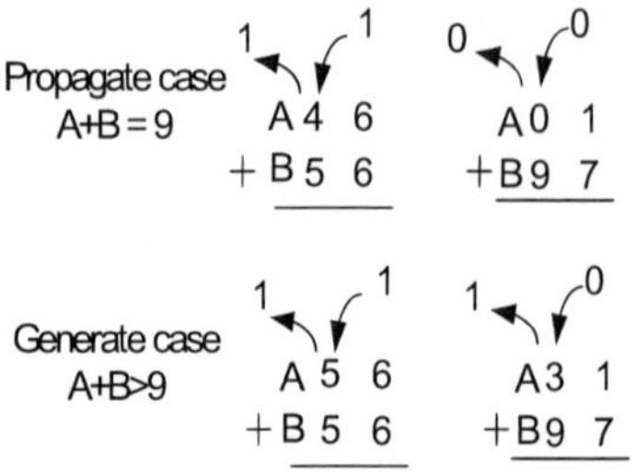

Fig 6. Examples of BCD addition illustrating the concept of propagate and generate

In the case of subtraction of BCD digits A and B (two digits at the kth stage) the following equations denote propagate and generate:

$P \Rightarrow A=B$

$G \Rightarrow A<B$

This operation can be better illustrated from the following example:

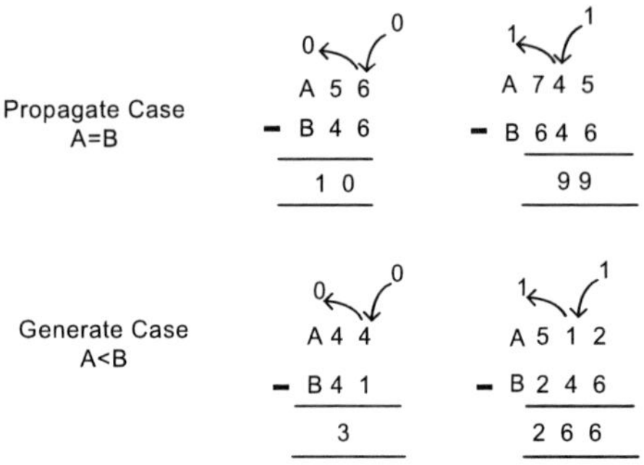

Fig 7. Examples of BCD subtraction illustrating the concept of propagate and generate

These propagate and generate bits are sent to the prefix network which has a network of blocks which calculates the group propagate and generate bits. The group $P_{k:0}$ and $G_{k:0}$ bits denote whether the first k stages propagate or generate the carry/borrow.

After the calculation of the group propagate and generate, the carry/borrow can be calculated based on the carry/borrow input based on the following equation.

$C_{out}=G + P. C_{in}$

Thus, the carry/borrow at every significant stage is obtained. These bits are sent to the post-computation blocks which compute the final sum/difference based on these bits. In case of binary, after obtaining the carry/borrow the sum/difference is the XOR of the carry/borrow and the propagate bit.

In case of BCD, it is a little complicated. As mentioned in Section 2, for the addition of two BCD digits if there is an overflow then a correction value of 0110(6) has to be added. For the subtraction of BCD digits, all the existing architectures are employing 10's complement subtraction. But computing 10's complement induces a very high latency in the operation.

Hence the proposed architecture uses propagate and generate bits defined specifically for subtraction to compute the borrow bits for every stage using the prefix network. After the borrow bit is computed, the individual digits at every significant stage are subtracted by using 2's complement using 4-bit binary adders. But then if a borrow output is generated then the output has to be corrected by adding 1010_2 to the difference. This is checked by inspecting the borrow input of the subsequent stage which has already been generated by the prefix network.

This operation is illustrated in the following example:

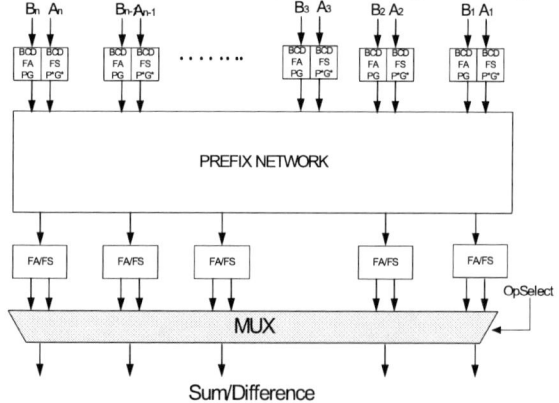

Fig 8. Examples of Binary and BCD operations illustrating the dataflow of the proposed architecture

5. Architecture of the proposed BCD and binary Adder/Subtractor

The proposed architecture performs both BCD and binary addition/subtraction including signed and unsigned numbers. This architecture can be divided into three major parts, the pre-computation stage, the prefix network and the post-computation stage. This architecture is illustrated by a block diagram in Fig. 9.

Fig 9. Block Diagram of the proposed architecture

For the case of BCD computation it can be observed from the diagram that the pre-computation block for every significant stage consists of logic to generate (P, G) and (P*, G*). Depending on whether addition or subtraction is selected the corresponding

550

978-1-4244-3039-0/08 $25.00 © 2008 IEEE

propagate and generate bits are sent to the prefix network using an array of multiplexers.

The selection of the prefix network can be made according to the requirements of area, power and delay from the wide range available in literature. For simulation purposes Sklansky network is used in the design. [19] Though Sklansky network is being used, any available prefix network [20] can be chosen depending upon the area, power and delay criterion.

This generates the group propagate and generate which when combined with the carry/borrow input generates carry/borrow for every stage.

These bits are taken by the final post-computation BCD Full Adder/Subtractor blocks shown in Fig 10.

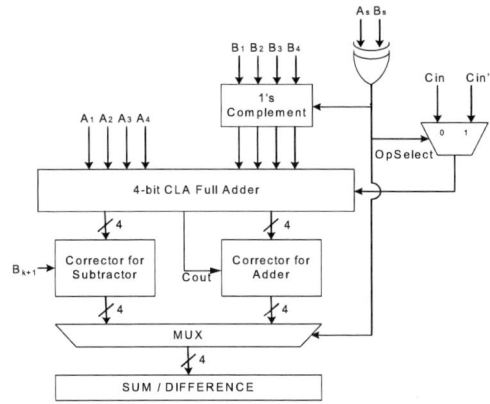

Fig 10. Block diagram of the 1-digit BCD adder/subtractor

The BCD Full Adder/Subtractor computes the sum if selected by the control signal. The addition operation is performed by adding the two BCD digits using the 4-bit binary carry look-ahead adder and the correction (mentioned in Sec. 2) block. This diagram is shown in Fig. 11.

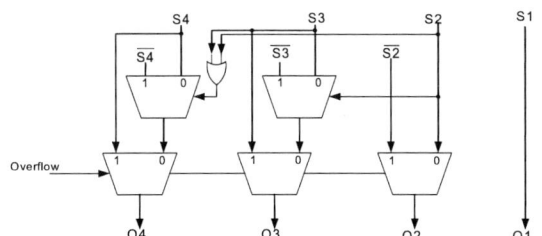

Fig 11. Gate-level diagram of the correction block for the addition operation

The subtraction operation is done by 2's complement addition. But the difference generated needs to be corrected. The borrow input for the subsequent stage that has already been generated by the prefix network is checked and the correction value (1010) is added. The correction block is shown in Fig.

12. The final set of multiplexers select the output based on the operation selected. Another thing that needs to be made sure in most of the existing architectures is that during subtraction the subtrahend is always the smaller number (in magnitude). One exception to this is the Humberto et. al. architecture. In the proposed architecture this is managed by using the outputs of the (P*, G*) blocks which already have comparators in their logic. Thus essentially there is no additional latency overhead due to this comparison.

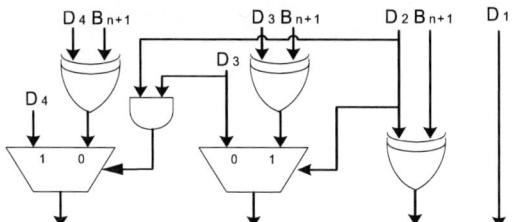

Fig 12. Gate level diagram of the correction block for the subtraction operation

As it comes to the binary operation of the architecture, the prefix network is the major block. The pre-computation of the binary numbers consists of an array of XOR gates which compute the 2's complement when subtraction needs to be performed. Then the propagate and generate bits are generated based on the equations mentioned in the previous section. These bits are sent to the prefix network which generates the carry at every stage. These individual carry bits are XORed to the corresponding propagate bits for every stage to compute the sum/difference.

Thus the prefix network that is the major block both in the BCD and binary computation is shared between the two operations to facilitate re-configurability. The signed numbers are taken care by the control logic at the beginning which take the two sign bits and OpSelect(Operation Select) as inputs to compute the control signal that selects the appropriate multiplexers depending on the operation.

6. Simulations and Results

6.1 Simulation Environment

The proposed architectures have been described using Verilog HDL and 7, 16 and 34 digits wide operands. The design was synthesized using Mentor Graphics' Leonardo spectrum targeting Xilinx Virtex-E v50ecs144 (speed grade-8) FPGA. The approximate values of area have also been mentioned. The inputs are given at a clock frequency of 500 MHz.

6.2 Results and Discussion

The proposed architectures have been simulated in the simulation environment mentioned above and the results for latency and area for 32-bits architectures are given in Table 1.

Table 1. Comparison of the existing and the proposed architectures in terms of latency and area

Design	Latency(ns)	Area (LUTs)
Hwang[13]	10.5	158
Haller[11]	10	584
Humberto[17]	12.1	495
Proposed	8.9	523

In terms of latency Haller turned out to be better of the all the existing designs. Compared to it, this design is 11% faster and uses 10% to 11% less hardware. This is due to the absence of 10's complement approach. In terms of area Hwang's proposal is by far the best one presented, as it uses only a single adder in its design.

7. Conclusion

Existing and proposed architectures for the BCD and binary reconfigurable adders are presented, simulated and compared. A novel way of implementing subtraction in Binary Coded Decimal without the use of 10's complement is explained. Though there is the necessity for the check of magnitude of the two numbers before subtraction it is implemented in a way as not to affect the latency. All the cases where correction is necessary and the logic correction blocks The proposed BCD and binary adder/subtractor is at least 11% faster than the fastest one till now while occupying a considerable amount of area.

8. References

[1] IEEE standard for floating–point arithmetic. IEEE SC, Oct. 2006 at http://754r.ucbtest.org/drafts/754r.pdf.

[2] Michael F. Cowlishaw, "Decimal Floating Point: Algorithm for Computers," *Proceedings of the 16th IEEE Symposium on Computer Arithmetic (ARITH '03)*, pp 104-111, June 2003.

[3] M.S .Schmookler and A.W. Weinderger, "High Speed Decimal Addition", *IEEE Transactions on. Computers*, vol. C-20, pp. 862-867, August 1971.

[4] W. Bultmann, W. Haller, H. Wetter, and A. Worner, "Binary and Decimal Adder Unit," U.S. Patent #6,292,819, September 2001.

[5] R.D. Kenney and M.J. Schulte, "Multioperand Decimal Addition," *Proc. IEEE CS Ann. Symp. VLSI*, pp. 251-253, Feb. 2004.

[6] M. J. Adiletta and V. C. Lamere. BCD Adder Circuit. *Digital Equipment Corporation, US patent 4805131*, pages 1 – 18, Jul 1989.

[7] J. L. Anderson. Binary or BCD Adder with Precorrected Result. Motorola, Inc., US patent 4172288, pages 1–8, Oct 1979.

[8] H. Fischer andW. Rohsaint. Circuit Arrangement for Adding or Subtracting Operands Coded in BCD-Code or Binary-Code. *Siemens Aktiengesellschaft, US patent 5146423*, pages 1 – 9, Sep 1992.

[9] M. J. Adiletta and V. C. Lamere. BCD Adder Circuit. *Digital Equipment Corporation, US patent 4805131*, pages 1 – 18, Jul 1989.

[10] M.S.Schmookler and A. Weinderger. Decimal Adder for Directly Implementing BCD Addition Utilizing Logic Circuitry. *International Business Machines Corporation, US patent 3629565*, pages 1 – 19, Dic 1971.

[11] W. Haller, U. Krauch, and H. Wetter. Combined Binary/Decimal Adder Unit. *International Business Machines Corporation, US patent 5928319*, pages 1 – 9, Jul 1999.

[12] W. Haller, W. H. Li, M. R. Kelly, and H. Wetter. Highly Parallel Structure for Fast Cycle Binary and Decimal Adder Unit. *International Business Machines Corporation, US patent 2006/0031289*, pages 1 – 8, Feb 2006.

[13] S. Hwang. High-Speed Binary and Decimal Arithmetic Logic Unit. *American Telephone and Telegraph Company, AT&T Bell Laboratories, US patent 4866656*, pages 1 – 11, Sep 1989.

[14] M. J. Adiletta and V. C. Lamere. BCD Adder Circuit. *Digital Equipment Corporation, US patent 4805131*, pages 1 – 18, Jul 1989.

[15] J. L. Anderson. Binary or BCD Adder with Precorrected Result. Motorola, Inc., US patent 4172288, pages 1–8, Oct 1979.

[16] S. R. Levine, S. Singh, and A. Weinberger. Integrated Binary-BCD Look-Ahead Adder. *International Business* Machines Corporation, US patent 4118786, pages 1–13, Oct 1978.

[17] U. Grupe. Decimal Adder. *Vereinigte Flugtechnische Werke-Fokker gmbH, US patent 3935438*, pages 1 – 11, Jan 1976.

[18] D.R.H. Calderón, G. N. Gaydadjiev, S. Vassiliadis, Reconfigurable Universal Adder, in Proceedings of the IEEE International Conference on Application-Specific Systems, Architectures, and Processors (ASAP 07), Montreal, Quebec, Canada, July 2007

[19] J. Sklansky, 'Conditional-sum addition logic," *IRE Trans. Electronic Computers*, vol. EC-9, pp. 226-231, June 1960

[20] Harris D, "A taxonomy of parallel prefix networks," in *Proc. 37th Asilomar Conf. Signals, Systems, and Computers*, Nov. 2003, Vol. 2, pp. 2213-2217

21st International Conference on VLSI Design

Memory Architecture Exploration Framework for Cache Based Embedded SoC

T.S. Rajesh Kumar C.P. Ravikumar
Texas Instruments India Ltd.
C.V. Raman Nagar
Bangalore, 560 091, India

{tsrk,ravikumar}@ti.com

R. Govindarajan
Supercomputer Education & Research Centre
Indian Institute of Science
Bangalore, 560 012, India

govind@serc.iisc.ernet.in

Abstract— Today's feature-rich multimedia products require embedded system solution with complex System-on-Chip (SoC) to meet market expectations of high performance at a low cost and lower energy consumption. The memory architecture of the embedded system strongly influences crtical system design objectives like area, power and performance. Hence the embedded system designer performs a complete memory architecture exploration to custom design a memory architecture for a given set of applications. Further, the designer would be interested in multiple optimal design points to address various market segments. However, tight time-to-market constraints enforces short design cycle time. In this paper we address the multi-level multi-objective memory architecture exploration problem through a combination of exhaustive-search based memory exploration at the outer level and a two step based integrated data layout for SPRAM-Cache based architectures at the inner level. We present a two step integrated approach for data layout for SPRAM-Cache based hybrid architectures with the first step as data-partitioning that partitions data between SPRAM and Cache, and the second step is the cache conscious data layout. We formulate the cache-conscious data layout as a graph partitioning problem and show that our approach gives up to 34% improvement over an existing approach and also optimizes the off-chip memory address space. We experimented our approach with 3 embedded multimedia applications and our approach explores several hundred memory configurations for each application, yielding several optimal design points in a few hours of computation on a standard desktop.

I. INTRODUCTION

In application-specific system on chip (SOC) designs, memory design is a critical step since system parameters such as area, power and performance show a direct dependence on the memory organization. Real-time embedded systems have heterogeneous on-chip memory architectures such as data cache, scratch-pad memory, and custom memory. Since the target application is known at compile time, the designer can use this knowledge to design memory organization to optimize the system parameters. However, this optimization is a daunting task since the search space of memory architectures is quite large. The evaluation of a memory architecture cannot be separated from the problem of *data layout*, which physically places the

application data in the memory. A non-optimal data layout will yield an inferior performance even on a very good memory architecture platform.

Designers of real-time embedded memories have typically preferred scratch-pad memories (SPRAM) over data caches since the latter lead to unpredictable run-times. Since the application code is known a priori, the data structures can be carefully placed in the SPRAM. This data layout is performed manually, since compilers cannot assume the code under compilation represents the entire system. Several efficient heuristic approaches for data layout have been published in the literature [2, 3, 1, 4].

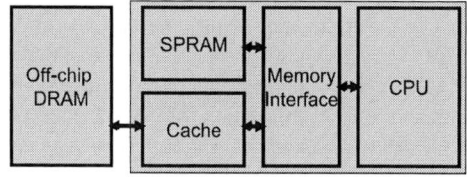

Fig. 1. Target Memory Architecture

In cache based architectures, data is placed in an off-chip RAM and copied at run-time from off-chip RAM to cache by a hardware cache controller. Cache controllers increase the silicon area, but eliminate the software overhead of data management, which mars the performance in dynamic data layout [5]. The mapping of data from off-chip RAM to L1-cache is dictated by the cache associativity scheme and can create potential side effects like thrashing. Careful analysis of data access characteristics and understanding of temporal access pattern of the data structures is required to eliminate thrashing.

In this paper our intention is to perform memory architecture exploration and data layout for real-time embedded SOC in an integrated manner, assuming a hybrid architecture which includes both on-chip SPRAM and data cache (see figure 1). We address this problem using a two step approach for each memory architecture: (a) data partitioning to divide the data between SPRAM and cache with the objective to improve overall memory sub-system performance and power and (b) Cache conscious data layout to minimize the number of cache misses

553

978-1-4244-3039-0/08 $25.00 © 2008 IEEE

IEEE
computer
society

within a given external memory address space.

Cache-conscious data layout has been studied in the literature [6, 7]. Authors of [2, 9, 8] propose *source level* transformations such as array tiling, re-ordering data structures and loop unrolling to improve cache performance. But we focus on optimizing *object module level* data placements without any code modifications. We emphasize that this is an important constraint since *application development flow typically involves integration of many IPs and the source may not be available*. Authors of [6, 7] propose data layout heuristics that aim at minimizing cache conflict misses. [7] formulates the problem as an Integer Linear Program (ILP) and proposes a heuristic to avoid the long run-times of ILP solvers. [6] uses a *Temporal Relationship Graph* (TRG) that captures the temporal access characteristics of data and proposes a greedy algorithm for cache conscious data layout.

While [7, 6] approaches only target the minimization of conflict misses *en masse*, our approach aims at minimizing conflict misses within a certain off-chip memory address space. The constraint of working within a certain external memory address space is very important for memory architecture exploration, since this makes the instruction-cache performance independent of data cache for architectures where the external memory address space is common for both data and instruction caches, and there by reducing the memory arch search space.

To the best of our knowledge, only [2] addresses data partitioning for hybrid architectures. Their data partitioning heuristic identifies the most conflicting data variable and places it in SPRAM. They also demonstrate memory exploration of hybrid architectures with their proposed data partitioning heuristic. But their memory exploration framework does not have an integrated cache-conscious data layout. They propose a model to estimate the number of cycles spent in cache access. Our approach proposes data partitioning based on three factors (i) profile, (ii) temporal access characteristics and (iii) spatial access characteristics. The proposed method is a comprehensive data layout approach for SPRAM-cache based architecture as we perform data partitioning followed by cache conscious data layout. Also our approach works on all the key system design objectives such as area, power and performance.

The major contributions in this paper are

- an efficient heuristic to partition data between SPRAM and caches based on access frequency, temporal and spatial access patterns

- a data layout heuristic for data caches that is independent of instruction caches, optimizes run-time and keeps the off-chip memory address space usage under check.

- hybrid memory architecture exploration with the objective to optimize run-time performance, power consumption and area.

The rest of the paper is organized as follows. In the following section we give an overview of the proposed method. In section III we explain our data partitioning heuristic. In section IV, the cache conscious data layout heuristic is explained.

In section V, we present the experimental results. Conclusions are presented in VI.

II. METHOD OVERVIEW

Figure 2 presents our memory architecture exploration framework, where the exploration phase targets the optimization of (a) cache size, (b) cache block-size, (c) cache associativity, and (d) SPRAM size. Our proposed memory exploration framework consists of two levels. The outer level explores various memory architectures while the inner level explores placement of data sections (data layout problem) to minimize memory stalls. We use an exhaustive search for memory architecture exploration by imposing certain practical constraints (such as, the memory bank size is always a power of 2) on the architectural parameters. These constraints limit the search space to include only "practical" architectures and limit the run-time. The exploration module takes the application's total data size as input and provides an instance of memory architecture by defining the above listed parameters. Based on the SPRAM size and the application access characteristics, the data partitioning heuristic identifies the data sections to be placed in SPRAM. The remaining data sections are placed in off-chip RAM.

The cache conscious data layout heuristic assigns addresses to the data sections placed in external RAM such that these data do not conflict in the cache. The data layout heuristic uses the temporal access information as input to find the optimal data placement. The objective is to minimize the number of cache misses. The data partitioning heuristic and data layout heuristic together place the application data in SPRAM and off-chip RAM respectively. From the temporal access information of data sections and access frequency information, the run-time performance is computed. Using the software eCacti [10] the power per cache read-hit, power per read-miss, power per write-hit and power per write-miss numbers are obtained. The SPRAM power/read access and power/write access are obtained from the semiconductor vendors memory library. The area for a given cache architecture is computed using eCacti [10] and the area value for SPRAM is obtained from the memory library.

The exploration process is repeated for all valid memory architectures and the area, power and performance are computed for each of these. The last step is to identify the list of "optimal" architectures. Since this is a multi-objective problem, all the solution points are evaluated according to the *Pareto optimality* conditions given by the following formulation [13]. Let $(M_{pow}^a, M_{cyc}^a, M_{area}^a)$ and $(M_{pow}^b, M_{cyc}^b, M_{area}^b)$ be the memory power, memory cycles and memory area for solution A and solution B respectively. We say A dominates B if the following expression is true.

$$(((M_{pow}^a < M_{pow}^b) \wedge (M_{cyc}^a \le M_{cyc}^b) \wedge (M_{area}^a \le M_{area}^b))$$
$$\vee((M_{cyc}^a < M_{cyc}^b) \wedge (M_{pow}^a \le M_{pow}^b) \wedge (M_{area}^a \le M_{area}^b))$$
$$\vee((M_{area}^a < M_{area}^b) \wedge (M_{cyc}^a \le M_{cyc}^b) \wedge (M_{pow}^a \le M_{pow}^b)))$$

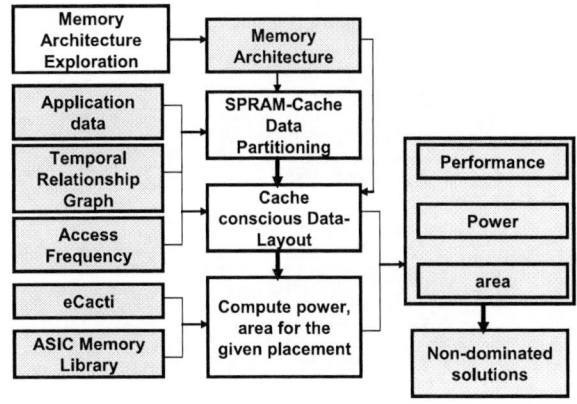

Fig. 2. Memory Exploration Framework

All the dominated solutions are identified and removed. The non-dominated solutions form the Pareto optimal set, which represents the set of good architectural solutions that can be individually examined by the designer.

A. Temporal Relationship Graph (TRG) and Input Parameters

In this section we explain some of the input parameters with an example. We refer to a set of data (one or more scalar variables or array variables) that are grouped together as one *data-section*. A data-section forms an atomic unit that will be assigned a memory address. All data that are part of a data section are placed in memory contiguously.

Let there be 4 data-sections a, b, c and d and the access pattern is given below.

$$a\,a\,a\,b\,c\,b\,c\,b\,c\,b\,c\,d\,d\,d\,d\,d\,a\,a\,a\,a\,a\,a\,a\,c\,a\,c\,a\,a\,c\,a\,c$$

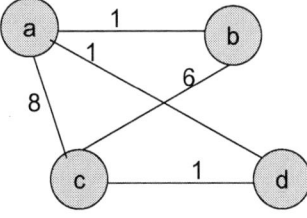

Fig. 3. Example: Temporal Relationship Graph

For this access pattern, the TRG is given in the Figure 3. Given a trace of data memory references, for any two data a and b, let $TRG(a,b)$ be the number of times that two successive occurrences of a are interleaved with at least one reference to b, or vice versa. As an example for the pattern $bcbcbcbc$, $TRG(b,c) = 6$. But for the pattern $ddddaaaa$, the $TRG(a,d) = 0$ as there are no interleaved accesses. TRG is computed for all the data sections from the address trace collected from an instruction set simulator.

We define $sumtrg(i)$ as the sum of all TRG weights on the edges connected to node i. As an example from Figure 3, $sumtrg(a) = 10$.

We define *non-uniform access factor* as the ratio between the number of the times a data-section is accessed non-uniformly to the size of the data-section that is actually accessed. Non-uniform access is computed from the address traces by simulating a single cache-block. A regularly accessed data-section will exhibit a very high spatial locality and there will be only cross-conflict misses. In non-uniformly accessed data variable, many self-conflict misses will be seen. This self-conflict misses represent the non-uniform data access. We also compute the actual size of the data-section that gets used by the application.

III. DATA PARTITIONING

The objective of data partitioning is to identify data sections that must be placed in SPRAM for best performance. SPRAM consumes much less area than caches on a per-bit basis. SPRAM memory accesses consume lesser power than a memory access that is a cache-hit. However, SPRAM space is assigned to data sections exclusively, if dynamic data layout is not used. As a result, SPRAM space can still be higher. There are three parameters that control the decision to keep a data section in an on-chip SPRAM.

- Access Frequency - Placing the most frequently accessed data section in SPRAM gives better power consumption and run-time performance

- Temporal Access Characteristics - Placing the most conflicting data section in SPRAM reduces the number of cache conflict misses and hence improves the overall memory subsystem performance.

- Data with non-uniform access - Data sections that are not uniformly accessed uses more cache lines simultaneously and thereby reduce the available cache space for other data.

A frequently accessed data that conflicts most with the rest of data and also exhibits lesser spatial locality is an ideal candidate to be placed in SPRAM as this gives the best performance from an overall memory subsystem perspective. Our data partitioning heuristic algorithm is explained in figure 4. For each of the data sections, a conflict index is computed using the three parameters mentioned above. The higher the conflict index, the more suitable the data section for SPRAM placement. The greedy heuristic assigns data sections based on the conflict index. The corresponding node is removed from the trg and conflict index for the remaining data sections are recomputed. Note that the above step is performed for every data section identified to be placed in SPRAM. This process is repeated either till the SPRAM space is full or until there are no more data sections to be placed.

555

978-1-4244-3039-0/08 $25.00 © 2008 IEEE

Algorithm: SPRAM-Cache Data Partitioning
Inputs:
  N = Number of data sections
  Access Frequency of all data sections
  Temporal Relationship Graph (TRG)
  Non-uniform Access Factor
  Data Section Sizes
Output:
   List of data sections to be placed in SPRAM
begin
1. compute access frequency per byte for all data sections
2. normalize the access frequency per byte information
3. **for** $i = 0$ to N-1
  3.1 compute sumtrg(i)
  3.2 sumtrg(i) is the sum of all edge weights of node i
4. compute normalized sumtrg(i)
5. compute normalized 'non-uniform access factor'
6. compute the conflict-index
  6.1 conflict-index(i) = normalized sumtrg(i)
      + normalized access-frequency-per-byte(i)
      + normalized non-uniform-access-factor(i)
7. sort thedata sections in descending order
    with respect to conflict-index
8. while (available space in SPRAM)
  8.1 identify the data section with the highest conflict index
  8.2 place it in SPRAM if it fits within the available space
  8.3 update SPRAM available space to account for the
    above placement
  8.4 remove from TRG
  8.5 recompute $sumtrg$ for the remaining nodes in TRG
  8.6 recompute conflict index with the newly updated $sumtrg$
9. exit
end

Fig. 4. Heuristic Algorithm for Data Partitioning

IV. CACHE CONSCIOUS DATA LAYOUT

A. Overview

The data partitioning step places the most conflicting data in SPRAM and there by reduces the possible conflict misses in cache. However, the SPRAM size typically is very small and only a few data-sections would have been placed in SPRAM. The remaining data placements still needs to be done carefully to further reduce the cache misses. In this section we will be discussing the cache conscious data layout.

The problem of cache-conscious data layout is to find optimal data placement in off-chip RAM with the following objectives: (a) to reduce the number of cache misses and (b) to reduce address space used in off-chip RAM. In other words, the objective is to reduce the "holes in off-chip RAM after placement. To our knowledge, reducing cache misses has been the only objective targeted by all the data layout approaches published in the literature [6, 7]. But it is very important to consider objective (a) in the context of objective (b) for the following reasons. Performing a cache conscious data layout within a certain memory size constraints (i.e, without too many holes) helps in isolating data placement from code placemnt and doesn't put additional constraints on code placement. Also, since this makes the data placement indepdenent of code placement, data cache architecture exploration can be

done independent of instruction cache architecture and there by reducing the memory design space.

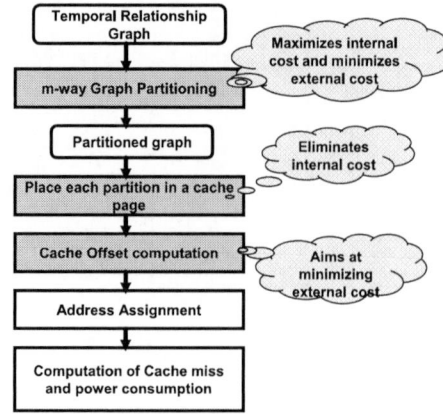

Fig. 5. Cache Conscious Data Layout

We formulate the cache conscious data layout problem as a graph partitioning problem. Inputs to the data layout algorithm are (i) data section sizes and (ii) Temporal Relationship Graph. The data layout algorithm is explained in a block diagram in figure 5. The first step is a k-way graph partitioning step that partitions the input TRG into k partitions, where $k = \lceil applicationdatasize/cache - size \rceil$. The idea is to assign highly conflicting data into the same partition with the constraint that the partition size is less than cache size. The output of graph partitioning step is k partitions with each partition having a set of data sections that conflicts among themselves the most. Since each of the k partitions is lesser than the cache size, each of these partitions can be mapped into off-chip RAM address space that corresponds to one cache page. This step eliminates all the conflicts between data-sections that are in the same partition. The graph partitioning method is discussed in detail in Section B. The last step in the data layout is to minimize the possible conflicts between data-sections that are in two different partitions. This is handled by the offset-computation step. The details of the offset computation is given in Section C. Once the offset-computation step assigns cache-block offsets for each of the data section, the address assignment step allocates unique off-chip addresses to all the data-sections. The following subsections details the graph partitioning heuristic and offset computation heuristic.

B. Graph Partitioning Formulation

In this section we explain our graph partitioning heuristic. Given a graph $G = \{V, E, s, w\}$, where V is the set of vertex representing data sections, E is the set of edges determining the connectivity between the nodes representing a temporal access conflict, s is the list of data section sizes, and w is the list of edge weights representing the number of temporal access conflicts between a pair of nodes. The graph partitioning problem aims at dividing G into m disjoint partitions. A m-way partition of G is a collection of subsets $G_i = \{V_i, E_i, s_i, w_i\}$, such that

- the subsets are disjoint,
- $\bigcup_{i=1}^m V_i = V$
- $\forall\, e_i$ that connects nodes $(u, v) \in G$, $e_i \in G_i$ iff $(u, v) \in G_i$.

The objective of the graph partitioning step is to maximize w_i with the constraint of $\Sigma s_i \leq cache - size$. An edge e_i is said to be an *external edge* for a partition G_i if one of the nodes connected by e_i is not in partition G_i. Similarly, an edge e_i is said to be an internal edge if both the nodes it connects are in the partition G_i. The sum of all the weights on the external edges is referred as external cost. Thus the objective of the partitioning problem is to find a partition with minimum external cost. The optimal partitioning problem is NP-Complete [11, 12]. There are a number of heuristic approaches to this problem, including the well known Kernigan Lin heuristic [11]. We extend the heuristic proposed in [11, 12] to model our problem. The Kernighan-Lin heuristic aims at finding a minimal external cost partition of a graph into two equally sized sub-graphs. The heuristic achieves this by starting with a random partition, and keeps swapping two nodes that gives the maximum gain. Gain is computed as the difference between internal and external costs. Let us consider two nodes a and b present in two different sub-graphs A and B respectively. We define external cost (ECost) of a as E_a and internal cost (ICost) of a as I_a for each $a \in A$. Similarly the ECost and ICost of b is defined as E_b and I_b respectively. Let $D_a = E_a - I_a$ be the difference between ECost and ICost for each $a \in A$. A result proved by Kernighan and Lin [11] shows that for any $a \in A$ and $b \in B$, if they are interchanged, the reduction in partitioning cost is given by $R_{ab} = D_a + D_b - 2 \times w_{ab}$. The nodes a and b are interchanged to partitions B and A respectively if $R_{ab} > 0$. [12] generalizes the graph partitioning heuristic to an m-way partition. [12] starts with a random set of partitions and picks any two of the partitions and applies the Kernighan-Lin heuristic. The two partitions are marked as pair-wise optimal and the algorithm then picks two other partitions to apply [11]'s heuristic. This process is repeated until all the partitions are pair-wise optimal.

We have adapted the algorithm of[12] and added additional constraints to make it work for our problem. The main constraints are as below:

- $\Sigma s_i <$ cache-size for all partitions
- if a data-section size $s_i >$ cache-size then this data-section is placed in a partition and marked optimal
- Nodes a and b are interchanged to partitions B and A respectively only if $R_{ab} > 0$ *and* if $\Sigma s_a <$ cache-size *and* if $\Sigma s_b <$ cache-size

The output of the graph partitioning step is a collection of sub-graphs that maximizes the internal cost and minimizes the

Algorithm: Offset Computation Heuristic
Inputs:
    Temporal Relationship Graph (TRG)
    External costs for all the partitions
    Internal costs for all the partitions
    Data Section Sizes
Output:
    Offsets assigned each of the data sections
begin
1. model two caches (a) main cache and (b) reference cache
   1.1   reference cache is used to find optimal offset
   1.2   main cache stores the data once the optimal offset is fixed
   1.3   these caches are used to compute the placement cost by
         simulating the cache behavior
2. Sort the partitions in the decreasing order of external cost
3. **for** i =0 to k partitions
   3.1   Pick the partition G_i with the highest external cost
   3.2   compute sum-trg for all data sections in partition G_i
   3.3   Sort the data sections in descending order
         with respect to sum-trg
   3.4   **for** alldata sections in G_i
       3.4.1 pick the data section i with highest sum-trg
       3.4.2 evaluate placement cost by placing i in each of the
             availablecache-line in the reference cache
       3.4.3 find the cache-line l that gives the minimal cost
       3.4.4 assign l as the starting point for i
       3.4.5 mark the cache lines from l to
             $l + size(i)/block - size$ as not available
   3.5   copy the content of reference cache to main-cache
4. placement complete
end

Fig. 6. Heuristic Algorithm for Offset Computation

external cost and ensures that no partition has a size larger than the cache size. Thus, each of the partition scan be placed in the off-chip RAM address space that maps to a cache page, such that none of the data sections that are part of the same partition will conflict in cache. Now we are left with optimizing the cache conflicts that might arise because of conflicts from data sections belonging to two different partitions. Since the external cost is already minimized, the number of such conflicts will already be very less. The offset computation step aims at reducing conflicts resulting from external cost.

C. Cache Offset Computation

The cache offset computation heuristic is explained in the Figure 6. As a first step the partitions are ordered based on the external cost. The data-sections part of partition p_1 with the highest external cost are first considered for offset assignment. The data-sections in $p1$ are ordered based on the $sumtrg$ and the data-section d_1 with the highest $sumtrg$ is taken for offset-computation.

To decide the offset that gives the lease conflicts, we adopt a cache simulation approach, a data-section is placed in all available cache block and the placement cost is computed for each instance of placement. To compute the placement cost, we simulate two caches (a) main cache and (b) reference cache. All the data in a partitions are placed temporarily in reference cache before fixing their offsets and once the offsets are decided, the data sections are moved to the main cache. While computing the cost, the data sections in the reference cache and main cache are considered together. The cache-block offset that gives the lease number of misses is taken as the final placement offset of the data-section and the main cache is updated by placing the data-section starting from the offset computed. At this point the reference cache contents are cleared and the cache-blocks that are assigned to data-section $d1$ are marked as not available. The next data-section from $p1$ is considered for offset-assignment. The same process is repeated till all the data-sections in $p1$ are handled.

V. EXPERIMENTAL RESULTS

A. Experimental Methodology

We have used Texas Instrument's TMS32064X processor for our experiments. This processor has 16K data cache and we have used Texas Instrument's Code Composer Studio (CCS) environment for obtaining profile data, data memory address traces and also for validating data-layout placements. We have used 3 different applications from the Media bench for performing the experiments. We compute the TRG, sumtrg, and non-uniform access factor from the data memory address traces obtained from the CCS. We have used [10] to obtain the area and power numbers for different cache configurations.

B. Cache-conscious Data Layout

In this section we present results on our cache conscious data layout and we compare our results with [6]'s approach. Table I presents the results of the data layout. We have used 3 applications and 4 different cache configurations. For all the cache sizes we have used a 32 byte cache-block size and direct mapped cache configuration. The third column in table I presents the number of cache-misses incurred when the data-layout of [6] is used and the column-4 gives the number of cache misses incurred when our data layout approach is applied. Note that there is an improvement of up to 34% and for most of the configurations our approach performs better. Also our approach consumes off-chip memory address space that is very close to the application data-size. This is by construction of the graph-partitioning approach as explained in section IV. Where as [6]'s approach consumes 1.5 to 2.6 times the application data-size as the off-chip address space to achieve the performance given in table I.

C. Memory Architecture Exploration

In this section we present the results from our memory architecture exploration. As mentioned we explore the Cache-SPRAM solution space with the following parameters: (a) cache-size, (b) cache block-size, (c) cache associativity and (d) SPRAM size. Again we have used the 3 applications. For each of the application, we use an exhaustive search method for memory exploration by varying the parameters. We start with

Appli-cation	Cache Size	Calder [6]	Graph-Partition (our approach)	improve-ment (%)
AAC	32K	0	0	0
	16K	14746	9711	34
	8K	155749	128322	17
	4K	446912	385795	14
MPEG	32K	17204	14574	15
	16K	275881	224278	19
	8K	2332008	2314398	1
	4K	11919814	11919814	0
JPEG	32K	0	0	0
	16K	0	0	0
	8K	2350	2112	10
	4K	10220	10294	-1

TABLE I
DATA LAYOUT COMPARISON

no SPRAM and a 4KB cache and keep increasing the cache sizes up to the application-data size. For each of the cache size explored, we then increase the SPRAM size from 0 to application data size with a 4KB step increase. Also for each of the cache configurations we also vary the block size from 8 bytes to 64 bytes with a 8-byte step increase and associativity is varied from 1 to 4 for each of the cache configurations. Based on the application data size, the number of memory configurations evaluated varies from 1200 to 2800. From the total memory configurations evaluated, we compute the non-dominated solutions based on the Pareto Optimal criteria explained in section II. Figures 7, 8, and 9 presents the non-dominated solutions for AAC, MPEG and JPEG respectively. In figure 7, the x-axis represents the number of memory stall cycles and the y-axis represents the power consumption. We have presented the power vs. performance graph for different area bands. We can observe from the figure 7 that as the area band increases, we get better power and performance. Note that the solution points are converging from the top-right portion of the graph (which is a high power and low performance region) to the lower left portion of the graph (which is the low power and high performance region) as the area is increased. In figure 8, the solution on the top right corner has the memory configuration of 4K cache size, direct mapped with 32 byte cache-block with no SPRAM. As we can observe this is a very conservative architecture giving very low performance and high power consumption. On the other hand, the solution in the lower left corner has the memory configuration of 8K cache with 2-way set-associativity with 16-byte cache-block and 128K of SPRAM. This is a very high end architecture consuming lot of area but gives the best performance and power consumption. Thus the set of Pareto Optimal design points presents a critical view to the designers to pick appropriate memory configurations that suit the application-system requirements.

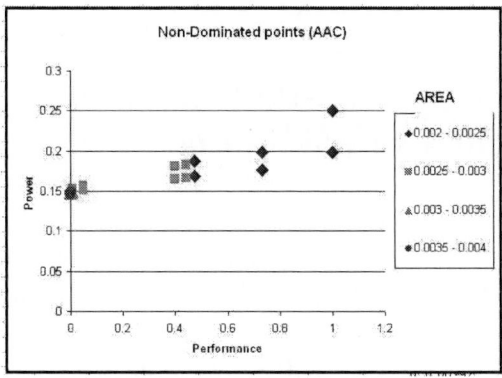

Fig. 7. AAC: Non-dominated Solutions

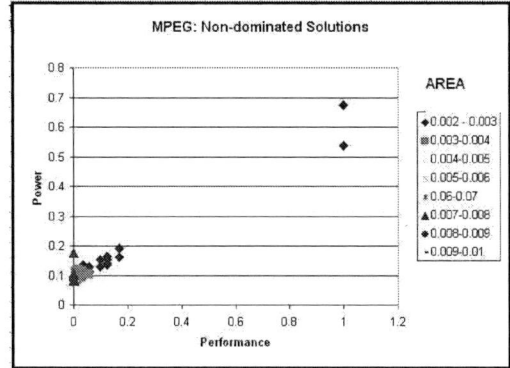

Fig. 8. MPEG: Non-dominated Solutions

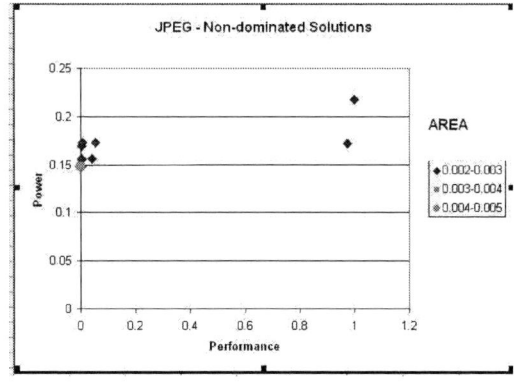

Fig. 9. JPEG: Non-dominated Solutions

VI. CONCLUSION

In this paper we have presented a memory architecture exploration framework for SPRAM-Cache based memory architectures for Real-Time Embedded SOC. Our framework integrates memory exploration, data partitioning between SPRAM

and Cache, and cache-conscious data layout to explore memory design space and presents a list of Pareto Optimal solutions. We have addressed three of the key system design objectives (i) memory area, (ii) performance and (iii) memory power. Our approach explores the memory design space and presents several Pareto Optimal solutions within a few hours on a standard desktop. Our solution is fully automated and meets the time-to-market requirements.

REFERENCES

[1] O. Avissar, R. Barua, D. Stewart. Heterogeneous Memory Management For Embedded Systems. CASES 2001, Nov 2001.

[2] P.R. Panda, N.D.Dutt, and A.Nicolau. Memory issues in Embedded Systems-on-chip:Optimizations and Exploration. Kluwer Academic Publishers, Norwell, Mass., 1998.

[3] J.Sjodin, and C.Platen. Storage Allocation for Embedded Processors. CASES 2001.

[4] P.R. Panda, N.D.Dutt, and A.Nicolau. On-chip vs. off-chip memory: The data partitioning problem in embedded processor-based systems. ACM Trans. Design Automation of Electronic Systems, 5(3):682–704, July 2000.

[5] Sumesh Udayakumaran, Andel Dominguez and Rajeev Barua. Dynamic Allocation for Scratch-Pad Memory Using Compile-Time Decisions. In ACM Tracsactions in Embedded Computing Systems, Vol V, Pages 1-33, 2005.

[6] Brad Calder, Chandra Krintz, Simmi John and Todd Austin. Cache-Conscious Data Placement. In the Eighth International Conference on Architectural Support for Programming Languages and Operating Systems, San Jose, California, October, 1998.

[7] C. Kulkarni, C.Ghez, M.Miranda, F.Catthoor, H.De Man. Cache conscious data layout organization for embedded multimedia applications. In Design, Automation and Test in Europe, 2001, Conference and Exhibition 2001, Proceedings , 2001 Page(s): 686 -691

[8] Mahmut Kandemir, J.Ramanujam and Alok Choudhary. Improving Cache Locality by a Combination of Loop and Data Transformations. In IEEE Transactions on Computers, 1999.

[9] Trishul M. Chilimbi, Mark D. Hill and James R.Larus. Cache Conscious Structure Layout. In the International Conference on Programming Languages Design and Implementation (PLDI99), May 1999.

[10] Mahesh Mamidipaka and Nikil Dutt. eCACTI: An Enhanced Power Estimation Model for On-chip Caches. Centre for Embedded Computer Systems, Technical Report 04-28, University of California, Irvine, California.

[11] B.W. Kernighan and S.Lin. An efficient heuristic procedure for partitioning graphs. The Bell System Technical Journal, 49(2):291-307, 1970.

[12] K.Shyam and R.Govindarajan. An Array Allocation Scheme for Energy Reduction in Partitioned Memory Architectures.

[13] Deb.K., Goel T. Controlled elitest non-dominated sorting genetic algorithms for better convergence. In Porceedings of Evolutionary Multi-Criterion Optimization. 67-81, 2001.

21st International Conference on VLSI Design

A 100MHz to 1GHz, 0.35V to 1.5V Supply 256 x 64 SRAM Block using Symmetrized 9T SRAM Cell with Controlled Read

Satish Anand Verkila
VLSI Circuits and Systems Lab, E.C.E,
IISc - Bangalore-560012
satishanandv@gmail.com

Siva Kumar Bondada
Nano Scale Device Research Lab,CEDT,
IISc - Bangalore-560012
sivakumar.bondada@gmail.com

Bharadwaj S. Amrutur
VLSI Circuits and Systems Lab, E.C.E,
IISc - Bangalore-560012.
amrutur@ece.iisc.ernet.in

Abstract

In this paper, we present Dynamic Voltage and Frequency Managed 256 x 64 SRAM block in 65nm technology, for frequency ranging from 100MHz to 1GHz. The total energy is minimized for any operating frequency in the above range and leakage energy is minimized during standby mode. Since noise margin of SRAM cell deteriorates at low voltages, we propose Static Noise Margin improvement circuitry, which symmetrizes the SRAM cell by controlling the body bias of pull down NMOS transistor. We used a 9T SRAM cell that isolates Read and Hold Noise Margin and has less leakage. We have implemented an efficient technique of pushing address decoder into zigzag-super-cut-off in stand-by mode without affecting its performance in active mode of operation. The Read Bit Line (RBL) voltage drop is controlled and pre-charge of bit lines is done only when needed for reducing power wastage.

1. Introduction

Dynamic voltage and frequency scaling has recently emerged as a very effective technique to trade off power and performance of a chip dynamically [8]. The supply voltage and substrate bias are adjusted to achieve the minimum power at any operating frequency [9, 10]. Most of the work in this area though has primarily focused on logic circuits. Since SRAMs consume a significant fraction of the total chip power, in this paper we look at the problem of designing SRAMs with a large frequency and voltage range of operation.

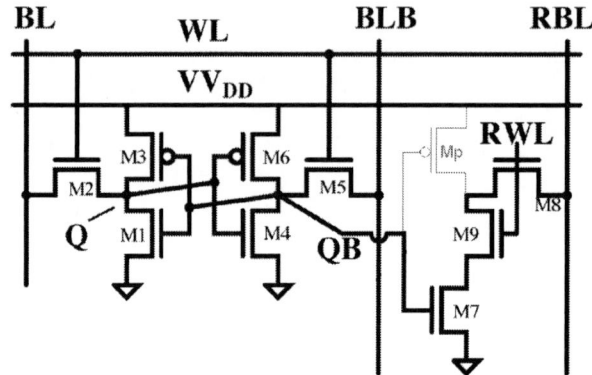

Fig.1 9 Transistor SRAM cell obtained by eliminating the PMOS of the 10T cell [1].

The main barrier to achieving a low-voltage SRAM is the increase in process variation with process scaling. Lower supply and higher variation in advanced processes makes the SRAM operation difficult, and some techniques must be used to expand its operating margin.

A standard 6 Transistor SRAM cell has significant problems working at low voltages due to its poor read and write noise margin. The read noise margin is typically only half of hold noise margin and it is difficult to robustly write into the cell at supplies less than 0.6V. The 6T cell also has large leakages which prevents having too many cells sharing a single bitline column. This implies a large partitioning overhead to accommodate buffers and muxes for each local bitline column.

The authors in [1, 2, 7] explore the use of cells with more than 6 transistors to reduce leakage and improve noise margins. Even though a single cell is larger in area compared to a 6 transistor cell, due to smaller

560

978-1-4244-3039-0/08 $25.00 © 2008 IEEE

partitioning overheads, larger arrays with these cells end up having smaller overall area.

The authors in [7] propose an 8 Transistor cell which is able to separate both read and write noise margins, but it too suffers from the bitline column height limitation due to large leakages.

The authors in [2] propose a 7 Transistor cell which has no read noise margin limitations. But during a read operation, a cell storage node is in dynamic state, which can lead to problems because of process variations, especially with long read times.

The authors in [1] use some of the ideas above to create a 10T SRAM cell shown in Fig. 1, mainly for sub-threshold operation. They separate the read and write ports and provide a read buffer consisting of 4 transistors (M7, M8, M9, Mp). This allows separation of read and hold noise margins. The read buffer has an extra series NMOS transistor, M9, which reduces bitline leakage drastically. They also solve the write noise margin problem by using the floating VDD technique [4].

We extend the ideas proposed in the above papers to design an SRAM with a large supply and frequency range of operation. We propose a 9T cell with threshold control for all the NMOS transistors to enable a very low supply operation. The threshold control enables symmetrizing the cell DC transfer curve and gives better noise margins at low supply voltages. The low leakage property of the cell also enables large bitline columns and hence less overall array area. Also we get good reduction in cell area compared to 10T in which we need some room in the layout to place PMOS Mp. Additionally we apply the Zigzag Super Cutoff CMOS scheme [11] for the decoder to achieve low standby power at very minimal performance loss.

In section 2 we discuss the 9T cell and reasons for going to 9T. Section 3 discusses how to symmetrize the cell. In section 4 we describe an efficient scheme of pushing the decoder into sleep mode without delay degradation. Section 5 gives results of our DVFM implementation. We conclude the paper in section 6.

2. 9T SRAM cell

The 9T cell is obtained by removing the PMOS Mp from the 10T cell (Fig.1). The main effect of this is that the node between M8 and M9 will be floating during the hold mode (RWL=0). We don't expect this floating node to cause any problems during hold as it will be disconnected from the large bitline capacitance. Even a 10T cell will have this node to be floating when QB=1. However during a read, if

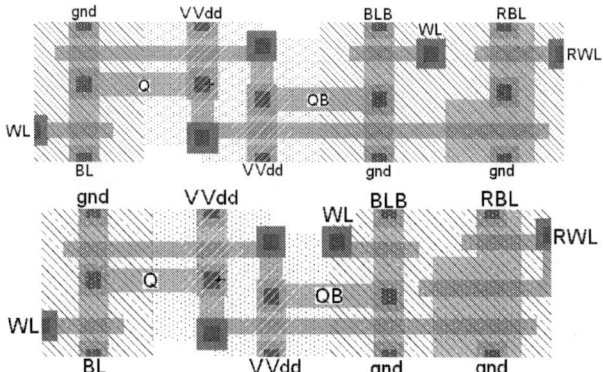

Fig. 2. Layout of a) 8T [2] and b) 9T SRAM cells

QB=0, the 9T cell connects a high impedance node to the bitline during the read (unlike the 10T cell). However if the bitline capacitance is large and it is precharged to Vdd, even under the worst case conditions, the bitline signal doesn't degrade by more than 10mV.

The 9T cell shares the same leakage reduction properties of the 10T cell due to the series stacking effect of the NMOS M7-9. Thus in a SRAM block having 128 cells per column using VDD = 0.35 (which is the minimum required supply at 100MHz operation) at 100C temperature, the ratio of Iread_cell to Itotal_leakage is 1.96 for a 9T design and is 1.06 for 8T design. Hence we can put double the number of 9T cells per column than in 8T design. This enables smaller bitline partitioning and reduced area overhead. A 9T SRAM cell has less area than 10T cell (due to missing PMOS) and is only a bit more area compared to the 8T SRAM cell (Fig. 2). Mosis scalable rules have been used to layout both the cells in the figure. The 6T SRAM cell area is 26X45 = 1170 and 8T SRAM cell area is 72X18 = 1296. So our 9T SRAM cell area is 72X19 = 1368. 8T has 10% more area compared to 6T and our 9T has 17% more area compared to 6T. 9T cell is having 16.5% less area than 10T cell reported in [1]. The area overhead is not so high, because in 8T and 9T the pull-down NMOS transistor size is not constrained by any condition unlike in 6T. We used large width for bottom two transistors in read buffer and small width for upper transistor, so that RBL capacitance is under control and circuit speed won't deteriorate much.

3. Symmetric Cell for Noise Margin Improvement

In a conventional SRAM cell, the PMOS is usually made very weak compared to the access transistor in the cell in order to satisfy the requirements of good

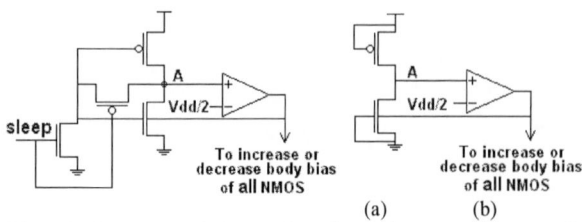

Fig. 3 Symmetrizing circuit using replica cell inverters with a) diode connected ON transistors; (b) OFF transistors

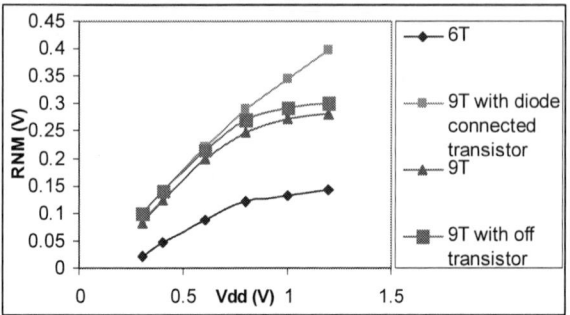

Fig.4 Comparison of read noise margin across supply range of 0.3V to 1.5V for 6T and 9T cells.

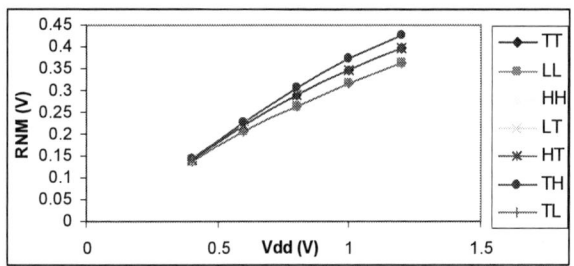

Fig.5 Change in RNM with process variations using symmetrizing circuit using diode connected transistor

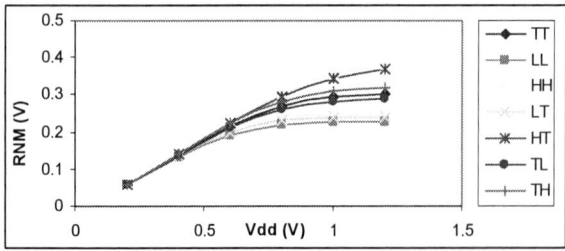

Fig.6 Change in RNM with process variations when using symmetrizing circuit using off transistor

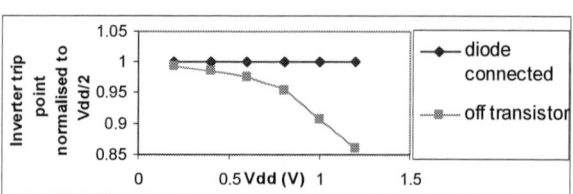

Fig.7 Comparison of adjusted inverter trip point using two symmetrizing circuits

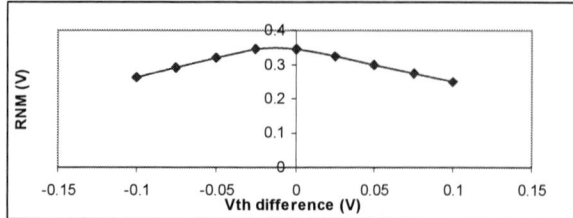

Fig.8(a) Degradation of read noise margin with variation effects for different •Vth between replica cell inverter and SRAM cell

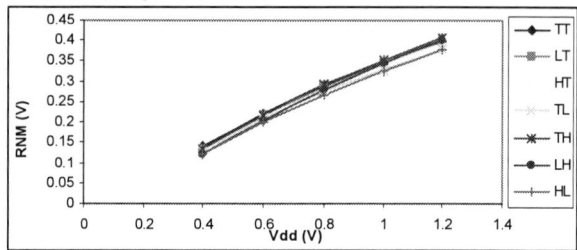

Fig.8(b) Degradation of read noise margin with supply for a •Vth=25mV between replica cell inverter and SRAM cell for different corners

write margin. Simultaneously, the access transistor is made weaker than the driver transistor to create good read margins. Thus the sizes of all the transistors are coupled and the DC transfer curve of the cell inverter is quite skewed as the PMOS becomes much weaker than the driver NMOS. However in the 9T cell, the write margins are improved by using the floating Vdd

technique [4]. Also the read buffer decouples the read and hold noise margins. This in turn allows the potential to have a symmetric DC transfer for the cell inverter to improve the cell noise margin.

Since the PMOS will be weaker than NMOS, their drive strengths can be equalized by raising the NMOS threshold voltage. This can be done for a triple well process with all the NMOS transistors sharing the same well. We have explored two possibilities for balancing the NMOS and PMOS drive strengths. In one, the ON currents are matched in a replica cell inverter by adjusting the body bias of the NMOS via a feedback amplifier (Fig 3a). In the other, the OFF currents are balanced in a similar way (Fig.3b).In either case, the body bias of the pull down NMOS is adjusted such that node A stays at Vdd/2. This body bias is applied to all SRAM cell NMOS's.

All our simulations are done using the PTM SPICE models [12] for the 65nm technology node. We emulate the process corners with fast (L), nominal (T) and slow (H) devices by creating transistor models with -0.1V and +0.1V threshold shift for the fast and slow cases compared to the nominal case.

562

978-1-4244-3039-0/08 $25.00 © 2008 IEEE

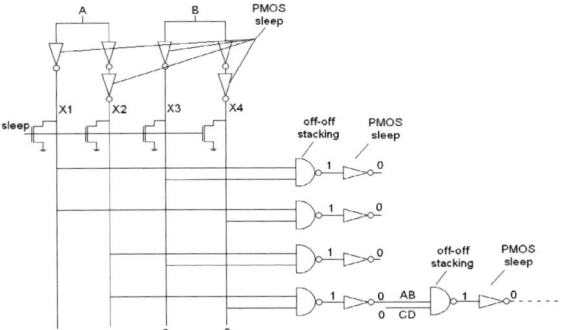

Fig. 9 Pushing decoder into low leakage stand by state

The NMOS is stronger for the process corners TT, LL and HH which are the most dominant process corners (the first letter applies to NMOS and the second to PMOS threshold). But for process corners such TL, LT, TH, HT we can adjust NMOS body bias from 0.3V to more negative voltage so as to make cell symmetric. But for cases like LH, HL, PMOS body bias has to be adjusted so as to make cell symmetric, which we don't do here.

Fig.4 shows read noise margins for the 6T, 9T, symmetrized 9T SRAM using Diode connected transistor (Fig. 3a) and 9T SRAM cell symmetrized using off transistor (Fig. 3b). 9T SRAM cell has 100% more SNM than 6T SRAM cell. 9T SRAM cell symmetrized using diode connected transistor has at least 15% more noise margin at lower Vdds and 40% more noise margin at higher Vdds compared to 9T SRAM cell. 9T SRAM cell symmetrized using OFF transistor has 6% to 20% more Noise Margin than 9T SRAM cell. This enables lower voltage operation with the symmetrized cell than is possible with the regular 9T SRAM cell.

Fig.5 reveals that the noise margin for the symmetrized 9T cell with diode connected transistor is high enough even across process corners. Fig.6 shows that noise margin for symmetrized 9T cell with OFF transistor feedback is less immune to process variations than that with diode connected transistor. This is merely a reflection of the fact that matching OFF currents doesn't imply a good match of the ON currents. Fig.7 gives the comparison of adjusted inverter trip point using diode connected symmetrizing circuit and off transistor symmetrizing circuit.

In reality, the replica cell inverter circuits in Fig. 3 will have a variation with respect to any cell in the array due to both systematic and random variations. We study the impact of this mismatch in figure 8 across supply voltages as well as the mismatch amount. Fig. 8(a) shows the degradation of the read noise margin as a function of the mismatch between

the replica cell inverter and the SRAM cell at 1V TT corner. Fig. 8(b) shows the change in noise margin across the Vdd for different corners for a 25mV Vth difference between the replica cell inverter and a SRAM cell.

4. Low Leakage High Speed Decoder

Decoders have very low activity since only one of 2^n output lines will be activated at a time. Additionally the gates in the decoder are sequentially activated from input to the output. Both of these properties can be used in combination with zigzag super cutoff technique [11] to reduce leakage in stand by mode without any delay degradation for active mode.

As an example consider the 2-to-4 decoder shown in Fig. 9. In stand by mode, by pushing output nodes x1, x2, x3, x4 to zero, all the NANDs will have off-off stacking and hence low leakage. The output of all NANDs will be '1' and output of all subsequent NOT gates will be '0'. Hence using PMOS sleep transistors in series for the inverters shown, even these will be in low leakage mode. This will in turn cause all the subsequent NAND gates of the next stage to get off-off stacked. Thus we can achieve low leakage without too much speed degradation. In each decode stage, only one NOT gate will be active at any time. Also the NOT gates across stages will be active at different times. Hence the PMOS sleep transistors for the NOT gates can all be merged into a single PMOS of the same size. So by using single PMOS sleep transistor and single NMOS sleep transistor (needed for buffer chain), the entire decoder can be pushed into a low leakage stand by mode. Sizing of the sleep transistors is done so as to not degrade speed by more than 2.5%(wire delay of decoder is also considered), yet achieve a leakage reduction of 80%.

5. Controlled Read and DVFM

We use a replica based control technique to monitor delay of our SRAM block and control the Read Bit Line voltage drop [6]. For the delay mimic we have used an extra column of SRAM cells.

Bit line sensing is done using a skewed inverter and this single ended reading is slower because one single buffer in SRAM cell has to discharge entire Read Bit Line. Hence it takes more time to read than to write into cell. In our SRAM block, reading is slower than writing (uses floating VDD technique). write operation is achieved by floating the cell's VDD prior to write [4]. In Fig.11 first cycle is write operation. As we can see that VVDD line is dropped down to 0.64V for

563

978-1-4244-3039-0/08 $25.00 © 2008 IEEE

1.15V operation, enabling easier writing of SRAM cell. Afterwards the QB signal recovered to 1.15V.

For DVFM implementation, we tap the replica bitline signal and delay it through buffers to provide margins. The output of each buffer is captured in flip flops [5]. The DVFM control loop can use this in its feedback to adjust the supply voltage so that required delay is met with some margin. We have used a flip flop chain to monitor the delay as shown in Fig. 10 and adjust for minimum VDD for that frequency of operation. In Fig.11 2nd and 3rd cycles are read operations. In Fig.11 we see the RBL line discharge is stopped after sensed_replica_output came (as internal word line (WL) is cut-off). Thus we can save major fraction of SRAM block power.

Fig. 12 shows the results of performing DVFM adjustment by using the replica bitline shown in Fig. 10 as the delay synthesizer. We have given a margin of 7.5% in terms of buffer delay after the replica bitline. The minimum supply voltages possible over clock frequency ranging from 100MHz to 1GHz are shown in the Fig.12. One problem of using symmetrizing technique for SRAM is we require separate well for read buffer because we should not apply body bias to it. So as to overcome this problem, what we do is use same P-well for all NMOSs as done routinely, but apply body bias for symmetrizing only at lower voltages, where noise margins are low. From fig.4 we can infer that, we need to improve RNM only for supply voltages less than 0.6V. So we use the following algorithm-

Step1: Get the minimum VDD required for the current frequency of operation.

Step2: If VDD < 0.6V apply body bias to symmetrize the SRAM cell (of course read buffer may become slow, as body bias is applied to it also).

Step3: Now take what ever VDD that is given by DVFM circuit.

In fig.12 we have also shown the increase in VDD after applying this algorithm. Supply voltage needed is maximum in HH corner (both PMOS and NMOS are weak) and minimum in LL corner (both PMOS and NMOS are strong). The maximum Vdd for 1GHz operation is 1.5V and minimum is 0.899V. The maximum Vdd for 100MHz operation is 0.751V and minimum is 0.35V. Thus having a replica bitline based delay column, local to each SRAM block, allows good tracking across process variations, as long as these are limited within a SRAM block.

While the replica bitline based delay mimic works well for tracking out correlated variations across the

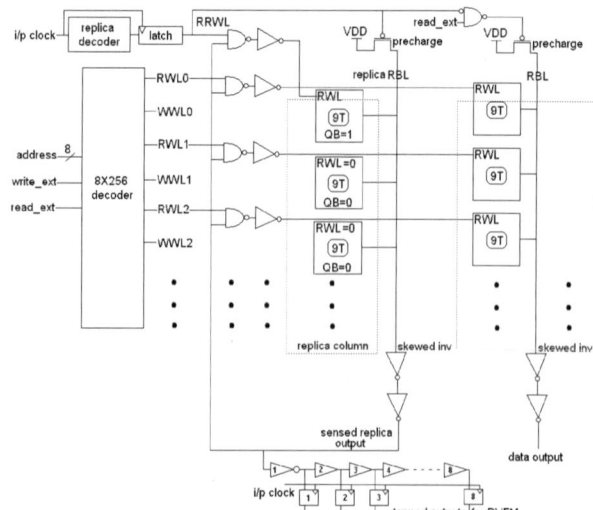

Fig.10 Model for implementing Read control and DVFM

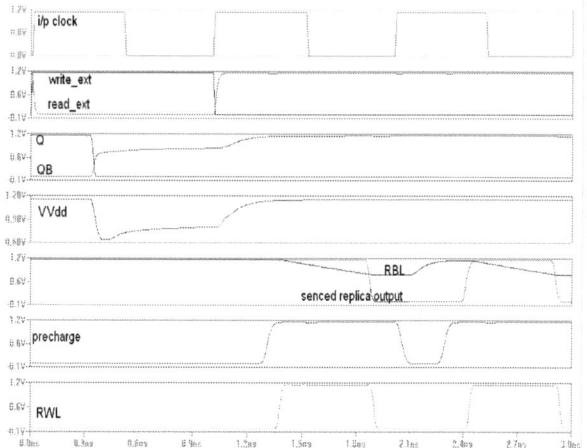

Fig.11 One cycle of write followed by two cycles of read

entire memory bank it will not track out cell to cell variations which are becoming significant as we scale the technology. In order to study this, we plot the delay difference between the replica path and the main bitline path in Fig. 13 as a function of the threshold voltage difference between the replica cell and the memory cell. We observe that for threshold voltage difference of 55mV and more the replica block is faster than SRAM block, thus turning off SRAM word line, leading to read failure. In the opposite direction of the mismatch with the replica cell having more Vth than the SRAM cells, power is wasted due to the delay in turning off the word line. A 200ps of delay margin for SRAM block is maintained at zero mismatch between replica and SRAM block. This extends allowable process variations.

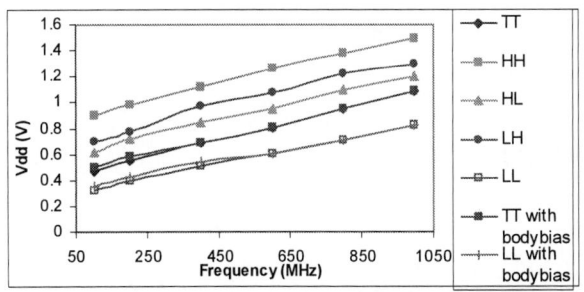

Fig.12 Minimum supply voltage for different frequencies at different process corners.

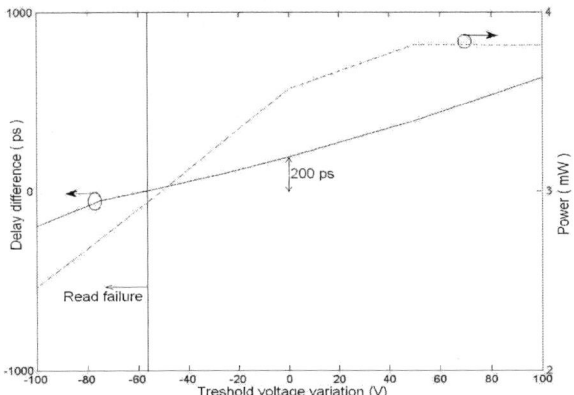

Fig.13 Delay and power changes with process variations

In stand by mode the decoder is completely cut off as described earlier and SRAM block leakage is reduced by reducing its supply to 0.3V where its hold noise margin is 0.117V. Overall, the 256X64 block in active mode for write operation consumes 2.624mW and the stand-by mode power is 6uW.

6. Conclusions

We have explored the design of a low-power high speed Dynamic Voltage and Frequency Managed 256 x 64 SRAM block in this paper. We have presented a 9T SRAM cell with improved the static noise margin and low leakage. The reduced leakage enables to have a larger number of cells in a bitline column, thus reducing overall memory area. The noise margins at very low supplies are further improved by symmetrizing the cell inverter's DC characteristics, via adjustment of the NMOS thresholds. A replica cell inverter is used in the feedback for this adjustment. This enables operation even into sub threshold region with a supply of about 0.3V. We reduce the stand-by power of the decoder by using the zigzag cut-off technique, with very little degradation of the decoder

speed. Replica bitline is used to provide both the delay mimic for DVFM control, as well as used to reduce the bitline power. Low voltage writes are achieved by adopting the floating-Vdd technique.

7. References

[1] B.Calhoun and A.Chandrakasan,"A 256-kb 65-nm Subthreshold SRAM Design for Ultra- Low-Voltage Operation". *IEEE JSSC , VOL. 42, NO. 3,* MAR. 2007

[2] K.Takeda et. al., "A Read-Static-Noise-Margin-Free SRAM Cell for Low- VDD and High-Speed Applications", *IEEE JSSC , VOL. 41, NO. 1,* JAN. 2006

[3] M.Nomura et. al "Delay and Power Monitoring Schemes for Minimizing Power Consumption by Means of Supply and Threshold Voltage Control in Active and Standby Modes" *IEEE JSSC, VOL. 41, NO. 4,* APRIL 2006

[4] M.Yamaoka, et. al. "90-nm Process-Variation Adaptive Embedded SRAM Modules With Power-Line-Floating Write Technique", *IEEE JSSC, VOL. 41, NO. 3,* MAR. 2006

[5] M.Nakai et. al., "Dynamic voltage and frequency management for a low-power embedded microprocessor," *IEEE JSSC, vol. 40, no. 1, pp. 28–35,* JAN. 2005.

[6] Bharadwaj S.Amrutur, and Mark A. Horowitz, "A Replica Technique for Wordline and Sense Control in Low-Power SRAM's" *IEEE JSSC, VOL. 33, NO. 8,* AUG. 1998

[7] L. Chang *et al.,* "Stable SRAM cell design for the 32 nm node and beyond," in *Symp. VLSI Tech. Dig.,* JUNE. 2005, pp. 128–129.

[8] M. Nakai et. al., "Dynamic voltage and frequency management for a low-power embedded microprocessor," *IEEE JSSC, vol. 40, no. 1,* JAN. 2005, pp. 28–35.

[9] K. Nose and T. Sakurai, "Optimization of VDD and VTH for lowpower and high-speed applications," *Proc. ASP-DAC,* Jan. 2000, pp. 469–474.

[10] Nomura et.al., "Delay and Power Monitoring Schemes for Minimizing Power Consumption by Supply and Threshold Voltage Control in Active and Standby Modes", *IEEE JSSC ,* April 2006.

[11] Min et. al., "ZigZag Super Cut-off CMOS (ZSSCMOS) Block Activation with self adaptive voltage level controller: An alternative to clock gating scheme in leakage dominant era", *ISSCC 2003.*

[12] PTM Spice models

21st International Conference on VLSI Design

A Novel Approach to Design BCD Adder and Carry Skip BCD Adder

Ashis Kumer Biswas, Md. Mahmudul Hasan, Moshaddek Hasan, Ahsan Raja Chowdhury and
Hafiz Md. Hasan Babu

Department of Computer Science and Engineering, University of Dhaka, Dhaka-1000, Bangladesh.
E-mails: ashis.csedu@gmail.com, nayeem81@gmail.com, moshaddekhasan@yahoo.com
farhan717@yahoo.com, hafizbabu@hotmail.com

Abstract

Reversible logic has become one of the most promising research areas in the past few decades and has found its applications in several technologies; such as low power CMOS, nanocomputing and optical computing. This paper presents improved and efficient reversible logic implementations for Binary Coded Decimal (BCD) adder as well as Carry Skip BCD adder. It has been shown that the modified designs outperform the existing ones in terms of number of gates, number of garbage output and delay.

1. Introduction

The advancement in higher-level integration and fabrication process has emerged in better logic circuits and energy loss has also been dramatically reduced over the last decades. This trend of reduction of heat in computation also has its physical limit according to Landauer [1,2] who proved that in logic computation every bit of information loss generates $kTln2$ *joules* of heat energy, where k is Boltzmann's constant of 1.38 x 10^{-23} *J/K,* and T is the absolute temperature of the environment. At room temperature the dissipating heat is around 2.9 x 10^{-21} *J.* Energy loss by Landauer limit is important because it is likely that the growth of heat generation due to information loss will be noticeable in future. Bennett [3] showed that zero energy dissipation would be possible if the network consists of reversible gates only. Reversible logic has also found its applications in several disciplines such as quantum computing [4], nanotechnology [5], DNA technology [6] and optical computing [7].

Due to inherent characteristics of floating point numbers and limitations on storing formats, not all floating-point numbers can be represented with desired precision [8]. Faster hardware for decimal floating-point arithmetic is also imminent as it has its importance in financial, Internet based applications. So, faster circuits for Binary Coded Decimal (BCD) numbers have great impact, as it is likely to be incorporated in more complex circuits like future mathematical processors.

Reversible logic implementations for BCD adder using 4-bit parallel adder is presented in [9], where the reversible Full-adder implementation of [10] is used. In [11], both BCD adder and Carry Skip BCD adder are implemented in reversible mode.

In this paper, improved design techniques of reversible logic implementation for BCD adder and Carry Skip BCD adder are presented. We have compared the proposed designs with the existing ones and found that modified designs are better than the existing ones in terms of number of gates, garbage output (output that is needed to maintain reversibility) and delay.

2. Basic Definitions and Literature Review

In this section, basic definitions and ideas related to reversible logic are presented.

Definition 2.1. *Reversible Gates* are circuits in which number of outputs is equal to the number of inputs and there is a one to one correspondence between the vector of inputs and outputs [10].

Example 2.1. Let the input vector be I_v, output vector O_v and they are defined as follows, $I_v = (I_i, I_{i+1}, I_{i+2} ... I_{k-1}, I_k)$ and $O_v = (O_i, O_{i+1}, O_{i+2} ... O_{k-1}, O_k)$. For each particular i, there exits the relationship $I_v \leftrightarrow O_v$.

Definition 2.2. Unwanted or unused output of reversible gate (or circuit) is known as *Garbage Output*.

Example 2.2. In Figure 2.1, P is a garbage output that propagates the primary input A.

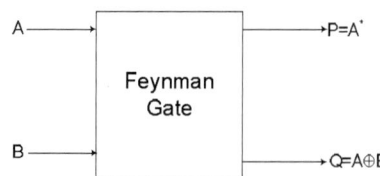

Figure 2.1. Garbage output

Definition 2.3. The *delay* of a logic circuit is the maximum number of gates in a path from any input line to any output line. This definition is based on the following assumptions:

   i. Each gate performs computation in one unit time

   ii. All inputs to the circuit are available before the computation begins.

A reversible circuit must incorporate reversible gates in it and the number of gates used and garbage output produced are always a good complexity measure for the circuit. Delay of a circuit should also be minimized.

3. Overview of the Existing Designs

The basic feature of the reversible BCD adder design in [9] is the use of combination of New gate [12] and Peres gate [13] as full-adder. The design techniques for both reversible BCD adder and Carry Skip BCD adder are presented in [11] and one of the basic characteristics of these designs is the use of TSG [11] gate as a full adder. For reversible BCD adder, one 4-bit parallel adder is used for binary addition of the numbers, a combinational circuit is used for detection of BCD overflow and another 4-bit parallel adder is used for error correction if overflow occurs. Apart from these basic components, in Carry Skip reversible BCD adder, carry skip logic is incorporated for faster carry generation, which is used in the overflow detection logic.

The general ideas of these designs are as follows: In the first 4-bit parallel adder, initial sum is produced by the binary addition of the two BCD numbers. In the combinational part, BCD overflow is detected. In the strict reversible sense, fan-out is restricted. Therefore, copying circuit is used. In the correction part, a 4-bit parallel adder is used to add the error correction value, i.e. in binary 0110, whenever overflow occurs. Otherwise, output produced by the first 4-bit parallel adder becomes the final output. However, there are scopes to improve the designs in terms of number of gates, garbage outputs and delay. We have designed reversible BCD adder and carry skip reversible BCD adder, which overcome the limitations of the existing designs.

4. Proposed Reversible BCD Adders

In this section, improved designs for reversible BCD and Carry Skip Reversible BCD adder have been presented with detail algorithms and figures.

4.1. Basic Properties

Definitions and necessary terminologies related to Binary Coded Decimal (BCD) number and adder are presented here.

Definition 4.1.1. A *full-adder* is a device that takes as input two input bits and a carry-in bit and produces as output the sum of the bits and the carry-out.

Theorem 4.1.1. *A reversible full adder can be realized by at least one gate.* □

Proof. The input and output vector, I_v and O_v for TSG [11] gate are $I_v = (A,B,C,D)$ and $O_v = (P = A, Q = \overline{A.C} \oplus \overline{B}, R = (\overline{A.C} \oplus \overline{B}) \oplus D, S = (\overline{A.C} \oplus \overline{B}).D \oplus (AB \oplus C)$ respectively. If we put input bit, $C = 0$ (constant) and other 3 inputs are for the 3 bits to be added; sum and carry are produced at output R and S respectively; and hence, a reversible full adder can be realized by at least one gate. ∎

Lemma 4.1.1. *A reversible 4-bit parallel adder can be realized by at least 4 reversible gates.*

Proof. Theorem 4.1.1 proves that a reversible full adder can be realized by at least one gate. As a reversible 4-bit parallel adder consists of 4 reversible full adders, a reversible 4-bit parallel adder can be realized by at least 4 x 1 = 4 gates. ∎

Definition 4.1.2. A *combinational circuit for BCD overflow detection* is a circuit that checks whether the result of the binary addition of the two BCD numbers overflows.

A BCD number overflow occurs if the resulting number is greater than 1001 (decimal 9). Let $A_3A_2A_1A_0$ and $B_3B_2B_1B_0$ be the two BCD numbers to be added and the resulting number is represented by $T_3T_2T_1T_0$. Carry out is represented by C_4. C_4 is set when the resulting number is greater than 1111, i.e. decimal 15. Six invalid BCD numbers can be detected by the condition $(T_2+T_1).T_3$. So, the expression for overflow detection bit, F is $(T_2+T_1).T_3 + C_4$. However, it is easy to note that $(T_2+T_1).T_3$ and C_4 cannot be set at the same time. Therefore, a revised expression for overflow detection bit is, $F = (T_2+T_1).T_3 \oplus C_4$.

If F is set, an overflow has been occurred. In error correction logic, 0110 (decimal 6) is added to the partial sum, $T_3T_2T_1T_0$ and any carry out from this addition is ignored. Carry out from the addition of two BCD numbers $A_3A_2A_1A_0$ and $B_3B_2B_1B_0$ is already computed along with F. If F is not set, no error correction is needed. The partial sum, $T_3T_2T_1T_0$ itself becomes the final result.

4.2. Proposed Reversible BCD Adder

A reversible BCD adder consists of three components: a 4-bit parallel adder, BCD adder overflow detection logic and BCD adder overflow correction logic. We will present these parts with proper algorithms and appropriate figures of the algorithms in this section.

978-1-4244-3039-0/08 $25.00 © 2008 IEEE

In order to propose the 1-digit BCD adder, we have proposed three algorithms. Algorithm 4.2.1, termed as Overflow_Detection_Algorithm (ODA), is used to detect the overflow produced by adding two BCD digits. Overflow_Correction_Algorithm (OCA), or Algorithm 4.2.2, is used to correct the error generated by adding two BCD digits. Finally, Algorithm 4.2.3, which is termed as BCD_Adder_Construction_Algorithm, is used to design the overall circuit.

Algorithm 4.2.1. ODA (T)

Input: T (C_4, T_3, T_2, T_1): a 4–bit vector which we mentioned as the partial sum received from the binary adder discussed in Sub-section 4.1.

Output: The vector $R=(T \cup F)$ would be the output from this algorithm, where F is the overflow detection bit (1 indicates overflow, 0 otherwise). The reversible logic design states that there must be no fan-out from any segment of the circuit. It is to be noted that, the T vector is required again for correction after overflow detection, but T was fed to this detection circuit. There are numerous ways of generating copies of T vector at any level, but we preferred this detection circuit to produce T vector as well.

begin

Overflow detection bit, $F = (T_2+T_1). T_3 \oplus C_4$. The expression shows that the resulting circuit may contain at least two blocks. The approach might be similar to the following -

Step 1: The first block will take T_1 and T_2 and output (T_2+T_1).

Step 2: The second block will take the T_3, C_4 and output from first block (T_2+T_1) and compute the result $F = (T_2+T_1). T_3 \oplus C_4$.

return $R:= T \cup F$;

end

Example 4.2.1. Figure 4.2.1 shows a direct implementation of Algorithm 4.2.1 where $T = (C_4, T_3, T_2, T_1)$.

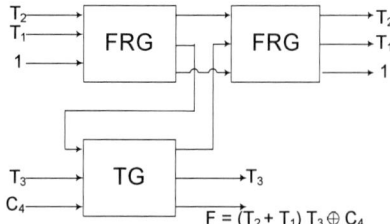

Figure 4.2.1. 1 digit BCD adder's overflow detection logic

Algorithm 4.2.2. OCA (R)

Input: $R = (T \cup F)$: a 4 –bit vector received from the overflow detection logic circuit.

Output: Final corrected BCD sum S $(C_{out,} S_3, S_2, S_1, S_0)$. As T vector that was fed to the detection logic does not include T_0, it is free and intact to use as S_0. It is not mandatory to wait for the final carry out, because if F is 1, we are sure that the final carry out $C_{out}= 1$, so we need not propagate further to compute this carry.

begin

Step 1: The first block will take T_1 and F from the overflow detection logic circuit and generate $S_1 = T_1 \oplus F$ and $carry\_out_1 = T_1.F$.

Step 2: The second block will take carry out of the first block, T_2 from the overflow detection circuit and F (this F can be duplicated using numerous techniques, in our circuit first block generates F again) and generate $S_2 = T_2 \oplus F \oplus carry\_out_1$. It will also generate $carry\_out_2 = (T_2 \oplus F). carry\_out_1 + T_2.F$.

Step 3: The third block will take carry out of the second block, T_3 from the overflow detection circuit and generate $S_3 = T_3 \oplus carry\_out_2$.

return S;

end

Example 4.2.2. Figure 4.2.2 shows a direct implementation of Algorithm 4.2.2.

Algorithm 4.2.3. BCD_ADDER_CONSTRUCTION _ ALGORITHM (A, B)

Input: $A = (A_3, A_2, A_1, A_0)$ and $B= (B_3, B_2, B_1, B_0)$ are two 4-bit input BCD vectors.

Output: Final corrected BCD sum S $(C_{out,} S_3, S_2, S_1, S_0)$.

begin

$T:=$ Binary Adder output(A, B);

$R:=$ ODA(T);

$S:=$ OCA(R);

return S;

end

Example 4.2.3. Figure 4.2.3 shows a direct implementation of Algorithm 4.2.3.

Table 1 shows the comparative analysis of the proposed reversible BCD adder with the designs presented in [9, 11] and it clearly shows that the proposed design outperforms both the existing designs in every metrics.

978-1-4244-3039-0/08 $25.00 © 2008 IEEE

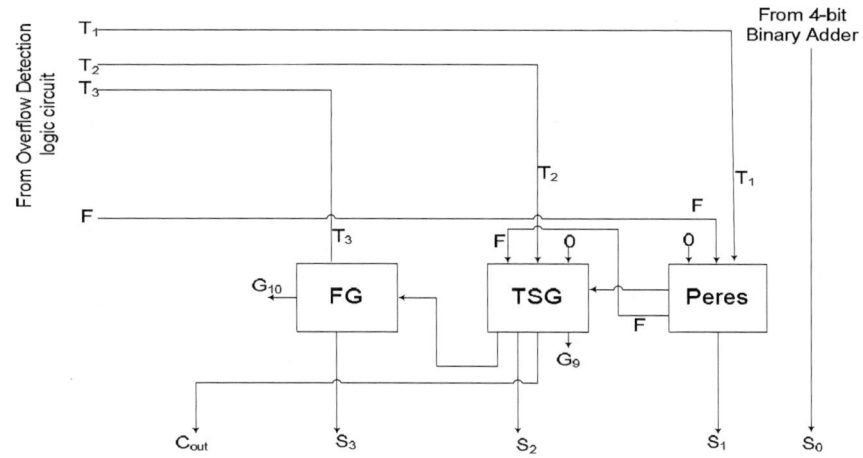

Figure 4.2.2. Designing a 1 bit BCD adder's correction logic circuit

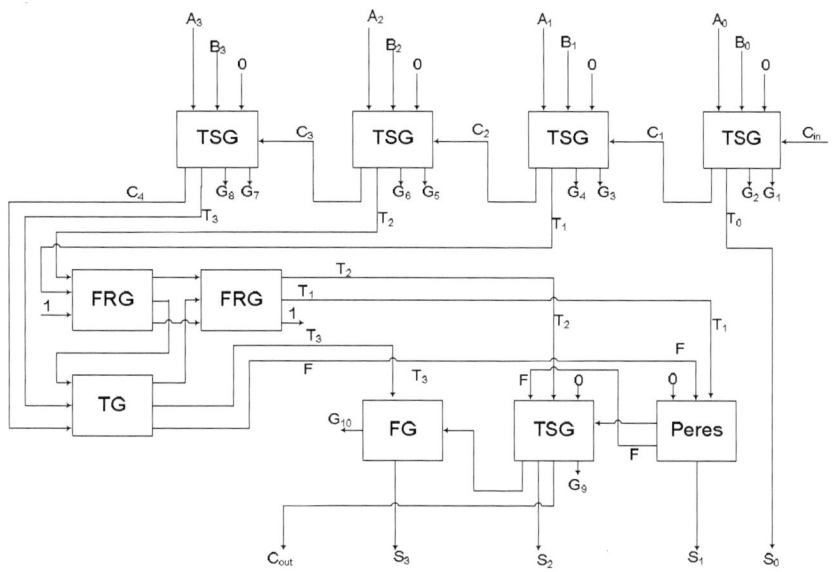

Figure 4.2.3. Design of a 1-digit BCD adder

Table 1. Comparison of different reversible BCD adders

	Existing Circuit 1 [11]	Existing Circuit 2 [9]	Proposed Circuit
Gates	11*	23	10
Garbage	22*	22	10
Delay	10*	13	10

The design in [11] contains multiple fan-outs, which are forbidden in strict reversible sense.

4.3 Proposed Carry Skip Reversible BCD Adder

A carry skip reversible BCD adder consists of the following components: a 4-bit parallel adder, Carry Skip logic, BCD adder overflow detection logic and BCD adder overflow correction logic. Carry skip logic may generate the carry out, C_{out} instantaneously. We will present these components with proper algorithms and appropriate figures. The proposed design is found to be much better than the existing one [11] in terms of number of gates, number of garbage and delay.

Carry skip logic circuit is the fundamental part to this design. We can propagate the carry in, C_{in} to the carry out, C_{out} of the block. Let, A_i and B_i be the inputs to i-th full-adder and either of them is set. Proper expression for this condition is: $P_i = A_i \oplus B_i$ and C_{in} to the block will propagate to the carry output of the block if the entire P_i's are set. In this way, we can generate C_{out} without waiting for it to be generated in ripple carry fashion. Let, the propagation signal for the block is denoted by P. Then, $P = P_3 \cdot P_2 \cdot P_1 \cdot P_0$. If P is set,

C_{in} will be propagated to the C_{out}. However, in the other case, C_{out} will be generated in the ripple carry fashion. So, carry skip logic bit of the block is $K = P.C_{in} + C_4$ where C_4 is the carry generated in the ripple carry fashion. The overall overflow detection bit, $F = (T_1+T_2).\ T_3 \oplus K$ is generated in the same way with Reversible BCD adder presented earlier in this paper. Overflow correction logic incorporated is the same as the Reversible BCD adder.

The following procedure (Algorithm 4.3.1) is used for the design of Carry Skip 1-digit BCD adder. This procedure is presented along with appropriate figure.

Algorithm 4.3.1. CARRY_SKIP_BCD_ADDER_ALGORITHM (A, B, C_i)

Input:　　　A (A_3, A_2, A_1, A_0) and B (B_3, B_2, B_1, B_0) are two input vectors and C_{in} is the carry in.

Output: A BCD adder capable of performing the sum $= A + B$. The buffer vector S $(C_{out}, S_3, S_2, S_1, S_0)$ will store the result.

begin

> **Step 1:** Compute P (propagate bit).
>> Initially $P := $ **true**
>> **for** all i in $\{0, 1 \dots 3\}$ **do**
>>> $P := P$ AND $(A_i \oplus B_i)$.

> **Step 2:**　　Compute $T := \{C_4, T_3, T_2, T_1, T_0\}$, where $T_i := A_i \oplus B_i \oplus C_i$ and C_i's are generated from each adder block.

> **Step 3:** Compute carry skip logic bit, $K := P.C_{in} + C_4$.

> **Step 4:** The overall overflow detection bit $F := (T_1+T_2)T_3 \oplus K$, which is **true** whenever a BCD overflow is detected.

> **Step 5:** Add binary 0110 to T if overflow detection bit F *is* **true**.

> **Step 6:** Compute $S := (C_{out}, S_3, S_2, S_1, S_0)$, the final sum of the addition process.

> **return** S;

end

Example 4.3.1. Figure 4.3.1 shows a direct implementation of Algorithm 4.3.1.

The Fredkin gates in the middle of the Figure 4.3.1 generate the block propagation, P and carry skip logic bit, K. Fredkin gates and Toffoli gate on the left side performs the BCD overflow detection same as for reversible BCD adder. BCD overflow correction logic is also like the reversible BCD adder. Table 2 shows the comparative analysis of the improved Carry Skip Reversible BCD adder with the one presented in [11] and it clearly shows that the proposed design outperforms the existing one in every metrics. Circuit presented in [11] allows multiple fan-outs that are prohibited in strict reversible sense.

Table 2. Comparison of different carry skip reversible BCD adders

		Existing Circuit [11]	Proposed Circuit (Without Fanout)
No. of gates	A	15	15
	B	21	
No. of garbage	A	27	14
	B	27	
Delay	A	10	10
	B	12	

 # A = With Fan out, B = Without Fan out
* fan outs in reversible design are forbidden. But as it was found in literature [11], the numbers are shown here only.

5. Conclusions

In this paper, reversible logic syntheses were carried out for both BCD adder and carry skip BCD adder. The designs have been done for ease of reversible logic implementation and it has been found that the proposed designs are far better than the existing ones [9,11] in terms of number of gates needed, number of garbage outputs produced and delay. Improved Carry Skip BCD adder can perform much faster than the BCD adder. If multiple BCD blocks are used in the carry skip adder, i.e. m-digit BCD numbers, then carry skip BCD adder has the potential to perform the desired operation much faster. BCD adders can be an important part of some other larger and more complex reversible circuits. Fast and improved BCD adders may also find its use in future quantum computers [4].

6. References

[1] Keyes R, Landauer R. Minimal Energy Dissipation in Logic. *IBM Journal of Research and Development* 1970; 14: 153-7.

[2] Landauer R. Irreversibility and heat generation in the computational process's. *IBM Journal of Research Development* 1961; 5: 183-91.

[3] Bennett CH. Logical reversibility of computation. *IBM Journal of Research and Development* 1973; 17: 525-32.

[4] Shende VV, Prasad AK, Markov IL, Hayes JP. Synthesis of reversible logic circuits. *IEEE Transaction on CAD* 2003; 22(6): 723-9.

[5] Moore GE. Cramming more components onto integrated circuits. *Journal of Electronics* 1965; 38(8).

[6] Frank M. Physical Limits of Computing. CIS 4930.1194X/6930.1078X, 2000.

[7] Perkowski M. Reversible Computation for Beginners. Lecture Series 2000. Portland State University. http://www.ee.pdx.edu/~mperkows.

[8] Hayes JP. *Computer Architecture and Organization*, 3rd ed. McGraw-Hill; 1998.

[9] Babu HMH, Chowdhury AR. Design of a Compact Reversible Binary Coded Decimal Adder Circuit. *Elsevier Journal of Systems Architecture* 2006, 52 (5): 272-82.

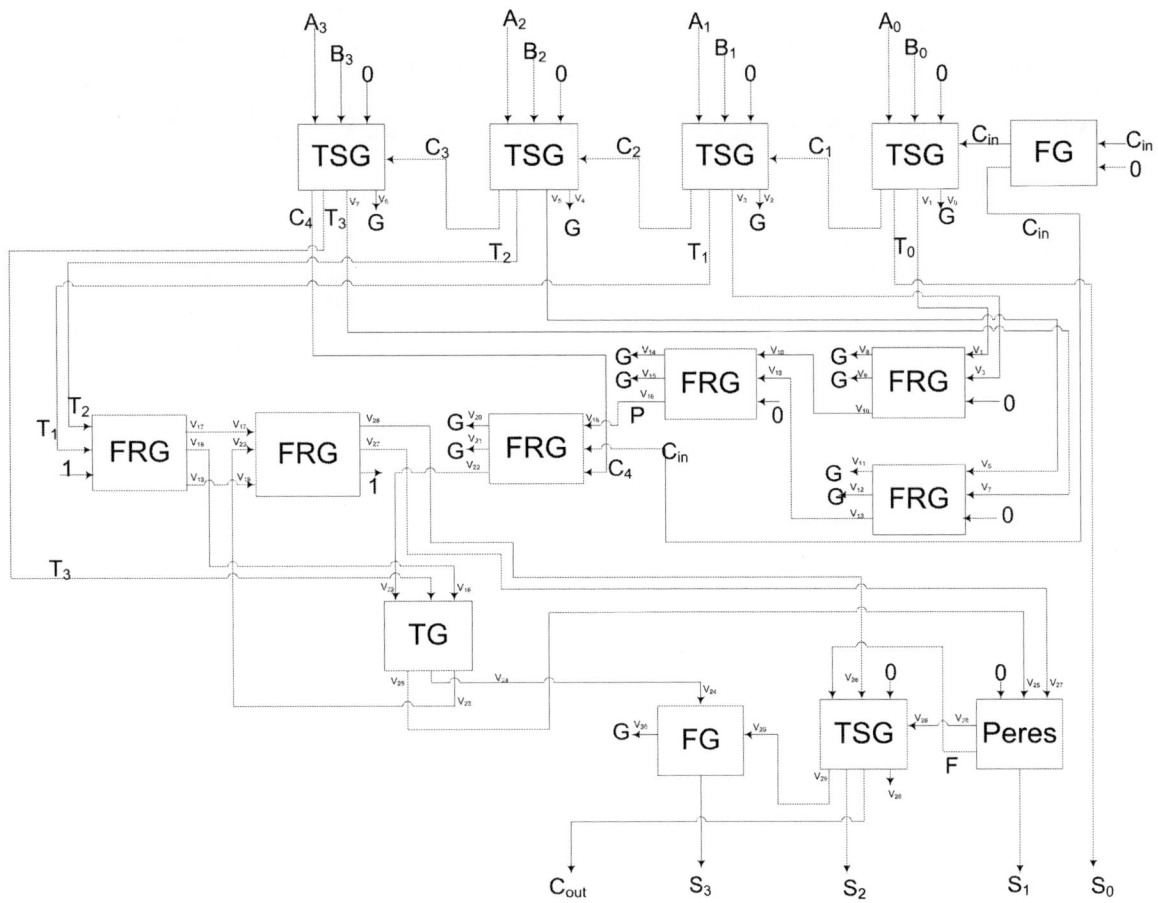

Figure 4.3.1. Design of a carry skip 1-digit BCD adder

[10] Babu HMH, Islam MR, Chowdhury AR, Chowdhury SMA. Synthesis of full-adder circuit using reversible logic. 17th International Conference on VLSI Design 2004; 757-60.

[11] Thapliyal H, Kotiyal S, Srinivas MB. Novel BCD Adders and their Reversible Logic Implementation for IEEE 754r Format. 19th International Conference on VLSI Design 2006; 387-92.

[12] Khan M. H. A and Perkowski M. Multi-output ESOP synthesis with cascades of new reversible families. 6th International Symposium on Representations and Methodology of Future Computing Technologies, March 2003, 144-153.

[13] Peres A. Reversible Logic and Quantum Computers. Physical Review 1985; 3266-76.

571

978-1-4244-3039-0/08 $25.00 © 2008 IEEE

21st International Conference on VLSI Design

A Merged Synthesis Technique for Fast Arithmetic Blocks involving Sum-of-Products and Shifters

Sabyasachi Das
Synplicity Inc
Sunnyvale, CA, USA
Email: sabya@synplicity.com

Sunil P. Khatri
Texas A&M University
College Station, TX, USA
Email: sunilkhatri@tamu.edu

Abstract— **In modern Digital Signal Processing (DSP) and Graphics applications, the arithmetic Sum-of-Products, Shifters and Adders are important modules, contributing a significant amount to the overall delay of the system. A datapath structure consisting of multiple arithmetic sum-of-product, shifter and adder blocks is often found in the timing-critical path of the chip. In this paper, we propose a new operator-level merging technique to synthesize this type of datapath structure. In our approach, we combine the shifting operation with the partial product reduction stage of the sum-of-product blocks. This enables us to implement the functionality of the original design by using only one carry-propagate adder block (instead of two carry-propagate adders). As a result, the timing-critical path of the design gets shortened by a significant percentage and the overall performance of the design improves. Our experimental data shows that the datapath block generated by our approach is significantly faster (13.28% on average) with a modest area penalty (3.24% on average) than the corresponding block generated by a commercially available best-in-class datapath synthesis tool. These improvements were verified on placed-and-routed designs as well.**

I. INTRODUCTION

The design complexity and performance requirements of datapath operations implemented in systems on chips has increased considerably over the years. This is especially true in ICs for communication, multimedia and graphic applications, which have highly parallel implementations of signal processing algorithms.

The arithmetic sum-of-products, shifters and adders are some of the most widely used arithmetic datapath operations in modern digital design. The block diagram of the Figure-1 is often seen in modern datapath designs. This design consists of multiple computationally expensive arithmetic sum-of-product (SOP) blocks. The outputs of some of the SOP blocks get shifted by different shift signals, followed by an addition of all the outputs of the shifters, remaining SOPs and some other additive input signals. The critical path of this design goes through an SOP, a shifter and an adder. Since this design requires intensive computations, they incur a significant amount of delay, and therefore tend to be typically found in the timing-critical path of the chip. Developing an efficient architecture for this design structure would reduce the delay of the individual blocks and thereby improve the performance of the IC. Hence there is great interest in generating timing-efficient architecture for this type of datapath structure.

In [1], [2], [3] and [4], the authors presented techniques to

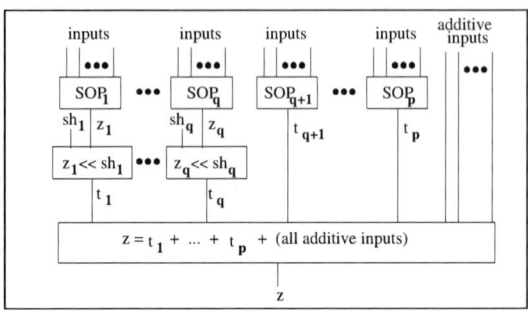

Fig. 1.   The generalized block-diagram of our problem statement

emphasize the usefulness of arithmetic sum-of-product (SOP) blocks over a collection of cascaded blocks performing unit operations (like additions, subtractions, multiplications etc). In [5], the critical path delays and hardware complexities of Multiplier-Accumulation units are explored to derive a high performance MAC. In [6], a technique to reduce the partial products in multiplication and sum-of-product units has been proposed. A hybrid compression technique to reduce the delay of SOP has been presented in [7]. The basic architecture for a barrel shifter was proposed in [8]. In [9], a timing-driven decomposition is introduced for fast shifter. The use of dynamic logic for shifter blocks was demonstrated in [10]. Timing-driven layout techniques of shifters were proposed in [11], [12]. A 32-bit rotator/shifter circuit design with short latency was discussed in [13]. Several architectures for performing fast timing-driven two-operand addition are explained in [14], [15] and [18]. A mix of these architectures can be used to synthesize the blocks involving sum-of-products and shifters.

In this paper, we propose an operator-level merging-based technique involving the sum-of-products (SOP) and shifters. In our approach, we combine the shifting operation with the partial product reduction stage of the SOP blocks. This enables us to implement the original design by using only one carry-propagate adder block. As a result, the timing-critical path of the design gets shortened by a significant percentage and the overall performance of the design improves. Our paper addresses a different datapath design issue than what was implemented in the references cited in this section.

We have organized the rest of the paper as follows: Some preliminary information is given in the Section II. In the Section III, we present the definition of the problem we are addressing in this paper. In the Section IV, we discuss our proposed approach in detail. The experimental setup is explained in the Section V. The Section VI presents the experimental results. Conclusions are drawn in the Section VII.

II. PRELIMINARIES

In this section, we briefly explain the concepts of an arithmetic *Sum-of-Product* (SOP) and a *Shifter* [18].

An example sum-of-product (SOP) block can be expressed by the following Verilog RTL:

assign $z = a * b + c * d + e * f + g + h$;

In this block, there are three product terms ($a*b$, $c*d$ and $e*f$) and two input sum terms (g and h). In general, the number of product and sum terms are arbitrary. After evaluating the product terms (or performing the multiplication operations for each product term), the results of each product term also get added with the input sum terms to produce the final result of the overall SOP block. A sum-of-product block can have any number (including zero) of product terms or sum terms. As a consequence, an SOP block is quite general. It can be used to implement a multiplier. MAC (multiply-accumulator), adder, subtractor, squarer, chain-of-adders or combinations thereof.

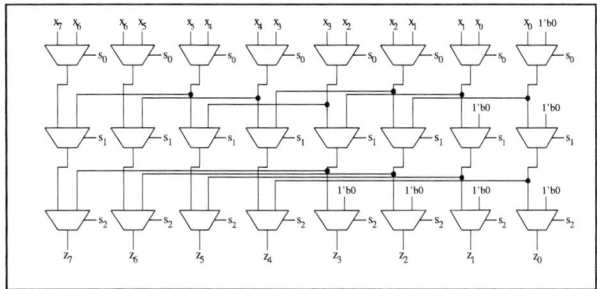

Fig. 2. A Traditional Barrel Shifter with 3-stages

A generalized left-shifter block can be expressed by the Verilog RTL: "assign $z = a << sh$"; where the input *data* signal (a) gets shifted by an input *shift* signal (sh) to produce the output (z) signal. If the data input signal is n-bits wide, then the shift signal is typically $\lceil log_2(n) \rceil$ bits wide. The width of the output (z) is also typically same as the input-width (n).

In a barrel shifter having an n bit wide data signal, the shifter is divided into $\lceil log_2(n) \rceil$ stages, where each stage (i) handles a single shift of 0 or 2^i bits. Each bit of the shift signal controls the functionality of exactly one barrel shifter stage. The input data will be shifted or not shifted by each of the stages in sequence. To implement this, multiplexers (or an equivalent logic circuit constituted using technology library cells) are used in each stage. Figure-2 shows the block-level diagram of a 3-stage barrel shifter. In this diagram, the data input signal (x) is 8-bit wide and the output signal (z) is also 8-bit wide. The shift signal has 3 bits ($\lceil log_2(8) \rceil = 3$) and the shifter consists of 3 stages only.

III. PROBLEM DEFINITION

In this paper, we propose an operator-level merging-driven technique to synthesize a fast arithmetic block involving sum-

of-product (SOP) and shifter blocks. The generalized description of our targeted datapath design block involving SOP and shifter blocks is as follows.

The datapath module consists of p arithmetic sum-of-product (SOP) blocks. Each SOP block takes any number of input vectors. The inputs to each of the p SOPs are arbitrary. The functionality of each of the p SOPs is arbitrary as well. Out of the p outputs of the p SOP blocks, q signals get shifted left by q shifter blocks. The q shift signals (sh_1, sh_2, ..., sh_q) are also arbitrary. Finally, the outputs of q shifters and the outputs of the remaining ($p-q$) sum-of-products (which do not go through the shifters) and r primary additive input signals (which do not go through any sum-of-products or shifters) get added together by a final adder.

The block diagrams shown in Figure-1 depicts the above-described datapath structure. This type of circuit topology is often seen in modern datapath designs (specially DSP ICs).

One specific example of the above-described generalized datapath design is shown in the Figure-3. In that specific example design, $p=2$, $q=1$ and $r=2$.

IV. OUR APPROACH

Our proposed synthesis approach has four steps. In the following sub-sections, we discuss each step in detail.

A. Generation of Partial Products

For every product term in an SOP block, partial products are generated by performing a bit-wise multiplication between the appropriate bits of the multiplicand and the multiplier. Each partial product is shifted by one or more bits, depending on the bit number of the multiplicand. If PP_i is the ith partial product of the product term $a*b$, and b has n bits, then partial products can be represented by the following expression:

$$PP_i = a * b_i * 2^i \qquad for \ i = 0, 1, ...(n-1)$$

After computing the set of all the partial products corresponding to the product terms, the r additive terms of the SOP expression get included in the set of partial products.

For an SOP with the expression $z = a * b + c$ (where a, b and c each are n-bits wide), there will be a total of $n+1$ partial products. Out of these, n partial products will be generated by the product term $a * b$. The remaining single partial product will be the sum-term c. Similarly, for a more complex SOP with the expression $z = a * b + c * d + e + f + g$ (where each input signal is n-bits wide), there will be a total of $2n+3$ partial products.

B. Shifting of Partial Products

As mentioned in the description of the datapath structure targeted in this paper, q of the p outputs of the sum-of-products need to be shifted. In the traditional synthesis approach, all the q SOP blocks get computed and then the q output values get shifted before being fed to the final adder module. In this case, the critical path traverses one SOP block (which in turn has a carry-propagate adder in it), one shifter block and one final carry propagate adder block. It is well-known that the carry propagate adder is one of the most delay-consuming

573

978-1-4244-3039-0/08 $25.00 © 2008 IEEE

arithmetic operations. Our experimental analysis suggests that the carry propagate adder inside a multiplier (which is one of the most heavily used sum-of-products block), typically contributes about 30% to the total delay of the multiplier [16]. Hence any reduction in the number and structure of the adder stages in our targeted datapath design plays a key role in determining the critical path delay.

In our approach, we implement the shifting of the q SOP outputs by performing the shift operation on *each* of the partial products corresponding to the q sum-of-product blocks. This eliminates the need for carry propagate adders in the q SOP blocks, leading to the improvement of about 30% delay of the individual SOP blocks. Note that, to achieve this improvement in speed of the individual SOP blocks, we require more shifter modules compared to the traditional approach.

To implement a fast shifter, we use the decomposition of the barrel shifter approach [9]. We describe this technique briefly, for completeness. In this approach, two or three stages of the shifter are merged into a single stage whenever feasible. If two stages are merged, the newly created stage is called a *dual merged* stage. On the other hand, if three stages are merged, then the new stage is called a *triple merged* stage.

In the case of *dual merged* stages, let us assume that the stages corresponding to the ith bit and the jth bit of the shift signal s were merged, where $0 \le i < n$, $0 \le j < n$ and $i \ne j$. Note that i and j do not require to be two consecutive bits of the shift signal. The newly created *dual merged* stage will perform one of the following four operations:

1) no shifting operation (if $s_i=0$ and $s_j=0$)
2) shift by 2^i bits (if $s_i=1$ and $s_j=0$)
3) shift by 2^j bits (if $s_i=0$ and $s_j=1$)
4) shift by (2^i+2^j) bits (if $s_i=1$ and $s_j=1$).

The functionality of each bit-slice of the *dual merged* stage for the left shifter is as follows (data signal is denoted as x):

$$out_q = \overline{\overline{(t_1)} \wedge \overline{(t_2)} \wedge \overline{(t_3)} \wedge \overline{(t_4)}}. \qquad \text{for } 0 \le q < n.$$
where $t_1 = x_q \wedge \overline{s_i} \wedge \overline{s_j}$
$\qquad t_2 = x_{(q-2^i)} \wedge s_i \wedge \overline{s_j}$
$\qquad t_3 = x_{(q-2^j)} \wedge \overline{s_i} \wedge s_j$
$\qquad t_4 = x_{(q-2^i-2^j)} \wedge s_i \wedge s_j$

In a similar manner, one can formulate the output equation of each bitslice for *triple merged* stages as well. Let us assume that the stages corresponding to the ith bit, j^{th} bit and the kth bit of the shift signal s are merged, where $0 \le i < n$, $0 \le j < n$, $0 \le k < n$, $i \ne j$, $j \ne k$ and $k \ne i$. The functionality of each bit-slice of *triple merged* stage for a left shifter is as follows:

$$out_q = \overline{\overline{(t_1)} \wedge \overline{(t_2)} \wedge \overline{(t_3)} \wedge \overline{(t_4)} \wedge \overline{(t_5)} \wedge \overline{(t_6)} \wedge \overline{(t_7)} \wedge \overline{(t_8)}}.$$
$$\text{for } 0 \le q < n.$$
where $t_1 = x_q \wedge \overline{s_i} \wedge \overline{s_j} \wedge \overline{s_k}$
$\qquad t_2 = x_{(q-2^i)} \wedge s_i \wedge \overline{s_j} \wedge \overline{s_k}$
$\qquad t_3 = x_{(q-2^j)} \wedge \overline{s_i} \wedge s_j \wedge \overline{s_k}$
$\qquad t_4 = x_{(q-2^k)} \wedge \overline{s_i} \wedge \overline{s_j} \wedge s_k$
$\qquad t_5 = x_{(q-2^i-2^j)} \wedge s_i \wedge s_j \wedge \overline{s_k}$
$\qquad t_6 = x_{(q-2^i-2^k)} \wedge s_i \wedge \overline{s_j} \wedge s_k$

$t_7 = x_{(q-2^j-2^k)} \wedge \overline{s_i} \wedge s_j \wedge s_k$
$t_8 = x_{(q-2^i-2^j-2^k)} \wedge s_i \wedge s_j \wedge s_k$

A general-purpose technology mapper should be able to identify the most efficient implementation of the traditional *unmerged* stage, *dual-merged* stage and *triple merged* stage of a shifter. The best possible delays of these three types of stages are denoted as Del_1, Del_2 and Del_3.

In addition to the design of the merged stages, the technique for identification of the *mergeable* stages plays a key role to determine the performance of the shifter architecture. Without an efficient algorithm to identify the *mergeable stages*, the design of merged stages would not be useful.

In the approach of [9], the following timing-driven analysis was done to find two or three stages for merging: Assume that the earliest arriving three shift signals are s_i, s_j and s_k. Let ts_i be the arrival time of the shift signal s_i. For the signals s_i and s_j, if a *dual merged* stage is constructed, then the output of the dual merged stage will be available at time

$$T_{dual} = ts_j + Del_2.$$
On the other hand, if two individual stages are constructed in cascade, then the output of the second stage will be available at time

$$T_{single2} = \text{Max} \left((ts_i + Del_1), ts_j \right) + Del_1.$$
Similarly, for the signals s_i, s_j and s_k, if a *triple merged* stage is constructed, then the output of the triple merged stage will be available at time

$$T_{triple} = ts_k + Del_3.$$
On the other hand, if three individual cascaded stages are constructed, then the output of the third stage will be available at time

$$T_{single3} = \text{Max} \left(T_{single2}, ts_k \right) + Del_1.$$
Now, if $(T_{triple} < T_{single3})$ and $(T_{triple} < (T_{dual} + (Del_2/2))$, then the three stages (i, j, and k) of the shifter are chosen as the *mergeable* stages. If the above conditions are not true and if $(T_{dual} < T_{single2})$, then the two stages (i and j) are selected as the *mergeable* stages. If both the above conditions are false (which means, $(T_{single2} \le T_{dual})$ and $(T_{single3} \le T_{triple})$), then i is not included into any merging combination and one single stage is utilized for the stage i. Next, the same analysis is performed with the three stages corresponding to the next three earliest arriving bits. This analysis and identification of *mergeable* stages continues until all the stages are analyzed. At the end of this algorithm, the list of all the *mergeable stages* is determined in this manner.

In terms of the execution of the flow, the mergeable stages are first identified. Once the configurations of all the *dual merged*, *triple merged* and *unmerged* stages are identified, then the merged stages as well as unmerged single-stages in the netlist are implemented with proper connectivity.

Note that during technology mapping in our approach, the mapper sizes the output of any node based on their load capacitance. Also, the delay analysis for each configuration considers actual capacitance of the output node, using a load-dependent delay model. Also, note that any of our nodes inside the shifter block do not have high fanouts.

C. Reduction of Partial Products

In our approach, all the shifted partial products (corresponding to the q sum-of-products), the non-shifted partial products (corresponding to the remaining $p-q$ sum-of-products) and the r additive terms are fed to a partial product reduction tree. The purpose of this partial product reduction tree is to perform column-wise reduction of the elements in each bitslice, such that each bitslice finally consists of 2 elements or less.

If all the sum-of-products in the Figure-1 are multipliers and each signal is n-bits wide, then a total of $(pn+r)$ partial products will be fed to this partial product reduction tree. On the other hand, if all the sum-of-products in the Figure-1 are multiply-accumulators (MAC) and each signal is n-bits wide, then a total of $(pn+p+r)$ partial products will be fed to this partial product reduction tree.

To perform the reduction of partial products, we use the technique presented in [7]. This technique uses the concept of *counters*. A *(p:q) counter* is a functional block, which adds p single-bit inputs and produces q single-bit outputs; where p and q satisfy the following equation: $q = \lfloor log_2 p + 1 \rfloor$

To reduce the partial products, we use the concept of (4:3) counter. This is defined as a functional block which accepts 4 single-bit signals in the ith bitslice and transform them into 3 different single-bit output signals (one for the ith bitslice; one for the $(i+1)^{th}$ bitslice and the third one for the $(i+2)^{th}$ bitslice). Let us assume that there is a (4:3) counter in bitslice$_i$, which takes 4 signals (a_i, b_i, c_i and d_i) as inputs and produces 3 outputs (x_{i+2} for the bitslice$_{i+2}$, x_{i+1} for the bitslice$_{i+1}$ and x_i for the bitslice$_i$). The functionality of the (4:3) counter is as follows:

- $x_i = (a_i \oplus b_i) \odot (c_i \odot d_i)$
- $x_{i+1} = \overline{((a_i \wedge b_i) \odot (c_i \wedge d_i)) \wedge \overline{((a_i \oplus b_i) \wedge (c_i \oplus d_i))}}$
- $x_{i+2} = a_i \wedge b_i \wedge c_i \wedge d_i$

The (3:2) counter is widely used in column-compression schemes. A (3:2) counter accepts 3 inputs signals (a_i, b_i and c_i) belonging to the ith column (bitslice) in the partial-products and would produce 1 output signal (x_i) for the ith column (bitslice) and 1 output signal (x_{i+1}) for the $(i+1)^{th}$ bitslice. The functionality of the (3:2) counter is:

- $x_i = a_i \oplus b_i \oplus c_i$
- $x_{i+1} = \overline{\overline{(a_i \wedge b_i)} \wedge \overline{(b_i \wedge c_i)} \wedge \overline{(c_i \wedge a_i)}}$

Similarly, the functionality of a (2:2) counter is:

- $x_i = a_i \oplus b_i$
- $x_{i+1} = (a_i \wedge b_i)$

In our approach, we use a timing-driven algorithm to design the partial product reduction tree by using a combination of (4:3), (3:2) and (2:2) counters. The key idea in our partial product reduction approach is to find the opportunity to use the (4:3) counters. Let us consider that a_i, b_i, c_i and d_i are four input signals (sorted in the ascending order of the arrival time) in bitslice$_i$. Our algorithm would use the following scheme to determine the type of counter to be instantiated:

- If a_i and b_i arrive at least a 2-input EXOR gate-delay before signal c_i arrives; then instantiate a (2:2) counter.

- If the above condition fails and a_i, b_i, c_i arrive at least two 2-input EXOR gate-delay before the signal d_i arrives; then instantiate a (3:2) counter.
- If both the above-mentioned conditions fail; then instantiate a (4:3) counter.

In other words, our algorithm instantiates a (4:3) counter, if the arrival times of all four signals at the bitslice$_i$ are *reasonably close* to each other. We continue to perform the reduction in all the bitslices until each of the bitslices contain ≤ 2 elements. With an instantiation of the (4:3) counter, four elements are reduced in every bit. In addition, due to the simple circuitry needed to generate x_{i+2} in a (4:3) counter; the arrival time of the signal (x_{i+2}) at the input of bitslice$_{i+2}$ is also low. This reduces timing-skew of the signals at the output of the reduction tree.

D. Computation of the Final Sum

After performing the column-wise reduction of the partial products, each column in the reduced element-set consists of a maximum of 2 elements. Hence, after the partial product reduction step, we effectively transform all the partial products into two operands. To produce the final output (z) of our targeted datapath design, the 2 addends of all the columns have to be added by a final carry propagate adder circuit. Therefore, a 2-operand addition is required to compute the final result of the arithmetic block.

To obtain faster performance, we need to use a fast addition technique. In high-frequency datapath designs, adders with parallel prefix computation methodologies [14] are very popular. The hybrid adder described in [16] exploits the skewed pattern of the input arrival-times and can be very effective for the sum-of-product computation. In our approach, we use a hybrid adder which consists of three subadders. The bit-width of each of the subadders in the hybrid adder are computed by using the approach discussed in [17]. The internal topology of our hybrid adder is as follows:

- *Ripple-Carry* for the few bits near the least significant bit (LSB).
- A fast *Kogge-Stone* adder for *several* bits in the middle.
- A carry-select adder based on the *Kogge-Stone* architecture for the *remaining* bits near the most significant bit (MSB).

V. EXPERIMENTAL SETUP

We have implemented our proposed approach in the C++ programming language. The experiments were performed with datapath RTL designs written in Verilog hardware description language. For all our experiments, we used a Linux workstation (RedHat 7.1) with dual-2.2GHz processors and 4GB memory.

To collect different data-points regarding the quality of results for the different types of datapath designs involving SOP and shifter blocks in the timing-critical portion of the design, we used the following variations:

- Multiple types of designs of different expressions and input bit-widths:

- Our first design has one SOP block (multiply-accumulator) driving a shifter. That means, $q=1$. The three Inputs to the SOP block are 16-bit wide. In addition, there is another SOP block (multiplier), whose output directly drives the adder. That means, $p=2$. The two inputs to the SOP block are 16-bit wide. Finally, there are two 16-bits wide additive signals which get fed to the final adder. That means, $r=2$. Figure-3 shows a block-diagram of this example. We referred to this design as Des-q1-p2-r2.
- Using this notation, the next design is referred to as Des-q3-p7-r6.
- We also experimented with the following 3 designs: Des-q2-p3-r0, Des-q4-p9-r5 and Des-q1-p6-r3.

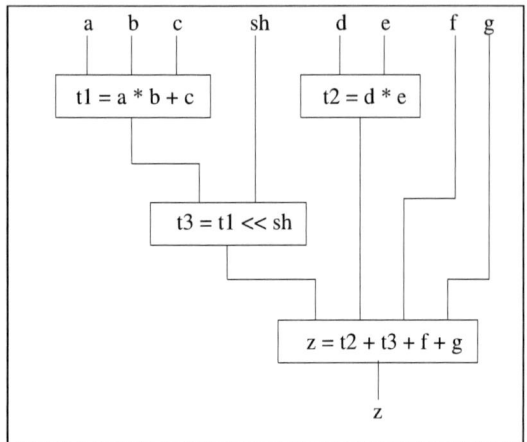

Fig. 3.   The block-diagram of one design (Des-q1-p2-r2)

- The different technologies and libraries we used are:
  - Two industrial libraries (L_1 and L_2) for a 0.13μ technology.
  - Two industrial libraries (L_3 and L_4) for a 0.09μ technology.
- Different input arrival time constraints:
  To facilitate the explanation, let us assume that there are four inputs to the datapath structure of the Figure-1 and each of the four input signals is n-bit wide. We have used the following types of input arrival time constraints:
  - All input bits of all the signals arrive at the same time. We refer to this constraint as Type-A. If we denote $Arr(a_i)$ as the arrival time of the bit a_i and if k is a constant number, then this Type-A constraint can be represented as:

$$Arr(a_i) = k; \qquad 0 \leq i < n$$
$$Arr(b_i) = k; \qquad 0 \leq i < n$$
$$Arr(c_i) = k; \qquad 0 \leq i < n$$
$$Arr(d_i) = k; \qquad 0 \leq i < n$$

This category represents the actual timing situations if the SOP blocks are placed immediately after a register-bank or the primary inputs of the design are fed to the SOP blocks.

- Different input bits arrive at different times. We refer to this category of timing constraints as Type-B. We believe that this category represents the actual timing situations for many sum-of-product blocks in real-life designs. Assuming that k is a constant number and δ is the delay of the fastest 2-input AND-gate in the given technology library, the following are some specific examples of the Type-B timing constraints. Here we have explained the arrival times for signal a_i. Similar expressions for arrival times of all the bits of signals b, c and d can be written as well.

1) $Arr(a_i) = i * k * \delta; \qquad 0 \leq i < n$

2) $Arr(a_i) = i^2 k\delta; \qquad 0 \leq i < n$

3) $Arr(a_i) = 0; \qquad 0 \leq i < \lceil n/2 \rceil$
   $Arr(a_i) = k\delta; \qquad \lceil n/2 \rceil \leq i < n$

4) $Arr(a_i) = 0; \qquad 0 \leq i < \lceil n/4 \rceil$
   $Arr(a_i) = k\delta; \qquad \lceil n/4 \rceil \leq i < \lceil n/2 \rceil$
   $Arr(a_i) = 2k\delta; \qquad \lceil n/2 \rceil \leq i < \lceil 3n/4 \rceil$
   $Arr(a_i) = 3k\delta; \qquad \lceil 3n/4 \rceil \leq i < n$

5) $Arr(a_i) = 0; \qquad 0 \leq i < \lceil n/4 \rceil$
   $Arr(a_i) = ik\delta; \qquad \lceil n/4 \rceil \leq i < \lceil n/2 \rceil$
   $Arr(a_i) = 2ik\delta; \qquad \lceil n/2 \rceil \leq i < \lceil 3n/4 \rceil$
   $Arr(a_i) = 3ik\delta; \qquad \lceil 3n/4 \rceil \leq i < n$

VI.   Experimental Results

We compared our approach against a commercially available datapath synthesis tool which is considered to be the best-in-class solution. The synthesis tool generates arithmetic-optimized architectures for all the arithmetic blocks (like sum-of-products, shifters, adders) and then it performs general-purpose operations like technology-independent optimizations, constant propagation, redundancy removal, technology mapping, timing-driven optimization, area-driven optimization, incremental optimization etc. While running the synthesis tool, we turned on all the above-mentioned optimizations. In the Table-I, we report the worst-case delay and the total area results obtained for the datapath block from the commercial synthesis tool and from our approach. In this table, we report 20 sets of data-points involving different combinations of datapath blocks and technology libraries.

If we compute the average of all the 20 data-points presented in the Table-I, then our approach results in about 13.28% faster implementation of the datapath block, with a 3.24% area penalty compared to the netlist generated by the commercial datapath synthesis tool. State-of-the-art designs have very strict timing goals, hence most designers would be willing to accept a 13.28% delay improvement at the expense of a 3.24% area penalty of the datapath block only.

To keep the size of the Table-I relatively brief, we do not report the results for different types of Type-B timing constraints. Note that the results in each of the combinations which are not reported here also support our conclusion that the proposed approach produces significantly faster netlist.

Design Name	Technology Library	Timing Constraint	Worst-case Delay (ps)			Area (μ^2)		
			Commercial Tool	Our Approach	(%) Improvement	Commercial Tool	Our Approach	(%) Penalty
Des-q1-p2-r2	Lib$_1$	Type-A	1572	1396	11.19%	6358	6469	1.76%
Des-q3-p7-r6	Lib$_1$	Type-A	1749	1523	12.92%	22413	23316	4.03%
Des-q2-p3-r0	Lib$_1$	Type-A	1527	1351	11.53%	9862	10125	2.67%
Des-q4-p9-r5	Lib$_1$	Type-A	1681	1447	13.87%	29571	31318	5.91%
Des-q1-p6-r3	Lib$_1$	Type-A	1603	1419	11.47%	20247	20814	2.80%
Des-q1-p2-r2	Lib$_2$	Type-A	1243	1068	14.08%	7416	7523	1.45%
Des-q3-p7-r6	Lib$_2$	Type-A	1419	1254	11.63%	25539	26514	3.82%
Des-q2-p3-r0	Lib$_2$	Type-A	1207	1045	13.42%	11282	11649	3.26%
Des-q4-p9-r5	Lib$_2$	Type-A	1356	1172	13.56%	34753	36480	4.97%
Des-q1-p6-r3	Lib$_2$	Type-A	1285	1093	14.94%	23927	24346	1.75%
Des-q1-p2-r2	Lib$_3$	Type-A	1847	1582	14.35%	4672	4779	2.31%
Des-q3-p7-r6	Lib$_3$	Type-A	2169	1814	16.37%	16461	17031	3.46%
Des-q2-p3-r0	Lib$_3$	Type-A	1785	1539	13.78%	7518	7786	3.58%
Des-q4-p9-r5	Lib$_3$	Type-A	2031	1763	13.19%	22085	23257	5.32%
Des-q1-p6-r3	Lib$_3$	Type-A	1914	1627	14.99%	14359	14652	2.04%
Des-q1-p2-r2	Lib$_4$	Type-A	1094	961	12.16%	6892	7019	1.85%
Des-q3-p7-r6	Lib$_4$	Type-A	1256	1089	13.29%	23641	24521	3.71%
Des-q2-p3-r0	Lib$_4$	Type-A	1061	913	13.95%	10367	10694	3.16%
Des-q4-p9-r5	Lib$_4$	Type-A	1183	1027	13.19%	32753	34286	4.69%
Des-q1-p6-r3	Lib$_4$	Type-A	1128	996	11.70%	21536	21863	1.52%
Average					13.28%			3.24%

TABLE I

AREA AND DELAY COMPARISON OF BLOCKS GENERATED BY A COMMERCIAL SYNTHESIS TOOL AND BY OUR APPROACH

To verify the correlation of the post-synthesis experimental data of the Table-I with the post place-and-route data, we performed placement and routing on Des-q1-p2-r2 and Des-q3-p7-r6. For these two testcases, the average improvement in the post-routing worst case delay of the datapath design generated by our proposed approach is 12% compared with the worst delay of the corresponding block generated by the commercial datapath synthesis tool (with the average 4% post-roting area penalty). The individual results for these testcases correlate closely with the post-synthesis numbers reported in the Table-I. These post-routing data confirm our conclusion about significant timing improvement of the netlist produced by using our approach (with a modest area penalty).

Our delay improvement is consistent across multiple types of designs, technology libraries and arrival time constraints. This underscores the strength of our approach. Since this type of datapath structure is frequently used in modern digital design, we believe that the timing-critical portions of many real-life designs can significantly benefit from our approach.

VII. CONCLUSION

In this paper, we have presented a new approach to implement a faster datapath block involving arithmetic sum-of-products, shifters and a final adder. Our approach would be very useful when the critical path of the design goes through such a block. Our approach to generate the timing-efficient architecture for this block works seamlessly with different types of datapath blocks, arrival timing constraints and across different technology domains (0.13μ, 0.09μ). The experimental results indicate that our implementation of the datapath block is significantly faster (with a modest area penalty) than the datapath block generated by a commercially available best-in-class datapath synthesis tool.

REFERENCES

[1] T. Kim, W. Jao, S. Jjiang. "Circuit optimization using carry-save-adder cells," in *IEEE Transactions on Computer-Aided Design of Integrated Circuits and Systems* CAD-17, pp. 974–984, 1998.

[2] A. Mathur, S. Saluja. "Improved merging of datapath operators using information content and required precision analysis," in *Proceedings of the 38th conference on Design Automation*, pp. 462-467, 2001.

[3] A. K. Verma, P. Ienne. "Improved Use of the Carry-Save Representation for the Synthesis of Complex Arithmetic Circuits," in *Proceedings of the 2004 IEEE/ACM International conference on Computer-aided design*, pp. 791-798, 2004.

[4] A. Fayed, W. Elgharbawy, M. Bayoumi, "A data merging technique for high-speed low-power multiply accumulate units," in *IEEE International Conference on Acoustics, Speech, and Signal Processing*, vol. 5, pp. 145–148, 2004.

[5] L. Chen, O. T. C. Wang, Y. C. Ma. "A multiplication accumulation computation unit with optimized compressors and minimized switching activities," in *IEEE International Symposium on Circuits and Systems*, vol. 6, pp. 6118–6121, 2005.

[6] C. S. Wallace, "A suggestion for a fast multiplier," in *IEEE Transactions on Electronic Computers*, EC-13(2):14-17, 1964.

[7] S. Das, S. P. Khatri, "A Timing-Driven Hybrid-Compression Algorithm for Faster Sum-of-Products", in *Proceedings of International Conference on Circuits, Signals and Systems*, 2007.

[8] R. S. Lim, "A Barrel Switch Design," in *Computer Design*, pp. 76-78, 1972.

[9] S. Das, S. P. Khatri, "Timing-Driven Decomposition of a Fast Barrel Shifter", in *Proceedings of IEEE MidWest Symposium on Circuits and Systems*, 2007.

[10] R. Rafati, S. M. Fakhraie, K. C. Smith, "A 16-Bit Barrel-Shifter Implemented in Data-Driven Dynamic Logic (D^3L)," in *IEEE Transactions on Circuits and Systems I*, vol 53, issue 10, pp. 2194-2202. 2006.

[11] P. M. Seidel, K. Fazel, "Two dimensional folding strategies for improved layouts of cyclic shifters," in *Proceedings of the IEEE Computer society Annual Symposium on VLSI*, pp. 277-278, 2004.

[12] M. A. Hillebrand, T. Schurger, P. M. Seidel, "How to half wire lengths in the layout of cyclic shifters," in *Proceedings of the IEEE International Conference on VLSI Design*, pp. 339-344, 2001.

[13] A. P. Singh, M. Barany, D. J. Deleganes, "A mixed signal rotator/shifter for 8GHz Intel/spl reg/ Pentium/spl reg/ 4 integer core," in *Proceedings of the Symposium on VLSI Circuits*, pp. 394-397, 2004.

[14] P. M. Kogge, H. S. Stone, "A parallel algorithm for the efficient solution of a general class of recurrence equations," in *IEEE Transactions on Computers*, C-22(8):783-91, 1973.

[15] R. P. Brent, H. T. Kung, "A regular layout for parallel adders," in *IEEE Transactions on Computers*, C-31(3):260-64, 1982.

[16] P. F. Stelling, V. G. Oklobdzija, "Design strategies for the final adder in a parallel multiplier," in 29th *Asilomar Conference on Signals, Systems and Computers*, pp. 591-595, vol. 1, 1995

[17] S. Das, S. P. Khatri, "Generation of the Optimal Bit-Width Topology of the Fast Hybrid Adder in a Parallel Multiplier", in *Proceedings of International Conference on Integrated Circuit Design and Technology*, 2007.

[18] M. D. Ercegovac, T. Lang, "Digital Arithmetic," *The Morgan Kaufmann Series in Computer Architecture and Design*, 2003

SESSION D4:
Design/MEMS/Optical

21st International Conference on VLSI Design

A Jitter Reduction Circuit Using Autocorrelation for Phase-Locked Loops and Serializer-Deserializer (SERDES) Circuits

Hari V. Venkatanarayanan   Michael L. Bushnell

ECE Dept. and CAIP Center, Rutgers University

96 Frelinghuysen Rd., Piscataway, NJ 08854-8088

harivijay@gmail.com   bushnell@caip.rutgers.edu

Abstract

A new jitter reduction circuit is proposed for reducing the timing jitter in a serializer-deserializer (SERDES). Instead of using elaborate hardware to calculate the jitter, we use the jittered signal's autocorrelation to remove the jitter. The motivation for this work was to provide a reduced jitter phase-locked loop (PLL), so that incorporating a built-in self-testing (BIST) mechanism for PLL's and SERDES would be simplified. The technique involves transmit and receive side jitter reducer pulse shaping circuits made of only 14 and 20 transistors, respectively. They reduce the jitter in the clock generated by the PLL at the transmit side, and the jitter between the recovered clock and the serial data at the receive side. The jitter reducers are designed in 70 nm Berkeley Predictive process models and tested with various types of input jitter. In the case of the transmit side, the peak-to-peak random jitter (RJ) is reduced, on average, by 45.51% and also the average transmit and receive side RMS jitter is reduced, on average, by 62.24% and 35.88%, respectively. The bit-error rate (BER) of the SERDES computed probabilistically is improved from 8.3×10^{-2} to 6.44×10^{-20}, for input RMS periodic jitter (PJ) of 71.77 ps. The BER for the PCI express bus must be $\leq 1 \times 10^{-12}$.

1 Introduction

SERDES is a high-speed serial data link [6] used in *integrated circuits* (ICs) to serialize the parallel data and transfer it at a much faster rate. A typical SERDES architecture resembles a communication set-up with a transmit and a receive side. A PLL at the transmit side generates the fast clock necessary to drive the serializer. At the receive side, a *clock and data recovery* (CDR) circuit recovers a clock from the transmitted serial data and retimes the data. One advantage of using SERDES is reduced clock skew, so data can be sent at the GHz rate. Also, it lowers pin count, chip area and power dissipation [1].

The main disadvantage in SERDES is timing jitter, the deviation of the actual signal transition from the expected transition in time. The SERDES jitter affects the BER [3], which is the ratio of the number of bit errors to the total number of bits transmitted. A bit error occurs if a bad bit is latched at the receive side. The clock generated by the PLL at the transmit side has timing jitter and when it drives the

serializer, jitter transfers to the data. The timing of the serial data is further distorted in magnitude and phase by the transmission path. Apart from that, at the receive side, the jitter in the serial data is not tracked properly by the CDR circuit due to the bandwidth limitation of the PLL [4], resulting in a lag between the recovered clock and the data. Jitter can be broadly classified into *deterministic* and *random* jitter [8], which has a Gaussian distribution. Deterministic jitter is further classified as *periodic*, with periodic signal transitions, and *duty cycle distortion* (DCD), with unequal pulse widths for HIGH and LOW logic values.

The BER of SERDES can be improved by reducing the jitter in the clock generated by the PLL at the transmit side, by designing a good transmission medium and by extracting an accurate clock from the serial data at the receive side. This can be achieved with a good low-jitter PLL design. Many techniques have been proposed [5, 10, 11] for reducing the jitter in the PLL. They involve on-chip jitter tracking mechanisms with huge hardware for measuring the jitter. The measured jitter is used to control the loop bandwidth of the PLL for better jitter performance.

In this work a new jitter reduction technique is proposed for reducing the jitter in the PLL and the CDR circuit and thereby improving the BER of the SERDES. We use external hardware that can reduce jitter without modifying the PLL circuitry. The jitter reducer is made of only 14 (transmit side) and 20 (receive side) transistors. The methodology involves generating a reference signal, and using this signal to correct the jittered clock signal. The amount of jitter reduced depends on how accurate the reference signal is and also on the process of jitter reduction. For the receive side, the serial data is used as the reference to reduce the jitter between the recovered clock and itself. The jitter reducers are tested extensively with various types of jitter. In the case of the transmit side jitter reducer, the random peak-to-peak jitter is reduced, on average, by 45.51%. Finally, a complete SERDES architecture is designed with the jitter reducers and tested for its jitter performance. For the case where the input RMS periodic jitter is 71.77 *ps* at the transmit side, the BER of the SERDES is reduced from 8.3×10^{-2} to 6.44×10^{-20}. The PCI express bus requires less than 1×10^{-12} jitter.

The primary advantage of this work is that it is the first method to improve the BER of the SERDES by reducing

978-1-4244-3039-0/08 $25.00 © 2008 IEEE

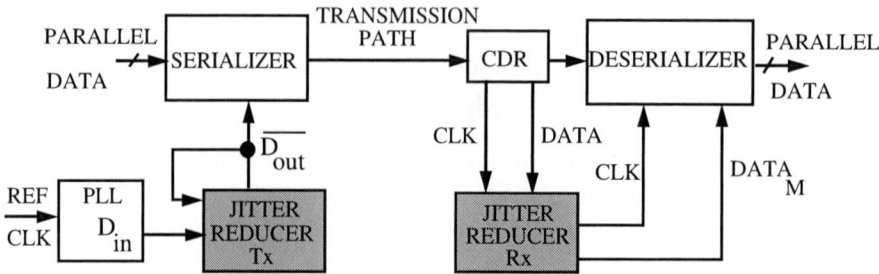

Figure 1. SERDES with Jitter Reduction Circuits

the jitter between the recovered clock and data externally without modifying the CDR circuit. The hardware proposed is relatively easier to control than the CDR circuit, in order to improve the BER of the SERDES.

Section 2 describes prior work. Section 3 explains the jitter reducer for both the transmit and the receive sides with results and circuit design issues are discussed in detail in Section 4. Finally, Section 5 concludes.

2 Prior Work

A jitter estimation algorithm is proposed and implemented by Vamvakos *et al.* [10] for reducing jitter in a PLL. They use two *voltage controlled delay lines* (VCDL) and counters along with edge comparison circuits to estimate jitter on-chip. The estimate is used to tune the loop parameters such as bandwidth and damping factor of the PLL. Telba *et al.* [9] use two PLLs; the first has a *voltage controlled crystal oscillator* (VCXO) and the second a wide loop bandwidth to decouple the dependency of the output clock jitter from the input reference clock jitter.

An adaptive PLL structure is proposed by Xia *et al.* [11]. It consists of a jitter test circuit, a locking indicator circuit and a control logic unit. The PLL is programmed to have wide bandwidth in the unlocked state by the control unit and when in the locked state, the jitter is tracked by the jitter test circuit and the loop bandwidth is accordingly adjusted based on the measured jitter. Hur *et al.* [5] proposed an analog adaptive PLL, where the charge pump is controlled depending on the lock status to adjust the loop bandwidth for minimizing the jitter. Mansuri *et al.* [7] proposed a method for measuring output jitter of a PLL on-chip and adjusting the loop bandwidth and peaking in the frequency response using a closed-loop control system. Dou *et al.* [2] proposed an on-chip jitter measurement technique using a dual-channel time digitizer.

3 SERDES with the Jitter Reducers

Figure 1 is a complete SERDES architecture with the *transmit* (Tx) and the *receive* (Rx) jitter reducer circuits (shaded blocks). The circuit has a PLL that generates a fast clock of $1\,GHz$, a Serializer (parallel to serial shifter) that converts 8-bit parallel data into serial data, a CDR circuit and a deserializer (serial to parallel shifter) that converts 1-bit serial data into 8-bit parallel data. The Type I PLL has

an XOR phase detector and a three stage ring oscillator. The CDR circuit has a similar PLL and a flip-flop for retiming the recovered data. A *low-pass filter* (LPF) of $1\,GHz$ cut-off frequency models the magnitude and frequency behavior of the transmission medium.

3.1 Jitter Reduction Technique for the Transmit Side Phase Locked Loop

Here, a new jitter reduction technique is proposed for the jitter present in the output of the PLL (D_{in}). A reference signal is used for determining the jitter that will be reduced from the jittered signal. The proposed jitter reducer (Figure 2) does: (1) inversion to generate the output signal $\overline{D_{out}}$, which has the opposite polarity to that of the input reference signal (D_{ref}), (2) jitter reduction and (3) generation of the reference signal D_{ref}.

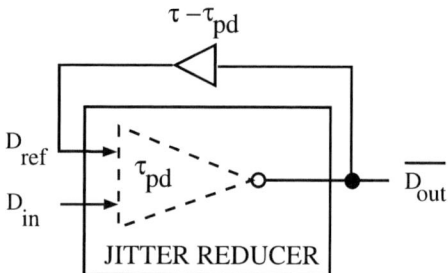

Figure 2. Jitter Reducer with Reference Signal $\overline{D_{out}}$

The jitter reducer behaves as an *inverter* that toggles the signal D_{ref} every τ *seconds*, where τ is the bit period (of either the logic '1' or '0' bit) for clock signal D_{in}. The buffer delays the signal from the inverter output to the input by $\tau - \tau_{pd}$ seconds, where τ_{pd} is the jitter reducer propagation delay. From the loop back signal D_{ref} (the modified input jittered signal) and the input jittered signal D_{in}, the circuit reduces the jitter in signal $\overline{D_{out}}$. The bit period of the loop back signal depends on the circuit delay and not on the input and, so, the jitter in the input is not transferred completely to the loop back signal $\overline{D_{out}}$. The loop back signal is not ideal and has some jitter in it that can be reduced by good circuit design. We use *auto-correlation*, where a signal correlates with its past, to determine the error. The correlating signals are the jittered input and the jitter reduced output signal. It is auto-correlation and not cross-correlation because the

582

978-1-4244-3039-0/08 $25.00 © 2008 IEEE

jitter reduced signal is nothing but the same delayed input signal but with less jitter in it.

3.1.1 Jitter Reduction Process

We obtain pulses giving the differences in time when the signals D_{ref} and D_{in} rise and fall to enable pulse shaping to reduce jitter.

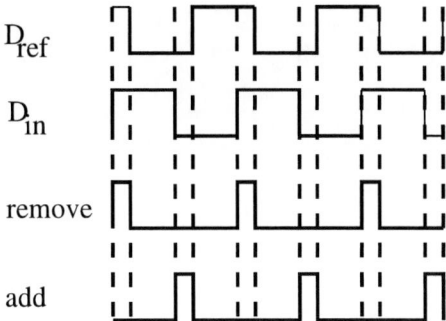

Figure 3. *Add* **and** *Remove* **Pulses**

The jitter reduction depends on how accurate these pulses are as compared to ideal reference pulses. In Figure 3, D_{in} leads D_{ref}. The *add* pulse adds a period of length τ_{add} to S_{PLL} and *remove* deletes a period of length τ_{remove} from D_{in}. If we add and remove pulses of period τ_{add} and τ_{remove} from D_{in} we get a pulse of period τ, which is the expected period of $\overline{D_{out}}$.

Theory: A mathematical proof is presented to show how jitter is reduced in the phase domain. D_{in} is the input jittered signal and D_{ref} be an ideal reference signal. We will represent these analog signals [12] as:

$$D_{in}(t) = \cos(2\pi f_o t + \triangle\phi_{in}(t)) \tag{1}$$
$$D_{ref}(t) = \cos(2\pi f_o t) \tag{2}$$

where f_o is the signal's fundamental frequency and $\triangle\phi_{in}(t)$ is the phase jitter. $R(t)$ and $A(t)$ are the functions that extract a phase jitter of opposite polarity from the input signals, i.e., if D_{in} has positive jitter then either $R(t)$ or $A(t)$ will extract a jitter of equal magnitude and negative polarity, so that the D_{in} jitter is nullified. $R(t)$ extracts $\triangle\phi_R(t)$ to be removed from D_{in}, and $A(t)$ extracts $\triangle\phi_A(t)$ to be added to signal D_{in} to transform it into D_{ref}.

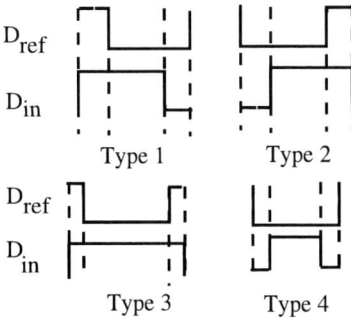

Figure 4. Four Types of Jitter Conditions

The jitter is positive when it advances a signal transition and negative when it delays a transition with respect to the reference signal. In Figure 4, Type 1 is where the jitter at the rising and falling transitions is positive, for Type 2 both are negative, for Type 3 the jitter at the rising transition is positive and at the falling one it is negative and, finally, for Type 4 it is negative and positive.

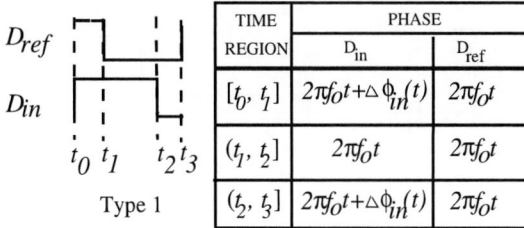

TIME	PHASE	
REGION	D_{in}	D_{ref}
$[t_0,\ t_1]$	$2\pi f_o t + \triangle\phi_{in}(t)$	$2\pi f_o t$
$(t_1,\ t_2]$	$2\pi f_o t$	$2\pi f_o t$
$(t_2,\ t_3]$	$2\pi f_o t + \triangle\phi_{in}(t)$	$2\pi f_o t$

Figure 5. D_{in} **and** D_{ref} **Phase in Three Time Regions**

Consider the case where D_{in} leads D_{ref} (Type 1). Figure 5 shows the phase of D_{in} and D_{ref} in three time regions. The general equation for jitter reduction is:

$$\overline{D_{out}} = \cos\left(((2\pi f_o t + \triangle\phi_{in} t)\ \pm\triangle\phi_R t)\ \pm\triangle\phi_A t\right) \tag{3}$$

In the time interval $[t_o, t_1]$, a negative jitter has to be removed since D_{in} has positive jitter. Therefore, $\triangle\phi_R t$ is $-\triangle\phi_{in} t$ and $\triangle\phi_A t$ is 0, since only removal is required.

$$\begin{aligned}\overline{D_{out}} &= \cos\left(((2\pi f_o t + \triangle\phi_{in} t) - \triangle\phi_{in} t) \pm 0\right) \\ &= \cos(2\pi f_o t)\end{aligned} \tag{4}$$

In the time interval $(t_1, t_2]$, $\triangle\phi_R t$ and $\triangle\phi_A t$ are zero, since there is no jitter in D_{in}.

$$\begin{aligned}\overline{D_{out}} &= \cos\left(((2\pi f_o t + 0) \pm 0) \pm 0\right) \\ &= \cos(2\pi f_o t)\end{aligned} \tag{5}$$

Finally in the time interval $(t_2, t_3]$, $\triangle\phi_A t$ is $-\triangle\phi_{in} t$ and $\triangle\phi_R t$ is 0, since only addition is required in this region.

$$\begin{aligned}\overline{D_{out}} &= \cos\left(((2\pi f_o t + \triangle\phi_{in} t) \pm 0) - \triangle\phi_{in} t\right) \\ &= \cos(2\pi f_o t)\end{aligned} \tag{6}$$

So in all the three regions the output signal is $2\pi f_o t$, which has the same phase as the jitter-free reference signal. Here, the reference signal used is the loop back signal, which will contain some jitter, therefore the jitter extracted by $R(t)$ and $D(t)$ will not be exactly the same as the jitter present in D_{in}. This implies that the signal at the output of the jitter reducer will still contain some jitter, but considerably less than the jitter in the input signal.

3.1.2 Architecture of Jitter Reducer

Figure 6 shows the jitter reducer. The part within dotted lines behaves as an inverter. It detects jitter and reduces it with reference to D_{ref}. Here τ is the desired bit period for the clock signal. The buffer G is used to delay the signal

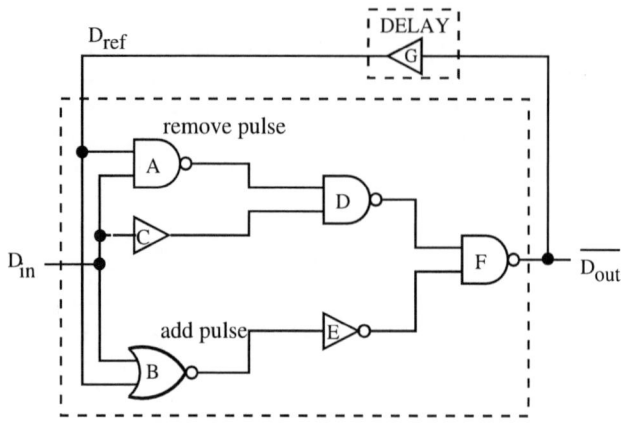

Figure 6. Jitter Reducer

$\overline{D_{out}}$ by $(\tau - \tau_{pd})$ *seconds*, before it reaches the input of the jitter reducer. The period τ_{pd} is the propagation delay of the circuit from D_{in} to $\overline{D_{out}}$. The inputs to this circuit are the signals D_{in} and D_{ref}. Signal D_{in} is the input jittered signal and D_{ref} is the loop back signal of the circuit. The BUFFER gate C delays the signal D_{in} by τ_A seconds,

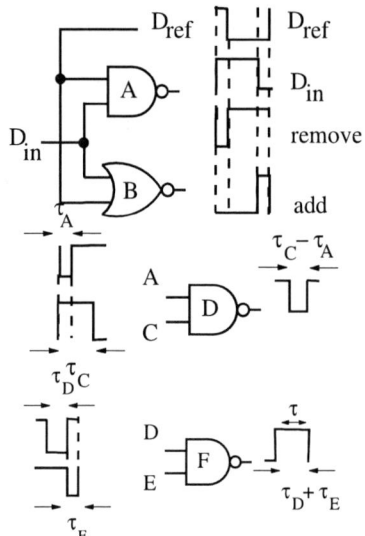

Figure 7. Output of Gates A, B, D and F

where τ_A is the propagation delay of gate A. The output of D (*add process*) is a pulse of period $\tau_C - \tau_A$ seconds and F (*remove process*) is a pulse of period $\tau_D + \tau_E$ seconds, which is equal to τ seconds (see Figure 7).

3.1.3 Optimized Jitter Reducer

In the original circuit, propagation delays of different paths must be identical for synchronizing various signals, which is impossible for both rising and falling transitions. This problem is overcome by optimizing the jitter reducer into one composite gate (transistors $5 - 14$) in Figure 8. Transistors $5 - 7$ of the *p*-tree and and transistors $10 - 12$ of the *n*-tree perform the *remove* operation. Transistors 8 and 9 of the *p*-tree and transistors 13 and 14 of the *n*-tree do the *add*

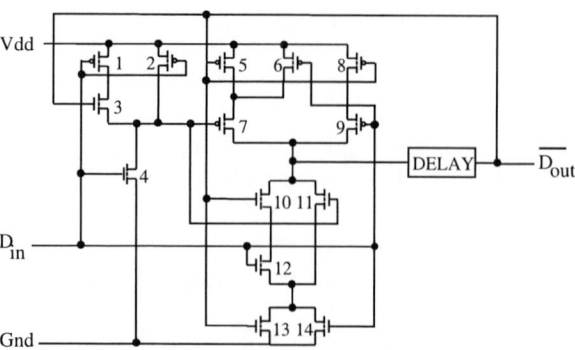

Figure 8. Optimized Jitter Reducer

operation. The delay is increased or decreased (depending on the signal transition direction) by changing the length L of each p and n transistor of each buffer. The delay is made of cascaded buffers to get the desired delay value. The main advantage of this circuit is that its different paths need not have identical path delays, avoiding a major circuit design problem.

Controlling the Jitter Reducer with an Inverter: We controlled the delay of D_{in} to reduce jitter. Transistors $1 - 4$ are a modified inverter that propagates the signal $\overline{D_{in}}$ with two different delays. The *p*-tree of the inverter provides two different delay paths. The *p*-transistor 2 is designed to have a longer delay than transistor 1. The two different paths are chosen by the *n*-transistor 3, which is controlled by D_{ref}. When *n*-transistor 3 is ON, the parallel combination of transistors 1 and 2 gives the shorter delay and when 3 is OFF, *p*-transistor 2 gives the longer delay.

3.1.4 Results for Tx Jitter Reducer

The Tx jitter reducer circuit is designed in the $70\,nm$ Berkeley Predictive process using CADENCE^{TM} and simulated using SPECTRE^{TM}. The circuit corrects a clock signal of frequency $1\,GHz$. The propagation delay τ_{pd} of the jitter reducer is $118.3\,ps$ for rising and $184.4\,ps$ for falling transitions. The delay element is designed using two buffers and their transistor lengths were chosen, such that the propagation delay (delay element) is $381.3\,ps$ for rising and $344.7\,ps$ for falling transitions. Now, the signal $\overline{D_{out}}$ will have a pulse width of $499.6\,ps$ for logic LOW and $529.1\,ps$ for logic HIGH, if the signal D_{in} is jitter free.

Table 1. Tx Jitter Reducer Experiment

Case #	Jitter Type	PJ ($ps\,RMS$)	RJ ($ps\,RMS$)
1	PJ ($60\,MHz$)	varies	-
2	PJ ($600\,MHz$)	varies	-
3	RJ	-	varies
4	PJ ($60\,MHz$), RJ	18.05	varies
5	PJ ($60\,MHz$), RJ	44.37	varies
6	PJ ($60\,MHz$), RJ	varies	10.47
7	PJ ($60\,MHz$), RJ	varies	18.23

584

978-1-4244-3039-0/08 $25.00 © 2008 IEEE

Testing the Tx Jitter Reducer with Input Jitter: Table 1 shows the types of input jitter for testing the Tx jitter reducer. Column 2 shows the jitter type, and columns 3 and 4 show whether periodic or random is varied or a fixed value is chosen. The input signal jitter is varied by changing the jitter amplitude. MATLAB is used to generate the input signals with various jitters and also to compute the RMS jitter in the output signal. The jitter reducer is simulated with these signals using SPECTRE. VERILOG modules are used to capture the time instants at which the signals transition for MATLAB to compute the jitter in the signals.

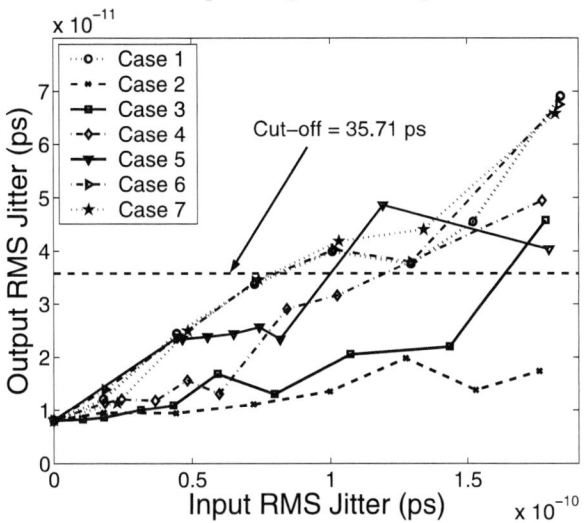

Figure 9. Testing the Tx Jitter Reducer

Analysis: For all cases the output RMS jitter is reduced compared to the input RMS jitter. An output RMS jitter of $35.71 ps$ is chosen as the cut-off, as it gives a BER of 10^{-12}. From Figure 9, we see that this cut-off is crossed only for input RMS jitter values greater than $35.71 ps$, the least being around $75 ps$ for Case 1. The transmit side jitter is reduced, on average, by 62.24%.

Table 2. Comparison of Peak-to-Peak Jitter Reduction

Method	Peak-to-Peak Jitter		%
	Before (ps)	After (ps)	Reduction
Adaptive 500 MHz PLL of Xia *et al.* [11]	29.2	17.5	40.06
New Jitter Reduction Technique at 1 GHz	21.5	16.5	23.26
	29.5	**17.6**	**40.34**
	37.2	21.7	41.67
	51.8	28.6	44.79
	78.5	32.0	59.24
	102.4	37.1	63.77

Comparison with Xia *et al.* [11]: The performance of the Tx jitter reducer is compared with the adaptive PLL technique proposed by Xia *et al.*. They have designed a PLL to generate a clock of frequency $500 MHz$ using IBM $180\ nm$ CMOS technology. Table 2 shows the results of random peak-to-peak jitter reduction by both the methods. Columns 2 and 3 give the peak-to-peak jitter before and after the jitter reducer is used. For the comparison, the result from the adaptive PLL technique is compared with the result shown in boldface for the new technique. We see that the results are comparable, but the difference is that the new technique is applied on a PLL that generates a clock of frequency $1\ GHz$ and the technique uses only 12 transistors as opposed to having logic with two counters and a comparator. The average peak-to-peak jitter reduction is 45.51%.

3.2 Jitter Reduction for Rx CDR Circuit

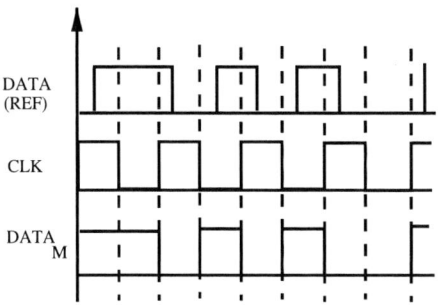

Figure 10. Waveforms for Rx Jitter Reduction

At the receive side of the SERDES, the timing jitter between the recovered clock and data from the CDR circuit results in erroneous data being latched by the flip-flop (a bit error). The solution is to reduce this jitter between the recovered clock and the data. Figure 10 shows the modified data signal $DATA_M$ that is aligned with the clock signal after the process of jitter reduction.

3.2.1 Rx Jitter Reducer Circuit

Figure 11 shows the receive side jitter reducer. The Rx jitter reduction process is similar to the Tx side as explained in Section 3.1.1, except that, instead of using a time delayed signal as a reference, the data signal from the CDR circuit is used. Since we now have an explicit reference signal, we use a different circuit for the jitter reducer because in the Tx jitter reducer, the feedback from $\overline{D_{out}}$ to transistor 3 was sufficient to control the delay when $\overline{D_{out}}$ leads D_{in}. However, on the Rx jitter reducer, there is no feedback, because there is a separate reference clock, and it is now necessary to add a subcircuit comprising transistors 24 and 30 to determine which delay to insert. Transistors 33 and 34 adjust the delay when D_{ref} leads D_{in}, while transistors 22 and 23 adjust the delay when D_{ref} lags D_{in}.

3.2.2 Controlling the Jitter Reducer

Apart from the inverter (transistors $15 - 18$) with two delay paths, the circuit performing the *add* operation is modified to have two delay paths, one with a slower *add* operation

585

978-1-4244-3039-0/08 $25.00 © 2008 IEEE

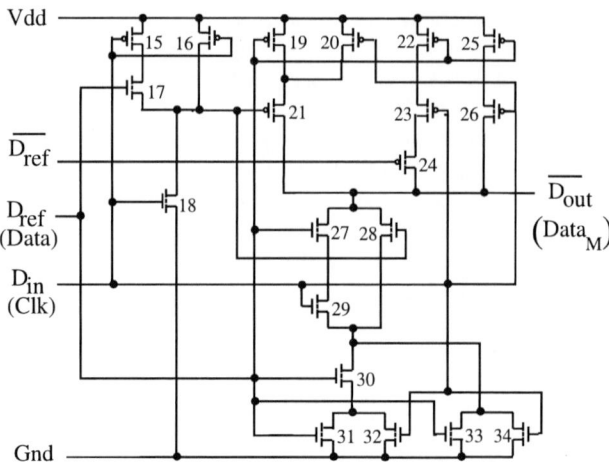

Figure 11. Receive Side Jitter Reducer

and the other faster, to give even more control in tuning the jitter reducer. The p-transistors 22 and 23 give shorter delay and transistors 25 and 26 (increased transistor lengths) give longer delay. The *add* operation delay is controlled by p-transistor 24, which is controlled by $\overline{D_{ref}}$. When transistor 24 is ON, the parallel combination of transistors 22, 23 and 25, 26 gives the shorter delay and when 24 is OFF, 25 and 26 in series give the longer delay. In the n-tree, transistors 31, 32 (shorter) and 33, 34 (longer) controlled by D_{ref} perform the *add* operation with two different delays.

3.2.3 Results for Rx Jitter Reducer

The circuit was designed in a $70\ nm$ Berkeley Predictive process using CADENCE^{TM} tools and simulated using SPECTRE. Table 3 shows the various jitter types on the D_{in} and D_{ref} signals during testing. Column 1 gives the type of jitter present either in the D_{in} (clock) or the D_{ref} (data) signal and columns $2-4$ show whether the jitter amplitude is varied or a fixed value is used.

Table 3. Input Jitter Types for Rx Jitter Reducer

#	Jitter Type	RMS PJ (ps)	RMS RJ (ps)	RMS DCD (ps)
1	PJ ($60\ MHz$) - D_{in}	varies	-	-
2	PJ ($60\ MHz$) - D_{ref}	varies	-	-
3	RJ - D_{in}, D_{ref}	-	varies	
4	PJ ($600\ MHz$) - D_{in} ($300\ MHz$)- D_{ref}	varies D_{in}	-	-
5	PJ ($600\ MHz$) - D_{in}, ($300\ MHz$) - D_{ref}	varies D_{in} & D_{ref}	-	-
6	DCD - D_{in}	-	-	varies
7	DCD - D_{ref}	-	-	varies
8	PJ ($60\ MHz$), RJ, DCD - D_{in} PJ ($60\ MHz$), RJ -D_{ref}	43.77 -D_{in} 42.44 -D_{ref}	5.15 -D_{in} 9.27 -D_{ref}	varies
9	PJ ($60\ MHz$), RJ - D_{in} PJ ($60\ MHz$), RJ, DCD -D_{ref}	43.77 -D_{in} 42.44 -D_{ref}	5.15 -D_{in} 9.27 -D_{ref}	varies

Analysis: Figure 12 shows the plot of input and output RMS jitter. The jitter between the clock and data signal is considerably reduced. The best performance is obtained for Case 7, where only DCD jitter is present in the data signal. When compared with the cut-off of $35.71\ ps$, in most cases, it is crossed only for input RMS jitter over $75\ ps$ except in Cases 3 and 8, where it is crossed around $40\ ps$. The Rx jitter is reduced, on average, by 35.88%.

3.3 Jitter Performance of the SERDES with the Tx and Rx Jitter Reducers

The SERDES described in Section 3 is integrated with the Tx and Rx jitter reducer blocks. The jitter performance of the SERDES is analyzed for three cases. The SERDES is tested for a PJ of $60\ MHz$ at the Tx side PLL. Figure 13 shows the plot of the RMS jitter between the recovered clock and the data signal and the input RMS periodic jitter at the transmit side PLL. In Case 1, where there are no jitter reducers, the jitter at the receive side is as high as the input jitter. For Case 2, where only the Rx jitter reducer is used, the jitter is considerably reduced, with the receive side jitter crossing the cut-off for input RMS jitter values greater than $50\ ps$. Finally, for Case 3, where both Tx and Rx jitter reducers are present, the receive jitter is drastically reduced. This is because the input periodic jitter is reduced by the Tx jitter reducer and hence its effect at the receive side is less pronounced.

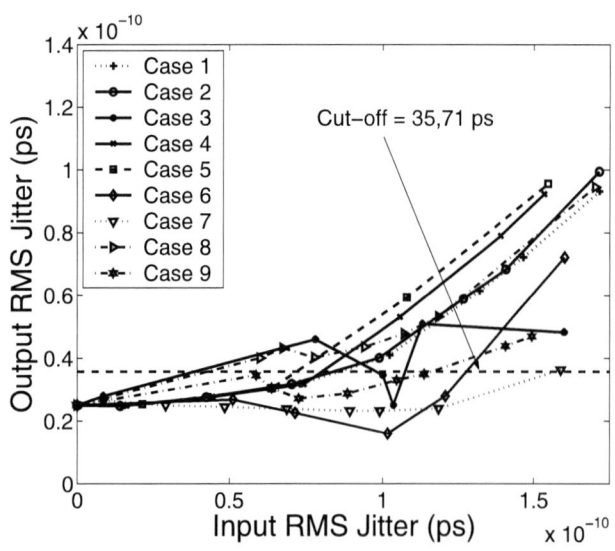

Figure 12. Testing the Rx Jitter Reducer

BER Analysis: The effect of the receive side jitter on the BER is determined probabilistically using the equations provided in the work of Hong *et al.* [4]. For the input RMS jitter of $71.77\,ps$, from Figure 13 we see that for Case 1, the receive side jitter is $144.44\,ps$, which gives a BER of 8.3×10^{-2}, for Case 2 it is $56.84\,ps$ with a BER of 1.09×10^{-5} and, finally, for Case 3, with both Tx and Rx jitter reducers, it is $27.59\,ps$ with a BER of 6.44×10^{-20}.

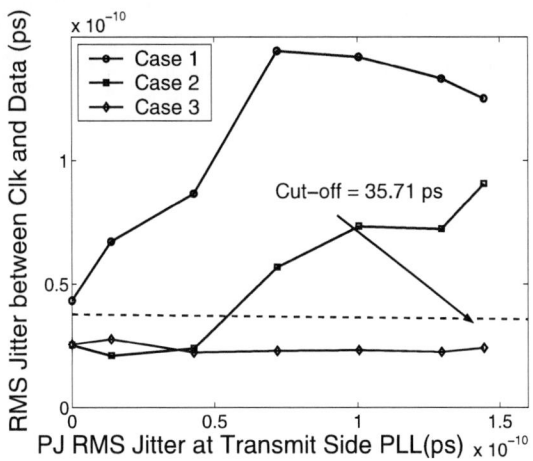

Figure 13. SERDES with the Tx and Rx Jitter Reducers

4 Circuit Design and Process Variations

The Tx and Rx jitter reducers are analyzed for *process variations* (PV) using Monte Carlo analysis with SPECTRE^{TM}. We vary these nMOS and pMOS transistor parameters of the Berkeley Predictive process: t_{ox} (gate oxide thickness), c_j (junction thickness), c_{jsw} (junction side wall capacitance), μ_0 (low-field surface mobility), v_{tho} (transistor threshold with 0 body bias) and p_{clm} (channel length modulation). All parameters have Gaussian distributions with their nominal values as mean and $\sigma = 20\%$ for c_j, c_{jsw} and p_{clm} and 10% for t_{ox}, v_{tho} and μ_0.

Optimal Transistor Width. For this experiment the Tx jitter reducer is tested with an input signal with no jitter in it. The transistor widths are varied and the output RMS jitter under process variations is determined.

Table 4. Optimal Transistor Width

W (nm)	Output RMS RJ (ps)	W (nm)	Output RMS RJ (ps)
100	39.08	500	37.13
160	39.14	700	46.09
200	38.63	1000	70.81
300	37.96		

Table 4 shows the output RMS random jitter versus the transistor width W. The RMS jitter decreases initially, but for widths over $700\,nm$ it increases. $W = 200\,nm$ is chosen, because the output RMS jitter is closer to the cut-off and it does not increase the jitter reducer chip area.

Tx and Rx Jitter Reducers under Process Variations. The Tx and Rx jitter reducers are tested for jitter under process variations. For the Tx jitter reducer, to simplify jitter measurement, instead of the loop back signal, an external reference signal is provided. It is tested with a periodic jitter of $60\,MHz$ on the input signal D_{in}. In the case of the Rx jitter reducer, it is tested with a periodic jitter of $60MHz$ on the D_{in} (clock) signal.

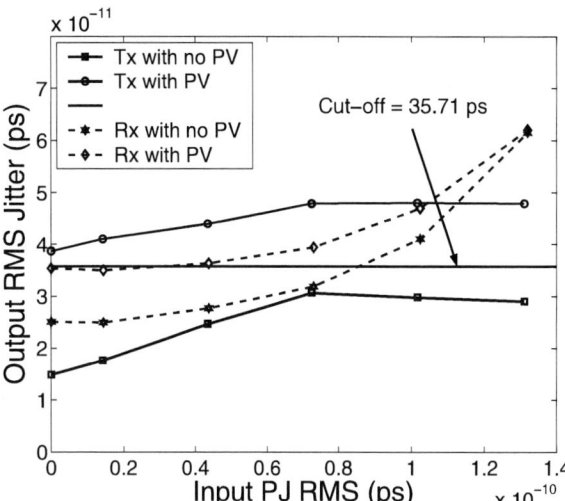

Figure 14. Rx and Tx Jitter Reducers under Process Variations

Figure 14 shows the performance of the jitter reducers. The solid lines and broken lines are for the Tx and Rx jitter reducers respectively. For the Tx jitter reducer the output RMS jitter under process variations is higher compared to the output RMS jitter without process variations, which can be improved with good circuit design and for the Rx jitter reducer, the output RMS jitter is initially higher but approaches the output RMS jitter without process variations for higher input RMS jitter values. But for both the Tx and Rx jitter reducers, even under process variations the output RMS jitter is considerably lower than the corresponding input RMS jitter. For the Tx jitter reducer an input RMS jitter of $72.59\,ps$ is reduced to $41.73\,ps$ (by 42.41%) and for the Rx jitter reducer an input RMS jitter of $73.14\,ps$ is reduced to $39.38\,ps$ (by 46.16%).

Layout and Parasitics. The Tx jitter reducer circuit was designed using the Layout Editor of CADENCE^{TM} and parasitic capacitances were extracted. The delay element was adjusted to account for the extra delay in the jitter reducer due to the parasitics. The extracted circuit was tested with a periodic jitter of $60\,MHz$ in Figure 15. The output RMS jitter is almost the same as for the jitter reducer with no parasitics, primarily due to the variable delay element.

978-1-4244-3039-0/08 $25.00 © 2008 IEEE

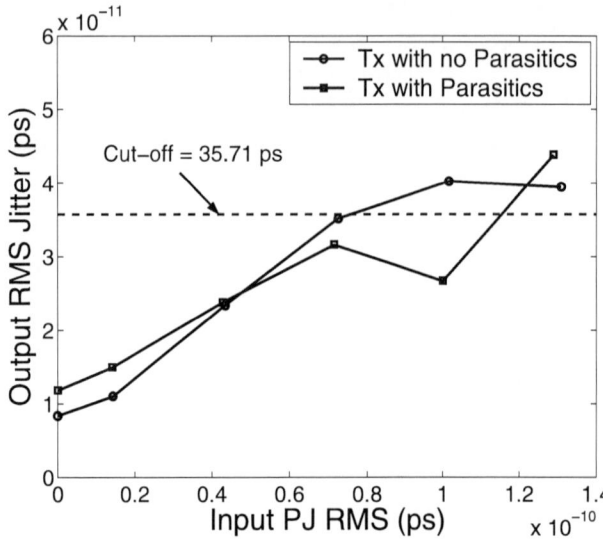

Figure 15. Tx Jitter Reducer with Parasitic Capacitances

Therefore, the delay constraint that the propagation delay of the delay element and the jitter reducer should be equal to τ (the bit period of logic HIGH and LOW) is not affected.

Pole-Zero Analysis. We analyzed the poles and zeroes of the Tx PLL, both with and without the jitter reducer, using SPECTRE^{TM}. The only change was to move a zero at $9.10407\ GHz$ to $9.10240\ GHz$, which is inconsequential for a $1\ GHz$ operating frequency. The Rx PLL has the same poles and zeroes as the Tx PLL, but the jitter reducer created three new poles and three new zeroes in the range of 14 to 52 GHz, which also is inconsequential.

Phase Noise Analysis. We analyzed the *phase noise* (PN) of the jitter reducer, which is mainly transistor thermal and flicker noise. From $1\ KHz$ to $1\ MHz$ the PN of the PLL is at $-103\ dB$, and reduces to $-126\ dB$ at $100\ MHz$. With the Tx jitter reducer, the PN drops to $-124\ dB$ from $1\ KHz$ to $1\ MHz$, and drops to $-132\ dB$ at $100\ MHz$. For the Rx side jitter reducer, the PN further drops to -146.1 to $-148.2\ dB$ from $1\ KHz$ to $100\ MHz$, so the jitter reducers significantly decrease PN.

5   Conclusion

The new jitter reduction technique effectively reduces the jitter at the transmit and receive side and improves the BER of the SERDES. The transmit and the receive side jitter is reduced, on average, by 42.41% and 46.16%, respectively, under process variations. The work demonstrates that an external hardware can be used to reduce jitter without changing the PLL circuitry. This method works for all types of PLLs, since it does not modify the PLL analog circuits. The method uses the autocorrelation of the jittered

signal in two adjacent clock periods to reduce jitter. This idea works, except when the nature of the jitter varies dramatically from clock period to clock period, which is most unlikely. This method allows for improved jitter reducer hardware on any type of PLL, so it is applicable to clock trees in microprocessors, and perhaps to communications circuits, although further research must determine the microwave frequency limits of the jitter reducer. Furthermore, this hardware can be modified for BIST for jitter testing.

Acknowledgment. We thank Tapan J. Chakraborty of Alcatel-Lucent for his valuable suggestions.

References

[1] A. Athavale and C. Christensen. *High-Speed Serial I/O Made Simple: A Designer's Guide with FPGA Applications.* Xilinx Connectivity Solutions, San Jose, CA, 2005.

[2] A. Dou and J. A. Abraham. On-Chip Jitter Measurement Using a Dual-Channel Undersampling Time Digitizer. In *Proc. of the Int'l. Mixed Signal Test Workshop*, pages 206–211, June 2006.

[3] Genesys Logic America, Inc. Multi-Gigabit SerDes: The Cornerstone of High Speed Serial Interconnects. Multi-Gigabit SerDes White Paper, Sept. 2003.

[4] D. Hong, C. K. Ong, and K. T. Cheng. BER Estimation for Serial Links Based on Jitter Spectrum and Clock and Recovery Characteristics. In *Proc. of the Int'l. Test Conf.*, pages 1138–1147, Oct. 2004.

[5] C. Hur, Y. Choi, H. Choi, and T. Kwon. A Low Jitter Phase-Locked Loop Based on a New Adaptive Bandwidth Controller. In *Proc. of the IEEE Asia-Pacific Conf. on Circuits and Systems*, pages 421–424, Dec. 2004.

[6] Lattice Semiconductor Corporation. Serdes Jitter. Technical Note TN1084, Hillsboro, OR, May 2006.

[7] M. Mansuri, A. Hadiashar, and C.-K. K. Yang. Methodology for On-Chip Adaptive Jitter Minimization in Phase-Locked Loops. *IEEE Trans. on Circuits and Systems*, 50(11):870–878, Nov. 2003.

[8] N. Ou, T. Farahmand, A. Kuo, S. Tabatabaei, and A. Ivanov. Jitter Models for the Design and Test of Gbps-Speed Serial Interconnects. *IEEE Design & Test of Computers*, 21(4):302–313, July 2004.

[9] A. Telba, J. M. Noras, M. Abou El Ela, and B. AlMashary. Jitter Minimzation in Digital Transmission using Dual Phase Locked Loops. In *Proc. of the Int'l. Conf. on Microelectronics*, pages 270–273, Dec. 2005.

[10] S. D. Vamvakos, C. Werner, and B. Nikolic. Phase-Locked Loop Architecture for Adaptive Jitter Optimization. In *Proc. of the Int'l. Symp. on Circuits and Systems*, pages 161–164, May 2004.

[11] T. Xia, S. Wyatt, and R. Ho. Automated Calibration of Phase Locked Loop with On-Chip Jitter Test. In *Proc. of the North Atlantic Test Workshop*, pages 140–147, May 2006.

[12] T. Yamaguchi, M. Soma, and M. Ishida. Extraction of Peak-to-Peak and RMS Sinusoidal Jitter Using an Analytic Signal Method. In *Proc. of the VLSI Test Symp.*, pages 394–402, May 2000.

978-1-4244-3039-0/08 $25.00 © 2008 IEEE

21st International Conference on VLSI Design

GyroCompiler: A Soft IP Model Synthesis and Analysis Framework for Design of MEMS based Gyroscopes

Jairam S
SDTC TI Bangalore India
(sjairam@ti.com)

Navakanta Bhat
ECE IISc Bangalore India
(navakant@ece.iisc.ernet.in)

Abstract

A model to create a simulation and a synthesis framework for design of Gyroscopes is proposed. The main motivation is to have a framwork for developing Gyroscope models in the form of soft intellectual properties (IPs) for their subsequent integration into mainstream VLSI systems. Synthesis targeting different performance classes of gyros is based on a simple table look-up. The next level of model refinement involving optimization of the different physical aspects of the gyro such as its shape is based on statistical design of experiments (DoE). Both FEM and Simulink based models have been used to build a custom DoE framework to estimate the parameters related to a desired gyro structure. A simple gyroscope structure is modeled and analysed with both FEM and Simulink based models. It is shown that DoE based framework can capture the parameters of a gyroscope structure, accurately and that it can be easily integrated with system level synthesis tools.

1 Introduction

MEMS based sensors like accelarometers, gyroscopes, gas sensors and varactors have become an integral part in the design of embedded systems for critical applications, spanning autmotative, aerospace, military avionics and related industrial sectors. With such penatration, efficient and automated design is becoming extremely critical [1]. This is a also very challenging task given the fact that design of MEMS structures are in general computationally intensive [2]. Given the safety critical nature of most embedded systems, achieving reduced error and tolerance margings in these designs makes this even harder. This also throws up the challenge of generating accurate and unambigous specifications by the system designer in the context of performance requirements. As an example consider MEMS based gyroscope for measuring angular velocity, which is integrated into an adaptive cruise controller modelled as a hybrid system. Gyroscope requirements given in terms of

bandwidth, sensitivity etc for various applications have significantly increased with each generation. Different classes of gyroscopes based on different applications and different performance requirements have been reported in literature. Most of these publications focus on specific gyroscope topology alongwith their design and optimization None of these publications report design procudures from a holistic system standpoint involving comprehensive model generation, analysis and integration in a single synthesis and design framework. Considering the availability of a family of large body of research publications targeting individual gyroscope structures and topologies, this can be easily leveraged to result in the above design framework for system integration and synthesis. It is also imperative to validate the correctness of the entire system with respect to the designed structure, keeping in mind the criticality of the application. We do not focus on this aspect in this paper.

The above mentioned model generations methods have been successfully deployed in the analog, digital and mixed-signal VLSI system domain. In the desired framework an ideal objective would mean: Firstly an automated generation of gyroscope structures based on the performance requirements of the target application in an embedded system. Secondly its inegration to the system and lastly the automated validation of the entire system vis-a-vis the generated MEMS structure. In this paper we focus of the first item. This involves automatically selecting the best possible topology followed by analysis of MEMS gyroscope structure, from the perspective of manufacturability based on its layout and area constraints imposed by processsing technology. Alternatively the designer can also provide area constraints as inputs to the synthesis framework so as to result in the best possible gyro structure. A very accurate model based on FEM based methods, may not always be required for first order system implementation. Under such circumstances a scalable first order model is needed which helps qualify the complete system design analysis, with acceptable error tolerances. Statistical methods based on DoE help provide such solutions. In a framwork based on the DoE approach one can easily take into account error tol-

978-1-4244-3039-0/08 $25.00 © 2008 IEEE

erances and bounded inaccuracies as these tend to average out, over a sample space. This renders the framework very easy to use. Statistical methods help model variations in the system design parameter values and also the fabricaion process parameter values. This enables its easy integration into a circuit simulation methodology.

In this paper we propose a GyroCompiler framework to address the above issues related to synthesis of MEMS based gyroscopes. The proposed approach is a combination of table look-up based topology selection built around a custom DOE methodology for various parametric optimizations of the gyroscope. The paper is divided into four sections. The Section 2 explains the construction and analytical modeling of gyroscopes. Section 3 explains the anatomy of the proposed GyroCompiler and some aspects related to its construction. Design of Experiments based modeling methodology is explained in Section 4. In section 5 we demonstrate the powerful advantages offered by our approach. In the final section we conclude with scope for further work.

2 Review of Gyro Functionality

2.1 Structure and Working Principle

Gyroscope is a mechanical construction used to measure the angular motion by sensing the Coriolis force on the structure driven to actuation. Figure 1 illustrates a simple gyroscope. The proof-mass is driven in the X direction. If the plane containing the structure rotates at an angular velocity $\Omega(t)$, the proof mass experiences a Coriolis force in the Y direction. If $\vec{v}$ is velocity in X direction, the Coriolis force is given by $2(\vec{\Omega} \times \vec{v})$. The structure then moves in a direction perpendicular to angular motion and the driven direction. This motion can be detected and measured by various means. Analysis techniques MEMS based gyroscope can be found in [3] [4]. Figure 2 illustrates a gyroscope diagram in one direction. The proof-mass is held along the drive and sense directions by a combination of stiff elements.

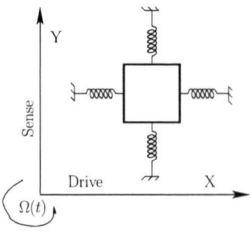

Figure.1 Gyroscope Functioning

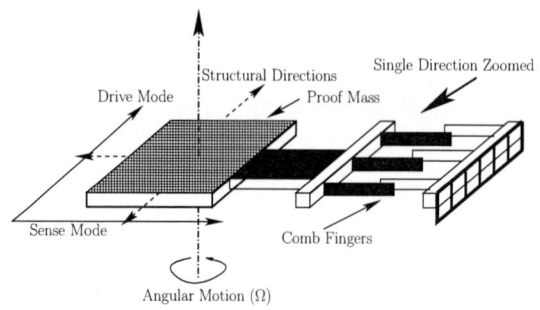

Figure. 2 A Gyroscope Structure

2.2 Modeling and Analysis

An analytical sketch of the gyroscope functionality is presented in this section. Dynamics of gyroscope can be modeled as a set of two second order coupled differential equations as shown below:

$$\ddot{x} + \omega_n^2 x - 2\Omega\dot{y} = 0 \quad (1)$$
$$\ddot{y} + \omega_n^2 y + 2\Omega\dot{x} = 0 \quad (2)$$

Equation (1) and (2)can be casted as a compact matrix representation of a system of linear second order differential equation of the following form:

$$\mathbf{A}\ddot{\mathbf{X}} + \mathbf{B}\dot{\mathbf{X}} + \mathbf{C}\mathbf{X} = \mathbf{0} \quad (3)$$

$\mathbf{X}$ is the vector (x, y), $\mathbf{A}$ and $\mathbf{B}$ being the respective matrices for the system describing the governing equations. The indvidual solutions for $x(t)$ and $y(t)$ can then be written as:

$$x(t) = A\,exp\{j\omega_n\lambda_1 t\}$$
$$y(t) = B\,exp\{j\omega_n\lambda_2 t\} \quad (4)$$

λ_1 and λ_2 are the eigenvalues of the system. For a first modeling approximation we decouple the differential equations as shown below:

$$M_d\ddot{x}_d + B_d\dot{x}_d + K_d x_d = F_{act} \quad (5)$$
$$M_s\ddot{y}_s + B_s\dot{y}_s + K_s y_s = 2\Omega\dot{x}_d \quad (6)$$

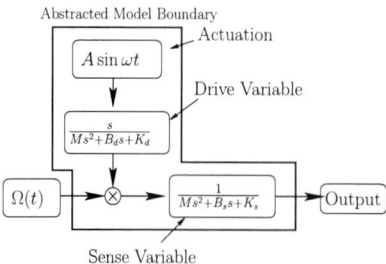

Figure. 3 Mathematical Model of a Gyroscope Structure

Figure 3 illustrates a suggested implementation of (6) and (7) [4]. The output is modeled as a capacitance, which

590

978-1-4244-3039-0/08 $25.00 © 2008 IEEE

is a function of input motion. The model transfer function $H(s)$ can be computed from figure 3 as shown below:

$$h_1(s) = \mathcal{K} \frac{A\omega^2}{s^2 + \omega^2} \frac{s}{(Ms^2 + B_d s + K_d)} \qquad (7)$$

$$h_2(s) = \frac{1}{(Ms^2 + B_s s + K_s)} \qquad (8)$$

where $\mathcal{K}$ is a constant, M is the proof-mass, $B_{s|d}$ and $K_{s|d}$ are the respective damping and stiffness co-efficients for drive and sense modes respectively. The output response for an input angular rate in time and frequency domain can be written down as:

$$O(s) = [\Omega(s) \star h_1(s)]h_2(s) \qquad (9)$$

$$O(t) = [\Omega(t) h_1(t)] \star h_2(t) \qquad (10)$$

2.3   Topological Requirements

The topological specifications needed for the design of a gyroscope translates into selection and optimization of multiple parameters. Figure 4 illustrates some simple topologies of gyroscopes.

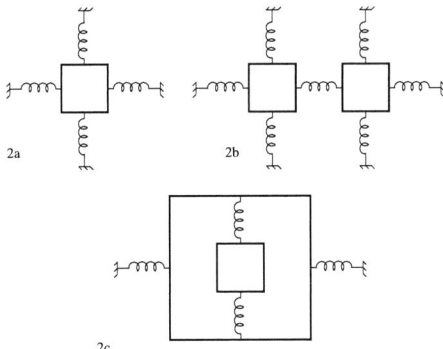

Figure. 4 A Sample set of Gyro Topologies [3]

The first level analysis requires the following properties to be considered:

- **Modal frequency computation:** This enables the designer to plan for the drive mode oscillation frequency.

- **Drive Mode Oscillation:** Given the physical parameters of the structure the designer should be able to explore if the structure would provide the required drive mode oscillation. Performing an FEM simulation for every explore cycle is expensive in computation.

- **Capacitance Computation:** Based on the physical phenomenon of the structure the capacitance variation can be computed. This variation either would be sensed by an electronic circuit or can be used for co-simulation of the system.

- **Gyro Sensitivity:** Lastly the sensitivity of the system to noise and bandwidth requirements also needs to be analysed based on the structural topology.

The physical dimension of the structure include mass, area of the cross-section of all the arms, drive mode rotation rate and derate factors. Based on these values, sense mode oscillation, maximum displacement and capacitance computation can be performed.

Once the correctness of the models have been established as explained earlier, a simple compiler based approach is adopted for synthesizing the gyro structure based on these models to enable system design at a much higher level of abstraction. A summary of some of the requirements tagtetted for such a compiler is given below:

- Correct prediction of topology for a given performance requirements.

- Enabling structural optimization vis-a-vis the chosen system based on performance.

- Allowing inclusion of area and process fabrication constraints for topological and structural optimisation.

It should be noted that mapping of a topology based on requirements can be partly automated and partly precompiled. For example vibratory type gyroscopes are bound by a performance class and hence can be used only to certain performance requirements. Similar can be the case for torsional type gyroscopes. Such requirements/classifications can also be directly pre-mapped.

3   Gyro Compiler Architecture

We now explain the architecture of the GyroCompiler. An analysis capability which is in-built in the backend of model generation, can be deployed for model synthesis also. It is on this premise that we propose the GyroCompiler. Given performance requirement, area constraints, material properties and performance class, a choice of gyroscope topology can be automatically made to meet the given specification. Figure 5 explains the architecture map for the proposed approach.

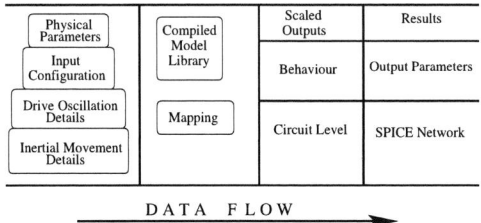

Figure. 5 GyroCompiler Architecture

Physical parameters, dynamic details and configuration are provided as inputs which are obatined based on system requirements. Configuration is an optional input, which is currently being worked out. In the first mode the gyroscope is modeled based on the physical parameters provided. Capacitance values and primary input oscillation (drive mode) related values can also be computed based on these details [4]. Most physical parameters would be replaced by performance related parametric inputs. These include available bandwidth utilisation, noise floor, maximum capacitance to be sensed and available Silicon floor area. In the current Simulink model, dynamic performance parameters are included.

Based on these requirements gyroscopes are mapped into various performance classes. Each class has its own topological selection criteria. Having chosen a class, one directly focuses on chossing the appropriate sub-topology that confirms to the performance specification. It is easy to note that the first level topological selection does not involve computationally intensive simulation. Rigorous FEM simulation may only be needed during validation of layout on Silicon. The synthesis architecture for GyroCompiler is shown in the Figure 6. The compiler is built on simple PERL based constructs. Further work includes expanding these inputs to a GUI based system. The selection optimization construct calls up a DoE based framework to arrive at a first order model for the gyroscope. We explain this methodology in the next section.

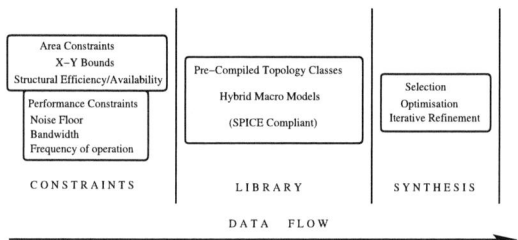

Figure. 6 GyroCompiler Synthesis Environment

4   DoE Based Gyro Model Generation

Traditional model generation methods for gyroscopes include running FEM simulations to capture the device modal frequency components which can then characterise the gyro behaviour at fairly abstracted level. Here we propose a method where this modal frequency behaviour can be obtained as a function of the primary input variables of the gyroscope. This is obtained through an approach based on *Design of Experiments* based methodology. Figure 7 illustrates the data flow diagram that we propose.

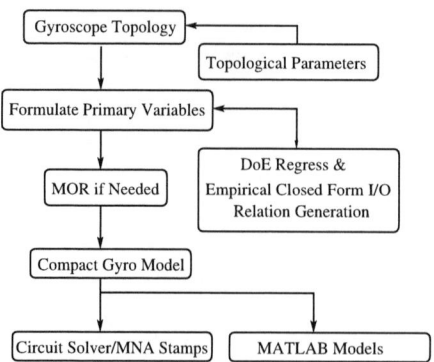

Figure. 7 Empirical Model Generation Flow

Primary input variables are selected based on groups. These groups can be subjective bases also. For example they can be a set of physical parameters, aspect ratio limiting factors and density variations. All the primary input variables are then mapped against targeted output variable set through a series of FEM simulations. However it should be noted that these FEM simulations are performed to characterise the gyro behaviour and are not actual FEM simulations for the model under consideration.

Design of experiments popularly called DOE [5], are statistical methods through which experiments can be planned and the resulting outputs can be analyzed in a methodical fashion. The maximal advantage of using such a methodology arises when there are multiple primary input variables and their relationship to output cannot be established in a quantitative manner. It is also important to establish, which of the given variables, is relatively more important. In this manner one can reduce the error bound variation of the output vis-a-vis a relative more important input variable. At the same time the region of operation in input variables also need to be ascertained. This also reduces the output error bound variation. The nature of experimentation in many cases can also be centered around dominant primary inputs. However at the same time the effects of the uncontrolled variables should also be minimised. Dependence on joint primary variables is also a possibility. Determining this interaction is also an objective of these experiments. We deploy this method to determine all the above properties and employ these relations for a first order model generation.

As an example, we consider the gyro structure shown in figure 1. Area of proof mass ($L * L$), Arm-Length variation and Arm-Width variation are the three parameters of the gyro considered for parametric variations. For a set of nominal values, modal frequency parameters of the structure are computed by FEM simulation in Coventorware [6]. As an ordered set of parametric variations about mean values, modal values are captured for a given range. These values are then fit through the response surface modeling method for a fully quadratic model [5]. Reference and cross-reference of multiple parameters are also computed. The output response y can be written for three input parameter in quadratic terms as :

$$
\begin{aligned}
y \ = \ & \beta_1 x_1 + \beta_2 x_2 + \beta_3 x_3 \\
& + \beta_{12} x_1 x_2 + \beta_{23} x_2 x_3 + \beta_{31} x_3 x_1 \\
& + \beta_{11} x_1^2 + \beta_{22} x_2^2 + \beta_{33} x_3^2 + \epsilon
\end{aligned}
$$

Here x_1, x_2 and x_3 are the three parameters mentioned before. The observations of factorial experiment can be casted as a simple model as opposed to the regression model mentioned above. This model can be written as:

$$
y_{ijk} \ = \ \mu + \tau_i + \beta_j + (\tau\beta)_{ij} + \epsilon_{ijk} \tag{11}
$$

Here i, j k can be represented as:

$$
\begin{cases}
i \ = \ 1,2,3\ldots,a \\
j \ = \ 1,2,3\ldots,b \\
k \ = \ 1,2,3\ldots,n
\end{cases}
$$

It should be noted that this model helps perform decision making for structural optimisation for maximising amplitude of the gyroscope. The amplitude still needs to be linked to the rotational aspects of the inputs to the gyroscope. These parameters can also be computed either based on simulations of analytical models or through direct FEM method. A list of these parameters are given below:

- Bandwidth, Noise Floor, ZRO, Scale Factor.

- Resolution, Drift.

- Angle random walk, Bias-drift.

- Maximum shock.

- Input referred thermo-mechanical noise.

We modeled the frequency response of the gyro by the analytical method described in section II based on Figure 2. A simulink model was developed and frequency response for various sets of M, $B_{d|s}$ and $K_{d|s}$ parameters. The frequency response was also casted as an output requirement to enable its function generation through a DOE model regression environment and co-efficients were obtained.

5 Results

The gyro shown in Figure 1 was simulated through a set of three input parameters. Fundamental modal frequency for the structure is shown in Table I as a variation of proof mass width, arm width and dielctric depth. The three input variables are coded -1, 0, and +1 for fitting a second order quadratic response surface model (RSM).

For a full factorial set of experiments response surface was generated. The optimised parameters for the equation shown above was computed for a full factorial model. The

model fit is shown in Figure 6. Since a 3-input, 1-output hyper surface cannot be represented, three separate graphs for the same relationship are shown. The percentage error bounds on model versus simulation for 20 random inputs are plotted in the Figure 7.

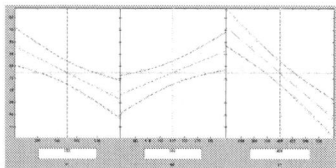

Figure 8 Response Surface: 3 Parameter Full Factorial Model

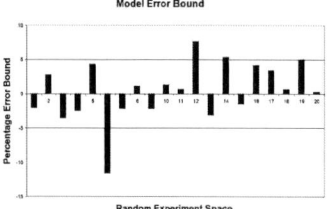

Figure. 9 Error Bounds for Random Sample Space

The Simulink model of the gyroscope is shown in Figure 10. A frequency sweep analysis of the model was performed for a given set of three second order system parameters. Another method tested was to check out the steady state response for a constant function for a given parameter.

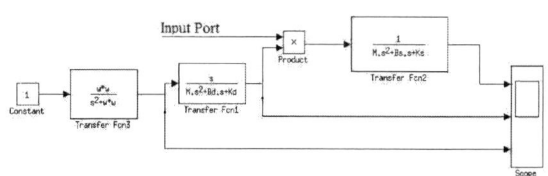

Figure. 10 Simulink Model of Gyroscope

The responses for different sets of inputs are shown in Figure 11. These Simulink snapshots clearly show that output response is an envelop riding over a Sine wave of the sense mode frequency. Frequency response curves for each these physical parameters were computed and these curves were sampled for frequency steps. The sampled responses were then fed through the same DOE framework mentioned in the previous section. In this case the primary input parameters included mass, damping and stiffness for drive and sense modes respectively. It can be seen that primary input variables are higher in number. Some of them again can be filtered based on the gyro topology or a DOE based methodology. For example, in plane and out of plane damping can be totally different. If the gyro topology is in-plane, a single primary input variable can be used as an input to the

DOE framework. Another conclusion can be drawn from this methodology can be drawn in the direction of system level parameter optimization. For example, if the model parameters obtained through Gyrocompiler are directly fed as inputs to system synthesis tools, an optimization loop can be formulated over this feature. A pertinent example is that of an adaptive cruise control problem. Based on requirements of traffic level parametric variations, parametric optimization of the gyro parameters can be achieved through this scheme.

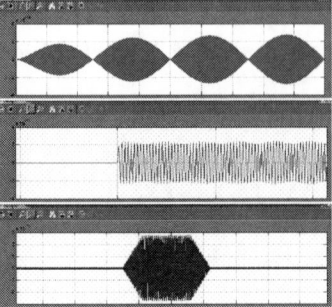

Figure. 11 Gyro Response to Sine, Step and Trapezoidal Inputs

We also developed an AWE (asymptotic waveform evaluation) [7] based model order reduction routine to collapse an nth order polynomial (either charatteristic equation of frequency response). The routine is function written in perl which integrates an AWE function with polynomial root finding solution in Octave™. Order reduction routine was tested with interconnect reduction benchmarks for accuracy of single point expansion. Reduced order fit agreed for step responses within 3% accuracy. This routine can be collapsed into the regular DoE generated first order model for nth order polynomials.

6 Conclusions

A solution for automatic soft IP generation approach *Gyrocompiler* is proposed. The solution produces based on a table look based solution performs topology selection and generates first order gyroscope models for these topologies. These scalability of these models depend their nature of integration (Matlab or circuit based). Further work includes part FEM integration requirements where these models can be exhaustively validated at the lower level of hierarchies.

References

[1] R. Alur and et. al. Hierarchical modeling and analysis of embedded systems. *Proc. IEEE*, 91(1):11–28, 2003.

[2] C. J. Tomlin, I. Mitchell, A. M. Bayen, and M. Oishi. Computational techniques for the verification of hybrid systems. *Proc. IEEE*, 91(7):,986–1001, 2003.

Table 1. Modal Frequency Beam Parameters

Proof Mass Width (μm)	Arm Width (μm)	Arm Length (μm)	Modal Frequency (kHz)
500	100	200	115.232
500	100	400	82.065
500	100	600	56.574
500	120	200	109.735
500	120	400	82.065
500	120	600	66.404
500	140	200	118.287
500	140	400	87.829
500	140	600	70.907
550	100	200	77.224
550	100	400	64.497
550	100	600	56.574
550	120	200	94.978
550	120	400	74.633
550	120	600	61.212
550	140	200	118.28
550	140	400	80.481
550	140	600	70.907
600	100	200	65.116
600	100	400	54.310
600	100	600	50.106
600	120	200	80.46
600	120	400	66.238
600	120	600	56.45
600	140	200	95.604
600	140	400	73.308
600	140	600	60.369

[3] H. Xie and G. K. Fedder. Integrated microelectromechanical gyroscopes. *Journal of Aerospace Engg.*, 6(2):65–75, 2002.

[4] S. Mohite, N. Patil, and R. Pratap. Design, modelling and simulation of vibratory micromachined gyroscopes. *Journal of Physics Conference Series*, 34:757–763, Apr. 2006.

[5] D. Montgomery. *Design and Analysis of Experiments*. John Wiley and Sons, 6st. edition, 2004.

[6] Coventorware. Coventor. Homepage. http://www.coventor.com/.

[7] P. Larry and R. Rohrer. Asymptotic waveform evaluation for timing analysis. *IEEE Trans. Computer-Aided Design*, 9(4):352–366, 1990.

21st International Conference on VLSI Design

Behavioral modeling of a CMOS compatible high precision MEMS based electron tunneling accelerometer

T.K. Bhattacharyya and Anandaroop Ghosh

Electronics and Electrical Communication Engineering Department, Indian Institute of Technology, Kharagpur

tkb@ece.iitkgp.ernet.in   anandaroopg@gmail.com

Abstract

The paper presents a comprehensive behavioral model of a high precision tunneling accelerometer. Design and optimization of the silicon based tunneling has also been reported in this work. The accelerometer is CMOS compatible and has actuation voltage within CMOS bias levels. The proposed structure uniquely combines the electron tunneling based sensing and capacitive actuation. A feedback controller is designed to measure the acceleration under constant gap mode of operation. The full dynamic range of operation is 1μg to 200μg with a resolution in the order of nano-g. The cross-axis sensitivity is less than 1% and the shock survivability is 10g for a 10 ms shock with 0.1 ms rise time. The Brownian noise floor of the system has also been studied and the squeeze film damping effects on the system has been shown.

1. Introduction

High precision accelerometers with wide dynamic range finds many applications such as satellite navigation and guidance system, geophones for oil exploration, seismology, under water acoustic measurements, etc. Micro-machined acceleration sensors have been developed with different working principles [1-3]. High sensitivity and low voltage operation of tunneling accelerometers have made it superior to other types of accelerometers [4]. The significant advantage of a simple constant thickness cantilever beam structure over a large bulk micro-machined proof mass design is low cross axis sensitivity. We have developed a high precision tunneling accelerometer having actuation voltage within CMOS bias levels. The accelerometer can be integrated with conventional CMOS circuits as SOC. The design parameters have been analytically optimized and validated by finite element analysis (FEA). The structure has been realized by surface micromachining techniques. The tunneling current density has been evaluated from the first principle of

quantum mechanics [14 and 15]. In electron tunneling transducers having a displacement resolution of around $10^{-4}\, \dot{A}\, Hz^{-\frac{1}{2}}$ [3], a 1% change in 1.5 n-A current between the tunneling electrodes corresponds to a displacement fluctuation of less than $0.1\dot{A}$. This high sensitivity is independent of the lateral size of the electrodes because the tunneling current occurs between two metal atoms located at opposite electrode surfaces. Due to their high sensitivity and miniature size it is possible to fabricate a high performance, small sized, light mass, inexpensive accelerometer. Attempts have been made to develop accelerometers based on electron tunneling phenomena [4 to 7] in recent years. The tunneling accelerometer structure is shown in figure (1). Figure (2) shows the dimensions of the beam and pads of the implemented tunneling sensor. In this work capacitive actuation has been employed along with a PD controller Fig (3) to obtain constant gap mode of operation.

MATLAB Simulink is chosen to model and simulate the function of the tunneling accelerometer.

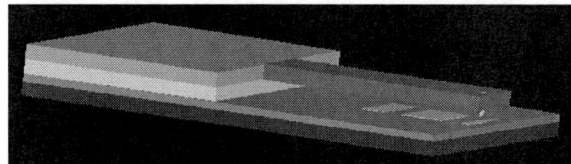

Figure-1:   Solid   Model   of   a   Tunneling Accelerometer

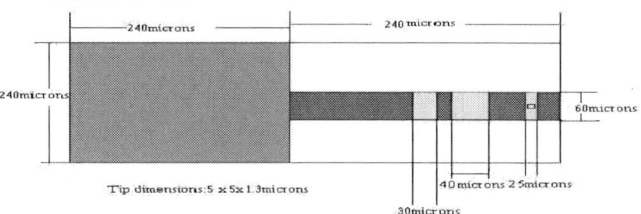

Figure-2: Dimensions of the structure

In this paper, section-2 deals with the description of the accelerometer structure. Section-3 describes the basic operating principle of the tunneling accelerometer. Section- 4 provides a detailed beam analysis and it primarily shows the action of electrostatic forces on the beam. In section-5 the

595

978-1-4244-3039-0/08 $25.00 © 2008 IEEE

non-linear phenomena in the system, namely the tunneling and electrostatic actuation phenomena are analyzed and linearized. Section-6 describes the feedback controller design while section-7 briefly mentions the sensor noise analysis followed by section-8 which describes the squeeze film damping effects on the system. Finally the paper is concluded at the end in section-9.

2. Accelerometer structure

The proposed tunneling accelerometer structure comprises of a cantilever beam integrated with a tunneling tip and electrodes assembly as shown in figure (1). The electrodes include a tunneling pad, an actuation pad and a testing pad. The metal of the electrode as well as the tip is a layer of Au film which is used because of its inert chemical characteristics and its relatively high work function. The metal is kept the same in both cases to avoid work function difference due to temperature variations. The fabricated structure is shown in fig (4).

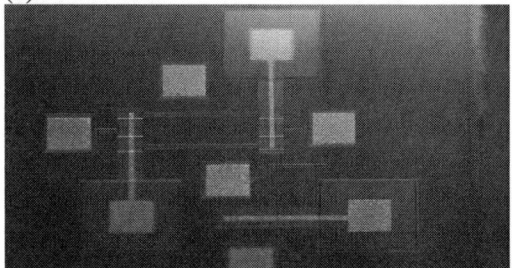

figure (4) Die picture of a multiple cantilever beam structure. Below each cantilever beam, a tunneling pad, an actuation pad and a testing pad are clearly visible.

The accelerometer has been designed and its performance has been simulated numerically using Coventorware software platform [8]. The optimized dimensions have been evaluated and tabulated as shown in Table-1.

Table-1
The Design Parameters

Beam	$240\,\mu m *60\,\mu m *5\,\mu m$
Tip	$5\,\mu m *5\,\mu m *1.3\,\mu m$
Beam Support	$240\,\mu m *240\,\mu m *5\,\mu m$
Tunneling Pad	$25\,\mu m *60\,\mu m *0.05\,\mu m$
Deflecting Pad	$40\,\mu m *60\,\mu m *0.05\,\mu m$
Self Test Pad	$30\,\mu m *60\,\mu m *0.05\,\mu m$

3. Operation principle

During the measurement, the accelerometer maintains a constant tip-to-tunneling pad distance by applying an electrostatic feedback force on cantilever beam. Usually at this constant gap mode of operation the separation between the tip and the tunneling electrode is about a few tens of $\dot{A}$.

A Matlab Simulink block diagram is constructed according to the accelerometer functional blocks as shown in figure (3). Due to an acceleration Δa the cantilever beam experiences an inertial force F_{ext} causing a displacement of ΔZ. Neglecting energy losses due to air damping, at force balance condition; $k\Delta Z = m\Delta a$, where m and k are the mass and stiffness constant of the cantilever beam respectively. The ratio k/m is inversely proportional to the sensitivity $\Delta Z/\Delta a$ and it describes the beam vibration frequency $\omega_n = \sqrt{k/m}$. In order to exhibit fast response time and large bandwidth, accelerometers require high natural frequencies. For better sensitivity and resolution, a smaller k/m ratio is necessary. Therefore, the most challenging task for an accelerometer design is to enhance the resolution while broadening the bandwidth.

When a force is applied on a beam, the cantilever beam frequency response is given by

$$T(s) = \frac{a}{s^2 + 2\zeta\omega_n s + \omega_n^2}$$

In present case we have analyzed the beam using the method of assumed modes. The separation between the tip and the pad ΔZ induces an exponential current given by $I = I_0 \exp[-\alpha\sqrt{\varphi}\Delta Z]$ (Simmon's & Binning's equation) where α is a constant and ϕ is the effective height of tunneling barrier. The tunneling current is converted into an equivalent voltage and compared with its steady state value. The error voltage is fedback to the electrostatic actuator through a PD controller producing a feedback force F_e given by $F_e = \dfrac{0.5\varepsilon_0 W w V^2}{g^2}$. This force F_e counter-balances the external force and keeps the tunneling tip at the operating point.

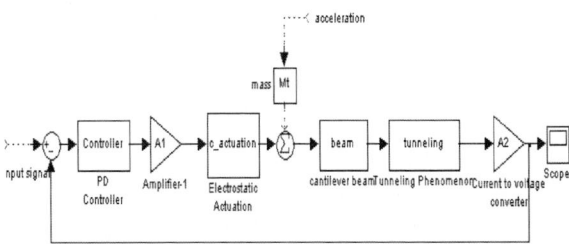

Figure –(3)　　Control block circuit

4. Beam transfer function and electrostatic actuation.

The method of assumed modes [16] has been used for the transfer function analysis of the beam. The solution of the beam deflection is given by

$$Z(x,t)=\sum_{i=1}^{\infty} \psi_i(x)q_i(t)$$

where $\psi_i(x)$ is the mode shape function and $q_i(t)$ is the temporal part of the solution. Here we assume that only the 1st mode is propagated through the cantilever beam. The transfer function of the beam is given by,

$$Z(s)=\frac{1}{m_i s^2 + b_i s + k_i}$$

where $m_i = \rho A \int_0^L \psi_i^2(x)dx$

$b_i = c\int_0^L \psi_i^2(x)dx$ and $k_i = EI\int_0^L \frac{\partial^4 \psi_i(x)}{\partial x^4}$; When the

cantilever beam is under the action of electrostatic forces the effective spring constant of the beam changes considerably [9]. L is taken to be the length of the cantilever beam, L_C is the length of application of external force on the cantilever beam and λ_r is a ratio which is defined as $\lambda_r = \frac{L_c}{L}$. Let y_{max} be the maximum displacement at the free end of the beam. The stiffness at the point of free end, i.e. at the point of maximum displacement is given by

$$K_{eff} = \frac{q_0 l_c}{y_{max}} = \frac{2}{3}\frac{Ebt^3}{l^3}[\frac{3}{8-6\lambda_r+\lambda_r^3}]$$

where q_0 is the distributed transverse load over the length L_C. The time response, root locus and the Bode plots of the proposed system are shown in figures (5a, b, c).

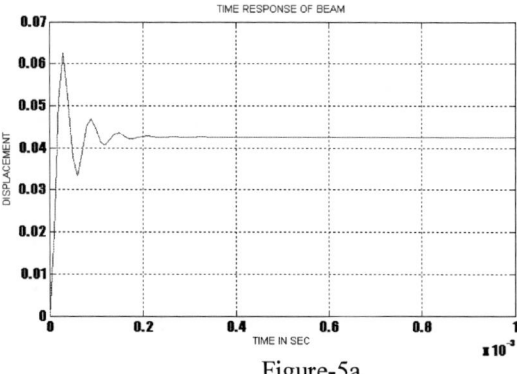

Figure-5a

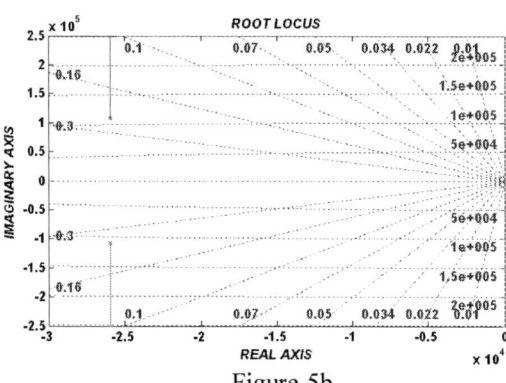

Figure-5b

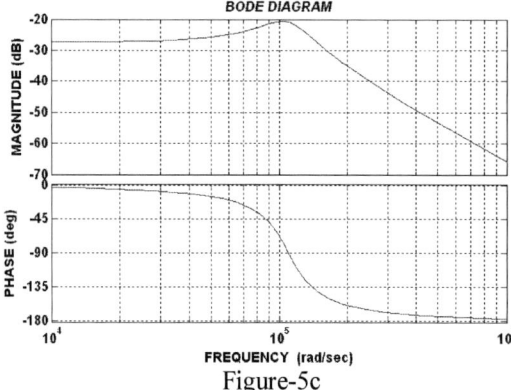

Figure-5c

It is clear from the root locus of the beam (fig5b) that the system is stable. The settling time of the open loop cantilever beam is around 0.2 ms as evident from the time response.

5. Small signal analysis

In the proposed system both the sensing mechanism and the actuation are nonlinear phenomenon.

Tunneling phenomenon-As the cantilever beam bends and the tip is brought sufficiently close to the tunneling pad (few tens of $\overset{\cdot}{A}$), using electrostatic actuation, a tunneling current is established and

597

978-1-4244-3039-0/08 $25.00 © 2008 IEEE

remains constant if the tunneling voltage and the distance between the tip and the counter-electrode are unchanged. The tunneling current I_{tun} is normally given by Simmons's and Binning equation; $I_{tun}=I_o\exp(-\alpha z\sqrt{\varphi})$ where I_{tun} =tunneling current, I_o = a constant, α =tunneling constant, z =tunneling gap and φ =effective work function. This result was also used in the seminal paper on Scanning Tunneling Microscope (STM) by Binning [10]. Nevertheless the result used by Binning is based on three approximations which though suitable for an STM are not compatible with the modeling of tunneling accelerometers. The approximations are as follows:

1) Simmon's and Binning's formula was derived for low voltage operation. This is correct for an STM. However tunneling accelerometers are often used at very high voltages.

2) In Simmon's original expression the barrier height present in the exponential term is an equivalent barrier height. Following Binning it is assumed to be the average barrier height for many applications. This assumption is valid for trapezoidal tunneling barriers only (as in STM). The tunneling accelerometer is normally operated at high voltages where the tunneling barrier is triangular; the slope of the triangle being dependent on the field. Thus in these situations the aforementioned approximations may induce gross errors.

3) Simmon's original expression contained an effective tunneling barrier thickness in the exponential term. Following Binning, most authors tend to assume this thickness to be physical tunneling barrier thickness. Again this is a valid approximation for trapezoidal barrier of STM [11]. But this assumption is quite unrealistic in the case of tunneling accelerometer. We have estimated the tunneling current using double box formalism [12]. Tunneling current density could be written from the first principle as

$$J=\frac{10e^3\sqrt{2m(V_0-\varphi)}}{\pi h_0^{\,2}\Delta E_p[1-\exp(-2pd)]}F^2\exp(-\frac{2p\varphi}{eF})$$

where m is the mass of the electron, $\hbar$ is the planck's constant, e is the charge of an electron, ϕ is the work function, V_0 is the barrier height, F is the field within the barrier (a function of the displacement z), d is the width of the barrier and ΔE_p is the difference between the successive energy levels at the transmitted end. The changes in the value of z during operation is about $10^{-1}\,\dot{A}$ to

$10^{-3}\,\dot{A}$ which is much small compared to the distance between the pad and the tip, typically a few $\dot{A}$ .In the present case the field within the barrier can be linearized by small displacement approximation. The denominator in the current density function can also be linearized by the same approximation. Thus the total current density function can be linearized.

Electrostatic actuation- The beam over the pull-down electrode is modeled as a parallel plate capacitor. Let the width of the beam be w & the width of the pull-down electrode is W (A=W*w)

So the parallel plate capacitance is given by $C=\dfrac{\varepsilon_0 A}{g}=\dfrac{\varepsilon_0 Ww}{g}$,where g is the height of the beam above the electrode. The electrostatic force applied to the beam is found by the power delivered to a time-dependent capacitance & is given by

$$F_e=0.5V^2\frac{dC(g)}{dg}=-\frac{0.5\varepsilon_0 WwV^2}{g^2},\text{where}\quad V\quad\text{is the}$$

applied voltage at the actuation electrode. The above calculated force is equal to the mechanical restoring force due to the stiffness of the beam (F=k*x). So we see that $\dfrac{0.5\varepsilon_0 WwV^2}{g^2}=k(g_0-g)$ where g_0 =reverse bias height. Solving the last equation, the electrostatic force F_e is given by

$F_e=\dfrac{\varepsilon_0 A_t V^2}{2g_0^{\,2}}$ where ε_0 is the permittivity of air or free space, A_t is the area of the actuation pad, g_0 is the gap between pad and tip, V is the total voltage acting on the actuation pad. The voltage V is a combination of a dc voltage $V_{d.c.}$ and a variable voltage $V_{a.c.}$ as shown.

$$V^2=(V_{d.c.}+V_{a.c.})^2=V_{d.c.}^{\,2}+2V_{d.c.}V_{a.c.}+V_{a.c.}^{\,2}$$

Here the dc voltage is used to set the tip at a fixed gap for tunneling; and for dynamic operation the effect of this term can be neglected. For dynamic operation only 2nd term of the above expression is considered. The transfer function for electrostatic actuation given by, $[\dfrac{F_a}{v_{a.c.}}]_s=\dfrac{(\varepsilon_0 A_t)}{g_0^{\,2}}$ .We take the area of the actuation pad as $60\mu m*20\mu m$ and $g_0=1.5\,\mu m$.

6. Feedback controller design

The primary objective of the controller design is to maintain a constant tunneling gap (typically a few $\overset{\cdot}{A}$) between the cantilever beam tip and the tunneling electrode to a very high degree of accuracy. This requirement results in an extension of dynamic range and linearity for micro-machined tunneling accelerometers. When considering accelerometer resolution, a smaller frequency results in higher resolution. The control system needs to enhance the system stability so as to protect the tunneling tip against parameter disturbances and signal impulses. The total closed loop control system obtained should have the characteristics as suggested in [5].

The displacement of the beam has been found to vary linearly with the variation in input acceleration (through Coventorware simulations) in the range of $0.1\ \mu g$ to $200\ \mu g$ (Fig7).

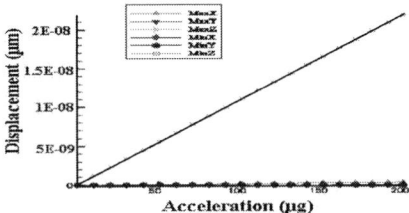

Figure: 7

The time response of the entire closed loop system is shown in figure-8. Stability is attained in less than 0.2 ms.

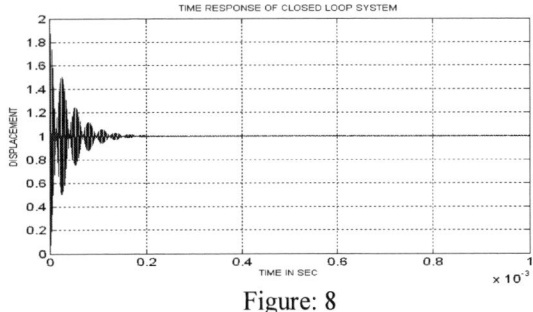

Figure: 8

7. Sensor noise analysis

Noise analysis is an important issue for a MEMS based accelerometer as it limits the sensitivity. Since the signal level is low, there is every possibility of the signal being masked by noise. Moreover the resolution of the accelerometer

is determined by the total noise of the system. The total noise equivalent acceleration [4] is given by $\text{TNEA} = \frac{1}{g}\sqrt{\frac{8\pi K_B T f_0}{MQ}}$ where K_B is the Boltzman constant, T is the temperature in Kelvin, f_0 is the resonant frequency, M is the mass of the beam , Q is the quality factor of the MEMS sensing element and g is the acceleration due to gravity. In the present tunneling structure the total noise equivalent acceleration is found to be $6.6\ \mu g/\sqrt{Hz}$.

8. Squeeze film damping

Whenever a body that travels through a fluid collides with the molecules of the fluid, it transmits some of its energy through those molecules. This impingement causes viscous damping as well as acoustic radiation When the beam approaches a stationary body in a perpendicular direction, squeeze film damping occurs. The energy loss due to squeeze film damping is significant compared to other sources of energy losses. Due to squeeze film damping [13] the effective damping constant b_{eff} is given by $b_{eff} = \frac{96\mu L W^3}{\pi^4 g_0{}^3}$ and the cut-off frequency is given by $\omega_c = \frac{\pi^2 g_0{}^2 P}{12\mu W^2}$ where μ is the air viscosity, W is the cantilever width, P is the normal pressure and g_0 is the nominal gap .Viscosity of the medium will change with temperature variations. This change in viscosity will affect the damping force and hence the damping ratio. Sutherland equation gives the temperature dependence of viscosity (μ) of air which can be expressed as $\mu = \mu_0 \frac{T_0 + T_s}{T + T_s}(\frac{T}{T_0})^{\frac{3}{2}}$ where $T_0 = 273 K$, μ_0 is the viscosity of air at T_0 and $T_s = 124$ K is a constant. In the present case $\mu_0 = 17.2 * 10^{-6}$ Pa-s. So μ $= 18.4 * 10^{-6}\ Pa - s$ for a temperature T=298 K. The viscosity μ is also dependent on the pressure. The pressure dependence of viscosity is given by: $\mu = \frac{\mu_{atm}}{1 + 9.638 K_n{}^{1.159}}$ where K_n is a ratio given by $K_n = \frac{\lambda}{h_0} = \frac{P_n \lambda_n}{P_0 h_0}$; where λ is the mean free path at the operating pressure, λ_n is the mean free path of air at known pressure P_n and P_0 is the operating pressure. Generally the mean free path is about 0.07-0.09 μm at normal atmospheric pressure. In the present case λ is assumed to be $0.08\,\mu m$. So $K_n = 0.053$ at 298 K and $\mu = 13.91*10^{-6}$ Pa-s. The effective damping constant b_{eff} and the cut-off frequency ω_0 comes out to be $2.11*10^{-4}\,N-s/m^2$ and 3.7 MHz respectively for the above mentioned surrounding temperature and pressure. From the above equations we get the effective quality factor

$$Q_{sq} = \frac{(1+9.638 K_n^{1.159})\pi^4 g_0^3 m_{eff}\omega_0}{96\mu_{atm}L_{eff}W_{eff}^3} = 33.97$$

9. Conclusions

By small signal and small displacement approximation the tunneling accelerometer model is analyzed and linearized. The closed loop system has improved the damping of the entire system while maintaining its high sensitivity. The model of the tunneling accelerometer sensor has been developed using an improved tunneling model. The actuation voltage has been kept within CMOS bias levels. It is integrable with the CMOS process as SOC. A detailed work will be published soon which would analyze the experimental results for the tunneling accelerometer.

10. References

1. T.W. Kenny et al., "Wide-bandwidth electromechanical actuators for tunneling displacement transducers" *IEEE J Microelectromech System*, vol3, pp 97-104, 1994
2. J. Grade et al., *Transducers97-International Conference on Solid-State Sensors and Actuators*, pp-1241-1244, June 1997.
3. R.L. Kubena et al, *IEEE Elec. Dev.Lett* vol-17, no-6, pp-306-308, June 1996.
4. Liu C-H and Kenny T W, "A high-precision wide-bandwidth micromachined tunneling accelerometer",

2001 *J. Microelectromech.Syst.***10** 425-33
5. Jian Wang, Yongjun Zhao, Tianhong Cui, and Kody Varahramyan, "Synthesis of the modeling and control systems of a tunneling accelerometer using MATLAB simulation", *Journal of Micromechanics and Micro-engineering.***12** (2002) 730-735
6. Waltman S B and Kaiser W J, "An electron tunneling sensor", 1989 *Sensors Actuators* **19** 201-10
7. Kenny T W, Waltman S B, Reynolds J K and Kaiser W J, "A Micro-machined silicon tunneling sensor for motion detection ",1991 *Appl.Phys.Lett.* **58** 100-2
8. http://www.coventor.com/coventorware
9. Sayanu Pamidighantam, Robert Puers, Kris Baert and Harrie A C Tilmans, " Pull-in Voltage Analysis of Electrostatically Actuated Beam Structures with fixed-fixed and fixed-free end conditions", 2002 *J. Micromech. Microengg.* **12** 458-464
10. 12. G. Binnig and H. Rohrer, "Scanning Tunneling Microscopy", *IBM J. Res. Develop.* Vol-30., pp-355, 1986.
11. G. Binning, H. Rohrer, C. Gerber and E. Weibel, "Surface Studies by Scanning Tunneling Microscopy", *Phys. Rev. Lett.* Vol-49, pp-5761 1982.
12. L. Larcher, A. Paccagnella and G. Ghidini, "Gate Current in Ultra-thin MOS capacitors: A New Model of Tunnel Current", *IEEE Trans. Electron Devices* , vol 48, no.2, pp 271-278, 2001.
13. Jay-Brotz, Gary Fedder and Tamal Mukherjee, "Damping in CMOS-MEMS Resonators" ,*Master's Project in ECE Carnegie Mellon University*
14. Angik Sarkar and T.K. Bhattacharyya, "Interpretation of Electron Tunneling from Uncertainty Principle", *arXiv: quant*-ph/0507293 v1 July 2005.
15. Debasish Paul and T.K. Bhattacharyya "Design & Fabrication of Silicon Based Microstructure for Tunneling Application to MEMS", M.S thesis *in E&ECE at IIT Kharagpur.June, 2006*
16. Shashikanth Suryanarayan and Amit Dixit " A Procedure for the Development of Control-Oriented Linear Models for Horizontal Axis Large Wind Turbines", *Journal of Dynamic Systems, Measurements and Control*, Volume 129/469, July 2007.

An optical reconfiguration system with four contexts

Naoki Yamaguchi and Minoru Watanabe
Electrical and Electronic Engineering
Shizuoka University
3-5-1 Jyohoku, Hamamatsu, Shizuoka, 432-8561 Japan
Email: tmwatan@ipc.shizuoka.ac.jp

Abstract

Optically reconfigurable gate arrays (ORGAs), which consist of a gate array VLSI, a holographic memory, and a laser diode array, are a type of programmable gate array that can achieve rapid reconfiguration and numerous reconfiguration contexts. The gate array of an ORGA is optically reconfigured using diffraction patterns from a holographic memory that is addressed using a laser diode array. It is noteworthy that ORGA-VLSIs which can be reconfigured in nanoseconds without any overhead have already been fabricated. However, to date, no multi-holographic reconfiguration system that is suitable for such rapidly reconfigurable ORGA-VLSIs without any overhead has ever been developed. As the first step toward realizing such a device, a four-context optical system is demonstrated experimentally using a liquid crystal spatial light modulator and a He-Ne laser. This paper describes those experimental results and plans for future work.

1 Introduction

Field Programmable Gate Arrays (FPGAs) have been used widely in recent years because of their flexible reconfiguration capabilities [1]-[3]. Moreover, demand for high-speed reconfigurable devices has been increasing. An idle circuit can be evacuated into memory if circuit information can be exchanged rapidly between the gate array and memory; other necessary circuits can be downloaded at that time from memory into the gate array, thereby increasing the gate array's activity. In addition, such devices offer the possibility of providing a virtual gate count that is much larger than those of currently available VLSIs. However, because the FPGA's reconfiguration requires more than several milliseconds, FP-GAs are unsuitable for dynamically reconfigurable devices.

On the other hand, high-speed reconfigurable devices have been developed: DAP/DNA chips, DRP chips, and multi-context FPGAs [4]-[9]. Those devices package reconfiguration memories and microprocessor arrays or gate arrays onto a chip. The in-ternal reconfiguration memory stores reconfiguration contexts of 4–16 banks. The banks can be changed from one to another on a clock. Thereby, the arithmetic logic unit or gate array of such devices can be reconfigured on every clock cycle in a few nanoseconds. However, increasing the internal reconfiguration memory while simultaneously maintaining the gate density is extremely difficult.

Mixed optical and electrical VLSIs [10]–[17] have been developed to address such situations. In particular, optically reconfigurable gate arrays (ORGAs) [15]–[17] have been developed to realize both capabilities of fast reconfiguration and numerous reconfiguration contexts. An ORGA consists of a gate-array VLSI, a holographic memory, and a laser diode array. The ORGA gate array is reconfigured optically using diffraction patterns from a holographic memory that is addressed using a laser diode array. However, the gate array cannot function during reconfiguration.

For those reasons, an optically differential reconfigurable gate array (ODRGA) VLSI has been developed: it provides more rapid nanosecond reconfiguration capability without any overhead [18]. In addition, a holographic reconfiguration system including a single context has been developed [19]. However, to date, a multi-holographic reconfiguration system that is suitable for such no-overhead and fast-reconfigurable ORGA-VLSIs has never been developed. Particularly, circuit implementation experiments have never been done using such a multi-holographic reconfiguration system. As the first step toward realization of such a system, a four-context optical system was demonstrated experimentally using a liquid crystal spatial light modulator and a He-Ne laser. This paper describes experimental results and plans for future work.

2 Reconfiguration context Generation

An ORGA is constructed as a three-layered structure, as shown in Fig. 1. The right layer is a gate array ORGA-VLSI, the middle layer is an optical holographic memory, and the left layer is a laser diode array. The holographic memory can store many reconfiguration contexts. The reconfiguration contexts

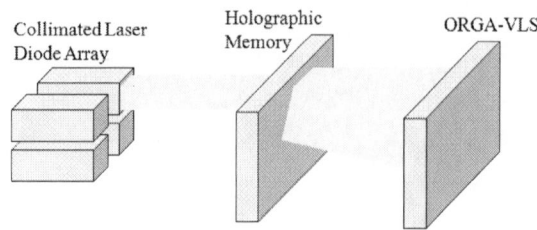

Figure 1. Experimental system setup of an ODRGA architecture.

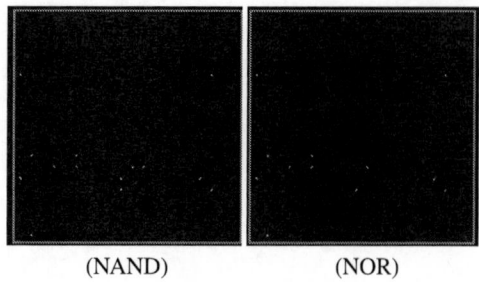

(NAND)   (NOR)

Figure 2. Sample contexts of a NAND circuit and a NOR circuit.

stored in the holographic memory are addressed using a laser diode array. The diffraction pattern from the holographic memory can be received as a reconfiguration context on a photodiode-array of an ORGA-VLSI.

2.1 Calculation method

Here, the calculation method of a holographic memory to generate reconfiguration contexts is described. A hologram for ORGAs is assumed as a thin holographic medium. A laser aperture plane, a holographic plane, and an ORGA-VLSI plane are parallelized. The laser beams are expanded, the sizes of which are fully wide for each reconfiguration context-recorded area of the holographic medium. Consequently, the laser beams can be considered as plane waves. Each reference wave from a laser propagates into the holographic medium. The holographic medium is divided into some areas. Each area includes a holographic pattern of a single context. The pixels are assumed to be analog values. On the other hand, the input object comprises small rectangular pixels, which can be modulated to be either on or off. The intensity distribution of each part of the holographic medium is calculable using the following equation.

$$H(x_1, y_1) \propto \int_{-\infty}^{\infty}\int_{-\infty}^{\infty} O(x_2, y_2)\sin(kr)dx_2dy_2,$$
$$r = \sqrt{Z_L^2 + (x_1 - x_2)^2 + (y_1 - y_2)^2}.$$

In that equation, $O(x_2, y_2)$ is a binary value of a reconfiguration context, k is the wave number, and Z_L is the distance between the holographic plane and the object plane. The value $H(x_1, y_1)$ is normalized as 0–1 for the minimum intensity H_{min} and maximum intensity H_{max} as the following.

$$H'(x_1, y_1) = \frac{H(x_1, y_1) - H_{min}}{H_{max} - H_{min}}. \quad (1)$$

Finally, the normalized image H' is used for implementing the holographic memory. The other areas on the holographic plane are opaque to the illumination.

3 Simulation results

Using previously explained formulations, simulations were executed. We assumed a four-context ORGA system. Each parameter was selected to fit an experimental system, as explained later. A holographic memory is presumed to be a liquid-crystal spatial light modulator (LC-SLM). The x-direction and y-direction pixel size of an ORGA-VLSI were defined respectively as 18 μm and 18 μm. Respective distances in the x-direction and y-direction between photodiodes of the fabricated ORGA-VLSI were 90 μm and 90 μm. A context consists of 20 × 17 bits. Next, the resolution of a holographic memory was designed as 14 μm × 14 μm, which is the same as the resolution of an LC-SLM. The number of pixels of a holographic memory of each part is 300 × 300. Each center of four regions of the four-holographic memories was placed with an x-direction offset of ±2,170 μm and a y-direction offset of ±2,170 μm from the center of the LC-SLM. Therefore, the total area occupied by the holographic memory is 8.5 × 8.5 mm^2. The target laser uses a wavelength of 632.8 nm. The distance Z_L between an object or an ORGA-VLSI and a holographic medium was designed as 100 mm. The intensity distribution of the holographic memory was normalized to 256 gradations, which is equal to that of an LC-SLM according to the calculated floating point value.

Figures 3 and 4 show simulation results for a NAND and a NOR circuit implementation shown in Fig. 2. Figures 3 (a-1), (a-2), (a-3), and (a-4) respectively portray the intensity distribution of an upper-left side holographic memory, an upper-right side holographic memory, a lower-left side holographic memory, and a lower-right side holographic memory. Figures 3 (b-1), (b-2), (b-3), and (b-4) respectively depict simulation results of context images reconstructed from the upper-left side holographic memory, the upper-right side holographic memory, the lower-left side holographic memory, and the lower-right side holographic memory. In addition, figures 4 (a-1), (a-2), (a-3), and (a-4) respectively por-

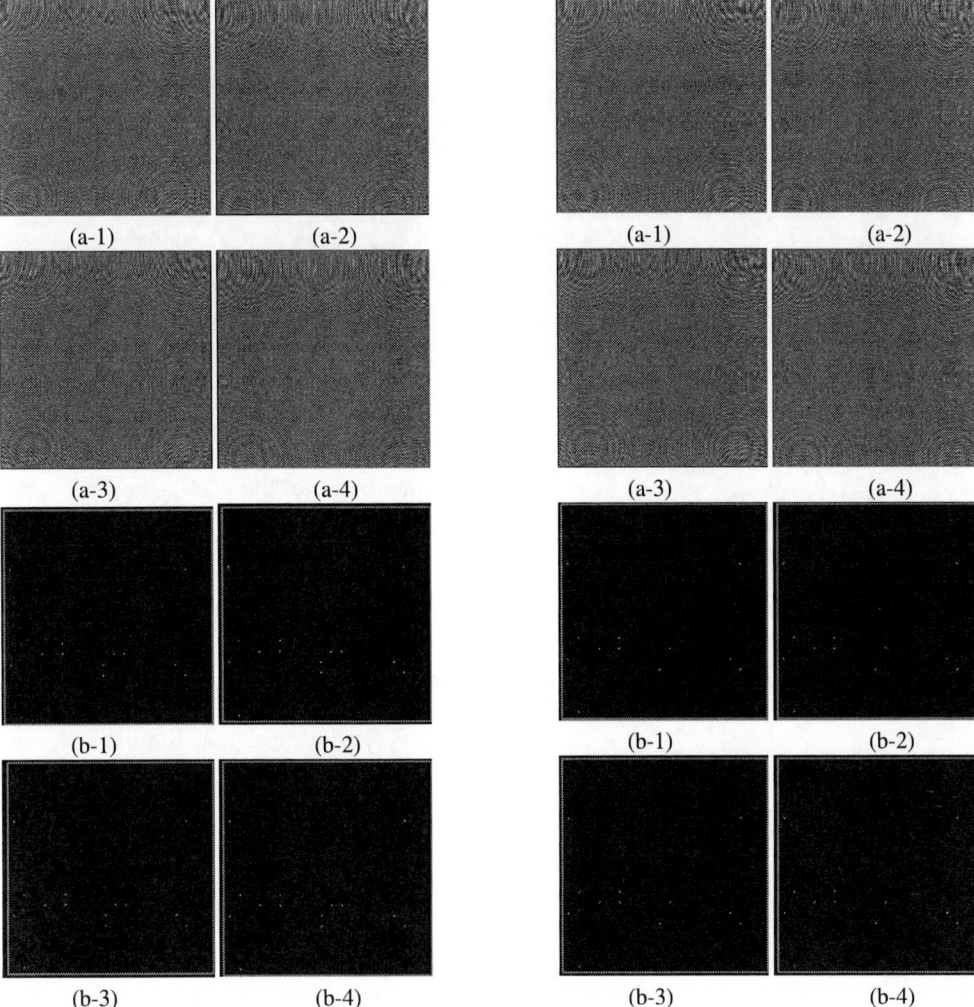

Figure 3. Simulation results for a NAND circuit implementation shown in Fig. 2. Figures (a-1), (a-2), (a-3), and (a-4) respectively depict the intensity distribution of an upper-left side holographic memory, an upper-right side holographic memory, a lower-left side holographic memory, and a lower-right side holographic memory. Figures (b-1), (b-2), (b-3), and (b-4) respectively portray the simulation results of context images reconstructed from the upper-left side holographic memory, the upper-right side holographic memory, the lower-left side holographic memory, and the lower-right side holographic memory.

Figure 4. Simulation results for a NOR circuit implementation shown in Fig. 2. Figures (a-1), (a-2), (a-3), and (a-4) respectively portray the intensity distribution of an upper-left side holographic memory, an upper-right side holographic memory, a lower-left side holographic memory, and a lower-right side holographic memory. Figures (b-1), (b-2), (b-3), and (b-4) respectively present the simulation results of context images reconstructed from the upper-left side holographic memory, the upper-right side holographic memory, the lower-left side holographic memory, and the lower-right side holographic memory.

tray the intensity distribution of an upper-left side holographic memory, an upper-right side holographic

memory, a lower-left side holographic memory, and a lower-right side holographic memory. Figures 4

Figure 5. Experimental system.

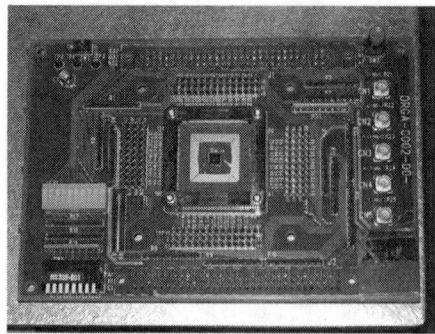

Figure 6. A photograph of an ODRGA-VLSI board with a 4.9 × 4.9 mm^2 ORGA-VLSI chip that was fabricated using a 0.35 μm three-metal CMOS process. The gate count of the gate array on the chip is 68. In all, 340 photodiodes were implemented and used for optical configurations.

(b-1), (b-2), (b-3), and (b-4) respectively present the simulation results of context images reconstructed from the upper-left side holographic memory, the upper-right side holographic memory, the lower-left side holographic memory, and the lower-right side holographic memory. The results show that the calculated holographic memories can generate good contrast contexts.

4 Experimental system and results

4.1 Experimental system

We have constructed a holographic memory system using a liquid crystal spatial light modulator (LC-SLM) as a holographic memory and a He-Ne laser to demonstrate some simulation results, as shown in Fig. 5. Four lasers are necessary to construct this experimental system correctly, but in this experiment, a 632.8 nm He-Ne laser is used as a common light source to emulate four laser sources. The LC-SLM used as a holographic memory including four contexts is a projection TV panel (L3P07X-31G0; Seiko Epson Corp.) which is a 90° twisted nematic device with a thin film transistor. The panel consists of 1,024 × 768 pixels, each having a size of 14 × 14 μm^2. The LC-SLM is connected to the evaluation board (L3B07X-E10A; Seiko Epson Corp.). The video input of the board is connected to the exter-

nal display terminal of a personal computer. Programming for the LC-SLM is executed by displaying a certain pattern of green with 256 gradation levels on the personal computer display. The ORGA-VLSI was placed at a distance of 100 mm from the LC-SLM. Figure 6 shows a chip photograph of the ORGA-VLSI. The ORGA-VLSI was fabricated using a 0.35 μm three-metal 4.9 × 4.9 mm^2 CMOS process chip. In this fabrication, the distance between each photodiode was designed as 90 μm; the photodiode is 25.5 × 25.5 μm^2 to ease the optical alignment. The total number of photodiodes is 340. The gate array's gate count is 68. Results of experiments confirmed that the ORGA-VLSI itself can be reconfigured in nanoseconds.

4.2 Experimental results

Two circuits, a NAND and a NOR circuit, were implemented respectively onto the LC-SLM using this experimental system. First, a NAND circuit was implemented for four regions of the LC-SLM, described above in Figs. 3 (b-1), (b-2), (b-3), and (b-4). Each holographic memory pattern was displayed on the LC-SLM with: an x-direction offset of -2,170 μm and a y-direction offset of +2,170 μm; an x-direction offset of +2,170 μm and a y-direction offset of +2,170 μm; an x-direction offset of -2,170 μm and a y-direction offset of -2,170 μm; and an x-direction offset of +2,170 μm and a y-direction offset of -2,170 μm. Then, in addition to the NAND implementation, a NOR circuit was also implemented for four regions of the LC-SLM with the same x-direction offsets and y-direction offsets. Experimental results of a NAND and a NOR circuit implementations are

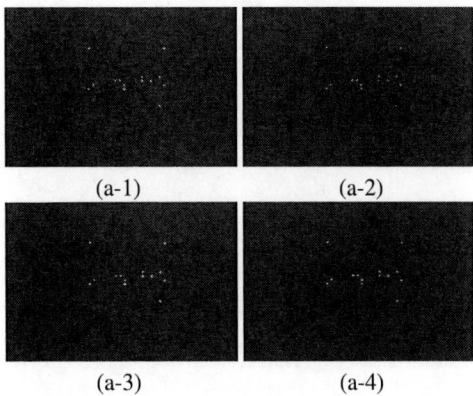

(a-1)　　　　　(a-2)

(a-3)　　　　　(a-4)

Figure 7. Experimental results of the NAND circuit implementation. Photographs (a-1), (a-2), (a-3), and (a-4) respectively portray CCD-received results at the ORGA-VLSI position, as reconstructed from the holographic memories of Figs. 3 (a-1), (a-2), (a-3), and (a-4).

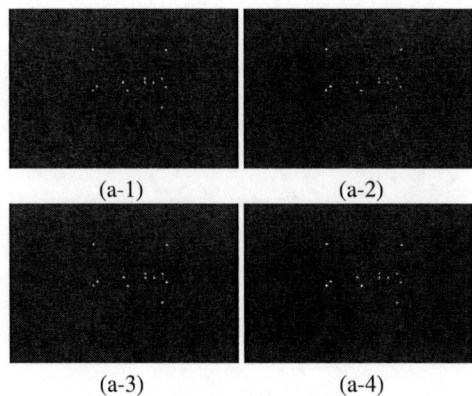

(a-1)　　　　　(a-2)

(a-3)　　　　　(a-4)

Figure 8. Experimental results of the NOR circuit implementation. Photographs (a-1), (a-2), (a-3), and (a-4) respectively present CCD-received results at the ORGA-VLSI position, as reconstructed from the holographic memories of Figs. 4 (a-1), (a-2), (a-3), and (a-4).

shown respectively in Figs. 7 and 8. The photographs presented in (a-1), (a-2), (a-3), and (a-4) respectively show CCD-received results at the ORGA-VLSI position, as reconstructed from the holographic memories of Figs. 3 (a-1), (a-2), (a-3), and (a-4). In addition, photographs (a-1), (a-2), (a-3), and (a-4) respectively show CCD-obtained results at the ORGA-VLSI position, as reconstructed from the holographic memories of Figs. 4 (a-1), (a-2), (a-3), and (a-4). The results indicate that contexts generated from holographic memories show good contrast. The reconfiguration procedure was executed using the diffraction patterns that indicate configuration contexts. As a result, reconfigurations from all regions of holographic memories of NAND and NOR circuits were successful, as depicted in Figs. 9 and 10. The reconfiguration times were measured as 2.50 ms to 3.99 ms. This experimental system demonstrated that a four-context-ORGA system can be realized. Moreover, by increasing the number of offset images, numerous reconfiguration contexts can be realized easily using this system.

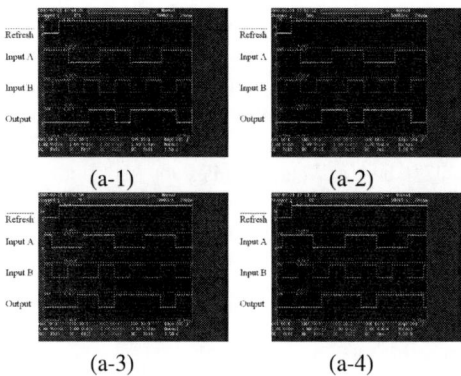

(a-1)　　　　　(a-2)

(a-3)　　　　　(a-4)

Figure 9. Waveforms of the NAND circuit implementations after a configuration procedure. Waveforms (a-1), (a-2), (a-3), and (a-4) respectively show results from holographic memories in Figs. 3 (a-1), (a-2), (a-3), and (a-4) on a LC-SLM, as measured using an oscilloscope.

5 Conclusion

This paper has presented a four-context holographic memory reconfigurable architecture. Simulation results showed that calculated holographic memories can generate good contrast contexts. Furthermore, experimental results showed that an LC-SLM storing four regions of the holographic contexts can generate very good contrast contexts. Using these contexts, the reconfiguration times with no overhead were measured as 2.50 ms to 3.99 ms. The current re-

configuration is not fast, but it can be improved easily by using high power lasers. Therefore, the reconfiguration speed problem need not be considered. Results obtained using this experimental system demonstrate that a four-context-ORGA system can be realized. Moreover, by increasing the number of offset images, numerous reconfiguration contexts can be realized easily using this system. This strategy for obtaining reconfiguration contexts remains as a goal of future work.

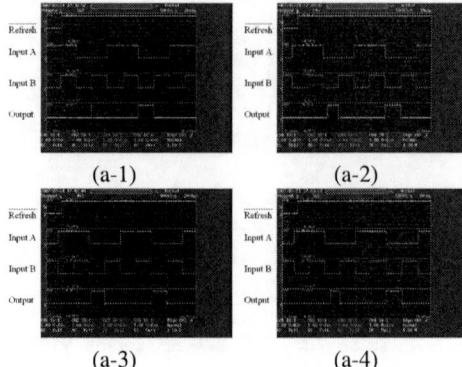

(a-1)   (a-2)

(a-3)   (a-4)

Figure 10. Waveforms of the NOR circuit implementations after a configuration procedure. Waveforms (a-1), (a-2), (a-3), and (a-4) respectively depict results from holographic memories, in Figs. 4 (a-1), (a-2), (a-3), and (a-4) on a LC-SLM, respectively, as measured using an oscilloscope.

6 Acknowledgments

This research was partially supported by the Ministry of Education, Science, Sports and Culture, Grant-in-Aid for Young Scientists (B), 18760256, 2007. The VLSI chip in this study was fabricated in the chip fabrication program of VLSI Design and Education Center (VDEC), the University of Tokyo in collaboration with Rohm Co. Ltd. and Toppan Printing Co. Ltd.

7. References

[1] Altera Corporation, "Altera Devices," http://www. altera.com.

[2] Xilinx Inc., "Xilinx Product Data Sheets," http://www. xilinx.com.

[3] Lattice Semiconductor Corporation, "LatticeECP and EC Family Data Sheet," http://www. latticesemi.co.jp/products, 2005.

[4] http://www.ipflex.co.jp

[5] H. Nakano, T. Shindo, T. Kazami, M. Motomura, "Development of dynamically reconfigurable processor LSI," NEC Tech. J. (Japan), vol. 56, no. 4, pp. 99–102, 2003.

[6] A. Dehon, "Dynamically Programmable Gate Arrays: A Step Toward Increased Computational Density," Fourth Canadian Workshop on Field Programmable Devices, pp. 47–54, 1996.

[7] S.M.Scalera and J.R.Vazquez, "The design and implementation of a context switching FPGA," IEEE symposium on FPGAs for Custom Computing Machines, pp. 78–85, 1998.

[8] S.Trimberger, et al. "A Time–Multiplexed FPGA," FCCM, pp. 22–28, 1997.

[9] D. Jones, D.M.Lewis, "A time–multiplexed FPGA architecture for logic emulation," Custom Integrated Circuits Conference, pp. 495 – 498, 1995.

[10] M. Vasilko and D. Ait-Boudaoud, "Optically Reconfigurable FPGAs: Is This a Future Trend ?," 6th International Workshop on Field-Programmable Logic and Applications, pp. 23–25, 1996.

[11] M.F. Sakr, S.P. Levitan, C.L. Giles, D.M. Chiarulli, "Reconfigurable Processor Employing Optical Channels," IEEE/OSA International Topical Meeting on Optics in Computing, SPIE Proceedings, Vol. 3490, pp. 564–567, 1998.

[12] L. Selavo, S.P. Levitan, D.M. Chiarulli, "An Optically Reconfigurable Field Programmable Gate Array," OSA Spring Topical Meeting on Optics in Computing, pp. 146-148, 1999.

[13] J. Depreitere, H. Neefs, H. V. Marck, J. V. Campenhout, R. Baets, B.Dhoedt, H. Thienpont, and I. Veretennicoff, "An optoelectronic 3-D field programmable gate array," FPL '94. Proc., pp.352–360, 1994.

[14] M. F. Sakr, S. P. Levitan, C. L. Giles, and D.M. Chiarulli, "Reconfigurable processor employing optical channels," Proc. SPIE - Int. Soc. Opt. Eng., vol. 3490, pp. 564–567, 1998.

[15] J. Mumbru, G. Panotopoulos, D. Psaltis, X. An, F. Mok, S. Ay, S. Barna, E. Fossum, "Optically Programmable Gate Array," SPIE of Optics in Computing 2000, Vol. 4089, pp. 763–771, 2000.

[16] J. Mumbru, G. Zhou, X. An, W. Liu, G. Panotopoulos, F. Mok, and D. Psaltis, "Optical memory for computing and information processing," SPIE on Algorithms, Devices, and Systems for Optical Information Processing III, Vol. 3804, pp. 14–24, 1999.

[17] J. Mumbru, G. Zhou, S. Ay, X. An, G. Panotopoulos, F. Mok, and D. Psaltis, "Optically Reconfigurable Processors," SPIE Critical Review 1999 Euro-American Workshop on Optoelectronic Information Processing, Vol. 74, pp. 265-288, 1999.

[18] M. Miyano, M. Watanabe, F. Kobayashi, "Rapid Reconfiguration of an Optically Differential Reconfigurable Gate Array with Pulse Lasers," IEEE International Conference on Field-Programmable Technology, pp. 287–288, 2005.

[19] M. Watanabe, M. Miyano, F. Kobayashi, "An optically differential reconfigurable gate array with a holographic memory," IEEE International parallel & Distributed Processing Symposium., 2006.

978-1-4244-3039-0/08 $25.00 © 2008 IEEE

21st International Conference on VLSI Design

An acceleration and optimization method for optical reconfiguration

Minoru Watanabe and Naoki Yamaguchi

Electrical and Electronic Engineering

Shizuoka University

3-5-1 Jyohoku, Hamamatsu, Shizuoka, 432-8561 Japan

Email: tmwatan@ipc.shizuoka.ac.jp

Abstract

Optically Reconfigurable Gate Arrays (ORGAs), by exploiting the large storage capacity of holographic memory, offer the possibility of providing a virtual gate count that is much larger than those of currently available VLSI circuits. Because circuits implemented on a gate array must often be changed using virtual circuits stored in a holographic memory, rapid reconfiguration is necessary to reduce the reconfiguration overhead. A simple means to realize a short reconfiguration time in ORGAs is to implement a high-power laser array. However, such an array presents the disadvantages of high power consumption, large implementation space, high cost, and so on. Therefore, this paper presents an acceleration method to increase ORGAs' reconfiguration frequency without the necessity for any increase of laser power. This technique also includes optimization between the number of reconfiguration contexts and the reconfiguration frequency. The description in this paper clarifies the advantages using simulation and experimental results.

1 Introduction

Field Programmable Gate Arrays (FPGAs) have been used widely in recent years because of their flexible reconfiguration capabilities [1]–[3]. Moreover, demand for high-speed reconfigurable devices has been increasing. An idle circuit could be evacuated into memory if circuit information could be exchanged rapidly between the gate array and memory; other necessary circuits might then be downloaded from memory into the gate array, thereby increasing the gate array's activity. Moreover, such devices offer the possibility of providing a virtual gate count that is much larger than those of currently available VLSIs. However, because the FPGA's reconfiguration requires more than several milliseconds, they are unsuitable for dynamically reconfigurable devices.

On the other hand, high-speed reconfigurable devices have been developed, e.g., DAP/DNA chips,

DRP chips, and multi-context FPGAs [4]-[9]. Those devices package reconfiguration memories and microprocessor arrays or gate arrays onto a chip. The internal reconfiguration memory stores reconfiguration contexts of 4-16 banks, which can be changed from one to another on a clock. Thereby, the arithmetic logic unit or gate array of such devices can be reconfigured on every clock cycle in a few nanoseconds. However, increasing the internal reconfiguration memory while maintaining the gate density is extremely difficult.

As with other rapidly reconfigurable devices, optically reconfigurable gate arrays (ORGAs) [10]–[27] that combine various optical and electrical techniques have been developed. Optically Programmable Gate Arrays (OPGAs) have become able to achieve 80 gate-count VLSI, a 16-20 μs reconfiguration period, and 50-100 reconfiguration contexts [18, 19, 20]. Furthermore, to date, a bit-by-bit optical reconfiguration capability [21, 22], with optical reconfigurations of less than 200 ns reconfiguration period [24] has been attainable. In addition, the possibility of a 51,272 gate-count ORGA-VLSI was demonstrated [25, 26, 27]. Optically Reconfigurable Gate Arrays (ORGAs), by exploiting the large storage capacity of holographic memory, offer the possibility of providing a virtual gate count that is much greater than those of currently available VLSI circuits. In such cases, because circuits implemented on a gate array must often be changed using virtual circuits stored in a holographic memory, rapid reconfiguration is necessary to reduce the reconfiguration overhead.

To realize a short reconfiguration time in ORGAs, the easiest way is to implement a high-power laser array. However, such an array presents the disadvantages of high power consumption, large implementation space, high cost, and so on. Therefore, this paper proposes an acceleration and optimization method of reconfiguration contexts to accelerate the reconfiguration speed of ORGAs without any increase of laser power.

Depending on the reconfigurable application, either the necessary number of contexts or the necessary reconfiguration frequency might be empha-

978-1-4244-3039-0/08 $25.00 © 2008 IEEE

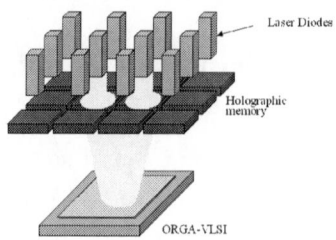

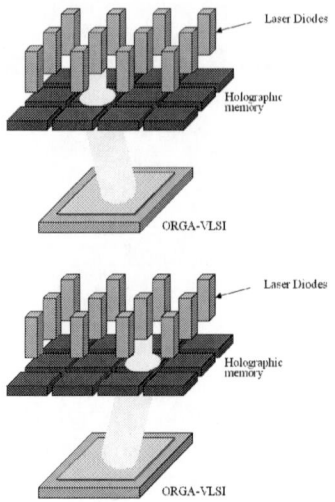

Figure 1. Block diagram of a reconfiguration method of conventional ORGAs. A single context was read out from a holographic memory by turning a single laser on.

Figure 2. Block diagram of an acceleration method of optical reconfigurations. This technique allows multiple laser diodes to be turned on when a single context is read out from a holographic memory onto an ORGA-VLSI.

2.1 Acceleration method

In previously proposed ORGAs, a single laser diode corresponded to a single configuration context. Figure 1 shows that only a single laser diode is used as light source when reconfiguring a gate array. The optical reconfiguration speed was designed to be constant. However, because the reconfiguration frequency is proportional to the light intensity received on photodiodes, if some laser can be turned on for reading a single context as shown in Fig. 2, the reconfiguration frequency can be accelerated without any increase of laser power.

In the acceleration method, first, the number of reconfiguration contexts consumed in an acceleration context must be determined depending on the necessary optical intensity or reconfiguration frequency for the acceleration context. Then, configuration information for the acceleration context is programmed onto configuration context areas selected above on a holographic memory. In other words, the same reconfiguration context is programmed onto a certain number of reconfiguration contexts in a holographic memory. All lasers corresponding to the acceleration reconfiguration contexts on which the same context information is programmed are turned on when the gate array of an ORGA-VLSI is reconfigured using the acceleration context. As a result, only the reconfiguration frequency of the target acceleration context can be increased. This method can be adapted to each context in an application on a gate array. Therefore, the method can be considered as a very flexible acceleration method.

2.2 Issue

As explained previously, in the acceleration method, the reconfiguration frequency of only a certain target context can be accelerated by consuming a certain number of reconfiguration context resources

sized. One application might require fewer contexts but rapid reconfiguration; other applications might require many contexts, but a slow reconfiguration frequency would be acceptable. Moreover, in a certain application, some procedures might require fast reconfigurations and the other procedures might require slow reconfiguration. Such requirements can be perfectly satisfied if the reconfiguration speed can be accelerated by using the resources of a number of contexts. This paper describes such an acceleration method of reconfiguration frequency by limiting the number of reconfiguration contexts and using an optimization method between the number of contexts and reconfiguration frequency. In addition, this paper discusses the effect of this method, which has been confirmed experimentally, and clarifies the advantages of this technique.

2 Acceleration and optimization method

A very simple technique to increase the power of a laser diode array is to increase the reconfiguration frequency of ORGAs. However, that technique presents the disadvantages of high power consumption, large implementation space, high cost, and so on. Therefore, it is a very important feature that, without any increase of laser power, the frequency of optical reconfiguration procedure can be accelerated.

in a holographic memory. Using the reconfiguration context resources seems to be a disadvantage of this method. However, ORGAs have numerous reconfiguration contexts. Some examples are OPGAs [18, 19, 20], which can have 50-100 reconfiguration contexts. Moreover, in the future, the number of reconfiguration contexts can be increased extremely by the advance of holographic memories. Therefore, if some fraction of the reconfiguration contexts is consumed to accelerate the reconfiguration frequency, the short-term and long-term effects on its development are expected to be negligible.

2.3 Optimization

In many reconfigurable applications, demands for the reconfiguration frequency and the number of contexts mutually differ. One application might require fewer contexts but fast reconfiguration, whereas other applications might require many contexts but slow reconfiguration frequency. Alternatively, one application might require both slow reconfiguration in a certain period and fast reconfiguration in the other period. At any one time, dynamically reconfigurable applications never require full reconfiguration contexts or the highest frequency for every reconfiguration. Therefore, optimization between the number of contexts and the reconfiguration frequency is a very useful technique for many applications. The following are some examples.

2.3.1 Twice or more acceleration example

The same context is written in two or more areas of a holographic memory in advance. During reconfiguration, all lasers corresponding to the areas turn on. The lasers are mutually incoherent. For that reason, the diffraction patterns are superimposed linearly. For example, we assume a two-reconfiguration-context superimposed case, as shown in Fig. 2. A certain context is written in two areas of a holographic memory. During reconfiguration, two laser diodes corresponding to the two areas turn on. Consequently, twice the light intensity distributions can be received on an ORGA-VLSI, thereby achieving twice the reconfiguration frequency.

2.3.2 More flexible example

Here, we assume that the basic reconfiguration frequency of a certain ORGA-VLSI is A MHz and that the ORGA-VLSI has 100 reconfiguration contexts. In addition to constant-frequency reconfiguration methods of conventional ORGAs, using only the basic reconfiguration frequency for all optical reconfigurations, we can use 100 reconfiguration contexts at A MHz reconfiguration frequency, as shown in the first line of Table 1. The second line of Table 1 shows an example that has a 20 times faster reconfiguration context ($20 \times A$ MHz) which consumes 20 re-

Table 1. Programming flexibility of the superimposing technique.

Programming Patterns	Number of Contexts	Reconfiguration Speed [MHz]
1(slow-many)	100	A
2(flexible)	30	A
	5	$10 \times A$
	1	$20 \times A$
3(flexible)	50	A
	5	$10 \times A$
4(fast-a few)	10	$10 \times A$

configuration context resources, 5–10 times faster reconfiguration contexts ($10 \times A$ MHz) which consumes 10 reconfiguration context resources, and 30 normal reconfiguration contexts (A MHz), which consumes only a reconfiguration context resource. The third line of Table 1 shows an example that has 5–10 times faster reconfiguration contexts ($10 \times A$ MHz) which consumes 10 reconfiguration context resources, and 50 normal reconfiguration contexts (A MHz), which consumes only a reconfiguration context resource. The fourth line of Table 1 shows an example that has all ten 10 times faster reconfiguration contexts ($10 \times A$ MHz), which consumes 10 reconfiguration context resources. Therefore, the optimization method allows mixed reconfigurations with a variety of frequencies. Therefore, this method is very useful for any reconfigurable applications.

3 Context generation

3.1 Calculation method

Here, a hologram for ORGAs is calculated as a thin holographic medium using the setup shown in Fig. 3. A collimated plane reference wave propagates in the holographic plane. A number of holographic patterns corresponding to reconfiguration contexts are programmed onto different positions of a single holographic medium and the other area of the holographic medium is assumed to be opaque. The thickness and pixels of the holographic medium are assumed respectively as negligible and analog values. The pixels of the input object can be modulated to be either on or off. The intensity distribution of a holographic medium is calculable using the following equation.

$$H(x_0, y_0) \quad \propto \quad \int_{-\infty}^{\infty} \int_{-\infty}^{\infty} O_n(x_1, y_1) \sin(kr) dx_1 dy_1,$$

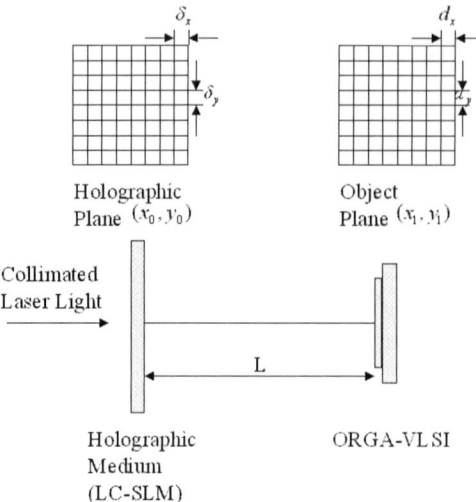

Figure 3. System configuration for a holographic memory.

Figure 4. Sample context pattern which can execute NOR circuit operation.

$$r = \sqrt{L^2 + (x_0 - x_1)^2 + (y_0 - y_1)^2}. \quad (1)$$

In that equation, $O_n(x_1, y_1)$ is a binary value of the n-th reconfiguration context, k is the wave number, and L is the distance between the holographic plane and the object plane. The value $H(x_0, y_0)$ is normalized as 0–1 for the minimum and maximum intensities as shown in the following.

$$H'(x_0, y_0) = \frac{H(x_0, y_0) - H_{min}}{H_{max} - H_{min}}. \quad (2)$$

Finally, the normalized image H' is used for implementing a holographic memory.

3.2 Context generation

A simulation was executed using previously explained formulations. To estimate the acceleration and optimization method, we have implemented four contexts. Each parameter was selected to fit an experimental system, as explained later. A holographic memory used for estimation is presumed to be a liquid crystal spatial light modulator (LC-SLM). The

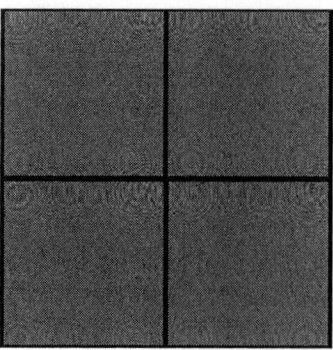

Figure 5. Recorded patterns of a holographic medium which includes four reconfiguration contexts. Each reconfiguration context is a NOR circuit implementation shown in Fig. 4.

resolution of a holographic memory was designed as 14 μm $\times$ 14 μm, which is the same as the resolution of an LC-SLM. The number of pixels of the holographic memory is 300 $\times$ 300. The center of each holographic region including a context information was placed with an x-direction offset of $\pm 2,170$ μm and a y-direction offset of $\pm 2,170$ μm from the center of the holographic memory. Therefore, the holographic memory size is 8.5 $\times$ 8.5 mm^2. The x-direction and y-direction spot size for photodiodes of an ORGA-VLSI were defined respectively as 18 μm and 18 μm. The distances of the x-direction and y-direction between spots were designed as 90 μm and 90 μm, respectively, as well as that of photodiodes of the fabricated ORGA-VLSI. The wavelength of a target laser was 632.8 nm. The distance L between an object or an ORGA-VLSI and a holographic medium was designed as 100 mm. The intensity distribution of the holographic memory was normalized to 256 gradations, which is identical to that of an LC-SLM after calculating the floating point value. A sample context which can execute NOR circuit operation shown in Fig. 4 was prepared for experiments. Different patterns are usually programmed on each region of a holographic medium. However, in this implementation, the sample context was programmed for four regions of the holographic medium to confirm the advantage of the acceleration and optimization method. Finally, a calculated holographic medium pattern including four contexts was generated, as shown in Fig. 5.

4 Experimental Results

Figure 6 shows that we constructed an ORGA architecture using a liquid crystal spatial light modulator (LC-SLM) as a holographic memory. In this

610

978-1-4244-3039-0/08 $25.00 © 2008 IEEE

Figure 6. Experimental system.

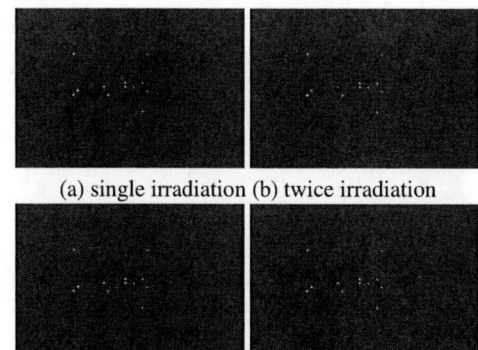

(a) single irradiation (b) twice irradiation

(c) three-times irradiation (d) four-times irradiation

Figure 7. Diffraction patterns from a holographic memory, as recorded using a CCD camera at an ORGA-VLSI position. As increase of the number of superimposition contexts; the intensity of diffraction pattern is also increased, thereby accelerating the reconfiguration frequency.

experiment, a He-Ne laser is shared for four reconfiguration context regions. The light of the He-Ne laser is incident to a LC-SLM after it is collimated. The LC-SLM used in this experiment is a projection TV panel (L3P07X-31G00; Seiko Epson) which is a 90° twisted nematic device with a thin-film transistor. The panel consists of 1024×768 pixels, each of which is $14 \times 14 \; \mu m^2$. The LC-SLM is connected to the evaluation board (L3B07X-E10A; Seiko Epson Corp.) and the board's video input is connected to the external display terminal of a personal computer. Programming for LC-SLM was executed by displaying a certain pattern of green with 256 gradation levels on the display of the personal computer. A holographic pattern shown in Fig. 5 was displayed on the LC-SLM. The ORGA-VLSI was placed at a distance of 100 mm from the LC-SLM. Finally, the diffraction pattern of a context image from the LC-SLM is received on the ORGA-VLSI.

The experimental result is shown in Fig. 7, which shows the diffraction patterns from a holographic memory at ORGA-VLSI position. It has been confirmed that, along with the increase of the number of superimposition contexts, the intensity of diffraction pattern also increases. In addition, reconfiguration times from a single context, two contexts, three contexts, and four contexts have been measured respectively as 3.894 ms, 2.995 ms, 2.097 ms, and 1.198 ms. Therefore, the acceleration method has been confirmed to be useful for accelerating the reconfiguration frequency. Even in the case of using only four contexts, about three-times-faster reconfiguration was achieved compared to conventional re-

configuration.

5 Conclusion

This paper has presented a new acceleration and optimization method for optical reconfiguration of ORGAs. For previously proposed ORGAs, the optical reconfiguration speed was designed as constant. However, using this method, the reconfiguration frequency could be accelerated flexibly. The reconfiguration times from a single context, two contexts, three contexts, and four contexts were measured respectively as 3.894 ms, 2.995 ms, 2.097 ms, and 1.198 ms. As increase of the number of superimposition contexts was observed; the intensity of diffraction pattern was also increased. Consequently, the acceleration method has been confirmed to be useful for accelerating the reconfiguration frequency. Even in the case of using only four contexts, about three-times-faster reconfiguration was achieved compared to conventional reconfiguration.

Moreover, when ORGA includes a number of reconfiguration contexts in the future, the effect of the acceleration and optimization will be further increased. We will adapt this method to future ORGAs.

6 Acknowledgments

This research was partially supported by the Ministry of Education, Science, Sports and Culture, Grant-in-Aid for Young Scientists (B), 18760256, 2007. The VLSI chip in this study was fabricated in the chip fabrication program of VLSI Design and Education Center (VDEC), the University of Tokyo in collaboration with Rohm Co. Ltd. and Toppan Printing Co. Ltd.

7. References

[1] Altera Corporation, "Altera Devices," http://www. altera.com.

[2] Xilinx Inc., "Xilinx Product Data Sheets," http://www. xilinx.com.

[3] Lattice Semiconductor Corporation, "LatticeECP and EC Family Data Sheet," http://www. latticesemi.co.jp/products, 2005.

[4] http://www.ipflex.co.jp

[5] H. Nakano, T. Shindo, T. Kazami, M. Motomura, "Development of dynamically reconfigurable processor LSI," NEC Tech. J. (Japan), vol. 56, no. 4, pp. 99–102, 2003.

[6] A. Dehon, "Dynamically Programmable Gate Arrays: A Step Toward Increased Computational Density," Fourth Canadian Workshop on Field Programmable Devices, pp. 47–54, 1996.

[7] S.M.Scalera and J.R.Vazquez, "The design and implementation of a context switching FPGA," IEEE symposium on FPGAs for Custom Computing Machines, pp. 78–85, 1998.

[8] S.Trimberger, et al. "A Time–Multiplexed FPGA," FCCM, pp.22–28, 1997.

[9] D. Jones, D.M.Lewis, "A time–multiplexed FPGA architecture for logic emulation," Custom Integrated Circuits Conference, pp. 495 – 498, 1995.

[10] M. Vasilko and D. Ait-Boudaoud, "Optically Reconfigurable FPGAs: Is This a Future Trend ?," 6th International Workshop on Field-Programmable Logic and Applications, pp. 23–25, 1996.

[11] M.F. Sakr, S.P. Levitan, C.L. Giles, D.M. Chiarulli, "Reconfigurable Processor Employing Optical Channels," IEEE/OSA International Topical Meeting on Optics in Computing, SPIE Proceedings, Vol. 3490, pp. 564–567, 1998.

[12] L. Selavo, S.P. Levitan, D.M. Chiarulli, "An Optically Reconfigurable Field Programmable Gate Array," OSA Spring Topical Meeting on Optics in Computing, pp. 146-148, 1999.

[13] J. Depreitere, H. Neefs, H. V. Marck, J. V. Campenhout, R. Baets, B.Dhoedt, H. Thienpont, and I. Veretennicoff, "An optoelectronic 3-D field programmable gate array," FPL '94. Proc., pp.352–360, 1994.

[14] J. V. Campenhout, H. V. Marck, J. Depreitere, and J. Dambre, "Optoelectronic FPGAs," IEEE J. Sel. Top. Quantum Electron, vol. 5, pp. 306–315, 1999.

[15] T. H. Szymanski, M. Saint-Laurent, V. Tyan, A. Au, and B. Supmonchai, "Field-programmable logic devices with optical input-output," Appl. Opt., vol. 39, pp.721–732, 2000.

[16] S.S. Sherif, S.K. Griebel, A. Au, D. Hui, T. H. Szymanski, and H.S. Hinton, "Field-programmable smart-pixel arrays: design, VLSI implementation, and applications," Appl. Opt., vol. 38, pp.838–846, 1999.

[17] M. F. Sakr, S. P. Levitan, C. L. Giles, and D.M. Chiarulli, "Reconfigurable processor employing optical channels," Proc. SPIE - Int. Soc. Opt. Eng., vol. 3490, pp. 564–567, 1998.

[18] J. Mumbru, G. Panotopoulos, D. Psaltis, X. An, F. Mok, S. Ay, S. Barna, E. Fossum, "Optically Programmable Gate Array," SPIE of Optics in Computing 2000, Vol. 4089, pp. 763–771, 2000.

[19] J. Mumbru, G. Zhou, X. An, W. Liu, G. Panotopoulos, F. Mok, and D. Psaltis, "Optical memory for computing and information processing," SPIE on Algorithms, Devices, and Systems for Optical Information Processing III, Vol. 3804, pp. 14–24, 1999.

[20] J. Mumbru, G. Zhou, S. Ay, X. An, G. Panotopoulos, F. Mok, and D. Psaltis, "Optically Reconfigurable Processors," SPIE Critical Review 1999 Euro-American Workshop on Optoelectronic Information Processing, Vol. 74, pp. 265-288, 1999.

[21] M. Watanabe, F. Kobayashi,"An optically differential reconfigurable gate array and its power consumption estimation," IEEE International Conference on Field-Programmable Technology, pp. 197–202, 2002.

[22] M. Watanabe, F. Kobayashi, "An Optically Differential Reconfigurable Gate Array using a 0.18 um CMOS process," IEEE International SOC Conference, pp. 281–284, 2004.

[23] M. Watanabe, F. Kobayashi, " An Optically Differential Reconfigurable Gate Array VLSI chip with a dynamic reconfiguration circuit," IEEE International Parallel & Distributed Processing Symposium, 2005.

[24] M. Miyano, M. Watanabe, F. Kobayashi, "Rapid Reconfiguration of an Optically Differential Reconfigurable Gate Array with Pulse Lasers," IEEE International Conference on Field-Programmable Technology, pp. 287–288, 2005.

[25] M. Watanabe, F. Kobayashi, "A high-density optically reconfigurable gate array using dynamic method," International conference on Field-Programmable Logic and its Applications, pp. 261–269, 2004.

[26] M. Watanabe, F. Kobayashi, "A dynamic optically reconfigurable gate array using dynamic method," International Workshop on Applied Reconfigurable Computing, pp. 50–58, 2005.

[27] M. Watanabe, F. Kobayashi, "A Dynamic Optically Reconfigurable Gate Array," Japanese Journal of Applied Physics, Vol. 45, No. 4B, pp. 3510–3515, 2006.

21st International Conference on VLSI Design

0.35μ, 1 GHz, CMOS Timing Generator Using Array of Digital Delay Lock Loops

S.Balaji
Programmer analyst trainee,
Cognizant Technology Solutions
Chennai, India
srinivasanbalaji@gmail.com

V.B Chandratre
Head, Microelectronics Section,
Electronics Division,
Bhabha Atomic Research Centre,
Mumbai, India

Menka Tewani
Scientist officer (D),
Microelectronics Section,
BARC, Mumbai, India

Abstract

This paper describes the architecture and performance of a 0.35μ, 1GHZ, CMOS Timing Generator Using Array of Delay Lock Loop. The timing generator is implemented as an array of delay locked loops. This architecture enables a timing generator with sub gate delay resolution to be implemented. The proposed Delay Lock Loops uses novel Multiplexer based Dual Phase and frequency Detector along with a charge pump where the injected charge approaches zero as the loop approaches lock on the leading edge and the trailing edge of an input clock reference. This greatly reduces the timing jitter, loop locks to both the leading and trailing clock edges as the dual phase and frequency detector along with charge pump converts the phase difference in to voltages. Test results show a timing jitter of less than 20 pS for the DLL (Delay Lock Loop) circuit .The DLL has a dead zone less than 0.01nS in the phase characteristics and has low phase sensitivity errors. The timing generator is implemented as an array of delay locked loops [1] which exponentially reduce the locking time. An experimental proto type was simulated at 0.7μ and 0.35μ technologies with a supply voltage of 5V and 3.3V respectively.

1. Introduction

The paper presented is originally motivated by the need of precise delay generation. Precise time measurements are required within most particle physics detectors. This may be, drift time of ionized tracks in a gas with an electrical drift field (drift chamber, wire chamber), propagation time of a signal on a wire to identify the origin of the signal along the wire or time of flight (TOF) of a particle. For drift time

measurements. a time resolution of the order of 1 nS is normally sufficient. For TOF detectors, an RMS error down to 10 pS may be desirable. Highly integrated multichannel time-to-digital converters for drift time measurements have been demonstrated and are already used in several experiments [2] .Very high resolution TDC's have been constructed at the board level and single channel IC's based on the current integration principle . These suffer from very low integration levels and are too expensive for future particle physics experiments with many hundreds of thousands of channels.TDC devices should not only be looked at from the point of view of resolution. Dynamic range, conversion speed, calibration procedure, data buffering, and read-out interface must also be taken into account. The mainstream IC technology today is submicron CMOS, which can contain very large amounts of digital functions, but it is not the optimal process for high resolution TDC's. Both phase-locked loops (PLLs) and Delay-locked loops (DLLs) are extensively used in many timing circuits. When either a DLL or a PLL can be used, a DLL is preferred in many cases because of its better stability, less jitter accumulation, and faster locking time compared to a PLL. We have identified that the best circuit to generate stable time delays on chip is a DLL. A DLL is widely used as a timing circuit in many systems for the purpose of clock generation, signal synchronization, and others. For example, a DLL is able to provide clock signals which are separated from each other by a well-controlled phase shift (delay). When appropriate logic, such as edge combining is used, a new clock signal which is of a different frequency can be generated by the DLL. Such an application of a DLL has been reported in for personal communication services. Another application of a DLL is for the purpose of clock deskewing in synchronous data transfer among communication chips. Reducing clock skew has become increasingly

613

978-1-4244-3039-0/08 $25.00 © 2008 IEEE

important with larger die size and higher clock frequency. A DLL for this application requires fast lock time and excellent phase alignment between the reference signal and the corrected output signal. Due to the wide range of applications of DLLs, the motivation of project is to design a high performance DLL with an accurate control which can be used for a variety of applications, including delay generation for testing purposes as well as for other timing applications, such as signal synchronization. The concept of using array of DLL's makes possible to generate high precision delay and fraction of a delay unit [3]. Many of the above applications share similar requirements for the DLL, such as a short locking time, low jitter, and a wide locking range. These common requirements have become our design goals.

In section II, the proposed dual loop DLL architecture is described. The array DLL is presented, and comparisons are made in section III & IV. Section V, draws a conclusion.

2. Dual Loop (Control) DLL Architecture

The Novel Dual loop (control) architecture ,shown in figure 1 ,has two, phase and frequency detector which detects both the rising and falling edges [4] of the signals unlike conventional system. There are two loops and correspondingly two control voltages generated, these control voltages control the rising and falling edges of the signal, this novelty helps in reducing the jitter and improves the stability of the system. The proposed dual loop DLL system increases the complexity of the circuit, also increase the area and power consumption of the circuit. The novel multiplexer based phase and frequency detector [5] uses only 9 CMOS transistor compared to the 42 CMOS transistor of the conventional phase and frequency detector [6]. The reduction in number of the transistor reduces the power consumption. Therefore the proposed Dual loop (control) DLL has reduced jitter, power consumption and area.

3. Array of Delay Locked Loops

The concept of using delay chains can be expanded to use an array of DLL's. We propose the use of several DLL's of the same configuration with a small phase difference between them. The problem here is to generate the small phase shift. A delay smaller than the delay unit in the DLL's cannot be made directly (otherwise that would have been used as the delay unit). It is, however, possible to generate with high precision a delay of one delay unit plus a fraction of a

delay unit. This is done by an additional DLL, with fewer delay elements in Fig. 2. Because of the symmetry of such an array it can be made to look like it is phase shifted by a fraction of the delay unit only. The advantage of using the dual control PFD is that it is able to control both leading and the trailing edge of the signal unlike the conventional PFD.Since the number of MOS transistor for the proposed novel PFD is less compared to the conventional PFD ,it reduces complexity and jitter exponentially.

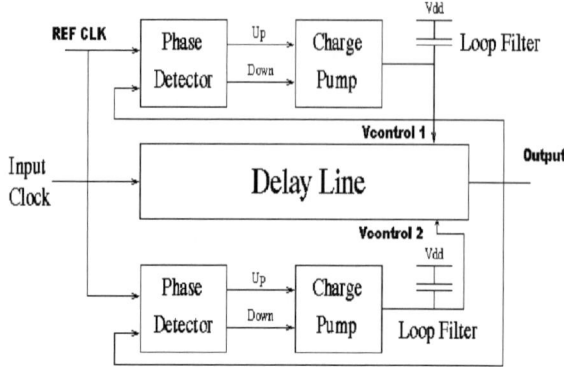

Fig.1. Dual Loop (Control) DLL Architecture

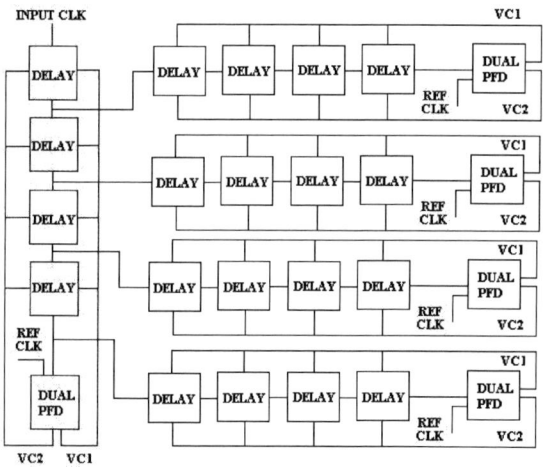

Fig. 2 . Dual Loop (Control)Array of DLL Architecture

3.1 Novel Multiplexer based Dual Edge Phase and Frequency Detector

The Phase Frequency Detector (PFD) [6] is the primary component of the entire architecture. The output of the

PFD depends on both the phase and frequency of the inputs. The conventional type of phase detector is also termed a sequential phase detector. It compares the leading edges of CLK and output of the Voltage Control Delay Line (VCDL). A VCDL (feedback) rising edge can not be present without a CLK rising edge. If the rising edge of the CLK leads the VCDL (feedback) rising edge, the "UP" output of the phase detector goes high while the "Down" output remains low. This causes the VCDL (feedback) frequency to increase and makes the edges move closer. If the VCDL (feedback) signal leads the CLK, "UP" remains low while the "DOWN" goes high. And we can find the phase difference between VCDL (feedback) and CLK. The architecture of our novel PFD [5], shown in figure 3, has one ,three input NAND gate and two 2:1 multiplexers. One of the input of the NAND gate is always made high and the other input is RESET(R) (output of the AND gate).The selection input of one multiplexer is CLK and of other is output of VCDL(output of the delay line).The output of the NAND gate is given to the input i0 of both the multiplexers and the input i1 of the both the multiplexer are grounded. The operation of the our novel PFD is if the CLK is active before VCDL the

becomes high and therefore the output value changes to low which depicts the reset state of the convention PFD [6]. This results in lower jitter and better linearity near steady state without dead zone and reduced power consumption.

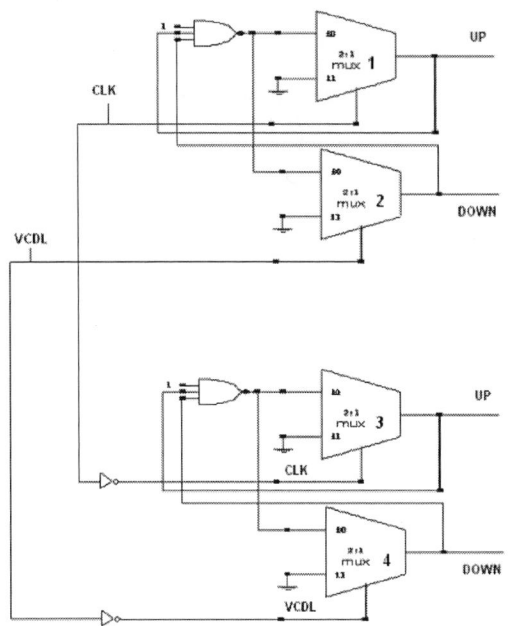

Fig. 4 Novel Dual PFD using Multiplexer

Table 1.Calculated power at 1GHz

	Conventional PFD	Novel PFD using MUX
Total number of Transistors	42	8
Power consumption	0.324mw	0.11mw

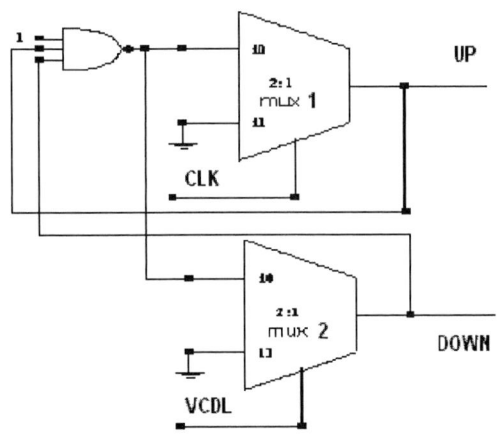

Fig. 3 Novel PFD using Multiplexer

UP output is generated because the CLK given to selection input of the Multiplexer (1) is activated before VCDL and the UP value is high so the reset value is low which in turn makes the input to be same. While DOWN is produced if VCDL is active before CLK because the VCDL given to selection input of the Multiplexer (2) is activated before CLK. As soon as the two clock signals are simultaneously active at the same time, ideally neither UP or DOWN has to be set. In such case, the reset signal R is given to the NAND gate

The conventional dual edge phase and frequency detector (DPFD) uses a four D Flip Flop structure. The DPFD compares both the falling and rising edge delay.If CLK is active before output of the Voltage control delay line (VCDL) the UP output is generated while DOWN is produced if VCDL(feedback) is active before CLK. As soon as the two clock signals are simultaneously active at the same time, ideally neither up or down has to be set.

In the design of our novel DPFD, have four multiplexer with two Multiplexer comparing the rise edge and other couple comparing falling edge. The design of the new DPFD has been constructed using two novel multiplexer based PFD .A basic schematic of DPFD is shown in figure 4 .This circuit's basic function is that the input signals which are to be compared are given to the select input of two multiplexers. If CLK is active before VCDL the UP output is generated while DOWN is produced if VCDL is active before CLK. When the two clock signals are simultaneously active at the same time, ideally neither up or down has to be set and the system is locked as there is perfect synchronization between CLK and VCDL (feedback).The objective of DPFD in this architecture is to compare the rising edge and the falling edge of the signal. so we invert the reference clock and the feed back and to similar PFD which will compare the falling edge of the input signals these two novel multiplexer based PFD will contribute DPFD, each PFD will have a charge pump there fore we have two control voltages VC1 and VC2.These control voltages as fed to the delay cell for appropriate delay generation corresponding to the reference clock.

3.2. Charge Pump

A charge pump [7]-[9], consists of two switched current sources that pump charge into or out of the loop filter according to two logical inputs. The circuit has three states .Charge pump consists of two PMOS and two NMOS, which are connected serially. Both of the NMOS are in the pull down section and both of the PMOS are in the pull up section. The gate of uppermost PMOS is connected to GND ,the gate of lowermost NMOS is connected to VDD. The gate terminals of the remaining NMOS and PMOS are connected to the "Down" and "Up" pin of the output of PFD. The Phase Difference is converted as control voltage.

3.3 Delay Cell

The Delay element ,shown in figure 5 ,of the Voltage Controlled Delay Line is made of current starved buffer, the transistor M0 and M3 receives the control voltages from the phase and frequency detector, these control voltages in turn adjust the delay by varying the voltages to the inverter. The current starved buffers are made of current starved inverters. The frequency of oscillation depends on the delay of each inverter, which is determined by the input capacity that each inverter

provides to its predecessor and the resistance that is inserted in the current path between the capacity and supply rails. In the simplest equivalent circuit, the inverter can be displayed as a switched resistor between Vdd, Gnd and the load capacity. The current starved buffer has two inverters , one of the inverters is a current starved inverter which is been controlled by the two control voltages ,VC1 and VC2 .These control voltages alters the delay according to the reference clock.The other inverter of the delay cell is a ordinary inverter .The transistor M4 and M5 are used to increase the speed of the delay cell.

3.4 Voltage Controlled Delay Line

In order to minimize the sensitivity to supply and substrate noise and to achieve a wide tuning range, the delay stage in [3] is used in the proposed DLL. This delay stage is built with a Current starved inverter (buffer) topology. This delay stage is built with a Current starved inverter (buffer) topology.

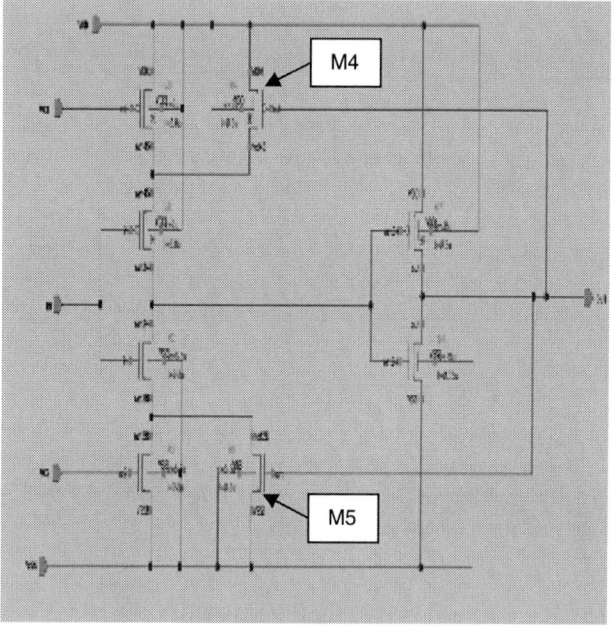

Fig. 5 Delay Cell

In the DLL clock generator, the accuracy of the generated setup clock and hold clock signals relies on the matching between the delay stages. In order to improve the matching, a shift-averaging VCDL [3] is used in the design. The shift-averaging technique can equalize the delay of each delay stage as well as improve the duty cycle of the generated clock signals [3]. This technique requires the VCDL to have an even number of delay stages. A design trade-off exists in the number of delay stage in the VCDL. More delay stages can enhance the phase resolution of the VCDL output signals, but at the same time more delay stages will increase the minimum VCDL delay and decrease the high end of the DLLs operating frequency range. On the other hand, fewer delay stages may boost the high end of the DLLs operating frequency range. Since the DLL can only generate clocks from a finite number of delay stages, the number of available time delays is limited by the number of delay stages in the VCDL. In other words, the target application influences the choice of the number of stages used in the VCDL.

4.Results and Discussion

In this section, we present the experimental results of the DLL. It has been designed using CADENCE VIRTUOSO & ORCAD PSPICE. This section will show the results of the simulations, and then present the measurements of the simulations. The Simulation is done Dual Loop DLL and Array DLL at 0.7μ and0.35μ CMOS technology.

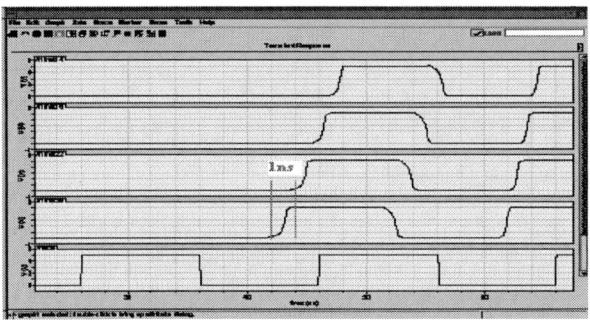

Figure 6: Output of Voltage controlled Delay Line

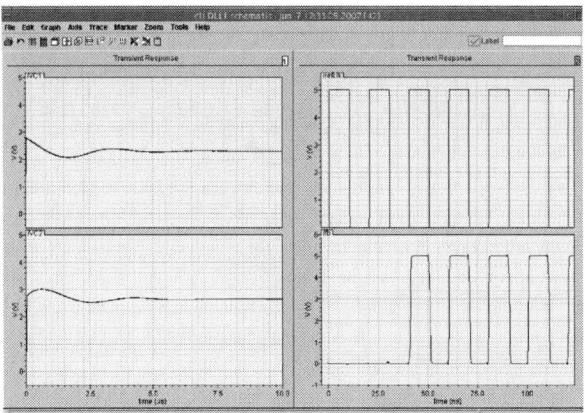

Figure 5: Simulation Output PFD and Charge Pump of Dual loop DLL

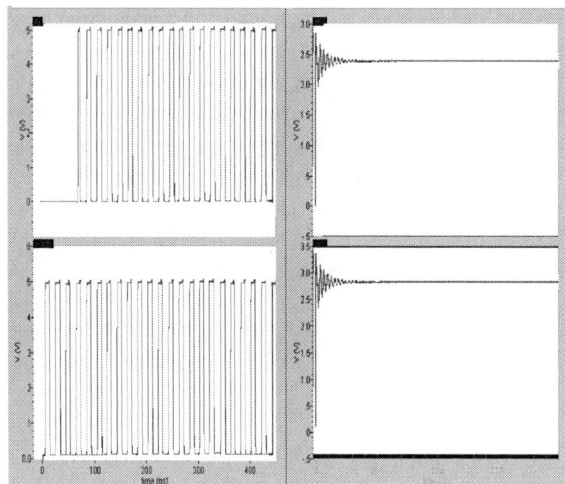

Figure 7: Output array of Dual (Control) loop Array Delay Locked

Table 2. Comparison of 0.7 μ and 0.35μ Delay Locked Loops

	Array Delay Locked Loops with dual control	Conventional Array Delay lock Loops
Locking Time (Array Dual Loop (control) DLL	≤2μs	≤10μs
Control Voltage	≤1.65V	≤2.5V
Jitter	≤25pS	≤80pS
Total number of transistor used in PFD	19	42
Total Power Consumed	1.12mW	2.15mw

5. Conclusion

A novel design of 1GHZ, CMOS Timing Generator Using Array of Delay Lock Loop has been implemented in both 0.7μ and 0.35μ technology. Various circuit and architecture parameter tradeoffs were evaluated. This new architecture outperforms comparable designs in performance, power dissipation with reduced jitter and using significantly less locking time. This architecture has a large degree of flexibility in design. The simulation results show that design is preferable for the next generation deep sub-micron CMOS DLL.The paper presents a novel architecture, using dual control voltages for SoC applications, the main advantage of this paper is that timing generator is implemented as an array of delay locked loops, this architecture enables a timing generator with sub gate delay resolution with high accuracy because both edges of the signal is controlled by dual control voltages, which reduces the jitter exponentially, this makes DLL ideal for the complete SoC implementation. The major trade off of complexity is reduced by novel multiplexer based PFD which uses reduced number of transistor there by reducing the power consumption. This design represents a drastic reduction of jitter compared to previously reported DLL implementations. An experimental prototype was simulated at 0.35 μm IMEC CMOS Technology with a supply voltage of 3.3V using Cadence Virtuoso. The Results shows exponential reduction in Jitter, power consumption and reduced locking time over a conventional DLL with comparable performance.

6.References

[1]J.Kostamovaara et al ., "An Integrated Digital CMOS Time-to-Digital Converter with Sub-Gate-Delay Resolution" Analog Integrated Circuits and Signal Processing,Springer, Netherlands,Jan 2000,pp.61-70

[2]J. Christiansen, C. Ljuslin, and A. Marchioro, "An integrated 16-channel CMOS time to digital converter," in Nuclear Science Symp. 1993, pp.625-629.

[3]J.Christiansen., "An Integrated High Resolution CMOS timing generator Based on an Array Of Delay Locked Loops" IEEE Journal, July 1996.

[4]Armin,"A Duty Cycle Control Circuit For High Speed Applications generation" in ISCAS '04

[5]S.Balaji., "Designing a low power multiplexer based phase and frequency detector for PLL" in 5th intl. conf .on Trends in Industrial Measurement and Automation,NITT,Jan 2007

[6]Von Kaenal , "A 320 Mhz ,1.5 mw @ 1.35V CMOS PLL for microprocessor clock generation" in ISSCC ''96

[7]Dinis M. Santos, "A CMOS Delay Locked Loop and Sub-Nanosecond Time- to-Digital Converter Chip" Vol.43,IEEE Trans.on Nuclear science June'96.

[8] Z. Wang, "An analysis of charge-pump phase-locked loops," IEEE Trans. On Circuits and Systems I, vol. 52, no. 10, Oct. 2005.

[9] P. K. Hanumolu, M. Brownlee, K. Mayaram, and U-K. Moon, "Analysis of charge pump phase-locked loops," IEEE Trans. on Circuits and Systems I, vol. 51, no. 9,Sep,2004.
[10] B. Razavi, Design of Analog CMOS Integrated Circuits, 2nd Ed., New York,McGraw-Hill, 2001.

[11] W. Chen. The VLSI Handbook. CRC Press, Boca Raton, FL, 2000

[12]R.J. Baker, H.W. Li, and D.E Boyce. CMOS Circuit Design, Layout, and Simulation.IEEE Press, New York, NY 1998

[13] R.E. Best. Phase-Locked Loops, 3rd Ed. McGraw Hill 1997

[14]Phillips Semiconductors, "An overview of Phase Locked Loops",AN177,Application Note,Dec. 1988.

[15]Phillip Allen, Douglas Holberg, CMOS Analog Circuit Design, Oxford Univ Press., Feb2002.

S.Balaji received Bachelors in Electrical and Electronics Engineering from Anna University and Masters in VLSI Design from Sathyabama University. His fields of interest are Analog and Digital Electronics, Low Power VLSI and Mixed Signal circuits. He was project trainee at Bhabha Atomic Research Centre, Mumbai, India and currently working in Cognizant technology solutions. Email:srinivasanbalaji@gmail.com

V.B Chandratre is Head, Microelectronics Section, electronics Division, Bhabha Atomic Research Centre, Mumbai, India. His fields of interest are Analog Electronics, Low Power VLSI and Mixed Signal circuits. He is currently working on high resolution time to digital converters, and high speed serial links

Menka Tewani is a Scientist officer (D), Microelectronics Section, electronics Division, Bhabha Atomic Research Centre, Mumbai, India. Her fields of interest are Analog Electronics and Mixed Signal circuits. She is currently working on high speed analog memory.

SESSION A5:
Synthesis

21st International Conference on VLSI Design

Variability-tolerant Register-transfer Level Synthesis

Anish Muttreja[†], Srivaths Ravi[‡], Niraj K.Jha[†]

†Dept. of Electrical Engineering, Princeton University, Princeton, NJ 08544

‡Texas Instruments India, Bangalore, Karnataka, India 560 093

muttreja@princeton.edu, srivaths.ravi@ti.com, jha@princeton.edu

Abstract

Variability in circuit delay is a significant challenge in the design and synthesis of digital circuits. While the challenge is being addressed at various levels of the design hierarchy, we argue that modern register-transfer level (RTL) synthesis tools can be enhanced to deal with this problem in an alternate, yet effective, manner.

Our solution involves the design of variability-tolerant, correct circuits assuming common-case, rather than worst-case, values for critical path delays. We propose a methodology to design variability-tolerant circuits that can, at runtime, detect and efficiently recover from delay errors, which would be inevitably introduced due to the use of common-case delay values. Variability-agnostic designs are automatically transformed into variability-tolerant circuits by the introduction of shadow logic to detect and recover from runtime errors, while exploiting data speculation to derive performance benefits. For various benchmark circuits, we show that the area overhead imposed by our scheme is only 11.4% on an average, while achieving upto 16.3% performance speedup over margined designs.

1 Introduction

Large variations in circuit delay are a significant bottleneck to the design and synthesis of digital circuits. Circuit delay depends on a number of factors, *viz.*, the values of present and past circuit inputs, operating conditions such as temperature and supply voltage, and for deeply scaled designs, process-induced device variations. Delay variability has become one of the most important design considerations because underestimating circuit delay during design can lead to errors at runtime. A straightforward variability-tolerant design technique is *worst-case* or margined design, wherein circuit delay is characterized over the space of the above variables [1] and worst-case delay values are used during design. Unfortunately, a worst-case design can severely limit opportunities to optimize circuit performance (reduce power consumption), for instance, by increasing clock frequency (reducing the supply voltage). A worst-case design point is particularly frustrating in situations where the critical delay distribution has a "long tail," *i.e.*, the worst-case delay, though improbable, is much greater in magnitude than the typical delay.

A common approach to escape the pessimism of a margined design is to design RTL circuits assuming a common-case operating point and process conditions. Undesigned-for variations in process conditions can be tackled by post-manufacturing delay testing, as is done for speed-binned circuits. Runtime variations might be addressed through the use of logic- or physical-level techniques where actual circuit speed is detected at runtime by the use of embedded test structures and the circuit operating point is adjusted in response. However, while synthesizing RTL components for use in larger systems, the availability of speed binning or lower-level closed-loop techniques to adjust the operating point cannot always be assumed. Furthermore, on-chip test structures may not be useful in predicting locality-sensitive variations accurately [2]. Such situations force RTL designers to stick to worst-case design.

In this work, we consider an RTL approach to variability-tolerant common-case design. We propose to "fix" a common-case design by directly detecting delay errors and recovering from them at runtime. First, we present previous research on variability-tolerant synthesis in Section 2. In Section 3, we introduce a design which is used as an example through the remainder of the paper. In Section 4, a scheme to catch delay errors, manifested as incorrect values being latched into registers, is considered. We then consider two schemes to recover from errors and demonstrate that both can deliver much higher throughput than worst-case design. The first scheme, considered in Section 5, is a simple dual clock frequency scheme that has small hardware overheads. The second, called RTL data speculation (Section 6) is based on the speculative use of critical path outputs and local recovery from misspeculation. Data speculation is more complex and requires both design and hardware support, but scales well to larger error probabilities. Experimental results are presented in Section 7, and conclusions in Section 8.

2 Related Work

Process variations had attracted the attention of researchers even before the start of the nanometer era [3]. The effect of process variations on circuit delay can be calibrated using statistical timing analysis tools [4]. From a synthesis perspective, a number of circuit-, logic- and architecture-level techniques have been proposed, some of which are described below.

Circuit-level techniques typically employ on-chip test structures to measure actual delay and leakage power [5]. Variations in delay/leakage are controlled through the adaptive control of power supply voltage [6], or transistor body bias voltages [7]. At the logic level, there has been recent work on statistical gate sizing to maximize timing yield [8] and minimize power consumption under yield constraints [9]. Statistical sizing techniques are used to obtain probabilistic delay constraints of the form:

$$P(delay \leq delay_{max}) \geq \eta \qquad (1)$$

978-1-4244-3039-0/08 $25.00 © 2008 IEEE

where $delay$ represents circuit delay, $delay_{max}$ the maximum allowed delay and η the confidence level of meeting this constraint. Statistical algorithms can often achieve lower circuit area and power for the same $delay_{max}$ than deterministic gate sizing, or when performing unconstrained minimization, yield much lower delays. Still, to ensure correctness, η is set to 99.9% to virtually guarantee no clock miss at runtime. We believe that further optimizations in the performance/power space are possible if the value of η can be relaxed. A framework to enable RTL design with relaxed delay constraints is the major contribution of this work.

Notable research at the system/architecture level includes a joint power/performance/variability metric to guide the microarchitecture [10] and variation-mitigating schemes for processor register file design [11, 12].

Our research is perhaps most closely related to the concept of circuit-level timing speculation for pipelined circuits [13], where cleverly modified latches, known as razor latches, were used to detect clock misses. While some circuit-level details and considerations in this work are therefore similar to the design in [13], we do not assume that the underlying circuit is pipelined. Delay errors in processor circuits, such as those examined in [13], can be recovered by flushing the pipeline. However, error recovery in non-pipelined circuits requires more extensive tool support and, to the best of our knowledge, has never been considered in previous works.

3 Example: Greatest Common Divisor

This section is devoted to the description of an RTL implementation for Euclid's greatest common divisor (GCD) algorithm, which is used as a running example to describe the challenges and demonstrate solutions throughout the paper. The design under consideration is concise and yet complex enough to capture most of the issues in variability-tolerant synthesis which we shall consider in this paper. The design was synthesized from a high-level description in [14], using a commercial behavioral compiler [15]. The pseudo-code for GCD, derived from the description in [14], is presented in Figure 1.

```
1 gcd(x, y){
2   while (x != y)
3     if (x > y)
4       x = x - y;
5     else
6       y = y - x;
7   return x;
8 }
```

Figure 1. High-level description of GCD

An RTL data path for our GCD implementation is presented in Figure 2(a), with the controller as specified in Figure 2(b). Parts of the data path shown shaded constitute additional hardware for error detection. Signals to control the shaded part are listed under *Error controls* in Figure 2(b). Further discussion of error detection is the subject of Section 4. A state transition graph (STG) for

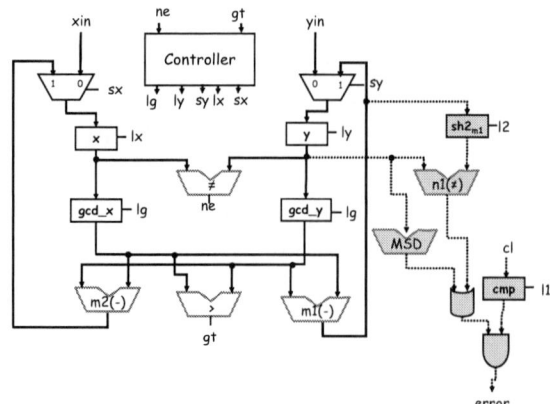

(a) GCD data path

States		Outputs					Error controls		
PS	NS	lg	ly	sy	lx	sx	$l1$	$l2$	cl
$st1$	$st2$	0	1	0	1	0	1	0	0
$st2$	$ne=0$ $st1$ / $ne=1$ $st3$	ne	0	0	0	0	0	1	0
$st3$	$st4$	0	0	0	0	0	1	0	0
$st4$	$st5$	0	0	0	0	0	0	0	0
$st5$	$st2$	0	gt	1	gt	1	1	1	1

(b) Controller specification for the data path

Figure 2. RTL implementation for GCD

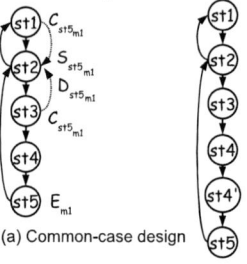

(a) Common-case design

(b) Worst-case design

Figure 3. State transition graphs for GCD

the controller finite-state machine (FSM) is presented in Figure 3. The controller was synthesized assuming nominal delay values for the critical paths.

As an illustration, we also synthesized another controller for GCD assuming worst-case delay values for all components. The STG for the worst-case controller, presented in Figure 3(b), requires an extra clock cycle compared to the common-case schedule in Figure 3(a). This may be understood by correlating the GCD algorithm in Figure 1 with the implementation in Figure 2. In state $st1$, new input values from signals xin and yin are loaded into registers x and y, respectively. In state $st2$, the not-equal-to comparison of line 2 is performed to check whether GCD computation for the current pair of input values has been completed. If so, the FSM jumps back to state $st1$ to load new input values. Otherwise, it continues with the execution of the while loop.

That computation should continue inside the while loop is indicated by signal ne being 1. Registers gcd_x and gcd_y are also loaded if $ne = 1$ in state $st2$. Note that subtract operations $x-y$ [1] and $y-x$, and the greater-

---

[1] Since registers gcd_x and gcd_y are loaded from registers x and y, respectively, it is convenient to transparently refer to the $gcd\_x - gcd\_y$ subtract operation as $x - y$.

than comparison $x > y$, are all computed in parallel in states $st3 \rightarrow st4 \rightarrow st5$. However, only one of the subtracter outputs is latched into register x or y, depending on the value of gt. With the given components, the schedule length is determined by the time required to complete the subtract operations. The data path of Figure 2(a) has two critical paths through subtracter $m1$ and multiplexer with select signal sy, and subtracter $m2$ and multiplexer with select signal sx, respectively, both of which are dominated by the delay of the subtracters. We modeled the subtracter delay, using Monte-Carlo simulation in SPICE, as a Gaussian distribution with mean $= 1ns$ and the 3σ point at $1.22ns$. A commercial 130nm CMOS process was used for obtaining the above estimate as well as all following experiments. The clock period was set to $0.35ns$. If a nominal delay is assumed for the subtracters, it can be seen that the three states $st3 \rightarrow st4 \rightarrow st5$ are sufficient for the above critical paths to complete execution. However, if a worst-case delay is assumed for $m1$, an extra state, labeled $st4'$ in Figure 3(b), is needed, leading to the six-state schedule in Figure 3(b).

The disadvantage of a common-case schedule is the possibility of delay errors at runtime. In the next section, we consider catching such errors.

4 Runtime Error Detection

Errors due to larger-than-expected delays in a combinational path p manifest themselves as incorrect values being latched into the register fed by p. Error detection can be based on the observation that a correct value is available at the output of p after a worst-case delay has been allowed to elapse. The error will then be indicated by the two output values of p, which are latched after the nominal and worst-case delays, being unequal. In Section 6, we propose a scheme to automatically modify RTL data paths in order to include error detection.

As an illustration, let us consider error detection in the GCD data path shown in Figure 2(a), under the scenario where $m1$ does not complete computation in time for the correct output to be latched into register y in state $st5$. The correct value of $m1$ is latched into register $sh2_{m1}$ in state $st2$. $sh2_{m1}$ and y are compared in the successor state of $st2$ to detect the error[2]. Since control flow diverges in state $st2$, the above comparison must be performed in both successor states, $st1$ and $st3$. The error is indicated in Figure 2(a) by signal $error$ (The metastability detector (MSD) block in Figure 2(a) is discussed below.)

The above comparison is, however, only valid if control arrived at states $st3$ or $st1$ through state $st5$ where a speculative assignment was made to register y. It would be erroneous to indicate an error otherwise. To remove this ambiguity in error detection, a control register cmp is added – we took the liberty of depicting it in the data path in Figure 2(a). The register is set in state $st5$ and cleared in states $st1$ and $st3$ to ensure that errors are only detected if control actually passed through state $st5$. Clearly, the presence of complex control flow makes error detection complex. This issue is revisited in Section 6.

Metastability: Another challenge in error detection for timing speculation schemes, such as ours, is metastability, which is to say that it is possible for the output of a register that contains speculatively-latched values to become *metastable*. A metastable signal is one that does not correctly resolve to logic 0 or logic 1, but instead hovers around $V_{dd}/2$. For instance, in the GCD data path, data might arrive at the input of register y at the same time as the clock, causing a hold-time violation and possibly leading to a metastable value at the output of register y. If the output of register y is metastable, it is impossible to determine whether a timing error has occurred using the scheme outlined above. It is thus crucial to reliably detect metastability at register outputs.

Below we outline a low-overhead metastability detection scheme, proposed in the Razor design [13], that was used in this work. An MSD block, constructed using two inverter gates with differently skewed P/N ratios, is included at the outputs of registers whose outputs might become metastable. Note that a detected metastability event can be treated the same way as a delay error. Therefore, since the output of the MSD block is logic 1 if metastability is detected, in Figure 2(a) it is ORed with the output of not-equal-to comparator $n1$ to indicate an error. However, a general principle is that complete system failure due to metastability cannot be ruled out, only its probability of occurrence can be reduced to negligible levels [16]. In the Razor design, as well as in this work, this manifests itself in a small, but finite, probability that the output of an MSD block can itself become metastable, causing a failure in error detection. An additional mechanism is, therefore, used to detect such failures, where the output of each MSD block is double-latched using two skewed flip-flops. The output of the second flip-flop is almost guaranteed to be stable, and a metastable event can, therefore, be detected with probability close to 1, by comparing the two flip-flop outputs. Under normal operation, both the flip-flops contain the same value. If, however, a difference is detected, error detection might have failed. Computation for the current input values must be then aborted and restarted. It should be noted that the probability of such an event is extremely low and restarted computations are very unlikely to face repeated restarts due to metastability.

5 Variability Tolerance with Dual Clocks

It is well known that logic delay is strongly dependent on input values – research in synthesis with variable-latency functional units [17] is based on this premise. A straightforward variability-tolerant synthesis technique based on the above premise and the ability to detect delay errors at runtime then suggests itself.

For example, under the assumption that delay misses are rare events, circuit schedules are designed using nominal delays for critical paths, such as in the GCD design in Figure 2(a). A pair of alternate clocks, fast and slow, are used. The system normally operates at the fast clock. However, when an error is caught, computation for the error-causing input values is restarted at the slower clock. Under the premise that delay errors only occur for some input values, the system can switch back to the faster clock on the next input value. This approach is evidently simple to implement and has small hardware overhead. It can also provide somewhat better performance than worst-case design. In Figure 4, we compare

---

[2]For ease of exposition, we only consider errors in $m1$ here. In our experiments, $m2$ was similarly instrumented.

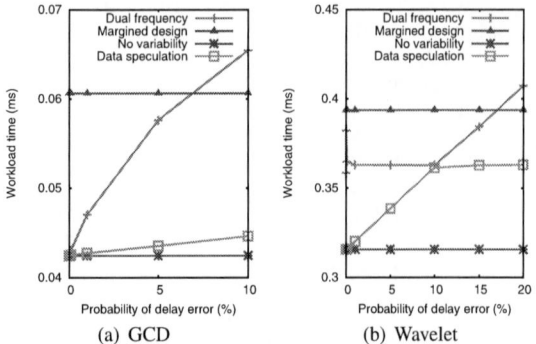

(a) GCD   (b) Wavelet

Figure 4. Performance of variability-tolerant synthesis techniques

the performance of this approach with the worst-case design approach. Results are presented for two benchmarks, our running example GCD, and a larger design, Wavelet, which implements a wavelet-based image filter. Each circuit was exercised with a large randomly-generated workload to measure its performance. GCD was simulated with 100 input pairs and Wavelet with a large dataset. The graphs plot the total time required to complete processing the given workloads against the probability of any critical path missing its deadline. A critical path which can miss its deadline is called a *non-robust critical (NRC)* path. The clock periods for GCD and Wavelet were set nominally at 0.35ns and 2.00ns, respectively. We wish to point out that in Figure 4, for the sake of illustration, the probability of a delay miss was varied independently of the clock period, which was held constant. In a more realistic scenario, such as the one considered in Section 7, the clock period would be optimized for each design causing the probability of a delay miss to vary in response to it.

Four synthesis scenarios are shown in Figure 4: the dual frequency approach considered here, a worst-case (margined design), a hypothetical ideal scenario which does not suffer from delay variability, and data-speculative designs synthesized using the approach considered in Section 6. GCD has two NRC paths, through the outputs of $m1$ and $m2$. Wavelet has four. For simplicity, all NRC paths are assumed to have the same error probability in this experiment. For GCD, the variability-tolerant design had a schedule length of five states compared to six for a margined design, as mentioned earlier. For Wavelet, the numbers were 38 and 55, respectively.

It is evident from Figure 4 that the dual-frequency approach can improve performance compared to a worst-case design if the probability of a delay error is modest. However, at larger probabilities, rising overheads make the scheme impractical as it performs worse than a margined design. Nevertheless, the performance of RTL data speculation, which is the subject of the next section, scales much better to high error probabilities.

6 RTL Data Speculation

As mentioned earlier, data speculation refers to the use of possibly incorrect logic values in dependent computations. The examples we considered in Section 5 demonstrate that this approach has low delay overheads and can yield large improvements even at high probabilities of delay errors. The high scalability of data speculation is actually attributable, not to speculation *per se*,

but to its ability to recover from misspeculation cheaply. Unlike frequency selection, the data speculation scheme considered here can recover from misspeculation by locally re-executing only the incorrect computations. Such local error recovery requires both hardware (runtime) and design support. In Section 6.1, we consider hardware support for data speculation. Design support, in the form of an RTL transformation to convert an unsafe speculative design to a safe one, is considered in Section 6.2. We wish to note that it is straightforward to obtain a design that uses data-speculative execution by simply allowing the use of nominal, rather than worst-case, delay values for logic paths during synthesis. The RTL transformation considered here is required in order to introduce the ability to *recover* from misspeculation. Thus, the five-state schedule for GCD in Figure 2(a) is speculative, but unsafe. We revisit our decision to treat the introduction of speculation and accompanying error recovery, respectively, as a process with two explicit steps (correcting unsafe designs *post-facto*), instead of a single-step process (designing safe speculative circuits), in Section 6.3.

6.1 Hardware Support for Data Speculation

To re-execute incorrect computations, we adopt the straightforward approach of restarting computation from a known correct state. A state is correct if no computation dependent on the misspeculated data value has been performed in its predecessors. Computation restarted from a correct state can be guaranteed to be correct. As an example, let us consider recovering from an error in the GCD design of Figure 2(a), where an incorrect value is latched into register y. The first computation that is dependent on the speculative y value is the not-equal-to comparison $x \neq y$ in state $st2$, which is, therefore, the last known correct state. To make the restarted operations execute correctly during recovery, the value of y is restored from the known-correct value $sh2_{m1}$, before restarting computation. There are, however, two sources of complexity in our scheme that were not present in [13]:

Loss of computation state: Computation state includes values in register and functional units. To ensure correctness, it is required that the computation state during error recovery be the same as it would have been during the initial speculative computation. Continuing our example, assume that speculative execution took the path $st5 \rightarrow st2 \rightarrow st1$. A misspeculation was discovered in $st1$ and control jumped back to $st2$. However, the value in register x, which is needed for computations in $st2$, is overwritten in $st1$.

Complex control flow: Control flow featuring branches and loops makes error detection and recovery more complex. For instance, error detection may need to be performed in state sti if a speculative register assignment was made in a previous state. In the presence of loops and branches, there might be more than one path that arrive at state sti. However, error detection should be performed only if control actually passed through a speculative assignment. In our running example, it is necessary to check for an error in state $st3$ if control arrived at it through the path $st5 \rightarrow st2 \rightarrow st3$. However, if the control path taken was $st1 \rightarrow st2 \rightarrow st3$, error detection in $st3$ would be spurious.

The above sources of complexity arise as a direct result of our desire to handle delay errors in their full-blown generality. For instance, loss of computation state is not a problem in [13] since a misspeculation is detected within half a clock cycle, before any values that are part of the state can be overwritten. In our approach, however, worst-case delay is allowed to exceed nominal delay by several cycles, so that by the time an error is detected, some of the values might have been overwritten. Also, our approach allows data speculation across control flow branches, unlike simplifying assumptions made in previous research on software data speculation [18].

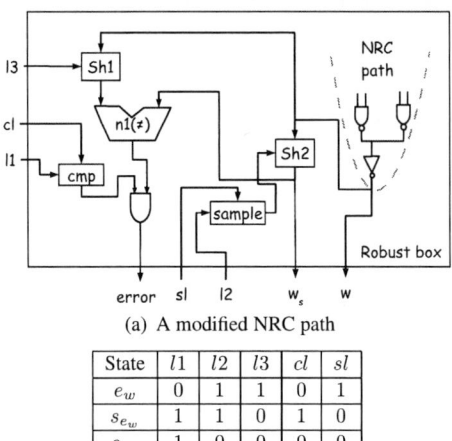

(a) A modified NRC path

State	$l1$	$l2$	$l3$	cl	sl
e_w	0	1	1	0	1
s_{e_w}	1	1	0	1	0
c_{e_w}	1	0	0	0	0

(b) Control signals for the NRC path in Figure 5(a)

Figure 5. A modified NRC path

Our choices necessitate the presence of guards around error detection code in some cases and extra registers to store the computation state that might be overwritten, but preserve the generality of the approach. These guards and other hardware support for error detection are introduced by modifying each NRC path. A modified NRC path is depicted in Figure 5(a) as enclosed inside a robust box that interacts with the outside world using output signals w, w_s and $error$ to provide the output logic values and indicate that an error has been detected, respectively, and input control signals $l1, l2, l3, cl$ and sl. w_s is guaranteed to be correct during recovery. To understand Figure 5(a), let us enumerate the stages involved in data speculation:

Early state, depicted as e_w, in Figure 5(b): In the early state, the NRC output w is speculatively latched, after a nominal delay has elapsed, for later comparison. For example, the output of $m1$ in Figure 2 is speculatively latched in register y in state $st5$. In Figure 5(a), w is also latched in a shadow register $sh1$, controlled by guard $l3$. The shadow register is used to guard against the possibility of y being overwritten by the time it is needed for error detection. Simultaneously, flip-flop $sample$ is set to 1. $sample$ is used to indicate that another sample of w must be taken in a later stage, after the worst-case delay has elapsed.

Sample state, s_{e_w}, in Figure 5(b): In the sample state, the correct value of w is stored in shadow register $sh2$. Note that $sh2$ is controlled by register $sample$, which would be 1 only if a speculative value was stored in $sh1$,

thus allowing control flow to be disambiguated. Also, in this cycle, flip-flop cmp is set to 1.

Compare state, c_{e_w}, in Figure 5(b): In the compare state, values in shadow registers $sh1$ and $sh2$ are compared to detect a delay error. An error detected by not-equal-to comparator $n1$ is only signaled to the outside world if $cmp = 1$, once again guarding against error due to complex control flow. It may not be evident from Figure 5(a), but the additional hardware in the form of added registers, flip-flops and comparator can be shared between NRC paths that are not simultaneously active, allowing much of the added area overhead to be recovered. Also, the additional registers are eliminated automatically unless required. As an example, error detection for GCD did not require register $sh1_{m1}$, since the y value was available in states $st3$ and $st1$[3]. Flip-flop $sample$ was also not needed. However, register cmp was required because control may arrive at states $st1$ and $st3$ through alternate paths in which a speculative value may or may not have been latched in register y.

6.2 RTL Transformation for Data Speculation

In this section, we consider an algorithm to automatically introduce error detection and recovery in speculative designs. Each NRC path is assumed to be modified as shown in Figure 5(a). We introduce some notation which will be useful in discussing the algorithm. We define $succ(e)$ as the set of states to which the controller might move in one clock cycle from state e. W is used to denote the set of wires that are outputs of NRC paths. For each wire $w \in W$, E_w is defined as the set of so-called *early* states during which w is speculatively latched into a register. For each state $e_w \in E_w$, we identify a set of w-dependent states in which the speculatively latched value of w is used for the first time.

$$D_{e_w} = \{s | \text{speculative value of } w \text{ is used in } s\} \quad (2)$$

Set D_{e_w} is typically a singleton containing the immediate successor of e_w. However, if control flow diverges in state e_w, it might have more than one member, each identifying a state d_{e_w} from which computation must be restarted for recovery. The following two sets of states are identified for each state $d_{e_w} \in D_{e_w}$:

$$S_{d_{e_w}} = \{s | w \text{ is sampled for the second time in} \atop \text{state } s\} \quad (3)$$

$$C_{d_{e_w}} = \{c | \text{a delay error for } w \text{ is detected in} \atop \text{state } c\} \quad (4)$$

Determining the above sets of states for each NRC path is the first step in our algorithm to introduce error detection and recovery (see Algorithm 1).

Next, we introduce some concepts to analyze dataflow. Define $defd(s)$ as the set of registers that are assigned in state s and $used(s)$ as the set of registers or wires that are used for some operation in s. The set of registers that constitute the computation state in

---

[3]Registers in our designs were loaded just before jumping to the next state, rather than being loaded immediately after jumping to it, as is perhaps more usually done. Thus, even though $ly = 1$ in state $st1$, the speculative value in register y is not overwritten until *after* comparator $n1$ has completed execution.

Algorithm 1: Transform-safe

input: FSM, W
1 **foreach** $w \in W$ **do**
2      $E_w = \text{identifyE}(w)$;
3   **foreach** $e_w \in E_w$ **do**
4          $D_{e_w} = \text{identifyD}(e_w, w)$;
5   **foreach** $d_{e_w} \in D_{e_w}$ **do**
6              $S_{d_{e_w}}, C_{d_{e_w}} = \text{identifySC}(d_{e_w}, w)$;
7   **foreach** $s \in d_{e_w} \cup S_{d_{e_w}} \cup C_{d_{e_w}}$ **do**
8   **foreach** $reg \in defd(s)$ **do**
9   **if** $reg \in live(d_{e_w})$ **then**
                            $\text{store}(reg, s)$;
10   **foreach** $fo \in used(s)$ **do**
11   **if** $\text{isTallFOp}(fo)$ **then**
                                $\text{store}(fo, s)$;
12                     $\text{addNRCcontrolsignals}(s, w)$;
13   **if** $s \in C_{d_{e_w}}$ **then**
14   restore lost state and jump to d_{e_w};
15   CleanUpRegisters;

state s can now be denoted as $live(s)$ and recursively defined [19], as follows:

$$live(s) = used(s) \setminus defd(s) \cup \bigcup_{s' \in succ(s)} live(s') \quad (5)$$

We are now in a position to explain the state annotations in Figure 3. The annotations depict sets E, D, S and C for our running example, considering a delay error due to $m1$. Set E_{m1} has only one member state $st5$, since the output of subtracter $m1$ is speculatively latched only in that state. $D_{st5_{m1}}$ is also a singleton, since state $st5$ has only one successor, state $st2$. Therefore, $D_{st5_{m1}} = \{st2\}$. Also, because the correct value of $m1$ is available in $st2$, $S_{st5_{m1}} = \{st2\}$. Assuming a comparator's delay is less than one cycle, $C_{st5_{m1}} = \{st1, st3\}$.

Let us examine Algorithm 1 now. It accepts the controller FSM and a set of NRC paths W in the circuit. It modifies the FSM to operate correctly in the presence of specified variations by sequentially introducing error detection for each wire $w \in W$. It begins by identifying the sets of states $E_w, D_{e_w}, S_{d_{e_w}}$ and $C_{d_{e_w}}$ in lines 2–6. Next, it determines if computation state might be lost during speculation. Since computation must restart from some state $d_{e_w} \in D_{e_w}$, we wish to ensure that the computation state associated with state d_{e_w} and its successors is preserved. Also, since speculative execution never visits a successor of any state $s \in S_{d_{e_w}}$, we need only concern ourselves with the set of states $d_{e_w} \cup S_{d_{e_w}} \cup C_{d_{e_w}}$[4]. Lost computation in these states may have two components:

1) Registers that are live in state d_{e_w} but would be overwritten before the computation is restarted. The loop in line 8 identifies such registers and stores them using the $\text{store}(r, s)$ function which introduces an extra register to store the value of r in state s.

---

[4]If the delay can vary across multiple cycles, there might be other states, denoted by set $F_{d_{e_w}}$ which are visited speculatively between states $d_{e_w} \in D_{e_w}$ and $S_{d_{e_w}}$. Algorithm 1 can be extended to handle this case by iterating over set $d_{e_w} \cup S_{d_{e_w}} \cup C_{d_{e_w}} \cup F_{d_{e_w}}$ in line 7.

2) Outputs of multi-cycled logic blocks which began computing before state e_w. Function isTallFOp checks if a wire is the output of such a "tall" logic block. Unless the inputs of a tall logic block were held constant until error detection and recovery had been completed, the output would be invalid during recovery. It is usually cheaper to store the output than hold all inputs of a logic block, but either solution could be employed.

Lost register values are restored by copying the stored values back into the register before jumping back for recovery. Logic block outputs are multiplexed to ensure that the stored value is used during re-execution. In line 12, the algorithm introduces control signals for error detection and recovery into sets $E_w, S_{d_{e_w}}$ and $C_{d_{e_w}}$ according to the specification in Figure 5(b). Finally, a cleanup phase in line 15 removes unneeded registers by sharing registers and eliminating unneeded hardware.

6.3 Implementing Error Recovery as a Post-scheduling Step

In the context of a synthesis system, the introduction of error recovery in a circuit can be considered an independent synthesis step. In this paper, error recovery is introduced using an RTL transformation after scheduling and binding have been performed.

In an alternative setup, data speculation and concomitant error recovery might be handled within the scheduler. This would allow data speculation to be considered along with other circuit transformations during design space exploration. In spite of its potential to improve performance, indiscriminate use of data speculation can actually degrade performance due to the overheads associated with recovery. Deciding where to employ data speculation during design space exploration is, therefore, an attractive way to limit potential overheads.

Instead, in our work, it is left to the designer to decide where data speculation should be employed. There are two main advantages to our approach. Firstly, a design constraint for this work was compatibility with a commercial behavioral compiler [15]. By using the RTL output of the compiler, as the input of our error-recovery transformation, this is achieved in a simple manner. In fact, manually-designed circuits or the RTL output of any other synthesis tool might also be used as an input to our transformation. On the other hand, integrating data speculation and error recovery directly into a scheduler would require significant changes in the given scheduler and may not always be possible. Secondly, we can fix potential clock misses that may only be discovered after detailed timing information from lower levels in the synthesis flow, *viz.* the logic or layout levels, is available. The designer can simply indicate such a clock miss and have the RTL description adjusted incrementally.

7 Experimental Methodology and Results

In this section, we evaluate the performance of RTL data speculation against a margined design approach for a number of benchmarks. Performance of both approaches is benchmarked against the hypothetical ideal scenario in which the critical paths do not suffer from any variation. All benchmarks, other than benchmark b04, were taken from the test suite in [14]. Benchmark b04 is a circuit that computes the minimum and maximum of a set of numbers and was taken from [20]. Benchmarks GCD and Wavelet were described in Section 5. Benchmarks Diffeq, Kalman and SOR imple-

626

978-1-4244-3039-0/08 $25.00 © 2008 IEEE

ment a differential equation solver, Kalman filter to track a system's state, and successive over-relaxation algorithm, respectively. All benchmarks were synthesized from their high-level descriptions using a commercial behavioral compiler [15]. An implementation of Algorithm 1 based on the CIL compilation framework [21] was used to post-process behavioral synthesis results in order to introduce data speculation.

Our criterion for performance is throughput, or else the time required by the circuit to complete execution of a large workload. A circuit's throughput is a function of the number of clock cycles consumed and the optimal clock period at which the circuit can operate. Determining the optimal clock period for each of the design approaches under consideration is the subject of Section 7.1. Experimental results are discussed in Section 7.2.

7.1 The Optimal Clock Period

It is straightforward to define the optimal clock period of a traditional design as the minimal clock period at which all critical paths meet their deadlines. On the other hand, operating at the minimal correct clock period may be suboptimal for a data-speculative design. This is because the probability of misspeculation and, therefore, the overhead imposed by it, typically increases with decreasing clock period. We propose the following analysis to estimate the expected time $T(p)$ required to compute a given benchmark at clock period p.

Using notation from Section 6, let E_w be the set of controller states, during which some NRC path w is activated. Assume an RTL execution trace, which lists the number of times $a(e_w)$ each state $e_w \in E_w$ is visited during the execution of the target benchmark, is available. Since it is possible to have more than one NRC path complete execution in a state, let $W_e = \{w_1, w_2, ..., w_n\}$ be the set of NRC wires that complete execution in state e. Denote by $m_i \in [0, 1]$ the event that wire w_i misses its deadline. Let $pm(m_1, m_2, ..., m_n, p)$ be the joint probability distribution of a combination of events m_i taking place. The probability distribution can be obtained as a function of clock period p using statistical timing analysis. Finally, let $o(m_1, m_2, ..., m_n)$ denote the overhead (in number of cycles) of events $m_1, m_2, ..., m_n$ being observed. Then, the following equation may be used to estimate T:

$$T(p) = p(N_o + N_v(p))$$
$$N_v(p) = \sum_{e \in E} a(e) \sum_{\overline{m} \in Z_2^n} pm(\overline{m}, p)o(\overline{m}) \qquad (6)$$

where E is the set of all early states associated with all NRC outputs, N_o is the number of cycles required by the benchmark if there was no misspeculation, $N_v(p)$ is the number of extra cycles because of misspeculation, and $\overline{m}$ is a vector that ranges over the space Z_2^n of n-dimensional binary numbers.

To illustrate the process, we consider the determination of the optimal clock period for our running example, GCD. As mentioned before, the two NRC paths in GCD, denoted w_1 and w_2, are composed, respectively, of subtracter $m1$ followed by a multiplexer feeding register y, and subtracter $m2$ followed by a multiplexer feeding register x (see Figure 2(a)). The worst-case delay for

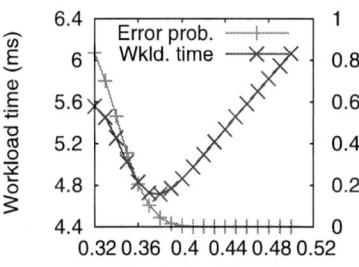

Figure 6. Probability of misspeculation and workload time for GCD

either path is the sum of the worst-case delays of a subtracter and a multiplexer, and register setup time. The total was estimated to be $1.245ns$. For error detection, we require that both w_1 and w_2 complete execution in four cycles, $st3 \rightarrow st4 \rightarrow st5 \rightarrow st2$. Thus, the minimal clock period at which the circuit can operate correctly is $1.245/4 = 0.312ns$. Accordingly, the range of allowed clock periods in this experiment was set as $0.32ns$–$0.50ns$.

To obtain the probability of misspeculation $pm(m_1, p)$, where m_1 is the event that w_1 misses its deadline, we conservatively modeled the delay distribution of w_1 as the sum of the probabilistic delay distribution for the subtracter and a worst-case delay value for the multiplexer. Assuming that the delay of the subtracter is modeled as a normal distribution with mean $1.00ns$ and the 3σ point at $1.22ns$, we obtain a normal distribution with mean $1.025ns$ and $3\sigma = 1.245ns$. An identical delay distribution is obtained for w_2.

Using the obtained delay distribution for w_1, probability $pm(m_1, p)$ can be obtained by straightforward integration and is plotted in Figure 6. We simulated the circuit to obtain visit count $a(st5)$ and used Equation (6) to estimate the workload time $T(p)$ as a function of the clock period. $T(p)$ is superimposed on $pm(m_1, p)$ to illustrate the tradeoff between clock period and misspeculation in Figure 6. The minimum workload time was found to be $47.1us$ with the clock period set at $0.38ns$.

In the ideal design scenario, it is assumed that the probability of misspeculation is zero. Therefore, the ideal design can not only utilize the speculative schedule, its optimal clock period is determined by the nominal delay of the critical paths in the speculative design. For instance, the optimal clock period for the ideal GCD implementation is determined by the restriction that the above-mentioned critical path through $m1$, with nominal delay $1.025ns$, completes execution in three cycles. Thus, we obtain $1.025ns/3 = 0.341ns$ for the minimal allowable clock period.

7.2 Experimental Results

For our experiments, we estimated the performance of each design at its optimal clock period. For margined designs, behavioral synthesis was performed using the worst-case delay of each functional unit. Speculative designs were derived from the corresponding margined designs: we first identified critical paths in the synthesized margined design where speculation could lead to a shorter nominal schedule length. Since the available behavioral synthesis system did not support data speculation, we synthesized unsafe common-case designs by redoing behavioral synthesis with nominal delay values

Table 1. Delay and area comparisons for variability-tolerant synthesis schemes

Bench.	N	Hypothetical ideal			Margined design				Data speculative design					
		L	Clock (ns)	T (μs)	L	Clock (ns)	T (μs)	slowdown	L	Clock (ns)	T (μs)	slowdown	speedup	ΔA
GCD	2	5	0.35	46.4	6	0.32	56.3	17.6%	5	0.28	47.1	1.5%	16.3%	29.1%
Wavelet	4	38	1.92	273.8	55	1.68	330.5	17.2%	38	2.01	293.6	7.0%	11.2%	10.3%
b04	3	6	0.26	16.6	7	0.26	19.2	13.5%	6	0.28	18.2	8.7%	5.2%	10.1%
Diffeq	1	56	0.94	11.2	60	0.95	12.8	12.5%	56	0.98	11.7	4.3%	8.6%	5.4%
Kalman	4	145	0.50	85.2	159	0.49	94.0	9.4%	145	0.49	88.6	3.8%	5.7%	6.1%
SOR	6	226	0.50	545.5	262	0.50	632.8	13.7%	226	0.51	580.7	6.1%	8.2%	7.1%

for some functional units in the identified critical paths, such as $m1$ in GCD. Algorithm 1 was used to introduce error detection and recovery into the obtained circuits.

Our results are tabulated in Table 1. We report the total time required and optimal clock period for the margined and speculative designs. The loss in performance, due to variability, of speculative and margined designs compared with the ideal scenario is reported in column *slowdown*. The speedup and area overhead (ΔA) observed with speculative designs over their margined counterparts are also reported. Area estimates for each design were obtained by performing logic synthesis using Synopsys Design Compiler [22]. A 130nm commercial CMOS process was used. The number of NRC paths that were modified is reported in column N. Column L lists the schedule length for each design. It is evident from Table 1 that improvements in performance can be obtained by speculative designs over their worst-case counterparts for a modest increase in area. Average reported speedup is 9.2% with the average increase in area being 11.4%. Also, more significantly, speculative designs achieve close to ideal performance. On an average, for our benchmarks, margined designs were 14.0% slower than the ideal scenario while data speculative designs achieved performance that was within 5.2% of it.

8 Conclusion

This paper demonstrated that RTL data speculation is a viable technique and, when guided by the designer, can be used to recover delay overheads imposed by variability. The paper provides a framework to integrate our methodology in a behavioral synthesis flow in a semi-automated manner. The next step for research efforts in this direction should be to integrate profile-guided data speculation into a high-level scheduler. It is our belief that such a development will expand both the applicability of data speculation as well as the benefits derivable from it.

References

[1] S. Nassif, A. Strojwas, and S. Director. A methodology for worst-case analysis of integrated circuits. *IEEE Trans. Computer-Aided Design*, 5(1):104–113, Jan. 1986.

[2] M. Orshansky, L. Milor, P. Chen, K. Keutzer, and C. Hu. Impact of spatial intrachip gate length variability on the performance of high-speed digital circuits. *IEEE Trans. Computer-Aided Design*, 21(5):544–553, May 2002.

[3] N. Herr and J. J. Barnes. Statistical circuit simulation modeling of CMOS VLSI. *IEEE Trans. Computer-Aided Design*, 5(1):15–22, Jan. 1986.

[4] C. Visweswariah, K. Ravindran, and K. Kalafala. First-order incremental block-based statistical timing analysis. In *Proc. Design Automation Conf.*, pages 331–336, June 2004.

[5] J. Tschanz, K. Bowman, and V. De. Variation-tolerant circuits: Circuit solutions and techniques. In *Proc. Design Automation Conf.*, pages 762–763, June 2005.

[6] S. Dhar, D. Maksimovi, and B. Kranzen. Closed-loop adaptive voltage scaling controller for standard-cell ASICs. In *Proc. Int. Symp. Low Power Electronics & Design*, pages 103–107, Aug. 2002.

[7] J. W. Tshanz et al. Adaptive body bias for reducing impacts of die-to-die and within-die variations on microprocessor frequency and leakage. *IEEE J. Solid-State Circuits*, 37(11):1396–1402, Nov. 2002.

[8] J. Luo, S. Sinha, Q. Su, J. Kawa, and C. Chiang. An IC manufacturing yield model considering intra-die variations. In *Proc. Design Automation Conf.*, pages 749–754, July 2006.

[9] M. Mani, A. Devgan, and M. Orshansky. An efficient algorithm for statistical minimization of total power under timing yield constraints. In *Proc. Design Automation Conf.*, pages 309–314, July 2005.

[10] D. Marculescu and E. Talpes. Variability and energy awareness: A microarchitecture-level perspective. In *Proc. Design Automation Conf.*, pages 11–16, June 2005.

[11] J. H. Tseng and K. Asanovic. A speculative control scheme for an energy-efficient banked register file. *IEEE Trans. Computers*, 54(6):741–751, June 2005.

[12] X. Liang and D. Brooks. Mitigating the impact of process variations on CPU register file and execution units. In *Proc. Int. Symp. Microarchitecture*, pages 504–514, Dec. 2006.

[13] D. Ernst et al. Razor: A low-power pipeline based on circuit-level timing speculation. In *Proc. Int. Symp. Microarchitecture*, pages 7–18, Dec. 2003.

[14] *High-Level Synthesis Benchmark Circuits*. http://www.ece.vt.edu/mhsiao/hlsyn.html.

[15] NEC Inc. *CyberWorkBench*. http://www.cyberworkbench.com/.

[16] N. Weste and D. Harris. *CMOS VLSI Design*. Addison Wesley, Boston, MA, 3rd edition, 2000.

[17] L. Benini, E. Macii, M. Poncino, and G. De Micheli. Telescopic units: A new paradigm for performance optimization of VLSI designs. *IEEE Trans. Computer-Aided Design*, 18(3):220–232, Mar. 1999.

[18] R. D.-C. Ju, K. Nomura, U. Mahadevan, and L.-C. Wu. A unified compiler framework for control and data speculation. In *Proc. Int. Conf. Parallel Architectures & Compilation Techniques*, pages 157–168, Oct. 2000.

[19] A. W. Appel. *Modern Compiler Implementation in ML*. Cambridge Univ. Press, Cambridge, U.K., 1998.

[20] *Int. Test Conf. '99 Benchmarks*. http://www.cerc.utexas.edu/itc99-benchmarks/bench.html.

[21] G. Necula, S. McPeak, S. Rahul, and W. Weimer. CIL: Intermediate language and tools for analysis and transformation. In *Proc. Conf. Compiler Construction*, pages 213–228, Apr. 2002.

[22] Synopsys Inc. *Design Compiler*. http://www.synopsys.com.

21st International Conference on VLSI Design

A Galois Field Based Logic Synthesis Approach with Testability

J. Mathew, H. Rahaman[a], A.K Singh[b], A. M. Jabir[c] and D.K Pradhan

Department of Computer Science, University of Bristol, UK
[b]School of Engineering, Curtin University of Technology, Malaysia
[a]Information Technology dept., Bengal Engineering and Science University, Shibpur, India.
[c]School of Technology, Oxford Brookes University, UK.

Abstract: In deep-submicron VLSI, efficient circuit testability is one of the most demanding requirements. Efficient testable logic synthesis is one way to tackle the problem. To this end, this paper introduces a new fast efficient graph-based decomposition technique for Boolean functions in finite fields, which utilizes the data structure of the Multiple-Output Decision Diagrams (MODD). In particular, the proposed technique is based on finite fields and can decompose any N valued arbitrary function F into N distinct sets conjunctively and N-1 distinct sets disjunctively. The proposed technique is capable of generating testable circuits. The experimental results show that the proposed method is more economical in terms of literal count compared to existing approaches. Furthermore, we have shown that the basic block can be tested with eight test vectors.

Key terms: Multiple-Output Decision Diagrams, testable, Galois Field, OBDD, Multipliers.

1. INTRODUCTION

Much progress has been seen in the automation of the design process for large integrated circuits. Tools for automatic synthesis play a crucial role in the progress of VLSI industry. With increasingly complex systems, an efficient testable synthesis process could help generating testable circuits. This paper will examine a Multiple-Output Decision Diagram (MODD) based logic synthesis process which produces more efficient testable circuits. Finite fields, also known as Galois fields, have gained wide spread uses in public-key cryptography, error detecting and correcting code [17], VLSI testing [24], digital signal processing [18].

In the design automation process, the role of Multi-valued functional representation is very important especially in the form of Multi-valued Decision Diagrams (MDD). Furthermore, word level diagrams can be useful in high level verification, and logic synthesis [1, 6, 12]. In this paper, the multi-valued functional representation using Galois field switching theory [10] is considered. A new decision diagram for representing multiple output binary functions was proposed in for the synthesis of Galois circuit, we decompose the circuit and re-express the logic expression in terms of a set of simple ones. The complete synthesis is shown in Fig. 1. In multi-level logic synthesis, a key step is to find the factored form for a logic function with fewer literals as much as possible. The number of literals in a factored form is directly related to the number of transistor pairs required to implement the function as a static CMOS gate. Plenty of research has been done in this area [13, 14, 21]. In particular algebraic logic minimization method such as MIS, are the most successful and relevant way to attain the minimization. This method is based on the cube set (or two level logic)

minimization and generates multilevel logic from cube sets by applying a weak division.

The advent of the Ordered Binary Decision Diagram (OBDD) decomposition produced strong impact on the logic synthesis methods and has been used in PTL synthesis [5], multilevel synthesis [1, 4, 6, 8]. In [1], the BDDs are used to obtain factored forms for logic functions. Boolean division and factorization based on BDD using internal cofactors were proposed in [8]. Yung et al. [7] presented an algorithm for disjunctive and non disjunctive decomposition of Boolean function and Boolean methods for identifying common sub functions from multiple Boolean functions using OBDD. A fast weak division method for cube set representation using Zero Suppressed Binary Decision Diagrams (ZBDD) is proposed in [6]. An algorithm for extracting a disjunctive decomposition from the BDD representation of function is presented in [4].

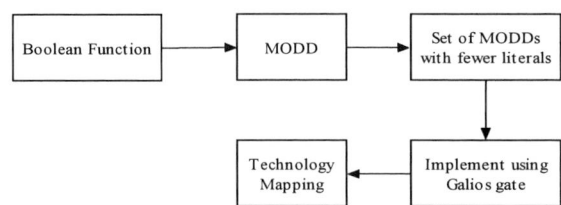

Fig. 1: Proposed Synthesis Flow

Yang [15] proposed a unified logic optimization method which handles both AND/OR intensive and XOR intensive function based on various dominators. Lang et al. [20] proposed algorithms that allow the realization of multi-valued functions as a multi-level network consisting of MIN and MAX gates. A method to decompose a multiple output circuit into two circuits with intermediate outputs using BDD for CF was presented in [25].

Considering the merits and demerits of existing techniques, this paper proposes a fast decomposition technique for function based on the MODD. The decomposition method presented in [7] is suitable for BDD based representations. The MODD is based on the literal forms of functions over the Galois fields.

2. PRELIMINARIES

Basically two types of decomposition can be performed on a function (i) Algebraic (ii) Boolean. In algebraic decomposition, the logic functions are treated as

978-1-4244-3039-0/08 $25.00 © 2008 IEEE

polynomials and based on division operation namely, rewriting a function F as $D \cdot Q + R$, where D, Q and R, are divisor, quotient and remainder, respectively. The theory of division was studied by Brayton et. al. and well developed in MIS package [2]. Algebraic methods can not use Boolean identities i.e. $x \cdot x = x, x \cdot \overline{x} = 0$ i.e. the don't care set; therefore fail to produce optimal factorization.

The MODD is a rooted directed acyclic graph (DAG) for representing multi-valued, multi-output function in literal form in GF(N) based on the following expansion [10],

$$f(x_1, x_2, ..., x_k, ..., x_n) = \sum_{e=0}^{N-1} g_e(x_k) f \big|_{x_k = \delta(e)}$$

where $g_e(x_k) = 1 - [x_k - \delta(e)]^{N-1}$.

The term $[1 - (x_k - \delta(e))]^{N-1}$ is a multiple valued literal in GF(N), where the Galois Field *GF(N)* is the set of N elements with two operators, '+' and '·'. The total number of elements of a finite field is a power of a prime number, i.e. $N = p^k$ where p is a prime number and k is an integer.

In an ordered MODD (OMODD) each variable is present at most once on each path from the root to a terminal node and if the variables are encountered in the same order of each path. A reduced ordered MODD (ROMODD) has no any non-terminal node which all the outgoing edge point to a same node and has no isomorphic sub graphs.

For example, Fig. 2 shows a 2×2 arithmetic multiplier using MODD in GF(4), where:

$$z_1 = \alpha \cdot X^\beta Y^\alpha + \alpha \cdot X \ Y^\alpha + X^\beta Y \ + \alpha \cdot X^\alpha Y \ + \beta \cdot X \ Y$$
$$\beta \cdot X^\beta Y^\beta + \alpha \cdot X^\alpha Y^\beta$$
$$z_2 = \beta \cdot X^\alpha Y^\alpha + \beta \cdot X^\alpha Y \ + \beta \cdot X \ Y^\alpha + \alpha \cdot X \ Y$$

The labeled edges 0, 1, α and β refer to the elements in *GF(4)*.

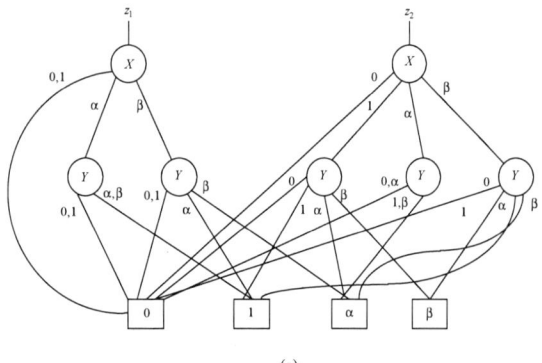

(a)

Fig. 2: MODD for the 2×2 arithmetic multiplier

An MODD can be seen as the tuple MODD = $[\phi, V, E, \{0, ..., N\text{-}1\}]$, where ϕ is the function node, V is the set of internal nodes representing the input variables. E is the set of edges and $\{0, ..., N\text{-}1\}$ are the terminal nodes.

Definition 1: The *leaf edge* is an edge $e \in E$, which is directly connected to a terminal node of the MODD. For example in *GF*(4), the set of leaf edge Σ can be partitioned into Σ_0, Σ_1, Σ_α and Σ_β, which are the sets of leaf edges $0, 1, \alpha, \beta$ respectively.

Definition 2: Zero, Non-Zero, and Don't Care Sets (F_i): A set of cubes for which $F_i = 0$ is called the zero set. A set of cubes for which $F_i \neq 0$ is called the non zero set and a set of cubes for which F_i is not specified is known as the don't care set.

3. DECOMPOSITION USING MODD

Decomposition is the process of representing a function $F(X,Y)$ as $F'(g(X),Y)$ such that the number of inputs of F' is smaller than that of F. The fundamental properties of Galois field allows any function to be decomposed in N different ways. In this paper we use the decomposition method based on horizontal cut proposed in [7] for BDD and extend to MODD with certain modification.

Definition 3: A *horizontal cut* $(D, V - D)$ of a MODD is a partitioning of its nodes into disjoint subsets (D) and $(V - D)$ such that $root \in D$, and $E \in (V - D)$, and their supports are disjoint. In other words, the horizontal cut can not cross a path more than once.

Definition 4: The *support* of a function is the variables on which the function depends.

Definition 5: The *dangling edges* are the edges, which are not leading to either terminal or non-terminal nodes.

Definition 6: Assuming a cut on an MODD splits it into two parts (1) the set of D and (2) the set of $(V - D)$, the upper part (D) is copied separately. The resulting graph is called *generalized dominator*, which decomposes the function conjunctively and disjunctively.

3.1. Conjunctive Decomposition

Definition 7: If an MODD of the function F can be decomposed into a quotient Q and divisor D such that $F = (D) \cdot (Q) = (D \cdot r_c) \cdot (Q \cdot r_c^{-1})$, where r_c is any non-zero element in GF(N), i.e. $r_c = \alpha, \alpha^2, ..., \alpha^{N-1}$ then F is said to be *Conjunctively Decomposable*.

Theorem 1: For a given generalized dominator of a function F, the D is obtained by assigning the dangling edge to r_c and connecting the terminal node by dividing them by r_c. The quotient (Q) is obtained by minimizing the original MODD by removing the OFF set path of D and connect the terminal node by multiplication of r_c.

Example 1: The procedure of conjunctive decomposition is shown in Fig. 3. For a function represented by MODD shown in Fig. 3(a), a horizontal cut is performed after the

630

978-1-4244-3039-0/08 $25.00 © 2008 IEEE

node x_2 at second level shown by dotted line. A generalized dominator is obtained by copying the upper part of graph as shown in Fig. 3(b). For generating the D part of function F we assign the non-terminal and dangling edge as $r = 1$. Therefore, $F = (1 \cdot D) \cdot (Q \cdot 1^{-1})$. Note that we are forcing to connect the non-zero terminal nodes to 1, even if it is not connected to 1 in original MODD. Since the function is in $GF(4)$, we can assign other values except 1 i.e. α, β and change the graph accordingly. Hence we can get three sets of $(D \cdot Q)$. In Figure 3(c) Q is generated by minimizing the original MODD using the off set path of D, which is shown by the arrow.

3.2. Disjunctive Decomposition

Definition 8: If the MODD can be decomposed as a sum of a divisor D and remainder R, such that $F = (D) + (R) = (D + r_d) + (R - r_d)$, then the function is said to be *Disjunctive Decomposable*.

Theorem 2: For a given generalized dominator of function F, the D is obtained by assigning the dangling edge to r_d, and connecting the terminal node by increment of r_d. The remainder (R) is obtained by minimizing the original MODD by removing the non-zero paths in D and decrementing the terminal nodes of the original MODD by r_d.

Example 2: The complete procedure of disjunctive decomposition is shown in Fig. 4. A horizontal cut is performed as in the previous example. The original MODD and the generalized dominator are shown in Fig. 4(a) and 4(b) respectively. For generation of D, the dangling edge is assigned to 0, and the terminal nodes are connected to the same terminal node as is in the original MODD, since $(D - 0) = D$. In Fig. 4(c), R is generated by minimizing the original MODD using the on set path of D, which is shown by the arrow and connected to the terminal node in the original MODD.

Note that the dangling edges can assigned to either 1, α, or β, but the remaining terminals have to be changed accordingly. Therefore, four sets of conjunctive decomposition is possible in *GF(4)*.

While performing a cut in an MODD a special case can arise where a node, apart from the root node, belongs to every path from the root node to one of the terminal nodes having the value k, where $k \in GF(N)$. Such a node is called a *k-dominator*. In that case we can decompose the MODD directly as shown in Fig. 5. In this figure node X_3 is an α-dominator, i.e. all the paths from the root node to the terminal node α go through X_3.

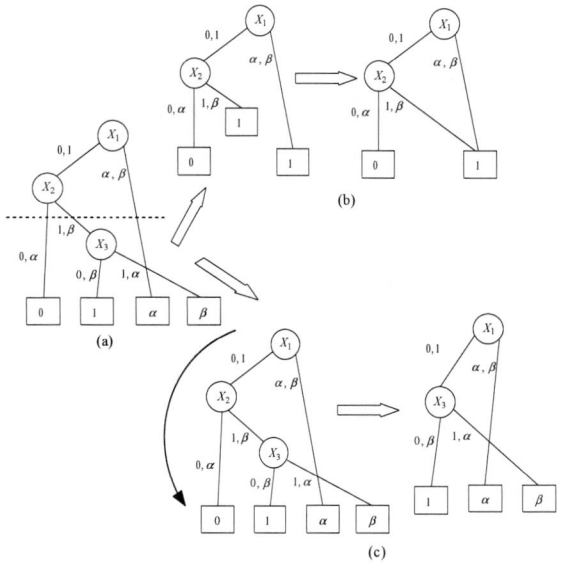

Fig. 3: Conjunctive Decomposition

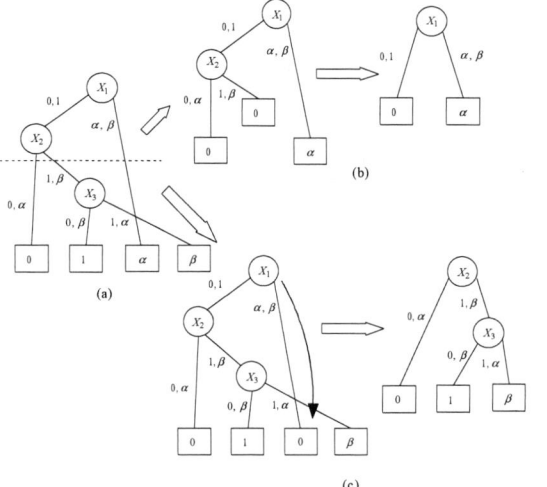

Fig. 4: Disjunctive Decomposition

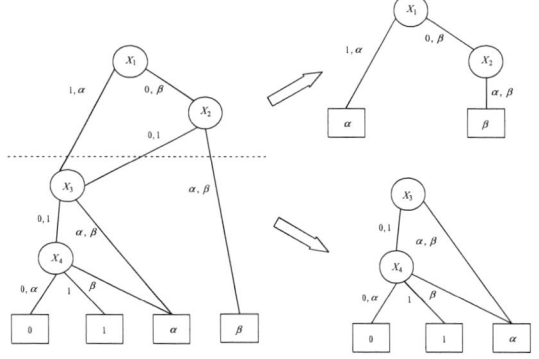

Fig. 5: An Example of α dominator

4. A FAST ALGORITHM FOR DECOMPOSITION

Fig 6 shows the outline of a greedy algorithm for decomposition a function F quickly into three components D, Q and R such that $F = D \cdot Q + R$, where the multiplication and addition are defined in GF(N).

```
1.  deque <int> NodeQ;
2.  Let 'root' be the root of the MODD;
3.  Let 'EtRoot' be the root of the ExprTree;
4.  Initialize NodeQ;
5.  Add root to ExprTree;
6.  Push root on to NodQ;
7.  do{
8.      ComputeParents(root);
9.      if(ParentList>0){
10.         Q = node with the maximum no. of parents.
11.         D = AssignPassAroundZ(Q);
12.         Assign all edges in D leading to Q to 1;
13.         R = AssignPassThroughZ(Q);
14.         D = reduce(D);
15.         R = reduce(R);
16.         Add Q, D, and R to ExprTree;
17.         Push Q, D, and R on to NodeQ;
18.     }
19.     NodeQ.pop_front();
20.     if(NodeQ is not empty) root = NodeQ[0];
21. } while(NodeQ is not empty);
22. EtRoot = EtReduce(EtRoot);
```

Fig. 6: A fast algorithm for decomposition

Line 10 constructs Q by selecting the node with the maximum number of parents. Line 11 assigns all paths *not* passing through Q to 0, while Line 13 assigns all paths passing through Q to 0. The decomposed expressions are stored as a shared binary expression DAG (ExprTree with root EtRoot), which is basically a shared netlist of 2-input GF(N) adders and multipliers. Line 22 reduces the netlist for sharing of resources. Simple optimization routines like $X + 0 = X$, $X \cdot 1 = X$, and $X \cdot 0 = 0$ are also performed on the netlist during this step. All basic operations, e.g., Lines 8, 11, 13-15, and 22 are $O(n)$ (n = number of nodes). It can be trivially argued that this algorithm ensures that the resulting netlist realizes disjoint expressions in GF(N).

5. EXPERIMENTAL RESULTS

Table 1 shows benchmark results from the IWLS'93 set. The results of the proposed technique appear in Column-6. For comparison results obtained by sis is also presented in Columns 4 and 5. In Column 4 the decomposition algorithm *fx* has been applied as follows: 1. simplify, 2. *fx*, 3. simplify. The total literal count is obtained in the factored form. The choice of *fx* has been made based on comparison with other techniques, such as *decomp*, *gcx*, and *gkx*, and it was concluded that *fx* was producing the smallest literal count in

the factored form. The results have also been compared with sis executed with *script.rugged*, perhaps somewhat unfairly. This is because *script.rugged* uses resynthesis and re-substitution in addition to decomposition, whereas the proposed technique is only a decomposition algorithm without any re-synthesis, re-substitution, or reordering. Clearly, for the majority of the cases the proposed technique has outperformed *fx* both in terms of speed and literal count. The proposed technique, despite being a decomposition technique without any resynthesis, has also outperformed sis with *script.rugged* for many benchmarks. Execution of sis with *script.rugged* had to be aborted for the benchmark alu4, because sis was taking too long.

Table 1: Comparison with SIS

BM	I/P	O/P	sis_decomp *fx*		sis *script.rugged*		Proposed decomposition		
			Lits	CPU (s)	Lits	Time (s)	Lits	Time (s)	Power μw
9sym	9	1	583	0.13	275	3.17	129	0.29	715
5xp1	7	10	350	0.07	132	0.43	146	0.21	820
9symml	9	1	341	0.11	227	2.32	129	0.29	715
rd53	5	3	166	0.07	34	0.11	55	0.05	266
rd73	7	3	663	0.31	189	1.36	125	0.20	452
rd84	8	4	1232	0.66	348	5.21	196	0.52	632
t481	16	1	2434	2.37	453	10.05	62	0.14	447
b12	15	9	1708	1.36	141	1.98	152	0.13	476
duke2	22	29	885	0.42	515	1.85	1648	10.88	1.64
alu2	10	8	709	0.41	372	17.99	308	0.36	583.33
f51m	8	8	302	0.10	117	0.26	179	0.21	623
table3	14	14	1812	1.46	1024	27.23	2790	21.4	3..931
misex1	8	7	100	0.03	60	0.09	75	0.07	327
misex2	25	18	186	0.06	106	0.13	167	0.12	382
b1	3	4	27	0.02	10	0.01	10	0.00	176
square5	5	8	209	0.06	67	0.22	70	0.04	345
alu4	14	8	5929	14.67	-	-	4199	67.63	6.622

Table 2 shows the proposed algorithm tested on adders (adr2-adr6) and multipliers (mul2-mul6) in addition to some of the IWLS'93 benchmarks in Table 1. In this case the field size has been varied from GF(2) to GF(64) (i.e.,

632

978-1-4244-3039-0/08 $25.00 © 2008 IEEE

1-bit to 6-bit word-sizes). The grouping has been done by placing adjacent bits together. Effort has been given to split the input and output bits into even word-sizes. Uneven words are padded with leading 0s to make the input and output word sizes the same. Columns with the heading "Basic Elements" represent total count of the 2-input GF(N) adders and multipliers. Although we have not reported the results on the gate count, the complexity of the circuits generated by the proposed technique can be reasoned about as follows: a 2-input GF(2^m) adder can be realized with m 2-input EXOR gates, while a 2-input multiplier can be realized in the worst case using m^2 2-input AND gates plus $3m$ 2-input EXOR gates [23]. Each 2-input EXOR gate requires roughly 1.5 times the chip area of 2-input AND gate. Using this formulation our decomposition technique requires fewer gates than the earlier reported techniques for many benchmarks. The power analysis is based on the 0.18 micron technology library.

Table 2: Results in different GF

Benchmark	GF(2)		GF(4)		GF(8)		GF(16)	
	Lits	Basic Elements	Lits	Basic Elements	Lits	Basic Elements	Lits	Basic Elements
9sym	129	128	-	-	70	61	-	-
b12	152	143	-	-	324	276	-	-
duke2	1648	1619	1940	1862	-	-	-	-
alu2	308	300	396	349	-	-	928	841
alu4	4199	4191	4679	4550	-	-	8038	7490
rd84	196	192	90	77	-	-	103	94
Adr2	26	23	33	23	60	51	-	-
Adr3	82	78	-	-	97	83	220	203
Adr4	216	211	232	199	-	-	321	299
adr5	504	598	-	-	-	-	-	-
mul2	22	18	22	15	-	-	-	-
mul3	108	103	-	-	103	88	-	-
mul4	467	459	-	-	-	-	446	415
mul5	1697	1687	-	-	-	-	-	-
mul6	-	-	4214	4030	6217	5596	-	-

6. DESIGN FOR TESTABILITY

The proposed design lends itself to admit easy testability. We

illustrate with an example over GF(2). Consider the MODD as shown in Fig 7 (a). Since for any given input combination only one path is activated, the design shown in Fig 7 (b) which is a direct mapping of the MODD is fully testable for single stuck at faults [24]. This principle can be extended to higher order fields. It may be noted that essentially what is proposed is a direct mapping of MODD to silicon.

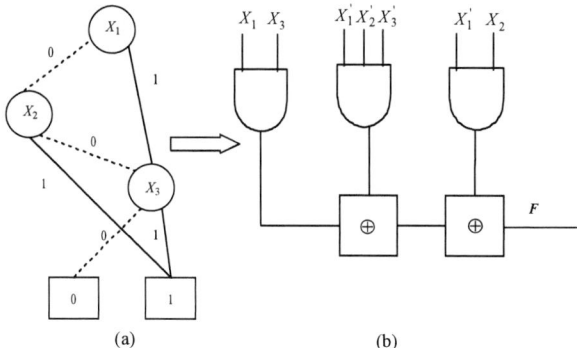

Fig. 7: MODD for *GF(2)* and its implementation

In this technique, any Boolean function can be implemented using only *GF* adders and multipliers. The low complexity Bit Parallel Polynomial Basis multipliers over $GF(2^m)$ proposed in [26] can effectively be used as the latency is very low. A GF adder is simply m two inputs EXOR gates. EXOR tree or EXOR cascade is easily testable for single stuck-at fault by 4 tests. In [24], a C-testable polynomial basis (PB) bit-parallel (BP) multiplier over $GF(2^m)$ for 100% coverage of stuck-at faults was presented. C-testability is achieved with three control inputs and approximately 6% additional hardware. Only 8 constant vectors are required irrespective of the size of the fields and the primitive polynomials to detect stuck-at faults in the multiplier circuit. The 4 tests which detect single stuck-at faults in the adder circuit are included in the 8 tests. In this implementation, the GF multiplier circuit outputs are EXORed to get the function outputs. As the GF multiplier circuit is AND-EXOR, the resultant circuit is also an AND-EXOR circuit. In [24], it has been proved that any AND-EXOR circuit is testable by only 8 tests. Therefore, this circuit is testable by the 8 tests.

Example 3: Consider a two-output four-variable function, where $y_1 = \Sigma$ (1, 2, 3, 9, 13, 15) and $y_2 = \Sigma$ (1, 2, 6, 7, 8, 9, 10, 11, 12, 14). The circuit is implemented as shown in Fig. 9 using GF adder and multiplier constructed by Irreducible Polynomial in $p(x) = x^2+x+1$ over GF(4). The GF multiplier circuit is shown in Fig. 8. IF $GF(2^m)$ multiplier is used in the realization, then number of horizontal EXOR cascade would be m. In an expression, if the output lines of a multiplier are feeding the input lines of another multiplier, then the output lines of the feeding multiplier will be partitioned from the input lines of the other multipliers using MUXs.

633

978-1-4244-3039-0/08 $25.00 © 2008 IEEE

The testable realization of the circuit of the Fig.9 is shown in the Fig. 10. When T=0, the testable circuit operates in the test mode. When T=1, the circuit will operate in the normal mode. The 8 test vectors derived in [24] is sufficient to detect all the single stuck-at faults in the circuit.

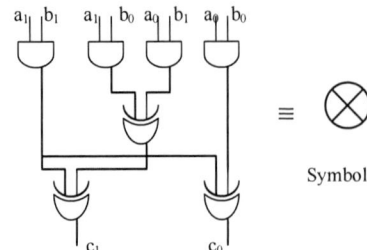

Fig. 8. $GF(2^2)$ multiplier

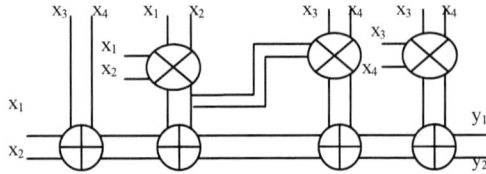

Fig. 9: Realization of the two-output function of Example 3.

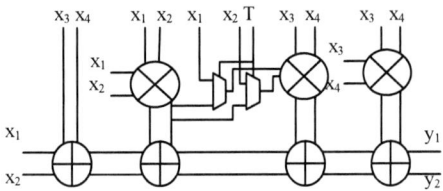

Fig. 10: Testable Realization of Example 3.

7. CONCLUSIONS

This paper presented a decomposition algorithm for highly-testable circuits in Galois fields. The 8 constant tests are sufficient to detect the faults in the circuit. The algorithm guarantees a network of disjoint products at each node thereby ensuring testability. For the time being the decomposition technique does not employ any re-substitution or reordering (e.g., reordering of the variables in the MODD) [7, 15] approaches. However, such approaches, both on the MODD and the netlist, can be applied to enhance the performance of the algorithm, which is mostly an implementation issue. The downside is that this would inevitably render the algorithm slow.

REFERENCES

[1] K. Karplus, "Representing Boolean functions with if-then-else DAGs", Board of studies in Computer Engineering", Univ. of Calif.-Santa Cruz, CA, Tech. Rep. USCS-CRL-88-29, Nov. 1988.

[2] R. K. Brayton, G. D. Hachtel and A. Sangiovanni-Vincentelli, "Multilevel logic synthesis," *IEEE Proc.*, pp. 264-300, Feb., 1990.

[3] Y. Ye and K. Roy, "A graph-based synthesis algorithm for AND/XOR networks," *Proc. DAC*, pp. 107-112, 1997.

[4] V. Bertacco and M. Damiani, "The disjunctive decomposition of logic functions", *Proc. ICCAD*," pp. 78-82, 1997.

[5] R. Chaudhry, T. Liu, A. Aziz and J. Burns, "Area-oriented synthesis for pass-transistor logic," *Proc. ICCAD*, pp. 160-167, 1998.

[6] S. Minato, "Fast Factorization Method for Implicit Cube Set Representation," *IEEE Trans. On CAD*, vol. 15, no. 4, pp. 377-384, April 1996.

[7] C. Yang, V. Singhal and M. Ciesielski, "BDD Decomposition for Efficient Logic Synthesis," *Proc. ICCD*, pp. 626-631, 1999.

[8] Ted Stanion and Carl Sechen "Boolean Division and Factorization Using Binary Decision Diagram," *IEEE Trans. on CAD*, vol. 13, no. 9, pp. 1179-1184, Sep. 1994.

[9] A.M. Jabir, and D.K. Pradhan "MODD: A New Decision Diagram and Representation for Multiple Output Binary Functions", Design Automation and Test in Europe 2004; Paris, France; Feb., 2004.

[10] D.K. Pradhan "A Theory of Galois Switching Functions", IEEE Transactions on Computers; vol. C- 27, no. 3, pp. 239 – 249; March 1978.

[11] Jabir A., Pradhan D., Mathew J. "An Efficient Technique for Synthesis and Optimization of Polynomials in GF(2m). IEEE International Conference on Computer Aided Design, November 2006, California, pp. 151–157.

[12] Randal E.Braynt, "Graph-based algorithms for Boolean manipulation," *IEEE Trans. on Computer*, vol. 35, no. 8, pp. 677-691, August 1986.

[13] U. Kebschull, E. Schubert and W. Rosenstiel, "Multilevel logic synthesis based on functional decision diagrams, "in *European Conference on Design Automation*, pp. 43-47, 1992.

[14] F. Mailhot and G. De Micheli "Algorithms for technology mapping based on binary decision diagrams and on Boolean operations," *IEEE Trans. on Computer*, vol. 12, no. 5, pp. 599-620, May 1993.

[15] C. Yang M. Ciesielski, "BDS: A BDD-Based logic optimization system" *IEEE Trans. on CAD*, vol. 21, no. 7 July 2002.

[16] W. Stallings, "Cryptography and Network Security", Englewood Cliffs, N. J.: Prentice Hall, 1999.

[17] S. B. Wicker, "Error Control Systems for Digital Communication and Storage" Englewood Cliffs, N. J.: Prentice Hall, 1995.

[18] R. E. Blahut, "Fast Algorithms for Digital signal Processing" Reading Mass: Addison-Wesley, 1984.

[19] T. Kam, T. Villa, R. K. Brayton and A. L. Vincentelli, "Multi-valued decision diagrams: Theory and Applications" *Multiple-valued Logic*, Vol.4, no.1-2, pp.9-62, 1998.

[20] Christian Lang and Bernd Steinbach, "Decomposition of Multi-Valued functions into min and max gates" Proceeding of ISMVL'01, pp. 173-178, 2001.

[21] Y. Jiang and B. K. Brayton, "Software synthesis from synchronous specifications using logic simulation techniques," *Design Automation Conference*, pp. 319-324, New Orleans, LA, U. S. A., June 10-14, 2002.

[22] C.-H. Wu, C.-M. Wu, M.-D. Shieh and Y.-T. Hwang, "High-Speed, Low-Complexity systolic design of novel iterative division algorithm in $GF(2^m)$" *IEEE Trans. Comp.*, vol. 53, pp. 375-380, March 2004.

[23] Peter Sweeney, "Error control coding from theory to practice" John Wiley & Sons Ltd, 2002.

[24] H. Rahaman, J. Mathew, and D. K. Pradhan, "Constant Function Independent Test Set for Fault Detection in Bit Parallel Multipliers over $GF(2^m)$", *VLSI Design 2007*, pp. 479-484.

[25] T. Sasao and M. Matsuura, "A method to decompose multiple-output logic functions," 41st Design Automation Conference, San Diego, CA, USA, June 2-6, pp.428-433, 2004.

[26] A. Reyhani-Masoleh, and M. A. Hasan, "Low Complexity Bit Parallel Architectures for Polynomial Basis Multiplication over GF(2m)", *IEEE Transactions on Computers*, Vol.53, No.8, pp.945-959, August 2004.

21st International Conference on VLSI Design

A Timing-Driven Synthesis Technique for Arithmetic Product-of-Sum Expressions

Sabyasachi Das
Synplicity Inc
Sunnyvale, CA, USA
Email: sabya@synplicity.com

Sunil P. Khatri
Texas A&M University
College Station, TX, USA
Email: sunilkhatri@tamu.edu

Abstract—The arithmetic Product-of-Sum (POS) is a frequently used datapath operation in modern integrated circuit designs, especially in Digital Signal Processing (DSP) and Graphics applications. Since POS blocks typically incur a significant amount of delay, these blocks often become critical in determining the performance of the entire chip. Hence, to improve the efficiency of this operation, it is desirable to use an architecture with good performance characteristics. This paper presents an architectural optimization approach to synthesize a faster POS block, which can be very useful in reducing the delay of the design without significantly impacting its area. Our architecture for the Product-of-Sum (POS) block is primarily based on the analysis of the corresponding Sum-of-Products (SOP) expression or a hybrid combination of the POS and the SOP expressions. In our technique, we extensively use arrival-times of the input signals of the Product-of-Sum block to determine the suitable architecture of the block. We have tested our approach using a variety of POS blocks implemented under varying timing constraints and technology libraries. Experimental results demonstrate that our proposed solution is 10.92% faster (and 3.54% larger) than the corresponding block generated by a commercially available best-in-class datapath synthesis tool. These improvements were verified on placed-and-routed designs as well.

I. INTRODUCTION

The complexity and the performance requirements of datapath operations implemented in systems-on-chips has increased considerably over the years. This is especially true in ICs used for communication, multimedia and graphic applications, because such circuits typically have highly parallel implementations of signal processing algorithms.

In modern digital designs, Product-of-Sum (POS) blocks are one of the most widely used arithmetic datapath operations. Since these blocks require intensive computations, they exhibit a significant amount of delay, and therefore they typically tend to be found in the timing-critical path of the chip. As a result, it is important to develop an efficient architecture that would reduce the delay of a POS block and thereby improve the performance of the chip.

A Product-of-Sum block can be implemented by using any adder/multiplier architecture. Carry lookahead adders based on parallel prefix computation methods are the fastest adders [1]. Kogge and Stone (KS) introduced a technique [2], where the prefix computation circuit is optimized for timing. In the Brent-Kung (BK) approach [3], the prefix computation is done in an area optimal way. A hybrid adder architecture based on

BK and KS techniques are introduced in [4]. Different adder families for prefix-computation are explained in [5].

Multiplication is a ubiquitous operation in digital signal processing (DSP) applications. In a multiplier, the reduction of partial products is the most timing-intensive operation. One of the popular techniques to implement the reduction phase was proposed by Wallace [6], where a column reduction technique is deployed, using a cascade of (3:2) counters. Dadda proposed a slightly different approach for the reduction of partial products [7], where the number of operands in each BitSlice are reduced less aggressively. An in-depth analysis of different column-compression techniques is presented in [8].

In this paper, we propose an approach to efficiently implement the arithmetic Product-of-Sum (POS) block. Our architecture is primarily based on the corresponding Sum-of-Products (SOP) expression or a hybrid combination of the POS and the SOP expressions. In our technique, we extensively use the arrival-times of input signals to the Product-of-Sum block, to determine the suitable architecture of the block.

We have organized the rest of the paper as follows: in Section II, we present background information and the definition of the problem we are addressing in this paper. In Section III, we discuss our proposed approach. Section IV presents experimental results. Conclusions are drawn in Section V.

II. PRELIMINARIES AND PROBLEM DEFINITION

In this section, we briefly explain the concept of the arithmetic Product-of-Sum (POS). A generalized Product-of-Sum block can be used to implement the multiplication of two sum terms representing an arbitrary additive (or subtractive) expression. As a consequence, a POS block is quite general. A generalized Product-of-Sum block can be used to implement arithmetic blocks having the following expressions (or combinations thereof): i) $z = (a + b) * (c + d)$; ii) $z = (a + b) * c$; iii) $z = (a + const_1) * c$; iv) $z = (a + const_1) * (c - d)$ v) $z = (a - const_1) * (c + const_2)$

Once a Product-of-Sum block is synthesized, the resulting netlist consists of the adders (to evaluate the sum terms) followed by the single multiplier (to evaluate the product term of the adder outputs). As described in the Section I, there are many well-known techniques to perform binary addition and multiplication, any of which can be used in the POS block.

978-1-4244-3039-0/08 $25.00 © 2008 IEEE

In any given adder architecture, the most expensive operation in terms of delay is *carry propagation*. On the other hand, the multiplier block also contains a binary carry propagate adder (CPA) inside it. As a result, the delay of the carry propagation through the final adder impacts the performance of the multiplier as well. In fact, experimental data shows that the delay of the Carry Propagate Adder contributes roughly 30%-40% to the total delay of the multiplier [9].

In the implementation of the POS block, since the outputs of the adders are multiplied by a binary multiplier, the resulting netlist has 2 carry propagate adders in the critical path. Since carry propagation is an expensive operation, this implementation contributes a significant amount of delay to the total delay of the POS block. Hence it becomes quite crucial to implement a Product-of-Sum in a manner that the critical path of the POS block does not contain 2 carry propagate adders.

In this paper, we propose an approach to efficiently implement the Product-of-Sum (POS) block. Our aim is to ensure that the critical path of the POS block does not traverse 2 binary carry propagate adders (CPAs) in cascade. In our proposed timing-driven architecture, we exploit the operator-merging based implementation [10] of the corresponding Sum-of-Products (SOP) expression. Depending on the arrival-times of the signals, our technique can also implement a hybrid combination of the POS and the SOP expressions.

A generalized Sum-of-Product block can be used to implement the addition of an arbitrary number (including zero) of product terms and sum terms. Many specialized techniques have been introduced to design arithmetic Sum-of-Product (SOP) blocks. Some of the well-known techniques are presented in [11], [12], [13] and [14].

III. OUR APPROACH

To facilitate the explanation, throughout the rest of the paper, we will use the example of a 4-input Product-of-Sum (POS) block, where each input is n-bits wide and output is $(2n+2)$-bits wide. The expression of the Product-of-Sum (POS) block is *assign z = (a + b) * (c + d)*

The expression of the equivalent Sum-of-Product (SOP) is: *assign z = (a * c) + (a * d) + (b * c) + (b * d)*

If the Product-of-Sum block is in the critical path of the design, we need to use a fast implementation for the block. The above-mentioned POS block requires two adders to implement $(a+b)$ and $(c+d)$ followed by a multiplier to multiply the outputs of the two adders. Similarly, to design the corresponding SOP block in the straight-forward way, we require four multipliers to implement $(a*c)$, $(a*d)$, $(b*c)$ and $(b*d)$ followed by an adder to add the outputs of the four multipliers. On the other hand, if we use the proposed hybrid implementation for the POS block, we again need to design one or more adders and one or more multipliers. The number of adder and multiplier modules used will depend on the input signal timing constraint and the characteristics of the technology cells available in the technology library. This indicates that for any type of implementation, to analyze the

performance of the POS/SOP/Hybrid block, we *estimate* or *model* the delays of the adder and the multiplier sub-blocks.

The final implementation will have one of the following three architectures:

- The Product-of-Sum (POS) block as-is.
- The equivalent Sum-of-Product (SOP) block.
- A hybrid implementation, where some portions of the functionality is implemented in the POS form and the remaining portion is implemented in an SOP form.

In the following two sub-sections, we discuss our analysis technique for each of these two sub-blocks (adder and multiplier) of the POS/SOP/Hybrid block. Finally, we discuss how we use these analysis to determine the timing-efficient architecture for the given POS block.

A. Design and Analysis of the Adder Sub-Block

Since carry lookahead adders based on parallel prefix computation methods are the fastest adders, we design the adder modules in each of our implementations (SOP/POS/Hybrid) by using the widely used Kogge-Stone adder approach [2]. Let us briefly explain the general idea behind the parallel prefix computation method and the Kogge-Stone approach.

Consider a binary addition block, which takes two n-bits wide inputs (a and b) and produces a $(n+1)$-bits wide output sum. In every bit (i) of the 2-operand adder block, the equation to produce the sum_i is (assuming $carry_i$ is the input carry for that bit): $sum_i = a_i \oplus b_i \oplus carry_i$.

Computation of the carry-in signals at every bit is the most critical and time-consuming operation. So, the primary focus of the carry-lookahead adder is to design a circuit which can efficiently compute the $(n-1)$ carry-in signals (c_1 to c_n) based on the $2n$ input bits (a_0, a_1, ..., a_{n-1} and b_0, b_1, ..., b_{n-1}). For any given bit-position, the *generate* (g_i) and *propagate* (p_i) signals are defined as follows: $g_i = a_i \wedge b_i$; $p_i = a_i \oplus b_i$

Let $B_{i,j+1}$ and $B_{j,k}$ be two adjacent blocks in an adder module. These two blocks consist of $(i-j)$ and $(j-k+1)$ bits respectively (the size of either one or both the blocks can also be equal to one) and $B_{i,j+1}$ consists of more significant bits than $B_{j,k}$. Now, if we combine these two adjacent blocks to form a single continuous block having $(i-k+1)$ bits, then the equations for computing the *generate* and *propagate* values of the combined block is as follows:

$$g_{i,k} = g_{i,j+1} \vee (p_{i,j+1} \wedge g_{j,k})$$
$$p_{i,k} = p_{i,j+1} \wedge p_{j,k}$$

The output of a parallel prefix computation tree is the set of all the $(g_{i,0}, p_{i,0})$ value pairs (for $i=0,1,...,(n-1)$). The equation to compute $carry_{i+1}$ is: $carry_{i+1} = g_{i,0} \vee (carry_0 \wedge p_{i,0})$

In [2], Kogge and Stone define an *"o"* operator, which performs the computation described in above equations (for any given *generate* and *propagate* value pairs $(g_{i,j+1}, p_{i,j+1})$ and $(g_{j,k}, p_{j,k})$). They suggest a divide and conquer tree-based computation approach, wherein pairs of adjacent blocks are merged together with the help of the *"o"* operator. This process is repeated until all the bits in the adder are consumed into a single block (in other words, until values of all $g_{i-1,0}$ (for $i=0, 1, ..., (n-1)$) are computed).

636

978-1-4244-3039-0/08 $25.00 © 2008 IEEE

To analyze the performance of this adder block, we use the technique presented in the Algorithm 1. In this algorithm, we assume that ta_i represents the arrival time of the ith bit of the signal a. This algorithm computes $tsum_i$ (for $i=0, 1, \ldots n-1$), which is the estimated time when the output at the ith bit will be available. In addition, Del_{pp} is the delay of the fastest prefix-computation cell ('o' operator) available in the technology library. Similarly, Del_{xor} is the delay of the fastest 2-input XOR cell and Del_{and} is the delay of the fastest 2-input AND cell present in the library.

Algorithm 1 : Analyze the performance of the Adder block

---

procedure $AnalyzeKSAdderDelay\ (a, b)$
// Loop-1: Find latest arriving signals (for each bit)
$Late_{(0,\ 0)} = \text{Max}(ta_0, tb_0)$
for i = 1 to $(n-1)$ **do**
  $Late_{(i,\ i)} = \text{Max}(ta_i, tb_i)$
  $Late_{(0,\ i)} = \text{Max}(Late_{(i,\ i)}, Late_{(0,\ i-1)})$
end for
// Loop-2: Compute the time when output bits will be ready
$tsum_0 = Del_{xor}$
for i = 1 to $(n-1)$ **do**
  $LogVal_i = \lceil log_2(i) \rceil$
  $tsum_i = (2 * Del_{xor}) + (LogVal_i * Del_{pp}) + Late_{(0,\ i)}$
end for
$tsum_n = tsum_{n-1} - Del_{xor}$
return all $tsum_i$ (for i = 0, 1, 2, ..., n)

In the algorithm, we first compute (in loop-1) the latest arrival time of signals in each bit. In addition, we compute $Late_{(0,\ i)}$, which is the latest arrival time among all the input signal across all bits starting from the 0th bit to the ith bit. This value ($Late_{(0,\ i)}$) is effectively the time when the execution of the parallel-prefix computation tree starts for the ith bit.

Next, in loop-2, we estimate the overall delay of the adder block designed with the Kogge-Stone architecture. We compute $LogVal_i$ to determine the number of levels of the 'o' operators needed for each bit (the number of levels required at each bit is different). Once we identify the delay of the fastest 'o' operator cell available in the provided technology library, this $LogVal_i$ is required to determine the delay of the prefix-computation tree for each bit. In addition, we consider two XOR-gate delays required to form the p_i values before the prefix-computation tree and the final sum output of the adder (sum_i) after the prefix-computation tree. In this way, our algorithm computes $tsum_i$ (for $i=0, 1, \ldots n$), which is the estimated time when the output at the ith bit of the adder will be available.

B. Design and Analysis of the Multiplier Sub-Block

To implement a multiplication between two vectors (a and b), we need to design the following three parts:

- Partial products Generator (PPGen)
- Partial Product Reduction Tree (PPRT)
- Final Carry propagate Adder (CPA)

Partial products are generated by performing a bit-wise multiplication between the appropriate bits of the multiplicand and the multiplier. Each partial product is shifted by one or more bits, depending on the bit number of the multiplicand. If PP_i is the ith partial product of the product term ($a*b$), and b has n bits, then partial products can be represented by the following expression: $PP_i = a*b_i*2^i \quad for\ i = 0, 1, \ldots(n-1)$

Algorithm 2 : Analyze the performance of the Multiplier

---

procedure $AnalyzeMultDelay\ (a, b)$
// Step-1: Compute the Timing at the end of the PPGen
for $i = 0$ to $(n-1)$ **do**
  for $j = 0$ to $(n-1)$ **do**
    $tppgen_{(i,j)} = \text{Max}(ta_i, tb_j) + Del_{and}$
  end for
end for
for $i = 0$ to $(2n-1)$ **do**
  Populate all the elements in $BsList_i$
end for
// Step-2: Compute the Timing at the end of the PPRT
for $i = 0$ to $(2n-1)$ **do**
  Sort $BsList_i$ in ascending order of the arrival time
  total = Elements in $BsList_i$
  while $total \geq 2$ **do**
    Select the three signals (s_{i1}, s_{i2} and s_{i3}) with the
      smallest arrival times
    Perform $(3:2)$ reduction of signals s_{i1}, s_{i2} and s_{i3}
    Compute arrival time of sum and add to $BsList_i$
    Compute arrival time of carry and add to $BsList_{i+1}$
    Remove signals s_{i1}, s_{i2} and s_{i3} from $BsList_i$
    $total = total - 3$
  end while
  if $total \leq 2$ **then**
    $v1_i$ = Remaining 0th element in the $BsList_i$
    $v2_i$ = Remaining 1st element in the $BsList_i$,
    t_{v1i} = Arrival Time of $v1_i$; t_{v2i} = Arrival Time of $v2_i$
  end if
end for
// Step-3: Compute the Timing at the end of the CPA
$finalDelayMult$ = AnalyzeKSAdderDelay($v1$, $v2$)
return all $finalDelayMult_i$ (for i = 0, 1, 2, ..., $(2n-1)$)

After the generation of the partial products, all these n partial products need to be reduced to two vectors. To perform this operation, a *Partial Product Reduction Tree* is used. To design the partial product reduction tree, we have used the Wallace tree column-compression technique [6]. This technique reduces the addition operations of any BitSlice into a cascade of full adder (3:2 counter) operations, yielding two addends per BitSlice.

After the reduction of partial products, the resulting two vectors get fed to a final carry propagate adder (CPA) which produces the final result of the Multiplier block. Since the functionality of this final carry propagate adder is identical to that of a stand-alone adder, we can use the technique to estimate the delay of an adder (explained in the previous subsection) in this situation as well.

To analyze the performance of the multiplier sub-block, which performs all three above mentioned steps, we use the technique presented in the Algorithm 2. In this algorithm, we assume that both the input signals (a and b) are n-bits wide and ta_i indicates the arrival time of the ith bit of the signal a. This algorithm computes two vectors each having a width of $(2n-1)$ bits. We denote $tppgen_{(i,j)}$ as the time when the element in the jth BitSlice of the ith partial product vector will be ready. After the generation of the partial products, we assume that the list of elements in BitSlice$_i$ is denoted by BsList$_i$ (for $i = 0, 1, ..., (2n-1)$). To maintain functional correctness in BsList$_i$, we perform appropriate shifting of the partial product elements. We denote $finalDelayMult_i$ (for $i=0, 1, ... n-1$) as the estimated time when the multiplier output at the ith bit will be available. Similarly, Del_{xor} is the delay of the fastest 2-input XOR cell and Del_{and} is the delay of the fastest 2-input AND cell present in the library. The following is a brief description of the Algorithm 2.

In the first step of the Algorithm 2, we compute the timing numbers at the end of the partial product generation phase. After that, we populate the BitSlice structure by putting all members of a given BitSlice (i) into a list called BsList$_i$. In step-2, we design the partial product reduction tree by using (3:2) reduction counters and compute the timing of each newly generated member element in each BitSlice, which get added to BsList$_i$ and BsList$_{i+1}$. At the end of this reduction phase, we create two vectors ($v1$ and $v2$) having a width of $(2n-1)$ bits each. The arrival time of a given bit of these two vectors are stored as t_{v1i} and t_{v2i}. In the third and final phase of adding $v1$ and $v2$, we use the $AnalyzeKSAdderDelay$ routine described in Algorithm 1. Finally, the output delay vector of $finalDelayMult_i$ (for $i = 0, 1, ..., (2n-1)$) is returned from Algorithm 2.

Note that the structure of a Sum-of-Products (SOP) block is quite similar to that of a multiplier. The only significant difference is that the Multiplier has one product term and no sum term, whereas the SOP can possibly have multiple product terms and multiple sum terms. That means that the number of partial products in the SOP could be more than that in a multiplier, leading to more elements in each BitSlice of the SOP. Hence for an SOP, we can use the similar estimation techniques in each of the three steps (Generation of partial Products, Reduction of Partial Products, Final Carry Propagation Addition) with minor modifications.

C. Design and Analysis of the Whole POS Block

Once we have modeled the delays of the stand-alone Adder, Multiplier and SOP blocks, we need to identify the efficient timing-driven implementation for a given Product-of-Sum expression. For this purpose, we use the technique presented in the Algorithm 3. Detailed comments are provided below:

In this algorithm, we first analyze if any of the sum terms present in the original product-of-sums expression need to be added with stand-alone adders. This is done in the Step-1 of

the algorithm, where we compute the output timing of the stand-alone adder and compare it with the arrival times of the other inputs of the Product-of-Sum block. If this analysis indicates that the outputs of the stand-alone adder will be available before the other input signals of the POS block arrive, then we implement this stand-alone adder. If a stand-alone adder is used, then the expressions of the POS gets simplified. For example, if the original POS expression is $z = (a + b) * (c + d)$ and the functionality of $(a + b)$ gets implemented as a stand-alone adder, then the modified expression of the POS becomes: $z = tmp1 * (c + d)$, where $tmp1$ is the output of $(a + b)$. This analysis (and the updating of the POS expression) is performed for each of the sum terms of the POS block.

Algorithm 3 :Implement the Timing-critical Product-of-Sum

procedure $ImplTimingCriticalBlock$ (input signals)
// Step-1: Any sum term needs stand-alone implementation
for each of the sum terms (like $a+b$) in the block **do**
  $PerformAdd$ = TRUE
  $delayAddAB$ = AnalyzeKSAdderDelay(a, b)
  for each of the remaining terms (like c etc.) in the block **do**
    for $i = 0$ to $(n-1)$ **do**
      if ($delayAddAB_i > tc_i$) **then**
        $PerformAdd$ = FALSE
      end if
    end for
  end for
  if $PerformAdd$ is TRUE **then**
    Implement the stand-alone adder ($a+b$)
    update the POS expression such that the output of ($a+b$)
      becomes a primary input to the modified POS
    Store $delayAddAB_i$ as the arrival-time of the new
      primary input of the modified POS
  end if
end for
// Step 2: Analyze POS implementation of the modified POS
for each of the remaining sum terms (like $c+d$) in the block **do**
  $delayAdderCD$ = AnalyzeKSAdderDelay(c, d)
end for
$delayFinalPOS$ = AnalyzeMultDelay(outputs of adders)
// Step 3: Analyze SOP implementation of the modified POS
Transform the modified POS to an SOP expression
$delayFinalSOP$ = AnalyzeMultDelay(inputs to SOP)
Compare $delayFinalSOP$ with $delayFinalSOP$ values
and return the faster implementation

At the completion of the Step-1, we have a potentially modified POS expression. In Step-2, we analyze the delay of the POS-based configuration. This is computed by using our previous approaches of modeling multiplier delays and the adder delays. In Step-3, we transform the Product-of-Sum (POS) expression into a Sum-of-Products (SOP) expression. After that, we analyze the delay of the SOP-based configuration. This analysis is performed by a simple extension of our multiplier delay modeling, as explained earlier.

Once the delays of the POS-based and the SOP-based configurations are computed, we compute the timing slacks at each output pin of both the configurations. The timing slack at a given output pin is defined as the difference between the required time and the time at which the data becomes available. From these two configurations, we finally select the configuration which produces a better worst case slack.

Note that during technology mapping in our approach, the mapper sizes the output of any node based on their load capacitance. Also, the delay analysis for each configuration considers actual capacitance of the output node, using a load-dependent delay model. Also, note that any of our nodes inside the POS block do not have high fanouts.

By using this algorithm, we can select either a fully-SOP based implementation, a fully-POS-based implementation or a hybrid implementation.

IV. Experimental Results

We have implemented our proposed approach in the C++ programming language. For all our experiments, we used a Linux workstation running on RedHat 7.1 with the dual-2.2 GHz processors and 4GB memory.

To test the effectiveness of our approach under varying design conditions, we used the following design constraints:

- Multiple types of Product-of-Sum designs of different expressions and input bit-widths: In the Table-I, we report 6 different configurations of the designs that have been used in our experiments.

Name of the Product-of-Sum (POS) Block	Functional Equation of the Product-of-Sum (POS) Block	Widths of the Input Signals of the POS Block
POS-A	$z = (a + b) * (c + d)$	16, 16, 16, 16
POS-B	$z = (a + b) * c$	18, 15, 21
POS-C	$z = (a + k_1) * (c + d)$	14, 23, 19
POS-D	$z = (a + b) * (a - b)$	24, 18, 14, 21
POS-E	$z = (a - k_1) * (c - k_2)$	32, 29
POS-F	$z = (a + b) * (c - d)$	36, 29, 33, 30

TABLE I

Characteristics of Different Product-of-Sum (POS) Blocks

- Different technologies and libraries: we used 3 technology libraries, provided by well-known commercial library vendors. One library (Lib-A) for 0.13μ technology, one library (Lib-B) for 0.09μ and one (Lib-C) for 0.065μ.
- Different input arrival time constraints:
  To facilitate the explanation, let us assume that the expression of the POS is $Z = (a * b) + (c * d)$ and each of the four input signals is n-bits wide. We have used the following types of input arrival time constraints:
  - All input bits of all the signals arrive at the same time. We refer to this constraint as Type-A. If we denote $Arr(a_i)$ as the arrival time of the bit a_i and if k is a constant, then this Type-A constraint can be represented as:
  $$Arr(a_i) = k; \qquad 0 \le i < n$$

$$Arr(b_i) = k; \qquad 0 \le i < n$$
$$Arr(c_i) = k; \qquad 0 \le i < n$$
$$Arr(d_i) = k; \qquad 0 \le i < n$$

This category represents the actual timing situation when the POS block is placed immediately after a register-bank, or if the primary inputs of the design are fed to the POS block.

- Different input bits arrive at different times. We refer to this category of timing constraints as Type-B. We believe that this category represents the actual timing situations in most of the Product-of-Sum blocks in real-life designs, assuming they are driven by other datapath blocks. Assuming that k is a constant and δ is the delay of the fastest 2-input AND-gate in the given technology library, the following are some specific examples of the Type-B timing constraints. We explain the Type-B constraints using the arrival times for signal a_i. Similar expressions for arrival times would apply to all the bits of all the input signals of the POS block.

1) $Arr(a_i) = i * k * \delta;$ $0 \le i < n$
2) $Arr(a_i) = i^2 k \delta;$ $0 \le i < n$
3) $Arr(a_i) = 0;$ $0 \le i < \lceil n/2 \rceil$
   $Arr(a_i) = k\delta;$ $\lceil n/2 \rceil \le i < n$
4) $Arr(a_i) = 0;$ $0 \le i < \lceil n/4 \rceil$
   $Arr(a_i) = k\delta;$ $\lceil n/4 \rceil \le i < \lceil n/2 \rceil$
   $Arr(a_i) = 2k\delta;$ $\lceil n/2 \rceil \le i < \lceil 3n/4 \rceil$
   $Arr(a_i) = 3k\delta;$ $\lceil 3n/4 \rceil \le i < n$
5) $Arr(a_i) = 0;$ $0 \le i < \lceil n/4 \rceil$
   $Arr(a_i) = ik\delta;$ $\lceil n/4 \rceil \le i < \lceil n/2 \rceil$
   $Arr(a_i) = 2ik\delta;$ $\lceil n/2 \rceil \le i < \lceil 3n/4 \rceil$
   $Arr(a_i) = 3ik\delta;$ $\lceil 3n/4 \rceil \le i < n$

We have compared our approach against a well-known commercially available datapath synthesis tool, which is considered to be the best-in-class solution. The synthesis tool generates arithmetic-optimized architectures for all the arithmetic blocks (like product-of-sum blocks) and then it performs general-purpose operations like technology-independent optimizations, constant propagation, redundancy removal, technology mapping, timing-driven optimization, area-driven optimization, incremental optimization etc. While running the synthesis tool, we turned on all the above-mentioned optimizations. Due to the licensing agreements, we are unable to mention the name of the commercial tool we used.

In Table-II, we report the worst-case delay and the total area results obtained for the POS blocks from the commercial synthesis tool and from our approach. In this table, we report 18 sets of data-points involving different combinations of POS blocks arrival timing constraints and technology libraries. If we compute the average of all the 18 data-points, then our approach results in about 10.92% faster implementations of the Product-of-Sum block, with a 3.54% area overhead. In addition, the total power consumption (dynamic and leakage power) of our netlist is quite comparable to the total power of the netlist generated by the commercial synthesis tool.

639

978-1-4244-3039-0/08 $25.00 © 2008 IEEE

Design	Library	Timing Constraint	Worst-case Delay (ps)			Area (μ^2)		
			Commercial Tool	Our Algorithm	(%) Improvement	Commercial Tool	Our Algorithm	(%) Penalty
POS-A	Lib-A	Type-A	3159	2897	8.29%	8372	8804	5.16%
POS-B	Lib-A	Type-B1	4374	3908	10.65%	8961	9296	3.74%
POS-C	Lib-A	Type-B2	4931	4362	11.53%	9703	10129	4.39%
POS-D	Lib-A	Type-B3	4461	3859	13.48%	9429	9578	1.58%
POS-E	Lib-A	Type-B4	6107	5391	11.72%	12305	12471	1.35%
POS-F	Lib-A	Type-B5	7483	6799	9.14%	17029	17869	4.93%
POS-A	Lib-B	Type-A	2837	2561	9.73%	9147	9586	4.81%
POS-B	Lib-B	Type-B1	3916	3492	10.81%	10142	10574	4.26%
POS-C	Lib-B	Type-B2	4531	3986	12.04%	10795	11236	4.09%
POS-D	Lib-B	Type-B3	3985	3475	12.79%	9963	10193	2.31%
POS-E	Lib-B	Type-B4	5347	4637	13.28%	13692	13944	1.84%
POS-F	Lib-B	Type-B5	6702	6148	8.26%	19247	20281	5.37%
POS-A	Lib-C	Type-A	2249	2043	9.13%	6597	6902	4.63%
POS-B	Lib-C	Type-B1	2981	2651	11.06%	7251	7539	3.97%
POS-C	Lib-C	Type-B2	3457	3048	11.81%	8062	8371	3.85%
POS-D	Lib-C	Type-B3	3274	2839	13.29%	7596	7675	1.04%
POS-E	Lib-C	Type-B4	4362	3861	11.48%	9971	10188	2.18%
POS-F	Lib-C	Type-B5	5240	4824	7.93%	13287	13870	4.39%
Average					10.92%			3.54%

TABLE II

DELAY AND AREA COMPARISON OF POS BLOCKS GENERATED BY A COMMERCIAL SYNTHESIS TOOL AND BY OUR APPROACH

To keep the sizes of the Table-II relatively brief, we did not report the results for other possible combinations of designs, timing constraints and technology libraries. Note that the results for each of the combinations which are not reported here also supported the conclusion of the Table-II (on an average, the improvement over all 108 combinations was 10.47% faster speed with 3.17% area overhead).

To verify the correlation of the post-synthesis experimental data of the Table-II with the post place-and-route data, we performed placement and routing on POS-A and POS-B. For these two testcases, the average improvement of the post-routing worst delay of the POS block generated by our proposed approach is 11.46% compared with the worst delay of the POS generated by the synthesis tool (with a post-routing 3.91% area penalty). In addition, the post-routing total power consumption of the POS generated by the synthesis tool and our techniques are comparable. These data confirm our conclusion about the delay and area efficiency of our approach.

Our proposed approach of synthesizing the faster Product-of-Sum block produces much faster results across different types of POS blocks, timing constraints and technology libraries. This underscores the strength and wide applicability of our algorithm. Since the POS is a highly compute intensive operation and plays a significant role in determining the overall performance of the design, we believe that many real-life designs can significantly benefit from our algorithm.

V. CONCLUSION

In this paper, we have presented a timing-driven approach of implementing a faster arithmetic Product-of-Sum (POS) block, which would be very useful when the critical path of the design goes through the POS block. This technique determines an efficient architecture of the POS block by analyzing the performance of the corresponding SOP block and a hybrid combination of the POS and the SOP expressions. Our synthesis technique works seamlessly with different types of POS blocks, different input arrival time constraints and

across different technology domains (0.13μ, 0.09μ, 0.065μ). The experimental results indicate that our implementation of the Product-of-Sum block is significantly faster (on an average, by 10.92%) and slightly larger (on an average, by 3.54%) than the POS block generated by a commercially available best-in-class datapath synthesis tool.

REFERENCES

[1] R. E. Ladner, M. J. Fischer, "Parallel prefix computation," in *Journal of the ACM* 27(4):831-838, 1980.

[2] P. M. Kogge, H. S. Stone, "A parallel algorithm for the efficient solution of a general class of recurrence equations," in *IEEE Transactions on Computers*, C-22(8):783-91, 1973.

[3] R. P. Brent, H. T. Kung, "A regular layout for parallel adders," in *IEEE Transactions on Computers*, C-31(3):260-64, 1982.

[4] T. Han, D. A. Carlson. "Fast area-efficient VLSI adders," in *8th Symposium on Computer Arithmetic*, 49-56, 1987.

[5] S. Knowles, "A Family of Adders," in Proceedings of the 15th IEEE Symposium on Computer Arithmetic (ARITH-15 '01), 277, 2001

[6] C. S. Wallace, "A suggestion for a fast multiplier," in *IEEE Transactions on Electronic Computers*, EC-13(2):14-17, 1964.

[7] L. Dadda, "Some schemes for parallel multipliers," in *Alta Frequenza*, vol. 34, pp. 349–356, 1965.

[8] K. C. Bickerstaff, E. E. Swartzlander, M. J. Schulte, "Analysis of column compression multipliers," in *Proceedings of 15th IEEE Symposium on Computer Arithmetic*, pp. 33–39, 2001.

[9] P. F. Stelling, V. G. Oklobdzija, "Design strategies for the final adder in a parallel multiplier," in 29th *Asilomar Conference on Signals, Systems and Computers*, pp. 591-595, vol. 1, 1995

[10] A. Mathur, S. Saluja, "Improved merging of datapath operators using information content and required precision analysis," in *IEEE 38th Conference on Design Automation (DAC, '01)*, pp. 462–467, 2001.

[11] T. Kim, W. Jao, S. Jjiang. "Circuit optimization using carry-save-adder cells," in *IEEE Transactions on Computer-Aided Design of Integrated Circuits and Systems CAD-17*, pp. 974–984, 1998.

[12] A. K. Verma, P. Ienne. "Improved Use of the Carry-Save Representation for the Synthesis of Complex Arithmetic Circuits," in *Proceedings of the 2004 IEEE/ACM International conference on Computer-aided design*, pp. 791-798, 2004.

[13] A. Fayed, W. Elgharbawy, M. Bayoumi, "A data merging technique for high-speed low-power multiply accumulate units," in *Proceedings of IEEE Internation Conference on Acoustics, Speech, and Signal Processing*, vol. 5, pp. 145–148, 2004.

[14] L. Chen, O. T. C. Wang, Y. C. Ma. "A multiplication-accumulation computation unit with optimized compressors and minimized switching activities," in *IEEE International Symposium on Circuits and Systems*, vol. 6, pp. 6118–6121, 2005.

21st International Conference on VLSI Design

Clock Period Minimization with Iterative Binding based on Stochastic Wirelength Estimation during High-Level Synthesis

Vyas Krishnan and Srinivas Katkoori

Department of Computer Science & Engineering,
University of South Florida, Tampa, Florida, USA
{krishnan, katkoori}@cse.usf.edu

Abstract

In this paper we present an iterative binding algorithm for high-level synthesis design space exploration, that simultaneously optimizes clock period and wirelength. Our algorithm uses a stochastic interconnect distribution model and a top-down partition-based global placement in a novel framework to provide fast and accurate estimates for wire length and wire delays during resource binding in high-level synthesis. The wirelength estimates used in our algorithm are within 15% of wirelengths in layouts created by commercial and academic placement tools. Experiments show that when compared to a clique-partitioning based binding technique, the proposed algorithm improves the clock period by an average of 18%, with minimal impact on the total wirelength. In addition, our algorithm is an order-of-magnitude faster than a traditional synthesis technique that uses a full place-and-route as part of the design space exploration process.

1. Introduction

Advances in scaling of process technology has increased the importance of interconnect-centric approaches to VLSI design. Interconnect capacitance now forms a significant portion of the total load capacitance of a gate [1]-[2], resulting in a corresponding increase in the overall delay and power dissipation due to interconnects [3].

Traditionally, the steps of high-level synthesis and physical synthesis are performed independently, and the design iterated through the behavioral and physical synthesis steps several times till all of design constraints such as timing, area, and power are met. Due to this separation of physical synthesis from behavioral synthesis, the impact of design decisions taken during HLS will only be known after physical synthesis, often necessitating multiple design re-spins.

Several researchers have proposed techniques to incorporate some measure of feedback from the physical level to the behavioral synthesis level. Techniques that use some form of physical synthesis within the HLS steps use a floorplanner to estimate the impact of high-level design decisions on the resulting layout [6]-[9]. The disadvantage of using a floorplanner is that it is computationally expensive. During design space exploration, many alternative designs are examined, necessitating the need for a way to quickly assess their wirelengths and achievable clock cycle times.

Stochastic wirelength estimation models [10]-[13] have been proposed as a suitable technique to estimate the wirelength distributions in logic gate netlists. These models are based on Rent's rule [11], which is an empirical law that relates the number of terminals in a logic block to the number of gates in the block. The models use Rent's rule to compute the wirelengths based on two empirically derived parameters, namely, *Rent exponent* (p) and *Rent coefficient* (k).

The stochastic wirelength estimation models proposed in earlier work assume some fixed values for a circuit's Rent parameters. However, during design space exploration, a variety of designs with different netlist structures are examined, often with significantly different Rent parameter values, making previous approaches to stochastic wirelength estimation inapplicable to the area of design space exploration. To handle this, we use a dynamic Rent-parameter estimation technique [14], to determine the wiring complexity of netlists examined during design space exploration.

This work addresses the problem of estimating the wiring complexity and clock period of standard-cell based designs examined during HLS design space exploration. To the best of our knowledge, this is the first work that has applied stochastic wirelength

978-1-4244-3039-0/08 $25.00 © 2008 IEEE

estimations to drive design space exploration during high-level synthesis.

The remainder of the paper is organized as follows. Section 2 provides a brief background on stochastic wirelength estimation techniques and describes related work. Section 3 provides an overview of the proposed iterative HLS binding algorithm for clock-period minimization that uses a stochastic RTL wirelength estimation technique. Section 4 presents experimental results from our implementation of the proposed framework, and Section 5 concludes the paper.

2. Related Work

Sutherland and Oestreicher were the first to derive an upper bound for the interconnect wirelength of a square array of gates [10]. Their models assumed a random placement of gates in a gate array, leading to overly pessimistic estimates. Donath [11] used Rent's rule [16] to improve on Sutherland's model, and derive a tighter upper bound for interconnect estimates for a hierarchical placement approach. Stroobandt [12] improved on Donath's model by introducing the concept of *occupation probability* to encapsulate the tendency of good placers to position strongly connected cells closer. Both, Donath's and Stroobandt's models assume a square array of gates. Dembre [18] extended Stroobandt's model to accommodate rectangular arrays of gates. Later, Davis [13] constructed a wirelength distribution model based on a non-hierarchical placement. This wirelength distribution is more accurate than the earlier models, especially for long wires. In this paper, we use the stochastic wirelength distribution model proposed by Davis, to rapidly estimate the wirelengths of RTL netlists during HLS. Since deriving stochastic wirelength estimates are faster than creating a layout, these estimates can be used to determine the effects of high-level design decisions, and use them to drive the high-level synthesis process.

The stochastic wirelength estimation used in this work is based on an empirical power law called Rent's rule [16]. Rent's rule relates the number of terminals (T) to the number of gates (N) in a gate-level netlist using a simple empirical formula, $T = k N^p$, where, p is known as the Rent exponent, and k as the Rent coefficient. The Rent exponent, whose values lie in the range $0<p<1$, provides an indication of the wiring complexity of a circuit [16], while the Rent coefficient (k) indicates the average number of terminals per gate. The Rent's constant is determined empirically by hierarchically partitioning a netlist, recording the number of terminals created in each partition, and then performing a linear regression on these data points on a log-log plot. The slope of the *log-log* plot is the Rent exponent.

Davis's wirelength distribution model assumes a square array of N uniformly tiled gates with $\sqrt{N}$ rows and $\sqrt{N}$ columns. Wirelengths are expressed in gate pitches, with the horizontal and vertical gate pitches being one unit. A continuous interconnect density function $F(l)$ is defined as the number of interconnects in this gate array, with l between gate pitches a and b as

$$\int_a^b F.dl \qquad (1)$$

The continuous interconnect density function $F(l)$ models the interconnections among the gates in the netlist. This interconnect density function is expressed in terms of a gate pair structural distribution function (which models the target layout fabric), and a net occupancy probability (which models the placement of connected modules on a layout). In Davis's model, the average interconnect length in a netlist can be derived as [13],

$$L_{avg} = \frac{L_{total}}{I_{total}} = \frac{\int_a^b l\,F.dl}{\int_a^b F.dl} \qquad (2)$$

where L_{total} represents the total estimated wirelength for the given netlist, and I_{total} is the total number of nets in the netlist. This closed form expression is used in our work to estimate the average wire length of a given topological netlist, while the total wirelength for the netlist is obtained by integrating the closed-form expression for the cumulative wirelength

$$L_{total} = \int_a^b l\,F.dl \qquad (3)$$

In equations (2) and (3), $a = 1$, and $b = 2\sqrt{N}$, where N is the number of gates.

3. Overview of the method

The proposed HLS design space exploration framework uses accurate estimates of the clock period and wirelength to guide HLS binding decisions. During design space exploration, often hundreds of candidate designs are examined, and any estimation technique used for design space exploration must be very fast, in addition to being reasonably accurate. The use of stochastic wirelength estimates and a partition-based global placer to provide wire delay bounds for clock period estimation, enables our technique to be both fast and accurate.

Algorithm 1 Evaluate solution cost

Inputs: (1) RTL netlist of datapath
(2) Cell library
(3) Scheduled dataflow graph

Outputs: (1) Solution cost
(2) Estimated clock period of datapath
(3) Estimated total wirelength

---

1. Flatten RTL netlist to a gate-level netlist.

2. Technology map gate-level netlist to target cell-library.

3. Use Algorithm-2 to estimate total wirelength and placement of cells in core cell-placement area.

4. Use Algorithm-3 to determine coordinates of the datapath RTL modules on the cell placement area.

5. Use Algorithm-4 to compute clock period of datapath RTL netlist.

6. Return the total wirelength and clock period.

Algorithm-1 illustrates the approach used to evaluate datapaths during design space exploration. The estimates provided by this algorithm are used to guide HLS binding decisions. The algorithm consists of three main steps: (1) estimating total wirelength of the gate-level netlist using a stochastic wirelength model, (2) determining cell locations on the core placement area through global placement, and (3) estimating the clock cycle of the RTL netlist. The following sub-sections elaborate on each of these steps. The inputs to the algorithm is a scheduled dataflow graph and an RTL netlist obtained from the allocation and binding sub-task of high-level synthesis. The algorithm assumes the availability of an RTL module library containing gate-level netlists of the RTL modules. The algorithm first flattens the input RTL netlist to its gate-level equivalent, and estimates the total layout area required to place the cells of the netlist. This gate-level netlist is then recursively bi-partitioned, using hMetis [15], with cut size minimization and area-balanced partitions, as the bi-partitioning criteria. Meanwhile, the layout area is also partitioned into placement regions, each of which contains a corresponding sub-circuit. The partitioned sub-circuits are then analyzed to estimate total wirelength and wire delays in the RTL netlist resulting from the HLS allocation and binding sub-task.

3.1. Stochastic wirelength estimation

Stochastic wirelength estimation models estimate wirelengths of gate-level netlists, based on (1) the number of gates, (2) the Rent exponent, and (3) the Rent coefficient. These parameters can be directly

extracted from a gate-level netlist. These parameters can differ significantly among designs examined by HLS design space exploration. Hence, any technique using stochastic wirelength estimation models must be able to dynamically determine these parameters from an RTL netlist *on-the-fly*. We dynamically compute the Rent parameters using the *partitioning*-tree method proposed in [14], and use them to estimate the wiring complexity of RTL datapaths. In addition, we enhance the algorithm of [14] to concurrently place the partitioned gate-level netlist on a 2-D plane, thus simulating the global cell placement process. This allows us to estimate the wirelengths between interconnected RTL modules. Wire delays for data transfers between datapath modules are computed using an accurate Elmore-based distributed wire delay model.

Algorithm 2 Stochastic wirelength estimation

Input: (1) Gate-level circuit netlist
(2) Core cell placement area

Output: (1) Total wirelength
(2) Placement of gates on core cell area

---

1. Recursively bi-partition original gate-level netlist, and the core cell area.

   At each recursive level,

   ☐ calculate the average number of cells per partition (N_i) and the average number of external nets (T_i) over all partitions. Save the data pair (N_i, T_i), where i is the depth of recursive partitioning. Partitioning stops when reaching a given depth k.

   ☐ assign the two partitioned sub-circuits to the two partitioned bins on the core cell area.

2. Apply linear regression on the log-log data pairs (N_i, T_i), (N_{i+1}, T_{i+1}), ... , (N_k, T_k), where k is a user-specified partitioning depth.

3. Determine the slope of the fitted line by linear regression, and its y-intercept.

4. Use equation (4) to calculate total wirelength.

5. Return the estimated wirelength placement of gates on the core cell area.

Algorithm-2 outlines the wirelength estimation technique used in our approach. To extract the Rent parameters from a given circuit netlist, the sub-circuits in the partition-tree, are analyzed to obtain their gate-count and terminal-count. The gate and terminal counts are used as data points on a log-log plot, to determine the Rent parameters of the netlist. A linear regression is applied to find the slope of the fitted line on this log-log plot. The slope represents the Rent exponent, while its y-intercept represents *log k,* where *k* is the Rent coefficient.

To compute the total wirelength of an RTL netlist, we use the stochastic interconnect distribution model proposed by Davis in [13], which provides a closed-form expression for the estimated wirelength of a gate-level netlist, in terms of the total number of gates, and the Rent parameters extracted from the netlist.

3.2. Estimating placement of RTL modules

To estimate the wire delays associated with data transfers between RTL modules in a datapath (and hence estimate the clock period), we need information on the relative locations of these modules on the chip floorplan. We obtain this by doing a top-down global placement of the gates in the netlist, concurrently while bi-partitioning the netlist for Rent parameter extraction.

Algorithm 3 Determine coordinates of RTL modules

Input: Placement of gates on core cell area
Output: Coordinates of datapath RTL modules

---

for each RTL resource R_k in the datapath
    1. Determine coordinates of all gates contained by R_k
    2. From gate coordinates, determine (X_{min}, Y_{min}) and (X_{max}, Y_{max}) of the bounding-box encompassing all gates contained by R_k
return (X_{min}, Y_{min}) and (X_{max}, Y_{max}) of all R_k

Algorithm-3 outlines the methodology used to determine the dimensions and locations of these RTL modules. We start with a core layout area whose dimensions are estimated based on the gate-count of the RTL netlist, and the layout area is recursively partitioned into placement bins. During partitioning, terminal propagation [19] is used so that connectivity of gates to areas outside each bin is properly handled. We assume that all gates assigned to a bin, are located at the center of the bin. An efficient cell placement tool will place all gates belonging to an RTL module in close proximity. With this assumption, we can approximately estimate the dimensions of the datapath RTL modules by recording the dimensions of a rectangular bounding box that spans all gates of the RTL module on the layout. After determining the locations and dimensions of the datapath modules, we can estimate the inter-module wire lengths and wire delays used for clock period estimation.

3.3. Clock period estimation

The process of determining the clock period for an RTL datapath is illustrated in Algorithm-4. It begins by

identifying each RTL module pair between which data flows when executing a DFG operation. To estimate the wire delays between a communicating module pair, we first determine the location and dimensions of these modules (Algorithm-3). The half-perimeter of the smallest rectangular bounding box that completely encloses the module-pair is used as an estimate for the wirelength between the module-pair. The widely used Elmore delay model is used to estimate wire-delays. Module delays are obtained from a pre-characterized RTL module library. The clock period is then computed by examining all register-to-register delays for data transfers in the scheduled dataflow graph.

Algorithm 4 Estimate clock period of datapath

Inputs: (1) Scheduled dataflow graph $G(V,E)$
        (2) HLS Binding of $G(V,E)$
        (3) Coordinates of datapath RTL modules

Output: Clock period CP of RTL datapath

---

for each DFG operation V_k in $G(V,E)$
    1. Determine the RTL module M_k bound to V_k
    2. Determine the register bindings $R1_k$ and $R2_k$ of the two inputs of V_k
    3. Determine the register binding $R3_k$ of the output of V_k
    4. Identify all input multiplexers MX_k to these modules.
    5. Look up the coordinates of the RTL modules M_k, $R1_k$, $R2_k$, $R3_k$, and MX_k.
    6. Compute path delay $d1$ from $R1_k$ to $R3_k$
    7. Compute path delay $d2$ from $R2_k$ to $R3_k$

Set CP = maximum path delay found in G
return CP

3.4. Iterative binding algorithm for high-level design space exploration

The design space exploration framework described in this paper is based on an iterative binding algorithm that uses the wirelength and clock period estimation techniques detailed in sections 3.1 to 3.3, to evaluate binding solutions examined by the algorithm. The algorithm starts with an initial binding and datapath, created using a constructive technique, and improves this solution through a greedy acceptance criterion.

The approach used in our iterative binding technique is shown in Algorithm-5. It accepts a scheduled dataflow graph and an initial binding as inputs, and returns a datapath optimized for clock period and wirelength. Starting with an initial solution, a move-based iterative improvement phase improves the datapath by reducing the wirelength and clock period while satisfying user constraints. A weighted cost function of the clock period and total wirelength is used to evaluate the quality of solutions. The iterative improvement stage is repeated over multiple passes, where a sequence of HLS binding moves is applied to the current solution in each pass. Each pass of the iterative binding algorithm starts by sorting all DFG operations of the scheduled dataflow graph in decreasing order of their register-to-register delay, and a series of HLS re-binding moves are tried. All binding moves that improve the cost over the best solution are accepted. The algorithm terminates when no improvement over the current best solution is found in one complete pass of the iterative improvement phase, or when a MAX_MOVES number of re-binding moves have been attempted.

Algorithm 5 Iterative Binding for Clock Period Optimization

Input: Scheduled Dataflow Graph $G(V,E)$
Output: Optimized RTL netlist of datapath

1. Read scheduled DFG
2. Use clique partitioning to bind RTL resources to G to create initial solution
3. Evaluate its cost function

while (cost improvement = TRUE
   OR number of binding moves < MAX_MOVES)
{
 sort DFG operations V_k of G by their delay
 for each V_k in G **do**
  i_1 = input operand 1 of V_k
  i_2 = input operand 2 of V_k
  o_k = output produced by V_k
 $list(R)$ = compatible RTL resources for V_k
 for each RTL resource r_k $list(R)$ **do**
  (1) tentatively bind V_k to r_k
  (2) evaluate cost of new solution $\mathbf{S}_k$
  (3) if cost($\mathbf{S}_k$)<cost of current solution
    then accept new binding for V_k
 $list(R)$ = compatible RTL resources for i_1
 for each RTL resource r_k $list(R)$ **do**
  (1) tentatively bind i_1 to r_k
  (2) evaluate cost of new solution $\mathbf{S}_k$
  (3) if cost($\mathbf{S}_k$)<cost of current solution
    then accept new binding for i_1
 $list(R)$ = compatible RTL resources for i_2
 for each RTL resource r_k $list(R)$ **do**
  (1) tentatively bind i_2 to r_k
  (2) evaluate cost of new solution $\mathbf{S}_k$
  (3) if cost($\mathbf{S}_k$)<cost of current solution
    then accept new binding for i_2
 $list(R)$ = compatible RTL resources for o_k
 for each RTL resource r_k $list(R)$ **do**
  (1) tentatively bind o_k to r_k
  (2) evaluate cost of new solution $\mathbf{S}_k$
  (3) if cost($\mathbf{S}_k$)<cost of current solution
    then accept new binding for o_k
}
return best solution in terms of clock period and total wirelength

---

4. Experimental results

All experiments with the algorithms described in this paper were performed on a Linux workstation using a 1.86 GHz Intel Core2 Duo CPU and 2GB RAM.

To assess the accuracy of stochastic wirelength estimation for HLS design space exploration, we used our dynamic stochastic wirelength estimation technique on RTL netlists implementing six benchmark instances drawn from DSP applications. The benchmark types studied were Infinite Impulse Response (IIR) filter, Elliptic Wave (EWF) filter, and an 8X8 Discrete Cosine Transform (DCT) filter. Each RTL netlist represents a different architecture.

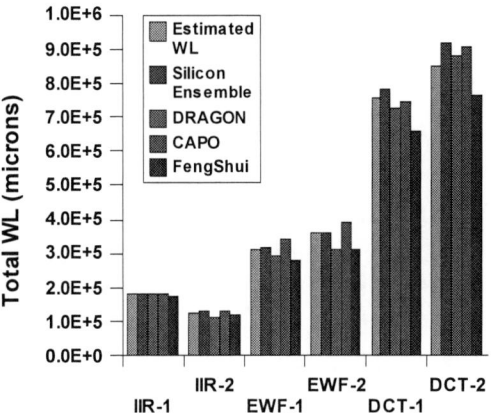

Figure 1. Comparison of wirelength estimates with post-placement wirelengths of DRAGON, CAPO, FengShui, and Silicon Ensemble

The wirelength estimates returned by the stochastic wirelength estimation technique were compared with the post-placement wirelengths from three state-of-the-art academic standard-cell placement tools, and Cadence Silicon Ensemble (a modern commercial tool). The academic cell placement tools used in our study were DRAGON [20], CAPO [21], and FengShui [22]. Figure 1 illustrates this comparison of the estimated and measured wirelengths from the academic placers DRAGON, CAPO, and FengShui, as well as Cadence Silicon Ensemble. From the figure, it is evident that the stochastic wirelength estimates closely agree with actual wirelengths obtained from four different placement tools.

Table 1 illustrates the results of our experiments on clock period optimization by the proposed iterative binding algorithm. The table lists the clock periods of the initial and best bindings for the benchmarks tested.

In the table, column 1 lists the benchmarks. Column 2 shows the estimated clock period of the initial solution provided to the iterative binding algorithm, while column 3 shows the clock period of the best binding solution found. Column 4 shows the percentage improvement in the clock period. The percentage improvements in the clock period vary from 13.8% for the IIR filter, to 22.7% for the EWF filter, with an average improvement of 18.2%. These improvements in the clock period were achieved without any significant overhead in total wirelength.

Table 1. Clock period improvement by the iterative binding algorithm

Benchmark	Initial CP (ps)	Best CP (ps)	% improvement
IIR	2146	1851	13.8
EWF	3152	2437	22.7
FIR	2991	2386	20.2
DCT	4165	3492	16.1
		Avg.	18.2

Table 2. Runtimes for iterative binding algorithm

HLS benchmark	Number of bindings moves	Execution time
IIR	200	9m:54s
EWF	500	14m:32s
FIR	400	33m:27s
DCT	500	51m:12s

Table 2 shows the runtime for design space exploration performed by our algorithm for the tested benchmarks. Column 2 lists the total number of datapath designs examined by the algorithm, and column 3 shows the CPU time in minutes and seconds. Each binding move attempted by the algorithm involves creating a new datapath architecture. In a typical physical synthesis step, the placement and routing of these datapaths would take several minutes. Hence, for a traditional synthesis flow, evaluating each binding move, by itself, would take several minutes to hours (depending on the design complexity), for physical synthesis and timing analysis. The iterative binding algorithm proposed in this paper performs the same task almost an order of magnitude faster than a traditional synthesis flow.

5. Conclusions

In this paper, we present an iterative binding algorithm for clock period optimization, that uses stochastic wirelength models to estimate the total wirelength of cell-based designs, and a top-down partitioned based RTL placement to estimate the clock period. Use of these estimates to guide HLS binding decisions enables our approach to achieve an order-of-magnitude improvement in the search time for HLS design space exploration. Our wirelength estimates are within 15% of Dragon, Capo, FengShui, and Cadence Silicon Ensemble. Experiments on dataflow intensive HLS benchmarks show that our iterative binding algorithm can improve the clock period of a datapath by an average of 18.2%, with minimal impact on the wirelength.

References

[1] A. Deutsch et al., "On-chip wiring design challenges for operation," *Proc. IEEE*, vol.89, no.4, pp. 529-555, Apr.2001.

[2] J. Cong, "An interconnect-centric design flow for nanometer technologies," *Proc. IEEE*, vol.89, no.4, pp. 505-528, Apr.2001.

[3] D. Sylvester and C. Hu, "Analytical modeling and characteriza -tion of deep-submicron interconnect," *Proc. IEEE*, May 2001.

[4] M. Xu and F.J. Kurdahi, "Layout-driven RTL binding techniques for high-level synthesis using accurate estimators," *ACM TODAES*, vol.2, no.4, pp.312-343, 1997.

[5] S. Tarafdar and M. Leeser, "A data-centric approach to high-level synthesis," *IEEE Trans. CAD*, Nov. 2000.

[6] J.P. Weng and A.C. Parker, "3D Scheduling: High-level synthesis with floorplanning," *Proc. DAC* 1992, pp. 668-673.

[7] Y.M. Fang and D.F. Wong, "Simultaneous functional-unit binding and floorplanning," *Proc. ICCAD* 1994, pp.317-321.

[8] P. Prabhakaran and P. Banerjee, "Simultaneous scheduling, allocation, binding, and floorplanning," *Int. Conf. VLSI* 1998.

[9] W.E. Dougherty and D.E. Thomas, "Unifying behavioral synthesis and physical design," *Proc. DAC* 2000, pp. 756-771.

[10] I.E. Sutherland and D. Oestriecher, "How big should a printed circuit board be?," *IEEE Trans. Computers*, pp.537-542, 1972.

[11] W.E. Donath, "Placement and average interconnect lengths of computer logic," *IEEE Trans. Circuits Syst.*, pp. 272-277, 1979.

[12] D. Stroobandt and J.V. Campenhout, "Accurate interconnect length estimations for predictions early in the design cycle," *VLSI Design*, vol. 10, no. 1, pp. 1-20, 1999.

[13] J.A. Davis, V.K. De, and J.D. Meindl, "A stochastic wirelength distribution for gigascale integration (GSI) – Part I: Derivation and validation," *IEEE TED*, vol.45, pp. 580-589, March 1998.

[14] V. Krishnan and S. Katkoori, "Design space exploration of RTL datapaths using Rent parameter based stochastic wirelength estimation," *Proc. ISQED* 2006.

[15] G. Karypis *et.al.*, "Multilevel hypergraph partitioning: Applications in VLSI domain," *IEEE Trans. VLSI*, March 1999.

[16] B.S. Landman and R.L. Russo, "On a pin versus block relationship for partitions of logic graphs," *IEEE Trans. Computers*, C-20:1469-1479, 1971.

[17] P. Christie and D. Stroobandt, "The interpretation and application of Rent's rule," *IEEE Trans. VLSI*, 2000.

[18] J. Dembre, *et.al.*, "On Rent's rule for rectangular regions," *Proc. Workshop System Level Interconnect Prediction*, 2001, pp. 49-56.

[19] A.E. Dunlap and B.W. Kernighan, "A procedure for placement of standard cell VLSI circuits," *IEEE Trans CAD*, April 1985.

[20] M. Wang, *et. al.*, "Dragon 2000: Standard-cell placement tool for large industry circuits," *ICCAD 2000*.

[21] J. Roy, *et. al.*, "Capo: Robust and scalable open-Source min-cut floorplacer," Proc. Intl. Symposium on Physical Design, 2005.

[22] A.R. Agnihotri, S. Ono, and P.H. Madden, "Recursive bisection placement: Feng Shui 5.0 Implementation details," *Proc. ISPD 2005*.

21st International Conference on VLSI Design

On the Use of Hash Tables for Efficient Analog Circuit Synthesis

Almitra Pradhan and Ranga Vemuri

Department of ECE, University of Cincinnati, Cincinnati OH 45221, USA

{pradhaa,ranga}@ececs.uc.edu

Abstract

Achieving accurate and speedy circuit sizing is a challenge in automated analog synthesis. System matrix model based estimators predict circuit performance accurately. In this paper we employ hashing in conjunction with matrix models for faster synthesis convergence. With hash tables some matrix element recomputations are avoided, thus improving synthesis time. Hashing is effectively performed by dividing matrix elements into classes and building classwise hash tables. Hash tables are updated over several synthesis runs which further expedites convergence. Experimental results show that the proposed method can provide 4x-6x speedup over that offered by synthesis approaches employing macromodels but no hashing.

1. Introduction

Circuit sizing is one of the most important steps in analog design. Given a topology, the aim of circuit sizing is to determine the device sizes and bias such that the specified performance goals are met. In optimization based analog CAD flows, a search algorithm such as Simulated Annealing (SA) is used to iteratively explore the design space and automate the sizing process. Optimization based flows typically perform two steps, (i) search and (ii) performance evaluation successively till the specified goal is achieved. In the search step the SA proposes a set of sizes and bias for circuit components. The evaluator subsequently calculates the performance to determine if the target has been met. The speed of the sizing process is dependent upon the speed of the evaluator. SPICE based evaluators are slow but very accurate [1]. Performance models can be used instead of SPICE which trade some accuracy for an improvement in speed. Performance models are constructed by sampling the entire design space of the circuit and gathering data for parameters of interest such

This work was supported in part by the National Science Foundation under award number CCF-0429717.

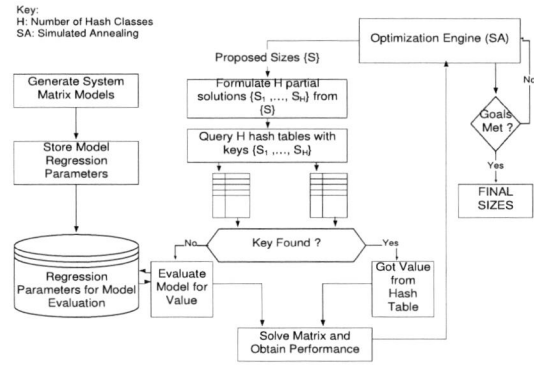

Fig. 1. Fast Analog Circuit Synthesis with Hashing

as gain, unity gain frequency, phase margin, bandwidth. Regression models, known as performance macromodels, are then chosen to accurately fit the data.

In recent years researchers have used a number of techniques for generating performance macromodels. In [2] Daems *et.al.* use geometric programming and posynomial response surfaces to model the performance parameters of analog circuits. Kiely *et.al.* [3] applied least square support vector machines for the same. Other modeling techniques applied to the performance modeling problem include neural networks [4], cubic-splines [5] and multivariate adaptive regression splines [6]. An important class of models based on symbolic analysis have also been extensively used for circuit synthesis [7], [8]. In [9], Mandal *et.al.* use a knowledge based method to model the pole zero information of analog circuits.

In our earlier work on performance estimation [10], we showed that the elements of the system matrix can be modeled with very high fidelity. Using these models in a synthesis environment enables fast and accurate sizing of analog circuits. The main contribution of this paper is the extensive use of hash tables to avoid the recomputation of circuit matrix elements and thus speed up the synthesis process. An average speedup of 4x-6x is achievable above the previous method which did not use hash tables. Figure 1 illustrates our proposed methodology

978-1-4244-3039-0/08 $25.00 © 2008 IEEE

TABLE I. Circuit Synthesis With Matrix Models

Circuit 1: Single Ended Opamp		
Performance Specification	Estimated	Actual
Gain $\geq$ 39.7 dB	39.64	39.65
UGF $\geq$ 13 MHz	13.02	13.00
Phase Margin $\geq$ 76 Deg	79.54	79.54
Bandwidth $\geq$ 123 KHz	126.9	126.5
Speedup Compared to Hspice: 10.13x		
Circuit 2: High Gain Differential Opamp		
Performance Specification	Estimated	Actual
Gain $\geq$ 80 dB	80.01	79.02
UGF $\geq$ 20 MHz	20.55	20.7
Phase Margin $\geq$ 90 Deg	89.5	89.33
Speedup Compared to Hspice: 6x		

of using hashing in conjunction with circuit matrix models for optimization based sizing of analog circuits.

This paper is organized as follows. Section 2 briefly reviews analog synthesis employing matrix models. Section 3 describes the method of generating and employing hash tables in the synthesis framework. Section 4 presents the experimental results of speedup offered by hashing for two benchmark circuits. Simple as well as reinforced hash tables are used for improved speedup. Section 5 concludes the paper.

2. Analog Circuit Synthesis Employing Matrix Models

This section presents a brief review of generation and use of circuit matrix models in a synthesis environment. A linear(ized) circuit can be represented in a matrix form by its Modified Nodal Analysis (MNA) formulation. The numerical values of the system matrix elements are a linear combination of active component small signal parameters and passive component values. Mosfet widths, passive components and bias are the design variables for analog synthesis. Thus, system matrix values are a function of the design variables. Performance parameters such as gain, bandwidth, phase margin are calculated from the circuit matrix.

In [10], we used pre-generated models of the matrix for performance estimation. Models are generated for all non-zero matrix elements. Device matching is used to reduce the number of models to be generated. The relation between matrix elements and design variables of the circuit is captured using polynomial response surface models.

The sizing operation is performed by an SA based optimization engine. In each sizing iteration, models for *all* matrix elements are evaluated for the proposed variable sizes. Once numerical values of the matrix elements are obtained, performance parameters can be estimated accurately. The results of using circuit matrix models in an SA based circuit synthesis environment for two benchmark circuits are shown in Table I.

In addition to accuracy, the advantage of using system matrix models instead of performance models is that any linear performance parameter can be calculated from the matrix. Performance macromodeling requires a separate model for each parameter. Data generation for the models is through simulation and is an expensive one-time process. Adding a new performance model to an existing set requires data for that parameter and is costly. However, with matrix models, the system matrix is available and the new parameter can be inexpensively computed.

3. Application of Hashing in an Analog Synthesis Environment

The primary contribution of this work is the extensive use of hash tables in conjunction with matrix models to speed circuit synthesis. The proposed method first divides matrix elements into hash classes based on the design variables they are dependent on. Storage and retrieval of matrix elements is then performed by building hash tables for each hash class. [11] used hash tables for obtaining symbolic solutions of circuits. No work to the best of our knowledge has implemented class based hashing to speed up analog circuit synthesis.

3.1. The Concept of Hash Classes

There are two primary reasons why hashing is useful in conjunction with matrix models:

- Each matrix element depends only on a *subset* of design variables.
- Multiple matrix elements are dependent on the *same* variable subset.

We explain this with the example of a simple two stage opamp in Fig. 2. This circuit has 8 transistors. The design variables of the circuit are mosfet widths w1-w4 and compensating capacitance Cc. Each element of the system matrix is a linear combination of small signal values. For ease of explanation, let us only consider the small signal parameter gm. The basic mosfet equations are:

$$I_d = \frac{1}{2} * \mu_n * C_{ox} * \left(\frac{W}{L}\right) * (V_{gs} - V_{th})^2 * (1 + \lambda V_{ds}) \quad (1)$$

$$gm = \sqrt{2 * \mu_n * C_{ox} * \left(\frac{W}{L}\right) * I_d} \quad (2)$$

Let us analyze the effect of change in w1. When w1 changes, V_{gs}(M1), gm(M1) change as per eq(1,2). The change in V_{gs}(M1) is effected by V(3), which in turn modifies V_{gs}(M5) and thus I_d(M5). Since I_d(M5) changes, the drain current of M1-M5 changes. Thus, w1 affects gm(M1)-gm(M5).

978-1-4244-3039-0/08 $25.00 © 2008 IEEE

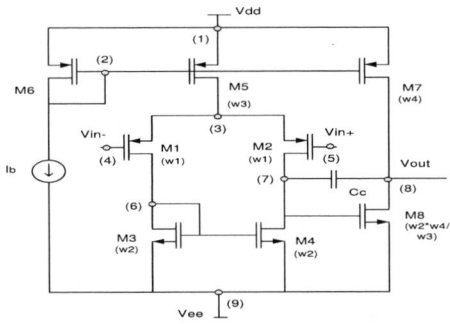

Fig. 2. A Simple Two Stage Op-Amp

TABLE II. Mapping Matrix Elements to Variables

Matrix Element	Dependency List	Matrix Element	Dependency List
C_{11}	{w3,w4}	C_{12}	{w3}
C_{17}	{w4}	C_{23}, C_{33}, C_{35}	{w1}
C_{56}	{w2}	C_{67}, C_{76}	{Cc}
C_{66}	{w2,w3,w4,Cc}	G_{55}, G_{65}	{w1,w2,w3}
C_{55}	{w1,w2}	G_{71}	{w1,w3,w4}
G_{77}, G_{76}	{w2,w3,w4}	$C_{22}, G_{21}, G_{22}, G_{23}, G_{26}, G_{52}, G_{66}$	{w1,w3}
C_{ii} (G_{ii}) are capacitive (conductive) elements in MNA matrix			

Now, consider a change in w2. A change in w2 affects voltage at nodes 6 and 7. V_{gs}(M5), I_d(M5) and gm(M5) are not affected. As I_d and width of M1 does not change, its gm is unaffected. Only gm(M3), gm(M4), gm(M8) vary because of change in w2.

From the above analysis it is seen that a design variable does not affect small signal parameters of all transistors. By a similar analysis it can be seen that each system matrix element is affected only by a *few* design variables. The dependent design variables for each matrix element are found as follows. The matrix is first symbolically generated from the circuit topology. Values for matrix elements are obtained by uniformly sampling the design space. Matrix elements are fit to the variables by a regression model. Student's t-test (p-value < 0.05) is used to determine the significant variables for each matrix element.

We group matrix elements that are dependent on the same subset of design variables in a class that we call a *hash class*. For each hash class this subset is referred to as the dependency list. The two stage op-amp has 12 hash classes as seen in Table II. The number of elements that get grouped in a hash class depends on the circuit topology and design variables.

3.2. Partial Solutions in SA based search

SA is commonly used in analog circuit synthesis due to its ability to traverse the search space without getting trapped in a local minima. In any SA based optimization engine the design space is explored by perturbing one variable at a time. If there are n design variables the set

of values $\{v1, v2, ..vn\}$ proposed by the SA at a given instance (iteration) forms the current solution.

A subset of the overall solution is called a partial solution. We are only interested in those partial solutions which form dependency lists of the hash classes. As the SA based search engine traverses the design space, it keeps proposing new values $\{v1, v2, ..vn\}$. An important observation is that although the solution changes in every SA iteration, all partial solutions do not change. Moreover, even though the SA is unlikely to generate the same solution multiple times, same partial solutions can be generated multiple times.

For example, consider a circuit having four variables v1,v2,v3,v4 being sized by an SA. Two solutions {v1=10, v2=10, v3=20, v4=20} and {v1=10, v2=30, v3=20, v4=40} are proposed by the SA in different iterations. The partial solution {v1=10,v3=20} is same in both cases. A hash class having dependency list {v1,v3} needs only these two variables for model evaluation. Thus, if matrix values corresponding to {v1=10,v3=20} are stored in the hash table, they can simply be retrieved and model computation is avoided.

3.3. Hash Table Organization

Having identified hash classes, one hash table is defined for each class. The hash table is accessed by a key that is a function of the partial solution. Each row of the hash table contains a key and numerical values of elements corresponding to that key. For example, a hash class that has a dependency list $\{v1, v2, ..vk\}$ and m hash class elements will have a hash table as shown in Fig. 3. As seen from the figure, when the table is queried with the key, all m elements can be retrieved *at once*. Thus, greater the number of elements in the hash class, more is the benefit from a hit in the hash table.

The benefit from hashing not only depends on the number of elements in a hash class but also the cost. Expensive hash classes have more elements, higher order models or both. More the cost, greater is the time saved when elements are obtained from hash table instead of model evaluation. The underlying assumption that hash class cost is more than the hash table access time is generally true.

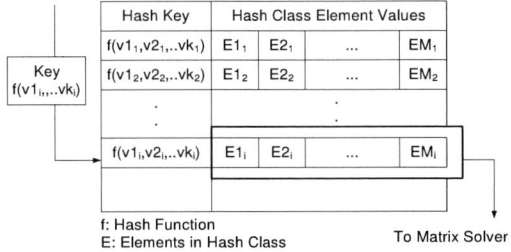

Fig. 3. Table for a Hash Class

649

978-1-4244-3039-0/08 $25.00 © 2008 IEEE

3.4. Estimation of Speedup by Hashing

In this section we provide an expression for time saved by using hash tables.

- Number of elements in system matrix: n,
- Number of hash classes: h ($h \leq n$),
- Iterations for SA to converge: $Iters$,
- Cost of hash class i: $class\_cost[i]$,
- Total hash class i entries not in table: $misses[i]$

When no hashing is used all n matrix elements need to be computed from their models during each iteration. Each of these n matrix elements belong to one of the h hash classes. Runtime for the entire sizing experiment is given by eq(3).

$$Synthesis\_Time_{no\_hash} = Iters * \sum_{i=1}^{h} class\_cost[i] \quad (3)$$

When hashing is used all the hash classes need not be computed in every iteration. If the hash key for the current solution matches the key for some previously visited solution, the evaluated matrix values exist in the hash table. Thus, the runtime depends on the number of misses for the hash classes. Runtime for the entire sizing experiment is given by eq(4). Expression (5) gives the time saved by use of hashing. Time for hash table access is negligible compared to model evaluation and is omitted from the expression.

$$Synthesis\_Time_{hash} = \sum_{i=1}^{h} class\_cost[i] * misses[i] \quad (4)$$

$$Iters * \sum_{i=1}^{h} class\_cost[i] - \sum_{i=1}^{h} class\_cost[i] * misses[i] \quad (5)$$

Thus, the speedup depends on

- Recomputations avoided by storing the previously visited solutions.
- The cost of the hash class whose computation was saved.

3.5. Integrating Hash Classes in the Synthesis Flow

The hash table contains values of matrix elements for a number of combinations of device sizes. To find if the element values corresponding to a particular set of sizes is present in the table, a hash key is used. For a hash class with k dependent variables, the hash key is a function $f(v1, .., vk)$. The key we use is a concatenated string of design variable values expressed as integers.

The hash tables themselves are implemented as Red-Black Trees [12]. Insert and find are the primary hash table operations. With R-B trees both insert and find are performed in $O(log(n))$ time where n is the table size. A separate hash table for each hash class also limits the maximum table size.

The sizing engine used in the synthesis flow is a SA based optimizer. Before the start of synthesis, empty hash tables are initiated for each hash class. As synthesis proceeds new solutions are proposed by the sizing engine. For each hash class, the key is formed and the hash table is queried with the key. If found, the element values can be directly obtained from the table otherwise hash class element values are computed from the model and a row comprising of the key and element values is added to the hash table. Algorithm 1 describes the procedure.

Algorithm 1 Analog Synthesis With Hash Classes

Input: Circuit Topology(SPICE), Design Variables, Performance Goals
Output: Sizing Solution for the Circuit
Identify unique matrix elements
Sample design space and generate data for these elements
$\forall$ Matrix Elements do:
  Form Dependency List.
  Identify Hash Classes.
  Distribute matrix elements into Hash Classes.
  Fit regression model from generated data.
while (! converged) **do**
  Get current SA solution.
  for $i = 1$ to Number_of_Hash_Classes **do**
    Form key$[i]$ from current solution
    if (key$[i]$ $\subseteq$ Hash_Table[i].keys) **then**
      Retrieve all element values from Hash_Table$[i]$
    else
      Compute elements values from models.
      Update Hash_Table$[i]$ with key$[i]$ and element values.
    end if
  end for
  Solve matrix to obtain performance parameters.
  if (SA_Cost is 0) **then**
    converged = true
  end if
end while

---

4. Experiments and Results

The synthesis framework incorporating sizing, hashing and matrix solver for performance estimation is developed as an integrated C++ package. All experiments are conducted on a Sunblade 1000 with Solaris (Sun OS), 2048 MB RAM and 2-750 MHz processors.

4.1. Single Ended Opamp

The first circuit on which we conduct our experiments is the Single Ended Opamp (SEO) of Fig. 4. The SEO is a 15 transistor benchmark. Mosfets $M0$ and $M5$, $M1, M3, M6$ and $M9$, $M15$ and $M19$, $M21$ and $M22$, $M20$ and $M23$, $M16$, $M17$ and $M18$ have matched device widths. Thus, this circuit has six device widths (w1-w6)

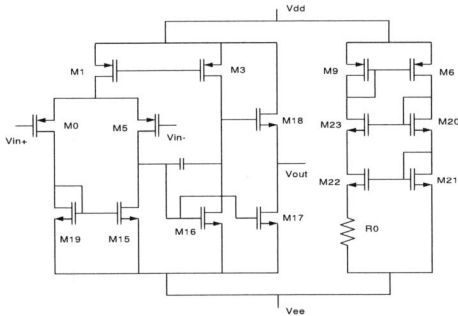

Fig. 4. The Single Ended OpAmp (SEO)

as the design variables. The system matrix of the SEO has 42 distinct elements. Based on the design variables that they are dependent on, these elements get divided into 13 hash classes. Details can be found in Table III. In order to demonstrate the effectiveness of hash tables in conjunction with analog circuit synthesis we perform the following 3 experiments.

Experiment A: Speedup by hashing

This experiment demonstrates the improvement in the convergence time of an SA based circuit sizer with the use of hashing. A set of ten performance goals (gain, ugf, phase margin and -3db bandwidth) is randomly generated for the SEO. Each performance goal is a group of specifications for the circuit. The performance goals are generated to lie within the maximum and minimum achievable performance parameters in the given design space. The goals vary from nominal to difficult.

In the first step, no hash tables are used. All matrix elements are evaluated from their models at every SA iteration. The CPU time for the SA to converge for each of the performance goals is noted. In the second step, the matrix elements are divided into hash classes and matrix elements are obtained from the hash table whenever possible as described in algorithm 1. The same set of ten performance goals are used to enable a runtime comparison. The random number generator of the SA engine is also seeded identically in both cases. Figure 5(i) compares the time for the SA based sizer to converge with and without the use of hashing.

Hashing provides speedup from 1.7x to 2.3x on these benchmarks. Difficult goals take longer time for synthesis

TABLE III. Hash Class Details for Benchmarks

	Single Ended Opamp	Differential Amplifier
Design Variables	6	5
Number of Hash Classes	13	9
$\sum_{i=1}^{h}$ Hash Class Cost (sec)	0.05	0.24
Max. Elements in Class	11	16
Max. Hash Class Cost (sec)	0.03	0.09
Perf. Calculation (sec)	0.01	0.026

than nominal ones. Hashing improves runtime for both. Longer running experiments have more hits in hash tables and are able to take more advantage from hashing.

Experiment B: Reinforced Hash Tables

Matrix models generated for a topology can be used to size the circuit for a number of performance goals. A lot of data is generated in each sizing operation. If the data generated in earlier synthesis operations is provided to the next one, it can converge faster.

In this experiment, the hash tables are successively updated (reinforced) with each sizing run. If a part of the design space that was visited in an earlier sizing operation is revisited in the current one, matrix element values can be obtained from the hash table and model evaluation is not required.

A set of 10 close performance goals is generated. Figure 5(ii) illustrates the time for synthesis with reinforced hash tables. The synthesis time with the competing approaches which use no hashing or simple hashing is also provided. The speedup with reinforced hashing improves upto 4.1x.

Experiment C: Average and Best Speedup by Reinforced Hashing

For the measurement of average speedup by reinforced hash tables, 300 performance goals are randomly generated and synthesized. As speedup of each sizing operation varies, the average speedup over a batch of 25 sizing runs is calculated and plotted in fig. 5(iii). The average speedup over the entire experiment is 4.3x and it improves with the number of runs.

The best speedup is achieved when a sizing operation obtains 95% or more matrix values from hash tables. As seen in fig. 5(iii), the best speedup for the SEO is 6x. Here, the second half of the 10 sizing runs are able to completely utilize hash tables filled by previous runs.

4.2. High Gain Differential Amplifier

The Differential Amplifier (DA) is a high gain amplifier designed with 33 transistors [13]. Its circuit matrix has 61 elements and there are 5 design variables [10]. Matrix elements are divided into 9 hash classes based on their dependency lists.

Similar to the SEO, experiments A, B and C demonstrate the effectiveness of hashing on synthesis of the DA. With simple hashing a speedup of 1.5x-2x is observed (expt. A). When reinforced hash tables are used the speedup obtained is 7.7x (expt. B). The average and best case speedup is 6.1x and 10.2x respectively (expt. C). Fig. 6 shows the results for speedup of the DA. The total number of hash table entries at the end of 300 synthesis runs is of the order of 10^5 for both SEO and DA.

651

978-1-4244-3039-0/08 $25.00 © 2008 IEEE

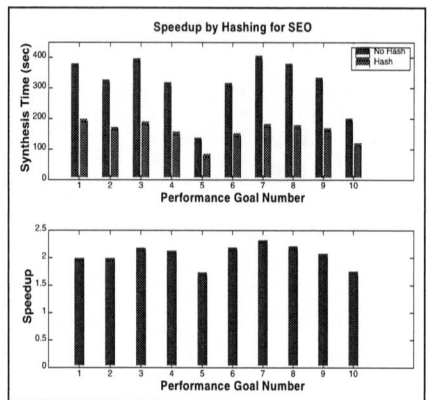 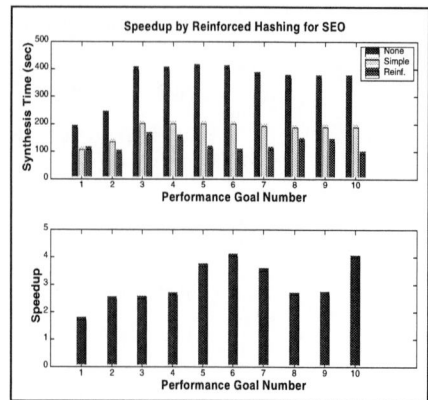 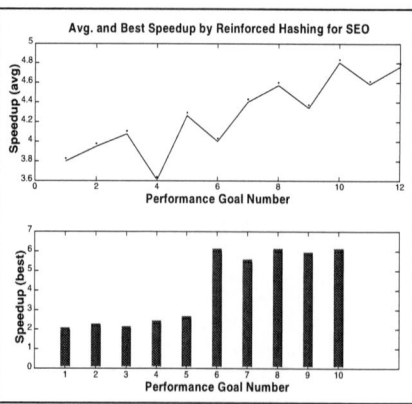

Fig. 5. SEO Speedup by (i) Expt A: Hashing; (ii) Expt B: Reinforced Hashing; (iii) Expt C: Average, Best Speedup

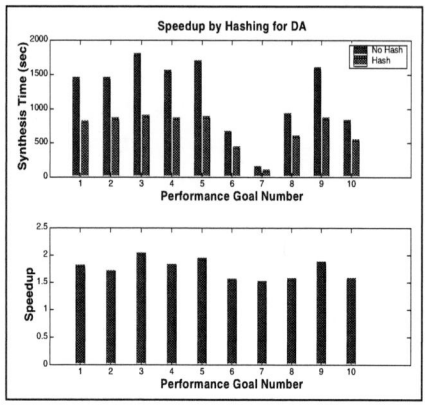 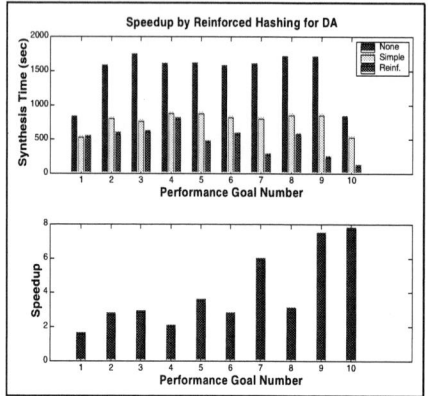 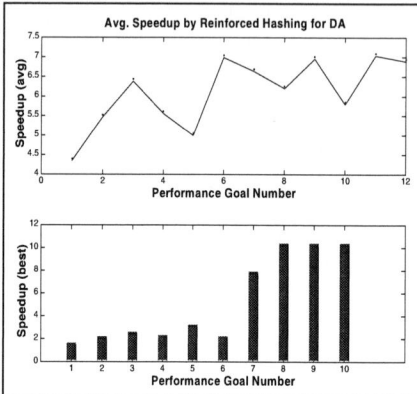

Fig. 6. DA Speedup by (i) Expt A: Hashing; (ii) Expt B: Reinforced Hashing; (iii) Expt C: Average, Best Speedup

5. Conclusion

This paper studied the application of hashing for improving the synthesis time of analog circuits. Hashing makes faster synthesis possible by avoiding matrix element recomputation. Separate tables for hash classes facilitates faster retrieval of hashed objects as well as limits the table size. Simple as well as continuously updated hash tables were used to increase the benefit from hashing. An average speedup of 4x-6x (400%-600%) demonstrates the effectiveness of the proposed approach.

References

[1] M. J. Krasnicki, R. Phelps, J. R. Hellums, M. McClung, R. A. Rutenbar, and L. R. Carley, "ASF: a practical simulation-based methodology for the synthesis of custom analog circuits," in *Proc. ICCAD*, 2001, pp. 350–357.

[2] W. Daems, G. Gielen, and W. Sansen, "Simulation-based automatic generation of signomial and posynomial performance models for the sizing of analog integrated circuit," in *IEEE Transactions on CAD*, vol. 22, no. 5, 2003, pp. 517–534.

[3] T. Kiely and G. Gielen, "Performance modeling of analog integrated circuits using least-squares support vector machines," in *Proc. Design, Automation and Test in Europe*, 2004, p. 10448.

[4] S. Doboli, G. Gothoskar, and A. Doboli, "Extraction of piecewise-linear analog circuit models from trained neural networks using hidden neuron clustering," in *Proc. Design, Automation and Test in Europe*, 2003, p. 11098.

[5] G. Wolfe and R. Vemuri, "Adaptive sampling and modeling of analog circuit performance parameters with pseudo-cubic splines," in *Proc. ICCAD*, 2004, pp. 931–938.

[6] D. Han and A. Chatterjee, "Adaptive response surface modeling-based method for analog circuit sizing," in *Proc. IEEE International SOC Conference*, 2004, pp. 109–112.

[7] T. McConaghy and G. Gielen, "Double-strength CAFFEINE: Fast template-free symbolic modeling of analog circuits via implicit canonical form functions and explicit introns," in *Proc. Design, Automation and Test in Europe*, 2006, pp. 269–274.

[8] A. Manthe and C. Shi, "Lower bound based DDD minimization for efficient symbolic circuit analysis," in *Proc. International Conference on Computer Design*, 2001, p. 374.

[9] P. Mandal and V. Visvanathan, "Macromodeling of the A.C. characteristics of CMOS op-amps," in *Proc. ICCAD*, vol. 7, no. 11, Nov. 1993, pp. 334–340.

[10] A. Pradhan and R. Vemuri, "Regression based circuit matrix models for accurate performance estimation of analog circuits," in *Proc. 15th IFIP Conference on VLSI SOC*, 2007, pp. 48–53.

[11] J.Harvey, M.Elmasry, and B.Leung, "STAIC: a synthesis tool for CMOS and BiCMOS analog integrated circuits," in *Proc. ISCAS*, 1991, pp. 2004–2007.

[12] T. H. Cormen, C. E. Leiserson, and R. L. Rivest, *Introduction to algorithms.* MIT Press, 2001.

[13] J. M. Cohn, *Analog Device-Level Layout Automation.* Kluwer Academic Publishers, 2000.

21st International Conference on VLSI Design

An Inversion-Based Synthesis Approach for Area and Power efficient Arithmetic Sum-of-Products

Sabyasachi Das
Synplicity Inc
Sunnyvale, CA, USA
Email: sabya@synplicity.com

Sunil P. Khatri
Texas A&M University
College Station, TX, USA
Email: sunilkhatri@tamu.edu

Abstract— In state-of-the-art Digital Signal Processing (DSP) and Graphics applications, the arithmetic Sum-of-Product (SOP) is an important and computationally intensive operation, consuming a significant amount of area, delay and power. This paper presents a new algorithmic approach to synthesize a non-timing critical SOP block in an area-efficient and power-efficient way, which can be very useful to reduce the size and power consumption of the non timing-critical portion in the design. We have divided the problem of generating the SOP into three parts: inversion-based creation of the BitClusters (sets of individual partial-product bits, which belong to the ith bitslice), propagation-based reduction of the BitClusters and selective-inversion based computation of the final sum result. Techniques used in these three steps help to reduce the implementation area and power consumption for the SOP block. Our experimental data shows that the SOP block generated by our approach is significantly smaller (8.59% on average) and marginally faster (0.42% on average) than the SOP block generated by a commercially available best-in-class datapath synthesis tool. In addition, our proposed SOP netlist consumes significantly less dynamic power (7.92% on average) and leakage power (5.65% on average) than the netlist generated by the synthesis tool. These improvements were verified on placed-and-routed designs as well.

I. INTRODUCTION

As we migrate toward ultra deep sub-micron feature sizes, designs are becoming increasingly complex, with very aggressive optimization goals. In all circuits, some portions of the design are timing-critical and other portions are not timing-critical. It is very important to use area-efficient and power-efficient architectures in the non timing-critical portion of the chip. This would reduce the overall size and power consumption of the design, with secondary improvement in circuit delay as well. In addition, it would also provide more options to the placement and routing of the timing-critical portions of the circuit, potentially leading to improved performance.

Sum-of-Product (SOP) blocks have been extensively used in DSP and Graphics algorithms. Some of the specific applications of SOP are Multiply-Accumulation (MAC), vector quantization, computation of the euclidean distance between two points, adaptive filtering, pattern recognition, image compression, decoding, etc. Hence, an area-efficient and power-efficient SOP architecture is becoming increasingly important.

There have been several techniques proposed, which can be used to improve the area of a Sum-of-Product block. In [1], [2], [3], [4], [5]; the authors have presented different ways to use carrysave arithmetic on multiple arithmetic blocks

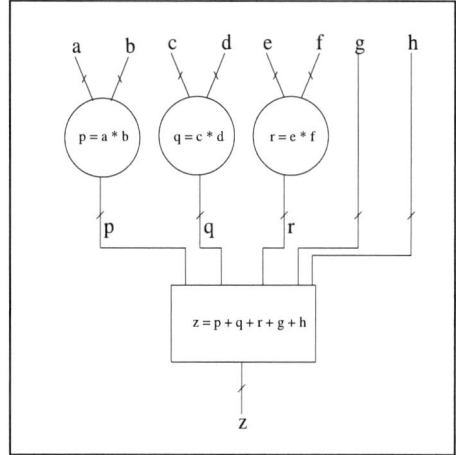

Fig. 1. Block Diagram of an 8-input Sum-of-Product (SOP) block

to design large SOP blocks. These techniques emphasizes the usefulness of SOP blocks over a collection of cascaded arithmetic blocks performing unit operations (like additions, subtractions, multiplications etc). There are several papers focusing on the generation of multiplication and addition units, which can be applied to the design of the SOP blocks also. In [6], a modified Wallace tree construction is discussed, to save most of the wasted area in the multiplier layout. In [7], a dependence graph and modified Booth algorithm is used to design a merged multiply-accumulate (MAC) hardware unit. The authors in [8] describe a split array multiplier organized in a left-to-right leapfrog structure, leading to less power consumption. In [9], [10], [11], different techniques have been proposed to reduce the partial products in multiplication and SOP units. Competitive analysis between different reduction approaches are presented in [12] and [13]. Among the adder architectures, Ripple carry adder is the smallest and it is widely used in non timing-critical path [14]. A mix of the above-mentioned architectures can be used to generate an SOP block.

In this paper, we propose a new scheme to synthesize SOP blocks in an area and power efficient manner. In our approach, we define the notion of a BitCluster for every bit in the SOP block. To generate the BitClusters for all bits, we use an inversion-based scheme. After the BitClusters are created, we perform a tree-reduction operation to reduce the BitClusters

653

978-1-4244-3039-0/08 $25.00 © 2008 IEEE

to two addends. In the third step, we add the two addends by using a selective-inversion based adder to produce the output.

We have organized the rest of the paper as follows: In Section II, we present some background information. We discuss our proposed approach in Section III. The experimental setup is explained in Section IV. Section V presents the experimental results. Conclusions are drawn in Section VI.

II. PRELIMINARIES

In this section, we briefly explain the concept of a generalized Sum-of-Product (SOP) block. The block diagram of an 8-input Sum-of-Product block is shown in the Figure 1. In this block, there are eight inputs (a, b, c, d, e, f, g and h), which produce the output z. In this SOP block, there are three product terms or multiplicative terms ($a*b$, $c*d$ and $e*f$) and two input sum terms or additive terms (g and h). A generalized Sum-of-Product block can be used to implement the addition of an arbitrary number of (including zero) product terms and sum terms. As a consequence, an SOP block is quite general.

Since a multiplier has only one product term ($a*b$) and no sum term, it can be considered as a special case of the Sum-of-Product block. On the other hand, a 2-input adder has only one sum term ($a+b$) and no product term, it can also be considered as a special case of the Sum-of-Product block. In addition to the multiplier and adder blocks, a generalized Sum-of-Product block can be used to implement the multiply-accumulator (MAC), subtractor, squarer, comparator, shared multiplier-adder, tree-of-adders or combinations thereof.

III. OUR APPROACH

Figure 2 describes our overall flow, which consists of three steps. In the following sub-sections, we discuss each step in detail.

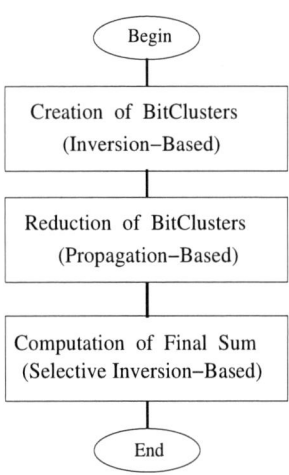

Fig. 2. Our Flow to Synthesize an Area and Power Efficient SOP Block

A. Creation of BitClusters

We define the BitCluster for the ith bit as the set of individual partial-product bits, which belong to the ith bitslice. To explain the creation of BitClusters, let us consider the

following Sum-of-Product (SOP) block: $Z = (a*b) + (c*d)$ where a, b are 4-bit wide and c, d are 2 bit-wide each. If we denote signal a by (a_3, a_2, a_1, a_0); signal b by (b_3, b_2, b_1, b_0); signal c by (c_1, c_0) and signal d by (d_1, d_0) then, the BitClusters are:

- $BitCluster_0 = \{a_0 \wedge b_0,\ c_0 \wedge d_0\}$
- $BitCluster_1 = \{a_1 \wedge b_0,\ a_0 \wedge b_1,\ c_1 \wedge d_0,\ c_0 \wedge d_1\}$
- $BitCluster_2 = \{a_2 \wedge b_0,\ a_1 \wedge b_1,\ a_0 \wedge b_2,\ c_1 \wedge d_1\}$
- $BitCluster_3 = \{a_3 \wedge b_0,\ a_2 \wedge b_1,\ a_1 \wedge b_2,\ a_0 \wedge b_3\}$
- $BitCluster_4 = \{a_3 \wedge b_1,\ a_2 \wedge b_2,\ a_1 \wedge b_3\}$
- $BitCluster_5 = \{a_3 \wedge b_2,\ a_2 \wedge b_3\}$
- $BitCluster_6 = \{a_3 \wedge b_3\}$

For any given (m-bit$*n$-bit) + (p-bit$*q$-bit) Sum-of-Product, we can compute all the BitClusters by performing 2-input NAND operations between the appropriate bits of the multiplicand and multiplier in each product expression. In such an approach, we need ($m*n + p*q$) number of 2-input NAND gates to generate max(m+n-1, p+q-1) BitClusters. In practice, due to the use of NAND gates, all the elements in the BitClusters contain a logical inversion. In CMOS technology, inverting functions (like NAND, NOR etc.) are typically more efficient in terms of area, delay and power than non-inverting functions (like AND, OR etc). We have found that all of the commercially available 0.13μ and 0.09μ technology libraries (that we have explored) have smaller and faster 2-input NAND gates than 2-input AND gates. Throughout the rest of this section, we denote the total number of BitClusters as N. In Algorithm 1, we present the way to create the BitClusters.

B. Reduction of BitClusters

After generating the BitClusters, most of the BitClusters contain more than two elements. For an SOP which implements the expression $a*b + c*d + e*f + g + h$ (and a, b, c, d, e, f, g and h all have the same bit-width); all the N BitClusters have more than two elements. In this step, we reduce each BitCluster to a maximum of two elements.

For the reduction of partial products, two techniques proposed in the context of multipliers are *Wallace Tree* [9] and *Dadda Tree* [10] reduction schemes. In these approaches, an n-input *Wallace Tree* or *Dadda Tree* reduces its k-bit inputs to two (k+log_2n-1)-bit outputs. In the *Wallace Tree*, the number of operands are reduced at the earliest opportunity. On the other hand, in the *Dadda Tree*, the number of operands are reduced in a more area-efficient way without any significant impact in the delay of the reduction tree. The authors of [12], [13] present a comparative study between different reduction techniques.

All these techniques use different types of *counters*. A *(p:q) counter* is defined as a functional block, which adds its p single-bit inputs and produces q single-bit outputs; where p and q satisfy the following equation:

$$q = \lfloor log_2 p + 1 \rfloor$$

In our approach for area and power efficient SOP block, we use the Dadda-tree reduction scheme with a modified (3:2) and (2:2) counters. In the *Reduction of BitClusters* phase, each BitCluster would possibly need multiple modified (3:2)

654

978-1-4244-3039-0/08 $25.00 © 2008 IEEE

Algorithm 1 : Creation of all the BitClusters

N = Total number of BitClusters in SOP

// Loop-1: Initialize All BitClusters to NULL
for i = 0 to $(N-1)$ **do**
   $BitCluster_i = \{NULL\}$
end for

// Loop-2: Compute All BitClusters
//   (for each multiplicative-term or product-term)
for (Each Multiplicative term in the SOP expression) **do**
   // (assume that the expression is $m1 * m2$.)
   $w1$ = Width($m1$)
   $w2$ = Width($m2$)
   for k = 0 to $(w1-1)$ **do**
      for l = 0 to $(w2-1)$ **do**
         $BitCluster_{(k+l)} = \{BitCluster_{k+l}$
                $\vee \ \overline{(m1_k \wedge m2_l)}\}$
      end for
   end for
end for

// Loop-3: Update All BitClusters
//   (for each additive-term or sum-term)
for (Each Additive term in the SOP expression) **do**
   // (assume that the additive expression is $m3$.)
   $w3$ = Width($m3$)
   for k = 0 to $(w3-1)$ **do**
      $BitCluster_{(k)} = \{BitCluster_k \vee \overline{m3_k}\}$
   end for
end for

return all (N) BitClusters

---

and (2:2) counters. After generating the BitClusters, all the elements in the BitClusters contain a logical inversion (due to the presence of the NAND gate in the BitCluster Creation phase). To ensure that the outputs at the end of the reduction of BitClusters also contain the logical inversion, we modify the functionality of the (3:2) and (2:2) counters. A modified (3:2) counter accepts 3 inputs signals (x_i, y_i and $carry_i$) belonging to i^{th} column (BitCluster) in the partial-products and produces 1 output signal (sum_i) for the ith column (BitCluster) and 1 more output signal ($carry_{i+1}$) for the $(i+1)^{th}$ BitCluster. The functionality of the sum_i and $carry_{i+1}$ of our modified (3:2) counter is written as:

- $sum_i = \overline{x_i \odot y_i \odot carry_i}$
     where $\odot$ represents a 2-input XNOR gate
- $carry_{i+1} = (x_i \vee y_i) \wedge (y_i \vee carry_i) \wedge (carry_i \vee x_i)$

Similarly, a modified (2:2) counter accepts 2 inputs signals (x_i and y_i) belonging to i^{th} column (BitCluster) in the partial-products and produces 1 output signal (sum_i) for the ith column (BitCluster) and 1 more output signal ($carry_{i+1}$) for the $(i+1)^{th}$ BitCluster. The functionality of the sum_i and

$carry_{i+1}$ of our modified (2:2) counter is written as:

- $sum_i = x_i \odot y_i$
- $carry_{i+1} = (x_i \vee y_i)$

In each BitCluster of the reduction-tree, we are able to use instantiations of the same counter structure. The sum_i output of each counter in $BitCluster_i$ gets fed to either another counter in the same $BitCluster_i$ or to the final adder stage (described in the next section of this paper). The $carry_{i+1}$ output of each counter in $BitCluster_i$ gets fed to either another counter in $BitCluster_{i+1}$, or to the final adder stage. At the output of the final level in each BitCluster, the inverted result is produced, and would get fed to the final adder stage. In this way, the reduction-tree structure *propagates* the inversion property to the final stage of the SOP block.

C. Computation of Final Sum

Since the Sum-of-Product circuit needs to present the final result in the single binary vector format, all the BitClusters have to be added (with the inversion property taken care of) by a final carry propagation adder circuit. After performing the reduction of the BitClusters, each BitCluster consists of ≤ 2 elements. Hence, we can view the N vertical BitClusters as two horizontal vectors or operands, each having N elements. Therefore, the problem of final addition of BitClusters gets converted to the problem of a specialized 2-operand addition. In this addition, the inputs are two inverted vectors of width N bits and the output is one non inverted vector of width $(N+1)$ bits. In this section, we refer to these two operands as

- vector x $(x_{N-1}, \ x_{N-2}, \ ..., \ x_1, \ x_0)$ and
- vector y $(y_{N-1}, \ y_{N-2}, \ ..., \ y_1, \ y_0)$

To have an area and power efficient SOP implementation in the non-timing-critical path of a design, we definitely need to use a low-area architecture for the final carry propagate addition. In datapath designs, the ripple-carry architecture is very useful in non-timing-critical portions of the design (if it can satisfy the timing requirement of the off-critical path). Our final adder is a modified version of the ripple-carry addition scheme. In our final adder, every bit (i) has a modified-full-adder cell; which takes 3 inputs (x_i, y_i and $carry_i$) and produces two outputs (sum_i and $carry_{i+1}$). The sum_i output from every modified-full-adder cell would have the correct polarity (non-inverted) and is the final result for the ith bit position. On the other hand, $carry_{i+1}$ remains in the *inverted* state. The Boolean expressions for the functionality of the modified-full-adder cell is the following:

- $sum_i = \overline{\overline{x_i} \oplus y_i \oplus carry_i}$
- $carry_{i+1} = \overline{\overline{(x_i \wedge y_i)} \wedge \overline{(y_i \wedge carry_i)} \wedge \overline{(carry_i \wedge x_i)}}$

Based on the De Morgan's law, we note that the above-mentioned equation for $carry_{i+1}$ is identical to the $carry_{i+1}$ output of a traditional (3:2) counter.

If a particular bit has less than three elements to be fed to the modified-full-adder cell, then the remaining input pins of the modified-full-adder cell need to be tied to the global logic-1 signal. This is due to the fact that all the inputs of the modified-full-adder are in inverted state. This situation could happen frequently in the least significant bit of the SOP.

The algorithm for the Computation of the Final Sum is presented in Algorithm 2.

Algorithm 2 :Computation of Final Sum

---

for $i = 0$ to $(N - 1)$ **do**
    // Handle all the non-existent elements
    if x_i does not exist **then**
        $x_i = 1'b1$
    end if
    if y_i does not exist **then**
        $y_i = 1'b1$
    end if
    if $carry_i$ does not exist **then**
        $carry_i = 1'b1$
    end if

    // Perform the addition in the ith bit
    Instantiate modified-full-adder cell with three inputs
    x_i, y_i, $carry_i$ and two outputs sum_i and $carry_{i+1}$
end for
$sum_N = carry_N$

return the $(N + 1)$-bit wide sum vector

IV. EXPERIMENTAL SETUP

We implemented our proposed approach in the C++ programming language. The experiments were performed with different Sum-of-Product RTL designs written in the Verilog hardware description language.

To collect different data-points regarding the quality of results for the Sum-of-Product blocks in the non timing-critical portion of the design, we used the following variations:

- Multiple types of Sum-of-Product designs of different expressions and input bit-widths:
  In the Table I, we report different configurations of the designs that are used in our experiments. Following is a brief description of the different SOP blocks presented in the Table I.
  - Two multiplier blocks ($Z = (a * b)$) having different bit-widths. We refer to these as Mult-1 and Mult-2.
  - Two Multiply-Accumulate blocks ($Z = (a * b) + c$). We refer to these designs as Mac-1 and Mac-2.
  - Two general SOP blocks. The block Sop-1 represents the functionality of $Z = (a*b)+(c*d)$ and the block Sop-2 represents the functionality of $Z = (a * b) + (c * d) + e$.
  - Two Squarer blocks ($Z = (a * a)$). We refer to these blocks as Sqr-1 and Sqr-2.
- The different technologies and libraries, we used are:
  - Two libraries (L_1 and L_2) for 0.13μ technology.
  - Two libraries (L_3 and L_4) for 0.09μ technology.

Name of the Sum-of-Product (SOP) Block	Widths of the Input Signals of the SOP Block	Width of the Output Signal of the SOP
Mult-1	16 , 16	32
Mult-2	24 , 31	55
Mac-1	32, 32, 32	64
Mac-2	28, 24, 32	52
Sop-1	34, 35, 23, 28	69
Sop-2	16, 23, 21, 17, 31	39
Sqr-1	25	50
Sqr-2	18	36

TABLE I

CHARACTERISTICS OF DIFFERENT SUM-OF-PRODUCT BLOCKS

- Different input arrival time constraints:
  To facilitate the explanation, let us assume that the expression of the SOP is $Z = (a * b) + (c * d)$ and each of the four input signals is n-bit wide. We have used the following types of input arrival time constraints:
  - All input bits of all the signals arrive at the same time. We refer to this constraint as Type-A. If we denote $Arr(a_i)$ as the arrival time of the bit a_i and if k is a constant number, then this Type-A constraint can be represented as:
    $$Arr(a_i) = k; \qquad 0 \leq i < n$$
    $$Arr(b_i) = k; \qquad 0 \leq i < n$$
    $$Arr(c_i) = k; \qquad 0 \leq i < n$$
    $$Arr(d_i) = k; \qquad 0 \leq i < n$$
    This category represents the actual timing situations if the SOP block is placed immediately after a register-bank or the primary inputs of the design are fed to the SOP block.
  - Different input bits arrive at different times. We refer to this category of timing constraints as Type-B. We believe that this category represents the actual timing situations in most of the Sum-of-Product blocks in real-life designs. Assuming that k is a constant number and δ is the delay of the fastest 2-input AND-gate in the given technology library, the following are some specific examples of the Type-B timing constraints. Here we have explained the arrival times for signal a_i. Similar expressions for arrival times applies to all the bits of signals b, c and d as well.
    1) $Arr(a_i) = i * k * \delta;$ $0 \leq i < n$
    2) $Arr(a_i) = i^2 k\delta;$ $0 \leq i < n$
    3) $Arr(a_i) = 0;$ $0 \leq i < \lceil n/2 \rceil$
       $Arr(a_i) = k\delta;$ $\lceil n/2 \rceil \leq i < n$
    4) $Arr(a_i) = 0;$ $0 \leq i < \lceil n/4 \rceil$
       $Arr(a_i) = k\delta;$ $\lceil n/4 \rceil \leq i < \lceil n/2 \rceil$
       $Arr(a_i) = 2k\delta;$ $\lceil n/2 \rceil \leq i < \lceil 3n/4 \rceil$
       $Arr(a_i) = 3k\delta;$ $\lceil 3n/4 \rceil \leq i < n$
    5) $Arr(a_i) = 0;$ $0 \leq i < \lceil n/4 \rceil$
       $Arr(a_i) = ik\delta;$ $\lceil n/4 \rceil \leq i < \lceil n/2 \rceil$
       $Arr(a_i) = 2ik\delta;$ $\lceil n/2 \rceil \leq i < \lceil 3n/4 \rceil$
       $Arr(a_i) = 3ik\delta;$ $\lceil 3n/4 \rceil \leq i < n$

Design Name	Technology Library	Timing Constraint	Area (μ^2)			Worst-case Delay (ps)		
			Commercial Tool	Our Approach	(%) Improvement	Commercial Tool	Our Approach	(%) Improvement
Mult-1	L_1	Type-A	2697	2472	8.34%	2148	2137	0.51%
Mult-2	L_1	Type-A	4619	4268	7.59%	2671	2646	0.94%
Mac-1	L_1	Type-A	5281	4983	5.63%	2836	2860	-0.84%
Mac-2	L_1	Type-A	4306	3959	8.07%	2792	2761	1.11%
Sop-1	L_1	Type-A	10743	9892	7.92%	3018	2982	1.19%
Sop-2	L_1	Type-A	6937	6343	8.56%	2675	2651	0.89%
Sqr-1	L_1	Type-A	2368	2151	9.16%	1864	1849	0.81%
Sqr-2	L_1	Type-A	1873	1710	8.65%	1753	1738	0.86%
Mult-1	L_3	Type-A	2953	2683	9.14%	1582	1573	0.69%
Mult-2	L_3	Type-A	5816	5372	7.62%	1739	1758	-1.09%
Mac-1	L_3	Type-A	6372	5826	8.57%	1951	1934	0.87%
Mac-2	L_3	Type-A	4901	4421	9.81%	1814	1829	-0.83%
Sop-1	L_3	Type-A	11694	10493	10.26%	2139	2126	0.61%
Sop-2	L_3	Type-A	8165	7401	9.36%	1863	1882	-1.02%
Sqr-1	L_3	Type-A	2687	2419	9.94%	1481	1453	1.89%
Sqr-2	L_3	Type-A	2342	2092	10.68%	1107	1084	2.07%
Mult-1	L_1	Type-B1	2103	1959	6.82%	2762	2735	0.97%
Mult-2	L_1	Type-B1	4592	4248	7.49%	2918	2874	1.51%
Mac-1	L_1	Type-B1	5286	4915	7.03%	3859	3841	0.46%
Mac-2	L_1	Type-B1	3741	3421	8.56%	3724	3759	-0.94%
Sop-1	L_1	Type-B1	9837	9036	8.14%	4206	4238	-0.76%
Sop-2	L_1	Type-B1	6372	5897	7.45%	3562	3540	0.61%
Sqr-1	L_1	Type-B1	1983	1812	8.62%	2844	2829	0.53%
Sqr-2	L_1	Type-B1	1562	1448	7.28%	2181	2164	0.78%
Mult-1	L_3	Type-B1	2972	2664	10.41%	1702	1737	-2.06%
Mult-2	L_3	Type-B1	4619	4213	8.79%	1867	1851	0.86%
Mac-1	L_3	Type-B1	5384	4937	8.30%	2914	2896	0.69%
Mac-2	L_3	Type-B1	4063	3652	10.12%	2876	2829	1.63%
Sop-1	L_3	Type-B1	8745	7946	9.14%	3285	3258	0.82%
Sop-2	L_3	Type-B1	5691	5209	8.47%	2973	2961	0.40%
Sqr-1	L_3	Type-B1	2284	2048	10.28%	2049	2084	-1.71%
Sqr-2	L_3	Type-B1	1956	1781	8.92%	1631	1612	1.16%
Average					8.59%			0.42%

TABLE II

AREA AND DELAY COMPARISON OF SUM-OF-PRODUCT BLOCKS GENERATED BY A COMMERCIAL SYNTHESIS TOOL AND BY OUR APPROACH

V. EXPERIMENTAL RESULTS

We compared the netlist produced by our approach against the output netlist of a commercially available best-in-class datapath synthesis tool. The synthesis tool generates arithmetic-optimized architectures for all the arithmetic blocks (like sum-of-products) and then it performs general-purpose operations like technology-independent optimizations, constant propagation, redundancy removal, technology mapping, timing-driven optimization, area-driven optimization, low-power optimization etc. While running the synthesis tool, we turned on all the above-mentioned optimizations. In the Table II and Table III, we report 32 sets of data-points (worst-case delay, total area, dynamic and leakage power consumption) involving SOPs having different widths and expressions, timing constraints and technology libraries. If we compute the average of all the 32 data-points in the Table II and compare our results with the results produced by the implementation of the commercial datapath synthesis tool, we see a 8.59% area savings in the SOP block (column 6 of Table II) with a marginal 0.42% speed improvement (column 9 of Table II). Similarly, the average dynamic and leakage power consumption of our SOP block is significantly less (7.92% for dynamic power and 5.65% for leakage power) than that of the SOP produced by the synthesis tool (columns 6 and 9 in the Table III).

We observe that in 8 cases, the delay of our SOP is slightly worse than the baseline. As expected, in each of these cases, the area and the power of our SOP is much better than the baseline. Since our approach is designed to be used in the area-critical portions of the design, savings in area and power are considered to be the primary goal and the blocks do not go though rigorous timing optimization phase. As a result, a marginal degradation in timing is not considered significant. Similarly, the slight improvement in timing in all the other 24 cases are also considered insignificant.

To keep the sizes of the Table II and Table III relatively brief, we did not report the results for all possible combinations of designs, timing constraints and technology libraries. Note that the results in each of the combinations, which are not reported here also supported our conclusion that, the proposed approach produces area and power efficient SOP blocks.

To verify the correlation of post-synthesis experimental data with the post place-and-route data, we performed placement and routing on Mult-1 and Mac-1. For these two testcases, the average post-routing total area of the SOP block generated by our proposed approach is 0.89 (normalized to the total area of the SOP generated by the commercial synthesis tool). Similarly, the post-routing total power consumption of the SOP block generated by our technique is 0.91 (normalized to the total power of the SOP generated by the synthesis tool). In addition, the post-routing worst delay of the SOP generated by the synthesis tool and our techniques are comparable. The individual results for the Mult-1 and Mac-1 designs correlate with the post-synthesis numbers reported in the Table II and Table III. These post-routing data confirm our conclusion about the area and power efficiency of our approach.

With this observation, we conclude that our area and power improvement is consistent across multiple types of SOPs,

Design Name	Technology Library	Timing Constraint	Dynamic Power (μW)			Leakage Power (μW)		
			Commercial Tool	Our Approach	(%) Improvement	Commercial Tool	Our Approach	(%) Improvement
Mult-1	L_1	Type-A	42	38	9.52%	9	9	0.00%
Mult-2	L_1	Type-A	76	71	6.57%	18	17	5.55%
Mac-1	L_1	Type-A	83	80	3.61%	19	17	10.53%
Mac-2	L_1	Type-A	68	64	5.88%	15	15	0.00%
Sop-1	L_1	Type-A	164	157	4.27%	34	32	5.89%
Sop-2	L_1	Type-A	107	99	7.48%	26	23	11.54%
Sqr-1	L_1	Type-A	30	30	0.00%	6	6	0.00%
Sqr-2	L_1	Type-A	29	27	6.90%	6	5	16.66%
Mult-1	L_3	Type-A	69	61	11.60%	21	20	4.76%
Mult-2	L_3	Type-A	114	103	9.65%	34	31	8.82%
Mac-1	L_3	Type-A	136	122	10.29%	40	37	7.50%
Mac-2	L_3	Type-A	97	94	3.08%	29	28	3.44%
Sop-1	L_3	Type-A	213	197	7.51%	63	63	0.00%
Sop-2	L_3	Type-A	165	146	11.52%	52	49	5.77%
Sqr-1	L_3	Type-A	51	45	11.76%	16	15	6.25%
Sqr-2	L_3	Type-A	42	39	7.14%	11	11	0.00%
Mult-1	L_1	Type-B1	37	34	8.11%	8	7	12.50%
Mult-2	L_1	Type-B1	68	63	7.35%	14	13	7.14%
Mac-1	L_1	Type-B1	71	66	7.04%	15	15	0.00%
Mac-2	L_1	Type-B1	67	59	11.94%	15	13	13.33%
Sop-1	L_1	Type-B1	146	138	5.48%	29	28	3.45%
Sop-2	L_1	Type-B1	95	90	5.26%	18	16	11.11%
Sqr-1	L_1	Type-B1	26	25	3.84%	5	5	0.00%
Sqr-2	L_1	Type-B1	21	19	9.52%	3	3	0.00%
Mult-1	L_3	Type-B1	62	55	11.29%	19	17	10.52%
Mult-2	L_3	Type-B1	99	91	8.08%	28	25	10.71%
Mac-1	L_3	Type-B1	138	121	12.32%	41	39	4.87%
Mac-2	L_3	Type-B1	94	85	9.57%	27	26	3.70%
Sop-1	L_3	Type-B1	205	187	8.78%	63	60	4.76%
Sop-2	L_3	Type-B1	147	136	7.53%	46	45	2.17%
Sqr-1	L_3	Type-B1	48	43	10.42%	11	11	0.00%
Sqr-2	L_3	Type-B1	39	35	10.25%	10	9	10.00%
Average					7.92%			5.65%

TABLE III

POWER COMPARISON OF SUM-OF-PRODUCT BLOCKS GENERATED BY A COMMERCIAL SYNTHESIS TOOL AND BY OUR APPROACH

timing constraints and technology libraries. This underscores the strength of our approach. Since the SOP is a very area and power intensive block, we believe that the non-timing critical portions of many real-life datapath designs can significantly benefit from our approach.

VI. CONCLUSION

In this paper, we have presented a new approach for implementing an area and power efficient sum-of-product (SOP) block, which would be very useful in the non timing-critical portion of the design. Our inversion and propagation based approach works seamlessly with different types of SOP blocks, and across different technology libraries (0.13μ, 0.09μ). Our experimental data shows that the SOP block generated by our approach is significantly smaller (8.59% on average) and marginally faster (0.42% on average) than the Sum-of-Product block generated by a commercially available best-in-class datapath synthesis tool. In addition, our proposed Sum-of-Product netlist consumes significantly less dynamic power (7.92% on average) and leakage power (5.65% on average) than the netlist generated by the datapath synthesis tool.

REFERENCES

[1] T. Kim, W. Jao, S. Jjiang. "Circuit optimization using carry-save-adder cells," in *IEEE Transactions on Computer-Aided Design of Integrated Circuits and Systems CAD-17*, pp. 974–984, 1998.

[2] A. Mathur, S. Saluja, "Improved merging of datapath operators using information content and required precision analysis," in *Proceedings of IEEE 38th Conference on Design Automation*, pp. 462–467, 2001.

[3] A. K. Verma, P. Ienne. "Improved Use of the Carry-Save Representation for the Synthesis of Complex Arithmetic Circuits," in *Proceedings of the 2004 IEEE/ACM International conference on Computer-aided design*, pp.791-798, 2004.

[4] A. Fayed, W. Elgharbawy, M. Bayoumi. "A data merging technique for high-speed low-power multiply accumulate units," in *Proceedings of IEEE Internation Conference on Acoustics, Speech, and Signal Processing*, vol. 5, pp. 145–148, 2004.

[5] P. F. Stelling, V. G. Oklobdzija, "Implementing Multiply-Accumulate Operation in Multiplication Time," in *Proceedings of 13th IEEE Symposium on Computer Arithmetic* pp. 99, 1997.

[6] N. Itoh, Y. Tsukamoto, T. Shibagaki, K. Nii, H. Takata, H. Makino, "A 32/spl times/24-bit multiplier-accumulator with advanced rectangular styled Wallace-tree structure," in *IEEE International Symposium on Circuits and Systems*, pp. 73-76 vol. 1, 2005.

[7] F. Elguibaly, "A fast parallel multiplier-accumulator using the modified Booth Algorithm," in *IEEE Transactions on Circuits and Systems II: Analog and Digital Signal Processing*, vol: 47(9), pp. 902–908, 2000.

[8] Z. Huang, M. D. Ercegovac, "High-performance low-power left-to-right array multiplier design," in *IEEE Transactions on Computers*, vol: 54, issue: 3, pp. 272-283, 2005

[9] C. S. Wallace, "A suggestion for a fast multiplier," in *IEEE Transactions on Electronic Computers*, EC-13(2):14-17, 1964.

[10] L. Dadda, "Some schemes for parallel multipliers," in *Alta Frequenza*, vol. 34, pp. 349–356, 1965.

[11] V. G. Oklobdzija, D. Villeger, "Improving multiplier design by using improved column compressiontree and optimized final adder in CMOS technology," in *IEEE Transactions on Very Large Scale Integration (VLSI) Systems*, vol. 3, issue. 2, pp. 292-301, 1995.

[12] T. Whitney, S. Earl, A. Jacob, "A comparison of Dadda and Wallace multiplier delays," in *Advanced Signal Processing Algorithms, Architectures, and Implementations XIII. Edited by Luk, Franklin T. Proceedings of the SPIE*, vol. 5205, pp. 552-560, 2003.

[13] K. C. Bickerstaff, E. E. Swartzlander, M. J. Schulte, "Analysis of column compression multipliers," in *Proceedings of 15th IEEE Symposium on Computer Arithmetic*, pp. 33–39, 2001.

[14] M. D. Ercegovac, T. Lang. "Digital Arithmetic," Morgan Kaufmann Publishers, San Francisco, 2004.

SESSION B5:
Low Power - III

21st International Conference on VLSI Design

A Low Voltage, Low Ripple, on Chip, Dual Switch-Capacitor Based Hybrid DC-DC Converter

Kaushik Bhattacharyya, Student Member, IEEE, Pradip Mandal, Member, IEEE

E&ECE Department, IIT-Kharagpur, India

{kaushik, pradip}@ece.iitkgp.ernet.in

Abstract

Here we propose a low voltage low ripple dual switch-capacitor based hybrid DC-DC converter which is suitable for high dropout embedded regulation. In the proposed topology, along with a linear regulator two switching capacitors are used to store and recycle the charge for better power efficiency. The linear regulator is used to reduce the amount of output voltage ripple that comes from the switching capacitors. The output ripple noise is further reduced by introducing a synthesized counter ripple through the linear regulator. With this noise reduction technique, for an acceptable output ripple noise, the switching capacitors are reduced to a value which can be implemented on chip. The proposed converter circuit is designed in 0.18μ process for 3.3V to 1.25V conversion. With two switching capacitors of 150pF each, for 10mA load current and 50pF load capacitor, peak-peak output voltage ripple is only 45 mV and the achieved power efficiency is 64%

1. Introduction

In recent years, low Voltage-Power converters are achieving a remarkable attention because the most effective ways to reduce the power of the active circuits is by operating at a lower power supply voltage, (vdd) [1]. There is a constant endeavor to build up a highly efficient-converter with a small size and weight for a long time. Buck converter [2] which uses inductor to step down the supply voltage, is the first choice to make a highly efficient converter. However, the main disadvantage of a buck converter is that it requires bulky inductor. In addition, it introduces high ripple at the output voltage. For this reason its application for analog and mixed signal is limited.

On the other hand, linear regulator is compact [3] and its output voltage is free from ripple. A fast dynamic response is also provided by this regulator. These features make linear regulator suitable for analog and embedded application. However its power efficiency reduces linearly with the increase of input and output voltage difference. This becomes major concern for battery operated system, where battery voltage remains same while the required internal voltage reduces with progress of technology.

Inductor-less switch-capacitor converter topologies are very much attractive to improve the efficiency [4]. It posses the advantages of reduced size and weight. A considerable amount of switching noise restricts its analog application while the switching noise can be reduced by using large capacitor(s) (may be of the order of nano-or micro Farad) [5] but at that time it is difficult to use it as an embedded regulator.

There is a recent trend of combining a switching regulator (inductor based or capacitor based) with a linear regulator to achieve good power efficiency with low output ripple noise. For instance, combining a switch-capacitor converter and a linear regulator [6] gives a high dropout converter with good power efficiency. In these hybrid converters the output ripple is kept low by using external capacitor. This manuscript discusses about a hybrid converter topology which can be implemented on-chip (without external capacitor) still maintaining its output ripple within acceptable level.

In the proposed topology, along with a linear regulator a switch able capacitor is used to recycle charge for improving power efficiency. On the other hand the linear regulator helps to reduce the output ripple-noise that is introduced by the switching circuit. Noise reduction which is provided by the linear regulator is analyzed and optimized. Here we Propose Dual-Switch capacitor Circuits instead of using the fixed charge storing capacitor. The second equal size flying capacitor is switching in opposite phase of that of the first one. This type of switching-capacitor configuration helps to reduce peak-peak ripple at the output of the Switching Circuit module. It also helps to maintain input supply current steady in contrast to rectangular profile with a single flying capacitor. The output ripple-noise is further reduced by introducing a synthesized ripple-noise in opposite phase through the linear regulator. A detailed analysis and the necessary condition to minimize the output ripple are provided. Finally, two switching circuits with appropriate phase difference are combined and fed to the linear regulator to provide a steady supply current to the linear regulator.

In Section 2, the operating principle of the proposed converter is discussed. The calculation of the power efficiency is depicted in Section3. Section 4 is dedicated to the analysis of the output ripple-noise. Implementation of the proposed topology in the circuit

978-1-4244-3039-0/08 $25.00 © 2008 IEEE

level of design is discussed in section 5. Simulation results and discussion is provided in the section 6. Finally concluding remarks are discussed.

2. Proposed Converter and its Working Principle

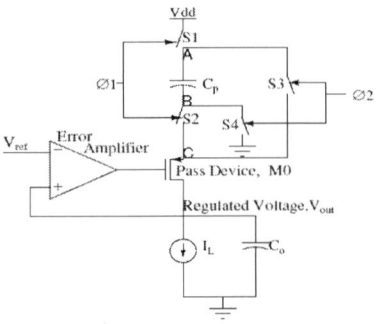

Fig. 1 Block Diagram of single capacitor based hybrid converter

The basic schematic diagram of a hybrid converter with single flying capacitor is shown in Fig.1. It comprises of an error amplifier followed by a pass-device, M0, a flying capacitor, Cp and four switches, S1-S4. I_L and Co are load current and load capacitance of the converter. The pass transistor and the error amplifier form a typical linear regulator except, the source of the pass transistor is connected to the charge recycling capacitor Cp rather than input supply. The four switches are controlled by two non-overlap signals Φ1 and Φ2 to connect Cp either in series or in parallel with the pass transistor of the linear regulator. In series configuration the capacitor stores charge which is then recycled to supply the output current in parallel configuration. In Fig2 operation of the converter is illustrated with sketch of different node voltages.

In *PHASE I*, the two switches, S1 and S2 are closed to connect the capacitor Cp in between the M0 and the input supply, *vdd*. With the progress of time, the capacitor is charged up at a rate of I_L/C_P. Hence, voltage at node B, V_B and the source voltage of M0, V_C decrease with time as shown in Fig.2. Ignoring voltage drop across switches, V_C follows V_B. Due to negative feedback at the gate of M0 through the error amplifier effect of variation in V_C on the output voltage is attenuated by power supply rejection ratio (PSRR) of the linear regulator [7]. For high PSRR the negative loop gain should be high and hence, M0 should be in saturation region. Therefore, this charging phase can continue as long as M0 is in saturation. After this charging phase the circuit moves into *PHASE II* at the phase transition point 'Y' in Fig. 2.

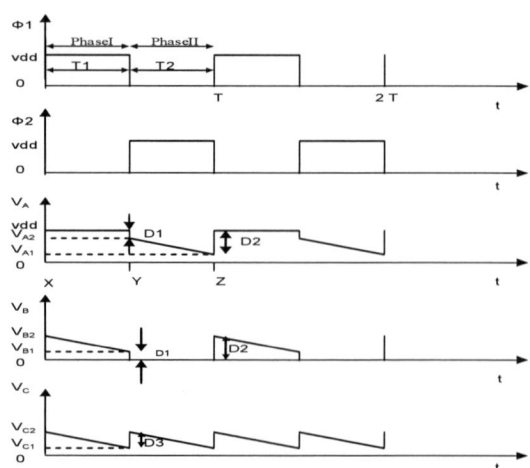

Fig. 2 Sketch of different node voltages of the proposed converter

In *PHASE II*, the two switches S1 and S2 are opened and the other two switches S3 and S4 are closed to connect the positive (A) and negative (B) terminals of Cp to the source node of M0 and the ground node respectively. As shown in Fig.2, at the transition phase point 'Y', V_B is made zero with a negative voltage step of $D1$. Since node 'A' is disconnected from the supply, this node also has same voltage step and it becomes (Vdd - $D1$) (denoted as V_{A2}) at this phase transition point. Ignoring voltage drop across S3, in this phase V_C follows V_A. Note that in *PHASE II*, Cp is the source of the output current. This phase we refer as charge recycle phase. With progress of time, voltage of Cp and hence, V_A and V_C reduce. Similar to charging phase, this phase may continue as long as M0 is in saturation. As shown in Fig. 2, at time point 'Z', *PHASE II* is terminated and *PHASE I* is resumed by opening S3 and S4 and subsequently closing S1 and S2. At this phase transition point, V_A goes to Vdd with a positive voltage step of D2 and V_B changes from 0V to V_{B2} with the same voltage step. This is how alternate charging and charge recycling phases continue and V_C, follows (with drop of a switch) V_B and V_A in consecutive phases. For simplicity, so far we have discussed the working principle of the hybrid converter with single flying capacitor [8]. However, in the final proposal two flying capacitors [9] are used. Unlike a typical switch capacitor circuit where a charge storing capacitor is used, we use second flying capacitor (of equal size) switching in opposite phase of that of the first one. This helps to maintain average of the switch capacitor output voltage (i.e. V_C) at $V_{dd}/2$.

Block diagram of the dual switch-capacitor based converter is shown in figure 3. In this circuit each of the flying capacitors supplies half of output current. So, for a steady load current, with equal duration of

662

978-1-4244-3039-0/08 $25.00 © 2008 IEEE

PHASE I and PHASE II V_C becomes a saw tooth waveform. Ignoring voltage drop across the switches it can be shown that the peak-peak ripple of the saw tooth waveform is $I_o.T / 4Cp$.

Use of dual switch-capacitors also helps to keep input supply current steady (around $I_o/2$) in contrast to rectangular profile with single flying capacitor. Note that, to avoid internal short circuit current loss through various switches two phases of capacitor switching are made non-overlap. However, due to non-overlap phases, there is a momentary inflow-current deficiency at node 'C' during phase transition. To supplement the current deficiency the transistor M_{DM} is turned on by a pulse generator.

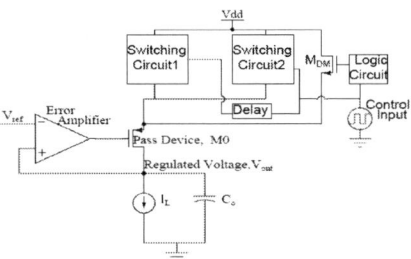

Fig. 3 Block Diagram of Dual switch- capacitor based hybrid converter

In the following two sections the converter is analyzed to find its power efficiency and attenuation of ripple noise that comes from the switch capacitor.

3. Analysis of Power Efficiency

For power efficiency, first we derive expression of output power of the converter [8]. For proper operation of the linear regulator, voltage at node 'C' should be sufficient to keep transistor M0 in saturation. So minimum voltage at node 'C',

$$V_{C1} \geq Vout + V_{DSAT\_M0} \quad (\approx V_{ref} + V_{DSAT\_M0}) \quad (1)$$

On the other hand, for an output current of I_o, expression of the minimum voltage is,

$$V_{C1} = \frac{Vdd}{2} - \frac{I_o.T}{8Cp} - 2I_oR_{SW}, \quad (2)$$

Where, R_{SW} is on resistance of each switch in the switch capacitor. For minimum switching power loss the switching period can be stretched so that minimum requirement of V_{c1} (Defined by equation (1)) is just meet. At the limiting situation,

$$Vout = V_{C1} - V_{DSAT\_M0}$$
$$= \frac{Vdd}{2} - \frac{I_o.T}{8Cp} - V_{DSAT\_M0} - 2I_oR_{SW} \quad (3)$$

So the output power is,

$$P_{out} = \left(\frac{Vdd}{2} - \frac{I_o.T}{8Cp} - V_{DSAT\_M0} - 2I_oR_{SW} \right).I_o \quad (4)$$

Now we find expression of total current drawn from the input supply. First we consider current drawn by the switch capacitors each one of which supply half of the load current. Note that, each of these capacitors draws $I_o/2$ current from Vdd only during their PHASE I operation. Considering dual operation of these capacitors, at any point of time (ignoring phase switchover duration) only one of them is in PHASE I. Hence, the total average current drawn by the two capacitors is $I_o/2$.

The total average current drawn from the input supply is summation of current through Cp, current to operate the eight switches and quiescent current of the error amplifier. Combining all the three current components total input power is,

$$P_{in} = Vdd\left(\frac{I_o}{2} + 8f.C_{ESW}.Vdd + I_{STAT} \right) \quad (5)$$

Where, C_{ESW} is the effective capacitance of a switch and Current in the error amplifier plus the bias circuit current is denote as. I_{STAT}

From equation (4) and (5) the maximum power efficiency of the converter is,

$$\eta = \frac{1 - \left(\left(\frac{I_o.T}{2Cp} + 2V_{DSAT\_M0} + 8I_oR_{SW} \right) \middle/ Vdd \right)}{\left(1 + 16\frac{f.C_{ESW}.Vdd}{I_o} + 2\frac{I_{STAT}}{I_o} \right)} X100\% \quad (6)$$

4. Analysis of Output Ripple-Noise and its Minimization

In section 2, it is shown that voltage at node C, Vc contains saw tooth ripple noise. This noise is attenuated by the linear regulator. The noise attenuation provided by the linear can be obtained by analyzing small signal equivalent circuit of the converter shown in Fig. 4. In this circuit, A_{v1} and r_{o1} are voltage gain and output impedance of the error amplifier respectively. Nodes G, D and S are gate, drain and source nodes of the pass transistor M0 respectively. Trans-conductance and drain-source resistance of M0 are indicated by g_m and r_{ds} respectively. C_t denotes total capacitance (inherent device capacitor plus extra capacitor) between gate and drain of M0. Similarly, C_{in} denotes total capacitance between gate and source of M0. The signal source Vs depicts the small signal noise which comes at the source node of M0 from the switching of Cp.

663

978-1-4244-3039-0/08 $25.00 © 2008 IEEE

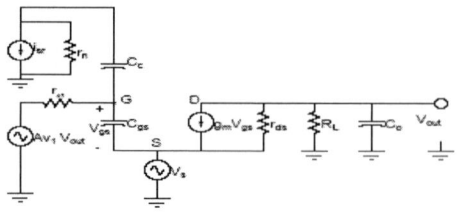

Fig. 4 Small signal model of the linear regulator for noise attenuation analysis

The final expression of V_s to Vout noise gain is,

$$A_{of} = \frac{(1 + g_m r_{ds})(1 + s/z_1)(1 + s/z_2)}{\{1 + r_{ds}/R_L + g_m r_{ds} A_{v1}\}(1 + s/p_1)(1 + s/p_2)}$$

$$\approx \frac{(1 + s/z_1)(1 + s/z_2)}{A_{v1}(1 + s/p_1)(1 + s/p_2)} \quad (7)$$

Where, the two zeros are (assuming, $z_2 \gg z_1$),

$$z_1 = \frac{(1 + g_m r_{ds})}{r_{01}(c_{in} + c_t g_m r_{ds})};$$

$$z_2 = \frac{r_{01}(c_{in} + c_t g_m r_{ds})}{c_{in} c_t r_{01} r_{ds}} \quad (8)$$

For higher supply-to-output noise attenuation the second zero should be pushed at higher frequency. On the other hand, $r_{01}(c_{in} + c_t g_m r_{ds})$ is the reciprocal of the first stage pole of the linear regulator. Position of this pole is decided by the output node pole and the low frequency gain of the linear regulator for phase margin requirement. Therefore, the second zero should be kept at high frequency with minimum value of C_t. In other words, for more noise attenuation, instead of Miller-effect capacitor, an extra capacitor between gate to source is suitable Intuitively, smaller value of C_t and higher value of C_{in} helps to couple source noise Vs to the gate node and hence, to keep $g_m v_{gs}$ small.

Apart from the voltage dependent current source, $g_m v_{gs}$, r_{ds} also provides path for noise to propagate at the output. For high frequency noise, assuming $r_{o1} \gg \dfrac{1}{sc_{in}}$ at ripple frequency and $c_t \approx 0$, the ripple noise goes to the output only through r_{ds}. The corresponding ripple current goes to the output node is,

$$i_{rds} = \frac{v_s}{r_{ds} + \left(\dfrac{1}{1/R_L + sc_o}\right)} \approx \frac{v_s}{r_{ds}} \quad (9)$$

To cancel this r_{ds} -coupled noise, a synchronizer ripple noise is synthesized and it is coupled at the gate of M0 through a capacitor. Small signal equivalent model of

the converter including a Norton equivalent model of ripple-noise synthesis circuit is shown in Fig.5.

Fig.5 Small Signal equivalent circuit of linear regulator with ripple synthesizer Block

Ignoring r_{o1} and r_n , the current (in Laplace domain) at the output node due to the synthesized noise source is,

$$g_m v_{gs} = \frac{g_m I_{sr}(s)}{sC_{gs}} \quad (10)$$

Total output noise can be minimized by making this current equal to the r_{ds} -coupled noise from equation (9) and (10). The required condition for minimum output noise is,

$$I_{sr}(s) = \frac{sC_{gs}}{g_m r_{ds}} \quad (11)$$

With Inverse Laplace transform the required condition becomes,

$$i_{sr}(t) = \frac{C_{gs}}{g_m r_{ds}} \frac{d v_s(t)}{dt} \quad (12)$$

Since Vs is a saw tooth waveform, the required profile of synthesized ripple current is a square wave with duty cycle close to 100% and with zero average.

5. Implementation of the Proposed Converter

While working principle and analysis of various building blocks are introduced in the previous sections, the consolidate transistor level topology of the proposed converter is shown in Figure 6. It has two switching capacitors working in dual mode, cascaded with a linear regulator consists of an error amplifier and a pass transistor M0. Two control signal generators provide controlling signals required for the two switch capacitors. Non-overlap-phase of the switch capacitors, for eliminating its short circuit current, is ensured by break-before- make mechanism [10]. Transistor M_{DM} is driven by a glitch generator to avoid momentary charge deficiency at the source node of M0 during phase transition. Duration of the glitches is adjusted to

978-1-4244-3039-0/08 $25.00 © 2008 IEEE

be equal to the duration of non-overlap phase of the switch capacitors. Output of the same glitch generator is also utilized to synthesize rectangular current which is then fed at the gate of the M0 through a capacitor.

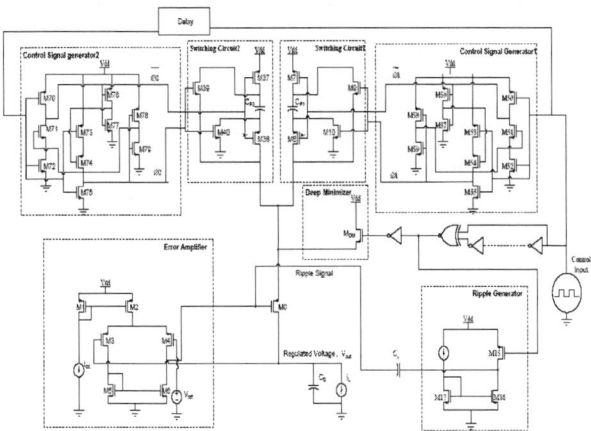

Fig. 6 Transistor level topology of the converter

6. Simulation Results and Discussion

The converter circuit (shown in figure 6) is designed in a 0.18μm CMOS process using dual oxide transistors for 3.3V to 1.25V conversion. The targeted specification parameters are shown in the second column of Table 1. Spice simulated performance of the converter at nominal process, voltage and temperature (PVT) is shown in the second column of the table. It is observed that the performance does not change remarkably at other process and temperature corners. However, the output ripple increases with input supply voltage less than 3.2V. For instance, with 3.0V input supply with 10mA load current the output ripple increases to 170 mV p-p. This is due to the pass transistor enters into linear region. Note that, for lower input supply voltage, the ripple can be brought back within the specified limit by adjusting the reference voltage to a lower value which might degrade power efficiency at nominal input supply.

The output voltage with 10.3mA steady load current is shown in Fig. 7. Transient behavior of the output voltage with load change is shown and Fig. 8. Dynamic range of load current is 1mA to 10mA

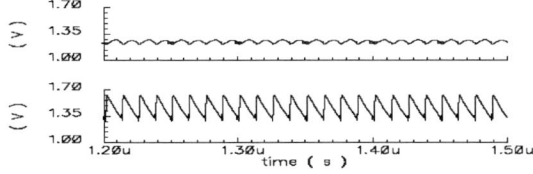

Fig. 7 Output voltage waveform of the Converter and output of the switching circuit (i.e. Vc)

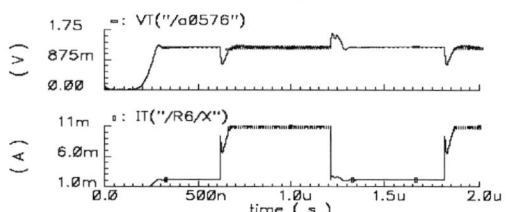

Fig. 8 Transient Response to load impedance variation

Table 1. Specification and Performance Summary

Performance Parameter	Requirement	Simulated value
Input voltage	>=3.3V	3.3 V
Output Voltage	1.25 V +/- 10%	1.25 V
Switch capacitor	Minimize	300pF (2 X150pF)
Max. Load Current	10 mA	10.3 mA
Load capacitor	50 pF	50 pF
Output Ripple (p-p)	< 60 mV	45 mV
Power Efficiency	Maximize	64%
Switching Frequency	< 50 MHz	40 MHz

In a charge pump based regulator [11] the ripple noise is reduced at the cost of external capacitor. However, we target embedded applications (without external capacitor) and hence, one of our key objectives is to minimize the capacitors. In a switch capacitor based converter, as shown in our analysis, the output ripple noise decreases with the capacitor (flying capacitor and/or load capacitor) and increases with the load current. Therefore, we consider

(ripple) X (capacitor) / (max. load current)

as a figure of metric to compare our topology with the existing charge pump based converter [11]. Value of this metric is 1.575×10^{-9} for the proposed converter whereas that of the existing converter [11] is 2625.13×10^{-9}.

The hybrid converter discussed in [6] is suitable for embedded applications and, compare to other literatures, it is the closest one to the present work. Therefore detail comparison with this literature is important. However, few important parameters such as value of load capacitor and ripple noise at final output are not available there. The remaining parameters are enlisted in Table 2.

665

978-1-4244-3039-0/08 $25.00 © 2008 IEEE

Table 2. Comparative Study

Performance Parameter	Reference [6]	This Work
Input Voltage	2.5 V	3.3 V
Output Voltage	0.65V to 2.5 V	1.25 V
Flying Capacitor	2 X 50 =100 pF	2 X 150 =300pF
Storing capacitor (at output of SC)	100pF	Not used
Load Capacitor	Not Reported	50 pF
Switching Frequency	125 MHz	40 MHz
Output Ripple (p-p)	Not Reported	45 mV
Power Efficiency	52 % (at 0.9V-30mA)	64%
Technology	0.25-μm CMOS	0.18-μm CMOS

7. Conclusion

A low voltage and low ripple dual switch-capacitor based DC-DC converter suitable for high drop-out and embedded application is proposed here. The converter consists of a linear regulator and two switching capacitors. The capacitors are used to achieve higher power efficiency by recycling charge. The linear regulator, on the other hand, reduces output ripple-noise. Use of dual flying capacitors helps to reduce peak-peak ripple at the output of switch capacitor. It also helps to maintain input supply current steady in contrast to rectangular profile with single flying capacitor. Noise reduction provided by the linear regulator is analyzed and minimized. The output ripple noise is further reduced by introducing a synthesized counter ripple noise through the linear regulator. Combination of the various techniques of ripple reduction helps to reduce the total capacitor value while keeping the ripple within acceptable limit.

The proposed converter circuit is designed in a 0.18μ CMOS process for 3.3V to 1.25V conversion with two flying capacitors of 150 pF each (which is possible to implement on-chip). For 10.3mA load current and with only 50pF load capacitor peak-peak output voltage ripple is 45 mV and power efficiency is 64%.

8. References

[1] K.Usami and M. Horowitz, "Clustered Voltage Scaling technique for low power design", Proc. of International Workshop on Low Power Design, April 1995, pp. 3-8

[2] Peter Hazucha et al, "A 233-MHz 80%-87% Efficient Four-Phase DC-DC Converter Utilizing Air-Core Inductors on Package", IEEE Journal of Solid-State Circuits, Vol. 40, No. 4, April 2005, pp. 838-845

[3] Gabriel A. Rincon-Mora and P.E. Allen, "A Low-Voltage, Low Quiescent Current, Low Drop-Out regulator", IEEE Journal of Solid-State Circuits, Vol. 33, No. 1, January 1998, pp. 36-44

[4] Marek S. Makowski and Dragan Maksimovic, "Performance Limits of Switched-Capacitor DC-DC Converters", 26th annual IEEE Power Electronics Specialists Conference, Atlanta, Ga, USA, Vol. 2, June 1995, pp. 1215-1221

[5] Hoi Lee and Philip K.T. Mok, " An SC Voltage Doubler with Pseudo-Continious output Regulation Using a Three – stage Switchable Opamp", IEEE Journal of Solid State Circuits, Vol. 42, No. 6, June 2007, pp. 1216-1229

[6] George Patounakis, Yee William Li, and Kenneth L. Shepard, "A Fully integrated On-Chip DC-DC Conversion and Power Management System", IEEE Journal of Solid-State Circuits, Vol. 39, No. 3, March 2004, pp. 443-451

[7] Vishal Gupta and Gabriel A. Rincon Mora, "A Low Dropout, CMOS Regulator with High PSR over Wideband Frequencies", IEEE International Symposium on Circuit and Systems, Vol.5, May 2005, pp. 4245-4248

[8] Pradip Mandal and Kaushik Bhattacharyya, "A Low voltage, Low ripple on Chip Hybrid DC-DC Converter", International Symposium on Integrated Circuits 2007(ISIC-2007), Singapore, 26-28 September 2007, pp. 442-445

[9] J.G. Ryan, K. J. Carroll, and B.D. Pless, "A four chip implantable defibrillator/pacemaker chipset", in proc. Custom Integrated Circuits Conference, 1989, pp. 7.6/1-7.6/4

[10] Hoi Lee and Philip K. T. Mok "Switching Noise and Shoot through Current Reduction Techniques for Switched-Capacitor Voltage Doubler ", IEEE Journal of Solid State Circuits, Vol. 40, No. 5, May 2005,pp. 1136-1146

[11] Jae-Youl Lee, et al, "A Regulated Charge Pump With Small Ripple Voltage and Fast Start-Up", IEEE Journal of Solid-State Circuits, Vol. 41, No. 2, February 2006, pp. 425-432

21st International Conference on VLSI Design

Voltage and Temperature Scalable Standard Cell Leakage Models Based On Stacks For Statistical Leakage Characterization

Janakiraman Viraraghavan
janakiram@ece.iisc.ernet.in

Bishnu Prasad Das
bpdas@cedt.iisc.ernet.in

Bharadwaj Amrutur
amrutur@ece.iisc.ernet.in

Indian Institute of Science. Bangalore, India - 560012

Abstract

With extensive use of Dynamic Voltage Scaling (DVS) there is increasing need for voltage scalable models. Similarly, leakage being very sensitive to temperature motivates the need for a temperature scalable model as well. We characterize standard cell libraries for statistical leakage analysis based on models for transistor stacks. Modeling stacks has the advantage of using a single model across many gates there by reducing the number of models that need to be characterized. Our experiments on 15 different gates show that we needed only 23 models to predict the leakage across 126 input vector combinations. We investigate the use of neural networks for the combined PVT model, for the stacks, which can capture the effect of inter die, intra gate variations, supply voltage(0.6-1.2V) and temperature($0 - 100^0C$) on leakage. Results show that neural network based stack models can predict the PDF of leakage current across supply voltage and temperature accurately with the average error in mean being less than 2% and that in standard deviation being less than 5% across a range of voltage, temperature.

1 Introduction

Statistical leakage analysis is gaining importance with scaling transistor dimensions. Leakage power will contribute approximately 50% of the total power in the 90nm technology node [1] and shrinking transistor sizes makes this leakage power more difficult to predict . Further, process variations i.e. variations in effective gate length, L_e, oxide thickness, T_{ox}, and threshold voltage V_{TH} can result in up to 20 $\times$ variations in the leakage of the manufactured chips [2]. Authors in [2] also show how environmental factors affect leakage.

Process Variations occur due to non uniformity in the manufacturing of the chips. These include Inter-die, Intra die and intra gate variations. Inter die variations [3] refer to variations that occur across dies, wafers or lots. The variation introduced is the same across the entire die. Intra-gate variations [7]refers to transistor to transistor variations within a gate. The impact of intra-gate variations on statistical analysis is a drastic increase in the number random variable needed to model them. Each transistor within a gate will now have one random variable per intra-gate process parameter. Intra-die variation, as the name suggests, refers to variation of a particular parameter within the die. Intra-die variations are usually spatially correlated and all transistors within a gate have the same value. In the past few years considerable amount of work has been done in building models that predict leakage accurately as function of process parameters. The empirical technique described in [3] captures the effect of inter die variations in gate length L_e and provides a simple analytical method to statistically analyze leakage but considers inter die variations only in L_e. It also does not take into account the effect of intra gate variations.

Currently industrial chips implement DVS in the range of $V_{DD}/2 - V_{DD}$ which drives the need for voltage scalable models. Further, depending on the location of a gate in the chip its operating temperature can vary from $25^0C - 120^0C$ [6] [2]. If the temperature profile is available a temperature scalable model enables more accurate analysis. We present one such model which uses Neural Networks to capture the effect of process, voltage and temperature (PVT). Authors in [8] modeled the leakage of a stack with a neural network but the extreme non-linearity in leakage due to temperature introduced too much error in the model when both voltage and temperature were included. Instead, we model *log* of the leakage current [3] and are able to include the effect of both voltage and temperature, along with process, in the same model.

Statistical leakage characterization, like static leakage characterization, involves characterizing every gate for every input vector. Each such characterization will involve a substantial number of SPICE runs. We look at the problem of reducing the number of such models by modeling different kinds of stacks present in the library.

The rest of the paper is organized as follows: section 2

978-1-4244-3039-0/08 $25.00 © 2008 IEEE

describes how modeling leakage through stacks can be used for statistical leakage characterization. In section 3 we explain how leakage through different gates are characterized with these stacks. We then present in section 4 a neural network based leakage current model for stacks. In section 5 we present the results obtained when we tested these stack models on different gates comparing them with accurate SPICE simulations across voltage and temperature and conclude in section 6.

2 Leakage modeling

2.1 Standard Cell Library Characterization for Statistical Analysis

The intent of statistical leakage characterization is to fit the leakage of every gate, for every input vector, to a convenient functional form which is usually an exponential of a quadratic polynomial [3]. If there are N gates in the library and the ith gate has M_i inputs we would require $\sum_{i=1}^{N} 2^{M_i}$ models. One way to reduce the number of models is to model the average leakage of a gate across all its input vectors. This assumes that each input vector is equally likely, which is too simplified an assumption as the probability of occurance of an input vector for a gate depends on the circuit structure. Thus, there is definitely a need for modeling each gate for every input vector. Stack modeling enables re-use of models which reduce the required number of models.

Authors in [3] model log of the leakage using a quadratic polynomial in L_e. However, they do not consider the effect of intra gate variations [7]. With intra gate variations it is expensive to consider the effect of all transistors in the gate for every input vector. Depending on the input vector being considered, some transistors may not not affect the leakage statistically. Stack modeling helps in this respect too.

2.2 Modeling stacks

It is well known that the leakage of a gate is primarily determined by the stacks through which the current flows. Considerable amount of work has been done in trying to find closed form expression for the leakage current through stacks [4]. The authors in [4] considers the case of an NMOS transistor being turned off only when its gate is grounded. But an NMOS transistor on the top of a stack trying to pass V_{DD} will turn off as soon as its source potential rises to $V_{DD} - V_{TH}$ hence we need to extend the idea in [4] to estimate the leakage across all input vectors. However, we use the ideas mentioned in [4] to identify the transistors on a stack that affect the leakage in a statistical sense. [8] uses this concept to model leakage for a few CMOS gates across all input vectors. The key ideas in [8]

for using the stack model to predict the leakage of any gate across all input vectors can be summarized as

- An NMOS/PMOS transistor trying to pass a 0/1 is completely turned on and hence can be treated as a short circuit [4]

- An NMOS/PMOS transistor trying to pass a 1/0 turns off as soon as its source potential increases to $(V_{DD} - V_{TH})/(V_{TH})$ and hence is turned off even though its gate is connected to (V_{DD}/GND)

- In an N transistor NMOS/PMOS stack, when there is exactly ONE NMOS/PMOS transistor whose gate is connected to GND/V_{DD} i.e. the gates of other $N - 1$ transistors are connected to V_{DD}/GND, the leakage will increase significantly as this transistor moves towards the output due to DIBL [8].

- When the leakage through a logic gate is predicted using a stack, the leakage obtained from the stack model has to be scaled by the ratio of effective widths of the transistors in the gate and those on the stack

- Leakage due to parallel stacks simply add up

Using the above mentioned stack rules we built models for the 18 commonly found stacks listed in the first column of Table 1. These stack models are the most basic stacks and are widely used. We assume that the standard cell library does not have gates with stack size greater than 4 as the increased logical effort will affect the delay of the circuit. For the gates tested in [8], any input combination will result in a set of stacks which are a subset of the list we have modeled in Table 1. In this paper we look at a few more gates that have different kinds of stacks. The naming convention for the commonly used stacks are as follows.

All model names in Table 1 are of the form, {Stack type}{Stack size}/{Input to the stack}

- Stack type indicates if it is an NMOS stack or a PMOS stack

- Stack size is the number of transistors on the stack

- Input is the decimal value of the input vector being applied to the stack with the LSB of the input vector applied to the transistor closest to the output

- The widths of the transistors on the stack are the unit inverters transistor widths scaled by the stack size

Consider an example of a three input majority gate shown in Fig 1. Apart from the stacks listed in Table 1 we need to model four more stacks to handle such gates. The leakage of this gate is determined by three different stacks, whose currents are denoted as I_1, I_2, I_3, and the total leakage is given by the sum of these three currents. Let us examine the input vectors that result in the use of these different stacks.

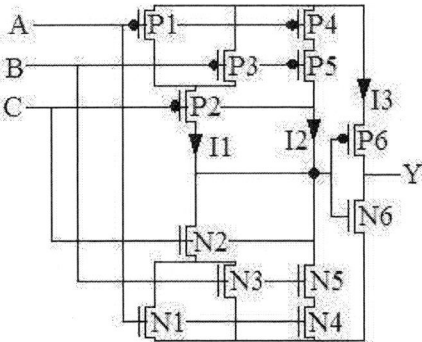

Figure 1. Three input majority gate

1. **Input vectors (000/111):** When the input is 000/111 PMOS/NMOS transistors {P1-P5}/{N1-N5} are completely turned on. N6/P6 is also turned on as the output of the majority gate is 0/1. I_2 is determined by the leakage through the stack {N4,N5}/{P4,P5} and can be estimated using the {n2/0}/{p2/3} model. Similarly I_3 can be estimated using the {p1/1}/{n1/0} model. I_1 is primarily determined by the stack {N1, N2, N3}/{P1,P2,P3}. The stack formed by {N1 N2 N3}/{P1, P2, P3} cannot be approximated by any of the stack models listed in Table 1 because of the parallel combination of N1/P1 and N3/P3 in series with N2/P2. Thus we need two more models.

Input vectors (001/110): The only difference between the previous input vector case and this case is that the input to C is 1/0. Thus I_2 and I_3 are identical to the previous case. I_1 is now determined by the stack {N1 N2 and N3}/{P1, P2, P3} with the input to N2/P2 being 1/0. As mentioned earlier, N2/P2 cannot be treated as short circuits as NMOS/PMOS transistors cannot pass 1/0 fully. This again requires two more stack models to model these two states.

Other Input vectors: Any of the other four input vectors will result in a set of stacks listed in Table 1. Similar to the explanation for the NAND4 gate in [8] we can deduce which of the models in Table 1 are needed for each input vector.

Similarly, we had to model one more PMOS stack for the two input XOR gate to accurately predict the leakage when the input vector 0. However for the inverters and AND/OR gates we could use the stacks models listed in Table 1 to predict the leakage of all their input states.

While static leakage characterization needs to be done for every gate for every input vector, it is more efficient to characterize different kinds of stacks[1] for statistical leakage

---
[1] The list of different stacks is specific to the library

Table 1. List of the common stack models used across all gates with Neural Network Training Details. Training time(**Time**), maximum testing error (**Err**), and number of hidden nodes(**H**)

Model	Size	Type	Input	Err(%)	Time(s)	H
n1/0	1	NMOS	0	7.03	3.16	9
n2/0	2	NMOS	00	4.39	9.70	13
n2/1	2	NMOS	01	4.02	9.36	13
n3/0	3	NMOS	000	7.76	16.17	17
n3/1	3	NMOS	001	5.43	21.92	17
n3/3	3	NMOS	011	7.20	9.45	17
n4/0	4	NMOS	0000	3.65	52.11	21
n4/1	4	NMOS	0001	4.61	49.26	21
n4/7	4	NMOS	0111	3.09	6.01	21
p1/1	1	PMOS	1	4.77	8.65	9
p2/3	2	PMOS	11	4.01	7.66	13
p2/2	2	PMOS	10	3.10	2.3	13
p3/7	3	PMOS	111	4.17	64.45	17
p3/6	3	PMOS	110	6.10	27.73	17
p3/4	3	PMOS	100	3.26	5.26	17
p4/15	4	PMOS	1111	4.37	65.70	21
p4/14	4	PMOS	1110	9.78	12.95	21
p4/8	4	PMOS	1000	6.31	5.42	21

characterization rather than characterize every gate for every input vector.

3 Deriving leakage of a logic gate from stack leakage

All static CMOS gates can be broken down into elementary stacks and hence characterization of all gates in the standard cell library will only involve mapping of their different leakage states to the corresponding stacks that cause the leakage. Thus, the leakage currents predicted by the elementary stack models are analogous to basis vectors in linear algebra. We have demonstrated the idea with a small subset of 15 gates in the standard cell library. However, by scanning the entire library and modeling the different stacks that appear in it, leakage current through any gate, for a given input vector can then be written as linear weighted sum of the currents through these stacks. If M is the total number of unique stack models present in the library and S_j is the leakage through the jth stack, the leakage through the ith gate in a circuit for a given input vector q_i can be written as

$$X_i^{q_i} = \sum_{j=1}^{M} \alpha_{ij}^{q_i} S_j \qquad (1)$$

Where S_j is the leakage through the jth stack model and $\alpha_{ij}^{q_i}$ is the scaling factor due to difference in effective widths between the stack used in the gate and the stack model. $\alpha_{ij}^{q_i}$ = 0 if the jth stack does not appear for that gate for the input vector q_i.

In certain applications the primary input is known apriori. For example, in the idle state, a combinational circuit is set to a particular input state which results in least leakage. Equation 1 can be used in such cases. However in many other applications, the input is not known and the state of the input becomes probabilistic [10]. In such cases, Eqn. 1 needs to be modified to accommodate the probability of an input vector occurring. This can be easily done in our framework. Let the probability of occurance of the input q_i for the ith gate be $p_i^{q_i}$. The average leakage of the ith gate can be written as [10]

$$X_i = \sum_{\forall q_i} p_i^{q_i} X_i^{q_i} \qquad (2)$$

Substituting Eqn. 1 we get

$$X_i = \sum_{j=1}^{M} \alpha_{ij}^{avg} S_j \qquad (3)$$

Where, $\alpha_{ij}^{avg} = \sum_{\forall q_i} p_i^{q_i} \alpha_{ij}^{q_i}$, is the average scaling across all input vectors. Thus, this formulation can be modified very easily for such applications as well. To test our formulation we need to use a model for these elementary stacks. In the next section we investigate the feasibility of using neural networks to model the leakage through these stacks.

4 Leakage modeling - Neural Networks

Existing leakage models capture the effect of process on leakage for a given supply voltage (V) and temperature (T) and these models are indexed by voltage and temperature. Also for (V, T) values not present in the look interpolation is used. We can do away with the look up table based approach, to account for voltage and temperature variations, if we have a unified PVT model. Thus, there is a clear need for a model which can capture the effect of Process (Inter die and Intra gate), Voltage and Temperature on leakage all in one model. Fitting log of the leakage to a second order polynomial, of process parameters alone, does work well when all possible cross terms are considered. A PVT quadratic polynomial model failed to fit the leakage when temperature was also considered and hence we tried out Artificial Neural Networks (ANN) for the combined model. Authors in [8] used ANN to model the sub-threshold leakage of a logic gate. But the exponential dependence of leakage on PVT introduces too much non-linearity and hence the model gives unacceptable error. Instead, in this paper we are able to model the effect of PVT on leakage by modeling *log* of the leakage current. The exact algorithm to train an ANN to model the leakage through a stack is shown in Fig2.

---

1. Generate N(1500) training samples for the P random input parameters from their distributions (Inter die and intra gate process parameters) - Produces an $N \times P$ matrix, $\mathbf{X_P}$

2. Divide the temperature range ($0 - 100^0 C$) and voltage range (0.6 - 1.2V) into N equally spaced samples

3. Randomly pair the ith temperature point with j^{th} voltage point - Produces an $N \times 2$ matrix $\mathbf{X_{VT}}$

4. Append matrix $\mathbf{X_{VT}}$ to the matrix $\mathbf{X_P}$ columnwise - Produces an $N \times (P+2)$ matrix $\mathbf{X = [X_P \quad X_{VT}]}$

5. Repeat steps 1 — 4 for another M(500) testing samples

6. Simulate the stack, to be modeled, using SPICE with the $N + M$ (2000) samples to obtain the corresponding leakage current values $\mathbf{Y_N^{SPICE}}$ and $\mathbf{Y_M^{SPICE}}$

7. Normalize the input and output according to equation $X_{NORM} = X/[max(X) - min(X)]$

8. Initialize the Neural network

9. Train the ANN with N *normalized* training samples according to the back propagation algorithm [5]

10. Feed the ANN with the M testing samples and obtain the outputs predicted by the ANN. Produces Y_M^{ANN}

11. Evaluate $\Delta = Max\{\frac{|Y_M^{ANN} - Y_M^{SPICE}|}{Y_M^{SPICE}}\}$

12. IF $\Delta < \delta$ (0.1) *Increase the number of hidden nodes by ONE and GO TO Step 8*

13. ELSE ANN model is trained

Figure 2. Algorithm to train the ANN model - In brackets we have indicated values used in this work

---

Steps 1 - 6 generate the necessary training and testing samples which are obtained through SPICE simulations. As indicated we had to do 2000 SPICE simulations to successfully train a stack. Step 1 generates all the process parameter values from their distribution. Step 2 and 3 generate the necessary temperature and voltage samples. Our ANN model is expected to predict the leakage accurately at any supply voltage and any temperature in the ranges we train the network with. Having considered a temperature range of $0 - 100^0 C$, with 1500 training samples, we divide the entire range in uniform steps of $0.067^0 C$. Similarly the voltage range of (0.6V - 1.2V) is divided in steps of 0.4mV. In step 3 we randomly pair these uniformly generated points in order to make sure that the training set contains as many different V,T pairs as possible. The testing set has a similar but smaller number of such points which in our case was 500. The reason we have to consider so many points is the extreme non linearity introduced by temperature. Step 6 creates the training and testing output data set through SPICE simulations. In step 7 we normalize both the input and output data set in order to improve the training. In step 8 we train the ANN through standard techniques as given in [5]. We used the Neural Network tool box provided by MATLAB for this purpose. We only had to choose the initialization parameters and train the network. The standard practice in Neural Networks is to train the network until the mean square error of one epoch is below the chosen threshold [5]. We then validate the trained network with a disjoint

670

978-1-4244-3039-0/08 $25.00 © 2008 IEEE

testing data set in step 9 and compute the maximum percentage error between the actual SPICE value and the one predicted by our *trained* ANN. If the error is above the chosen threshold, which in our case was 10%, we have to increase the number of hidden nodes and re-train the network. However, among the stacks listed in Table 1 we did not have to do this retraining even once. This kind of re-training is a common procedure with neural networks [5].

5 Results

We trained the ANN to model the leakage through the different stacks listed in col 1 of Table 1 and the 5 extra models for the majority gate and the xor gate. We then used them to predict the PDF of the different gates listed in Table 2. Each model captures the effect of Process (inter die[1] and intra gate variations) in L, T_{OX} and V_{TH}, supply voltage (0.6 - 1.2V) and temperature ($0 - 100^0C$). All process parameters were sampled from Gaussian distributions with $3\sigma = 10\%$ of their mean. Table 1 shows the details of training the network. From the table we observe that the maximum training time amongst all models is around 66 seconds. For each gate in col 1 of Table 2 we have listed the mean and standard deviation error, compared to SPICE, for the input vector 0 at a supply voltage of 0.9V and an operating temperature of 50^0C. All SPICE simulations were done with HSPICE using an industrial 130nm model file. From Table 2 we see that the error for most gates is less than 1% while the error for the the xor gate is 5%. This is a result of the stacking approximation. The maximum standard deviation error across all the gates across all input vectors is around 20%(not shown in Table 2) for the NAND4 gate with its input vector set to $14(1110_2)$. We see that the model used to predict the leakage of a NAND4 gate with an input vector 14 is the n1/0 model, which is nothing but a single NMOS OFF transistor. As per the stack approximation the bottom three transistors are treated as short circuits. This assumption is not completely true. The voltage drop across the three ON transistors, albeit very small, does affect the leakage of the top most OFF transistor due to body effect. Similarly the error is maximum for the NAND3 and NAND2 gates when the n1/0 model is used to predict their leakage. Since the number of transistors on the stack progressively decrease from NAND4 to NAND2 the error drops from 20% for a NAND4 gate to 10% for a NAND2 gate. Similar results are observed for NOR gates as well. However, the average mean error and average standard deviation error, for a gate, across all input vectors, are less than 2% and 5% respectively. A library consisting of an inverter, 2-4 input NAND-NOR-AND-OR gates, an XOR2 gate and

---

[1]Intra gate variations make sense only when a gate is placed in a circuit. For an isolated gate level model, both inter die and intra die variations are one and the same

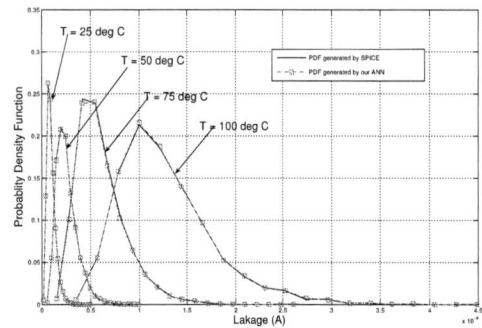

Figure 3. Temperature scalable plots: PDF from our model, for NAND4 gate, follows SPICE closely across different temperatures

a three input majority gate, needed just 23 stack models to predict the leakage across 126 leakage states. The idea of using stacks as elementary models being independent of the model used, we tried out the conventional exponential polynomial approach before using the ANN.

We first tried to model log of the leakage with a quadratic polynomial, in the process parameters alone, for a four transistor NMOS stack which has 3 inter die parameters and $3 \times 4 = 12$ intra gate parameters. The model did work well but quadratic polynomial will have total of 15 first order terms and $15 \times 8 = 120$ second order terms. To fit this polynomial with 135 coefficients we require at least 135 SPICE simulations. We actually needed 150 SPICE simulations to fit the leakage accurately, using the least square fit algorithm, and 50 disjoint SPICE simulations to test it. Thus if we want to use this exponential polynomial model and index it by voltage and temperature, in voltage steps of 50mV and temperature steps of 25^0C, we would need 48 different exponential quadratic models in a voltage range of 0.6-1.2V and $0 - 100^0C$. This requires $150 \times 48 = 7200$ SPICE simulations for a four transistor NMOS stack. However, our ANN model required just 2000 SPICE simulations. Further, the natural ability of the ANN to interpolate accurately also gives us the freedom to use it any (V, T). Since our ANN model is both voltage and temperature scalable, we also show that it can predict the leakage across different voltages and temperatures. The plot in Fig 3 shows how accurately the ANN model is able to predict the PDF of the leakage across a range of temperatures ($25 - 100^0C$) for a NAND4 gate. Similarly, Fig 4 shows a voltage scalable plot for an XOR2 gate at 0.6V and 1.2V. Both figures also show the exact PDF obtained through MC SPICE simulation.

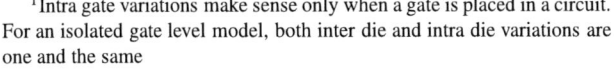

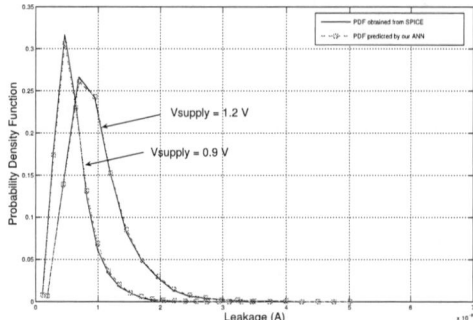

Figure 4. Voltage scalable plots: PDF from our model, for an XOR2 gate, follows SPICE closely across different voltages

6 Conclusion

Modeling stacks greatly helps in reducing the number of models required to characterize a standard cell library for statistical analysis. Once all kinds of stacks present in the library have been modeled, characterization of a gate, across all its input vectors, only involves mapping each leakage state to the corresponding stack models that cause the leakage. With 23 stack models we were able to predict the leakage for 15 gates in the standard cell library across 126 input vector combinations. We used Neural Networks to model the leakage through stacks as conventional techniques failed to capture the effect of process, temperature and voltage together. It was found that neural networks were able to model the leakage through stacks taking into account variations in process (both inter die and intra gate), voltage (0.6-1.2V) and temperature($0 - 100^0C$) making it a suitable voltage and temperature scalable model. The average mean error across all gates across all input vectors was less than 2% and the average standard deviation error was less than 5%.

References

[1] Intel Corp. http://www.intel.com/cd/ids/developer/asmona/eng/strategy/182440.htm?page=2.

[2] Sherkar Borkar et al. *Parametric variations and impact on circuits and microacrchitecture. In Proc. DAC 2003.*

[3] R. Rao, A. Srivastava, D. Blaauw and D. Sylvester. *Statistical analysis of sub-threshold leakage current for VLSI circuits IEEE. Trans. Very Large Scale, 12(2);131-139, 2004.*

[4] Z. Chen, M. Johnson, L. Wei and K Roy. *Estimation of standby leakage power in CMOS circuits consider-*

Table 2. Mean and Std Dev Error between SPICE and our ANN model for different gates. V = 0.9V and T = 50^0C Input = 0.

Gate	$\mu_{SPICE}{}^a$ (pA)	$\Delta\mu$ (%)	σ_{SPICE} (pA)	$\Delta\sigma$ (%)
nand4	263.08	0.1	118.87	0.1
nand3	272.90	0.0	129.96	0.1
nand2	316.88	0.2	168.02	0.7
nor4	3234.38	0.1	1505.08	0.2
nor3	2402.75	0.1	1161.32	0.2
nor2	1605.88	0.1	825.47	0.3
and4	1029.09	0.1	537.64	0.3
and3	1036.30	0.1	579.48	0.4
and2	1104.27	0.1	611.00	0.4
or4	3999.14	0.1	1710.71	0.1
or3	3149.84	0.0	1372.48	0.1
or2	2364.97	0.0	1072.01	0.0
xor2	6127.25	3.1	2889.81	5.1
inv	815.60	0.1	492.99	0.2
maj3	1605.56	0.0	724.25	0.2

[a]$\mu_{SPICE}/\sigma_{SPICE}$ - Actual mean/std dev from SPICE. $\Delta\mu/\Delta\sigma$ % error in mean/ std dev between SPICE and our ANN

ing accurate modeling of transistor stacks In ISLPED '98: Proceedings of the 1998 international symposium on Low power electronics and design, pages 239-244. New York, NY, USA, 1998. ACM Press.

[5] Simon Haykin *Neural networks a comprehensive foundation, 2 edition, Prentice-Hall Engineering, USA, 1998*

[6] H.Su, E. Acar and S. R. Nassif. *Full chip leakage estimation considering power supply and temperature variations In ISLPED '03: Proceedings of the 2003 international symposium on Low power electronics and design, pages 78-83. New York, NY, USA, 2003. ACM Press*

[7] K. Okada, K.Yamaoka and H.Onodera (2003). *A Statistical Gate-Delay Model considering Intra-gate Variability,. ICCAD, pp.908-913, 2003*

[8] Janakiraman. V et al. *Leakage modelling of logic gates considering the effect of input vectors. In press VDAT, 2007.*

[9] Hongliang Chang and Sachin Sapatnekar. *Full chip analysis of leakage power under process variations, including spatial correlations In Proc of DAC 2005*

[10] E. Acar et al. *Leakage and Leakage Sensitivity Computation for Combinational Circuits In ISLPED '03: pages 96-99. New York, NY, USA, 2003. ACM Press*

21st International Conference on VLSI Design

Self-Sleep Buffer for Distributed MTCMOS Design

Charbel J. Akl and Magdy A. Bayoumi

The Center for Advanced Computer Studies (CACS)
University of Louisiana at Lafayette, LA 70504, USA
{cja3455, mab}@cacs.louisiana.edu

Abstract

Leakage power is considered as major concern in deep sub-micrometer VLSI designs. MTCMOS technology was introduced to provide considerable power reduction in standby mode, while maintaining high performance in active mode. However, MTCMOS presents new challenges that require extra design effort. This paper targets the challenges and complexities related to sleep signal distribution in a distributed MTCMOS design. We propose synchronized dual-V_{th} self-sleep buffer method that eliminates the need for sleep signal distribution and allows easy implementation of MTCMOS wakeup scheduling. Guidelines for designing and sizing the self-sleep buffer circuit are provided. In a 90-nm technology and 2-GHz clock frequency, the self-sleep buffer consumes only 1.46-uW in active mode, while eliminating the sleep distribution network overheads and providing fast, low-energy active-to-standby-to-active transitions.

1. Introduction

Scaling down of transistor dimensions and increasing the integration density of VLSI chips, have made leakage power a significant contributor to the overall chip power. Below 90-*nm* technology node, leakage power becomes comparable, or even higher, than switching power [1]. Fabrication Processes with dual or multi threshold voltage (V_{th}) became standard at 0.13-*um* technology and below, to contend with the fast increasing rate of leakage power. MTCMOS (also known as power/ground gating) is a common design methodology that exploits the dual/multi V_{th} technology to cutoff leakage currents in standby mode, while achieving high performance in active mode. MTCMOS uses high-speed low-V_{th} devices for logic cells to achieve high performance in active mode, and low-leakage high-V_{th} sleep devices to reduce power in standby mode [2]. *NMOS* and/or *PMOS* devices can be used for ground and/or power gating, respectively. However, *NMOS* is preferred due to its reduced on-resistance compared to same size *PMOS*. During standby mode, the sleep transistors are turned off to disconnect the logic cells from the actual power/ground lines. During active mode, the sleep transistors are turned on to maintain the functionality of the circuit.

Although MTCMOS is a very effective technique, it introduces some overheads and design challenges. Several solutions have been proposed in the literature to deal with these challenges [3]-[10]. The area penalty due to the sleep transistors is one of MTCMOS overheads. Correct sizing of sleep transistor is required since over-sizing leads to an extra unnecessary area penalty and under-sizing leads to performance degradation and reduced noise margin [3, 4]. Signal integrity is another issue that should be considered during standby-to-active transition [5]-[8]. Also, MTCMOS latches and flip-flops must retain the data in sleep mode [9]. The energy and performance overheads of going from active-to-standby and standby-to-active should be minimized [10]. Also, MTCMOS requires extra routing due to the virtual power/ground lines, and sleep signal distribution. The latter, which is a major issue in distributed MTCMOS, and its accompanied concerns are the target of this work. Section 2 reviews MTCMOS design styles and presents the tradeoffs between centralized versus distributed sleep devices. Section 3 presents the control signal overheads in distributed MTCMOS design. The self-sleep buffer is proposed and analyzed in Section 4. Section 5 provides the self-sleep buffer simulation and characteristics, and Section 6 concludes this paper.

2. MTCMOS design styles

In this paper, we define a *block* as a circuit whose elements share the same sleep signal. Depending on the system-level design, a block can be a whole chip, a core, a clock/V_{DD} domain, a data path, and it can be a global bus (MTCMOS repeater) as well. Applying MTCMOS design to a block can be done in several ways. Global MTCMOS design (Figure 1(a)) controls the whole block via a large centralized single sleep transistor [2]. Whereas, distributed MTCMOS design (Figure 1(b)) employs multiple sleep

978-1-4244-3039-0/08 $25.00 © 2008 IEEE

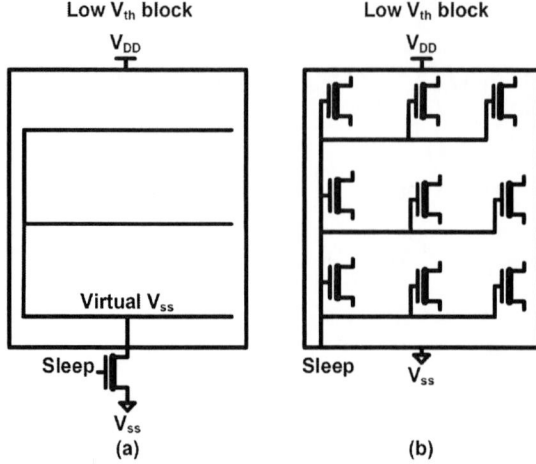

Figure 1. (a) Global MTCMOS design, (b) Distributed MTCMOS design.

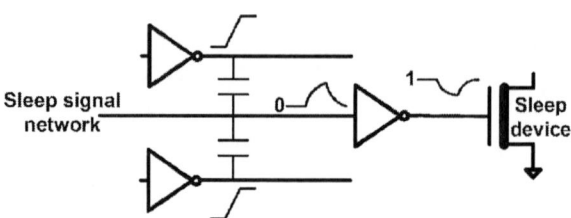

Figure 2.Sleep signal crosstalk noise during active mode.

transistors for a single block. Distributed MTCMOS, in turn, can be applied in two ways. One way is cluster-based where a block is divided into clusters and each cluster has its own virtual power/ground and sleep device [11]. Clustering is done based on the switching behavior of the gates to minimize the total sleep transistor area. Another way for distributed MTCMOS design is network-based (also known as coarse-grain) where many distributed sleep transistors are inserted between the actual and virtual power/ground networks inside the block, and these sleep transistors share the charge/discharge currents [12, 13]. Distributed sleep transistor network was shown to be better than distributed clusters in terms of sleep transistor area and performance. Another MTCMOS design style is known as fine-grain where each gate has its own sleep device [14].

Different MTCMOS styles present different design tradeoffs [15]. Global MTCMOS style has the lowest optimal total sleep transistor area. However, determining the optimal size of the global sleep transistor is hard, and impractical for large blocks. This complicates the design and leads, in most cases, to an over-sized sleep transistor, which in turn reduces the efficiency of global MTCMOS in terms of area. Determining the optimal sleep transistor size for fine-grain MTCMOS is easy. However, the area penalty is large. Distributed MTCMOS styles simplify sleep transistor sizing compared to global MTCMOS, and reduce the total sleep transistor area compared to fine-grain MTCMOS. Regarding signal integrity, global MTCMOS is also impractical since it suffers from degraded noise margin and large ground bounce in power/ground networks. Cluster and network based MTCMOS offers better signal integrity than global MTCMOS, and fine-grain MTCMOS has the best signal integrity since the virtual power/ground are embedded within the gates. Global MTCMOS does not require intra-block sleep signal routing but it suffers from high virtual power/ground lines sizing and routing complexity. Fine-

grain MTCMOS does not require virtual power/ground traces, but the sleep signal has to be delivered to all gates. In deep submicron technologies, the increased power density coupled with reduced supply voltage and increased interconnect resistance make global MTCMOS not feasible. Fine-grained MTCMOS offers many desirable advantages in terms of signal integrity and sleep transistor sizing complexity. However, it suffers from a large area penalty, and it can only be considered when the sleep transistor area penalty can be tolerated. Distributed Cluster and network based MTCMOS are the most commonly used in industrial designs since they combine the advantages of both global and fine-grain MTCMOS. Moreover, unlike fine-grain style, distributed MTCMOS styles use the same standard-cells provided by library vendors and ASIC foundries.

In this work, we target the distributed MTCMOS challenges related to intra-block sleep signal distribution. The sleep network grows as the block size increases and more sleep transistors are employed, and it presents several concerns that are discussed in the next section. We propose a distributed MTCMOS self-sleep buffer that is capable of generating a sleep signal based on the clock behavior. The self-sleep buffer has low overhead, and it can be simultaneously applied with the previously proposed methods that deal with different MTCMOS issues.

3. Sleep distribution network overhead in distributed MTCMOS design

In distributed MTCMOS (cluster-based or network-based), the sleep distribution network presents an overhead since the sleep signal has to be routed to all the sleep devices within the block. The sleep network is a multi-sink network which adds considerable routing complexity. Regular sleep transistor placement reduces the sleep signal routing complexity. However, the sleep signal still consumes precious routing resources and increases the total intra-block wirelength, which in turn increases design cost.

Maintaining good signal integrity in the sleep distribution network is essential, especially for cluster-based design where all the charging/discharging currents flow through a single sleep device. Sleep distribution network noise during active mode, as shown in Figure 2,

974

978-1-4244-3039-0/08 $25.00 © 2008 IEEE

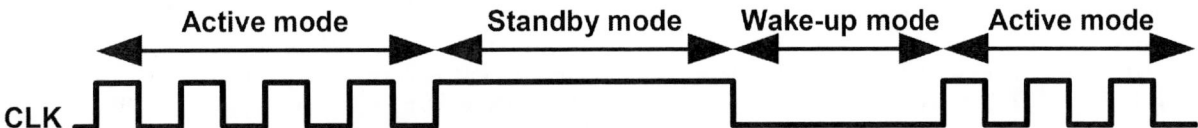

Figure 3. Operation modes based on the clock signal.

increases the on-resistance of the sleep device and reduces its overdrive, which in turn degrades circuit performance and noise margin. This noise issue is critical for cluster based design since a single sleep transistor is shared by all the gates in the cluster. Therefore, the increased on-resistance of the sleep device affects all the gates performance. If such noise occurs during the peak current switching of the cluster, functional failure might occur. Buffering and shielding the sleep signal greatly helps in this issue. However, shielding increases the wiring cost and area. Whereas, buffering increases area and active-to-standby-to-active power overhead. Moreover, sleep signal buffers consume leakage power in active and standby modes, and complicate buffer floor-planning and sleep signal routing. Previous projection indicates that repeaters and buffers required for intra-block signaling will take 70% of the total block cell count at the 32-*nm* node [16]. Therefore, adding extra intra-block signals and buffers is not desirable.

Important metrics that greatly affect MTCMOS efficiency include active-to-sleep and sleep-to-active energy overheads. These metrics determine the minimum standby period that achieves overall power saving. Charging/discharging the sleep distribution network and its associated buffers introduce energy overhead comparable to the energy overhead of charging/discharging the virtual power/ground networks and turning-on the sleep transistors during standby-to-active mode, and is the major contributor in the active-to-standby energy overhead. Also, sleep signal network delay affects performance since it is also a major contributor in the total standby-to-active delay. Moreover, sleep signal network requires careful timing analysis to accurately determine the wake-up delay.

With all these complexities related to sleep signal distribution in distributed MTCMOS design styles, and as the integration density of a single block continues to increase, design methods that take sleep signal distribution into account should be considered.

4. Self-sleep buffer

In order to fully reduce power during standby mode, when a block goes into standby, its sleep transistors, as well as its clock, are turned off. If the block is a whole chip or if its clock is independent from the other blocks clock, clock gating can be done from the clock source. In the case of a multi-block system with global clock, clock-gating can be done locally from the node that distributes the clock signal to the idle block. Our method uses the relation

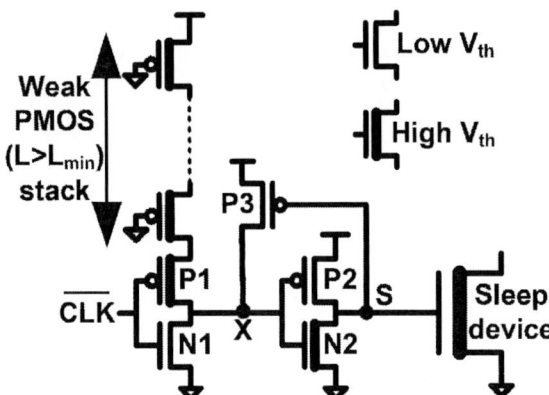

Figure 4. Two stages dual-V_{th} self-sleep buffer.

between sleep signal and the clock in order to eliminate the sleep distribution network. We assume that the clock is gated high, as shown in Figure 3. Before the block switches back to active mode, the clock goes low, which indicates a wake-up period. The wake-up time depends on the distributed MTCMOS style being used. Also, several MTCMOS techniques that were previously proposed can be employed during this mode to achieve better signal integrity and lower energy. After the wake-up period, the block resumes its normal operation.

Figure 4 presents a two stages dual-V_{th} self-sleep buffer. The self-sleep buffer outputs the sleep signal based on the clock. In the active mode and during the low-phase of the clock, node X is low and node S (sleep) is high. During the high-phase of the clock, node X starts charging. The self-sleep buffer should be designed such that at the worst corner, the voltage at node X can not reach a level that causes a glitch at node S. Therefore, node S remains high during all the active period. Transistors $N2$ and $P2$ have high and low V_{th}, respectively. This increases the high-to-low switching threshold of the second stage inverter in the buffer, which helps in maintaining a glitch-free node S. Transistors $N1$ and $P1$ have minimum size channel length and width, in order to maintain the minimum possible clock loading. In the case of a large sleep device, extra stages can be added to the buffer to maintain the minimum possible clock load at the input. $P1$ and the weak *PMOS* stack have high V_{th}, and the channel length of the transistors in the weak *PMOS* stack equals to the maximum channel length (L_{max}) allowed by the design/manufacturing rules. The number of transistors in the weak *PMOS* stack is based on the desired peak voltage at node X during active

675

978-1-4244-3039-0/08 $25.00 © 2008 IEEE

mode $V_X(t_{duty})$, where t_{duty} is the duty cycle of the clock since node X is charged during the entire high-phase of the clock. The active-mode peak voltage at node X equals

$$V_X(t_{duty}) = V_{DD}(1 - e^{-\frac{t_{duty}}{R_P C_X}}) \tag{1}$$

where V_{DD} is the supply voltage, C_X is the total capacitance at node X, and R_P is the sum of the weak $PMOS$ stack resistance and transistor $P1$ on-resistance.

We define δ as the peak percentage swing at node X during the active mode, therefore

$$\delta = \frac{V_X(t_{duty})}{V_{DD}} \tag{2}$$

δ should be larger than zero, therefore

$$I_P(on) > I_{N1}(off) \tag{3}$$

where $I_P(on)$ is the current flowing through the weak $PMOS$ stack and $P1$ during the high-phase of the clock, and $I_{N1}(off)$ is the drain leakage current of $N1$. Equation (3) is very easy to satisfy and it should be automatically satisfied at all corners. The upper bound on δ is

$$\delta < \beta_S(P2, N2) \tag{4}$$

where $\beta_S(P2, N2)$ is the ratio of high-to-low switching threshold of the second stage inverter ($P2, N2$) to V_{DD}, as discussed earlier in this section.

Based on equation (1), t_{duty}, V_{DD} are known, and C_X is determined based on the size of the second stage inverter, which in turn is determined based on the sleep device size. The number of weak transistors in the weak $PMOS$ stack can be adjusted to achieve the desired δ, since

$$R_P = -\frac{t_{duty}}{C_X \log_e(1 - \delta)} \tag{5}$$

δ presents a tradeoff between active-mode power and active-to-standby delay and energy. Using a small δ achieves low active-mode power since the swing at node X is reduced, but it increases the active-to-standby delay of the buffer, as well as the active-to-standby short circuit current due to the slow slew rate at node X. However, the active-to-standby delay is not important as the standby-to-active delay which affects the overall performance. The self-sleep buffer offers fast sleep-to-active transition through transistor $N1$ and $P2$. Transistor $P3$ acts as a booster which reduces the transition time at node X during active-to-standby transition, in order to reduce the short circuit power at the second stage inverter. However, $P3$ should have minimum size since it has to be weaker than $N1$.

4.1 Advantages of self-sleep buffer

For distributed cluster-based and network-based MTCMOS, a self-sleep buffer is assigned to each sleep transistor. The number of stages and the transistor sizes of the self-sleep buffer are based on the size of the sleep

device. The self-sleep buffer eliminates the need for distributing a sleep signal to all the sleep devices since it uses the available clock signal as its input. As a result, the total intra-block wirelength, routing complexity, and interconnect buffers are reduced.

Sleep signal integrity during active mode is improved by the self-sleep buffers. The first reason for that is the very high resistance of the weak $PMOS$ stack and $P1$, and the skewed second stage inverter. This makes the self-sleep buffer a noise filter during active mode, which helps in maintaining a glitch free sleep signal. The second reason is that the self-sleep buffer is driven by the clock signal which is the most well designed signal on the chip. Moreover, the delay and power characteristics of the clock network are known, which eliminates the need for extra design effort in characterizing the delay and energy of the sleep network and maintaining its signal integrity.

The intrinsic sleep-to-active delay of the buffer is small, but since the buffer is driven by the clock, the source-to-sinks delay of the clock should be added to the overall wake-up delay. The synchronization of the self-sleep buffers helps in several ways. Scheduling the sleep signals inside the block to reduce ground bounce during sleep-to-active transition can be done easily by adding various delay elements to the self-sleep buffers. Also, the timing relation between the sleep signal and the clock is well determined, which helps in timing synchronization when going from standby to active mode.

The active-to-sleep and sleep-to-active MTCMOS energy overheads are reduced with the self-sleep buffer due to the elimination of the charging/discharging of the sleep signal interconnect and its associated buffers. The clock signal is gated in the presence of a sleep distribution network as well, thus clock switching during active-to-standby and standby-to-active transitions can not be considered as overhead for self-sleep buffer method. The reduction in MTCMOS energy overhead during mode transitions reduces the minimum standby period that is required to achieve overall power reduction, and helps in switching more frequently between active and standby modes.

5. Self-sleep buffer simulation and characteristics

A self-sleep buffer cell was designed in a 90-*nm* CMOS technology with 1V supply voltage. The buffer design has two stages in order to drive a 9-*um* wide *NMOS* sleep device (60× minimum device width) with minimal clock load at the input. Four minimum width high-V_{th} devices with a channel length of 0.3-*um* are used to build the weak $PMOS$ stack to achieve a δ ratio of 0.2 at 2-GHz clock frequency. Table 1 presents the channel length and width of the self-sleep buffer transistors. Figure 5 shows a standard-cell layout of the self-sleep buffer routed with

Table 1. Transistor sizes of the two-stages self-sleep buffer (60× sleep device load).

	4 HVT transistors in the weak *PMOS* stack	N1	N2	P1	P2	P3
W (um)	0.15	0.15	0.6	0.15	1.6	0.15
L (um)	0.3	0.1	0.1	0.1	0.1	0.1

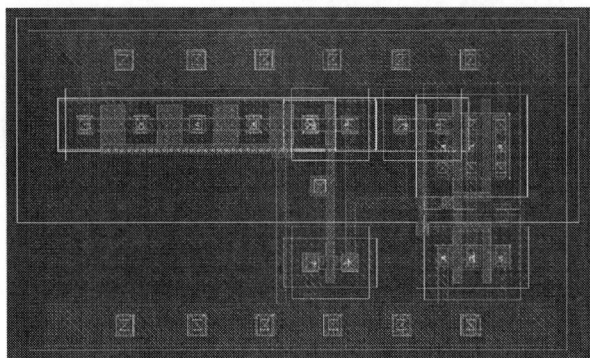

Figure 5. Self-sleep buffer layout.

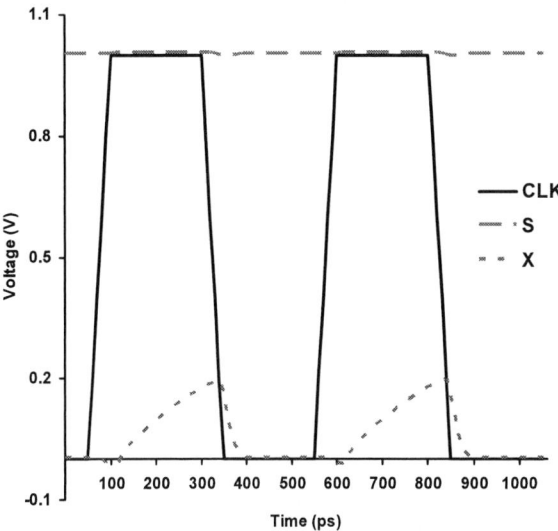

Figure 6. Active-mode simulation waveform.

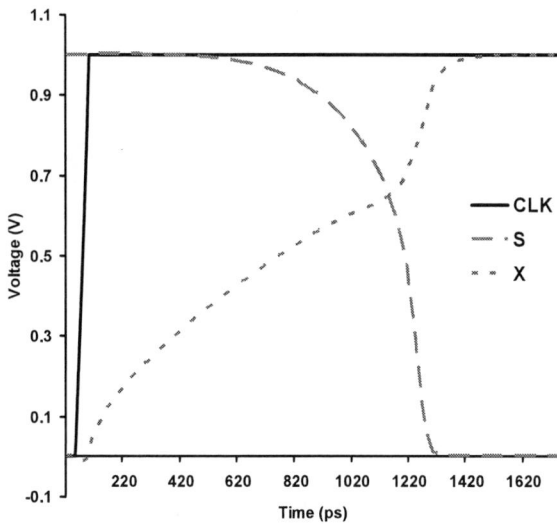

Figure 7. Active-to-standby simulation waveform.

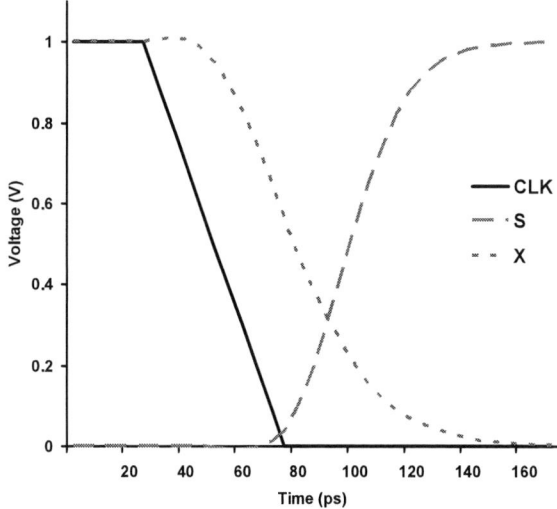

Figure 8. Standby-to-active simulation waveform.

poly and METAL1 with a total area of 21.95-um^2. The *PMOS* stack causes a considerable loss in the *N*-region area due to the long channel of its *PMOS* devices, and the small number of *NMOS* devices in the self-sleep buffer compared to the number of *PMOS* devices. However, the self-sleep buffer layout can be further optimized based on the layout style and its context. For example, the cell can have two rows of *PMOS* devices, where the weak stack occupies one row. This considerably reduces the area, but it might require an extra METAL2 line to be used for intra-cell routing. Another way that can be used to avoid the area loss in the *N*-region is by embedding the multi-finger *NMOS* sleep device inside the self-sleep buffer cell. Figure 6 shows the simulation waveform during the active mode for a 2-GHz clock frequency. Node *X* has a peak voltage of 0.2V which does not affect node *S* that remains high during

the entire active period. The active-to-standby delay of the buffer is considerably larger than its standby-to-active delay, as shown in Figures 7 and 8. However, active-to-standby delay does not affect the design performance, unlike standby-to-active delay which is critical for wake-up performance.

Table 2 presents the characteristics of the self-sleep buffer. The CLK-to-S active-to-standby delay is about 2.25 clock cycles, and most of this delay is in driving node *X*. However, the performance critical CLK-to-S standby-to-active delay is only 48.42-*ps*. The active-to-standby energy

Table 2.Self-sleep buffer characteristics (2-GHz clock frequency, $\delta = 0.2$, 60× minimum device width sleep transistor).

Area	21.95-um^2
CLK-to-S active-to-standby delay	1.125-ns
Active-to-standby energy	30.86-fJ
CLK-to-S standby-to-active delay	48.52-ps
Standby-to-active energy	8.478-fJ
Active power	1.467-uW
Standby power	0.4536-uW

is larger than standby-to-active energy due to the slow low-to-high transition of node X which causes some short circuit current until the booster $P3$ is turned on. Although node X is being charged/discharged during active mode, the active power of the self-sleep buffer is similar to the active-power of any other cell in the library at a typical switching activity, due to the very low swing and large charging time of node X. The active power of the self-sleep buffer is around 3× its standby power. The load that the self-sleep buffer presents on the clock is small and equals to the gate capacitance of two minimum size transistors, which is equivalent to 0.55 the input capacitance of a 1× inverter cell with equal rise and fall delays.

6. Conclusion

We have proposed a self-sleep buffer for distributed MTCMOS design with gated clock. The self-sleep buffer eliminates the need for a sleep distribution network, while maintaining good sleep signal integrity in active mode and low energy overhead during active-to-standby-to-active transitions. Wakeup scheduling can be easily and reliability implemented with the proposed self-sleep buffer. The energy overhead due to charging/discharging sleep signal interconnect and buffers during active-to-standby-to-active is eliminated. One disadvantage of the self-sleep buffer is that it consumes power in active mode; however its active power is reasonable and similar to the power consumed by any other cell in the circuit. The self-sleep buffer does not pose a restriction on applying the previously proposed MTCMOS techniques, and it can be applied simultaneously with them to reduce the MTCMOS overheads while achieving low standby leakage.

7. Acknowledgment

This work was supported in part by the US Department of Energy (DoE), DE97ER12220, and the Governor's Information Technology Initiative.
The authors would like to thank Dr. Soumik Ghosh from University of Louisiana at Lafayette for his help with the design tools.

8. References

[1] T. Karnik, S. Borkar, V. De, "Sub-90nm technologies-challenges and opportunities for CAD," in *Proc. IEEE/ACM Int. Conference on CAD*, pp. 203-206, 2002.

[2] S. Mutoh *et al.*, "1-V power supply high-speed digital circuit technology with multithreshold-voltage CMOS," *Jour. Solid-State Circuits*, vol. 30, pp. 847-854, Aug. 1995.

[3] J. Kao, A. Chandrakasan, D. Antoniadis, "Transistor sizing issues and tool for multi-threshold CMOS technology," in *Proc. IEEE/ACM Design Automation Conf.*, pp. 409-414, June 1997.

[4] J. Kao, S. Narendra, A. Chandrakasan, "MTCMOS hierarchical sizing based on mutual exclusive discharge pattern," in *Proc. IEEE/ACM Design Automation Conf.*, pp. 495-500, June 1998.

[5] S. Kim, S. V. Kosonocky, D. R. Knebel, and S. Stawiasz, "Experimental measurement of a novel power gating structure with intermediate power saving mode," in *Proc. IEEE Int. Symp. On Low Power Elec. Design*, pp. 20-25, August 2004.

[6] S. Kim, S. V. Kosonocky, D. R. Knebel, "Understanding and minimizing ground bounce during mode transition of power gating structures," in *Proc. IEEE Int. Symp. On Low Power Elec. Design*, pp. 22-25, August 2003.

[7] A. Ramalingam, A. Devgan, and D. Z. Pan, "Wakeup scheduling in MTCMOS circuits using successive relaxation to minimize ground bounce," *Jour. Of Low Power Electronics*, vol. 3, no. 1, 2007.

[8] A. Abdollahi, F. Fallah, and M. Pedram, "A robust power gating structure and power mode transition strategy for MTCMOS design," *IEEE Trans. VLSI Syst.*, vol. 15, no. 1, Jan. 2007.

[9] S. Shigematsu *et al.*, "A 1-V high-speed MTCMOS circuit scheme for power-down applications," in *Symp. VLSI Circuits Dig. Tech. Papers*, pp. 125-126, 1995.

[10] E. Pakbaznia, F. Fallah, M. Pedram, "Charge recycling in MTCMOS circuits: concept and analysis," in *Proc. IEEE/ACM Design Automation Conf.*, pp. 97-102, June 2006.

[11] M. Anis, S. Areibi, and M. Elmasry, "Design and optimization of multithreshold CMOS (MTCMOS) circuits," *IEEE Trans. CAD of Integ. Circuits and Syst.*, vol. 22, no. 10, pp. 1324-1342, Oct. 2003.

[12] C. Long, and L. He, "Distributed sleep transistor network for power reduction," *IEEE Trans. VLSI Syst.*, vol. 12, no. 9, Sep. 2004.

[13] K. Shi, D. Howard, "Sleep transistor design and implementation – simple concepts yet challenges to be optimum," in *Proc. Int. Symp. VLSI Design, Automation and Test*, pp. 1-4, April 2006.

[14] V. Khandelwal and A. Srivastava, "Leakage control through fine-grained placement and sizing of sleep transistors," in *Proc. IEEE/ACM Int. Conference on CAD*, pp. 533-536, 2004.

[15] B. H. Calhoun, F. A. Honore, and A. P. Chandrakasan, "A leakage reduction methodology for distributed MTCMOS," *Jour. Solid-State Circuits*, vol. 39, no. 5, pp. 818-826, May 2004.

[16] P. Saxena, N. Menezes, P. Cocchini, D. A. Kirkpatrick, "Repeater scaling and its impact on CAD," *IEEE Trans. CAD of Integ. Circuits and Syst.*, vol. 23, no. 4, pp. 451-463, Apr. 2004.

21st International Conference on VLSI Design

Power Management of Interactive 3D Games using Frame Structures

Yan Gu   Samarjit Chakraborty

Department of Computer Science, National University of Singapore

E-mail: {guyan, samarjit}@comp.nus.edu.sg

Abstract

We propose a novel dynamic voltage scaling (DVS) scheme that is specifically directed towards 3D graphics-intensive interactive game applications running on battery-operated portable devices. The key to this DVS scheme lies in parsing each game frame to estimate its rendering workload and then using such an estimate to scale the voltage/frequency of the underlying processor. The main novelty of this scheme stems from the fact that game frames offer a rich variety of "structural" information (e.g. number of brush and alias models, texture information and light maps) which can be exploited to estimate their processing workload. Although DVS has been extensively applied to video decoding applications, compressed video frames do not offer any information (beyond the frame types – I, B or P) that can be used in a similar manner to estimate their processing workload. As a result, DVS algorithms designed for video decoding mostly rely on control-theoretic feedback mechanisms, where the workload of a frame is predicted from the workloads of the previously-rendered frames. We show that compared to such history-based predictors, our proposed scheme performs significantly better for game applications. Our experimental results, based on the Quake II game engine running on Windows XP, show that for the same energy consumption our scheme results in more than 50% improvement in quality (measured in terms of number of frames meeting their deadlines) compared to history-based prediction schemes.

1 Introduction

Graphics-intensive game applications are now increasingly spilling over from high-end desktops to mobile devices (e.g. PDAs, cell phones and portable game consoles) for which battery-life is a major concern. This has resulted in a growing interest in power management schemes specifically directed towards 3D graphics and game applications [5, 9, 10]. It is now well-established that such applications exhibit sufficient variability in their workload for dynamic voltage scaling (DVS) algorithms to be meaningfully applied [5, 7, 12]. Over the last few years, such algorithms have been very successfully applied to video encoding/decoding applications which are also computationally expensive and where the workload associated with processing different frames can vary significantly (for example, see

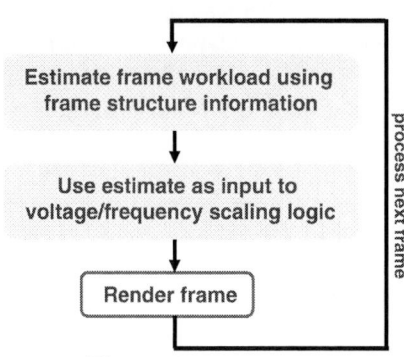

Figure 1. DVS in a game loop.

[1, 3, 6, 14]). The basic principle behind most of these algorithms is to predict the workload associated with processing a video frame from the workloads of the previously decoded frames. The voltage/frequency of the underlying processor is then scaled based on such history-based workload predictions. This basic scheme has also been refined using control-theoretic feedback mechanisms, where previous prediction errors are taken into account while estimating the workload of a current frame [8, 11].

In this paper we propose a workload prediction and DVS scheme specifically directed towards 3D game applications. Instead of predicting the workload associated with processing a game frame from the history of previously-processed frames — as done with video decoding applications — we estimate this workload by *parsing* the game frame. The main novelty of our scheme stems from the observation that game frames offer a rich variety of *structural* information which can be used to predict their workload or processor cycle requirements. The workload involved in processing a frame largely depends on the scene that the frame is depicting. In other words, the workload depends on the structure/contents of the frame, or the constituting *objects* that need to be processed. Such structural information can be efficiently obtained by parsing a frame, prior to it being actually processed, which can then be used to estimate the frame's processing workload.

An overview of this scheme is shown in Figure 1. The "voltage/frequency scaling logic" is used to decide whether the voltage/frequency of the processor is to be changed from its current level based on the workload estimation. Since scaling the voltage/frequency of a processor involves a certain overhead — which depends on the processor's architecture as well as the underlying operating system — it might not be meaningful to switch the voltage and frequency in response to every workload change. This is explained in further detail later in this paper. Note that this scheme of parsing a frame to estimate its processing workload cannot be applied to video frames, which offer no structural information beyond the frame type (i.e. I, B or P).

We have evaluated our proposed scheme using the Quake II game engine running on a notebook with a 1400 MHz Intel Pentium Mobile processor with Windows XP. All the power measurements were conducted by connecting this notebook to a National Instruments PXI-4071 $7\frac{1}{2}$-digit Digital Multimeter, using which the instantaneous voltage and current drawn by the notebook was recorded. Our estimated energy consumptions therefore refer to those incurred by the *entire* notebook (i.e. all its components, including processor, display, disk drive, etc.) and not the processor alone. When compared with history-based workload predictors, for the same amount of energy consumption, our scheme results in more than 50% improvement in output quality (which is measured in terms of the number of frames meeting their deadlines for a prespecified frame rate). Compared to running the processor at a fixed frequency (i.e. no voltage/frequency scaling), our scheme achieved 22% *system-wide* energy savings. The savings for the processor alone would certainly be much higher. However, we did not have any mechanism for measuring the power consumption of the processor alone.

The rest of this paper is organized as follows. In the next section we give a brief overview of a game pipeline. In Section 3 we describe our workload prediction scheme using frame structures, which is followed by an outline of our DVS algorithm based on this predictor. Section 5 contains our experimental results. Finally we conclude in Section 6 by outlining some directions for future work.

2 Game Pipeline

Similar to what is shown in Figure 1, a game application runs in an infinite loop, where the body of the loop consists of tasks responsible for processing a single game frame. Such tasks can be broadly classified into those responsible for *computing* and those for *rendering*. Examples of computing tasks are collision detection, AI, and simulation of game physics and particle systems. Rendering tasks implement algorithms to generate an image or a frame from a model, which is then displayed on the screen. Such models are typically descriptions of several 3D-objects (e.g. characters in the game, weapons, buildings, etc.) using a predefined language or data structure. These descriptions consist of geometry, viewpoint, texture and lighting information. A more detailed discussion of the game pipeline may be found in [2, 13].

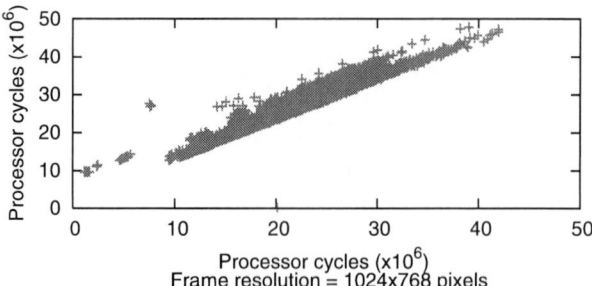

Figure 2. Linear correlation between rasterization and total processing workload of a frame.

The rendering algorithms typically transform vertices of 3D/solid objects to the 2D screen space, delete invisible pixels by clipping, perform rasterization, delete occluded pixels and interpolate various parameters. These algorithms are computationally expensive and are typically mapped onto a graphics processing unit (GPU) in desktop computers or high-end notebooks. However, most mobile devices (e.g. PDAs and cell phones) currently do not have GPUs and instead implement the rendering algorithms in software. We believe that this trend will continue at least for the next few years. Hence, for this work we have assumed all rendering tasks to be implemented in software running on the common voltage/frequency-scalable processor along with the computing tasks of the game engine.

3 Workload Prediction via Frame Structures

In this section we describe the frame structure-based workload prediction scheme that forms the basis of our DVS algorithm. It may be noted that a significant component of the rendering task involves *rasterizing* objects on the screen. Our experimental results suggest that the total workload generated from processing a frame is almost linearly correlated with its rasterization workload. Hence, we predict the total workload by estimating the rasterization workload of a frame. Figure 2 shows the correlation between these two workloads, with the horizontal axis denoting the rasterization workload.

3.1 Exploiting the Frame Structure

Figure 3 shows an overview of the proposed frame structure-based workload prediction scheme. Note that it primarily consists of estimating the rasterization workload for each frame. Towards this, we compute the number of occurrences of the different primitives in a frame (e.g. brush models, alias models and particles) and multiply these with the workload involved in processing each of these primitives. This is possible because, once again, the workload involved in processing all primitives of the same type almost linearly scales with the number of primitives occurring in the frame. The workload corresponding to a single primitive of any given type is computed in an offline fashion. We have experimentally verified this (nearly) linear correlation for most of the primitives such as brush models, alias

978-1-4244-3039-0/08 $25.00 © 2008 IEEE

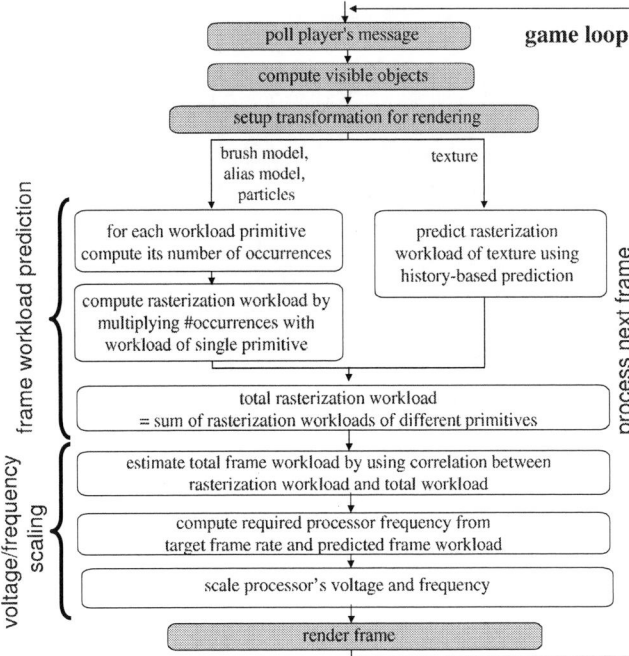

Figure 3. Overview of the frame structure based workload prediction scheme.

models, textures and particles. It may also be noted that the workload corresponding to each of these primitives exhibits sufficient short-term variability, as shown in Figure 4. Hence, our proposed scheme performs significantly better than history-based predictors even if they are applied individually to the different primitive-types. In what follows, we describe our rasterization workload estimation in further details.

For each frame, once the current *view frustum* is computed based on the user input, the number of occurrences of the different primitives in the frame is estimated (e.g. the number of brush models, alias models, etc.). Further, for each of these primitives, their detailed constitution is also computed. For each brush model, this amounts to computing its number of constituent polygons. For each alias model it amounts to computing its number of pixels, for each texture model its number of surfaces, and for each particle its number of pixels. Based on offline simulation, the workload associated with each of the different primitives parameterized by their constitution is stored in a table. For example, this table contains the workload associated with processing a brush model with n polygons for different values of n. Let us assume that $c(n)$ is the number of processor cycles requires to rasterize any *one* brush model with n polygons. Then $c(n)$ is stored in the above-mentioned table. To compute the rasterization workload for *all* brush models in a frame, let us assume that $B(n)$ is the number of brush models in this frame with n polygons. Then the total rasterization workload for all the brush models in the frame is equal to $\sum_{n=1,...,\infty} c(n) \times B(n)$ processor cycles

(where ∞ is the maximum possible number of polygons in any brush model). This procedure is followed for all the different primitives, with the exception of texture models (which is explained below).

The abovementioned estimation process clearly has a computational overhead, which turns out to be prohibitive in the case of textures. Textures are drawn on brush models and hence its number of constituent surfaces cannot be determined unless the associated brush models are rasterized. However, rasterizing brush models solely for workload estimation purposes is prohibitively expensive. Hence, as shown in Figure 3, the workload associated with rasterizing texture models in a frame is estimated using a history-based prediction scheme. Although this is not as accurate as frame structure-based predictions, this loss of accuracy is unavoidable. However, using frame structure-based estimation, at least for the other primitives like brush models, alias models and particles, reduces the overall error compared to using history-based predictors for all the primitives. Further, this mix of two different estimation schemes results in a good tradeoff between prediction accuracy and computational overhead.

3.2 Prediction Accuracy

Figure 5 shows a comparison of a history-based predictor and our proposed frame structure-based predictor against the actual workload. The excerpt shown in this figure was generated from a four-second demo file of Quake II (massive1.dm2[1]). The history-based predictor estimated the total processing workload of a frame by averaging the actual workload of a number of previously processed frames. Each point in Figures 5(a) and 5(b) corresponds to a frame and the horizontal axis refers to the time stamp (in milliseconds) associated with each frame. The vertical axis refers to the total rasterization workload of a frame in terms of the number of processor cycles. It may be noted that the this workload varies between 15 million cycles to 45 million cycles per frame.

It it clear from this figure that our proposed scheme matches the profile of the actual rasterization workload more closely than history-based predictors, especially because of the high variability exhibited by game workloads. To measure the prediction errors incurred by the two schemes, we used two metrics: (i) the *absolute prediction error* is defined as the absolute difference in processor cycles between the actual and predicted workloads, (ii) the *relative prediction error* is defined as the ratio between the absolute prediction error and the actual workload. The absolute and relative prediction errors in Figure 5(a) (i.e. using the history-based predictor) turn out to be 3.6 million cycles and 0.15 respectively. These errors drop to 2.1 million cycles and 0.09 respectively for our proposed predic-

---

[1]http://cure.gamepoint.net/files/massive1.zip

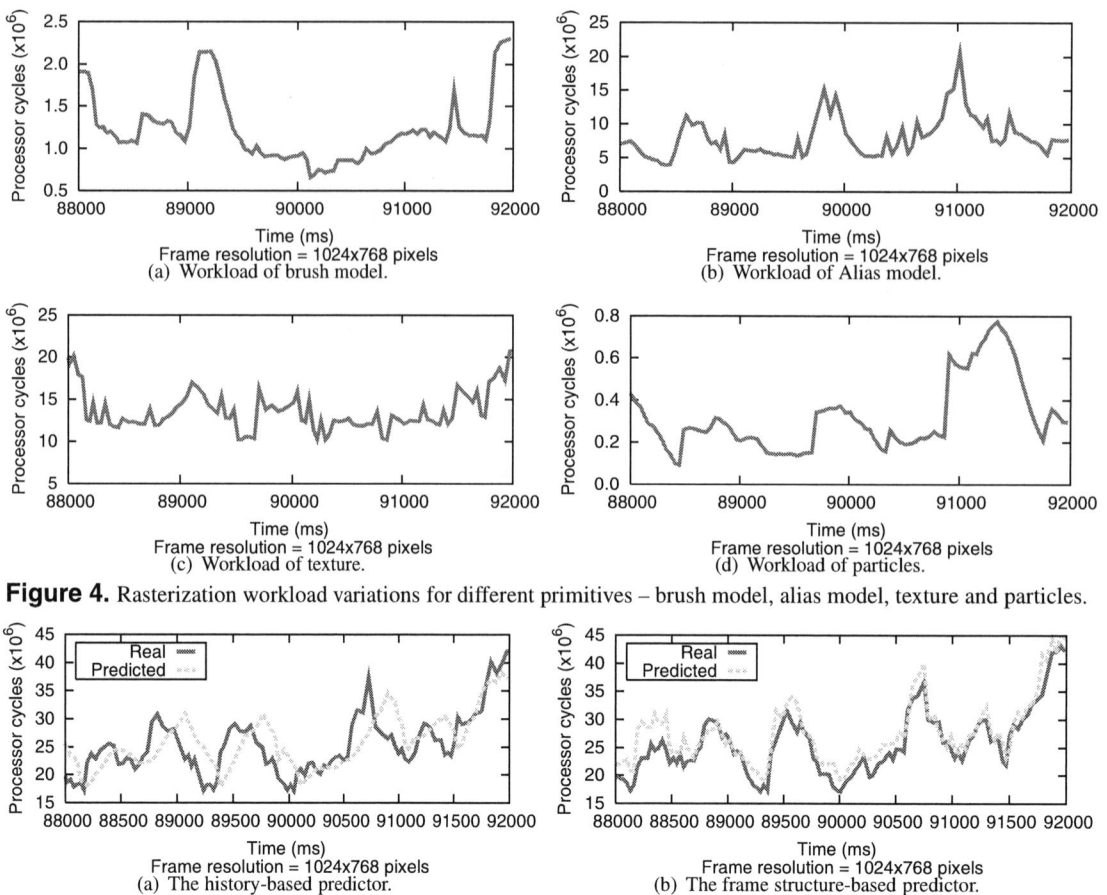

Figure 4. Rasterization workload variations for different primitives – brush model, alias model, texture and particles.

Figure 5. A comparison of history-based and frame structure-based workload predictors.

tion scheme (Figure 5(b)). Hence, our proposed scheme results in more than 40% improvement in prediction accuracy over history-based predictors.

4 Frame Structure-based DVS

In this section we discuss how the above workload prediction scheme is integrated with a DVS algorithm. As shown in Figure 1, the predicted workload for each frame is fed into a voltage/frequency scaling logic, which takes into account several hardware and systems issues to decide whether the voltage/frequency should be switched from its current level. Since scaling a processor's voltage/frequency is associated with a certain overhead — which depends on the processor's microarchitecture and the OS running on top of it — it might not be meaningful to switch the clock frequency at every possible game frame or workload change. This decision is handled by the voltage/frequency scaling logic. Further, most processors support a fixed number of discrete operating frequency (and associated voltage) levels. From predicted workload of a game frame and the target frame/display rate, the optimum operating frequency of the processor may be calculated. This calculated frequency needs to be mapped onto the discrete frequency levels available on the processor in a conservative manner. Below we describe these steps in further detail.

4.1 Frequency Mapping

A number of previously-proposed algorithms for DVS have assumed the processor's frequency range to be continuous (e.g. see [8]). However, most voltage/frequency-scalable processors only support a fixed number of discrete frequency levels (which is often five). Hence, in this work we have assumed that only a fixed number of frequency levels are available and the computed optimum frequency is mapped onto the next available higher frequency level. Such a conservative mapping satisfies the workload demands of the game application, at the cost of less than ideal energy savings. However, we have also conducted simulations where we assumed that the processor's frequency is continuously scalable. In Section 5 we present a comparison of the energy savings obtained with such ideal settings and where the frequency can only be set to five discrete levels.

4.2 Frequency Transition

As mentioned before, switching the frequency of a processor is associated with an overhead which depends on the processor's microarchitecture as well as the OS running on top of it. Our experimental results suggest that for the same processor, this overhead is higher in Windows XP compared to Linux. The average transition overhead in Windows XP running on an Intel Pentium Mobile processor is 20 million cycles, i.e., the overhead is 14 milliseconds with the operating frequency set to 1400 MHz.

Hence, to skip unnecessary frequency switches, we have used a *lazy* transition mechanism. Instead of immediately switching the processor frequency whenever the predicted workload of a game frame changes, we defer the switch to the immediate next frame. For instance, if the estimated quantized frequency (i.e. the computed frequency mapped to the frequency level available on the processor) for the current frame i is different from the frequency associated with the previous frame $i-1$, then the switching decision is deferred to the next frame (i.e. frame $i+1$). If the computed quantized frequency of the $i+1$th frame is also different from the frequency of the $i-1$th frame, then the frequency of the processor is changed to the frequency computed for frame $i+1$, otherwise the operating frequency is kept unchanged. Such a *lazy* frequency scaling is resilient to frequent frequency adaptations which might be unnecessary and expensive. Note that we defer the frequency scaling decision by only one frame. Our experimental results suggest that for our setup this provides satisfactory results. However, if in a different setting, the switching overhead is even higher, then it might be meaningful to defer the switching decision by multiple frames.

5   Experimental Evaluation

We have evaluated our proposed DVS algorithm on the Quake II game engine in three different settings: (i) On a notebook with an Intel Pentium Mobile processor running Windows XP, (ii) Using a discrete event simulator where the processor has the same power consumption characteristics as in the previous case, but the frequency transition overhead is assumed to be zero. However, only five discrete frequency levels are available, and (iii) Same as (ii) with the exception that the frequency of the processor is assumed to be continuously scalable. Settings (ii) and (iii) are referred to as *simu-disc* (i.e. simulation with discrete frequency levels) and *simu-cont* (simulation with continuous frequency levels) respectively.

Our motivation behind using the Quake II game engine primarily stems from the fact that it is a popular game that can be played on a variety of mobile devices such as PDAs, mobile phones and laptops without additional graphics hardware. Further, this game engine forms the core of a number of other First Person Shooter games (e.g. Hexen II) and its software architecture is representative of those in many other commercially-available games. Finally, the source code of Quake II is freely available, which allows for experimentation and appropriate modification. To ensure reproducibility, we have used pre-recorded demo files

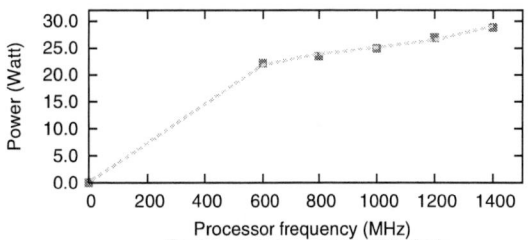

Figure 6. Processor frequency versus total system power consumption.

(.dm2[2]) of Quake II in all our experiments. Since these demo files keep pre-recorded states and therefore they are not computed during playback, there is some difference in workload when compared to a real-time game play. However, we have verified that these differences are negligible and do not affect the conclusions derived from this study. Finally, the game resolution was set to 1024×768 pixels, running in full-screen mode. Again, the conclusions derived from this setting also hold for other resolutions, as verified by our initial experiments with running Quake II on a PDA. To ensure that the game process is not preempted by other processes, it was set to the highest priority.

The notebook used for the experiments (with the 1400 MHz Intel Pentium Mobile processor) was equipped with Speedstep technology and had an ATI Radeon Mobility Video card. The processor supported five different operating points with clock frequencies of 1400, 1200, 1000, 800 and 600 MHz. All processor cycle counts were measured using the RDTSC (read time-stamp counter) instruction that was inserted into Quake II source code.

Recall, that all our power measurements refer to the full system and not the processor alone. Figure 6 shows the total system power consumption for the five different processor frequency levels. It may be noted that this varies between 28.8 Watts and 22.1 Watts, which corresponds to the processor frequencies of 1400 MHz and 600 MHz respectively. Hence, the maximum possible reduction in power consumption is upper bounded by 23%.

5.1   Results

We have compared our proposed scheme against a history-based predictor that is commonly used in DVS algorithms for video decoding. Towards this, we have defined two quality metrics that have been motivated by a study in [4]. This study concluded that while frame rates higher than a pre-defined constant target frame rate do not improve the overall gaming experience, lower than target frame rates severely degrade the game quality. For all our experiments, we have set the target frame rate to be 20 frames/second. Hence, each frame has to be processed within 1/20th of a second, which is set as the *frame deadline*.

---

[2]http://demospecs.planetquake.gamespy.com/dm2/

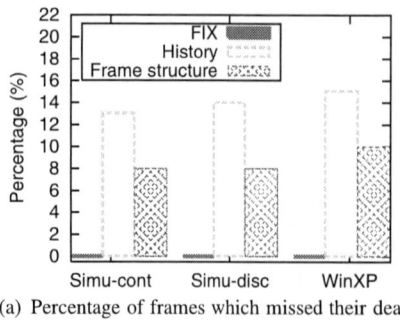
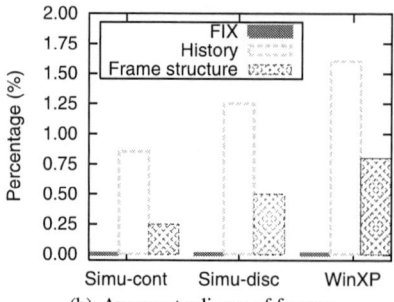

(a) Percentage of frames which missed their deadlines.

(b) Average tardiness of frames.

Figure 7. A comparison of game quality using frame structure-based and history-based prediction on three different settings.

Our first metric only measures the percentage of frames missing their deadlines. The second metric also takes into account the magnitude of the missed deadline (or tardiness). Figure 7 shows the game quality under these two metrics for three different DVS schemes: FIX (where the processor is run at a constant frequency of 1400 MHz, i.e. no frequency scaling), History (DVS with a history-based predictor), and Frame structure (DVS using our proposed workload prediction scheme). From this figure, it may be noted that under the average tardiness metric, our proposed scheme results in more than 50% improvement in game quality in the Windows XP setting for the same amount of energy consumption. The results under the simulation setting are even more attractive. In terms of power savings, compared to the FIX scheme, our scheme achieves up to 22% power savings, where the upper bound on the savings — as mentioned before — is 23%. Note that to match the target frame deadline, most of the frequencies computed for the estimated frame workload approach to the lowest possible frequencies (i.e. 600 MHz) on the laptop.

When the target frame deadline is reduced to 1/30th of a second (i.e. 30 frames/sec), more processor cycles are required to speed up the game play. Therefore, the power saving from Frame structure drops to 13% comparing with FIX, while the game quality obtained using Frame structure is consistently better than that obtained using History.

6   Concluding Remarks

In this paper we have proposed a novel DVS scheme specifically targeted towards graphics-intensive game applications. Our results indicate attractive energy savings and output quality compared to known history-based DVS algorithm that was developed for video decoding applications. We have implemented our scheme on a Intel Pentium based notebook running Windows XP. Currently we are in the process of implementing it on a PDA with a voltage/frequency-scalable processor. Further, we are also exploring possibilities of designing hybrid workload predictors that combine history-based (which has low prediction overhead) and frame-structure-based predictors.

References

[1] A. Acquaviva, L. Benini, and B. Riccó. An adaptive algorithm for low-power streaming multimedia processing. In *Design, Automation and Test in Europe (DATE)*, 2001.

[2] L. Bishop, D. Eberly, T. Whitted, M. Finch, and M. Shantz. Designing a PC game engine. *IEEE Computer Graphics and Applications*, 18(1), 1998.

[3] K. Choi, K. Dantu, W.-C. Cheng, and M. Pedram. Frame-based dynamic voltage and frequency scaling for a MPEG decoder. In *International Conference on Computer-Aided Design (ICCAD)*, 2002.

[4] M. Claypool, K. Claypool, and F. Damaa. The effects of frame rate and resolution on users playing First Person Shooter games. In *Multimedia Computing and Networking Conference (MMCN)*, 2006.

[5] Y. Gu, S. Chakraborty, and W. T. Ooi. Games are up for DVFS. In *Design Automation Conference (DAC)*, 2006.

[6] C. J. Hughes and S. V. Adve. A formal approach to frequent energy adaptations for multimedia applications. In *International Symposium on Computer Architecture (ISCA)*, 2004.

[7] G. Lafruit, L. Nachtergaele, K. Denolf, and J. Bormans. 3D computational graceful degradation. In *ISCAS - Workshop and Exhibition on MPEG-4, Vol. III*, 2000.

[8] Z. Lu, J. Hein, M. Humphrey, M. Stan, J. Lach, and K. Skadron. Control-theoretic dynamic frequency and voltage scaling for multimedia workloads. In *International Conference on Compilers, Architecture and Synthesis for Embedded Systems (CASES)*, 2002.

[9] B. Mochocki, K. Lahiri, and S. Cadambi. Power analysis of mobile 3D graphics. In *Design, Automation, and Test in Europe (DATE)*, 2006.

[10] B. Mochocki, K. Lahiri, S. Cadambi, and X. S. Hu. Signature-based workload estimation for mobile 3d graphics. In *Design Automation Conference (DAC)*, 2006.

[11] C. Poellabauer, L. Singleton, and K. Schwan. Feedback-based dynamic frequency scaling for memory-bound real-time applications. In *RTAS*, 2005.

[12] N. Tack, F. Morán, G. Lafruit, and R. Lauwereins. 3D graphics rendering time modeling and control for mobile terminals. In *9th International Conference on 3D Web Technology*, 2004.

[13] A. Watt and F. Policarpo. *3D Games: Real-time Rendering and Software Technology, Volume 1.* Addison-Wesley, 2001.

[14] W. Yuan and K. Nahrstedt. Practical voltage scaling for mobile multimedia devices. In *ACM Multimedia (MM)*, 2004.

978-1-4244-3039-0/08 $25.00 © 2008 IEEE

21st International Conference on VLSI Design

Voltage and Temperature Scalable Gate Delay and Slew Models Including Intra-Gate Variations

Bishnu Prasad Das[1]*, Janakiraman V.[2], Bharadwaj Amrutur[2], H.S. Jamadagni[1], N.V. Arvind[3]

1 C.E.D.T., Indian Institute of Science, Bangalore, India.
2 Dept of E.C.E., Indian Institute of Science, Bangalore, India.
3 Texas Instruments, Bangalore, India

Abstract

We investigate the feasibility of developing a comprehensive gate delay and slew models which incorporates output load, input edge slew, supply voltage, temperature, global process variations and local process variations all in the same model. We find that the standard polynomial models cannot handle such a large heterogeneous set of input variables. We instead use neural networks, which are well known for their ability to approximate any arbitrary continuous function. Our initial experiments with a small subset of standard cell gates of an industrial 65nm library show promising results with error in mean less than 1%, error in standard deviation less than 3% and maximum error less than 11% as compared to SPICE for models covering 0.9-1.1V of supply, -40^0C to 125^0C of temperature, load, slew and global and local process parameters. Enhancing the conventional libraries to be voltage and temperature scalable with similar accuracy requires on an average 4x more SPICE characterization runs.

1. Introduction

In today's complex industrial designs, both temperature and supply voltage have strong location dependency, i.e. they are non-uniform across the chip. The current practice is to do a worst case design using the extreme values of supply and temperature for timing and power analysis. The authors in [6] argue that this can lead to unnecessarily large margins being included in the design and hence advocate a voltage and temperature aware timing analysis. Such an analysis is now feasible due to the emergence of sophisticated power grid analysis tools, which can predict the supply voltage at any point in the power grid. Similarly, a chip's thermal profile can also be estimated as a function of the computational activity and the heat removal capability of the cooling system. Thus it is possible in principle to use a better estimate of the local supply and temperature [11] at any gate and hence provide more accurate bounds on the

gate's and hence the chip's speed.

Due to technology scaling, a large number of process related effects force a wide spread in process parameters, in turn causing a large variation in the chip's delay and power [1] [10] [11]. The existing corner based models leads to over design of the chip. Some of the design margins can be recovered by using dynamic voltage and frequency management (DVFM) as demonstrated by the authors in [7]. Here the supply voltage to the chip is adjusted on a chip by chip basis, as well as for different performance targets over the course of the chip's operations. Thus the chip no longer runs at a given supply, but instead uses a dynamic range of supplies. Timing analysis for such applications provides another motivation for voltage aware gate delay models.

On the analysis side, Statistical Timing Analysis (SSTA) is being advocated for better insights into the chip's delay spread instead of worst case corner based analysis [2] [5]. Most of the existing literature on SSTA [2] [5] [10], only considers gate level variations, in effect having one random process parameter per gate. But the authors in [8] and [9] have shown that ignoring intra-gate variations can introduce substantial errors. Dealing with intra-gate variations involves considering the effect of process variations on each transistor in the gate. For a complex gate with N transistors and M parameters per transistor, we would need NM parameters as input to the delay model. The authors in [9] propose a sensitivity-based enhancement to the linear delay model.

Existing delay models for both static and statistical timing analysis are table based. Usually one delay table per process, supply and temperature corner is created for every gate. The table is indexed by the gate's load and the input edge slew and contains the gate's delay in the corresponding table entry . Similarly, statistical delay models, which are usually linear [2] or quadratic [5], use such tables to store the delay sensitivity coefficients. Extending this approach to cover multiple supply and temperature points at finer granularity for voltage and temperature analysis will be quite expensive in terms of characterization effort (for

---

*Corresponding Author, Email: bpdas@cedt.iisc.ernet.in

978-1-4244-3039-0/08 $25.00 © 2008 IEEE

e.g., [7] uses 5mV steps for dynamic supply control from 0.9V to 1.6V).

In this research, we investigate the feasibility of a comprehensive gate delay model which includes load, input slew, supply, temperature, global (inter-die) and local (intra-gate) process variations. However, the delay dependence on supply and temperature is quite non-linear as shown in Fig. 1a. Due to the contrary movements of the mobility and threshold voltage with temperature, the delay actually degrades for lower temperature at low supply voltages. Similarly delay is non-linear with some process parameters like the flat band voltage (Fig 1b). The figure shows the normalized rise delay (for cell 1xNOR2 at 65nm) at 0.9V and 25^0C versus the variations of the flat band voltage from the nominal (of 0), in units of 1σ . The $+ 3\sigma$ points are also highlighted. With increased variations in future technologies, the non-linearity within $+ 3\sigma$ region will increase.

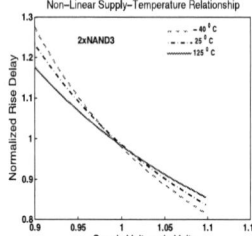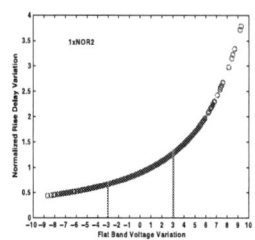

(a) Non-linear relation of delay on temperature and supply

(b) Non-linear dependence on flat band voltage

Figure 1.

Standard polynomial models don't work well for these requirements and we need to use a more sophisticated non-linear model. We have investigated neural networks as a modeling template in this work. In the next section, we provide a brief background of neural networks and explain our methodology of characterizing the library cells over large process and environmental ranges. We also present model complexity in this section. In Section 3 we provide some experimental results comparing the accuracy of neural network based models with SPICE. In Section 4 we motivate a few applications of such voltage and temperature models and finally conclude in Section 5.

2. Gate delay modeling using neural networks

Neural networks have been used for over a decade in pattern recognition applications [3] [4]. The ability of a neural network to model complex systems, which are dependent on a number of parameters, makes it a promising candidate for a comprehensive delay model. Thus any continuous function f(x), where x is an input vector, can be modeled very well by a neural network with a single hidden layer [3]. The exact equation for delay of a gate, though complicated to evaluate in closed form, is a smooth, continuous, but non linear, function of the underlying process parameters, load,

input slew, supply voltage and temperature and hence it is expected that the delay of a gate can be modeled well by a neural network with a single hidden layer.

Mathematically, the neural network can be described as a real valued multi-variable function of input variables X_1, ..., X_n, as :

$$\hat{F}(X) = \sum_{j=1}^{M} \phi_j \left(\sum_{k=1}^{N} X_k W_{jk} + b_j \right) a_j + b_0 \quad (1)$$

Where, W_{jk} is the weight between the kth input and j^{th} hidden node
a_j is the weight between the jth hidden node and the output layer
X is the vector of inputs $X_1, \ldots, X_n$ to the network
b_j is the bias to the jth hidden node
b_0 is the output layer bias
The activation function $\phi(t)$ used in our work is the tan sigmoid function given by

$$\phi(t) = \tanh(t) \quad (2)$$

The model is fit to sample delay data by adjusting the weights and bias values to give the minimum fitting error [3]. For timing analysis, one needs four delay parameters per gate: rise delay, fall delay, output rise slew and output fall slew. These are functions of the load, input slews, supply, temperature and process parameters. Hence we create four different neural networks, one for each of the above delay parameter.

$$D = F(G, L, Load, Slew, Supply, Temperature) \quad (3)$$

We consider only the global process variations (G) and local process variations (L) for the model. Global variations are also known as inter-die or Die to Die variations. Local variations are also called intra-gate or Mismatch variations. The other set of variations are the intra-die (or within die) variations and these are spatially correlated. These can be handled by partitioning the chip into grids and can be decor-related using the technique in [2]. After decorrelation, they become like inter-die parameters and can be handled within our model. The number of global process parameters for our process is 8 and there are 2 local parameters per transistor in each gate.

The procedure for gate delay model creation using neural network is described as follows. The first step is to generate samples from SPICE simulations of the gate to use for training and testing the gate's neural network model. For our examples, we have used the layout extracted library cells (using STAR-RCXT) for the SPICE simulations. All the sample data is normalized prior to use for model training or testing as follows. For any input parameter X, its normalized version X_{nom}, is scaled to lie between 0 and 1 as follows

$$X_{nom} = \frac{X - min(X)}{max(X) - min(X)} \quad (4)$$

The output of the network is scaled as follows:

$$D_{nom} = \frac{D}{max(D)} \quad (5)$$

Data from 1000 SPICE runs for the cell is used to train the network. Data from a further 900 SPICE runs is used to test the error between SPICE output and that predicted by the created model. If the error is unacceptable, then either the number of hidden nodes or the number of epochs or both is increased and retraining with the original 1000 SPICE samples is done [11]. In our experiments with 11 different standard library cells, we have found a successful model within 5 iterations, with the most complicated model having 26 hidden nodes (see Table I). We have used MATLAB's built in toolbox to train the network and no gate took more than 1000 seconds to train.

2.1 Sample selection for Modeling

Selecting the appropriate samples from the feasible space of input parameters for modeling is a crucial task. Unfortunately, finding an optimal training sample set is an open problem. Here, we have employed both uniform and random sampling. The range of input slew taken is 1ps-500ps. Parameters like load and input slew are sampled uniformly within their bounds to create about 36 sample points. These are combined with five sample points for supply (0.9, 0.95, 1.0, 1.05, 1.1) and four sample points for temperature (-40^0C, 0^0C, 100^0C, 125^0C), to result in about 720 combinations for the load, slew, voltage and temperature. Another 280 samples are taken from uniform distribution between their bounds to obtain a total of 1000 samples needed for training. All the global and local process parameters are samples from a zero mean unit variance Gaussian distribution and are used to generate samples for Monte Carlo SPICE simulations of the standard cell gate. An additional set of 900 SPICE simulations have been performed to test the models. This is used during the model training phase. These include 9 corners for three temperatures (-40^0C, 25^0C and 125^0C) and three supply voltages (0.9v, 1.0v and 1.1v) constituting nine combinations. In each combination of 100 samples, load and slew are taken from uniform distribution from their bounds and process parameters are varied from Gaussian distribution with mean 0 and standard deviation 1. Hence a total of 1900 SPICE characterizations need to be done per gate, to create a NN model for that gate.

Our standard cell library contains cells with different drive strengths e.g. 1xNAND2, 2xNAND2, etc. If the implementation of the larger drive strength is done using more segmented transistors, it increases the number of local parameters for the gate. In our case, we have eight (4*2) local process parameters in 1xNAND2 and sixteen (8*2) for 2xNAND2.

We reduce the number of local process parameters by about half by only considering the transistors that get sensi-

Table 1. Number of hidden node used by the model for different gates

Gate Types	RD Hidden node	FD Hidden node	RS Hidden node	FS Hidden node
1xNAND2	19	18	19	18
2xNAND2	19	18	19	24
2xNAND3	24	26	19	25
1xNAND4	19	21	19	19
1xNOR2	19	19	19	20
2xNOR2	18	18	17	20
1xNOR3	19	19	19	20
2xNOR3	21	18	19	20
1xNOR4	20	21	18	19
1xXOR	22	20	22	18
1xINV	20	13	20	13

tized for a particular transition. For example, in the case of an inverter, we use local process parameters of PMOS for rise delay/slew modeling, and local process parameters of NMOS for fall delay/slew modeling.

Table 1 shows the number of hidden nodes (a measure of the model complexity) for 11 gates from the std. cell library, for rise delay (RD), fall delay (FD), rise slew (RS) and fall slew (FS).

2.2 Model evaluation complexity

The gate delay model is meant to be used for static and statistical timing analysis, during which the most fundamental operation is model evaluation for a given set of input parameters. With N inputs and H hidden nodes, the number of double precision multiplications, additions and tanh evaluations is (N+1)*H, (N+1)*H and H respectively. A single evaluation of tanh function is equivalent to 39 floating point operations. The two most complex models from the table above are the 2xNAND3 and 1xXOR. A single evaluation of each is equivalent to 1612 and 1276 floating point multiplications respectively. The additional cost of two double precision additions and divisions (Equations 4,5) needed to be carried out on the input and output parameters for normalization and renormalization can be amortized over a large number of model evaluations. While the NN model evaluation is at least three orders of magnitude faster than SPICE, further reduction of this computational cost is a topic of ongoing research.

3 EXPERIMENTAL RESULTS

The quadratic polynomial model is created as follows:

$$D = \sum_{i=1}^{N} \sum_{j=1}^{N} c(i,j) X_{nom}^i X_{nom}^j + \sum_{i=1}^{N} d(i) X_{nom}^i + e \quad (6)$$

The $c(i, j), d(i)$ and e are constants determined using the MATLAB's lsqcurvefit function. We find that for a given output load, input slew, supply and temperature, the quadratic model gives good accuracy for global and local process variations. However, the model fails to work well when the load, slew, voltage and temperature are incorporated, due to the inherently large non linearity. Fig.2 shows the maximum % error in the model by NN and the quadratic polynomial [5] for different gates in the library when compared to SPICE when tested over data from an additional 8000 sample points which are different from the 1900 used for the model creation. As expected the polynomial model shows a large error of about 100% for this application. The maximum error with NN model is less than 11% over the 11 gates shown in the Fig.2. Characterization of the other gates in the library is on going and we expect the neural network model to work as well for those based on the theory [4] and our experience with the gates shown in the figures.

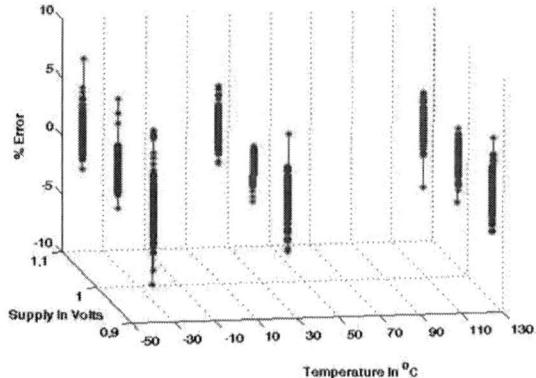

Figure 3. % error at nine corners of voltage and temperature for 2xNOR2 cell

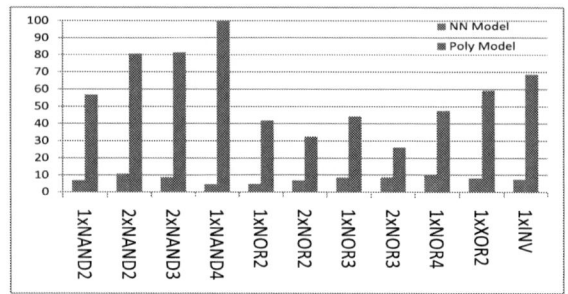

Figure 2. Comparison of maximum % error by NN and Polynomial model

Fig.3 shows the % error at extreme voltage and temperature corners for fall slew of 2xNOR2 gate. We have chosen nine corners with voltage at (0.9v, 1v and 1.1v) and temperature at $(-40^0C, 25^0C, 125^0C)$. At each corner point, 1000 samples are generated with different output load and input slew with uniform distributions between their bounds, global and local process parameters are taken from Gaussian distribution. The figure shows that the maximum error is within 8% . Since the model works with acceptable error bounds at all the nine corners, we expect it to work well at any intermediate point based on the interpolation properties of the network [4].

Fig.4 shows the voltage scalable PDFs generated by SPICE, NN model and quadratic model at 0.9v and 1.1v for fall slew of 2xNAND3 cell. The figure clearly depicts the closeness of SPICE and NN model. The quadratic model is unable to generate accurate PDF. As expected, the spread in PDF is more at 0.9v as compared to spread at 1.1v due to the larger impact of process variation at reduced gate overdrive (supply voltage minus threshold voltage). Fig.5 shows the temperature scalable PDFs generated by SPICE, NN and the quadratic polynomial model for fall slew of 2xNOR2

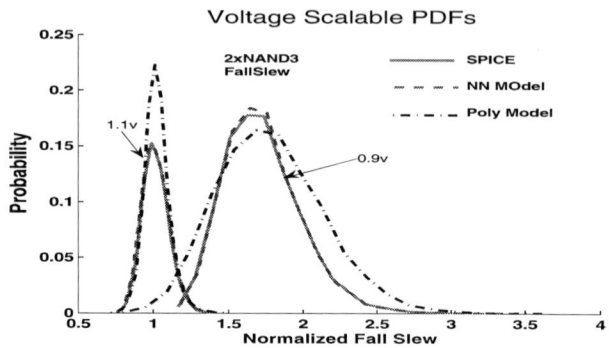

Figure 4. Comparison of voltage scalable PDFs

cell at supply voltage of 1.1v. The PDF from the NN model matches well with SPICE, unlike the quadratic model.

4. Applications

4.1 Voltage and temperature aware static timing analysis

The delay models can be used for voltage and temperature aware timing analysis [6] as well as analysis for dynamic voltage scaling applications. They can replace the corner based delay tables currently being used for static timing analysis. With such tables, interpolations need to be done to obtain values at intermediate parameter points. A small interpolation error necessitates more SPICE characterizations at finer input parameter granularities. Whereas the neural network naturally does good interpolation for arbitrary cell supply voltage and temperature. Such evaluations can done either real time during the analysis or alternatively, the model can be used to generate delay tables at arbitrary voltage and temperature points, thus saving costly SPICE characterization time.

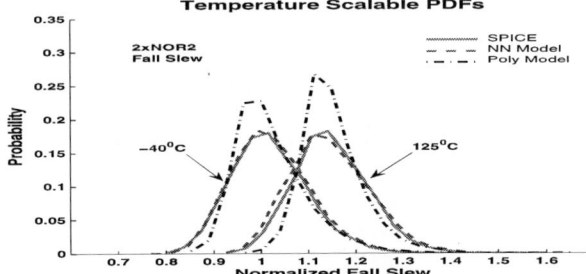

Figure 5. Comparison of temperature scalable PDFs

4.2 Voltage and temperature aware statistical timing analysis

We can extend the work in [6] to perform statistical timing analysis using the IR drop profile and/or the thermal profile in the chip. Fig.6 shows the Cumulative Distribution Function (CDF) of the rise delay for a path containing ten 1xNAND2 gates for three different voltages obtained via Monte Carlo evaluation of the NN model. The results match closely with the Monte Carlo evaluations done using SPICE. We can see that at the normalized rise delay of 2, one can get 100% yield at 1.1v, 90% yield at 1.05v and 40% yield at 1.0v. Thus one can do yield analysis for dynamic voltage scaling applications.

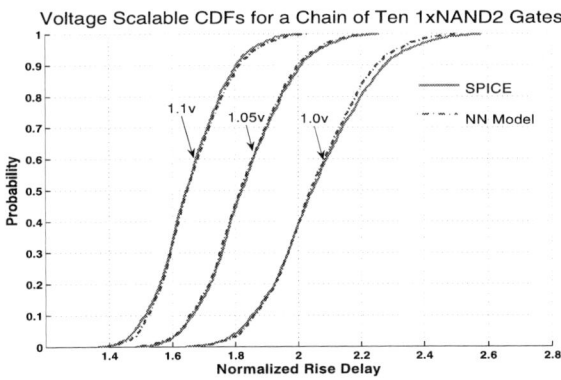

Figure 6. Voltage Scalable CDFs Comparison between NN and SPICE

One of the main arguments in [6] for voltage aware static timing analysis was that valuable excess margin put in by designers can be reduced. We will next show that doing voltage aware statistical analysis can help reduce excess margin further. We demonstrate this on a critical path of ISCAS c7552 consisting of 50 stages using the profile from [6] (Fig.7). In this profile, the voltage is assumed to be constant for five stages of gates and decreases in the steps of 0.05v till the middle of the critical path and increases in the steps of 0.05v till end of the path. We have done four dif-

ferent kinds of analyses on this critical path:

(1)Voltage is worst and process is worst (conventional worst case).

(2) Voltage is taken from profile and process is worst ([6]).

(3) Voltage is worst and process is statistical (conventional statistical).

(4) Voltage is taken from the profile and process is statistical (proposed).

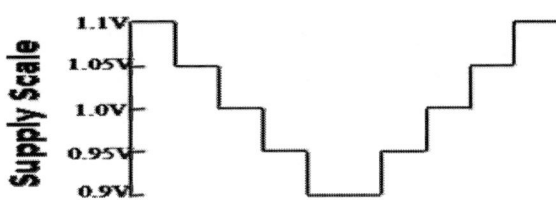

Figure 7. Voltage profile

Fig.8 shows normalized delay (in FO4 inverter delay) of the critical path of ISCAS c7552 benchmark for different types of analyses. Conventional worst case (case 1) has 47% excess margin compared to the proposed case 4. The approach in [6] (case 2) helps reduce the excess margin to 33% while the conventional statistical analysis (case 3) reduces the excess margin but still leaves a valuable 17% of margin on the table, compared to the more accurate voltage aware statistical analysis (case 4). We expect further savings in excess margins when accurate temperature profile is also considered. This clearly motivates the need for voltage and temperature scalable models to enable such statistical analysis

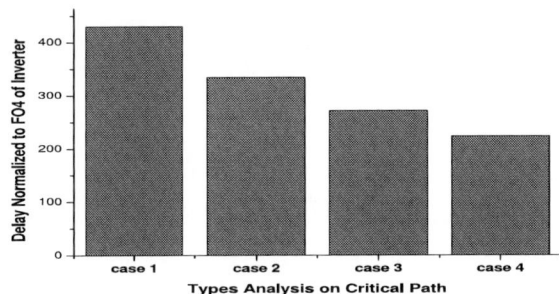

Figure 8. Different types of analysis on ISCAS c7552 critical path

A linear or quadratic sensitivity based delay model is quite accurate for a given supply, temperature, load and input slew [2] [5]. A table based approach for storing these sensitivities across a range of these conditions leads to large interpolations errors, unless fine gridding of these parameters are used. But this would in turn increase the SPICE characterization time to obtain these tables if one wanted to cover large ranges of voltage and temperature. Instead these sensitivities can be accurately generated by the neu-

978-1-4244-3039-0/08 $25.00 © 2008 IEEE

ral network model using just one model evaluation per linear sensitivity calculation. Fig.9 shows delay sensitivity to flat band voltage generated by NN model and SPICE across a range of voltages at three different temperatures for the 1xNOR2 cell. The sensitivity values match quite well with SPICE. We can also see that the delay is more sensitive to temperature at lower supply voltage due to reduced gate overdrive.

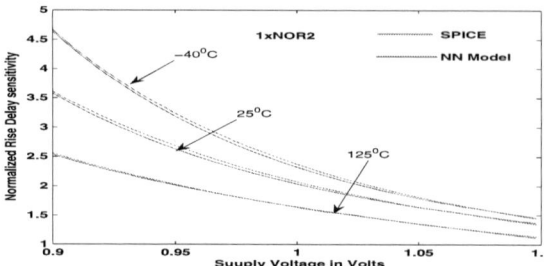

Figure 9. Delay Sensitivity comparison between NN and SPICE

Table 2 compares the number of SPICE runs needed to characterize the eleven cells used in this study for standard delay library format (.lib) and the NN model. NN model requires 1900 SPICE runs independent of types of gates. But .lib characterization needs different SPICE runs for different gates because of the intra-gate process parameters. We have taken a table of 6x6 for input slew and output load. One nominal simulation is needed at each slew and load point. There are 8 global process parameters and 2 local process parameters per transistor. There will be 9 corners due to three supply voltage points (0.9v, 1.0v, 1.1v) and three temperature points (-40^0C, 25^0C, 125^0C). In case of inverter, the number of SPICE runs required is (1+8+4)*6*6*9=4212. There is a minimum saving of 2.21x SPICE runs in case of inverter and maximum saving of 5.62x in case of 2xNAND3 cell.

5 Conclusion

A single delay model which is a function of supply voltage, temperature, inter-die and intra-gate process parameters will enable voltage and temperature aware static and statistical timing analysis. A neural network is a good model template for delay modeling. An initial experiments on a subset of 11 gates from a 65nm standard cell library showing maximum error of less than 11% compared to SPICE, over a large voltage, temperature, load and slew range. A quadratic polynomial model shows errors of up to 100% for the same range.

The neural network models can be created with less number of SPICE runs than that required for table lookup based models. Hence these can be used to efficiently generate table based models for delays as well sensitivities for the

Table 2. The SPICE run required for .lib creation vs NN model creation

Gate Types	SPICE Run for .lib creation	SPICE run for NN model creation	Saving in SPICE Run
Inverter	4212	1900	2.21
1xNAND2	5508	1900	2.89
2xNAND2	8100	1900	4.26
2xNAND3	10692	1900	5.62
1xNOR2	5508	1900	2.89
2xNOR2	8100	1900	4.26
1xNOR3	6804	1900	3.58
2xNOR3	10692	1900	5.62
1xNOR4	8100	1900	4.26
1xNAND4	8100	1900	4.26
1xXOR2	8100	1900	4.26

linear and quadratic statistical delay models for any supply, temperature, load and slew conditions. This will enable efficient voltage and temperature aware statistical timing analysis as well as yield analysis under dynamic voltage scaling framework.

References

[1] S. Borkar et al. Parameter variations and impact on circuits and microarchitecture. In *DAC*, pages 338–342, 2003.
[2] H. Chang and S. Sapatnekar. Statistical timing analysis under spatial correlations. *IEEE Transcation CAD*, 24(9):1467–1482, Sept. 2005.
[3] S. Haykin. *Neural Network A comprehensive foundation.* PHI,, New Delhi, India, 1999.
[4] K. Hornik. Approximation capabilities of multilayer feedforward networks. *Neural Networks*, (4):251257, 1991.
[5] V. Khandalwal and A. Srivastava. A general framework for accurate statistical timing analysis considering correlation. In *DAC*, pages 89–94, 2005.
[6] B. Lasbouygues, R. Wilson, N. Azemard, and P. Maurine. Temperature and voltage aware timing analysis. *IEEE Transcation CAD*, 26(4):801–815, April 2007.
[7] H. Mahmoodi, S. Mukhopadhyay, and K. Roy. Dynamic voltage and frequency management for a low-power embedded microprocessor. *IEEE JSSC*, 40(1):28–35, Jan 2005.
[8] H. Mahmoodi, S. Mukhopadhyay, and K. Roy. Estimation of delay variation due to random-dopant fluctuations in nanoscale cmos circuits. *IEEE JSSC*, 40(9):1787–1796, Sept. 2005.
[9] K. Okada, K.Yamaoka, and H.Onodera. A statistical gate-delay model considering intra-gate variability. In *ICCAD*, pages 908–913, 2003.
[10] A. Srivastava, D. Sylvester, and D. Blaauw. *Statistical analysis and optimization for VLSI: Timing and Power.* Springer,, USA, 2005.
[11] H. Su et al. Full chip leakage estimation considering power supply and temperature variations. In *ISLPED*, 2003.

SESSION C5:
Security

21st International Conference on VLSI Design

Single Chip Encryptor/Decryptor Core Implementation of AES Algorithm

Monjur Alam   Santosh Ghosh   Dipanwita RoyChowdhury   Indranil Sengupta

Department of Computer Science and Engineering

Indian Institute of Technology, Kharagpur

{monjur, santosh, drc, isg}@cse.iitkgp.ernet.in

Abstract—This paper presents a single chip encryptor/decryptor core implementation of Advanced Encryption Standard (AES-Rijndael) cryptosystem. The suggested architecture is capable of handling all possible combinations of standard bit lengths (128,192,256) of data and key. The fully rolled inner-pipelined architecture ensures lesser hardware complexity. The architecture does reutilize precomputed blocks, in the sense that the same hardware is shared during encryption and decryption as much as possible. The design has been implemented on Xilinx XCVe1000-8bg560 device. The performance of the architecture has been compared with existing results in the literature and has been found to be the most efficient (throughput/area) implementation of the AES algorithm.

Index Terms—**Reconfigurable Architecture, AES, Rijndael, S-box, Composite fields.**

I. INTRODUCTION

The Rijndael[1] block cipher [4] was chosen as the Advanced Encryption Standard (AES) by the National Institute for Standard and Technology (NIST) in 2000. Compared to their relatively slower software [2] counterparts, VLSI (FPGA and ASIC) implementations of AES-Rijndael have become attractive. Although the popularly used AES-Rijndael is of 128-bit, originally the algorithm was proposed for 128, 192 and 256-bit. To keep up the market demand it is absolutely incumbent that the proposed AES architecture should work under any possible (128, 192 and 256-bit) key or data bit frame sizes, so that they can be used in multifarious fields (Viz: E-banking, Identity card, SIM card). Thus the issue of reconfigurability is of immense practical importance over the last few years.

Over the recent years many FPGA [9], [10], [14], [16], [17], [21] and ASIC [3], [5], [6], [7], [8], [11], [18] implementations for Rijndael has been reported. Most of them have used look-up tables to implement S-Boxes. However, none of them are able to offer the reconfigurability feature. The advent of composite field $GF((2^4)^2)$ arithmetic in S-box operation was first noted in the works of Rijmen [4] and Rudra *et al.* [6]. Among the designers who tried to produce an area optimized implementation using composite field arithmetic, the works [3], [7], [8] are of importance. Feldhofer *et al.* [7] implemented 128-bit AES on a grain of sand. It is a 8-bit

architecture which exploits composite field $GF((2^4)^2)$ for S-box optimization. To our knowledge, it is the most compact AES implementation so far. Single chip implementation of Rijndael with hardware sharing was introduced by Zhang et al. [19]. It did rearrange the *MixColumn* operation and shared maximum hardware. All the above mentioned approaches are capable of handling fixed bit length (128-bit) of data and key.

Rijndael encryptor capable of handling all possible data and key lengths was first introduced by Verbauwhede *et al.* [18]. Although it supports all possible combination of data and key lengths (128, 192 and 256-bit), it does not consider the design of decryption unit. A highly regular approach was presented by Mangard *et al.* [13]. This 32-bit architecture performs encryption and decryption for various key sizes, but fixed block size (128-bit). The most area optimized reconfigurable AES-Rijndael implementation to date was demonstrated by Monjur *et al.* [3]. This work developed a FSMD model based controller which is ideal for such iterative implementation of AES. But, it has also not explored the design at decryption unit. In a follow-up article the authors came up with a latency optimized version of their architecture [21]. The architecture can process data during both edges of clock transition. At each clock cycle, the key scheduler generates a round key, which ensures enhanced performance. Also the fully rolled encryption unit can complete a round transformation in a single clock cycle with more combinational path delay.

The reconfigurable AES-Rijndael architecture proposed in this paper is the enhanced version of the works [3] and [21]. The architecture can process all 9 possible combinations of key and data lengths. All computations are done in composite field $GF(((2^2)^2)^2)$ rather than $GF(2^8)$. The effective use of composite field $GF(((2^2)^2)^2)$ helps to reduce the hardware complexity. It reutilizes precomputed blocks and the same hardware is shared during encryption and decryption as much as possible. Due to its rolling technique the design supports all standard modes of encryption operation including CBC, OFB etc.

The remainder of this paper is organized as follows. In Section II, the AES algorithm is briefly described. Section III discusses our proposed AES architecture with several sub-unit optimizations. Section IV presents the implementation results and compares our results with existing ones, followed by concluding remarks in section V.

---

[1]The terms AES, Rijndael and AES-Rijndael are used in the same meaning through out the paper

II. AES-Rijndael Algorithm

The AES is a symmetric block cipher [1]. It operates on 128-bit blocks of data. The algorithm can encrypt and decrypt blocks using secret keys of size either 128-bit, 192-bit, or 256-bit. The Rijndael block cipher [4] was designed by John Daemen and Vincent Rijmen. It operates on various blocks (128, 192, or 256-bit) with the combination of different key and data length (128, 192 or 256-bit).

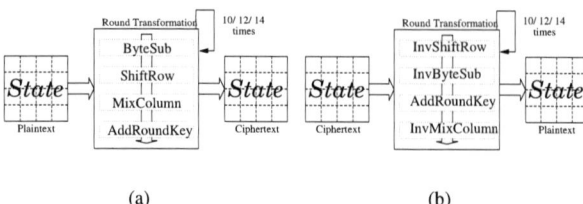

(a)   (b)

Fig. 1.   Architecture of (a) Encryption and (b) Decryption datapath

Figure 1 outlines the basic structure (128-bit) of the algorithm. The round transformation consists of four different steps: *ByteSub, ShiftRow, MixColumn and AddRoundKey*. They are performed in this order with the exception of the final round which is slightly different. All transformations are based on byte-oriented operations. *AddRoundkey* consists of bitwise XOR operations. The transformations operate on the intermediate result, which is called the *State*. The initial *State* is the input plaintext and the final *State*, after the round transformations, is the output ciphertext. The state is organized as a 4×4 (for 128-bit block) or 4×8 (for 256-bit block) matrix of bytes. The round transformation scrambles the bytes of the *State* either individually, row-wise, or column-wise by applying the functions *ByteSub, ShiftRow, MixColumn and AddRoundKey* sequentially.

The *ByteSub*[2] transformation is a non-linear byte substitution, also called S-box. The S-box is invertible and consists of the following two operations:

- Inversion in the $GF(2^8)$ field, modulo the irreducible polynomial m(x) = $x^8 + x^4 + x^3 + x + 1$.
- Affine transformation defined as: Y = AX^{-1} + B, where A is an 8×8 fixed matrix and B is an 8×1 vector-matrix. The matrix A is defined in [4].

The inverse ByteSub called *InvByteSub* (InvS-box), operates upon the bytes in the reverse order, that is, first an inverse affine block followed by the inverse operation, and the field polynomial is same as m(x).

In *ShiftRow*, the rows of the *State* are cyclically shifted over different offsets (Table I [4]); row 0 is not shifted. The inverse *ShiftRow* performs the circular shift in the opposite direction. *Shiftrow* implementations do not require FPGA resources as they can be implemented by rewiring.

---

[2]The terms ByteSub(s), SubBytes(s), S-box(es) carry same meaning throughout this paper

---

TABLE I
SHIFT OFFSETS FOR DIFFERENT BLOCKS

DataBlock	Row1	Row2	Row3
128	1-Byte	2-Byte	3-Byte
192	1-Byte	2-Byte	3-Byte
256	1-Byte	3-Byte	4-Byte

The *MixColumn* transformation operates on each column of *State* (X) individually. If $F_m(X)$ is defined as a function of the transformation of *MixColumn* that operates on *State* X, then:

$$F_m(X_{ij}) = (m_1 X_{1j} + m_2 X_{2j} + m_3 X_{3j} + m_4 X_{4j}) \text{ modulo}$$

$x^4 + 1$, where (m_1, m_2, m_3, m_4) is a permutation of $(\{01\}_{16}, \{01\}_{16}, \{02\}_{16}, \{03\}_{16})$ and X_{ij} is the $(i,j)^{th}$ element of the *State* matrix. In matrix form the *MixColumn* transformation can be expressed as

$$\begin{bmatrix} X'_{0,j} \\ X'_{1,j} \\ X'_{2,j} \\ X'_{3,j} \end{bmatrix} = \begin{bmatrix} \{02\}_{16} & \{03\}_{16} & \{01\}_{16} & \{01\}_{16} \\ \{01\}_{16} & \{02\}_{16} & \{03\}_{16} & \{01\}_{16} \\ \{01\}_{16} & \{01\}_{16} & \{02\}_{16} & \{03\}_{16} \\ \{03\}_{16} & \{01\}_{16} & \{01\}_{16} & \{02\}_{16} \end{bmatrix} \begin{bmatrix} X_{0,j} \\ X_{1,j} \\ X_{2,j} \\ X_{3,j} \end{bmatrix}$$

(1)

In case of inverse *MixColumn* (*InvMixColumn*), the same polynomial is used. If $F_m^{-1}(X)$ is defined as a function of the transformation of *InvMixColumn* that operates on *State* X then:

$$F_m^{-1}(X_{ij}) = (m_1^{-1} X_{1j} + m_2^{-1} X_{2j} + m_3^{-1} X_{3j} + m_4^{-1} X_{4j}) \text{ modulo}$$

$x^4 + 1$, where $(m_1^{-1}, m_2^{-1}, m_3^{-1}, m_4^{-1})$ is a permutation of $(\{0E\}_{16}, \{0B\}_{16}, \{0D\}_{16}, \{09\}_{16})$ and X_{ij} is the $(i,j)^{th}$ element of the *State* matrix.

In matrix form the *InvMixColumn* transformation can be expressed as

$$\begin{bmatrix} X''_{0,j} \\ X''_{1,j} \\ X''_{2,j} \\ X''_{3,j} \end{bmatrix} = \begin{bmatrix} \{0E\}_{16} & \{0B\}_{16} & \{0D\}_{16} & \{09\}_{16} \\ \{09\}_{16} & \{0E\}_{16} & \{0B\}_{16} & \{0D\}_{16} \\ \{0D\}_{16} & \{09\}_{16} & \{0E\}_{16} & \{0B\}_{16} \\ \{0B\}_{16} & \{0D\}_{16} & \{09\}_{16} & \{0E\}_{16} \end{bmatrix} \begin{bmatrix} X_{0,j} \\ X_{1,j} \\ X_{2,j} \\ X_{3,j} \end{bmatrix}$$

(2)

III. PROPOSED AES ARCHITECTURE

During encryption, the data are organized conceptually in an 4×8 matrix of bytes (Figure 2(a)). This organization is used for data block sizes of 256 bits. For smaller data block sizes (128 or 192 bits), the leftmost columns of the matrix are unused. The encryption data path processes a full 32-byte block in parallel. A complete round transformation executes in four clock cycles.

The decryption structure can be derived by inverting the encryption structure directly. However, the sequence of the sub-units operations will be different from that in encryption (see Fig. 1). This feature prohibits resource sharing between encryptor and decryptor. To make maximum resource sharing we need some technique and some sort of re-arrangement of sub-units operations.

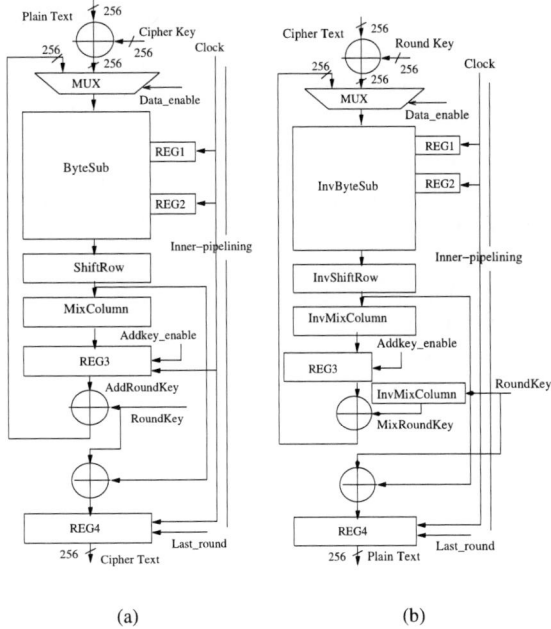

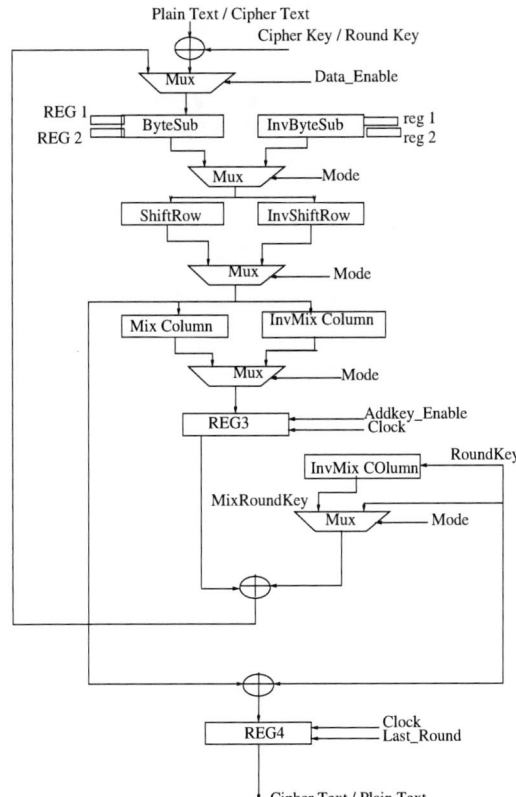

Fig. 2. (a) Architecture of Encryption Datapath, (b) Architecture of Decryption Datapath

It can be observed from the operations involved in the decryption transformations that, the inverse ShiftRow (InvShiftRow) and the InvByteSub can be exchanged without affecting the decryption process. The InvMixcolumn can be moved before the *AddRoundKey*, provided that the InvMix-Column is applied to the roundkeys before it is added. Taking these into consideration, an equivalent decryption structure can be used (Figure 2(b)). In this figure, the *MixRoundkey* is the modified roundkey resulting from applying InvMixColumn to the *AddRoundKey*. The equivalent decryption structure has the same sequence of transformations as that in the encryption structure, and thus, resource sharing between encryptor and decryptor are possible. Figure 3 shows the block diagram of Encryption/Decryption core into a single chip. The operation either of encryption or decryption is selected by user defined control signal called *Mode*. In the Figure 3, *REG1 and REG2* are used for sub-pipelined S-box or InvS-box implementations and is discussed in Section III-A. Hardware sharing of each sub-unit for Encryption/Decryption core implementation is shown in the respective sections.

A. S-box/InvS-box Design

The major computation inside the S-box is to find out the multiplicative inverse of an element in the finite field $GF(2^8)$. The Inverse S-box (InvS-box) operates upon the bytes in the reverse order, that is, first an inverse affine block followed by the inverse operation . For encryptor/decryptor core design, the S-box or Inverse S-box can share multiplicative inversion unit. In our design, the *Finding Inverse* unit is same as described in [21]. S-box/InvS-box can be implemented according to

Fig. 3. Single Chip Encryptor/Decryptor Core Design

the schematic illustrated in Figure 4. S-box or Inverse S-box operation is selected by a user specified control signal called *Mode*.

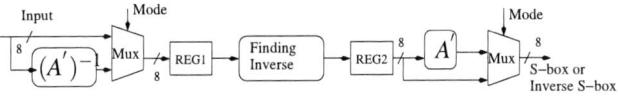

Fig. 4. Block diagram of 2-stage sub-pipelined S-box. The block "Finding Inverse" is similar as described in [21]

A feature that can be exploited to gain higher throughput is pipelining. Pipelining is a technique which subdivides the critical path by insertion of storing elements (flipflops). Subdividing the S-box functionality into a number of stages is easy to accomplish since flip flops can be inserted nearly anywhere when S-boxes are implemented with combinational logic. Pipelining introduces latency but the additional clock cycles are made up by an increased clock frequency. To reduce the critical path delay, let us insert buffers at appropriate points of S-box such that the delay is equally distributed. We have implemented 2-stage pipelined S-box (Figure 4).

B. Implementation of MixColumn/InvMixColumn Transformation

Various architectures have been proposed for the implementation of the *MixColumn/InvMixColumn* transformation [6],

695

978-1-4244-3039-0/08 $25.00 © 2008 IEEE

[19]. The following observations are useful in the implementation [6]:

- If $x \in GF(((2^2)^2)^2)$ then $F_m(01) \times x = x$ as the identity element is mapped to the identity element in a homomorphism.
- $F_m(03) = F_m(02) + F_m(01)$.

By applying these technique, equation 1 can be rewritten as

$$\begin{cases} X'_{0,j} = \{02\}_{16}(X_{0,j}+X_{1,j}) + (X_{2,j}+X_{3,j}) + X_{1,j} \\ X'_{1,j} = \{02\}_{16}(X_{1,j}+X_{2,j}) + (X_{3,j}+X_{0,j}) + X_{2,j} \\ X'_{2,j} = \{02\}_{16}(X_{2,j}+X_{3,j}) + (X_{0,j}+X_{1,j}) + X_{3,j} \\ X'_{3,j} = \{02\}_{16}(X_{3,j}+X_{0,j}) + (X_{1,j}+X_{2,j}) + X_{0,j} \end{cases} \quad (3)$$

Similarly, in the *InvMixColumn* transformation, equation 2 can be rewritten as

$$\begin{cases} X''_{0,j} = (\{02\}_{16}(X_{0,j}+X_{1,j}) + (X_{2,j}+X_{3,j}) + X_{1,j}) \\ \qquad + (\{02\}_{16}(\{04\}_{16}(X_{0,j}+X_{2,j}) \\ \qquad + \{04\}_{16}(X_{1,j}+X_{3,j})) + \{04\}_{16}(X_{0,j}+X_{2,j})) \\ X''_{1,j} = \{02\}_{16}(X_{1,j}+X_{2,j}) + (X_{3,j}+X_{0,j}) + X_{2,j} \\ \qquad + (\{02\}_{16}(\{04\}_{16}(X_{0,j}+X_{2,j}) \\ \qquad + \{04\}_{16}(X_{1,j}+X_{3,j})) + \{04\}_{16}(X_{1,j}+X_{3,j})) \\ X''_{2,j} = \{02\}_{16}(X_{2,j}+X_{3,j}) + (X_{0,j}+X_{1,j}) + X_{3,j} \\ \qquad + (\{02\}_{16}(\{04\}_{16}(X_{0,j}+X_{2,j}) \\ \qquad + \{04\}_{16}(X_{1,j}+X_{3,j})) + \{04\}_{16}(X_{0,j}+X_{2,j})) \\ X''_{3,j} = \{02\}_{16}(X_{3,j}+X_{0,j}) + (X_{1,j}+X_{2,j}) + X_{0,j} \\ \qquad + (\{02\}_{16}(\{04\}_{16}(X_{0,j}+X_{2,j}) \\ \qquad + \{04\}_{16}(X_{1,j}+X_{3,j})) + \{04\}_{16}(X_{1,j}+X_{3,j})) \end{cases} \quad (4)$$

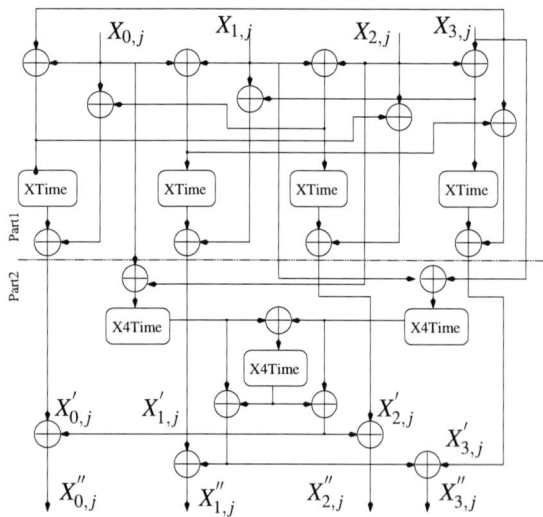

Fig. 5.   Block diagram of the MixColumn/InvMixColumn

Using substructure sharing, equations 3 and 4 can be implemented by the architecture illustrated in Fig. 5. The function of block *XTime* is to compute constant multiplication by $\{02\}_{16}$. The block *X4Time* computes the constant multiplication of $\{04\}_{16}$ and can be implemented by two serially concatenated

XTime blocks. For encryption operation only *Part1* of the Fig. 5 is required and combination of *Part1* and *part2* are used for decryption operation.

C. Key Scheduler

The *RoundKeys* are derived from the Cipher Key by means of the key schedule. This consists of two components: the Key Expansion and the Round Key Selection. The basic principle is the following:

- The total number of Round Key bits is equal to the block length multiplied by the number of rounds plus 1. For example, for a block length of 128 bits and 10 rounds, 1408 Round Key bits are needed.
- The Cipher Key is expanded into an Expanded Key.
- *RoundKeys* are taken from this Expanded Key in the following way: the first Round Key consists of the first N_b words where N_b denotes the length of the data block divided by 32, the second one of the following N_b words, and so on.

Algorithm 1 KeyExpansion(byte Key[$4 * N_k$] word W[$N_b * (N_r + 1)$])

---

for $i = 0$ to $N_k - 1$ **do**
    W[i] = (key[4*i], key[4*i+1], key[4*i+2], key[4*i+3])
end for
for $i = N_k$ to $N_b * (N_r + 1)$ **do**
    temp = W[i - 1]
    if (i mod N_k is 0) **then**
        temp = SubByte(RotByte(temp)) $\oplus$ Rcon[i/N_k]
    end if
    if (i mod N_k is 4) **then**
        temp = SubByte(temp)
    end if
    W[i] = W[i - N_k] $\oplus$ temp
end for

The *Key Expansion* algorithm is given by Algorithm 1. The Expanded Key is a linear array of 4-byte words and is denoted by $W[N_b * (N_r + 1)]$ where N_r is the number of round. The values of N_r are determined from the Table II [4]. The first N_k words contain the Cipher Key where N_k denotes the length of the key divided by 32. All other words are defined recursively in terms of words with smaller indices. In this description, SubByte(W) is a function that returns a 4-byte word in which each byte is the result of applying the Rijndael S-box to the byte at the corresponding position in the input word. The function RotByte(W) returns a word in which the bytes are a cyclic permutation of those in its input such that the input word (a,b,c,d) produces the output word (b,c,d,a).

The KeyExpan unit is shown in Figure 7. W's are the 32-bit shift registers. R is a 256-bit register to store initial key. The architecture is almost same as the KeyExpan unit described in [3]. P/P^{-1} block performs the operation of S-box/InvS-box with maximum hardware sharing as discussed in Section III-A.

696

978-1-4244-3039-0/08 $25.00 © 2008 IEEE

TABLE II
NUMBERS OF ROUNDS (N_r) AS A FUNCTION OF THE BLOCK AND KEY

N_r	$N_b=4$	$N_b=6$	$N_b=8$
$N_k=4$	10	12	14
$N_k=6$	12	12	14
$N_k=8$	14	14	14

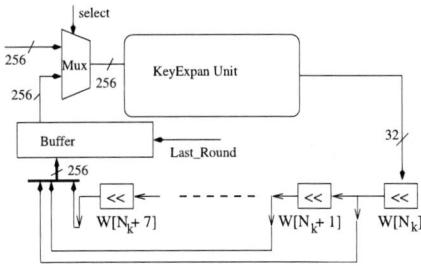

Fig. 6. The KeyScheduler

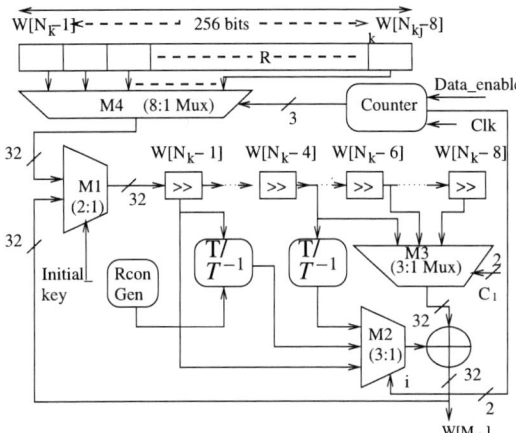

Fig. 7. The KeyExpan Unit

For decryption operation the *KeyScheduler* should generate last *RoundKey* first using initial cipher key. This last *RoundKey* key may be the round keys of either 10, 12 or 14 rounds. For example, in case of 128-bit decryption operation, first round key should be the keys of W[43], W[42], W[41] and W[40]. And these keys are stored to generate all round keys, i.e., $W[39], \cdots, W[0]$. In our design we store the round keys $W[43], \cdots, W[40]$ (in case of 128-bit block) using the *Buffer* shown in the Figure 6. This *Buffer* changes its values when Last_round signal is reset to value 1. The *Mux* (Figure 6) selects either 256-bit initial keys or last round keys depending upon the control signal *select*. The signal *select* is set to 1 when *Last_round* is set to 1 for decryption mode. One point worth mentioning is that the status (either set or reset) of the *Last_round* signal depends on another signal *Mode* which is user specified. All the other control signals (like *Last_round, Data_enable, Key_enable*, etc.) are generated by a control unit. The P or P^{-1} are same like S-box or InvS-box.

The architecture takes N_b clock cycles to generate single round key. A maximum of $8 \times 14 + 8 = 120$ clock cycles are required to generate complete round key for 256 bits data and

key. In case of decryption, it takes extra $N_b \times N_r$ clock cycles for the operation on the initial data input. From the operation on the next data input, it requires the same clock cycles as for encryption.

IV. EXPERIMENTAL RESULTS

The proposed design has been implemented on Xilinx XCVe10002-8bg560 device and simulated by ModelSim8.1i. The performances (throughput and frequency) are shown at Table III. Throughput ($\square$) is calculated as:
$\square = (\square \times f)/(\square)$, where $\square$, f and $\square$ stand for block length, clock frequency and number of clock cycles respectively. Latency is defined as the time required to encrypt a single block of data. It can be measured in terms of total clock cycles. In our approach $N_b \times N_r$ clock cycles are needed to generate a block of cipher text.
$\square = (N_b \times 32 \times f)/(N_b \times N_r) = (32 \times f)/N_r$. For example (in FPGA), $\square = 32 \times 135/14 = 432$ Mb/s for 256 bits block length, as the clock frequency is 135 MHz and N_r is 14.

TABLE III
THROUGHPUT IN FPGA (XCV1000E-8)

Clock Frequency = 135 MHz			
Results shown using 2-Stage Sub-Pipelined S-box are shown			
Throughput (Mb/s)	$N_b=4$	$N_b=6$	$N_b=8$
$N_k=4$	432	360	308.6
$N_k=6$	360	360	308.6
$N_k=8$	308.6	308.6	308.6

A. Comparison with Other Designs

The simulated or synthesized results for a particular design may vary if targeted devices or technology change. Different modes of operation (CBC, OFB, ECB, etc.) or different choices of operation (like encryption/decryption, different block or key length) may lead to have different throughput and hardware overheads of a particular design in a particular device. So it is quite difficult to have a fair comparison of the existing designs with our suggested design. Still, we have explicitly mentioned the targeted devices, different modes of operation and different choices of operation of the existing designs along with our suggested design to make relatively fair comparison.

In Table IV we illustrate the comparative analysis of different existing AES architectures along with our suggested one. The symbol (∗) means Mb/s per Slice. The implementation by Zhang *at el.* [19] (Table IV) seems to be the better one in terms of throughput/slice. But it does not support the all combinations of key and block lengths. As it is pipelined architecture, to support 256-bit AES it should need 40 percent more hardware (as there is 14 round for 256-bit block length). It means throughput/area ratio becomes 0.57 ($0.95 \times 100/140$) keeping throughput same.

V. CONCLUSIONS

In this paper we have presented a single chip encryptor/decryptor of reconfigurable AES algorithm. The design

TABLE IV
PERFORMANCES OF COMPARED CORES IN FPGA

Design	Device	Area		□ (Mb/s)	□ /Slice (*)	E/D	Data Length (128/192/ 256-bit)	Key Length (128/192/ 256-bit)	Mode
		Slice	RAM						
[9]	Spartan II	606	3	166	0.27	E	128	128	C,O
[10]	Vertex-E	13416	0	3136	0.24	E	128	128	E
[16]	Spartan II	264	2	20.2	0.1	E	128	128	C,O,Cf
[12]	Vertex-E	4389	4	1019	0.27	E	all	all	C,O
[19]	XCV1000e-8	11022	0	21556	0.95	all	128	128	C,O
[15]	XCV2000e	20300	100	6810	0.79	all	128	128	C,O
[20]	XCV4000	1780	0	1000	0.56	all	128	128	C,O
[17]	XC3S50-4	547	0	208	0.38	all	128	128	C,O
[3]	XCV1000	520	0	120.74	0.23	E	all	all	C,O
[21]	SXC3S5000	1760	0	600	0.34	E	all	all	C,O
Our (with sub-pipelined S-box)									
0-Stage	XCV1000e-8	480	0	320	0.67	E	all	all	C,O,
2-Stage	XCV1000e-8	510	0	432	0.84	E	all	all	C,O
Encryptor/ Decryptor	XCV1000e-8	622	0	432	0.70	all	all	all	C,O

□ stands for throughput. E, C, O, Cf sequentially stand for electronic code book, cipher block chaining, output feedback and cipher feedback

exploits the theory of composite field arithmetic $GF(((2^2)^2)^2)$ to compute all nonlinear operations of S-boxes and thus optimizes the hardware complexity. It does reutilize precomputed blocks. The same hardware is shared in encryption and decryption as much as possible. After exhaustive survey in literature we have seen that this is the first work of single chip encryptor/decryptor core implementation of AES-Rijndael which can work under any possible (128, 192 and 256-bit) key or data bit frames.

Acknowledgement

I would like to give special thanks to Dr. Debdeep Mukhopadhyay, Assistant Professor at IIT Madras, for constant source of inspiration and suggestions behind this work.

REFERENCES

[1] National Institute of Standards and Technology (NIST). FIPS-197: Advanced Encryption Standard, November 2001. http://www.itl.nist.gov/fipspubs/

[2] Guido Bertoni et al, "Efficient Software Implementation of AES on 32-bits Platforms", in *Cryptographic Hardware and Embedded System (CHES 2002)*, Revised Papers, LNCS Vol. 2523, pp.159-171.

[3] M. Alam, S. Ray, D. Mukhopadhyay, S. Ghosh, D. Roychowdhury and I. Sengupta: "An Area Optimized Reconfigurable Encryptor for AES-Rijndael", in the proceeding of *Design, Automation and Test in Europe (DATE 2007)*, pp.1116-1121, April 16-21, Nice, France.

[4] J.Daemen and V.Rijmen, "The Design of Rijndael: AES-The Advanced Encryption Standard", Springer-Verlag, Berlin, New York, 2002.

[5] D. Mukhopadhyay, D. RoyChowdhury, "An Efficient End to End Design of Rijndael Cryptosystem in 0.18μ CMOS", *The 18th International Conference on VLSI Design and The 4th International Conference on Embedded Systems (VLSID 2005)*, pp.405-410, Kolkata.

[6] A. Rudra, P. K. Dubey, C. S. Jutla, V. Kumar, J. R. Rao and P. Rohatgi, "fficient Rijndael Encryption Implementation with Composite Field Arithmetic", in *Cryptographic Hardware and Embedded System (CHES 2001)*, LNCS Vol. 2162, pp.171-184.

[7] M. Feldhofer, J. Wolkerstorfer, J. Rijmen, "AES implementation on a grain of sand," in *IEE Procidings in Information Security*, July, 2005.

[8] A. Satoh, S. Morioka, K. Takona and S. Munetoh, "A Compact Rijndael Hardware Architecture with S-Box optimization", in *Advances in Cryptography-ASIACRYPT 2001*, LNCS Vol. 2248, pp.239-254.

[9] P. Chodowiec and K. Gaj, "Very Compact FPGA Implementation of the AES Algorithm", in *Cryptographic Hardware and Embeded System (CHES 2003)*, LNCS Vol. 2779, pp.319-333.

[10] N. Saqib, F. Henriquez, A. Perez, "Two Approaches for a Single-Chip FPGA Implementation of an Encryptor/Decryptor AES Core," in *Cryptographic Hardware and Embedded System (CHES 2005)*, LNCS Vol. 2779, pp.319-333.

[11] R. Sever, A. Neslin, Y. Tekmen, M. Asker, "A High Speed ASIC Implementation of the Rijndael Algorithm," in *International Symphosium of Circuit and System (ISCAS-2004)*, IEEE Vol.2, pp.541-4.

[12] R. Sever, A. Neslin, Y. Tekmen, M. Asker, B. Okcan, "A High Speed FPGA Implementation of the Rijndael Algorithm" in *Proc. EUROMICRO Systems on Digital System Design (DSD 2004)*, IEEE Vol.2, pp.541-554.

[13] S. Mangard, M. Aigner and S. Dominikus, "A highly regular and scalable AES hardware architecture", *IEEE Trans. Comput., 2003*, Vol. 52 (4), pp.483-491

[14] N. Praustaller, S. Mangard, S, Dominikus, J. Wolkerstorfer, "Efficient AES implementation on ASIC's and FPGA's", in *Fourth Workshop on the Advanced Encryption Standard (AES 2004)*, LNCS Vol.3373, pp.98-112.

[15] G. Saggese, A. Mazzeo, N. Mazocca, A. Strollo, "An FPGA based performance analysis of the unrolling, tiling and pipelining of the AES algorithm" in *Field Programmable Logic (FPL 2003)*, pp.292-302, Portugal, 2003.

[16] T. Good, M. Benaissa, "AES on FPGA from the Fastest to the Smallest", in *Cryptographic Hardware and Embedded System (CHES 2005)*, LNCS Vol. 3659, pp.427-440, Springer 2005.

[17] N. Pramstaller, J. Wolkerstorfer, "A Universal and Efficient AES Co-processor for Field Prograble Logic Arrays," in Field Prograble Logic (FPL 2004), LNCS Vol. 3203, pp.565-574, Springer 2004.

[18] I. Verbauwhede, P. Schaumont, H. Kuo, "Design and Performance Testing of a 2.29-GB/s Rijndael Processor", in *IEEE Journal of Solid State Circuit*, Vol. 38, No. 3, pp.569-572, March 2003.

[19] X. Zhang, K. Parhi, "igh-Speed VLSI Architectures for the AES Algorithm", in *IEEE Transactions on Very Large Scale Integration (VLSI) Systems*, Vol. 12, No. 3, pp.957-967, 2004

[20] J. Zambreno, D. Nguyen, and A. Choudhary, "Exploring Area/Delay Tradeoffs in an AES FPGA Implementation" in *Field Programmable Logic (FPL 2004)*, LNCS Vol. 3203, pp.575-585.

[21] M. Alam, S. Ghosh, D. Mukhopadhyay, D. Roychowdhury and I. Sengupta: "Latency Optimized AES-Rijndael with Flexible Mode of Operation", *11th IEEE VLSI Design And Test Symposium (VDAT 2007)*, pp.413-420, August 8-11, Kolkata, India.

978-1-4244-3039-0/08 $25.00 © 2008 IEEE

21st International Conference on VLSI Design

Reduced Complementary Dynamic and Differential Logic: A CMOS Logic Style for DPA-resistant Secure IC Design

Srividhya Rammohan, Vijay Sundaresan and Ranga Vemuri

Department of ECE, University of Cincinnati, Cincinnati, Ohio 45221

{rammohs, sundarvy, ranga}@ececs.uc.edu

Abstract

In recent years, Differential Power Analysis (DPA) attack has become a major threat to the security of embedded cryptographic ICs (secure ICs) like smart cards. DPA attack is a powerful side-channel attack. During a DPA attack, the attacker uses power consumption measurements from the secure IC and statistical techniques to correlate the power consumption information leaked with the secret key stored in the secure IC, thus retrieving the secret key, and effectively breaking the secure IC. In this paper, we present a Reduced Complementary Dynamic and Differential Logic (RCDDL) style to design DPA-resistant, secure ICs. RCDDL style ensures that the power consumption of the secure IC remains invariant, *and hence, uncorrelated to the input data (secret key). As opposed to existing DDL styles that complement every gate in the uncomplementary logic to generate the differential output, RCDDL style proposes reuse of gates, thus ensuring that a reduced number of gates in the uncomplementary logic are complemented to generate the differential output. Further, we present an analysis of how reduced complementation is achieved while maintaining the capacitance and switching requirements for power invariance. To evaluate the proposed logic style, we built a set of logic gates typically used to design secure ICs. Experiments on a set of circuits, designed using the set of RCDDL gates, show significant improvements in security strength, power consumption and area.*

I. Introduction

In recent years, embedded systems like smart cards (used in credit card readers, identification, pay phones, and many others), personal computers, mobile phones and PDAs have become an integral part of our everyday life. Many of these embedded systems need to access, process and communicate confidential data. The price of insecurity in these embedded systems is very steep. The 2006 CSI/FBI Computer Crime and Security Survey [1] reports that almost 40% of the financial losses in 2006 (approximately US $38 million) were due to unauthorized access to secret and sensitive information, theft of proprietary information, financial fraud and system penetration by an outsider. According to a global survey by the Economist Intelligence Unit and sponsored by Symantec Corp [2], security concerns are the biggest obstacle to the widespread adoption of wireless and remote computing in businesses worldwide. Hence, increase in security and proof of validity measures of security for all these embedded systems is of utmost importance, not only for financial purposes, but also for complete adoption of all technological advances.

Security of embedded systems has been a subject of intensive research in recent years. Security of embedded systems is, typically, ensured using embedded cryptographic ICs (secure ICs). Secure ICs are hardware implementations of cryptographic algorithms [3] like DES [4] and AES [5], which typically use a *secret key* to ensure security of confidential data. Security attacks on embedded systems can be classified as mathematical attacks and implementation attacks [6] [7]. While mathematical attacks target the functionality of cryptographic algorithms, implementation attacks attempt to exploit the weaknesses of the hardware implementation of the cryptographic algorithms.

In recent years, several side-channel implementation attacks have been proposed [7]. The most popular and powerful side-channel attack [8] is the Differential Power Analysis (DPA) attack [9]. DPA attacks are powerful because they can be mounted using cheap, easily available equipments. They require no kind of physical intrusion into the secure IC. DPA attacks can be carried out by any attacker who has sufficient knowledge on the internal workings (cryptographic algorithm) of the secure IC, with little or no information on the target implementation. More importantly, DPA attacks can be automated, while obtaining highly accurate results. DPA attacks have been successfully carried out on a wide variety of cryptographic products [9]. During a DPA attack, the attacker records power consumption information from the secure IC. Then, by analyzing the power consumption using statistical techniques, identifies the correlation between power consumed by the secure IC and the secret key stored in the secure IC, thereby, retrieving the secret key. DPA exploits a basic property of secure IC implementations. Secure ICs are, typically, implemented using Static-CMOS (sCMOS) logic style [10]. In sCMOS hardware implementations, power is consumed only when the output transitions from $0 \rightarrow 1$. This property makes the power consumption of the secure IC *correlated* to the secret key. While most cryptographic algorithms are functionally strong, they are weak against side-channel attacks, like DPA, that attempt to extract the secret key at the hardware implementation level (secure ICs).

A. DPA Countermeasures - Related Work

DPA uses two important weaknesses of hardware implementations of cryptographic algorithms (secure ICs) to attack them:

1) The output of the algorithm is directly related to whole or parts of the secret key and the secret key can be traced back from the output (*algorithmic weakness*).

2) Hardware implementations of cryptographic algorithms

978-1-4244-3039-0/08 $25.00 © 2008 IEEE

have visible characteristics such as power consumption, which depend on the input data *(circuit level weakness)*.

Existing DPA countermeasures can be classified as: 1) *Algorithmic countermeasures* and 2) *Circuit level countermeasures*. Algorithmic countermeasures attempt to conceal power variations at the architectural level, and are not effective against DPA or its derivatives, as the variations originate at circuit level. Circuit level countermeasures involve changing circuit structure such that the power traces are not directly correlated to the data being manipulated. Numerous circuit level countermeasures have been proposed that use dedicated CMOS circuit and logic design techniques to counter DPA, by making the power consumption of the circuit independent of the input data (secret key). We restrict comparisons to a few relevant ones. Many circuit level countermeasures introduce random power consuming operations causing random power consumption to mask power variations [11] [12]. Analysis of *masking* techniques indicates that masking merely lowers power information leaked, but does not eliminate it, and can be overcome using variants of DPA [13].

Recently, several circuit-level countermeasures have been proposed that completely eliminate power variations. These *power invariant* techniques attack the very foundation of DPA by not creating any side channel power information. These techniques implement cryptographic algorithms in CMOS logic styles with constant power consumption for all input data. These include Sense Amplifier Based Logic (SABL) [14], Simple Dynamic and Differential Logic (SDDL) [15] and Wave Dynamic Differential Logic (WDDL) [15]. Of these, WDDL style has been the more widely analyzed circuit-level countermeasure. Hence, in this paper, we compare RCDDL results with WDDL as well as non-secure sCMOS logic styles. All existing circuit level power invariant countermeasures suffer from reduced security strength, increased area or power consumption penalty, and in some techniques, design issues like cascading and signal integrity. This is because all existing power invariant countermeasures complement every gate in the uncomplementary logic to generate the differential output. This is primarily done to match the capacitance and switching requirements of the circuit for each input data. To our knowledge, the proposed work is the first to incorporate *reuse* in dynamic and differential logic style.

In this paper, we present Reduced Complementary Dynamic and Differential Logic (RCDDL) style. RCDDL is a circuit level power invariant countermeasure against DPA attacks. Unlike all existing circuit level DDL countermeasures that complement every gate in the uncomplementary logic to generate the differential output, RCDDL proposes *reuse* of gates in the uncomplementary logic to generate the differential output. *Reuse* is achieved while ensuring that a constant capacitance is charged during every input data cycle (cycle). Further, we present an analysis of how reuse is achieved while maintaining the capacitance and switching requirements for constant power consumption. *Reuse* reduces power and area consumption as well as improves power variation characteristics of the circuit designed. Also, unlike existing static-CMOS DDL styles, RCDDL style encourages use of both negative (NAND/NOR) as well as positive (AND/OR) logic gates to design each RCDDL gate, thereby, making RCDDL gate design conducive to logic synthesis and optimization strate-

gies. RCDDL is a static-CMOS implementation of dynamic and differential logic. Static-CMOS implementation eliminates any cascading or signal integrity issues, as faced by dynamic logic implementations. RCDDL uses wave (of '0's) propagation [15] to mimic dynamic behavior. However, unlike the restrictions in existing DDL style, RCDDL achieves precharge wave propagation without restricting to use of positive gates. The RCDDL style is formulated as a standard cell methodology where each cell has been characterized to show power invariance. Formulation as a standard cell methodology ensures possibility of automation and optimization of RCDDL cells as well as circuits designed using RCDDL cells, thereby fitting well into existing design flow. Cell and Circuit-level design methodology development and associated issues are presented in [16]. To evaluate the proposed RCDDL style, we built a set of logic gates necessary to build secure ICs, and a set of test circuits. When compared with WDDL [15] and non-secure sCMOS implementations, significant improvements in security strength, power consumption and area were obtained.

This paper is organized as follows. Section II motivates the need for DDL style to achieve power invariance and presents the RCDDL gate architecture. Further, it presents the modes of operation of an RCDDL gate and analyzes how constant current draw (power consumption) is achieved independent of the input data. Section III examines the various types and implementations of RCDDL gates. Section IV discusses first set of experimental results and analyzes the RCDDL circuit design issues. Section V discusses the second set of experimental results by performing actual DPA attacks on implementations of a subset of DES cryptographic algorithm. Section VI concludes the paper.

II. Reduced Complementary Dynamic and Differential Logic (RCDDL) style [17]

A. Motivation

To achieve power invariance, and hence DPA-resistance, the circuit must charge/discharge a constant value of capacitance in every cycle. For any CMOS logic style to satisfy this requirement, there are two conditions [14]:

(1) **Exactly ONE output transition must occur in every cycle.**

This condition can be satisfied by designing the secure IC using Dynamic and Differential CMOS Logic (DDL) [14]. Fig. 1 illustrates the need for DDL with a two input XOR logic function. In sCMOS implementation, Fig. 1(a), only one set of input data results in $0 \rightarrow 1$ output transition. In DDL implementation, Fig. 1(b), *all inputs result in exactly ONE output transition (charging/discharging)*.

(2) **The total capacitance charged/discharged in every cycle must remain constant.**

All the existing DPA countermeasures using DDL styles, complement every gate in the uncomplemented logic to generate the differential output and to equalize the capacitances. By default, the power consumption and area of these implementations would increase by at least 2X, when compared with a non-secure sCMOS implementation.

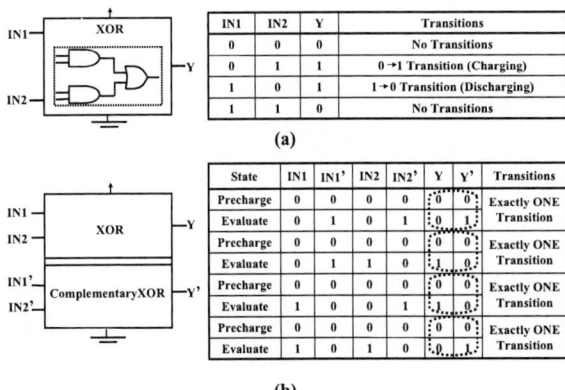

IN1	IN2	Y	Transitions
0	0	0	No Transitions
0	1	1	0 →1 Transition (Charging)
1	0	1	1→0 Transition (Discharging)
1	1	0	No Transitions

(a)

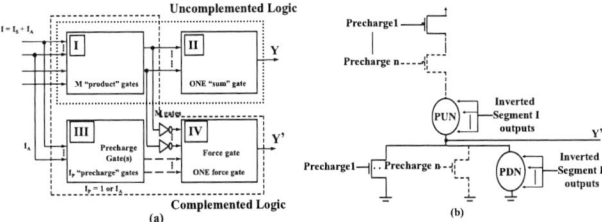

State	IN1	IN1'	IN2	IN2'	Y	Y'	Transitions
Precharge	0	0	0	0	0	0	Exactly ONE Transition
Evaluate	0	1	0	1	0	1	
Precharge	0	0	0	0	0	0	Exactly ONE Transition
Evaluate	0	1	1	0	1	0	
Precharge	0	0	0	0	0	0	Exactly ONE Transition
Evaluate	1	0	0	1	1	0	
Precharge	0	0	0	0	0	0	Exactly ONE Transition
Evaluate	1	0	1	0	0	1	

(b)

Fig. 1. (a) sCMOS XOR (b) Dual-rail DDL XOR

B. RCDDL - Gate Architecture

The RCDDL gate architecture is shown in Fig. 2 (a). Each RCDDL gate consists of four Segments. Any logic function can be designed as an RCDDL gate using the two-level Sum-Of-Products (SOP) expression of the logic function to be implemented. Let the two-level SOP expression contain M product terms and I literals (I_S is the number of symmetric inputs (i.e. both A and A' are present in the expression) and I_A is the number of asymmetric inputs (i.e. only A or A' is present in the expression).

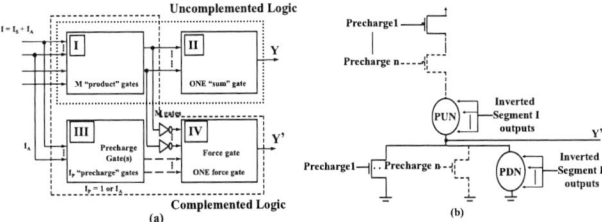

Fig. 2. (a) RCDDL - Gate Architecture (b) Force Gate Architecture

Segment I: Segment I contains M "product" gates implementing the product terms in the SOP expression and M NOT gates generating inverted signals for Segment IV. Therefore, Segment I outputs M uninverted signals connected to Segment II and M inverted signals connected to Segment IV.

Segment II: Segment II contains ONE gate implementing the sum term in the SOP expression and generates the uncomplemented output Y.

Segment III: Segment III generates the precharge signal(s) (I_P). Each precharge signal is generated using a NOR gate that takes a set of differential inputs to the RCDDL gate as inputs (e.g. A and A'). The number of precharge signals generated (I_P) is dependent on the nature of inputs of the SOP expression of the logic function. For SOP expressions that have all differential inputs available (i.e. $I_A = 0$), a single precharge signal is generated using only one differential input from the set of inputs to the gate, to generate ONE precharge signal (i.e. $I_P = 1$). E.g. XOR gate ($Y = A'.B + A.B'$), the SOP expression contains both

uncomplementary and complementary inputs for all inputs. For SOP expressions that do not have all differential inputs available, precharge signals are generated for all asymmetric (I_A) inputs (i.e. $I_P = I_A$). E.g. MUX gate ($Y = S'.A + S.B$), the SOP expression does not contain A' and B' signals. Hence, precharge signals are generated using both A, A' and B, B' differential inputs. Although the RCDDL gate will function without the redundant precharge signals, they are included only to account for all differential inputs.

Segment IV: Segment IV contains the Force Gate (Fig. 2 (b)). The force gate, generates the complemented output Y'. The force gate takes the precharge signal(s) (I_P) from Segment III and the inverted outputs (M) from Segment I. In the force gate, PUN/PDN sub-circuits comprises of the complement of the logic gate in Segment II. The precharge signal(s), are connected to PMOS transistors in series with PUN, and to NMOS transistors in parallel with PDN.

Segments I and II constitute the *Uncomplemented Logic*. The gates in Segments I and II can be implemented using both negative (NAND/NOR) and positive (AND/OR) gates. Care must be taken to ensure that the uncomplemented output Y must be '0' during the precharge state i.e. when all the differential inputs (e.g. both A and A') are '0's. Segments I, III and IV constitute the *Complemented Logic*.

C. RCDDL - Modes of Operation and Switching Analysis

RCDDL gates are dynamic and differential logic gates. The gates are designed in sCMOS logic.

Precharge state is achieved by propagating a precharge wave (of '0's) through all the differential inputs of the circuit [15]. During precharge state, M "product" gates in Segment I are switched ON and M Inverters are switched OFF. Therefore, in Segment I exactly M gates draw current from the source. In Segment III, all precharge signal(s) become '1'. Therefore, exactly I_P gates draw current from the source. By logic, the uncomplemented output Y is '0'. The precharge signals force the complemented output Y' to '0'. Hence, the name, "force" gate. Thus, Segments II and IV are switched OFF. Therefore, exactly $M + I_P$ gates draw current from the source during every precharge mode of operation. Thus, when all the differential inputs are '0's, all differential outputs of each RCDDL gate become '0's, so that all subsequent RCDDL gates in the circuit can be precharged. The precharge '0's pass through the circuit like a wave, forcing all the outputs in the circuit to '0's. [15] details precharge wave generation circuitry and strategies required for seamless circuit operation.

Evaluation state is achieved when complemented differential inputs are provided to the gate. During evaluation state, in Segment I, m number of "product" gates switch OFF and corresponding m inverters switch ON. m could vary from $0...M$, depending on the input data. However, if the "product" gates and their corresponding inverters are sized to equal capacitances, we can conclude that in Segment I exactly M number of gates ("product" gates and inverters) draw current from the source. Segment III is switched OFF and all precharge signal(s) become '0's. Thereby, Segment IV behaves as a complement of Segment

II. Thus, based on the input provided either Segment II (output Y) *or* Segment IV (output Y') draws current from the source. If the gate in Segment II and PUN/PDN part of Segment IV are sized to equate their capacitances, we can conclude that exactly $M + 1$ number of gates draw current from the source.

From the above analysis, by placing the Segments at minimum distances from each other to minimize the interconnect capacitances and using differential routing to equate the interconnect capacitances, we can conclude that during both modes of operation, the number of gates drawing current from the source is a function of M and I_P, which are independent of the input data. Thus, ensuring constant current draw from the source (power consumption).

D. RCDDL - Sizing Requirements

After analyzing the switching of an RCDDL gate in both modes of operations, assuming there are no timing requirements on inputs to various segments, the following additional sizing requirement is necessary for proper functionality of the gate.

1) Force gate sizing strategy: The structure of the force gate imposes two restrictions on the arrival times of signals:

(i) *During precharge state, the precharge signal(s) must arrive earlier than the signals from Segment I* - This is inherently ensured, due to the presence of inverters connected to the inputs from Segment I. This requirement can further be ensured by increasing the widths of NMOS transistor(s) connected to the precharge signal(s).

(ii) *During evaluation state, the precharge signal(s) must arrive later than the signals from Segment I* - This can be ensured by decreasing the widths (or increasing the lengths) of PMOS transistor(s) connected to the precharge signal(s).

Now, based on the sizing changes made to the force gate, the transistors in the remaining Segments are sized to ensure that the supply currents drawn are within less than 10% of each other in every cycle.

It is important to note that improvements in RCDDL implementations are due to *reuse* of part of the RCDDL gate (Segment I). Thus, building one-level gates (AND/OR) using RCDDL would prove to be inefficient. However, cryptographic algorithms (secure ICs) predominantly use operations such as permutation, substitution and other operations based on Galois Field arithmetic [3] [14], which are designed mostly using XOR and MUX logic functions. Hence, when designing secure ICs, the proposed RCDDL style would be an efficient alternative to existing logic styles.

III. RCDDL Gates - Implementation

In our experiments, we designed many RCDDL gates. In this paper, for explanation and analysis, and due to space restrictions, we discuss three RCDDL gates (XOR, MUX and XorPlus). XOR and MUX gates are predominantly used in embedded cryptographic IC (secure IC) implementations. To further highlight the capabilities of RCDDL style, we present XorPlus gate (random logic operation).

A. RCDDL XOR Gate - Architecture

The logic diagram of the RCDDL XOR gate ($Y = A'.B + A.B'$) and the layout-level implementation are shown in Fig. 3. Note that we have used negative gates (NAND) to design the uncomplemented logic, in contrast with the design restrictions existing in WDDL style [15].

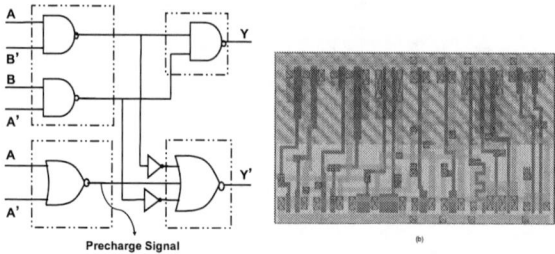

Fig. 3. (a) RCDDL XOR gate (b) RCDDL XOR Layout

1) RCDDL XOR - Operation: Precharge state: All inputs (A, A', B, and B') are set to '0'. In Segment II, output Y generates '0'. The precharge signal from Segment III becomes 1. In Segment IV, the force gate forces the output Y' to '0'. Hence, all differential outputs are '0' when all differential inputs are '0's. *Evaluation state:* Normal complemented differential inputs are provided to the gate. Segments I and II operate normally, producing output Y. In Segment III, the precharge signal becomes '0'. Thus, the force gate (Segment IV) works normally, producing the complemented output Y'.

B. RCDDL MUX Gate - Architecture

The logic diagram of the RCDDL MUX gate ($Y = S'.A + S.B$) and the layout-level implementation are shown in Fig. 4. In MUX gate, the *asymmetric* differential inputs, i.e. inputs whose complementary or uncomplementary input is not part of the SOP expression, are connected by introducing redundant precharge logic in the gate. Two asymmetric differential inputs A' and B' are missing from the SOP expression. These are connected in the RCDDL MUX gate, by generating a precharge signal from both asymmetric inputs.

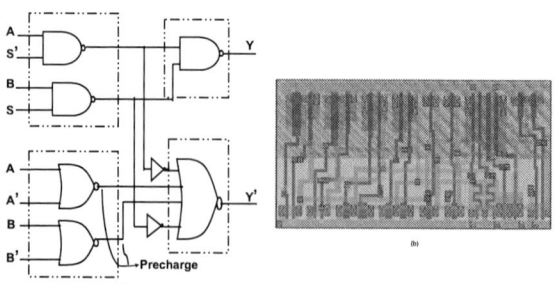

Fig. 4. (a) RCDDL MUX gate (b) RCDDL MUX Layout

1) RCDDL MUX - Operation: Precharge state: All differential inputs (A, A', B, B', S and S') are set to '0'. In Segment II, output Y generates '0'. In Segment III, all precharge signals generated using asymmetric differential inputs (A and B) become

978-1-4244-3039-0/08 $25.00 © 2008 IEEE

'1'. In Segment IV, the force gate forces the output Y' to '0'. Hence, all differential outputs are '0' when all differential inputs are '0's. *Evaluation state:* Normal differential (complemented) inputs are provided to the gate. Segments I and II operate normally, producing output Y. In Segment III, all precharge signals become '0'. Thus, in Segment IV, the force gate works normally, producing the complemented output Y'.

C. RCDDL XorPlus - Gate Architecture

The logic diagram of the RCDDL XorPlus gate ($Y = A'.B + A.B' + B.C$) and the layout-level implementation are shown in Fig. 5.

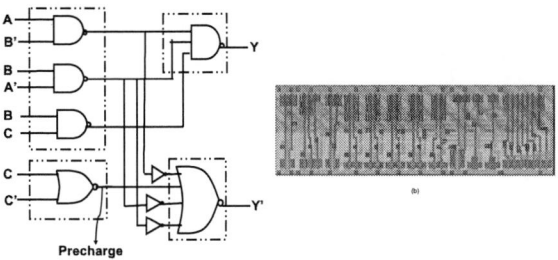

Fig. 5. (a) RCDDL XorPlus gate (b) RCDDL XorPlus Layout

1) RCDDL XorPlus - Operation: Precharge state: All differential inputs (A, A', B, B', C and C') are set to '0'. In Segment II, output Y generates '0'. In Segment III, the precharge signal generated using C and C' (asymmetric input) become '1'. In Segment IV, the force gate forces the output Y' to '0'. Hence, all differential outputs are '0' when all differential inputs are '0's. *Evaluation state:* Normal differential (complemented) inputs are provided to the gate. Segments I and II operate normally, producing output Y. In Segment III, the precharge signal becomes '0'. Thus, in Segment IV, the force gate works normally, producing the complemented output Y'.

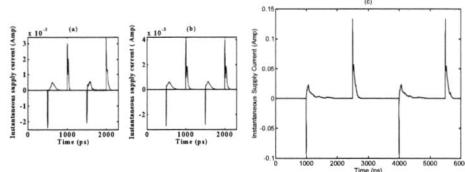

Fig. 6. Inst. Supply Current of (a) RCDDL XOR (b) RCDDL MUX (c) RCDDL XorPlus

Gates	SCMOS				WDDL				RCDDL			
	Max. I_{inst} Variance (Amp)	Avg. Power (mW)	Area (λ^2)	Prop. Delay (ns)	Max. I_{inst} Variance (Amp)	Avg. Power (mW)	Area (λ^2)	Prop. Delay (ns)	Max. I_{inst} Variance (Amp)	Avg. Power (mW)	Area (λ^2)	Prop. Delay (ns)
XOR	9.52xe-6	0.1327	7020	0.2768	0.234xe-6	0.4039	22248	0.5070	0.0808xe-6	0.2944	19548	0.8325
MUX	2.05xe-6	0.0087	6264	0.1609	0.2042xe-6	0.4033	22356	0.5554	0.0169xe-6	0.3981	22140	0.7612
XorPlus	1.261xe-6	0.162	15120	0.776	0.342xe-6	0.577	31644	2.645	0.179xe-6	0.493	28296	3.013

Fig. 7. Max. Inst. Current Variation, Avg. Power, Area and Delay of sCMOS, WDDL and RCDDL XOR, MUX and XorPlus Gates

Fig. 6 shows the Instantaneous supply current curves for RCDDL XOR, RCDDL MUX and RCDDL XorPlus respectively. Table in Fig. 7 shows the Max. Instantaneous Current Variation, Average Power and Area improvements (with delay penalty) of RCDDL XOR, RCDDL MUX and RCDDL XorPlus gates over sCMOS and WDDL implementations.

IV. Experimental Results I - Circuit Design, Characteristics & Analysis

All experiments were performed on a Sun Blade 100 workstation with 512MB RAM. All gates were implemented in 0.25u technology, with 5V supply voltage. All gates were designed and extracted using Magic layout editor, and simulated using Synopsys Hspice transistor-level simulator. *All circuit experiments included layout parasitics.* Instantaneous power (supply current) measurements from Hspice were taken every 10ps, equaling a sampling rate of 100GHz. DPA attacks and analysis were performed using MATLAB/C++. The RCDDL library built consisted of XOR, MUX and XorPlus cells; combined with AND/OR gate implemented using WDDL logic; two sCMOS buffers with propagation delays equal to 0.3ns and 0.5ns, and a current-matched sCMOS D-FlipFlops. We compared results for RCDDL circuits with the circuit implementations using WDDL and sCMOS logic styles.

The *security strength* of any secure IC implementation can be measured using the maximum instantaneous power (supply current) variation ($Var(I_{inst})$) as the metric [8]. In this paper, maximum instantaneous supply current variation along with average power consumption, area and delay characteristics are analyzed.

Any circuit can be designed using RCDDL cells. Cryptographic algorithms predominantly use Galois Field (GF) operations [3] such as permutations and substitutions. Additionally, Galois Field arithmetic operations like GF addition, GF multiplication and GF inversion are used to implement cryptographic algorithms. These operations are designed predominantly using XORs and MUXes. Due to space restrictions, the authors request the readers to refer to [3], [18] and [19] for more information on the architectures of each GF operation.

Our first task was to analyze the efficiency of the proposed RCDDL style by building Galois Field arithmetic operations and other realistic circuits. We designed 8 Galois Field Arithmetic operations, namely, $GF(2^3)Plus$, $GF(2^3)Multiplier$, $GF(2^3)Inversion$, $GF(2^4)Plus$, $GF(2^4)Multiplier$, $GF(2^4)Inversion$, $GF(2^5)Plus$ and $GF(2^5)Multiplier$ operations. Further, to analyze impact of varying circuit sizes and configurations, we performed experiments on 6 synthetic circuits. The synthetic benchmarks were generated as balanced random-trees, where each node in the tree contains any random cell from the cell library. Experiments were conducted by simulating the circuits for 800 random input vectors, supplied at the rate of 100 MHz. Fig. 8 shows the maximum $Var(I_{inst})$, avg. power consumption, area from the 14 circuit implementations.

When compared with WDDL implementations, on average, we can see that using RCDDL style, reduced $Var(I_{inst})$ by 62%,

703

978-1-4244-3039-0/08 $25.00 © 2008 IEEE

Test Circuits	SCMOS			WDDL			RCDDL		
	Max.Var.	Avg.Pwr. (mW)	Area (λ²)	Max.Var.	Avg.Pwr. (mW)	Area (λ²)	Max.Var.	Avg.Pwr. (mW)	Area (λ²)
GF(2³) Plus	38.55xe⁻⁶	0.58	21060	0.871xe⁻⁶	2.60	66744	0.413xe⁻⁶	1.52	58644
GF(2³) Multiplier	88.64xe⁻⁶	7.75	115020	20.51xe⁻⁶	17.58	303912	15.36xe⁻⁶	10.20	279612
GF(2³) Inversion	87.06xe⁻⁶	5.34	55080	8.54xe⁻⁶	9.65	159408	5.11xe⁻⁶	4.04	143208
GF(2⁴) Plus	64.74xe⁻⁶	0.73	28080	1.137xe⁻⁶	1.76	88992	0.560xe⁻⁶	1.24	78192
GF(2⁴) Multiplier	252.79xe⁻⁶	3.70	202500	9.27xe⁻⁶	11.56	560952	6.16xe⁻⁶	10.16	509652
GF(2⁴) Inversion	802.11xe⁻⁶	33.47	263520	324.66xe⁻⁶	60.71	724032	79.13xe⁻⁶	25.33	659232
GF(2⁵) Plus	80.72xe⁻⁶	0.93	35100	1.482xe⁻⁶	2.21	111240	0.61xe⁻⁶	1.57	97740
GF(2⁵) Multiplier	422.92xe⁻⁶	25.32	254880	81.02xe⁻⁶	46.52	706752	59.41xe⁻⁶	20.51	641952
Synth1	19.3xe⁻⁶	0.39	20736	0.612xe⁻⁶	0.98	53136	0.249xe⁻⁶	0.82	47952
Synth2	15.4xe⁻⁶	0.55	21816	0.47xe⁻⁶	1.39	66852	0.41xe⁻⁶	1.10	61452
Synth3	39.7xe⁻⁶	1.28	48384	1.23xe⁻⁶	3.43	156168	0.85xe⁻⁶	2.90	147528
Synth4	27.0xe⁻⁶	0.84	59832	4.78xe⁻⁶	3.90	187164	3.72xe⁻⁶	3.50	178308
Synth5	15.1xe⁻⁶	6.82	235008	3.37xe⁻⁶	15.01	690120	1.856xe⁻⁶	11.32	619272
Synth6	41.4xe⁻⁶	3.37	146448	2.51xe⁻⁶	9.03	426492	1.66xe⁻⁶	7.64	394632

Fig. 8. Experimental Results - Galois Field Arithmetic Circuits and Synthetic Circuits.

reduced average power consumption by 45% and reduced area by 9%. When compared with sCMOS implementations, $Var(I_{inst})$ reduced by 11.37X (average), thus drastically improving the security strength of the RCDDL implementations. However, due to use of DDL style, we see an increase in average power consumption (1.19X) and area (2.6X).

A. Buffering Analysis

At circuit-level to satisfy the force gate timing requirements (refer section II-D), the slower arriving differential input signal(s) are used to generate the precharge signal(s). However, when several RCDDL cells are connected in series, there is a differential delay introduced between the inputs to the force gate during evaluation state. Further analysis about the implication of differential delay is explained in [16]. This differential delay problem does not pose any functionality issues and can be overcome by using buffering to equalize input arrival times.

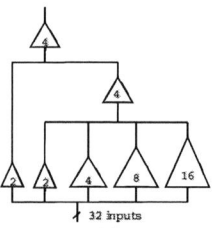

Test Circuits	WDDL		RCDDL	
	Max.Curr. Var.	AvgPwr (mW)	Max.Curr. Var.	AvgPwr (mW)
Xunbal1	1.24xe⁻⁶	1.16	0.69xe⁻⁶	0.88
Xunbal2	22.2xe⁻⁶	2.49	17.6xe⁻⁶	2.08
Xunbal3	34.5xe⁻⁶	5.17	27.7xe⁻⁶	4.04

(a)   (b)

Fig. 9. (a) Sample Unbalanced XOR-Tree (b) Experimental Results

Our second experiment was performed to analyze the RCDDL style by building *realistic* circuits with buffering requirements due to unequal logic depths and wiring. Our test circuits consisted of three unbalanced XOR-trees containing logic depths from 2 to 5. An example of our test circuit is shown in Figure 9 (a). The results are presented in Fig. 9 (b). We can see that

the Xunbal (RCDDL) has improved I_{inst} variance and average power consumption over WDDL implementations.

In RCDDL, the propagation delay of RCDDL implementations is on average 40% more than WDDL implementations. However, since the sizes of secure IC designs are small, the delay penalty incurred does not drastically affect the operation of the embedded systems using the secure IC. Also, when considering the price of insecurity in embedded systems, from all experimental results, we can conclude that RCDDL style implementations consistently out-perform WDDL implementations by providing better security strength with improvements in average power consumption and area.

V. Experimental Results II - DPA on DES Cryptographic Algorithm

A. DES Cryptographic Algorithm

While DPA can be applied to many (types) cryptographic algorithms including RSA [20] and AES [5], in this paper we use DES algorithm as an example. Data Encryption Standard [4] algorithm one of the most widely used and studied cryptographic algorithm [9]. DES encrypts/decrypts 64-bits of data (plaintext/ciphertext). It performs 16 rounds of encryption on the 64-bit data. Each encryption round involves permutation, shift operations, substitution and parity checks. DES uses a 64-bit key containing 8 parity bits. The resultant 56-bit actual key is used to generate 16 48-bit subkeys ($K_i | i = 1, 2, ..16$) rounds, one subkey for each round of DES encryption. Each round of DES encryption can be summarized as follows. The plaintext is first split into two 32-bit L and R subtexts (L_0 and R_0). The encrypted output from each round L_i and R_i can be represented as shown below.

$$L_i = R_{i-1}$$
$$R_i = L_{i-1} \oplus f(R_{i-1}, K_i)$$

Cipher function $f()$ includes S-box table lookup selection function that takes a 6-bit input to yield a 4-bit output. i denotes the DES round ranging from 1 to 16. $R_{16}L_{16}$ is the encrypted ciphertext preoutput from the 16th round of the DES algorithm. For *decryption*, the cipher text is given as the input to the DES algorithm, and the subkeys are applied in the reverse order to obtain the unencrypted plaintext as output.

B. DPA attack on DES [9]

The attack on DES begins by recording the supply current (instantaneous power) traces (T_j) from the secure IC, sampled at discrete intervals ($j = 1..k$ intervals), during encryption of a large set of plaintexts ($M_i | i = 1..N$). In addition, the attacker also records the ciphertexts (C) obtained from the DES for each of the plaintexts (M_i) encrypted.

The attacker attempts to decipher the correct key, typically, during the last round of DES encryption. DPA on the last round is performed on 8, 6-bit subsets of the DES round, independently, to obtain the 48-bit subkey for the last round. Using this subkey, upon further analysis, the correct secret key could be obtained.

For each subset in the last round of the DES, the attacker makes a 6-bit key guess (K). The total possible key guesses for each subset are $K = 2^6$. A DPA selection function (D) is then evaluated for each R_i selection bit (b), $i = 1..4$ in each subset. Based on the results of the DPA selection function (D), the set of power traces for all N ciphertexts is split into two sets, $T_{D=0}$ and $T_{D=1}$. A differential power trace (T_Δ) is generated by averaging the power traces in each set over each instant sampled ($j = 1..k$).

For an incorrect key guess, the trace components will remain uncorrelated to the key and behave as noise, causing the differential trace to become (close to) flat. *For a correct key guess, however, the trace components will be correlated to the key, and the differential trace will become equal to a correlated value ($\neq 0$). Hence, the correct key guesses will result in a peak (spike) in the differential power trace.* The peak of the correct key becomes larger as the number of ciphertexts considered increases. This is because of the randomness in the uncorrelated differential trace, for an incorrect key, when $N \to \infty$, $T_\Delta \to 0$.

C. DPA attack on DES - Results

Our DES implementation circuit consisted of one round [16] and four rounds of a subset of a round of DES encryption. This subset has been shown to be a sufficient subset of the DES cryptographic algorithm [8]. We performed simulations by encrypting 2048 random input vectors, with a fixed secret key. We implemented the DES test circuits in all three logic styles. The secret key stored in DES (sCMOS) was retrieved, and in DES (WDDL) and DES (RCDDL), we were unable to recover the secret key even when using the exhaustive set of input texts. From the table in Fig. 10, DES (RCDDL) exhibits lower max. $Var(I_{inst})$ (better security strength) and avg. power consumption when compared with DES (WDDL).

Test Circuits	WDDL		RCDDL	
	Max.Curr. Var.	Avg Pwr (mW)	Max.Curr. Var.	Avg Pwr (mW)
DES [16]	0.849xe^{-4}	4.53	0.532xe^{-4}	2.99
DES-4S	3.212xe^{-4}	17.79	2.616xe^{-4}	9.76

Fig. 10. DES Experimental Results [16].

VI. Conclusions

Security of embedded cryptographic ICs (secure ICs) has become a critical issue since the advent of powerful side-channel attacks like DPA. In this paper, we presented a novel Reduced Complementary Dynamic and Differential Logic (RCDDL) style to design DPA-resistant, secure ICs. RCDDL style ensures that the power consumption of the secure IC remains *invariant*, and hence, uncorrelated to the input data (secret key). Also, RCDDL style proposes reuse of gates, thus ensuring that a reduced number of gates in the uncomplementary logic are complemented to generate the differential output. This is achieved while maintaining the capacitance and switching requirements for power invariance. Experimental results on all 19 test circuits, including DES, Xunbal, GF arithmetic circuits and synthetic circuits, show that circuits implemented using RCDDL style have significant improvements in security strength (approx. 42%

reduction in max. instantaneous current variation) when compared with WDDL and 11.37X improvement when compared with sCMOS implementations. Using RCDDL style also shows improvements in average power consumption and area, but, with a delay penalty when compared with WDDL style. From the results, we can conclude that secure ICs designed using RCDDL style would lead to more secure embedded systems, leading to better protection of confidential data.

References

[1] "2006 CSI/FBI Computer Crime and Security Survey." [Online]. Available: http//www.gocsi.com/forms/fbi/csi_fbi_survey.jhtml

[2] "Security Concerns Threaten Enterprise Rollout of Mobile Technology- Symantec Corporation," article appeared on 04-04-06. [Online]. Available: http://www.symantec.com/about/news/release/index.jsp

[3] A. J. Menezes, P. C. van Oorschot and S. A. Vanstone, *Handbook of Applied Cryptography*. CRC, October 1996, no. 978-0849385230.

[4] "NIST - FIPS 46-3: Data Encryption Standard," National Institute of Standards and Technology (NIST), Tech. Rep., October 1999.

[5] "NIST - FIPS 197: Advanced Encryption Standard," National Institute of Standards and Technology (NIST), Tech. Rep., 2001.

[6] S. Ravi, A. Raghunathan, P. Kocher, and S. Hattangady, "Security in embedded systems: Design challenges," *Trans. on Embedded Computing Sys.*, vol. 3, no. 3, pp. 461–491, 2004.

[7] KULRD and SCARD Consortium, "Intermediate report side-channel attacks," SCARD, Tech. Rep., January 2005.

[8] K. Tiri et al, "Securing Encryption Algorithms against DPA at Logic Level: Next Generation Smart Card Technology," *Lecture Notes in Computer Science*, vol. 2779, pp. 125–136, 2003.

[9] P. Kocher, J. Jaffe, and B. Jun, "Differential Power Analysis," *Lecture Notes in Computer Science*, vol. 1666, pp. 388–397, 1999.

[10] J. Rabaey et al, *Digital Integrated Circuits: A Design Perspective (2nd Edition)*. Prentice Hall, 2002.

[11] D. Suzuki et al, "Random Switching Logic: A New Countermeasure against DPA and Second-Order DPA at the Logic Level," *IEICE Transactions on Fundamentals of Electronics, Communications and Computer Sciences*, vol. E90-A, no. 1, pp. 160–168, 2007.

[12] S. Yang et al, "Power Attack Resistant Cryptosystem Design: A Dynamic Voltage and Frequency Switching Approach," in *Proc. of Design Automation and Test in Europe Conference (DATE 2005)*.

[13] T. S. Messerges, "Using Second-Order Power Analysis to Attack DPA Resistant Software," in *Cryptographic Hardware and Embedded Systems (CHES '00)*, vol. LNCS 1965, 2000, pp. 238 – 251.

[14] K. Tiri et al, "A Dynamic and Differential CMOS Logic with Signal Independent Power Consumption to Withstand Differential Power Analysis on Smart Cards," in *Proc. of European Solid-State Circuits Conference (ESSCIRC 2002)*, 2002, pp. 403–406.

[15] K. Tiri and et al, "A Logic Level Design Methodology for a Secure DPA Resistant ASIC or FPGA Implementation," in *Proc. of Design Automation and Test in Europe Conference (DATE 2004)*.

[16] V. Sundaresan, S. Rammohan, and R. Vemuri, "Power Invariant Secure IC Design Methodology using Reduced Complementary Dynamic and Differential Logic," in *Proc. of the 2007 IFIP Intnl. Conf. on VLSI (VLSI-SoC '07)*, Atlanta, USA, 2007, pp. 1–6.

[17] S. Rammohan, "Reduced Complementary Dynamic and Differential Logic: a Circuit Design Methodology for DPA-resistant Cryptographic ICs," Master's thesis, University of Cincinnati, May 2007.

[18] C. Paar, "Efficient VLSI Architectures for Bit Parallel Computation in Galois Fields," Ph.D. dissertation, Inst. for Exp. Math., 1994.

[19] E. D. Mastrovito, "VLSI Designs for Multiplication over Finite Fields GF (2m)," in *Proc. of the 6th Intnl. Conf. on App. Algebra, Algebraic Alg. and Error-Correcting Codes*, 1989, pp. 297–309.

[20] RSA Laboratories, "PKCS: RSA Encryption Standard, version 2.1," RSA Security Inc., Tech. Rep., 2002.

978-1-4244-3039-0/08 $25.00 © 2008 IEEE

21st International Conference on VLSI Design

Power Attack Resistant Efficient FPGA Architecture for Karatsuba Multiplier

Chester Rebeiro[1] and Debdeep Mukhopadhyay[2]
[1] MS Scholar, Dept. of Computer Science and Engineering
[2] Assistant Professor, Dept. of Computer Science and Engineering
Indian Institute of Technology Madras, India
{rebeiro, debdeep}@cse.iitm.ernet.in

Abstract

The paper presents an architecture to implement Karatsuba Multiplier on an FPGA platform. Detailed analysis has been carried out on how existing algorithms utilize FPGA resources. Based on the observations the work develops a hybrid technique which has a better area delay product compared to the known algorithms. The results have been practically demonstrated through a large number of experiments. Subsequently, the work develops a masking strategy to prevent power based side channel attacks on the multiplier. It has been found that the proposed masked Hybrid Karatsuba multiplier is more compact compared to existing designs.

1 Introduction

Elliptic Curve Cryptography (ECC) is becoming increasing popular for public key cryptosystems because it offers more security per key bit than any other crypto algorithm. This results in shorter keys and smaller implementations. ECC implementation follows a layered hierarchical scheme. The lower layers in the hierarchy are the finite field arithmetic operations, while the top layers are the scalar multiplications in the group formed by the Elliptic Curves. The performance of the top layers of the hierarchy is greatly influenced by the performance of the underlying layers. It is therefore important to have efficient implementations of finite field arithmetic. In this paper we discuss an important component of the ECC implementation: the finite field multiplier.

There are several methods to implement finite field multiplication, however the Karatsuba multiplication [4] algorithm is the most efficient. The basic Karatsuba algorithm is ideally suited when the number of bits (n) of the multiplicands is a power of 2. i.e. $n = 2^k$. If $n \neq 2^k$, as in the finite fields used for ECC, an adaption of the Karatsuba algorithm has to be used. There have been several research

works [6][11][3][9] for Karatsuba multipliers in such fields.

In this paper we present a novel Hybrid Karatsuba multiplier for elliptic curves. Our multiplier is designed specifically for FPGA platforms and requires least resources as compared to other Karatsuba multipliers used for elliptic curve arithmetic. We selected FPGAs as our development platform because of the low non recurring costs, simpler design cycles, faster time to market, greater performance per unit area and the rapid growth in FPGA usage. Besides these, FPGAs for cryptography applications have several advantages over ASICs[12]. However there has been little work done on the security aspects of FPGAs. It has been shown in [8] that Side Channel Attacks (SCAs) on FPGAs is a reality, therefore cryptography implementations on the FPGA have to be made resistant to such attacks. In this paper we develop an efficient technique to prevent Differential Power Analysis (DPA) [5] on our multiplier.

In section 2 we present the background for finite field multiplication and side channel attacks. Section 3 discusses how the existing Karatsuba algorithms get mapped on to an FPGA. Section 4 presents our proposed Karatsuba multiplier. Section 5 shows how our multiplier can be made resistant to DPA attacks. Section 6 has the conclusion.

2 Preliminaries

In this section we present a background on the topic of finite field multiplication and the Karatsuba method to multiply efficiently. We also briefly state about Side Channel Attacks (SCA) on cryptographic algorithms.

2.1 Finite Field Multiplications and the Karatsuba Algorithm

Finite field multiplication of two elements in the field $GF(2^n)$ is defined as

$$C(x) = A(x)B(x) mod P(x) \qquad (1)$$

where $A(x)$, $B(x)$ and $C(x) \in GF(2^n)$, and $P(x)$ is the irreducible polynomial of degree n which generates the field

706

978-1-4244-3039-0/08 $25.00 © 2008 IEEE

$GF(2^n)$. Implementing the multiplication requires two steps. First, the polynomial product $C'(x) = A(x)B(x)$ is determined, then, the modulo operation is done on $C'(x)$. The Karatsuba Algorithm is used for the polynomial multiplication. The Karatsuba multiplier achieves its efficiency by splitting the n bit multiplicands into two 2-term polynomials as shown in equation (2).

$$A(x) = A_h x^{n/2} + A_l \qquad B(x) = B_h x^{n/2} + B_l \qquad (2)$$

The multiplication is then done using three $n/2$ bit multiplications as shown in equation (3). The three $n/2$ bit multiplications are implemented recursively using the Karatsuba algorithm.

$$
\begin{aligned}
C'(x) &= (A_h x^{n/2} + A_l)(B_h x^{n/2} + B_l) \\
&= A_h B_h x^n + (A_h B_l + A_l B_h)x^{n/2} + A_l B_l \\
&= A_h B_h x^n \\
&\quad + ((A_h + A_l)(B_h + B_l) + A_h B_h + A_l B_l)x^{n/2} \\
&\quad + A_l B_l
\end{aligned}
\qquad (3)
$$

The gate requirements for an n bit recursive Karatsuba multiplier is given below.

$\#AND$ gates : $n^{log_2 3}$

$\#XOR$ gates : $\sum_{r=0}^{log_2 n} 3^r \left(4n/2^r - 4\right)$

2.2 Karatsuba Multiplication for Elliptic Curves

The basic recursive Karatsuba multiplier cannot be applied directly to ECC, because the binary extension fields used in standards such as in [10] have a degree which is prime. There have been several published works which implement a modified Karatsuba algorithm for use in elliptic curves. The easiest method to modify the Karatsuba algorithm for elliptic curves is by padding. The *Padded Karatsuba Multiplier* extends the n bit multiplicands to $2^{\lceil log_2 n \rceil}$ bits by padding the most significant bits with zeroes. This allows the use of the basic recursive Karatsuba algorithm. The obvious drawback of this method is the extra arithmetic introduced due to the padding.

In [6], a *Binary Karatsuba Multiplier* was proposed to handle multiplications in any field of the form $GF(2^n)$, where $n = 2^k + d$, and k is the largest integer such that $2^k < n$. The Binary Karatsuba multiplier splits the n bit multiplicands (A and B) into two terms. The lower terms (A_l and B_l) have 2^k bits while the higher terms (A_h and B_h) have d bits. Two 2^k bit multipliers are required to obtain the partial products $A_l B_l$ and $(A_h + A_l)(B_h + B_l)$. For the latter multiplication, the A_h and B_h terms have to be padded by $2^k - d$ bits. $A_h B_h$ product is determined using a d bit Binary Karatsuba multiplier.

In [9], a *Recursively Applied Iterative Karatsuba* implementation was presented. In this architecture, each of the $n/2$ bit multiplications is done serially using the same hardware. This has the advantage of a reduction in the area required at the cost of increased timing requirements. However this technique still relies on padding when n is not a

power of two. Although there is a marginal reduction in the number of XOR gates, the number of AND gates remains the same.

The *Simple Karatsuba Multiplier* [11] is the basic recursive Karatsuba multiplier with a small modification. If an n bit multiplication is needed to be done, n being any integer, it is split into two polynomials as in equation (3). The A_l and B_l terms have $\lceil n/2 \rceil$ bits and the A_h and B_h terms have $\lfloor n/2 \rfloor$ bits. The Karatsuba multiplication can then be done with two $\lceil n/2 \rceil$ bit multiplications and one $\lfloor n/2 \rfloor$ bit multiplication. The upper bound for the number of AND gates and XOR gates required for the Simple Karatsuba multiplier is the same as that of a $2^{\lceil log_2 n \rceil}$ bit basic recursive Karatsuba multiplier. The maximum number of gates required and the time delay for an n bit Simple Karatsuba multiplier is given below.

$\#AND$ gates : $3^{\lceil log_2 n \rceil}$

$\#XOR$ gates : $\sum_{r=0}^{\lceil log_2 n \rceil} 3^r \left(4\lceil n/2^r \rceil - 4\right)$

In the *General Karatsuba Multiplier* [11], the multiplicands are split into more than two terms. For example an n term multiplier is split into n different terms. The number of gates required is given below.

$\#AND$ gates : $n(n+1)/2$

$\#XOR$ gates : $\frac{5}{2}n^2 - \frac{7}{2}n + 1$

2.3 Side Channel Resistant Multiplication

An important aspect of the finite field multiplier used for elliptic curve cryptosystems is that they have to be resistant to SCAs. SCAs are the biggest threat to modern cryptosystems. In these attacks knowledge is gathered about the key by exploiting the information that leaks from various sources in the device. Among all side channel attacks, SCA based on power consumption of the device is the most practical and threatening. There are two techniques for power analysis: Simple Power Analysis (SPA) and Differential Power Analysis (DPA). DPA exploits the fact that power consumption of a chip depends on intermediate results of the algorithm.

One of the most common techniques to counter DPA in multipliers is by using masking [2]. The main idea behind a masked multiplier is to make all intermediate values of the multiplier independent of the multiplicands. Such multipliers are secure against power attacks, if the underlying CMOS gates switch once per clock cycle [7].

3 Karatsuba Implementation on FPGA

In this section we discuss the mapping of the recursive Karatsuba algorithm and the General Karatsuba algorithm on an FPGA. We estimate the amount of FPGA resources that is required for the Karatsuba implementations. This demands a special section because of the unique architecture of the FPGA compared with other programmable devices. The Xilinx FPGA [14] is made up of Configurable Logic Blocks (CLBs). Each CLB on a Xilinx Virtex 4 FPGA

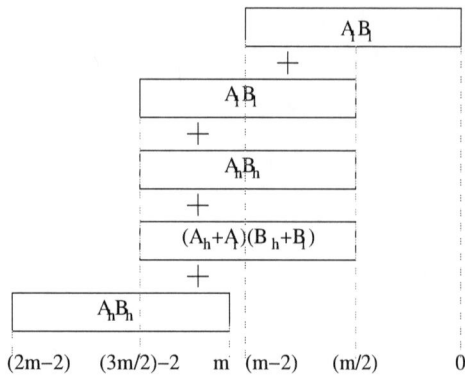

A_lB_l	
+	

Figure 1. Combining the Partial Products

contains two slices. Each slice contains two lookup tables (LUTs). Each LUT has four inputs and can be configured for any logic function having a maximum of four inputs. The LUT can also be used to implement logic functions having less than four inputs, two for example. In this case, only half the LUT is utilized the remaining part is not utilized. Such a LUT which has less than four inputs is an under utilized LUT. For example, the logic function $y = x_1 + x_2$ under utilizes the LUT as it has only two inputs. Most compact implementations are obtained when the utilization of each LUT is maximized. From the above fact it may be derived that the minimum number of LUTs required for a q bit input is given by the equation (4):

$$\#LUT(q) = \begin{cases} 0 & \text{if } q = 1 \\ 1 & \text{if } 1 < q \leq 4 \\ \lceil q/3 \rceil & \text{if } q > 4 \text{ and } q \bmod 3 = 2 \\ \lfloor q/3 \rfloor & \text{if } q > 4 \text{ and } q \bmod 3 \neq 2 \end{cases} \quad (4)$$

The mapping of the algorithm on the FPGA depends on several factors in addition to the FPGA architecture such as the synthesis tool used, the synthesis options used etc. In our analysis we consider only the algorithm and the FPGA architecture. We have verified our analysis with Xilinx ISE synthesis tool [13] and found that for the Recursive Karatsuba multiplier our estimation of the LUTs used match the tool results. For the General Karatsuba multiplier our estimation for the LUTs was marginally lower than the results produced by the tool.

Recursive Karatsuba Multiplier : In an n bit Recursive Karatsuba Multiplier, each recursion reduces the size of the input by half while tripling the number of multiplications required. At each recursion, except the final, only XOR operations are involved. Let $m = 2^{(log_2 n)-k}$ be the size of the inputs, A and B, for the kth recursion of an n bit multiplier. There are 3^k such m bit multipliers required. The A and B inputs are split into two: A_h, A_l and B_h, B_l respectively with each term having $m/2$ bits. $m/2$ two input XORs are required for the computation of $A_h + A_l$ and $B_h + B_l$ respectively (Equation :3). Each two input XOR requires one LUT on the FPGA. Combining the partial products as shown in the figure (1) is the last step of the recursion. Determining the output bits $m - 2$ to $m/2$

and $3m/2 - 2$ to m requires $3(m/2 - 1)$ two input XORs each. The output bit $m - 1$ requires 2 two input XORs. In all $(3m - 4)$ two input $XORs$ are required to add the partial products. The number of LUTs required to combine the partial products is much lower. This is because each LUT implements a four input XOR. Each output bit $m/2$ to $3m/2 - 2$ requires one LUT, therefore $(m - 1)$ LUTs are required for the purpose. In total, $2m - 1$ LUTs are required for each recursion on the FPGA. The final recursion has $3^{(log_2 n)-1}$ two bit Karatsuba multipliers. The equation for the two bit Karatsuba multiplier is shown in equation (5).

$$\begin{aligned} C_0 &= A_0 B_0 \\ C_1 &= A_0 B_0 + A_1 B_1 + (A_0 + A_1)(B_0 + B_1) \\ C_2 &= A_1 B_1 \end{aligned} \quad (5)$$

This requires three LUTs on the FPGA; one for each of the output bits (C_0, C_1, C_2).

The total number of LUTs required for the n bit recursive Karatsuba multiplication is given by equation (6).

$$\begin{aligned} \#LUTS &= 3 * 3^{log_2 n - 1} + \sum_{k=0}^{log_2 n - 2} 3^k (2 * 2^{log_2 n - k} - 1) \\ &= \sum_{k=0}^{log_2 n - 1} 3^k (2^{log_2 n - k + 1} - 1) \end{aligned} \quad (6)$$

The FPGA usage for the Simple Karatsuba multiplier [11] is similar to that of the recursive Karatsuba multiplier.

Algorithm 1: gkmul (*General Karatsuba Multiplier*)

Input: A, B are multiplicands of n bits
Output: C of length $2n - 1$ bits

```
/*  Define :   M_x → A_x B_x                              */
/*  Define :   M_(x,y) → (A_x + A_y)(B_x + B_y)           */
```

1 **begin**
2   **for** $i = 0$ *to* $n - 2$ **do**
3        $C_i = C_{2n-2-i} = 0$
4   **for** $j = 0$ *to* $\lfloor i/2 \rfloor$ **do**
5   **if** $i = 2j$ **then**
6              $C_i = C_i + M_j$
7              $C_{2n-2-i} = C_{2n-2-i} + M_{n-1-j}$
8   **else**
9              $C_i = C_i + M_j + M_{i-j} + M_{(j,i-j)}$
10             $C_{2n-2-i} = C_{2n-2-i} + M_{n-1-j}$
11                $+ M_{n-1-i+j} + M_{(n-1-j,n-1-i+j)}$
12   **end**
13   **end**
14   **end**
15    $C_{n-1} = 0$
16   **for** $j = 0$ *to* $\lfloor (n-1)/2 \rfloor$ **do**
17   **if** $n - 1 = 2j$ **then**
18          $C_{n-1} = C_{n-1} + M_j$
19   **else**
20          $C_{n-1} = C_{n-1} + M_j + M_{n-1-j} + M_{(j,n-1-j)}$
21   **end**
22   **end**
23 **end**

General Karatsuba Multiplier : The n bit General Karatsuba algorithm [11] is shown in Algorithm 1. The '+' in the algorithm denotes XOR operation as the multiplication is in the field $GF(2^n)$. Each iteration of i computes two output bits C_i and C_{2n-2-i}. Computing the two output bits require same amount of resources on an FPGA. The line 6 in the algorithm is executed once for every even iteration of i, and is not executed for odd iterations of i. The term $M_j + M_{i-j} + M_{(j,i-j)}$ is computed with the four inputs A_j, A_{i-j}, B_j and B_{i-j}, therefore, on an FPGA, computing the term would require one LUT. For an odd i, the C_i would have $\lceil i/2 \rceil$ such LUTs whose outputs have to be added. The LUTs required for this can be obtained from equation (4).An even value of i would have an additional two inputs corresponding to $M_{i/2}$ which have to be added. The LUTs required for computing C_i ($0 \le i \le n-1$) is given by equation (7).

$$
\#LUT_{C_i} = \begin{cases} 1 & \text{if } i = 0 \\ \lceil i/2 \rceil + \#LUT(\lceil i/2 \rceil) & \text{if } i \text{ is odd} \\ i/2 + \#LUT(i/2 + 2) & \text{if } i \text{ is even} \end{cases} \quad (7)
$$

The total number of LUTs required for the General Karatsuba multiplier is given by equation(8).

$$
\#LUTS = 2\left(\sum_{i=0}^{n-2} LUT_{C_i} \right) + LUT_{C_{n-1}} \quad (8)
$$

4 Hybrid Karatsuba Multiplier

In this section we present our Hybrid Karatsuba multiplier. We show how we combine techniques to maximize the utilization of each LUT resulting in minimum area. We also show that the proposed multiplier has minimum area delay product on the FPGA.

The table (1) compares the General and Simple Karatsuba algorithms for gate counts (two input XOR and AND gates), LUTs required on a Xilinx Virtex 4 FPGA and the percentage of LUTs under utilized. The percentage of under utilized LUTs is determined using the equation (9). LUT_k is the number of LUTs having k inputs.

$$
\%UnderUtilizedLUTs = \frac{LUT_2 + LUT_3}{LUT_2 + LUT_3 + LUT_4} * 100 \quad (9)
$$

n	General			Simple		
	Gates	LUTs	LUTs Under Utilized	Gates	LUTs	LUTs Under Utilized
2	7	3	66.6%	7	3	66.6%
4	37	11	45.5%	33	16	68.7%
8	169	53	20.7%	127	63	66.6%
16	721	188	17.0%	441	220	65.0%
29	2437	670	10.7%	1339	669	65.4%
32	2977	799	11.3%	1447	723	63.9%

Table 1: *Multiplication Comparison*

The Simple Karatsuba multiplier is not efficient for FPGA platforms as the number of under utilized LUTs is about 65%. For an n bit simple recursive multiplier the two bit multipliers take up approximately a third of the area (for $n = 256$). In a two bit multiplier, two out of three LUTs required, are under utilized (In equation (5), C_0 and C_2 result in under utilized LUTs). In addition to this, around half the LUTs used for each recursion is under utilized. The under utilized LUTs results in a bloated area requirement on the FPGA.

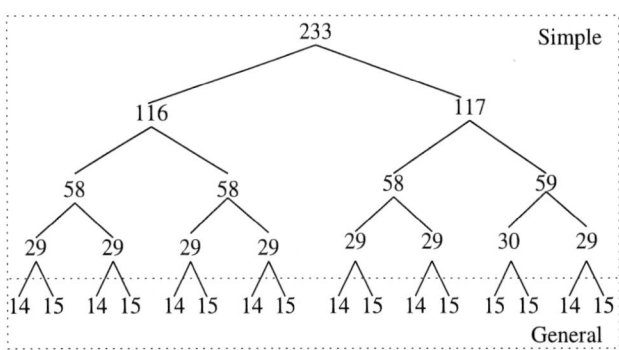

Figure 2. 233 Bit Hybrid Karatsuba Multiplier

The n-term General Karatsuba is more efficient on the FPGA for small values on n (Table 1) even though the gate count is significantly higher. This is because a large number of operations can be grouped in fours which fully utilizes the LUT. For small values of n ($n < 29$) the compactness obtained by the fully utilized LUTs is more prominent than the large gate count, resulting in low footprints on the FPGA. For $n \ge 29$, the gate count far exceeds the efficiency obtained by the fully utilized LUTs, resulting in larger footprints with respect to the Simple Karatsuba implementation.

In our proposed Hybrid Karatsuba multiplier, shown in the algorithm (2), the n bit multiplicands are split into two parts when the number of bits is greater than or equal to the threshold 29. The higher term has $\lfloor n/2 \rfloor$ bits while the lower term has $\lceil n/2 \rceil$ bits. If the number of bits of the multiplicand is less than 29 the General Karatsuba algorithm is invoked. The General Karatsuba algorithm ensures maximum utilization of the LUTs for the smaller bit multiplications, while the Simple Karatsuba algorithm ensures least gate count for the larger bit multiplications.

Algorithm 2: hmul *(Hybrid Karatsuba Multiplier)*

Input: The multiplicands A, B and their length n
Output: C of length $2n - 1$ bits

1 **begin**
2   **if** $n < 29$ **then**
3   **return** $gkmul(A, B, n)$
4   **else**
5        $l = \lceil n/2 \rceil$
6        $A' = A_{[n-1 \ldots l]} + A_{[l-1 \ldots 0]}$
7        $B' = B_{[n-1 \ldots l]} + B_{[l-1 \ldots 0]}$
8        $C_{p1} = hmul(A_{[l-1 \ldots 0]}, B_{[l-1 \ldots 0]}, l)$
9        $C_{p2} = hmul(A', B', l)$
10       $C_{p3} = hmul(A_{[n-1 \ldots l]}, B_{[n-1 \ldots l]}, n - l)$
11   **return**
      $(C_{p3} << 2l) + (C_{p1} + C_{p2} + C_{p3}) << l + C_{p1}$
      ;   /* $<<$ *indicates left shift* */
12
13   **end**
14 **end**

The ith recursion ($0 \le i < r$) of the n bit multiplier has 3^i multiplications. The multipliers in this recursion have bit lengths $\lceil n/2^i \rceil$ and $\lfloor n/2^i \rfloor$. For simplicity we assume

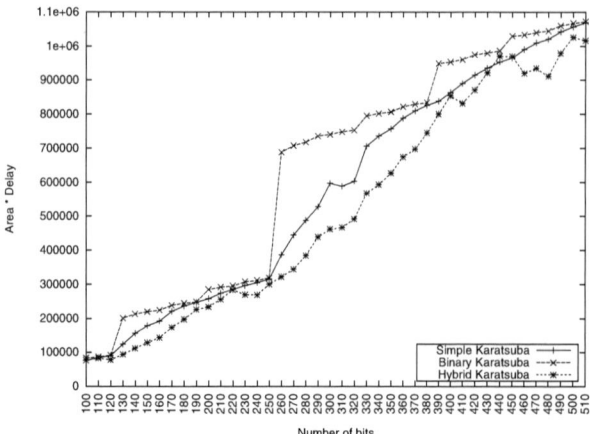

Figure 3. n bit multiplication vs Area X Delay

the number of gates required for the $\lceil n/2^i \rceil$ and $\lfloor n/2^i \rfloor$ bit multipliers is equal to that of a $\lceil n/2^i \rceil$ bit multiplier. The total number of AND gates required is the AND gates for the multiplier in the last recursion (i.e. $\lceil n/2^{r-1} \rceil$ bit multiplier) times the number of $\lceil n/2^{r-1} \rceil$ multipliers present.

$$\#AND = 3^{r-1} \lceil \frac{n}{2^r} \rceil \left(\lceil \frac{n}{2^{r-1}} \rceil + 1 \right) \qquad (10)$$

The number of XOR gates required for the ith recursion is given by $4\lceil \frac{n}{2^i} \rceil - 4$. The total number of two input $XORs$ is the sum of the $XORs$ required for last recursion, $\#XOR_{g_{r-1}}$, and the $XORs$ required for the other recursions, $\#XOR_{s_i}$.

$$\#XOR = 3^{r-1}\#XOR_{g_{r-1}} + \sum_{i=0}^{r-2} 3^i \#XOR_{s_i}$$

$$= 3^{r-1} \left(10\lceil \frac{n}{2^r} \rceil^2 - 7\lceil \frac{n}{2^r} \rceil + 1 \right) + \sum_{i=0}^{r-2} 3^i \left(4\lceil \frac{n}{2^i} \rceil - 4 \right)$$

$$(11)$$

4.1 Performance Results

The graph in figure (3) compares the area delay product for the Hybrid Karatsuba multiplier with the Simple Karatsuba Multiplier and the Binary Karatsuba multipliers for increasing values of n. The area and delay was obtained by synthesizing each multiplier using Xilinx's ISE[13] for a Virtex 4 FPGA. The area was determined by the number of LUTs required for the multiplier. The delay is in nanoseconds. The graph shows that the area delay product for the Hybrid Karatsuba multiplier is lesser compared to the other multipliers.

5 DPA Resistant Multiplier

DPA attacks monitor the power consumption of the multiplier for several different inputs and then use statistical analysis to extract critical information about the key. One

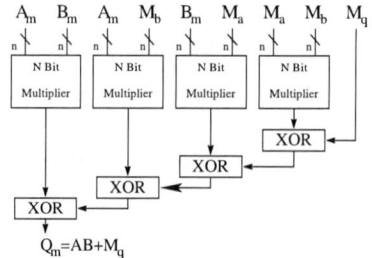

Figure 4. Generic Masked Multiplier

way to avoid DPA based attacks is to make the power consumption independent of the inputs. Masking the inputs with random values can achieve this[2]. However the masking should be in such a way that the required result can be inferred from the result obtained by the masked inputs.

For an n bit masked multiplier the inputs A and B are masked with random values M_a and M_b.

$$A_m = A + M_a \qquad B_m = B + M_b \qquad (12)$$

The above equations can be rewritten as

$$A = A_m + M_a \qquad B = B_m + M_b \qquad (13)$$

The generic way to build a masked multiplier is as follows. The masked inputs A_m, B_m and masks M_a, M_b are fed to the masked multiplier as shown in figure (4). The output of the masked multiplier is the sum of $(A * B)$ and another mask M_q as shown in the equation (14).

$$Q_m = A_m B_m + A_m M_b + B_m M_a + M_a M_b + M_q$$
$$= (A_m + M_a)(B_m + M_b) + M_q \qquad (14)$$
$$= AB + M_q$$

The four N bit multipliers are implemented with any of the multiplication techniques described in the previous sections. Let MUL_A and MUL_X be the number of AND and XOR gates required for each multiplier respectively. The gates required by the masked multiplier is shown below. The $4(2n - 1)$ extra $XORs$ are required to sum the outputs of the individual multiplications.

$$\#AND : 4MUL_A(n)$$
$$\#XOR : 4MUL_X(n) + 4(2n - 1) \qquad (15)$$

Masking a multiplier requires more than four times the

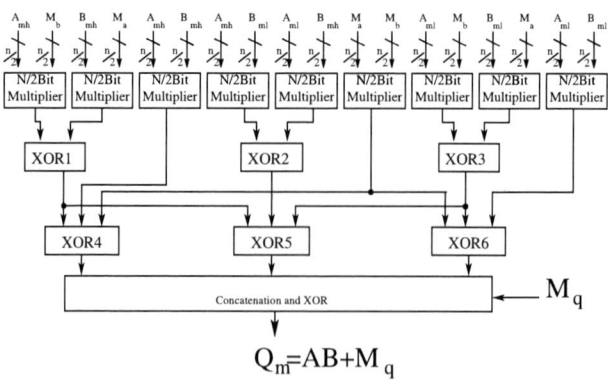

Figure 5. Proposed Masked Multiplier

gates as that of an unmasked multiplier. It is therefore important to search for masked techniques with the number of gates minimized. In this perspective we propose a masked multiplier (Figure 5) with lesser number of gates as compared with the generic masked multiplier. The multiplicands A and B of our masked multiplier are split into two terms just as in the Karatsuba algorithm. We then use an $n/2$ bit mask, M_a, to mask both the A_h and A_l terms. Similarly an $n/2$ bit mask, M_b, masks B_h and B_l as shown in equation (16).

$$A_{mh} = A_h + M_a \qquad A_{ml} = A_l + M_a$$
$$B_{mh} = B_h + M_b \qquad B_{ml} = B_l + M_b \qquad (16)$$

The masked inputs A_{mh}, A_{ml}, B_{mh}, B_{ml}, and the masks M_a, M_b and M_q are fed to the masked multiplier. The output (Equation 17) is the product of $(A * B)$ and the $2n - 1$ bit mask M_q.

$$
\begin{aligned}
Q_m &= AB + M_q \\
&= [A_h x^{\frac{n}{2}} + A_l][B_h x^{\frac{n}{2}} + B_l] + M_q \\
&= [(A_{mh} + M_a)x^{\frac{n}{2}} + (A_{ml} + M_a)] \\
&\quad [(B_{mh} + M_b)x^{\frac{n}{2}} + (B_{ml} + M_b)] + M_q \\
&= (A_{mh}B_{mh} + A_{mh}M_b + B_{mh}M_a + M_aM_b)x^n \\
&\quad + (A_{mh}B_{ml} + A_{mh}M_b + B_{ml}M_a \\
&\quad + A_{ml}B_{mh} + A_{ml}M_b + B_{mh}M_a)x^{\frac{n}{2}} \\
&\quad + (A_{ml}B_{ml} + A_{ml}M_b + B_{ml}M_a + M_aM_b) + M_q
\end{aligned}
\tag{17}
$$

The number of AND gates required is nine times that of an $n/2$ bit multiplier.

$$\#AND = 9MUL_A(n/2) \tag{18}$$

In addition to the $XORs$ required for the $n/2$ bit multipliers, XOR gates are required for combining the products, adding the mask M_q and for the gates $XOR1$ to $XOR6$. The $XOR1$, $XOR2$ and $XOR3$ require $(n - 1)$ $XORs$ each. The $XOR4$, $XOR5$ and $XOR6$ require $2(n - 1)$ two input $XORs$ each. Combining and adding the mask M_q requires $3(n - 1)$ $XORs$.

$$\#XOR = 9MUL_X(n/2) + 12(n - 1) \tag{19}$$

The graph in figure (6) shows the number of gates ($AND + XOR$) required verses the size of the multiplicands for the generic and proposed masked multipliers. The underlying multiplier in both cases is the Hybrid Karatsuba multiplier proposed earlier. From the graph the proposed masked multiplier has lesser gates requirements as compared to the generic masked multiplier and hence will require lesser area on the FPGA. The critical delay for $n = 233$ for the generic masked multiplier and the proposed masked multiplier was found to be almost the same (around 17 ns on a Xilinx Virtex 4 FPGA).

6 Conclusion

In this paper we proposed a hybrid technique of implementing the Karatsuba multiplier. Our proposed design results in best area $\times$ delay products on an FPGA compared

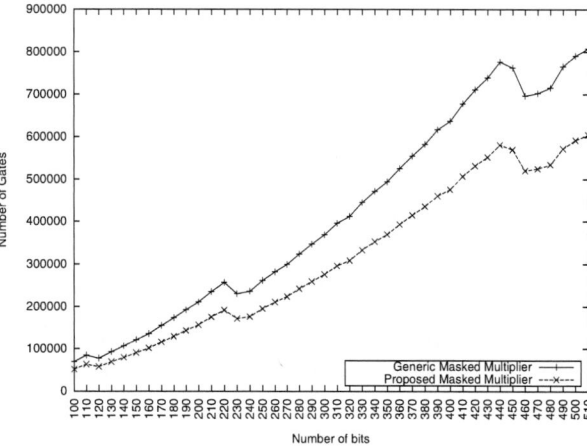

Figure 6. Gates required for n bit masked multiplication

to other Karatsuba implementations. We then proposed a new technique based on masking to prevent DPA attacks on the multiplier. Our technique is more efficient than conventional masking techniques. This multiplier forms an ideal base for a side channel resistant Elliptic Curve Crypto implementations on FPGA platforms.

References

[1] A. DeHon, *The Density Advantage of Configurable Computing*, IEEE Computer Society, 2000.

[2] J.D. Golic, *Techniques for Random Masking in Hardware*, IEEE Transactions on Circuits and Systems, vol 54, No 2, Pg 291-300, 2007.

[3] C. Grabbe, et. al., *FPGA Designs of Parallel High Performance $GF(2^{233})$ Multipliers*, International Symposium on Circuits and Systems, ISCS 2003.

[4] A. Karatsuba and Y. Ofman, *Multiplication of Multidigit Numbers on Automata*, Soviet Phys. Doklady (English Translation), vol 7, no 7, pg 595-596, 1963.

[5] P. Kocher, et. al., *Differential Power Analysis*, Advances in Cryptology-CRYPTO'99, 1999.

[6] F. R. Henriquez, et. al., *On Fully Parallel Karatsuba Multipliers for $GF(2^m)$*, Proceedings of the International Conference on Computer Science and Technology, CST 2003, 2003.

[7] S. Mangard, et. al., *Pinpointing the Side-Channel Leakage of Masked CMOS Implementations*, Workshop on Cryptographic Hardware and Embedded Systems 2006, CHES 2006.

[8] S.B. Ors, et. al., *Power-Analysis Attacks on an FPGA - First Experimental Results*, Workshop of Cryptographic Hardware and Embedded Systems - CHES 2003.

[9] S. Peter, et. al., *An Efficient Polynomial Multiplier in $GF(2^m)$ and its Applications to ECC Design*, Design, Automation and Test in Europe, DATE 2007, DATE.

[10] U.S. Department of Commerce/National Institute of Standards and Technology, *Digital Signature Standards*, Federal Information Processing Standards, Publication 186.

[11] A. Weimerskirch, et. al., *Generalizations of the Karatsuba Algorithm for Efficient Implementations*, http://eprint.iacr.org/2006/224.pdf, 2006.

[12] T. Wollenger. et. al., *Security on FPGAs: State of the Art Implementations and Attacks*, ACM Transactions in Embedded Computing Systems, Vol 3, No. 3, 2004.

[13] Xilinx, ISE Foundation, *http://www.xilinx.com/ise/logic_design_prod/foundation.htm*

[14] Xilinx, Virtex 4 User Guide, *http://direct.xilinx.com/bvdocs/user guides/ug070.pdf*

21st International Conference on VLSI Design

Watermarking Video Clips with Workload Information for DVS

Yicheng Huang   Samarjit Chakraborty   Ye Wang

Department of Computer Science, National University of Singapore

E-mail: {huangyic, samarjit, wangye}@comp.nus.edu.sg

Abstract

We present a lightweight scheme for watermarking or annotating video clips with information describing the workload that would be incurred while decoding the clip. This information can be used at run time to scale the operating voltage/frequency of the processor on which the video clip is to be decoded. Our main contribution is a fast, low-cost bitstream analysis technique for estimating the decoding workload of a video clip. Using this technique the workload information can be inserted into a clip while it is being downloaded onto a battery-powered portable device from a desktop computer or a server, for later playback. In contrast to control-theoretic feedback techniques that have been traditionally used for predicting video decoding workload at runtime for dynamic voltage/frequency scaling, we show that our scheme performs better in terms of energy savings and has significantly lower buffer requirements.

1 Introduction

Energy efficiency is today one of the most critical issues in the design of battery-powered portable devices such as mobile phones, PDAs and audio/video players. The predominant workload running on most of these devices are now generated by multimedia processing applications (e.g. audio/video decoders). This has resulted in a considerable interest in power management schemes for portable devices running multimedia applications [2, 3, 4, 9, 10].

In this paper we propose a new approach for dynamic voltage scaling (DVS) in the context of multimedia applications. In what follows, we will only be concerned with MPEG-2 video decoding. However, the proposed scheme is very general and can be applied to both, other types of video, as well as audio processing applications. The scheme relies on a lightweight offline bitstream analysis of a video clip to predict the workload that will be generated while decoding the clip. Based on this analysis, the video clip is watermarked with the predicted workload information. Such watermarking can either consist of metadata information being inserted into the video clip or such information being saved as a separate file. At runtime, the decoder reads this metadata information and controls (or scales) the voltage and frequency of the processor. The metadata information will typically consist of the frequency at which the

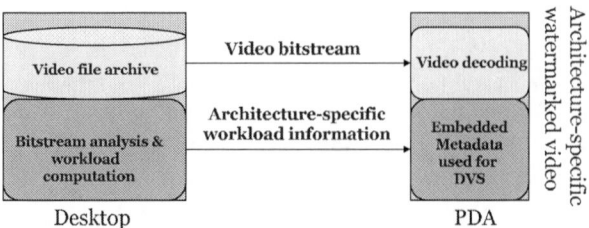

Figure 1. Overview of the proposed scheme.

processor needs to be run at any point in time. However, the metadata might also consist of workload information (such as processor cycle demands), from which the required processor frequency is computed at runtime. The amount of metadata that needs to be inserted depends on the granularity, or how often the frequency of the processor needs to be changed. If the amount of metadata allowed is large, then potentially higher amounts of energy can be saved.

Figure 1 illustrates the key idea behind our scheme. It shows a setup where a desktop computer or a multimedia server stores video clips. When such clips are being downloaded into a portable device, a lightweight bitstream analysis scheme runs on the desktop computer and watermarks the video clip with workload information. The watermarked video clips are then stored in the portable device. At runtime, the workload information is read out and used for dynamic voltage and frequency scaling. It may be noted here that downloading audio/video clips from a desktop into a portable device is currently the most common use scenario. The other possible scenario where such clips are directly downloaded into the portable device over the network is still fairly uncommon and many portable media players currently don't even support network interfaces; however that might change over the next few years.

The key point to note here is that all the previously known techniques predict at *runtime* the processor frequency f with which a segment of the video clip s needs to be decoded *without* looking into s. In contrast to such techniques, we perform an offline analysis of the compressed bitstream corresponding to s and insert the metadata f *before* the start of s. The runtime system simply reads f and sets the processor frequency to this value. Also, note that the metadata information need not be equally spaced out within the video clip. If the computational workload of a clip is highly variable and irregular, then this might require

712

978-1-4244-3039-0/08 $25.00 © 2008 IEEE

more metadata. Whereas certain portions of the clip might not exhibit any variation, in which case it might suffice to run the processor at a constant frequency (and hence only this constant frequency value needs to be inserted once). The inserted metadata information might consist of frequency as well as voltage values, depending on the type of the underlying processor. Again, in contrast to this, many current approaches attempt to scale the processor frequency at regular intervals (e.g. at all video frame boundaries). We show that our scheme leads to energy savings that are comparable and often better than those achieved by known DVS schemes based on runtime (control-theoretic) prediction techniques. More importantly, the buffer requirement in our scheme is significantly smaller.

Main Challenge: The main challenge in implementing the proposed scheme lies in the metadata computation process. Clearly, the exact values of the metadata inserted will depend on the architecture of the processor in the portable device (e.g. its instruction set architecture, voltage/frequency range and the steps in which they can be changed) and also on the decoder application running on this processor. The metadata computation task running on the desktop is aware of the architecture of the portable device. However, the generated metadata is transparent to the user; it only resides inside the portable device.

One possibility is to insert this metadata information directly during the encoding process. However, this would assume that the details of the decoder and also the processor on which the decoder would run are already known at the time of encoding. It would also amount to generating video clips which can only be played on certain devices or on devices manufactured by the same company, which are all based on the same or on similar processor architectures. Although this is a feasible option (e.g. the Windows Media format is only targeted towards Windows platforms), it is clearly very restrictive.

We therefore propose a scheme where the metadata information is directly inserted into a video clip based on the architecture of the portable device. Towards this, we assume the following scenario. To download a video file into such a device, it would be connected to a desktop computer on which an application program specialized for this device would run. This program would perform a bitstream analysis of the video file being downloaded, calculate the appropriate metadata information and insert this information into the file. Since the program is specialized for this device, the metadata computed is specific to its processor architecture and also to the decoder application running on the device. Each such device would therefore be shipped with an application program (that would run on the desktop computer) that is specific to the device. This scheme has two main advantages: (i) It is flexible, i.e. the portable device can play video files encoded in standard formats such as MPEG-2

and the metadata-inserted files are not visible to the external world; they only exist inside the portable device. (ii) The bitstream analysis process, which might be involved, can run on a desktop computer and not on the portable device, which would typically be resource constrained.

Metadata Computation: The only remaining question that needs to be answered is, given a video file, how is the metadata or the watermark information exactly computed? What follows in this paper will mostly be concerned with answering this question.

The most straightforward answer to this question is, simulate the decoding of the given video file on a software model of the processor's architecture. This would result in a trace of the file's processor cycle requirements, e.g. the number of processor cycles required to decode each macroblock of the video file. From this trace, the clock frequency with which the processor should be run while decoding any segment of the file can be computed. The computed frequencies will constitute the metadata information to be inserted into the video file. Towards this, it would be possible to use processor instruction set simulators like SimpleScalar [1] to compute the trace of processor cycle requirements of a video file. However, a cycle-accurate simulation of the execution of a processor is extremely expensive in terms of the simulation time involved. For example, simulating the decoding of a 30 seconds long MPEG-2 video clip requires more than half an hour using SimpleScalar. Hence, this scheme is not feasible if the metadata computation needs to be done while downloading a video file from a desktop computer into a portable device, without any perceptible delay.

We therefore propose an alternative lightweight scheme where we do not simulate the execution/decoding of the video clip. Instead we perform a bitstream analysis to *predict* the processor cycle requirements of each macroblock. The scheme has to be lightweight since it has to be fast enough and not delay the downloading process from the desktop computer to the portable device. We would again like to point out that in contrast to this, runtime prediction schemes predict the processor cycle requirement of a video segment *without* looking into the segment. Our scheme allows for the bitstream analysis because it is done offline (i.e. not at runtime) while the video file is being downloaded into the device. The prediction scheme we propose is based on classifying the video decoding tasks into two groups—those that are CPU-bound, such as motion compensation, and others which are memory-bound such as those responsible for dithering. The processor cycle requirements of memory-bound tasks are almost constant and are hence easy to predict. Hence, we shall mostly be concerned with predicting the processor cycle requirements of CPU-bound tasks. As already mentioned, in this paper we will use MPEG-2 for the sake of illustration.

Paper Organization: The rest of this paper is organized as follows. In the next section we present our analysis scheme for MPEG-2. This consists of estimating the workload incurred by the different subtasks of an MPEG-2 decoder. In Section 3 we show how our approach compares with known DVS schemes that use control-theoretic techniques for online/runtime workload prediction. Finally, we conclude in Section 4 by outlining some directions for future work.

2 MPEG-2 Bitstream Analysis

An MPEG-2 video sequence is made up of a number of *frames*, where each frame contains several *slices*. Each slice in turn consists of a number of *macroblocks* (MBs). Decoding an MPEG-2 video can therefore be considered as decoding a sequence of MBs. This involves executing the following tasks for each MB: variable length decoding (VLD), inverse discrete cosine transformation (IDCT) and motion compensation (MC). Other tasks, such as inverse quantization (IQ) involves a negligible amount of computational workload and hence we ignore them for the purpose of our analysis. The analysis we present here can be used for voltage/frequency scaling at the MB granularity (clearly, the same analysis can be used at the slice or frame granularity as well). Given a sequence of MBs, in this section we describe how to predict the processor cycle requirements corresponding to the tasks VLD, IDCT and MC for each of these MBs. We compare our predicted results with those obtained from simulating the execution of these tasks using the SimpleScalar [1] instruction set simulator (with the Sim-Profile configuration), with the same sequence of MBs as input. Since we envisage the decoder to run on a general-purpose processor (such as those found on a PDA), we choose our processor to be a RISC processor (similar to a MIPS3000) without any MPEG-specific instructions. We use Test Model 5 (TM5) [7] as our MPEG-2 decoder application. Although not an optimized decoder, it is acceptable for our analysis since all MPEG-2 decoders have a similar code structure. We experimented with five different commonly used benchmark video clips, encoded with a 4M/s bitrate: (i) Flwr (has moderate motion), (ii) Tennis (still background with moving foreground), (iii) Susi (very low motion), (iv) V700 (still image) and (v) Football (very fast motion).

The Variable Length Decoding Task: The IDCT coefficients in MPEG-2 are encoded using variable length encoding, which involves Run-Length Coding followed by Huffman Coding. Some run-length codes are coded using longer Huffman codes compared to the others. The number of processor cycles required for the Huffman decoding depends on the length of the Huffman codes used. Therefore, the number of processor cycles required by the VLD task

for any input MB is expected to depend on the number of non-zero IDCT coefficients in it. Our simulations confirm that this is indeed the case and the relationship is a linear one. Hence, we use the following function as an estimate of the number of processor cycles required by the VLD task for any MB: $n_{\text{vld}} = a \times n_{\text{coeff}} + b$. Here, n_{vld} is the estimated number of processor cycles, n_{coeff} is the number of non-zero coefficients in the MB and a and b are constants which depend on the processor architecture and the VLD code. From our experiments we determined the values of a and b to be 140 and 3000 respectively for our setup. For the Flwr video we noted that for around 36% of the MBs, the processor cycle requirements were predicted with an error of less than 2%. For *all* MBs, the prediction error was less than 10% (in the range of -1000 to $+2000$ processor cycles). Other clips also had similar error distributions. It should be noted that the values of a and b in this case capture the characteristics of one specific architecture. Recall that our watermarking is architecture-specific. For a different processor architecture and decoder implementation, these values will change and they are hardcoded in the metadata insertion task running on the desktop (see Figure 1).

The Motion Compensation Task: MBs constituting an MPEG-2 clip may be classified into three categories: those involving no motion compensation (I-type), those involving only forward motion compensation (P-type) and those involving both forward and backward motion compensation (B-type). Therefore, the MC task for P-type MBs incur about half the number of processor cycles compared to B-type MBs and I-type MBs do not incur any computational workload.

We used SimpleScalar simulations to obtain the processor cycle distribution for the MC task for each of our five MPEG-2 video clips. As expected, with the exception of the V700 clip (still image), the number of processor cycles for all of these clips were distributed into three distinct clusters. The first cluster (located around 0 processor cycles) correspond to the I-type MBs, the second (around 3000 - 7000 cycles) correspond to the P-type MBs, and finally the third cluster (around 9000 - 17000 cycles) correspond to the B-type MBs. In the V700 clip, almost all the MBs use the same type of motion compensation, thereby resulting in a single cluster.

Since the processor cycle distribution within each cluster is reasonably large, a prediction solely based on MB type will not be accurate enough. The variability within each cluster results from factors like whether the MC task is frame- or field-based and whether the motion vectors are half- or one-pixel accurate. We account for these as follows.

The code for the MC task may be considered to be composed of a number of subroutines, each of which is essentially the same function, but called with different parame-

714

978-1-4244-3039-0/08 $25.00 © 2008 IEEE

ters. Let us denote this function by $\mathcal{F}$. The number of processor cycles required to execute $\mathcal{F}$ depends only on its input parameters. Depending on the input MB, these parameters include whether (i) Y^1 component's x-dimension is HALF-PIXEL, (ii) Y component's y-dimension is HALF-PIXEL, (iii) U or V component's x-dimension is HALF-PIXEL, (iv) U or V component's y-dimension is HALF-PIXEL, (v) forward or backward motion compensation is required, and (vi) the motion compensation window size is 16×8 or 16×16. Different MBs call $\mathcal{F}$ different number of times and with different values of the above boolean parameters. For example, a P-type non-progressively coded MB, which uses frame-based motion compensation, will call $\mathcal{F}$ twice. Both of these calls are with the same list of parameters $(0, 0, 0, 0, 1, 16 \times 8)$. Similarly, a B-type, progressively coded MB, which uses field-based motion compensation, will also call $\mathcal{F}$ twice, but with the parameters $(1, 1, 1, 1, 1, 16 \times 16)$ and $(1, 1, 1, 1, 0, 16 \times 16)$.

Based on this observation, we predict the processor cycle requirement of the MC task by first simulating the execution of $\mathcal{F}$ with all possible input parameter values. Since there are six boolean parameters, they result in a total of $2^6 = 64$ possible input values. The processor cycle requirement of $\mathcal{F}$ corresponding to each of these 64 possible inputs is stored in a table. Now, given a sequence of MBs, by parsing each MB, we determine the number of times $\mathcal{F}$ is called and with what input parameter values. Using these and our precomputed table of cycle requirements we are able to predict the cycle requirements for each of the MBs. For approximately 40% of the MBs the error incurred is less than 2%. Further, none of the MBs incur an error of more than 4%. Again, it should be noted that the contents of the precomputed table is architecture-specific.

The Inverse Discrete Cosine Transform Task: Usually each MB in MPEG-2 contains four Y blocks, one U block and one V block. Each of these blocks are of 8×8 pixels size. Hence, the input data size to the IDCT task is the same for all MBs, which results in the same computational workload being incurred for all MBs. However, an optimized implementation of the IDCT task takes into account that several IDCT coefficients might be zero and exploits this fact to save some computation. In spite of this, it is a reasonably good approximation to assume that the number of processor cycles incurred by the IDCT task for any MB is a constant, as is confirmed by our experimental results. We selected $2 \times 10^4 + 4000$ as the processor cycle requirement for any macroblock (where 4000 cycles is an architecture-specific "safety margin"). With this prediction, around 61% of the MBs incur an error of less than 2% and 91% of the MBs incur an error of less than 10%.

---

[1]Each frame in MPEG-2 is represented in the YUV color space. See the ISO MPEG-2 standard for details.

Total Cycle Requirements: The total number of processor cycles required to decode a MB may be predicted by summing up the predicted values for the VLD, MC and IDCT tasks and adding a safety margin of 500 cycles (this value may again be obtained from simulations and would depend on the processor architecture and the decoder code).

3 Experimental Results

To evaluate our scheme, we conducted several experiments using an MPEG-2 decoder application. For estimating the power consumption, we adopted the model from [5]: $P \propto \sum(v_{dd,i})^2 f_i$, where $v_{dd,i}$ and f_i are voltage and frequency values set by the scheduler at ith adaptation point. We assumed that $f \propto v_{dd,i}$. We also assumed that the processor can be scaled continuously. The experiments were repeated on five different video sequences that we listed at the beginning of Section 2. Further, we used three different adaptation intervals: every one macroblock of the bitstream, every ten macroblocks and every twenty macroblocks.

We compared our energy savings with those achieved by the DVS scheme published in Wu et al. [8]. Wu et al. uses a PID controller which tracks changes in the buffer fill level (at the input of the processor) and accordingly regulates the processor's speed and voltage. The reason we selected this scheme is because (i) it can handle the stream burstiness and the data-dependent variability in the decoder's execution demand; (ii) it is suitable for buffer-constrained architectures; and (iii) it also uses fixed adaptation intervals as we do in our scheme. Furthermore, to the best of our knowledge, the scheme of Wu et al. represents on of the most advanced DVS techniques recently published.

Comparative results: Here, we would like to point out the contrasts between our approach and that of Wu et al. We predict the workload incurred by the decoder in an offline fashion by analyzing the video clips. Wu et al. on the other hand uses control theoretic techniques to predict the future workload in an online fashion. As we show in this section, the energy savings obtained by both these schemes are similar, with our scheme performing marginally better for most clips. However, the main difference is in terms of the quality of the decoded video, especially in cases where the playout buffer is small. Using the PID controller to scale the processor frequency results in many decoded macroblocks to miss their deadlines (i.e. the output display device is occasionally unable to read a decoded macroblock from the playout buffer). This problem is especially acute when the playout buffer size is small (≤ 1.5 kB). On the other hand, our scheme consistently performs well even for very small buffer sizes and achieves better energy savings. Lastly, as already mentioned, because our scheme relies on an offline

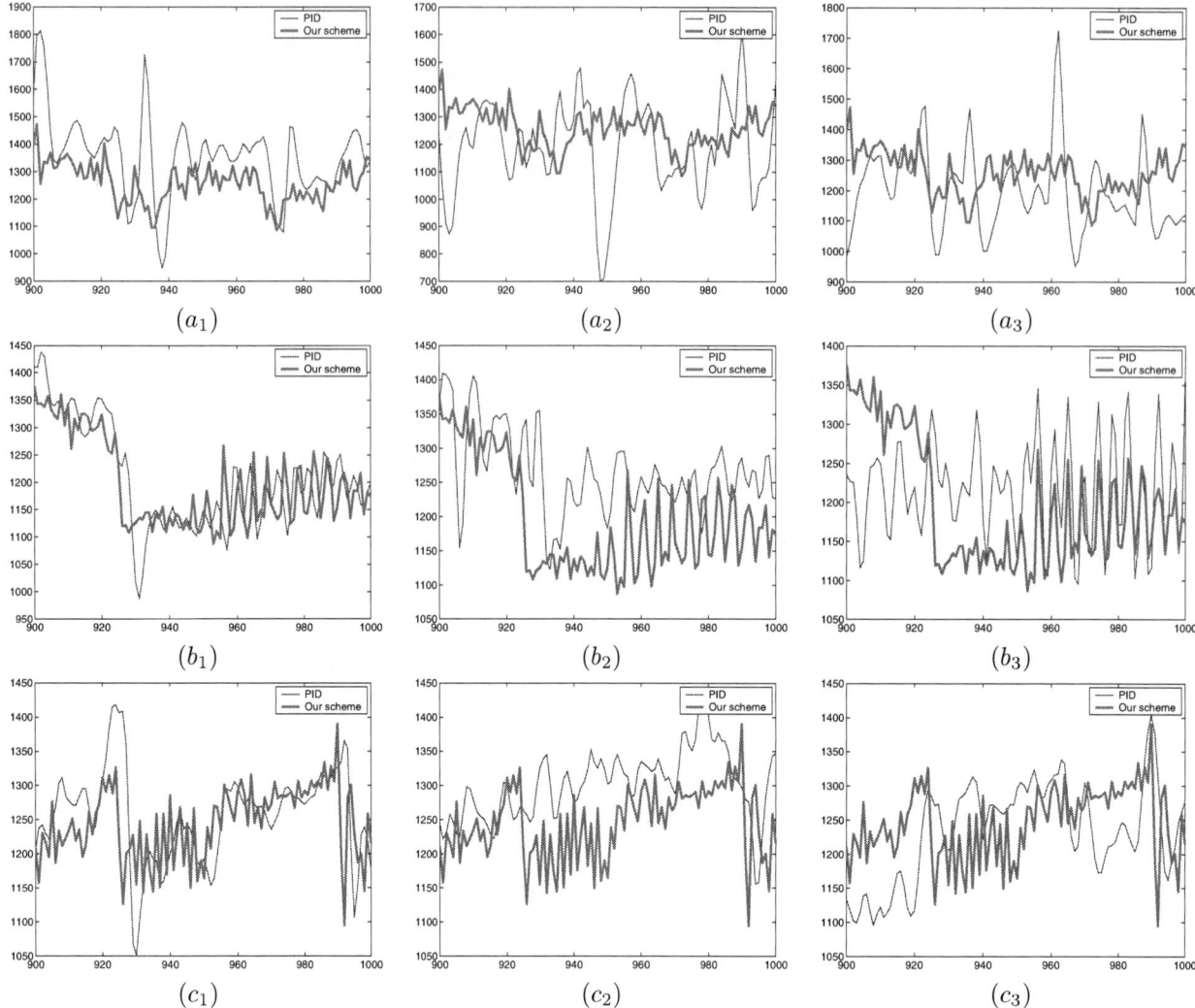

Figure 2. Excerpts from processor frequency schedules generated by our scheme and the PID controller. The horizontal axis represents the adaptation points (adaptation interval specified in terms of no. of macroblocks) and the vertical axis represents the processor frequency (in Hz). (a), (b) and (c) represent the video clips Flwr, Susi and Football respectively. Indices 1 - 3 represent the adaptation intervals of 1, 10 and 20 macroblocks respectively.

workload prediction, it incurs a smaller runtime overhead compared to online schemes.

Figure 2 shows how the processor's frequency is scaled using the PID controller and using our scheme. Each row in this figure represents a specific video clip. The first row shows the results obtained with the clip Flwr. Recall that this clip contains moderate amounts of motion. The second row shows the results for the clip Susi, which has very low motion. Finally, the third row shows the results for Football which has very high motion. All the clips have a frame resolution of 704×480 pixels and are displayed at the rate of 30 frames/second. Each column in this figure shows the results for a specific adaptation interval. Column 1 shows the results when the adaption interval was set to one macroblock, i.e. the processor's voltage and fre-

quency are changed after the decoding of each macroblock. Column 2 shows the results for an adaptation interval of 10 macroblocks and finally Column 3 shows the results for an adaptation interval of 20 macroblocks. While an adaptation interval of one macroblock might incur too much of an overhead, we believe that adaptation intervals of 10 and 20 macroblocks are quite realistic.

The setup we used consists of a playout buffer which stores the decoded macroblocks. This buffer is read out by output device at a constant frame rate. For the PID controller, the expected buffer fill level was always set to half the size of this playout buffer. The baseline case—for comparing the energy savings obtained by both the schemes—was set as a constant processor frequency. This frequency was computed assuming that *all* macroblocks

Buffer size (kB)	Watermarking	PID
0.5	0.1987	42.8906
1.0	0.1852	26.8384
1.5	0.1768	10.3906
2.0	0.1700	2.5253
5.0	0.1077	0.0101

Table 1. Percentage of macroblocks missing their deadlines for different playout buffer sizes.

have the worst-case execution requirement and an output rate of 30 frames/sec will have to be sustained. With this base case, and a one-macroblock adaptation interval, both our scheme and the PID controller-based scheme achieved energy savings in the range of 89.1% and 96.3%. For all the five clips that we experimented with (each having a varying degree of motion), our scheme obtained larger energy savings except for the V700 clip where our scheme obtained a savings of 94.6% and the PID controller-based scheme obtained a savings of 96.3%.

When the adaptation interval was set to 10 macroblocks, the energy savings for both the schemes were in the range of 83% and 96.1%. Again, for all the clips, our scheme performed better than the PID controller except in the case of the V700 clip. Finally, with an adaptation interval of 20 macroblocks, the energy savings were again in the range of 83% and 96.3% with our scheme performing worse than the PID controller for the V700 clip.

Our results show that both the schemes are fairly robust in terms of the adaptation interval and are hence practical to implement. However, the buffer requirements for the two schemes are quite different. For the one macroblock adaptation interval, the buffer requirement of our scheme was between 19% and 50% of that of the PID controller-based scheme (when the buffer size was not constrained). In particular, for the V700 clip, the buffer requirement of our scheme was only 19% of what was required by the PID controller scheme. For adaptation intervals of 10 and 20 macroblocks, the buffer requirements of our scheme varied between 22% and 88% of that required by the PID controller.

When the buffer size was restricted, the PID controller-based scheme suffered from severe deterioration in the output video quality because of several macroblocks missing their deadlines. Table 1 lists the percentage of macroblocks missing their deadlines for different playout buffer sizes. Note that for relatively small buffer sizes, more than 20% of the macroblocks can miss their deadlines when the PID controller-based scheme is used. Our proposed scheme, on the other hand, consistently performs well even for small buffers. However, with relative large buffers, almost all the macroblocks meet their deadlines with the PID controller.

In summary, the energy savings achieved by both the schemes are comparable. However, the buffer requirements

for the PID controller scheme, on an average, was almost double of what was required by our scheme. Hence, our scheme is particularly interesting given the current interest in DVS schemes for buffer-constrained architectures [6]. Further, as mentioned before, our scheme does not involve any runtime overhead, whereas online workload prediction schemes such as the PID controller often incurs a considerable runtime overhead and also requires constant monitoring of the buffer fill level.

4 Concluding Remarks

In this paper we presented a novel scheme for watermarking video clips with workload information, which can be extracted at runtime for dynamic voltage/frequency scaling. The inserted metadata is not visible to the external world, in contrast to previous proposals for the video *encoder* to generate the workload information, which we believe is less flexible because it restricts the platforms on which the resulting video clips can be decoded. Our main contribution is a fast bitstream analysis technique to predict the computational workload involved in decoding a video clip. We believe that the basic scheme that we presented in this paper can be further refined to more accurately estimate the decoding workload. In particular, the cache/memory organization of the target device (e.g. PDA) may be taken into account in conjunction with the bitstream analysis technique that we presented here.

References

[1] T. Austin, E. Larson, and D. Ernst. SimpleScalar: An infrastructure for computer system modeling. *IEEE Computer*, 35(2):59–67, 2002.

[2] M. Buss, T. Givargis, and N. Dutt. Exploring efficient operating points for voltage scaled embedded processor cores. In *Real-Time Systems Symposium (RTSS)*, 2003.

[3] H. V. A. et al. Energy-aware system design for wireless multimedia. In *DATE*, 2004.

[4] S. M. et al. Integrated power management for video streaming to mobile handheld devices. In *ACM Multimedia (MM)*, 2003.

[5] Y.-H. Lu, L. Benini, and G. D. Micheli. Dynamic frequency scaling with buffer insertion for mixed workloads. *IEEE Trans. on CAD of Integrated Circuits andd System*, 2002.

[6] A. Maxiaguine, S. Chakraborty, and L. Thiele:. DVS for buffer-constrained architectures with predictable QoS-energy tradeoffs. In *CODES+ISSS*, 2005.

[7] http://www.tns.lcs.mit.edu/manuals/mpeg2/.

[8] Q. Wu, P. Juang, M. Martonosi, and D. W. Clark. Formal online methods for voltage/frequency control in multiple clock domain microprocessors. *ASPLOS*, 2004.

[9] W. Yuan and K. Nahrstedt. Energy-efficient soft real-time CPU scheduling for mobile multimedia systems. In *ACM Symposium on Operating Systems Principles (SOSP)*, 2003.

[10] W. Yuan and K. Nahrstedt. Practical voltage scaling for mobile multimedia devices. In *ACM Multimedia (MM)*, 2004.

978-1-4244-3039-0/08 $25.00 © 2008 IEEE

21st International Conference on VLSI Design

Throughput efficient Parallel Implementation of SPIHT algorithm

Anilkumar V. Nandi[*], Dr. R.M.Banakar[**]
[*] *Assistant Professor,* [**] *Professor and Member - IEEE*
Electronics and Communication Engineering Department, B.V.B. College of Engineering and Technology, Vidyanagar, HUBLI-580031, Karnataka, INDIA
E-mail: anilnandy@bvb.edu, banakar@bvb.edu

Abstract

We present a throughput efficient FPGA implementation of the 'Set Partitioning in Hierarchical Trees' (SPIHT) algorithm for compression of images. The SPIHT uses inherent redundancy among wavelet coefficients and suited for both gray and color images. The SPIHT algorithm uses dynamic data structures which hinders hardware realization. In our FPGA implementation we have modified basic SPIHT in two ways, one by using static (fixed) mappings which represent significant information and the other by interchanging the sorting and refinement passes. A hardware realization is done in a Xilinx XC2S30 device. Significant compression ratio and throughput is obtained for a sample image of size 128 x128 pixels

1. Introduction

DWT of an image leads to its representation in different scales in the form of Spatial Orientation Trees (SOT). The coefficients in SOT of an image DWT are encoded using embedded zero-tree coding which is proved to be the best image compression technique [1], [2]. After the transform, the lowest frequency coefficients concentrate most of the energy of the transformed image. The high frequency coefficients of different scales and orientations indicate the strong self similarity among themselves. These properties are exploited in SPIHT [1]. After EBCOT in JPEG2000, SPIHT a sophisticated coding technique belongs to next generation of encoders for wavelet transformed images. The basic SPIHT uses dynamic data structures for exploiting self similarities mentioned above. These dynamic data structures impose practical limitation on hardware implementation of SPIHT, unlike software implementation where dynamic data structures can be implemented conveniently using linked lists [3]. Hence we have modified the basic SPIHT algorithm to overcome dynamic allocation problem. The FPGA implementation of modified version has resulted in

significant optimization in memory requirements and speed.

The paper is organized as follows. In Section 2, a brief description of the basic *SPIHT* algorithm and the modified SPIHT algorithm are described. Section 3 presents comparison of basic SPIHT and modified SPIHT. The hardware implementation in a Xilinx XC2S30 device is presented in Section 4.

2. SPIHT Algorithm

A. Introduction: The SPIHT algorithm uses a partitioning of the spatial orientation trees in a manner that tends to keep insignificant coefficients together in larger subsets. The partitioning decisions are binary decisions that are transmitted to the decoder, providing a significance map encoding. The thresholds used for checking significance are powers of two, so in essence the SPIHT algorithm sends the binary representation of the integer value of the wavelet coefficients. The significance map encoding or set partitioning and ordering step is followed by a refinement step in which the representations of the significant coefficients are refined. The SPIHT algorithm can be applied to both grey-scale and colored images. SPIHT displays exceptional characteristics over several properties like good image quality, fast coding and decoding, a fully progressive bit stream, application in lossless compression, error protection and ability to code for exact bit rate.

The SPIHT process represents a very effective form of coding. A straightforward consequence of the compression simplicity is the greater coding/decoding speed. The SPIHT algorithm is nearly symmetric, i.e., the time to encode is nearly equal to the time to decode. SPIHT codes the individual bits of the image wavelet transform coefficients following a bit plane sequence. Thus, it is capable of recovering the image perfectly by coding all bits of the transform. In practice it is frequently possible to recover the image perfectly using rounding after recovery, but this is not the most efficient approach.

SPIHT due to its embedded coding property is much easier to design efficient error-resilient schemes. This happens because with embedded coding the information is sorted according to its importance, and the requirement for powerful error correction codes decreases from the beginning to the end of the compressed file. If an error is detected, but not corrected, the decoder can discard the data after point and still display the image obtained with the bits received before the error. Another reason is that SPIHT generates two types of data. The first is sorting information, which needs error protection. The second consists of uncompressed sign and refinement bits, which do not need special protection because they affect only one pixel.

B. Algorithm:

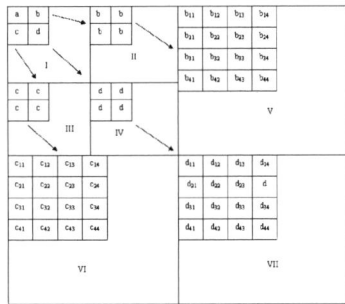

Figure 1: Data Structure used in SPIHT

In SPIHT algorithm, the wavelet coefficients are divided into trees originating from the lowest resolution band. The coefficients are grouped into 2-by-2 arrays that, except for the coefficients in band 1, are offspring of a coefficient of a lower resolution band. The coefficients in the lowest resolution band are also divided into 2-by-2 arrays. The coefficient in the top-left corner of the array does not have any offspring and is known as the root node. This data structure is shown pictorially in Figure 1 for seven-band decomposition.

The trees are further partitioned into four types of sets, which are sets of coordinates of the coefficients:

- $O(i, j)$ – Set of coordinates of the off springs of the wavelet coefficient at location (i, j). As each node can have either four off springs or none, the size of $O(i, j)$ is either zero or four. For example, in Figure 1 the set $O(0, 1)$ consists of the coordinates of the coefficients b_1, b_2, b_3 and b_4.
- $D(i, j)$ – Set of all descendants of the coefficient at location (i, j). Descendants include the offspring,

the offspring of the offspring, and so on. For example, in Figure 1 the set $D(0,1)$ consists of the coordinates of the coefficients $b_1, \ldots, b_4, b_{11}, \ldots, b_{14}, \ldots, b_{44}$. The number of offspring can either be zero or four, the size of $D(i, j)$ is either zero or a sum of powers of four.

- H – Set of all root nodes – essentially band I in the case of Figure 1.
- $L(i, j)$ – Set of coordinates of all descendants of the coefficient at location (i, j) except for the immediate offspring of the coefficient at location (i, j). In other words,

$$L(i, j) = D(i, j) - O(i, j)$$

In Figure 1, the set $L(0, 1)$ consists of the coordinates of the coefficients $b_{11}, \ldots, b_{14}, \ldots, b_{44}$.

A set $D(i, j)$ of $L(i, j)$ is said to be significant if any coefficient in the set has a magnitude greater than the threshold. Finally, thresholds used for checking significance are powers of two, so in essence the SPIHT algorithm sends the binary representation of the integer value of the wavelet coefficients. The bits are numbered with the least significant bit being the zero[th] bit, the next bit being the first significant bit, and the kth bit being referred to as the k-1 most significant bit.

The SPIHT algorithm makes use of 3 lists namely, List of Insignificant Pixels (LIP), List of Significant Pixels (LSP) and List of Insignificant Sets (LIS). The LIP and LSP lists will contain the coordinates of the coefficients, while the LIS will contain the coordinates of the roots of sets of type D or L. We start by determining the initial value of the threshold. It can be done by calculating $n = |\log_2 C_{max}|$, where, c_{max} is the maximum magnitude of the coefficients to be needed. The LIP list is initialized with the set H. Those elements of H that have descendants are also placed in LIS as type D entries. The LSP list is initially empty. In each pass, the members of LIP are processed first, then the members of LIS. This is essentially the significance map encoding step. Later the elements of LSP are processed in the refinement step.

We begin by examining each coordinate contained in LIP. If the coefficient at that coordinate is significant (that is, it is greater than 2^n), we transmit a 1 followed by a bit representing the sign of the coefficient. We consider a 0 for positive and a 1 for negative sign. This coordinate is later moved to LSP list. If the coefficient at that coordinate is not significant, a 0 is transmitted.

After examining each coordinate in LIP, we begin examining the sets in LIS. If the set at coordinate (i, j) is not significant, we transmit a 0. If the set is significant, a 1 is transmitted. The next steps depend on whether the set is of type D or L.

If the set is of type D, we check each of the offspring of the coefficient at that coordinate. In other words, we check the four coefficients whose coordinates are in O (i,j). For each coefficient that is significant, a 1 is transmitted, with the sign of the coefficient, and then the coefficient is moved to the LSP. For the rest a 0 is transmitted and their coefficients are added to LIP. Now that the coordinates of O (i, j) are removed from the set, what is left is simply the set L (i, j). If the set is not empty, it is moved to the end of LIS and marked to be of type L. Note that this new entry into the LIS has to be examined during this pass. If the set is empty, the coordinate is removed from the list.

If the set is of type L, each coordinate in O (i, j) is added to the end of the LIS as the root of a set of type D. Note that these new entries in the LIS have to be examined during this pass. (i, j) is then removed from the LIS.

Once each of the sets of LIS have been processed (including the newly formed ones), we proceed to the refinement step. In the refinement step, each coefficient that was in the LSP prior to the current pass is examined and the nth most significant bit of $|c_{i, j}|$ is output. The coefficients that have been added to the list in this pass are ignored because, by declaring them significant at this particular level, the decoder of the value of the nth most significant bit has already been informed.

This completes one pass. Depending on the availability of more bits or external factors, if we decide to continue with the coding process, n is decremented by one and the process is continued.

3. Modified SPIHT Algorithm

The SPIHT algorithm computes the SOT wavelet coefficients progressively in dynamic order. The work by Singh et. al. [4] uses content addressable memories to keep track of the order in which the coefficients are scanned. Algorithm is implemented directly without modification not taking into account hardware optimization. If the bit streams end within the middle of the bit-plane, then the quality of the image will be better for such a scheme. Since every image has a unique order of coefficients determined by their values. The generation of bit-stream depends on 2x2 block of coefficients, their children and maximum value within

that sub-tree. So, every block of coefficients can be operated independent of others and also in parallel to one another. However the order in which they are to be operated is not known in advance, due to uniqueness of every image.

The basic SPIHT algorithm determines the ordering of coefficients in sequential manner. The computation traverses the coefficients of an image many times in each bit plane and inserts or deletes the coefficients in the lists. Such a scheme is not suitable for hardware implementation using parallelism and limits the throughput.

Our scheme uses fixed order SPIHT [5], in which the order in which the block of coefficients are transmitted is fixed in advance. The blocks of coefficients are inserted in the predetermined order. The order is based upon Morton Scan ordering [6]. This eliminates the overhead of computing the order of coefficients in each bit-plane. The parallel units of SPIHT can be built with this technique so that throughput can be increased.

Another modification is in the refinement pass. In basic SPIHT, the information regarding status for the elements of LSP, whether they have been added in the current iteration of sorting pass, need to be maintained. In the worst case if all the coefficients become significant in the same iteration, we need large memory requirement to store this information. If we exchange the sorting pass and refinement pass we need not store this information. The data stream is still decodable and does not increase in size. We need to consider the reordering of the transmitted bits during the iteration in the decoding process.

A. The modified SPIHT algorithm is as follows:

Initialization
        n = |log2 (max |coeff|)|
        LIP = All elements in **H**
        LSP = Empty
        LIS = **D**'s of Roots
Refinement Pass
        Process LSP
for each element (i, j) in LSP – *except* those just added above.
Output the nth most significant bit of coefficient
End loop over LSP

Significance Map Encoding ("Sorting Pass")
        Process LIP

```
for each coeff (i, j) in LIP
        Output Sn (i, j)
        If Sn (i, j)=1
Output sign of coeff (i, j): 0/1 = -/+
Move (i, j) to the LSP
End if
        End loop over LIP

Process LIS
        for each set (i, j) in LIS
                if type D
                        Send Sn(D (i, j))
                        If Sn (D (i, j))=1
                for each (k, l)☐ O (i, j)
        output Sn (k, l)
        if Sn (k, l)=1, then
    add (k,l) to the LSP and output sign of coeff:0/1+/-
    if Sn (k,l)=0, then add (k,l) to the end of the LIP
    end for
    end if
    else (type L )
    Send Sn (L(i,j))
    If Sn (L(i,j))=1
    add each (k,l)☐O (i,j) to the end of LIS as an entry of
    type D
    remove (i,j) from the LIS
    end if on type
    End loop over LIS
    Update
    Decrement n by 1
    Go to Significance Map Encoding Step
```

4. Architecture

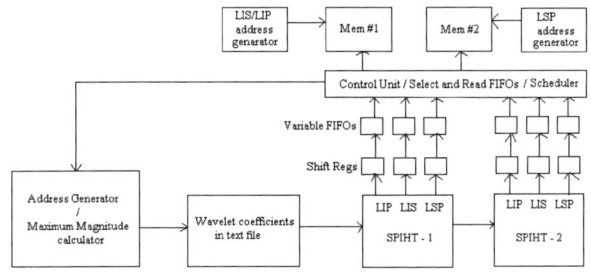

Figure 2 Fixed Order Parallel SPIHT block diagram

The Fixed order Parallel architecture is shown in Figure 2.

A. Maximum magnitude calculator / Address generator block:

The hierarchical tree structure obtained after wavelet decomposition has a particular pattern for accessing the coefficients. The relation between the row and the column values of a parent co-ordinate and its four descendants are as follows. Suppose the parent be 'a' and its row and column address be X[0] and Y[0] respectively. Then the row and column addresses for its descendants a1, a2, a3, a4 are as given below.

a1:	$X[1]= 2X[0]$,	$Y[1]= 2Y[0]$
a2:	$X[2]= X[1]$,	$Y[2]= Y[1]+1$
a3:	$X[3]= X[1]+1$,	$Y[3]= Y[2]-1$
a4:	$X[4]= X[3]$,	$Y[4]= Y[3]+1$

Using the same pattern, the tree grows on for each of the nodes and it stops at the last element of the matrix. The address generator has been designed to implement the above mentioned pattern in hardware.

Thus the random access of coefficients is converted into sequential access. Maximum magnitude coefficient is computed by searching all the coefficients of the image matrix using sequential addressing for the wavelet coefficients. This process of finding the maximum value of coefficient slows down the system performance. By using Depth First Search order algorithm [7] performance can be improved. Maximum magnitude calculation phase calculates the maximum value of coefficients and rearranges the following information for SPIHT block. a) Maximum magnitude of four child trees b) Current maximum magnitude c) Threshold and sign data of each of the 16 child coefficients 4) Reorder the coefficients into sequential order.

The address generator also generates the addresses fixed order addresses and picks the data (wavelet coefficients) from text files. We have used text file storage for wavelet coefficients. In real time operation, they are fed from the wavelet transform block. The coefficients are fed to the SPIHT blocks (SPIHT-1 and SPIHT-2) in the predefined order.

B. SPIHT block:

We have implemented SPIHT block of stages which does coding in parallel based on data from magnitude calculator/Address generator block. Coefficient blocks are read from higher level to lower level. The fixed point numerical representation is used for each wavelet level. When both blocks receive data in common format, parallel version of SPIHT operates to generate code corresponding to information in each block that contributes to each bit-plane. Each SPIHT block generates bit stream which are added and grouped before sending to the variable FIFOs for each bit-plane. The data received in FIFOs vary in size depending on the coefficients values in each bit-plane. Maximum size of FIFO is kept at 20 bits after verification by software program [3] for a class of 128x128 pixel images. Variable FIFOs arrange the block of data into regular size of 16 bit words for memory access. Scheduler is used to take care by stalling the algorithm operation if one of the FIFOs becomes full. The scheduler determines the filled FIFO and writes the filled data into memory.

C. Platform for Implementation:

Hardware is modeled using VHDL under Xilinx 7.1i tools with modelsim simulator. XST is used to compile VHDL code and generate the net list. Xilinx 7.1i is used to place and route the design on Spartan II FPGA reconfigurable system with XC2S30 device consisting of 972 logic cells, 30,000 gates and 384 CLBs.

5. Results and Discussion

A. Memory requirement of SPIHT: The size of required working memory at any instant of time depends on the number of entries in the lists. The memory requirements at the end of each pass (bit-plane) and maximum (worst case) memory requirement are considered for memory requirement. The three lists LIP, LIS and LSP are used in SPIHT. Each entry in the lists uses co-ordinates of wavelet coefficients.

The memory requirements are calculated as follows:

The total memory required for SPIHT be M_{SPIHT}.
$M_{SPIHT} = C N_{LIP} + (c-1) N_{LIS} + c N_{LSP}$
where N_{LIP}, N_{LIS} and N_{LSP} are number of elements in the lists LIP, LIS and LSP respectively.
Worst case is when
$N_{LIP} + N_{LSP} = M.N.$ (for M x N image).
$N_{LIS} = M.N / 4$

Considering the worst case when High frequency sub-bands will never enter into LIS.

$M_{SPIHT (max)} = (5C-1) (M.N) / 4$.
For 8 x 8 image and c=9 bits, $M_{SPIHT (max)}$ will be 704 bits.

B. Clock rates and Throughput:
The clock cycles for SPIHT encoding for 128x128 test image is 1030 cycles. It does not include the cycles for Maximum Magnitude calculation and address generation block. The maximum clock rate available on the board (TKBase XC2S30) is 10MHz. The average number of bits generated at the output of Parallel SPIHT block is 3 bits per clock cycle.

However the throughput can be increased to a greater extend if the higher end FPGA board is used. The Table 1 shows the results

Phase	Clock cycles / 8 x 8 image	Clock rate	Throughput
SPIHT	1030	10MHz	30 MPixels/Second

Table 1. Results for SPIHT block

C. PSNR measurement:
At the end of each bit-plane same bit streams will be obtained as that of basic SPIHT (without fixed order) but in different order. Hence PSNR at the end of each bit-plane will be closely matching to that of regular SPIHT as shown in the Figure 3. The slight difference is observed due to short length of each bit stream in fixed order scheme. Maximum loss is found to be 0.18 dB over the shown bit rates.

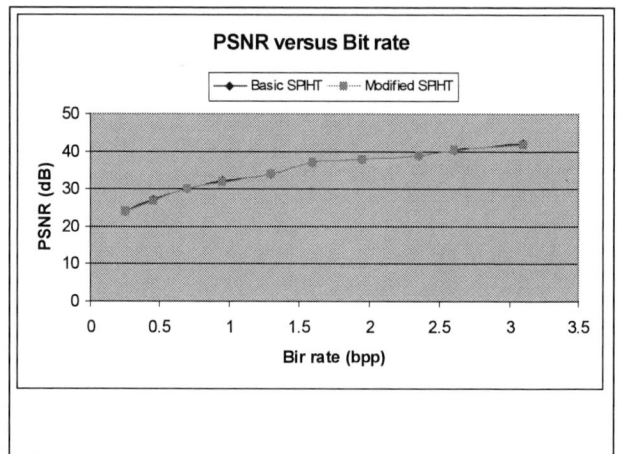

Figure 3 PSNR versus Bit rate

6. References

1. A. Said and W. A. Pearlman. A new fast and efficient image codec based on set partitioning in hierarchical trees. In *Trans. Signal Processing*, volume 5, no.9, pages 1303–1310. IEEE, 1996.

2. I. Daubechies. *Ten lectures of wavelets*. SIAM, Philadelphia PA, 1992.

3. Priyanka G,Shalini Neeli,Trupti R,Kavya P, Anil V. Nandi, R.M.Banakar, *ImageCodec using Wavelet/Set Partitioning In Hierarchical Tree approach*, National Conference on Signal Processing and Communications-2006 (NCSPC-2006), at JNNCE, Shimoga.

4 J Singh, A. Antoniou, D. J. Shpak, "Hardware Implementation of a Wavelet based Image Compression

Coder," *IEEE Symposium on Advances in Digital Filtering and Signal Processing*, pp 169 – 173, 1998.

5 Thomas W. Fry, Scott Hauck, "SPIHT image compression on FPGAs", *IEEE Transactions on circuits and systems for video technology*, Vol 15, No.9, September 2005 pg 1138-1147.

6. V. R. Algazi, R. R. Estes. "Analysis based coding of image transform and sub band coefficients," *Applications of 64 of SPIE Proceedings*, pages 11-21, 1995.

7. T. Cormen, C. Leiserson, R. Rivest, *Introduction to Algorithms*, MIT Press, Cambridge, Massachusetts, 1997.

978-1-4244-3039-0/08 $25.00 © 2008 IEEE

SESSION D5:
Invited Special Session:
Standards in EDA

21st International Conference on VLSI Design

Invited Special Session in VLSI Design 2008:

Session D5 – Jan 8, 2008 (11.00 am – 1.00 pm)

"Standards in EDA"

Organizer & Chair : Nagi Naganathan, LSI Corp

Industry Standards from Accellera

Shrenik Mehta, SUN Microsystems

Abstract:

Accellera's (www.accellera.org) mission is to drive worldwide development and use of standards required by systems, semiconductors and design tools companies, which enhance a language based design automation process.

Overview of different standards from Accellera and how they fit into the design flow process will be presented. Status of various technical sub-committees like Open Compression Interface (OCI), Unified Power Format (UPF), Unified Coverage Interoperatbility (UCIS) and Open Verification Library(OVL) will also be covered.

Speaker Bio:

Shrenik Mehta is Senior Director of Frontend Technologies and OpenSPARC program at Sun Microsystems. Shrenik has been driving the details of the OpenSPARC project from its infancy to the public release. Shrenik is also currently the Chair of Accellera, an industry standards organization and was the Vice Chair of the IEEE 1800 SystemVerilog Working Group. He holds eight US patents and one patent in Taiwan.

He has a Master'd degree in Computer Engineering from University of Michigan, Ann Arbor and a Bachelor's degree from Institute of Technology - BHU, Varansasi (India).

[Editorial note: The Verilog-AMS standard from Accellera will be covered by Sri Chandra of Freescale and fellow speaker]

IEEE Market-Oriented Standards Process and the EDA Industry

Dennis Brophy, Mentor Graphics

Abstract:

The IEEE has collaborated with numerous consortia to develop EDA standards for more than a decade. The recent success of new and emerging standards have borrowed from the market-oriented approaches to ensure immediate suppliers of tools and technology that embrace IEEE standards that give consumers confidence they should plan for their immediate use. The SystemVerilog success will be explored and it will be demonstrated how it can apply to other work in the IEEE.

Speaker Bio:

Dennis B. Brophy is director of strategic business development at Mentor Graphic Corporation. He is also volunteer vice-chair and past chair of Accellera, an electronic design automation standardization group. Dennis has been in the electronic design automation industry for the past 27 years. He was first with Hewlett-Packard for five years, and then joined Mentor Graphics where he has held several positions the past 22 years. He is a member of the IEEE Standards Association (SA) Board of Governors (BOG) and a member of the IEEE SA Corporate Advisory Group (CAG). He is secretary of the IEEE P1800 SystemVerilog Working Group, secretary of the IEEE 1666 SystemC Working Group and participates in the IEEE P1801 Low Power Working Group. Dennis received a Bachelor of Science degree from the University of California at Davis in electrical engineering and computer engineering in 1980.

Design Automation Standards the IP providers perspective

Dr. John Goodenough, ARM

Abstract:

Design chain standards are all ultimately aimed to make the task of the Design Integration and Manufacture of System on Chip Products more efficient, improving turn Around Time, and effective improving quality and yield. Dr Goodenough will outline the standardization activities in which ARM is currently involved {including those managed by Si2 Accellera, SPIRIT, JEDEC, Eclipse, OpenMax OpenGL} and their relevance to the issues of the IP supply chain. He will discuss types of standards an their impact on IP. The presentation will also focus on some of the challenges in managing viable standards to broad market acceptance and the consequent need for an integrated roadmap between various standardization activities to give maximum benefit and leverage to the final end customers

Speaker Bio:

John Goodenough is Worldwide Director of Design Technology at ARM, responsible for all aspects of design methodology including support for ARM's internal production flows and IP deployment and integration. In the latter role he works extensively with ARM's design chain partners and customers. John began his career in electronics as a Sheffield University research fellow investigating novel VLSI signal processing architectures. He then co-founded Infinite Designs, a consulting house specializing in advanced ASIC and embedded system design methodologies. As a present board member of Si2 consortium, past board member of Accellera, OSCI and the VSIA alliance and a founding board member of the SPIRIT consortium John has a passionate interest in finding pragmatic approaches to improving quality and turn around time for System-SoC designs. Dr Goodenough has strategic oversight of all of ARMs design standards activities. Dr. Goodenough holds a B.Sc in Engineering Science from Durham University and a Ph.D in VLSI Design from Sheffield University.

Dr. Goodenough is currently active board member of Accellera, SPIRIT, Si2 and have accountability for ARMs activities in OSCI.Jedec

21st International Conference on VLSI Design

Driving Analog Mixed Signal Verification through Verilog-AMS

Sri Chandra, Freescale

Abstract:

The complexity of today's SoCs and applications are driving the need for faster and more accurate mixed signal verification. Additionally the percentage of analog content in mixed-signal designs is increasing rapidly. This requires a change in mindset: no longer can the analog and digital modules be verified independantly. For these reasons Accellera has been leading the development of the Verilog-AMS standard, to enable accurate mixed signal design verification of systems containing thousands of analog/digital interface connections. The presentation will discuss the recent language enhancements that have been driven by the Verilog-AMS technical committee, to make system level analysis of analog and mixed signal designs much more efficient and accurate.

Speaker Bio:

Sri has been associated with Freescale's analog & mixed signal internal tool development for the past 7 years. Currently he manages the internal tool development activities for Freescale Semiconductor in Noida, and his interests include power estimation and analysis at the system level. He has been actively driving the Verilog-AMS language development efforts for the past 4 years through his role as the technical chairperson of the AMS committee. Sri has worked for Freescale Semiconductor (formerly Motorola Inc) for 13 years, starting his career with MIEL Bangalore and has worked at the Austin and Adelaide development centers. Sri holds a Master of Engineering degree from the Indian Institute of Science, Bangalore.

978-1-4244-3039-0/08 $25.00 © 2008 IEEE

21st International Conference on VLSI Design

VSI Standards, Current Status and Future Work

Kathy Werner, Freescale

Abstract:

The VSI Alliance has been involved in SoC standards and documents for 11 years and laid the groundwork for the IP industry addressing the issues associated with reusing IP, both from the technical as well as the business perspectives. VSIA brought together the EDA, Electronics and Semiconductor industries to enable a dramatic design paradigm shift. The increasing proliferation of SoCs and virtual platforms in 2007 is testimony that VSIA has succeeded in making that paradigm shift. Additionally, much of the work and documents created are the basis for the great work happening in other groups today. With the planned closing of the organization, the presentation will discuss the status and disposition of VSI's work.

Speaker Bio:

Kathy Werner is the manager of the corporate standards, intellectual property and writers recognition programs at Freescale Semiconductor. She has led the industry in its quest for IP standards and been involved in design, NPI and management at several companies including wireless, processor development, consumer and computing and military, in both direct and consulting roles . She also has long experience in standards development, including President of the VSIA industry consortium. She also chaired the VSI Alliance Quality IP Pillar and led the development and introduction of the Quality IP (QIP) metric. She had published in Electronic Business, EE Times, Chip Design and others, as well as being the co-executive editor and contributor of an upcoming book on Intellectual Property.

978-1-4244-3039-0/08 $25.00 © 2008 IEEE

Author Index

Abelé, N. 119

Abraham, Jacob A. 521

Afzali-Kusha, Ali.................................. 541

Agarwal, Sundeepkumar......................... 371

Agrawal, Banit.............................. 59, 354

Agrawal, Vishwani D. 429, 527

Akl, Charbel J. 195, 673

Alaghi, Armin....................................... 409

Alam, Monjur....................................... 693

Almeida, Manuel................................... 85

Amiri-Kamalabad, Mojtaba..................... 21

Amrutur, Bharadwaj..................... 667, 503

Amrutur, Bharadwaj S. 509, 560

Amrutur, Janakiraman V. Bharadwaj........ 685

Anbugeetha, K. 317

Andrei, Alexandru............................... 103

Arslan, Tughrul.................................... 389

Arvind, N.V. 685

Ashouei, Maryam................................. 27

Babu, Hafiz Md. Hasan......................... 566

Bagan, K. Boopathy.............................. 77

Bahari, Asral....................................... 389

Balajee, S. 3

Balaji, S. 613

Baldawa, Sandeep................................ 279

Banakar, R.M. 718

Bandyopadhyay, Saurav......................... 208

Banerjee, Soumitro.............................. 305

Banerjee, Swapna................................ 323

Bansal, Aditya.................................... 125

Basu, Anupam..................................... 111

Basu, Samik.. 492

Basu, Shubhankar................................ 287

Bayoumi, Magdy A. 195, 673

Bennett, Terrell.................................... 267

Bhardwaj, Sarvesh......................... 533, 515

Bhat, Navakanta.................................. 589

Bhatia, Sandeep................................... 187

Bhattacharya, Bhargab B. 163

Bhattacharyya, Kaushik......................... 661

Bhattacharyya, T.K............................... 595

Bhattacharyya, Tarun Kanti.................... 311

Bhunia, Swarup................................... 441

Biswas, Ashis Kumar............................ 566

Bondada, Siva Kumar............................ 560

Boppana, Dr. Vamsi.............................. 3

Brien, Cameron................................... 461

Brophy, Dennis.................................... 727

Brown, Richard B. 143

Bushnell, Michael Lee........................... 581

Cadambi, Srihari.................................. 435

Calhoun, Benton H. 131

Carvajal, Ramon Gonzalez..................... 294

Catthoor, Francky................................. 201

Chakraborty, Rajat Subhra..................... 441

Chakraborty, Samarjit.............. 679, 712, 3

Chakraborty, Tapan Jyoti........................ 39

Chandra, Sri....................................... 727

Chandrachoodan, Nitin.......................... 169

Chandratre, Vinay B. 613

Chatterjee, Abhijit................... 65, 27, 71

Chatterjee, Subho................................ 311

Chauhan, Rajat.................................... 383

Chauhan, Yogesh Singh.......................... 119

Chong, Frederic T. 59

Chowdhury, Ahsan Raja......................... 566

Author Index

Chuang, Ching-Te............... 125, 143

Ciesielski, Maciej..................... 487

Costas, A. 33

Culpepper, Barry..................... 331

Das, Bishnu Prasad........... 667, 685

Das, Debesh K. 163

Das, Sabyasachi........... 653, 635, 572

Datta, Abhishek. 475

Declercq, M. 119

Dev, Nilabha......................... 187

Dey, Soumyajit....................... 111

Dhanasekaran, D. 77

Dutt, Nikil.......................... 363, 3

Dutt, Nikil D. 421

Eggimann, C. 119

Eles, Petru.............................. 103

Eltawil, Ahmed........................... 3

Erdogan, Ahmet T. 389

Erraguntla, Vasantha............... 273

Fathy, Mahmood..................... 409

Fazeli, Mahdi............................ 21

Fernando, Pradeep.................. 337

Gandhi, Kaushal R. 45

Garimella, Annajirao................. 294

Garimella, Sri Raga Sudha......... 294, 300

Genova, Sue........................... 187

Ghosh, Amlan......................... 143

Ghosh, Anandaroop................. 595

Ghosh, Jyotirmoy.................... 331

Ghosh, Santosh....................... 693

Goodenough, John................... 727

Govindarajan, R. 553

Gu, Yan................................. 679

Gueron, Shay......................... 273

Gupta, Ankur......................... 383

Gupta, Aseem......................... 421

H.M., Roopashree.................... 383

Harame, D. 3

Hasan, Md. Mahmudul............. 566

Hasan, Moshaddek................... 566

Hessabi, Shaahin.................... 415

Huang, Yicheng...................... 712

I, Mohammed Shareef............. 503

Ionescu, A.M. 119

Isoaho, Jouni......................... 228

Ivancic, Franjo....................... 435

Iyer, Subramanian S. 3

Jabir, A.M. 629, 33

Jagirdar, Aditya....................... 39

Jain, Jalaj............................... 97

Jamadagni, H.S. 685

Janarthanan, Arun.................. 397

Jayapala, Murali..................... 201

Jha, Niraj K. 435, 621, 220

Jose, Babita R. 453

Joshi, Rajiv............................... 3

Jr., John E. Barth........................ 3

Jung, Hwisung....................... 249

K, Pavankumar V..................... 371

Kadam, Vinayak..................... 187

Kalyan, T. Venkata................. 235

Kalyani-Garimella, Lalitha Mohana.......... 294

Kannan, Deepa............. 533, 515, 421

Kathail, Vinod............................ 3

Katkoori, Srinivas.............. 641, 337

Kedia, Monu........................... 111

Author Index

Khatri, Sunil P. 653, 635, 572

Kim, Jae-Joon................................. 125

Kim, Jae-joon................................. 143

Kim, Keunwoo................................ 125

Kitamichi, Junji............................... 91

Kokrady, Aman................................ 169

Kole, Dipak K. 163

Kommineni, Balaji............................ 287

Koohi, Somayyeh.............................. 415

Koopahi, Elnaz................................ 409

Krishnamurthy, Ram.......................... 273

Krishnan, Vyas................................ 641

Krishnapura, Nagendra........................ 3

Kumar, Anshul........................... 348, 261

Kumar, Nidhir................................... 3

Kumar, T.S. Rajesh........................... 553

Kundu, Amal Kumar.......................... 311

Kundu, Sandip................................ 151

Kurdahi, Fadi.................................... 3

Kurdahi, Fadi J. 363, 421

Kuroda, Kenichi............................... 91

Lambrechts, Andy............................ 201

Lee, Dwayne.................................... 3

Lei, Chi-Un.................................... 469

Lopez-Martin, Antonio....................... 294

Lotfi-Kamran, Pejman.................... 541, 481

Lu, Yuanlin................................... 527

M, Kirthi Krishna............................ 547

Mahapatra, Nihar R. 45

Mahapatra, Rabi.............................. 243

Mahapatra, Santanu.......................... 447

Malik, Sharad................................ 461

Mandal, Pradip.......................... 208, 661

Massoumi, Mehran............................ 481

Mathew, J. 629

Mathew, Jimson.......................... 453, 33

Mathew, Sanu................................ 273

Mehta, Shrenik.............................. 727

Mei, Tawen.................................. 331

Menezes, Vinod.............................. 383

Miller, Tom.................................... 3

Miremadi, Seyed Ghassem.................... 21

Mirza-Aghatabar, Mohammad................ 415

Mirzaei, Mohammad.......................... 481

Mishra, Prateek.............................. 220

Mohammad, Mohammad Gh. 157

Mohan, Vipin................................ 515

Mueller, Jeff................................. 214

Mukherjee, Rupam........................... 305

Mukherjee, Subhasish........................ 187

Mukhopadhyay, Saibal........................ 125

Mukhopadhyay, Siddhartha.................... 331

Mukhpodhyay, Debdeep...................... 706

Muttreja, Anish.......................... 621, 220

Mutyam, Madhu.............................. 235

Mysore, Shashidhar........................... 59

N.S., Sreeram................................ 509

Nair, Pradeep................................ 503

Nandi, Anilkumar V. 718

Narang, Vikas................................ 383

Nassif, Sani.................................... 3

Navabi, Zainalabedin................. 541, 481, 409

Nigussie, Ethiopia........................... 228

Nisar, Muhammad Mudassar................... 71

Oliveira, Roystein............................ 39

Park, Young-Hwan............................ 363

Author Index

Pasricha, Sudeep.................................. 363
Patra, Amit................................... 305, 331
Paul, Kolin................................... 348, 261
Paul, Somnath...................................... 441
Pavan, Shanthi....................................... 3
Pedram, Masoud................................... 415
Pedram, Massoud................................. 249
Pedrotti, Kenneth................................ 208
Peng, Zebo...................................... 103
Pétrot, Frédéric................................. 403
Pimentel, Bruno................................... 85
Pomeranz, Irith.................. 151, 181, 175
Pothineni, Nagaraju............... 348, 261
Pradhan, Almitra................................. 647
Pradhan, D.K..................................... 629
Pradhan, D.K...................................... 33
Pradhan, Dhiraj K. 453
Prasad, Shashank................................. 343
R, Yokesh....................................... 371
Raghavan, Praveen............................. 201
Rahaman, H. 629, 33
Rahaman, Hafizur....................... 453, 163
Rahmani, Amir-Mohammad..................... 541
RahoulVarma...................................... 3
Ralph, Stephen E. 208
Ramanarayanan, Rajaraman..................... 273
Ramasamy, S. 317
Ramesh, S. 3
Ramirez-Angulo, Jaime....................... 294
Rammohan, Srividhya....................... 699
Rao, P. Vijaya Sankara....................... 235
Rao, Rahul M. 143
Ravi, Srivaths.................................. 621

Ravikumar, C.P. 553, 169
Ray, Biswajit.................................... 447
Rebeiro, Chester.................................. 706
Reddy, Sudhakar M. 151, 181, 175
Rizwan, Shahid..................................... 53
Roetteler, Martin................................. 435
Roop, Partha S. 492
Rosen, Jakob...................................... 103
Roy, Kaushik.................................... 125
Roy, Sounak.................................... 323
RoyChowdhury, Dipanwita..................... 693
Rutenbar, Rob A. 131
S, Bharat...................................... 547
S, Jairam...................................... 589
S, Subroto.................................... 547
S.A., Kannan................................... 509
Saleh, Resve................................... 214
Salehpour, Ali-Asghar........................ 541
Saluja, Kewal K. 157, 137
Samanta, Rupak................................. 243
Sambamurthy, Sriram........................ 521
Sangireddy, Rama........................ 267, 279
Sedghi, Mahshid................................ 409
Sen, Shreyas..................................... 65
Sengupta, Indranil............................. 693
Senguttuvan, Rajarajan..................... 65, 71
Shen, Hao...................................... 403
Sherwood, Timothy........................ 59, 354
Shim, Kyuho................................... 487
Shin, Chulho.................................. 354
Shrivastava, Aviral.............. 533, 515, 421
Simsir, Muzaffer O. 435
Singh, A.K.................................... 629

Author Index

Singh, Adit D. 27
Singh, Dr Adit D. 3
Singhal, Vigyan............................ 475
Singhee, Amith............................. 131
Sinha, Roopak 492
Skliarova, Iouliia......................... 255, 85
Sklyarov, Valery........................... 255, 85
Sootkaneung, Warin........................ 137
Srinivas, M.B. 547
Sundaresan, Vijay.......................... 699
Surprise, Jason............................ 243
Talupuru, Kesava........................... 487
Tamarapalli, Nagesh........................ 3
Tenhunen, Hannu............................ 228
Tewani, Menka 613
Tomko, Karen A. 397
Tsamados, D. 119
Tupuri, Raghuram S. 521
Tuuna, Sampo............................... 228
Ueda, Koji 91
V, Prateek G............................... 547
Varma, Rahoul.............................. 3
Veeramachaneni, Sreehari................... 547
Velu, Senthil N. 3
Vemuri, Ranga............................ 647, 287, 699

Venkatanarayanan, Hari Vijay............... 581
Venkataraman, Srikanth..................... 3
Venkataramani, B. 317
Verkest, Diederik.......................... 201
Verkila, Satish Anand...................... 560
Verma, Rajan............................... 3
Vijayakumar, Shriram....................... 137
Viraraghavan, Janakiraman.................. 667
Vrudhula, Sarma........................... 533, 515
Wang, Fan.................................. 429
Wang, Hui.................................. 279
Wang, Jiajing.............................. 131
Wang, Jinxiang............................. 377
Wang, Ye................................... 712
Watanabe, Minoru......................... 607, 601
Watts, Josef S. 3
Wong, Ngai................................. 469
Yamaguchi, Naoki......................... 607, 601
Yang, Seiyang.............................. 487
Yang, Xaingning............................ 137
Ye, Yizheng................................ 377
Yoon, Simon................................ 354
Yu, Yinlei................................. 461
Zhang, Qingli.............................. 377

9781424430390